Precalculus

Version 3.0

Edward B. Burger

AF864801

FlatWorld

Precalculus Version 3.0

Edward B. Burger

Published by:

FlatWorld
175 Portland Street
Boston, MA 02114

About the Author

Edward B. Burger

Dr. Edward Burger is President of Southwestern University as well as a professor of mathematics and an educational leader on thinking, innovation, and creativity. Previously, he was the Francis Christopher Oakley Third Century Professor of Mathematics at Williams College. He has delivered over 700 addresses worldwide at venues including Berkeley, Harvard, Princeton, and Johns Hopkins as well as at the Smithsonian Institution, Microsoft Corporation, the World Bank, the International Monetary Fund, the U.S. Department of the Interior, the New York Public Library, and the National Academy of Sciences. He is the author of over 70 research articles, books, and video series (starring in over 4,000 online videos), including the book *The 5 Elements of Effective Thinking*, published by Princeton University Press and translated into over a dozen languages worldwide. His book, *Making Up Your Mind: Thinking Effectively Through Creative Puzzle-Solving*, also published by Princeton University Press, was on several of Amazon's Hot New Releases lists.

In 2006, *Reader's Digest* listed Burger in their annual "America's 100 Best" list as *America's Best Math Teacher*. In 2010, he was named the winner of the Robert Foster Cherry Award for Great Teaching—the largest prize in higher education teaching across all disciplines in the English-speaking world. Also in 2010, he starred in a mathematics segment for NBC-TV on the *Today Show*; that appearance earned him a 2010 Telly Award. The *Huffington Post* named him one of their 2010 Game Changers: "HuffPost's Game Changers salutes 100 innovators, visionaries, mavericks, and leaders who are reshaping their fields and changing the world." In 2012, Microsoft Worldwide Education selected him as one of their "Global Heroes in Education." In 2013, Burger was inducted as an inaugural Fellow of the American Mathematical Society. In 2014, Burger was elected to the Philosophical Society of Texas. He is now in his fourth season of his weekly program on thinking and higher education produced by NPR's Austin affiliate KUT. The series, *Higher ED*, is available at kut.org/topic/higher-ed/ and on iTunes.

Preface

About FlatWorld's Precalculus

Modern calculus has its roots in the 17th century Europe, and is credited with revolutionizing the way we understand concepts such as motion, force, and time, along with many other facets of nature. Calculus provides a means of explaining how the world works. In any situation where change occurs, there is likely to be an application of calculus.

FlatWorld's *Precalculus* offers the subjects, skills, and insights needed to understand calculus. It is the gateway to dynamic and advanced mathematics and will provide you with the tools you'll need for success in precalculus and beyond.

This textbook is designed to provide you with a variety of engaging approaches to precalculus concepts. Video lectures by the award-winning professor Edward B. Burger provide compelling instruction for the core concepts of precalculus.

Your Path to Success

Keep It Simple

Don't try to tackle a complex problem all in one step. Instead, first identify what you know and what you don't know. Establishing what you do know gives you a solid foundation on which to build your skills and knowledge. Identifying what you do no know will help you to focus more clearly on what you still need to learn.

Make Mistakes

Making mistakes is a great way to learn. Don't be afraid to take a wrong turn, or to explore the uncharted territory. Reaching a wrong answer or winding up at a dead end is actually useful information. Learn to see that every "mistake" is actually a means of getting to what is truly important—the "ah-ha" moment that happens to you when you learn something new.

Ask Questions

Take a few minutes as you begin learning each new concept to ask yourself one or more questions about what it is you need to learn. This will help you focus your thinking as you engage in the material.

... and, be sure to make use of all of the textbook features and online resources available to you.

Acknowledgments

Thank you to the following reviewers, contributors, and colleagues, whose commitment to excellence made this book possible.

Cristina Caputo, *University of Texas*
Julie Hess, *Grand Rapids Community College*
Patrick Jones, *Patrick JMT.com*
Eric Mann, *The Khabele School*
Allison M. Pacelli, *Williams College*
Douglas Quinney, *Keele University*
Danielle Rivard, *Post University*
Mike Smith, *Ohio Valley University*

PRECALCULUS AT A GLANCE

P PREREQUISITES: BASIC ALGEBRA REVIEW

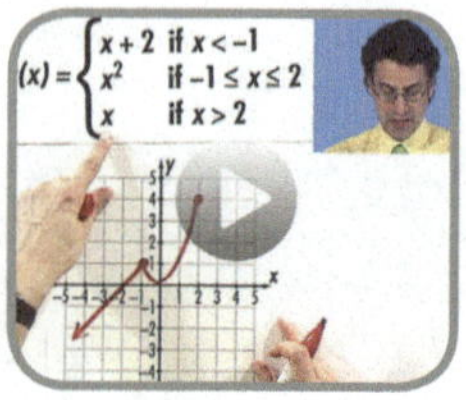

1 FUNCTIONS AND THEIR GRAPHS

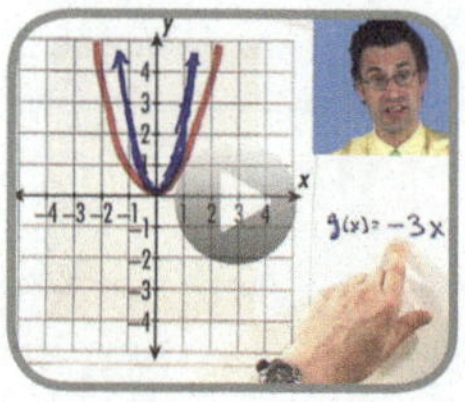

2 POLYNOMIAL FUNCTIONS

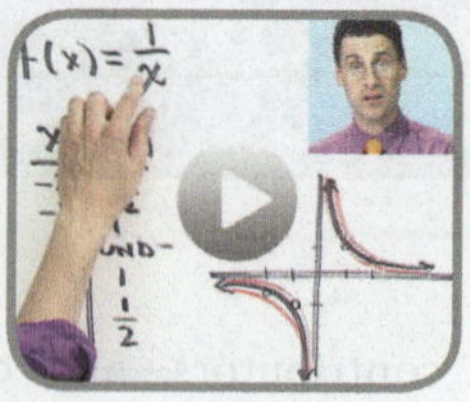

3 RATIONAL FUNCTIONS AND CONICS

4 EXPONENTIAL AND LOGARITHMIC FUNCTIONS

5 TRIGONOMETRIC FUNCTIONS

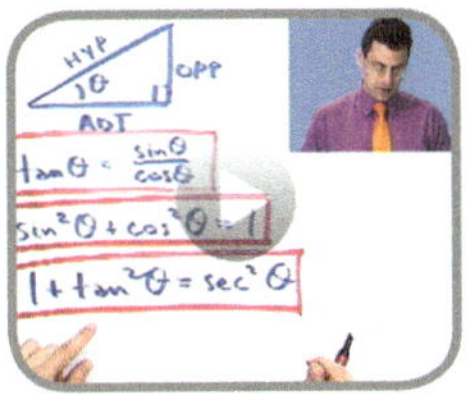

6 ANALYTIC TRIGONOMETRY

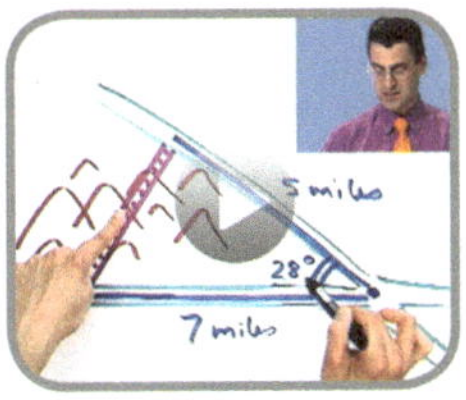

7 APPLICATIONS OF TRIGONOMETRY

8 SYSTEMS OF EQUATIONS AND INEQUALITIES

9 MATRICES AND DETERMINANTS

10 ANALYTIC GEOMETRY

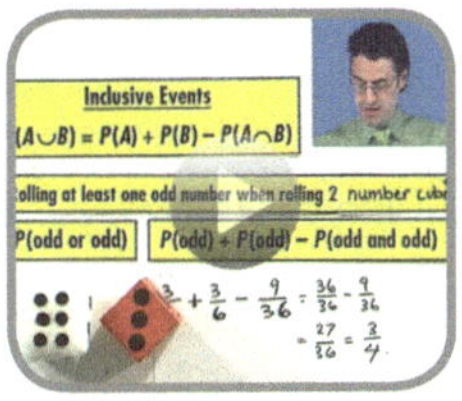

11 SEQUENCES, SERIES, AND PROBABILITY

TABLE OF CONTENTS

1 FUNCTIONS AND THEIR GRAPHS 115

2 POLYNOMIAL FUNCTIONS 225

4 EXPONENTIAL AND LOGARITHMIC FUNCTIONS 337

6 ANALYTIC TRIGONOMETRY 497

7 APPLICATIONS OF TRIGONOMETRY 547

9 MATRICES AND DETERMINANTS 649

10 ANALYTIC GEOMETRY 691

11 SEQUENCES, SERIES, AND PROBABILITY 737

ANSWERS 785

INDEX 877

PRECALCULUS

PREREQUISITES: BASIC ALGEBRA REVIEW

CHAPTER

P

P.1 REAL NUMBERS

OBJECTIVES

- Compare, order, and classify real numbers.
- Use properties of real numbers to simplify expressions.
- Evaluate algebraic expressions.
- Write sets of numbers using different notations: interval notation, set-builder notation, and roster notation.
- Evaluate absolute value expressions.

PREREQUISITE VOCABULARY TERMS

algebraic expression
coefficient
inequality
numeric expression
variable

P.1.1 Ordering and Classifying Real Numbers

Anything that can be quantified or measured, such as temperature, distance, speed, or volume, can be described by a **real number.** The set of all real numbers is divided into several overlapping subsets. One subset of the real numbers, called the **natural numbers**, is made up of the counting numbers starting with 1, 2, 3, . . . (and continuing to infinity). The subset of the real numbers that includes the natural numbers and 0 is called the **whole numbers**. The set of numbers called the **integers** is the subset of the real numbers that includes all positive whole numbers, negative whole numbers, and 0.

Natural Numbers
1, 2, 3, 4, 5, . . .

Whole Numbers
0, 1, 2, 3, 4, . . .

Integers
. . . , −2, −1, 0, 1, 2, . . .

These subsets of the real numbers are displayed in the Venn diagram in Figure P.1a.

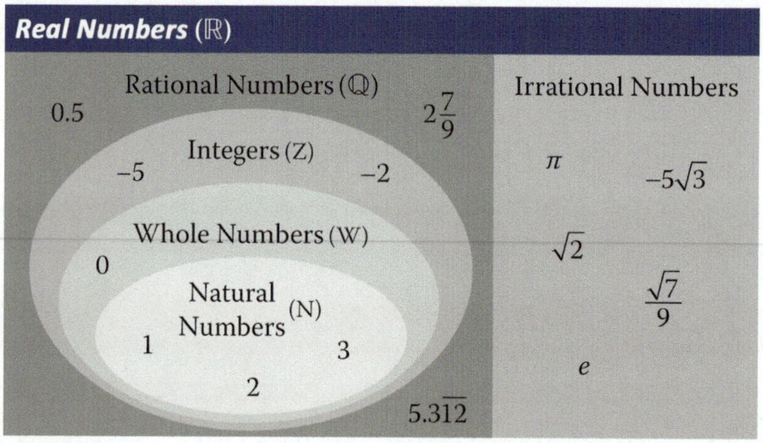

Figure P.1a

IMPORTANT

Every natural number is also a whole number, an integer, and a rational number. Every real number can be classified as either a rational or an irrational number.

Two nonoverlapping subsets of the real numbers are the **rational numbers** and the **irrational numbers**. A rational number is a ratio of integers a to b where $b \neq 0$. So, $\frac{2}{3}$, $-\frac{10}{7}$, and 5 (since 5 can be written as $\frac{5}{1}$) are all examples of rational numbers. Notice that all integers, whole numbers, and natural numbers are rational numbers. All rational numbers can be expressed as either a **terminating decimal** ($\frac{3}{4} = 0.75$) or as a **repeating decimal** ($\frac{16}{3} = 5.3\overline{3}$).

TIP

Irrational numbers and repeating decimals can be written as decimal approximations. Use the ≈ symbol to indicate a decimal approximation.
$\pi \approx 3.14$

Some decimal numbers do not terminate or repeat. Any real number that is equivalent to a nonterminating, nonrepeating decimal is an **irrational number**. For example, the decimal equivalent to the irrational number $\sqrt{2}$ is 1.4142135623730950488016887242097 . . . , where the decimal-place digits continue infinitely without repeating. The most familiar irrational number is π. Every real number (rational and irrational) can be expressed as a decimal that is either terminating, repeating, or nonterminating and nonrepeating.

Ordering Real Numbers

Real numbers can be represented (plotted) and ordered on a number line, where any number is greater than all real numbers to its left and less than all real numbers to its right (i.e., left to right is least to greatest). For example, 2, –4, and $\frac{3}{4}$ are plotted on the number line in Figure P.1b.

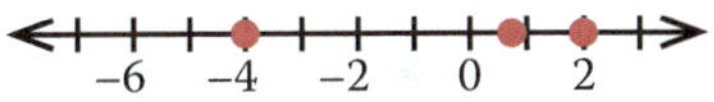

Figure P.1b

The fraction $\frac{3}{4}$ is plotted between 0 and 1.

IMPORTANT

For any real numbers a and b, if $a < b$, then a is to the left of b on a number line. If $a = b$, then the location of a must be the same as the location of b on a number line.

Clearly, $-4 < \frac{3}{4} < 2$, and this fact is confirmed by the order of the points on the number line, since the numbers are plotted in that order from left to right.

When ordering real numbers, it is often helpful to write fractions, mixed numbers, or irrational numbers as equivalent or approximate decimals, as demonstrated in **Example 1**.

EXAMPLE 1 Classifying and Ordering a List of Real Numbers

Consider the numbers –0.061, $\sqrt{3}$, 1.07, $\frac{4}{3}$, $\sqrt{9}$, and –1. Order the numbers from least to greatest, then classify each number by the subsets of the real numbers to which it belongs.

TIP

Write real numbers as decimals to plot them on a number line and to compare them. Use a calculator to find the decimal approximation of an irrational number.

SOLUTION

Write each nondecimal, noninteger number as a decimal equivalent or approximation.

$\sqrt{3} = 1.732\ldots \approx 1.73$ *Use a calculator and round to the nearest hundredth.*

$\frac{4}{3} = 1.3\overline{3} \approx 1.33$ *Divide 4 by 3 and round to the nearest hundredth.*

$\sqrt{9} = 3$ *9 is a perfect square.*

Use the decimal approximations to order the numbers.

$$-1 < -0.061 < 1.07 < 1.33 < 1.73 < 3$$

Therefore, $-1 < -0.061 < 1.07 < \frac{4}{3} < \sqrt{3} < \sqrt{9}$.

The numbers can be plotted on a number line.

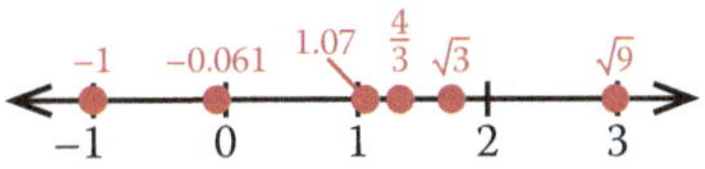

So, from least to greatest, the numbers are: –1, –0.061, 1.07, $\frac{4}{3}$, $\sqrt{3}$, and $\sqrt{9}$.

Use the chart to classify each number.

	Real	Rational	Integer	Whole	Natural	Irrational
–0.061	✓	✓				
$\sqrt{3}$	✓					✓
1.07	✓	✓				
$\frac{4}{3}$	✓	✓				
$\sqrt{9}$	✓	✓	✓	✓	✓	
–1	✓	✓	✓			

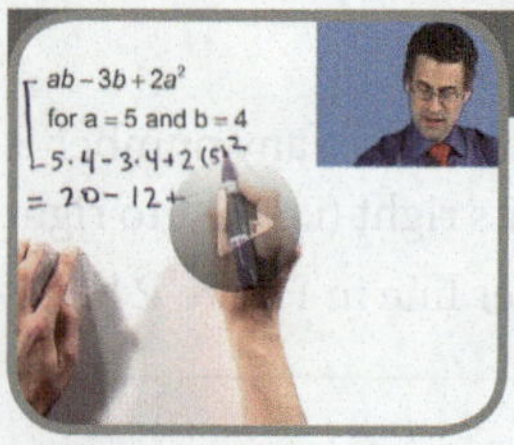

P.1.2 Evaluating an Algebraic Expression

An **algebraic expression** is a mathematical expression that includes at least one variable. An algebraic expression also typically includes at least one operation, such as multiplication or subtraction.

Examples of Algebraic Expressions	
x	*an unknown value represented by the letter x*
$2p + 1$	*the product of 2 and p increased by 1*
$\frac{m^2 - 1}{n}$	*m squared decreased by 1, divided by n*
$a(b + c)$	*the product of a and the sum of b and c*

IMPORTANT

A variable is an unknown value that is typically represented by a letter, such as x or n.

To **evaluate** an algebraic expression, substitute a given value for the variable and then simplify the expression by following the order of operations.

IMPORTANT

Order of Operations
1. Complete operations within grouping symbols (e.g., parentheses).
2. Simplify powers (i.e., exponents).
3. Multiply and divide from left to right.
4. Add and subtract from left to right.

EXAMPLE 2 Evaluating an Algebraic Expression

Evaluate each expression using the given value of each variable.

A. $\frac{2(k-1)}{10-k+2}$ for $k = 4$

B. $pq - 3p^2 + 5(q - p)$ for $p = -3$ and $q = 2$

SOLUTION

A. Substitute 4 for k in the expression. Then simplify the numeric expression by following the order of operations.

$$\frac{2(k-1)}{10-k+2} = \frac{2(4-1)}{10-4+2} = \frac{2(3)}{10-4+2} = \frac{6}{8} = \frac{3}{4}$$

B. Substitute −3 for p and 2 for q, then simplify.

Since the p-value is negative, use parentheses to prevent calculation errors.

$$pq - 3p^2 + 5(q - p) = (-3)(2) - 3(-3)^2 + 5(2 - (-3))$$ *Substitute p = −3 and q = 2.*

$$= (-3)(2) - 3(-3)^2 + 5(5)$$ *Subtract within parentheses.*

$$= (-3)(2) - 3(9) + 5(5)$$ *Simplify the power.*

$$= -6 - 27 + 25$$ *Multiply.*

$$= -8$$ *Add and subtract.*

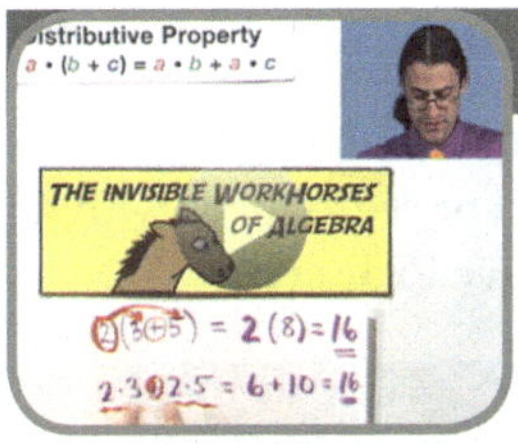

P.1.3 Using Properties of Real Numbers

Additive and Multiplicative Identities and Inverses

When an **identity** is combined (using some operation) with a quantity, the quantity does not change. The **additive identity** is 0 because adding 0 to any real number does not change the value of that number. Similarly, the **multiplicative identity** is 1 because multiplying any real number by 1 does not change the value of that number.

When an **inverse** is combined (using some operation) with some quantity, the result is the operation's identity. The **additive inverse** of any real number is its **opposite** (i.e., the differently signed number that is the same distance from 0 on a number line) because for any real number a, $a + (-a) = 0$ (the additive identity). For example, the additive inverse of 9 is -9 because $9 + (-9) = 0$. Every real number has an additive inverse.

The **multiplicative inverse** of a nonzero real number a is equal to $\frac{1}{a}$, since $a \cdot \frac{1}{a} = 1$, and 1 is the multiplicative identity. The multiplicative inverse of a real number is also known as the number's **reciprocal**. Every real number except 0 has a multiplicative inverse.

NOTICE THAT

A fraction is written as an equivalent fraction using the multiplicative identity. For example, $\frac{2}{3}$ is written as $\frac{10}{15}$ by multiplying by $\frac{5}{5}$ (which is equivalent to 1, the multiplicative identity).

$$\frac{2}{3} = \frac{2}{3} \cdot 1 = \frac{2}{3} \cdot \frac{5}{5} = \frac{10}{15}$$

Identities and Inverses of Real Numbers

	Additive		Multiplicative	
Identity	0	*If a is any real number, then* $a + 0 = a$.	1	*If a is any real number, then* $a \cdot 1 = a$.
Inverse of *a*	$-a$	*If a is any real number, then* $a + (-a) = 0$.	$\frac{1}{a}$ (when $a \neq 0$)	*If a is any real number, except 0, then* $a \cdot \frac{1}{a} = 1$

IMPORTANT

$-a$: the opposite of a

$\frac{1}{a}$: the reciprocal of a

EXAMPLE 3 Finding Inverses of Real Numbers

State the additive and multiplicative inverses of -20, $\frac{9}{10}$, 0.75, and $-\frac{1}{n}$. Assume that $n \neq 0$.

SOLUTION

The additive inverse of a number is its opposite.
The multiplicative inverse of a number is its reciprocal.

	Additive Inverse		Multiplicative Inverse	
-20	20	$-20 + 20 = 0$	$-\frac{1}{20}$	$-20 \cdot -\frac{1}{20} = 1$
$\frac{9}{10}$	$-\frac{9}{10}$	$\frac{9}{10} + \left(-\frac{9}{10}\right) = 0$	$\frac{10}{9}$	$\frac{9}{10} \cdot \frac{10}{9} = 1$
0.75	-0.75	$0.75 + -0.75 = 0$	$\frac{4}{3}$	$0.75 = \frac{3}{4}$ and $\frac{3}{4} \cdot \frac{4}{3} = 1$
$-\frac{1}{n}$	$\frac{1}{n}$	$-\frac{1}{n} + \frac{1}{n} = 0$	$-n$	$-\frac{1}{n} \cdot (-n) = 1$

The Commutative, Associative, and Distributive Properties

There are several properties which allow an expression to be manipulated and simplified without changing its value. Sums and products of real numbers can be rearranged or regrouped (without changing the value of the expression) by the **Commutative** and **Associative Properties.**

a, b, and c represent any real number.

Properties of Real Numbers: Commutative and Associative		
	Addition	**Multiplication**
Commutative Property	$a + b = b + a$	$ab = ba$
Associative Property	$(a + b) + c = a + (b + c)$	$(a \cdot b) \cdot c = a \cdot (b \cdot c)$

These properties are demonstrated in the following equations.

$5 + 4 = 4 + 5$ *Commutative Property of Addition*

$3 \cdot 4 = 4 \cdot 3$ *Commutative Property of Multiplication*

$(1 + 2) + 8 = 1 + (2 + 8)$ *Associative Property of Addition*

$(6 \cdot 2) \cdot 3 = 6 \cdot (2 \cdot 3)$ *Associative Property of Multiplication*

By the **Distributive Property**, a real number can be multiplied by each **number** in a sum or difference of real numbers without changing the value of the expression.

a, b, c, and d represent any real number.

Properties of Real Numbers: Distributive	
$a(b + c) = ab + ac$	$a(b - c) = ab - ac$

For example, consider the expression $3(2 + 5)$. Applying the order of operations to simplify the expression yields $3(2 + 5) = 3(7) = 21$. By the Distributive Property, multiplying 3 by each number in the sum first yields the same result: $3(2 + 5) = 3 \cdot 2 + 3 \cdot 5 = 6 + 15 = 21$.

EXAMPLE 4 Identifying Properties of Real Numbers

Identify the properties demonstrated in each equation.

IMPORTANT

The Commutative, Associative, and Distributive Properties can be applied to expressions that are algebraic as well as numeric.

A. $\frac{2}{3} + \left(\frac{1}{8} + \frac{1}{3}\right) = \left(\frac{2}{3} + \frac{1}{3}\right) + \frac{1}{8}$ **B.** $2(x + y) = 2y + 2x$

SOLUTION

A. The numbers are reordered, as well as regrouped.

$\frac{2}{3} + \left(\frac{1}{8} + \frac{1}{3}\right) = \frac{2}{3} + \left(\frac{1}{3} + \frac{1}{8}\right)$ *Reorder the fractions within parentheses by the Commutative Property of Addition.*

$= \left(\frac{2}{3} + \frac{1}{3}\right) + \frac{1}{8}$ *Regroup the fractions by the Associative Property of Addition.*

The Commutative and Associative Properties of Addition are demonstrated in the equation.

IMPORTANT

Assume that variables in algebraic expressions represent real numbers unless otherwise stated.

B. By the Distributive Property, 2 can be multiplied by each term in the sum $x + y$.

$2(x + y) = 2x + 2y$ *Multiply 2 by each term in the sum by the Distributive Property.*

$= 2y + 2x$ *Reorder the terms by the Commutative Property of Addition.*

Alternatively, the Commutative Property of Addition can be applied first.

$2(x + y) = 2(y + x)$ *Reorder the terms by the Commutative Property of Addition.*

$= 2y + 2x$ *Multiply 2 by each term in the sum by the Distributive Property.*

The Distributive Property and the Commutative Property of Addition are demonstrated in the equation.

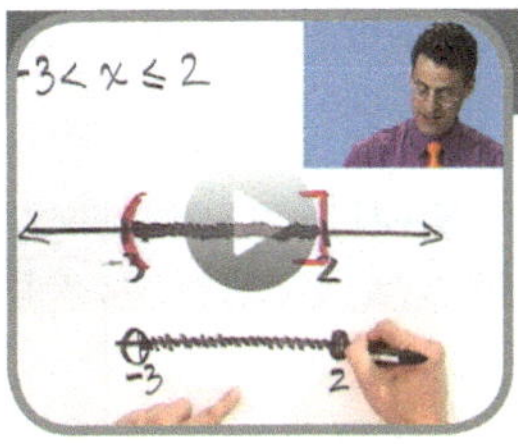

P.1.4 Inequalities and Interval Notation

An inequality expresses an ordered relationship between two quantities, such as between two numbers, two variables, or between a variable and a number, as in $x < 6$, $n \geq 0$, or $1 < z \leq 5$. The **solution set** for each of these inequalities (i.e., the set of all real numbers that are solutions to the inequality) is a subset of the real numbers called an **interval**.

DEFINITION

An **interval** is the set of all real numbers between two real numbers a and b, where a and b are called the interval's endpoints.

An interval has two endpoints, but those endpoints are not necessarily included in the interval. For example, the solution set of the inequality $-1 < x \leq 3$ includes all real x-values that are greater than 1 and less than or equal to 3. Therefore, the solution set is the interval with endpoints -1 and 3, where 3 is included in the interval (because x is less than *or equal to* 3), but -1 is not included in the interval (because x is greater than *but not equal to* -1).

Consider the inequality $x \geq 2$. The solution set includes all real x-values greater than or equal to 2, which is the interval with an endpoint at 2 that extends infinitely from 2 in a positive direction. When an interval extends infinitely in a positive direction, the endpoint is said to be positive infinity (∞). When an interval extends infinitely in a negative direction, the endpoint is said to be negative infinity ($-\infty$). Note that ∞ and $-\infty$ are never included in the interval.

A **closed interval** includes both endpoints and an **open interval** includes neither endpoint. (An interval that includes exactly one endpoint is neither open nor closed.) Examples of the different types of intervals are given in the following table.

Interval Endpoints and Inequalities

Inequality	Solution Set Description	Endpoints of Interval	Type of Interval	
$x < 6$	all real numbers less than 6	$-\infty$ and 6	open	*The interval includes neither endpoint.*
$n \geq 0$	all real numbers greater than or equal to 0	0 and ∞	neither open nor closed	*The interval includes only one endpoint.*
$1 \leq z \leq 5$	all real numbers greater than or equal to 1 and less than or equal to 5	1 and 5	closed	*The interval includes both endpoints.*

Graphs of Intervals and Interval Notation

An interval can be expressed as a graph or by using **interval notation**. To express an interval using interval notation, write the endpoints separated by a comma and enclose the endpoints with parentheses, brackets, or one of each. A parenthesis indicates that the endpoint is not included in the interval, and a bracket indicates that the endpoint is included in the interval. A parenthesis is always used with $-\infty$ and ∞, because $-\infty$ and ∞ are never included in the interval.

Alternative Method
On a number line, a bracket can be used instead of a closed point, and a parenthesis can be used instead of an open point.

To graph an interval, plot the interval's numeric endpoint(s) as points on a number line. Use an open point (empty circle) when plotting an endpoint that is not included in the interval, and a closed point (filled circle) when plotting an endpoint that is included in the interval. Shade the region between the endpoints. Shade the number line to the right end to indicate that the interval extends infinitely in the positive direction (∞), or to the left end to indicate that the interval extends infinitely in the negative direction ($-\infty$).

All possible notations for a single interval are listed in the following table, where the endpoints a and b represent real numbers such that $a < b$.

Interval Notation, Inequalities, and Graphs of Intervals

	Notation	Inequality	Graph
The interval includes b.	$(-\infty, b]$	$x \le b$	
The open interval does not include b.	$(-\infty, b)$	$x < b$	
The interval includes a.	$[a, \infty)$	$x \ge a$	
The open interval does not include a.	(a, ∞)	$x > a$	
The open interval does not include a or b.	(a, b)	$a < x < b$	
The closed interval includes a and b.	$[a, b]$	$a \le x \le b$	
The interval includes a, but not b.	$[a, b)$	$a \le x < b$	
The interval includes b, but not a.	$(a, b]$	$a < x \le b$	
The interval includes all real numbers.	$(-\infty, \infty)$	$\mathbb{R}$	

EXAMPLE 5 Inequalities and Interval Notation

Express the solution set for each inequality using interval notation.

A. $x < 10$ **B.** $-5 < x \le 0$

SOLUTION

IMPORTANT

Brackets are never used with $-\infty$ or ∞.

A. $(-\infty, 10)$ *The interval includes all real numbers less than 10 (10 is not included).*

Note that 10 is followed by a parenthesis since 10 is *not* included in the interval.

B. $(-5, 0]$ *The interval includes all real numbers between −5 and 0, not including −5.*

Note that a parenthesis is used before −5 to indicate that −5 is *not* included in the interval, and 0 is followed by a bracket to indicate that 0 *is* included in the interval.

EXAMPLE 6 Inequalities and Interval Notation

Write an inequality for each solution set.

A. $[0.15, \infty)$ **B.** $\left(-\frac{1}{3}, 0\right]$

SOLUTION

A. $x \ge 0.15$ *The interval includes all real numbers greater than or equal to 0.15.*

B. $-\frac{1}{3} < x \le 0$ *The interval includes all real numbers between $-\frac{1}{3}$ and 0, not including $-\frac{1}{3}$.*

P.1.5 Interval Notation: Another Example

Two or more intervals can be combined by writing the union symbol, ∪, between the intervals.

EXAMPLE 7 Writing Intervals Represented by Graphs

Write the interval described by each graph using interval notation.

A.

B.

SOLUTION

A. $[0, 3)$ *The interval includes all real numbers between 0 and 3, not including 3.*

B. $(-\infty, -4) \cup [-2, 3]$ *The interval includes all real numbers less than –4 and all real numbers between –2 and 3 inclusive.*

IMPORTANT

"All real numbers between –2 and 3 inclusive" means that the endpoints, –2 and 3, are included in the interval.

WHAT'S THE BIG IDEA?

Explain how the endpoints of an interval are expressed on a number line and in interval notation.

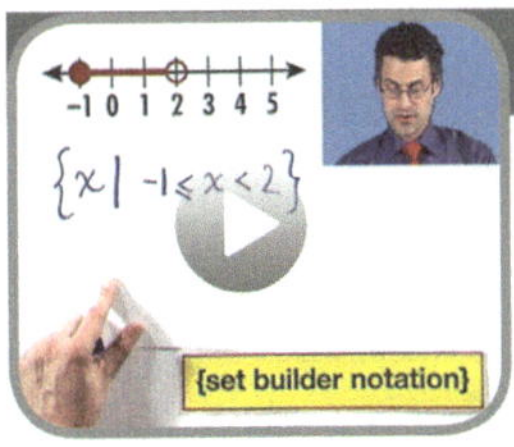

P.1.6 Set Notation

Another notation used to express an interval is **set-builder notation**. For the interval (1, 3), the related set-builder notation is $\{x|\ 1 < x < 3\}$, which is read as "the set of all real numbers x such that x is greater than 1 and less than 3." Braces (curly brackets) are always used with set-builder notation.

EXAMPLE 8 Representing Intervals Using Set-Builder Notation

Write the interval using set-builder notation.

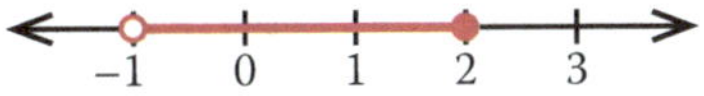

SOLUTION

The solution set of the inequality $-1 < x \le 2$ is represented on the graph.
Use this inequality to write the set-builder notation.

$$\{x|\ -1 < x \le 2\}$$

The set contains all real numbers x such that x is greater than –1 and less than or equal to 2.

A set containing distinct values (not an interval of values) can be represented using **roster notation**. Roster notation shows the set's **elements** (the items in the set) listed between braces, such as {2, 4, 6}. An ellipsis (" . . . ") can be used to indicate that the set continues infinitely in the same manner. For example, {2, 4, 6, 8, . . . } is the set of all positive multiples of 2. Additionally, an ellipsis can be used to indicate that the set continues in the same manner between two values. For example, {1, 2, 3, . . . , 10} is the set of natural number from 1 to 10.

Natural Numbers ▶
$\{1, 2, 3, 4, 5, \ldots\}$

Whole Numbers
$\{0, 1, 2, 3, 4, \ldots\}$

Integers
$\{\ldots, -2, -1, 0, 1, 2, \ldots\}$

Since the sets of natural numbers, whole numbers, and integers are sets of distinct values, roster notation can be used to represent these sets. The following table shows examples of sets in roster notation.

Examples of Sets in Roster Notation	
$\{1, 2, 3, 4\}$	*The set includes only 1, 2, 3, and 4.*
$\{2, 3, 4, \ldots\}$	*The set includes all whole numbers greater than or equal to 2.*
$\{3, 6, 9, \ldots\}$	*The set includes all positive multiples of 3.*
$\{-3, -2, -1, \ldots, 5\}$	*The set includes all integers between −3 and 5.*

EXAMPLE 9 Representing a Set of Numbers Using Roster Notation

Write the set using roster notation: "all positive integers evenly divisible by 5."

SOLUTION

The set of all positive integers that are evenly divisible by 5 begins with 5 and continues infinitely in a positive direction, including all multiples of 5. List at least the first three elements of the set, and then add an ellipsis to indicate that the set continues infinitely.

$$\{5, 10, 15, 20, \ldots\}$$

WHAT'S THE BIG IDEA?

Roster notation cannot be used to express the solution set for $x > 0$. Explain.

P.1.7 Evaluating Absolute Value Expressions

The distance between a real number, a, and 0 on a number line is called the **absolute value** of a, denoted $|a|$. Since distance is always nonnegative, the absolute value of a number is always nonnegative, and so $|a| \geq 0$.

The following properties can be applied to the absolute value for any real number.

Properties of Absolute Value	
$\lvert a\rvert \geq 0$	*The absolute value of a number is always positive or 0 (i.e., nonnegative).*
$\lvert a\rvert = \lvert -a\rvert$	*A number and its opposite have the same absolute value.*
$\lvert ab\rvert = \lvert a\rvert \cdot \lvert b\rvert$	*The absolute value of a product is equal to the product of the absolute values of the factors.*
$\left\lvert \frac{a}{b} \right\rvert = \frac{\lvert a\rvert}{\lvert b\rvert}$ (when $b \neq 0$)	*The absolute value of a quotient is equal to the quotient of the absolute values of the numerator and denominator.*

Use these properties and follow the order of operations to simplify a numeric expression containing absolute value. Absolute value behaves like parentheses (grouping symbols) in the order of operations, so simplify any operations within absolute value symbols and then take the absolute value of the number. Additionally, when a number is written immediately before (or after) an absolute value, the operation is understood to be multiplication. For example, $a|b|$ means "the product of a and the absolute value of b."

EXAMPLE 10 Simplifying a Numeric Absolute Value Expression

Simplify. $10 + 5|7 - 12| - |3^2|$

SOLUTION

Follow the order of operations.

$10 + 5\|7 - 12\| - \|3^2\| = 10 + 5\|-5\| - \|9\|$	*Simplify expressions within the absolute value.*
$= 10 + 5(5) - 9$	*Take the absolute value of −5 and 9.*
$= 10 + 25 - 9$	*Multiply.*
$= 35 - 9$	*Add.*
$= 26$	*Subtract.*

Simplifying the Absolute Value of an Algebraic Expression

Consider the relationship between the absolute value of a number and the number itself. For any real number a that is

- greater than or equal to 0, the absolute value of a is equal to a.
- less than 0, the absolute value of a is equal to the opposite of a, or $-a$.

These facts are represented symbolically in the following table.

Absolute Value	
For all real numbers a, $\|a\| = a$ when $a \geq 0$.	For all real numbers a, $\|a\| = -a$ when $a < 0$.
The absolute value of a is equal to a when a is greater than or equal to 0 (nonnegative).	*The absolute value of a is equal to the opposite of a (i.e., −a) when a is negative.*

CAUTION

The statement "$|a| = -a$ when $a < 0$" does not mean that the absolute value of a number is equal to a negative number. Instead, the statement means that when a number is negative, the absolute value of that number is equal to the opposite of that number.

These two facts can be used to write an algebraic absolute value expression (an expression that contains the absolute value of at least one variable or number) as an equivalent expression that does not include absolute value. Specifically, when $a < 0$, any instance of $|a|$ in an expression can be replaced by $-a$, and when $a \geq 0$, any instance of $|a|$ in an expression can be replaced by a.

EXAMPLE 11 Simplifying an Algebraic Absolute Value Expression

Simplify $8n - |n|$ when $n \geq 0$, and when $n < 0$.

SOLUTION

If $n \geq 0$, then $|n| = n$. So, substitute n for $|n|$ and simplify.

$$8n - |n| = 8n - n = 7n, \text{ when } n \geq 0.$$

If $n < 0$, then $|n| = -n$. So, substitute $-n$ for $|n|$ and simplify.

$$8n - |n| = 8n - (-n) = 8n + n = 9n, \text{ when } n < 0.$$

Therefore, when $n \geq 0$, $8n - |n| = 7n$, and when $n < 0$, $8n - |n| = 9n$.

WHAT'S THE BIG IDEA?

Can the expression $x - (3|x|)^2$ be simplified if x is any real number? Explain.
Can the expression $x - (3 + |x|)^2$ be simplified if x is any real number? Explain.

The Distance between Two Points on a Number Line

Recall that the absolute value of a number is equal to the distance between that number and 0 on a number line. So, absolute value can be used to find the distance between a number and 0, or to find the distance between two numbers, or between two points on a number line.

TIP

The distance between a and b on a number line is equal to either $|b - a|$ or $|a - b|$, since the distance between a and b is equal to the distance between b and a.

Distance between Two Points on a Number Line

For all real numbers a and b, the distance between a and b on a number line is given by $|b - a|$.

EXAMPLE 12 Finding the Distance between Two Points on a Number Line

Find the distance between p and q if $p = 10$ and $q = -7$.

SOLUTION

The distance between 10 and –7 is given by $|10 - (-7)|$. Simplify.

$$|10 - (-7)| = |10 + 7| = |17| = 17$$

Alternatively, the distance between 10 and –7 is also given by $|-7 - 10|$. Simplify.

$$|-7 - 10| = |-17| = 17$$

SECTION P.1 EXERCISES

Warm Up

1. Simplify. $-\frac{5}{8}-\left(-\frac{1}{6}\right)$
2. Simplify. $7k-4+k^2+k-7k^2$
3. Describe the solution set for $-10<x\leq 3$.

Just the Facts

Fill in the blanks.

4. Any real number is in the set of either rational numbers or ____.
5. If 0 is removed from the ____, the resulting set is the same as the natural numbers.
6. ____ numbers can be written as $\frac{a}{b}$, where a and b are ____, and b ____.
7. The multiplicative inverse of ____ is −0.01.
8. The difference between the intervals $(a, b]$ and $[a, b)$ is that $(a, b]$ ____, but $[a, b)$ ____.
9. The number line showing the interval [4, 5) has a point at 4 that is ____ and a point at 5 that is ____.
10. The absolute value of a number is equal to its ____ from ____ on a number line.

Essential Skills

In Exercises 1–4, order the numbers from least to greatest.

1. 11/3, 2.09, $\sqrt{10}$, 3.72, and −3
2. $\sqrt{7}$, 2.41, 7/3, 0.7, and 0.07
3. −4.1, 12/5, −4.11, 1/4, and $\sqrt{13}$
4. 2.83, $\sqrt{2}$, 0.29, 1, and 7/2

In Exercises 5–10, classify each number by the subsets of the real numbers to which it belongs.

5. 0
6. $\sqrt{5}$
7. 15.333 . . .
8. −3
9. $\sqrt{49}$
10. $\frac{9}{4}$

In Exercises 11–18, evaluate each expression.

11. $10+2x^2$ for $x=5$
12. $\frac{9(n+3)}{3-n}$ for $n=-6$
13. $7-6(2-y)+2y+5y^2$ for $y=-2$
14. $1+2p-3(p+6)^2$ for $p=-8$
15. $\frac{-w}{z+w}-zw^2$ for $w=-1$ and $z=-3$
16. $4xy-3x^2+2(1-y)+y^2$ for $x=5$ and $y=-3$
17. $-m+4(n-m)+mn^2$ for $m=-4$ and $n=6$
18. $\frac{-2r+3s}{r}+r^2s-2$ for $r=-1$ and $s=3$

In Exercises 19–22, find the additive inverse and multiplicative inverse of each number.

19. 3/4
20. −7
21. −8.1
22. −7/6

In Exercises 23–26, identify the property demonstrated in each equation.

23. $(p+(q-r))+st=p+((q-r)+st)$
24. $5(3a+2)=15a+10$
25. $-\frac{1}{k}\cdot -k=1$
26. $(x)(3)=3x$

In Exercises 27–34, express the solution set for each inequality using interval notation.

27. $x>0$
28. $x\leq -4$
29. $10\geq x$
30. $3>x$
31. $0<x\leq 3.2$
32. $-4\leq x<-1$
33. $7<x\leq 20$
34. $2\leq x\leq 11$

In Exercises 35–42, write an inequality for each solution set.

35. $(-\infty, 4]$

36. $\left(-\dfrac{2}{3}, \infty\right)$

37. $[-1.5, \infty)$

38. $(-\infty, -7)$

39. $[5, 10]$

40. $(-0.7, 0]$

41. $[-3.8, 12)$

42. $(2, 5)$

In Exercises 43–48, write the solution set displayed on each number line using the indicated notation(s).

43. 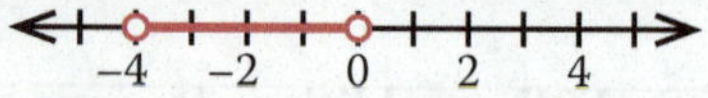

inequality and interval notation

44. 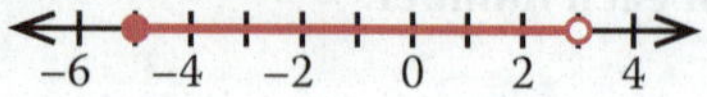

inequality and interval notation

45. 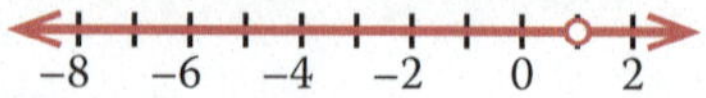

interval notation

46. 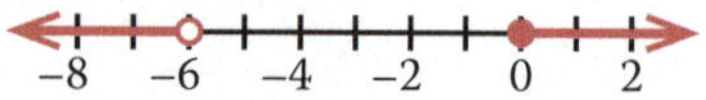

interval notation

47. 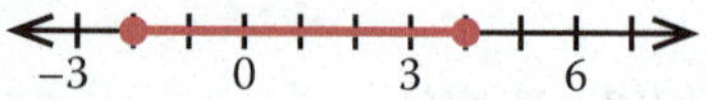

inequality, interval notation, and set-builder notation

48. 

set-builder notation

In Exercises 49–52, write each set using roster notation.

49. all positive integers

50. all positive multiples of 3

51. all even natural numbers

52. all positive powers of 0.1

In Exercises 53–56, simplify.

53. $|1-7| - |7-1|$

54. $2|-3+4| - |-8|$

55. $-4|6-9| + 3|-5|$

56. $|-12| - 3|14-18|$

In Exercises 57–60, simplify each expression for when $x \geq 0$ and when $x < 0$.

57. $-|x|$

58. $|x| + 2x$

59. $-4x + |x|$

60. $-7|x| - 3x$

In Exercises 61–64, find the distance between m and n.

61. $m = 12$ and $n = -3$

62. $m = -7$ and $n = 4$

63. $m = -3/4$ and $n = 5$

64. $m = -15$ and $n = -3/5$

Extensions

In Exercises 65–70, give three numbers that fit each description, if possible. If there are fewer than three possible numbers, list all possibilities.

65. a rational number that is not an integer

66. a natural number that is not an integer

67. an integer that is not a natural number

68. an integer that is also a whole number and a natural number

69. a whole number that is not a natural number

70. a real number that is not a rational or irrational number

71. A student was asked to evaluate the expression $\dfrac{15-2(n+3)+p}{p+m-n} + np$ for $n = -3$, $m = 1$, and $p = 5$. They completed the first step in the process (the substitution) as shown below. Correct any errors.

$$\frac{15-2(-3+3)+5}{5+1-3} + (-3)(5)$$

In Exercises 72–77, use the following information.

If A and B are sets, then the notation $A \cup B$ (the union of A and B) denotes the set of all elements from A combined with all elements from B. The notation $A \cap B$ (the intersection of A and B) denotes the set of the elements that A and B have in common. When A and B have no elements in common, then their intersection is the empty set, denoted $A \cap B = \emptyset$. Using interval notation, write $A \cup B$ and $A \cap B$ for each pair of intervals.

72. $A = (0, 2]$, $B = [1, 9)$

73. $A = [-5, 0]$, $B = [-1, \infty)$

74. $A = (-\infty, 2]$, $B = [1, 9)$

75. $A = (-\infty, -4)$, $B = (-3, \infty)$

76. $A = (-\infty, 0)$, $B = (-5, 1)$

77. $A = (-\infty, 10]$, $B = [10, \infty)$

In Exercises 78–81, use the properties of absolute value to determine whether each statement is true or false for all real values of *x* and *y*.

78. $|xy| = |(-x)(-y)|$

79. $|xy| \geq 0$

80. $|x + y| = |x| + |y|$

81. $|x| - |x| = 0$

82. Suppose line segment $\overline{AB}$ has endpoints at $A(-5, 0)$ and $B(7, 0)$, and line segment $\overline{CD}$ has endpoints at $C(3, 10)$ and $D(3, -4)$. Is the length of line segment $\overline{AB}$ equal to the length of line segment $\overline{CD}$? Explain.

83. Suppose parallelogram $ABCD$ has vertices at $A(-6, 7)$, $B(5, 7)$, $C(5, -4)$, and $D(-6, -4)$. Is parallelogram $ABCD$ a rhombus? Explain.

P.2 INTEGER EXPONENTS

OBJECTIVES

- Use the properties of exponents to evaluate and simplify expressions.
- Use scientific notation to represent real numbers.
- Perform operations with numbers written in scientific notation.

PREREQUISITE VOCABULARY TERMS

algebraic expression
integer
real number

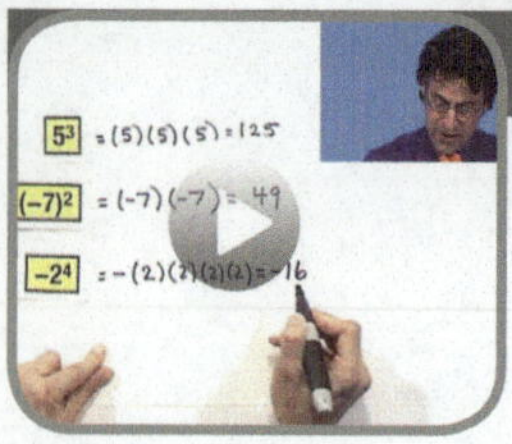

P.2.1 An Introduction to Exponents

Exponents provide a shorthand for writing repeated multiplication. For example, $3 \cdot 3 \cdot 3 \cdot 3$ can be written as 3^4.

DEFINITION

If a is a nonzero real number and n is an integer, then $a^n = \overbrace{a \cdot a \cdot a \cdot \ldots \cdot a}^{n \text{ times}}$, where a is a factor n times. In the exponential expression a^n (called a **power**), a is the **base**, and n is the **exponent**.

TIP

The power a^n is read as "a to the nth power" or "a to the power of n." The exponent n is also sometimes called a power.

Exponential notation can be used for repeated multiplication of a number or an expression. Examples of powers and their equivalent expanded forms are shown in the following table.

	Exponential Form	Expanded Form
The base is 5.	5^4	$(5)(5)(5)(5)$
The base is xy.	$(xy)^5$	$(xy)(xy)(xy)(xy)(xy)$
The base is y.	xy^5	$x(y)(y)(y)(y)(y)$
The base is −2.	$(-2)^3$	$(-2)(-2)(-2)$
The base is 2.	-2^3	$-(2)(2)(2)$

TIP

The base of -2^n is 2, not −2.
$-2^n = (-1)2^n$

When simplifying expressions with exponents, pay close attention when identifying the base of the power. Be sure that only the power's base is used as the repeated factor. The base must be enclosed within parentheses except when the base is a single variable or positive number.

EXAMPLE 1 Simplifying a Power

Simplify each expression.

A. $(-10)^3$

B. -15^2

TIP

An odd power of a negative base is always negative, and an even power of a negative base is always positive.

SOLUTION

A. The base −10 is a factor 3 times.
$(-10)^3 = (-10)(-10)(-10) = -1000$

B. The base 15 is a factor 2 times.
$-15^2 = -(15)(15) = -225$

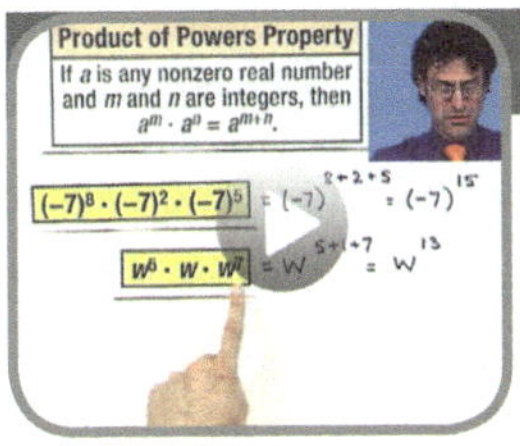

P.2.2 Product of Powers Property

There are several properties of powers (also called properties of exponents) that allow expressions with powers to be simplified. The first of these concerns products of powers, where each power has the same base. Consider the product of 4^2 and 4^3. Simplify the product by expanding and then writing repeated multiplication as a power.

$$4^2 \cdot 4^3 = (4)(4) \cdot (4)(4)(4) \quad \textit{4 is a factor a total of 5 times.}$$
$$= 4^5 \quad \textit{Use exponential notation to write as a single power.}$$

Notice that the exponent in 4^5 is equal to the *sum* of the exponents in the equivalent expression $4^2 \cdot 4^3$. This is not a coincidence. A *product* of powers with *like bases* can be written as a single power using the like base and the sum of the exponents by the **Product of Powers Property**.

IMPORTANT

*The Product of Powers Property can only be applied when powers **with like bases** are **multiplied**.*

Product of Powers Property
If a is a nonzero real number and m and n are integers, then $a^m \cdot a^n = a^{m+n}$.

The Product of Powers Property can be applied to products of powers where the bases are nonzero real numbers or algebraic expressions. When an exponential expression contains a variable as its base, assume the base represents a nonzero real number.

EXAMPLE 2 Using the Product of Powers Property

Simplify each expression.

A. k^3k^{10} **B.** $t^2 + t^4$ **C.** b^2b^9b **D.** $x^2y^3z^3$

SOLUTION

Product of Powers Property $a^m \cdot a^n = a^{m+n}$

A. The powers in the product have like bases.

$$k^3k^{10} = k^{3+10} = k^{13}$$

B. The expression is simplified.

C. The powers in the product have like bases.

$$b^2b^9b = b^2b^9b^1 = b^{2+9+1} = b^{12}$$

The exponent for the factor b is 1, since $b = b^1$.

D. The expression is simplified.

TIP

A nonzero number or expression can be written as a power with an exponent equal to 1.

$x = x^1$

To simplify expressions such as $5x^3(2x^7y^4)$, use the Commutative Property of Multiplication to group the coefficients and the powers with like bases: $5x^3(2x^7y^4) = (5 \cdot 2)(x^3 \cdot x^7)(y^4)$. Then multiply the coefficients and use the Product of Powers Property: $(5 \cdot 2)(x^3 \cdot x^7)(y^4) = 10x^{10}y^4$.

EXAMPLE 3 Using the Product of Powers Property

Simplify each expression.

A. $n^4p(5n^2p^4)$ **B.** $(-12a^5bc^2)(3b^5c)$

SOLUTION

Group the coefficients and the powers with like bases, then use the Product of Powers Property.

A. $n^4p(5n^2p^4) = (5)(n^4n^2)(pp^4) = 5n^6p^5$

B. $(-12a^5bc^2)(3b^5c) = (-12 \cdot 3)(a^5)(bb^5)(c^2c) = -36a^5b^6c^3$

TIP

Group the coefficients first and then the variables (in alphabetical order).

P.2.3 Power of a Power Property

The next property concerns a power of a power, or, in other words, a power where the base is itself a power, as in $(4^2)^3$. In this expression, notice that the base of the power with exponent 3 is 4^2. Simplify the power by expanding and then writing repeated multiplication as a power.

$$(4^2)^3 = 4^2 \cdot 4^2 \cdot 4^2 \qquad \textit{4}^2 \textit{ is a factor 3 times.}$$
$$= (4)(4) \cdot (4)(4) \cdot (4)(4) \qquad \textit{4 is a factor a total of 6 times.}$$
$$= 4^6 \qquad \textit{Use exponential notation to write as a single power.}$$

Notice that the exponent in 4^6 is equal to the *product* of the exponents in the equivalent expression $(4^2)^3$. Again, this is not a coincidence. A power of a power can be written as a single power using that base and the *product* of the exponents by the **Power of a Power Property**.

TIP

The expression $(a^m)^n$ is read as "a to the m to the nth power" or as "a to the m to the power of n."

Power of a Power Property

If a is a nonzero real number and m and n are integers, then $(a^m)^n = a^{m \cdot n}$.

As with the Product of Powers Property, the Power of a Power Property can be applied when the base of a power is a nonzero real number or an algebraic expression.

EXAMPLE 4 Using the Power of a Power Property to Simplify an Expression

Simplify each expression.

A. $(y^5)^4$

B. $(a^3)^2(a^4)^6$

SOLUTION

Each expression includes a power of a power, so use the Power of a Power Property to simplify.

Power of a Power Property $(a^m)^n = a^{m \cdot n}$

A. $(y^5)^4 = y^{5 \cdot 4} = y^{20}$

B. $(a^3)^2(a^4)^6 = (a^{3 \cdot 2})(a^{4 \cdot 6}) = (a^6)(a^{24}) = a^{6+24} = a^{30}$

Use the Product of Powers Property to simplify $(a^6)(a^{24})$.

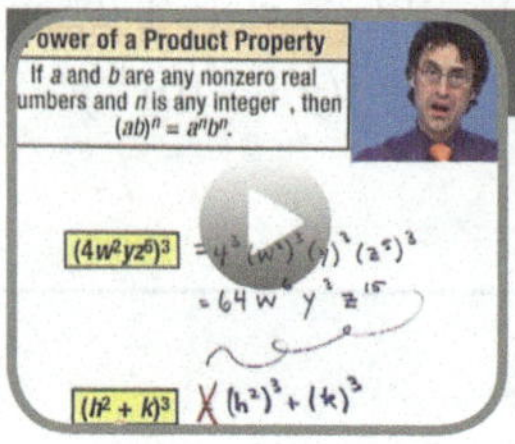

P.2.4 Power of a Product Property

Another property of powers can be applied when the base of a power is a product (which can include numbers, variables, powers, or algebraic expressions). Consider the power $(2 \cdot 7^4 \cdot 5^2)^3$. Notice that the base is the product of 2, 7^4, and 5^2. Simplify the power by expanding and then writing repeated multiplication using exponents.

$$(2 \cdot 7^4 \cdot 5^2)^3 = (2 \cdot 7^4 \cdot 5^2) \cdot (2 \cdot 7^4 \cdot 5^2) \cdot (2 \cdot 7^4 \cdot 5^2) \qquad \textit{(2} \cdot 7^4 \cdot 5^2\textit{) is a factor 3 times.}$$
$$= (2)(2)(2) \cdot (7^4)(7^4)(7^4) \cdot (5^2)(5^2)(5^2) \qquad \textit{Commutative Property of Multiplication}$$
$$= (2)^3 \cdot (7^4)^3 \cdot (5^2)^3 \qquad \textit{Use exponential notation.}$$

So, $(2 \cdot 7^4 \cdot 5^2)^3$ can be written equivalently as $(2)^3 \cdot (7^4)^3 \cdot (5^2)^3$. Notice that in $(2)^3 \cdot (7^4)^3 \cdot (5^2)^3$, each of the original base's factors, 2, 7^4, and 5^2, is the base of a power where the exponent is 3 (the original exponent). This is an example of the **Power of a Product Property**, which states that a power of a product can be written as a product of powers.

TIP

It is easy to confuse the Power of a Product Property with the Product of Powers Property. When considering the properties, keep in mind that the Power of a Product Property is used when the base is a product.

Power of a Product Property

If a is a nonzero real number and n is an integer, then $(ab)^n = a^n b^n$.

Note that when the product in the power of a product contains factors that are powers, the Power of a Power Property will be applied to fully simplify the expression.

EXAMPLE 5 Using the Power of a Product Property to Simplify an Expression

Simplify. $(-3x^2yz^5)^4$

SOLUTION

Power of a Product Property $(ab)^n = a^n b^n$

The expression is a power in which the base is a product, so use the Power of a Product Property to simplify.

$$(-3x^2yz^5)^4 = (-3)^4(x^2)^4(y)^4(z^5)^4 \quad \text{Power of a Product Property}$$

$$= (-3)^4(x^8)(y^4)(z^{20}) \quad \text{Power of a Power Property}$$

$$= 81x^8y^4z^{20} \quad \text{Simplify.}$$

WHAT'S THE BIG IDEA?

Write an expression that can be simplified by using the Product of Powers, Power of a Power, and Power of a Product Properties.

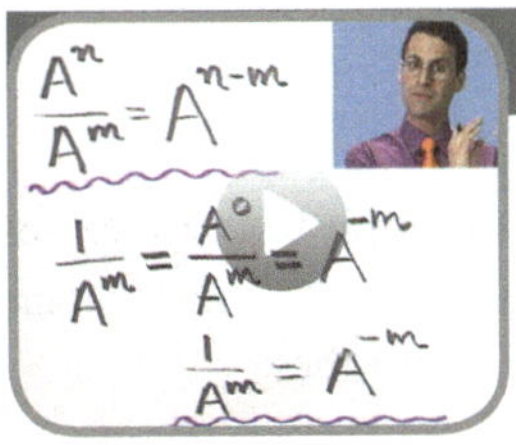

P.2.5 Quotient Properties and Negative Exponents

As with products of powers, quotients of powers can be simplified when the powers have like bases. Consider the fraction $\frac{3^6}{3^2}$. Simplify by expanding the power in the numerator and denominator and then writing repeated multiplication as a power.

$$\frac{3^6}{3^2} = \frac{(3)(3)(3)(3)(3)(3)}{(3)(3)} \quad \text{Expand the powers.}$$

$$= (3)(3)(3)(3) \quad \text{Remove the common factors.}$$

$$= 3^4 \quad \text{Use exponential notation to write as a single power.}$$

Notice that the exponent in 3^4 is equal to the *difference* of the exponents in $\frac{3^6}{3^2}$. A *quotient* of powers with *like bases* can be written as a single power using that like base and the difference of the exponents by the **Quotient of Powers Property**.

IMPORTANT

*The Quotient of Powers Property can only be applied when powers **with like bases** are **divided**.*

Quotient of Powers Property

If a is a nonzero real number and m and n are integers, then $\frac{a^m}{a^n} = a^{m-n}$.

Two additional properties are consequences of the Quotient of Powers Property. First, consider the case of a quotient of powers with like bases where the exponent in the numerator is *less than* the exponent in the denominator, such as $\frac{3^2}{3^6}$. Expanding, removing common factors, and writing in exponential notation gives the result $\frac{1}{3^4}$. However, notice that applying the Quotient of Powers Property yields $\frac{3^2}{3^6} = 3^{2-6} = 3^{-4}$. Therefore, $3^{-4} = \frac{1}{3^4}$, which is an example of the **Negative Exponent Property.**

Negative Exponent Property
If a is a nonzero real number and m is an integer, then $a^{-m} = \frac{1}{a^m}$.

Now consider a quotient of powers with like bases where the exponents in the numerator and denominator are *equal*, such as $\frac{3^6}{3^6}$. Expanding, removing common factors, and writing in exponential notation gives the result 1 (since the numerator and denominator are equal). However, notice that applying the Quotient of Powers Property yields $\frac{3^6}{3^6} = 3^{6-6} = 3^0$. Therefore, $3^0 = 1$, which is an example of the **Zero Exponent Property.**

Zero Exponent Property
If a is a nonzero real number, then $a^0 = 1$.

The last property of powers introduced in this section is the **Power of a Quotient Property**, which states that the power of a quotient can be written as a quotient of powers.

Power of a Quotient Property
If a and b are nonzero real numbers and n is an integer, then $\left(\frac{a}{b}\right)^n = \frac{a^n}{b^n}$.

The properties of powers are summarized in the following table. Note that all apply to exponential expressions where the base is a nonzero real number or an algebraic expression. Assume that an algebraic expression used as the base of a power represents a nonzero real number.

IMPORTANT

In each of these properties, a and b are nonzero real numbers, and m and n are integers.

Properties of Powers	
Product of Powers	$a^m a^n = a^{m+n}$
Power of a Power	$(a^m)^n = a^{m \cdot n}$
Power of a Product	$(ab)^n = a^n b^n$
Quotient of Powers	$\frac{a^m}{a^n} = a^{m-n}$
Power of a Quotient	$\left(\frac{a}{b}\right)^n = \frac{a^n}{b^n}$
Negative Exponent	$a^{-m} = \frac{1}{a^m}$
Zero Exponent	$a^0 = 1$

EXAMPLE 6 Using Properties of Powers to Simplify Expressions

Simplify. Write each expression using only positive exponents.

A. $\frac{b^9}{b^3}$ B. $\frac{x^2}{x^{10}}$

SOLUTION

Each expression is a quotient of powers with like bases, so use the Quotient of Powers Property.

Quotient of Powers Property $\frac{a^m}{a^n} = a^{m-n}$

A. $\frac{b^9}{b^3} = b^{9-3} = b^6$ B. $\frac{x^2}{x^{10}} = x^{2-10} = x^{-8} = \frac{1}{x^8}$ *Use the Negative Exponent Property to write x^{-8} with a positive exponent.*

EXAMPLE 7 Using Properties of Powers to Simplify Expressions

Simplify. Write each expression using only positive exponents.

A. 4^{-4} B. $(-2)^{-5}$ C. $\left(\frac{1}{2}\right)^{-3}$ D. $\left(-\frac{7}{9}\right)^{-2}$

SOLUTION

Negative Exponent Property $a^{-m} = \frac{1}{a^m}$

Each power has a negative exponent, so use the Negative Exponent Property to simplify.

A. $4^{-4} = \frac{1}{4^4} = \frac{1}{256}$

B. $(-2)^{-5} = \frac{1}{(-2)^5} = -\frac{1}{32}$

TIP

Applying the Negative Exponent Property does not change the sign of the base.

C. $\left(\frac{1}{2}\right)^{-3} = \frac{1}{\left(\frac{1}{2}\right)^3}$ *Negative Exponent Property*

$= \frac{1}{\frac{1^3}{2^3}}$ *Power of a Quotient Property*

$= \frac{1}{\frac{1}{8}}$ *Simplify.*

$= 8$ $1 \div \frac{1}{8} = 8$

D. $\left(-\frac{7}{9}\right)^{-2} = \frac{1}{\left(-\frac{7}{9}\right)^2}$ *Negative Exponent Property*

$= \frac{1}{\frac{(-7)^2}{9^2}}$ *Power of a Quotient Property*

$= \frac{1}{\frac{49}{81}}$ *Simplify.*

$= \frac{81}{49}$ $1 \div \frac{49}{81} = \frac{81}{49}$

ALTERNATIVE METHOD

Using the Power of a Power Property, write each expression a^{-m} as $\left(a^{-1}\right)^m$, then simplify.

A. $4^{-4} = (4^{-1})^4$ *Power of a Power Property*

$= \left(\frac{1}{4}\right)^4$ *Negative Exponent Property*

$= \frac{1^4}{4^4}$ *Power of a Quotient Property*

$= \frac{1}{256}$ *Simplify.*

B. $(-2)^{-5} = \left((-2)^{-1}\right)^5$ *Power of a Power Property*

$= \left(-\frac{1}{2}\right)^5$ *Negative Exponent Property*

$= \frac{(-1)^5}{2^5}$ *Power of a Quotient Property*

$= -\frac{1}{32}$ *Simplify.*

TIP

The reciprocal of x is $\frac{1}{x}$.

The reciprocal of $\frac{x}{y}$ is $\frac{y}{x}$.

C. $\left(\frac{1}{2}\right)^{-3} = \left(\left(\frac{1}{2}\right)^{-1}\right)^{3}$ *Power of a Power Property*

$= 2^3$ *Negative Exponent Property*

$= 8$ *Simplify.*

D. $\left(-\frac{7}{9}\right)^{-2} = \left(\left(-\frac{7}{9}\right)^{-1}\right)^{2}$ *Power of a Power Property*

$= \left(-\frac{9}{7}\right)^{2}$ *Negative Exponent Property*

$= \frac{(-9)^2}{7^2}$ *Power of a Quotient Property*

$= \frac{81}{49}$ *Simplify.*

The alternative method demonstrated in **Example 7** can be generalized as $a^{-m} = \left(a^{-1}\right)^{m} = \left(\frac{1}{a}\right)^{m}$ or $\left(\frac{a}{b}\right)^{-m} = \left(\left(\frac{a}{b}\right)^{-1}\right)^{m} = \left(\frac{b}{a}\right)^{m}$, where a and b are nonzero real numbers. Any exponential expression can be rewritten as its reciprocal base to the negative exponent, but it is particularly helpful when the base is a fraction (as demonstrated in **Example 7C** and **Example 7D**).

WHAT'S THE BIG IDEA?

Write $\left(\frac{b}{c}\right)^{-d}$ as an equivalent expression that meets each of the following criteria:

A. an expression with one positive exponent

B. an expression with two negative exponents

C. an expression with two positive exponents

D. an expression with one positive and one negative exponent

(Assume that b and c represent nonzero real numbers and d represents an integer.)

P.2.6 Scientific Notation

Scientific notation is a method for writing a very large or very small positive real number as an equivalent number. A number written in scientific notation is the product of two factors where one factor has exactly one whole number digit and the other factor is a power of 10.

DEFINITION

A number written in **scientific notation** is the product of a real number a and a power of 10 where a is greater than or equal to 1 but less than 10, and the exponent in the power of 10 is an integer.

$a \times 10^n$, where $1 \leq a < 10$ and n is an integer

Scientific notation is a standard way of writing real numbers. Very large and very small numbers are easier to compare, order, and manipulate when written using scientific notation.

Converting from Decimal Notation to Scientific Notation

Consider ways in which the number 12,000 (which is in decimal notation) can be written as the product of a number and a power of 10.

$$12{,}000 = 1200 \times 10 = 120 \times 10^2 = 12 \times 10^3 = 1.2 \times 10^4 = 0.12 \times 10^5 = \ldots$$

Each is the product of a positive real number and 10^n (where n is an integer), but only 1.2×10^4 is in scientific notation because $1 \le 1.2 < 10$.

Steps for Writing a Number in Scientific Notation

❶ Move the number's decimal point so that the resulting number is greater than or equal to 1 but less than 10 (i.e., so the number has exactly one nonzero digit to the left of the decimal point).

❷ Determine the exponent for the power of 10 by counting the number of places that the decimal point was moved in step Σ.

- For numbers greater than 10, the exponent will be a positive integer.
- For numbers between 0 and 1, the exponent will be a negative integer.

❸ Write the product of the number from step ❶ and the power of 10 from step ❷.

For example, to write 18,000,000 in scientific notation, place the decimal point after 1 (since 1 is the first nonzero digit). The decimal point was moved 7 places and the original number is greater than 10, so the exponent in the power of 10 is 7.

$$18{,}000{,}000 = 1.8 \times 10^7$$

To write 0.0000402 in scientific notation, place the decimal point after 4 (since 4 is the first nonzero digit). The decimal point was moved 5 places and the original number is 0 and 1, so the exponent in the power of 10 is −5.

$$0.0000402 = 4.02 \times 10^{-5}$$

EXAMPLE 8 Writing a Number in Scientific Notation

Write each number in scientific notation.

A. 82,040,000,000 **B.** 0.000000006

SOLUTION

A. $82{,}040{,}000{,}000 = 8.204 \times 10^{10}$ *number > 1 ⇒ exponent > 0*

B. $0.000000006 = 6.0 \times 10^{-9}$ *number < 1 ⇒ exponent < 0*

Converting from Scientific Notation to Decimal Notation

To write a number that is in scientific notation in its expanded form, multiply the decimal number a by the power of 10. The product of a and the power of 10 can be found by moving the decimal point the number of places indicated by the exponent in the power of 10, and inserting placeholder 0s as needed. If the exponent is positive, then the decimal point is moved to the right (resulting in a number greater than 10), and if the exponent is negative, then the decimal point is moved to the left (resulting in a number that is less than 1).

EXAMPLE 9 Converting from Scientific Notation

Write each number in decimal notation.

A. 3×10^6 B. 1.002×10^{-3}

SOLUTION

A. $3 \times 10^6 = 3{,}000{,}000$ *exponent > 0 ⇒ number > 1*

B. $1.002 \times 10^{-3} = 0.001002$ *exponent < 0 ⇒ number < 1*

Operations with Numbers in Scientific Notation

Arithmetic can be performed with numbers written in scientific notation. Numbers in scientific notation follow the multiplicative properties of real numbers, such as the Associative and Commutative Properties of Multiplication. When finding products and quotients of the powers of 10, use the properties of powers, such as the Product of Powers and the Quotient of Powers Properties.

EXAMPLE 10 Calculating with Numbers in Scientific Notation

Simplify. Write each answer in scientific notation.

A. $(9.6 \times 10^4)(8.71 \times 10^{-8})$ B. $\dfrac{1.08\times10^{-3}}{1.2\times10^{-7}}$ C. $\dfrac{(7\times10^{-6})(1.1\times10^{9})}{8\times10^{11}}$

SOLUTION

Use the Commutative and Associative Properties of Multiplication to group the decimal numbers and the powers of 10, then simplify.

A. $(9.6 \times 10^4)(8.71 \times 10^{-8}) = (9.6)(8.71) \times (10^4)(10^{-8})$

$= 83.616 \times 10^{-4}$ *Product of Powers Property*

$= (8.3616 \times 10^1) \times 10^{-4}$ *Write 83.616 in scientific notation.*

$= 8.3616 \times 10^{-3}$ *Product of Powers Property*

B. $\dfrac{1.08\times10^{-3}}{1.2\times10^{-7}} = \dfrac{1.08}{1.2}\times\dfrac{10^{-3}}{10^{-7}}$

$= 0.9\times10^{4}$ *Quotient of Powers Property*

$= (9.0\times10^{-1})\times10^{4}$ *Write 0.9 in scientific notation.*

$= 9.0\times10^{3}$ *Product of Powers Property*

C. $\dfrac{(7\times10^{-6})(1.1\times10^{9})}{8\times10^{11}} = \dfrac{(7)(1.1)}{8}\times\dfrac{(10^{-6})(10^{9})}{10^{11}}$

$= 0.9625\times10^{-8}$ *Product and Quotient of Powers Properties*

$= (9.625\times10^{-1})\times10^{-8}$ *Write 0.9625 in scientific notation.*

$= 9.625\times10^{-9}$ *Product of Powers Property*

SECTION P.2 EXERCISES

Warm Up

1. Simplify. 5^3
2. Simplify. $\left(\frac{3}{7}\right)^2$
3. Write $8 \cdot 8 \cdot 8 \cdot 8$ as a power.

Just the Facts

Fill in the blanks.

4. A power is an expression written as a^n where a is a(n) ____ real number called the ____ and n is an integer called the ____ or ____.
5. When the Product of Powers Property is applied to powers with ____ that are multiplied, the exponents are ____ to simplify.
6. The Power of a Power Property states that when the ____ of a power is a power, then the exponents are ____ to simplify.
7. When the ____ Property is applied to powers with ____ that are divided, the exponents are ____ to simplify.
8. The ____ Property states that any nonzero real number to the ____ power is equal to 1.
9. A number written in scientific notation is a product of a number ____ 1 but ____ 10, and a(n) ____ of 10.
10. The decimal notation equivalent of a number in scientific notation where the power of 10 is ____ must be between 0 and ____.

Essential Skills

In Exercises 1–40, simplify each expression. Include only positive exponents in simplified expressions.

1. $(-7)^2$
2. $(-5)^3$
3. $(-2)^5$
4. $(-3)^6$
5. -2^5
6. -6^4
7. -1^6
8. -3^3
9. x^2x^6
10. p^7p^4p
11. $z^8 + z^9$
12. a^6b^{10}
13. m^7m^3m
14. $x^5 - x^4$
15. $w^3x^2(w^{10}x^4)$
16. $b^4c^5(12bc^2)$
17. $(-cd^3)(-2b^5c^2d^{11})$
18. $(10m^4p)(-7m^3n^8p^2)$
19. $(-4g^4h^2k)(4g^8h^3k^7)$
20. $(-5r^2s^2p)(11r^3s^4p^2)$
21. $(d^5)^2$
22. $(k^7)^4$
23. $(h^2)(h^{10})^3$
24. $(w^3)^2(w^6)^4$
25. $(5p^8q)^2$
26. $(-7xy^2z^5)^3$
27. $(-2g^3h)^2(3g^4h^6k)$
28. $(4w^2x^2y)^3(w^3xy^2)^4$
29. $\frac{h^9}{h}$
30. $\frac{z^7}{z^3}$
31. $\frac{m}{m^8}$
32. $\frac{k^5}{k^{12}}$
33. $\frac{j^{11}}{j^{12}}$
34. $\frac{q^4}{q^8}$
35. $\left(\frac{1}{4}\right)^{-2}$
36. $\left(-\frac{5}{3}\right)^{-3}$
37. 9^{-3}
38. $(-5)^{-4}$
39. $\left(-\frac{2}{5}\right)^{-3}$
40. $\left(\frac{3}{8}\right)^{-2}$

In Exercises 41–46, write each number in scientific notation.

41. 602,000
42. 90,005,000
43. 0.00003
44. 0.0000801
45. 4,690,000,000
46. 0.00356

In Exercises 47–52, write each number in decimal notation.

47. 7.1×10^5
48. 2.59×10^4
49. 5.2×10^{-4}
50. 9.8014×10^{-6}
51. 1.8×10^{-7}
52. 1.07378×10^3

In Exercises 53–60, simplify. Write each answer in scientific notation.

53. $(3.2 \times 10^3)(1.5 \times 10^7)$
54. $(8.5 \times 10^6)(4 \times 10^{-2})$
55. $\frac{9.6 \times 10^5}{3.2 \times 10^{12}}$
56. $\frac{(5 \times 10^{16})(8.2 \times 10^{-10})}{2.5 \times 10^{-8}}$
57. $\frac{(4.2 \times 10^{-10})(3.5 \times 10^6)}{1.5 \times 10^{-8}}$
58. $(2.7 \times 10^{-5})(3.4 \times 10^{12})$
59. $(1.8 \times 10^7)(5.4 \times 10^{-11})$
60. $\frac{(3.5 \times 10^{13})(7.2 \times 10^{-9})}{2.4 \times 10^{-9}}$

Extensions

In Exercises 61–63, simplify each expression and state the properties of powers that are applied.

61. $\dfrac{(3d^6)^2}{18d(d^5)}$

62. $\dfrac{10w^{-3}}{(2w)^3}$

63. $(-8m)^{-2}(-n)^0\left(\dfrac{-3n^2}{4}\right)^{-3}(m^2)$

In Exercises 64–69, write two expressions that use the given properties and are equal to the given number or expression.

64. Zero Exponent; 1

65. Zero Exponent; 3

66. Quotient of Powers; x^2

67. Quotient of Powers; $\dfrac{2}{x^2}$

68. Product of Powers; $x^{10}y^9z$

69. Power of a Product and Power of a Power; $16x^{12}y^8z^4$

In Exercises 70–72, find and correct the error, if any.

70. $\dfrac{(s^3)(s^6)}{s^{-2}} = s^{3+6-2} = s^7$

71. $(2c^{-5})^{-3} = 2^{-3}c^{-5(-3)} = -8c^{15}$

72. $\left(\dfrac{h^2}{5}\right)^{-3} = \left(\dfrac{5}{h^2}\right)^3 = \dfrac{5^3}{(h^2)^3} = \dfrac{125}{h^6}$

P.3 RATIONAL EXPONENTS AND RADICALS

OBJECTIVES

- Use properties of radicals to evaluate and simplify expressions.
- Use properties of rational exponents to evaluate and simplify expressions.
- Rationalize denominators.

PREREQUISITE VOCABULARY TERMS

irrational number
power
rational number

P.3.1 Finding Real Roots

Addition and subtraction are **inverse operations** because one operation *undoes* the other. For example, for any numbers m and n, $m + n - n = m$. In other words, the value of a number does not change when some number is added to and subtracted from that number. Another example of inverse operations is multiplication and division. Powers also have an inverse operation. The inverse of a number to the nth power is the ***n*th root** of the number.

DEFINITION

If a and b are real numbers and n is an integer greater than or equal to 2 such that $a = b^n$, then b is an ***n*th root** of a.

IMPORTANT

Some numbers have nth roots that are not real numbers, but for now, our discussion is limited to real nth roots.

So, an nth root of a real number a is a number such that its nth power is a. Some numbers have more than one nth root. For example, 2 is a sixth root of 64 because $2^6 = 64$. Additionally, -2 is a sixth root of 64 because $(-2)^6 = 64$.

TIP

$n = 2$: square root
$n = 3$: cube root
$n = 4$: fourth root
. . .

When $n = 2$, the second root is called the **square root**. A third root is referred to as a **cube root**. After $n = 3$, roots are named by the number of the root (e.g., the fourth root and fifth root).

EXAMPLE 1 Identifying the Real Roots of a Number

Identify each statement as true or false.

A. 10 and -10 are real square roots of 100.

B. -9 has one real square root: -3.

C. 125 has two real cube roots: 5 and -5.

D. -11 is a real cube root of -1331.

E. 2 and -2 are real fourth roots of 16.

F. -16 has no real fourth roots.

SOLUTION

A. True. Since $10^2 = 100$ and $(-10)^2 = 100$, 10 and -10 are real square roots of 100.

B. False. Since $(-3)^2 = 9$, not -9, -3 is not a square root of -9.

IMPORTANT

There is no real number whose square is -9, so -9 has no real square roots.

C. False. Since $(-5)^3 = -125$, not 125, -5 is not a real cube root of 125. However, since $5^3 = 125$, 5 is a real cube root of 125.

D. True. Since $(-11)^3 = -1331$, -11 is a real cube root of -1331.

E. True. Since $2^4 = 16$ and $(-2)^4 = 16$, 2 and -2 are real fourth roots of 16.

F. True. There is no real number whose fourth power is -16, so -16 has no real fourth roots.

IMPORTANT

Negative real numbers have no real even roots because a real number to an even power is always positive.

P.3.2 Simplifying Radical Expressions

The radical symbol, $\sqrt[n]{\ }$, is used to indicate the **principal *n*th root** of a number.

Principal *n*th Root of a Number

The principal *n*th root of *a*, denoted by the **radical** $\sqrt[n]{a}$, is the *n*th root of *a* with the same sign as *a*. The positive integer *n* is the radical's **index**, and the number *a* is the **radicand**.

IMPORTANT

The expression $\sqrt[n]{a}$ is called a radical.

Recall from **Example 1** that there are two square roots of 100: 10 and −10. A principal *n*th root of 100 must have the same sign as 100 (which is positive), so there is only one principal square root of 100: $\sqrt{100} = 10$. The value of a simplified radical is always one value, not two, since the radical symbol indicates only the principal *n*th root.

The square root is a special case of the principal *n*th root because the index value ($n = 2$) is not shown in the radical. The symbol $\sqrt{\ }$ over a positive number a means "the positive square root of *a*."

A **perfect square** is the square of a whole number. For example, 9 and 25 are perfect squares since $9 = 3^2$ and $25 = 5^2$. Principal square roots are easy to simplify when the radicand is a perfect square because the radical simplifies to a whole number. For example, $\sqrt{9} = 3$ and $\sqrt{25} = 5$.

TIP

To simplify calculations, it can be helpful to memorize the first 15 perfect squares and the first 6 perfect cubes.

First 15 Perfect Squares
1, 4, 9, 16, 25, 36, 49, 64, 81, 100, 121, 144, 169, 196, 225

First 6 Perfect Cubes
1, 8, 27, 64, 125, 216

Similarly, **perfect cubes** are cubes of whole numbers. When the radicand of a principal cube root is a perfect cube, the radical simplifies to a whole number. For example, $\sqrt[3]{8} = 2$ (because $2^3 = 8$) and $\sqrt[3]{64} = 4$ (because $4^3 = 64$).

The following properties of radicals can be applied to simplify radical expressions (assuming that the roots exist).

Properties of Radicals

Product Property $\sqrt[n]{ab} = \sqrt[n]{a} \cdot \sqrt[n]{b}$ **Quotient Property** $\sqrt[n]{\frac{a}{b}} = \frac{\sqrt[n]{a}}{\sqrt[n]{b}}$

Clearly, $\sqrt{a}$ can be simplified when *a* is a perfect square. Additionally, by using the Product Property of radicals, $\sqrt{a}$ can be simplified even when *a* is not a perfect square, but only when *a* contains a factor that is a perfect square. (If *a* does not contain a perfect-square factor, then $\sqrt{a}$ is in simplest form.) For example, consider $\sqrt{75}$. The radicand, 75, is not a perfect square, but 75 can be written as a product where one of the factors is a perfect square. After factoring the radicand, the Product Property of radicals can be applied to simplify the radical.

$\sqrt{75} = \sqrt{25 \cdot 3}$ *Since 25 is a perfect-square factor of 75, write 75 as $25 \cdot 3$.*

$= \sqrt{25} \cdot \sqrt{3}$ *Product Property of Radicals*

$= 5\sqrt{3}$ *Simplify the radical containing the perfect square: $\sqrt{25} = 5$.*

So, $5\sqrt{3}$ is the exact value of $\sqrt{75}$.

Note that when simplifying radicals, a real number coefficient (in this case, 5) is written in front of the radical factor. The expression $5\sqrt{3}$ can be read as "5 times the square root of 3," or as "5 square root of 3."

Similarly, $\sqrt[3]{a}$ can be simplified if the radicand contains a factor that is a perfect cube, and so on. The process of using the Product Property to simplify radicals is demonstrated in **Examples 2** and **3**.

EXAMPLE 2 Simplifying Radical Expressions

Simplify.

A. $\sqrt{225}$ **B.** $\sqrt{32}$ **C.** $\sqrt[3]{16}$

SOLUTION

A. $\sqrt{225} = 15$ *The radical is a square root and the radicand is a perfect square ($15^2 = 225$).*

B. The radical is a square root, but the radicand, 32, is not a perfect square. So, simplify the radical by writing the radicand as a product where at least one factor is a *perfect square*. Then apply the Product Property of radicals to simplify.

TIP

To simplify the calculations, use the radicand's greatest perfect-square factor.

$\sqrt{32} = \sqrt{16 \cdot 2}$ *Since 16 is a perfect-square factor of 32, write 32 as $16 \cdot 2$.*

$= \sqrt{16} \cdot \sqrt{2}$ *Product Property of Radicals*

$= 4\sqrt{2}$ *Simplify the radical containing the perfect square: $\sqrt{16} = 4$.*

So, the exact value of $\sqrt{32}$ is $4\sqrt{2}$.

ALTERNATIVE METHOD

Notice that 4 is also a perfect-square factor of 32, and the radicand can be written as $4 \cdot 8$. However, 8 also contains a perfect-square factor: 4. So, write the radicand as $4 \cdot 4 \cdot 2$, then apply the Product Property of radicals to simplify.

$\sqrt{32} = \sqrt{4 \cdot 4 \cdot 2}$ *Write 32 as $4 \cdot 4 \cdot 2$.*

$= \sqrt{4} \cdot \sqrt{4} \cdot \sqrt{2}$ *Product Property of Radicals*

$= 2 \cdot 2\sqrt{2}$ *Simplify the radicals containing perfect squares: $\sqrt{4} = 2$.*

$= 4\sqrt{2}$ *Multiply.*

CHECK

Using a calculator, find decimal approximations of $\sqrt{32}$ and $4\sqrt{2}$. $\sqrt{32} \approx 5.66 \approx 4\sqrt{2}$ ✓

C. The radical is a cube root, but the radicand, 16, is not a perfect cube. So, simplify the radical by writing the radicand as a product where at least one factor is a *perfect cube*. Then apply the Product Property of radicals to simplify.

$\sqrt[3]{16} = \sqrt[3]{8 \cdot 2}$ *Since 8 is a perfect-cube factor of 16, write 16 as $8 \cdot 2$.*

$= \sqrt[3]{8} \cdot \sqrt[3]{2}$ *Product Property of Radicals*

$= 2\sqrt[3]{2}$ *Simplify the radical containing the perfect cube: $\sqrt[3]{8} = 2$.*

So, the exact value of $\sqrt[3]{16}$ is $2\sqrt[3]{2}$, which can be read as "2 times the cube root of 2."

A quotient of radicals is simplified in **Example 3**. When simplifying quotients of radicals, pay close attention to whether the radical contains the numerator *and* the denominator, or just the numerator (or just the denominator). For example, consider the expressions $\sqrt{\frac{3}{3}}$ and $\frac{\sqrt{3}}{3}$. It may appear that both expressions are equal to 1, but this is not true.

$\sqrt{\frac{3}{3}} = \sqrt{1} = 1$ *The radical contains the numerator and the denominator.*

However, $\frac{\sqrt{3}}{3} \neq 1$ because $\sqrt{3}$ in the numerator cannot be canceled with 3 in the denominator.

EXAMPLE 3 Simplifying Radical Expressions

Simplify.

A. $3\sqrt{20}\cdot\sqrt{3}$ **B.** $8\sqrt{\frac{25}{16}}$

SOLUTION

A. The Product Property of radicals can be applied to write the expression using a single radical symbol because the radicals have the same index and they are being multiplied.

$$3\sqrt{20}\cdot\sqrt{3} = 3\sqrt{20\cdot 3} = 3\sqrt{60}$$

The radical is not in simplest form because the radicand, 60, contains a perfect-square factor. Simplify the radical.

$3\sqrt{60} = 3\sqrt{4\cdot 15}$ *Since 4 is a perfect-square factor of 60, write 60 as 4 · 15.*

$= 3\sqrt{4}\sqrt{15}$ *Product Property of Radicals*

$= 3\cdot 2\sqrt{15}$ *Simplify the radical of the perfect square:* $\sqrt{4}=2$.

$= 6\sqrt{15}$ *Multiply.*

ALTERNATIVE METHOD

Simplify $\sqrt{20}$ first, then use the Product Property to write the expression using a single radical symbol.

$3\sqrt{20}\cdot\sqrt{3} = 3\sqrt{4\cdot 5}\cdot\sqrt{3}$ *Since 4 is a perfect-square factor of 20, write 20 as 4 · 5.*

$= 3\sqrt{4}\cdot\sqrt{5}\cdot\sqrt{3}$ *Product Property of Radicals*

$= 3\cdot 2\sqrt{5}\cdot\sqrt{3}$ *Simplify the radical of the perfect square:* $\sqrt{4}=2$.

$= 3\cdot 2\sqrt{5\cdot 3}$ *Product Property of Radicals*

$= 6\sqrt{15}$ *Multiply.*

B. Notice that the radical contains a fraction, and the numerator and denominator are both perfect squares. So, use the Quotient Property of radicals to simplify the expression.

$8\sqrt{\frac{25}{16}} = 8\left(\frac{\sqrt{25}}{\sqrt{16}}\right)$ *Quotient Property of Radicals*

$= 8\left(\frac{5}{4}\right)$ *Simplify the radicals.*

$= 10$ *Multiply.*

P.3.3 Adding and Subtracting Radical Expressions

Just as terms in an algebraic expression can be combined (added or subtracted) when they are like terms, radical terms can be combined when the terms contain **like radicals**. Like radicals are radicals that have the same index and the same radicand. As with like terms with a variable, combine like radical terms by adding or subtracting the coefficients, while keeping the radical part of the terms the same. When simplifying an expression with radical terms, simplify each radical first, then look for like radical terms to combine.

EXAMPLE 4 Combining Radical Terms

Simplify each expression.

A. $10\sqrt{3}+2\sqrt{3}$ **B.** $8\sqrt[3]{7}-\sqrt[3]{7}$ **C.** $-4\sqrt{18}+5\sqrt{2}$ **D.** $\sqrt{500}-16\sqrt{20}+\sqrt{2}$

SOLUTION

A. $10\sqrt{3}+2\sqrt{3}=12\sqrt{3}$ *The terms are like radical terms, so add the coefficients.*

B. $8\sqrt[3]{7}-\sqrt[3]{7}=7\sqrt[3]{7}$ *Subtract the coefficients. The coefficient of $\sqrt[3]{7}$ is 1.*

In **C** and **D**, the terms are not like radical terms, but the radicals are not in simplest form. Simplify the radicals and then combine the terms if the radicands are the same.

C.
$$
\begin{aligned}
-4\sqrt{18}+5\sqrt{2} &= -4\sqrt{9\cdot 2}+5\sqrt{2} && \textit{Write 18 as } 9\cdot 2.\\
&= -4\cdot 3\sqrt{2}+5\sqrt{2} && \textit{Simplify the radical.}\\
&= -12\sqrt{2}+5\sqrt{2} && \textit{Multiply.}\\
&= -7\sqrt{2} && \textit{Subtract the coefficients.}
\end{aligned}
$$

D.
$$
\begin{aligned}
\sqrt{500}-16\sqrt{20}+\sqrt{2} &= \sqrt{100\cdot 5}-16\sqrt{4\cdot 5}+\sqrt{2} && \textit{Write 500 as } 100\cdot 5 \textit{ and 20 as } 4\cdot 5.\\
&= 10\sqrt{5}-16\cdot 2\sqrt{5}+\sqrt{2} && \textit{Simplify the radicals.}\\
&= 10\sqrt{5}-32\sqrt{5}+\sqrt{2} && \textit{Multiply.}\\
&= -22\sqrt{5}+\sqrt{2} && \textit{Subtract the coefficients.}
\end{aligned}
$$

IMPORTANT

$\sqrt{5}$ and $\sqrt{2}$ are not like radicals because their radicands are not the same.

P.3.4 Rationalizing a Denominator

Typically, a **fractional expression** (a quotient of algebraic expressions) that contains a radical in the denominator will be written as an equivalent expression in which the denominator is a rational number (so it does not contain a radical), which is called **rationalizing the denominator**. To rationalize a denominator, multiply the fractional expression by a number (or expression) equivalent to 1 that will remove the denominator's radical. The following table generalizes the process for several types of denominators. (Assume that all expressions exist.)

Rationalizing Denominators

Irrational Denominator	Rationalizing the Denominator	Rational Denominator
$\dfrac{c}{b\sqrt{n}}$	$\dfrac{c}{b\sqrt{n}}\cdot\dfrac{\sqrt{n}}{\sqrt{n}}$	$\dfrac{c\sqrt{n}}{bn}$
$\dfrac{c}{a+b\sqrt{n}}$	$\dfrac{c}{a+b\sqrt{n}}\cdot\dfrac{a-b\sqrt{n}}{a-b\sqrt{n}}$	$\dfrac{c\left(a-b\sqrt{n}\right)}{a^2-b^2n}$
$\dfrac{c}{a-b\sqrt{n}}$	$\dfrac{c}{a-b\sqrt{n}}\cdot\dfrac{a+b\sqrt{n}}{a+b\sqrt{n}}$	$\dfrac{c\left(a+b\sqrt{n}\right)}{a^2-b^2n}$
$\dfrac{c}{b\sqrt[3]{n}}$	$\dfrac{c}{b\sqrt[3]{n}}\cdot\dfrac{\sqrt[3]{n^2}}{\sqrt[3]{n^2}}$	$\dfrac{c\sqrt[3]{n^2}}{bn}$

For example, if the denominator is $b\sqrt{n}$, where n is any real positive number, then multiplying the fractional expression by $\frac{\sqrt{n}}{\sqrt{n}}$ (which is equivalent to 1) will remove the radical from the denominator because the denominator will then be $b\sqrt{n} \cdot \sqrt{n}$, which is equal to just bn.

Notice that if the denominator is a sum of terms where one term contains a square root, multiplication by the difference of those same terms will rationalize the denominator. These expressions, $a + b\sqrt{n}$ and $a - b\sqrt{n}$, are called **conjugates**. Multiplying an expression by the denominator's conjugate over itself will rationalize a denominator that is a sum (or difference) of terms.

Examples of Conjugate Pairs ▶

$1+2\sqrt{3}$ and $1-2\sqrt{3}$

$\sqrt{5}-4$ and $\sqrt{5}+4$

EXAMPLE 5 Rationalizing the Denominator

Rationalize the denominator of each expression.

A. $\dfrac{8}{\sqrt{18}}$ **B.** $-\dfrac{1}{\sqrt[3]{4}}$ **C.** $\dfrac{3}{5+\sqrt{2}}$

SOLUTION

A. The square root in the denominator can be simplified.

$$\frac{8}{\sqrt{18}} = \frac{8}{3\sqrt{2}}$$

Multiply by $\frac{\sqrt{2}}{\sqrt{2}}$ to rationalize the denominator.

$$\frac{8}{3\sqrt{2}} = \frac{8}{3\sqrt{2}} \cdot \frac{\sqrt{2}}{\sqrt{2}} = \frac{8\sqrt{2}}{3 \cdot 2} = \frac{4\sqrt{2}}{3}$$

TIP

Use the Product Property of radicals to simplify the denominator.

$3\sqrt{2} \cdot \sqrt{2} = 3\sqrt{4} = 3 \cdot 2$

B. Notice that the radical in the denominator is a cube root. Multiplying the denominator by $\sqrt[3]{4^2}$ will remove the radical. So, multiply by $\frac{\sqrt[3]{16}}{\sqrt[3]{16}}$ to rationalize the denominator.

$$-\frac{1}{\sqrt[3]{4}} = -\frac{1}{\sqrt[3]{4}} \cdot \frac{\sqrt[3]{16}}{\sqrt[3]{16}} = -\frac{\sqrt[3]{16}}{\sqrt[3]{64}} = -\frac{2\sqrt[3]{2}}{4} = -\frac{\sqrt[3]{2}}{2}$$

TIP

Use the Product Property of radicals to simplify the numerator.

$\sqrt[3]{16} = \sqrt[3]{8} \cdot \sqrt[3]{2} = 2\sqrt[3]{2}$

C. The denominator contains a radical and is a sum of terms, so multiply by the conjugate over itself to rationalize the denominator.

$$\frac{3}{5+\sqrt{2}} = \frac{3}{5+\sqrt{2}} \cdot \frac{5-\sqrt{2}}{5-\sqrt{2}}$$ *The conjugate of $5+\sqrt{2}$ is $5-\sqrt{2}$.*

$$= \frac{3(5-\sqrt{2})}{(5+\sqrt{2})(5-\sqrt{2})}$$ *Multiply the numerators and multiply the denominators.*

$$= \frac{3(5-\sqrt{2})}{5(5-\sqrt{2})+\sqrt{2}(5-\sqrt{2})}$$ *Apply the Distributive Property twice in the denominator.*

$$= \frac{15-3\sqrt{2}}{25-5\sqrt{2}+5\sqrt{2}-2}$$ *Distribute.*

$$= \frac{15-3\sqrt{2}}{23}$$ *Combine like terms.*

WHAT'S THE BIG IDEA?

It has been shown that multiplying a fractional expression by $\frac{\sqrt{n}}{\sqrt{n}}$ will rationalize the denominator $b\sqrt{n}$, and that multiplying a fractional expression by $\frac{\sqrt[3]{n^2}}{\sqrt[3]{n^2}}$ will rationalize the denominator $b\sqrt[3]{n}$. What fraction can be used to rationalize the denominator $b\sqrt[4]{n}$? Generalize this pattern for the denominator $b\sqrt[x]{n}$, where x is any positive whole number greater than 1.

P.3.5 Converting Rational Exponents and Radicals

Whole number exponents, zero exponents, and negative integer exponents have been discussed previously in this chapter. Exponents can also be rational numbers, known as **rational exponents**. A power with a rational exponent can be written as a radical, as shown in the following table. (Assume that all expressions exist.)

Rational Exponents	
$a^{\frac{1}{n}} = \sqrt[n]{a}$	*The power's base is the radicand.* *The rational exponent's denominator is the radical's index.*
$a^{\frac{m}{n}} = \left(\sqrt[n]{a}\right)^m$	*The power's base is the radicand.* *The rational exponent's denominator is the radical's index.* *The rational exponent's numerator is the exponent of the radical.*
$a^{\frac{m}{n}} = \sqrt[n]{a^m}$	*The power's base is the radicand.* *The rational exponent's denominator is the radical's index.* *The rational exponent's numerator is the exponent of the radicand.*

TIP

There are two ways to write a power with exponent m/n as a radical. The difference is that this exponent is outside the radical in one form and inside the radical in the other form.

Notice that $a^{\frac{m}{n}} = \left(\sqrt[n]{a}\right)^m = \sqrt[n]{a^m}$. Therefore, the rational exponent's numerator m can be either the power of the radical or the power of the radicand. The difference between these two expressions is the order in which the operations are completed.

$\left(\sqrt[n]{a}\right)^m$ The radical is evaluated first, and then the power is applied to the value.

$\sqrt[n]{a^m}$ The power is evaluated first, and then the root of the power is taken.

Choose the option that makes simplification easier when simplifying a power with a rational exponent, $a^{\frac{m}{n}}$. Typically, $\left(\sqrt[n]{a}\right)^m$ (i.e., simplifying the radical first and then applying the power) is the easier option, because simplifying the radical first results in a smaller number (as opposed to evaluating the power first, which results in a larger number). If the radical cannot be simplified, then try writing the expression as $\sqrt[n]{a^m}$.

EXAMPLE 6 Simplifying a Power with a Rational Exponent

Simplify each expression.

A. $16^{\frac{1}{2}}$ **B.** $4^{\frac{3}{2}}$ **C.** $27^{\frac{2}{3}}$

TIP

A radical can be evaluated using a calculator by writing the radical as a rational exponent. For example, to find $\sqrt[4]{625}$, enter 625^(1/4). Note that the ^ symbol indicates that the number (or expression) that follows is an exponent.

SOLUTION

Write each power as a radical where the denominator of the rational exponent is the radical's index and the numerator is either the power of the radicand or the power of the radical. Then simplify.

A. $16^{\frac{1}{2}} = \sqrt{16} = 4$

B. $4^{\frac{3}{2}} = \left(\sqrt{4}\right)^3 = 2^3 = 8$ *Alternatively, the power could be written as* $\sqrt{4^3} = \sqrt{64} = 8$.

C. $27^{\frac{2}{3}} = \left(\sqrt[3]{27}\right)^2 = 3^2 = 9$ *Alternatively, the power could be written as* $\sqrt[3]{27^2} = \sqrt[3]{729} = 9$.

WHAT'S THE BIG IDEA?

Use properties of powers and properties of real numbers to show that $a^{\frac{m}{n}} = \left(\sqrt[n]{a}\right)^m$ and that $a^{\frac{m}{n}} = \sqrt[n]{a^m}$.

Properties of exponents, such as the Product of Powers Property, $a^m \cdot a^n = a^{m+n}$, can be used to simplify radical expressions and expressions with rational exponents. Recall that the Product of Powers Property is applied only to powers with like bases.

EXAMPLE 7 Simplifying a Product of Radicals

Write each product as a single radical, if possible.

A. $\sqrt{3}\left(\sqrt[5]{3}\right)$ **B.** $\sqrt[4]{7^3}\left(\sqrt{7}\right)$

SOLUTION

Write each radical as a power with a rational exponent.

A.

$\sqrt{3}\left(\sqrt[5]{3}\right)$

$= 3^{\frac{1}{2}}(3^{\frac{1}{5}})$ *Write each radical using a rational exponent.*

$= 3^{\frac{1}{2}+\frac{1}{5}}$ *Product of Powers*

$= 3^{\frac{7}{10}}$ *Simplify the exponent.*

$= \left(\sqrt[10]{3}\right)^7$ *Convert to radical form.*

The radical can be written equivalently as $\sqrt[10]{3^7}$.

B.

$\sqrt[4]{7^3}\left(\sqrt{7}\right)$

$= 7^{\frac{3}{4}}(7^{\frac{1}{2}})$ *Write each radical using a rational exponent.*

$= 7^{\frac{3}{4}+\frac{1}{2}}$ *Product of Powers*

$= 7^{\frac{5}{4}}$ *Simplify the exponent.*

$= 7^1 7^{\frac{1}{4}}$ *Properties of Exponents*

$= 7\sqrt[4]{7}$ *Convert to radical form.*

Negative Rational Exponents

When a rational exponent is negative, apply the Negative Exponent Property, $a^{-m} = \frac{1}{a^m}$, to write the power with a positive exponent.

EXAMPLE 8 Simplifying an Expression with a Negative Rational Exponent

Simplify each expression.

A. $25^{-\frac{1}{2}}$ **B.** $27^{-\frac{4}{3}}$

SOLUTION

Use the Negative Exponent Property to write each negative rational exponent as a positive rational exponent. Then convert the rational exponent to a radical and simplify.

A. $25^{-\frac{1}{2}} = \frac{1}{25^{\frac{1}{2}}} = \frac{1}{\sqrt{25}} = \frac{1}{5}$

B. $27^{-\frac{4}{3}} = \frac{1}{27^{\frac{4}{3}}} = \frac{1}{\left(\sqrt[3]{27}\right)^4} = \frac{1}{3^4} = \frac{1}{81}$

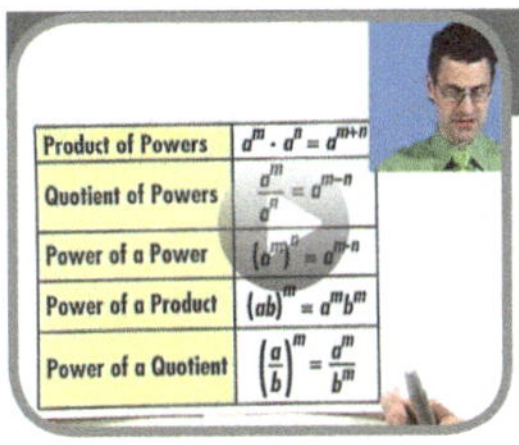

P.3.6 Simplifying Expressions with Rational Exponents

All of the properties of exponents can be applied to rational exponents as they have been applied previously to integer exponents. When simplifying a product or quotient of powers with rational exponents, it is often best to multiply or divide the powers first (by applying a property of powers). Then convert the simplified power to radical notation.

EXAMPLE 9 Simplifying an Expression with Rational Exponents

Simplify each expression.

A. $9^{\frac{1}{6}} \cdot 9^{\frac{1}{3}}$ **B.** $\frac{16^{-\frac{5}{4}}}{16^{-\frac{1}{2}}}$ **C.** $\frac{243^{\frac{1}{5}}}{243^{\frac{7}{10}} \cdot 243^{\frac{1}{10}}}$

SOLUTION

Use the properties of exponents to write each expression as a single power, then simplify.

A. $9^{\frac{1}{6}} \cdot 9^{\frac{1}{3}} = 9^{\frac{1}{6}+\frac{1}{3}} = 9^{\frac{1}{2}} = \sqrt{9} = 3$

B. $\frac{16^{-\frac{5}{4}}}{16^{-\frac{1}{2}}} = 16^{-\frac{5}{4}-\left(-\frac{1}{2}\right)} = 16^{-\frac{3}{4}} = \frac{1}{16^{\frac{3}{4}}} = \frac{1}{\left(\sqrt[4]{16}\right)^3} = \frac{1}{2^3} = \frac{1}{8}$

C. $\frac{243^{\frac{1}{5}}}{243^{\frac{7}{10}} \cdot 243^{\frac{1}{10}}} = 243^{\frac{1}{5}-\left(\frac{7}{10}+\frac{1}{10}\right)} = 243^{-\frac{3}{5}} = \frac{1}{243^{\frac{3}{5}}} = \frac{1}{\left(\sqrt[5]{243}\right)^3} = \frac{1}{3^3} = \frac{1}{27}$

P.3.7 Simplifying Radical Expressions with Variables

Consider the radical $\sqrt{x^6}$, where x is assumed to be a positive real number. Since $(x^3)^2 = x^6$, the radical simplifies to x^3. So, $\sqrt{x^6} = x^3$. The radical could also be simplified by converting it to a rational exponent and then applying the Power of a Power Property, $(a^m)^n = a^{mn}$.

$$\sqrt{x^6} = (x^6)^{\frac{1}{2}} = x^{6\cdot\frac{1}{2}} = x^3$$

Radicals in which the radicand is a product of powers can also be simplified. For example, consider the radical $\sqrt[3]{x^{12}y^3}$. By the Product Property of radicals, $\sqrt[3]{x^{12}y^3} = \sqrt[3]{x^{12}} \cdot \sqrt[3]{y^3}$. Since $(x^4)^3 = x^{12}$ and $(y^1)^3 = y^3$, the radical simplifies to x^4y^1 or x^4y.

Alternatively, the radical can be simplified in the following way: convert it to a rational exponent and then apply the Power of a Product Property, $(ab)^n = a^nb^n$ (since the radicand is a product of powers). Finally, apply the Power of a Power Property, $(a^m)^n = a^{mn}$.

$$\sqrt[3]{x^{12}y^3} = (x^{12}y^3)^{\frac{1}{3}} = (x^{12})^{\frac{1}{3}}(y^3)^{\frac{1}{3}} = x^{12\cdot\frac{1}{3}}y^{3\cdot\frac{1}{3}} = x^4y^1 = x^4y$$

When the exponents in the radicand are multiples of the index (as was the case in the two previous examples), the simplified expression will no longer contain a radical. However, if the exponents in the radicand are not multiples of the index, the simplified expression will contain a radical. For example, consider the radical $\sqrt{x^5}$. There is no power of x that can be squared to give x^5, since 5 is not a multiple of 2 (the radical's index). If the radicand x^5 is written as x^4x, then it can be simplified, but the simplified expression will still contain a radical factor. By the Product Property of radicals, $\sqrt{x^5} = \sqrt{x^4x} = \sqrt{x^4} \cdot \sqrt{x}$. Since $\sqrt{x^4} = x^2$, it follows that $\sqrt{x^5} = x^2\sqrt{x}$. Again, the alternative method is to convert the radical to a rational exponent and then apply the properties of exponents.

$$\sqrt{x^5} = (x^5)^{\frac{1}{2}} = (x^4x)^{\frac{1}{2}} = x^{4\cdot\frac{1}{2}}x^{\frac{1}{2}} = x^2\sqrt{x}$$

In this example, the radicand x^5 was written as x^4x. Another possibility is to just multiply the exponents, 5 and 1/2, and then write that product as a mixed number. Then apply the Product of Powers Property, $a^m \cdot a^n = a^{m+n}$, to simplify. (Note that here the Product of Powers Property is used to write an expression of the form a^{m+n} as $a^m \cdot a^n$.)

$$\sqrt{x^5} = (x^5)^{\frac{1}{2}} = x^{\frac{5}{2}} = x^{2+\frac{1}{2}} = x^2x^{\frac{1}{2}} = x^2\sqrt{x}$$

While there are typically several different ways to simplify a radical expression that contains powers with variable bases in the radicand, the following steps can be applied in *most* cases.

Steps for Simplifying a Radical Expression with Variables in the Radicand

1. Convert the radical to a rational exponent.
2. Apply the Power of a Product Property, $(ab)^n = a^nb^n$.
3. Apply the Power of a Power Property, $(a^m)^n = a^{mn}$.
4. For any powers with rational exponents that are improper, write those rational exponents as the sum of their whole and fractional parts, then apply the Product of Powers Property, $a^m \cdot a^n = a^{m+n}$.
5. Write any powers with rational exponents that are proper fractions in radical notation.

EXAMPLE 10 Using a Rational Exponent to Simplify a Radical Expression with Variables

Simplify each expression. Assume that all variables represent positive numbers.

A. $\sqrt[3]{8p^6q^{21}}$ **B.** $\sqrt{20a^8bc^3}$

SOLUTION

A. The coefficient 8 is a perfect cube, and the exponents are both multiples of the index 3, so the simplified expression will not contain a radical. Simplify the coefficient and then write the remaining factors in the radicand with a the rational exponent.

$\sqrt[3]{8p^6q^{21}} = \sqrt[3]{8}\left(\sqrt[3]{p^6q^{21}}\right)$	*Product Property of Radicals*
$= 2(p^6q^{21})^{\frac{1}{3}}$	*Simplify the coefficient. Write the radical as a rational exponent for the remaining factors.*
$= 2(p^6)^{\frac{1}{3}}(q^{21})^{\frac{1}{3}}$	*Power of a Product Property*
$= 2(p^{6\cdot\frac{1}{3}})(q^{21\cdot\frac{1}{3}})$	*Power of a Power Property*
$= 2p^2q^7$	*Simplify the exponents.*

TIP

It is often helpful to simplify any coefficient first and keep that factor separate from the rest of the calculations.

B. The coefficient 20 is not a perfect square, and the exponents are not all multiples of the index 2, so the simplified expression will contain a radical.

$\sqrt{20a^8bc^3} = \sqrt{20}\sqrt{a^8bc^3}$	*Product Property of Radicals*
$= 2\sqrt{5}(a^8bc^3)^{\frac{1}{2}}$	*Simplify the coefficient. Write the radical as a rational exponent for the remaining factors.*
$= 2\sqrt{5}(a^4b^{\frac{1}{2}}c^{\frac{3}{2}})$	*Apply the Power of a Product and the Power of a Power Properties.*
$= 2\sqrt{5}(a^4b^{\frac{1}{2}}c^{1+\frac{1}{2}})$	*Write the improper fraction as the sum of the whole and fraction parts.*
$= 2\sqrt{5}(a^4b^{\frac{1}{2}}c^1c^{\frac{1}{2}})$	*Apply the Product of Powers Property.*
$= 2a^4c\sqrt{5}\left(\sqrt{b}\right)\left(\sqrt{c}\right)$	*Write the powers with rational exponents in radical notation.*
$= 2a^4c\sqrt{5bc}$	*Product Property of Radicals*

Simplifying Quotients of Radical Expressions with the Same Index

When simplifying quotients (or products) of radical expressions with the same index, it is often helpful to first combine the radicals by applying the Quotient Property of radicals (or the Product Property of radicals, as applicable). After the radicands are written as a single radicand, apply properties of exponents to simplify the radicand.

EXAMPLE 11 **Using Rational Exponents to Simplify a Product or Quotient of Radicals**

Simplify. Assume all variables represent positive numbers. $\dfrac{\sqrt{3b^2c}\sqrt{6ac^3}}{\sqrt{2a^5b^8c^{-1}}}$

SOLUTION

Each radical is a square root. So, by the Product and Quotient Properties of radicals, the expression can be written as a single radical. Then multiply to simplify the numerator.

$$\frac{\sqrt{3b^2c}\sqrt{6ac^3}}{\sqrt{2a^5b^8c^{-1}}} = \sqrt{\frac{3b^2c\left(6ac^3\right)}{2a^5b^8c^{-1}}} = \sqrt{\frac{18ab^2c^4}{2a^5b^8c^{-1}}}$$

Divide the coefficients and subtract the exponents with like bases to simplify the radicand.

$$\sqrt{\frac{18ab^2c^4}{2a^5b^8c^{-1}}} = \sqrt{9a^{1-5}b^{2-8}c^{4-(-1)}} = \sqrt{9a^{-4}b^{-6}c^5}$$

Find the square root of 9. Write the radical as a rational exponent, then apply the Power of a Product and the Power of a Power Properties.

$$\sqrt{9a^{-4}b^{-6}c^5} = 3(a^{-4}b^{-6}c^5)^{\frac{1}{2}} = 3(a^{-2}b^{-3}c^{\frac{5}{2}})$$

Write the improper fraction as the sum of the whole and fraction parts, then apply the Product of Powers Property.

$$3(a^{-2}b^{-3}c^{\frac{5}{2}}) = 3(a^{-2}b^{-3}c^{2+\frac{1}{2}}) = 3(a^{-2}b^{-3}c^2c^{\frac{1}{2}})$$

Write the negative exponents as positive exponents, and write the power with the rational exponent in radical notation.

$$3(a^{-2}b^{-3}c^2c^{\frac{1}{2}}) = \frac{3c^2}{a^2b^3}\sqrt{c}$$

SECTION P.3 EXERCISES

Warm Up

1. Evaluate. 2^6
2. Simplify. $5x - 7 - y + x + 3y + 1$
3. Simplify. $k^3(k^4)^2$

Just the Facts

Fill in the blanks.

4. If p is the nth root of q, then ____ $= q$.
5. Given $\sqrt[n]{a}$, the case where $n = 2$ is called the ____ root of a number, and the case where $n =$ ____ is called the cube root of a number.
6. ____ roots of negative numbers do not exist.
7. ____ has one real cube root: -8.
8. 1, 4, 9, 16, and 25 are the first five ____.
9. Radical terms can be combined if the terms contain ____, which have the same ____ and ____.
10. If the denominator of an expression is ____, then multiplying the expression by the denominator's ____, $p + q\sqrt{n}$, will rationalize the denominator.

Essential Skills

In Exercises 1–8, determine whether each statement is true or false.

1. The square roots of 225 are 15 and -15.
2. The fourth roots of 6561 are 9 and -9.
3. 14 is a cube root of 2744.
4. The fifth roots of 100,000 are 10 and -10.
5. -8 is a square root of -64.
6. $-10{,}000$ has no real fourth roots.
7. -3 is a cube root of -27.
8. The real fifth roots of -3125 are 5 and -5.

In Exercises 9–74, simplify. Do not use a calculator. Rationalize the denominator as necessary. Assume that all variables represent positive numbers.

9. $\sqrt{175}$
10. $-5\sqrt{800}$
11. $\sqrt[3]{\frac{128}{8}}$
12. $12\sqrt{\frac{75}{16}}$
13. $-6\sqrt{32}$
14. $7\sqrt{40} \cdot 3\sqrt{75}$
15. $4\sqrt[3]{\frac{24}{27}}$
16. $15\sqrt{\frac{18}{125}}$
17. $3\sqrt[3]{6} + \sqrt[3]{6} + \sqrt[3]{48}$
18. $\sqrt{45} - 7\sqrt{5}$
19. $4\sqrt{8} - \sqrt{200} + 6\sqrt{12}$
20. $-\sqrt[3]{54} + 4\sqrt[3]{24} - \sqrt[3]{21}$
21. $3\sqrt{27} - 5\sqrt{12} + \sqrt{48}$
22. $-8\sqrt{32} + \sqrt{72} + 9\sqrt{18}$
23. $\frac{2\sqrt{3}}{\sqrt{6}}$
24. $-\frac{2}{\sqrt{13}}$
25. $-\frac{5}{\sqrt{75}}$
26. $\frac{3\sqrt{6}}{\sqrt{10}}$
27. $\frac{4\sqrt{2}}{\sqrt{7}}$
28. $-\frac{6}{\sqrt{48}}$
29. $\sqrt[3]{\frac{3}{5}}$
30. $-\frac{1}{\sqrt[3]{-10}}$
31. $-\frac{5}{\sqrt[3]{-3}}$
32. $\frac{4}{\sqrt[3]{6}}$
33. $\frac{4}{\sqrt{7}+3}$
34. $\frac{1}{\sqrt{5}-12}$
35. $\frac{2}{5-\sqrt{3}}$
36. $\frac{3}{2+\sqrt{6}}$
37. $\frac{6}{8+\sqrt{10}}$
38. $\frac{7}{4-\sqrt{11}}$
39. $25^{\frac{3}{2}}$
40. $8^{\frac{7}{3}}$
41. $27^{\frac{4}{3}}$
42. $4^{\frac{3}{2}}$
43. $16^{\frac{5}{4}}$
44. $32^{\frac{2}{5}}$
45. $x^{\frac{2}{7}}x^{\frac{3}{14}}$
46. $\sqrt[4]{5}\left(\sqrt{5}\right)^3$
47. $\sqrt[3]{2}\left(\sqrt{2^3}\right)$
48. $m^{\frac{3}{4}}m^{\frac{5}{8}}$
49. $y^{\frac{2}{3}}y^{\frac{5}{6}}$
50. $\sqrt[6]{7}\left(\sqrt{7}\right)^2$
51. $64^{-\frac{4}{3}}$
52. $81^{-\frac{3}{4}}$
53. $49^{-\frac{3}{2}}$
54. $32^{-\frac{2}{5}}$
55. $16^{-\frac{3}{4}}$
56. $78{,}125^{-\frac{3}{7}}$
57. $\frac{27^{\frac{2}{3}}}{27^{\frac{1}{6}}}$
58. $\frac{25^{\frac{1}{8}} \cdot 25^{\frac{1}{8}}}{25^{\frac{3}{4}}}$
59. $\frac{32^{\frac{4}{5}}}{32^{\frac{1}{2}} \cdot 32^{\frac{1}{10}}}$
60. $\frac{36^{-\frac{3}{4}}}{36^{-\frac{1}{4}}}$
61. $\frac{27^{-\frac{2}{3}}}{27^{-\frac{4}{3}}}$
62. $\frac{49^{\frac{1}{6}} \cdot 49^{\frac{5}{6}}}{49^{\frac{18}{12}}}$
63. $\sqrt{9u^7v^2z}$
64. $\sqrt{28p^8qr^3}$
65. $\sqrt{72g^4h^6j^3}$
66. $\sqrt[3]{27a^9bc^{24}}$
67. $\sqrt[3]{16mn^{12}p^{16}}$
68. $\sqrt{275s^4tp^9}$

69. $\sqrt{\dfrac{20u^4v^2w^5}{80u^6v^3w^3}}$

70. $\dfrac{\sqrt[3]{40x^8y^3z}}{\sqrt[3]{5x^2y^2z}\cdot\sqrt[3]{yz^3}}$

71. $\dfrac{\sqrt[3]{256t^4x^7y}}{\sqrt[3]{4t^{-5}x^4y^7}\cdot\sqrt[3]{x^3y^3}}$

72. $\sqrt{\dfrac{36ab^7c^4}{72a^5b^3c}}$

73. $\dfrac{\sqrt{14m^2n^5p}\cdot\sqrt{3mn^4p^2}}{\sqrt{6mn^3p}}$

74. $\dfrac{\sqrt[3]{108z^4s^7t}}{\sqrt[3]{4z^{-2}s^4t^2}\cdot\sqrt[3]{s^3t^5}}$

Extensions

In Exercises 75–78, identify the two integers that each root is between. Then use a calculator to approximate the root, rounding to the nearest tenth as needed.

75. $\sqrt{31}$

76. $100^{\frac{1}{4}}$

77. $\sqrt[3]{20}$

78. $\sqrt[5]{\dfrac{7}{4}}$

In Exercises 79–81, write each expression in radical form using only positive exponents. Rationalize the denominator as needed. Assume that all variables represent positive numbers.

79. $5\left(\dfrac{x}{2}\right)^{\frac{1}{2}}$

80. $-\left(\dfrac{2}{y}\right)^{\frac{1}{3}}$

81. $\left(\dfrac{4}{1+y}\right)^{\frac{1}{2}}$

82. A student simplified the expression $3^{\frac{1}{2}}+3^{\frac{1}{4}}$ to $\sqrt[4]{3}^{\,3}$, which is incorrect. Explain the student's misconception.

83. Which of the following numbers are *not* equal to $\sqrt{2}$?

A. $\dfrac{2}{\sqrt{2}}$ B. $\dfrac{\sqrt{6}}{\sqrt{3}}$ C. $\dfrac{2\sqrt{3}}{\sqrt{6}}$

D. $\dfrac{\sqrt{2}+\sqrt{6}}{1+\sqrt{3}}$ E. $\dfrac{20-3\sqrt{2}}{5\sqrt{2}-3}$ F. $\dfrac{8\sqrt{3}-2}{4\sqrt{6}-\sqrt{2}}$

In Exercises 84–86, use the Power of a Power Property to write each expression as a single radical. Then simplify.

84. $\sqrt{\sqrt{64}}$

85. $\sqrt{\sqrt{32}}$

86. $\sqrt[3]{\sqrt{x^{12}y^{10}z^{-3}}}$

87. Explain two ways that $\sqrt[4]{4}$ can be simplified to $\sqrt{2}$ by using the Power of a Power Property.

P.4 POLYNOMIALS

OBJECTIVES

- Classify polynomials by degree and number of terms.
- Write polynomials in standard form.
- Add, subtract, and multiply polynomials.
- Use formulas to multiply special products of polynomials.

PREREQUISITE VOCABULARY TERMS

coefficient
cube (of a number)
power
square (of a number)
variable

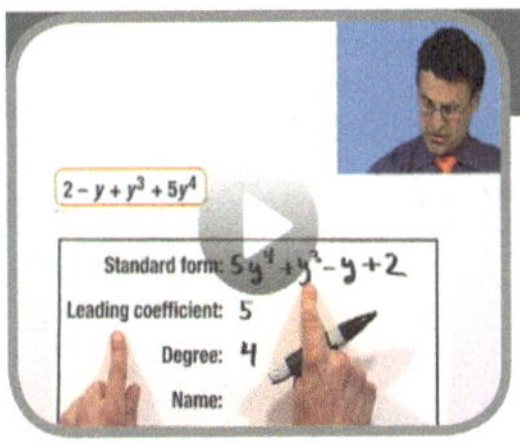

P.4.1 Introduction to Polynomials

A **monomial** is an algebraic expression of the form ax^n where a is a real number and n is a nonnegative integer (i.e., a whole number). A monomial can be a number or a product of numbers and variables with whole number exponents, such as 4, $3x$, or $-a^2b$. A **polynomial** is a monomial or a sum or difference of monomials, such as w, $2p + 4$, or $c^3 - 5c + 10$.

DEFINITION

A **polynomial** in terms of x is an expression of the form

$$a_n x^n + a_{n-1}x^{n-1} + \cdots + a_1 x + a_0$$

where $a_0, a_1, a_2, \ldots, a_n$ are real numbers and n is a nonnegative integer.

IMPORTANT

Since the exponents of the variables in a polynomial must be whole numbers, a variable with a negative exponent or a rational exponent cannot be a part of a polynomial. It follows that a polynomial cannot include variables in the denominator of any term, nor can a polynomial contain a radical with a variable in a radicand.

Classifying Polynomials

Polynomials are classified by their **degree** and number of terms. The **degree of a monomial** is given by the sum of the exponents of the variables. For example, the degree of the monomial $-a^2b$ is 3 because the exponents of the variables are 2 (from a^2) and 1 (from b). The **degree of a polynomial** is equal to the degree of the term with the greatest degree. For example, the degree of $c^3 - 5c + 10$ is 3 since the term with the greatest degree is c^3, which has degree 3. The following table shows the names of polynomials of degrees 0 through 5.

Classifications of Polynomials (by Degree)

Degree	Polynomial Name	Example Polynomials
0	Constant	10, 0
1	Linear	$2a + 2$, $-x$, $6 - w$
2	Quadratic	$7z^2 + 3z - 2$, d^2, $5 + x^2$
3	Cubic	p^3, $b^3 - b$, $y^3 + y^2 + y + 1$
4	Quartic	$3v^2 + v^4$, $9a^4$
5	Quintic	$4m^5 + 2m^4 + 3$, $10 + k^5$

A polynomial with degree greater than 5 does not have a specific name, and the names for polynomials with degrees 4 and 5 are rarely used. The names for polynomials with degrees 1, 2, and 3 (linear, quadratic, and cubic, respectively) will be seen often in this course.

IMPORTANT

monomial: one term
binomial: two terms
trinomial: three terms

A polynomial can have any number of terms, but polynomials with one, two, and three terms have specific names. From the definition, a polynomial with only one term is a monomial. A polynomial with two terms is called a **binomial**, and a polynomial with three terms is called a **trinomial**.

A polynomial is often classified and named by its degree *and* its number of terms. For example, $b^3 - b$ is a cubic binomial, $7z^2 + 3z - 2$ is a quadratic trinomial, and $2a + 2$ is a linear binomial.

IMPORTANT

A simplified polynomial should always be written in standard form, unless otherwise stated.

Polynomials are typically written in **standard form**, where the terms are written in descending order by degree. For example, the quadratic binomial $h^2 - 5$ is in standard form because the first term's degree is greater than the second term's degree. The trinomial $a^2 - a^4 + 2$ is not in standard form because the degree of the first term is less than the degree of the second term. The trinomial can be written in standard form by rearranging the terms so that they are in descending order by degree: $-a^4 + a^2 + 2$. The coefficient of the term with the greatest degree in a polynomial is called the **leading coefficient**. When a polynomial is written in standard form, the leading coefficient is written first.

EXAMPLE 1 Writing a Polynomial in Standard Form and Classifying the Polynomial

Write each polynomial in standard form. Then identify its degree and leading coefficient, and classify it according to its degree and number of terms.

A. $w + 3 - w^2$

B. $1 + 7n^2 - 5n^3$

SOLUTION

A. The polynomial has 3 terms.

- standard form: $-w^2 + w + 3$
- degree: 2
- leading coefficient: –1
- classification: quadratic trinomial

B. The polynomial has 3 terms.

- standard form: $-5n^3 + 7n^2 + 1$
- degree: 3
- leading coefficient: –5
- classification: cubic trinomial

WHAT'S THE BIG IDEA?

What information can be deduced about each of the following polynomials? (Assume that each polynomial contains only one variable.)

- a quartic binomial
- a quadratic monomial
- a quintic polynomial

P.4.2 Adding, Subtracting, and Multiplying Polynomials

The Associative, Commutative, and Distributive Properties are applied when adding, subtracting, and multiplying polynomials. Each process includes applying appropriate operations or properties to remove any parentheses and then combining any like terms, as was done previously when simplifying an expression. (Division of polynomials will be discussed in Chapter 2.)

Subtracting Polynomials

When one polynomial is subtracted from another, the subtracted polynomial must be enclosed within parentheses.

$(a + b) - (c + d - e)$ *Trinomial $c + d - e$ is subtracted from binomial $a + b$.*

Subtraction of a polynomial can be rewritten (or just thought of) as addition of the negative of the polynomial.

$$(a + b) - (c + d - e) = (a + b) + (-1)(c + d - e)$$

Distributive Property ▶ $a(b + c) = ab + ac$; $a(b - c) = ab - ac$

To simplify this type of expression, distribute the -1 to *each term* within the parentheses. Multiplying each term by -1 changes the sign of the term (i.e., changes addition to subtraction and subtraction to addition).

$$(a + b) + (-1)(c + d - e) = (a + b) - c - d + e$$

Grouping symbols that are preceded by plus signs, or nothing at all, are unnecessary because they can be removed without changing any of the enclosed terms.

$$(a + b) - c - d + e = a + b - c - d + e$$

Therefore, $(a + b) - (c + d - e) = a + b - c - d + e$. The expression $a + b - c - d + e$ contains no like terms or parentheses, so the expression is simplified.

EXAMPLE 2 Adding and Subtracting Polynomials

Simplify.

A. $x^3 + (x^2 + 3x - 4) + 3(2x^3 - x + 1)$ **B.** $[(2x + 3x^2) - x^2] - [x^3 - 3(5x^3 - x) + 4x^2]$

SOLUTION

A. Use the Distributive Property, then combine the like terms.

$x^3 + (x^2 + 3x - 4) + 3(2x^3 - x + 1)$

$= x^3 + x^2 + 3x - 4 + 3(2x^3 - x + 1)$ *Remove the parentheses.*

$= x^3 + x^2 + 3x - 4 + 6x^3 - 3x + 3$ *Distribute the 3.*

$= x^3 + 6x^3 + x^2 + 3x - 3x - 4 + 3$ *Group the like terms.*

$= 7x^3 + x^2 - 1$ *Combine the like terms.*

TIP

The parentheses around the polynomial $x^2 + 3x - 4$ can be removed without changing any of the enclosed terms because only a plus sign immediately precedes the opening parenthesis.

B. Start by simplifying within the brackets. Next, use the Distributive Property to remove the parentheses, and then combine the like terms. Remember, the grouping symbols that are preceded by a plus sign, or nothing at all, can be removed without changing any of the enclosed terms.

$[(2x + 3x^2) - x^2] - [x^3 - 3(5x^3 - x) + 4x^2]$

$= 2x + 3x^2 - x^2 - [x^3 - 3(5x^3 - x) + 4x^2]$ *Remove the unnecessary grouping symbols.*

$= 2x + 3x^2 - x^2 - [x^3 - 15x^3 + 3x + 4x^2]$ *Distribute the -3.*

$= 2x + 3x^2 - x^2 - x^3 + 15x^3 - 3x - 4x^2$ *Distribute the -1.*

$= 2x - 3x + 3x^2 - x^2 - 4x^2 - x^3 + 15x^3$ *Group the like terms.*

$= -x - 2x^2 + 14x^3$ *Combine the like terms.*

$= 14x^3 - 2x^2 - x$ *Write the polynomial in standard form.*

TIP

Multiplying a polynomial by -1 changes the sign of each term.

$- [x^3 - 15x^3 + 3x + 4x^2]$

$= (-1)[x^3 - 15x^3 + 3x + 4x^2]$

$= -x^3 + 15x^3 - 3x - 4x^2$

The Product of a Monomial and a Polynomial

Polynomials can be multiplied using the properties of real numbers. Since polynomials typically contain powers, the Product of Powers Property of exponents is often used when multiplying polynomials.

Product of Powers Property

$a^m \cdot a^n = a^{m+n}$

To multiply a monomial by a monomial, such as $3x^2(5x^4)$, multiply the coefficients and use the Product of Powers Property to multiply any powers: $3x^2(5x^4) = 15x^6$. Multiplying a monomial by a polynomial is just a repetition of this process (using the Distributive Property) as many times as there are terms in the polynomial. For example, to simplify $3x^2(5x^4 + 2x^2 - x)$, multiply $3x^2$ by each term in the trinomial.

IMPORTANT

Multiplication of polynomials is also referred to as expanding.

$$3x^2(5x^4 + 2x^2 - x) = 3x^2(5x^4) + 3x^2(2x^2) - 3x^2(x)$$
$$= 15x^6 + 6x^4 - 3x^3$$

The Product of Two Binomials

A special case of multiplying polynomials is the case of a binomial times a binomial. The Distributive Property can be used twice, as was previously discussed. Alternatively, the **FOIL method** can be used. The FOIL method is a way of applying the distributing steps in a particular order that makes the steps easier to remember. Each letter in the word "FOIL" represents a multiplication step used in the process.

F – multiply the *first* terms from each binomial	$(\underline{a} + b)(\underline{c} + d) = ac +$
O – multiply the *outer* terms from each binomial	$(\underline{a} + b)(c + \underline{d}) = ac + ad +$
I – multiply the *inner* terms from each binomial	$(a + \underline{b})(\underline{c} + d) = ac + ad + bc +$
L – multiply the *last* terms from each binomial	$(a + \underline{b})(c + \underline{d}) = ac + ad + bc + bd$

When "FOILing" (i.e., using the FOIL method to multiply two binomials), consider a term that follows a subtraction symbol to be negative. For example, to FOIL $(a - b)(c - d)$, think of the last term as $-b$ and $-d$.

$$(a - b)(c - d) = ac + (-ad) + (-bc) + bd$$
$$= ac - ad - bc + bd$$

EXAMPLE 3 Multiplying Two Binomials

Simplify.

A. $(5a + 2)(3a - 1)$ **B.** $(y^3 - 4)(2y^3 - 3)$ **C.** $(5x + 3)(x^2 - 8)$

SOLUTION

Using the Distributive Property twice will yield the same result as the FOIL method. Here, the FOIL method is used. After FOILing, combine any like terms.

A. $(5a + 2)(3a - 1)$
$= 15a^2 - 5a + 6a - 2$
$= 15a^2 + a - 2$

B. $(y^3 - 4)(2y^3 - 3)$
$= 2y^6 - 3y^3 - 8y^3 + 12$
$= 2y^6 - 11y^3 + 12$

C. $(5x + 3)(x^2 - 8)$
$= 5x^3 - 40x + 3x^2 - 24$
$= 5x^3 + 3x^2 - 40x - 24$

Notice that the expansions of the binomials in **Examples 3A** and **B** are trinomials after the like terms are combined. However, the expansion of the binomials in **Example 3C** is a polynomial with four terms, not a trinomial, because none of the terms produced by the FOIL steps are like terms.

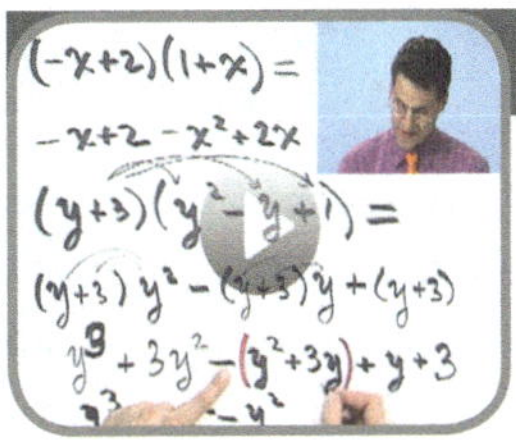

P.4.3 Multiplying Polynomials by Polynomials

To multiply a polynomial by a polynomial, use the Distributive Property and multiply the first polynomial by each term in the second polynomial. The Distributive Property will then be used as many times as there are terms in the second polynomial. For example, consider the product $(x + 3)(x^2 - 5x + 2)$. Distribute the first polynomial, $x + 3$, to each term in the second polynomial, $x^2 - 5x + 2$.

$$(x + 3)(x^2 - 5x + 2) = (x + 3)(x^2) - (x + 3)(5x) + (x + 3)(2)$$

Notice that the second polynomial is a trinomial, so the Distributive Property is used three times to simplify. Combine the like terms to complete the simplification process.

$$\begin{aligned}(x + 3)(x^2 - 5x + 2) &= (x + 3)(x^2) - (x + 3)(5x) + (x + 3)(2)\\ &= x^3 + 3x^2 - (5x^2 + 15x) + 2x + 6\\ &= x^3 + 3x^2 - 5x^2 - 15x + 2x + 6\\ &= x^3 - 2x^2 - 13x + 6\end{aligned}$$

IMPORTANT

The parentheses are added to indicate that the entire quantity is being subtracted.

Since multiplication is commutative, each term in the first polynomial can be distributed to the second polynomial, or vice versa.

$$(a + b)(c + d + e) = (a)(c + d + e) + (b)(c + d + e)$$

Each term in $(a + b)$ is distributed to $(c + d + e)$.

$$(a + b)(c + d + e) = (a + b)(c) + (a + b)(d) + (a + b)(e)$$

$(a + b)$ is distributed to each term in $(c + d + e)$.

TIP

Be sure that each polynomial is simplified before multiplying.

EXAMPLE 4 Multiplying Two Polynomials

Simplify.

A. $(x - 3x^2)(x^3 + x^2 - 5x + 4 + x^2)$

B. $(3y^2 + xy - x^3)^2$

SOLUTION

A. $(x - 3x^2)(x^3 + x^2 - 5x + 4 + x^2)$

$= (x - 3x^2)(x^3 + 2x^2 - 5x + 4)$ *Simplify within parentheses.*

$= (x)(x^3 + 2x^2 - 5x + 4) - (3x^2)(x^3 + 2x^2 - 5x + 4)$ *Use the Distributive Property.*

$= (x^4 + 2x^3 - 5x^2 + 4x) - (3x^5 + 6x^4 - 15x^3 + 12x^2)$ *Distribute twice.*

$= x^4 + 2x^3 - 5x^2 + 4x - 3x^5 - 6x^4 + 15x^3 - 12x^2$ *Subtract (distribute the -1).*

$= -5x^4 + 17x^3 - 17x^2 + 4x - 3x^5$ *Combine the like terms.*

$= -3x^5 - 5x^4 + 17x^3 - 17x^2 + 4x$ *Write in standard form.*

B. $(3y^2 + xy - x^3)^2$

$= (3y^2 + xy - x^3)(3y^2 + xy - x^3)$ *Expand the power.*

$= (3y^2)(3y^2 + xy - x^3) + (xy)(3y^2 + xy - x^3) - (x^3)(3y^2 + xy - x^3)$ *Distributive Property*

$= (9y^4 + 3xy^3 - 3x^3y^2) + (3xy^3 + x^2y^2 - x^4y) - (3x^3y^2 + x^4y - x^6)$ *Distribute three times.*

$= 9y^4 + 3xy^3 - 3x^3y^2 + 3xy^3 + x^2y^2 - x^4y - 3x^3y^2 - x^4y + x^6$ *Subtract (distribute -1).*

$= 9y^4 + 6xy^3 - 6x^3y^2 + x^2y^2 - 2x^4y + x^6$ *Combine the like terms.*

$= x^6 - 2x^4y - 6x^3y^2 + x^2y^2 + 6xy^3 + 9y^4$ *Write in standard form.*

P.4.4 Special Products of Binomials: Squares and Cubes

The following table summarizes the formulas that can be used to find products of binomials when the binomials contain the same terms (with the same or opposite signs), such as squares and cubes of binomials. These formulas can be helpful, but they are not necessary. You can multiply by using the Distributive Property or FOIL instead of using a formula.

Formulas for Special Products of Binomials

Sum and Difference of Same Terms	$(A+B)(A-B)=A^2-B^2$
Square of a Sum	$(A+B)^2=A^2+2AB+B^2$
Square of a Difference	$(A-B)^2=A^2-2AB+B^2$
Cube of a Sum	$(A+B)^3=A^3+3A^2B+3AB^2+B^3$
Cube of a Difference	$(A-B)^3=A^3-3A^2B+3AB^2-B^3$

TIP

The product of two binomials is a binomial only when the two binomials have the same terms with opposite signs, as in $(a+b)(a-b)$. This case is called a ***Sum and Difference of Same Terms.***

In these formulas, A and B can be real numbers, powers of variables, or any kind of polynomial. Note that A and B do *not* represent the coefficients of the variables.

EXAMPLE 5 Squares and Cubes of Binomials

Simplify.

A. $(x+5)^2$ **B.** $(3p^3-4r)^2$ **C.** $(3+5z^2)^3$ **D.** $(4a^5-\sqrt{2}b)^3$

SOLUTION

A. Since the squared expression is a binomial sum, simplify by using the Square of a Sum Formula, $(A+B)^2=A^2+2AB+B^2$, where $A=x$ and $B=5$.

$$(x+5)^2=x^2+2(x)(5)+5^2 \quad \textit{Substitute.}$$
$$=x^2+10x+25 \quad \textit{Simplify.}$$

Check by FOILing
$(x+5)^2$
$=(x+5)(x+5)$
$=x^2+5x+5x+25$
$=x^2+10x+25$ ✓

B. Since the squared expression is a binomial difference, simplify by using the Square of a Difference Formula, $(A-B)^2=A^2-2AB+B^2$, where $A=3p^3$ and $B=4r$.

$$(3p^3-4r)^2=(3p^3)^2-2(3p^3)(4r)+(4r)^2 \quad \textit{Substitute.}$$
$$=(3)^2(p^3)^2-2(3p^3)(4r)+(4)^2(r)^2 \quad \textit{Power of a Product Property}$$
$$=9(p^6)-2(3p^3)(4r)+16(r^2) \quad \textit{Power of a Power Property}$$
$$=9p^6-24p^3r+16r^2 \quad \textit{Multiply.}$$

TIP

When A or B represents a product or sum, be sure to enclose the expression for A or B within parentheses. Otherwise, the powers of products may be evaluated incorrectly.

C. Since the cubed expression is a binomial sum, simplify by using the Cube of a Sum Formula, $(A+B)^3=A^3+3A^2B+3AB^2+B^3$, where $A=3$ and $B=5z^2$.

$$(3+5z^2)^3$$
$$=(3)^3+3(3)^2(5z^2)+3(3)(5z^2)^2+(5z^2)^3 \quad \textit{Substitute.}$$
$$=(3)^3+3(3)^2(5z^2)+3(3)(5)^2(z^2)^2+(5)^3(z^2)^3 \quad \textit{Power of a Product Property}$$
$$=(3)^3+3(3)^2(5z^2)+3(3)(5)^2(z^4)+(5)^3(z^6) \quad \textit{Power of a Power Property}$$
$$=27+3(9)(5z^2)+3(3)(25)(z^4)+(125)(z^6) \quad \textit{Evaluate the powers.}$$
$$=27+135z^2+225z^4+125z^6 \quad \textit{Multiply.}$$
$$=125z^6+225z^4+135z^2+27 \quad \textit{Write the polynomial in standard form.}$$

Power of a Product Property
$(ab)^n=a^nb^n$

D. Since the cubed expression is a binomial difference, simplify by using the Cube of a Difference Formula, $(A - B)^3 = A^3 - 3A^2B + 3AB^2 - B^3$, where $A = 4a^5$ and $B = \sqrt{2}b$.

$(4a^5 - \sqrt{2}b)^3$

$= (4a^5)^3 - 3(4a^5)^2(\sqrt{2}b) + 3(4a^5)(\sqrt{2}b)^2 - (\sqrt{2}b)^3$ *Substitute.*

$= (4)^3(a^5)^3 - 3(4)^2(a^5)^2(\sqrt{2}b) + 3(4a^5)(\sqrt{2})^2(b)^2 - (\sqrt{2})^3(b)^3$ *Power of a Product Property*

$= (4)^3(a^{15}) - 3(4)^2(a^{10})(\sqrt{2}b) + 3(4a^5)(\sqrt{2})^2(b^2) - (\sqrt{2})^3(b^3)$ *Power of a Power Property*

$= 64(a^{15}) - 3(16)(a^{10})(\sqrt{2}b) + 3(4a^5)(2)(b^2) - (2\sqrt{2})(b^3)$ *Evaluate the powers.*

$= 64a^{15} - 48\sqrt{2}a^{10}b + 24a^5b^2 - 2\sqrt{2}b^3$ *Multiply.*

Power of a Power Property

$(a^m)^n = a^{mn}$

REMEMBER

$(\sqrt{2})^3 = \sqrt{2}\cdot\sqrt{2}\cdot\sqrt{2}$
$= \sqrt{4}\cdot\sqrt{2}$
$= 2\sqrt{2}$

SECTION P.4 EXERCISES

Warm Up

Simplify.

1. $5(3x^2 - 7x + 2)$
2. $2(4y + 1) + 10y$
3. $-6p^5qr^2(3p^4r)^2$

Just the Facts

Fill in the blanks.

4. An algebraic expression of the form ax^n is called a ____.
5. Any ____ in a monomial cannot be negative or a fraction.
6. A polynomial is a sum or ____ of ____.
7. A cubic binomial has ____ terms and degree ____.
8. A simplified ____ polynomial in one variable can have at most three terms.
9. The coefficient of the term with the greatest ____ is called the ____.
10. FOIL is a process used for ____ two ____.

Essential Skills

In Exercises 1–6, write each polynomial in standard form. Then identify the leading coefficient and classify it according to its degree and number of terms.

1. $12 + k^2 - 5k$
2. $3x^2 - 8x - 5x^3 + 6$
3. $16m - 6m^3 + 3m^2 - 5m^5$
4. $-4x^3 - 2x^5 - 3$
5. $4b^3 - b + 2b^2 + b^4$
6. $2a - 4a^2 + 2a^3 + 2$

In Exercises 7–32, simplify each expression.

7. $(x^2 - 2x^3 + 16) - (4x^3 + x^2 - x + 10)$
8. $[2a^4 + (5a + 6) - a^2] - [7a - (6a^3 - 2a^2 + 5a) - 4(a + 6)]$
9. $(y^3 + 3y^2 + 5 - 6y) - 4(y + 1) + 3(y^3 - 2y^2 + y)$
10. $(4b^5 + 7b^2 + 2 - b) - (3b^3 - 5b^2 + 6b) - 3(2b + 1)$
11. $[(5m + 6) - 3m^2 + m] - [8m - 2(m^3 + 4m^2 - 5m) - 5m^3]$
12. $5x^4 + 3(5x + 6) - x^2 - [4(x + 9) - (x^3 - 4x^2 + 3x)]$
13. $(5x - 8)(6 - x)$
14. $(4x^2 - 5)(3x^2 + 7)$
15. $(3a^2 + 4)(2a - 5)$
16. $(4h^3 - 1)(5h^3 - 4)$
17. $(2 - 7m^2)(m^2 + 3)$
18. $(b^2 - 6)(2 - 7b^2)$
19. $(x - 1)(3x^3 - 5x^2 + 6x - 1)$
20. $(x + 3)(x^2 + 2x + 4 + x)$
21. $(x^2 + 2x + 5)(2x^2 - 3x + 1)$
22. $(2x^2 + 3xy - 5y^2)^2$
23. $(6y^3 - 2yz - 2z^2)^2$
24. $(2b + 3)(b^3 + 4b^2 - 3b - 7)$
25. $(3x + 7)^2$
26. $(4x - 3y)^2$
27. $(5a^2 + 2bc)^3$
28. $\left(\sqrt{5}p - 2r^4\right)^3$
29. $(2x - 5)(2x + 5)$
30. $(4 + 3m^2)^3$
31. $\left(\sqrt{2}y - 7z^3\right)^3$
32. $(3a^3 - 2b)^2$

Extensions

In Exercises 33–37, simplify each expression.

33. $(9n^2 + n)(9n^2 - n)$
34. $\left(\frac{2}{3}x^4 + 6\right)\left(\frac{2}{3}x^4 - 6\right)$
35. $((1 + 3b^2) + 5b)((1 + 3b^2) - 5b)$
36. $((9 + 5a) - a^3)^2$
37. $((4r^2 + 2) - r)^3$
38. Write six pairs of binomials that meet these criteria:
    - the product is a quadratic trinomial in terms of x,
    - the product is a sum of terms,
    - the leading coefficient of the product is 8, and
    - the constant term of the product is 12.
39. Write six pairs of binomials that meet these criteria:
    - the product is a quadratic trinomial in terms of x,
    - the product is a sum and difference of terms,
    - the leading coefficient of the product is 6, and
    - the constant term of the product is 10 or -10.
40. Write a pair of binomials whose product is a quadratic binomial in terms of x, where the leading coefficient is 1 and the constant term is -25.

41. Write a pair of binomials whose product is a quadratic binomial in terms of x, where the leading coefficient is 9 and the constant term is -100.

42. Write a pair of binomials whose product is a quartic binomial in terms of x, where the leading coefficient is 4 and the constant term is -64.

43. Write six pairs of binomials that meet these criteria:
- the product is a cubic polynomial in terms of x,
- the product is a sum of 4 terms,
- the leading coefficient of the product is 20, and
- the constant term of the product is 1.

44. Suppose the lengths of the sides of a rectangle are represented by the expressions $4x^2 + 2$ and $5x + 7$. Write expressions for the perimeter and area of the rectangle.

45. A square pool is surrounded by a rectangular patio. The lengths of the sides of the patio are represented by the expressions $3x + 1$ and $x + 10$. The length of a side of the pool is represented by $x + 2$. Write an expression for the area of the patio.

46. The length of a side of a cube can be represented by the expression $3x + 8$. Write an expression for the surface area of the cube and an expression for the volume of the cube.

P.5 FACTORING

OBJECTIVES

- Factor the GCF (greatest common factor) from polynomials.
- Factor 4-term polynomials using grouping.
- Factor trinomials using trial and error.
- Factor trinomials using the product-and-sum method.
- Use special factoring formulas.

PREREQUISITE VOCABULARY TERMS

binomial
monomial
polynomial
trinomial

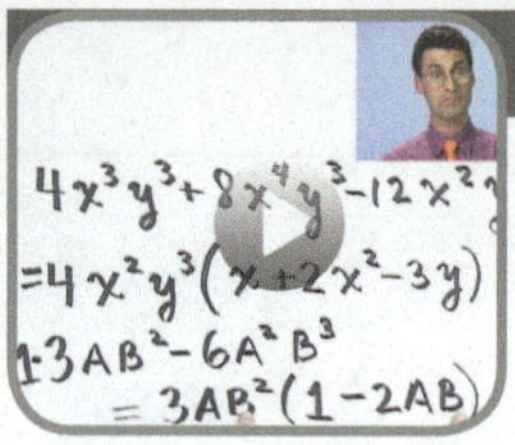

P.5.1 Factoring Using the Greatest Common Factor

Introduction to Factoring

Factors of a number n are the numbers that divide evenly into n. For example, the factors of 12 are 1, 2, 3, 4, 6, and 12. There are several ways to write 12 as a product of two factors.

$$12 = 1 \cdot 12$$
$$12 = 2 \cdot 6$$
$$12 = 3 \cdot 4$$

The factors in each pair can be reversed, since multiplication is commutative.

Just as a number can be written as a product of its numeric factors, a polynomial can be written as a product of its **polynomial factors**. For example, consider the monomial $12x^2$. There are several ways to write $12x^2$ as a product of two monomial factors.

$12x^2 = 1 \cdot 12x^2$	$12x^2 = 4 \cdot 3x^2$	$12x^2 = x \cdot 12x$	*Again, the factors in each*
$12x^2 = 2 \cdot 6x^2$	$12x^2 = 6 \cdot 2x^2$	$12x^2 = 2x \cdot 6x$	*pair can be reversed, since*
$12x^2 = 3 \cdot 4x^2$	$12x^2 = 12 \cdot x^2$	$12x^2 = 3x \cdot 4x$	*multiplication is commutative.*

A polynomial with more than one term may also be written as a product of its factors by applying the Distributive Property. For example, consider the polynomial $10x + 15$. This can be written as $5(2x) + 5(3)$, because $5(2x) = 10x$ and $5(3) = 15$. Recall that the Distributive Property states that $ab + ac = a(b + c)$. So, by the Distributive Property, $5(2x) + 5(3) = 5(2x + 3)$. Therefore, $10x + 15 = 5(2x + 3)$, and so 5 and $2x + 3$ are factors of the polynomial $10x + 15$.

TIP

When a polynomial is factored, the value of the polynomial does not change, only the appearance.

This process of writing a polynomial as a product of factors, called **factoring a polynomial**, can be thought of as *undoing* multiplication. Some polynomials are **irreducible**, meaning that they cannot be factored because they have no factors other than 1 and the polynomial itself.

Factoring Out the GCF

IMPORTANT

For some polynomials, the GCF of the terms is 1.

Several methods for factoring polynomials will be discussed in this section, beginning with factoring the greatest common factor (GCF) of a polynomial's terms from each term.

For example, consider the binomial $12x^3yz^2 + 8x^2y^4$. The GCF of its terms is the product of

- 4 (the GCF of the coefficients 12 and 8),
- x^2 (the power with the greatest exponent common to x^3 and x^2), and
- y (the power with the greatest exponent common to y and y^4).

So, $4x^2y$ is the GCF of the terms in the binomial $12x^3yz^2 + 8x^2y^4$. Note that the variable z is not included in the GCF because z is not common to both terms in the binomial.

To complete the factorization of $12x^3yz^2 + 8x^2y^4$, factor the GCF from each term in the original polynomial, then use the Distributive Property to write the product of the factors.

$$12x^3yz^2 + 8x^2y^4 = (4x^2y)(3xz^2) + (4x^2y)(2y^3) \quad \textit{Write each term as a product using the GCF.}$$
$$= (4x^2y)(3xz^2 + 2y^3) \quad \textit{Distributive Property}$$

A factorization can always be checked by multiplying to confirm that the product is equal to the original polynomial. By the Distributive Property, $(4x^2y)(3xz^2 + 2y^3) = 12x^3yz^2 + 8x^2y^4$, so the factorization is correct.

EXAMPLE 1 Factoring a Monomial from a Polynomial

Identify the GCF and then factor each polynomial.

A. $15x^4 + 35x^7$ **B.** $8p^5q^4 + 12p^3q^3 - 4pq^2$ **C.** $4a^3 + 9b^2 + 6$

SOLUTION

A. Identify the GCF of the binomial's terms.

- The GCF of the coefficients is 5.
- The GCF of x^4 and x^7 is x^4. *The greatest exponent common to x^4 and x^7 is 4.*

So, the GCF of the binomial's terms is $5x^4$.

Factor $5x^4$ from each term in the binomial.

$$15x^4 + 35x^7 = 5x^4(3 + 7x^3)$$

CHECK

Check the factorization by multiplying.

$$5x^4(3 + 7x^3) = 5x^4(3) + 5x^4(7x^3) = 15x^4 + 35x^7 \checkmark$$

Product of Powers ▶ $a^m \cdot a^n = a^{m+n}$

B. Identify the GCF of the trinomial's terms.

- The GCF of the coefficients is 4.
- The GCF of p^5, p^3, and p is p. *The greatest exponent common to p^5, p^3, and p is 1.*
- The GCF of q^4, q^3, and q^2 is q^2. *The greatest exponent common to q^4, q^3, and q^2 is 2.*

So, the GCF of the trinomial's terms is $4pq^2$.

Factor $4pq^2$ from each term in the trinomial.

$$8p^5q^4 + 12p^3q^3 - 4pq^2 = 4pq^2(2p^4q^2 + 3p^2q - 1)$$

TIP

$4pq^2(2p^4q^2 + 3p^2q - 1)$ is equivalent to $(2p^4q^2 + 3p^2q - 1)4pq^2$, since multiplication is commutative.

CHECK

The factorization can be checked by multiplying.

$$4pq^2(2p^4q^2 + 3p^2q - 1) = 4pq^2(2p^4q^2) + 4pq^2(3p^2q) - 4pq^2(1)$$
$$= 8p^5q^4 + 12p^3q^3 - 4pq^2 \checkmark$$

C. The GCF of the coefficients is 1, and the terms have no common variable. So, the GCF of the trinomial's terms is 1. The polynomial is irreducible.

IMPORTANT

If the GCF of a polynomial's terms is 1, the polynomial is not necessarily irreducible.

The GCF of a polynomial's terms is not always a monomial. If each term in a polynomial contains a common binomial factor (or common polynomial with any number of terms), then that common factor must be included in the GCF of its terms, as demonstrated in **Example 2**.

EXAMPLE 2 Factoring a Binomial Factor from a Polynomial

Identify the GCF and then factor each polynomial.

A. $18n^5m^4(p + 4) + 6n^3(p + 4) - 12n^3m^2(p + 4)$ **B.** $15w^4(z - 3) + 10w^3(3 - z) - 20w(z - 3)$

SOLUTION

> **IMPORTANT**
> *Factorizations are typically written with the factor with the least number of terms first. For example, if a factorization is a monomial and a binomial, the monomial is typically written first. However, any order is correct.*

A. Notice that the polynomial has three terms.

$$18n^5m^4(p + 4) + 6n^3(p + 4) - 12n^3m^2(p + 4)$$

The first term is the product of 18, n^5, m^4, and the binomial $(p + 4)$. *The second term is the product of 6, n^3, and the binomial $(p + 4)$.* *The third term is the product of 12, n^3, m^2, and the binomial $(p + 4)$.*

Identify the GCF of the trinomial's terms.

- The GCF of the coefficients is 6.
- The GCF of n^5 and n^3 is n^3. *The greatest exponent common to n^5 and n^3 is 3.*
- Each term contains the factor $(p + 4)$.

> **NOTICE THAT**
> *The second term does not contain a power of m, so the variable m is not included in the GCF.*

The GCF of the trinomial's terms is $6n^3(p + 4)$.

Factor $6n^3(p + 4)$ from each term in the trinomial.

$$18n^5m^4(p + 4) + 6n^3(p + 4) - 12n^3m^2(p + 4) = 6n^3(p + 4)(3n^2m^4 + 1 - 2m^2)$$

CHECK

Check by multiplying. Since each term in the original polynomial contains a factor of $(p + 4)$, do not expand $(p + 4)$.

$$6n^3(p + 4)(3n^2m^4 + 1 - 2m^2)$$
$$= 6n^3(p + 4)(3n^2m^4) + 6n^3(p + 4)(1) - 6n^3(p + 4)(2m^2)$$
$$= 18n^5m^4(p + 4) + 6n^3(p + 4) - 12n^3m^2(p + 4) \checkmark$$

B. This polynomial also has three terms.

$$15w^4(z - 3) + 10w^3(3 - z) - 20w(z - 3)$$

Notice that the middle term contains the binomial factor $(3 - z)$, but the other two terms contain $(z - 3)$. Factoring -1 from $(3 - z)$ yields $(z - 3)$.

$$15w^4(z - 3) + 10w^3(3 - z) - 20w(z - 3)$$
$$= 15w^4(z - 3) + 10w^3(-1)(z - 3) - 20w(z - 3)$$
$$= 15w^4(z - 3) - 10w^3(z - 3) - 20w(z - 3)$$

The -1 changes the preceding operation to subtraction.

The GCF of the trinomial's terms is $5w(z - 3)$.

Factor $5w(z - 3)$ from each term.

> **TIP**
> *When multiplying these polynomials to check, do not expand the factor $(z - 3)$.*

$$15w^4(z - 3) - 10w^3(z - 3) - 20w(z - 3)$$
$$= 5w(z - 3)(3w^3 - 2w^2 - 4)$$

P.5.2 Factoring Polynomials by Grouping

A polynomial that does not contain a common factor for all terms, but does contain a common factor for some pairs of terms, can sometimes be factored by grouping.

EXAMPLE 3 Factoring by Grouping

Factor each polynomial.

A. $5x^4 - 7x^3 + 5x - 7$ **B.** $8yz - 12y - 15 + 10z$

SOLUTION

A. The terms of the polynomial do not share a common factor. However, notice that the first two terms do have a common factor, and their GCF is x^3.

$$5x^4 - 7x^3 + 5x - 7 = x^3(5x - 7) + 5x - 7 \quad \textit{Factor } x^3 \textit{ from the first two terms.}$$
$$= x^3(5x - 7) + (5x - 7) \quad \textit{Associative Property}$$
$$= (5x - 7)(x^3 + 1) \quad \textit{Factor } (5x - 7) \textit{ from each term.}$$

TIP

This form of the polynomial has two terms.

$x^3(5x - 7) + (5x - 7)$

The coefficient of the second term is 1.

$x^3(5x - 7) + 1(5x - 7)$

CHECK

Check by multiplying.

$$(5x - 7)(x^3 + 1) = 5x^4 + 5x - 7x^3 - 7 = 5x^4 - 7x^3 + 5x - 7 \checkmark$$

B. The terms of the polynomial do not share a common factor. However, notice that the GCF of the first two terms is $4y$, and the GCF of the last two terms is 5. Rearrange the last two terms, and then factor the GCF from each pair of terms.

$$8yz - 12y - 15 + 10z = 8yz - 12y + 10z - 15 \quad \textit{Rearrange the last two terms.}$$
$$= 4y(2z - 3) + 5(2z - 3) \quad \textit{Factor the GCF from each pair of terms.}$$
$$= (2z - 3)(4y + 5) \quad \textit{Factor the GCF } (2z - 3) \textit{ from each term.}$$

Notice that there are several ways to group the terms in **Example 3B**, but the factorization remains the same. For example, rearranging the terms as $8yz + 10z - 12y - 15$ gives $2z(4y + 5) - 3(4y + 5)$. Factoring out $(4y + 5)$ yields $(4y + 5)(2z - 3)$, which is equivalent to the factorization from **Example 3B**, $(2z - 3)(4y + 5)$, because multiplication is commutative.

P.5.3 Factoring Trinomials Using Trial and Error

Recall that a product of two binomials can be found by using the FOIL method.

$$(a + 2)(a + 1) = a^2 + a + 2a + 2 = a^2 + 3a + 2$$

Notice that the first term of the trinomial, a^2, is the product of the binomials' first terms. Also, the last term of the trinomial, 2, is the product of the binomials' last terms.

product of the first terms

$$(a + 2)(a + 1) = a^2 + 3a + 2$$

product of the last terms

The product of two binomials that have two pairs of like terms between them (for example, like first terms and like last terms) is always a trinomial.

$$(a + 2)(a + 1) = a^2 + 3a + 2$$

like terms *like terms*

It follows that a reducible (i.e., factorable) trinomial can be written as the product of two binomials that have two pairs of like terms between them. The product of the binomials' first terms will be the first term of the trinomial, and the product of the binomials' last terms will be the last term of the trinomial.

These facts can be used to find a list of pairs of possible binomial factors for a trinomial. Each pair of binomial factors must be multiplied to determine if the factorization is correct.

Steps for Factoring a Trinomial ($ax^2 + bx + c$) Using Trial and Error

❶ Find two like-term factors of the trinomial's first term.

❷ Find two like-term factors of the trinomial's last term.

❸ Write these factors as the first and last terms in two binomials and use the FOIL method with the binomials to determine if the factors are correct.

When writing the binomials in step ❸, you must decide whether to include addition or subtraction between the terms in each binomial. Assuming the sign of the trinomial's first term is positive, if the trinomial's last term is *positive*, then the binomials will both use the *same operation*, and if the trinomial's last term is *negative*, then the binomials will use *different operations.*

More specifically, if the trinomial's

- last and middle term are both positive, then both binomials will be sums.
- last term is positive and the middle term is negative, then both binomials will be differences.
- last term is negative, then one binomial will be a sum and the other will be a difference.

The trinomials in **Example 4** will be of the form $ax^2 + bx + c$, where $a = 1$. In the case where $a = 1$, the binomial's first terms will always be the same (e.g., x and x).

EXAMPLE 4 Factoring a Trinomial ($a = 1$) by Trial and Error

Factor.

A. $x^2 + 9x + 8$ **B.** $n^2 + 4n - 12$

SOLUTION

A. The trinomial's last term and middle term are both positive, so the binomial factors will be sums.

❶ Like-term factors of x^2 are x and x.

❷ Like-term factors of 8 are 2 and 4 *or* 1 and 8.

❸ The first term in each binomial factor will be x. The last terms in the binomial factors will be either 2 and 4 *or* 1 and 8. Write the possible binomial pairs and then use the FOIL method to determine the correct factorization.

Possible Factorizations	Product	
$(x + 2)(x + 4)$	$x^2 + 6x + 8$	✗
$(x + 1)(x + 8)$	$x^2 + 9x + 8$	✓

Therefore, $x^2 + 9x + 8 = (x + 1)(x + 8)$.

B. The trinomial's last term is negative. So, one binomial factor will be a sum, and the other will be a difference.

❶ Like-term factors of n^2 are n and n.

❷ Like-term factors of 12 are 3 and 4, 2 and 6, or 1 and 12.

❸ The first term in each binomial factor will be n. The last terms in the binomial factors will be either 3 and 4, 2 and 6, *or* 1 and 12. Write the possible binomial pairs, then use the FOIL method to determine the correct factorization.

Possible Factorizations	Product	
$(n + 3)(n - 4)$	$n^2 - n - 12$	☒
$(n - 3)(n + 4)$	$n^2 + n - 12$	☒
$(n + 2)(n - 6)$	$n^2 - 4n - 12$	☒
$(n - 2)(n + 6)$	$n^2 + 4n - 12$	✓

Therefore, $n^2 + 4n - 12 = (n - 2)(n + 6)$.

IMPORTANT

When the operations in the binomial factors are different, there are two options for each pair of factors, for example, $(n + 3)(n - 4)$ and $(n - 3)(n + 4)$.

TIP

There is no need to write the binomial factors using 1 and 12 since the correct factorization has already been found.

In the next example, trinomials of the form $ax^2 + bx + c$ where a is *not* 1 will be factored (e.g., $5x^2 + 8x - 3$). Having an a-value not equal to 1 complicates the factoring process because the binomials' first terms will not always be the same. This creates many additional possibilities for the two binomial factors.

EXAMPLE 5 Factoring a Trinomial ($a \neq 1$) by Trial and Error

Factor.

A. $5x^2 - 8x + 3$

B. $8p^2 + 31p - 4$

SOLUTION

A. The trinomial's last term is positive, and its middle term is negative. So, the binomial factors will both be differences.

❶ Like-term factors of $5x^2$ are $5x$ and x.

❷ Like-term factors of 3 are 1 and 3.

❸ The first terms in the binomial factors will be $5x$ and x. The last terms will be 1 and 3. Write the possible binomial pairs, then use the FOIL method to determine the correct factorization.

Possible Factorizations	Product	
$(5x - 1)(x - 3)$	$5x^2 - 16x + 3$	☒
$(5x - 3)(x - 1)$	$5x^2 - 8x + 3$	✓

Therefore, $5x^2 - 8x + 3 = (5x - 3)(x - 1)$.

TIP

Once the correct factors are found, there is no need to continue multiplying possible factorizations.

B. The trinomial's last term is negative. So, one binomial factor will be a sum and the other will be a difference.

❶ Like-term factors of $8p^2$ are $2p$ and $4p$, p and $8p$.

❷ Like-term factors of 4 are 2 and 2, 1 and 4.

❸ The first terms in the binomial factors will be either $2p$ and $4p$ or p and $8p$. The last terms will be either 2 and 2 *or* 1 and 4. Write the possible binomial pairs, then use the FOIL method to determine the correct factorization.

TIP

The operations in the binomial factors are different, so there are two options for each pairs of factors.

Binomials with 2p and 4p as the first terms

Possible Factorizations	Product	
$(2p + 2)(4p - 2)$	$8p^2 + 4p - 4$	✗
$(2p - 2)(4p + 2)$	$8p^2 - 4p - 4$	✗
$(2p + 1)(4p - 4)$	$8p^2 - 4p - 4$	✗
$(2p - 1)(4p + 4)$	$8p^2 + 4p - 4$	✗
$(2p + 4)(4p - 1)$	$8p^2 + 14p - 4$	✗
$(2p - 4)(4p + 1)$	$8p^2 - 14p - 4$	✗

Binomials with p and 8p as the first terms

Possible Factorizations	Product	
$(p + 2)(8p - 2)$	$8p^2 + 14p - 4$	✗
$(p - 2)(8p + 2)$	$8p^2 - 14p - 4$	✗
$(p + 1)(8p - 4)$	$8p^2 + 4p - 4$	✗
$(p - 1)(8p + 4)$	$8p^2 - 4p - 4$	✗
$(p + 4)(8p - 1)$	$8p^2 + 31p - 4$	✓

Therefore, $8p^2 + 31p - 4 = (p + 4)(8p - 1)$.

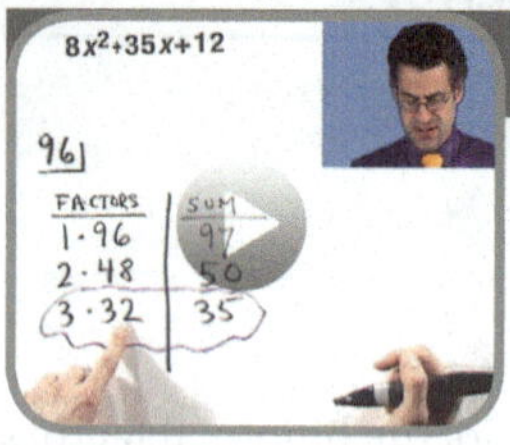

P.5.4 Factoring Trinomials Using the Product-and-Sum Method

Trinomials can be factored into two binomials using a systematic method which is called the **product-and-sum method**. Consider the trinomial $x^2 + 9x + 8$ from **Example 4**. The factorization of this trinomial is $(x + 1)(x + 8)$. Notice that the binomials' last terms, 1 and 8, have a product of 8 and a sum of 9. Furthermore, notice that this product is equal to the trinomial's last term and that the sum is equal to the coefficient of the trinomial's middle term, $9x$. This pattern can be seen in all factorizations of trinomials of the form $ax^2 + bx + c$, when $a = 1$. It follows that for all trinomials of the form $x^2 + bx + c$, the factors of c whose sum is b will be the last terms in the binomial factors.

Factoring $ax^2 + bx + c$ When $a = 1$

If a, b, c, m, and n are numbers such that $a = 1$, $mn = c$, and $m + n = b$, then $ax^2 + bx + c = (x + m)(x + n)$.

If m and n are factors of c such that $m + n = b$, then $x^2 + bx + c = (x + m)(x + n)$.

Note that terms following a subtraction symbol are considered to be negative. For example, $x^2 + bx - c$ is equal to $x^2 + bx + (-c)$. In the case where c is negative, the factors of c must have opposite signs, as in $(x + m)(x + (-n))$ or $(x + m)(x - n)$.

EXAMPLE 6 Factoring a Trinomial ($a = 1$) Using the Product-and-Sum Method

Factor.

A. $x^2 + 7x + 12$

B. $2z^2 - 26z - 28$

SOLUTION

A. Find the factors of c, 12, whose sum is b, 7.

Factors of c	Sum	
(1)(12)	13 *(1)(12) = 12, but 1 + 12 ≠ 7.*	☒
(2)(6)	8 *(2)(6) = 12, but 2 + 6 ≠ 7.*	☒
(3)(4)	7 *(3)(4) = 12, and 3 + 4 = 7.*	✓

Since $3(4) = 12$ and $3 + 4 = 7$, the last terms of the binomial factors will be 3 and 4. Therefore, $x^2 + 7x + 12 = (x + 3)(x + 4)$.

CHECK

Multiply to check the factors.

$(x + 3)(x + 4)$
$= x^2 + 4x + 3x + 12$
$= x^2 + 7x + 12$

So, the factorization is correct. ✓

B. Notice that the a-value (the coefficient of the squared term) in $2z^2 - 26z - 28$ is not 1. Furthermore, notice that the trinomial has a common factor. So, factor out the GCF.

$$2z^2 - 26z - 28 = 2(z^2 - 13z - 14)$$

Factor the trinomial $z^2 - 13z - 14$. (Note that a is now 1.)

Find the factors of c, -14, whose sum is b, -13. Since c is negative, the two factors of c must have opposite signs (i.e., one positive and one negative factor). And since b is also negative, the factor of c with the greater absolute value must be negative.

Factors of c	Sum	
(1)(−14)	−13 *(1)(−14) = −14, and 1 + (−14) = −13.*	✓

Since $1(-14) = -14$ and $1 + (-14) = -13$, the last terms of the binomial factors will be 1 and -14. Therefore, $2z^2 - 26z - 28 = 2(z + 1)(z + (-14)) = 2(z + 1)(z - 14)$.

CHECK

Multiply to check the factors.

$2(z + 1)(z - 14)$
$= 2(z^2 - 14z + z - 14)$
$= 2(z^2 - 13z - 14)$
$= 2z^2 - 26z - 28$

So, the factorization is correct. ✓

When factoring a trinomial of the form $ax^2 + bx + c$, where $a \neq 1$, the factors of a and c must be considered. In this case, the factors of ac whose sum is b will be used in the factorization.

Factoring $ax^2 + bx + c$ When $a \neq 1$

If a, b, c, m, and n are numbers such that $pq = a$, $mn = c$, and $pn + qm = b$, then $ax^2 + bx + c = (px + m)(qx + n)$.

If p and q are factors of a, and m and n are factors of c, such that $pn + qm = b$, then $x^2 + bx + c = (px + m)(qx + n)$.

To factor a trinomial of the form $ax^2 + bx + c$, write the trinomial and express b as the sum of the factors of ac whose sum is b. Then distribute and factor the resulting 4-term polynomial by grouping.

EXAMPLE 7 Factoring a Trinomial ($a \neq 1$) Using the Product-and-Sum Method

Factor.

A. $3x^2 + 11x + 10$ **B.** $10r^2 + 25r - 35$

SOLUTION

A. Identify a, b, and c from the trinomial. $a = 3, b = 11, c = 10$

Find factors of ac, 30, whose sum is b, 11.

Factors of ac	Sum	
(1)(30)	31 *(1)(30) = 30, but 1 + 30 ≠ 11.*	☒
(2)(15)	17 *(2)(15) = 30, but 2 + 15 ≠ 11.*	☒
(3)(10)	13 *(3)(10) = 30, but 3 + 10 ≠ 11.*	☒
(5)(6)	11 *(5)(6) = 30 and 5 + 6 = 11.*	✓

Write the trinomial, and express b as the sum of 5 and 6. Then distribute and factor the 4-term polynomial by grouping.

$3x^2 + 11x + 10 = 3x^2 + (6 + 5)x + 10$ *Write 11 as the sum of 5 and 6.*

$= 3x^2 + 6x + 5x + 10$ *Distribute x.*

$= 3x(x + 2) + 5(x + 2)$ *Factor the GCF from the first two terms and from the last two terms.*

$= (x + 2)(3x + 5)$ *Factor (x + 2) from each of the two terms.*

Therefore, $3x^2 + 11x + 10 = (x + 2)(3x + 5)$.

CHECK

Multiply to check the factors.

$(x + 2)(3x + 5)$

$= 3x^2 + 5x + 6x + 10$

$= 3x^2 + 11x + 10$

So, the factorization is correct. ✓

B. Notice that the terms have a common factor. So, factor out the GCF.

$10r^2 + 25r - 35 = 5(2r^2 + 5r - 7)$

Identify a, b, and c from the trinomial. $a = 2, b = 5, c = -7$

Find factors of ac, −14, whose sum is b, 5. Since ac is negative, the two factors of ac must have opposite signs (i.e., one positive and one negative factor). And since b is positive, the factor of ac with the greater absolute value must be positive.

Factors of ac	Sum	
(−1)(14)	13 *(−1)(14) = −14, but −1 + 14 ≠ 5.*	☒
(−2)(7)	5 *(−2)(7) = −14, and −2 + 7 = 5.*	✓

Write the trinomial, and express b as the sum of −2 and 7. Then distribute and factor the 4-term polynomial by grouping.

$2r^2 + 5r - 7 = 2r^2 + (-2 + 7)r - 7$ *Write 5 as the sum of −2 and 7.*

$= 2r^2 - 2r + 7r - 7$ *Distribute r.*

$= 2r(r - 1) + 7(r - 1)$ *Factor the GCF from the first two terms and from the last two terms.*

$= (r - 1)(2r + 7)$ *Factor (r − 1) from each of the two terms.*

Therefore, $10r^2 + 25r - 35 = 5(r - 1)(2r + 7)$.

P.5.5 Factoring Perfect-Square Trinomials

Consider the following examples of the square of a binomial.

$$(x+3)^2 = (x+3)(x+3) = x^2 + 6x + 9 \qquad (x-5)^2 = (x-5)(x-5) = x^2 - 10x + 25 \qquad (4x+3)^2 = (4x+3)(4x+3) = 16x^2 + 24x + 9$$

Notice that in each trinomial product,
- the first term is the square of the binomial's first term,
- the middle term is twice the product of the binomial's terms, and
- the last term is the square of the binomial's last term.

To generalize this pattern, consider the squares of the binomials $(A + B)$ and $(A - B)$.

$$(A+B)^2 = (A+B)(A+B) = A^2 + 2AB + B^2$$
$$(A-B)^2 = (A-B)(A-B) = A^2 - 2AB + B^2$$

A **perfect-square trinomial** is any trinomial that can be written in the form $A^2 + 2AB + B^2$ or $A^2 - 2AB + B^2$. It follows that all perfect-square trinomials can be written as the square of a binomial. Use these formulas to factor perfect-square trinomials.

Factoring Perfect-Square Trinomials
$A^2 + 2AB + B^2 = (A+B)^2$
$A^2 - 2AB + B^2 = (A-B)^2$

EXAMPLE 8 Factoring a Perfect-Square Trinomial

Factor.

A. $49x^2 + 28x + 4$

B. $36a^2 - 60ab + 25b^2$

SOLUTION

Notice that the first and last terms of each trinomial are perfect squares, and that each trinomial can be written in the form of a perfect-square trinomial: $A^2 + 2AB + B^2$ or $A^2 - 2AB + B^2$.

	$49x^2 + 28x + 4$	$36a^2 - 60ab + 25b^2$
Perfect-Square Terms	$49x^2 = (7x)^2$ and $4 = (2)^2$	$36a^2 = (6a)^2$ and $25b^2 = (5b)^2$
Perfect-Square Trinomial	$(7x)^2 + 2(7x)(2) + 2^2$ $A^2 + 2AB + B^2$	$(6a)^2 - 2(6a)(5b) + (5b)^2$ $A^2 - 2AB + B^2$
A- and B-values	$A = 7x$ and $B = 2$	$A = 6a$ and $B = 5b$

A. Substitute $A = 7x$ and $B = 2$ into $(A + B)^2$ to find the factorization.

$$49x^2 + 28x + 4 = (7x + 2)^2$$

B. Substitute $A = 6a$ and $B = 5b$ into $(A - B)^2$ to find the factorization.

$$6a^2 - 60ab + 25b^2 = (6a - 5b)^2$$

CHECK

Multiply to check each factorization.

A. $(7x+2)^2$
$= (7x+2)(7x+2)$
$= 49x^2 + 28x + 4$

B. $(6a-5b)^2$
$= (6a-5b)(6a-5b)$
$= 36a^2 - 60ab + 25b^2$

So, the factorizations are correct. ✓

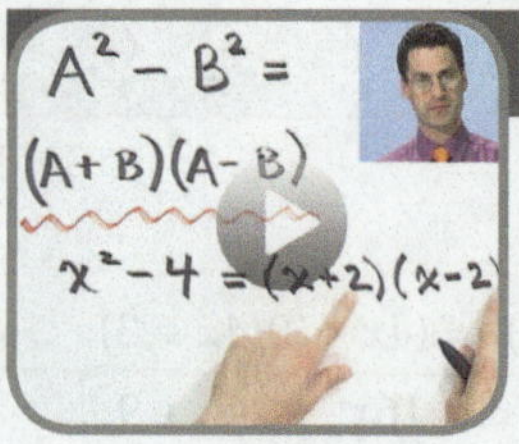

P.5.6 Factoring the Difference of Two Squares

Recall that the product of a *sum and difference of same terms* is a special case in which the product of two binomials is also a binomial.

$$(x + 3)(x - 3) = x^2 - 3x + 3x - 9 = x^2 - 9$$
$$(5w + 2y)(5w - 2y) = 25w^2 - 10wy + 10wy - 4y^2 = 25w^2 - 4y^2$$

This pattern can be generalized using the product of $(A + B)$ and $(A - B)$.

$$(A + B)(A - B) = A^2 - B^2$$

It follows that a binomial of the form $A^2 - B^2$, called a **difference of two squares**, can be factored into two binomials that are the sum and difference of same terms.

Factoring Difference-of-Two-Squares Binomials

$$A^2 - B^2 = (A + B)(A - B)$$

EXAMPLE 9 Factoring a Difference of Two Squares

Factor.

A. $1 - 64m^2$ **B.** $25a^6 - b^4$

SOLUTION

Notice that each binomial is a difference of perfect squares, and that each binomial can be written as a difference of two squares: $A^2 - B^2$.

	$1 - 64m^2$	**$25a^6 - b^4$**
Perfect-Square Terms	$1 = (1)^2$ and $64m^2 = (8m)^2$	$25a^6 = (5a^3)^2$ and $b^4 = (b^2)^2$
Difference of Two Squares	$(1)^2 - (8m)^2$ $A^2 - B^2$	$(5a^3)^2 - (b^2)^2$ $A^2 - B^2$
A- and B-values	$A = 1$ and $B = 8m$	$A = 5a^3$ and $B = b^2$

A. Substitute $A = 1$ and $B = 8m$ into $(A + B)(A - B)$ to find the factorization.

$$1 - 64m^2 = (1 + 8m)(1 - 8m)$$

B. Substitute $A = 5a^3$ and $B = b^2$ into $(A + B)(A - B)$ to find the factorization.

$$25a^6 - b^4 = (5a^3 + b^2)(5a^3 - b^2)$$

CHECK

Multiply to check each factorization.

A. $(1 + 8m)(1 - 8m)$
$= 1 - 8m + 8m - 64m^2$
$= 1 - 64m^2$

B. $(5a^3 + b^2)(5a^3 - b^2)$
$= 25a^6 - 5a^3b^2 + 5a^3b^2 - b^4$
$= 25a^6 - b^4$

So, the factorizations are correct. ✓

WHAT'S THE BIG IDEA?

Explain how to identify and factor a perfect-square trinomial and a difference of squares.

P.5.7 Factoring Sums and Differences of Cubes

In the previous topic, it was shown that a difference of squares can be factored into two binomials that are the sum and difference of the same terms. Note that a *sum* of squares cannot be factored into two binomials.

Sums *and* differences of cubes can be factored, but not into two binomials. A **sum of cubes** is a binomial that can be written in the form $A^3 + B^3$. Similarly, a **difference of cubes** is a binomial that can be written in the form $A^3 - B^3$. The following formulas can be used to factor a sum or difference of cubes.

Factoring Sums and Differences of Two Cubes
$A^3 + B^3 = (A + B)(A^2 - AB + B^2)$
$A^3 - B^3 = (A - B)(A^2 + AB + B^2)$

EXAMPLE 10 Factoring a Sum or Difference of Cubes

Factor.

A. $1 + 8c^3$ **B.** $64w^6 - 125y^3z^{12}$ **C.** $d^3 - (1 + g)^6$

SOLUTION

A. The binomial is a sum of perfect cubes. Since $1 = (1)^3$ and $8c^3 = (2c)^3$, it follows that $A = 1$ and $B = 2c$. Substitute $A = 1$ and $B = 2c$ into $(A + B)(A^2 - AB + B^2)$, and simplify to find the factorization.

$$1 + 8c^3 = (1 + 2c)((1)^2 - (1)(2c) + (2c)^2) = (1 + 2c)(1 - 2c + 4c^2)$$

TIP

Note that the trinomial factor $1 - 2c + 4c^2$ is irreducible, so it cannot be factored further.

B. The binomial is a difference of perfect cubes. Since $64w^6 = (4w^2)^3$ and $125y^3z^{12} = (5yz^4)^3$, it follows that $A = 4w^2$ and $B = 5yz^4$. Substitute $A = 4w^2$ and $B = 5yz^4$ into $(A - B)(A^2 + AB + B^2)$, and simplify to find the factorization.

$$\begin{aligned} 64w^6 - 125y^3z^{12} &= (4w^2 - 5yz^4)((4w^2)^2 + (4w^2)(5yz^4) + (5yz^4)^2) \\ &= (4w^2 - 5yz^4)(16w^4 + 20w^2yz^4 + 25y^2z^8) \end{aligned}$$

C. The binomial is a difference of perfect cubes. Since $d^3 = (d)^3$ and $(1 + g)^6 = ((1 + g)^2)^3$, it follows that $A = d$ and $B = (1 + g)^2$. Substitute $A = d$ and $B = (1 + g)^2$ into $(A - B)(A^2 + AB + B^2)$, and simplify to find the factorization.

$$\begin{aligned} d^3 - (1 + g)^6 &= (d - (1 + g)^2)((d)^2 + (d)((1 + g)^2) + ((1 + g)^2)^2) \\ &= (d - (1 + g)^2)(d^2 + d(1 + g)^2 + (1 + g)^4) \end{aligned}$$

WHAT'S THE BIG IDEA?

Explain how to identify and factor a sum of cubes and a difference of cubes.

SECTION P.5 EXERCISES

Warm Up

Simplify.

1. $2m^2(8m^2 + m - 4)$
2. $(b + 2)(b + 9)$
3. $(2p + 5)(2p - 5)$

Just the Facts

Fill in the blanks.

4. A factored polynomial is written as the ____ of two or more polynomial factors.
5. $xy^2 + 2x^2y =$ (____)$(y + 2x)$
6. $18mn^7 + 24m^2np =$ (____)$(3n^6 + 4mp)$
7. $20y^3 + 15y^2 - 30y =$ (____)$(4y^2 + 3y - 6)$
8. The factorization of a polynomial can be checked by ____ the polynomial factors.
9. If one factor of a difference of two squares is ____, then the other factor is $(x + y)$.
10. The ____ of two cubes factors into a binomial difference and a trinomial sum.

Essential Skills

In Exercises 1–62, factor each polynomial completely.

1. $2x^3y^8 + 6x^4y^2 + 10x^5y^{10}$
2. $16a^3b^2 + 4a^2b^2 - 12ab^3 - 4ab^2$
3. $3x^2(2x + 5y) + 7y^2(2x + 5y)$
4. $6x^3(2 - y) - 8x^2(2 - y) + 4x(y - 2)$
5. $21b^3(c + 4) + 42b^2(c + 4) - 14b(c + 4)$
6. $15x^3(y + 3) - 40x^2(-3 - y) - 10x(y + 3)$
7. $8a^3(b + 5) - 2a^2(-5 - b) - 22a(b + 5)$
8. $8m^6n^4 - 14m^3n^2 + 2mn^3 - 7mn^2$
9. $10x^2 + 2xy + 15xy + 3y^2$
10. $2y^3 + 3y^2 - 4y - 6$
11. $4a^2 + 8ab + 48b + 24a$
12. $7y^3 + 3y^2 - 56y - 24$
13. $5x^2 - 15xy + 45y - 15x$
14. $6m^2 + 8mn - 12n - 9m$
15. $k^2 + 14k + 24$
16. $z^2 - 9z + 20$
17. $b^2 + 13b - 30$
18. $m^2 - 18m - 40$
19. $a^2 - 5a - 6$
20. $x^2 + x - 20$
21. $3h^2 + 19h + 6$
22. $2x^2 + 11x - 6$
23. $6x^2 - 11x + 3$
24. $7x^2 - 43x - 42$
25. $3b^2 + 7b - 6$
26. $6y^2 - 31y - 30$
27. $14x^2 + 15xy + 4y^2$
28. $3y^4 - 8y^2z + 5z^2$
29. $3p^4 + 4p^2q - 4q^2$
30. $4s^4 - 13s^2t + 12t^2$
31. $5w^6 + 12w^3x - 9x^2$
32. $4t^4 + 5t^2x - 6x^2$
33. $8c^2 - 52c + 60$
34. $40x^2 - 70x + 15$
35. $60m^2 + 51m - 30$
36. $36y^2 + 21y - 30$
37. $80x^2 - 132x + 40$
38. $80s^2 + 68s - 40$
39. $n^2 + 20n + 100$
40. $25x^2 + 30x + 9$
41. $1 - 10z^2 + 25z^4$
42. $16x^2 - 72xy + 81y^2$
43. $49x^2 - 70xy + 25y^2$
44. $4s^2 + 12st + 9t^2$
45. $9x^2 - 49$
46. $121m^2 - n^6$
47. $4p^{10} + 9w^2$
48. $16q^8 - 25r^2$
49. $144x^4 - 100y^2$
50. $9p^4 - 16w^2$
51. $27p^3 + 8$
52. $216x^3 + 1$
53. $u^6 - 8$
54. $64 - 27t^3$

55. $343 - 64y^3$

56. $27w^3 + 1$

57. $q^9 + r^6$

58. $s^6t^3 - 125u^{12}$

59. $(x + 2y)^6 - 8$

60. $(2r - 3s)^3 + t^9$

61. $(m + n)^9 - 27$

62. $343t^3z^6 - 64s^3$

Extensions

In Exercises 63–72, factor.

63. $\pi r^2h + \frac{4}{3}\pi r^3h^2 + \frac{1}{3}\pi r^2h^3$

64. $8m^4 - 20m^2n - 12n^2$

65. $-84np^2 - 21npz + 28npz^2 + 7nz^3$

66. $30x^2 + 87xy + 30y^2\, 20y^{2n} + 16y^n + 3$

67. $20y^{2n} + 16y^n + 3$

68. $y^5 - 2y^4 - 35y^3$

69. $x^2 + 2ax + a^2 - y^2 + 2by - b^2$

70. $7x^2 + 10\sqrt{7}x + 25$

71. $(8(x - 5y))^3 + 27z^3$

72. $64x^9 - 8y^9$

73. A student factored $2m^4 - am^4 + 6 - 3a$ by grouping and obtained $(m^4 + 3)(2 - a)$. Another student factored the same polynomial and gave an answer of $(a - 2)(3 - m^4)$. Which answer is correct?

74. The area of a rectangle is represented by $9k^2 - 24k + 16$. Find expressions for the dimensions of the rectangle in terms of k.

75. Suppose $cx^2 - 81$ is a difference of two squares and c is a natural number. What must be true about c?

76. Factor $4x^2y^2 - 7z^2$ into a product of two binomials that are a sum and a difference of same terms.

P.6 RATIONAL EXPRESSIONS

OBJECTIVES

- Simplify and find the domain of rational expressions.
- Multiply, divide, add, and subtract rational expressions.
- Simplify complex fractions.

PREREQUISITE VOCABULARY TERMS

factor (of a polynomial)
fractional expression
polynomial

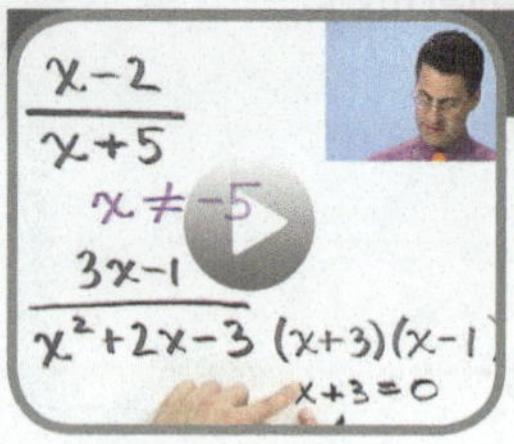

P.6.1 Rational Expressions and Domain

Consider any polynomial, such as $3x^2 + 5x - 1$. Substituting any real number for x results in a real number. Therefore, all real numbers are *allowable values* for x in $3x^2 + 5x - 1$. The set of allowable values for the variable in an expression is called the expression's **domain**. So, the domain of $3x^2 + 5x - 1$, or of any polynomial, is all real numbers, $(-\infty, \infty)$.

When any value of the variable in an expression results in an undefined value, then this expression's domain is not all real numbers. One type of expression where the domain may not be all real numbers is a **rational expression**.

DEFINITION

A fractional expression where the numerator and denominator are polynomials (and the denominator is not 0) is called a **rational expression**.

REMEMBER

The numerator and denominator of a fractional expression can be any algebraic expression.

For example, consider the rational expression $\frac{2}{x}$. Since fractions are undefined when the denominator is 0, the domain of $\frac{2}{x}$ is the set of all real numbers except $x = 0$. The domain of $\frac{2}{x}$ can be expressed using interval notation as $(-\infty, 0) \cup (0, \infty)$, or using set-builder notation as $\{x \mid x \neq 0\}$.

Generally, to find the domain of a rational expression, find all values of the variable that result in a 0 denominator, and then exclude those values from the domain. Find the values that result in a 0 denominator by setting the expression in the denominator equal to 0 and solving.

EXAMPLE 1 Finding the Domain of a Rational Expression

Write the domain of each expression using set-builder notation or interval notation.

A. $\dfrac{x+10}{4-x}$ **B.** $\dfrac{3}{x^2+3x+2}$

SOLUTION

Find the values of x that result in a 0 denominator, then exclude those values from the domain.

A. $4 - x = 0 \Rightarrow x = 4$ Domain: $\{x \mid x \neq 4\}$ *The domain is all real numbers except 4.*

B. $x^2 + 3x + 2 = 0 \Rightarrow (x+2)(x+1) = 0 \Rightarrow x = -2$ or -1

Domain: $\{x \mid x \neq -2$ and $x \neq -1\}$ *The domain is all real numbers except −2 and −1.*

P.6.2 Simplifying Rational Expressions

Recall that numeric fractions are simplified by removing (canceling) any factors common to the numerator and denominator (other than 1). For example, the fraction $\frac{10}{15}$ can be simplified to $\frac{2}{3}$ by removing the common factor 5 from the numerator and denominator.

$$\frac{10}{15}=\frac{2\cancel{(5)}}{3\cancel{(5)}}=\frac{2}{3}$$

A fraction is in simplest form when the numerator and denominator have no common factors other than 1.

The same process is used for simplifying rational expressions.

Simplifying Rational Expressions

To simplify a rational expression, remove the factors common to the numerator and denominator. If A, B, and C are polynomials such that $B \neq 0$ and $C \neq 0$, then $\dfrac{AC}{BC}=\dfrac{A}{B}$.

A rational expression is in **simplest form** when the numerator and denominator have no common factors other than 1.

IMPORTANT

Always use the original expression to determine the values excluded from the domain.

Unless otherwise directed, the domain of a rational expression needs to be stated along with the simplified form.

Steps for Simplifying a Rational Expression

1. Factor the polynomials in the numerator and denominator completely.
2. Remove all factors common to the numerator and denominator.
3. Write the remaining factors and multiply.
4. State the domain of the original expression (unless otherwise directed).

EXAMPLE 2 Simplifying a Rational Expression

Simplify. State the values excluded from the domain, if any.

A. $\dfrac{(2x+3)(x+4)}{2x+8}$ **B.** $\dfrac{24x^2-6}{48x^2+8x-16}$

SOLUTION

Factor each polynomial (as needed), then remove the common factors. Exclude from the domain the x-values that result in a 0 denominator in the original expression.

IMPORTANT

The value excluded from the domain is -4 because $2x+8$ is 0 when $x=-4$.

A. $\dfrac{(2x+3)(x+4)}{2x+8}=\dfrac{(2x+3)\cancel{(x+4)}}{2\cancel{(x+4)}}=\dfrac{2x+3}{2},\ x\neq -4$

The expression is in simplest form because 2 (the denominator) is not a factor of $2x+3$ (the numerator).

B. $\dfrac{24x^2-6}{48x^2+8x-16}=\dfrac{6(4x^2-1)}{8(6x^2+x-2)}$ *Factor out the GCF in the numerator and denominator.*

$=\dfrac{6(2x+1)\cancel{(2x-1)}}{8\cancel{(2x-1)}(3x+2)}$ *Factor the difference of two squares (numerator) and the trinomial (denominator).*

$=\dfrac{3(2x+1)}{4(3x+2)}$ *Remove the common factors.*

$=\dfrac{6x+3}{12x+8},\ x\neq\dfrac{1}{2}$ and $x\neq-\dfrac{2}{3}$ *Distribute and write the excluded values.*

P.6.3 Factoring –1 from a Rational Expression to Simplify

When the numerator of a rational expression contains a factor of the form $(a + b)$ and the denominator contains a factor of the form $(b + a)$, the factors can be removed because addition is commutative: $(a + b) = (b + a)$.

However, if the numerator of a rational expression contains a factor of the form $(a - b)$ and the denominator contains a factor of the form $(b - a)$, then the factors cannot be removed because subtraction is not commutative: $(a - b) \neq (b - a)$. The factors can be removed only after factoring –1 from either (but not both) of the factors.

IMPORTANT

$b - a$ is 0 when $a = b$.

$$\frac{a-b}{b-a} = \frac{(-1)(-a+b)}{b-a} = \frac{(-1)(b+(-a))}{b-a} = \frac{(-1)\cancel{(b-a)}}{\cancel{b-a}} = \frac{-1}{1} = -1\,,\; a \neq b$$

EXAMPLE 3 Simplifying a Rational Expression

Simplify $\dfrac{4x-3x^2}{9x^2-16}$. State the values excluded from the domain, if any.

SOLUTION

Factor each polynomial, then remove the common factors. Exclude from the domain the x-values that result in a 0 denominator in the original expression.

IMPORTANT

$9x^2 - 16$ is 0 when $x = -4/3$ and $x = 4/3$.

$$\frac{4x-3x^2}{9x^2-16} = \frac{x(4-3x)}{(3x+4)(3x-4)} = \frac{x(-1)\cancel{(3x-4)}}{(3x+4)\cancel{(3x-4)}} = \frac{-x}{3x+4}\,,\; x \neq \frac{4}{3} \text{ and } x \neq -\frac{4}{3}$$

P.6.4 Multiplying and Dividing Rational Expressions

Recall that numeric fractions are multiplied by multiplying the numerators and multiplying the denominators. The product can then be simplified (reduced) by removing (canceling) any common factors from the numerator and denominator. Common factors can be removed before or after multiplying the numerators and denominators, but removing the common factors before multiplying is typically easier. For example, consider the product of $\frac{4}{9}$ and $\frac{12}{5}$.

$$\frac{4}{9}\cdot\frac{12}{5} = \frac{4(12)}{9(5)} = \frac{4\cancel{(3)}(4)}{\cancel{(3)}(3)(5)} = \frac{4(4)}{(3)(5)} = \frac{16}{15}$$

The common factor 3 is removed before multiplying.

The same process is used for multiplying rational expressions.

Multiplying Rational Expressions

Rational expressions can be multiplied by multiplying the numerators and multiplying the denominators.

If A, B, C, and D are polynomials such that $B \neq 0$ and $D \neq 0$, then $\dfrac{A}{B}\cdot\dfrac{C}{D} = \dfrac{AC}{BD}$.

Common factors can be removed before or after multiplying.

Recall that the process used for dividing fractions is the same as the process for multiplying fractions, but with one additional step: the division is written as multiplication, and the divisor is written as its reciprocal. Rational expressions can be divided using the same process.

Dividing Rational Expressions

If A, B, C, and D are polynomials such that B, C, and $D \neq 0$, then $\frac{A}{B} \div \frac{C}{D} = \frac{A}{B} \cdot \frac{D}{C} = \frac{AD}{BC}$.

The steps for dividing rational expressions are the same as the steps for multiplying rational expressions, after the division expression has been written as a multiplication expression.

From here forward, we will assume that all variables result in nonzero denominators, and so we will no longer state the domain of an expression when simplifying.

EXAMPLE 4 Multiplying and Dividing Rational Expressions

Simplify. Assume that all variables result in nonzero denominators.

A. $\frac{10c}{3c^2d^2 - 2cd^3} \cdot \frac{6d^3 - 9cd^2}{2 - 8c}$ **B.** $\frac{6x^2 + 3x}{x + 1} \div \frac{2x^2 - 3x - 2}{x + 4}$

SOLUTION

A. $$\frac{10c}{3c^2d^2 - 2cd^3} \cdot \frac{6d^3 - 9cd^2}{2 - 8c} = \frac{(10c)(6d^3 - 9cd^2)}{(3c^2d^2 - 2cd^3)(2 - 8c)}$$ *Multiply numerators and denominators.*

Notice that expressions in the numerator and denominator can be factored.

- Factor the GCF $3d^2$ from $6d^3 - 9cd^2$ in the numerator.
- Factor the GCF cd^2 from $3c^2d^2 - 2cd^3$ in the denominator.
- Factor the GCF 2 from $2 - 8c$ in the denominator.

$$\frac{(10c)(6d^3 - 9cd^2)}{(3c^2d^2 - 2cd^3)(2 - 8c)} = \frac{(10c)(3d^2)(2d - 3c)}{(cd^2)(3c - 2d)(2)(1 - 4c)}$$ *Factor.*

The numerator and denominator both contain the factors c and d^2, so those factors can be removed. Additionally, 10 in the numerator can be written as (2)(5), and then the common factor 2 can be removed.

$$\frac{(10c)(3d^2)(2d - 3c)}{(cd^2)(3c - 2d)(2)(1 - 4c)} = \frac{\cancel{(2)}(5)(\cancel{c})(3\cancel{d^2})(2d - 3c)}{(\cancel{c}\,\cancel{d^2})(3c - 2d)\cancel{(2)}(1 - 4c)}$$ *Remove common factors.*

$$= \frac{5(3)(2d - 3c)}{(3c - 2d)(1 - 4c)}$$ *Write the remaining factors.*

Notice that the numerator contains $2d - 3c$ and the denominator contains $3c - 2d$. So, factor -1 from $3c - 2d$ in the denominator (or from $2d - 3c$ in the numerator, but not both), and then remove the common factor $2d - 3c$.

$$\frac{5(3)(2d - 3c)}{(3c - 2d)(1 - 4c)} = \frac{5(3)(2d - 3c)}{(-1)(2d - 3c)(1 - 4c)}$$ *Factor.*

$$= \frac{5(3)\cancel{(2d - 3c)}}{(-1)\cancel{(2d - 3c)}(1 - 4c)}$$ *Remove the common factor.*

$$= \frac{5(3)}{(-1)(1 - 4c)}$$ *Write the remaining factors.*

$$= \frac{15}{4c - 1}$$ *Multiply.*

TIP

Factoring -1 from $3c - 2d$ results in $2d - 3c$.

$3c - 2d = (-1)(-3c + 2d)$
$= (-1)(2d - 3c)$

B. $\dfrac{6x^2+3x}{x+1} \div \dfrac{2x^2-3x-2}{x+4} = \dfrac{6x^2+3x}{x+1} \cdot \dfrac{x+4}{2x^2-3x-2}$ *Multiply by the reciprocal of the divisor.*

$= \dfrac{(6x^2+3x)(x+4)}{(x+1)(2x^2-3x-2)}$ *Multiply numerators and denominators.*

$= \dfrac{(3x)(2x+1)(x+4)}{(x+1)(2x+1)(x-2)}$ *Factor and remove common factors.*

$= \dfrac{(3x)(x+4)}{(x+1)(x-2)}$ *Write the remaining factors.*

$= \dfrac{3x^2+12x}{x^2-x-2}$ *Multiply.*

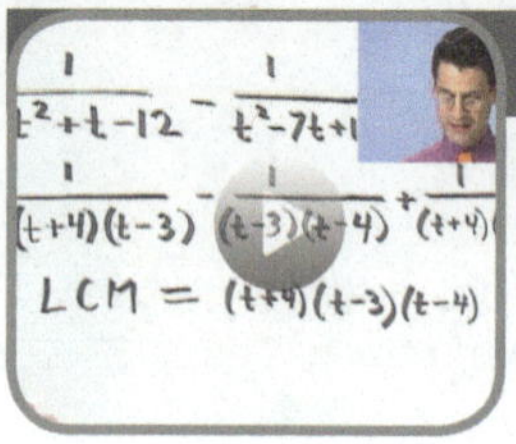

P.6.5 Adding and Subtracting Rational Expressions

As with numeric fractions, rational expressions can be added or subtracted when the expressions have a common denominator. For rational expressions with unlike denominators, the common denominator is a common multiple of the denominators.

Adding and Subtracting Rational Expressions

Like Denominators
If A, B, and C are polynomials such that $B \neq 0$, then

$$\frac{A}{B}+\frac{C}{B}=\frac{A+C}{B} \text{ and } \frac{A}{B}-\frac{C}{B}=\frac{A-C}{B}.$$

Unlike Denominators
If A, B, C, and D are polynomials such that B and $D \neq 0$, then

$$\frac{A}{B}+\frac{C}{D}=\frac{AD+BC}{BD} \text{ and } \frac{A}{B}-\frac{C}{D}=\frac{AD-BC}{BD}.$$

TIP

In $\frac{A-C}{B}$, if C is a polynomial with more than one term, be sure to distribute the negative to each term in C.

Addition (and subtraction) of rational expressions is easiest when the LCM (least common multiple) of the denominators is used for the common denominator. The LCM of the denominators is the product of the greatest power of the denominators' factors. For example, if the denominators of three rational expressions are xy, x^2, and xz, then the LCM of the denominators is x^2yz.

EXAMPLE 5 Adding and Subtracting Rational Expressions

Simplify. Assume that all variables result in nonzero denominators.

A. $\dfrac{7x}{y^4z^3}+\dfrac{3}{y^5z}$ **B.** $\dfrac{7ab}{a^2-b^2}-\dfrac{a-b}{a+b}$

SOLUTION

A. The factors in the denominators are powers of y and z. The greatest power of y is y^5, and the greatest power of z is z^3, so the LCM of the denominators is y^5z^3. Since $y^4z^3(y) = y^5z^3$, multiply the first term by $\frac{y}{y}$. Since $y^5z(z^2) = y^5z^3$, multiply the second term by $\frac{z^2}{z^2}$.

$\dfrac{7x}{y^4z^3}+\dfrac{3}{y^5z} = \dfrac{7x}{y^4z^3}\cdot\dfrac{y}{y}+\dfrac{3}{y^5z}\cdot\dfrac{z^2}{z^2}$ *Write the terms with a common denominator.*

$= \dfrac{7xy}{y^5z^3}+\dfrac{3z^2}{y^5z^3}$ *Multiply.*

$= \dfrac{7xy+3z^2}{y^5z^3}$ *Add the numerators.*

B. Factor the first term's denominator, and then identify the LCM of the denominators.

$$\frac{7ab}{a^2-b^2}-\frac{a-b}{a+b}=\frac{7ab}{(a+b)(a-b)}-\frac{a-b}{a+b}$$

The factors in the denominators are $(a + b)$ and $(a - b)$, where each is to the power of 1. So, the LCM of the denominators is $(a + b)(a - b)$. The first term's denominator is already $(a+b)(a-b)$. Multiply the second term by $\frac{a-b}{a-b}$, so that its denominator will be $(a+b)(a-b)$.

TIP

The denominator of the first expression is equal to the LCM.

$$\frac{7ab}{(a+b)(a-b)}-\frac{a-b}{a+b}=\frac{7ab}{(a+b)(a-b)}-\frac{(a-b)(a-b)}{(a+b)(a-b)}$$ *Write the terms with a common denominator.*

$$=\frac{7ab-(a-b)(a-b)}{(a+b)(a-b)}$$ *Subtract the numerators.*

$$=\frac{7ab-(a^2-2ab+b^2)}{(a+b)(a-b)}$$ *FOIL in the numerator.*

$$=\frac{7ab-a^2+2ab-b^2}{(a+b)(a-b)}$$ *Distribute.*

$$=\frac{-a^2+9ab-b^2}{a^2-b^2}$$ *Simplify.*

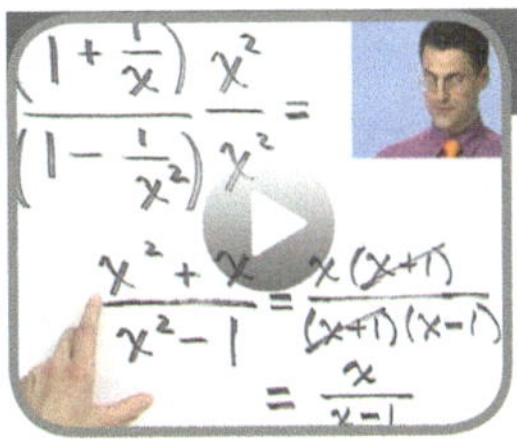

P.6.6 Rewriting Complex Fractions

A rational expression that contains a rational expression in the numerator or denominator is called a **complex fraction** (i.e., a complex rational expression). Two methods for simplifying complex fractions will be discussed in this topic.

IMPORTANT

A sum or difference of fractions should be combined using a common denominator before writing as division.

Simplifying Complex Fractions

Method 1: Multiply by the Reciprocal
Write the fraction as division, then multiply the dividend (the numerator) by the reciprocal of the divisor (the denominator).

Method 2: Use the LCM
Find the LCM of the denominators, then multiply the expression by the $\frac{\text{LCM}}{\text{LCM}}$.

The methods are demonstrated here using two example complex fractions.

$\dfrac{\frac{2}{3}}{\frac{6}{7}}$ *Equivalent to* $\frac{2}{3}\div\frac{6}{7}$ $\qquad$ $\dfrac{\frac{3}{x}}{\frac{x+1}{x^2}}$ *Equivalent to* $\frac{3}{x}\div\frac{x+1}{x^2}$

	Method 1: Multiply by the Reciprocal	**Method 2: Use the LCM**	
$\dfrac{\frac{2}{3}}{\frac{6}{7}}$	$\dfrac{\frac{2}{3}}{\frac{6}{7}}=\frac{2}{3}\div\frac{6}{7}=\frac{2}{3}\cdot\frac{7}{6}=\frac{7}{9}$	$\dfrac{\frac{2}{3}}{\frac{6}{7}}\cdot\frac{21}{21}=\dfrac{\frac{2}{3}\cdot 21}{\frac{6}{7}\cdot 21}=\frac{14}{18}=\frac{7}{9}$	*LCM(3 and 7) = 21*
$\dfrac{\frac{3}{x}}{\frac{x+1}{x^2}}$	$\dfrac{\frac{3}{x}}{\frac{x+1}{x^2}}=\frac{3}{x}\div\frac{x+1}{x^2}=\frac{3}{x}\cdot\frac{x^2}{x+1}=\frac{3x}{x+1}$	$\dfrac{\frac{3}{x}}{\frac{x+1}{x^2}}\cdot\frac{x^2}{x^2}=\dfrac{\frac{3}{x}\cdot x^2}{\frac{x+1}{x^2}\cdot x^2}=\frac{3x}{x+1}$	*LCM(x and x^2) = x^2*

EXAMPLE 6 Simplifying a Complex Fraction

Simplify. Assume that all variables result in nonzero denominators.

A. $\dfrac{4+\frac{8}{x}}{1+\frac{x}{2}}$ **B.** $\dfrac{\frac{5}{x+5}}{\frac{1}{x^2+5x}-\frac{2}{x}}$

SOLUTION

IMPORTANT

Combine each sum (in the numerator and in the denominator) before writing as division.

A. Method 1: Multiply by the Reciprocal

$$\frac{4+\frac{8}{x}}{1+\frac{x}{2}}=\frac{\frac{4x}{x}+\frac{8}{x}}{\frac{2}{2}+\frac{x}{2}}$$ *Write the terms with a common denominator.*

$$=\frac{\frac{4x+8}{x}}{\frac{2+x}{2}}$$ *Add the numerators.*

$$=\frac{4x+8}{x}\div\frac{2+x}{2}$$ *Write as division.*

$$=\frac{4x+8}{x}\cdot\frac{2}{2+x}$$ *Multiply by the reciprocal.*

$$=\frac{(4x+8)(2)}{x(2+x)}$$ *Multiply the numerators and denominators.*

$$=\frac{4\cancel{(x+2)}(2)}{x\cancel{(2+x)}}$$ *Factor and remove the common factors.*

$$=\frac{8}{x}$$ *Simplify.*

TIP

x + 2 and 2 + x are equivalent expressions.

B. Method 2: Use the LCM

Find the LCM of the denominators in $\frac{5}{x+5}$, $\frac{1}{x^2+5x}$, and $\frac{2}{x}$.

$\text{LCM}(x+5, x^2+5x, \text{ and } x) = x(x+5)$ *x^2+5x can be factored as $x(x+5)$.*

Multiply the expression by $\frac{x(x+5)}{x(x+5)}$, then factor and simplify.

$$\frac{\frac{5}{x+5}}{\frac{1}{x^2+5x}-\frac{2}{x}}\cdot\frac{x(x+5)}{x(x+5)}=\frac{\left(\frac{5}{x+5}\right)(x(x+5))}{\left(\frac{1}{x^2+5x}-\frac{2}{x}\right)(x(x+5))}$$ *Multiply by the LCM/LCM.*

$$=\frac{\frac{5x\cancel{(x+5)}}{\cancel{x+5}}}{\frac{\cancel{x}\cancel{(x+5)}}{\cancel{x}\cancel{(x+5)}}-\frac{2\cancel{x}(x+5)}{\cancel{x}}}$$ *Distribute, factor, and remove the common factors.*

$$=\frac{5x}{1-2(x+5)}$$ *Simplify.*

$$=\frac{5x}{-2x-9}$$ *Simplify the denominator.*

The simplified expression in **Example 6B** can be written as $-\dfrac{5x}{2x+9}$ by factoring -1 from $-2x-9$.

$$\frac{5x}{-2x-9}=\frac{5x}{-(2x+9)}=-\frac{5x}{2x+9}$$

SECTION P.6 EXERCISES

Warm Up

Simplify.

1. $\dfrac{24}{40}$
2. $\dfrac{9}{10} \div \dfrac{15}{2}$
3. $\dfrac{5}{8} + \left(\dfrac{2}{3} - \dfrac{1}{6}\right)$

Just the Facts

Fill in the blanks.

4. Fractional expressions where the numerator and denominator are _____ are called _____.
5. The domain of an expression is the set of all _____ values of the variable.
6. The domain of an expression with denominator $x^2 + 1$ is _____.
7. If the denominator of a rational expression contains the factor $x - 3$, then _____ must be excluded from the _____.
8. A rational expression can be simplified by removing _____.
9. A rational expression where the numerator is $x^2 + 6x + 5$ and the denominator is _____ simplifies to $x + 5$.
10. The LCM of $x^3(x + 1)^2$, x^6, and $x^2(x + 1)(x + 2)$ is _____.

Essential Skills

In Exercises 1–6, write the domain of each expression using set-builder notation or interval notation.

1. $x^3 + 4$
2. $\dfrac{3b}{2 - b}$
3. $\dfrac{x + 5}{6x^2 + 8x^3}$
4. $\dfrac{p - 5}{p^2 - 2p - 63}$
5. $\dfrac{6x}{x^2 - 5x + 6}$
6. $\dfrac{6n}{3 + n^2}$

In Exercises 7–24, simplify and state any values excluded from the domain of each expression.

7. $\dfrac{(x + 3)(x - 4)}{(x - 4)(x - 1)}$
8. $\dfrac{4x - 4}{12x - 12}$
9. $\dfrac{(a - 6)(a + 5)}{(a + 3)(a - 6)}$
10. $\dfrac{6p - 24}{12p - 48}$
11. $\dfrac{20 - 4x}{(x + 8)(5 - x)}$
12. $\dfrac{(3x + 1)(2x - 5)}{4(2x - 5)(x + 7)}$
13. $\dfrac{2x^2 + 6x}{x^2 - 9}$
14. $\dfrac{x^2 - 10x + 24}{x^2 - 12x + 32}$
15. $\dfrac{5a^2 - 20}{a^2 + 6a - 16}$
16. $\dfrac{q^2 - 8q + 15}{q^2 - 7q + 12}$
17. $\dfrac{b^2 + 3b - 18}{6b^2 + 36b}$
18. $\dfrac{m^2 - 8m + 12}{m^2 - 7m + 10}$
19. $\dfrac{1 - x^2}{x^2 - 3x + 2}$
20. $\dfrac{5x - x^2}{x^2 - 8x + 15}$
21. $\dfrac{5x - 2x^2}{4x^2 - 25}$
22. $\dfrac{2y - y^2}{y^2 - 5y + 6}$
23. $\dfrac{36 - a^2}{a^2 - a - 30}$
24. $\dfrac{7x - x^2}{x^2 - 10x + 21}$

In Exercises 25–60, simplify. Assume that all variables result in nonzero denominators.

25. $\dfrac{5x}{2x + 2} \cdot \dfrac{(3x - 2)(x + 1)}{x}$
26. $\dfrac{8x^2y - 4xy}{3y} \cdot \dfrac{6}{2x - 1}$
27. $\dfrac{36}{9x^2 + 18x} \cdot \dfrac{x + 2}{3}$
28. $\dfrac{3xy + 3x^2}{y} \cdot \dfrac{1}{3x}$
29. $\dfrac{z + 1}{z - 1} \cdot \dfrac{8z - 8}{z^2 + 1}$
30. $\dfrac{5q^3y - 20q^2y}{3y^4} \cdot \dfrac{9}{q - 4}$
31. $\dfrac{36 - x^2}{8x + 8} \div \dfrac{2x - 12}{x + 1}$
32. $\dfrac{12j^2 + 5j - 2}{4j^2 + 13j + 3} \div \dfrac{4 - 9j^2}{8j^2 + 2j}$
33. $\dfrac{10b^2 - b - 3}{5b^2 + 28b + 15} \div \dfrac{1 - 4b^2}{20b^2 + 12b}$
34. $\dfrac{4q^2 + 7q - 15}{4q^2 - 5q - 6} \div \dfrac{q^2 - 9}{12q^2 + 9q}$
35. $\dfrac{4k^2 - k - 3}{k^2 + 6k + 5} \div \dfrac{9 - 16k^2}{2k^2 + 2k}$
36. $\dfrac{3n^2 + 2n - 8}{3n^2 - 10n - 8} \div \dfrac{n^2 - 4}{6n^2 + 4n}$
37. $\dfrac{4}{2x^2 + x} + \dfrac{3}{5 + 10x}$
38. $\dfrac{3}{x^2 - 3x - 18} - \dfrac{4}{x^2 - 36}$
39. $\dfrac{1}{x^2 - 3x + 2} + \dfrac{2x}{x^2 - 1}$

40. $\dfrac{x}{x^2+x-2}+\dfrac{2}{x^2-5x+4}$

41. $\dfrac{x}{x^2+2x+1}-\dfrac{3}{x+1}$

42. $\dfrac{14}{z^2+14z-147}-\dfrac{7}{z^2-49}$

43. $\dfrac{x+1}{3x}+\dfrac{5}{x}-\dfrac{2x}{3x^2-3x}$

44. $\dfrac{2x}{x+1}-\dfrac{x-4}{x}+\dfrac{2}{x+1}$

45. $\dfrac{2}{a^3}-\dfrac{1}{a}+\dfrac{4}{a^2}$

46. $\dfrac{3}{x^2-x}+\dfrac{4}{x}-\dfrac{5}{x-1}$

47. $\dfrac{2m+1}{4m^2+6m}-\dfrac{3}{2m}+\dfrac{7m}{2m+3}$

48. $\dfrac{5y}{y+1}-\dfrac{y-7}{y}+\dfrac{5}{y+1}$

49. $\dfrac{\frac{3}{x}+1}{1+\frac{x}{3}}$

50. $\dfrac{\frac{4}{2x}-\frac{3}{3x}}{\frac{2}{x}-\frac{3}{2x}}$

51. $\dfrac{\frac{5}{x}-\frac{1}{3x}}{\frac{1}{3x}-\frac{3}{2x}}$

52. $\dfrac{1+\frac{9}{x}}{1+\frac{x}{9}}$

53. $\dfrac{\frac{2}{x}-\frac{1}{2x}}{\frac{4}{3x}-\frac{1}{4x}}$

54. $\dfrac{\frac{6}{2x}-\frac{3}{5x}}{\frac{2}{x}-\frac{2}{2x}}$

55. $\dfrac{n+\frac{2}{n}}{n^2-\frac{4}{n^2}}$

56. $\dfrac{\frac{5}{x+5}}{\frac{1}{x}+2}$

57. $\dfrac{\frac{2}{x^2+3x}}{\frac{2}{x}+\frac{4}{x+3}}$

58. $\dfrac{\frac{5}{2x^2-6x}}{\frac{2}{x}-\frac{1}{x-3}}$

59. $\dfrac{\frac{3}{x-2}}{\frac{2}{x^2-2x}-\frac{1}{x}}$

60. $\dfrac{\frac{7}{7+y}}{\frac{1}{y}+4}$

Extensions

In Exercises 61–63, simplify and state the domain of each expression using set-builder notation or interval notation.

61. $\dfrac{3x^3-2x^2-3x+2}{15x^2-5x-20}$

62. $\dfrac{3x^3+6x^2}{x^2-x-12}\cdot\dfrac{x^2+5x+6}{x^2}\div\dfrac{x^2-x}{x^2-2x-8}$

63. $\dfrac{5}{3-2y}+\dfrac{3}{2y-3}-\dfrac{y-3}{2y^2-y-3}$

In Exercises 64–69, simplify. Assume that all variables result in nonzero denominators.

64. $\left(\dfrac{\frac{y+1}{y-1}+1}{\frac{y+1}{y-1}-1}\right)^5$

65. $1+\dfrac{1}{1+\dfrac{1}{1+\dfrac{1}{1+\frac{1}{x}}}}$

66. $\dfrac{\frac{1}{(x+h)^2+9}-\frac{1}{x^2+9}}{h}$

67. $\dfrac{\sqrt{y+b}}{\sqrt{y-b}}-\dfrac{\sqrt{y-b}}{\sqrt{y+b}}$

68. $\dfrac{\left(b+\frac{1}{a}\right)^b\left(b-\frac{1}{a}\right)^a}{\left(a+\frac{1}{b}\right)^b\left(a-\frac{1}{b}\right)^a}$

69. $\dfrac{\frac{1}{m^2-5m-14}+\frac{1}{m+1}}{\frac{1}{m+2}-\frac{1}{m^2-6m-7}}$

70. True or False? The domain of a rational expression can never be the set of all real numbers.

In Exercises 71–72, write expressions for the perimeter and area of each rectangle with the given length and width.

71. $\dfrac{w+4}{3}$ and $\dfrac{w-2}{5}$

72. $\dfrac{3}{l+4}$ and $\dfrac{2}{l-5}$

P.7 SOLVING LINEAR AND QUADRATIC EQUATIONS

OBJECTIVES

- Solve linear equations.
- Solve rational equations.
- Solve quadratic equations by factoring and completing the square.
- Write the square root of a negative number as a complex number.
- Use the Quadratic Formula to solve quadratic equations.
- Model geometric situations with quadratic equations.

PREREQUISITE VOCABULARY TERMS

Distributive Property
FOIL
linear expression
perfect-square trinomial
quadratic expression
rational expression

P.7.1 Solving a Linear Equation

A **linear equation** in one variable is an equation that can be written in the form $ax + b = 0$, where x is a variable and a and b are real numbers. To solve a linear equation, apply operations to both sides of the equation, with the goal of isolating the variable on one side of the equation. Note that a linear equation may have one solution, no solutions, or all real numbers as solutions.

To solve a linear equation with the variable on both sides (i.e., $ax + b = cx + d$, where a, b, c, and d are numbers), isolate the x-terms on one side and the constant terms on the other, then divide both sides by the coefficient of x. If either side of the equation contains like terms or parentheses, then simplify each side before applying these solving steps.

EXAMPLE 1 Simplifying and Solving a Linear Equation

Solve. $1 + 3(y + 8) = 7(3 - y) + 14$

SOLUTION

Begin by simplifying each side of the equation. Apply inverse operations to move the variable terms to the left side and the constant terms to the right side only after each side is fully simplified.

$1 + 3(y + 8) = 7(3 - y) + 14$	
$1 + 3y + 24 = 21 - 7y + 14$	*Distribute.*
$3y + 25 = 35 - 7y$	*Combine the like terms on each side.*
$10y + 25 = 35$	*Add 7y to each side.*
$10y = 10$	*Subtract 25 from each side.*
$y = 1$	*Divide each side by 10.*

REMEMBER

Distributive Property
$a(b + c) = ab + ac$
$a(b - c) = ab - ac$

CHECK

Check the solution by substituting $y = 1$ in the original equation.

$1 + 3(1 + 8) \stackrel{?}{=} 7(3 - 1) + 14$	*Substitute y = 1.*
$1 + 3(9) \stackrel{?}{=} 7(2) + 14$	*Simplify within parentheses.*
$1 + 27 \stackrel{?}{=} 14 + 14$	*Multiply.*
$28 = 28$ ✓	*Add.*

Solving Linear Equations with Rational Terms

TIP

Applying the Cross-Products Property is more commonly referred to as "cross multiplying."

Equations containing rational terms (i.e., terms that are rational expressions) can be simplified by multiplying the equation by the LCM of the denominators. Alternatively, the **Cross-Products Property** can be used when the equation is a proportion.

Cross-Products Property

If A, B, C, and D are algebraic expressions such that B and $D \neq 0$ and $\frac{A}{B} = \frac{C}{D}$, then $AD = BC$.

EXAMPLE 2 **Solving a Linear Equation with Rational Terms**

Solve each equation.

A. $\frac{3x-1}{4} = \frac{10-x}{5}$ **B.** $\frac{4(x-1)}{5} = 1 + \frac{x}{2}$

SOLUTION

A. Use the Cross-Products Property (since the equation has one term on each side).

$$\frac{3x-1}{4} = \frac{10-x}{5}$$

$$5(3x-1) = 4(10-x) \quad \textit{Cross multiply.}$$

$$15x - 5 = 40 - 4x \quad \textit{Distribute.}$$

$$19x = 45 \quad \textit{Add 4x and 5 to each side.}$$

$$x = \frac{45}{19} \quad \textit{Divide each side by 19.}$$

IMPORTANT

The cross products are not used because the equation is not a proportion (there is more than one term on one side).

B. Multiply the equation by the LCM of the denominators, 10.

$$10\left(\frac{4(x-1)}{5}\right) = \left(1 + \frac{x}{2}\right)10 \quad \textit{Multiply the equation by 10.}$$

$$\frac{40(x-1)}{5} = 10 + \frac{10x}{2} \quad \textit{Multiply on the left side and distribute on the right side.}$$

$$8(x-1) = 10 + 5x \quad \textit{Simplify the fractions.}$$

$$8x - 8 = 10 + 5x \quad \textit{Distribute.}$$

$$3x - 8 = 10 \quad \textit{Subtract 5x from each side.}$$

$$3x = 18 \quad \textit{Add 8 to each side.}$$

$$x = 6 \quad \textit{Divide each side by 3.}$$

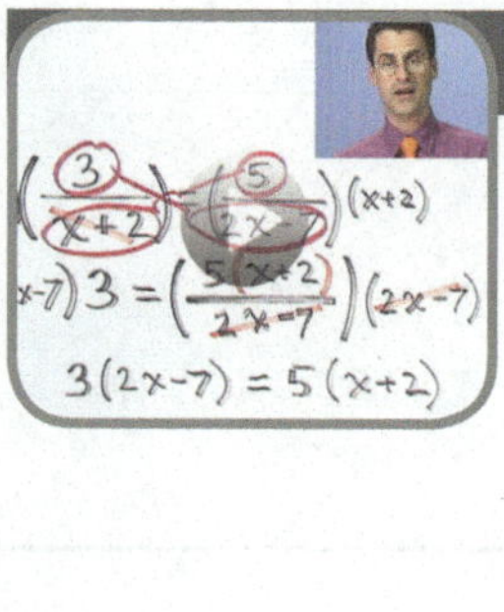

P.7.2 Solving a Linear Equation with Rationals

An equation where the domain is not all real numbers may have an **extraneous solution**, which is a solution not in the equation's domain. When solving an equation where the domain is not all real numbers, be sure to confirm that any solutions are in the equation's domain (and therefore not extraneous).

For example, recall that the domain of a rational expression includes all real numbers except those that result in a 0 denominator. Likewise, the domain of an equation with rational terms must also exclude values that result in a 0 denominator. Therefore, when solving an equation with rational terms, be sure to verify that any solutions are not extraneous.

REMEMBER

Set the denominator equal to 0 and solve to identify values excluded from the domain of a rational expression.

Some equations with rational terms that contain a variable in the denominator will simplify to linear form after cross multiplying, or after multiplying the equation by the LCM (least common multiple) of the denominators. These equations can be solved using the same process as used for linear equations with a numeric denominator (as in **Example 2**), with one additional step: check for extraneous solutions, as demonstrated in **Example 3**.

EXAMPLE 3 Solving a Linear Equation with Rational Terms

Solve each equation.

A. $\frac{4}{3x-1} = \frac{2}{x-5}$ **B.** $\frac{7y}{y+9} - 2 = \frac{2}{y+9}$

SOLUTION

REMEMBER

Values of the variable that make a denominator 0 must be excluded from the domain of the equation.

A. Use the Cross-Products Property (since the equation is a proportion).

$$\frac{4}{3x-1} = \frac{2}{x-5}$$

$4(x-5) = 2(3x-1)$	*Cross multiply.*
$4x - 20 = 6x - 2$	*Distribute.*
$-2x = 18$	*Subtract 6x from each side and add 20 to each side.*
$x = -9$	*Divide each side by –2.*

IMPORTANT

Alternatively, extraneous solutions can be identified by substituting the solution into the original equation. If the result is a false statement, then the solution is extraneous.

The values excluded from the equation's domain are $x = \frac{1}{3}$ and $x = 5$ because the equation is undefined when x is either of these values. Since –9 is in the equation's domain, $x = -9$ is not an extraneous solution.

B. Multiply the equation by the denominator, $(y + 9)$.

$(y+9)\left(\frac{7y}{y+9} - 2\right) = \left(\frac{2}{y+9}\right)(y+9)$	*Multiply the equation by (y + 9).*
$7y - 2(y+9) = 2$	*Distribute.*
$7y - 2y - 18 = 2$	*Distribute.*
$5y - 18 = 2$	*Combine the like terms.*
$5y = 20$	*Add 18 to each side.*
$y = 4$	*Divide each side by 5.*

The equation's domain is all real numbers except $y = -9$. Since 4 is in the equation's domain, $y = 4$ is not an extraneous solution.

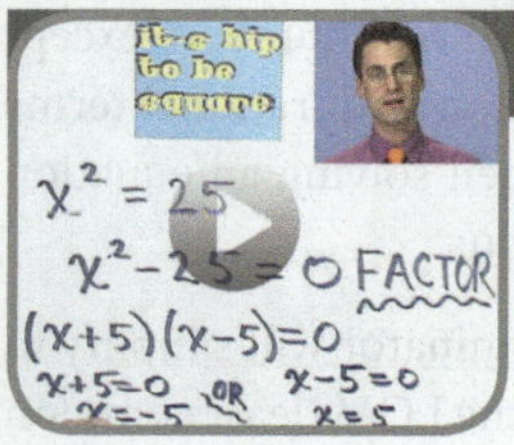

P.7.3 Solving Quadratics by Factoring

A **quadratic equation** in one variable can be written in the form $ax^2 + bx + c = 0$, where x is a variable and a, b, and c are real numbers such that $a \neq 0$. Some quadratic equations can be solved by using the factoring method, which uses this basic property of real numbers.

Zero-Product Property

If A and B are algebraic expressions, and $AB = 0$, then $A = 0$ or $B = 0$.

Steps for Solving a Quadratic Equation by Factoring

1. Write the equation in general form, $ax^2 + bx + c = 0$ (i.e., where one side is 0).
2. Factor the quadratic expression.
3. Set each factor equal to 0, then solve each equation.

It follows from the Zero-Product Property that if a product of two algebraic expressions is known to be equal to 0, then it can be assumed that either one or both of those factors is also equal to 0. For example, if $(x + 2)(x - 5) = 0$, then $x + 2 = 0$ or $x - 5 = 0$. Solving each linear equation gives $x = -2$ or $x = 5$. So, the solutions for the equation $(x + 2)(x - 5) = 0$ are $x = -2$ and $x = 5$.

NOTICE THAT

The equation $(x + 2)(x - 5) = 0$ is quadratic since FOILing the factors gives a quadratic expression: $x^2 - 3x - 10 = 0$.

EXAMPLE 4 Solving a Quadratic Equation by Factoring

Solve. $2x^2 - x = 25 - 6x$

SOLUTION

The equation is quadratic, so follow the steps for solving a quadratic equation by factoring.

$$2x^2 - x = 25 - 6x$$

$$2x^2 + 5x - 25 = 0 \qquad \textit{Write the equation in general form.}$$

$$(2x - 5)(x + 5) = 0 \qquad \textit{Factor.}$$

$$2x - 5 = 0 \quad \text{or} \quad x + 5 = 0 \qquad \textit{Set each factor equal to 0.}$$

$$x = \frac{5}{2} \qquad x = -5 \qquad \textit{Solve each equation.}$$

TIP

FOIL to check the factors.

$$(2x - 5)(x + 5) = 2x^2 + 10x - 5x - 25 = 2x^2 + 5x - 25$$

CHECK

Check each solution by substituting into the original equation.

Check $x = \frac{5}{2}$:

$$2\left(\frac{5}{2}\right)^2 - \frac{5}{2} \stackrel{?}{=} 25 - 6\left(\frac{5}{2}\right)$$

$$2\left(\frac{25}{4}\right) - \frac{5}{2} \stackrel{?}{=} 25 - 6\left(\frac{5}{2}\right)$$

$$\frac{25}{2} - \frac{5}{2} \stackrel{?}{=} 25 - 15$$

$$10 = 10 \checkmark$$

Check $x = -5$:

$$2(-5)^2 - (-5) \stackrel{?}{=} 25 - 6(-5)$$

$$2(25) + 5 \stackrel{?}{=} 25 - 6(-5)$$

$$50 + 5 \stackrel{?}{=} 25 + 30$$

$$55 = 55 \checkmark$$

Therefore, the solutions are $x = -5$ and $x = \frac{5}{2}$.

Remember that some quadratic expressions cannot be factored (i.e., they are irreducible), so there are some quadratic equations that cannot be solved by factoring. For example, the equation $x^2 + 8x + 5 = 0$ is quadratic and in general form, but it cannot be solved by factoring because $x^2 + 8x + 5$ is irreducible (because there are no factors of 5 whose sum is 8). However, this does not necessarily mean that the equation has no real solution; it just must be solved using some other method. Methods for solving an irreducible quadratic equation will be discussed later in this section.

P.7.4 Introducing Complex Numbers

The Square Roots Method for Solving a Quadratic Equation

If a quadratic equation can be written in the form $ax^2 + c = 0$ (i.e., the quadratic equation has no x-term), then its solutions can be found by using the **square roots method**, which involves isolating the x^2-term on one side of the equation and then taking ± the square root of both sides of the equation. For example, to solve $x^2 = 9$, take ± the square root of each side.

IMPORTANT

The ± symbol must be written before the square root to account for the two possible solutions to the quadratic equation.

$$x^2 = 9 \Rightarrow x = \pm\sqrt{9} \Rightarrow x = \pm 3$$

Complex Numbers

The square roots method can be used even when the equation has a squared variable (or linear expression) equal to a *negative* number. For example, taking ± the square root of both sides of $x^2 = -9$ yields $x = \pm\sqrt{-9}$. However, the square root of a negative number is not a real number.

In this case, each of the equation's solutions is a **complex number**. Complex numbers are based on the **imaginary unit *i***.

$$i = \sqrt{-1}$$ *Note that i is not a variable; instead, i represents a number: $\sqrt{-1}$.*

DEFINITION

A **complex number** is an expression that can be written in the form

$$a + bi,$$

where a and b are real numbers and ***i*** is the **imaginary unit**. The **real part** of $a + bi$ is a, and the **imaginary part** is b. Two complex numbers are equal when their real parts are equal and their imaginary parts are equal. A complex number of the form bi (where $a = 0$) is called a **purely imaginary number**.

IMPORTANT

Additional operations with complex numbers will be discussed in Chapter 2.

Purely imaginary numbers can be operated on like real numbers. For example, $i + i = 2i$ and $i \cdot i = i^2$. From the definition of i, $i = \sqrt{-1}$, it follows that $i^2 = -1$. The properties of exponents can be applied to powers of purely imaginary numbers.

A square root of a negative number can be expressed by a complex number that is a purely imaginary number. For example, consider the square root expression $\sqrt{-9}$. The expression can be simplified using properties of radicals and the fact that $i = \sqrt{-1}$.

$$\sqrt{-9} = \sqrt{9(-1)} = \sqrt{9} \cdot \sqrt{-1} = 3i$$

Since complex numbers can be operated on like real numbers, it follows that expressions with complex numbers can be simplified.

EXAMPLE 5 Simplifying Square Root Expressions with Negative Radicands

Simplify. Write imaginary expressions in terms of i.

A. $\sqrt{-300}$ **B.** $\sqrt{-20}+\sqrt{-500}-\sqrt{45}$

SOLUTION

A. $\sqrt{-300}=\sqrt{100(-1)(3)}=\sqrt{100}\cdot\sqrt{-1}\cdot\sqrt{3}=10i\sqrt{3}$

B. $\sqrt{-20}+\sqrt{-500}-\sqrt{45}$

$=\sqrt{(4)(-1)(5)}+\sqrt{(100)(-1)(5)}-\sqrt{(9)(5)}$ *Factor each radicand.*

$=2i\sqrt{5}+10i\sqrt{5}-3\sqrt{5}$ *Simplify each radical.*

$=12i\sqrt{5}-3\sqrt{5}$ *Combine the like terms.*

$=-3\sqrt{5}+12i\sqrt{5}$ *Write the expression in the standard form of a complex number.*

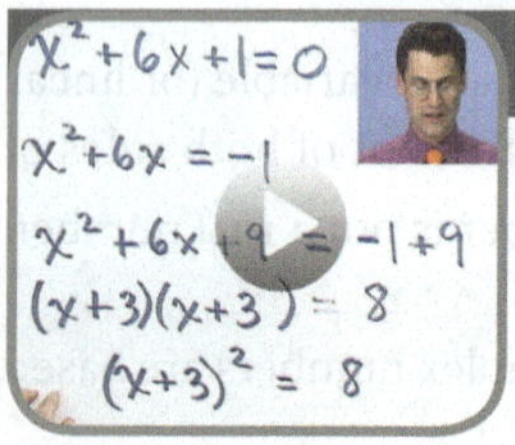

P.7.5 Solving Quadratics by Completing the Square

Now that complex numbers have been defined, quadratic equations with complex solutions can be solved. The methods for solving quadratic equations discussed in the next three topics can be used to solve quadratic equations with real or complex solutions.

Any quadratic equation that can be written with a perfect-square quadratic trinomial on one side and a constant value on the other side can be solved by taking square roots. For example, $x^2+12x+36=12$ can be solved by factoring the perfect-square quadratic trinomial $x^2+12x+36$ and then taking $\pm$ the square root of both sides.

$$x^2+12x+36=12$$

$$(x+6)^2=12 \quad \textit{Factor.}$$

$$x+6=\pm\sqrt{12} \quad \textit{Take} \pm \textit{the square root of each side.}$$

$$x+6=\pm2\sqrt{3} \quad \textit{Simplify the radical.}$$

$$x=-6\pm2\sqrt{3} \quad \textit{Subtract 6 from each side.}$$

TIP

Perfect-Square Trinomial

$a^2+2ab+b^2=(a+b)^2$

If the quadratic equation is not written with one side as a perfect-square quadratic trinomial and the other side as a constant value, this method of taking square roots can still be applied to solve the equation, but only after some algebraic manipulation.

Consider the equation $x^2+4x+3=8$. Notice that x^2+4x+3 is not a perfect-square quadratic trinomial. However, adding 1 to both sides will result in a perfect-square quadratic trinomial on the left side. Then the equation can be solved by taking square roots.

$$x^2+4x+3=8$$

$$x^2+4x+4=9 \quad \textit{Add 1 to each side (making the left side a perfect-square).}$$

$$(x+2)^2=9 \quad \textit{Factor.}$$

$$x+2=\pm3 \quad \textit{Take} \pm \textit{the square root of each side.}$$

$$x=1 \text{ or } x=-5 \quad \textit{Subtract 2 from each side.}$$

The process of manipulating an expression into the form of a perfect-square trinomial is called **completing the square**. The complicated part of the process can be identifying the value that needs to be added to make the perfect-square trinomial. Examining perfect-square trinomials will give some insight into this process. The following table shows three examples of perfect-square trinomials (where the coefficient of x^2 is 1). Consider the relationship between the coefficient of x and the constant term in each.

Perfect-Square Trinomial	Coefficient of x and the Constant Term	Relationship
$x^2 + 12x + 36$	12 and 36	$\left(\frac{12}{2}\right)^2 = 6^2 = 36$
$x^2 + 6x + 9$	6 and 9	$\left(\frac{6}{2}\right)^2 = 3^2 = 9$
$x^2 + 8x + 16$	8 and 16	$\left(\frac{8}{2}\right)^2 = 4^2 = 16$

The constant term in each perfect-square quadratic trinomial is equal to the square of half the linear term's coefficient (i.e., the coefficient of x). Because this is always true when the coefficient of x^2 is 1 in a perfect-square trinomial, we can use this fact to write a quadratic equation of the form $x^2 + bx = c$ with a perfect-square quadratic trinomial on one side.

IMPORTANT

The purpose of completing the square is to write the equation in a form that can be solved by taking square roots.

Completing the Square

To write $x^2 + bx = c$ as an equation with a perfect-square quadratic trinomial on one side, add $\left(\frac{b}{2}\right)^2$ to both sides of the equation.

$$x^2 + bx = c$$

$$x^2 + bx + \left(\frac{b}{2}\right)^2 = c + \left(\frac{b}{2}\right)^2$$

$$\left(x + \frac{b}{2}\right)^2 = c + \left(\frac{b}{2}\right)^2$$

For example, the equation $x^2 + 10x = 7$ can be written with a perfect-square trinomial on the left side by adding 25 to both sides (because $(10/2)^2 = 25$).

$$x^2 + 10x = 7$$

$$x^2 + 10x + 25 = 7 + 25 \quad \textit{Complete the square by adding 25 to both sides.}$$

$$(x + 5)^2 = 32 \quad \textit{Factor and simplify.}$$

This equation could now be solved by taking ± the square root of both sides, simplifying the radical, and then subtracting 5 from both sides.

Steps for Solving a Quadratic Equation by Completing the Square

❶ Write the equation in the form $x^2 + bx = c$, where b and c are real numbers.

❷ Add $\left(\frac{b}{2}\right)^2$ to both sides of the equation.

❸ Factor the perfect-square trinomial into the form $\left(x+\frac{b}{2}\right)^2$.

❹ Take ± the square root of each side.

❺ Simplify the radical.

❻ Isolate x.

EXAMPLE 6 Solving a Quadratic Equation by Completing the Square

Solve. $x^2 + 7 = 26 + 2x$

SOLUTION

$$x^2 + 7 = 26 + 2x$$

$$x^2 - 2x = 19 \quad \textit{Subtract 2x and 7 from each side.}$$

$$x^2 - 2x + 1 = 19 + 1 \quad \textit{Add } (-2/2)^2 \textit{ to each side.}$$

$$(x-1)^2 = 20 \quad \textit{Factor and simplify.}$$

$$x - 1 = \pm\sqrt{20} \quad \textit{Take} \pm \textit{the square root of each side.}$$

$$x - 1 = \pm 2\sqrt{5} \quad \textit{Simplify the radical.}$$

$$x = 1 \pm 2\sqrt{5} \quad \textit{Add 1 to each side.}$$

Therefore, the solutions are $x = 1 - 2\sqrt{5}$ and $x = 1 + 2\sqrt{5}$.

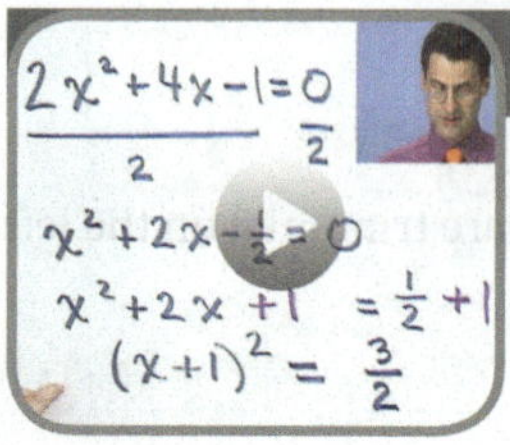

P.7.6 Completing the Square: Another Example

So far, all of the equations solved by completing the square have had an x^2-term coefficient equal to 1 (i.e., $a = 1$). To solve a quadratic equation where $a \neq 1$ by completing the square, divide both sides of the equation by a and then complete the square (see steps for solving by completing the square in the previous topic). This process is demonstrated in **Example 7**.

To solve a quadratic equation $ax^2 + bx = c$ where $a \neq 1$ by completing the square, **divide both sides of the equation by a** (the coefficient of x^2) before proceeding with the steps for solving by completing the square.

EXAMPLE 7 Solving a Quadratic Equation by Completing the Square

Solve. $4x^2 + 12x = 5$

SOLUTION

$$4x^2 + 12x = 5$$

$$x^2 + 3x = \frac{5}{4}$$ *Divide each side by 4.*

$$x^2 + 3x + \frac{9}{4} = \frac{14}{4}$$ *Add $(3/2)^2$ to each side.*

$$\left(x + \frac{3}{2}\right)^2 = \frac{14}{4}$$ *Factor.*

$$x + \frac{3}{2} = \pm\sqrt{\frac{14}{4}}$$ *Take ± the square root of each side.*

$$x = -\frac{3}{2} \pm \frac{\sqrt{14}}{2}$$ *Simplify the radical and subtract 3/2 from each side.*

$$x = \frac{-3 \pm \sqrt{14}}{2}$$ *Combine the terms. (The preceding form of the answer is also acceptable.)*

Therefore, the solutions are $x = \dfrac{-3 - \sqrt{14}}{2}$ and $x = \dfrac{-3 + \sqrt{14}}{2}$.

NOTICE THAT

The fraction 14/4 is not reduced to 7/2 because the next step is to take ± the square root of both sides, so it is better to leave 4 as a perfect square. If 14/4 is reduced to 7/2, then the denominator must be rationalized later, resulting in the same answer, but with additional steps.

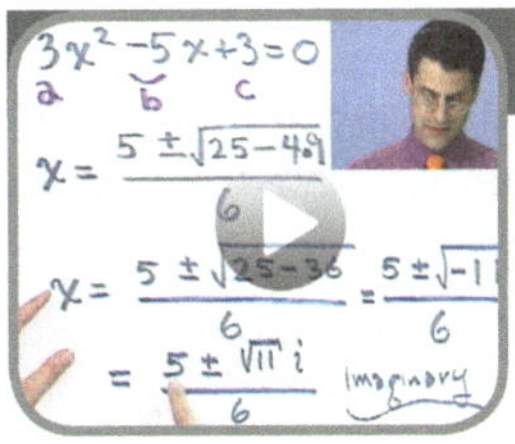

P.7.7 Using the Quadratic Formula

The **Quadratic Formula** can also be used to solve a quadratic equation.

The Quadratic Formula

The solutions of a quadratic equation of the form $ax^2 + bx + c = 0$ are given by the formula

$$x = \frac{-b \pm \sqrt{b^2 - 4ac}}{2a}.$$

When using the Quadratic Formula to solve a quadratic equation, be sure to write the quadratic equation in general form first, and then identify the values of a, b, and c. When identifying a, b, and c, note that the value will be negative when a subtraction symbol precedes the term.

Steps for Solving a Quadratic Equation by Using the Quadratic Formula

1. Write the equation in general form ($ax^2 + bx + c = 0$).
2. Identify the values of a, b, and c from the general form equation.
3. Substitute the values of a, b, and c into the Quadratic Formula, $x = \dfrac{-b \pm \sqrt{b^2 - 4ac}}{2a}$, and simplify, starting with the radicand.

TIP

Enclose the substituted values within parentheses to prevent calculation errors.

EXAMPLE 8 Using the Quadratic Formula to Solve a Quadratic Equation

Solve. $2m^2 + 3 = m$

SOLUTION

❶ Subtract m from both sides to write the equation in the general form.

$$2m^2 + 3 = m \Rightarrow 2m^2 - m + 3 = 0$$

❷ From the general form equation, $a = 2$, $b = -1$, and $c = 3$.

❸ Substitute $a = 2$, $b = -1$, and $c = 3$ into the Quadratic Formula and simplify.

$$m = \frac{-(-1) \pm \sqrt{(-1)^2 - 4(2)(3)}}{2(2)}$$ *Substitute.*

$$= \frac{1 \pm \sqrt{1 - 24}}{4}$$ *Evaluate the power and multiply within the radicand. Multiply in the denominator.*

$$= \frac{1 \pm \sqrt{-23}}{4}$$ *Subtract in the radicand.*

$$= \frac{1 \pm i\sqrt{23}}{4}$$ *Simplify the radical.*

Therefore, the solutions are $m = \frac{1 + i\sqrt{23}}{4}$ and $m = \frac{1 - i\sqrt{23}}{4}$.

P.7.8 Modeling with Quadratic Geometric Equations

There are many real-life questions that can be answered by writing and solving equations. This process of writing an equation to fit a specific situation is called **modeling**. An unknown quantity is represented by a variable (or algebraic expression) in the model.

Steps for Modeling with an Equation

❶ Identify the unknown value(s).

❷ Translate the given information into algebraic parts.

❸ Set up the model (i.e., write the equation), and solve. State the answer.

❹ Confirm the answer.

Modeling geometric situations with equations usually involves substituting given values into some known formula and then solving for the unknown variable. Many common geometric formulas are quadratic equations, such as the formula for the area of a circle, $A = \pi r^2$. See *Appendix A: Common Formulas and Properties* for additional common formulas.

Using a Relationship between Two Unknown Values

A situation involving two unknown values can be modeled using an equation in two variables. However, if there is a relationship between the two unknown values, then one unknown value can be expressed in terms of the other unknown value. In this case, the situation involving two unknown values can be modeled using an equation in one variable (which can then be solved).

For example, suppose there is a rectangle where the length is 10 feet greater than the width. There are two unknown values, the length of the rectangle and the width of the rectangle, but we can express both unknowns in terms of the same variable because a relationship between the two unknown values is given: the length is 10 feet greater than the width.

If we let the width be w, then the length must be $w + 10$. Thus, the unknown length is expressed in terms of the unknown width. Alternatively, if the length is identified as l, then the width must be $l - 10$.

Note that an additional piece of information regarding the rectangle must be given (such as the perimeter or area of the rectangle) in order to model the situation with an equation in one variable, as demonstrated in **Example 9**.

EXAMPLE 9 Modeling a Quadratic Equation for a Geometric Situation

The rectangular picture inside a frame of the same shape is 3 inches longer than it is wide. The frame extends 2 inches on each side of the picture. The area of the picture and the frame is 108 square inches. Find the dimensions of the frame.

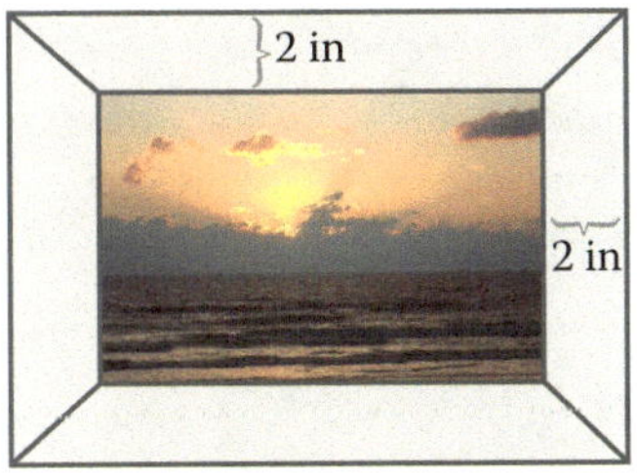

SOLUTION

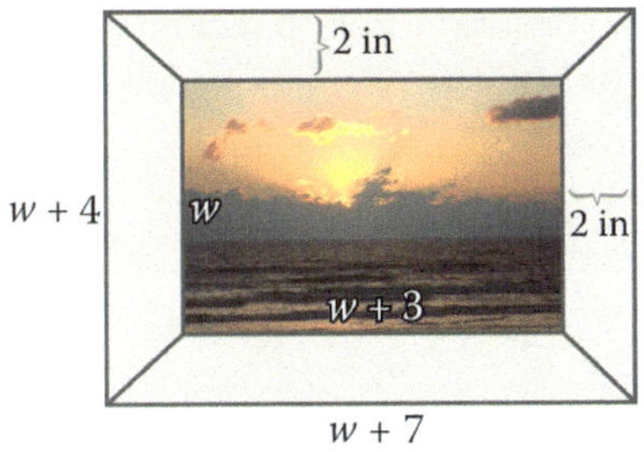

❶ Since the picture is 3 inches longer than it is wide, let w represent the width of the picture. Then $w + 3$ represents the length of the picture.

❷ The area of the picture and the frame is 108 square inches. Since the frame extends 2 inches on each side of the picture, the length and width of the frame are each 4 inches longer than the length and width of the picture.

Total area	108
Frame width	$w + 4$
Frame length	$(w + 3) + 4 = w + 7$

❸ The formula for the area of a rectangle is $A = lw$.

Total area = frame length · frame width

$$108 = (w + 7)(w + 4)$$

Solve the equation for w.

$$108 = (w + 7)(w + 4)$$

$$108 = w^2 + 11w + 28 \qquad \textit{FOIL.}$$

$$0 = w^2 + 11w - 80 \qquad \textit{Subtract 108 from each side.}$$

$$0 = (w + 16)(w - 5) \qquad \textit{Factor.}$$

$$w + 16 = 0 \quad \text{or} \quad w - 5 = 0 \qquad \textit{Set each factor equal to 0.}$$

$$w = -16 \qquad w = 5 \qquad \textit{Solve each equation.}$$

TIP

This quadratic equation could be solved by factoring, completing the square, or using the Quadratic Formula.

The width of a rectangle cannot be negative, so the only solution is $w = 5$.

The length and width of the frame are represented by $w + 7$ and $w + 4$, so the length of the frame is $5 + 7 = 12$ inches, and the width of the frame is $5 + 4 = 9$ inches. Therefore, the frame's dimensions are 12 inches by 9 inches.

CHECK

❹ Confirm that the area of a 12 inch by 9 inch rectangle is 108 square inches.

$$A = 12(9) = 108 \text{ in}^2 \checkmark$$

EXAMPLE 10 Modeling a Geometric Situation with an Equation

A gardener wants to fence in a rectangular plot in two rectangular sections. If she has 54 yards of fence and the area of the entire plot is 108 square yards, find the possible dimensions of the entire rectangular plot, in yards.

SOLUTION

TIP

The relationship between the unknowns is given in the equation for the total amount of fencing, which is the perimeter of the rectangular plot, $2l + 2w$, plus one extra side.

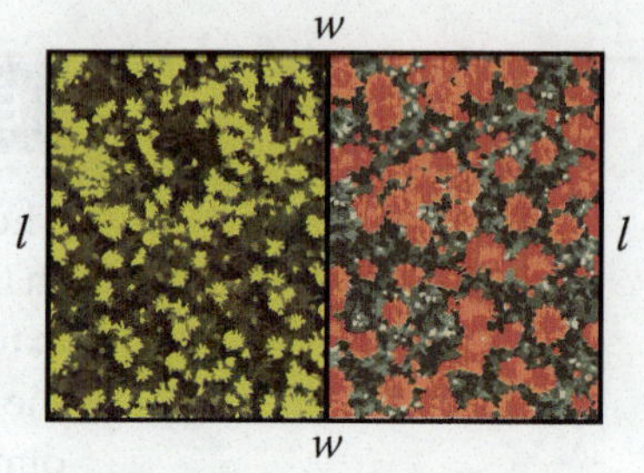

❶ Let l represent the length and let w represent the width of the rectangular plot. Let the additional piece of fence (dividing the plot into two rectangular parts) be parallel to the length. Then $3l + 2w$ yards of fencing are needed for the plot. She has 54 feet of fence, so $54 = 3l + 2w$. This equation gives the relationship between the length and the width.

Solve the equation for w. $\quad 54 = 3l + 2w \Rightarrow w = \dfrac{54 - 3l}{2}$

❷ The area of the entire rectangular plot is 108 square yards.

Total area	108
Plot's width	$\dfrac{54-3l}{2}$
Plot's length	l

❸ The formula for the area of a rectangle is $A = lw$. $\quad$ Total area = plot's length · plot's width

$$108 = l\left(\frac{54-3l}{2}\right)$$

Solve the equation for l.

$$108 = \frac{54l - 3l^2}{2} \qquad \textit{Distribute.}$$

$$216 = 54l - 3l^2 \qquad \textit{Multiply each side by 2.}$$

$$3l^2 - 54l + 216 = 0 \qquad \textit{Write the equation in general form.}$$

$$3(l^2 - 18l + 72) = 0 \qquad \textit{Factor out the GCF of the terms.}$$

$$3(l - 12)(l - 6) = 0 \qquad \textit{Factor the trinomial.}$$

$$l - 12 = 0 \quad \text{or} \quad l - 6 = 0 \qquad \textit{Set each factor equal to 0.}$$

$$l = 12 \qquad l = 6 \qquad \textit{Solve each equation.}$$

Use 12 yards and 6 yards to find the possible widths of the rectangle.

$$w = \frac{54 - 3(12)}{2} = 9 \quad \text{and} \quad w = \frac{54 - 3(6)}{2} = 18$$

The dimensions of the rectangular plot are 12 yards by 9 yards or 6 yards by 18 yards.

CHECK

❹ Confirm that the area is 108 square yards for a 12 × 9-yard rectangular plot and for a 6 × 18-yard rectangular plot, and confirm that both plots will require 54 yards of fencing.

$A = 12(9) = 108\text{ yd}^2$ ✓ $\qquad A = 6(18) = 108\text{ yd}^2$ ✓

Three 12-yard sides and two 9-yard sides = $3(12) + 2(9) = 54$ yd ✓

Three 6-yard sides and two 18-yard sides = $3(6) + 2(18) = 54$ yd ✓

SECTION P.7 EXERCISES

Warm Up

Simplify.

1. $10 - 5(x + 3)$
2. $8m + 11 - (4 - 2m) + 7m$
3. $\sqrt{125} + \sqrt{45}$

Just the Facts

4. The inverse operation of _____ is subtraction and of multiplication is _____.
5. An equation with fractional terms can be simplified by multiplying both sides by the _____ of the denominators.
6. An equation where the domain is not all real numbers may have an _____ solution.
7. _____ and using the _____ are solving methods that can be used to solve any quadratic equation.
8. The _____ method cannot be used to solve a quadratic equation in the form $ax^2 + bx + c = 0$ unless the quadratic expression is a perfect-square trinomial.
9. _____ are based on the imaginary unit i.
10. An equation of the form $x^2 = c$ has complex solutions when c is a _____ real number.

Essential Skills

In Exercises 1–30, solve each equation.

1. $10 + 2p = 2(p + 10)$
2. $3(x - 2) = 12 - x$
3. $s + 3 - 4s = 12 + 9s - s + 1$
4. $7(4 - 3z) + 3 = -6(z - 4) - 8$
5. $-(-3 + 2x) - (7 - x) - 3 = -(2 - x) + 1$
6. $-(-2 + x) - (8 - x) + 11 = -(-4 - x) + (-x) + 1$
7. $\frac{3x}{8} = \frac{x+1}{6}$
8. $\frac{7-2x}{4} = \frac{3(x+5)}{9}$
9. $\frac{2x-3}{4} = \frac{3x-4}{5}$
10. $\frac{5x-1}{2} = -\frac{2x-7}{4}$
11. $\frac{3(b-1)}{4} = \frac{2b+1}{8}$
12. $\frac{10-4w}{7} = \frac{7(w+2)}{6}$
13. $\frac{d}{9} = 2 + \frac{5d}{6}$
14. $5 - \frac{x}{3} = \frac{3(x+1)}{7}$
15. $\frac{5(g+2)}{3} = 5 - \frac{g}{5}$
16. $\frac{a}{4} + 3 = \frac{6a}{11}$
17. $\frac{2(d-1)}{3} + 4 = \frac{d}{2}$
18. $6 - \frac{w}{7} = \frac{7(w+2)}{6}$
19. $\frac{8}{x+3} = \frac{5}{x-2}$
20. $-\frac{5}{3x-7} = \frac{3}{3x+1}$
21. $\frac{2}{5m-1} = \frac{1}{2m-3}$
22. $\frac{4}{w-1} = \frac{7}{2w+5}$
23. $\frac{3}{10n+10} = -\frac{10}{10n-3}$
24. $-\frac{8}{3k+4} = \frac{4}{3k-3}$
25. $\frac{15}{u} - 1 = \frac{4}{u}$
26. $-\frac{x}{x-4} + 9 = \frac{6x}{x-4}$
27. $3 - \frac{2x}{x-3} = \frac{2x}{x-3}$
28. $\frac{11}{10w-11} = 4 - \frac{3w}{10w-11}$
29. $\frac{5b}{b-2} = -\frac{4b}{b-2} + 11$
30. $-\frac{4q}{5q-4} + 3 = \frac{4}{5q-4}$

In Exercises 31–42, solve each equation by factoring.

31. $m^2 + 2m - 35 = 0$
32. $x^2 = 3x + 18$
33. $d^2 - 5d = 0$
34. $h^2 - h = 56$
35. $a^2 = 10a - 9$
36. $x^2 = -7x - 12$
37. $2p^2 + p - 21 = 0$
38. $6x^2 + 13x = 5$
39. $8v^2 + 10v = 25$
40. $4y^2 - 15y - 25 = 0$

41. $2k^2 + 3k - 35 = 0$

42. $9x^2 - 14x = 8$

In Exercises 43–48, simplify each expression. Write imaginary expressions in terms of *i*.

43. $\sqrt{27} + \sqrt{-48}$

44. $-\sqrt{12} - \sqrt{-72}$

45. $-\sqrt{112} - \sqrt{-45}$

46. $-\sqrt{176} - \sqrt{-28}$

47. $\sqrt{-99} - \sqrt{32} + \sqrt{50}$

48. $\sqrt{63} + \sqrt{-44} - \sqrt{125}$

In Exercises 49–60, solve each equation by completing the square.

49. $p^2 + 6p = 18$

50. $x^2 - 8 = -8x$

51. $a^2 - 29 = -8a$

52. $r^2 + 10r - 1 = 22$

53. $y^2 + 182 = -6y - 2$

54. $t^2 + 224 = -14t$

55. $4v^2 + 48v = 32$

56. $-3x^2 - 6x = -1$

57. $-7b^2 - 84b = 216$

58. $5r^2 + 70r = -261$

59. $4d^2 + 45 = -24d$

60. $-5z^2 - 229 = 70z$

In Exercises 61–66, solve each equation by using the Quadratic Formula.

61. $11x^2 + 12x + 1 = 0$

62. $3w^2 + 2w = -1$

63. $9m^2 + 9m - 2 = 0$

64. $3r^2 + 2r - 3 = 0$

65. $4d^2 - 3d = -2$

66. $2x^2 + 3x = 1$

67. A rectangle has an area of 84 square inches, and the length is 8 inches longer than the width. What are the dimensions of the rectangle?

68. At a tennis club, a 15,000-square-foot rectangular area is partitioned into three rectangular courts of equal size. A total of 800 feet of fencing is used to enclose the three courts, including the interior sides. What are the possible dimensions, in feet, of the entire rectangular area?

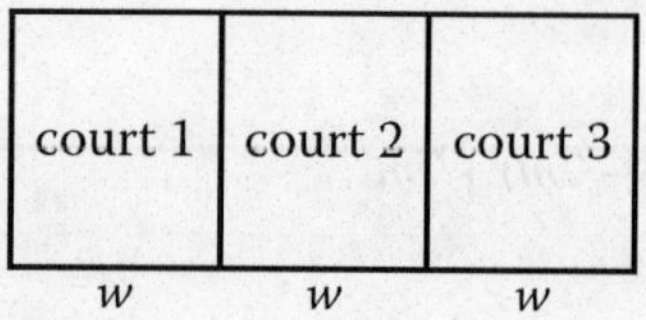

69. A rectangular field is enclosed on three sides by a fence, while the fourth side is a stream. The fence opposite the stream is 9 meters longer than the other sides. If the area of the field is 112 square meters, find its dimensions.

70. At a tennis club, a 16,500-square-foot rectangular area is partitioned into three rectangular courts of equal size. A total of 820 feet of fencing is used to enclose the three courts, including the interior sides. What are the possible dimensions, in feet, of the entire rectangular area?

Extensions

In Exercises 71–76, use the following information.

The perimeter *P* of a square can be found using the equation $P = 4s$, where *s* represents the length of a side of the square.

71. Find P when $s = 4$ and when $s = 8$ units.

72. Find s when $P = 6$ and when $P = 35$ units.

73. Find s when $P = 10$ and when $P = 20$ units.

74. Find P when $s = 1/2$ and when $s = 6.2$ units.

75. When the length of a side of a square doubles, does the square's perimeter always double?

76. When the perimeter of a square triples, does the square's side length always triple?

In Exercises 77–82, use the following information.

The equation $P - dP = S$ relates the original price *P* of an item to the sale price *S*, where *d* is the percent of the discount.

77. Find the sale prices of items with original prices \$459, \$720, and \$58 when the discount is 10%.

78. Find the sale prices of items with original prices \$18, \$150, and \$625 when the discount is 18%.

79. Find the percent discount when the original price of an item is \$180 and its sale price is \$135.

80. Find the percent discount when the original price of an item is \$25 and its sale price is \$18.

81. Find the original price of an item when the sale price is \$93 and the discount is 15%.

82. Find the original price of an item when the sale price is \$152 and the discount is 25%.

83. 172 meters of wire is cut into three pieces, where the second piece is 4 meters longer than the first piece, and the third piece is 4/5 as long as the first piece. Use the equation $f + (f + 4) + 0.8f = 172$, where f represents the length of the first piece, to find the length of each piece.

In Exercises 84–86, solve each equation.

84. $\frac{2}{3}\left(\frac{7}{8} - 4x\right) - \frac{5}{8} = \frac{3}{8}$

85. $0.008 + 9.62x - 42.8 = 0.944x + 0.0083 - x$

Round the answer to the nearest tenth, if necessary.

86. $\frac{4 - 3x}{7} = \frac{2 + 5x}{49} - \frac{x}{14}$

87. Two cars leave an intersection at the same time. One drives east, while the other travels south at 20 miles per hour faster than the first. After 3 hours, the cars are 300 miles apart. How fast is the southbound car driving?

88. Two buses make a 1680-mile trip. One bus makes the trip in 4 hours fewer than the other, traveling 10 miles per hour faster than the other. How long did it take the slower bus to complete the trip?

89. To solve the equation $4x^2 + 48x = 32x$, a student divides each side by $4x$ and solves the equation $x + 12 = 8$. The resulting solution, $x = -4$, satisfies the original equation. Is there an error? Explain.

90. Solve $ax^2 + bx = 0$ for x, given that a and b are not 0.

91. Write a quadratic equation that has $2 + \sqrt{3}$ and $2 - \sqrt{3}$ as solutions.

92. Two consecutive odd integers have squares that differ by 80. Find the integers.

P.8 OTHER TYPES OF EQUATIONS

OBJECTIVES

- Solve basic polynomial equations by factoring.
- Solve radical equations with one or two radical expressions.
- Solve equations with rational exponents.
- Solve rational equations.
- Use quadratic solving techniques to solve equations of quadratic type.
- Solve absolute value equations with one or two absolute value expressions.

PREREQUISITE VOCABULARY TERMS

absolute value
domain
polynomial
quadratic equation
rational expression
Zero-Product Property

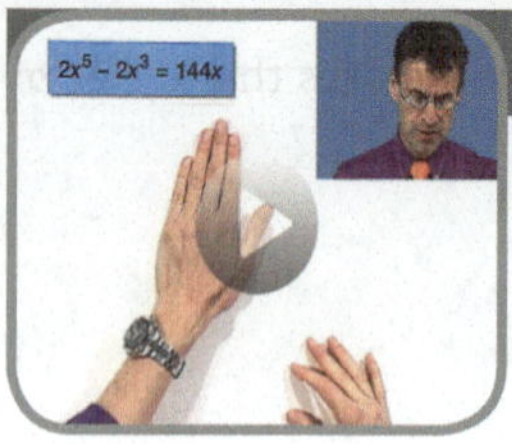

P.8.1 Solving a Polynomial Equation by Factoring

Recall that a polynomial is an expression in terms of x that can be written in the form $a_n x^n + a_{n-1} x^{n-1} + \cdots + a_1 x + a_0$, where the coefficients are real numbers, and n is nonnegative. A **polynomial equation** is an equation that can be expressed as a polynomial equal to 0.

DEFINITION

A **polynomial equation** can be written in the general form

$$a_n x^n + a_{n-1} x^{n-1} + \cdots + a_2 x^2 + a_1 x + a_0 = 0,$$

where $a_0, a_1, a_2, \ldots, a_n$ are real numbers, and n is a nonnegative integer.

TIP

A polynomial equation with degree 2 is a quadratic equation, and a polynomial equation with degree 1 is a linear equation.

A quadratic equation is a polynomial equation where $n = 2$ (i.e., the degree is 2). One method for solving a quadratic equation discussed in the previous section was solving by factoring. Recall that by the Zero-Product Property, if a product of algebraic expressions is equal to 0, then at least one of those expressions must be equal to 0. This property can also be applied to cases where there are three or more algebraic expressions in the product. It follows that if a polynomial equation can be written as a product of factors, then the factoring method can also be used to solve a polynomial equation.

Steps for Solving a Polynomial Equation by Factoring

1. Write the polynomial equation in general form.
2. Factor the polynomial.
3. Set each factor equal to 0, and solve each equation.

Note that the quadratic equations solved previously by factoring were always factored into a product of linear expressions. A polynomial equation can be factored into a product of linear expressions, a product of linear and quadratic expressions, or even a product of factors that are higher-degree expressions. The factoring method can be used to solve a polynomial equation as long as the polynomial's factors form equations (when set equal to 0) that can be solved.

EXAMPLE 1 Solving a Polynomial Equation by Factoring

Solve each equation.

A. $5x^5 = 5x^3$ **B.** $x^3 + 8x^2 = 13x^2 - 5x + 25$

SOLUTION

Write each equation in general form, and then factor.

CAUTION

Do not divide both sides of the equation by x^3. If you do so, you will not find the solution $x = 0$. Do not divide an equation by an expression containing a variable unless you know the expression cannot equal 0.

A.

$5x^5 = 5x^3$	
$5x^5 - 5x^3 = 0$	*Write the equation in general form.*
$5x^3(x^2 - 1) = 0$	*Factor the GCF from each term.*
$5x^3(x + 1)(x - 1) = 0$	*Factor the difference of squares.*
$5x^3 = 0$ or $x + 1 = 0$ or $x - 1 = 0$	*Set each factor equal to 0.*
$x = 0$ $\quad x = -1$ $\quad x = 1$	*Solve each equation.*

CHECK

Each solution can be checked by substituting back into the original equation.

$x = 0$: $5(0)^5 \stackrel{?}{=} 5(0)^3$, $0 = 0$ ✓ $\quad$ $x = -1$: $5(-1)^5 \stackrel{?}{=} 5(-1)^3$, $-5 = -5$ ✓ $\quad$ $x = 1$: $5(1)^5 \stackrel{?}{=} 5(1)^3$, $5 = 5$ ✓

Therefore, the solutions are $x = 0$, $x = -1$, and $x = 1$.

B.

$x^3 + 8x^2 = 13x^2 - 5x + 25$	
$x^3 - 5x^2 + 5x - 25 = 0$	*Write the equation in general form.*
$x^2(x - 5) + 5(x - 5) = 0$	*Factor by grouping.*
$(x - 5)(x^2 + 5) = 0$	*Factor the binomial $(x - 5)$ from each term.*
$x - 5 = 0$ or $x^2 + 5 = 0$	*Set each factor equal to 0.*
$x = 5$ $\quad x = \pm i\sqrt{5}$	*Solve each equation.*

Therefore, the solutions are $x = -i\sqrt{5}$, $x = i\sqrt{5}$, and $x = 5$.

TIP

To solve $x^2 + 5 = 0$, subtract 5 from each side then take ± the square root of each side.

$$x^2 + 5 = 0$$
$$x = \pm i\sqrt{5}$$

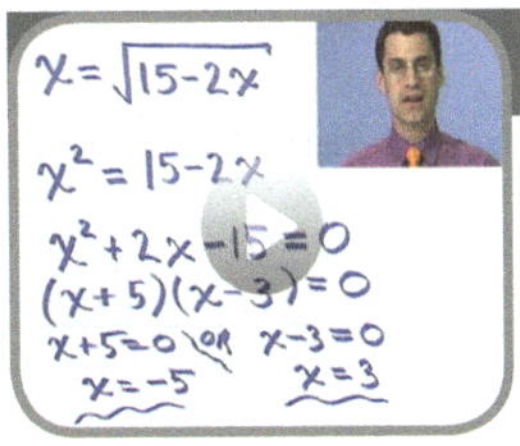

P.8.2 Solving an Equation Containing a Radical

An equation that contains a radical where the radicand includes a variable is called a **radical equation**. Some radical equations can be solved by isolating the radical expression and then applying a power to each side of the equation to eliminate the radical. The power is equal to the index of the radical. For example, squaring both sides of an equation containing a square root expression will cancel out a square root (assuming the square root expression is isolated on one side). To solve an equation containing a cube root expression, isolate the cube root expression and then raise each side to the third power. After the radical is eliminated, solve the resulting equation using any of the techniques covered previously.

Extraneous solutions can occur when both sides of an equation are raised to some power, so be sure to check each solution in the original equation to confirm that it is not extraneous.

Steps for Solving a Radical Equation with One Radical Expression

❶ Isolate the radical expression on one side of the equation.

❷ Apply a power equal to the radical's index to each side of the equation.

❸ Solve the resulting equation.

❹ Check for extraneous solutions.

EXAMPLE 2 Solving a Radical Equation

Solve each equation.

A. $\sqrt{z+10}+2=z$ **B.** $1=\sqrt{x-5}+2$

SOLUTION

A. Isolate the radical expression. Then square both sides, since the radical is a square root (i.e., index = 2), and simplify.

$$\sqrt{z+10}+2=z$$

$$\sqrt{z+10}=z-2 \qquad \textit{Subtract 2 from each side.}$$

$$z+10=(z-2)^2 \qquad \textit{Square both sides.}$$

$$z+10=z^2-4z+4 \qquad \textit{FOIL}$$

The resulting equation is quadratic, so manipulate the equation so that it is in general form, then use one of the quadratic solving techniques.

$$0=z^2-5z-6 \qquad \textit{Write the equation in general form.}$$

$$0=(z-6)(z+1) \qquad \textit{Factor.}$$

$$z=6 \quad \text{or} \quad z=-1 \qquad \textit{Set each factor equal to 0 and solve each equation.}$$

IMPORTANT

Always check for extraneous solutions when the solving steps involve squaring both sides of an equation or taking both sides to any power.

Check for extraneous solutions.

$$\sqrt{6+10}+2\stackrel{?}{=}6 \qquad \sqrt{-1+10}+2\stackrel{?}{=}-1$$

$$\sqrt{16}+2\stackrel{?}{=}6 \qquad \sqrt{9}+2\stackrel{?}{=}-1$$

$$6=6\ \checkmark \qquad 5\neq-1$$

Substituting $z=-1$ results in a false statement, so this solution is extraneous.

Therefore, the only solution is $z=6$.

B. Isolating the radical expression gives $-1=\sqrt{x-5}$. There is no number such that its square root is equal to a negative number. Therefore, there is no solution.

P.8.3 Solving an Equation with Two Radicals

When a radical equation contains two radical expressions A and B such that the index of A is equal to the index of B, and the equation contains no additional terms, then the equation can be solved by simply taking both sides to a power equal to the radical's index (assuming the equation has a solution). Doing so will eliminate the radical from both sides of the equation, leaving an equation where the radicand of A is equal to the radicand of B. Examples of such equations where the index is 2, 3, and 4 follow.

Index = 2 (square root)	Index = 3 (cube root)	Index = 4
Square each side.	Cube each side.	Take each side to the 4th power.
$\sqrt{2x+1} = \sqrt{x+4}$	$\sqrt[3]{x} = \sqrt[3]{2x-5}$	$\sqrt[4]{3x-1} = \sqrt[4]{2x}$
$\left(\sqrt{2x+1}\right)^2 = \left(\sqrt{x+4}\right)^2$	$\left(\sqrt[3]{x}\right)^3 = \left(\sqrt[3]{2x-5}\right)^3$	$\left(\sqrt[4]{3x-1}\right)^4 = \left(\sqrt[4]{2x}\right)^4$
$2x+1 = x+4$	$x = 2x-5$	$3x-1 = 2x$

Additionally, this technique may be applied when one or both of the radical expressions has a coefficient. In the case of a coefficient, be sure to apply the power to the coefficient as well as to the radical expression (by the Power of a Product Property).

In the case where a radical equation contains two square root expressions A and B, and at least one additional term, each side of the equation must be squared twice to find solutions.

Index = 2
Square each side. Be sure to square the coefficients.
$3\sqrt{2x+1} = 5\sqrt{x+4}$
$\left(3\sqrt{2x+1}\right)^2 = \left(5\sqrt{x+4}\right)^2$
$3^2\left(\sqrt{2x+1}\right)^2 = 5^2\left(\sqrt{x+4}\right)^2$
$9(2x+1) = 25(x+4)$

Steps for Solving a Square Root Equation with Two Radical Expressions and Additional Terms

1. Isolate one of the radical expressions on one side of the equation.
2. Square both sides of the equation and simplify.
3. Isolate the remaining radical expression on one side of the equation.
4. Square both sides of the equation and simplify.
5. Solve the resulting equation.
6. Check for extraneous solutions.

EXAMPLE 3 Solving a Radical Equation

Solve. $\sqrt{3x+10}+2 = \sqrt{2x+5}+3$

SOLUTION

Isolate one of the radical expressions by subtracting either 2 or 3 from both sides. Then square both sides and simplify.

$$\sqrt{3x+10}+2 = \sqrt{2x+5}+3$$

$$\sqrt{3x+10} = \sqrt{2x+5}+1 \qquad \textit{Subtract 2 from each side.}$$

$$\left(\sqrt{3x+10}\right)^2 = \left(\sqrt{2x+5}+1\right)^2 \qquad \textit{Square both sides.}$$

$$3x+10 = \left(\sqrt{2x+5}+1\right)\left(\sqrt{2x+5}+1\right) \qquad \textit{Simplify and expand the power.}$$

$$3x+10 = \left(\sqrt{2x+5}\right)^2 + 2\sqrt{2x+5}+1 \qquad \textit{FOIL}$$

$$3x+10 = 2x+5+2\sqrt{2x+5}+1 \qquad \textit{Simplify the power.}$$

$$3x+10 = 2x+6+2\sqrt{2x+5} \qquad \textit{Combine the like terms.}$$

Now isolate the remaining radical, and then square both sides again.

$$\frac{x+4}{2} = \sqrt{2x+5}$$ *Subtract 2x and 6 from each side, and divide each side by 2.*

$$\left(\frac{x+4}{2}\right)^2 = \left(\sqrt{2x+5}\right)^2$$ *Square both sides.*

$$\frac{(x+4)^2}{2^2} = 2x+5$$ *Simplify the powers.*

$$\frac{x^2+8x+16}{4} = 2x+5$$ *FOIL*

$$x^2+8x+16 = 8x+20$$ *Multiply both sides by 4.*

The resulting equation is quadratic, so manipulate the equation so that it is in general form, then use one of the quadratic solving techniques.

$$x^2-4=0$$ *Write the equation in general form.*

$$(x-2)(x+2)=0$$ *Factor.*

$$x=2 \quad \text{or} \quad x=-2$$ *Set each factor equal to 0 and solve each equation.*

Check for extraneous solutions.

$$\sqrt{3(2)+10}+2 \stackrel{?}{=} \sqrt{2(2)+5}+3 \qquad \sqrt{3(-2)+10}+2 \stackrel{?}{=} \sqrt{2(-2)+5}+3$$

$$\sqrt{16}+2 \stackrel{?}{=} \sqrt{9}+3 \qquad \sqrt{4}+2 \stackrel{?}{=} \sqrt{1}+3$$

$$6=6\ \checkmark \qquad 4=4\ \checkmark$$

Both solutions result in true statements, so neither solution is extraneous.

Therefore, the solutions are $x = -2$ and $x = 2$.

P.8.4 Solving an Equation with Rational Exponents

In Section P.3, we saw that an expression with a rational exponent can be written as a radical where the denominator of the rational exponent is the index of the radical.

$$a^{\frac{m}{n}} = \left(\sqrt[n]{a}\right)^m = \sqrt[n]{a^m}$$

An equation with a rational exponent can be solved by writing the expression with the rational exponent as a radical and then using the radical equation solving techniques.

To solve an equation that contains two terms with rational exponents where the denominators are not equal and there are no additional terms, write each term on opposite sides of the equation, and then raise each side to the power equal to the LCM of the denominators. Again, if either expression with a rational exponent has a coefficient, be sure to apply the power to the coefficient as well.

EXAMPLE 4 Solving an Equation with Rational Exponents

Solve each equation.

A. $(x+1)^{\frac{2}{3}} = 25$ **B.** $3(y^2+y-1)^{\frac{1}{4}} = (9y)^{\frac{1}{2}}$

SOLUTION

A. Convert the rational exponent to a cube root, and then raise each side to the 3rd power.

$$(x+1)^{\frac{2}{3}} = 25$$

$$\sqrt[3]{(x+1)^2} = 25 \quad \textit{Convert to a radical.}$$

$$\left(\sqrt[3]{(x+1)^2}\right)^3 = (25)^3 \quad \textit{Cube both sides.}$$

$$(x+1)^2 = 15{,}625 \quad \textit{Simplify.}$$

$$x+1 = \pm 125 \quad \textit{Take } \pm \textit{ the square root of each side.}$$

$$x = 124 \text{ or } x = -126 \quad \textit{Subtract 1 from each side.}$$

IMPORTANT

Be sure to check for extraneous solutions.

Check for extraneous solutions.

$$(124+1)^{\frac{2}{3}} = (125)^{\frac{2}{3}} = \left(\sqrt[3]{125}\right)^2 = 5^2 = 25 \checkmark$$

$$(-126+1)^{\frac{2}{3}} = (-125)^{\frac{2}{3}} = \left(\sqrt[3]{-125}\right)^2 = (-5)^2 = 25 \checkmark$$

Substituting $x = 124$ and $x = -126$ into the original equation results in true statements, so neither solution is extraneous. Therefore, the solutions are $x = -126$ and $x = 124$.

ALTERNATIVE METHOD

Raise each side to the 3/2 power.

$$((x+1)^{\frac{2}{3}})^{\frac{3}{2}} = (25)^{\frac{3}{2}} \quad \textit{Raise each side to the 3/2 power.}$$

$$x+1 = \left(\pm\sqrt{25}\right)^3 \quad \textit{Simplify.}$$

$$x+1 = (\pm 5)^3 \quad \textit{Simplify the radical.}$$

$$x+1 = \pm 125 \quad \textit{Simplify.}$$

$$x = 124 \text{ or } x = -126 \quad \textit{Subtract 1 from each side.}$$

B. Since the rational exponents are not the same, raise both sides to the power equal to the LCM of the denominators, which is 4.

$$3(y^2+y-1)^{\frac{1}{4}} = (9y)^{\frac{1}{2}}$$

$$(3(y^2+y-1)^{\frac{1}{4}})^4 = ((9y)^{\frac{1}{2}})^4 \quad \textit{Raise each side to the 4th power.}$$

$$(3)^4((y^2+y-1)^{\frac{1}{4}})^4 = ((9y)^{\frac{1}{2}})^4 \quad \textit{Power of a Product Property}$$

$$(3)^4(y^2+y-1)^{\frac{1}{4}\cdot 4} = (9y)^{\frac{1}{2}\cdot 4} \quad \textit{Power of a Power Property}$$

$$(3)^4(y^2+y-1)^1 = (9y)^2 \quad \textit{Multiply the exponents.}$$

$$(3)^4(y^2+y-1) = (9^2)(y^2) \quad \textit{Power of a Product Property}$$

$$81(y^2+y-1) = 81y^2 \quad \textit{Evaluate the powers.}$$

$$y^2+y-1 = y^2 \quad \textit{Divide each side by 81.}$$

$$y-1 = 0 \quad \textit{Subtract } y^2 \textit{ from each side.}$$

$$y = 1 \quad \textit{Add 1 to each side.}$$

Check for extraneous solutions. Substituting $y = 1$ into the original equation results in a true statement, so the solution is $y = 1$.

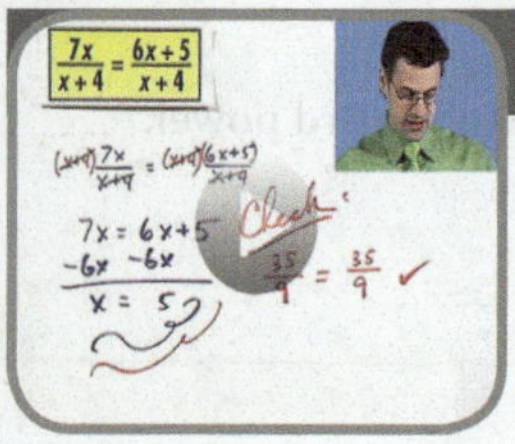

P.8.5 Solving Rational Equations

Two techniques for solving rational equations were discussed previously. One method is to use the Cross-Product Property. This method can be used when the equation is an equality between two rational expressions (i.e., the equation is a proportion). The second method is to multiply the equation by a common denominator. This method will work for any type of rational equation. If the terms in the equation contain multiple denominators, then multiply the equation by the LCD (least common denominator), which is the LCM of the denominators.

EXAMPLE 5 Solving a Rational Equation

Solve. $\dfrac{1}{w-2}+\dfrac{2}{w+3}=\dfrac{8}{w^2+w-6}$

SOLUTION

The equation contains unlike denominators, so multiply the equation by the LCD (the LCM of $w-2$, $w+3$, and w^2+w-6). The factorization of w^2+w-6 is $(w-2)(w+3)$. Since the other denominators are $w-2$ and $w+3$, the LCD is $(w-2)(w+3)$.

$$\frac{1}{w-2}+\frac{2}{w+3}=\frac{8}{w^2+w-6}$$

$$\frac{1}{w-2}+\frac{2}{w+3}=\frac{8}{(w-2)(w+3)}$$ *Factor.*

$$(w-2)(w+3)\left(\frac{1}{w-2}+\frac{2}{w+3}\right)=\left(\frac{8}{(w-2)(w+3)}\right)(w-2)(w+3)$$ *Multiply both sides by the LCD.*

$$\frac{\cancel{(w-2)}(w+3)}{\cancel{w-2}}+\frac{2(w-2)\cancel{(w+3)}}{\cancel{w+3}}=\frac{8\cancel{(w-2)}\cancel{(w+3)}}{\cancel{(w-2)}\cancel{(w+3)}}$$ *Distribute and remove common factors.*

$$(w+3)+2(w-2)=8$$ *Write the remaining factors.*

$$w+3+2w-4=8$$ *Distribute.*

$$3w-1=8$$ *Simplify.*

$$w=3$$ *Solve for w.*

Check for extraneous solutions. Substituting $w = 3$ into the original equation results in a true statement, so the solution is $w = 3$.

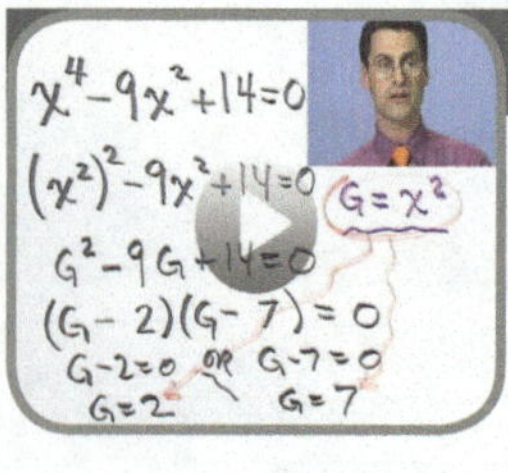

P.8.6 Solving Equations of Quadratic Type

Expressions that are not quadratic can sometimes be written in quadratic form by substituting a variable for an expression that is a factor in one or two of the terms. For example, consider the expression x^6-5x^3+6. This expression is not quadratic. However, notice that if the power in the middle term, x^3, is squared, the result is the power in the first term, x^6: $(x^3)^2=x^6$. So, the expression can be written as $(x^3)^2-5(x^3)+6$. If we let $G=x^3$, then the expression can be written as G^2-5G+6, which is a quadratic expression. Using this substitution method, a nonquadratic expression can be written in quadratic form.

This method can also be used to write a nonquadratic equation in quadratic form. Once the equation is in quadratic form, any of the quadratic solving techniques may be applied. After the quadratic form of the equation is solved, the last step is to substitute those solutions into the original G-substitution equation. For example, solving $G^2-5G+6=0$ by factoring gives $G=3$ or $G=2$.

However, $G = 3$ or $G = 2$ are not the solutions of $x^6 - 5x^3 + 6 = 0$. To find the solutions of $x^6 - 5x^3 + 6 = 0$, substitute $G = 3$ and $G = 2$ back into the original G-substitution equation, which is $G = x^3$: $3 = x^3$ or $2 = x^3$. Take the cube root of both sides to solve for x: $x = \sqrt[3]{3}$ or $x = \sqrt[3]{2}$.

EXAMPLE 6 Solving an Equation by Writing as a Quadratic

Solve each equation.

A. $q^4 + 900 = 61q^2$ **B.** $(2y-1)^{\frac{2}{5}} + 1 = 2(2y-1)^{\frac{1}{5}}$

SOLUTION

NOTICE THAT

G is substituted for a monomial with a whole number exponent: q^2.

A. In general form, the equation is $q^4 - 61q^2 + 900 = 0$. Notice that the equation can be written in the form of a quadratic where q^2 is the variable: $(q^2)^2 - 61(q^2) + 900 = 0$. So, substitute using $G = q^2$ to write the equation in quadratic form.

$$(q^2)^2 - 61(q^2) + 900 = 0$$
$$G^2 - 61G + 900 = 0 \quad \textit{Let } q^2 = G.$$
$$(G - 36)(G - 25) = 0 \quad \textit{Factor.}$$
$$G = 36 \text{ or } G = 25 \quad \textit{Set each factor equal to 0 and solve each equation.}$$
$$q^2 = 36 \quad q^2 = 25 \quad \textit{Substitute } q^2 \textit{ for } G.$$
$$q = \pm 6 \quad q = \pm 5 \quad \textit{Take} \pm \textit{the square root of each side.}$$

Substitute to check for extraneous solutions.

$$(-6)^4 + 900 \stackrel{?}{=} 61(-6)^2 \qquad (-5)^4 + 900 \stackrel{?}{=} 61(-5)^2 \qquad (5)^4 + 900 \stackrel{?}{=} 61(5)^2 \qquad (6)^4 + 900 \stackrel{?}{=} 61(6)^2$$
$$1296 + 900 \stackrel{?}{=} 61(36) \qquad 625 + 900 \stackrel{?}{=} 61(25) \qquad 625 + 900 \stackrel{?}{=} 61(25) \qquad 1296 + 900 \stackrel{?}{=} 61(36)$$
$$2196 = 2196 \checkmark \qquad 1525 = 1525 \checkmark \qquad 1525 = 1525 \checkmark \qquad 2196 = 2196 \checkmark$$

Therefore, the solutions are $q = -6$, $q = -5$, $q = 5$, and $q = 6$.

IMPORTANT

Raising both sides to the 5th power complicates the equation. So, solving by raising both sides to the 5th power is not a useful option.

B. In general form, the equation is $(2y-1)^{\frac{2}{5}} - 2(2y-1)^{\frac{1}{5}} + 1 = 0$. Notice that the binomial $(2y - 1)$ is used in each term with a power. Additionally, notice that the equation can be written in quadratic form where $(2y-1)^{\frac{1}{5}}$ is the variable: $((2y-1)^{\frac{1}{5}})^2 - 2(2y-1)^{\frac{1}{5}} + 1 = 0$.

So, substitute using $G = (2y-1)^{\frac{1}{5}}$ to write the equation in quadratic form.

NOTICE THAT

G is substituted for a binomial with a rational exponent: $(2y-1)^{\frac{1}{5}}$.

$$((2y-1)^{\frac{1}{5}})^2 - 2(2y-1)^{\frac{1}{5}} + 1 = 0$$
$$G^2 - 2G + 1 = 0 \quad \textit{Let } (2y-1)^{\frac{1}{5}} = G.$$
$$(G-1)(G-1) = 0 \quad \textit{Factor.}$$

Solving $G - 1 = 0$ yields $G = 1$. Substitute $(2y-1)^{\frac{1}{5}}$ for G and solve.

$$(2y-1)^{\frac{1}{5}} = 1$$
$$((2y-1)^{\frac{1}{5}})^5 = 1^5 \quad \textit{Raise each side to the 5th power.}$$
$$2y - 1 = 1 \quad \textit{Simplify.}$$
$$y = 1 \quad \textit{Solve for } y.$$

Check for extraneous solutions. Substituting $y = 1$ into the original equation results in a true statement, so the solution is $y = 1$.

P.8.7 Solving Absolute Value Equations

Recall that the absolute value of a number is the distance between that number and 0 on a number line. Since absolute value is a distance, the absolute value of a number is always positive. Consider the equation $|A| = B$, where A and B are some numbers. If $|A| = B$, then A must be equal to either B or $-B$. The following property follows from this statement.

Absolute Value Equations

If A and B are some numbers or algebraic expressions such that $|A| = B$, then $A = \pm B$. So, the equation $|A| = B$ can be written as the two equivalent equations $A = B$ and $A = -B$.

This property is used to solve equations in which a variable is contained within one or more absolute value expressions, called **absolute value equations**. The domain of an absolute value equation may not be all real numbers. Therefore, solving an absolute value equation may produce extraneous solutions.

Steps for Solving an Absolute Value Equation

1. Isolate an absolute value expression on one side of the equation: $|A| = B$.
2. Write the equation as the two equivalent equations $A = B$ and $A = -B$.
3. Solve each equation.
4. Check for extraneous solutions.

CAUTION

When B is a multiterm expression, be sure to distribute the negative in $A = -B$ to each term in B.

EXAMPLE 7 Solving an Absolute Value Equation

Solve each equation.

A. $|4x - 2| - 6 = 0$ **B.** $1 = \left|\frac{x-7}{5}\right| + 4$ **C.** $\left|\frac{6}{x+4}\right| - 9 = -6$

SOLUTION

A. Add 6 to both sides to isolate the absolute value expression $|4x - 2|$. Then write and solve the two equations.

$$|4x - 2| - 6 = 0$$

$$|4x - 2| = 6 \qquad \textit{Isolate the absolute value expression.}$$

$$4x - 2 = 6 \quad \text{or} \quad 4x - 2 = -6 \qquad \textit{Write } |A| = B \textit{ as } A = B \textit{ or as } A = -B.$$

$$x = 2 \qquad\qquad x = -1 \qquad \textit{Solve each equation.}$$

Substitute to check for extraneous solutions.

$$|4(2) - 2| - 6 = |6| - 6 = 0 \checkmark$$

$$|4(-1) - 2| - 6 = |-6| - 6 = 0 \checkmark$$

Therefore, the solutions are $x = -1$ and $x = 2$.

B. Subtracting 4 from both sides to isolate the absolute value expression gives $-3 = \left|\frac{x-7}{5}\right|$. The absolute value of a number cannot equal a negative number, so there is no solution.

C. Add 9 to both sides to isolate the absolute value expression, then write and solve the two equations. Notice that the resulting equations will be rational and therefore should be solved using one of the techniques for solving rational equations.

$$\left|\frac{6}{x+4}\right| - 9 = -6$$

$$\left|\frac{6}{x+4}\right| = 3 \qquad \textit{Isolate the absolute value expression.}$$

$$\frac{6}{x+4} = 3 \quad \text{or} \quad \frac{6}{x+4} = -3 \qquad \textit{Write } |A| = B \textit{ as } A = B \textit{ or as } A = -B.$$

$$6 = 3(x+4) \qquad 6 = -3(x+4) \qquad \textit{Multiply both sides by } x+4.$$

$$6 = 3x + 12 \qquad 6 = -3x - 12 \qquad \textit{Distribute.}$$

$$-2 = x \qquad -6 = x \qquad \textit{Solve each equation for } x.$$

Check for extraneous solutions. Substituting $x = -2$ into the original equation results in a true statement, as does substituting $x = -6$, so the solutions are $x = -6$ and $x = -2$.

SECTION P.8 EXERCISES

Warm Up

1. Simplify. $-|8 - 3(-5 + 2)|$
2. Evaluate $\frac{3x+4}{x+1} + \frac{x}{5}$ for $x = 1$.
3. Factor. $6x^2 - 5x - 4$

Just the Facts

Fill in the blanks.

4. The equation $a_n x^n + a_{n-1} x^{n-1} + \cdots + a_2 x^2 + a_1 x + a_0 = 0$ is a(n) _____ equation written in general form.
5. If a radical equation contains two square root expressions and at least one additional term, then the equation can be _____ twice to find solutions.
6. An equation that contains a radical where the radicand includes a variable is called a(n) _____ equation.
7. Always check for _____ solutions when solving a radical equation because they can occur when the equation's domain is not all real numbers.
8. An expression with a(n) _____ exponent can be written as a radical where the _____ is the index of the radical.
9. When solving an absolute value equation, isolate the _____ expression on one side of the equation, then write the two equations $A = B$ and _____.
10. The denominators in a rational equation can be cleared by _____.

Essential Skills

In Exercises 1–60, solve each equation.

1. $2x^6 + 3x^5 - 7x^4 - 2x^6 + 15x^5 + 9x^4 = 0$
2. $6x^5 + 40x^4 + 20x^3 = 4x^4 - 28x^3$
3. $6x^6 + 33x^5 - 133x^4 = 9x^5 - 7x^4$
4. $3t^5 + 3t^4 - 79t^3 - 9t^4 = -7t^3$
5. $5r^7 + 40r^6 + 35r^5 - 10r^6 = -5r^5$
6. $3r^9 + 15r^8 - 60r^7 = 6r^8 - 6r^7$
7. $5x^3 + 7x^2 - 27x + 15 = 2x^3 + 6x^2 + 24$
8. $2x^3 - x^2 - x = 2x^2 + x - 3$
9. $3a^3 + 4a^2 - 80a + 50 = 6a^2 - 5a$
10. $5b^3 + 3b^2 - 130b = 5b^2 - 5b - 50$
11. $4m^3 + 2m^2 - 102m + 75 - 5m^2 = -2m$
12. $2x^3 + 3x^2 - 22x + 27 = 6x^2 - 4x$
13. $\sqrt{z+10} = 2$
14. $\sqrt{3w+14} + 6 = 3w$
15. $\sqrt{15a+10} = 5a - 6$
16. $6 - 7w = -\sqrt{21w+10}$
17. $4y + 1 = \sqrt{4y+7}$
18. $\sqrt{14x+12} + 6 = 7x$
19. $\sqrt{x} = \sqrt{x+5} - 1$
20. $\sqrt{8x+4} = \sqrt{2x+1} + 3$
21. $\sqrt{3z} + 1 = 3 + \sqrt{2z-8}$
22. $\sqrt{a+1} = 3 - \sqrt{4-a}$
23. $\sqrt{2x-11} - \sqrt{3x+6} = -3$
24. $\sqrt{22w+4} - 1 = \sqrt{4w+1} + 2$
25. $(x^2 - 4x + 3)^{\frac{1}{3}} = x^{\frac{2}{3}}$
26. $(y+3)^{\frac{1}{2}} = 2y^{\frac{1}{4}}$
27. $(2k-1)^{\frac{1}{5}} = k^{\frac{2}{5}}$
28. $(17d^2 - 16)^{\frac{1}{4}} = d$
29. $(w-1)^{\frac{2}{3}} = (w^2 - 4w)^{\frac{1}{3}}$
30. $(z+8)^{\frac{1}{2}} = 3z^{\frac{1}{4}}$
31. $\frac{x}{4} + \frac{2x+3}{x-2} = \frac{3x+1}{x-2}$
32. $\frac{x+4}{x+1} + \frac{x}{5} = \frac{2x+5}{x+1}$
33. $\frac{1}{2} - \frac{2}{x-1} = -\frac{3}{2x-2}$
34. $\frac{5}{4} - \frac{1}{t+2} = \frac{1}{t-1}$
35. $\frac{5}{2g+6} + \frac{1}{g-2} = \frac{3}{g+3}$
36. $\frac{32}{d^2-25} = \frac{2}{d-5} + \frac{4}{d+5}$
37. $y^6 + 16y^3 + 64 = 0$
38. $z^{10} - 6z^5 = 2z^5 - 15$
39. $81b^8 + 80b^4 + 64 = 65$
40. $27b^6 + 19b^3 = 8$
41. $25c^4 + 10 = -46c^2 + 18$
42. $a^6 - 4a^3 + 20 = 5a^3$
43. $2x^{\frac{2}{5}} - x^{\frac{1}{5}} - 1 = 0$

44. $(x+2)^{\frac{1}{2}}+4=5(x+2)^{\frac{1}{4}}$

45. $(n-3)^{\frac{2}{3}}-3(n-3)^{\frac{1}{3}}+2=0$

46. $(2x-1)^{\frac{2}{5}}-2(2x-1)^{\frac{1}{5}}+1=0$

47. $(b+2)^{\frac{1}{2}}-5(b+2)^{\frac{1}{4}}=-4$

48. $(w+6)^{\frac{1}{2}}+8=6(w+6)^{\frac{1}{4}}$

49. $|5a+2|=-1$

50. $5=|6x+4|$

51. $7=|2p+5|$

52. $|9t+4|=8$

53. $|5m-14|-6=0$

54. $8+|-3x+9|=0$

55. $\left|\dfrac{3}{b+1}\right|=2$

56. $\left|\dfrac{5}{x+2}\right|=7$

57. $10+\left|\dfrac{4}{p+6}\right|=16$

58. $6-\left|\dfrac{2}{k-5}\right|=1$

59. $\left|\dfrac{w+5}{2}\right|-7=9$

60. $4-\left|\dfrac{a-2}{3}\right|=10$

Extensions

In Exercises 61–64, use techniques for solving quadratic equations to solve each equation for the indicated variable.

61. $(y+2)^2=3x-5$ for y

62. $0=(-3+x)^2-8z$ for x

63. $5cp-6ap^2+cp=0$ for p

64. $x+2hm=7rm^2+hm$ for m

In Exercises 65–69, solve each equation.

65. $\sqrt{1-\sqrt{x-1}}+x-1=1$

66. $\dfrac{1}{x-1}+\dfrac{2}{x-5}=\dfrac{4}{x+2}$

67. $|3x-4|=|5x+2|$

68. $9x^4+3\sqrt{7}x^3+4x^2=2x^4-11\sqrt{7}x^3-45x^2$

69. $6x^{-2}+11x^{-1}=10$

70. In still water, a barge travels at an average of 8 miles per hour. The barge travels 60 miles up the Mississippi River and 60 miles down the river in 16.5 hours. What is the average speed of the current in miles per hour? Round to the nearest tenth of a mile per hour.

71. A woman spends her morning kayaking on a river. She travels 6 miles upstream and 6 miles downstream in 4.5 hours. In still water, she averages 3 miles per hour. What is the average speed of the current in miles per hour? Round to the nearest mile per hour.

P.9 INEQUALITIES

OBJECTIVES

- Solve linear inequalities.
- Solve compound inequalities.
- Solve absolute value inequalities.
- Use a sign chart to solve quadratic, polynomial, and rational inequalities.

PREREQUISITE VOCABULARY TERMS

absolute value equation
interval notation
polynomial equation
quadratic equation
rational equation

P.9.1 An Introduction to Solving Inequalities

A **linear inequality** is almost identical to a linear equation. However, a linear inequality contains an inequality symbol instead of an equals sign. For example, $2x + 1 > 9$ is a linear inequality in terms of x.

Solving linear inequalities is similar to solving linear equations: numbers are added, subtracted, multiplied, or divided on both sides of the inequality to isolate the variable. However, if an inequality is *multiplied* or *divided* by a *negative* number, the inequality symbol must be reversed.

Solving an inequality gives its solution set (the set of all real numbers that satisfy the inequality), which is typically expressed in interval notation. The **empty set** ø represents the solution set of an inequality that has no real solution.

REMEMBER

A bracket is used to indicate that the endpoint is included in the interval. A parenthesis is used to indicate that the endpoint is not included in the interval. A parenthesis is always used for the opening symbol with −∞ and the closing symbol with ∞.

Review of Inequalities and Interval Notation

Inequality	Interval Notation	Description
$x \leq 4$	$(-\infty, 4]$	*all numbers less than or equal to 4*
$0 < y \leq 1$	$(0, 1]$	*all numbers between 0 and 1, including 1*
$z < -5$ or $z \geq 2$	$(-\infty, -5) \cup [2, \infty)$	*all numbers less than −5 and all numbers greater than or equal to 2*

EXAMPLE 1 Solving a Linear Inequality in One Variable

Solve each inequality. Express each solution set in interval notation.

A. $2x + 3 \leq -15$ **B.** $4(x - 8) > -6(7 - x)$

SOLUTION

A. $2x + 3 \leq -15$

$2x \leq -18$ *Subtract 3 from each side.*

$x \leq -9$ *Divide each side by 2.*   The solution set is the interval $(-\infty, -9]$.

B. $4(x - 8) > -6(7 - x)$

$4x - 32 > -42 + 6x$ *Distribute.*

$-2x > -10$ *Move the variables to one side and the constant terms to the other.*

$x < 5$ *Divide each side by −2 and reverse the inequality symbol.*

The solution set is the interval $(-\infty, 5)$.

NOTICE THAT

The inequality symbol is not reversed when −6 is distributed on the right side because the entire inequality is not multiplied by a negative number at that point.

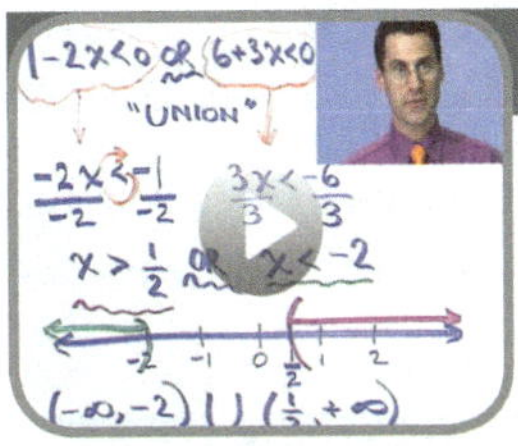

P.9.2 Solving Compound Inequalities

Two inequalities connected by the word "and" or the word "or" form a **compound inequality**.

An "and" compound inequality, called a **conjunction**, indicates an intersection of intervals. For example, the solution set of the conjunction "$x > 2$ and $x < 5$" is all values greater than 2 that are also less than 5, which is all values in the interval (2, 5). Note that a conjunction can sometimes be expressed as a double inequality. For example, "$x > 2$ and $x < 5$" can be expressed as "$2 < x < 5$."

An "or" compound inequality is called a **disjunction**. The solution set of a disjunction is the union of the solution set of each inequality in the disjunction. In other words, a disjunction's solution set includes all values from the solution set of each inequality in the disjunction. For example, the solution set of the disjunction "$x > 2$ or $x < 5$" is all values greater than 2 combined with all values less than 5, which is the interval of all real numbers: $(-\infty, \infty)$.

The solution set of a disjunction can be expressed as a single interval only when the solution sets of the individual inequalities are intervals that overlap. Otherwise, the solution set of a disjunction is the union of two intervals (expressed using the union symbol ∪ between the intervals).

A **linear inequality** is similar to a linear equation, but with an inequality symbol replacing the equals sign. To solve a linear inequality (disjunction or conjunction), isolate the variable just as you would when solving a linear equation. However, if the inequality is multiplied or divided by a negative number, the inequality symbol must be reversed.

IMPORTANT

If the intervals for the two inequalities in a conjunction do not overlap, then the conjunction's solution set is the empty set.

EXAMPLE 2 Solving a Linear Inequality

Solve each inequality. Express each solution set in interval notation.

A. $10 - 4x \le -18$ or $3x + 7 \le 4$ B. $-9 < 6 - 3x \le 0$

SOLUTION

A. Solve each inequality individually.

$10 - 4x \le -18$

$-4x \le -28$ *Subtract 10 from each side.*

$x \ge 7$ *Divide each side by −4 and reverse the inequality symbol.*

$3x + 7 \le 4$

$3x \le -3$ *Subtract 7 from each side.*

$x \le -1$ *Divide each side by 3.*

The solution set of $x \ge 7$ or $x \le -1$ is $(-\infty, -1] \cup [7, \infty)$.

B. Solve the double inequality by applying each solving step to all three sides.

$-9 < 6 - 3x \le 0$

$-15 < -3x \le -6$ *Subtract 6 from each side.*

$5 > x \ge 2$ *Divide each side by −3 and reverse both inequality symbols.*

The inequality $5 > x \ge 2$ should be written as $2 \le x < 5$.

The solution set is [2, 5).

IMPORTANT

Reverse the inequality symbol when the inequality is multiplied or divided by a negative number. For a double inequality, reverse both inequality symbols.

P.9.3 Solving Absolute Value Inequalities

Recall that an absolute value equation is solved by isolating the absolute value expression and then writing $|A| = B$ as two equivalent equations: $A = B$ and $A = -B$. A similar procedure is used for solving **absolute value inequalities.**

CAUTION

If B represents a number such that $B < 0$ (i.e., B is negative), then $|A| < B$ and $|A| \leq B$ have no solutions.

Absolute Value Inequalities

If A and B are some numbers or algebraic expressions such that

- $|A| < B$, then $-B < A < B$,
- $|A| \leq B$, then $-B \leq A \leq B$,
- $|A| > B$, then $A > B$ or $A < -B$, and
- $|A| \geq B$, then $A \geq B$ or $A \leq -B$.

So, a *less than* absolute value inequality ($|A| < B$ or $|A| \leq B$) can be written as an "and" compound inequality, and a *greater than* absolute value inequality ($|A| > B$ or $|A| \geq B$) can be written as an "or" compound inequality. In each case, the resulting inequality (which does not contain an absolute value expression) can be solved using the techniques for solving compound inequalities.

Steps for Solving an Absolute Value Inequality

❶ Isolate the absolute value expression on one side of the inequality.

❷ Write the absolute value inequality as the equivalent compound inequality.

❸ Solve the compound inequality.

EXAMPLE 3 Solving an Absolute Value Inequality

Solve each inequality. Express each solution set in interval notation.

A. $12 > |9 - 3x|$ **B.** $|x - 7| - 1 > 5$

SOLUTION

A. Reverse the inequality so that the absolute value expression is on the left side.

$$12 > |9 - 3x| \;\Rightarrow\; |9 - 3x| < 12$$

Express the absolute value inequality as a compound inequality using $|A| < B \Rightarrow -B < A < B$, then solve the double inequality by applying each solving step to all three sides.

$$-12 < 9 - 3x < 12$$

$$-21 < -3x < 3 \qquad \textit{Subtract 9 from each side.}$$

$$7 > x > -1 \qquad \textit{Divide each side by } -3 \textit{ and reverse both inequality symbols.}$$

The inequality $7 > x > -1$ should be written as $-1 < x < 7$. The solution set is $(-1, 7)$.

B. Add 1 to each side to isolate the absolute value expression on the left side.

$$|x - 7| - 1 > 5 \;\Rightarrow\; |x - 7| > 6$$

Express the absolute value inequality as a compound inequality using $|A| > B \;\Rightarrow\; A < B$ or $A < -B$, and then solve each inequality individually.

$$x - 7 > 6 \;\Rightarrow\; x > 13 \qquad \text{or} \qquad x - 7 < -6 \;\Rightarrow\; x < 1$$

The solution set of $x > 13$ or $x < 1$ is $(-\infty, 1) \cup (13, \infty)$.

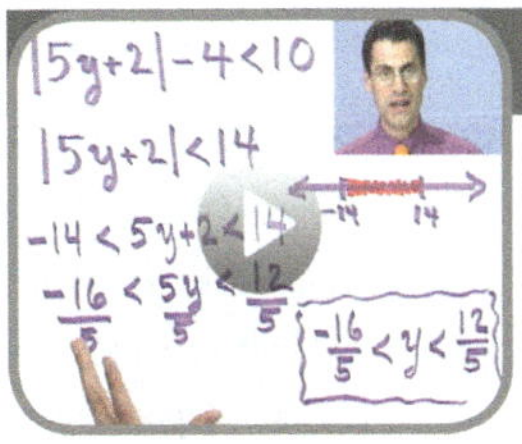

P.9.4 Solving Absolute Value Inequalities: Another Example

When solving an absolute value inequality, be sure to isolate the absolute value expression on one side of the inequality (typically the left side) and simplify the absolute value expression before writing the compound inequality.

EXAMPLE 4 Solving an Absolute Value Inequality

Solve $18 - 5|2 - (3x - 5)| \geq -37$. Express the solution set in interval notation.

SOLUTION

Simplify the absolute value expression, then isolate the expression on the left side.

$18 - 5|2 - (3x - 5)| \geq -37$

$18 - 5|2 - 3x + 5| \geq -37$ *Distribute.*

$18 - 5|-3x + 7| \geq -37$ *Combine the like terms.*

$-5|-3x + 7| \geq -55$ *Subtract 18 from each side.*

$|-3x + 7| \leq 11$ *Divide each side by –5 and reverse the inequality symbol.*

Express the absolute value inequality as a compound inequality using $|A| \leq B \Rightarrow -B \leq A \leq B$, and then solve the double inequality by applying each solving step to all three sides.

$-11 \leq -3x + 7 \leq 11$

$-18 \leq -3x \leq 4$ *Subtract 7 from each side.*

$6 \geq x \geq -\dfrac{4}{3}$ *Divide each side by –3 and reverse both inequality symbols.*

The inequality $6 \geq x \geq -\frac{4}{3}$ should be written as $-\frac{4}{3} \leq x \leq 6$. The solution set is $\left[-\frac{4}{3}, 6\right]$.

WHAT'S THE BIG IDEA?

Explain why $|A| < B$ can be written equivalently as $-B < A < B$.

P.9.5 Solving Quadratic Inequalities

When an inequality is linear and has a solution set, its solution set is a single interval of real numbers. For example, the solution set of the linear inequality $x - 1 > 0$ is all real numbers x in the interval $(1, \infty)$. Notice that the solution of the *related equation* $x - 1 = 0$ (i.e., $x = 1$) gives the interval's endpoint.

If a solution set exists for a **quadratic inequality,** it can consist of one or two intervals of real numbers where each endpoint (other than $-\infty$ and ∞) is a solution of the *related equation* (i.e., the equation given by replacing the inequality symbol with an equals sign).

$x^2 - 4x + 3 = 0$
$(x - 1)(x - 3) = 0$
$x - 1 = 0$ or $x - 3 = 0$
$x = 1$ $\quad$ $x = 3$

For example, consider the quadratic inequality $x^2 - 4x + 3 > 0$. The related equation $x^2 - 4x + 3 = 0$ has solutions $x = 1$ and $x = 3$. So, the solution set for $x^2 - 4x + 3 > 0$ (if one exists) is all real numbers x in one or more of the intervals $(-\infty, 1)$, $(1, 3)$, and $(3, \infty)$.

IMPORTANT

For a quadratic inequality, the endpoints of the test intervals are the solutions of its related equation.

To determine which of these intervals are in the solution set of an inequality, we can use a **sign chart**. A sign chart is a number line (representing all real numbers) divided into intervals called *test intervals*. For example, consider the sign chart for $x^2 - 4x + 3$ shown in Figure P.9a. The number line is divided into the three test intervals, $(-\infty, 1)$, $(1, 3)$, and $(3, \infty)$, because the solutions of the equation $x^2 - 4x + 3 = 0$ are $x = 1$ and $x = 3$. The sign of $x^2 - 4x + 3$ within each test interval is found by substituting any value from that interval into the expression $x^2 - 4x + 3$. This sign is then noted above the number line to complete the sign chart.

Sign Chart for
$x^2 - 4x + 3$

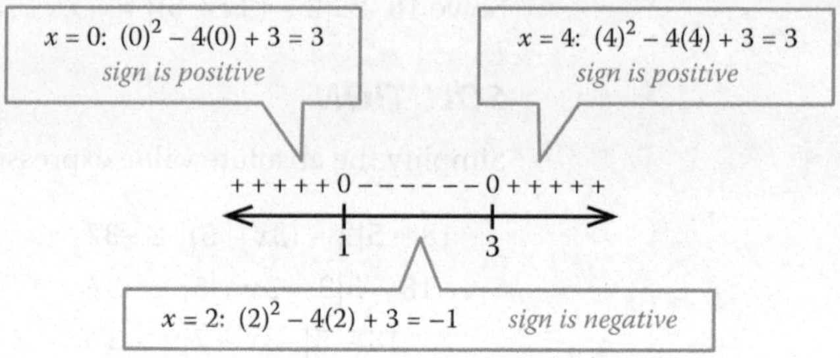

Figure P.9a

IMPORTANT

If one real number in a test interval satisfies the inequality, then we can assume that the inequality is satisfied for all real numbers within that test interval.

From the sign chart, we see that for all x in the interval
- $(-\infty, 1)$, the sign of $x^2 - 4x + 3$ is positive,
- $(1, 3)$, the sign of $x^2 - 4x + 3$ is negative, and
- $(3, \infty)$, the sign of $x^2 - 4x + 3$ is positive.

The information from the sign chart can be used to find the solution set of any quadratic inequality where $x^2 - 4x + 3$ is on one side and 0 is on the other side (i.e., $x^2 - 4x + 3 < 0$, $x^2 - 4x + 3 \le 0$, $x^2 - 4x + 3 > 0$, or $x^2 - 4x + 3 \ge 0$).

The solution sets of both $x^2 - 4x + 3 < 0$ and $x^2 - 4x + 3 \le 0$ include all x-values that make $x^2 - 4x + 3$ negative (i.e., *less than* 0). Additionally, the solution set of $x^2 - 4x + 3 \le 0$ contains the x-values that make $x^2 - 4x + 3$ equal to 0 (i.e., $x = 1$ and $x = 3$). From the sign chart, we see that $x^2 - 4x + 3$ is negative for all x in the interval $(1, 3)$. So, the solution set of $x^2 - 4x + 3 < 0$ is $(1, 3)$, and the solution set of $x^2 - 4x + 3 \le 0$ is $[1, 3]$.

REMEMBER

The brackets in [1, 3] indicate that 1 and 3 are included in the intervals.

The solution sets of both $x^2 - 4x + 3 > 0$ and $x^2 - 4x + 3 \ge 0$ include all x-values that make $x^2 - 4x + 3$ positive (i.e., *greater than* 0), that is, the intervals $(-\infty, 1)$ and $(3, \infty)$. Additionally, the solution set of $x^2 - 4x + 3 \ge 0$ contains the x-values that make $x^2 - 4x + 3$ equal to 0 (i.e., $x = 1$ and $x = 3$). So, the solution set of $x^2 - 4x + 3 > 0$ is $(-\infty, 1) \cup (3, \infty)$, and the solution set of $x^2 - 4x + 3 \ge 0$ is $(-\infty, 1] \cup [3, \infty)$.

REMEMBER

The brackets in $(-\infty, 1] \cup [3, \infty)$ indicate that 1 and 3 are included in the intervals.

Steps for Solving a Quadratic Inequality

❶ Write the inequality in general form (i.e., with 0 on one side) and then find the solutions of the related equation.

❷ Begin a sign chart: divide a number line into test intervals where the endpoints are the solutions of the related equation.

❸ Complete the sign chart: substitute a value from each test interval into the quadratic expression, and note the sign of the resulting value.

❹ Identify the test intervals where the inequality is satisfied. State the solution interval(s), including the solutions of the related equation when the inequality symbol is ≥ or ≤.

EXAMPLE 5 Solving a Quadratic Inequality

Solve each inequality. Express each solution set in interval notation.

A. $x^2 - 8x + 7 \le 0$ **B.** $x^2 - 8x + 7 < 0$ **C.** $x^2 - 8x + 7 > 0$ **D.** $x^2 - 8x + 7 \ge 0$

SOLUTION

The same quadratic expression is seen in each of these inequalities. The only difference between the four inequalities is the inequality symbol itself. So, the related equation for all four inequalities is $x^2 - 8x + 7 = 0$.

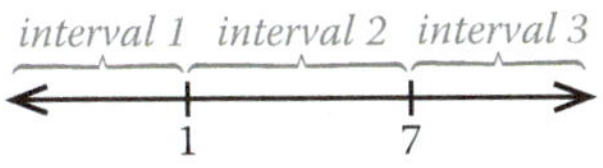

Sign Chart for $x^2 - 8x + 7$

❶ The solutions of $x^2 - 8x + 7 = 0$ are $x = 1$ and $x = 7$. $x^2 - 8x + 7 = (x - 1)(x - 7)$

❷ Note the solutions on a sign chart as the endpoints of the test intervals.

❸ Substitute a value from each test interval into $x^2 - 8x + 7$ and complete the sign chart.

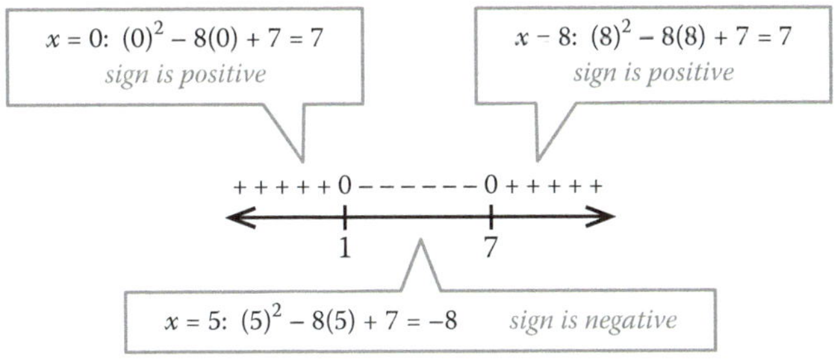

NOTICE THAT

The sign chart and the table apply to all four inequalities (A-D) because the related equation is the same for each inequality.

Test Interval	x-value	$x^2 - 8x + 7$	Sign
$(-\infty, 1)$	$x = 0$	$(0)^2 - 8(0) + 7 = 7$	positive
$(1, 7)$	$x = 5$	$(5)^2 - 8(5) + 7 = -8$	negative
$(7, \infty)$	$x = 8$	$(8)^2 - 8(8) + 7 = 7$	positive

❹ Only step ❹ is specific to the inequality. Identify the solution set of each inequality.

A. The solution set of $x^2 - 8x + 7 \le 0$ includes all values where $x^2 - 8x + 7$ is *less than or equal to* 0. Since $x^2 - 8x + 7$ is negative in $(1, 7)$, and $x^2 - 8x + 7 = 0$ when $x = 1$ and 7, the solution set is $[1, 7]$.

B. The solution set of $x^2 - 8x + 7 < 0$ includes all values where $x^2 - 8x + 7$ is *less than* 0 (but *not* equal to 0). Since $x^2 - 8x + 7$ is negative in $(1, 7)$, the solution set is $(1, 7)$.

IMPORTANT

The solutions are not included in the solution set when the inequality symbol is < or >.

C. The solution set of $x^2 - 8x + 7 > 0$ includes all values where $x^2 - 8x + 7$ is *greater than* 0 (but *not* equal to 0). Since $x^2 - 8x + 7$ is positive in $(-\infty, 1)$ and $(7, \infty)$, the solution set is $(-\infty, 1) \cup (7, \infty)$.

D. The solution set of $x^2 - 8x + 7 \ge 0$ includes all values where $x^2 - 8x + 7$ is *greater than or equal to* 0. Since $x^2 - 8x + 7$ is positive in $(-\infty, 1)$ and $(7, \infty)$, and $x^2 - 8x + 7 = 0$ when $x = 1$ and 7, the solution set is $(-\infty, 1] \cup [7, \infty)$.

P.9.6 Solving Quadratic Inequalities: Another Example

The first step in solving the four inequalities in **Example 5** was to find the solutions of the related equation. Notice that each inequality was given in general form. If the inequality is not given in general form, begin by rewriting the inequality so that one side is 0.

EXAMPLE 6 Solving a Quadratic Inequality

Solve $84 \le 4x^2 + 16x$. Express the solution set in interval notation.

SOLUTION

NOTICE THAT

Dividing by −1 gives $4x^2 + 16x - 84 \ge 0$. Either form of the inequality can be used to find its solution set.

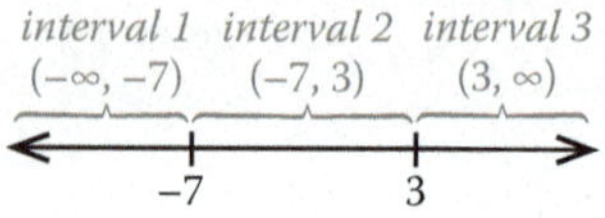

Sign Chart for $-4x^2 - 16x + 84$

❶ Rewriting the inequality gives $-4x^2 - 16x + 84 \le 0$.

The related equation is $-4x^2 - 16x + 84 = 0$. Solve by factoring.

$$-4(x^2 + 4x - 21) = 0 \qquad \textit{Factor out } -4.$$

$$-4(x + 7)(x - 3) = 0 \qquad \textit{Factor the trinomial.}$$

The solutions of $-4x^2 - 16x + 84 = 0$ are $x = -7$ and $x = 3$.

❷ Note the solutions on a sign chart as the endpoints of the test intervals.

❸ Substitute a value from each test interval into $-4x^2 - 16x + 84$ and complete the sign chart.

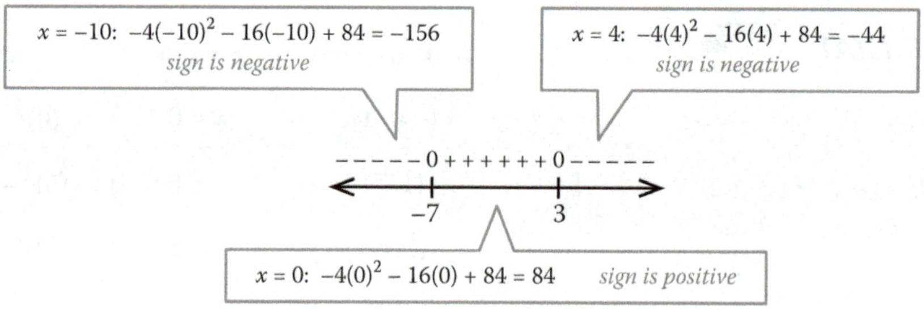

TIP

Substituting into the factored form of the expression, $-4(x + 7)(x - 3)$, is allowed as well.

Test Interval	x-value	$-4x^2 - 16x + 84$	Sign
$(-\infty, -7)$	$x = -10$	$-4(-10)^2 - 16(-10) + 84 = -156$	negative
$(-7, 3)$	$x = 0$	$-4(0)^2 - 16(0) + 84 = 84$	positive
$(3, \infty)$	$x = 4$	$-4(4)^2 - 16(4) + 84 = -44$	negative

❹ The solution set of $-4x^2 - 16x + 84 \le 0$ includes all x-values where $-4x^2 - 16x + 84$ is *less than or equal to* 0. Since $-4x^2 - 16x + 84$ is negative in $(-\infty, -7)$ and $(3, \infty)$, and $-4x^2 - 16x + 84 = 0$ when $x = -7$ and 3, the solution set is $(-\infty, -7] \cup [3, \infty)$.

IMPORTANT

The inequality symbol includes "or equal to," so the solution will include the related equation's solutions.

WHAT'S THE BIG IDEA?

Explain the purpose of the sign chart when solving a quadratic inequality.

P.9.7 Solving a Polynomial Inequality

A sign chart can be used to find the solution set of a **polynomial inequality**. However, if the degree of the polynomial is greater than 2 (i.e., the polynomial inequality is not linear or quadratic), then the sign chart may have more than three test intervals. When substituting a value from each test interval into the polynomial expression to determine its sign, keep in mind that the value can be substituted into the factored form of the polynomial. Doing so almost always makes the calculations easier.

EXAMPLE 7 Solving a Polynomial Inequality

Solve $x^3 + 2x^2 > 18 + 9x$. Express the solution set in interval notation.

SOLUTION

❶ Rewriting the inequality gives $x^3 + 2x^2 - 9x - 18 > 0$.

The related equation is $x^3 + 2x^2 - 9x - 18 = 0$. Solve by factoring.

$$x^3 + 2x^2 - 9x - 18 = 0$$

$$x^2(x+2) - 9(x+2) = 0 \quad \textit{Factor by grouping.}$$

$$(x+2)(x^2-9) = 0 \quad \textit{Factor } (x+2) \textit{ from each term.}$$

$$(x+2)(x-3)(x+3) = 0 \quad \textit{Factor the difference of two squares.}$$

The solutions of $x^3 + 2x^2 - 9x - 18 = 0$ are $x = -3$, $x = -2$, and $x = 3$.

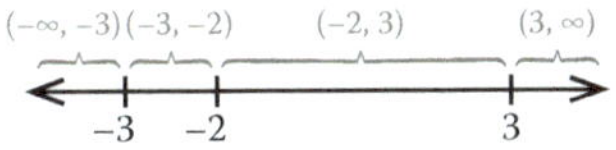

Sign Chart for
$x^3 + 2x^2 - 9x - 18$

❷ Note the solutions on a sign chart as the endpoints of the test intervals.

❸ Substitute a value from each test interval into $x^3 + 2x^2 - 9x - 18$. Complete the sign chart.

Here, the factored form, $(x+2)(x-3)(x+3)$, is used to simplify the calculations.

Test Interval	x-value	$(x+2)(x-3)(x+3)$	Sign
$(-\infty, -3)$	$x = -4$	$(-4+2)(-4-3)(-4+3)$ $(-)\ (-)\ (-) =$ *negative*	negative
$(-3, -2)$	$x = -\frac{5}{2}$	$\left(-\frac{5}{2}+2\right)\left(-\frac{5}{2}-3\right)\left(-\frac{5}{2}+3\right)$ $(-)\ (-)\ (+) =$ *positive*	positive
$(-2, 3)$	$x = 0$	$(0+2)(0-3)(0+3)$ $(+)\ (-)\ (+) =$ *negative*	negative
$(3, \infty)$	$x = 4$	$(4+2)(4-3)(4+3)$ $(+)\ (+)\ (+) =$ *positive*	positive

❹ The solution set of $x^3 + 2x^2 - 9x - 18 > 0$ includes all values where $x^3 + 2x^2 - 9x - 18$ is *greater than* 0 (but not equal to 0). Since $(x+2)(x-3)(x+3)$ is positive in $(-3, -2)$ and $(3, \infty)$, the solution set of $x^3 + 2x^2 > 18 + 9x$ is $(-3, -2) \cup (3, \infty)$.

TIP

The actual value of the expression is not important; only the sign is needed. So, determine whether each factor is positive or negative, and then use those signs to determine the sign of the expression, instead of simplifying and finding the exact value.

WHAT'S THE BIG IDEA?

How would the solution set of the inequality in **Example** 7 have changed if the original inequality was $x^3 + 2x^2 \leq 18 + 9x$?

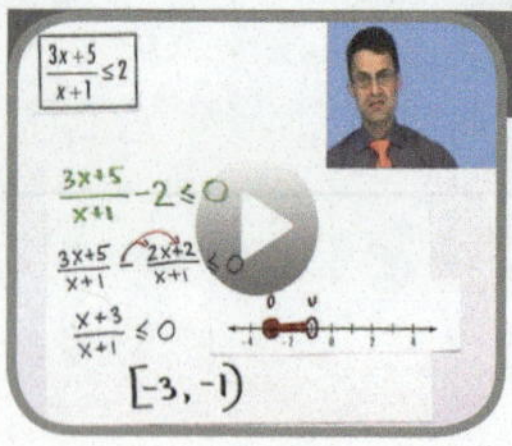

P.9.8 Solving Rational Inequalities

As with polynomial inequalities, a sign chart can also be used to find the solution set of a **rational inequality** (i.e., an inequality that can be expressed as a quotient of polynomials in relation to 0), where the values that make the related rational equation equal to 0 are endpoints for the test intervals.

However, the values that make the related rational expression undefined must also be included as endpoints for the test intervals.

REMEMBER

A rational expression is equal to 0 when its numerator is 0.

A rational expression is undefined when its denominator is 0.

Steps for Solving a Rational Inequality

❶ Write the inequality in the general form (where one side of the inequality is a rational expression and the other side of the inequality is 0). Then find the solutions of the related equation and the values where the rational expression is undefined.

❷ Begin a sign chart: divide a number line into test intervals where the values from step ❶ are the endpoints.

❸ Complete the sign chart: substitute a value from each test interval into the rational expression, and note the sign of the expression.

❹ Identify the test intervals where the inequality is satisfied. State the solution interval(s), including the solutions of the related equation when the inequality symbol is $\geq$ or $\leq$.

CAUTION

A value that makes the rational expression undefined is never included in the inequality's solution set. A value that makes the rational expression equal to 0 is included in the inequality's solution set, but only when the inequality symbol is $\geq$ or $\leq$.

Finding the Endpoints of the Test Intervals for a Rational Inequality

The related equation for a rational inequality is an equation with a rational expression on one side and 0 on the other side. So, a solution of this related equation is a value of the variable that makes the rational expression equal 0. Thus, the endpoints of the test intervals for a rational inequality are given by

a) the value(s) where the rational expression is 0, and

b) the value(s) where the rational expression is undefined.

To find the values where the rational expression is 0, set the rational expression's *numerator* (a polynomial) equal to 0 and solve for the variable.

To find the values where the rational expression is undefined, set the rational expression's *denominator* (a polynomial) equal to 0 and solve for the variable.

EXAMPLE 8 Solving a Rational Inequality

Solve $\dfrac{4x+2}{x+2} \geq 3$. Express the solution set in interval notation.

SOLUTION

❶ Rewrite the inequality by subtracting 3 from each side. Then use a common denominator to combine the terms.

$$\frac{4x+2}{x+2} - 3 \geq 0 \qquad \textit{Subtract 3 from each side.}$$

$$\frac{4x+2}{x+2} - \frac{3(x+2)}{x+2} \geq 0 \qquad \textit{Get the common denominator.}$$

$$\frac{4x+2-3(x+2)}{x+2} \geq 0 \qquad \textit{Combine the numerators.}$$

$$\frac{x-4}{x+2} \geq 0 \qquad \textit{Simplify the numerator.}$$

Determine the values where the rational expression $\dfrac{x-4}{x+2}$ is a) 0, and b) undefined.

a) $x - 4 = 0 \Rightarrow x = 4$
The numerator is 0 when x is 4.

b) $x + 2 = 0 \Rightarrow x = -2$
The denominator is 0 when x is −2.

So, the rational expression is 0 when $x = 4$ and undefined when $x = -2$.

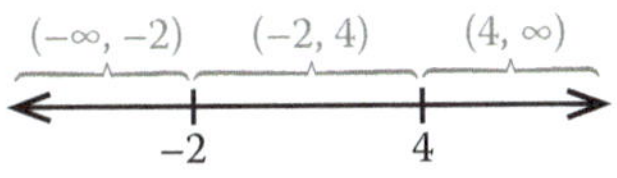

Sign Chart for $\dfrac{x-4}{x+2}$

❷ Note the values on a sign chart as the endpoints of the test intervals.

❸ Substitute a value from each test interval into $\dfrac{x-4}{x+2}$ and complete the sign chart.

Test Interval	x-value	$\dfrac{x-4}{x+2}$	Sign
$(-\infty, -2)$	$x = -3$	$\dfrac{-3-4}{-3+2} = \dfrac{neg}{neg} = positive$	positive
$(-2, 4)$	$x = 0$	$\dfrac{0-4}{0+2} = \dfrac{neg}{pos} = negative$	negative
$(4, \infty)$	$x = 5$	$\dfrac{5-4}{5+2} = \dfrac{pos}{pos} = positive$	positive

❹ The solution set of $\dfrac{x-4}{x+2} \geq 0$ includes all values where $\dfrac{x-4}{x+2}$ is *greater than or equal to* 0. Since $\dfrac{x-4}{x+2}$ is positive in $(-\infty, -2)$ and $(4, \infty)$, and 0 when $x = 4$, the solution set is $(-\infty, -2) \cup [4, \infty)$.

WHAT'S THE BIG IDEA?

Compare solving a polynomial inequality with solving a rational inequality.

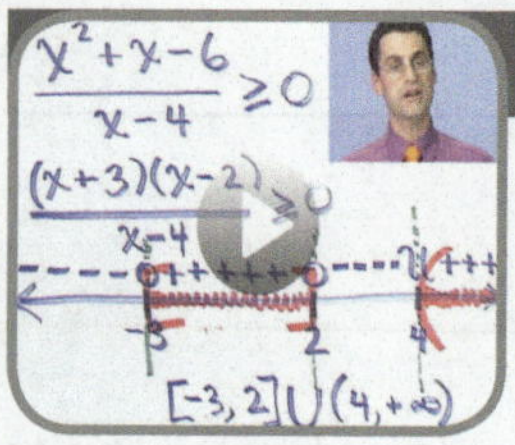

P.9.9 Solving Rational Inequalities: Another Example

In **Example 8**, the rational expression had a linear numerator and a linear denominator. Therefore, there was one value that made the expression 0 and one value that made the expression undefined, resulting in three possible solution intervals on the sign chart. However, a sign chart for a rational inequality may be divided into more than three test intervals when the degree of the polynomial in the numerator or denominator is greater than 1.

EXAMPLE 9 Solving a Rational Inequality with Common Denominators

Solve $\frac{x^2-7}{x-6} < \frac{9}{x-6}$. Express the solution set in interval notation.

SOLUTION

❶ Rewrite the rational inequality.

$$\frac{x^2-7}{x-6} - \frac{9}{x-6} < 0 \qquad \textit{Subtract } \frac{9}{x-6} \textit{ from each side.}$$

$$\frac{x^2-16}{x-6} < 0 \qquad \textit{Combine the numerators.}$$

Determine the values where the rational expression $\frac{x^2-16}{x-6}$ is a) 0, and b) undefined.

a) $x^2 - 16 = 0$
$(x+4)(x-4) = 0$
$x = -4$ or $x = 4$
The numerator is 0 when x is –4 or 4.

b) $x - 6 = 0$
$x = 6$
The denominator is 0 when x is 6.

So, the rational expression is 0 when $x = -4$ or 4 and undefined when $x = 6$.

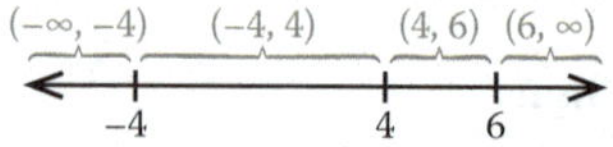

Sign Chart for $\frac{x^2-16}{x-6}$

❷ Note the values on a sign chart to denote the test intervals.

❸ Substitute a value from each test interval into $\frac{x^2-16}{x-6}$ and complete the sign chart.

Test Interval	x-value	$\frac{x^2-16}{x-6}$	Sign
$(-\infty, -4)$	$x = -5$	$\frac{(-5)^2-16}{-5-6} = \frac{pos}{neg}$ = *negative*	negative
$(-4, 4)$	$x = 0$	$\frac{(0)^2-16}{0-6} = \frac{neg}{neg}$ = *positive*	positive
$(4, 6)$	$x = 5$	$\frac{(5)^2-16}{5-6} = \frac{pos}{neg}$ = *negative*	negative
$(6, \infty)$	$x = 7$	$\frac{(7)^2-16}{7-6} = \frac{pos}{pos}$ = *positive*	positive

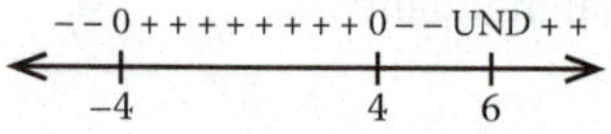

❹ The solution set of $\frac{x^2-16}{x-6} < 0$ includes all values where $\frac{x^2-16}{x-6}$ is *less than 0*. Since $\frac{x^2-16}{x-6}$ is negative in $(-\infty, -4)$ and $(4, 6)$, the solution set is $(-\infty, -4) \cup (4, 6)$.

SECTION P.9 EXERCISES

Warm Up

Write each set using interval notation.

1. all numbers greater than 5
2. all numbers excluding 0
3. $\{x| -2 \leq x \leq 70\}$

Just the Facts

Fill in the blanks.

4. The _____ symbol must be reversed when an inequality is _____ or _____ by a _____ number.
5. The phrase "x is at least y" can be represented by the inequality _____.
6. The phrase "x is no more than y" can be represented by the inequality _____.
7. When two inequalities are connected by the word "and" or "or", the inequality is called a _____.
8. A *less than* absolute value inequality can be rewritten as a compound inequality using the word _____. A *greater than* absolute value inequality can be rewritten as a compound inequality using the word _____.
9. The solution sets for the inequalities $ax^2 + bx + c$ _____ 0 and $ax^2 + bx + c$ _____ 0 do not include the solutions of the related equation $ax^2 + bx + c = 0$.
10. The solution sets for the inequalities $ax^2 + bx + c$ _____ 0 and $ax^2 + bx + c$ _____ 0 include the solutions of the related equation $ax^2 + bx + c = 0$.

Essential Skills

In Exercises 1–6, write the solution set of each inequality using interval notation.

1. $8h + 3 \leq 4h - 21$
2. $-5x + 3 \geq 6(x - 5)$
3. $3(1 + 2b) > 2 - b$
4. $6x - 14 < 4(x - 6)$
5. $5(d - 2) \geq -(8d - 3)$
6. $-2(5a + 1) > 7(4 - 3a)$

In Exercises 7–66, write the solution set of each inequality using interval notation. If there is no solution, write the empty set, ø.

7. $3x \leq 6$ or $3x \geq 21$
8. $5 - 3x \leq -7$ or $x + 8 \leq -1$
9. $-2a + 1 \geq 9$ or $5a + 2 \geq -28$
10. $4 - 10x \leq 10$ or $x + 16 \leq 14$
11. $9 - 7h \leq -12$ or $h + 9 < 4$
12. $2d - 6 < -8$ or $-5d + 7 \leq -13$
13. $-9 \leq -3x + 6 \leq 0$
14. $-7 \leq -13 - 2x \leq 15$
15. $16 < -2 + 3d \leq 34$
16. $12 \leq 7w - 2 < 40$
17. $-4 \leq -3a - 16 < 5$
18. $-5 \leq -9 - 2y \leq 3$
19. $|3 - 2y| > 5$
20. $8 > |3 - x|$
21. $|22 + 4q| < 2$
22. $|36 + 8p| \leq 4$
23. $6 \geq |6b + 24|$
24. $9 > |12 - 2z| - 1$
25. $4 > |8 - 2f|$
26. $|2 - 4x| \geq 6$
27. $1 + |12 + 2c| \geq 5$
28. $2 < |14 + 2y| - 2$
29. $|-2v + 7| - 2 > 1$
30. $|-8z + 44| + 1 \geq 13$
31. $12 \leq 3|2 - (3x - 6)|$
32. $30 - 4|2 - (3d - 4)| \geq -6$
33. $27 - 9|1 - 2(c - 1)| < 108$
34. $40 + 4|3 - 3(p - 3)| < 100$
35. $18 - 5|2 - (3w - 5)| \geq -37$
36. $15 - 3|-3(x + 2) + 2| \geq -12$
37. $3k^2 - 14k + 8 < 0$
38. $5x^2 + 7x - 6 > 0$
39. $5a^2 + 39a - 8 \leq 0$
40. $2z^2 - 3z - 14 \geq 0$
41. $6m^2 + 7m - 20 \geq 0$
42. $3p^2 + 2p - 8 \leq 0$
43. $-10x + 12 \geq -4x^2 + 9x$
44. $6z^2 - 1 \leq z^2 - 17z - 7$
45. $x^2 - 27 \geq -3x - x^2$

46. $7x^2 - 28 > x^2 - 17x$

47. $5b^2 - 13b + 12 \geq b^2 + 6b$

48. $7a^2 - 6a + 4 < a^2 + 10 - a$

49. $x^3 - 3x^2 < 9x - 27$

50. $x^3 + 7x^2 > 49x + 343$

51. $a^3 + 2a^2 + 3 \leq 4a + 11$

52. $p^3 + 8p^2 - 16p \leq 4p^2 + 64$

53. $c^3 + 6c^2 > -2c^2 + 64c + 512$

54. $y^3 + 4y^2 - 4y \leq 2y^2 + 8$

55. $\frac{x+1}{2x-7} \geq 0$

56. $\frac{4x-3}{x+4} < 1$

57. $\frac{x+3}{2x-5} \leq 0$

58. $\frac{6x-6}{x+6} \geq 1$

59. $\frac{3x+7}{5x-4} > 0$

60. $\frac{7y-5}{y+7} < 2$

61. $\frac{3x-2}{x-1} < \frac{6}{x-1}$

62. $\frac{x^2-3}{x+6} \geq \frac{1}{x+6}$

63. $\frac{3}{d+5} \leq \frac{d^2-1}{d+5}$

64. $\frac{y^2-2}{y+4} \geq \frac{2}{y+4}$

65. $\frac{7x-2}{x+2} \geq \frac{3}{x+2}$

66. $\frac{t^2-1}{t+7} < \frac{24}{t+7}$

Extensions

In Exercises 67–70, write the solution set of each inequality using interval notation. If there is no solution, write the empty set, ø.

67. $5 - |x - 12| + |2x + 4| > 0$

68. $\frac{2x^3 + 4x^2 - 30x}{x^2 + 3x - 4} < 0$

69. $\frac{3+x}{3-x} - \frac{3-x}{3+x} < -6$

70. $\frac{2x^2 + 5x - 3}{x^2 + 8x + 15} > 0$

71. The two shorter sides of a right triangle are x and $3 - x$, where $x > 0$. For which values of x will the hypotenuse be less than $\sqrt{15}$?

72. The sum of the first n natural numbers is given by $1+2+3+\ldots+n = \frac{n(n+1)}{2}$. For which values of n will the sum be less than 36?

73. A college student earns \$20 per lawn mowing lawns in his neighborhood. If he pays \$350 per month in rent, what is the minimum number of lawns that he must mow each month to be able to pay the rent?

74. At a rental car company, the total price for a rental car is either \$100 per week plus \$0.20 per mile driven or \$50 per week plus \$0.30 per mile driven. For what number of miles per week is the total price for the first option less than the total price for the second option?

75. The high-school marching band is selling boxes of candy bars to earn money for new uniforms. The uniforms will cost \$12,000, but the band already has \$4800. If they profit \$4 from each box of candy sold, what is the minimum number of boxes they must sell in order to have enough money to buy the uniforms?

76. A yo-yo factory has fixed operating costs that add up to \$450,000 per year. In addition to the fixed costs, the cost to produce one yo-yo is \$1.20. The yo-yo company sells the yo-yos to a distributor for \$4.80 each. How many yo-yos must the factory sell to the distributor in a year in order to earn a profit of at least \$100,000?

CHAPTER P REVIEW EXERCISES

Section P.1 Real Numbers

1. Evaluate $3a - 4ab^2 - 6(10 + b) - a^2$ for $a = 3$ and $b = -2$.
2. Write the solution set for $-10 < x \le -7$ using interval notation.
3. Write the inequality with solution set $[-1, \infty)$.
4. Express the interval $(-\infty, 2] \cup (3, 5)$ on a number line.
5. Simplify. $-4|7 - 12| + |-18|$

Section P.2 Integer Exponents

In Exercises 6–8, simplify each expression. Include only positive exponents in simplified expressions.

6. $(-5a^2bc)(3a^4b)^2$
7. $\dfrac{(-2g^3)^2}{16g^2(g^5)}$
8. $(2^{-3})\left(\dfrac{4}{3}\right)^{-2}$
9. Write 0.0000047 in scientific notation.
10. Simplify. Write the answer in scientific notation. $\dfrac{(3\times10^7)(6.4\times10^{-4})}{7.5\times10^{-3}}$

Section P.3 Rational Exponents and Radicals

In Exercises 11–15, simplify. Do not use a calculator. Rationalize the denominator as necessary. Assume all variables represent positive numbers.

11. $2\sqrt{18}+\sqrt{60}+\sqrt{72}-\sqrt{135}$
12. $\dfrac{-5}{\sqrt{6}+7}$
13. $\left(\dfrac{27}{64}\right)^{-\frac{2}{3}}$
14. $\sqrt{12b^3c^2d^{11}}$
15. $\sqrt{\dfrac{100x^9y^2z^6}{4x^6y^3}}$

Section P.4 Polynomials

In Exercises 16–18, simplify each expression. Write the polynomial answer in standard form. Then identify the leading coefficient and classify it according to its degree and number of terms.

16. $[2x^2 - (7x + 6)] + [-(5x^3 - 8x^2) - 14x]$
17. $(3x + 5)(2x - 3)$
18. $(x - 5)(x^2 + 3x - 1 + x)$

In Exercises 19–20, expand each expression.

19. $(c^2 + 2d)(c^2 - 2d)$
20. $(6x - 5y)^3$

Section P.5 Factoring

In Exercises 21–25, factor each polynomial completely.

21. $6a^2 + 4a - 10$
22. $25n^6 - 1$
23. $3x^3 - 8y - 2x^2y + 12x$
24. $8c^3d^{12} + 1$
25. $4z^5 - 4z^3 - 4z^2 + 4$

Section P.6 Rational Expressions

In exercises 26–27, simplify each expression and state any excluded values of the variable.

26. $\dfrac{-x^2+2x+35}{7-x}$
27. $\dfrac{x^3+27}{x^2-x-12}$

In Exercises 28–30, simplify. Assume that all variables result in nonzero denominators.

28. $\dfrac{4k^2-k-3}{k^2+6k+5} \div \dfrac{9-16k^2}{2k^2+2k}$
29. $\dfrac{1}{x^2-4x+3}-\dfrac{3}{x^2-9}$
30. $\dfrac{\dfrac{5}{2x^2-6x}}{\dfrac{2}{x}-\dfrac{1}{x-3}}$

Section P.7 Solving Linear and Quadratic Equations

In Exercises 31–37, solve each equation.

31. $9(x-1)+3x=-9x+5$

32. $4-\dfrac{x}{2}=\dfrac{2(x+1)}{5}$

33. $\dfrac{2}{5x-1}=\dfrac{6}{2x-3}$

34. $\dfrac{z-4}{z+1}=-\dfrac{24}{11}$

35. $8x^2-17x-14=21+x$

36. $x^2-6x=2$

37. $6x^2-\sqrt{7}x+3=0$

38. Two cars leave an intersection at the same time. One travels east $5(x+1)$ miles while the other travels south $6(x-1)$ miles. At that point, the cars are $2(4x-3)$ miles apart. How many miles did the eastbound car travel?

Section P.8 Other Types of Equations

In Exercises 39–42, solve each equation. Rationalize the denominator when necessary.

39. $\left|\dfrac{8}{5x-3}\right|=\left|\dfrac{6}{2x+7}\right|$

40. $(y-3)^{\frac{1}{2}}=(-2y+5)^{\frac{1}{4}}$

41. $\sqrt{2x+10}=\sqrt{2x-6}+4$

42. $2x^3+2x^2=x+1$

Section P.9 Inequalities

In Exercises 43–46, write the solution set of each inequality using interval notation. If there is no solution, write the empty set, ø.

43. $-9\le\dfrac{2x+6}{4}<4$

44. $\left|\dfrac{1-3x}{8}\right|\ge 3$

45. $3x^2-7x-6\le 0$

46. $\dfrac{x-7}{5x-1}+4\le\dfrac{9x+5}{5x-1}$

Functions and Their Graphs

Chapter

1

1.1 COORDINATES AND GRAPHS

OBJECTIVES

- Graph points on a coordinate plane.
- Use the Distance and Midpoint Formulas.
- Graph equations by plotting points.
- Find and plot the x- and y-intercepts of an equation.
- Graph circles in a coordinate plane.
- Write the equation of a circle.
- Determine types of symmetry from equations.

PREREQUISITE VOCABULARY TERMS

absolute value equation
completing the square
inequality
linear equation
perfect-square trinomial
Pythagorean Theorem
quadratic equation
square root

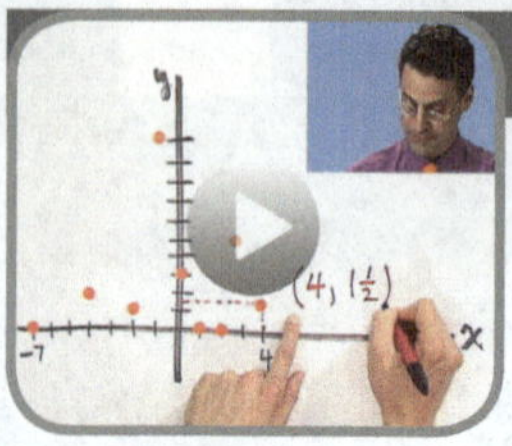

1.1.1 Using the Cartesian System

Introduction to the Coordinate Plane

A **Cartesian plane** or **coordinate plane** is formed by two perpendicular number lines (one horizontal and the other vertical), intersecting at 0 on each. A coordinate plane has seven basic parts: the two axes, the origin, and four quadrants. The number lines are the **axes**, and the point at which the axes intersect is the **origin**, O. Specifically, the horizontal axis is called the ***x*-axis**, and the vertical axis is called the ***y*-axis**. The axes divide a coordinate plane into four **quadrants**, labeled I, II, III, and IV (Figure 1.1a).

TIP

Ordered pairs are also referred to as points or coordinate pairs.

Recall that each point on a real number line corresponds to a unique real number. Similarly, each point P on a coordinate plane corresponds to a unique pair of real numbers called an **ordered pair**. In any ordered pair (a, b), the first number, a, is the ***x*-coordinate**, corresponding to a location indicated by the x-axis, and the second number, b, is the ***y*-coordinate**, corresponding to a location indicated by the y-axis. Therefore, ordered pairs are often referred to generally as (x, y). For example, the ordered pair $(1, 6)$ corresponds to 1 on the x-axis and 6 on the y-axis.

Many aspects of algebra, such as ordered pairs (Figure 1.1b), equations (Figure 1.1c), and inequalities (Figure 1.1d), can be visualized by representations on the coordinate plane.

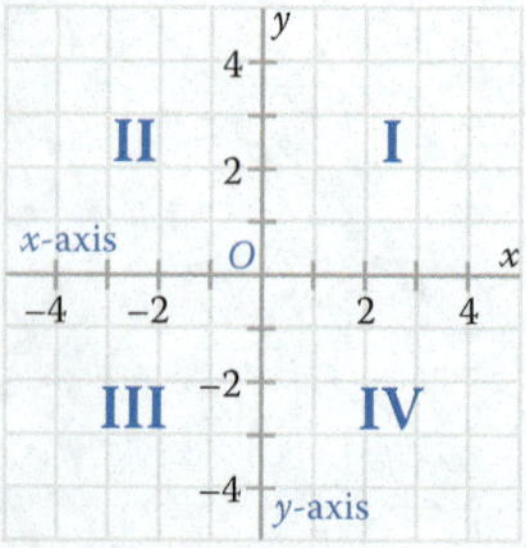

Figure 1.1a

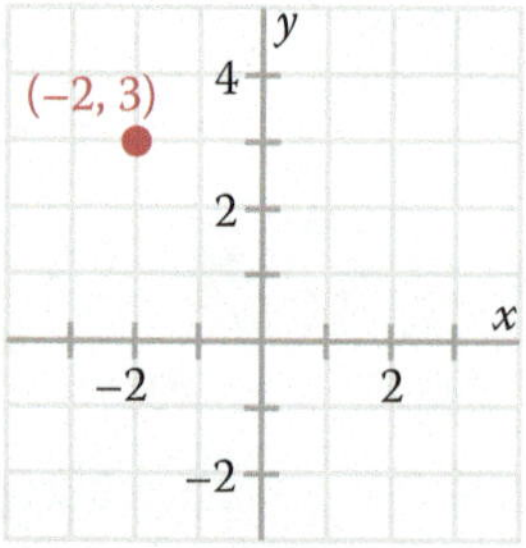

Figure 1.1b

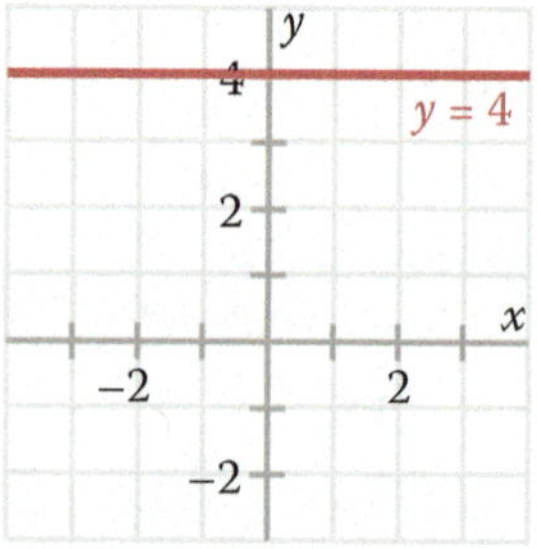

Figure 1.1c

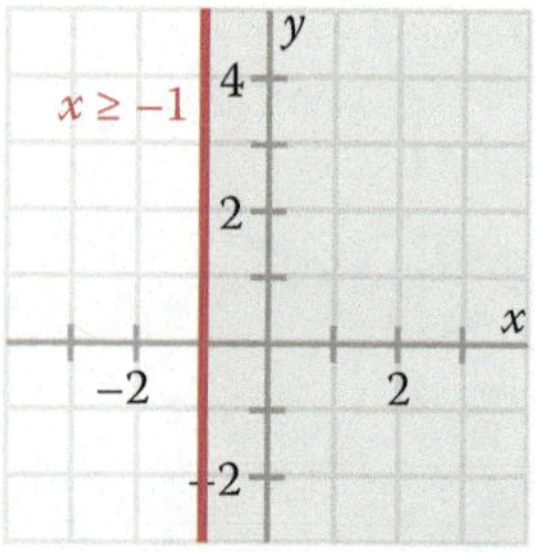

Figure 1.1d

Plotting a Point on a Coordinate Plane

To graph (or plot) an ordered pair on a coordinate plane, start at the position on the x-axis indicated by the x-coordinate. From that position on the x-axis, move up or down the number of units indicated by the y-coordinate. Move up if the y-coordinate is positive, or down if the y-coordinate is negative. Note that if the ordered pair's x-coordinate is 0, then the point is located on the y-axis. Similarly, if the ordered pair's y-coordinate is 0, then the point is located on the x-axis. The ordered pair (0, 0) is the origin.

EXAMPLE 1 Plotting Points

Identify the location of $A(3, -1)$, $B(0, 2)$, and $C(-4.5, 0)$.

SOLUTION

Plot each point on a coordinate plane.

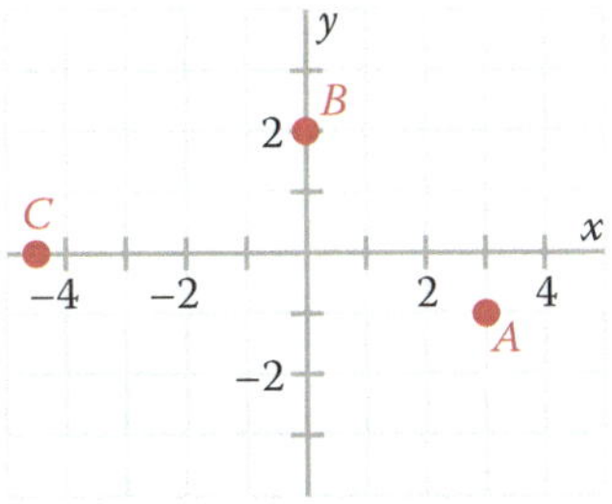

The x-coordinate of point C is -4.5, so the position on the x-axis is halfway between -4 and -5.

Point A is in quadrant IV, point B is on the y-axis, and point C is on the x-axis.

Tables and Ordered Pairs

Lists of ordered pairs can be written in a table that has either two rows or two columns. For a two-column table (a vertical table), the first column contains the x-coordinates, and the second column contains the y-coordinates (Figure 1.1e). Similarly, for a two-row table (a horizontal table), the first row contains the x-coordinates, and the second row contains the y-coordinates (Figure 1.1f).

x	y
−3	1
−1	0
0	−2
2	4

Figure 1.1e

x	−3	−1	0	2
y	1	0	−2	4

Figure 1.1f

Both tables list the points (−3, 1), (−1, 0), (0, −2), and (2, 4).

EXAMPLE 2 Plotting Points from a Table

NOTICE THAT

This two-row table could also be written as a two-column table.

x	y
0	4
−1	1
−3	−2
3	−1

Plot the points from the table on a coordinate plane.

x	0	−1	−3	3
y	4	1	−2	−1

SOLUTION

The table lists four ordered pairs: (0, 4), (−1, 1), (−3, −2), and (3, −1).

Plot each point on a coordinate plane.

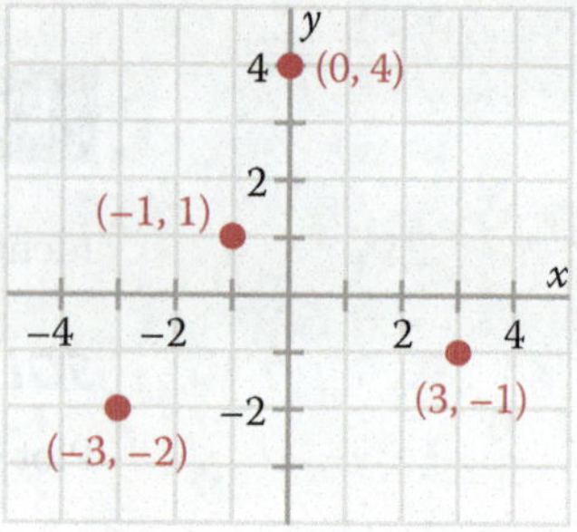

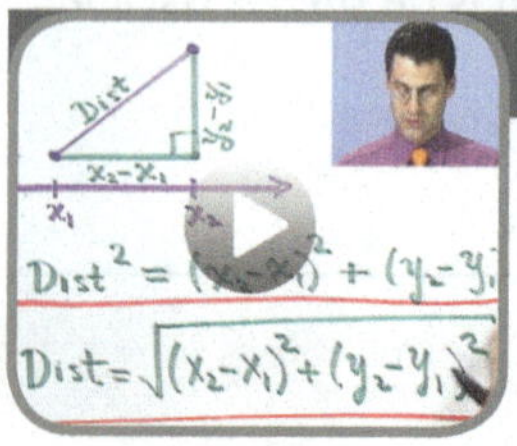

1.1.2 The Distance and Midpoint Formulas

Finding the Distance between Two Points

The distance between two points on a coordinate plane can be found using the **Distance Formula**.

The Distance Formula

The distance d between two points (x_1, y_1) and (x_2, y_2) on a coordinate plane is $d = \sqrt{(x_2 - x_1)^2 + (y_2 - y_1)^2}$.

CAUTION

The notation (x_1, y_1) and (x_2, y_2) is used to mean the first point (x_1, y_1) and the second point (x_2, y_2). The subscripts 1 and 2 do not indicate any sort of mathematical operation.

The Distance Formula is derived by using the Pythagorean Theorem, $a^2 + b^2 = c^2$, with two general points on a coordinate plane, (x_1, y_1) and (x_2, y_2), where the distance between the two points is the length of the hypotenuse of a right triangle.

IMPORTANT

Either one of the two points can be identified as (x_1, y_1), because the distance between point 1 and point 2 is the same as the distance between point 2 and point 1.

Steps for Finding the Distance between Two Points

❶ Decide which of the two points will be the first point, (x_1, y_1), and which will be the second point, (x_2, y_2).

❷ Substitute the coordinates from the two points, x_1, x_2, y_1, and y_2, into the Distance Formula.

❸ Simplify the radicand, and then find the square root. If the radicand is not a perfect square, write the square root as a decimal (using a calculator) or in simplest radical form.

EXAMPLE 3 Finding the Distance between Two Points

Determine the distance between $A(8, -10)$ and $B(0, 5)$.

SOLUTION

TIP

The same distance will be found if $A(8, -10)$ is (x_1, y_1) and $B(0, 5)$ is (x_2, y_2).

Let $B(0, 5)$ be (x_1, y_1), and let $A(8, -10)$ be (x_2, y_2). *$x_1 = 0, x_2 = 8, y_1 = 5, y_2 = -10$*

Substitute the coordinates into the Distance Formula and simplify.

$$d = \sqrt{(8-0)^2 + (-10-5)^2}$$ *Substitute.*

$$= \sqrt{(8)^2 + (-15)^2}$$ *Subtract within the parentheses.*

$$= \sqrt{64 + 225}$$ *Evaluate the powers.*

$$= \sqrt{289}$$ *Add.*

$$= 17$$ *Evaluate the square root.*

The distance between points A and B is 17 units.

ALTERNATIVE METHOD

Graph the two points, and then sketch a triangle where the distance between the two points is the hypotenuse.

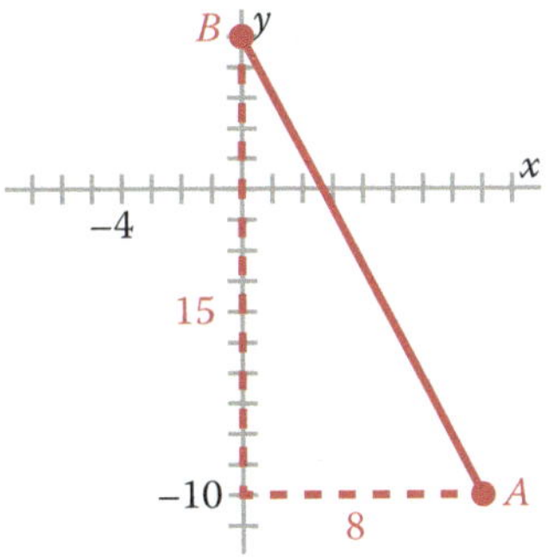

The length of the horizontal leg is 8 units since $|8 - 0| = 8$.

The length of the vertical leg is 15 units since $|-10 - 5| = 15$.

IMPORTANT

When the square root of both sides is taken, only the positive square root is needed, because c is a distance, and distance must be positive.

Use the Pythagorean Theorem to find the length of the hypotenuse AB.

$$a^2 + b^2 = c^2$$

$$8^2 + 15^2 = c^2$$ *Substitute the lengths of the legs.*

$$64 + 225 = c^2$$ *Evaluate the powers.*

$$289 = c^2$$ *Add.*

$$17 = c$$ *Take the square root of each side.*

The length of the hypotenuse AB (the distance between points A and B) is 17 units.

WHAT'S THE BIG IDEA?

Plot the points (0, 0), (x_1, y_1), and (x_2, y_2), where (x_1, y_1) and (x_2, y_2) are in quadrant I. Then use the distance between these points to explain the relationship between the Pythagorean Theorem and the Distance Formula.

Finding the Midpoint between Two Points

The midpoint between two points on a coordinate plane can be found using the **Midpoint Formula.**

TIP

As with the Distance Formula, either of the two points can be designated as (x_1, y_1) when using the Midpoint Formula. Furthermore, the x-coordinates and the y-coordinates can be added in any order in each coordinate's numerator, because addition is commutative.

The Midpoint Formula

The midpoint between two points (x_1, y_1) and (x_2, y_2) on a coordinate plane is $M = \left(\frac{x_1 + x_2}{2}, \frac{y_1 + y_2}{2}\right)$.

Steps for Finding the Midpoint between Two Points

1. Decide which of the two points will be the first point, (x_1, y_1), and which will be the second point, (x_2, y_2).
2. Substitute the coordinates from the two points, x_1, x_2, y_1, and y_2, into the Midpoint Formula.
3. Simplify each coordinate.

EXAMPLE 4 Finding the Midpoint between Two Points

Determine the midpoint between $A(8, -10)$ and $B(0, 5)$.

SOLUTION

TIP

The same midpoint will be found if $A(8, -10)$ is (x_1, y_1) and $B(0, 5)$ is (x_2, y_2).

Let $B(0, 5)$ be (x_1, y_1), and let $A(8, -10)$ be (x_2, y_2). *$x_1 = 0, x_2 = 8, y_1 = 5, y_2 = -10$*

Substitute the coordinates into the Midpoint Formula and simplify.

$$\left(\frac{0+8}{2}, \frac{5+(-10)}{2}\right) \quad \textit{Substitute.}$$

$$= \left(\frac{8}{2}, \frac{-5}{2}\right) \quad \textit{Simplify the numerators.}$$

$$= \left(4, -\frac{5}{2}\right) \quad \textit{Simplify the fraction.}$$

The midpoint between points A and B is $\left(4, -\frac{5}{2}\right)$.

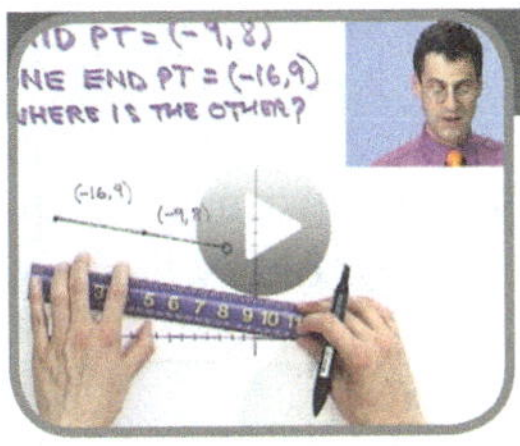

1.1.3 Finding the Second Endpoint of a Segment

If one endpoint and the midpoint of a segment are known, the midpoint formula can be used to find the segment's other endpoint.

Equation for the Midpoint's *x*-coordinate	Equation for the Midpoint's *y*-coordinate
$x_M = \frac{x_1 + x_2}{2}$ *x_1 and x_2 are the x-coordinates of the endpoints, and x_M is the x-coordinate of the midpoint.*	$y_M = \frac{y_1 + y_2}{2}$ *y_1 and y_2 are the y-coordinates of the endpoints, and y_M is the y-coordinate of the midpoint.*

EXAMPLE 5 Finding the Coordinates of the Second Endpoint

$C(4, 3)$ is the midpoint between $A(x, y)$ and $B(11, 7)$. Find the coordinates of A.

SOLUTION

Substitute the x-coordinate from the midpoint and the x-coordinate from B into $x_M = \frac{x_1 + x_2}{2}$, and solve to find the x-coordinate of A.

$4 = \frac{x + 11}{2}$ *Substitute.*

$8 = x + 11$ *Multiply both sides by 2.*

$-3 = x$ *Subtract 11 from each side.*

TIP

The x-coordinate from the midpoint must be substituted for x_M, but the x-coordinate of the endpoint can be substituted for either x_1 or x_2.

Substitute the y-coordinate from the midpoint and the y-coordinate from B into $y_M = \frac{y_1 + y_2}{2}$, and solve to find the y-coordinate of A.

$3 = \frac{y + 7}{2}$ *Substitute.*

$6 = y + 7$ *Multiply both sides by 2.*

$-1 = y$ *Subtract 7 from each side.*

TIP

The y-coordinate from the midpoint must be substituted for y_M, but the y-coordinate of the endpoint can be substituted for either y_1 or y_2.

Therefore, the coordinates of A are $(-3, -1)$.

TIP

The answer can be checked by finding the midpoint between $(-3, -1)$ and $(11, 7)$.

WHAT'S THE BIG IDEA?

Explain why each coordinate in a midpoint is found by adding the coordinates from the endpoints and dividing by 2.

1.1.4 Collinearity and Distance

Points contained on the same line on a coordinate plane are **collinear**. Since a line can be drawn between any two points, any two points are collinear. Three points are collinear only when one line can be drawn containing the three points.

Suppose A, B, and C are collinear points such that the distance between A and C is greater than the distance between A and B, and also greater than the distance between B and C. Since the three points are collinear, the sum of the distance between A and B and the distance between B and C must be equal to the distance between A and C. This fact can be used to determine whether three points are collinear.

For example, suppose there are three collinear points E, F, and G on a coordinate plane, such that the distance between E and F is 4 units and the distance between F and G is 10 units. If F is between E and G, then $EF + FG = 14 = EG$ (Figure 1.1g). Furthermore, if E is between F and G, then $FE + EG = 4 + 6 = 10 = FG$ (Figure 1.1h).

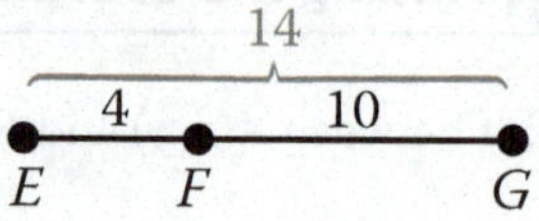

Figure 1.1g

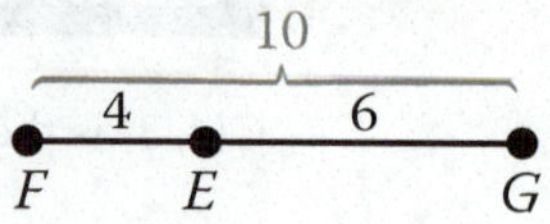

Figure 1.1h

IMPORTANT

Any three points that are not collinear form a triangle.

Steps for Determining the Collinearity of Three Points

❶ Find the distance between each of the three pairs of points.

❷ Add the two lesser lengths. The points are collinear when this sum is equal to the third length.

EXAMPLE 6 Determining When Three Points Are Collinear

True or false? $A(20, -15)$, $B(-13, 18)$, and $C(0, 5)$ are collinear.

SOLUTION

❶ Use the Distance Formula to find the distance between each pair of points, AB, BC, and AC.

$$AB = \sqrt{(20-(-13))^2 + (-15-18)^2} = \sqrt{2178} = 33\sqrt{2}$$

$$BC = \sqrt{(-13-0)^2 + (18-5)^2} = \sqrt{338} = 13\sqrt{2}$$

$$AC = \sqrt{(20-0)^2 + (-15-5)^2} = \sqrt{800} = 20\sqrt{2}$$

❷ Since AB is greater than BC and AC, check whether the sum of BC and AC is equal to AB.

$$BC + AC = 13\sqrt{2} + 20\sqrt{2} = 33\sqrt{2} = AB$$

Therefore, A, B, and C are collinear.

1.1.5 Triangles and Distance

Three points on a coordinate plane that are *not* collinear form a triangle. In this case, the Distance Formula can be used to find the length of each of the triangle's sides, and then the triangle can be classified as equilateral, isosceles, or scalene. Furthermore, if the side lengths satisfy the Pythagorean Theorem, then the triangle can be classified as a right isosceles triangle or a right scalene triangle.

Triangle Classifications		
Equilateral		*All three side lengths are equal.*
Isosceles	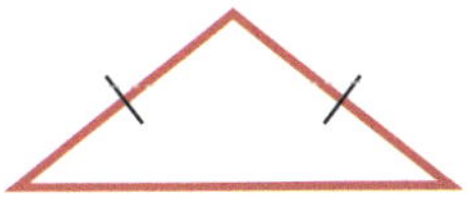	*Exactly two side lengths are equal.*
Scalene	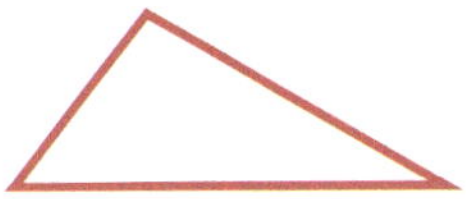	*No side lengths are equal.*
Right Isosceles	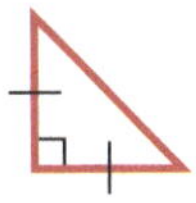	*Exactly two side lengths are equal, and there is one right angle.*
Right Scalene	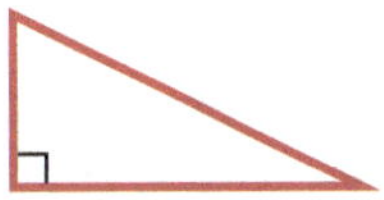	*No side lengths are equal, and there is one right angle.*

EXAMPLE 7 Classifying a Triangle

Classify the figure formed by $A(1, 2)$, $B(5, 3)$, and $C(4, 7)$.

SOLUTION

Use the Distance Formula to find the distance between each pair of points.

$$AB = \sqrt{(5-1)^2 + (3-2)^2} = \sqrt{17}$$

$$BC = \sqrt{(4-5)^2 + (7-3)^2} = \sqrt{17}$$

$$AC = \sqrt{(4-1)^2 + (7-2)^2} = \sqrt{34}$$

Since $AB + BC \neq AC$, the points are not collinear, and therefore, they form a triangle.

Since $AB = BC$, the triangle is isosceles.

Since $\sqrt{17}^2 + \sqrt{17}^2 = \sqrt{34}^2 \Rightarrow AB^2 + BC^2 = AC^2$, the side lengths satisfy the Pythagorean Theorem.

Therefore, triangle ABC is a right isosceles triangle.

REMEMBER

The hypotenuse of a right triangle is always the longest side.

1.1.6 Graphing Equations by Plotting Points

A solution to an equation in one variable is a value of the variable that makes the equation true. For example, a solution to the equation $3 = x + 1$ is $x = 2$, since $3 = 2 + 1$. Equations in two variables also have solutions, but the solutions are ordered pairs. For example, consider the equation $y = x + 1$, which mean that each y-value is equal to each x-value increased by 1. One solution to this equation is $(5, 6)$ since $5 + 1 = 6$.

REMEMBER

In an ordered pair, the first coordinate is the x-value, and the second coordinate is the y-value.

There are an infinite number of solutions to $y = x + 1$ including $(1, 2)$, $(10, 11)$, $(0, 1)$, and $(-4, -3)$. Solutions to this equation can be found by choosing any allowable value for x (i.e., any x-value in the domain) and substituting this x-value into the equation to find the corresponding y-value. These solutions are ordered pairs, which can be graphed on a coordinate plane.

When the solutions to an equation in two variables are plotted on a coordinate plane, the points form a pattern. A line or curve can be sketched through these points, resulting in the **graph of the equation**. The line or curve can be thought of as extending infinitely through all of the solutions to the equation.

DEFINITION

The line or curve on a coordinate plane that represents all of the ordered pair solutions for an equation is called the **graph of the equation**.

General Types of Equations and Their Graphs

The ordered pair solutions for equations of similar type form similar patterns (shapes) on the coordinate plane, and therefore have similar graphs. For example, the graph of any linear equation is a line, and the graph of any quadratic equation is a U-shaped graph called a parabola. The following table shows three common types of equations and the general shape of their graphs. (These types of equations will be discussed in greater detail in later sections.)

REMEMBER

The greatest power of x is 2 in a quadratic equation, and the greatest power of x is 1 (so no exponent is shown) in a linear equation. For example, $y = x^2 + 3x + 5$ is a quadratic equation, and $y = 2x + 1$ is a linear equation.

Three Types of Equations and Their Graphs

	Example Equations	Example Graphs	
Linear *Graphs are lines.*	$y = x - 2$ $-x = y$ $y = 4x + 1$	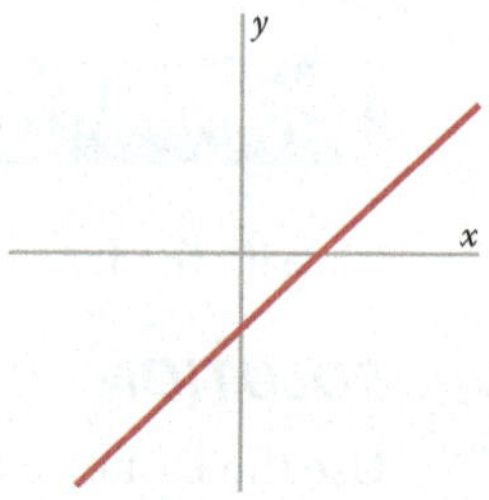	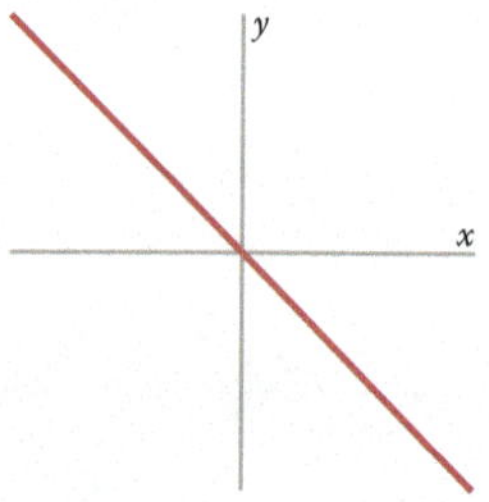
Quadratic *Graphs are parabolas (U-shaped).*	$y = -x^2 + 2$ $x = y^2 - 1$ $y = (x + 1)^2 - 3$	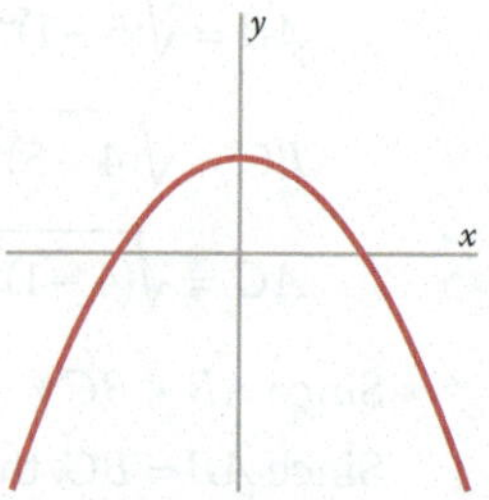	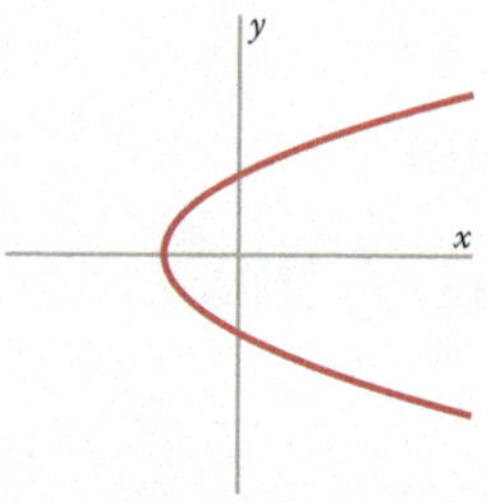
Absolute Value *Graphs are V-shaped.*	$y = \|0.5x\|$ $y = 1 - \|x\|$ $x = \|3y - 5\|$	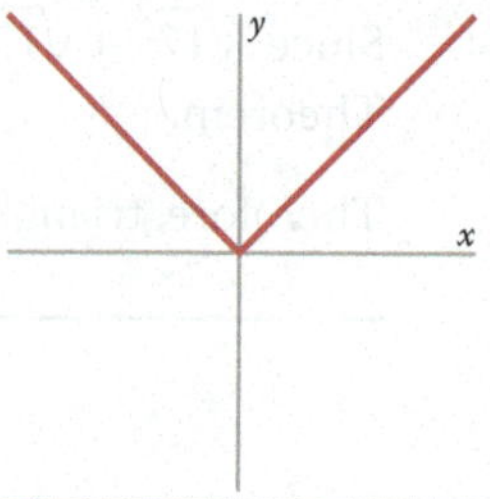	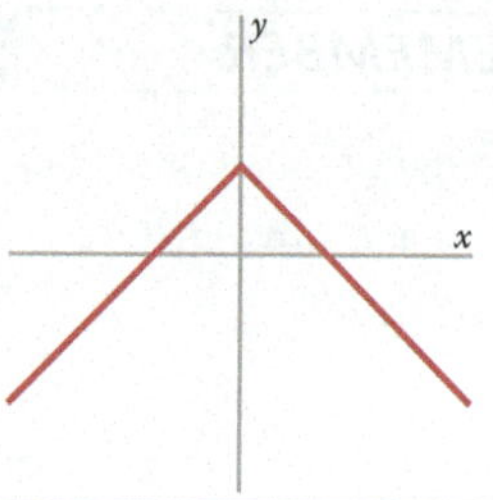

This method for graphing an equation by finding solutions and plotting those solutions (points) on a coordinate plane is called **graphing an equation by plotting points**.

Steps for Graphing an Equation by Plotting Points

1. Choose several allowable values for x.
2. Substitute each x-value into the equation and find the corresponding y-value, resulting in an ordered pair solution (x, y) to the equation. List these ordered pair solutions in a two-column (or two-row) table.
3. Plot each ordered pair (x, y) on a coordinate plane.
4. Connect the points with a straight line or smooth curve (using the shape of the equation type as a guide if known).

The number of x-values chosen in step 1 gives the number of ordered pair solutions that will be used to sketch the graph of the equation. Only a few (two to four) ordered pair solutions need to be found when the general shape of the equation's graph is known. However, when the general shape of the graph is not known, additional ordered pair solutions are needed in order to clearly reveal the shape of the equation's graph.

CAUTION

When y-values are chosen and corresponding x-values are found, be sure to list those chosen y-values in the table's second column, not the first column.

When graphing an equation by plotting points, several y-values can be chosen for the table instead of x-values. In this case, the chosen y-values are substituted into the given equation, and then the corresponding x-values are found. Choosing y-values (instead of x-values) makes finding the ordered pair solutions easier when

- the equation is solved for x,
- the y-variable in the equations is squared (or taken to any exponent), or
- the equation includes $|y|$.

EXAMPLE 8 Graphing an Equation by Plotting Points

Sketch the graph of each equation by making a table of values.

A. $y = 3 - 2x$ **B.** $y = 1 + x^2$ **C.** $\left|\frac{x}{2}\right| = y - 1$ **D.** $y^2 + y = x$

SOLUTION

A. Make a table for the equation. The equation's domain is all real numbers, so any x-value can be substituted into the equation. Here, −1, 0, 1, and 2 are chosen for the x-values.

x	y
−1	5
0	3
1	1
2	−1

$y = 3 - 2x$

$y = 3 - 2(-1) = 3 + 2 = 5$

$y = 3 - 2(0) = 3 - 0 = 3$

$y = 3 - 2(1) = 3 - 2 = 1$

$y = 3 - 2(2) = 3 - 4 = -1$

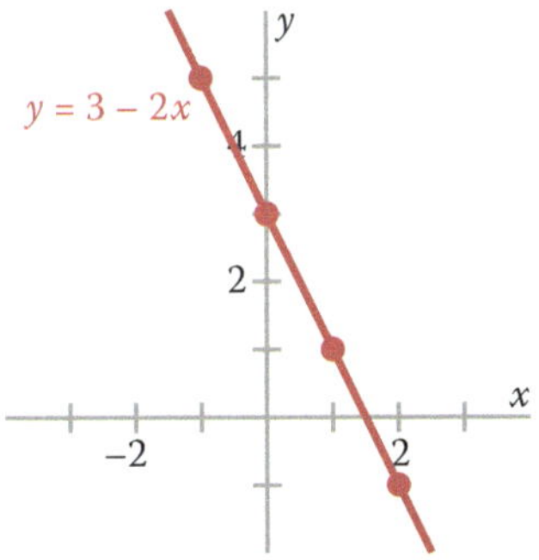

Plot each ordered pair solution from the table, and then draw a line (because the equation is linear) through the points.

IMPORTANT

Regardless of which x-values are used for a linear equation, the general pattern formed by the ordered pair solutions will not change, and so the graph is also the same (when the line is extended far enough). Furthermore, the x-values chosen for the table do not have to be in any particular order.

B. Make a table for the equation. The equation's domain is all real numbers, so any x-value can be substituted into the equation. Again, –1, 0, 1, and 2 are chosen here for the x-values.

x	y
–1	2
0	1
1	2
2	5

$y = 1 + x^2$

$y = 1 + (-1)^2 = 1 + 1 = 2$

$y = 1 + (0)^2 = 1 + 0 = 1$

$y = 1 + (1)^2 = 1 + 1 = 2$

$y = 1 + (2)^2 = 1 + 4 = 5$

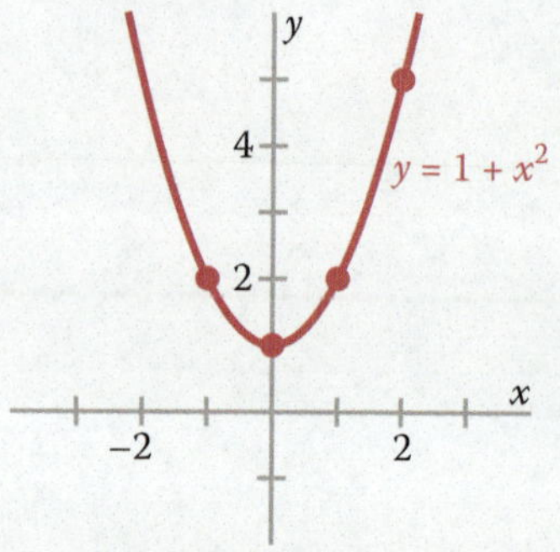

Plot each ordered pair solution from the table, and then draw a U-shaped graph (because the equation is quadratic) through the points.

C. This equation is not solved for x or y. Finding the ordered pair solutions is easier when the equation is solved for one of the variables, so solve the equation for y first, and then make the table.

$$\left|\frac{x}{2}\right| = y - 1 \Rightarrow y = \left|\frac{x}{2}\right| + 1$$

IMPORTANT

Even x-values are chosen for the table so that the resulting y-values will be integers, making the points easier to plot.

Here, –2, 0, 2, and 4 are chosen for the x-values. Plot each ordered pair solution from the table, and then draw a V-shaped graph (because the equation is an absolute value equation) through the points.

x	y
–2	2
0	1
2	2
4	3

$y = \left|\frac{x}{2}\right| + 1$

$y = \left|\frac{-2}{2}\right| + 1 = \left|-1\right| + 1 = 2$

$y = \left|\frac{0}{2}\right| + 1 = \left|0\right| + 1 = 1$

$y = \left|\frac{2}{2}\right| + 1 = \left|1\right| + 1 = 2$

$y = \left|\frac{4}{2}\right| + 1 = \left|2\right| + 1 = 3$

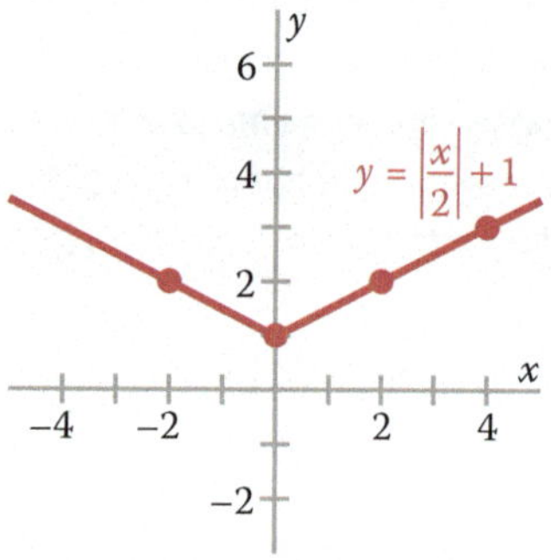

D. This equation is solved for x, so choose y-values for the table. Here, –1, 0, 1, and 2 are chosen for the y-values.

IMPORTANT

Since y is the squared variable, y-values are chosen and listed in the second column of the table. Next, each y-value is substituted into the equation for y to find the corresponding x-value in each ordered pair solution. Those x-values are then listed in the first column of the table.

x	y
0	–1
0	0
2	1
6	2

$x = y^2 + y$

$x = (-1)^2 + (-1) = 1 - 1 = 0$

$x = (0)^2 + 0 = 0$

$x = (1)^2 + 1 = 1 + 1 = 2$

$x = (2)^2 + 2 = 4 + 2 = 6$

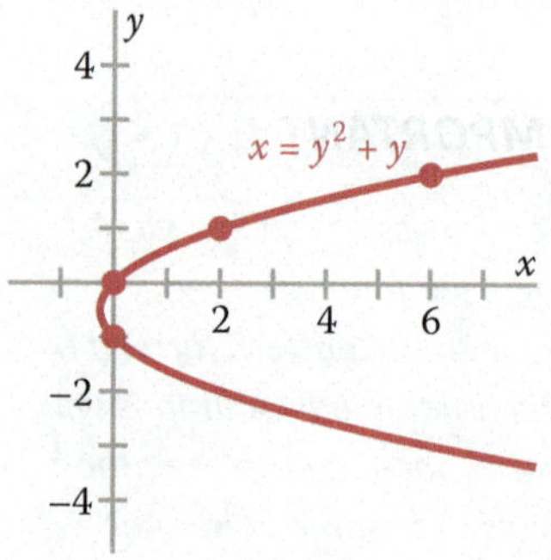

Plot each ordered pair solution from the table, and then draw a U-shaped graph (because the equation is quadratic) through the points.

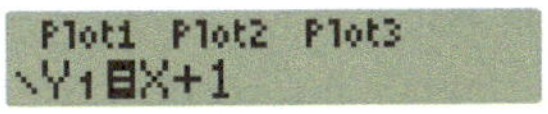

Figure 1.1i

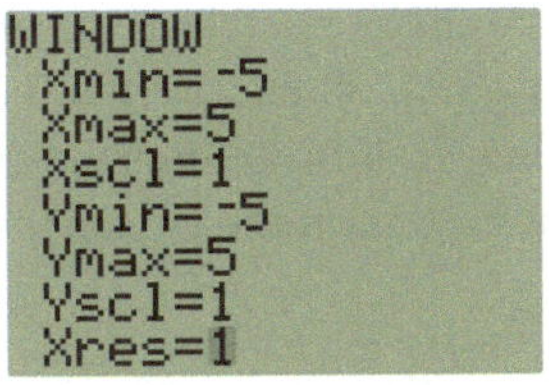

Figure 1.1j

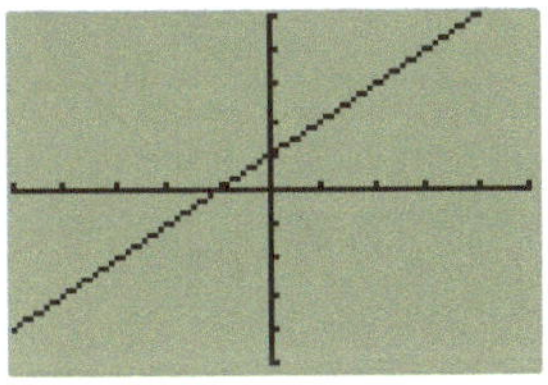

Figure 1.1k

WHAT'S THE BIG IDEA?

Describe the relationship between an equation in two variables, a table of ordered pairs defined by the equation, and the graph of the equation.

Using a Graphing Calculator to Graph an Equation

Once an equation is solved for y, the equation can be graphed using a graphing calculator. Use the following steps to graph an equation using a graphing calculator.

Steps for Graphing an Equation Using a Graphing Calculator

1. Solve the equation for y.
2. Using a graphing calculator, go to the **Y=** window and enter the expression for y.
3. Adjust the viewing window (**WINDOW**) as needed.
4. View the graph (**GRAPH**).

For example, to graph the equation $y = x + 1$ (the equation is already solved for y), enter "$x + 1$" in the **Y=** window, as shown in Figure 1.1i. Then adjust the **WINDOW** as needed. The window in Figure 1.1j is set to ±5 on the x- and y-axes, with a scale of 1 on both axes. Then view the graph, as shown in Figure 1.1k.

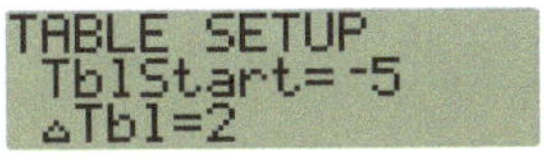

Figure 1.1l

Table for $y = x + 1$

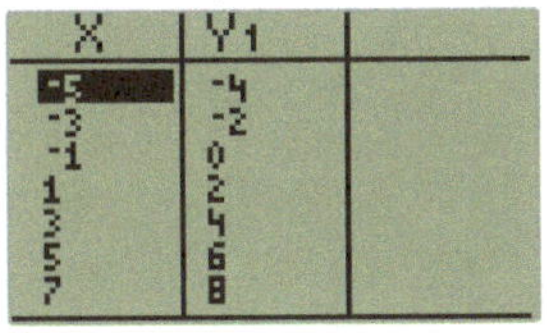

Figure 1.1m

After an equation is entered into a graphing calculator, a corresponding table of ordered pairs can be viewed by using the table function (**TABLE**). Most graphing calculators also have a function for setting the first value seen in the table, as well as the interval for the change in the x-values. For example, to view a table starting with $x = -5$ where the x-values increase by 2 (i.e., the x-values are $-5, -3, -1, 1, 3 \ldots$), set the start value to -5 and the interval (often denoted by Δ) to 2, as shown in Figures 1.1l and 1.1m.

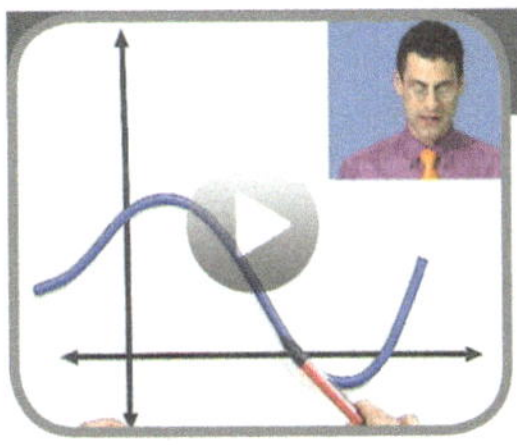

1.1.7 Finding the x- and y-Intercepts of an Equation

Another method for graphing an equation in two variables is to find and plot the ***x*- and *y*-intercepts** (the points at which the equation's graph crosses the x- and y-axes). The equation's graph can be sketched by drawing a line or curve through the x- and y-intercepts, keeping in mind the general shape of the graph (e.g., a line for a linear equation, a parabola for a quadratic equation, etc.). Some additional points on the graph are often needed to reveal the shape of the graph.

TIP

An x-intercept is a point at which a graph crosses the x-axis, and a y-intercept is a point at which a graph crosses the y-axis.

DEFINITION

An ***x*-intercept** of an equation is a point at which the equation's graph intersects the x-axis. The y-coordinate of an x-intercept is always 0.

A ***y*-intercept** of an equation is a point at which the equation's graph intersects the y-axis. The x-coordinate of a y-intercept is always 0.

x-intercept: $(x, 0)$ $\qquad$ y-intercept: $(0, y)$

On a coordinate plane, every line that is not horizontal or vertical has exactly one x-intercept and one y-intercept. All horizontal lines (except the line $y = 0$) have one y-intercept but no x-intercept, and all vertical lines (except the line $x = 0$) have one x-intercept but no y-intercept. The lines $y = 0$ and $x = 0$ are exceptions because these lines are the axes themselves (i.e., the horizontal line $y = 0$ is the x-axis and the vertical line $x = 0$ is the y-axis). Since $y = 0$ is the x-axis, and thus every point on $y = 0$ is on the x-axis, the line $y = 0$ has an infinite number of x-intercepts. Similarly, since $x = 0$ is the y-axis, and thus every point on $x = 0$ is on the y-axis, the line $x = 0$ has an infinite number of y-intercepts.

Other types of graphs may have any number of x- and y-intercepts, including no x- and y-intercepts, as shown in the following table.

TIP

When a graph passes through the origin, the graph has an x- and y-intercept at the same point.

One x-intercept **One y-intercept**	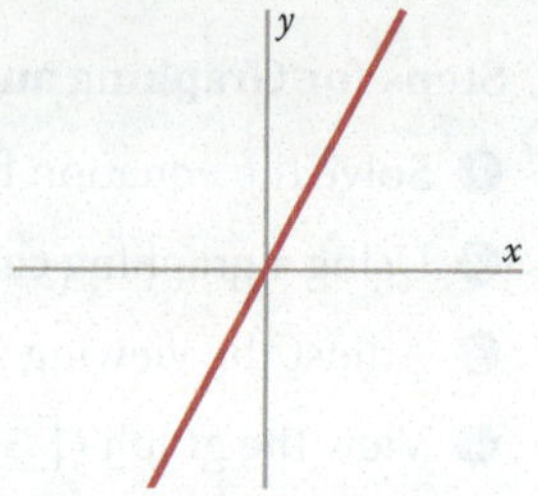	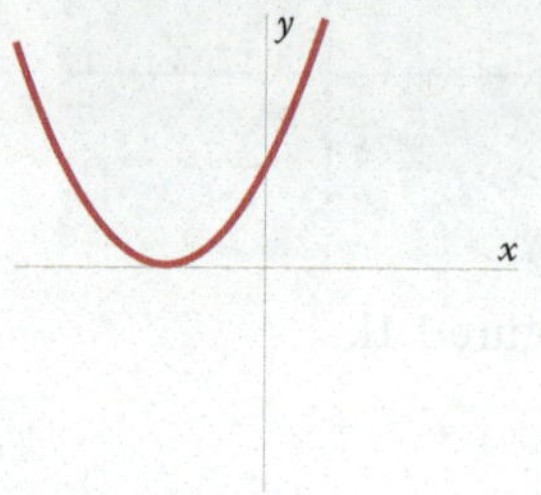
Two x-intercepts **One y-intercept**	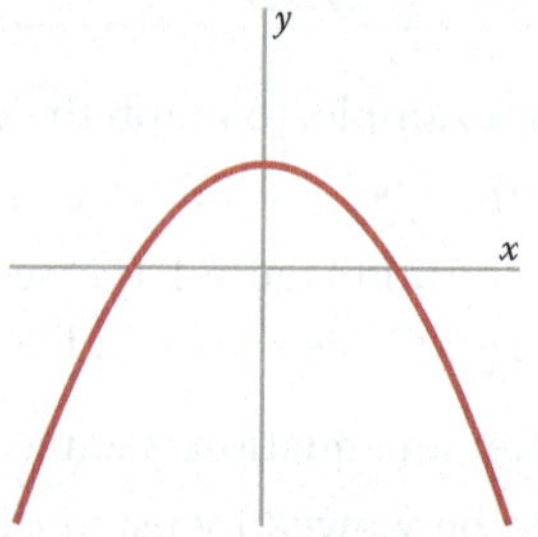	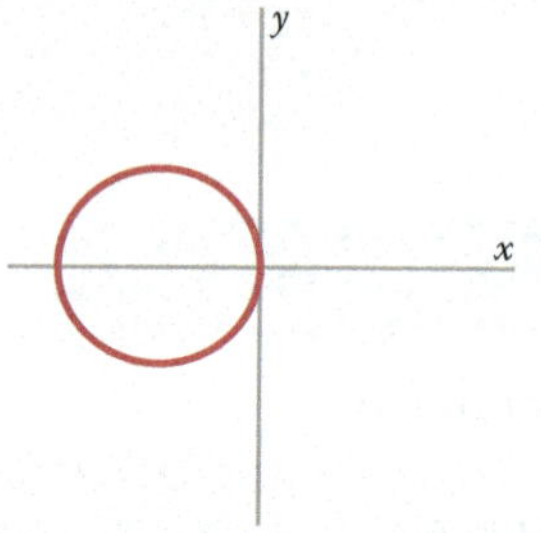
One x-intercept **Two y-intercepts**	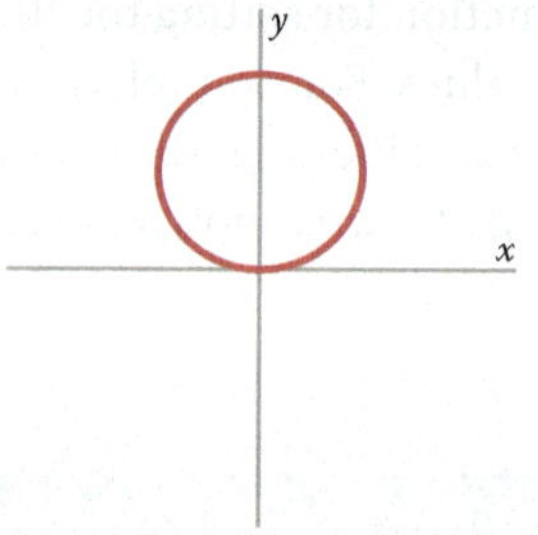	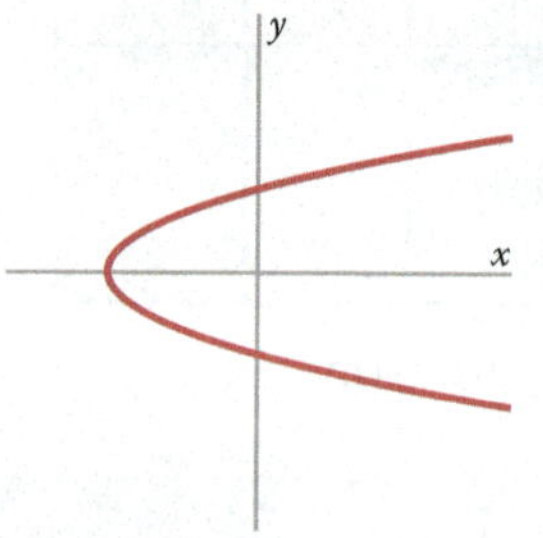
Four x-intercepts **One y-intercept**	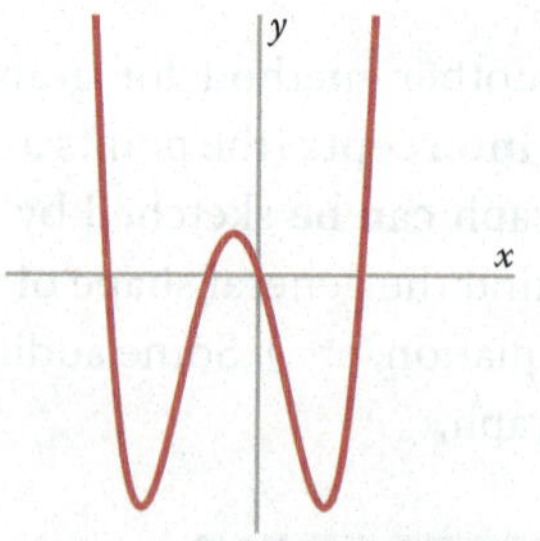	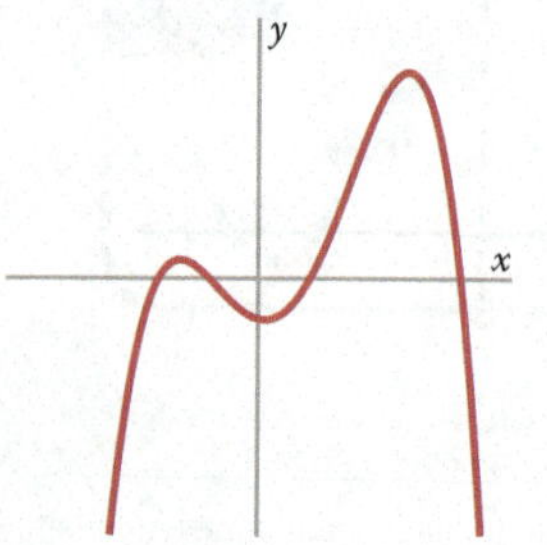
No intercepts	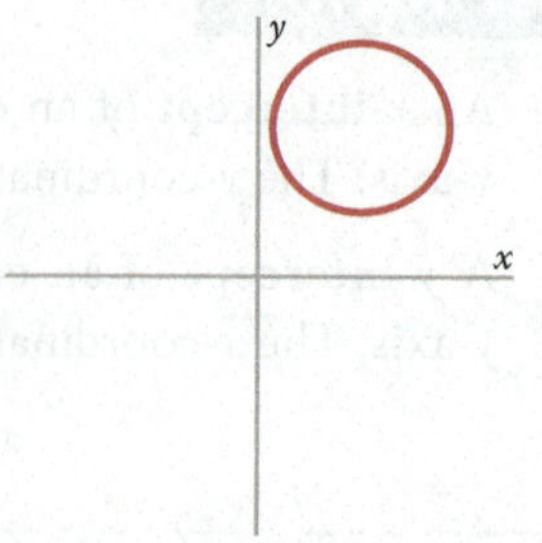	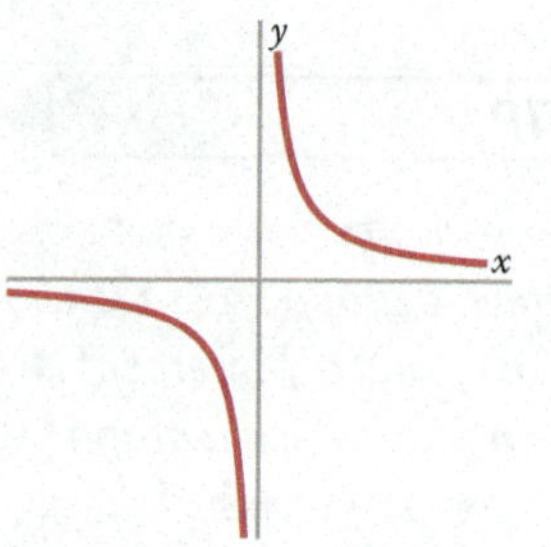

IMPORTANT

Many equations cannot be graphed using this method.

Steps for Graphing an Equation Using Intercepts

1. Find the x-intercept(s). (Substitute 0 for y in the equation and solve for x.)
2. Find the y-intercept(s). (Substitute 0 for x in the equation and solve for y.)
3. Plot the x- and y-intercepts on a coordinate plane.
4. Sketch the graph using the intercepts and the general shape of the graph, given the equation type.

EXAMPLE 9 Graphing an Equation Using Intercepts

Use the intercepts to sketch the graph of each equation.

A. $4x - y = 8$ **B.** $y = x^2 - 4$

SOLUTION

A. Find the x- and y-intercepts.

To find the x-intercept, substitute 0 for y and solve for x.

$$4x - 0 = 8 \Rightarrow x = 2$$

The x-intercept is at (2, 0).

To find the y-intercept, substitute 0 for x and solve for y.

$$4(0) - y = 8 \Rightarrow y = -8$$

The y-intercept is at (0, −8).

The equation is linear, so the graph forms a line. Plot the x- and y-intercepts, and then sketch the line through the intercepts.

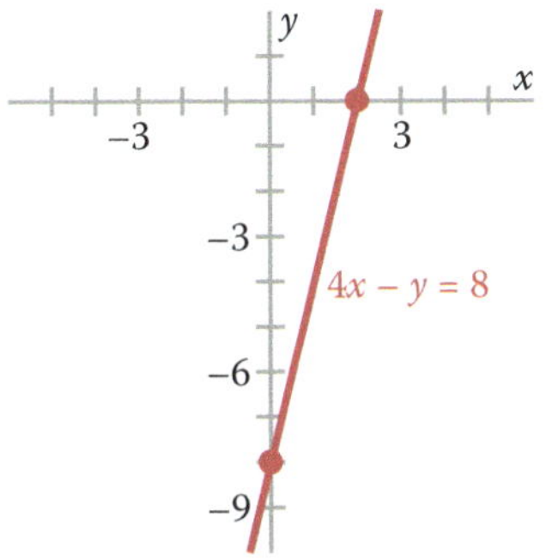

B. Find the x- and y-intercepts.

To find the x-intercept, substitute 0 for y and solve for x.

$$0 = x^2 - 4 \Rightarrow x = 2 \text{ or } x = -2$$

There are two x-intercepts, (2, 0) and (−2, 0).

To find the y-intercept, substitute 0 for x and solve for y.

$$y = (0)^2 - 4 = -4$$

The y-intercept is at (0, −4).

The equation is quadratic, so the graph forms a U-shape (parabola). Plot the x- and y-intercepts, and then sketch the parabola through the intercepts.

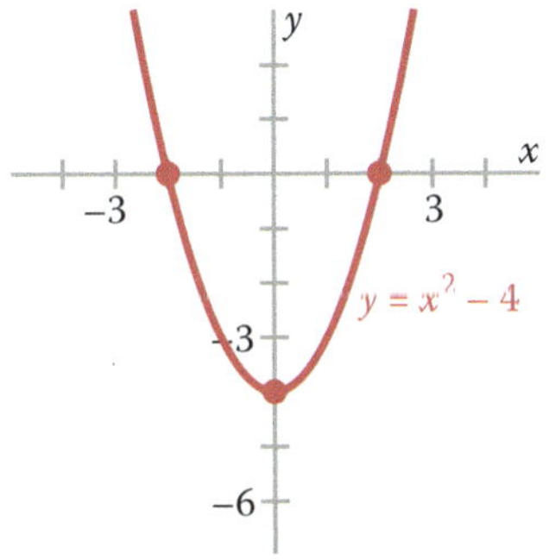

TIP

Find additional points on a graph (by making a table) to confirm the shape of a graph.

1.1.8 Introduction to the Equation of a Circle

On a coordinate plane, a **circle** is the set of all points that are a set distance from a specific point, the circle's **center**. This set distance is the circle's **radius**. Suppose the center of a circle on a coordinate plane is at (h, k). Then the standard form for the equation of a circle can be derived by using the Distance Formula to find the set of all points (x, y) that are r units from a specific point, the circle's center at (h, k).

The Standard Form Equation of a Circle

The standard form equation of a circle with radius r and center (h, k) is $(x - h)^2 + (y - k)^2 = r^2$.

The standard form equation of a circle with radius r and center at the origin (so $h = 0$ and $k = 0$) is $x^2 + y^2 = r^2$.

When the center and radius of a circle on a coordinate plane are known, then the values can be substituted into the standard form equation of a circle to write the equation of a circle.

EXAMPLE 10 Writing the Equation of a Circle Given the Center and Radius

Write the equation of each circle with the following properties.

A. $r = 6$ and center at the origin **B.** $r = 1$ and center at $(2, -6)$

SOLUTION

A. Substitute $r = 6$ into the equation of a circle with center at the origin, $x^2 + y^2 = r^2$, and simplify.

$$x^2 + y^2 = (6)^2 \Rightarrow x^2 + y^2 = 36 \text{ (Figure 1.1n)}$$

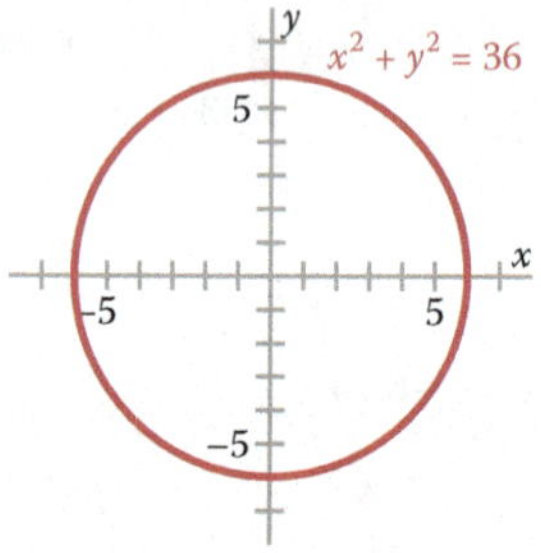

Figure 1.1n

B. Substitute $r = 1$, $h = 2$, and $k = -6$ into the equation of a circle with center at (h, k), $(x - h)^2 + (y - k)^2 = r^2$, and simplify.

$$(x - 2)^2 + (y - (-6))^2 = (1)^2 \Rightarrow (x - 2)^2 + (y + 6)^2 = 1 \text{ (Figure 1.1o)}$$

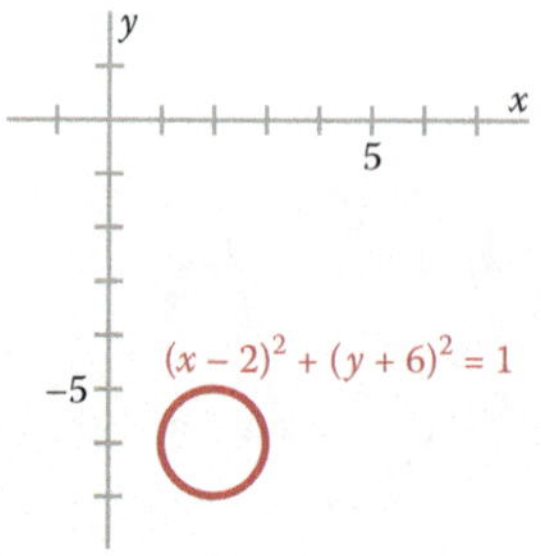

Figure 1.1o

WHAT'S THE BIG IDEA?

Use the Distance Formula to derive the equation of each circle with the given center when the radius is r.

- Center at (h, k)
- Center on the x-axis with x-coordinate h
- Center on the y-axis with y-coordinate k

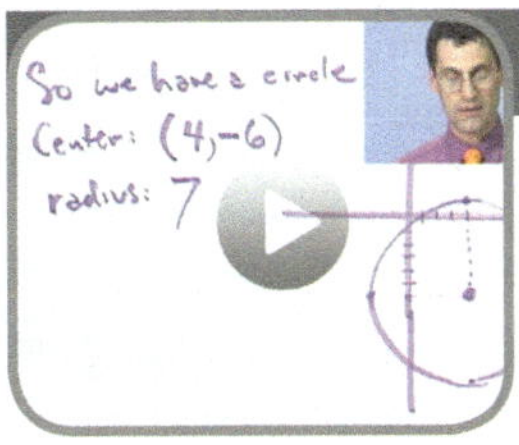

1.1.9 Graphing a Circle

To graph a circle when the center and radius are known, first plot the circle's center. Then, using the radius r, identify four points on the circle: the points r units above, below, to the left, and to the right of the center. Plot these four points, and sketch a circle that passes through each point. (Remember, the center is *not* a point on the circle.)

Steps for Graphing a Circle from Its Equation

1. Identify h, k, and r from the equation.
2. Identify (h, k), the center of the circle.
3. Plot the points r units above, below, to the left, and to the right of (h, k).
4. Sketch the circle through these four points.

Note that in the equation of a circle, the constant value on one side of the equation is the square of the radius. So, take the square root of that constant value to find the circle's radius.

EXAMPLE 11 Graphing a Circle Using Its Equation

Sketch the graph of $(x + 5)^2 + y^2 = 4$.

SOLUTION

IMPORTANT

$(x + 5)^2 + y^2 = 4 \Rightarrow$
$(x - (-5))^2 + (y - 0)^2 = 2^2$

From the equation, $h = -5$, $k = 0$, and $r^2 = 4$, so $r = 2$. So, the circle's center is at $(-5, 0)$, and the radius is 2.

Plot the points 2 units above, below, to the left, and to the right of $(-5, 0)$, and then sketch the circle through those points.

2 units above $(-5, 0) \Rightarrow (-5, 0 + 2) \Rightarrow (-5, 2)$
2 units below $(-5, 0) \Rightarrow (-5, 0 - 2) \Rightarrow (-5, -2)$
2 units left of $(-5, 0) \Rightarrow (-5 - 2, 0) \Rightarrow (-7, 0)$
2 units right of $(-5, 0) \Rightarrow (-5 + 2, 0) \Rightarrow (-3, 0)$

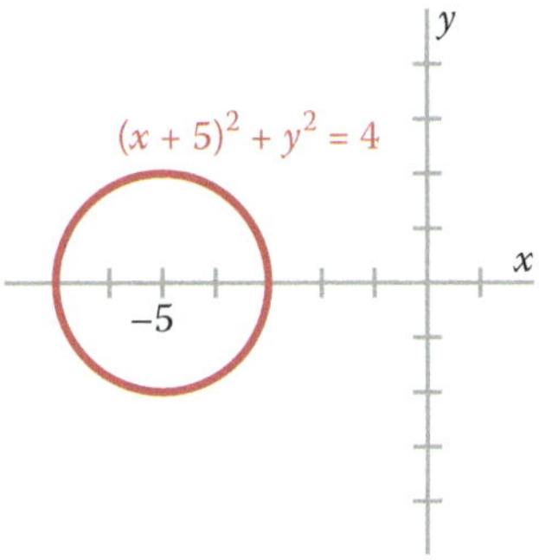

1.1.10 Writing the Equation of a Circle

Notice that the equation of a circle with center (h, k), $(x - h)^2 + (y - k)^2 = r^2$, contains two terms that are squares of binomials. Recall that completing the square was a process used to write a quadratic equation as the square of a binomial equal to a constant. Completing the square can also be applied to the equation of circle. An equation of a circle in general form (expanded form) can be written in standard form, $(x - h)^2 + (y - k)^2 = r^2$, by completing the square for the x-terms, for the y-terms, or for both the x-terms and the y-terms.

Steps for Writing the Equation of a Circle in Standard Form

❶ Group the x-terms together and the y-terms together, both on one side of the equation, and move the constant term to the other side.

❷ Complete the square as needed to write a perfect-square trinomial with the x-terms and with the y-terms.

❸ Factor each trinomial.

EXAMPLE 12 Writing the Equation of a Circle in Standard Form

Write $x^2 + y^2 - 8x + 12y + 34 = 0$ in the standard form of a circle. Then identify the circle's center and radius.

SOLUTION

REMEMBER

To complete the square for $x^2 + bx$, add the square of half of b to each side of the equation. Then the perfect-square trinomial $x^2 + bx + \left(\frac{b}{2}\right)^2$ can be written as $\left(x + \frac{b}{2}\right)^2$.

Grouping the x-terms and the y-terms and moving the constant term to the right side gives $(x^2 - 8x) + (y^2 + 12y) = -34$. Complete the square twice to write the x-terms as a perfect-square trinomial and the y-terms as a perfect-square trinomial.

$$(x^2 - 8x) + (y^2 + 12y) = -34$$

$$(x^2 - 8x + 16) + (y^2 + 12y + 36) = -34 + 16 + 36$$ *Complete the square twice.*

$$(x - 4)^2 + (y + 6)^2 = 18$$ *Factor each perfect-square trinomial and simplify the right side.*

The standard form equation is $(x - 4)^2 + (y + 6)^2 = 18$.

From the equation, $h = 4$, $k = -6$, and $r^2 = 18$, so $r = 3\sqrt{2}$. So, the circle's center is at $(4, -6)$, and the radius is $3\sqrt{2}$ units.

1.1.11 Writing the Equation of a Circle: Another Example

Any segment that passes through the center of a circle with endpoints on the circle is called a **diameter** of the circle. The length of a diameter is equal to twice the length of the circle's radius, and the midpoint of any diameter is the center of the circle, as shown in Figure 1.1p.

When a circle is graphed on a coordinate plane, the Distance Formula can be used to find the length of the circle's diameter (the distance between a diameter's endpoints, which are two specific points on the circle). The Distance Formula can also be used to find the length of the circle's radius (the distance between the circle's center and any point on the circle). The Midpoint Formula can be used to find the coordinates of the circle's center, which is the midpoint of a diameter.

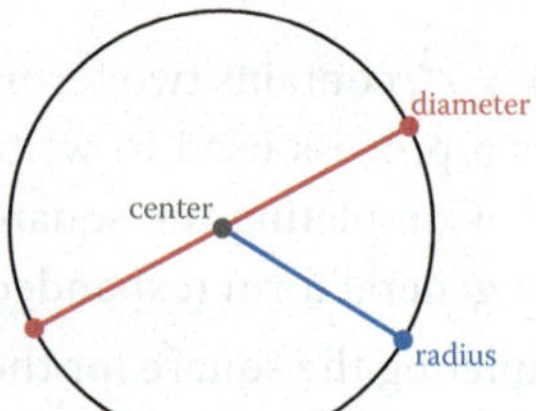

Figure 1.1p

EXAMPLE 13 Writing the Equation of a Circle Given the Endpoints of a Diameter

Write the equation of the circle where $A(0, 0)$ and $B(4, 6)$ are the endpoints of a diameter.

SOLUTION

To write the equation of a circle, find the radius r and the coordinates of the center (h, k), then substitute those values into the standard form equation of a circle, $(x - h)^2 + (y - k)^2 = r^2$.

Find the radius.
The line segment AB is a diameter, so the length of the circle's radius is half of the length of AB.

Use the Distance Formula to find the length of AB.

Let $A(0, 0)$ be (x_1, y_1), and let $B(4, 6)$ be (x_2, y_2). *$x_1 = 0, x_2 = 4, y_1 = 0, y_2 = 6$*

Distance Formula ▶ $d = \sqrt{(x_2 - x_1)^2 + (y_2 - y_1)^2}$

Substitute the coordinates into the Distance Formula and simplify.

$$AB = \sqrt{(4-0)^2 + (6-0)^2} \quad \text{Substitute.}$$
$$= \sqrt{(4)^2 + (6)^2} \quad \text{Subtract within the parentheses.}$$
$$= \sqrt{16+36} \quad \text{Evaluate the powers.}$$
$$= \sqrt{52} \quad \text{Add.}$$
$$= 2\sqrt{13} \quad \text{Simplify the square root.}$$

The length of the circle's diameter is $2\sqrt{13}$ units, so the length of its radius is $\frac{2\sqrt{13}}{2} = \sqrt{13}$ units.

Find the center.
Since the line segment AB is a diameter, the midpoint of AB is the circle's center.

Use the Midpoint Formula to find the midpoint of AB.

Again, let $A(0, 0)$ be (x_1, y_1), and let $B(4, 6)$ be (x_2, y_2). *$x_1 = 0, x_2 = 4, y_1 = 0, y_2 = 6$*

Midpoint Formula ▶ $M = \left(\frac{x_1 + x_2}{2}, \frac{y_1 + y_2}{2}\right)$

Substitute the coordinates into the Midpoint Formula and simplify.

$$\left(\frac{0+4}{2}, \frac{0+6}{2}\right) = (2, 3)$$

The midpoint between points A and B is $(2, 3)$, so the circle's center is at $(2, 3)$.

Alternative Method ▶ *Find the circle's center first (the midpoint between A and B), and then find the circle's radius (the distance between the center and either A or B).*

Write the equation.

Substitute $r = \sqrt{13}$, $h = 2$, and $k = 3$ into the equation of a circle with center at (h, k), $(x - h)^2 + (y - k)^2 = r^2$, and simplify.

$$(x-2)^2 + (y-3)^2 = (\sqrt{13})^2 \Rightarrow (x-2)^2 + (y-3)^2 = 13$$

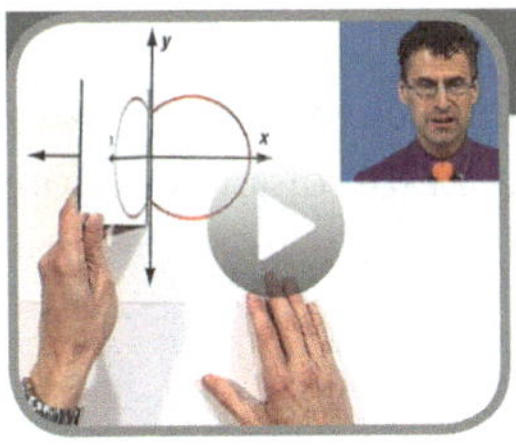

1.1.12 Testing for Symmetry

A circle is a figure with perfect **symmetry**, because any line drawn through the center of the circle divides the circle into two parts that are **reflections** of each other. A line that divides a figure so that one side is a reflection of the other is called a **line of symmetry**. So, a circle has an infinite number of lines of symmetry.

On a coordinate plane, if the x-axis is a line of symmetry for a figure, then the figure is **symmetric with respect to the x-axis**. Similarly, if the y-axis is a line of symmetry for a figure, then the figure is **symmetric with respect to the y-axis**. A figure formed by a reflection over the y-axis and then a reflection over the x-axis is **symmetric with respect to the origin**.

For example, a circle on a coordinate plane can be symmetric with respect to the x-axis, the y-axis, or the origin, depending on the location of the circle's center.

TIP

A circle centered at the origin is symmetric with respect to the x-axis, the y-axis, and the origin.

Symmetry of Circles

Symmetric with respect to the origin	Symmetric with respect to the x-axis	Symmetric with respect to the y-axis
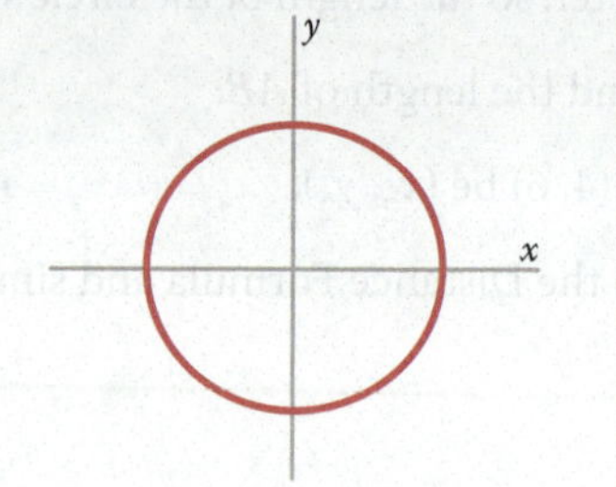	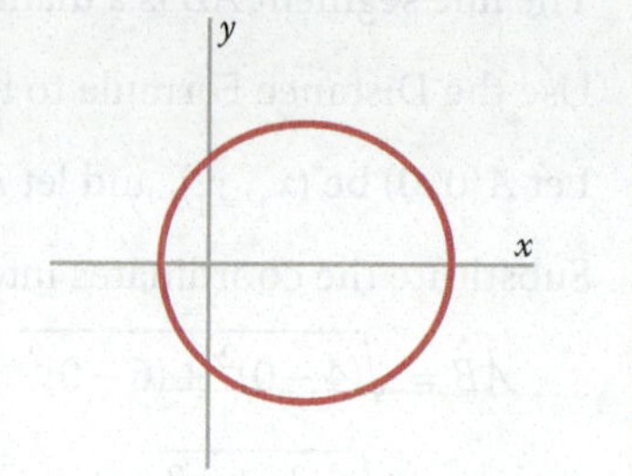	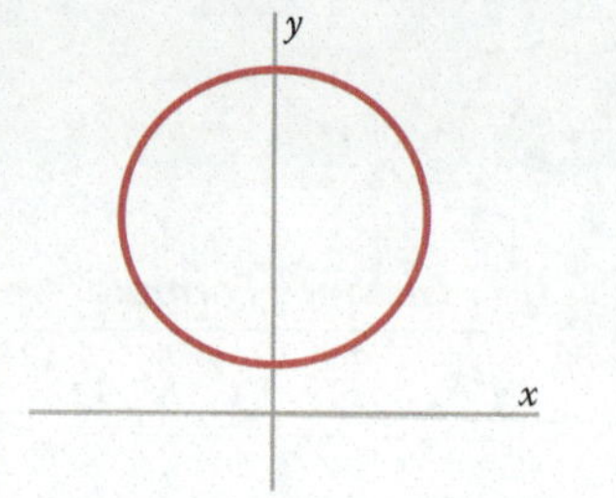
The circle's center is at the origin.	*The circle's center is on the x-axis, not at the origin.*	*The circle's center is on the y-axis, not at the origin.*

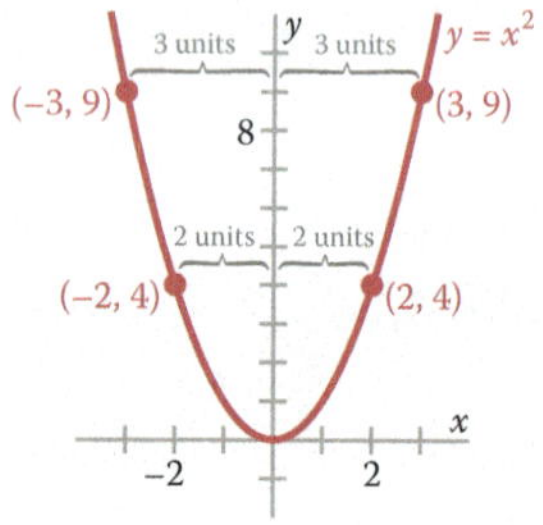

Figure 1.1q

Consider the graph of some equation that is symmetric to the y-axis, such as the graph of $y = x^2$ (Figure 1.1q). Notice that each point on $y = x^2$ that is to the left of the y-axis (the line of symmetry) has a reflection point on the right of the y-axis that is also on the graph. For example, $(-2, 4)$ and $(2, 4)$ are reflections of one another across the y-axis, as are $(-3, 9)$ and $(3, 9)$. These points are reflections of one another across the y-axis because they have the same y-coordinate and opposite x-coordinates. It follows that if (x, y) is on a graph that is symmetric with respect to the y-axis, then $(-x, y)$ must also be on this graph.

Similarly, if a graph is symmetric with respect to the x-axis, and (x, y) is on the graph, then $(x, -y)$ must also be on this graph. And if a graph is symmetric with respect to the origin, and (x, y) is on the graph, then $(-x, -y)$ must also be on this graph.

IMPORTANT

A graph that is symmetric with respect to the origin may or may not be symmetric with respect to the x- or y-axis.

Symmetry and Points on a Graph

Symmetric with respect to the origin	Symmetric with respect to the x-axis	Symmetric with respect to the y-axis
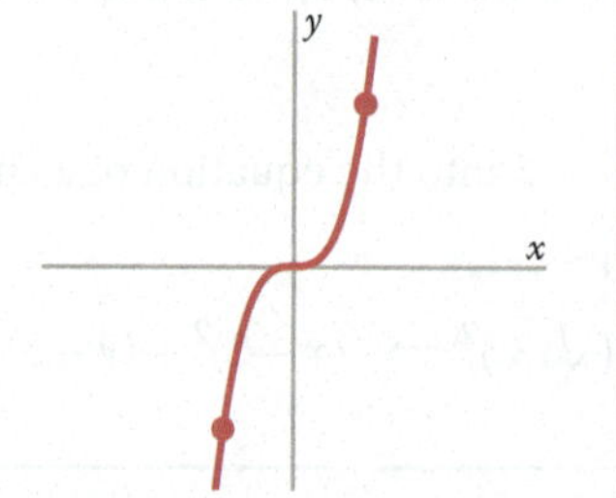	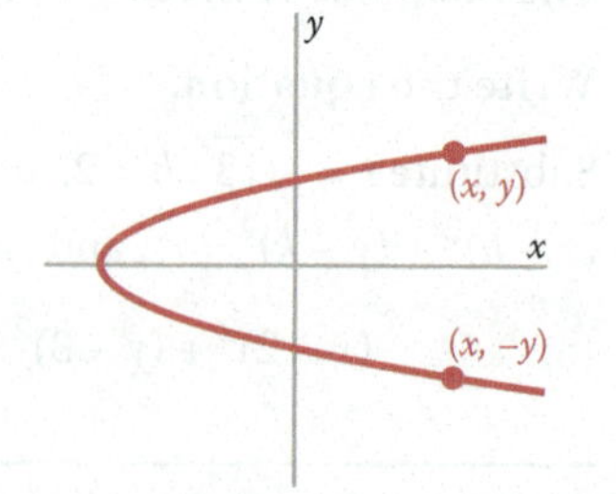	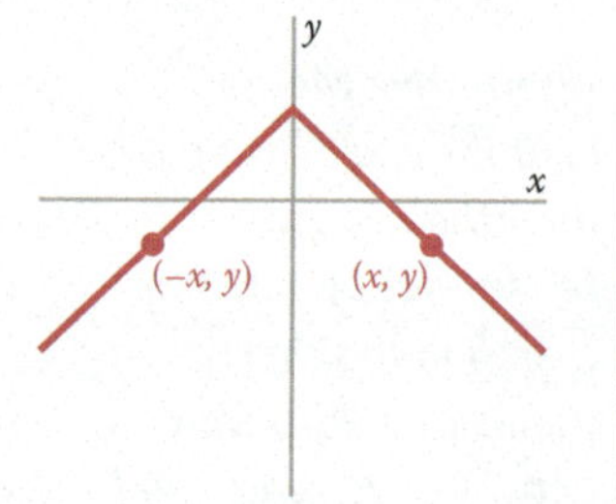
For every (x, y) on the graph, $(-x, -y)$ is also on the graph.	*For every (x, y) on the graph, $(x, -y)$ is also on the graph.*	*For every (x, y) on the graph, $(-x, y)$ is also on the graph.*

The graphs of many types of equations are symmetric with respect to one or both of the axes, or the origin. Knowing the type of symmetry is helpful when sketching the graph of an equation. The relationship between the points on a symmetric graph leads to an algebraic test for the type of symmetry that can be applied to any equation.

Tests for Symmetry	
Symmetric with respect to the origin	Substituting $-x$ for x and $-y$ for y yields an equivalent equation.
Symmetric with respect to the x-axis	Substituting $-y$ for y yields an equivalent equation.
Symmetric with respect to the y-axis	Substituting $-x$ for x yields an equivalent equation.

EXAMPLE 14 Determining the Type of Symmetry from an Equation

Identify the type of symmetry for the graph of $y = x - 5x^3$.

SOLUTION

Test the equation for each type of symmetry.

With Respect to the x-Axis	With Respect to the y-Axis	With Respect to the Origin
Substitute $-y$ for y. $-y = x - 5x^3$ $y = -x + 5x^3$	Substitute $-x$ for x. $y = (-x) - 5(-x)^3$ $y = -x + 5x^3$	Substitute $-x$ for x and $-y$ for y. $-y = (-x) - 5(-x)^3$ $-y = -x + 5x^3$ $y = x - 5x^3$

The original equation, $y = x - 5x^3$, is not equivalent to $y = -x + 5x^3$, so the graph of $y = x - 5x^3$ is not symmetric with respect to the x-axis or the y-axis. The graph of $y = x - 5x^3$ is symmetric with respect to the origin, because substituting $-x$ for x and $-y$ for y results in an equivalent equation.

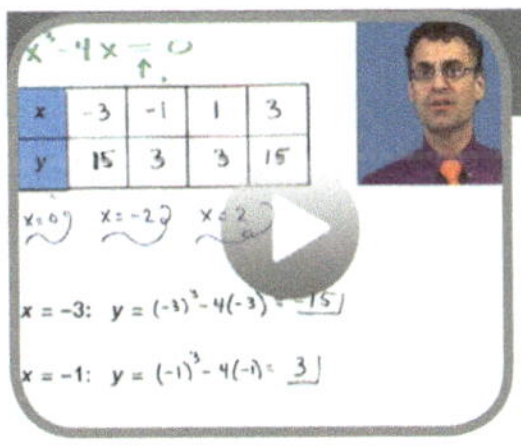

1.1.13 Using Symmetry as a Sketching Aid

Before sketching the graph of an equation, use the algebraic tests for symmetry to determine the type of symmetry, if any, for an equation's graph. Then use the type of symmetry to find additional points on the graph.

Steps for Graphing an Equation Using Intercepts and Symmetry

1. Test the equation for possible types of symmetry.
2. Find and plot the x- and y-intercepts on a coordinate plane.
3. Find and plot some additional points on one side of the line of symmetry (i.e., make a table).
4. Use the type of symmetry to plot a few additional points on the other side of the line of symmetry.

EXAMPLE 15 Using Symmetry to Graph an Equation

Use symmetry and intercepts to sketch the graph of $y = |x| - 2$.

SOLUTION

Test the equation for each type of symmetry.

With Respect to the x-Axis	With Respect to the y-Axis	With Respect to the Origin
$-y = \|x\| - 2$ $y = -\|x\| + 2$ *Not equivalent to* $y = \|x\| - 2$	$y = \|-x\| - 2$ $y = \|x\| - 2$ *Equivalent to* $y = \|x\| - 2$	$-y = \|-x\| - 2$ $y = -\|x\| + 2$ *Not equivalent to* $y = \|x\| - 2$

So, the graph $y = |x| - 2$ is symmetric with respect to the y-axis.

Find the intercepts.

x-intercept: $0 = |x| - 2 \Rightarrow x = 2$ or $x = -2$ *Substitute 0 for y and solve for x.*

y-intercept: $y = |0| - 2 = -2$ *Substitute 0 for x and simplify to find y.*

TIP

The equation $0 = |x| - 2$ is an absolute value equation, so solve using $A = |B| \Rightarrow A = B$ or $A = -B$.

So, the graph $y = |x| - 2$ crosses the x-axis at (2, 0) and (−2, 0), and the y-axis at (0, −2).

Find some additional points on the graph on one side of the y-axis (the line of symmetry). Organize these points in a table.

x	1	3	4	5
y	−1	1	2	3

Choose four x-values. Substitute each into the equation to find the corresponding y-value.

The graph is symmetric with respect to the y-axis, so for each (x, y) on the graph, $(-x, y)$ is also on the graph. Use this fact to find additional points on the graph. Organize these points in a table.

x	−1	−3	−4	−5
y	−1	1	2	3

Replace x with −x for each value in the table.

TIP

The x-value 2 is not included in the table because the point on the graph where x = 2 is already known; it is the x-intercept.

Plot the intercepts and all of the points from the tables on a coordinate plane, and sketch the graph.

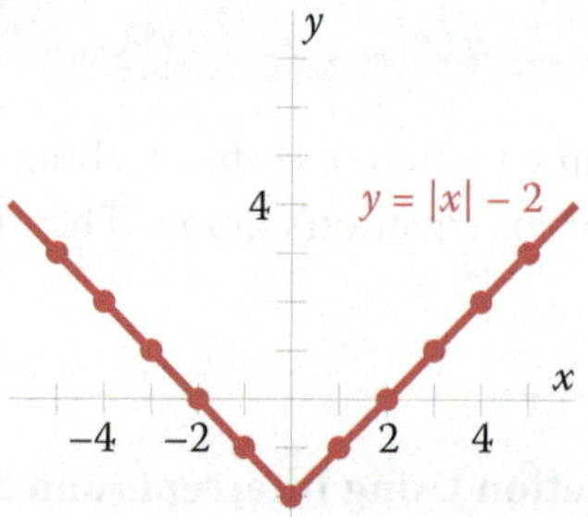

The graph is symmetric with respect to the y-axis, so each of the graph's points on the left side of the y-axis is a reflection of a point on the graph on the right side of the y-axis.

WHAT'S THE BIG IDEA?

How does the fact that the graph of an equation is symmetric with respect to the origin help when graphing the equation?

SECTION 1.1 EXERCISES

Warm Up

1. Simplify. $\sqrt{(4-(-3))^2+(-5-3)^2}$
2. Simplify. $\left(\frac{-6+4}{2}, \frac{3-9}{2}\right)$
3. Factor. $x^2 - 4x + 4$

Just the Facts

Fill in the blanks.

4. The Cartesian plane has these basic parts: the two _____, the _____ quadrants, and the _____.
5. The point (2, 4) is located in quadrant _____ , the point (–3, 1) is located in quadrant _____, and the point (–5, –8) is located in quadrant _____.
6. The formula $d = \sqrt{(x_2-x_1)^2+(y_2-y_1)^2}$ represents the _____ between two points _____ and _____.
7. If A is (x_1, y_1), B is (x_2, y_2), and $C = \left(\frac{x_1+x_2}{2}, \frac{y_1+y_2}{2}\right)$, then C is the _____ between _____ and _____.
8. If point G is between points A and E and $AG + GE = AE$, then the three points are _____.
9. To find the x-intercept of $y = ax + b$ (where a and b are real numbers), substitute 0 for _____ and solve for _____. Similarly, to find the y-intercept, substitute _____ for _____ and solve for _____.
10. The graph of the equation $x^2 + y^2 = 100$ is a(n) _____ with center _____ and radius _____.

Essential Skills

In Exercises 1–6, identify the quadrant or axis that contains each point.

1. (8, 6)
2. (–7, –3)
3. (0, –9)
4. (–2, 4)
5. (2.5, –4)
6. (–3, 0)

In Exercises 7–8, graph the points from each table in a coordinate plane.

7.

x	1	5	–4	–3
y	–1	1	2	–3

8.

x	y
0	4
3	1
–2	0
1	–1

In Exercises 9–14, determine the distance between each pair of points. Round to the nearest hundredth, as needed.

9. (–2, 5) and (–7, –3)
10. (11, –4) and (5, 12)
11. (–10, 14) and (3, –8)
12. (–2, –3) and (3, 4)
13. (3.1, 1.6) and (–1.4, 5.3)
14. (–2.5, 3.8) and (–1.2, –7.2)

In Exercises 15–20, determine the midpoint between each pair of points.

15. $G(4, 7)$ and $Q(-6, 6)$
16. $A(-8, 6)$ and $B(2, -3)$
17. $Y(-2, -5)$ and $Z(4, -8)$
18. $E(-1, 3)$ and $F(7, 8)$
19. $\left(-\frac{1}{10}, \frac{3}{8}\right)$ and $\left(\frac{1}{4}, -\frac{1}{4}\right)$
20. $\left(-\frac{3}{8}, -\frac{1}{4}\right)$ and $\left(\frac{2}{3}, -\frac{5}{6}\right)$

In Exercises 21–26, determine the unknown endpoint of each segment.

21. $P(-6, 3)$ is the midpoint between $F(x, y)$ and $T(-2, 5)$. Find the coordinates of F.
22. $M(3, -4)$ is the midpoint between $R(x, y)$ and $S(6, 0)$. Find the coordinates of R.
23. $K(-7, -5)$ is the midpoint between $M(x, y)$ and $P(-1, -3)$. Find the coordinates of M.
24. $T(-1, 2)$ is the midpoint between $S(x, y)$ and $U(4, 10)$. Find the coordinates of S.
25. $D(-3.6, -7.1)$ is the midpoint between $F(x, y)$ and $B(2.8, -2.7)$. Find the coordinates of F.
26. $G(8.3, -1.4)$ is the midpoint between $B(x, y)$ and $L(-5.9, 8)$. Find the coordinates of B.

In Exercises 27–30, determine whether each set of points is collinear.

27. $M(-2, 3)$, $N(-3, 8)$, and $P(1, -6)$
28. $A(-5, 20)$, $B(6, -13)$, and $C(-2, 11)$
29. $J(3, -10)$, $K(-1, 2)$, and $L(-7, 20)$
30. $A(4, 8)$, $B(7, 13)$, and $C(16, 23)$

In Exercises 31–38, classify the figure formed by each set of points.

31. $A(6, -2)$, $B(4, 3)$, and $C(-1, 5)$
32. $X(-3, 1)$, $Y(5, 3)$, and $Z(6, -1)$
33. $E(3, 17)$, $F(-16, 5)$, and $G(4, 4)$
34. $K(-2, -1)$, $L(12, -1)$, and $M(-2, 13)$
35. $A(-1, 0)$, $B(2, -3\sqrt{3})$, and $C(5, 0)$
36. $P(-1, 3)$, $Q(13, -3)$, and $R(-1, 24)$
37. $U(-15, 12)$, $V(-13, 6)$, and $W(-12, 3)$
38. $A(-6, 4)$, $B(-2, 4)$, and $C(-4, 4 - 2\sqrt{3})$

In Exercises 39–42, sketch the graph of each equation by plotting points.

39. $y = |x - 3| - 2$
40. $y = 1 + 2x$
41. $y^2 = 4x + 1$
42. $y - 5 = -x^2$

In Exercises 43–46, use the intercepts to sketch the graph of each equation.

43. $-4 = -x + y$
44. $2x + y = 2$
45. $y - 9 = -x^2$
46. $y = 2x^2 + x - 15$

In Exercises 47–52, write the equation of each circle.

47. radius 7 and center at the origin
48. radius 11 and center at the origin
49. radius 6 and center at $(-4, -3)$
50. radius 15 and center at $(0, -9)$
51. radius 1/2 and center at $(7, 2)$
52. radius 5 and center $(-5, 3)$

In Exercises 53–54, graph each circle, and identify its center and radius.

53. $x^2 + (y + 4)^2 = 25$
54. $(x + 3)^2 + (y - 3)^2 = 144$

In Exercises 55–60, write each equation in the standard form of a circle. Then identify the circle's center and radius.

55. $x^2 - 2x + y^2 + 10y - 170 = 0$
56. $x^2 + y^2 - 8x + 4y - 205 = 0$
57. $x^2 + 12x + y^2 - 14y - 204 = 0$
58. $x^2 + y^2 - 6x - 2y - 26 = 0$
59. $x^2 + y^2 - 8x + 16y + 16 = 0$
60. $x^2 - 4x + y^2 + 4y - 28 = 0$

In Exercises 61–66, write the equation of each circle where the given coordinates are the endpoints of a diameter.

61. $A(3, 4)$ and $B(-5, 6)$
62. $Q(1, -4)$ and $R(-7, 8)$
63. $C(-2, -6)$ and $D(8, 10)$
64. $P(6, 4)$ and $Q(-4, 8)$
65. $H(-3, -2)$ and $F(7, 4)$
66. $S(9, 10)$ and $V(-1, 4)$

In Exercises 67–70, identify the type of symmetry, if any, for the graph of each equation.

67. $y = 3x^2 - 4$
68. $y = 2x^3$
69. $y = -x^6 + 4x^4 + 5$
70. $y = 2x^4 + 3x$

In Exercises 71–72, use symmetry and intercepts to sketch each graph.

71. $y^2 = x + 4$
72. $y = 2|x| + 4$

Extensions

73. Complete the square for both x and y, and write the equation in the standard form of a circle.

$x^2 + 6\sqrt{2}x + y^2 + 4\sqrt{2}y = 2$

74. Use the Midpoint Formula to find the points that divide the line segment with endpoints (10, 2) and (−2, 6) into 8 equal parts.

75. True or False? The points in the table form a regular hexagon.

x	5	−5	2.5	−2.5	2.5	−2.5
y	0	0	$2.5\sqrt{3}$	$2.5\sqrt{3}$	$-2.5\sqrt{3}$	$-2.5\sqrt{3}$

In Exercises 76–78, find m and n so that the graph of $y = mx^3 - nx^2 + mx$ has the type of symmetry indicated.

76. symmetry with respect to the x-axis

77. symmetry with respect to the y-axis

78. symmetry with respect to the origin

79. True or False? The Midpoint Formula must be used 12 times to divide a line segment into 12 equal parts.

80. Solve the Distance Formula for x_1.

81. Write the equation in the standard form of a circle.

$16x^2 + 16y^2 - 32x + 64y + 64 = 80$

1.2 LINES

OBJECTIVES

- Find the slope of a line.
- Graph lines using the slope and a point on the line.
- Graph lines given point-slope, slope-intercept, and standard form equations.
- Write the equations of lines in point-slope and slope-intercept form.
- Write the equations of parallel and perpendicular lines.
- Model with linear equations.

PREREQUISITE VOCABULARY TERMS

coordinate plane
linear equation
ordered pair
***x*-axis and *y*-axis**
***x*-intercept**
***y*-intercept**

1.2.1 Finding the Slope of a Line from Two Points

The **slope** of a line between two points (a measure of the line's steepness) is equal to the vertical change between the points divided by the corresponding horizontal change. Slope is a constant value for any line, because the steepness of a line does not change between any two points on the line. It follows that the slope of a line can be determined between any two points on the line.

In a coordinate plane, the slope of a line between two points (x_1, y_1) and (x_2, y_2) is the change in the y-coordinates divided by the change in the x-coordinates. The change in the y-coordinates between two points (the vertical change) is called the "rise" or "change in y" (Δy), and the change in the x-coordinates between two points is called the "run" or "change in x" (Δx).

DEFINITION

The **slope** m of a line that passes through the points (x_1, y_1) and (x_2, y_2), where $x_1 \neq x_2$, is the quotient of the vertical change (rise) and the horizontal change (run) between those two points.

$$m = \frac{\text{rise}}{\text{run}} = \frac{y}{x} = \frac{y_2 - y_1}{x_2 - x_1}$$

When $x_1 = x_2$, the line is vertical and the slope is undefined.

A line that increases (heads upward from left to right) has a positive slope (Figure 1.2a), and a line that decreases (heads downward from left to right) has a negative slope (Figure 1.2b).

Line with $m > 0$

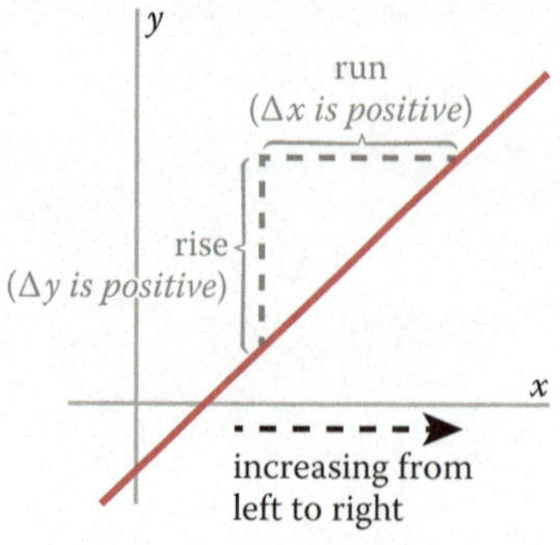

Figure 1.2a

Line with $m < 0$

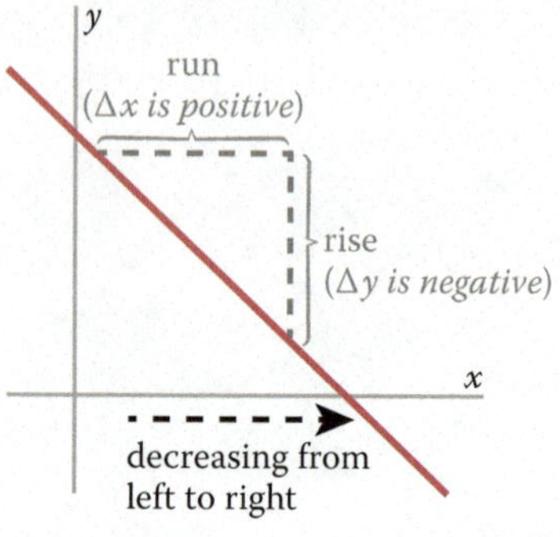

Figure 1.2b

Steps for Finding the Slope of a Line

1. Identify any two points on the line. Designate one point (either point) to be (x_1, y_1) and the other to be (x_2, y_2).
2. Substitute the coordinates from the two points into $m = \dfrac{y_2 - y_1}{x_2 - x_1}$ and simplify.

IMPORTANT

If A and B are two points on a line, then the slope from A to B is the same as the slope from B to A. So, when using the slope formula to find the slope of a line, it doesn't matter which of the two points is (x_1, y_1).

EXAMPLE 1 Finding the Slope of a Line

Find the slope of the line.

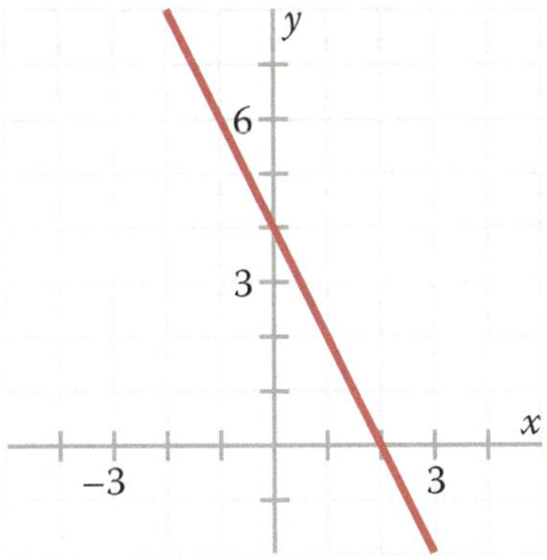

SOLUTION

1. Identify any two points on the line, such as (0, 4) and (2, 0) (the intercepts). Let (0, 4) be (x_1, y_1) and (2, 0) be (x_2, y_2).
2. Substitute the coordinates into the slope formula and simplify.

$$m = \frac{0-4}{2-0} = \frac{-4}{2} = -2$$

The line's slope is −2.

IMPORTANT

The slope is −2 regardless of which two points on the line are used to find the slope. For example, substitute (−1, 6) and (2, 0) into the slope formula.

$$m = \frac{0-6}{2-(-1)} = \frac{-6}{3} = -2$$

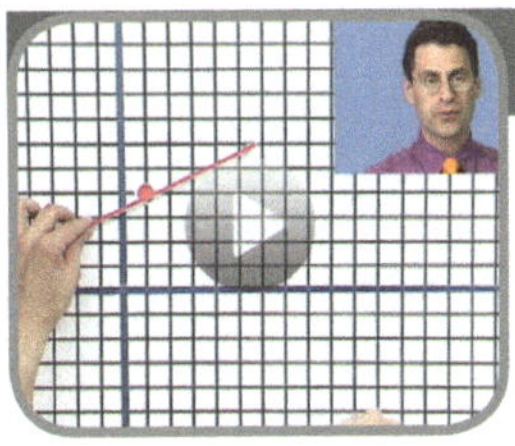

1.2.2 Graphing a Line Using a Point and the Slope

There are an infinite number of lines that have any one particular slope. For example, each of the lines in Figure 1.2c has a slope equal to 2. So, knowing the slope of a line is not sufficient information for sketching the graph of the line. However, there is only one line with a specific slope that passes through a specific point. So, when the slope of a line is given, along with a point that the line passes through, the graph of the line can be sketched.

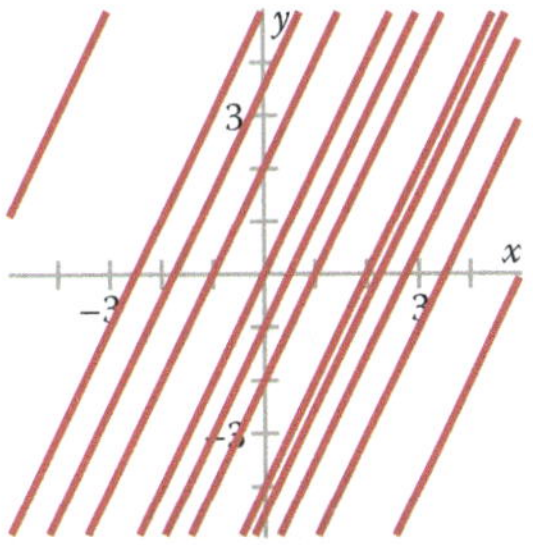

Figure 1.2c

Steps for Graphing a Line Using a Point and the Slope

1. Plot the point.
2. Using the slope, identify the rise (numerator) and run (denominator) values.
3. Starting at the point, move up (for positive slope) or down (for negative slope) the number of units indicated by the slope's rise value.
4. Move to the right the number of units indicated by the slope's run value (run is always to the right when using this method) and plot a second point.
5. Sketch the line through the two points.

EXAMPLE 2 Graphing a Line Using the Slope and a Point on the Line

Graph the line with slope –3 that passes through the point (2, –1).

SOLUTION

❶ Plot the point (2, –1).

❷ The line's slope is –3, which is equivalent to $\frac{-3}{1}$, so the rise value is –3 and the run value is 1.

❸ From (2, –1), move down (since the slope is negative) 3 units. (Do not plot the point yet.)

❹ From 3 units below (2, –1), move to the right 1 unit, which is the point (3, –4). Plot the second point at (3, –4).

❺ Sketch the line through the two points, (2, –1) and (3, –4).

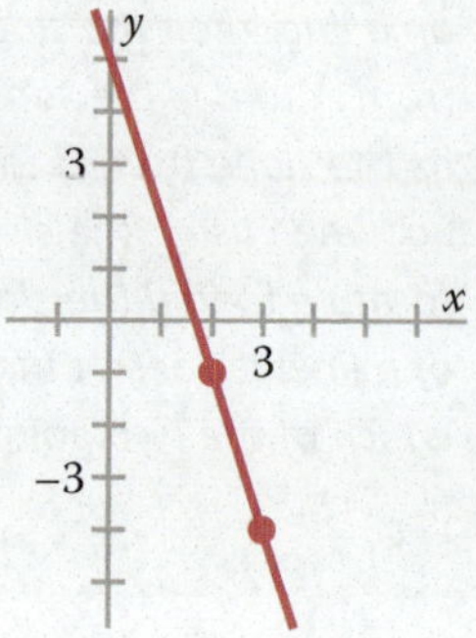

TIP

3 units down and 1 unit to the right of (2, –1) is (2 + 1, –1 – 3), or (3, –4).

WHAT'S THE BIG IDEA?

When graphing a line using its slope and a given point, why is the point plotted before the slope is used?

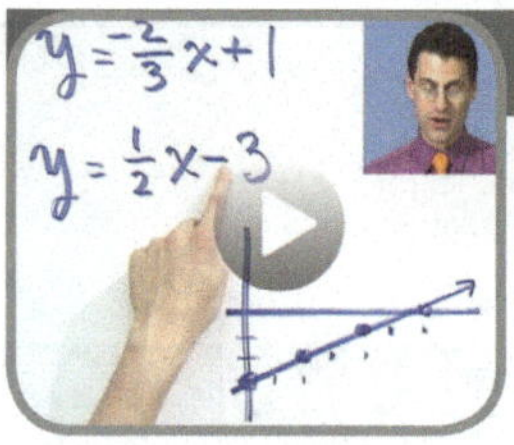

1.2.3 Introduction to Slope-Intercept Form

Previously, a linear equation in one variable was defined to be any equation that can be written in the form $ax + b = 0$, where a and b are real numbers. The definition of a linear equation will now be extended to include equations with two variables.

DEFINITION

A **linear equation** is an equation that can be written in the general form $Ax + By + C = 0$, where A, B, and C are real numbers, and A and B cannot both be 0.

The graph of every linear equation is a line, and the equation of every line is a linear equation.

TIP

In a linear equation, the term Ax is referred to as the "linear x-term," the term By as the "linear y-term," and the term C as the "constant term."

There are several forms of a linear equation from which information about the equation's graph (a line) can be deduced and then used to graph the line. These forms of a linear equation can be used to write the equation of a given line. One form of a linear equation (the equation of a line) is **slope-intercept form**. The "intercept" in slope-intercept form refers to the line's y-intercept.

Slope-Intercept Form of the Equation of a Line

The equation of the line with slope m and y-intercept b is $y = mx + b$.

When a linear equation is written in slope-intercept form, the slope and y-intercept can be deduced from the equation (and then used to graph the equation). Conversely, when the slope and y-intercept of a line are known, the corresponding linear equation can be written by substituting those values into slope-intercept form.

Steps for Writing the Equation of a Line in Slope-Intercept Form

1. Identify the line's slope m and y-intercept b.
2. Substitute the m- and b-values into $y = mx + b$, and simplify.

EXAMPLE 3 Writing the Equation of a Line

Write the slope-intercept form equation of the line with slope 2 that passes through (0, 0).

SOLUTION

1. A line that passes through the origin has y-intercept at (0, 0). So, $m = 2$ and $b = 0$.
2. Substitute these values into the slope-intercept form of the equation of a line, $y = mx + b$, and simplify.

$$y = mx + b \;\Rightarrow\; y = 2x + 0 \;\Rightarrow\; y = 2x$$

So, the equation of the line with slope 2 that passes through the origin is $y = 2x$ (Figure 1.2d).

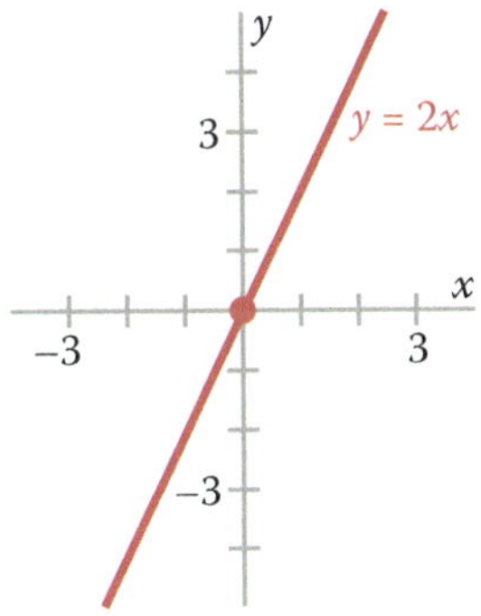

Figure 1.2d

Recall that the line through some point with some slope can be graphed by plotting the point and then using the slope to find a second point on the line. It follows that since the slope and y-intercept (a point) can be deduced from an equation in slope-intercept form, a slope-intercept equation can be graphed without making a table of values or using any of the other techniques previously used for graphing an equation.

Steps for Graphing an Equation in Slope-Intercept Form

1. Identify the slope m and the y-intercept b from the equation.
2. Plot the y-intercept (on the y-axis).
3. Use the rise and run values from the slope to find a second point on the line.
4. Sketch the line through the two points.

EXAMPLE 4 Graphing a Slope-Intercept Equation

Graph $4 - \frac{7}{3}x = y$.

SOLUTION

1. From the equation, $m = \frac{-7}{3}$ and $b = 4$.
2. Plot the y-intercept, (0, 4).
3. The line's slope is $\frac{-7}{3}$, so the rise value is –7 and the run value is 3. So, from (0, 4), move down (since the slope is negative) 7 units, and then to the right 3 units, and plot the second point at (3, –3).
4. Sketch the line through the two points, (0, 4) and (3, –3).

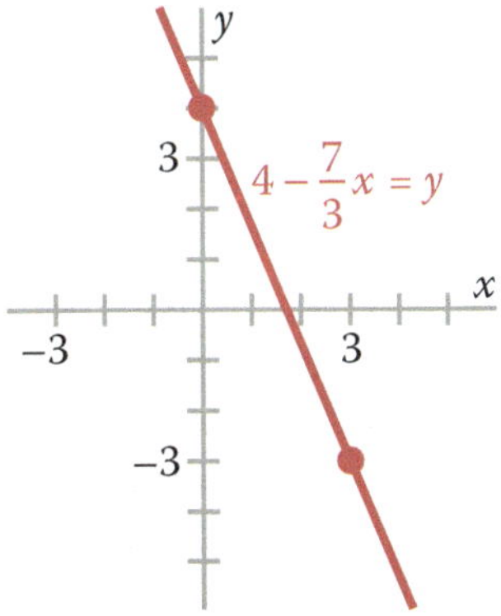

IMPORTANT

When using the slope to find a second point on a line, remember that this method defines the rise to be up or down (depending on the sign of the slope), but the run is always to the right.

1.2.4 Writing the Equation of a Line

It is easy to write the equation of a line when the slope and y-intercept are known: just substitute the m- and b-values into the slope-intercept form of a linear equation, $y = mx + b$, and simplify. The equation of any line can also be written when the slope and y-intercept are *not* known, as long as two points on the line are known.

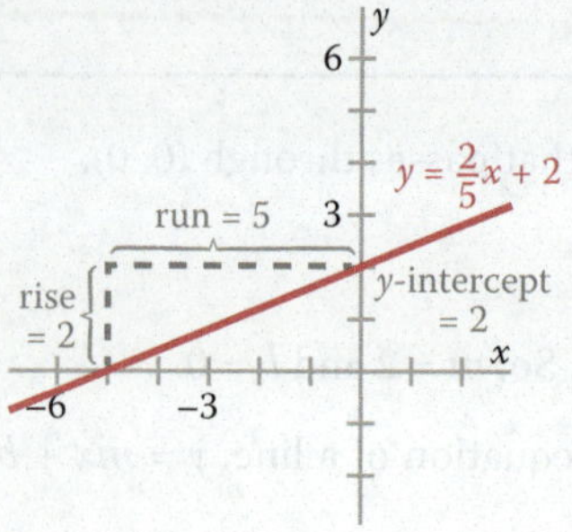

Figure 1.2e

One method for writing the equation of a line on a coordinate plane is to identify the slope and y-intercept by looking at the graph, and then substitute those m- and b-values into $y = mx + b$. For example, the slope of the line shown in Figure 1.2e is observed to be $\frac{2}{5}$ (by counting the rise and the run), and the y-intercept is at 2, so the equation of the line is $y = \frac{2}{5}x + 2$.

This method works well as long as the y-intercept is an integer. However, if the y-intercept is not an integer, an algebraic method for finding the equation of the line should be used to ensure that the exact slope and y-intercept are found.

The algebraic method for writing the equation of a line involves using the slope formula to find the slope of the line and then algebraically finding the b-value by solving an equation. This method can be used to find the equation of any line when at least two points on the line are known.

Steps for Writing the Equation of a Line in Slope-Intercept Form Algebraically

❶ Use the slope formula to find the slope m.

❷ To find b, substitute the x- and y-coordinates from *any* point on the line and the m-value into $y = mx + b$, and solve for b.

❸ Substitute the m- and b-values into the slope-intercept form equation, $y = mx + b$, and simplify.

EXAMPLE 5 Writing the Equation of a Line Algebraically

Write the slope-intercept form equation of the line that passes through (2, 10) and (4, −4).

SOLUTION

TIP

It doesn't matter which point is used as (x_1, y_1) or (x_2, y_2).

❶ Substitute the coordinates from each point into the slope formula and simplify.

$$m = \frac{-4-10}{4-2} = \frac{-14}{2} = -7 \quad \text{The line's slope is } -7.$$

❷ Substitute the m-value and the x- and y-coordinates from *either* one of the points into $y = mx + b$, and solve for b. Here, (2, 10) is used, so 2 is substituted for x and 10 is substituted for y.

$$y = mx + b$$

$$10 = (-7)(2) + b \qquad \textit{Substitute.}$$

$$10 = -14 + b \qquad \textit{Multiply.}$$

$$24 = b \qquad \textit{Add 14 to each side to solve for } b.$$

The line's y-intercept is at (0, 24).

❸ Substitute $m = -7$ and $b = 24$ into $y = mx + b$.

$$y = mx + b \;\Rightarrow\; y = -7x + 24$$

So, the equation of the line that passes through (2, 10) and (4, −4) is $y = -7x + 24$ (Figure 1.2f).

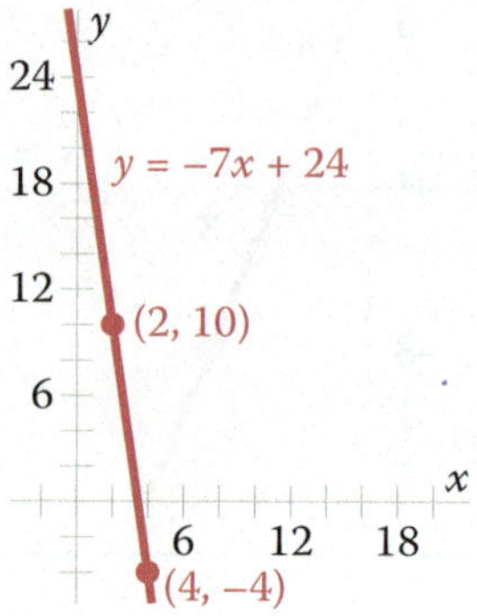

Figure 1.2f

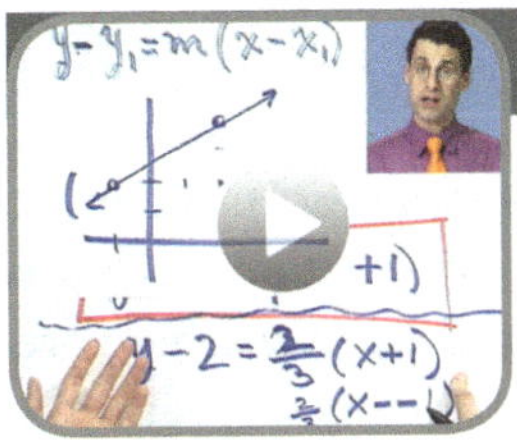

1.2.5 Introduction to Point-Slope Form

Another form of a linear equation is **point-slope form**.

Point-Slope Form of the Equation of a Line

The equation of the line with slope m that passes through (x_1, y_1) is $y - y_1 = m(x - x_1)$.

When a linear equation is written in point-slope form, the slope and a point on the line can be deduced from the equation, which can then be used to graph the equation. Conversely, when the slope and the coordinates of a point on the line are known, then the corresponding linear equation can be written by substituting those values into point-slope form.

Steps for Writing the Equation of a Line in Point-Slope Form

1. Find the slope m and a point on the line (x_1, y_1).
2. Substitute the values of m, x_1, and y_1 into the point-slope form equation, $y - y_1 = m(x - x1)$.

To convert an equation that is in point-slope form to slope-intercept form, distribute m to $(x - x_1)$ and then add y_1 to each side to solve for y. For example, to write the point-slope form equation $y - 2 = 3(x - 5)$ in slope-intercept form, distribute 3 to $(x - 5)$ and then add 2 to each side to solve for y.

$$\underbrace{y - 2 = 3(x - 5)}_{\text{point-slope form}} \Rightarrow y - 2 = 3x - 15 \Rightarrow \underbrace{y = 3x - 13}_{\text{slope-intercept form}}$$

EXAMPLE 6 Writing the Equation of a Line Using the Slope and a Point

Write the equation of the line that passes through (–2, 3) with slope 5 in point-slope form and in slope-intercept form.

SOLUTION

Substitute $m = 5$, $x_1 = -2$, and $y_1 = 3$ into the point-slope form of the equation of a line.

$$y - y_1 = m(x - x_1) \Rightarrow y - 3 = 5(x - (-2)) \Rightarrow y - 3 = 5(x + 2)$$

Simplify and solve the equation for y to write the equation in slope-intercept form, $y = mx + b$.

$$\begin{aligned} y - 3 &= 5x + 10 && \textit{Distribute.} \\ y &= 5x + 13 && \textit{Add 3 to both sides.} \end{aligned}$$

So, the point-slope equation is $y - 3 = 5(x + 2)$, and the slope-intercept equation is $y = 5x + 13$.

WHAT'S THE BIG IDEA?

Explain the relationship between the point-slope form of a linear equation and the slope formula.

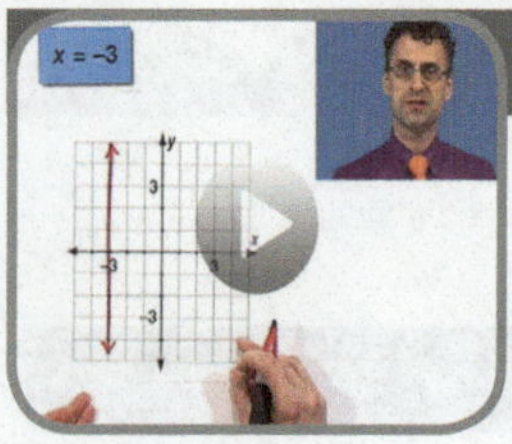

1.2.6 Vertical and Horizontal Lines

All points on any horizontal line on a coordinate plane have equal y-coordinates. For example, consider the horizontal line shown in Figure 1.2g. The y-coordinate of every point on this line is 2. The x-coordinate of any point on the line in Figure 1.2g could be any real number. Therefore, either of the following tables is a corresponding table of ordered pairs for the line given in Figure 1.2g.

x	−2	−1	0	1
y	2	2	2	2

x	1	2	3	4
y	2	2	2	2

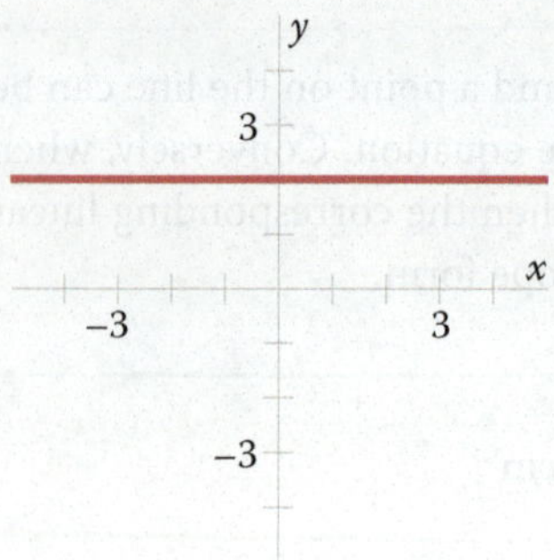

Figure 1.2g

The slope of a horizontal line is 0 because the change in y between any two points on a horizontal line is 0. Substituting $m = 0$ into slope-intercept form gives the general equation of a horizontal line.

$$y = mx + b \Rightarrow y = (0)x + b \Rightarrow y = b$$

So, the equation of any horizontal line is $y = b$, where b is the line's y-intercept. Therefore, the equation of the line in Figure 1.2g is $y = 2$.

Similarly, all points on a vertical line have equal x-coordinates. The horizontal change between points with equal x-coordinates is 0. So, the slope of any line through points with equal x-coordinates must be undefined (since the change in x, the denominator of slope, is 0). Therefore, the slope of any vertical line is undefined. The equation of any vertical line is $x = c$, where c is the line's x-intercept.

Horizontal and Vertical Lines

	Horizontal Line	Vertical Line
Slope	0	undefined
Equation	$y = b$ *b is the y-intercept.*	$x = c$ *c is the x-intercept.*

TIP

On a coordinate plane, horizontal lines do not intersect the x-axis, so they have no x-intercept; vertical lines do not intersect the y-axis, so they have no y-intercept. The only exceptions are the axes. The equation of the line formed by the y-axis is $x = 0$, and the equation of the line formed by the x-axis is $y = 0$.

EXAMPLE 7 Graphing Horizontal and Vertical Lines

Graph $y = -1$ and $x = 4$ on a coordinate plane.

SOLUTION

The graph of $y = -1$ is a horizontal line with y-intercept at $(0, -1)$.

The graph of $x = 4$ is a vertical line with x-intercept at $(4, 0)$.

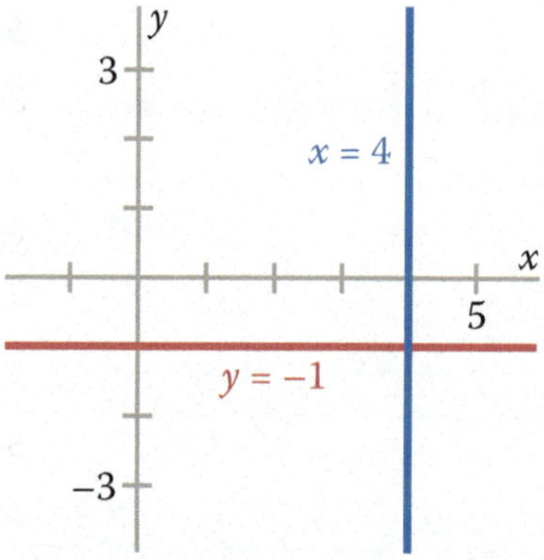

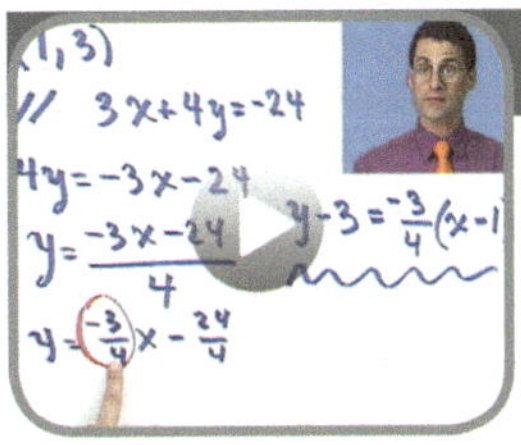

1.2.7 Graphing a Standard Form Linear Equation

Recall the two forms of the equation of a line discussed so far in this section.

Slope-Intercept Form: $y = mx + b$ *The line's slope and y-intercept are m and b.*

Point-Slope Form: $y - y_1 = m(x - x_1)$ *The line has slope m and passes through* (x_1, y_1).

A third form of a linear equation (the equation of a line) is the **standard form**.

Standard Form of the Equation of a Line

The equation of the line can be written in the form $Ax + By = C$, where A, B, and C are real numbers, and A and B are not both 0.

There are many ways to graph a linear equation in standard form, including the two methods described here.

Graphing a Linear Equation in Standard Form

Method 1: Use the Intercepts
Find and plot the x- and y-intercepts. Then draw a line through those two points.

Method 2: Use Slope-Intercept Form
Write the equation in slope-intercept form (by solving for y), and then use the slope and y-intercept from the equation to draw the line.

EXAMPLE 8 Graphing a Standard Form Linear Equation

Graph $3x - 4y = -8$ on a coordinate plane.

SOLUTION

Solve the equation for y to write the equation in slope-intercept form.

$$3x - 4y = -8$$
$$-4y = -8 - 3x$$
$$y = \frac{-8 - 3x}{-4}$$
$$y = 2 + \frac{3}{4}x$$

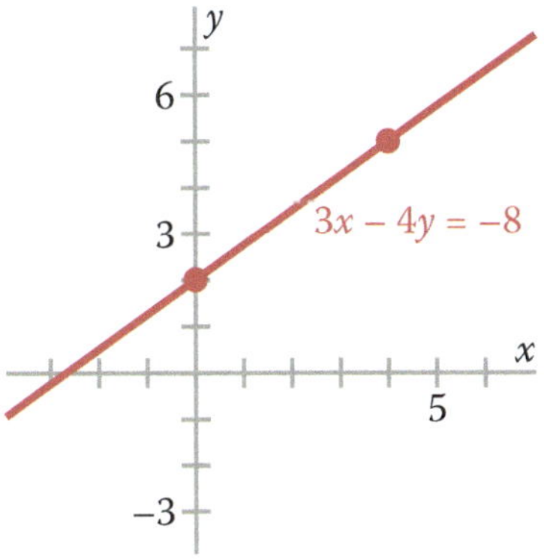

From the equation, $m = \frac{3}{4}$ and $b = 2$.

Plot the y-intercept, then use the slope to find a second point on the line.

Sketch the line through the two points.

ALTERNATIVE METHOD

Find the x- and y-intercepts. Substitute 0 for y and solve for x to find the x-intercept.

$$3x - 4(0) = -8$$
$$3x = -8$$
$$x = -\frac{8}{3}$$

Substitute 0 for x and solve for y to find the y-intercept.

$$3(0) - 4y = -8$$
$$-4y = -8$$
$$y = 2$$

Plot the x-intercept at $-\frac{8}{3}$ *(i.e.,* $-2\frac{2}{3}$*) and the y-intercept at 2, then sketch the line through the two intercepts.*

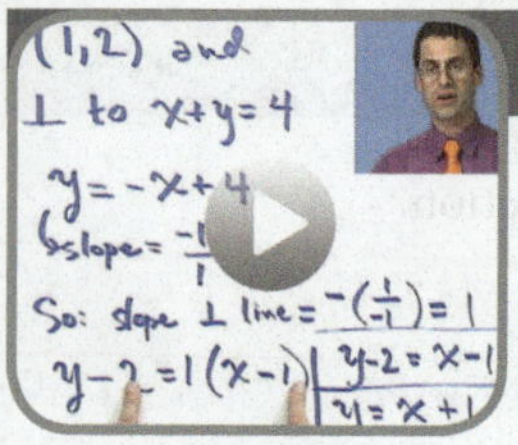

1.2.8 Equations of Parallel and Perpendicular Lines

Parallel lines never intersect, and **perpendicular lines** intersect at a 90° angle. There are many facts regarding the slopes of parallel and perpendicular lines that can be applied when writing the equations of or graphing lines that are parallel or perpendicular to each other.

IMPORTANT

Lines with the same slope and the same y-intercept are not parallel, because they form the same line.

Parallel and Perpendicular Lines

- Parallel lines have equal slopes.
- Perpendicular lines have slopes that are opposite reciprocals of one another.
- The product of the slopes of two perpendicular lines is equal to –1.
- All vertical lines are parallel.
- All horizontal lines are parallel.
- A vertical line is perpendicular to any horizontal line.
- A horizontal line is perpendicular to any vertical line.

EXAMPLE 9 Writing the Equations of Parallel Lines

IMPORTANT

The notation $l \parallel p$ means that line l is parallel to line p.

Line l passes through (4, 5), and line p is the graph of $y - 3 = 2(x + 7)$. If $l \parallel p$, what is the slope-intercept form equation of l?

SOLUTION

Since line l is parallel to line p, the two lines must have the same slope. So, the slope of line l is 2, because the slope of line p is 2 (from the equation of line p).

Substitute the slope of line l, 2, and the coordinates from the point on line l, (4, 5), into the point-slope form equation to write the equation of line l.

$$y - y_1 = m(x - x_1) \Rightarrow y - 5 = 2(x - 4)$$

Simplify and solve for y to write the equation of line l in slope-intercept form.

$$y - 5 = 2(x - 4) \Rightarrow y - 5 = 2x - 8 \Rightarrow y = 2x - 3$$

So, the slope-intercept form equation of line l is $y = 2x - 3$.

CHECK

The answer can be checked graphically or algebraically.

Graphically

Graph the equations for l and p. Visually confirm that the lines are parallel and that l passes through (4, 5) (Figure 1.2h).

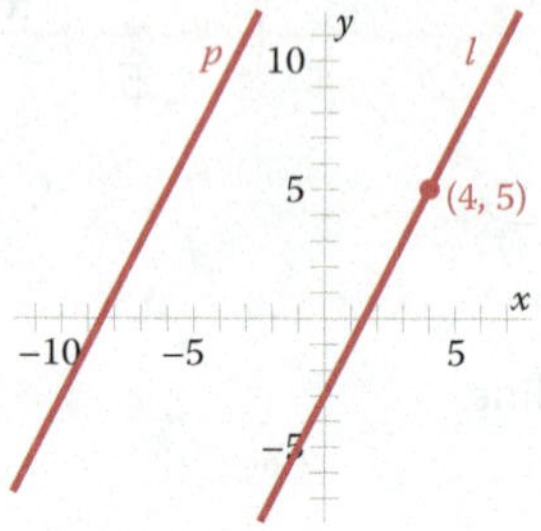

Figure 1.2h

Algebraically

Verify that l and p are parallel using the equations, and then verify that (4, 5) is on l algebraically.

From the equations for l and p, $y = 2x - 3$ and $y - 3 = 2(x + 7)$, the slopes of l and p are equal (both 2).

Lines l and p are parallel. ✓

Substituting $x = 4$ and $y = 5$ into the equation of l results in $5 = 2(4) - 3$, which is a true statement. So, (4, 5) is a solution to $y = 2x - 3$.

Line l contains (4, 5). ✓

EXAMPLE 10 Writing the Equations of Perpendicular Lines

IMPORTANT

The notation $l \perp p$ means that line l is perpendicular to line p.

Line l passes through (–5, 2), and line p is the graph of $2x - y = 1$. If $l \perp p$, what is the slope-intercept form equation of l?

SOLUTION

Since line l is perpendicular to line p, the two lines must have slopes that are opposite reciprocals of one another. Find the slope of line p by writing the equation of p in slope-intercept form.

$$2x - y = 1 \;\Rightarrow\; -y = -2x + 1 \;\Rightarrow\; y = 2x - 1 \qquad \text{So, the slope of line } p \text{ is } 2.$$

The opposite reciprocal of 2 is $-\frac{1}{2}$, so the slope of l is $-\frac{1}{2}$.

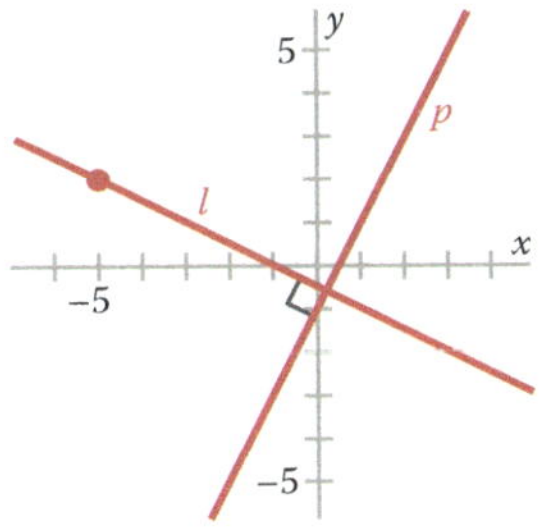

Figure 1.2i

To write the equation of line l, substitute the slope of l, $-\frac{1}{2}$, and the coordinate from the point on line l, (–5, 2), into the point-slope form equation.

$$y - y_1 = m(x - x_1) \;\Rightarrow\; y - 2 = -\frac{1}{2}(x - (-5)) \;\Rightarrow\; y - 2 = -\frac{1}{2}(x + 5)$$

Simplify and solve for y to write the equation of line l in slope-intercept form.

$$y - 2 = -\frac{1}{2}(x + 5) \;\Rightarrow\; y - 2 = -\frac{1}{2}x - \frac{5}{2} \;\Rightarrow\; y = -\frac{1}{2}x - \frac{1}{2}$$

So, the slope-intercept form equation of line l is $y = -\frac{1}{2}x - \frac{1}{2}$.

As in **Example 9**, the answer can be checked graphically (Figure 1.2i) or algebraically.

1.2.9 Modeling Rate of Change with Linear Equations

A linear equation (and its corresponding line) can model a relationship between two real quantities, such as the relationship between the number of hours traveled and the total distance traveled, or the relationship between the number of days worked and the total pay earned. The two quantities are represented in the equation by the two variables. The slope of the equation can be interpreted as the **rate of change** of one quantity in relation to the other—specifically, the rate of change in the y-variable in relation to the change in the x-variable (since slope is $\frac{\Delta y}{\Delta x}$).

TIP

The graph of a linear equation will involve only quadrant I of the coordinate plane (commonly referred to as an L-graph) when the quantities represented by the two variables are only positive numbers.

For example, the line shown in Figure 1.2j models the relationship between the total pay (in dollars) and time worked (in hours) for a person paid by the hour. The equation of the line is $y = 12x$, so the line's slope is 12. Considered as a rate of change, a slope of 12 in this situation means that the rate of pay is 12 dollars per hour (\$12/h).

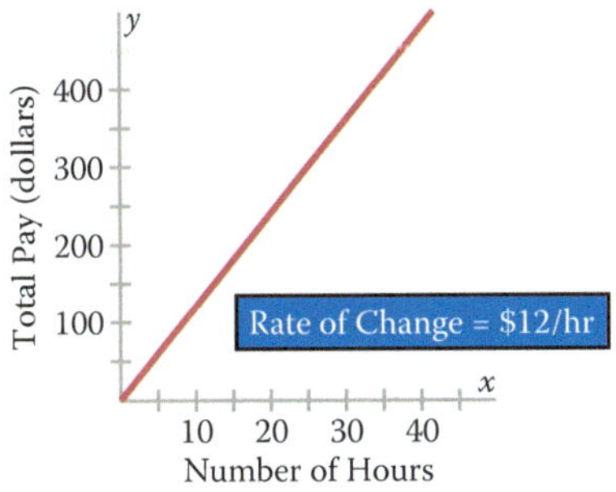

Figure 1.2j

The Units Used with Slope

IMPORTANT

When modeling a linear equation to represent the relationship between two quantities, the x-variable often represents time, such as hours, months, or years.

When slope is interpreted as a rate of change, the units are the y-variable's units divided by the x-variable's units (because slope is the ratio of the change in y to the change in x). For example, in the linear equation $y = 60x + 500$, if x represents the number of hours (time) traveled and y represents the number of miles (distance) traveled, then the slope, 60, can be interpreted as the rate of change "60 miles per hour."

If the units of x and y are the same, then the units for slope cancel out, and the slope is just a number with no units.

EXAMPLE 11 Modeling Change in Value with a Linear Equation

The estimated value of a used car x years after purchase can be represented by $y = 15{,}000 - 750x$. Explain the significance of the line's slope, y-intercept, and x-intercept within the context of the situation.

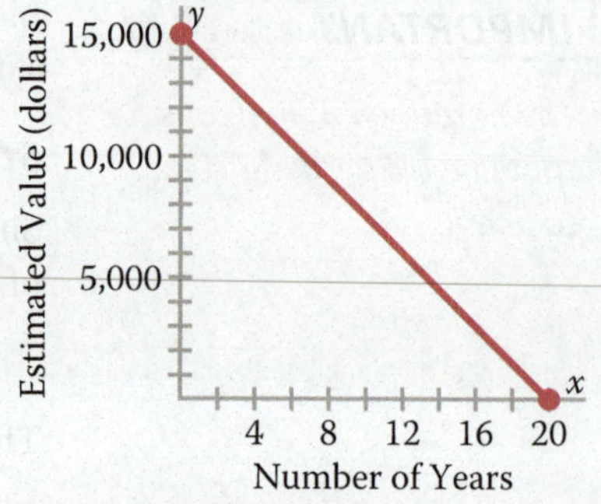

SOLUTION

From the equation, the slope is −750 and the y-intercept is 15,000. The y-intercept is the value of the car at year 0. So, the y-intercept indicates that the initial value of the car (the value at time of purchase) was $15,000.

TIP

The slope is negative, so the value of the car is decreasing. In finance, a decrease in value is called "depreciation."

The slope is the rate of change in the value of the car (y) per year (x), which means that the value of the car is depreciating at a rate of $750 per year. Substitute 0 into the equation for y and solve for x to find the x-intercept.

$$0 = 15{,}000 - 750x \Rightarrow 750x = 15{,}000 \Rightarrow x = 20$$

So, the x-intercept is 20 years, which means that after 20 years, the value of the car will be $0.

EXAMPLE 12 Modeling Change in Temperature with a Linear Equation

After a constant heat source was applied to it for 3 minutes, the temperature of a substance was found to be 170° F. After 7 minutes, the temperature was 210° F. Assuming a linear equation can be used to model the change in heat over time between $x = 3$ minutes and $x = 7$ minutes, write this linear model in slope-intercept form. Explain the significance of the slope and y-intercept in the context of this situation. Then predict the temperature of the substance after 15 minutes and the number of minutes it will take for the substance to reach 400° F using this linear model.

SOLUTION

IMPORTANT

When an equation is modeled involving time and some other quantity, time is usually the x-variable (independent variable), and the other quantity is the y-variable (dependent variable).

This situation involves two variables: the time (in minutes) that the substance has been heated, and the temperature of the substance (in ° F). Let x be the number of minutes that the substance is heated, and let y be the temperature of the substance after x minutes.

The two given pairs of corresponding x- and y-values can be written as ordered pairs: (3, 170) and (7, 210). Write the equation of the line through these two points.

Find the slope.

$$m = \frac{210 - 170}{7 - 3} = \frac{40}{4} = 10$$

Write the equation of the line.

$$y - y_1 = m(x - x_1)$$ *Substitute $m = 10$, $x_1 = 3$, and $y_1 = 170$*

$$y - 170 = 10(x - 3)$$ *into the point-slope form equation.*

So, the equation of the line is $y - 170 = 10(x - 3)$, or $y = 10x + 140$ in slope-intercept form.

Slope is the rate of change, so the temperature is increasing at a rate of 10° F per minute. Since the y-intercept (140) is the temperature that corresponds to time 0, the starting temperature of the substance was 140° F.

TIP

Substitute 15 for x and simplify to predict the temperature after 15 minutes.

Use the slope-intercept form equation, $y = 10x + 140$, to make the predictions.

According to the model, the substance will be $10(15) + 140 = 290$° F after 15 minutes.

Substitute 400 for y and solve for x to predict the time it takes for the substance to reach 400° F.

$$400 = 10x + 140 \Rightarrow 260 = 10x \Rightarrow 26 = x$$

According to the model, 26 minutes after the heat source is applied, the substance will be 400° F.

SECTION 1.2 EXERCISES

Warm Up

1. Simplify. $\dfrac{-4-6}{7-(-1)}$
2. Solve for y. $y - 20 + 8x = -5x + 6$
3. What is the opposite reciprocal of 2/3?

Just the Facts

Fill in the blanks.

4. The _____ of a line is a measure of the line's steepness. It is equal to the _____ change divided by the _____ change.
5. To find the slope of the line that passes through (2, 6) and (0, 5), use the formula _____. It doesn't matter which point is chosen for _____ or _____.
6. $y = mx + b$ is the _____ form of the equation of a line. $y - y_1 = m(x - x_1)$ is the _____ form of the equation of a line, where m is the _____, b is the _____, and (x_1, y_1) is a(n) _____ on the line.
7. Vertical lines have a slope that is _____, because the horizontal change between any two y-coordinates is _____; _____ lines have slope equal to 0, since the _____ change between any two x-coordinates is 0.
8. _____ is the standard form equation of a line.
9. Parallel lines have the _____ slope, while _____ lines have slopes that are _____. The symbol for _____ is ||, and the symbol for perpendicular is _____.
10. If the equation of a line is $y - p = q(x - r)$, then the slope of the line is _____, and the line passes through the point _____.

Essential Skills

In Exercises 1–8, find the slope of each line.

1. passes through (3, 5) and (–3, –11)
2.

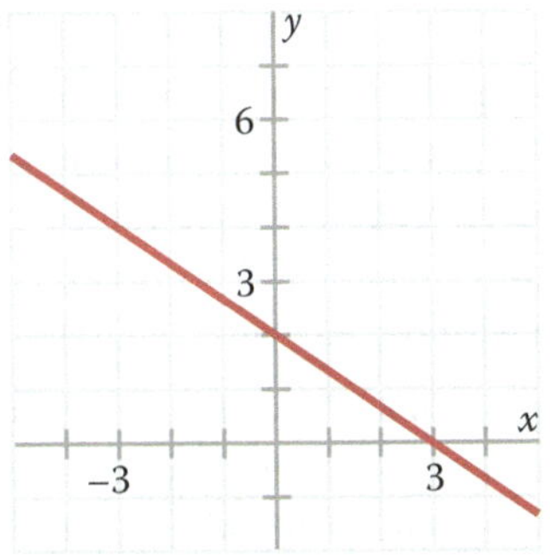

3. passes through (–5, 8) and (4, 2)
4. passes through (–2, 7) and (–2, –6)
5. passes through (–13, 9) and (–4, 3)
6. passes through (1, –2) and (–3, 6)
7. passes through (–5, –6) and (0, –6)
8. passes through (–7, 3) and (8, 15)

In Exercises 9–10, graph each line using the given slope and point.

9. slope 2/3, passes through (0, –2)
10. slope –1, passes through (–1, 2)

In Exercises 11–16, write the slope-intercept form equation of each line.

11. slope 2, passes through (0, 1)
12. slope –1, passes through (0, –5)
13. slope 3/5, passes through (0, 4)
14. slope –7/6, passes through (0, –2)
15. slope 9/4, passes through (0, 0)
16. slope –8/3, passes through (0, 7)

In Exercises 17–18, graph each equation.

17. $-x + 1 = y$
18. $y = 3x - 4$

In Exercises 19–28, write the slope-intercept form equation of each line.

19. passes through (5, –6) and (9, –2)
20. passes through (1, –8) and (–5, –20)
21. passes through (4, 7) and (0, 5)
22. passes through (7, –2) and (0, 2)
23. passes through (–2, –15) and (–1, 7)
24. passes through (–5, 8) and (–2, –4)
25. passes through (3, –5) and (–1, 6)
26. passes through (–1, 4) and (2, 0)
27. passes through (2, 5) and (8, 3)
28. passes through (–6, 8) and (9, –3)

In Exercises 29–36, write the point-slope form and the slope-intercept form equation of each line.

29. slope 4, passes through (5, –3)
30. slope –2/5, passes through (–5, 7)
31. slope 1/2, passes through (–6, 5)
32. slope –5, passes through (–1/2, 1)
33. slope 6, passes through (8, 3)
34. slope 7/8, passes through (–4, 1)

35. slope –4/5, passes through (10, 7)

36. slope –3/2, passes through (–8, –6)

In Exercises 37–42, write the equation of each line.

37. passes through (4, –5) and (10, –5)

38. passes through (–1, 6) and (–1, –4)

39. passes through (–3, 9) and (–8, 9)

40. passes through (6, –5) and (6, 3)

41. passes through (2, –8) and (2, 7)

42. passes through (–5, –4) and (–6, –4)

In Exercises 43–48, find the slope and *y*-intercept of the graph of each equation.

43. $8 = 3x - 2y$

44. $5x - 2y = 14$

45. $3x + 4y = 3$

46. $-11 = 8x + 2y$

47. $5 = 2x - 3y$

48. $-2x - 7y = -6$

In Exercises 49–56, if $l \parallel p$, what is the slope-intercept form equation of *l*?

49. Line *l* passes through (9, 4), and line *p* is the graph of $y + 5 = -(x + 2)$.

50. Line *l* passes through (1, –5), and line *p* is the graph of $y - 2 = -3(x + 6)$.

51. Line *l* passes through (4, 6), and line *p* is the graph of $y - 13 = -5(x - 1)$.

52. Line *l* passes through (–3, –1), and line *p* is the graph of $y + 4 = -2(x - 9)$.

53. Line *l* passes through (7, –2), and line *p* is the graph of $y - 1 = \frac{3}{4}(x + 5)$.

54. Line *l* passes through (2, 3), and line *p* is the graph of $y + 8 = \frac{4}{9}(x - 2)$.

55. Line *l* passes through (–4, 1), and line *p* is the graph of $y + 3 = -\frac{2}{3}(x - 7)$.

56. Line *l* passes through (3, –5), and line *p* is the graph of $y - 4 = -\frac{7}{2}(x + 4)$.

In Exercises 57–64, if $l \perp p$, what is the slope-intercept form equation of *l*?

57. Line *l* passes through (–5, 2), and line *p* is the graph of $y = 2x - 1$.

58. Line *l* passes through (–4, –3), and line *p* is the graph of $5x - 6y = 24$.

59. Line *l* passes through (6, –1), and line *p* is the graph of $y = -3x + 5$.

60. Line *l* passes through (–3, 8), and line *p* is the graph of $y = 4x - 7$.

61. Line *l* passes through (2, 7), and line *p* is the graph of $y = -\frac{1}{5}x - 9$.

62. Line *l* passes through (–4, –2), and line *p* is the graph of $y = \frac{8}{3}x + 2$.

63. Line *l* passes through (6, 1), and line *p* is the graph of $5x - 6y = 24$.

64. Line *l* passes through (9, –5), and line *p* is the graph of $3x - 5y = 6$.

In Exercises 65–67, use the following information.

A woman drove 800 miles to Dallas. She traveled 292 miles in 4 hours, and after 9 hours, she had traveled 657 miles.

65. Assuming a linear equation can be used to model the change in distance traveled over time between $x = 4$ hours and $x = 9$ hours, write this linear model in slope-intercept form.

66. Explain the significance of the slope in the context of this situation.

67. Determine how long it took to complete the trip to Dallas. Round to the nearest minute, if necessary.

In Exercises 68–70, use the following information.

After 2 minutes, an ant had traveled 100 meters, and after 6 minutes, the ant had traveled 220 meters.

68. Assuming a linear equation can be used to model the change in distance traveled over time between $x = 2$ minutes and $x = 6$ minutes, write this linear model in slope-intercept form.

69. Explain the significance of the slope in the context of this situation.

70. Using the linear model, predict the distance that the ant will travel after 10 minutes and the number of minutes it will take for the ant to travel 400 meters.

Extensions

71. A local restaurant installed a wheelchair ramp that rose 20 inches over a horizontal length of 22 feet. The Americans with Disabilities Act (ADA) requires that ramps have at least a 1:12 slope for wheelchairs and scooters. Is the restaurant in compliance with the ADA?

72. The HSBC Arena in Buffalo paid $225,000 for a new model 700 Zamboni ice resurfacer. It has a useful life of 10 years, after which time the salvage value is half its purchase price. Write a linear equation that describes the value of the Zamboni each year. What is the value after 3 years?

73. A computer salesman earns a monthly salary of \$2250 plus 5% commission when his total monthly sales are under \$10,000, 10% commission when his total monthly sales are between \$10,000 and \$14,999, and 12% commission when his total monthly sales are between \$15,000 and \$17,999. Write three separate linear equations representing the salesman's monthly wage W in terms of the monthly sales S. What is his monthly salary if he sold 16,000 dollars' worth of computer equipment in the month of July?

74. Show that the slope between the points $(x + 2, (x + 2)^2)$ and $(2, 4)$ is $x + 4$.

In Exercises 75–77, use the following information.

In 2007, Apple opened its 200th store, and in 2011, they had 357 stores in ten countries.

75. Assuming a linear equation can be used to model the change in stores over time between 2007 and 2011, write this linear model in slope-intercept form. Use $t = 7$ to represent the year 2007.

76. Use the linear model to predict the number of stores that Apple had in 1998. Is the answer reasonable? Explain.

77. Use the linear model to predict the number of stores that Apple had in 2009. Is the answer reasonable? Explain.

In Exercises 78–81, use the following information.

A landscaping contractor purchased a dump truck for \$21,900. The contractor pays a dump truck driver \$10.75 per operating hour and the truck costs the contractor an average of \$4.20 per operating hour for fuel and maintenance.

78. Write a linear equation giving the average total cost C of operating the dump truck for t hours, including the purchase price.

79. The contractor charges customers \$23.75 per hour for the services of the truck and driver. Write an equation for the revenue R for t hours of use.

80. Use $P = R - C$, the formula for profit, to write an equation for the profit earned from t hours of use.

81. After how many hours will the contractor break even?

1.3 VARIATION

OBJECTIVES	**PREREQUISITE VOCABULARY TERMS**
• Model equations for direct variation.	linear equation
• Model equations for inverse variation.	proportion
• Model equations for joint variation.	rational equation
• Model equations for combined variation.	slope

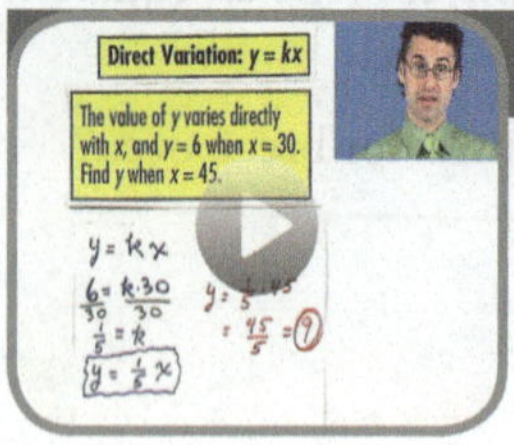

1.3.1 Direct Variation

A situation involving two variables x and y that can be modeled by an equation of the form $y = kx$, where k is some nonzero constant, is called a **direct variation**.

DEFINITION

If two quantities x and y are related by the equation $y = kx$, where k, the **constant of variation,** is some constant such that $k \neq 0$, then the relationship between x and y is a **direct variation** where y varies directly as x.

A direct variation is also called a **direct proportion**. If y varies directly as x, then it can also be said that y is directly proportional to x or that y is proportional to x. The k-value in a direct variation equation is also called the **constant of proportionality**.

Recall that the equation of any line can be written as $y = mx + b$ (slope-intercept form), where m and b are the line's slope and y-intercept, respectively. The direct variation equation, $y = kx$, is a linear equation in slope-intercept form where $m = k$ and $b = 0$. Any line with a y-intercept equal to 0 passes through the origin. Therefore, a direct variation equation, $y = kx$, describes a line with slope k that passes through the origin.

IMPORTANT

The x- and y-intercepts of $y = kx$ are both 0.

Steps for Writing and Using an Equation When y Varies Directly as x

❶ Substitute the given corresponding x- and y-values into $y = kx$ and solve for k.

❷ Write the direct variation equation by substituting the k-value into $y = kx$. (Be sure to leave x and y as variables.)

❸ Find an unknown value by substituting the corresponding known value into the direct variation equation (from step ❷) and solving for the unknown.

EXAMPLE 1 Writing a Direct Variation Equation

Write the variation equation for x and y, if y varies directly as x, and $y = 8$ when $x = 12$. Then find y when $x = 45$ and find x when $y = 50$.

SOLUTION

Since y varies directly as x, the relationship between x and y is a direct variation, and the equation $y = kx$ models the situation.

❶ Substitute the given corresponding x- and y-values, $x = 12$ and $y = 8$, into $y = kx$ and solve for k.

$y = kx \Rightarrow 8 = k(12) \Rightarrow k = \frac{2}{3}$ *The constant of variation is 2/3.*

❷ Substitute $k = \frac{2}{3}$ into $y = kx$. The direct variation equation is $y = \frac{2}{3}x$.

❸ Use $y = \frac{2}{3}x$ to find y when $x = 45$ and x when $y = 50$.

Find y when $x = 45$.

$y = \frac{2}{3}(45) = 30$

Find x when $y = 50$.

$50 = \frac{2}{3}x \Rightarrow x = \frac{3}{2}(50) = 75$

Direct Variation as a Proportion

Solving $y = kx$ for k gives $k = \frac{y}{x}$. It follows that y/x will be equal for all corresponding x- and y-values in a direct variation. Therefore, two pairs of x- and y-values from a direct variation can be written as a proportion: $\frac{y_1}{x_1} = \frac{y_2}{x_2}$.

TIP

From this proportion, it follows that $x_2y_1 = x_1y_2$ for any two ordered pairs in a direct variation.

The relationship given in **Example 1** ("y varies directly as x, and $y = 8$ when $x = 12$") can be written as a proportion to find y when $x = 45$ and to find x when $y = 50$.

Find y when $x = 45$.

$\frac{8}{12} = \frac{y}{45} \Rightarrow 12y = 8(45) \Rightarrow y = 30$

Find x when $y = 50$.

$\frac{8}{12} = \frac{50}{x} \Rightarrow 8x = 50(12) \Rightarrow x = 75$

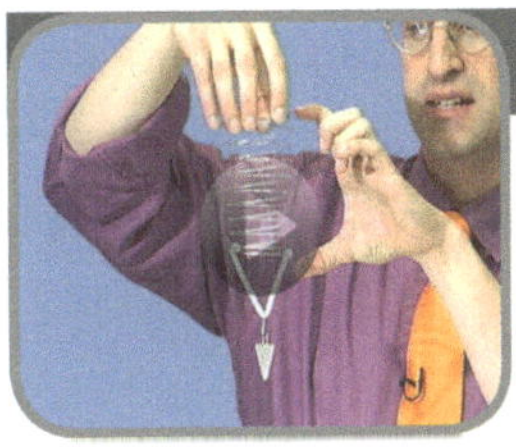

1.3.2 Modeling with Direct Variation

EXAMPLE 2 Modeling a Direct Variation

The dosage of a medication in milligrams varies directly as the weight of the patient in kilograms. A 50-kilogram patient receives 2000 milligrams of the medication. Find the constant of variation, write the variation equation, and find the amount of medication that a patient weighing 60 kilograms should receive.

SOLUTION

❶ Since the dosage (milligrams) varies directly as the patient's weight (kilograms), let x be the weight of the patient in kilograms, and let y be the dosage of the medication in milligrams. (So, $y = kx$.)

Find the constant of variation using $x = 50$ and $y = 2000$.

$2000 = k(50) \Rightarrow k = 40$ *The constant of variation is 40.*

❷ So, the direct variation equation is $y = 40x$.

❸ Use the equation to find the amount of medication needed for a 60-kilogram patient.

$y = 40(60) = 2400$ mg The 60 kg patient requires 2400 mg of the medication.

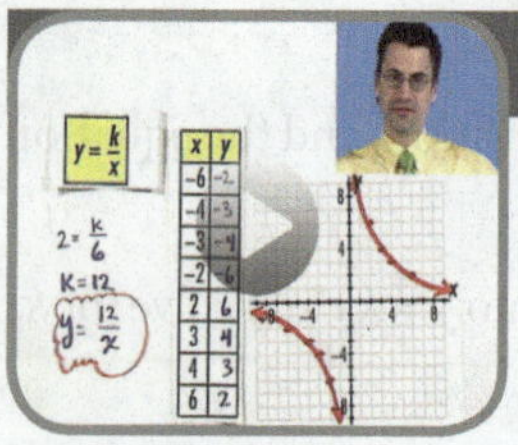

1.3.3 Inverse Variation

A situation involving two variables x and y that can be modeled by an equation of the form $y = \frac{k}{x}$, where k is some nonzero constant and $x \neq 0$, is called an **inverse variation**.

IMPORTANT

An inverse variation is also called an inverse proportion.

DEFINITION

If two quantities x and y are related by the equation $y = \frac{k}{x}$, where $x \neq 0$ and k is some constant such that $k \neq 0$, then the relationship between x and y is an **inverse variation** where y varies inversely as x.

NOTICE THAT

Since $x \neq 0$ in an inverse variation, the graph has no x-intercept. Furthermore, since $k \neq 0$ in an inverse variation, the graph has no y-intercept.

The general shape of a rational equation of the form $y = \frac{k}{x}$ is two curves.

- When $k > 0$, one curve is in quadrant I and the other is in quadrant III (Figure 1.3a).
- When $k < 0$, one curve is in quadrant II and the other is in quadrant IV (Figure 1.3b).

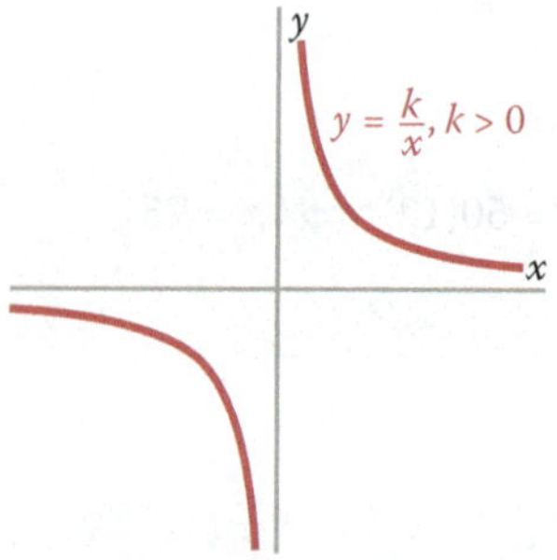

Figure 1.3a

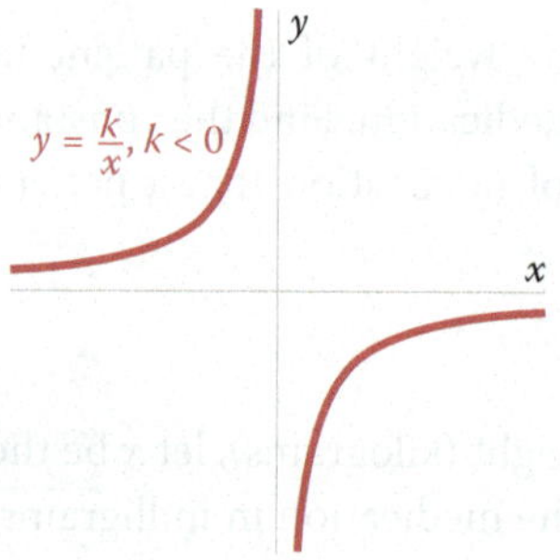

Figure 1.3b

Steps for Writing and Using an Equation When y Varies Inversely as x

❶ Substitute the given corresponding x- and y-values into $y = \frac{k}{x}$ and solve for k.

❷ Write the inverse variation equation by substituting the k-value into $y = \frac{k}{x}$. (Be sure to leave x and y as variables.)

❸ Find an unknown value by substituting the corresponding known value into the inverse variation equation (from step ❷) and solving for the unknown.

EXAMPLE 3 Writing an Inverse Variation Equation

Write the variation equation for x and y, if y varies inversely as x, and $y = 1$ when $x = 10$. Then find y when $x = 5$, and find x when $y = -10$.

SOLUTION

Since y varies inversely as x, the relationship between x and y is an inverse variation, and the equation $y = \frac{k}{x}$ models the situation.

❶ Substitute $x = 10$ and $y = 1$, into $y = \frac{k}{x}$, then solve for k.

$$y = \frac{k}{x} \Rightarrow 1 = \frac{k}{10} \Rightarrow k = 10$$ *The constant of variation is 10.*

❷ Substitute $k = 10$ into $y = \frac{k}{x}$. The inverse variation equation is $y = \frac{10}{x}$.

❸ Use $y = \frac{10}{x}$ to find y when $x = 5$, and to find x when $y = -10$.

Find y when $x = 5$.

$$y = \frac{10}{5} = 2$$

Find x when $y = -10$.

$$-10 = \frac{10}{x} \Rightarrow x = -1$$

Alternative Method: Inverse Variation as Products of Coordinates

Solving the general inverse variation equation $y = \frac{k}{x}$ for k gives $k = xy$. It follows that for all corresponding x- and y-values in an inverse variation, the value of xy must be the same (and equal to k). Therefore, for any two pairs of x- and y-values from an inverse variation, $x_1y_1 = x_2y_2$.

Consider the relationship given in **Example 3**: y varies inversely as x, and $y = 1$ when $x = 10$. Since the relationship between x and y is an inverse variation, the fact that $x_1y_1 = x_2y_2$ can be used to find y when $x = 5$, as well as to find x when $y = -10$.

Find y when $x = 5$.

$10(1) = 5y$ *$x_1y_1 = x_2y_2$*

$y = 2$ *Solve for y.*

Find x when $y = -10$.

$10(1) = x(-10)$ *$x_1y_1 = x_2y_2$*

$x = -1$ *Solve for x.*

1.3.4 Modeling with Inverse Variation

Variation equations may involve powers of the quantities in the relationship. For example, if y varies inversely as the cube of x, then $y = \frac{k}{x^3}$.

EXAMPLE 4 Modeling an Inverse Variation

In order for all of the cylindrical cans produced by a manufacturer to have a constant volume, the height of each can must vary inversely as the square of the diameter of the can's base. One type of can produced has a diameter of 3 inches and a height of 4 inches. Find the constant of variation, write the variation equation, and find the diameter for a can with height 9 inches.

SOLUTION

Since a can's height varies inversely as the square of the base's diameter, let x be the diameter and let y be the height. Then $y = \frac{k}{x^2}$.

Use $x = 3$ and $y = 4$ to find k. $4 = \frac{k}{3^2} \Rightarrow k = 4(3^2) = 36$ *The constant of variation is 36.*

So, the inverse variation equation is $y = \frac{36}{x^2}$.

Use the equation to find the diameter of a can with height 9 inches.

$$9 = \frac{36}{x^2} \Rightarrow 9x^2 = 36 \Rightarrow x^2 = 4 \Rightarrow x = 2$$

The diameter of a can must be 2 inches when the height is 9 inches.

IMPORTANT

Since x represents a length, x must be positive. So, only the positive square root is needed.

A variation equation can also be used to determine how a specific change in one variable, such as doubling the variable, affects the other variable. In this case, specific x- and y-values are not substituted into the variation equation. Instead, an expression for one of the variables is substituted into the general variation equation. For example, to determine how y is affected if x is doubled when x and y are inversely related, substitute $2x$ into the general inverse variation equation, $y = \frac{k}{x}$. This procedure is demonstrated in **Example 5**.

EXAMPLE 5 Modeling an Inverse Variation

NOTICE THAT

No specific P- or k-values are given, so the value of k cannot be found.

At some constant temperature, the volume V of a gas varies inversely as the pressure P. How does the volume of the gas change if the pressure is tripled?

SOLUTION

Since the volume V varies inversely as the pressure P, $V_1 = \frac{k}{P_1}$, where V_1 and P_1 represent the original volume and pressure (i.e., the volume and pressure before the pressure is tripled).

To find the effect of tripling the pressure on the volume, substitute $3P_1$ for P_1 in the variation equation, and let V represent the volume of the gas *after the pressure is tripled*.

$$V = \frac{k}{3P_1}$$

From the original variation equation, $V_1 = \frac{k}{P_1}$. So, substitute V_1 into $V = \frac{k}{3P_1}$ for $\frac{k}{P_1}$.

$$V = \frac{k}{3P_1} = \frac{1}{3} \cdot \frac{k}{P_1} = \frac{1}{3}V_1$$

After the pressure is tripled, the volume, V, is equal to one-third of the original volume, $\frac{1}{3}V_1$. In other words, tripling the pressure results in one-third of the original volume.

1.3.5 Joint Variation

A situation involving three variables, x, y, and z, which can be modeled by an equation of the form $y = kxz$, where k is some nonzero constant, is called a **joint variation**.

IMPORTANT

A joint variation is also called a joint proportion.

DEFINITION

If three quantities x, y, and z are related by the equation $y = kxz$, where k is some constant such that $k \neq 0$, then the relationship between x, y, and z is a **joint variation** where y varies jointly as x and z.

A joint variation may involve powers of x or z. Specifically, x, y, and z are related by an equation of the form $y = kx^n z^m$ when y varies jointly as x to the nth power and z to the mth power. For example, the equation $y = kxz^2$ describes a variation where y varies jointly as x and the square of z.

The steps for writing a joint variation equation are the same as those for writing any other type of variation equation, except that the values of the variables are substituted into $y = kxz$ for a joint variation.

Steps for Writing and Using an Equation When y Varies Jointly as x and z

❶ Substitute the given corresponding x-, y-, and z-values into $y = kxz$ and solve for k.

❷ Write the joint variation equation by substituting the k-value into $y = kxz$. (Be sure to leave x, y, and z as variables.)

❸ Find an unknown value by substituting the corresponding known values into the joint variation equation (from step ❷) and solving for the unknown.

EXAMPLE 6 Writing a Joint Variation Equation

The volume V of a rectangular pyramid varies jointly as the area of the base B and the height h. The volume of a rectangular pyramid is 56 cubic meters when the area of its base is 24 square meters and its height is 7 meters. Find the constant of variation, write the variation equation, and find the area of a rectangular pyramid's base when its volume is 81 cubic meters and its height is 9 meters.

SOLUTION

Since V varies jointly as B and h, the relationship between V, B, and h is a joint variation, and the equation $V = kBh$ models the situation.

❶ Substitute the given corresponding V-, B-, and h-values, $V = 56$, $B = 24$, and $h = 7$, into $V = kBh$, and solve for k.

$$V = kBh \ \Rightarrow\ 56 = k(24)(7) \ \Rightarrow\ k = \frac{1}{3}$$ *The constant of variation is 1/3.*

❷ Substitute $k = \frac{1}{3}$ into $V = kBh$. The direct variation equation is $V = \frac{1}{3}Bh$, or $V = \frac{Bh}{3}$.

❸ Use $V = \frac{Bh}{3}$ to find B when $V = 81$ and $h = 9$.

When $V = 81$ and $h = 9$, $81 = \frac{B(9)}{3}$, and so $B = \frac{81(3)}{9} = 27$.

So, the area of a rectangular pyramid's base is 27 square meters when its volume is 81 cubic meters and its height is 9 meters.

WHAT'S THE BIG IDEA?

How is a joint variation similar to a direct variation?

1.3.6 Modeling with Combined Variation

A **combined variation** is a relationship between three or more variables that combines both direct and inverse variation. In a combined variation, the constant of variation k is multiplied by a ratio of variables where quantities that vary directly are in the numerator of the ratio and quantities that vary inversely are in the denominator of the ratio.

DEFINITION

If three (or more) quantities x, y, and z are related by the equation $y = \frac{kx}{z}$, where $z \neq 0$ and k is some constant such that $k \neq 0$, then the relationship between x, y, and z is a **combined variation** where y varies directly as x and inversely as z.

IMPORTANT

A combined variation is also called a combined proportion.

The steps for writing a combined variation equation are the same as those for writing any other type of variation equation, except that the values of the variables are substituted into $y = \frac{kx}{z}$ for a combined variation.

Steps for Writing and Using an Equation When y Varies Directly as x and Inversely as z

❶ Substitute the given corresponding x-, y-, and z-values (where x represents the variable(s) that vary directly as y, and z represents the variable(s) that vary inversely as y) into $y = \frac{kx}{z}$ and solve for k.

❷ Write the combined variation equation by substituting the k-value into $y = \frac{kx}{z}$. (Be sure to leave x, y, and z as variables.)

❸ Find an unknown value by substituting the corresponding known values into the combined variation equation (from step ❷) and solving for the unknown.

EXAMPLE 7 Modeling a Combined Variation Equation

The pressure P of a gas varies directly as the temperature T and inversely as the volume V, and 100 liters (L) of the gas exerts a pressure of 33.2 kilopascal (kPa) when its temperature is 400 Kelvins (K). Find the constant of variation, write the variation equation, and find the pressure for 80 L of the gas at 500 K.

SOLUTION

Since P varies directly as T and inversely as V, the relationship between P, T, and V is a combined variation, and the equation $P = \frac{kT}{V}$ models the situation.

❶ Substitute $P = 33.2$, $T = 400$, and $V = 100$ into $P = \frac{kT}{V}$, and solve for k.

$$P = \frac{kT}{V} \quad \Rightarrow \quad 33.2 = \frac{k(400)}{100} \quad \Rightarrow \quad k = 8.3$$ *The constant of variation is 8.3.*

❷ Substitute $k = 8.3$ into $P = \frac{kT}{V}$. The combined variation equation is $P = \frac{8.3T}{V}$.

❸ Use $P = \frac{8.3T}{V}$ to find P when $V = 80$ and $T = 500$. $P = \frac{8.3(500)}{80} = 51.875$

So, the pressure of the gas is 51.875 kPa when its volume is 80 L and its temperature is 500 K.

SECTION 1.3 EXERCISES

Warm Up

Solve each equation.

1. $5y = (10)(8)$
2. $(7)(6)(3)k = 20$
3. $75 = \dfrac{A(5)}{3}$

Just the Facts

Fill in the blanks.

4. For $k > 0$, if y varies directly with x, then when x _____, y decreases, and when x increases, y _____.
5. The four types of variation situations are _____, _____, _____, and _____.
6. $y = kx$ is a(n) _____ variation equation where the graph passes through the _____ and the slope is _____.
7. An inverse variation where y varies inversely as x is written _____, where _____ $\neq 0$.
8. "P is jointly proportional to r and the third power of m" is written as $P =$ _____.
9. The equation $y = kmn$ is a(n) _____ variation equation where y varies _____ with _____, _____, and _____, which is some nonzero constant.
10. Mathematical models that involve both direct and inverse variation are said to have a(n) _____ variation.

Essential Skills

In Exercises 1–10, write the direct variation equation and find the specified values.

1. Two values, d and g, are directly proportional. If $d = 16$ when $g = 14$, what is the value of g when d is 25?
2. Suppose y varies directly with x, and $y = 18$ when $x = 30$. Find y when $x = 60$, and find x when $y = 3$.
3. Suppose t varies directly with s, and $t = 8$ when $s = 14$. Find t when $s = 21$, and find s when $t = 20$.
4. Suppose z varies directly with g, and $z = 36$ when $g = 30$. Find z when $g = 54$, and find g when $z = 60$.
5. Two values, p and q, are directly proportional. If $p = 4.8$ when $q = 2.8$, what is the value of q when p is 3.6?
6. Suppose b varies directly with a, and $b = 6.2$ when $a = 1.5$. Find b when $a = 7.4$, and find a when $b = 9.6$.
7. Suppose u varies directly with v, and $u = -12$ when $v = 8.4$. Find v when $u = 9.2$ and find u when $v = 4.9$.
8. Suppose m varies directly with n, and $m = 11.6$ when $n = -2.4$. Find n when $m = -29$, and find m when $n = 22.5$.
9. Suppose w varies directly with r, and $w = 14/3$ when $r = -5/6$. Find w when $r = -10/3$, and find r when $w = 7/4$.
10. Suppose p varies directly with l, and $p = -7/8$ when $l = -5/2$. Find p when $l = 2/3$, and find l when $p = -4/5$.

In Exercises 11–13, use the following information.

A kite company produces kites whose lengths vary directly with their widths.

11. If a kite with length 80 centimeters has width 65 centimeters, what is the width of a 96-centimeter-long kite?
12. If a kite with length 63 inches has width 48 inches, what is the length of a 40-inch-wide kite?
13. If a kite with length 3.4 feet has width 2.5 feet, what is the width of a 5.1-foot-long kite?

In Exercises 14–16, use the following information.

If income tax varies directly with income and a person earning $30,000 per year pays $8,400 in taxes, how much will a person earning each of the following amounts pay?

14. $42,000 per year
15. $136,000 per year
16. $78,500 per year

In Exercises 17–24, write the inverse variation equation and find the specified values.

17. Suppose $x = 8.4$ when $y = 1$, and x varies inversely with y. Find x when $y = 4$.
18. Suppose y varies inversely with x, and $y = 1/2$ when $x = 8$. Find y when $x = -1$, and find x when $y = 1$.
19. Suppose a varies inversely with b, and $b = 3/8$ when $a = 61$. Find b when $a = 1/3$, and find a when $b = 15$.
20. Suppose t varies inversely with u, and $u = 3.7$ when $t = 14$. Find u when $t = 7.5$, and find t when $u = 40$.
21. Suppose y varies inversely with x, and $y = 12$ when $x = -3/4$. Find y when $x = 6/11$, and find x when $y = -3/2$.
22. Suppose $n = -15/16$ when $p = 8$, and n varies inversely with p. Find n when $p = -4/3$, and find p when $n = 1/6$.
23. Suppose c varies inversely with d, and $c = -7/4$ when $d = -6/5$. Find d when $c = -9/10$, and find c when $d = 3/8$.
24. Suppose w varies inversely with z, and $w = 3/10$ when $z = -5/9$. Find z when $w = -2/7$, and find w when $z = -13/2$.

In Exercises 25–27, use the following information.

The number of days needed to harvest a wheat field varies inversely with the number of people harvesting the field.

25. If it takes 13 days for 12 people to harvest the field, how many days would it take for 4 people to harvest the field?

26. If it takes 24 days for 8 people to harvest the field, how many days would it take for 10 people to harvest the field?

27. If it takes 9 days for 27 people to harvest the field, how many days would it take for 45 people to harvest the field?

In Exercises 28–30, use the following information.

Power in an electric circuit varies inversely with the resistance.

28. If the power is 1200 watts when the resistance is 12 ohms, find the resistance when the power is 2400 watts.

29. If the power is 1700 watts when the resistance is 25 ohms, find the resistance when the power is 5000 watts.

30. If the power is 625 watts when the resistance is 15 ohms, find the power when the resistance is 12 ohms.

In Exercises 31–33, use the following information.

The illumination, measured in candelas, from a light bulb varies inversely with the square of the distance *d* from the light bulb. Round to the nearest tenth, as needed.

31. If the illumination is 85 candelas at 5 meters, what is the illumination at 15 meters?

32. If the illumination is 78 candelas at 6 meters, what is the illumination at 20 meters?

33. If the illumination is 56 candelas at 12 meters, what is the distance away from the light bulb if the illumination is 42 candelas?

In Exercises 34–36, use the following information.

The frequency of the oscillations of a pendulum varies inversely with the square root of the length of the pendulum.

34. How does the frequency change if the pendulum's length is increased to 4 times the original length?

35. How does the frequency change if the pendulum's length is increased to 9 times the original length?

36. How does the frequency change if the pendulum's length is decreased to one-sixteenth of the original length?

In Exercises 37–39, use the following information.

The area *A* of a triangle varies jointly with the base *b* and height *h*.

37. Suppose $A = 24$ square inches when $b = 6$ inches, and $h = 8$ inches. Find h when $A = 100$ square inches and $b = 10$ inches.

38. Suppose $A = 18$ square inches when $b = 2$ inches, and $h = 3$ inches. Find b when $A = 72$ square inches and $h = 8$ inches.

39. Suppose $A = 124$ square feet when $b = 9$ feet, and $h = 12$ feet. Find A when $b = 5$ feet and $h = 27$ feet.

In Exercises 40–42, use the following information.

The volume *V* of a cone varies jointly with the area of the base *B* and the height *h*.

40. If $V = 32$ cubic centimeters when $B = 16$ square centimeters, and $h = 6$ centimeters, what is h when $V = 60$ cubic centimeters and $B = 20$ square centimeters?

41. If $V = 138$ cubic inches when $B = 23$ square inches, and $h = 18$ inches, what is h when $V = 57$ cubic inches and $B = 19$ square inches?

42. If $V = 144$ cubic millimeters when $B = 12$ square millimeters, and $h = 36$ millimeters, what is h when $V = 240$ cubic millimeters and $B = 64$ square millimeters?

43. The amount of light E provided by a light bulb is inversely proportional to the square of the distance d in meters from the bulb. When a person is 3 meters from the bulb, the amount of light is 5.9 lux. Write the variation equation.

44. The resistance R of a wire varies directly with its length L and inversely with the square of its diameter D. Write an equation for the constant of variation in terms of R, L, and D.

In Exercises 45–47, use the following information.

The number of vibrations *n* per second of a nylon guitar string varies directly with the square root of the tension *T* and inversely with the length *L* of the string. Round to the nearest kilogram, as needed.

45. If the tension is 400 kilograms when the number of vibrations per second is 12 and the length is 0.7 meters, find the tension when the length is 0.4 meters and the number of vibrations is 8.

46. If the tension is 256 kilograms when the number of vibrations per second is 15 and the length is 0.6 meters, find the tension when the length is 0.3 meters and the number of vibrations is 12.

47. If the tension is 324 kilograms when the number of vibrations per second is 14 and the length is 0.8 meters, find the tension when the length is 0.2 meters and the number of vibrations is 12.

In Exercises 48–50, use the following information.

The power P that must be delivered by a car's engine varies directly with the distance d that the car moves and inversely with the time t required to move that distance. Round to the nearest foot, as needed.

48. To move the car 2000 feet in 75 seconds, the engine must deliver 152 kilowatts of power. Find the distance (in feet) the car moves when 189 kilowatts of power is delivered for 90 seconds.

49. To move the car 600 feet in 60 seconds, the engine must deliver 56 kilowatts of power. Find the distance (in feet) the car moves when 79 kilowatts of power is delivered for 80 seconds.

50. To move the car 1800 feet in 90 seconds, the engine must deliver 114 kilowatts of power. Find the distance (in feet) the car moves when 140 kilowatts of power is delivered for 110 seconds.

Extensions

In Exercises 51–53, use the following information.

At Chapel Hill Tubal Reversal Center, patients are required to have a body mass index (BMI) of less than 37 at least three weeks prior to the date of any scheduled tubal reversal surgery. This policy is in place to ensure optimal patient safety when undergoing this elective, outpatient surgery. A person's BMI is directly proportional to their weight (w) in pounds, and inversely proportional to the square of their height (h) in inches. A person who is 65 inches tall and weighs 130 pounds has a BMI of 21.6 pounds per square inch.

51. Would a person 5 feet 3 inches tall weighing 180 pounds be allowed to have the surgery? Depending upon the result, what amount is the person over/under? Round to the nearest tenth.

52. Would a person 6 feet 4 inches tall weighing 290 pounds be allowed to have the surgery? Depending upon the result, what amount is the person over/under? Round to the nearest tenth.

53. What is the most that a person 5 feet 8 inches tall could weigh and still have the surgery? Round to the nearest pound.

54. In photography, the luminance L varies directly as the square of the f-stop f and inversely as the product of the film ISO number s and the exposure time t. An ISO 100 film has a luminance of 14 candelas per square meter when shot for 1 second at an f-stop of 8. Write the equation. What is the f-stop of an ISO 400 film with a luminance of 22 candelas per square meter when shot for 1.3 seconds?

55. Suppose b varies directly as the cube of w and inversely as the square of h. If w is doubled and h is halved, what happens to b?

56. If x varies directly as a^2 and h^4, and a doubles while h triples, what happens to x?

57. The force needed to keep a car from skidding on a curve varies inversely as the radius r of the curve and jointly as the weight w of the car and the square of the speed s. If an 1800-pound car traveling 20 miles per hour takes a force of 3600 pounds to keep from skidding on a curve of radius 600 feet, what force would be required to prevent the same car from skidding at the same curve traveling 15 miles per hour faster? Write the variation equation.

58. Write a sentence using variation terminology to describe the formula $E = mc^2$ where E is energy, m is mass, and c is the speed of light.

1.4 FUNCTIONS

OBJECTIVES

- Identify functions graphically.
- Identify functions algebraically.
- Understand and use function notation.
- Evaluate a piecewise-defined function.
- Model with functions.
- Find the domain of a function.

PREREQUISITE VOCABULARY TERMS

coordinate plane
domain
equation
formula
ordered pair

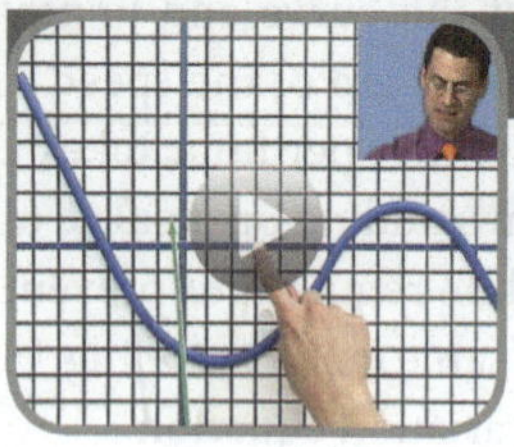

1.4.1 Functions and the Vertical Line Test

A **relation** is any relationship between two quantities (variables, such as x and y) or sets. A special kind of relation where each item from one set, the **domain**, corresponds to exactly one item from a second set, the **range**, is called a **function**.

DEFINITION

A **relation** that assigns each element in a set A to exactly one element in a set B is a **function** from A to B. The set A is the **domain**, or set of allowable input values, and the set B is the **range**, or set of corresponding output values.

There are many ways to express a function, including as a set of ordered pairs. For example, the set of ordered pairs {(1, 2), (2, 4), (3, 6), (4, 8)} expresses a function from A to B where set A is {1, 2, 3, 4} and set B is {2, 4, 6, 8}. Notice that each element in A is assigned to exactly one element in B.

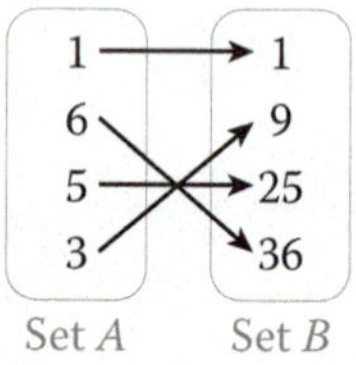

Figure 1.4a

Another way to express a function is a **mapping diagram** (also known as an **arrow diagram**). The mapping diagram in Figure 1.4a expresses the function {(1, 1), (6, 36), (5, 25), (3, 9)}. The domain is the set A, {1, 3, 5, 6}, and the range is the set B, {1, 9, 25, 36}. In a mapping diagram, a function will show one arrow from each element in set A pointing to exactly one element in set B. The following mapping diagrams show examples of relations that are functions and relations that are not functions.

Mapping Diagrams

Functions

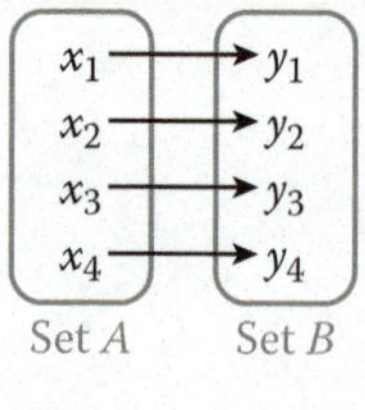

Each x is mapped to exactly one y.

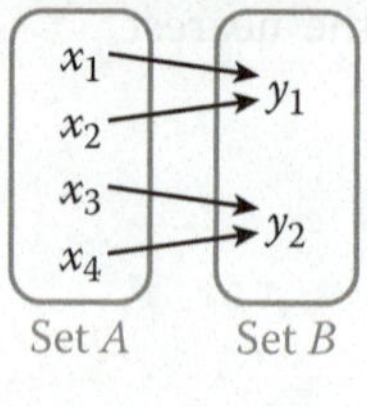

Each x is mapped to exactly one y. More than one x can be mapped to the same y.

Not Functions

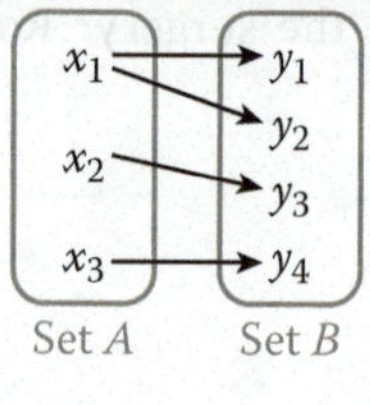

Not a function because x_1 is mapped to more than one y.

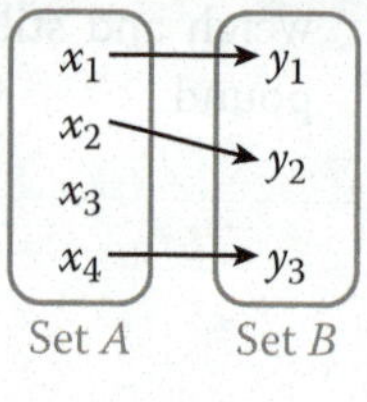

Not a function because x_3 is not mapped to any y.

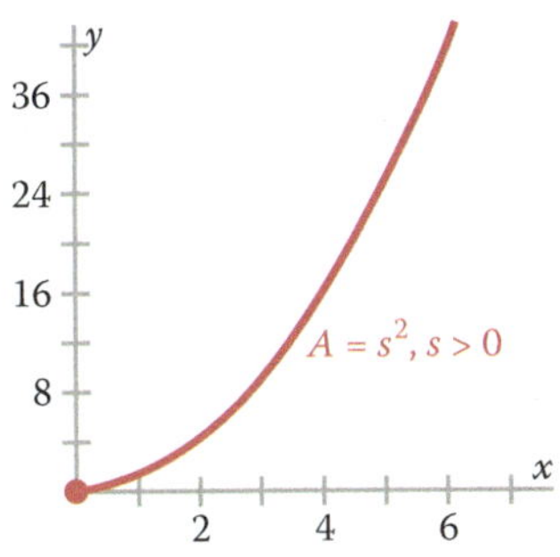

Figure 1.4b

A function can also be represented by an equation (or formula), or by a graph in a coordinate plane. For example, the relationship between the length s of the side of a square and the area A of the square is a function, where the area of the square *is a function of* the length of its side. In this case, the s-values are the input (domain) and the A-values are the output (range). This function can be expressed by the equation $A = s^2$, or by the graph in Figure 1.4b.

Determining Whether a Graph Represents a Function

A function must assign each value in its domain to exactly one value in its range. Therefore, if a graph on a coordinate plane is a function, then the graph contains points where each x-coordinate (domain value) is assigned to exactly one y-coordinate (range value). In other words, a graph that is a function *cannot* contain two points with the same x-value and different y-values. A test for determining whether a graph is a function follows from this statement.

The Vertical Line Test

If any vertical line intersects a graph more than once, then the graph is not a function.

For example, consider a graph that contains the points (3, 2) and (3, 5). We know that this graph cannot be a function because the x-coordinate 3 is assigned to two y-coordinates, 2 and 5. Additionally, by the Vertical Line Test, this graph cannot be a function because the vertical line $x = 3$ intersects the two points (3, 5) and (3, 2).

EXAMPLE 1 Using the Vertical Line Test

Use the Vertical Line Test to determine whether each graph represents a function.

A.

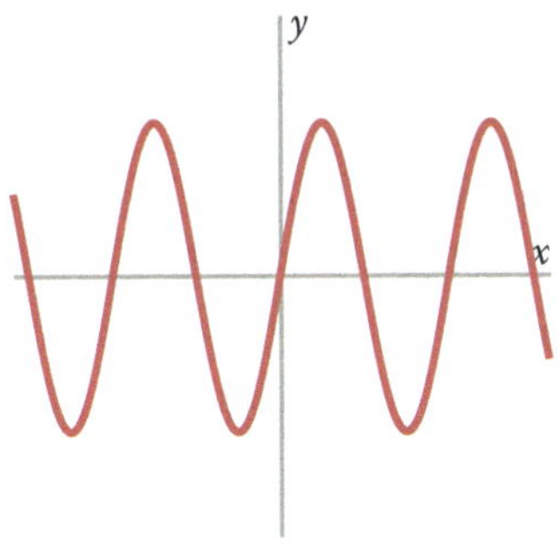

B.

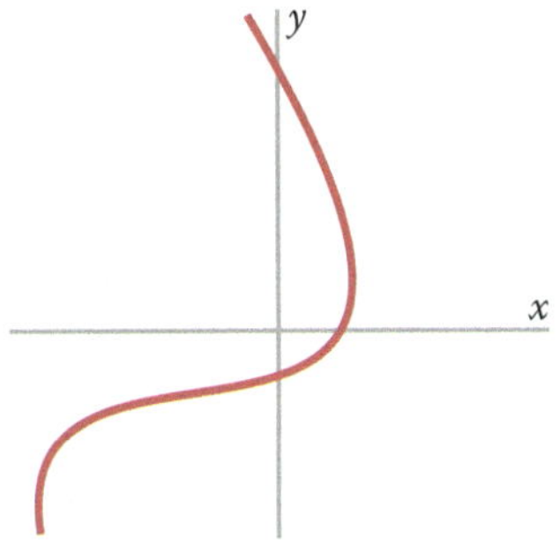

SOLUTION

A. The graph passes the Vertical Line Test because no vertical line intersects the graph more than once. Therefore, the graph is a function.

B. The graph fails the Vertical Line Test because there are many vertical lines that will intersect the graph more than once. One such vertical line is shown in Figure 1.4c. Therefore, the graph is not a function.

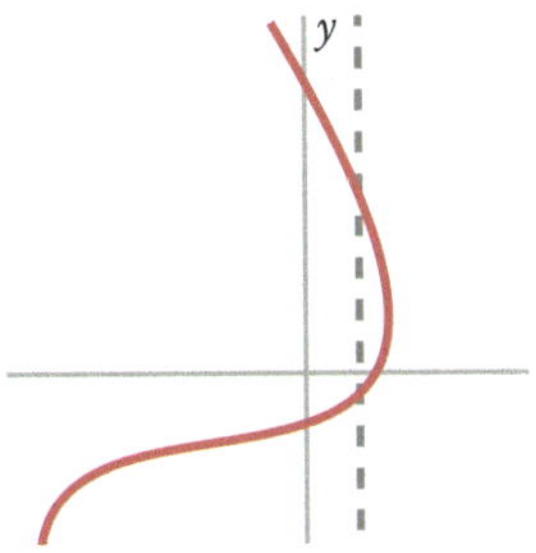

Figure 1.4c

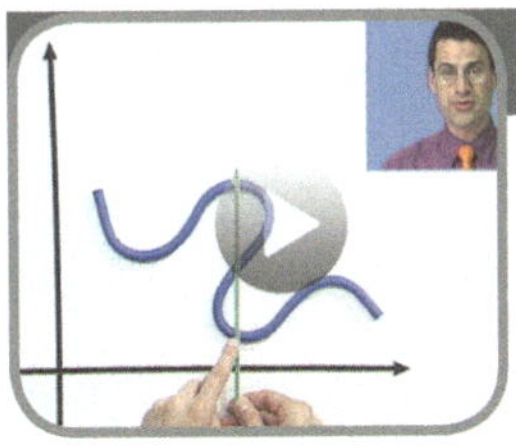

1.4.2 Identifying Functions Algebraically

To determine algebraically whether an equation in x and y is a function, solve the equation for y and then analyze the expression for y to determine whether there could possibly be more than one y-value for any x-value in the equation's domain.

For example, the equation $2x^2 + y = 1$ is a function. Solving for y gives $y = 1 - 2x^2$. Notice that any substituted x-value will result in only one y-value.

REMEMBER

To undo the nth power when n is even, take plus or minus the nth root of both sides of the equation.

However, the equation $2x + y^2 = 1$ is not a function. Solving for y gives $y = \pm\sqrt{1-2x}$. Notice that when $x = 0$, the corresponding y-values are -1 and 1. So, there is at least one x-value in the equation's domain with more than one corresponding y-value.

EXAMPLE 2 Identifying Functions Algebraically

Determine algebraically whether each equation represents a function.

A. $x + 2y = 0$ **B.** $x^2 + y^2 = 16$

SOLUTION

Solve each equation for y.

A. $x + 2y = 0 \Rightarrow y = -\frac{x}{2}$

Each x-value has one corresponding y-value: $-\frac{x}{2}$. So, $x + 2y = 0$ is a function.

B. $x^2 + y^2 = 16 \Rightarrow y = \pm\sqrt{16-x^2}$

Each x-value has two corresponding y-values: $\sqrt{16-x^2}$ and $-\sqrt{16-x^2}$. So, $x^2 + y^2 = 16$ is not a function.

NOTICE THAT

The graph of $x + 2y = 0$ is a nonvertical line. Since any nonvertical line passes the Vertical Line Test, this equation is a function.

The graph of $x^2 + y^2 = 16$ is a circle. Since any circle fails the Vertical Line Test, this equation is not a function.

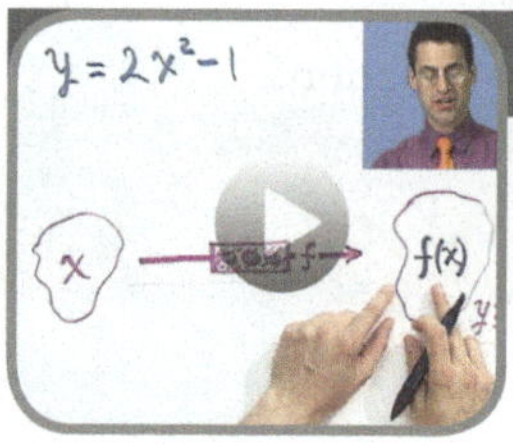

1.4.3 Function Notation and Finding Function Values

Function notation can be used to write the equation of a function where the output value, y, is replaced with $f(x)$, which is read as "f of x". This notation

- names the function with a letter (usually f, but other letters can be used),
- shows the input value (usually x, but other letters can be used), and
- shows the rule described by the function.

IMPORTANT

The notation $f(x)$ does not mean a variable f is multiplied by a variable x. Instead, $f(x)$ means that the function is named f and the two variables in the function are x and $f(x)$.

For example, $f(x) = 2x + 1$ is the function notation for $y = 2x + 1$, where $y = f(x)$. This function could also be written as $g(x) = 2x + 1$ or $p(x) = 2x + 1$ (all are equivalent functions). The two variables in $f(x) = 2x + 1$ are x and $f(x)$, where x is the input and $f(x)$ is the output. Two additional examples of functions written using function notation follow.

Example function: $f(x) = x^2$

- The function is named f.
- The input is x.
- The rule is x^2.
- The output is $f(x)$.
- Equivalent to $y = x^2$.
- f is a quadratic function.

Example function: $r(a) = |4 - a|$

- The function is named r.
- The input is a.
- The rule is $|4 - a|$.
- The output is $r(a)$.
- Equivalent to $y = |4 - a|$.
- r is an absolute value function.

TIP

When using function notation, remember that "$f(x)$" is just another way of writing "y."

When the variable is replaced by a number (or by another variable or expression) in function notation, it denotes evaluation. For example, if $f(x) = 2x + 1$, then $f(5)$ means to evaluate the function f for $x = 5$. So, $f(5) = 2(5) + 1 = 11$.

Additionally, $f(b + 1)$ means to evaluate the function f for $x = b + 1$.

$$f(b + 1) = 2(b + 1) + 1 = 2b + 3$$

EXAMPLE 3 Evaluating a Function

If $h(x) = x + 3x^2$, find $h(4)$, $h(-1)$, and $h(a + 1)$.

SOLUTION

Substitute each given input value into the function and simplify.

$$h(4) = (4) + 3(4)^2 = 4 + 3(16) = 4 + 48 = 52$$
$$h(-1) = (-1) + 3(-1)^2 = -1 + 3(1) = -1 + 3 = 2$$

$h(a + 1) = (a + 1) + 3(a + 1)^2$		*Substitute a + 1 for x.*
	$= a + 1 + 3(a + 1)(a + 1)$	*Expand the power.*
	$= a + 1 + 3(a^2 + 2a + 1)$	*FOIL*
	$= a + 1 + 3a^2 + 6a + 3$	*Distribute.*
	$= 3a^2 + 7a + 4$	*Combine the like terms.*

So, $h(4) = 52$, $h(-1) = 2$, and $h(a + 1) = 3a^2 + 7a + 4$.

TIP

Finding $h(4)$ when $h(x) = x + 3x^2$ is equivalent to evaluating $y = x + 3x^2$ for $x = 4$.

1.4.4 Evaluating Piecewise-Defined Functions

A **piecewise-defined function** is a combination of two or more functions. There are specific values of the domain that correspond to each piece (part) of the function. The values from the domain that correspond to each of the function's pieces are written in an inequality (or equation) to the right of that piece. For example, $f(x)$ is a piecewise-defined function with the domain divided into three parts: values less than 0, values between 0 and 4 (including 0 and 4), and values greater than 4.

$$f(x) = \begin{cases} 5 - x & \text{if } x < 0 \\ 2x & \text{if } 0 \le x \le 4 \\ (3 + x)^2 & \text{if } x > 4 \end{cases}$$

Piece 1: For domain values less than 0, $f(x) = 5 - x$.
Piece 2: For domain values from 0 to 4, $f(x) = 2x$.
Piece 3: For domain values greater than 4, $f(x) = (3 + x)^2$.

To evaluate a piecewise-defined function for some domain value x, first determine which piece of the function corresponds to that domain value. Then substitute the x-value into just that piece of the function. For example, to evaluate $f(3)$ using $f(x)$ given above, substitute 3 into $f(x) = 2x$ (since $x = 3$ is a solution to the inequality $0 \le x \le 4$): $f(3) = 2(3) = 6$.

IMPORTANT

A piecewise-defined function is not necessarily defined for all real numbers.

EXAMPLE 4 Evaluating a Piecewise-Defined Function

Given $g(x) = \begin{cases} x^2 - 3x & \text{if } x \le -3 \\ 7 & \text{if } -3 < x \le 1 \\ 5x & \text{if } x > 1 \end{cases}$, find $g(-5)$ and $g(1)$.

SOLUTION

Identify the piece of the function that corresponds to each given x-value. Then substitute that x-value into that piece of the function and simplify.

Evaluate $g(-5)$ by substituting -5 into $g(x) = x^2 - 3x$ (since $-5 \le -3$).

$$g(-5) = (-5)^2 - 3(-5) = 25 - (-15) = 40$$

Evaluate $g(1)$ by substituting 1 into $g(x) = 7$ (since $-3 < 1 \le 1$).

$$g(1) = 7$$

So, $g(-5) = 40$ and $g(1) = 7$.

IMPORTANT

The domain is divided into three parts.
Domain values in $(-\infty, -3]$ correspond to $g(x) = x^2 - 3x$.
Domain values in $(-3, 1]$ correspond to $g(x) = 7$.
Domain values in $(1, \infty)$ correspond to $g(x) = 5x$.

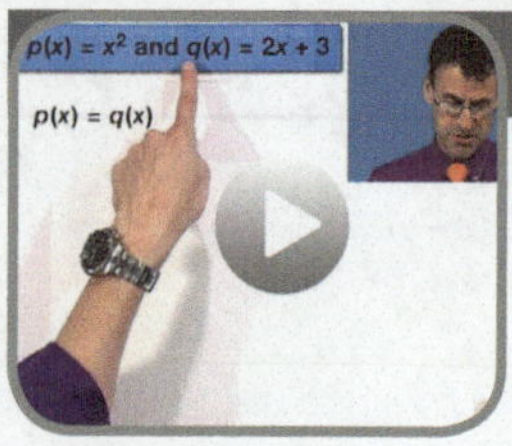

1.4.5 Finding Specific Function Values

So far in this section, all functions have been evaluated for some particular x-value (domain value). So, the corresponding range value has been found for some specific domain value. It is also possible to find a corresponding domain value when a range value is given. For example, suppose $f(x) = 2x + 1$. To find the domain value that corresponds to the range value 9, substitute 9 for $f(x)$ and solve for x. So, if $f(x) = 2x + 1$ and $f(x) = 9$, then $9 = 2x + 1$, and thus $x = 4$.

IMPORTANT

$f(x) = 9$ means that the function f is equal to 9.
$f(9)$ means that the function f is evaluated for $x = 9$.

When a function is set equal to 0 (i.e., $f(x) = 0$) and solved for x, the resulting x-values (the solutions) are called the **roots of the function** or the **zeros of the function**. Since these x-values correspond to where $y = 0$, these x-values are also the x-intercepts of the function's graph.

EXAMPLE 5 Finding Values for Which a Function is Equal to 0

Find all real x-values such that $r(x) = 0$.

A. $r(x) = 5x + 10$ **B.** $r(x) = x^2 - 13x + 30$

SOLUTION

For each function, substitute 0 for $r(x)$ and then solve for x.

REMEMBER

The equation $0 = x^2 - 13x + 30$ is quadratic, so solve by factoring, completing the square, or using the Quadratic Formula.

A.
$0 = 5x + 10$
$-10 = 5x$ *Subtract 10 from each side.*
$-2 = x$ *Divide each side by 5.*

So, if $r(x) = 5x + 10$, then $r(x) = 0$ when $x = -2$.

B.
$0 = x^2 - 13x + 30$
$0 = (x - 3)(x - 10)$ *Factor.*
$x - 3 = 0$ or $x - 10 = 0$ *Set each factor equal to 0.*
$x = 3$ $\quad x = 10$ *Solve each equation.*

So, if $r(x) = x^2 - 13x + 30$, then $r(x) = 0$ when $x = 3$ or $x = 10$.

The x-values found in **Example 5** correspond to where $r(x) = 0$, so these x-values are the roots, zeros, or solutions of the functions. These x-values are also the x-intercepts of the function, as shown in Figures 1.4d and 1.4e.

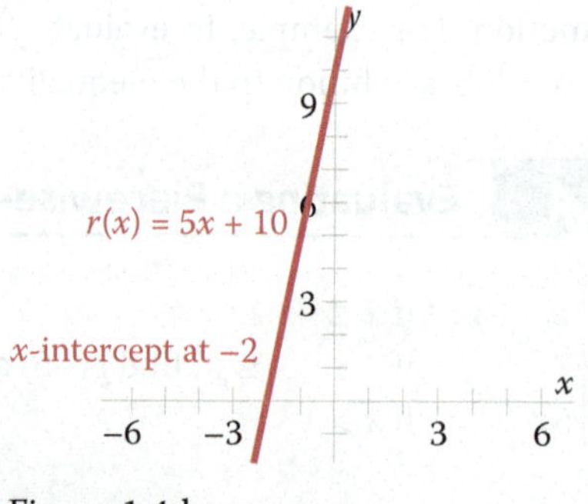

Figure 1.4d

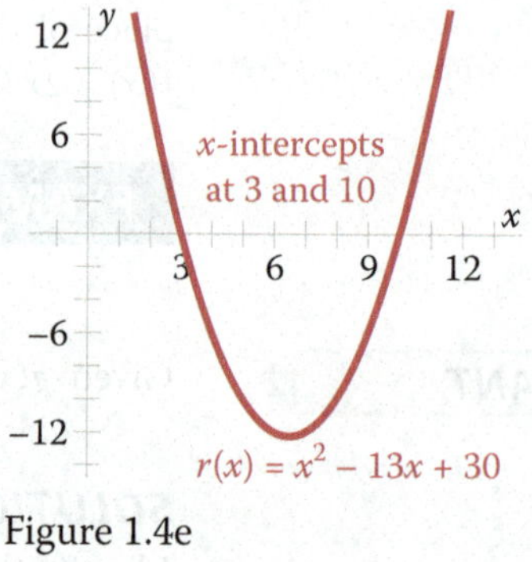

Figure 1.4e

WHAT'S THE BIG IDEA?

If $p(a)$ is some function, explain the meaning of $p(0)$ and $p(a) = 0$.

Values of x for Which Two Functions are Equal

Two functions f and g can be set equal to each other to find the x-values where the functions are equal. So, these x-values correspond to the points (x, y) that are the same on both functions. Graphically, these are the points at which the two functions intersect each other.

EXAMPLE 6 Finding Values for Which Two Functions are Equal

If $g(x) = x + 12$ and $h(x) = x^2$, find all real x-values such that $g(x) = h(x)$.

SOLUTION

> **TIP**
>
> *Any x-value that makes $g(x) = h(x)$ is an x-value that corresponds to the same y-value for both functions.*

To find the x-value(s) for which $g(x) = h(x)$, set the expression for $g(x)$, $x + 12$, equal to the expression for $h(x)$, x^2, and then solve for x.

$g(x) = h(x)$	
$x + 12 = x^2$	*Substitute $x + 12$ for $g(x)$ and x^2 for $h(x)$.*
$0 = x^2 - x - 12$	*Subtract x and 12 from each side.*
$0 = (x - 4)(x + 3)$	*Factor.*
$x - 4 = 0$ or $x + 3 = 0$	*Set each factor equal to 0.*
$x = 4$ $x = -3$	*Solve each equation.*

So, if $g(x) = x + 12$ and $h(x) = x^2$, then $g(x) = h(x)$ when $x = 4$ or $x = -3$.

CHECK

Check the answer by evaluating $g(4)$ and $h(4)$, and then evaluating $g(-3)$ and $h(-3)$.

$$g(4) = 4 + 12 = 16 \qquad g(-3) = -3 + 12 = 9$$
$$h(4) = 4^2 = 16 \qquad h(-3) = (-3)^2 = 9$$

So, when $x = 4$, $g(x) = 16$ and $h(x) = 16$, and when $x = -3$, $g(x) = 9$ and $h(x) = 9$. ✓

Figure 1.4f shows the graphs of functions g and h from **Example 6**. Notice that the functions intersect at the points (4, 16) and (–3, 9).

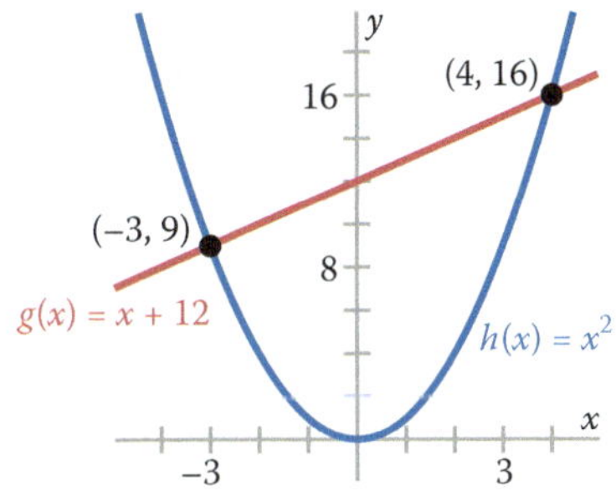

Figure 1.4f

1.4.6 Modeling with Functions

Independent and Dependent Variables

A function's x-variable is also called the **independent variable,** and its y-variable is also called the **dependent variable**, since the y-value depends on the x-value. The function's domain and range are the sets of all values for the independent and dependent variables, respectively.

When modeling, the relationship between two variables A and B can be stated as "A is a function of B." From this statement, it can be determined that B is the independent variable (x) and A is the dependent variable (y).

$$A \text{ is a function of } B \Rightarrow f(B) = A \qquad y \text{ is a function of } x \Rightarrow f(x) = y$$

Four Ways to Express a Function

A function can be expressed in four ways:

- verbally, using words to describe the rule,
- algebraically, using an equation or formula,
- visually, using a graph, or
- numerically, using a table or a list of ordered pairs.

For example, a description of a function rule is "To convert from feet to inches, multiply the number of feet by 12." In this rule, the number of inches is a function of the number of feet. If n represents the number of feet, then $f(n)$ represents the number of inches. Thus the corresponding equation is $f(n) = 12n$. The function's corresponding graph is shown in Figure 1.4g, and a table corresponding to the function is shown in Figure 1.4h.

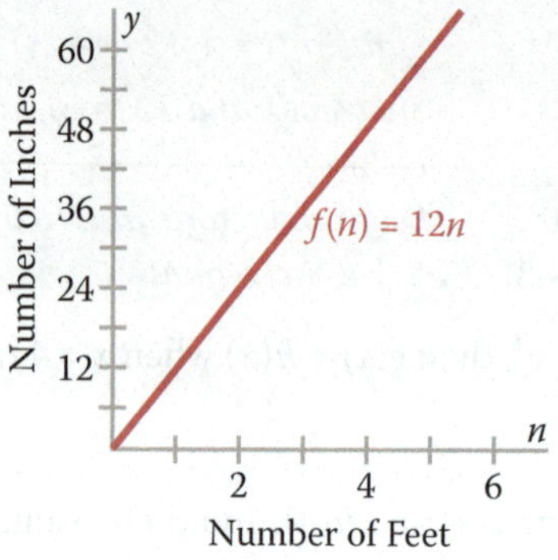

Figure 1.4g

n	$f(n)$
1	12
2	24
3	36
4	48

Figure 1.4h

EXAMPLE 7 Modeling with a Linear Function

Suppose a person runs a 200 meter race at a constant rate of 4 meters per second. The distance between the person and the finish line can be described as a function of t, where t is the elapsed time (in seconds) since the person started running. Express this function as an equation and a graph.

SOLUTION

REMEMBER

$d = rt$, where d is the distance traveled, r is the rate (speed), and t is the time traveled.

The distance that the person has traveled after t seconds is given by $4t$. So, the person's distance from the finish line, which is 200 meters at $t = 0$, is given by $200 - 4t$. The slope is negative because as the elapsed time increases, the distance between the person and the finish line decreases.

Naming the function f gives the equation $f(t) = 200 - 4t$.

Sketch a graph of this equation.

The t-variable acts as the x-value, and the $f(t)$-variable acts as the y-value in the graph.

Since t represents time, only positive t-values are needed for the domain. The equation $f(t) = 200 - 4t$ describes a line with slope -4 and y-intercept 200.

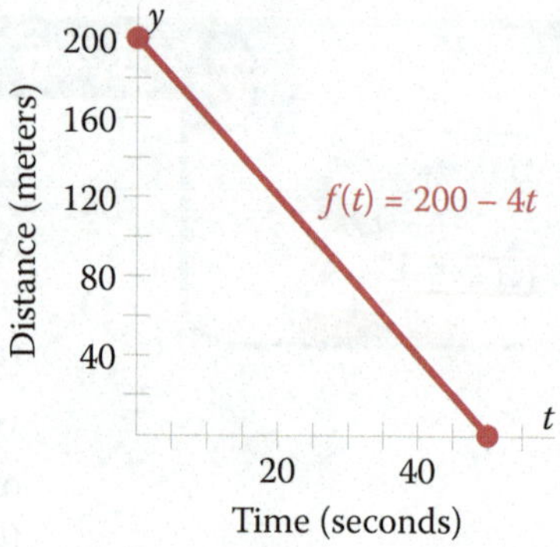

EXAMPLE 8 Modeling with a Function

IMPORTANT

Writing an equation where y is a function of x means to write an equation for y in terms of x.

Use function notation to write an equation that describes the area of a circle as a function of its circumference. Then find the area of a circle with circumference 20 centimeters, and the circumference of a circle with area 50π square centimeters.

SOLUTION

This equation will involve the formulas for the area A and circumference C of a circle.

$A = \pi r^2$ *area of a circle as a function of its radius*

$C = 2\pi r$ *circumference of a circle as a function of its radius*

To write an equation for the area of a circle as a function of its circumference, the variable r must be replaced by an expression in terms of C. So, solve the circumference equation for r.

$$C = 2\pi r \Rightarrow r = \frac{C}{2\pi}$$

Now substitute $C/2\pi$ for r in the area equation and simplify.

$$A = \pi r^2 = \pi\left(\frac{C}{2\pi}\right)^2 = \pi\left(\frac{C^2}{4\pi^2}\right) = \frac{C^2}{4\pi}$$

To use function notation, name the function using any letter (e.g, A or f). Here, the function is named f. The input value is C, regardless of the function's name.

TIP

Area of the circle = $f(C)$

The area of a circle as a function of its circumference is given by the equation $f(C) = \dfrac{C^2}{4\pi}$.

Use the equation to find the area of a circle with circumference 20 centimeters.

$f(20) = \dfrac{20^2}{4\pi} = \dfrac{400}{4\pi} = \dfrac{100}{\pi}$ *Substitute 20 for C and simplify.*

Use the equation to find the circumference of a circle with area 50π square centimeters. In other words, find C when $f(C) = 50\pi$.

$f(C) = \dfrac{C^2}{4\pi}$

$50\pi = \dfrac{C^2}{4\pi}$ *Substitute 50π for $f(C)$.*

$50\pi(4\pi) = C^2$ *Multiply both sides by 4π.*

$200\pi^2 = C^2$ *Simplify.*

$\sqrt{200\pi^2} = C$ *Take the square root of each side.*

$10\pi\sqrt{2} = C$ *Simplify the radical.*

IMPORTANT

When solving a quadratic equation using square roots where the variable cannot be negative, only the positive square root of both sides is taken.

So, the area of a circle with circumference 20 centimeters is $\dfrac{100}{\pi}$ square centimeters, and the circumference of a circle with area 50π square centimeters is $10\pi\sqrt{2}$ centimeters.

1.4.7 Satisfying the Domain of a Function

Recall that the domain of an expression in one variable is all allowable values for that variable (i.e., values that result in a real number when substituted into the expression). For example, the domain of $5 + x^2$ is all real numbers, because the result is always a real number when any real number is substituted into the expression. The domain of $\sqrt{x+3}$ is all real x-values such that $x \geq -3$ (i.e., $[-3, \infty)$), because substituting any number that is greater than or equal to -3 results in a real number, but substituting numbers less than -3 results in a negative radicand, which is not a real number.

REMEMBER

The set of all real numbers can be represented using the symbol $\mathbb{R}$ or in interval notation as $(-\infty, \infty)$.

The domain of a function is the domain of the expression used to define the function. For example, the domain of $f(x) = 5 + x^2$ is all real numbers, because the domain of $5 + x^2$ is all real numbers. So, to find the domain of a function, simply find the domain of the related expression. As with expressions, the domains of functions do not include values that result in a 0 denominator or in a negative radicand.

EXAMPLE 9 Finding the Domain of a Function

Find the domain of each function.

A. f: $\{(-2, -3), (0, -1), (1, 0), (5, 4)\}$ **B.** $g(x) = \sqrt{4 - x}$ **C.** $p(x) = \dfrac{x^2}{4x^2 - 1}$

SOLUTION

A. The domain of f is the set of all x, which is the first coordinate in each ordered pair.

Domain of f: $\{-2, 0, 1, 5\}$

B. The domain of g is all real numbers except those x-values that make the radicand < 0. So, set the expression in the radicand ≥ 0 and solve for x.

$4 - x \geq 0 \Rightarrow x \leq 4$ *Reverse the inequality symbol when dividing by –1.*

Domain of g: $(-\infty, 4]$ *The domain is all real numbers less than or equal to 4.*

C. The domain of p is all real numbers except those x-values that make the denominator 0. So, set the expression in the denominator equal to 0 and solve for x.

$$4x^2 - 1 = 0 \Rightarrow (2x - 1)(2x + 1) = 0 \Rightarrow x = \frac{1}{2} \text{ or } x = -\frac{1}{2}$$

Domain of p: $\{x \mid x \neq \frac{1}{2}, x \neq -\frac{1}{2}\}$

IMPORTANT

The set of "all real numbers except $\frac{1}{2}$ or $-\frac{1}{2}$" can be expressed as $\mathbb{R}\backslash\{\frac{1}{2}, -\frac{1}{2}\}$.

WHAT'S THE BIG IDEA?

Explain the differences between $y = 3x + 1$, $f(x) = 3x + 1$, and $3x + 1$.

SECTION 1.4 EXERCISES

Warm Up

1. Given $x - 5 = y$, find y when $x = -4$.

2. Given $m = 3n^2 - n + 2$, find m when $n = 3$.

3. Given $y = 4x^3 + 9x$, find y when $x = 2$.

Just the Facts

Fill in the blanks.

4. The set of ordered pairs {(2, −3), (4, −7), (−3, 7), (0, 1)} represent a(n) _____ because each x-value corresponds to only one _____.

5. To determine graphically whether a relation is a function, draw a(n) _____. If it intersects the graph _____, it is a(n) _____.

6. $f(x)$ is just another name for _____, the dependent variable. The equation $y = 2x + 1$ would be rewritten using function notation as _____.

7. To find where two functions intersect, set the functions _____ to one another and solve.

8. The domain of $g(x) = 6x + 1$ is _____.

9. The four ways to express a function are _____, _____, _____, and _____. A list of ordered pairs is an example of expressing a function _____.

10. A combination of two or more functions in which each part has its own domain is a(n) _____-defined function. To evaluate, identify the part that corresponds to that _____ value, substitute that value, and _____.

Essential Skills

In Exercises 1–2, determine whether each graph represents a function.

1.

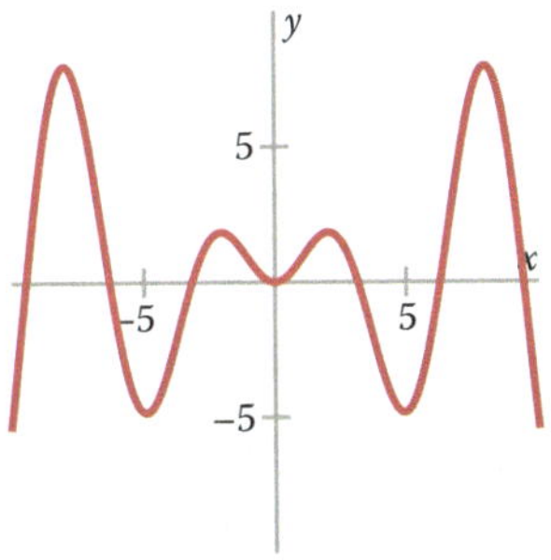

2.

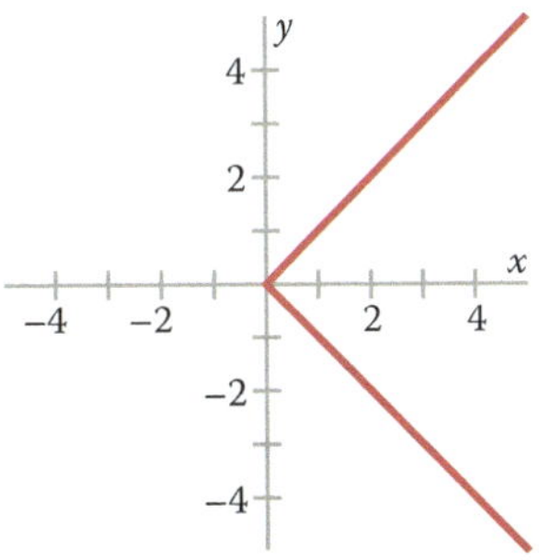

In Exercises 3–10, determine whether each equation represents a function.

3. $3 = |x - 6| + y$

4. $x^2 + 5 - y = 1$

5. $x^2 + 5 - y = 1$

6. $x - 6 + 8y^2 = 9$

7. $-4x^3 + x^2 + 3x = y$

8. $x^2 + 4y^2 = 36$

9. $y^2 = x - 2$

10. $y = 1$

In Exercises 11–28, evaluate.

11. If $f(x) = 2x^2 - x - 8$, find $f(-4)$ and $f(1)$.

12. If $g(x) = 2x - |5 + x|$, find $g(10)$ and $g(-7)$.

13. If $w(x) = 3x - |4 + x|$, find $w(3)$ and $w(-3)$.

14. If $r(x) = 6x^2 - 6x + 4$, find $r(2)$ and $r(-6)$.

15. If $t(x) = -x^3 + 2x$, find $t(-1)$ and $t(-4)$.

16. If $q(x) = x^4 - 5x^3 - 3x$, find $q(-3)$ and $q(1)$.

17. If $p(x) = |4x - 3|$, find $p(m - 7)$.

18. If $h(x) = x^2 - 2x$, find $h(3 + c)$.

19. If $j(x) = -3x^2 + 2x$, find $j(r - 5)$.

20. If $d(x) = -x^2 + 4x - 9$, find $d(s - 2)$.

21. If $b(x) = 2x^2 - 3x + 5$, find $b(4 + z)$.

22. If $v(x) = 4x^2 - 2x + 1$, find $v(1 - t)$.

23. If $f(x) = \begin{cases} -2x + 7 \text{ if } x < -8 \\ 3x^2 + x \text{ if } -8 \le x < -1 \\ -6x - 1 \text{ if } x \ge -1 \end{cases}$, find $f(-9)$ and $f(0)$.

24. If $p(x) = \begin{cases} 1 \text{ if } x \le -5 \\ |7 - 2x| \text{ if } -5 < x < 0 \\ x^2 \text{ if } x \ge 0 \end{cases}$, find $p(0)$ and $p(-12)$.

25. If $u(x) = \begin{cases} (2 - x)^2 \text{ if } x < -9 \\ -x^2 \text{ if } -9 \le x \le -2 \\ 8 - x \text{ if } x > -2 \end{cases}$, find $u(-10)$ and $u(-2)$.

26. If $k(x) = \begin{cases} 4 - 3x \text{ if } x \le 1 \\ -(3 - x)^2 \text{ if } 1 < x \le 4 \\ 5x \text{ if } x > 4 \end{cases}$, find $k(1)$ and $k(4)$.

27. If $c(x) = \begin{cases} 2 \text{ if } x \le -11 \\ |6x - 2| \text{ if } -11 < x < 5 \\ 6x^2 + 6 \text{ if } x \ge 5 \end{cases}$, find $c(6)$ and $c(-14)$.

28. If $n(x) = \begin{cases} 3x - 7 \text{ if } x < -4 \\ -x^2 + 5x \text{ if } -4 \le x < 3 \\ -4x - 2 \text{ if } x \ge 3 \end{cases}$, find $n(-4)$ and $n(3)$.

In Exercises 29–36, find the real x-value(s) such that the function equals 0.

29. $q(x) = 7x - 8$

30. $f(x) = 3x^2 + 13x - 10$

31. $v(x) = -\frac{2}{3}x - 14$

32. $h(x) = 16 - 9x$

33. $m(x) = 4x^2 - 17x - 15$

34. $w(x) = 18 - \frac{12}{5}x$

35. $p(x) = 5x^3 - 7x^2 - 6x$

36. $g(x) = 2x^3 + 3x^2 - 20x$

In Exercises 37–44, find the real x-value(s) so that the two functions are equal.

37. $g(x) = 10x - 9$ and $f(x) = -8x + 6$

38. $p(x) = 2 + 40x - x^2$ and $q(x) = 4x^2 + 6x - 5$

39. $m(x) = 4x + 13$ and $n(x) = x^2 - 8$

40. $d(x) = x^2 - 2x + 7$ and $c(x) = 29 - 14x + x^2$

41. $s(x) = 3x^2 - 9x$ and $t(x) = -7x - 2x + 3x^2$

42. $h(x) = 2x^2 - x - 15$ and $k(x) = 21 - x - 2x^2$

43. $j(x) = x^3 - 11x^2 + 3$ and $l(x) = -8x^2 + 18x + 3$

44. $u(x) = -2x^3 + 23x^2 - 8x$ and $v(x) = 2x^3 - 3x^2 + 4x$

In Exercises 45–46, express each function with an equation, table, and graph.

45. At the local department store, the average pay is $9 per hour. The total pay p an employee earns is a function of the number of hours h they work.

46. If a person drives 400 miles at an average of 50 miles per hour, then their distance d from the destination (in miles) is a function of the number of hours h driven.

In Exercises 47–49, use the following information.

Use a function to find the perimeter of a rectangle with the given area and length.

47. area: 20 square feet; length: 5 feet

48. area: 36 square feet; length: 4 feet

49. area: 313.5 square feet; length: 19 feet

In Exercises 50–52, use the following information.

A piece of rope is cut into two pieces. One piece is used to form a circle, and the other is used to form a square. Use the given length to write a function f representing the area of the circle as a function of the length of one side of the square s. Then write a function g representing the area of the square as a function of the radius of the circle r. Hint: If C is the circumference of the circle and P is the perimeter of the square, then the length equals $C + P$.

50. length: 20 feet

51. length: 12 feet

52. length: 36 feet

In Exercises 53–60, find the domain of each function.

53. $\{(2, -4), (4, 0), (-3, -14), (0, -8)\}$

54. $f(x) = \sqrt{-2x+8}$

55. $v(x) = \sqrt{63-7x}$

56. $w(x) = \sqrt{11+5x}$

57. $p(x) = \dfrac{x}{x^2 - 10x - 24}$

58. $q(x) = \dfrac{x-3}{x^2 + x - 12}$

59. $h(x) = \dfrac{x+6}{x^2 - 36}$

60. $d(x) = \dfrac{x-4}{x^2 + 3x - 40}$

Extensions

61. Find the domain of $f(x) = \dfrac{\sqrt{x^2-9}}{x^2 - 5x - 6}$.

62. True or False? The following graph is a function.

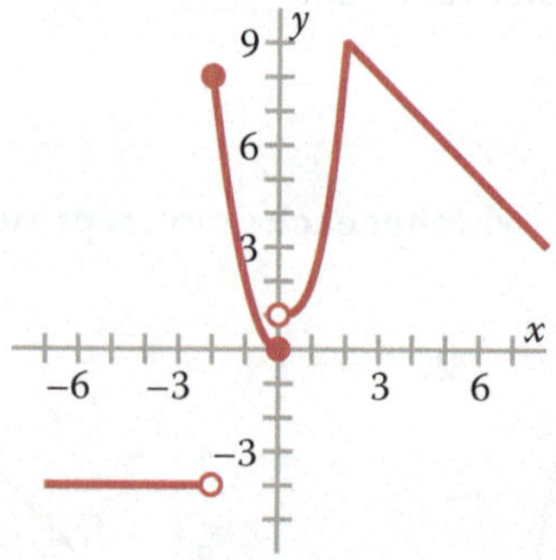

63. For $f(x) = x^2 + 3x - 10$, find $\dfrac{f(x+h) - f(x)}{h}$.

64. If $g(x) = -5x^2 - 3kx + 15$ and $g(2) = 10$, find k.

65. The height y (in feet) of a ball thrown by a child is $y = -0.083x^2 + 2x + 4$, where x is the horizontal distance in feet from the point at which the ball is thrown. Will the ball fly over the head of a 4-foot-tall child trying to catch the ball 24 feet away?

66. Will the graphs of $f(x) = \sqrt{2x^2 + x - 6}$ and $g(x) = x - 2$ intersect?

1.5 GRAPHS OF FUNCTIONS

OBJECTIVES

- Identify the domain and range of functions.
- Graph common types of functions.
- Graph piecewise-defined functions.
- Graph and use greatest integer (step) functions.
- Find the real zeros of functions.
- Identify intervals where functions are strictly increasing or decreasing.
- Find the maximum or minimum value(s) of functions.

PREREQUISITE VOCABULARY TERMS

absolute value equation
domain
function
linear equation
piecewise-defined function
quadratic equation
range

1.5.1 Finding the Domain and Range

Recall that the domain of a function is all allowable input values (x-values) and that the range of a function is all possible output values (y-values). The domain and range of a function can be identified visually from a graph of the function. A function's graph is the collection of ordered pairs (x, y) such that x is in the function's domain and y is in the function's range.

If function notation is used, then the ordered pairs can be expressed as $(x, f(x))$ or as (x, y). Since $f(x) = y$, each point on a function's graph is located x units from the y-axis and $f(x)$ units from the x-axis, as shown in Figure 1.5a.

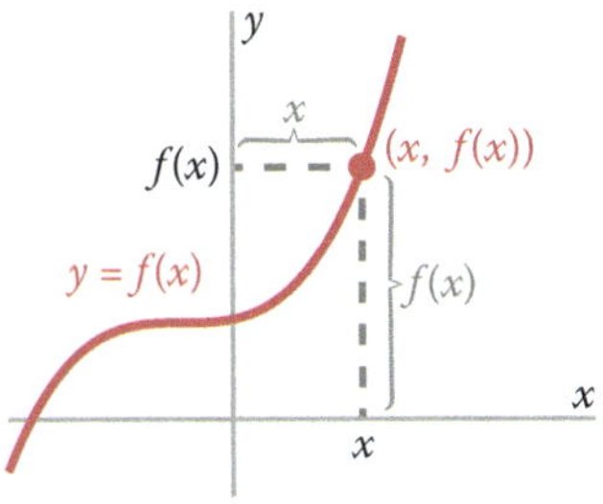

Figure 1.5a

Both the domain and range of the function shown in Figure 1.5a are all real numbers. It is assumed that the graph extends infinitely at both ends, unless the graph shows an open or closed point at one end.

Recall that an open point (open circle) indicates that the point is *not* included on a graph, and a closed point (closed circle) indicates that the point is included on a graph.

Evaluating a Function Graphically

Recall that a function can be evaluated algebraically for x by substituting the given x-value into the equation and simplifying to find the corresponding y-value (i.e., $f(x)$-value). For example, to evaluate $f(x) = 2x - 3$ for $x = 1$, substitute 1 into the function for x and simplify.

$$f(1) = 2(1) - 3 = -1$$

So, when $x = 1, f(x) = -1$ for $f(x) = 2x - 3$.

> **TIP**
>
> *The statement $f(x) = -1$ is equivalent to $y = -1$.*

These corresponding x and $f(x)$ values are an ordered pair on the graph of f. Specifically, the ordered pair $(1, -1)$ is a point on the graph of $f(x) = 2x - 3$. It follows that a function can also be evaluated by using the graph of the function. To evaluate a function for x using the graph, identify the point on the graph with this x-coordinate. The corresponding y-value from this point on the graph is the $f(x)$-value. This method for evaluating a function is demonstrated in **Example 1**.

EXAMPLE 1 Finding the Domain and Range of a Function

Identify the domain and range of g. Then, using the graph, find $g(0)$, $g(3)$, and all x such that $g(x) = 0$.

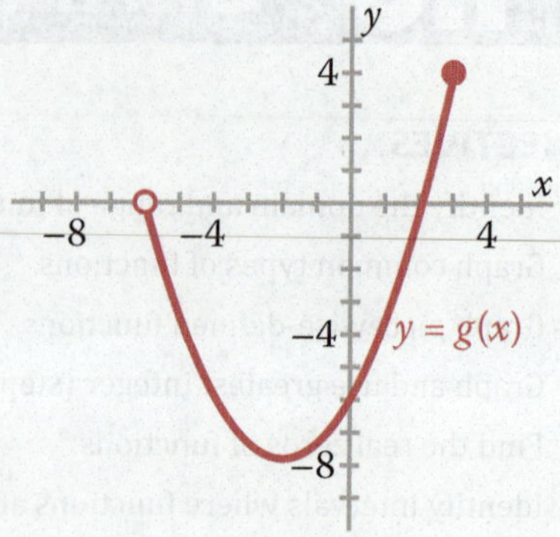

SOLUTION

The graph of g has a point at each end. Therefore, it does not continue infinitely on either end (i.e., to ∞ or −∞). Because the graph contains an open point at (−6, 0), this point is not included on the graph. Because the graph contains a closed point at (3, 4), this point is included on the graph.

Domain of g

The points on the graph includee all x-values between −6 and 3, not including −6 (because there is an open point where $x = -6$). So, the domain is all x-values such that $-6 < x \leq 3$.

Domain of g: (−6, 3] *The domain is all x-values between −6 and 3, not including −6.*

CAUTION

There is an open point at (−6, 0), but 0 is still in the function's range because the graph contains another point where y = 0: (2, 0). If the graph did not contain a point where y = 0, then 0 would not be in the function's range.

Range of g

The points on the graph include all y-values between −8 and 4. So, the range is all y-values such that $-8 \leq y \leq 4$.

Range of g: [−8, 4] *The range is all y-values between −8 and 4.*

Find $g(0)$, $g(3)$, and x such that $g(x) = 0$

Use the graph to find the coordinates that correspond to $x = 0$, $x = 3$, and $y = 0$.

$g(0) = -6$ *The graph passes through (0, −6), so when $x = 0$, $g(x) = -6$.*

$g(3) = 4$ *The graph passes through (3, 4), so when $x = 3$, $g(x) = 4$.*

The graph includes the point (2, 0), and this is the only point on the graph where $y = 0$. So, $g(2) = 0$. Therefore, $g(x) = 0$ only when $x = 2$.

1.5.2 Graphing Some Important Functions

Several general types of equations and their graphs were demonstrated previously, including linear, quadratic, and absolute value equations. The graph of a function can be sketched using the general shape of a type of function and a few points on the function. Recall that an equation in two variables can be graphed by making a table (i.e., listing several chosen x-values and the corresponding y-value for each in a table) and then plotting the points from that table. Additionally, recall that the x- and y-intercepts (the points on the graph where $y = 0$ and $x = 0$, respectively) are also good points to find when graphing.

The following table shows the basic function for several function types. All functions of the same type will have a similarly shaped graph. For example, all absolute value functions will have the general V-shape, but the V may point up or down (or be wider or more narrow than the graph of $f(x) = |x|$), depending on the specific equation.

TIP

A constant function is a type of linear function.

Types of Functions

Constant: $f(x) = c$	Linear: $f(x) = x$	Quadratic: $f(x) = x^2$
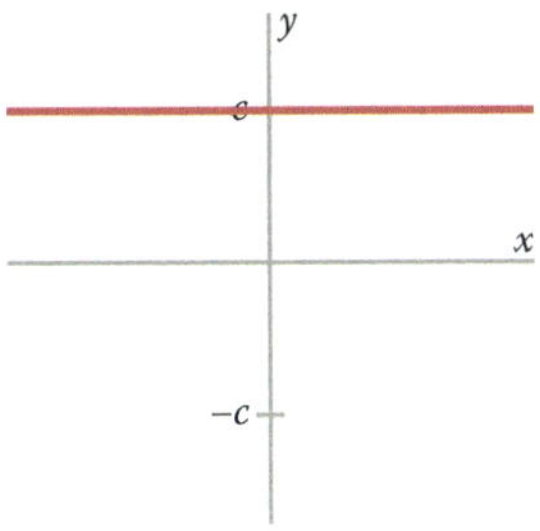	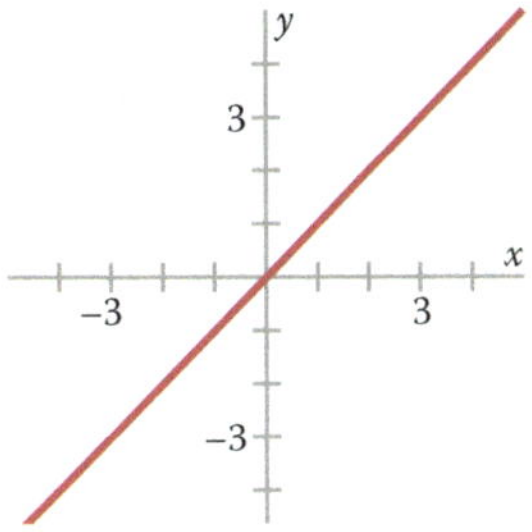	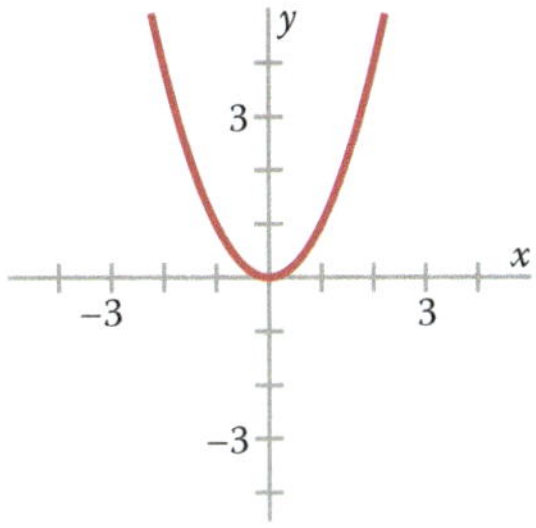
Cubic: $f(x) = x^3$	**Square Root: $f(x) = \sqrt{x}$**	**Absolute Value: $f(x) = \lvert x \rvert$**
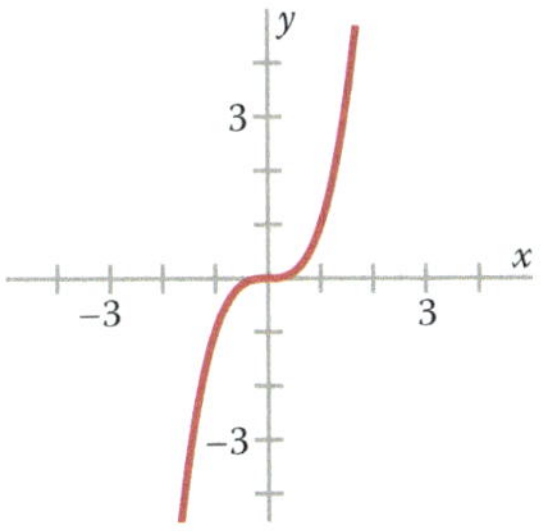	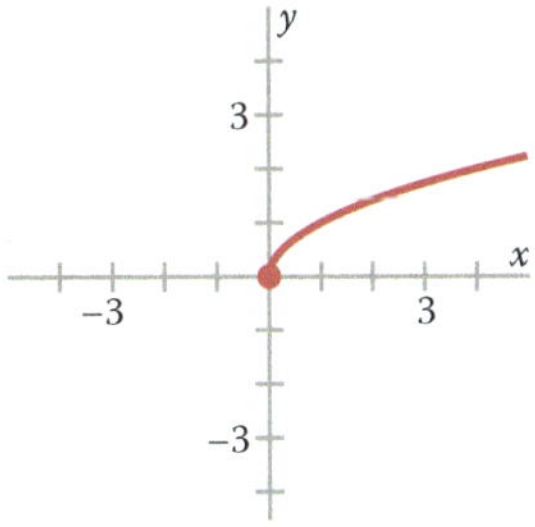	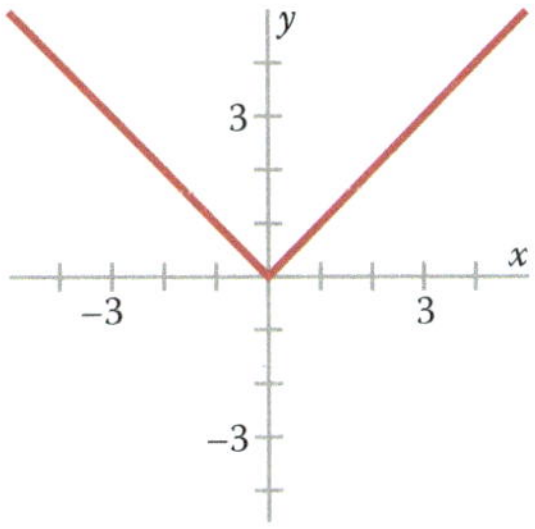

Steps for Graphing a Function

❶ Identify the type of function and the general shape of the function's graph.

❷ Find the y-intercept(s) by evaluating $f(0)$, and the x-intercept(s) by solving $f(x) = 0$. Plot these points.

❸ Make a table of points by choosing several allowable values for x and finding the corresponding $f(x)$-value. Plot each ordered pair $(x, f(x))$ on a coordinate plane.

❹ Connect the points with a straight line or smooth curve, corresponding to the general shape of the graph defined by the function type.

EXAMPLE 2 Graphing Functions by Plotting Points

Find the intercepts, make a table, and sketch a graph for each function.

A. $f(x) = x^3 - 1$ **B.** $f(x) = -|x| + 2$

SOLUTION

A. Notice that the function type is cubic. The basic cubic function is $f(x) = x^3$.

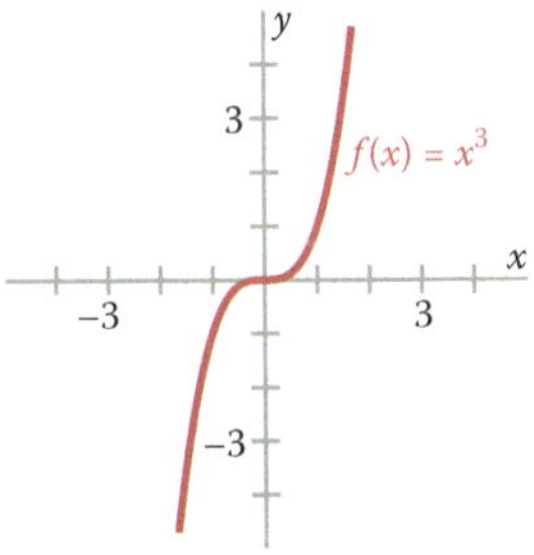

Figure 1.5b

❶ The function's graph will be in the general shape of a cubic function (Figure 1.5b).

❷ Find the x- and y-intercepts.

y-intercept: $f(0) = (0)^3 - 1 = -1$ x-intercept: $f(x) = 0 \Rightarrow 0 = x^3 - 1 \Rightarrow x = 1$

There is one y-intercept, $(0, -1)$, and one x-intercept, $(1, 0)$.

TIP

Solve $0 = x^3 - 1$ by adding 1 to each side, then taking the cube root of each side.

$x^3 = 1$

$x = \sqrt[3]{1} = 1$

❸ Choose x-values near the x-intercept for the table, such as –2, –1, and 2.

x	$f(x)$
–2	–9
–1	–2
2	7

$f(-2) = (-2)^3 - 1 = -9$

$f(-1) = (-1)^3 - 1 = -2$

$f(2) = (2)^3 - 1 = 7$

❹ Plot the intercepts and the points from the table. Then sketch a curve through the points in the general shape of a cubic function.

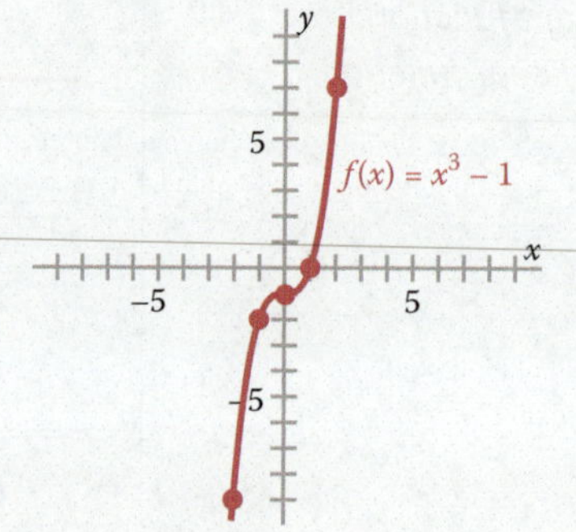

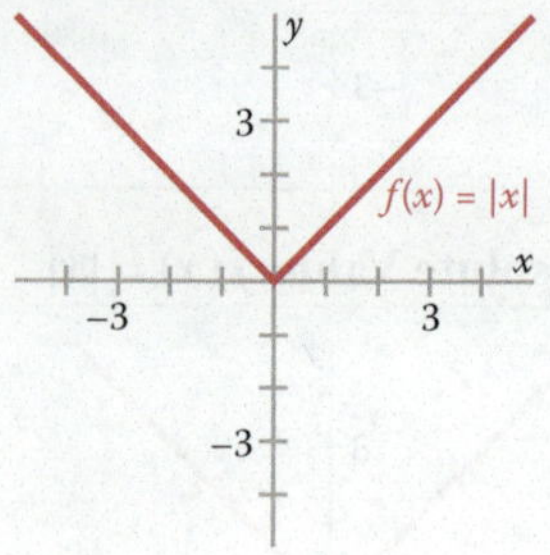

Figure 1.5c

B. Notice that the function type is absolute value. The basic absolute value function is $f(x) = |x|$.

❶ The function's graph will be in the general shape of an absolute value function (Figure 1.5c).

❷ Find the x- and y-intercepts.

y-intercept: $f(0) = -|0| + 2 = 2$ x-intercept: $f(x) = 0 \Rightarrow 0 = -|x| + 2 \Rightarrow x = 2$ or $x = -2$

There is one y-intercept, (0, 2), and two x-intercepts, (–2, 0) and (2, 0).

❸ Choose x-values near the x-intercepts for the table, such as –3, –1, 1, and 3.

x	$f(x)$
–3	–1
–1	1
1	1
3	–1

$f(-3) = -|-3| + 2 = -1$

$f(-1) = -|-1| + 2 = 1$

$f(1) = -|1| + 2 = 1$

$f(3) = -|3| + 2 = -1$

❹ Plot the intercepts and the points from the table. Then sketch the graph through the points in the general shape of an absolute value function.

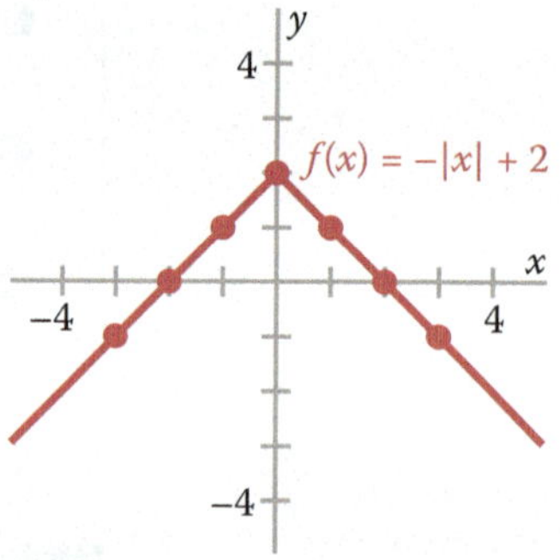

TIP

Solve $0 = -|x| + 2$ by isolating the absolute value expression, then writing the two equations. Remember, if $|A| = B$, then $A = B$ or $A = -B$.

$$0 = -|x| + 2$$
$$-|x| + 2 = 0$$
$$-|x| = -2$$
$$|x| = 2$$
$$x = 2 \text{ or } x = -2$$

NOTICE THAT

The graph is symmetric with respect to the y-axis.

1.5.3 Graphing Piecewise-Defined Functions

Recall that a piecewise-defined function is a combination of two or more functions where a specific interval from the domain corresponds to each of the function's pieces. The graph of a piecewise-defined function is also divided into separate pieces.

EXAMPLE 3 Graphing a Piecewise-Defined Function

Sketch the graph of $f(x)=\begin{cases}1-x & \text{if } x<-3\\ 3 & \text{if } -3\le x<0\\ 2x+1 & \text{if } x\ge 0\end{cases}$.

SOLUTION

Graph the three pieces for the specific corresponding domain values. Notice that all three pieces are linear functions, so the graph of each piece is a line.

Function Piece	Domain	Graph Description
$f(x)=1-x$	$\{x \mid x<-3\}$	line with slope -1 and y-intercept 1
$f(x)=3$	$\{x \mid -3\le x<0\}$	horizontal line with y-intercept 3
$f(x)=2x+1$	$\{x \mid x\ge 0\}$	line with slope 2 and y-intercept 1

IMPORTANT

When graphing each piece of a piecewise-defined function, be sure to use an open point (open circle) for any endpoint when the corresponding x-value is not included in the domain.

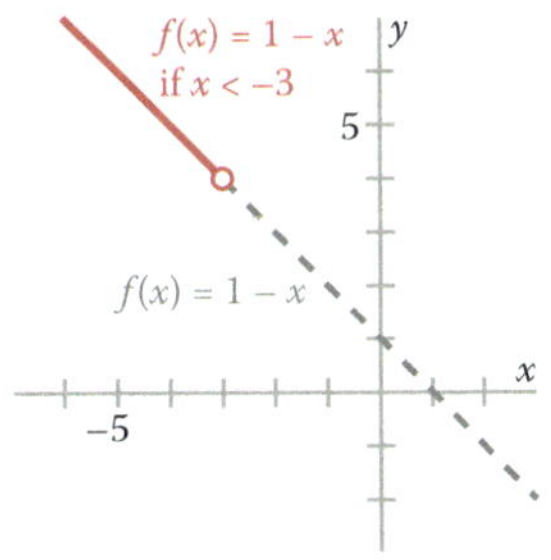

Figure 1.5d

$f(x)=1-x$ if $x<-3$

Sketch the line $y=1-x$, then remove the section of the line where $x\ge -3$. Since $x=-3$ is not included in the domain of this piece, use an open point where $x=-3$, which is $(-3, 4)$, as shown in Figure 1.5d.

$f(x)=3$ if $-3\le x<0$

Sketch the horizontal line $y=3$ starting where $x=-3$ and ending where $x=0$. Since -3 is included in the domain of this piece, use a closed point at $(-3, 3)$. Since 0 is not included in the domain of this piece, use an open point at $(0, 3)$.

$f(x)=2x+1$ if $x\ge 0$

Sketch the line $y=2x+1$, then remove the section of the line where $x<0$. Since $x=0$ is included in the domain of this piece, use a closed point where $x=0$, which is $(0, 1)$.

NOTICE THAT

If both points where $x=-3$ were closed points, then the graph would not be a function. Similarly, if both points where $x=0$ were closed points, then the graph would not be a function.

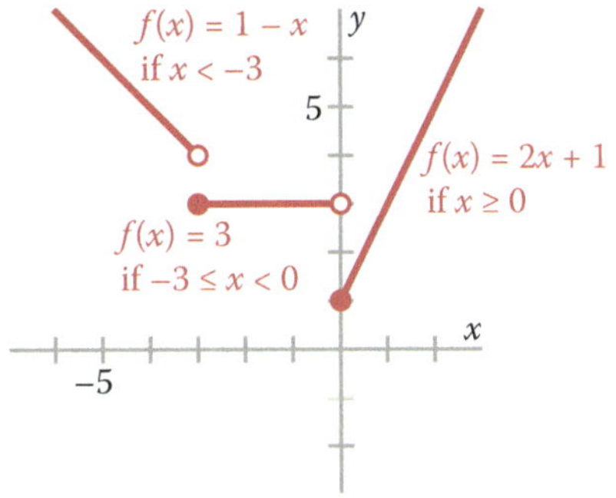

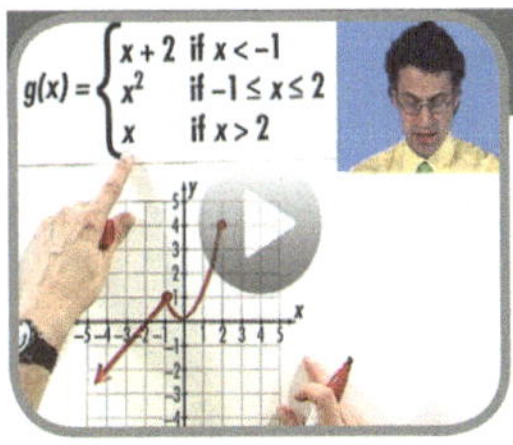

1.5.4 Using a Table to Graph Piecewise-Defined Functions

The piecewise-defined function in the preceding example contained three pieces where each piece was a linear function, and so the graph of each piece was a line. Any type of function can be included in a piecewise-defined function. In the case where a piece is not linear, graph the piece by making a table of values and noting the general shape of the graph, as demonstrated in **Example 4**.

EXAMPLE 4 Graphing a Piecewise-Defined Function by Using a Table

Sketch the graph of $f(x) = \begin{cases} 0.5x^2 & \text{if } x \le 0 \\ -x & \text{if } 0 < x < 2 \\ |x| + 1 & \text{if } x \ge 2 \end{cases}$.

SOLUTION

Graph the three pieces for the specific corresponding domain values.

Function Piece	Domain	Graph Description
$f(x) = 0.5x^2$	$\{x \mid x \le 0\}$	parabola with y-intercept 0
$f(x) = -x$	$\{x \mid 0 < x < 2\}$	line with slope -1 and y-intercept 0
$f(x) = \lvert x \rvert + 1$	$\{x \mid x \ge 2\}$	V-shaped graph with y-intercept 1

Since each piece is not a line, use a table to graph the piecewise-defined function. Include the endpoints of the domain intervals in the x-values for the table.

TIP

The entire table does not need to be completed. Substitute the x-values from the table into only the corresponding piece of the function. If a domain interval's endpoint is included for more than one piece (as is the case with $x = 0$ and $x = 2$), evaluate that endpoint value in both applicable functions.

x	$f(x) = 0.5x^2$	$f(x) = -x$	$f(x) = \lvert x \rvert + 1$
−4	8		
−2	2		
0	0	0	
1		−1	
2		−2	3
3			4
4			5

Plot the points from the table, keeping in mind the general shape of each piece.

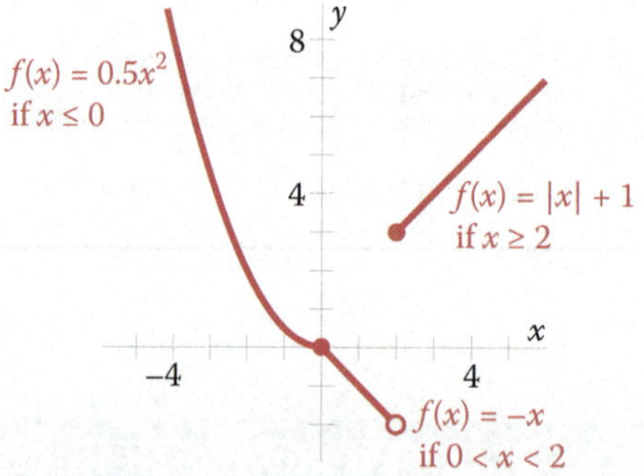

(0, 0) is included in the graph of $f(x) = 0.5x^2$, so a closed circle is used for the point.

WHAT'S THE BIG IDEA?

Explain the instances where open and closed points are used in the graph of a piecewise-defined function.

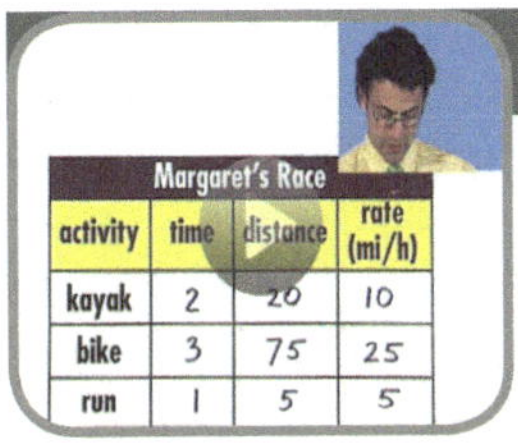

1.5.5 Modeling with Piecewise-Defined Functions

A piecewise-defined function will be used to model a situation when the independent variable (domain) is divided into parts. For example, suppose a person earns \$30 for mowing a yard if it takes him between 0 and 2 hours. He earns \$50 if it takes him longer than 2 hours. This situation's independent variable (the domain), the amount of time spent mowing a yard, is divided into two parts: 1) between 0 and 2 hours, and 2) more than 2 hours.

This situation is modeled by the piecewise-defined function shown in Figure 1.5e, where $f(x)$, the total pay, is a function of x, the number of hours spent mowing.

$$f(x) = \begin{cases} 30 & \text{if } 0 < x \leq 2 \\ 50 & \text{if } x > 2 \end{cases}$$

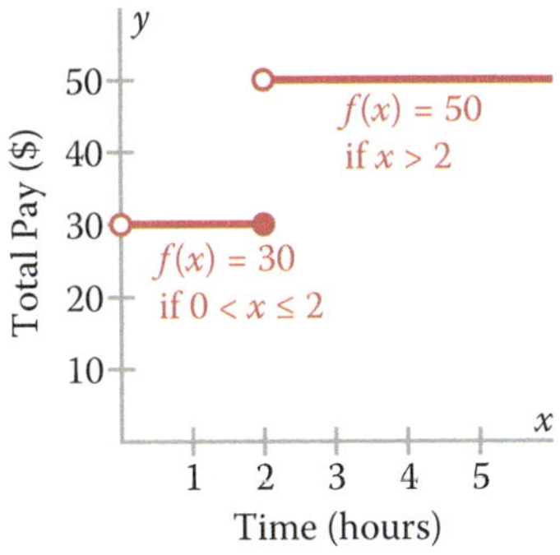

Figure 1.5e

EXAMPLE 5 Modeling a Piecewise-Defined Function

NOTICE THAT

The pay here is a rate. For example, "\$200 per project for the first 5 projects" means that he is paid \$200 for 1 project, \$400 for 2 projects, \$600 for 3 projects, and so on.

A freelance designer earns \$200 per project for the first 5 projects, \$250 per project for the next 10 projects, and \$350 per project for any additional projects he completes. Write a piecewise-defined function for his total pay.

SOLUTION

The domain is the number of projects completed. So, write a function for the total pay, $f(x)$, as a function of the number of projects completed, x (where x is a whole number).

The domain is divided into 3 parts: the first 5 projects, the next 10 projects (the 6th through the 15th projects), and all additional projects (all projects after the 15th project).

Domain Parts: $x \leq 5$, $5 < x \leq 15$, and $x > 15$

Determine how the total pay is calculated for each piece.

Domain	Pay Description	Function Piece
$\{x \mid x \leq 5\}$	\$200 per project for projects 1 through 5	$f(x) = 200x$
$\{x \mid 5 < x \leq 15\}$	\$250 per project for projects 6 though 15	$f(x) = 250x - 250$
$\{x \mid x > 15\}$	\$350 per project for projects 16 and greater	$f(x) = 350x - 1750$

The piecewise-defined function for his total pay is $f(x) = \begin{cases} 200x & \text{if } x \leq 5 \\ 250x - 250 & \text{if } 5 < x \leq 15 \\ 350x - 1750 & \text{if } x > 15 \end{cases}$.

CAUTION

It may seem that the function for projects 6 through 15 should be $f(x) = 250x$ (since the pay for projects 6 through 15 is \$250/project). However, if the designer completed 8 projects, he would not earn \$250(8) = \$2000. Instead, he would earn \$200(5) + \$250(3) = \$1750 (\$200(5) for the first 5 projects and \$250(3) for projects 6, 7, and 8).

The function corresponding to projects 6 through 15 must take into account that the pay rate is \$200 per project for projects 1 through 5. So, if he completes between 6 and 15 projects, the total pay can be calculated by adding \$200(5) and \$250(x – 5), where \$200(5) is the pay for the first 5 projects and \$250(x – 5) is the pay for projects over 5.

$$f(x) = 200(5) + 250(x - 5) = 1000 + 250x - 1250 = 250x - 250$$

Similarly, the function for 16 or more projects must take into account that the pay rate for projects 1 through 5 and for projects 6 through 15 was not \$350 per project. The function for 16 or more projects can be found by adding \$200(5), \$250(10), and \$350(x – 15), where \$200(5) is the pay for the first 5 projects, \$250(10) is the pay for projects 6 through 15, and \$350(x – 15) is the pay for projects over 15.

$$f(x) = 200(5) + 250(10) + 350(x - 15) = 1000 + 2500 + 350x - 5250 = 350x - 1750$$

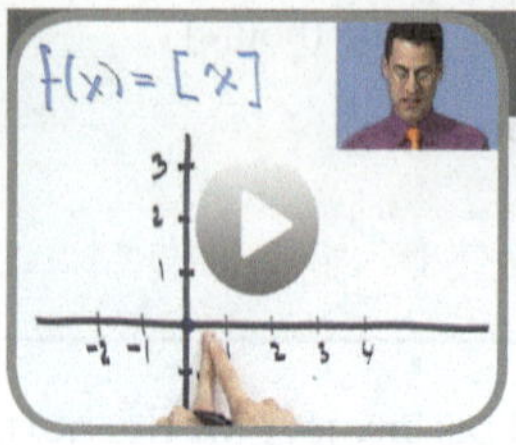

1.5.6 The Greatest Integer Function

The **greatest integer function**, denoted as [[x]], is a piecewise-defined function. The output of the greatest integer function is the greatest integer that is less than or equal to the input value. So, [[x]] represents the greatest integer that is less than or equal to x. The greatest integer function is an example of a **step function**, which is a function where the graph is a series of disconnected horizontal line segments resembling a staircase.

EXAMPLE 6 Evaluating the Greatest Integer Function

Evaluate [[16.9]], [[405.111]], [[3]], [[–5.7]], and [[–9(2)]] – [[0.4]] + [[–0.82]].

SOLUTION

Determine the greatest integer that is less than or equal to each input value. If the input value x is an integer, then the output is the same integer. If the input value x is not an integer, then [[x]] is the greatest integer less than x.

[[16.9]] = 16	*16 is the greatest integer less than 16.9.*
[[405.111]] = 405	*405 is the greatest integer less than 405.111.*
[[3]] = 3	*3 is equal to 3.*
[[–5.7]] = –6	*–6 is the greatest integer less than –5.7.*

The next expression involves addition and subtraction. The greatest integer function acts like grouping symbols in the order of operations.

[[–9(2)]] – [[0.4]] + [[–0.82]] = [[–18]] – [[0.4]] + [[–0.82]]	*Multiply.*
= –18 – 0 + (–1)	*Evaluate each function.*
= –19	*Simplify.*

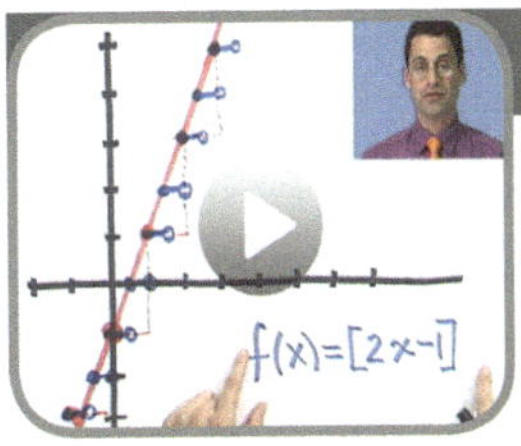

1.5.7 Graphing the Greatest Integer Function

The domain of the greatest integer function is all real numbers. The range of the greatest integer function is only integers because the output of the function is always an integer. Therefore, non-integer y-values are not represented on the graph of $f(x) = [[x]]$, and so the graph is a series of horizontal line segments.

EXAMPLE 7 Graphing the Greatest Integer Function

Sketch the graph of $y = [[x]]$.

SOLUTION

Make a table to locate some points on the graph. Use some x-values that are not integers in the table.

x	−2.5	−2	−1.5	−1	−0.5	0	0.5	1	1.5	2	2.5
$f(x)$	−3	−2	−2	−1	−1	0	0	1	1	2	2

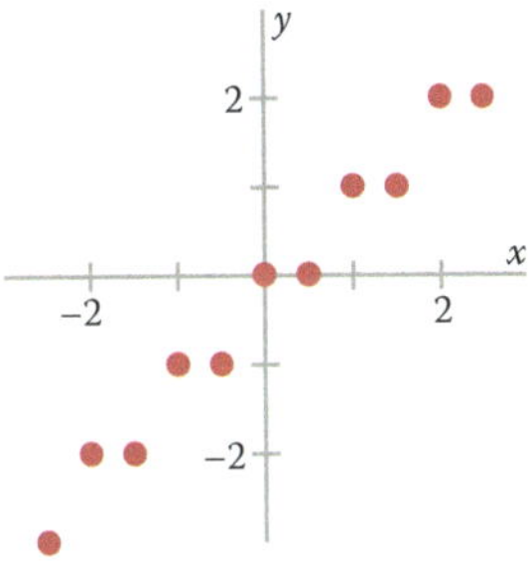

Figure 1.5f

Plot these points (Figure 1.5f) and look for a pattern.

Notice the pattern with the x-values ≥ 0.

$0 \leq x < 1 \Rightarrow y = 0$

$1 \leq x < 2 \Rightarrow y = 1$

$2 \leq x < 3 \Rightarrow y = 2$

Similarly, for x-values < 0, the pattern continues.

$-1 \leq x < 0 \Rightarrow y = -1$

$-2 \leq x < -1 \Rightarrow y = -2$

$-3 \leq x < -2 \Rightarrow y = -3$

TIP

The x-intervals and corresponding y-values can be written in a table.

x	y
$-3 \leq x < -2$	-3
$-2 \leq x < -1$	-2
$-1 \leq x < 0$	-1
$0 \leq x < 1$	0
$1 \leq x < 2$	1
$2 \leq x < 3$	2

The pattern forms a piecewise-defined function. Since all of the x-values between two consecutive integer x-values correspond to a single range value, the graph is a series of horizontal line segments. Since only the smaller of the two consecutive integer x-values is included in the domain of that interval, the graph includes a closed point on the left end and an open point on the right end of each horizontal line segment.

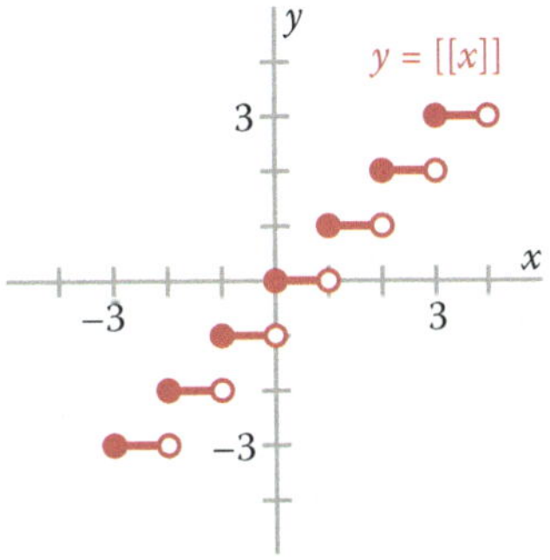

$y = 3$ if $3 \leq x < 4$
$y = 2$ if $2 \leq x < 3$
$y = 1$ if $1 \leq x < 2$
$y = 0$ if $0 \leq x < 1$
$y = -1$ if $-1 \leq x < 0$
$y = -2$ if $-2 \leq x < -1$
$y = -3$ if $-3 \leq x < -2$

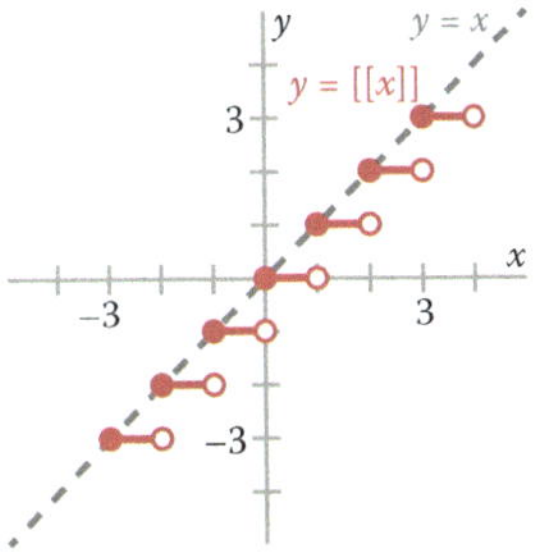

Figure 1.5g

The horizontal line segments in the graph of $y = [[x]]$ occur where the range values are integers. Notice that the line $y = x$ (the related line to $y = [[x]]$) passes through the left endpoint of each line segment in the graph of $y = [[x]]$, as shown in Figure 1.5g. This is not a coincidence. For all greatest integer functions where the expression is linear, the related line passes through the endpoints of the line segments. This fact can be used to graph a greatest integer function where the expression is linear.

EXAMPLE 8 Graphing the Greatest Integer Function

Sketch the graph of $y = [[2x + 1]]$.

SOLUTION

Sketch the related line, $y = 2x + 1$ (slope 2 and y-intercept 1), using a dashed line.

> **TIP**
>
> *Make a table of points if this process is difficult to visualize.*

Then draw a horizontal line segment starting at each point on the line with an integer y-value. End each line segment at the x-value that corresponds to the next point with an integer y-value.

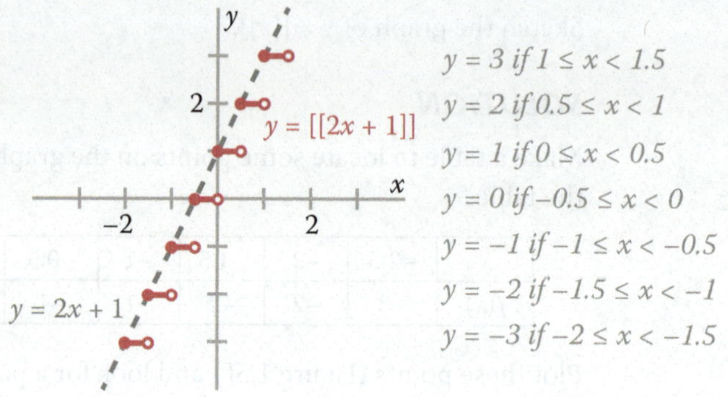

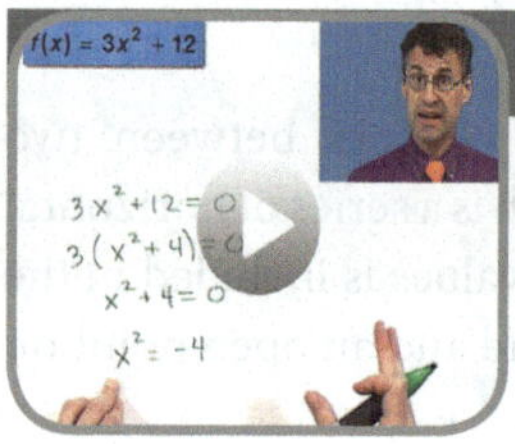

1.5.8 Finding Zeros of a Function

Recall that a zero of a function is an x-value such that $f(x) = 0$. Furthermore, a zero of a function is the x-coordinate of an x-intercept of the function.

> **DEFINITION**
>
> The **zeros** of a function $f(x)$ are the x-values where $f(x) = 0$.
>
> If the graph of f has an x-intercept at $(a, 0)$, then a is a zero of f.

It follows that a function's zeros can be found algebraically or graphically. To find a function's zeros algebraically, substitute 0 for $f(x)$ and solve for x. (The method used for solving depends upon the type of function, e.g., linear, quadratic, cubic.) To find a function's zeros graphically, sketch a graph of the function and then identify the x-coordinate of any x-intercepts. (The method used for graphing again depends upon the type of function.)

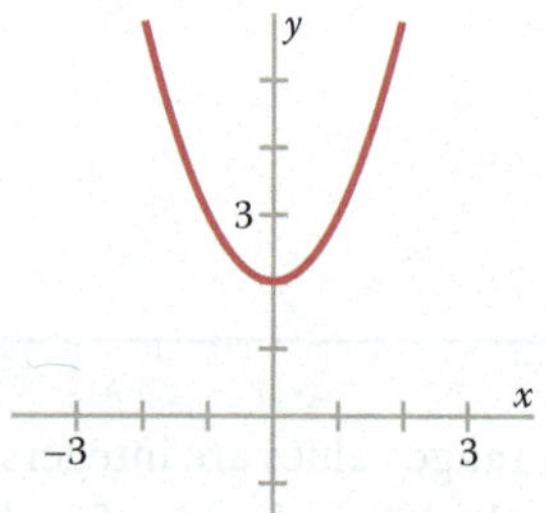

Figure 1.5h

Just as the graph of a function can have any number of x-intercepts, a function can have any number of zeros, including no zeros. Since it is possible for a graph of a function to not intersect the x-axis, as is the case in Figure 1.5h, the graph of a function can have no x-intercepts, and will then also have no zeros.

EXAMPLE 9 Finding the Real Zeros of a Function

Find the real zeros of each function.

A. $f(x) = 2x^3 + 5x^2 - 2x - 5$ **B.** $g(x) = 2\sqrt{x^2 + 9}$ **C.** $h(x) = \dfrac{2 - x}{3x + 5}$

SOLUTION

To find a function's zero(s), substitute 0 for the range value and find the corresponding domain value(s), if any.

A. The function is a polynomial function, so solve by factoring.

$f(x) = 2x^3 + 5x^2 - 2x - 5$	
$0 = 2x^3 + 5x^2 - 2x - 5$	*Substitute 0 for $f(x)$.*
$0 = x^2(2x + 5) - (2x + 5)$	*Factor by grouping.*
$0 = (2x + 5)(x^2 - 1)$	*Factor $(2x + 5)$ from each term.*
$2x + 5 = 0$ or $x^2 - 1 = 0$	*Set each factor equal to 0.*
$2x = -5$ $\quad x^2 = 1$	*Solve each equation for x.*
$x = -\dfrac{5}{2}$ $\quad x = \pm 1$	

REMEMBER

Take ± the square root of each side to solve $x^2 = 1$.

So, $f(x) = 2x^3 + 5x^2 - 2x - 5$ has zeros at $x = -\dfrac{5}{2}$, -1, and 1.

Use a graphing calculator to graph the function and check the answer.

The function's graph has x-intercepts at $\left(-\frac{5}{2}, 0\right)$, $(-1, 0)$, and $(1, 0)$.

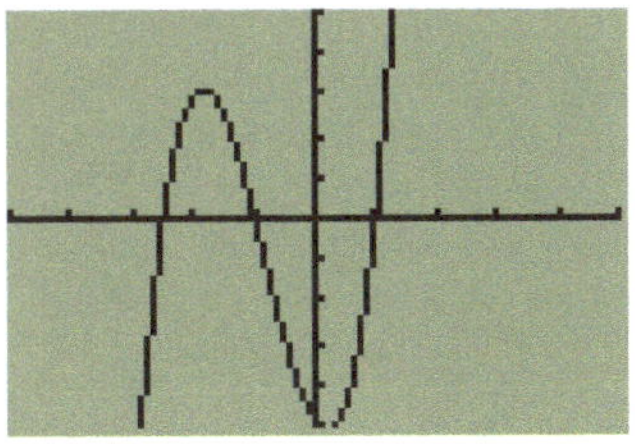

B. The function is a square root function, so solve by isolating the radical expression on one side of the equation and then squaring both sides.

$g(x) = 2\sqrt{x^2 + 9}$	
$0 = 2\sqrt{x^2 + 9}$	*Substitute 0 for $g(x)$.*
$0 = \sqrt{x^2 + 9}$	*Divide each side by 2 to isolate the radical.*
$(0)^2 = \left(\sqrt{x^2 + 9}\right)^2$	*Square each side.*
$0 = x^2 + 9$	*Simplify.*
$-9 = x^2$	*Subtract 9 from each side.*
$\pm\sqrt{-9} = x$	*Take ± the square root of each side.*

The x-values are not real numbers, so the function has no real zeros. Notice that the graph of the function, as shown in Figure 1.5i, does not intersect the x-axis.

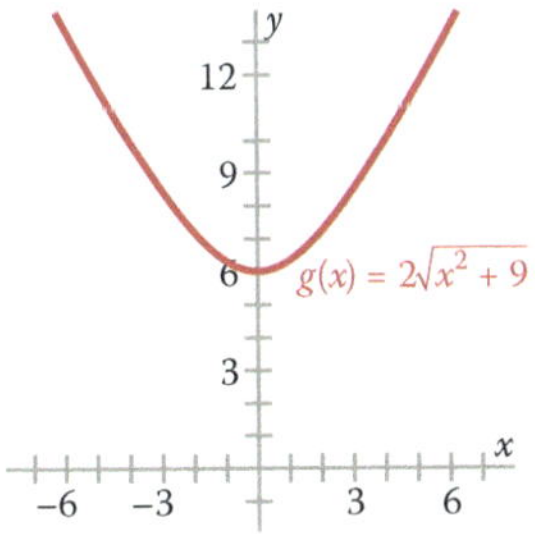

Figure 1.5i

C. The function is a rational function, so solve by setting the expression in the numerator equal to zero and solving for x.

$h(x) = \dfrac{2 - x}{3x + 5}$	
$0 = \dfrac{2 - x}{3x + 5}$	*Substitute 0 for $h(x)$.*
$2 - x = 0$	*Set the expression in the numerator equal to 0.*
$x = 2$	*Solve for x.*

So, $h(x) = \dfrac{2 - x}{3x + 5}$ has one zero at $x = 2$. Notice that the graph of the function, as shown in Figure 1.5j, intersects the x-axis only at $(2, 0)$.

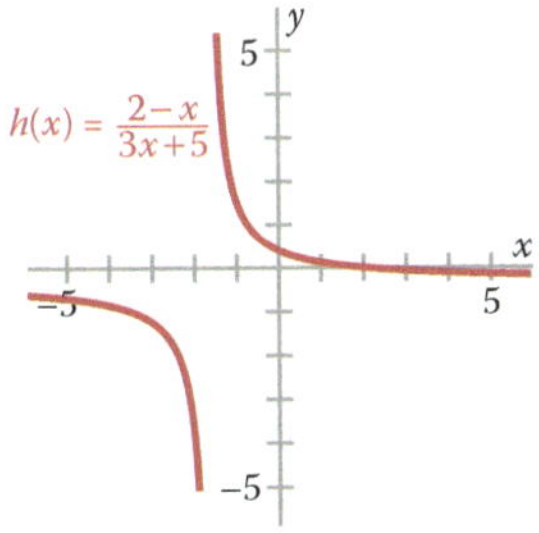

Figure 1.5j

WHAT'S THE BIG IDEA?

Explain the relationship between a function's zero and the graph of a function, as well as the relationship between a function's zero and the equation of a function.

1.5.9 Determining Intervals Over Which a Function Is Increasing

Recall that a line with positive slope increases from left to right, as shown in Figure 1.5k, and a line with negative slope decreases from left to right, as shown in Figure 1.5l. So, we say that a linear function with a positive slope is **increasing** for all domain values, and a linear function with a negative slope is **decreasing** for all domain values.

Linear Function with Positive Slope

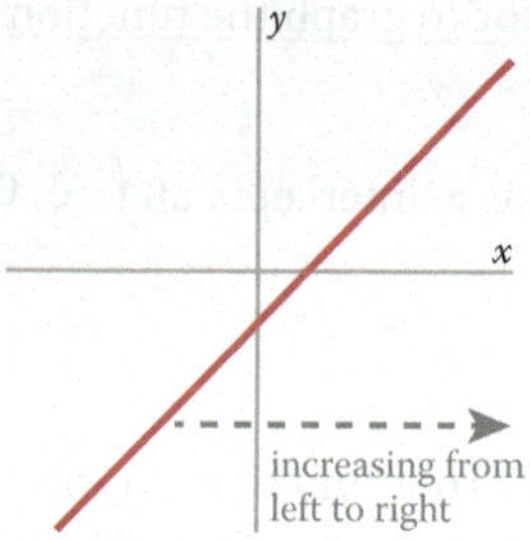

Figure 1.5k

Linear Function with Negative Slope

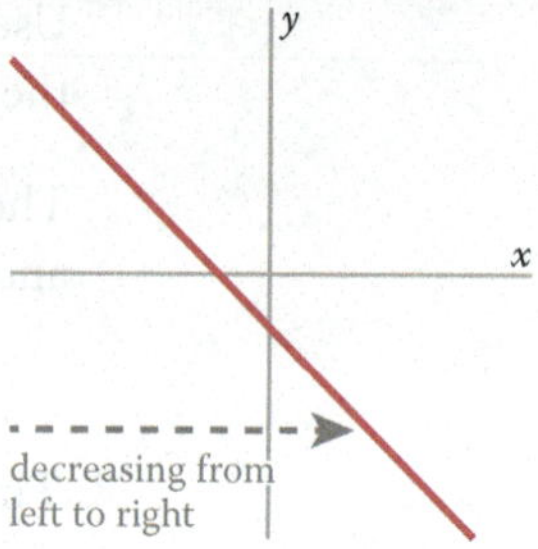

Figure 1.5l

The third possibility for a linear function is the horizontal line, where the slope is 0. A linear function with slope 0 (a constant function) is **constant** for all domain values. Since a line has a constant rate of change, a linear function is either only increasing, only decreasing, or only constant for all domain values.

Graphs of nonlinear functions may not have a constant slope. In the case where the slope is not constant, the function may be **strictly increasing** on some intervals (i.e., some interval of domain values), **strictly decreasing** on some other intervals, and **constant** on some intervals.

For example, consider the graph of the quadratic function $f(x) = x^2$ (Figure 1.5m). This function is strictly decreasing on the interval $(-\infty, 0)$ and strictly increasing on the interval $(0, \infty)$.

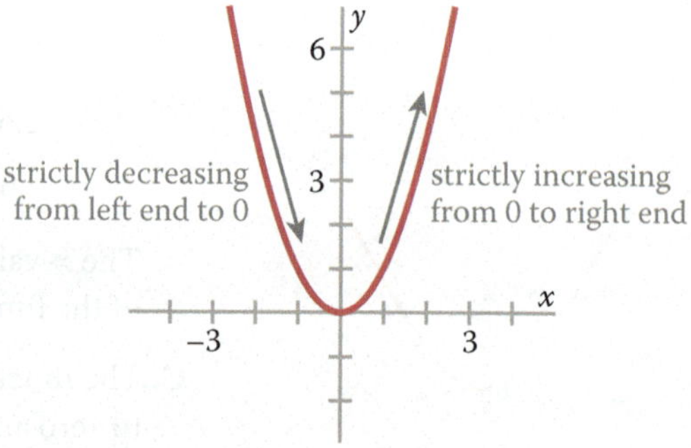

Figure 1.5m

Increasing, Decreasing, and Constant Functions

A function f is **strictly increasing** on an open interval when for any x_1 and x_2 in the interval where $x_1 < x_2$, it is also true that $f(x_1) < f(x_2)$.

A function f is **strictly decreasing** on an open interval when for any x_1 and x_2 in the interval where $x_1 < x_2$, it is also true that $f(x_1) > f(x_2)$.

A function f is **constant** on an open interval when for any x_1 and x_2 in the interval, it is true that $f(x_1) = f(x_2)$.

EXAMPLE 10 Identifying a Function's Strictly Increasing and Strictly Decreasing Intervals

Identify the intervals on which each function is strictly increasing, strictly decreasing, or constant.

A.

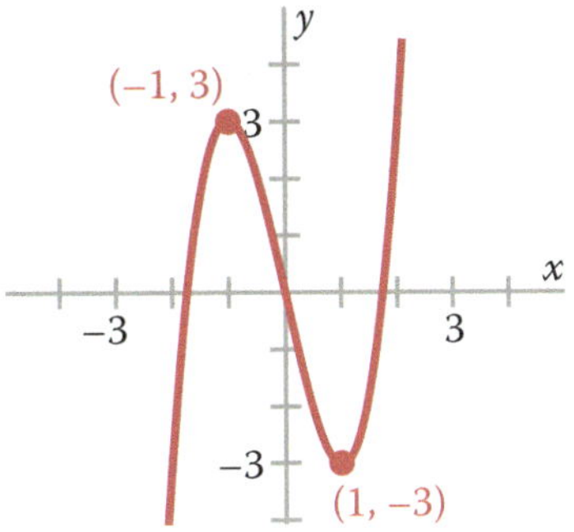

B.

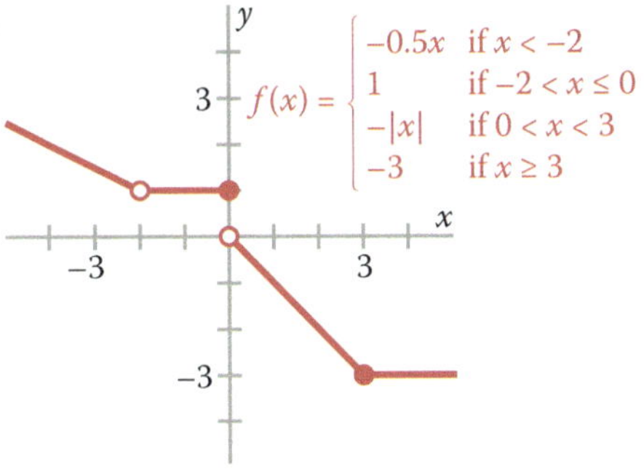

SOLUTION

A. Starting on the left end, the function is strictly increasing on the interval $(-\infty, -1)$, strictly decreasing on the interval $(-1, -1)$, and strictly increasing again on the interval $(1, \infty)$.

B. Starting on the left end, the function is strictly decreasing on the interval $(-\infty, -2)$, constant on the interval $(-2, 0)$, strictly decreasing again on the interval $(0, 3)$, and constant on the interval $(3, \infty)$.

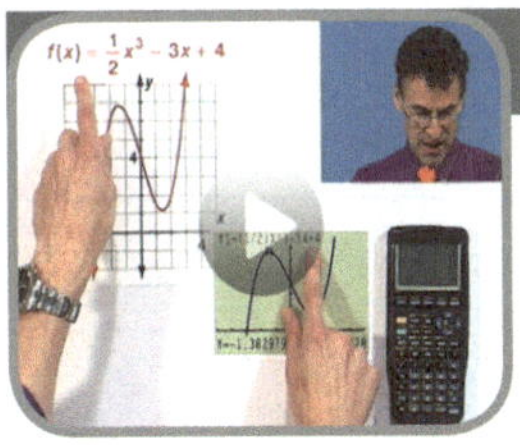

1.5.10 Relative Minimums and Maximums

Some functions have a **minimum** y-value, a **maximum** y-value, or both a minimum and a maximum y-value, but not all functions do. A minimum is the y-value of the point on a function with the least y-value, and a maximum is the y-value of the point on a function with the greatest y-value. Consider the graph of a linear function that is not horizontal. Regardless of the slope, this function has no maximum or minimum y-value because the line continues infinitely in both directions. Consider the graph of the parabola shown in Figure 1.5m. This graph has a minimum y-value at $(0, 0)$. However, it has no maximum y-value because the graph extends infinitely in an upward direction.

A point on the graph of a function such that the graph is increasing on one side and decreasing on the other is a **relative maximum** or **relative minimum** (also called a **local maximum** or **minimum**). For example, the graph in **Example 10A** contains one relative maximum point at $(-1, 3)$, so the relative maximum y-value is 3. This graph also contains one relative minimum point at $(1, -3)$, so the relative minimum y-value is -3.

A function may have any number of relative maximum or minimum y-values. These values can be identified by graphing the function. If the relative maximum or minimum points do not have integer coordinates, then they can be approximated by using the TRACE function on a graphing calculator.

EXAMPLE 11 Identifying Relative Maximum and Minimum Values

Identify the relative maximum and minimum values of $y = -x^3 + 5x + 1$, if any.

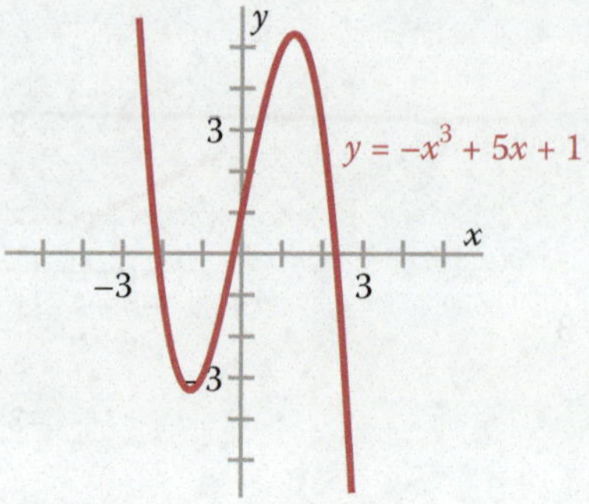

SOLUTION

From the graph, the function has a relative minimum between $x = -2$ and $x = 0$, and a relative maximum between $x = 0$ and $x = 2$.

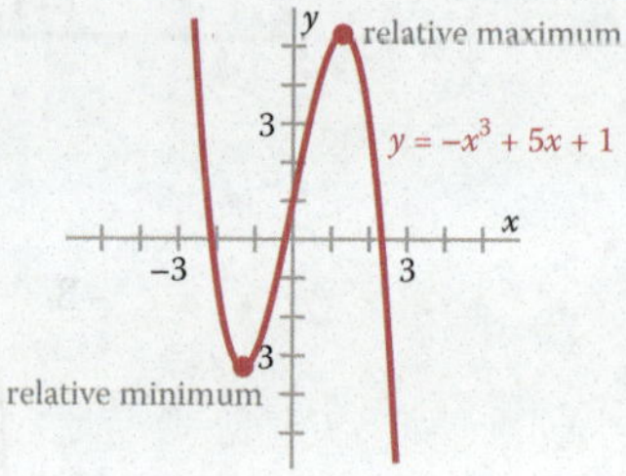

The exact coordinates of the minimum and maximum points cannot be determined from this graph. So, enter the function into a graphing calculator and use the TRACE function to approximate the coordinates of the minimum and maximum points.

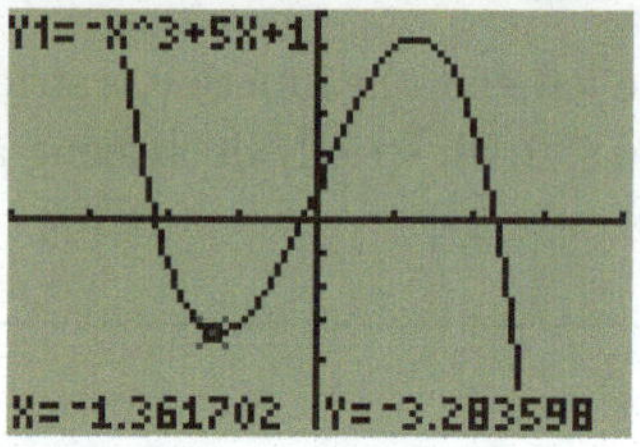

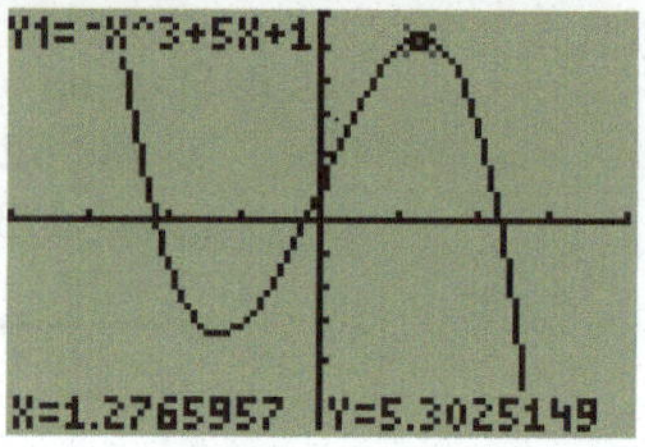

From the calculator, the function has a relative minimum at approximately (−1.4, −3.3), and a relative maximum at approximately (1.3, 5.3).

SECTION 1.5 EXERCISES

Warm Up

Describe the shape of the graph of each function.

1. $y = 2|x + 6|$
2. $x^2 = y$
3. $y = 2x + 9$

Just the Facts

Fill in the blanks.

4. To indicate that the point (a, b) is not included on the graph of a function, use a(n) _____ point at (a, b).
5. The greatest integer function, denoted as _____, is an example of a(n) _____ function.
6. The _____ of a function f are the values of x for which $f(x) = 0$. Graphically, they are the _____.
7. A function f is strictly decreasing on an interval when, for any x_1 and x_2 in the interval where $x_1 < x_2$, it is also true that $f(x_1)$ _____ $f(x_2)$.
8. If a function is increasing on one side of a point and decreasing on the other side of the point, then the y-coordinate of that point is a relative _____ or _____ value.
9. If the domain of a function f is $[3, 5)$, then the graph of f will contain a(n) _____ point at $(3, f(5))$ and a(n) _____ point at $(5, f(5))$.
10. An absolute value function will always have one interval where the function is _____ and one interval where the function is _____, because its graph is _____-shaped.

Essential Skills

In Exercises 1–8, write the domain and range of each function. Use interval notation when possible; otherwise, use set-builder notation.

1.

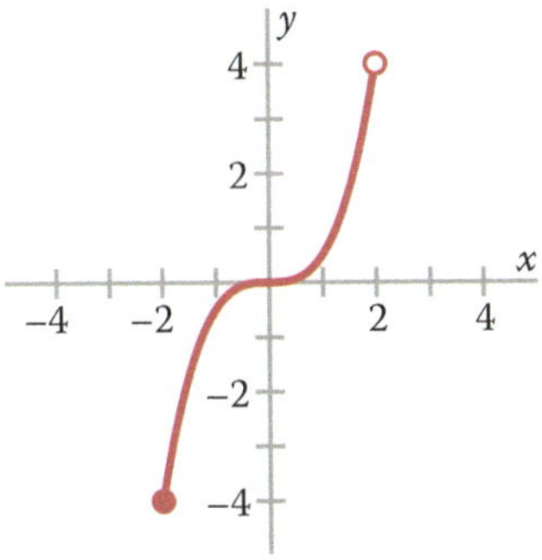

2.

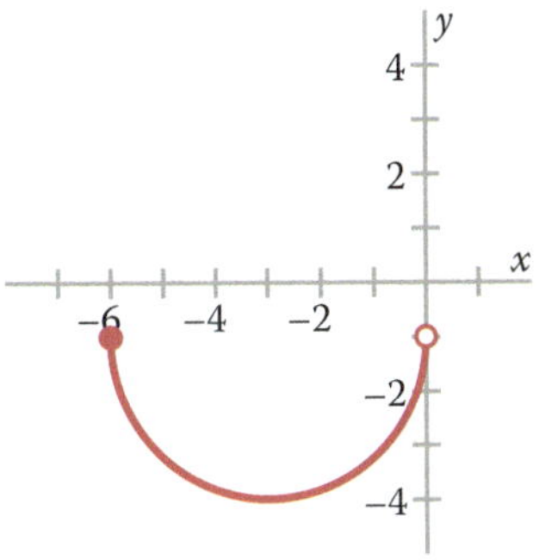

3.

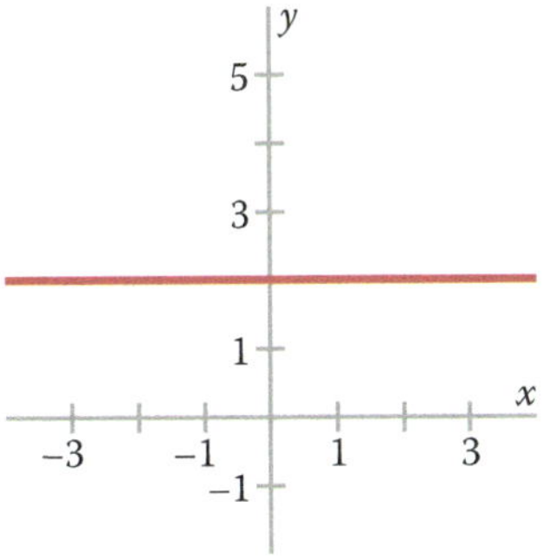

4.

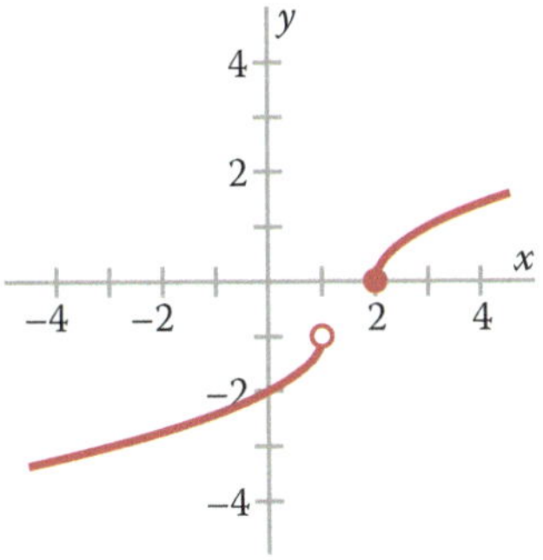

5.

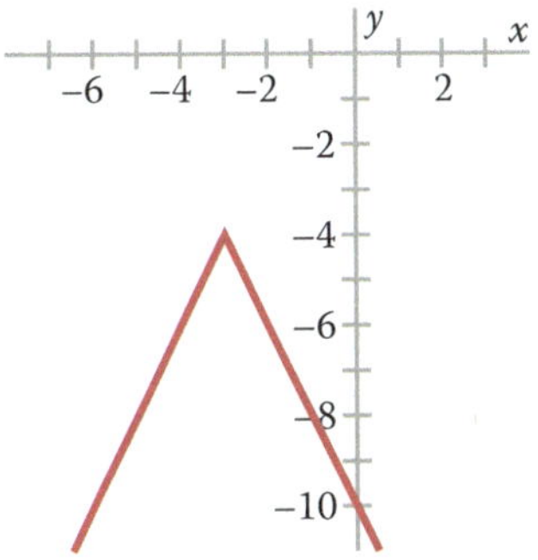

6.

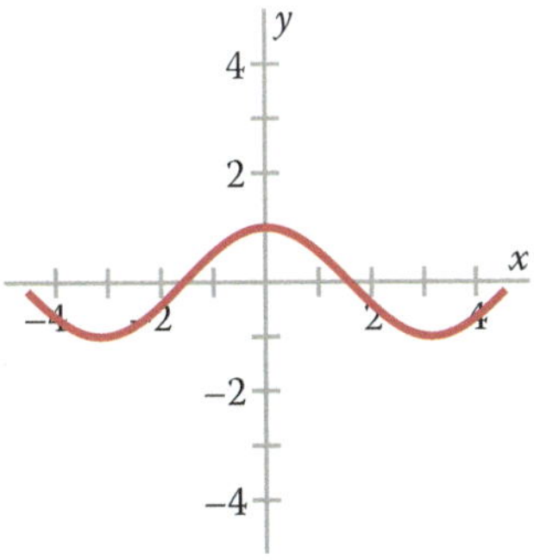

7.

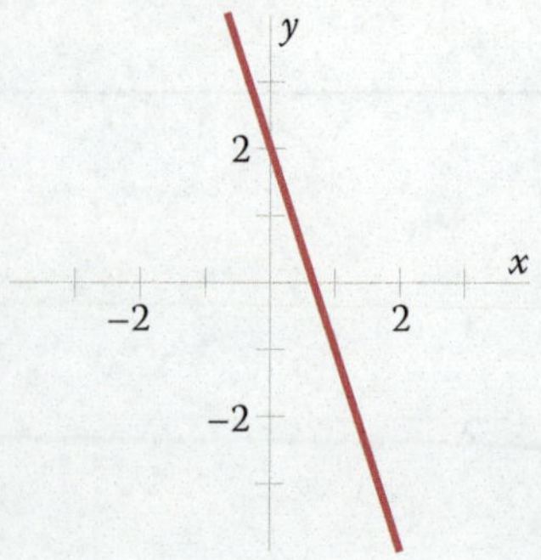

8.

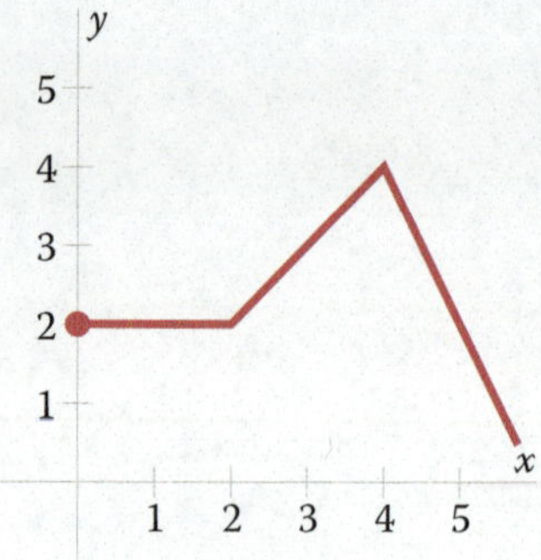

9. Using the graph, evaluate $f(4)$ and find all real values of x such that $f(x) = 2$.

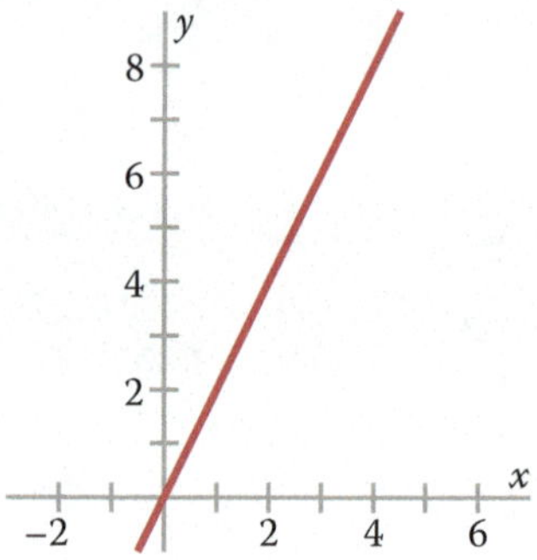

10. Using the graph, evaluate $f(0)$ and find all real value of x such that $f(x) = -5$.

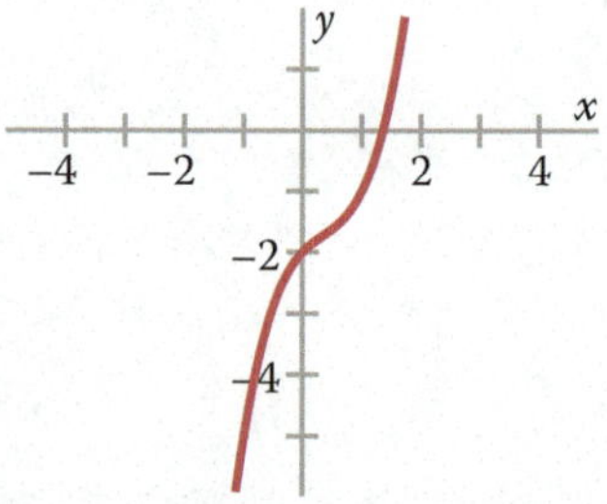

11. Using the graph, evaluate $f(2)$ and find all real value of x such that $f(x) = 0$.

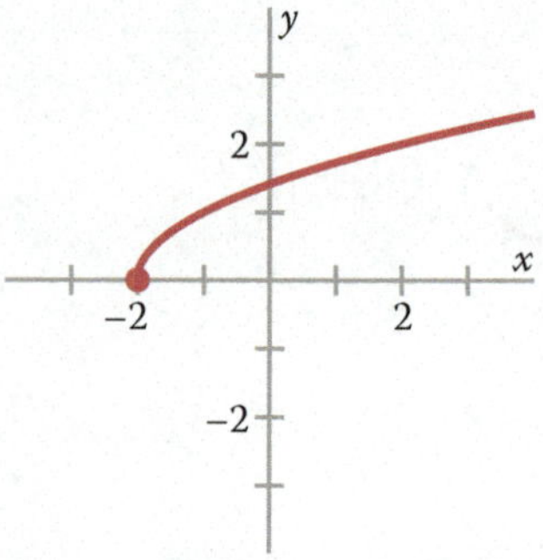

12. Using the graph, evaluate $f(0)$ and find all real value of x such that $f(x) = -1$.

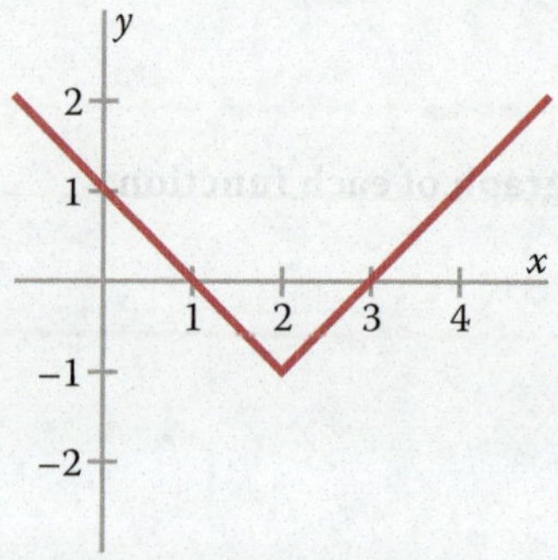

13. Using the graph, evaluate $f(2)$ and find all real value of x such that $f(x) = 1$.

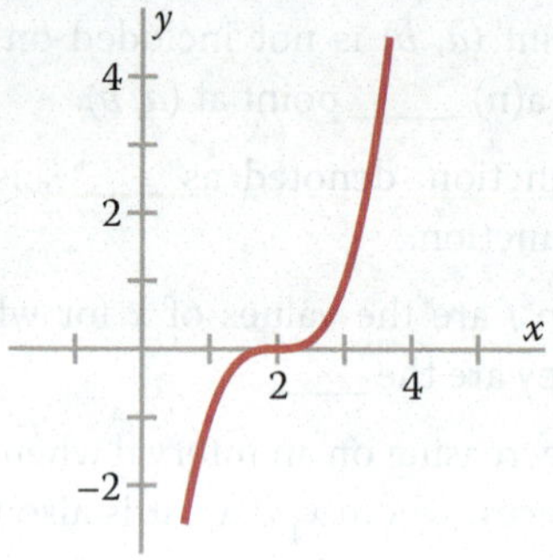

14. Using the graph, evaluate $f(4)$ and find all real value of x such that $f(x) = -3$.

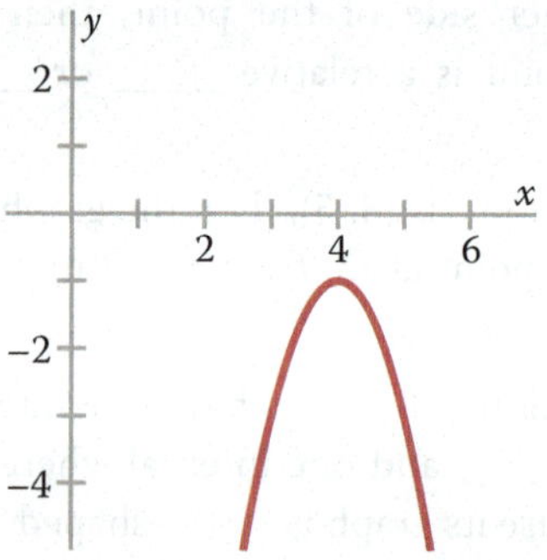

In Exercises 15–26, graph each function.

15. $g(x) = -3x + 2$

16. $h(x) = -x^2 - 4$

17. $m(x) = |x + 2| - 2$

18. $f(x) = 1 + \sqrt{x}$

19. $f(x) = x^2 + 1$

20. $k(x) = 4 - x^3$

21. $f(x) = -2x^2 + 1$

22. $v(x) = 4 - \sqrt{x}$

23. $p(x) = \begin{cases} x & \text{if } x < 0 \\ 3 & \text{if } 0 \le x < 2 \\ -x + 4 & \text{if } x \ge 2 \end{cases}$

24. $f(x)=\begin{cases}-2 & \text{if } x<-3\\ 5+3x & \text{if } -3\le x<0\\ -x & \text{if } x\ge 0\end{cases}$

25. $g(x)=\begin{cases}x^2-2 & \text{if } x\le 0\\ -|x| & \text{if } x>0\end{cases}$

26. $f(x)=\begin{cases}x+1 & \text{if } x<0\\ \sqrt{x}-3 & \text{if } x\ge 0\end{cases}$

In Exercises 27–29, use the following information.

A popular cell-phone company offers a package based on the minutes m used per month. Write a piecewise-defined function to model each situation.

27. For 450 minutes, they charge \$40; for 900 minutes, they charge \$60; and for anything ovcr 900 minutes, they charge \$100.

28. For 300 minutes, they charge \$35; for 600 minutes, they charge \$50; and for anything over 600 minutes, they charge \$90.

29. For 200 minutes, they charge \$25; for 500 minutes, they charge \$35; and for anything over 500 minutes, they charge \$60.

In Exercises 30–32, use the following information.

A fishing company offers to rent fishing boats based on the number of hours x the boat is used. Write a piecewise-defined function for each situation.

30. The cost of renting a fishing boat is \$25 per hour for the first 7 hours, \$10 per hour for the next 10 hours, and \$5 per hour for any additional hours.

31. The cost of renting a fishing boat is \$20 per hour for the first 6 hours, \$10 per hour for the next 16 hours, and \$5 per hour for any additional hours.

32. The cost of renting a fishing boat is \$23 per hour for the first 5 hours, \$12 per hour for the next 6 hours, and \$8 per hour for any additional hours.

In Exercises 33–40, simplify each expression.

33. $3-[[-5.62]]+[[9.4-3.6]]$

34. $[[4.08]]-[[-0.6]]+[[4.6+5.1]]$

35. $[[15.62]]-[[-7.16]]+[[7.04+9.3]]$

36. $[[-9.1]]-[[19.81]]+[[9.4-6.3]]$

37. $-18-[[7.9-8.2]]+[[16.5-21]]$

38. $-[[-5.6+3.5]]+[[2.9-4.3]]-[[4.7-7]]$

39. $-[[7.1-12.9]]-[[6.8-8]]+[[-4.+0.9]]$

40. $[[-0.6+6.1]]-[[7.9-13.5]]-[[8.2-11]]$

In Exercises 41–42, graph each function.

41. $y=[[x]]+2$

42. $y=[[x-5]]$

In Exercises 43–62, find the real zeros of each function.

43. $g(x)=x^2+7x+6$

44. $f(x)=25-x^2$

45. $h(x)=-7x^3+35x^2$

46. $u(x)=5x^2+15x+12$

47. $v(x)=6x^2+19x-20$

48. $k(x)=2x^2+9x-18$

49. $w(x)=-3x^2-4x-9$

50. $g(x)=x-x^3$

51. $s(x)=3x^3-12x$

52. $p(x)=x^3-4x^2+6x-24$

53. $r(x)=x^3-2x^2+3x-6$

54. $q(x)=x^3+2x^2-9x-18$

55. $f(x)=4+\sqrt{x-9}$

56. $f(x)=-4\sqrt{16-x^2}$

57. $g(x)=\dfrac{2}{3}\sqrt{x^2-81}$

58. $r(x)=3\sqrt{24+x^2}$

59. $f(x)=\dfrac{5x-4}{x+2}$

60. $g(x)=\dfrac{6x^2+7x-3}{x+1}$

61. $c(x)=\dfrac{15x^2+7x-4}{x+3}$

62. $d(x)=\dfrac{3x^2+19x+31}{x+5}$

In Exercises 63–70, identify the intervals on which each function is strictly increasing, strictly decreasing, or constant.

63.

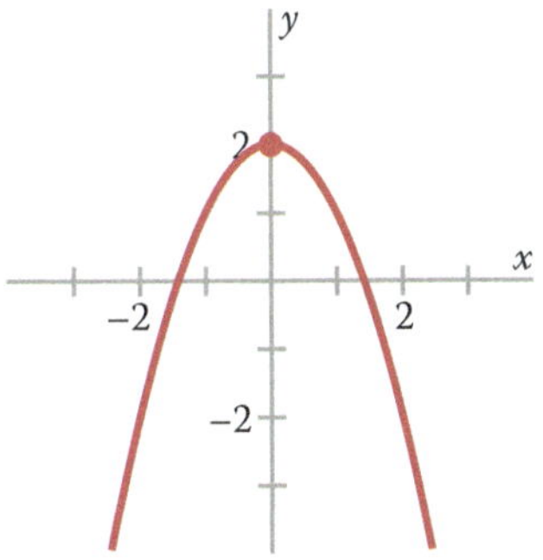

64.

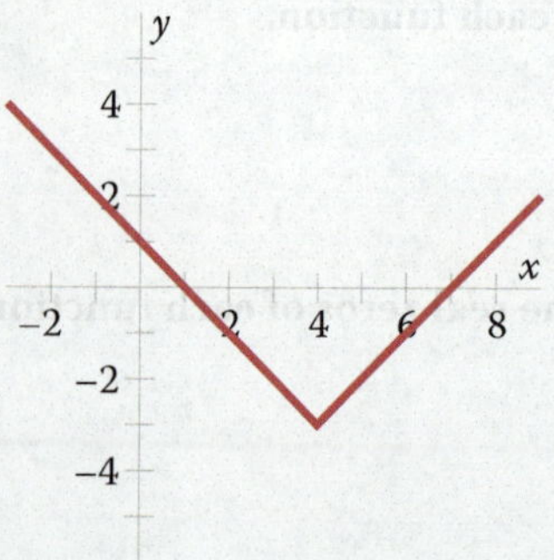

65.

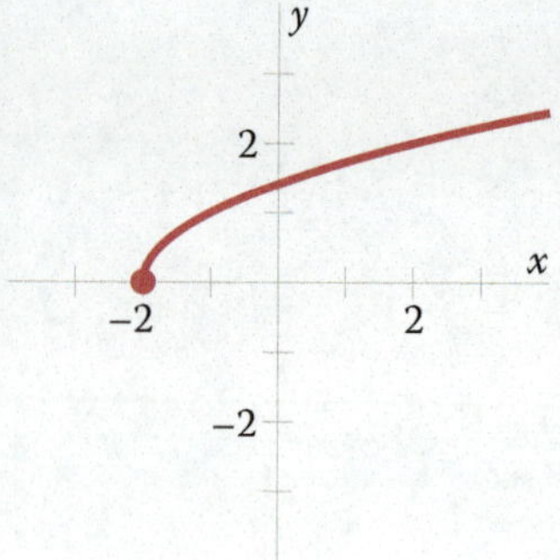

66.

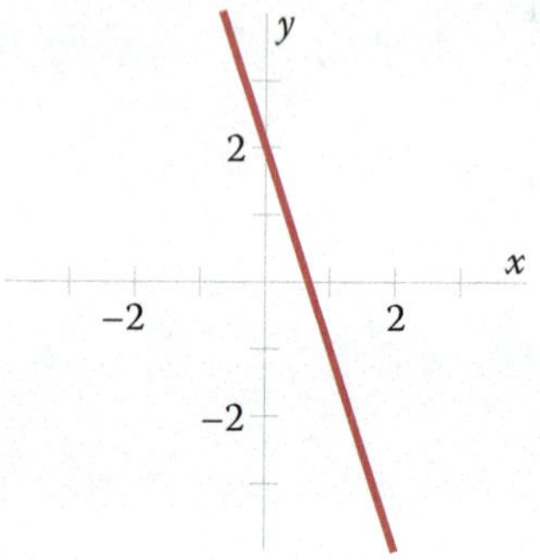

67.

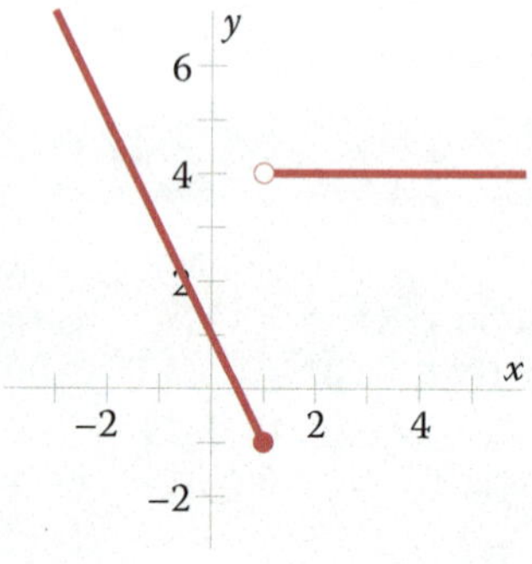

68.

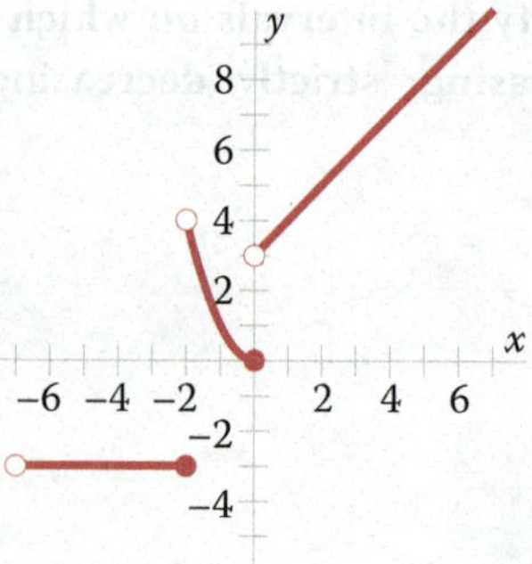

69.

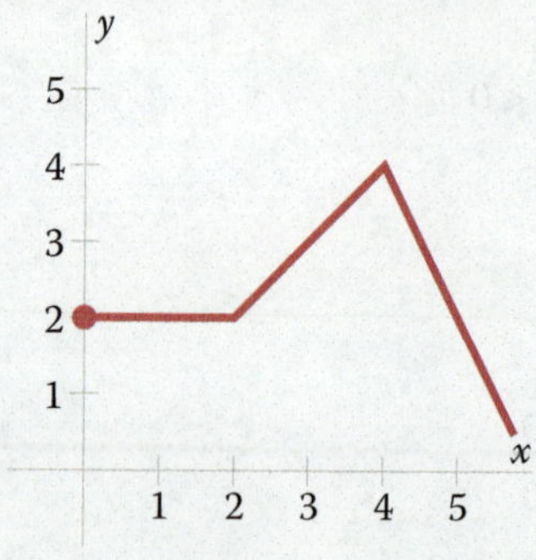

70.

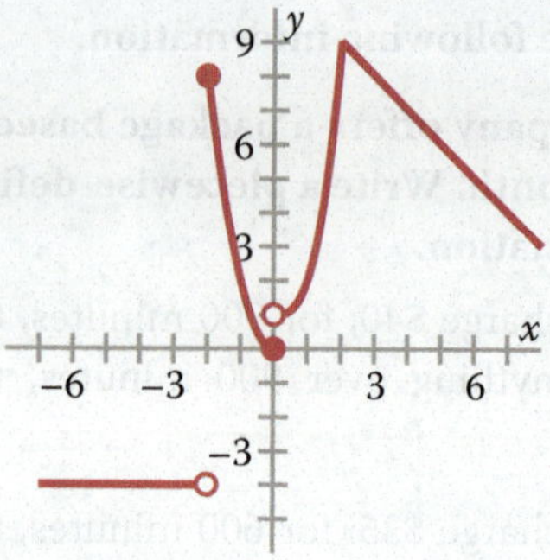

In Exercises 71–72, identify the relative maximum and relative minimum points on each graph, if any. Round to the nearest tenth.

71.

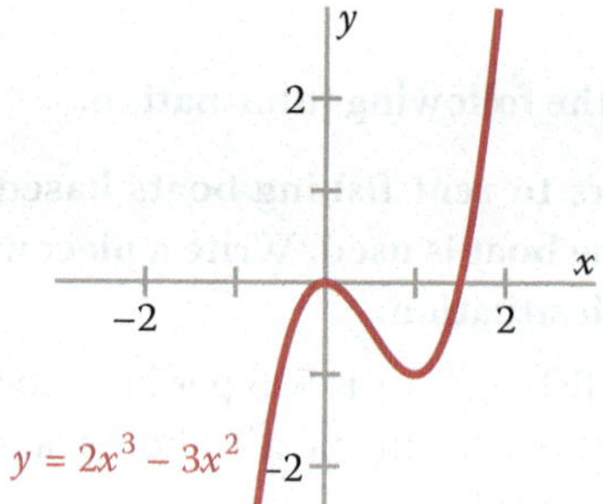

72.

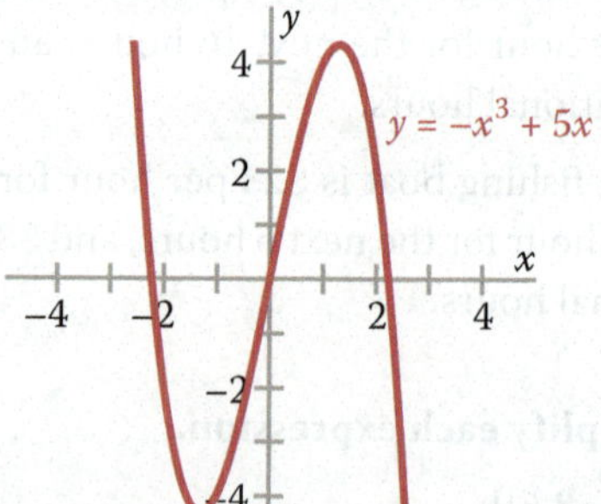

Extensions

In Exercises 73–76, identify the domain and range of the following functions. Then identify the intervals on which each function is strictly increasing, strictly decreasing, or constant.

73. $f(x) = \frac{1}{10}x - 6$

74. $h(x) = -2x^2 + 5$

75. $t(x) = |4x - 3| - 2$

76. $q(x) = -\sqrt{x - 5}$

77. The cost of sending an overnight package from Texas to Massachusetts is \$100.42 for a package weighing up to and including 1 pound, plus an additional \$6.42 for each additional pound or portion of a pound. Use the greatest integer function to create a model for the cost c of overnight delivery of a package weighing p pounds. Then graph the function.

78. Write the piecewise-defined function for the graph.

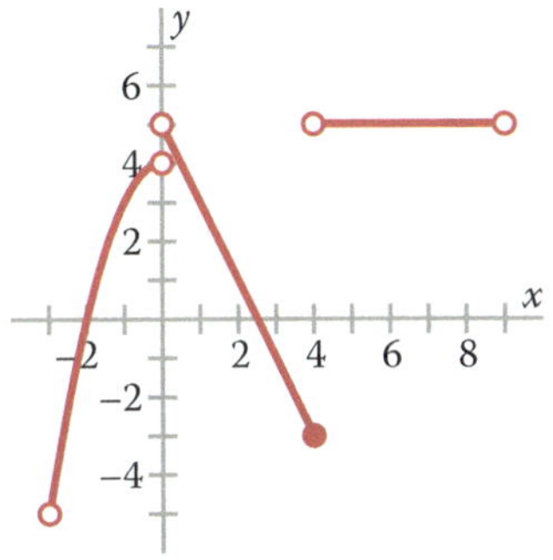

79. Credit card companies base the minimum payment due on the amount owed. The following rules apply: For a bill less than \$20, the entire amount is due. For a bill of at least \$20 but less than \$1000, the minimum payment due is \$20. On a bill of at least \$1000 but less than \$3000, a minimum of \$50 is due. Additionally, there is a minimum of \$100 due on bills of \$3000 and over. Write the function $p(x)$ that describes the minimum payment due on a bill of x dollars (where x is a whole number).

80. A car rental company offers a midsize car for \$130 per week. Extra days cost \$30 per day until the rate exceeds the quoted weekly rate, in which case the weekly rate applies. Write a piecewise-defined function for the cost c of renting a midsize car as a function of the number of days x, where $0 < x \leq 20$.

81. Solve $h(x) - -7x^3 + 35x^2 - 5x + 25$. How many real zeros does the function have?

82. True or False? A function that has three relative maximums must have three relative minimums.

83. Sketch the graph with the following properties: zeros at 2 and −4, y-intercept at 4, a relative maximum at (−2, 8), a relative minimum at (2, 0), strictly increasing from (−∞, −2) and (2, ∞), strictly decreasing from (−2, 2), with a domain and range of all real numbers.

1.6 TRANSFORMATIONS OF FUNCTIONS

OBJECTIVES

- Graph functions using translations.
- Graph functions using reflections.
- Stretch graphs of functions.
- Determine whether functions are even or odd.
- Identify symmetry of functions.

PREREQUISITE VOCABULARY TERMS

domain
function
range
reflection
symmetry

A **transformation** is an operation that changes the graph of an equation. The original graph is called the **pre-image**, and the transformed graph is called the **image**.

One type of transformation is called a **translation**, which is a shift of a graph horizontally or vertically. For example, Figure 1.6a shows the graphs of $y = x$ and $y = x + 5$. Notice that shifting the graph of $y = x$ upward 5 units forms the graph of $y = x + 5$. So, we can say that the graph of $y = x + 5$ is the result of translating the graph of $y = x$ upward 5 units. In this case, the graph of $y = x$ is the pre-image and the graph of $y = x + 5$ is the image.

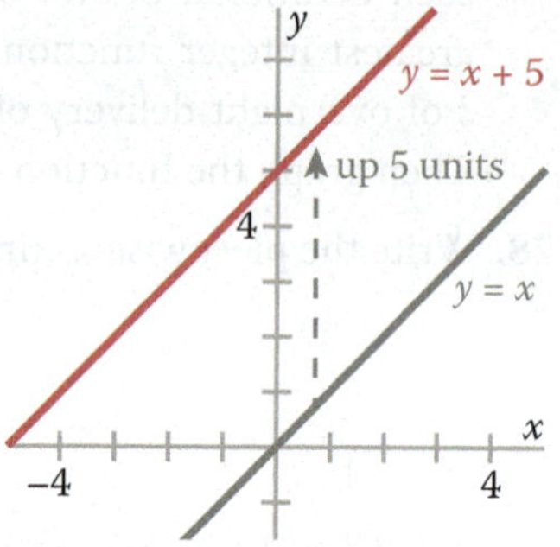

Figure 1.6a

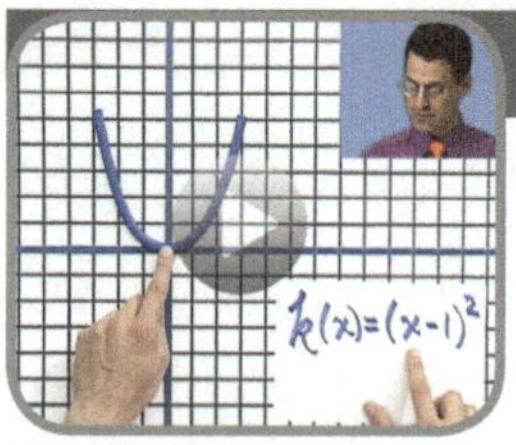

1.6.1 Translating Functions

The graphs of six basic functions are shown in the following table.

<table>
<tr><th colspan="3">Basic Functions</th></tr>
<tr><td>Constant: $f(x) = c$</td><td>Linear: $f(x) = x$</td><td>Quadratic: $f(x) = x^2$</td></tr>
<tr><td>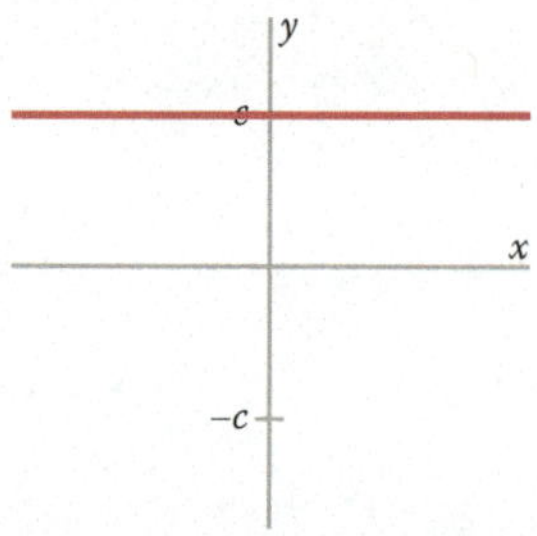</td><td>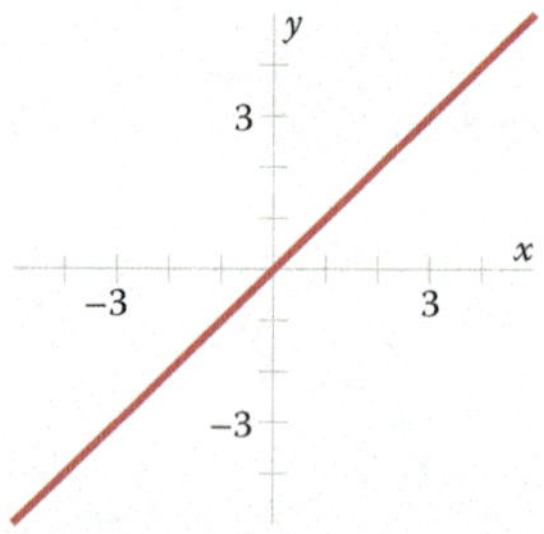</td><td>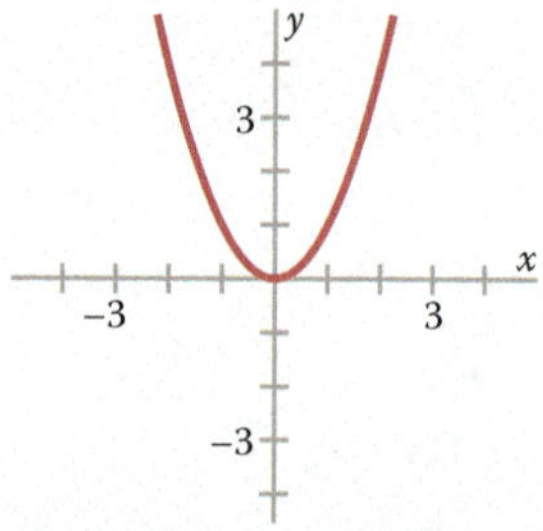</td></tr>
<tr><td>Cubic: $f(x) = x^3$</td><td>Square Root: $f(x) = \sqrt{x}$</td><td>Absolute Value: $f(x) = |x|$</td></tr>
<tr><td>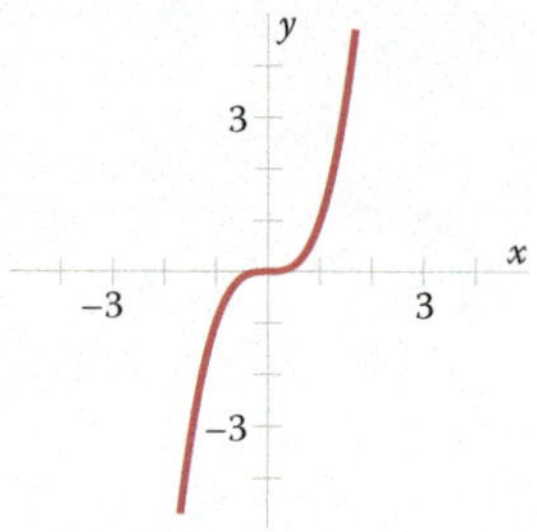</td><td>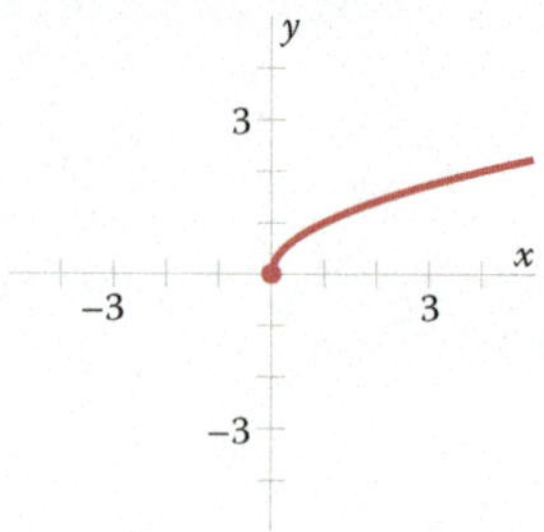</td><td>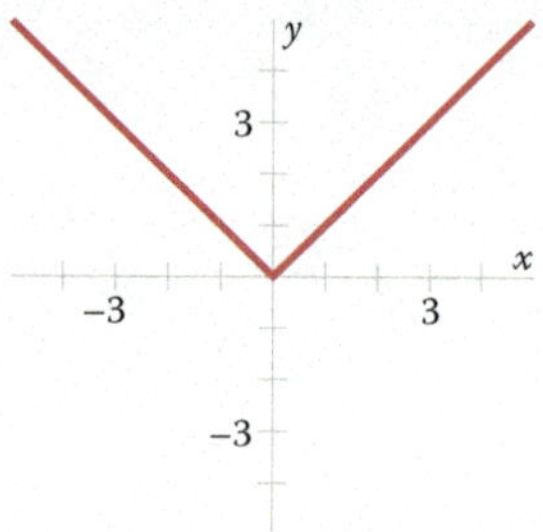</td></tr>
</table>

When the equation of one of the basic functions is modified by adding or subtracting a number, the graph of that basic function is translated either horizontally or vertically.

Generally, suppose $g(x)$ and $f(x)$ are some functions and $c > 0$.

- If $g(x) = f(x) + c$, then the graph of $g(x)$ is the image of $f(x)$ after a vertical translation of c units upward. So, every point (x, y) on $f(x)$ corresponds to a point $(x, y + c)$ on $g(x)$.
- If $g(x) = f(x) - c$, then the graph of $g(x)$ is the image of $f(x)$ after a vertical translation of c units downward. So, every point (x, y) on $f(x)$ corresponds to a point $(x, y - c)$ on $g(x)$.
- If $g(x) = f(x + c)$, then the graph of $g(x)$ is the image of $f(x)$ after a horizontal translation of c units to the left. So, every point (x, y) on $f(x)$ corresponds to a point $(x - c, y)$ on $g(x)$.
- If $g(x) = f(x - c)$, then the graph of $g(x)$ is the image of $f(x)$ after a horizontal translation of c units to the right. So, every point (x, y) on $f(x)$ corresponds to a point $(x + c, y)$ on $g(x)$.

Steps for Graphing a Function $g(x)$ by Using Translations Where $c > 0$

❶ Identify the basic type of function, $f(x)$, that $g(x)$ is based on.

❷ Identify any horizontal and/or vertical translations from the equation of $g(x)$.
- $g(x) = f(x) + c \Rightarrow f(x)$ is shifted c units upward $\Rightarrow (x, y) \to (x, y + c)$
- $g(x) = f(x) - c \Rightarrow f(x)$ is shifted c units downward $\Rightarrow (x, y) \to (x, y - c)$
- $g(x) = f(x + c) \Rightarrow f(x)$ is shifted c units to the left $\Rightarrow (x, y) \to (x - c, y)$
- $g(x) = f(x - c) \Rightarrow f(x)$ is shifted c units to the right $\Rightarrow (x, y) \to (x + c, y)$

❸ Apply these translations to the graph of $f(x)$, resulting in the graph of $g(x)$.

EXAMPLE 1 Graphing a Function by Using Vertical and Horizontal Translations

Sketch the graph of each function by using a translation of the corresponding basic function.

A. $g(x) = x^2 - 3$ **B.** $h(x) = (x + 1)^3 + 2$

SOLUTION

A. The function $g(x) = x^2 - 3$ corresponds to the basic quadratic function $f(x) = x^2$.

So, $g(x) = f(x) - 3$, where $f(x) = x^2$.

Therefore, the graph of $g(x)$ is the image of $f(x) = x^2$ after a vertical translation downward 3 units.

To complete this translation of $f(x) = x^2$, shift each point (x, y) on $f(x) = x^2$ to $(x, y - 3)$.

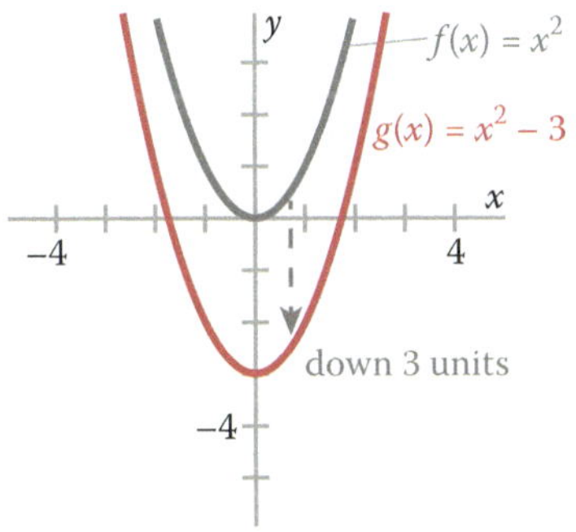

For example, the points $(-1, 1)$, $(0, 0)$, and $(1, 1)$ are on the graph of $f(x) = x^2$. So, the points $(-1, -2)$, $(0, -3)$, and $(1, -2)$ are on the graph of $g(x) = x^2 - 3$.

$f(x) \to g(x)$
$(-1, 1) \to (-1, -2)$
$(0, 0) \to (0, -3)$
$(1, 1) \to (1, -2)$

B. The function $h(x) = (x + 1)^3 + 2$ corresponds to the basic cubic function $f(x) = x^3$.

So, $h(x) = f(x + 1) + 2$, where $f(x) = x^3$.

Therefore, the graph of $h(x)$ is the image of $f(x) = x^3$ after a vertical translation upward 2 units and a horizontal translation to the left 1 unit.

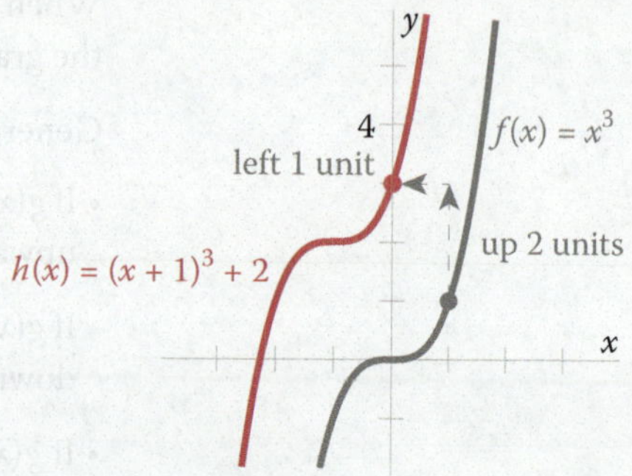

To complete this translation of $f(x) = x^3$, shift each point (x, y) on $f(x) = x^3$ to $(x - 1, y + 2)$.

For example, the points $(-1, -1)$, $(0, 0)$, and $(1, 1)$ are on the graph of $f(x) = x^3$. So, the points $(-2, 1)$, $(-1, 2)$, and $(0, 3)$ are on the graph of $h(x) = (x + 1)^3 + 2$.

$f(x) \to h(x)$
$(-1, -1) \to (-2, 1)$
$(0, 0) \to (-1, 2)$
$(1, 1) \to (0, 3)$

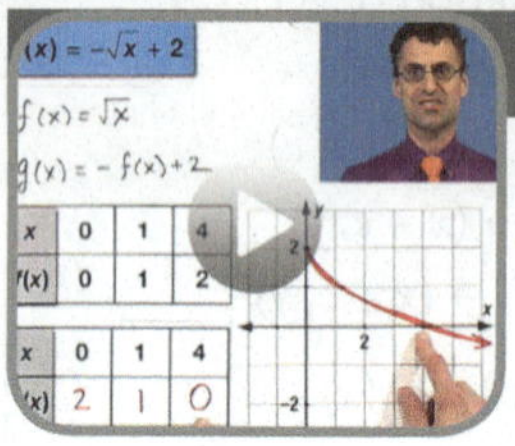

1.6.2 Reflecting Functions

A reflection is another type of transformation. In Figure 1.6b, the graph of $g(x) = -\sqrt{x}$ is a reflection of the graph of $f(x) = \sqrt{x}$ over the x-axis. Generally, if $g(x) = -f(x)$, then the graph of $g(x)$ is a reflection of the graph of $f(x)$ over the x-axis. Thus, every point (x, y) on $f(x)$ corresponds to a point $(x, -y)$ on $g(x)$.

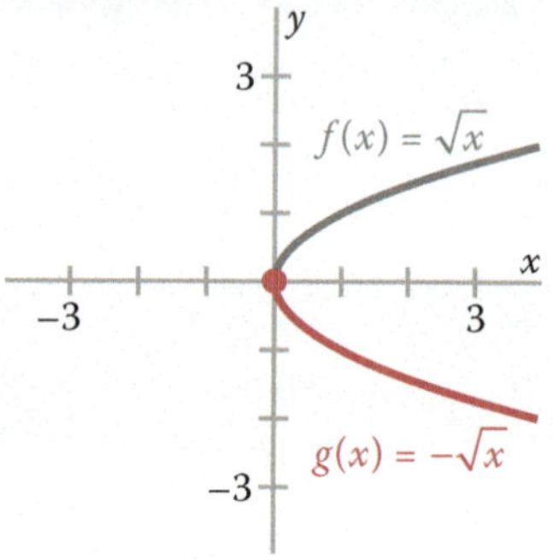

Figure 1.6b

In Figure 1.6c, notice that the graph of $g(x) = \sqrt{-x}$ is a reflection of the graph of $f(x) = \sqrt{x}$ over the y-axis. Generally, if $g(x) = f(-x)$, then the graph of $g(x)$ is a reflection of the graph of $f(x)$ over the y-axis. So, every point (x, y) on $f(x)$ corresponds to a point $(-x, y)$ on $g(x)$.

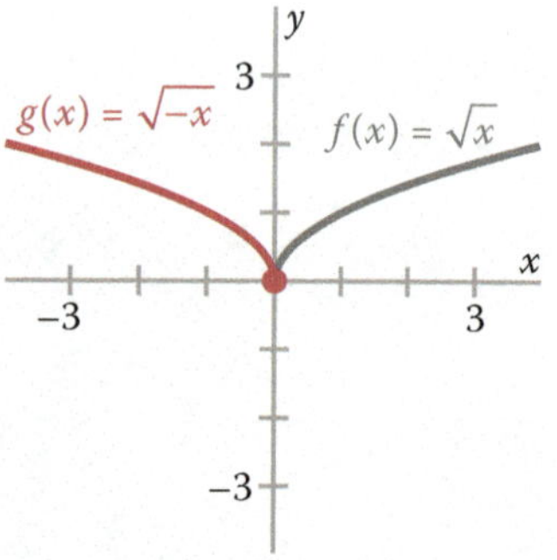

Figure 1.6c

Steps for Graphing a Function $g(x)$ by Using Reflections

1. Identify the basic type of function, $f(x)$, that $g(x)$ is based on.
2. Identify any reflections from the equation of $g(x)$.
   - $g(x) = -f(x) \Rightarrow f(x)$ is reflected over the x-axis $\Rightarrow (x, y) \to (x, -y)$
   - $g(x) = f(-x) \Rightarrow f(x)$ is reflected over the y-axis $\Rightarrow (x, y) \to (-x, y)$
3. Apply the reflection to the graph of $f(x)$, resulting in the graph of $g(x)$.

Reflections can also be combined with translations to graph a function.

EXAMPLE 2 Graphing a Function by Using a Reflection

Sketch the graph of each function by using a reflection or translation of the corresponding basic function.

A. $g(x) = -x^2$ **B.** $h(x) = (-x)^3 - 2$

SOLUTION

A. The function $g(x) = -x^2$ corresponds to the basic quadratic function $f(x) = x^2$.

So, $g(x) = -f(x)$, where $f(x) = x^2$.

Therefore, the graph of $g(x)$ is the image of $f(x) = x^2$ after a reflection over the x-axis.

To complete this reflection of $f(x) = x^2$, reflect each point (x, y) on $f(x) = x^2$ to $(x, -y)$.

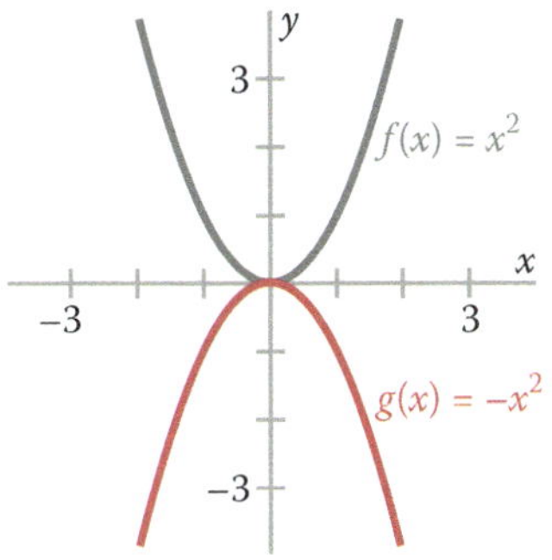

$f(x) \to g(x)$
$(-1, 1) \to (-1, -1)$
$(0, 0) \to (0, 0)$
$(1, 1) \to (1, -1)$

▶ For example, the points $(-1, 1)$, $(0, 0)$, and $(1, 1)$ are on the graph of $f(x) = x^2$. So, the points $(-1, -1)$, $(0, 0)$, and $(1, -1)$ are on the graph of $g(x) = -x^2$.

B. The function $h(x) = (-x)^3 - 2$ corresponds to the basic cubic function $f(x) = x^3$.

So, $h(x) = f(-x) - 2$, where $f(x) = x^3$.

Therefore, the graph of $h(x)$ is the image of $f(x) = x^3$ after a reflection over the y-axis and a vertical translation downward 2 units.

To complete this translation of $f(x) = x^3$, shift each point (x, y) on $f(x) = x^3$ to $(-x, y - 2)$.

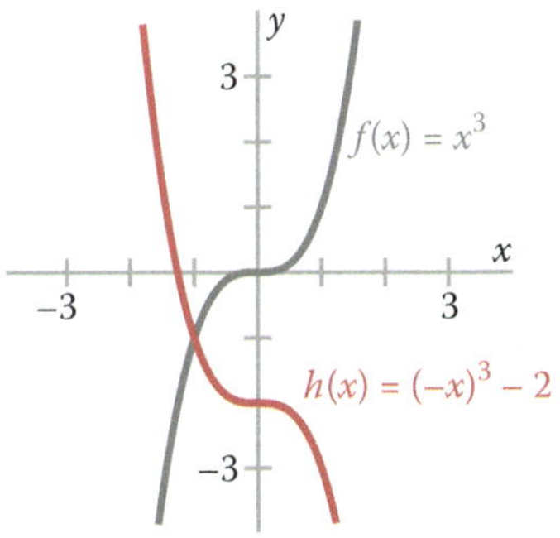

$f(x) \to h(x)$
$(-1, -1) \to (1, -3)$
$(0, 0) \to (0, -2)$
$(1, 1) \to (-1, -1)$

▶ For example, the points $(-1, -1)$, $(0, 0)$, and $(1, 1)$ are on the graph of $f(x) = x^3$. So, the points $(1, -3)$, $(0, -2)$, and $(-1, -1)$ are on the graph of $h(x) = (-x)^3 - 2$.

WHAT'S THE BIG IDEA?

Describe the graph of $g(x) = -(x - h)^2 + k$, where $h > 0$ and $k > 0$, as compared to the graph of $f(x) = x^2$.

1.6.3 Stretching Functions

When the graph of a function is translated or reflected, the position of the graph changes, but the shape of the graph remains the same. So, translations and reflections are **rigid transformations**. A **nonrigid transformation** is a change to the graph that results in a change in the shape of the graph. One example of a nonrigid transformation is a vertical **stretch**, where the graph of a function either widens or narrows.

IMPORTANT

The notation $cf(x)$ means that a constant value c and a function $f(x)$ are multiplied.

If $g(x)$ and $f(x)$ are some functions such that $g(x) = cf(x)$, where c is some positive number, then the graph of $g(x)$ is the image of $f(x)$ after a vertical stretch by a factor of c.

- If $c > 1$, then the vertical stretch narrows the graph of $f(x)$.
- If $0 < c < 1$, then the vertical stretch widens the graph of $f(x)$.

In either case, every point (x, y) on $f(x)$ corresponds to a point (x, cy) on $g(x)$.

Steps for Graphing a Function $g(x)$ by Stretching

❶ Identify the basic type of function, $f(x)$, that $g(x)$ is based on.

❷ Identify any stretch from the equation of $g(x)$.

❸ Apply the stretch to the graph of $f(x)$, resulting in the graph of $g(x)$.

Reflections and translations can be combined with a stretch to graph a function.

EXAMPLE 3 Graphing a Function by Using a Stretch

Sketch the graph of $g(x) = 2|x|$ by stretching the corresponding basic function.

SOLUTION

The function $g(x) = 2|x|$ corresponds to the basic absolute value function $f(x) = |x|$.

So, $g(x) = 2f(x)$ where $f(x) = |x|$.

Therefore, the graph of $g(x)$ is the image of $f(x) = |x|$ after a vertical stretch by a factor of 2.

To complete this stretch of $f(x) = |x|$, change each point (x, y) on $f(x) = |x|$ to $(x, 2y)$.

For example, the points $(-1, 1)$, $(0, 0)$, and $(1, 1)$ are on the graph of $f(x) = |x|$. So, the points $(-1, 2)$, $(0, 0)$, and $(1, 2)$ are on the graph of $g(x) = 2|x|$.

$f(x) \rightarrow g(x)$
$(-1, 1) \rightarrow (-1, 2)$
$(0, 0) \rightarrow (0, 0)$
$(1, 1) \rightarrow (1, 2)$

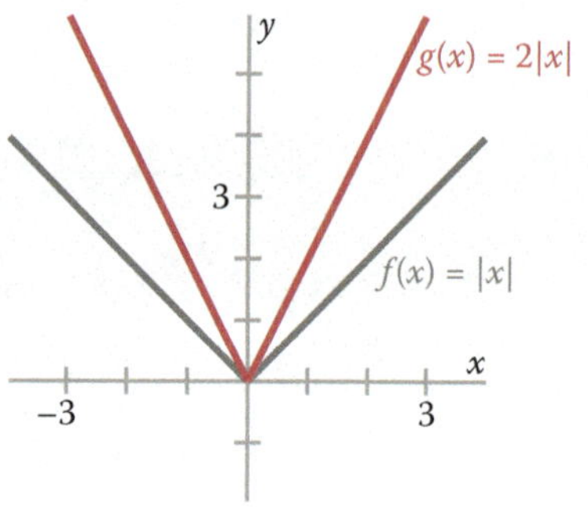

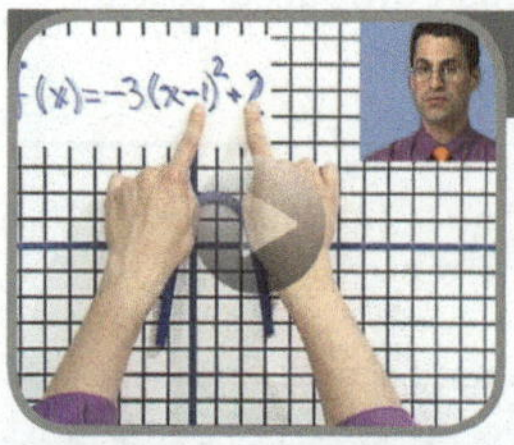

1.6.4 Using Patterns to Graph Functions

Translations, reflections, and stretches can be combined to sketch the graph of a function.

Steps for Graphing a Function $g(x)$ by Translating, Reflecting, and Stretching

❶ Identify the basic type of function, $f(x)$, that $g(x)$ is based on.

❷ Identify any reflection from the equation of $g(x)$.

❸ Identify any stretch from the equation of $g(x)$.

❹ Identify any translation(s) (shifts) from the equation of $g(x)$.

❺ Apply the transformations to the graph of $f(x)$, resulting in the graph of $g(x)$.

EXAMPLE 4 Graphing a Function by Using Patterns

Sketch the graph of $g(x) = -\frac{1}{2}|x+1| - 4$.

SOLUTION

The function $g(x) = -\frac{1}{2}|x+1| - 4$ corresponds to the basic absolute value function $f(x) = |x|$. Identify any reflections, stretches, or translations to the graph of $f(x) = |x|$ from the equation of $g(x)$.

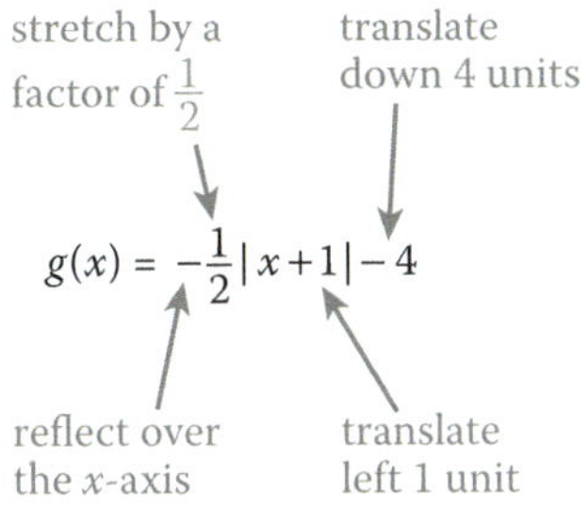

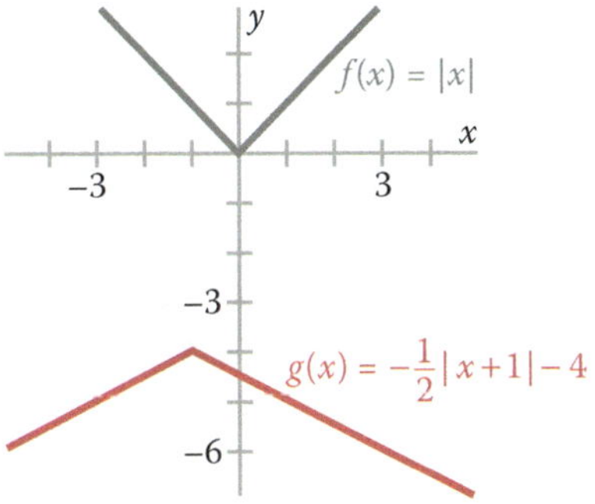

Apply these changes to the graph of $f(x) = |x|$. The changes are applied to three points on $f(x)$ in the following table.

$f(x)$	→	reflect over x-axis	→	stretch by factor of $\frac{1}{2}$	→	translate left 1 unit	→	translate downward 4 units
(x, y)	→	$(x, -y)$	→	$\left(x, -\frac{1}{2}y\right)$	→	$\left(x-1, -\frac{1}{2}y\right)$	→	$\left(x-1, -\frac{1}{2}y-4\right)$
$(-2, 2)$	→	$(-2, -2)$	→	$(-2, -1)$	→	$(-3, -1)$	→	$(-3, -5)$
$(0, 0)$	→	$(0, 0)$	→	$(0, 0)$	→	$(-1, 0)$	→	$(-1, -4)$
$(2, 2)$	→	$(2, -2)$	→	$(2, -1)$	→	$(1, -1)$	→	$(1, -5)$

1.6.5 Even and Odd Functions

Recall that the graph of an equation (which may or may not be a function) can be symmetric with respect to the x-axis, the y-axis, or the origin, as summarized in the following table.

Symmetry and Points on a Graph

Symmetric with respect to the origin	Symmetric with respect to the x-axis	Symmetric with respect to the y-axis
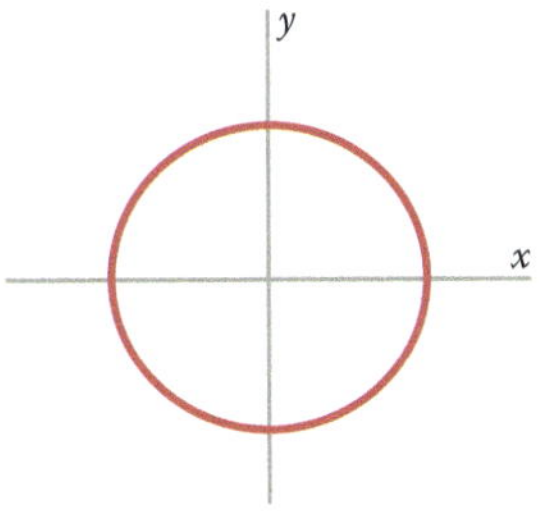	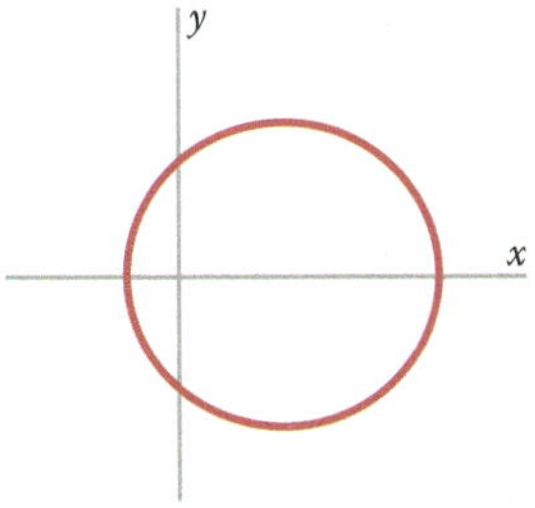	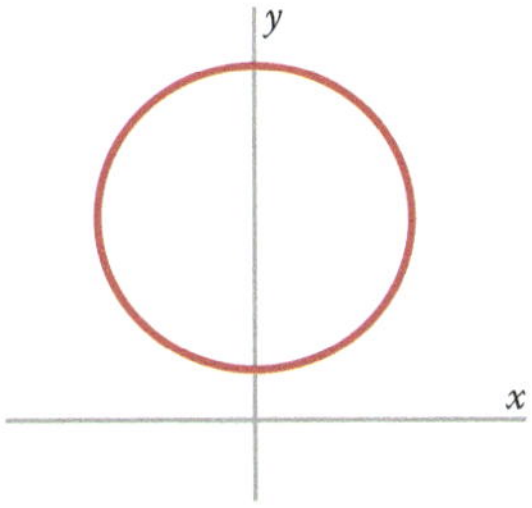
For every (x, y) on the graph, $(-x, -y)$ is also on the graph.	*For every (x, y) on the graph, $(x, -y)$ is also on the graph.*	*For every (x, y) on the graph, $(-x, y)$ is also on the graph.*

Functions that are symmetric with respect to the y-axis are called **even functions**. Functions that are symmetric with respect to the origin are called **odd functions**. So, whether a function is even or odd can be determined by analyzing the graph of the function. This can also be determined algebraically by substituting $-x$ for x in the function, as was done previously to determine the type of symmetry. When $-x$ is substituted for x in an even function $f(x)$, the result is $f(x)$. When $-x$ is substituted for x in an odd function $f(x)$, the result is $-f(x)$.

IMPORTANT

Since many functions are not symmetric with respect to the origin or the y-axis, many functions are neither even nor odd.

Even and Odd Functions

If $f(x)$ is a function, then

- $f(x)$ is even when $f(-x)$ is equal to $f(x)$ for every x in the domain of $f(x)$, and
- $f(x)$ is odd when $f(-x)$ is equal to $-f(x)$ for every x in the domain of $f(x)$.

EXAMPLE 5 Determining Whether a Function Is Even or Odd

Determine whether $p(x) = x - x^3$ is even, odd, or neither.

SOLUTION

Substitute $-x$ for x and simplify.

$$p(-x) = (-x) - (-x)^3$$

$$= (-x) - (-x^3) \qquad (-x)^3 = -x^3$$

$$= -x + x^3 \qquad \text{Subtracting a negative can be written as addition.}$$

$$= -(x - x^3) \qquad \text{Factor } -1 \text{ from each term.}$$

$$= -p(x) \qquad \text{Substitute } p(x) \text{ for } x - x^3.$$

Since $p(-x) = -p(x)$, $p(x)$ is odd. (The graph of $p(x)$ is shown in Figure 1.6d.)

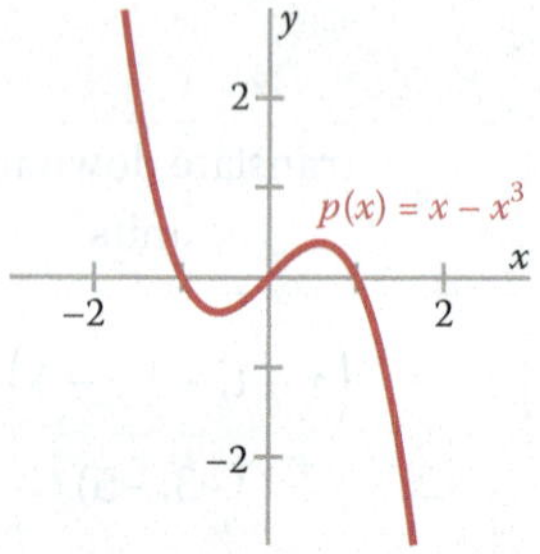

Figure 1.6d

The function is odd, so the graph is symmetric with respect to the origin.

WHAT'S THE BIG IDEA?

How can knowing that a function is even or odd be useful when graphing a function?

SECTION 1.6 EXERCISES

Warm Up

True or False?

1. The graph of $f(x) = x^2 + c$, where c is any real number, is always symmetric with respect to the y-axis.
2. A square root function will sometimes have a domain restriction.
3. Zeros of a function are also called y-intercepts.

Just the Facts

Fill in the blanks.

4. _____ and _____ are rigid transformations in which the shape of the graph _____ change, whereas transformations that cause a distortion of shape are called _____ transformations.
5. If the point (x, y) on the pre-image corresponds to the point $(x - c, y)$ on the image, then the coordinate has been shifted _____.
6. If $g(x) = -f(x)$, then the graph of $g(x)$ is a(n) _____ of the graph of $f(x)$ over the _____.
7. When $g(x) = cf(x)$, where $0 < c < 1$, the graph of $g(x)$ is _____ than the graph of $f(x)$.
8. The image of the point (2, 0) after a reflection across the y-axis and a shift down 4 units is _____.
9. $g(x) = (x + 5)^2 - 3$ is a shift of $f(x) =$ _____ to the _____ 5 units and _____ 3 units.
10. If for all (a, b) on the graph of $f(x)$, $(-a, b)$ is also on the graph of $f(x)$, then $f(x)$ is a(n) _____ function.

Essential Skills

In Exercises 1–30, sketch the graph of each function.

1. $f(x) = x^2 + 1$
2. $g(x) = (x - 2)^3$
3. $h(x) = |x - 4|$
4. $q(x) = \sqrt{x-3}$
5. $p(x) = (x - 1)^2 - 2$
6. $g(x) = \sqrt{x+5} - 7$
7. $w(x) = |x - 3| - 2$
8. $s(x) = (x + 4)^3 + 5$
9. $f(x) = (x + 1)^4 - 1$
10. $c(x) = \sqrt{x-6} - 3$
11. $f(x) = (x - 5)^2 + 8$
12. $w(x) = |x + 3| + 7$
13. $p(x) = -x^3 - 2$
14. $g(x) = -|x + 6|$
15. $k(x) = -x^2 + 9$
16. $d(x) = -\sqrt{x} - 4$
17. $h(x) = -|x| + 3$
18. $b(x) = -(x - 1)^2$
19. $m(x) = -x^5 + 2$
20. $t(x) = -\sqrt{x+3}$
21. $g(x) = 2\sqrt{x}$
22. $g(x) = \frac{1}{2}x^3$
23. $f(x) = 3x^2$
24. $v(x) = \frac{1}{4}|x|$
25. $f(x) = (x + 3)^2 + 1$
26. $p(x) = -2(x - 10)^3 + 9$
27. $g(x) = 2\sqrt{x+3} - 4$
28. $h(x) = -|x - 5| - 2$
29. $q(x) = \frac{1}{3}(x+6)^2 + 1$
30. $c(x) = -4\sqrt{x-1} + 2$

In Exercises 31–36, determine whether each function is even, odd, or neither.

31. $f(x) = x^2 - 1$
32. $r(x) = 2x^5 - x^3$
33. $k(x) = 7x^6 - x^4 + 3$
34. $m(x) = -3|x| + 9$
35. $p(x) = 4x^5 + 2x^4 - x^3 - 2$
36. $h(x) = -x^3 + x + 5$

Extensions

In Exercises 37–41, use the following information.

Given that f is an odd function, determine whether g is even, odd, or neither.

37. $g(x) = f(-x)$
38. $g(x) = -f(-x)$
39. $g(x) = f(x + 3)$
40. $g(x) = -f(x)$
41. $g(x) = (f(x))^2$

42. Write the equation of $g(x)$ if g is the image of the basic cubic function after a reflection across the x-axis and translations right 2 units, down 5 units.

43. Write the equation of $h(x)$ if h is the image of the basic absolute value function after a reflection across the x-axis and translations left 1 unit, up 2 units.

44. Write the equation of the graph.

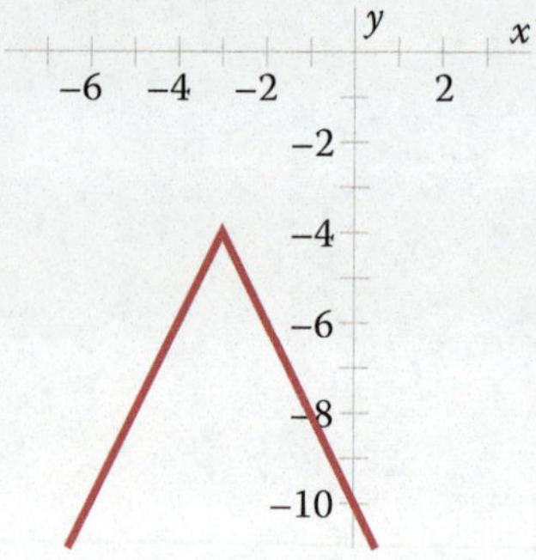

45. True or False? It is possible to have an even function with a domain of $[0, \infty)$.

46. Complete the graph of h given that it is an odd function.

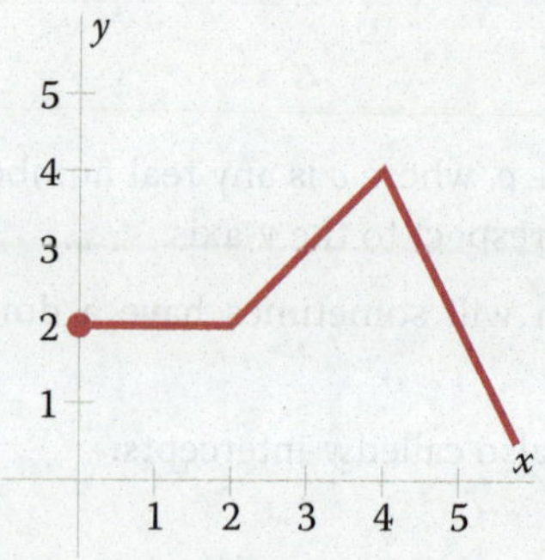

1.7 COMBINING FUNCTIONS

OBJECTIVES

- Perform arithmetic with functions.
- Calculate the difference quotient for functions and average rates of change.
- Compose functions and evaluate compositions of functions.
- Express functions as a composite of functions.
- Model with compositions of functions.

PREREQUISITE VOCABULARY TERMS

domain
function
rate of change
slope

1.7.1 Using Operations with Functions

Two functions can be combined by adding, subtracting, multiplying, or dividing to form a new function. There are no new rules or processes for combining functions because the same properties used for performing operations with real numbers and algebraic expressions are utilized. However, there is some new notation for combining functions, as summarized in the following table.

Arithmetic with Functions $f(x)$ and $g(x)$	
Sum of Functions $(f+g)(x) = f(x) + g(x)$	Difference of Functions $(f-g)(x) = f(x) - g(x)$
Product of Functions $(fg)(x) = f(x)g(x)$	Quotient of Functions $\left(\frac{f}{g}\right)(x) = \frac{f(x)}{g(x)}$ (when $g(x) \neq 0$)

The Domain of a Combination of Functions

The domain of a combination of functions f and g is all real numbers that are in both the domain of f and the domain of g. For example, if the domain of f is all real numbers x such that $x > 2$, and the domain of g is all real numbers x such that $x < 5$, then the domain of f combined with g (by using addition, subtraction, or multiplication) is the intersection of the domain of f and g, or $\{x \mid 2 < x < 5\}$. Additionally, in the case of a quotient of functions f/g, the domain is further restricted so that $g(x) \neq 0$.

EXAMPLE 1 Combining Functions

Given $p(x) = 1 - 5x^2$ and $r(x) = 2 - x$, find $(p + r)(x)$, $(r - p)(x)$, and $(pr)(x)$.

SOLUTION

Find the sum $(p + r)(x)$ by adding the two functions.

$(p + r)(x) = p(x) + r(x) = 1 - 5x^2 + 2 - x = -5x^2 - x + 3$ *Combine the like terms.*

Find the difference $(r - p)(x)$ by subtracting $p(x)$ from $r(x)$.

$(r - p)(x) = r(x) - p(x) = 2 - x - (1 - 5x^2) = 2 - x - 1 + 5x^2 = 5x^2 - x + 1$

IMPORTANT

Place the subtracted function, $p(x)$, within parentheses.

Find the product $(pr)(x)$ by multiplying the two functions.

$(pr)(x) = p(x)r(x) = (1 - 5x^2)(2 - x) = 2 - x - 10x^2 + 5x^3 = 5x^3 - 10x^2 - x + 2$ *FOIL the binomials.*

Evaluating a Combination of Functions

A combination of two or more functions can be evaluated for a given number by either

- evaluating each function for the given number and then combining the results, or
- combining the functions and then evaluating the combination for the given number.

For example, consider $(f+g)(2)$ when $f(x) = 5x$ and $g(x) = 3x$. One method for evaluating $(f+g)(2)$ is to evaluate each function for $x = 2$ first, and then combine (in this case add) those values.

$$(f+g)(2) = f(2) + g(2) = 5(2) + 3(2) = 16$$

The other method for evaluating $(f+g)(2)$ is to combine (add) the functions first, and then evaluate the combination for $x = 2$.

$$(f+g)(x) = f(x) + g(x) = 5x + 3x = 8x \quad \textit{Add the functions.}$$
$$(f+g)(2) = 8(2) = 16 \quad \textit{Evaluate the combination for } x = 2.$$

EXAMPLE 2 Evaluating a Combination of Functions

Given $p(x) = 1 - 5x^2$ and $r(x) = 2 - x$, find $\left(\frac{p}{r}\right)(-1)$.

SOLUTION

Find the quotient by dividing $p(-1)$ by $r(-1)$.

$$\left(\frac{p}{r}\right)(-1) = \frac{p(-1)}{r(-1)} = \frac{1-5(-1)^2}{2-(-1)} = \frac{1-5(1)}{2-(-1)} = \frac{-4}{3} = -\frac{4}{3}$$

ALTERNATIVE METHOD

Find the quotient $\left(\frac{p}{r}\right)(x)$ first, then evaluate the quotient for $x = -1$.

$$\left(\frac{p}{r}\right)(x) = \frac{p(x)}{r(x)} = \frac{1-5x^2}{2-x},\ x \neq 2 \qquad \left(\frac{p}{r}\right)(-1) = \frac{1-5(-1)^2}{2-(-1)} = \frac{1-5(1)}{2-(-1)} = \frac{-4}{3} = -\frac{4}{3}$$

IMPORTANT

Exclude 2 from the rational expression's domain because the denominator is 0 when $x = 2$.

1.7.2 The Difference Quotient

Recall that the slope of a line is the ratio of the change in y to the change in x, $\frac{y_2 - y_1}{x_2 - x_1}$, and that slope is also the average rate of change of y with respect to x.

Similarly with functions, the average rate of change in the value of a function f between two domain values x_1 and x_2 is given by the ratio of

- the change in f between $f(x_1)$ and $f(x_2)$, and
- the change in x between x_1 and x_2.

Let h be the difference between the two domain values x_1 and x_2 (where $x_1 < x_2$). Then, if $x_1 = a$, it follows that $x_2 = a + h$. In this case, the change in f between $f(x_1)$ and $f(x_2)$ is $f(a+h) - f(a)$. So, the ratio of the change in f to the change in x is given by $\frac{f(a+h) - f(a)}{h}$, which is known as the **difference quotient**.

DEFINITION

The **difference quotient,**

$$\frac{f(a+h)-f(a)}{h},$$

gives the average rate of change in f between the two domain values a and $a + h$.

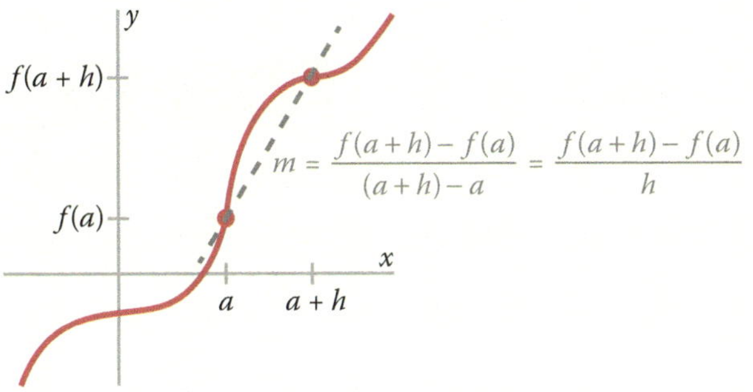

EXAMPLE 3 Simplifying the Difference Quotient

Simplify the difference quotient for each function.

A. $f(x) = 2 - 5x$ **B.** $f(x) = x^2 + 6x + 2$

SOLUTION

The difference quotient $\frac{f(a+h)-f(a)}{h}$ is the difference of $f(a + h)$ and $f(a)$ divided by h. So, use each given function to evaluate $f(a + h)$ and $f(a)$, subtract, and then divide by h.

TIP

$f(a + h) = 2 - 5(a + h)$
$= 2 - 5a - 5h$

$f(a) = 2 - 5(a)$
$= 2 - 5a$

A.

$$\frac{f(a+h)-f(a)}{h} = \frac{2-5(a+h)-(2-5(a))}{h}$$ *Evaluate $f(a + h)$ and $f(a)$.*

$$= \frac{2-5a-5h-2+5a}{h}$$ *Distribute.*

$$= \frac{-5h}{h}$$ *Combine the like terms.*

$$= -5$$ *Divide $-5h$ by h.*

TIP

$f(a + h) =$
$(a + h)^2 + 6(a + h) + 2 =$
$a^2 + 2ah + h^2 + 6a + 6h + 2$

$f(a) = (a)^2 + 6(a) + 2$
$= a^2 + 6a + 2$

B.

$$\frac{f(a+h)-f(a)}{h} = \frac{(a+h)^2+6(a+h)+2-((a)^2+6(a)+2)}{h}$$ *Evaluate $f(a + h)$ and $f(a)$.*

$$= \frac{a^2+2ah+h^2+6a+6h+2-a^2-6a-2}{h}$$ *FOIL and distribute.*

$$= \frac{2ah+h^2+6h}{h}$$ *Combine the like terms.*

$$= 2a+h+6$$ *Divide each term in the numerator by h.*

Examining the Difference Quotients (Rates of Change) in Example 3

Notice that the difference quotient for $f(x) = 2 - 5x$ is the constant -5. So, the rate of change between any two domain values for $f(x) = 2 - 5x$ will always be -5. This is to be expected since the rate of change (slope) of any line is constant and the slope of $f(x) = 2 - 5x$ is -5. Generally, since the rate of change for any linear function $f(x) = mx + b$ is the constant value m, the difference quotient of any linear function $f(x) = mx + b$ is also the constant m.

IMPORTANT

The difference quotient is equal to the slope of the line for any linear function.

The difference quotient for $f(x) = x^2 + 6x + 2$ is $2a + h + 6$, which is not a constant value. This is also to be expected, because the graph of $f(x)$ is not a line. The graph of $f(x)$ is a parabola, which does not have a constant rate of change between any two domain values on the parabola. In other words, the rate of change between domain values x_1 and x_2 is *not* equal to the rate of change between domain values x_3 and x_4 for all domain values. Instead, the rate of change between any two domain values on $f(x) = x^2 + 6x + 2$ is given by $2a + h + 6$, where a is the first domain value and h is the difference between the two domain values.

In the following example, the difference quotient will be used to find the average rate of change between two domain values on a nonlinear function.

EXAMPLE 4 Finding the Average Rate of Change

Suppose a ball is released from the top of a hill. The distance (in feet) that the ball rolls from the top of the hill t seconds after it is let go is given by the function $d(t) = 5.7t^2$. Find the ball's average speed between 7 and 12 seconds.

SOLUTION

REMEMBER

Since distance is equal to rate · time ($d = rt$), rate (speed) is equal to d/t.

The average rate of change in the function is the ratio of the change in d, the distance, to the change in t, the time, or d/t. Since $d/t = r$ (where r is the average rate or speed), the average rate of change in the function gives the ball's average speed. So, use the difference quotient to find the average rate of change in d.

$$\frac{d(a+h)-d(a)}{h} = \text{the ball's average speed between } t = a \text{ and } t = a + h$$

TIP

The variable h represents the difference between the domain values and the variable a represents the first domain value.

The domain values are 7 and 12. Since $12 = 7 + 5$, it follows that $a = 7$, $a + h = 12$, and $h = 5$.

$$\frac{d(12)-d(7)}{5} = \frac{5.7(12)^2 - 5.7(7)^2}{5} = \frac{820.8 - 279.3}{5} = \frac{541.5}{5} = 108.3$$

So, the ball's average speed between 7 and 12 seconds is 108.3 feet per second.

WHAT'S THE BIG IDEA?

How is the difference quotient related to slope?

1.7.3 Composition of Functions

Another way of combining two functions to form a new function is to **compose** the two functions (i.e., to form the **composition** of the two functions). Two functions are composed when one function is evaluated for another function.

Recall that the notation $f(3)$ means to evaluate f for 3, or to substitute 3 into f for x. It follows that $f(g(3))$ (read as "f of g of 3") means to evaluate f for $g(3)$, or to substitute the value of $g(3)$ into f for x.

For example, suppose $f(x) = 5x + 3$ and $g(x) = x^2$. To find $f(g(3))$, first evaluate $g(3)$, then use that number to evaluate f.

$f(g(3)) = f(9)$	*Evaluate $g(3)$: $g(3) = 3^2 = 9$.*
$= 5(9) + 3$	*Evaluate f for $g(3)$, or evaluate $f(9)$.*
$= 48$	*Simplify.*

DEFINITION

The **composition** of the function f with the function g, $f(g(x))$, is the result of evaluating f for g.

$$f(g(x)) = (f \circ g)(x)$$

The domain of $f \circ g$ is the set of all x in the domain of g such that $g(x)$ is in the domain of f.

TIP

The expression $f \circ g$ is read as "f composed with g."

EXAMPLE 5 Composition of Functions

Given $p(x) = 1 - 5x^2$ and $r(x) = 2 - x$, find $(p \circ r)(3)$ and $(r \circ p)(3)$.

SOLUTION

To find $(p \circ r)(3)$, evaluate $r(3)$, and then evaluate p for $r(3)$.

$(p \circ r)(3)$	$= p(r(3))$	*Write p composed with r(3).*
	$= p(2 - 3)$	*Evaluate r(3) by substituting 3 into $r(x) = 2 - x$ for x.*
	$= p(-1)$	*Subtract.*
	$= 1 - 5(-1)^2$	*Evaluate p(−1) by substituting −1 into $p(x) = 1 - 5x^2$ for x.*
	$= 1 - 5(1)$	*Evaluate the power.*
	$= -4$	*Simplify.*

ALTERNATIVE METHOD

Evaluate r(3) first.

$r(3) = 2 - 3 = -1$

Now evaluate p(−1).

$p(-1) = 1 - 5(-1)^2$

$= 1 - 5(1)$

$= -4$

To find $(r \circ p)(3)$, evaluate $p(3)$, and then evaluate r for $p(3)$.

$(r \circ p)(3)$	$= r(p(3))$	*Write r composed with p(3).*
	$= r(1 - 5(3)^2)$	*Evaluate p(3) by substituting 3 into $p(x) = 1 - 5x^2$ for x.*
	$= r(-44)$	*Simplify: $1 - 5(3)^2 = 1 - 5(9) = 1 - 45 = -44$.*
	$= 2 - (-44)$	*Evaluate r(−44) by substituting −44 into $r(x) = 2 - x$ for x.*
	$= 46$	*Simplify.*

So, $(p \circ r)(3) = -4$ and $(r \circ p)(3) = 46$.

NOTICE THAT

Since $(p \circ r)(3) = -4$ and $(r \circ p)(3) = 46$, $(p \circ r)(3) \neq (r \circ p)(3)$. It follows that for any functions f and g, $(f \circ g)$ is not necessarily equivalent to $(g \circ f)$.

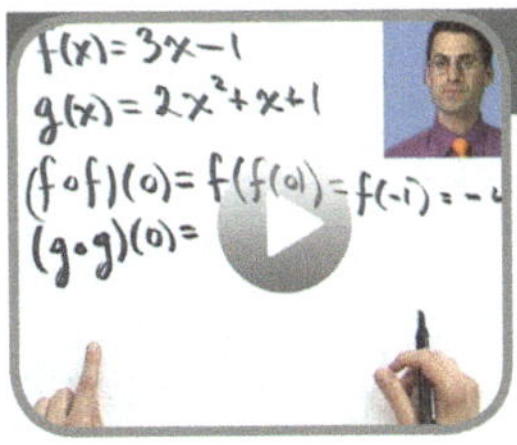

1.7.4 Composition of Functions: Another Example

In the previous topic, a composition of two functions was evaluated for a given number, for example, $(p \circ r)(3)$. In this topic, a function will be composed with itself, and then evaluated for some given number, for example, $(f \circ f)(5)$.

Additionally, a function will be composed with a function that is not evaluated for a number, for example, $(f \circ g)(x)$. The notation $(f \circ g)(x)$ means to evaluate f for g, or to substitute $g(x)$ into the function f for x. For example, suppose $f(x) = 5x + 3$ and $g(x) = x^2$. Since $g(x) = x^2$, the x in $f(x) = 5x + 3$ is replaced by x^2 when evaluating $(f \circ g)(x)$.

$$\text{Evaluate } f \text{ for } g \;\Rightarrow\; (f \circ g)(x) = f(g(x)) = 5(g(x)) + 3 = 5x^2 + 3$$

So, the new function formed by the composition of f and g is $(f \circ g)(x) = 5x^2 + 3$.

EXAMPLE 6 Composition of Functions

Given $r(x) = 2 - x$ and $p(x) = 1 - 5x^2$, find $(r \circ r)(3)$ and $(p \circ r)(x)$.

SOLUTION

To find $(r \circ r)(3)$, evaluate $r(3)$, and then evaluate r for $r(3)$.

$(r \circ r)(3)$	$= r(r(3))$	*Write r composed with r(3).*
	$= r(2 - 3)$	*Evaluate r(3) by substituting 3 into $r(x) = 2 - x$ for x.*
	$= r(-1)$	*Subtract.*
	$= 2 - (-1)$	*Evaluate r(−1) by substituting −1 into $r(x) = 2 - x$ for x.*
	$= 3$	*Simplify.*

To find $(p \circ r)(x)$, evaluate p for $r(x)$.

$(p \circ r)(x) = p(r(x))$	*Write p composed with r(x).*
$= p(2 - x)$	*Substitute 2 – x for r(x).*
$= 1 - 5(2 - x)^2$	*Substitute 2 – x into $p(x) = 1 - 5x^2$ for x.*
$= 1 - 5(2 - x)(2 - x)$	*Evaluate the power.*
$= 1 - 5(4 - 4x + x^2)$	*FOIL the binomials.*
$= 1 - 20 + 20x - 5x^2$	*Distribute –5.*
$= -5x^2 + 20x - 19$	*Combine the like terms.*

WHAT'S THE BIG IDEA?

Explain the difference between $(f \circ g)(x)$ and $(g \circ f)(x)$.

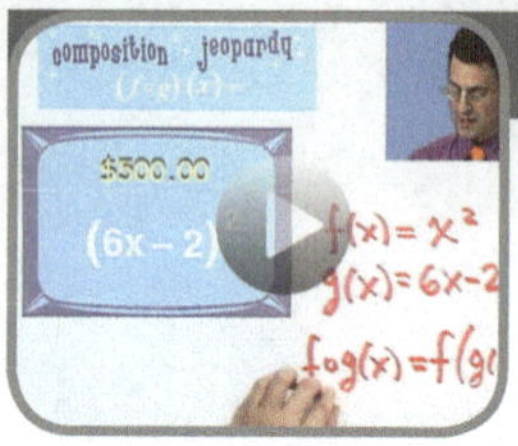

1.7.5 Finding Functions That Form a Given Composite

Sometimes a given function can be written as a composition of two functions. The process of finding two functions that can be composed to form a given function is called **decomposing**.

To decompose a function, analyze the function, considering its parts, until a possibility for two functions has been determined. Then compose these two functions to check that they do form the given composite. There is often more than one way to decompose a composite function.

EXAMPLE 7 Decomposing a Composite Function

Write $p(x) = \dfrac{9}{2(1-x)^3}$ as a composition of two functions in two different ways.

SOLUTION

Notice that the expression $(1 - x)$ is cubed and multiplied by 2, and that 9 is divided by that result. Breaking this into parts reveals that if $f(x) = 1 - x$ and $g(x) = \dfrac{9}{2x^3}$, then $g(f(x)) = p(x)$. Compose g with f to confirm.

$$g(f(x)) = g(1-x) = \frac{9}{2(1-x)^3} = p(x)$$

Again, breaking the expression for $p(x)$, $\dfrac{9}{2(1-x)^3}$, into parts reveals that if $f(x) = 2(1 - x)^3$ and $g(x) = \dfrac{9}{x}$, then $g(f(x)) = p(x)$. Compose g with f to confirm.

$$g(f(x)) = g(2(1-x)^3) = \frac{9}{2(1-x)^3} = p(x)$$

IMPORTANT

There are other ways to write p(x) as a composite of two functions.

CAUTION

Notice that in the previous example, writing $p(x)$ as the composition $g(f(x))$ where $f(x) = 2(1 - x)$ and $g(x) = \frac{9}{x^3}$ would not be correct, because the coefficient of 2 in $f(x) = 2(1 - x)$ must also be cubed when evaluating the composite.

$$g(f(x)) = g(2(1-x)) = \frac{9}{(2(1-x))^3} = \frac{9}{8(1-x)^3} \neq p(x)$$

Therefore, $p(x) \neq g(f(x))$ when $f(x) = 2(1 - x)$ and $g(x) = \frac{9}{x^3}$.

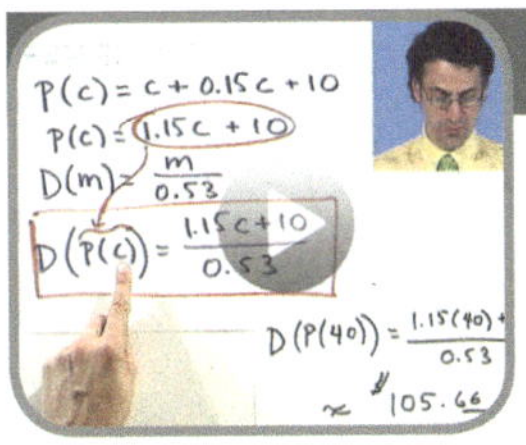

1.7.6 Modeling with Composite Functions

A composite function can be used to model a situation where one quantity is dependent on two different quantities (i.e., one quantity is a function of two different quantities). For example, suppose there are two discounts on the price of an item. In this case, the sale price of the item is a function of the price after the second discount, which is a function of the price after the first discount. A multiple-discount situation is modeled with a composition of functions in the following example.

EXAMPLE 8 Modeling with a Composite Function

A cell phone company offers a \$100 discount on the price of a phone for customers signing a two-year contract. The same company is offering a 15% discount on their phones for customers during a holiday weekend. Let x represent the regular price of a phone. Write two composite functions: one to represent the sale price of the phone if the 15% discount is applied after the \$100 discount, and one to represent the sale price of the phone if the 15% discount is applied before the \$100 discount. Which is the better deal for the customer?

SOLUTION

Write a function for each discount. Let the 15% discount be represented by the function f, and let the \$100 discount be represented by the function g.

$$f(x) = x - 0.15x = 0.85x \qquad g(x) = x - 100$$

15% discount applied *after* the \$100 discount	15% discount applied *before* the \$100 discount
$f(g(x)) = 0.85(x - 100) = 0.85x - 85$	$g(f(x)) = 0.85x - 100$

Notice that both compositions start with $0.85x$, or 85% of the original price. When the \$100 discount is applied first (as in $(f \circ g)(x)$), the sale price is \$85 less than 85% of the original price. When the \$100 discount is applied second (as in $(g \circ f)(x)$), the sale price is \$100 less than 85% of the original price. So, applying the 15% discount first and the \$100 discount second results in the better deal for the customer.

SECTION 1.7 EXERCISES

Warm Up

Find the x- and y-intercepts of each function.

1. $y = 2x - 1$
2. $g(x) = 3x^2 - 27$
3. $p(x) = \sqrt{x+1}$

Just the Facts

Fill in the blanks.

4. Two functions g and h can be combined by the arithmetic operations of _____, _____, _____, and _____ to form new functions.
5. The _____ of the function g with f is $(g \circ f)(x) = (g(f(x))$.
6. The domain of $(g + f)$ is all real numbers that are in both the domain of _____ and the domain of _____.
7. The difference quotient of any linear function $y = mx + b$ is always _____.
8. The _____ gives the average rate of change in f between two domain values such that the difference in the domain values is _____.
9. _____ is the process of finding two functions that can be composed to form a given function.
10. The quotient $\left(\frac{g}{h}\right)(x)$ is defined only when _____ does not equal 0.

Essential Skills

In Exercises 1–32, use the given functions to complete each operation.

1. $p(x) = -2x$ and $r(x) = x + 4$; $(p + r)(x)$
2. $f(x) = 1 + x^2$ and $g(x) = 3x^2 + 5$; $(f + g)(x)$
3. $g(x) = x^2 + 1$ and $h(x) = -x^2 - 5$; $(g - h)(x)$
4. $r(x) = 6x - 2$ and $t(x) = 4x^2 + 23$; $(r - t)(x)$
5. $s(x) = 11 + 4x^2$ and $t(x) = 3x^2 - 11x + 4$; $(s + t)(x)$
6. $r(x) = 9x^2 + 8x - 5$ and $w(x) = x^2 - 7x + 3$; $(r - w)(x)$
7. $c(x) = 3 - x - 6x^2$ and $d(x) = -2x^2 + x - 4$; $(c - d)(x)$
8. $p(x) = -4x^2 - 15x$ and $q(x) = 8x^2 + 3x - 9$; $(p + q)(x)$
9. $u(x) = \frac{2}{x}$ and $v(x) = \frac{3}{x-1}$; $(u + v)(x)$
10. $b(x) = -\frac{4}{x+3}$ and $n(x) = \frac{x}{x-2}$; $(b - n)(x)$
11. $k(x) = \frac{x+1}{x-5}$ and $h(x) = -\frac{2x+3}{x^2-25}$; $(k + h)(x)$
12. $f(x) = \frac{3x-5}{x^2+x-20}$ and $g(x) = \frac{2x-1}{x+5}$; $(f - g)(x)$
13. $h(x) = 2x + 1$ and $g(x) = -x + 8$; $(hg)(x)$
14. $f(x) = 5 - x^2$ and $g(x) = 6 - x$; $(fg)(x)$
15. $p(x) = x + 7$ and $r(x) = x^2 + 14x + 49$; $\left(\frac{p}{r}\right)(x)$
16. $h(x) = 5 + x$ and $g(x) = 25 - x^2$; $\left(\frac{g}{h}\right)(x)$
17. $k(x) = 3x - 8$ and $l(x) = 8 + 3x^2$; $(kl)(x)$
18. $m(x) = x^2 - 19x + 78$ and $n(x) = x - 6$; $\left(\frac{m}{n}\right)(x)$
19. $u(x) = 5x^2 - 20$ and $v(x) = 2 + x$; $\left(\frac{u}{v}\right)(x)$
20. $s(x) = 2x^2 - x$ and $t(x) = x^2 - 4$; $(st)(x)$
21. $q(x) = -\frac{5x}{x+4}$ and $r(x) = \frac{x+2}{x-3}$; $(qr)(x)$
22. $g(x) = \frac{4x-1}{x-6}$ and $t(x) = \frac{x+7}{2x-3}$; $(gt)(x)$
23. $f(x) = \frac{2x^2-2}{x+3}$ and $h(x) = \frac{x^2+5x+4}{x^2-9}$; $\left(\frac{f}{h}\right)(x)$
24. $w(x) = \frac{x^2-7x-8}{x^2+3x-18}$ and $z(x) = \frac{x^2-6x-16}{x^2+2x-24}$; $\left(\frac{w}{z}\right)(x)$
25. $g(x) = 3x + 2$ and $h(x) = -2x$; $(gh)(4)$
26. $f(x) = x^2$ and $g(x) = x + 1$; $(fg)(-3)$
27. $h(x) = -x^2 - 6$ and $k(x) = 9 - 4x$; $(h + k)(-2)$
28. $p(x) = 7 - 2x^2$ and $q(x) = -5x + 8$; $(p - q)(3)$
29. $b(x) = -4x^2 - 2$ and $c(x) = 2x + 5$; $(bc)(-1)$
30. $k(x) = 3x^2 - 5x$ and $n(x) = x - 9$; $(kn)(-6)$
31. $d(x) = \sqrt{1-x}$ and $p(x) = \frac{2}{x-7}$; $\left(\frac{d}{p}\right)(-3)$
32. $t(x) = \frac{3x}{x+11}$ and $w(x) = \sqrt{x^2-7}$; $\left(\frac{t}{w}\right)(4)$

In Exercises 33–40, simplify the difference quotient for each function.

33. $g(x) = -3x + 8$
34. $f(x) = -x^2 + 7x + 3$
35. $r(x) = 9 - 2x^2$
36. $c(x) = 14 - 5x$
37. $k(x) = 6x^2 - 4x + 1$
38. $d(x) = -3x^2 - 2x + 7$
39. $m(x) = x^3 - 5x$
40. $n(x) = 4x^3 - x^2 + 1$

In Exercises 41–60, use the given functions to find each composite.

41. $(f \circ g)(9)$ for $f(x) = x + 4$ and $g(x) = 9x - 3$

42. $(g \circ f)(3)$ for $f(x) = -2x^2 + 5x + 7$ and $g(x) = 4x - 3$

43. $(g \circ f)(-2)$ for $f(x) = -4x + 1$ and $g(x) = -x^2 + 3x + 2$

44. $(f \circ g)(-1)$ for $f(x) = -3x^2 + x + 9$ and $g(x) = 2x^2 + x - 8$

45. $(g \circ f)(-4)$ for $f(x) = 8 - x^2$ and $g(x) = \dfrac{x}{x+1}$

46. $(f \circ g)(-4)$ for $f(x) = 8 - x^2$ and $g(x) = \dfrac{x}{x+1}$

47. $(g \circ g)(2)$ for $g(x) = 2x + 1$

48. $(f \circ f)(-2)$ for $f(x) = x^2 + 4x$

49. $(f \circ f)(-1)$ for $f(x) = 3x^2 + 4x - 2$

50. $(g \circ g)(3)$ for $g(x) = -x^2 - 2x + 5$

51. $(f \circ f)(8)$ for $f(x) = \dfrac{x-1}{2}$

52. $(g \circ g)(-5)$ for $g(x) = \dfrac{x+2}{x}$

53. $(g \circ f)(x)$ and $(f \circ g)(x)$ for $f(x) = 6 + 5x$ and $g(x) = 4x - 4$

54. $(f \circ g)(x)$ and $(g \circ f)(x)$ for $f(x) = 5 - 4x$ and $g(x) = x^2 - 4$

55. $(f \circ g)(x)$ and $(g \circ f)(x)$ for $f(x) = 2x - 4$ and $g(x) = \dfrac{1}{2}x + 2$.

56. $(g \circ f)(x)$ and $(f \circ g)(x)$ for $f(x) = \dfrac{3}{2}x - 6$ and $g(x) = \dfrac{2}{3}x + 4$.

57. $(f \circ g)(x)$ and $(g \circ f)(x)$ for $f(x) = 5x - 4$ and $g(x) = x^2 - 5x + 6$.

58. $(g \circ f)(x)$ and $(f \circ g)(x)$ for $f(x) = 3 - x$ and $g(x) = 2x^2 - x + 4$.

59. $(g \circ f)(x)$ and $(f \circ g)(x)$ for $f(x) = 2\sqrt{x}$ and $g(x) = x^2 + 3$.

60. $(f \circ g)(x)$ and $(g \circ f)(x)$ for $f(x) = 2x^2 + 1$ and $g(x) = \sqrt{x - 4}$.

In Exercises 61–62, find the function $f(x)$ that forms each given composite function with $g(x) = x^2$.

61. $(g \circ f)(x) = (-3x - 10)^2$

62. $(f \circ g)(x) = x^4 - 5x^2$

63. Find $f(x)$ such that $(f \circ g)(x) = 6x + 10$ and $g(x) = 3x + 2$.

64. Find $f(x)$ such that $(f \circ g)(x) = 8x + 22$ and $g(x) = 2x + 4$.

65. Find $g(x)$ such that $(f \circ g)(x) = 12x - 5$ and $f(x) = 4x - 1$.

66. Find $g(x)$ such that $(f \circ g)(x) = -7x + 9$ and $f(x) = -x - 3$.

67. Find $g(x)$ such that $(f \circ g)(x) = 2x^6 - 5x^3$ and $f(x) = 2x^2 - 5x$.

68. Find $f(x)$ such that $(f \circ g)(x) = 3x^8 - 6x^4$ and $g(x) = x^4$.

69. A department store offers their employees a 20% discount on all of their merchandise m. An employee has a coupon from the local paper for $10 off any purchase. Write a function p for the cost of the merchandise after receiving only the employee discount. Write a function c for the cost of the merchandise after receiving only the coupon discount. Evaluate $(p \circ c)(m)$ and explain what the composition represents.

70. A car dealership offers a $1500 factory rebate and a 10% discount off the price of a new car c. Write a function r for the cost of the car after receiving only the factory rebate. Write a function p for the cost of the car after receiving only the dealership discount. Evaluate $(r \circ p)(c)$ and explain what the composition represents.

Extensions

In Exercises 71–79, use the graphs of $f(x)$ and $g(x)$ to evaluate each expression.

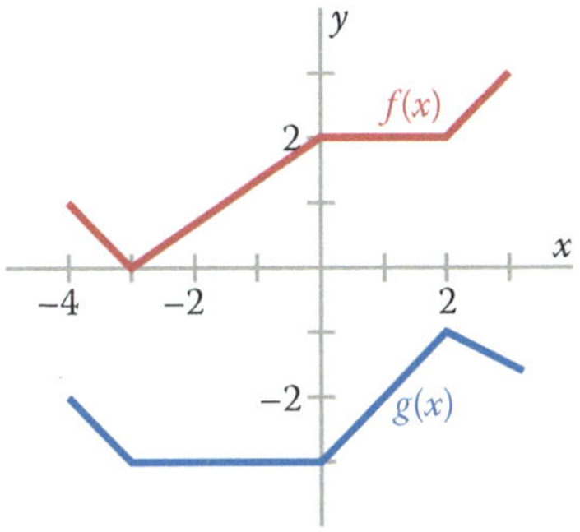

71. $(f + g)(0)$

72. $(f - g)(2)$

73. $(gf)(0)$

74. $(fg)(-3)$

75. $\left(\dfrac{f}{g}\right)(-4)$

76. $\left(\dfrac{g}{f}\right)(2)$

77. $(f \circ g)(0)$

78. $(g \circ f)(1)$

79. $(f \circ f)(-3)$

In Exercises 80–82, use the function $f(x) = 3x^2 - 2x + 1$ to find the average rate of change between each pair of values.

80. 4 and 7

81. 0 and 2

82. a and b

83. Let $f(m)$ be the number of full-time employees in a company in month m, and let $p(m)$ be the number of part-time employees in month m. Let $m = 1$ correspond to January. Find the function $t(m)$ that represents the total number of employees in the company. Then interpret the value $t(7)$.

84. A car salesman is paid an annual salary plus a bonus of 5% for sales over \$750,000. Given the two functions $g(x) = x - 750{,}000$ and $f(x) = 0.05x$, where x represents the sales over \$750,000, does $f(g(x))$ or $g(f(x))$ represent the bonus?

85. Given $h(x) = \dfrac{2}{x-5}$ and $g(x) = -3x + 8$, find $(h \circ g)(x)$ and $(g \circ h)(x)$. State the domain of each composite function.

In Exercises 86–92, given $f(x) = 3x^2$ and $g(x) = x^2$, describe the graph of the function.

86. $(f + g)(x)$

87. $(f - g)(x)$

88. $(g - f)(x)$

89. $(fg)(x)$

90. $\left(\dfrac{f}{g}\right)(x)$

91. $\left(\dfrac{g}{f}\right)(x)$

92. $(g \circ g)(x)$

1.8 INVERSE FUNCTIONS

OBJECTIVES

- Understand inverse functions and one-to-one functions.
- Use the Horizontal Line Test.
- Verify that two functions are inverses.
- Find the graph of an inverse function.
- Find the inverse function of a function algebraically.

PREREQUISITE VOCABULARY TERMS

domain
function
function composition
Vertical Line Test

1.8.1 Introduction to Inverse Functions

Recall that inverse operations are two operations that undo one another. For example, the inverse operation of multiplication is division, because when a number a is multiplied by a nonzero number b and then divided by b, the result is a ($a \cdot b \div b = a$). The same result occurs when a is divided by a nonzero number b and then multiplied by b ($a \div b \cdot b = a$).

Two functions can be **inverses** as well. If f is a function with domain A and range B, then the inverse function of f is a function g with range A and domain B such that for each point (c, d) on the graph of f there is a point (d, c) on the graph of g. In this way, the inverse function of a function *undoes* the original function.

DEFINITION

If two functions f and g are **inverse functions** then

- for every x in the domain of f, $g(f(x)) = x$, and
- for every x in the domain of g, $f(g(x)) = x$.

The domain of f must be equal to the range of g, and the domain of g must be equal to the range of f.

The relationship between the domain and range of two inverse functions f and g is shown in Figure 1.8a.

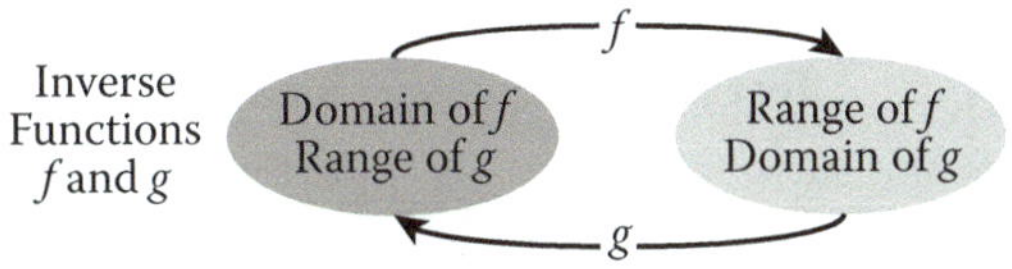

Figure 1.8a

Consider the function $f(x) = x - 3$. When the input is 5, the output is 2, since $f(5) = 5 - 3 = 2$ (i.e., (5, 2) is a point on the graph of f). So, the inverse function of $f(x) = x - 3$ must take 2 as an input value and return 5 as the output value (i.e., (2, 5) must be a point on the inverse of f). The inverse function of $f(x) = x - 3$ must be a function that *undoes* subtracting 3 from the domain value. Since addition and subtraction are inverse operations, it seems reasonable to expect that the inverse function of $f(x) = x - 3$ is $g(x) = x + 3$.

By the definition of inverse functions, we know that if f and g are inverse functions, then $g(f(x)) = x$ and $f(g(x)) = x$. Notice that composing the functions $f(x) = x - 3$ and $g(x) = x + 3$ results in x.

$$g(f(x)) = (x - 3) + 3 = x \qquad\qquad f(g(x)) = (x + 3) - 3 = x$$

Therefore, $f(x) = x - 3$ and $g(x) = x + 3$ are inverse functions, as expected.

EXAMPLE 1 Verifying Inverse Functions

Use $(f \circ g)$ and $(g \circ f)$ to confirm that $f(x) = 5x$ and $g(x) = \dfrac{x}{5}$ are inverse functions.

SOLUTION

Evaluate each composition.

$$(f \circ g)(x) = f(g(x)) = f\left(\frac{x}{5}\right) = 5 \cdot \frac{x}{5} = x$$

$$(g \circ f)(x) = g(f(x)) = g(5x) = \frac{5x}{5} = x$$

So, since $(f \circ g)(x) = x$ and $(g \circ f)(x) = x$, f and g are inverse functions.

EXAMPLE 2 Evaluating with Inverse Functions

Suppose that g and h are inverse functions, and that $h(1) = 3$, $g(-4) = 10$, and $h(0) = -2$. Use the given information to evaluate $h(10)$, $g(-2)$, and $g(1)$, if possible.

SOLUTION

Since g and h are inverse functions, we know that if $g(a) = b$, then $h(b) = a$. So, $h(10) = -4$ because $g(-4) = 10$.

Additionally, since g and h are inverse functions, we know that if $h(a) = b$, then $g(b) = a$. So, $g(-2) = 0$ because $h(0) = -2$.

We are not given a domain value for h that results in a range value of 1. Therefore, $g(1)$ cannot be evaluated using the given information.

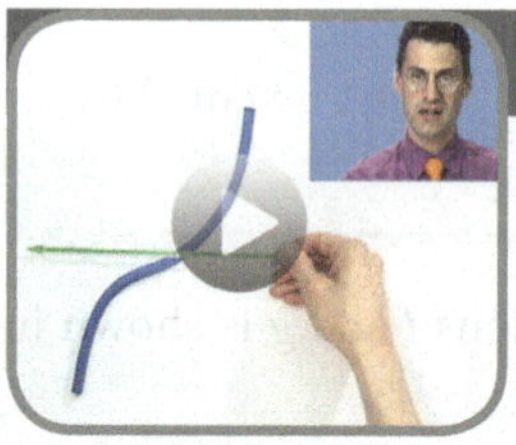

1.8.2 The Horizontal Line Test

Recall that there are some graphs that are not functions. The Vertical Line Test is used to determine whether a graph is a function. When any vertical line intersects a graph no more than one time, then the graph passes the Vertical Line Test and therefore is a function.

Similarly, there are some functions that do not have inverse functions. The **Horizontal Line Test** is used to determine whether a function has an inverse function.

Horizontal Line Test

A function f has an inverse function if any possible horizontal line can intersect the graph of f at most once.

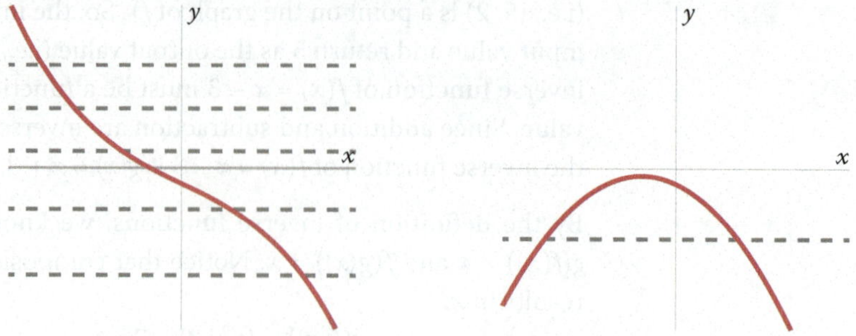

Passes the Horizontal Line Test   Fails the Horizontal Line Test

When a function passes the Horizontal Line Test, each of its domain values corresponds to exactly one range value.

DEFINITION

A function f that has exactly one range value corresponding to each of its domain values is a **one-to-one function**. In other words, $f(x_1) = f(x_2)$ if and only if $x_1 = x_2$.

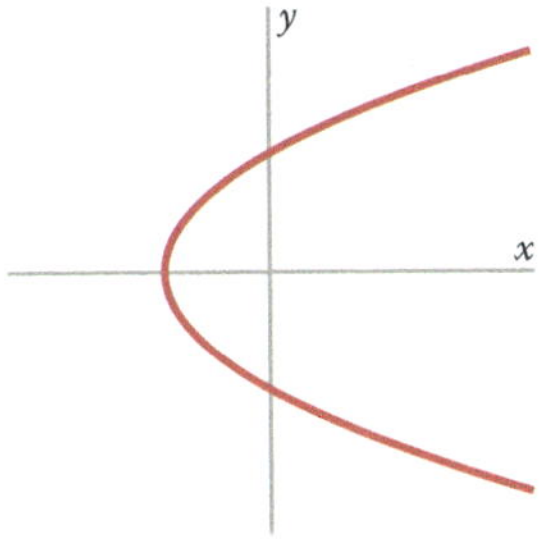

Figure 1.8b

A one-to-one function will always pass the Horizontal Line Test because it has no more than one point on its graph for any y-value in its range. Therefore, a one-to-one function will always have an inverse function.

CAUTION

Figure 1.8b shows a graph that no horizontal line will intersect more than once, so it may appear to pass the Horizontal Line Test. However, this is not the graph of a one-to-one function because it is not the graph of a function (i.e., it does not pass the Vertical Line Test).

EXAMPLE 3 Using the Horizontal Line Test

Use the Horizontal Line Test to determine whether each graph is a one-to-one function with an inverse function.

A.

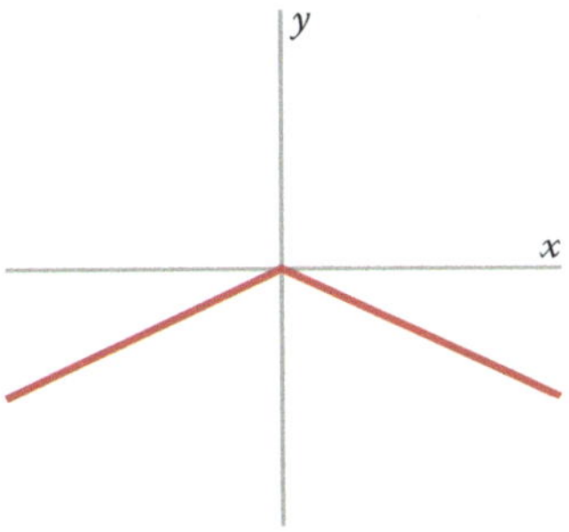

B.

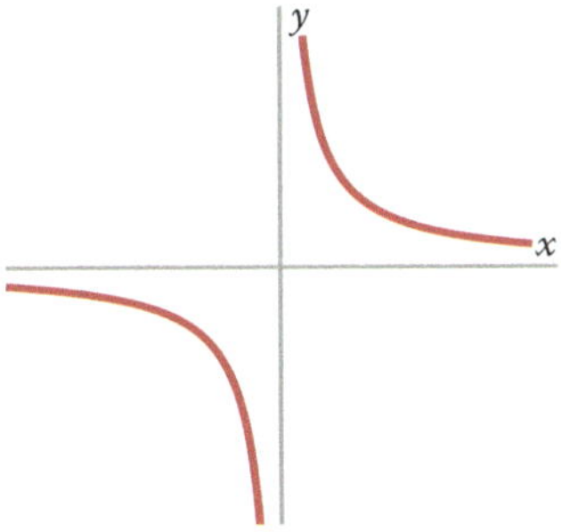

SOLUTION

REMEMBER

The graph passes the Vertical Line Test and is therefore a function.

A. The graph fails the Horizontal Line Test because at least one horizontal line will intersect the graph at more than one point. Therefore, the graph is not a one-to-one function and does not have an inverse function.

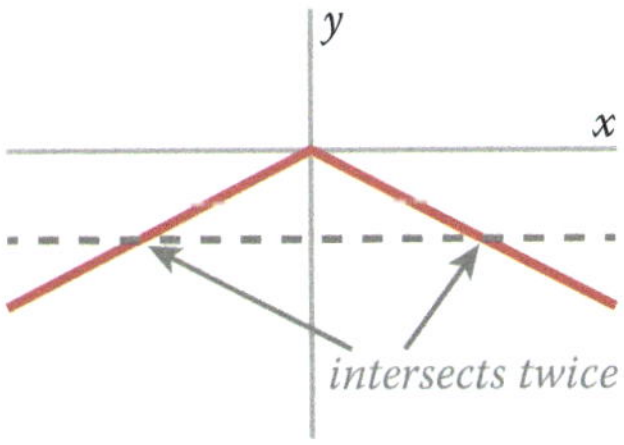

B. The graph passes the Horizontal Line Test because any horizontal line will intersect the graph at most once. Therefore, the graph is a one-to-one function, and the function has an inverse function.

1.8.3 Verifying That Functions Are Inverses

Inverse Function Notation

If f is a one-to-one function, then the inverse function of f is denoted f^{-1} (read as "f-inverse") or $f^{-1}(x)$.

CAUTION

There are two common misconceptions with inverse function notation. First, the inverse function of f is *not* denoted as $f(x)^{-1}$. Instead, the superscript −1 must follow the f, not the x. Second, the superscript −1 in inverse function notation is not an exponent. So, $f^{-1}(x) \neq \dfrac{1}{f(x)}$.

Verifying Inverse Functions

There are two methods for verifying that two functions are inverses: algebraically and graphically. The algebraic method, which was demonstrated previously, uses the compositions of the two functions. When two functions are inverses, their compositions are equal to x.

$$f^{-1}(f(x)) = x \quad \text{and} \quad f(f^{-1}(x)) = x$$

The other method for verifying that two functions are inverses utilizes the graphs of the functions. By the definition of an inverse function, (a, b) is a point on f if and only if (b, a) is a point on f^{-1}. Recall that reflecting a point (x, y) over the line $y = x$ gives the point (y, x). Therefore, the graph of a function and its inverse function are reflections over the line $y = x$, as shown in Figure 1.8c.

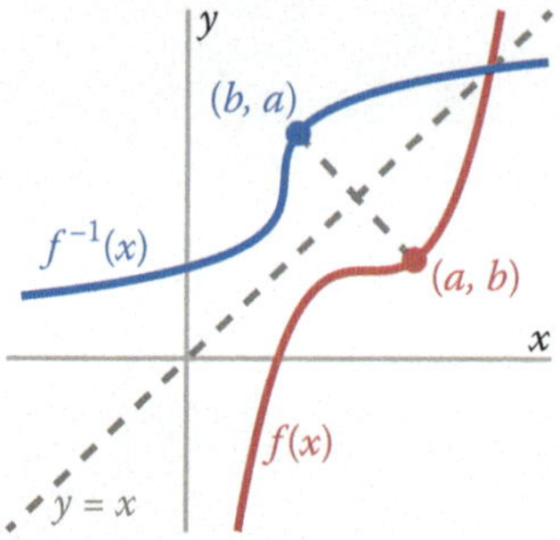

Figure 1.8c

EXAMPLE 4 Verifying That Two Functions Are Inverse Functions

Show graphically and algebraically that $f(x) = 3x + 6$ and $g(x) = \dfrac{x}{3} - 2$ are inverse functions.

SOLUTION

<u>Graphically</u>

$f(x) = 3x + 6$ is a line with slope 3 and y-intercept 6.

$g(x) = \dfrac{x}{3} - 2$ is a line with slope $\dfrac{1}{3}$ and y-intercept −2.

The lines are reflections over the line $y = x$.

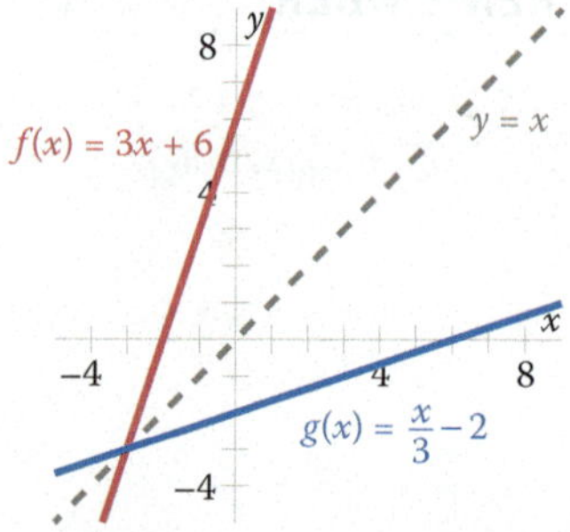

REMEMBER

When two graphs are reflections over the line y = x, then for each point (a, b) on one graph, the point (b, a) is on the other graph.

<u>Algebraically</u>

Verify that $g(f(x)) = x$ and $f(g(x)) = x$.

$$g(f(x)) = \frac{3x+6}{3} - 2 = (x+2) - 2 = x \qquad f(g(x)) = 3\left(\frac{x}{3} - 2\right) + 6 = (x-6) + 6 = x$$

Since $g(f(x)) = x$ and $f(g(x)) = x$, the functions are inverses.

1.8.4 Finding the Inverse of a Function Graphically

The graph of a function can be used to determine whether the function has an inverse function and to find the graph of that inverse function when it exists. Recall that if the graph of a function passes the Horizontal Line Test, then it is a one-to-one function and it does have an inverse function. So, by applying the Horizontal Line Test, the existence of an inverse function can be determined. Also recall that functions are inverses when their graphs are reflections over the line $y = x$. So, when the graph of a function passes the Horizontal Line Test, then the inverse function of this function can be graphed by reflecting the function's graph over the line $y = x$.

EXAMPLE 5 Finding the Graph of an Inverse Function

Graph the inverse function of f, if it exists.

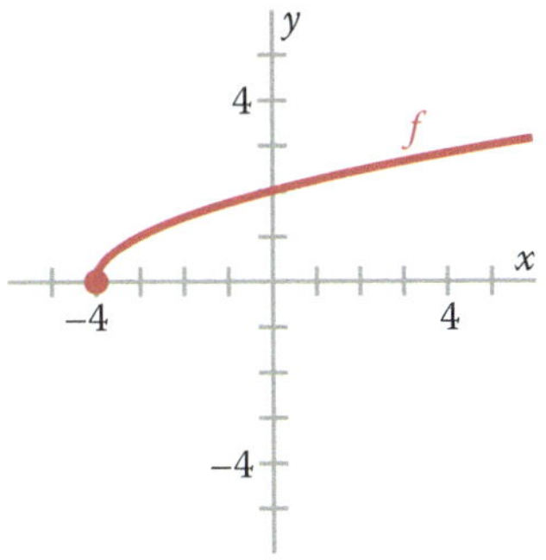

SOLUTION

The function f passes the Horizontal Line Test, so it is a one-to-one function and has an inverse function.

Reflect f over the line $y = x$ to graph the inverse function of f.

Since the graph of f passes through the line $y = x$, the graph of f^{-1} must also pass through this same point on $y = x$. Find additional points on the graph of f^{-1} by identifying points on f and then reversing the coordinates. For example, since $(-4, 0)$ and $(0, 2)$ are points on f, $(0, -4)$ and $(2, 0)$ must be points on f^{-1}.

So, to graph f^{-1}, sketch a curve that passes through $(0, -4)$, $(2, 0)$, and the point at which f intersects the line $y = x$.

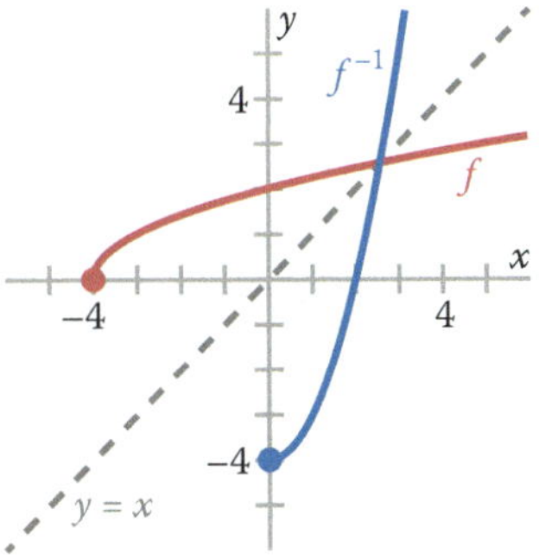

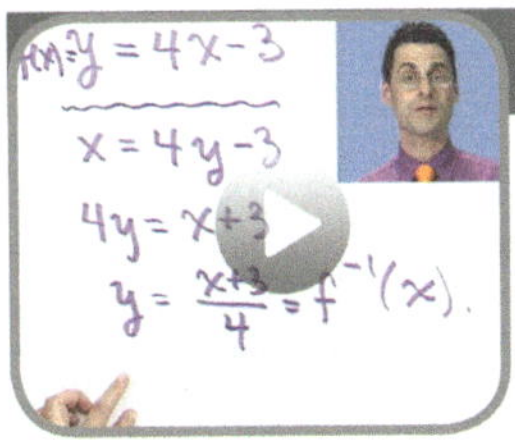

1.8.5 Finding the Inverse of a Function Algebraically

The inverse function of simple functions can be determined visually or by considering how the operations in the function can be undone. A more systematic algebraic method should be used to find the inverse function of a complicated function.

Steps for Finding the Inverse Function of a Function *f*

❶ Use the Horizontal Line Test to verify that the inverse function of f exists.

❷ If the equation is in function notation, replace $f(x)$ with y.

❸ Interchange x and y in the equation, and then solve for y.

❹ Replace y with f^{-1} in the equation from step ❸.

EXAMPLE 6 Finding the Equation of an Inverse Function Algebraically

Find the inverse function of each function, if it exists.

A. $f(x) = \dfrac{3x-2}{5}$ **B.** $g(x) = \sqrt{4x+1}$

SOLUTION

A. Since f is a linear equation, the graph of f is a line. The line is not horizontal, because the equation contains x. Any nonhorizontal line passes the Horizontal Line Test, so f is a one-to-one function and has an inverse function.

To find the inverse function of f, replace $f(x)$ with y, interchange x and y, then solve for y.

$$f(x) = \frac{3x-2}{5} \quad \textit{Write the function.}$$

$$y = \frac{3x-2}{5} \quad \textit{Replace } f(x) \textit{ with } y.$$

$$x = \frac{3y-2}{5} \quad \textit{Interchange } x \textit{ and } y.$$

$$5x = 3y-2 \quad \textit{Multiply both sides by 5.}$$

$$5x+2 = 3y \quad \textit{Add 2 to each side.}$$

$$\frac{5x+2}{3} = y \quad \textit{Divide each side by 3.}$$

Replace y with f^{-1} to write the inverse function's equation: $f^{-1}(x) = \dfrac{5x+2}{3}$.

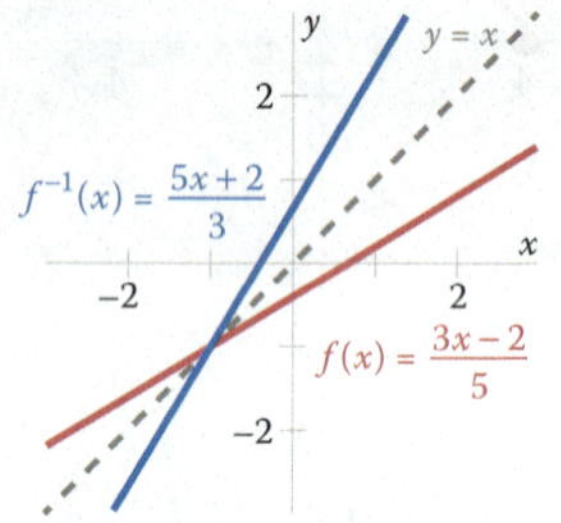

Figure 1.8d

The inverse function can be checked by confirming that $f^{-1}(f(x)) = x$ and $f(f^{-1}(x)) = x$, or by graphing the functions and verifying that the graphs are reflections over the line $y = x$, as shown in Figure 1.8d.

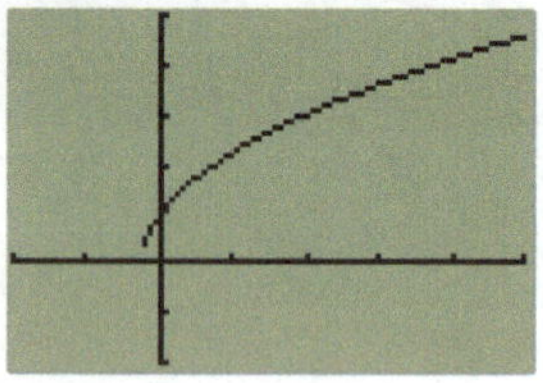

Figure 1.8e

B. Graph g with a graphing calculator, as shown in Figure 1.8e. The graph passes the Horizontal Line Test, so the function is one-to-one and has an inverse function.

The domain of g is the interval $\left[-\frac{1}{4}, \infty\right)$, so the range of g^{-1} must also be $\left[-\frac{1}{4}, \infty\right)$. Furthermore, the range of g is the interval $[0, \infty)$, so the domain of g^{-1} must also be $[0, \infty)$.

To find the inverse function of g, replace $g(x)$ with y, interchange x and y, then solve for y.

$$g(x) = \sqrt{4x+1} \quad \textit{Write the function.}$$

$$y = \sqrt{4x+1} \quad \textit{Replace } g(x) \textit{ with } y.$$

$$x = \sqrt{4y+1} \quad \textit{Interchange } x \textit{ and } y.$$

$$x^2 = 4y+1 \quad \textit{Square both sides.}$$

$$\frac{x^2-1}{4} = y \quad \textit{Subtract 1 from both sides and divide each side by 4 to solve for } y.$$

Replace y with g^{-1} to write the inverse function's equation: $g^{-1}(x) = \dfrac{x^2-1}{4}, x \geq 0$.

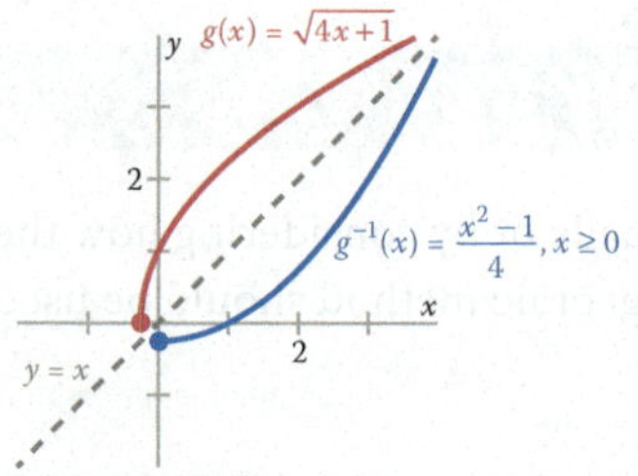

Figure 1.8f

The graphs of g and g^{-1} are shown in Figure 1.8f.

WHAT'S THE BIG IDEA?

Explain the relationship between a function, its inverse function, and the line $y = x$.

SECTION 1.8 EXERCISES

Warm Up

Write the domain of each function using interval notation.

1. $g(x) = -3x + 2$
2. $f(x) = \sqrt{9 - x^2}$
3. $p(x) = \dfrac{1}{x+3}$

Just the Facts

Fill in the blanks.

4. A function and its inverse function, when graphed, will always be symmetric with respect to the line _____.
5. If (x, y) is a point on the graph of $f(x)$, then _____ is a point on the graph of $f^{-1}(x)$.
6. Functions that pass the _____ are one-to-one functions and they have _____ functions.
7. If $x > 9$ is the domain of f, then the range of f^{-1} is _____.
8. If $(f \circ g)(x) = x$ and $(g \circ f)(x) = x$, then f and g are _____ functions.
9. The function $g(x) = -3|x + 2|$ does not have an inverse function because the graph of g does not pass the _____.
10. If $g(x) = x + 1$, its inverse function will be $g^{-1}(x) = x - 1$, since the addition is undone by the _____

Essential Skills

In Exercises 1–6, evaluate, if possible, given that p and q are inverse functions.

1. If $q(3) = 5$, what is $p(5)$?
2. If $p(9) = -4$, what is $q(-4)$?
3. If $q(-1) = 8$, what is $p(-8)$?
4. If $q(2) = -7$, what is $p(-7)$?
5. If $p(-0.6) = -1.3$, what is $q(-1.3)$?
6. If $p(1.8) = 0.5$, what is $q(1.8)$?

In Exercises 7–12, use the Horizontal Line Test to determine whether each graph is a one-to-one function with an inverse function.

7.

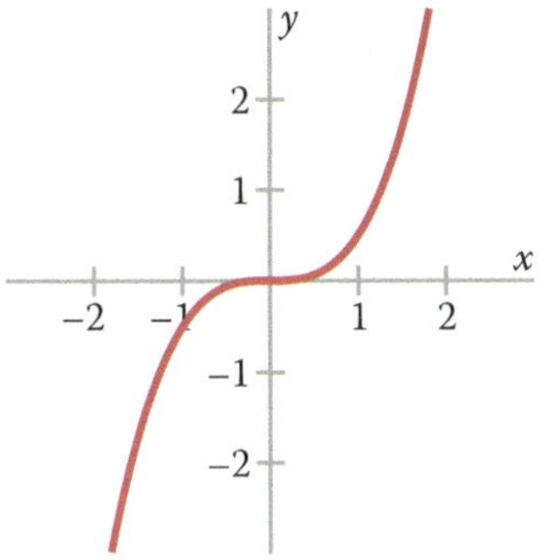

8.

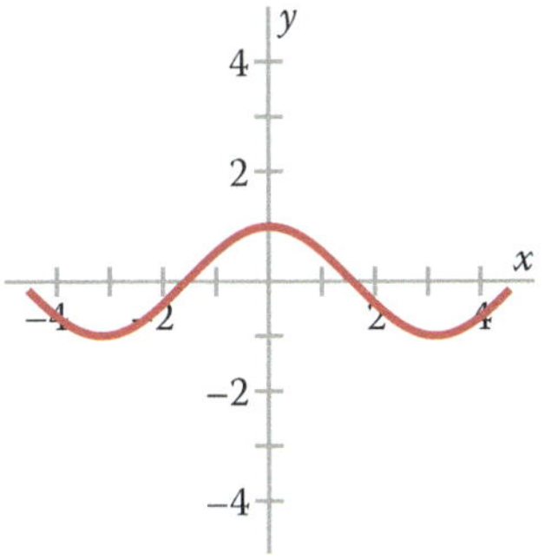

9.

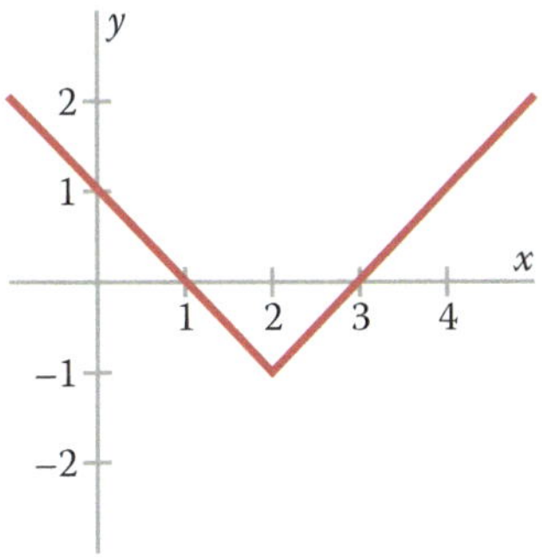

10.

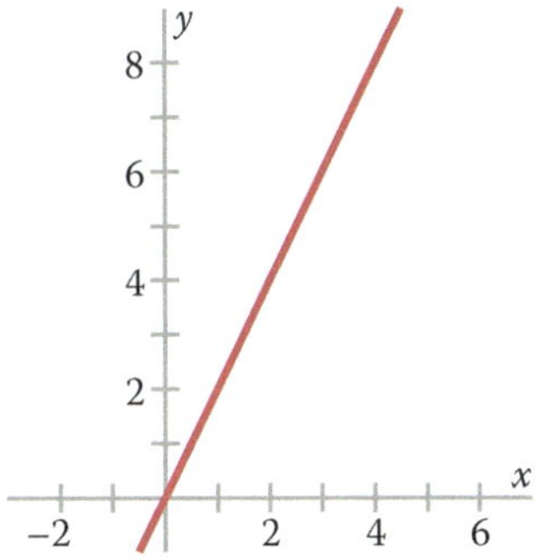

11.

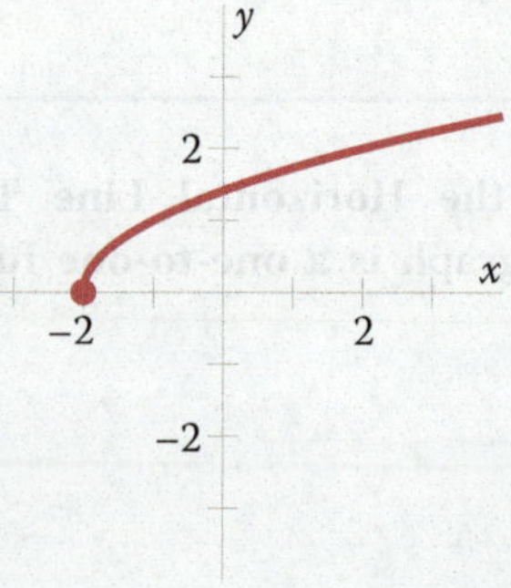

12.

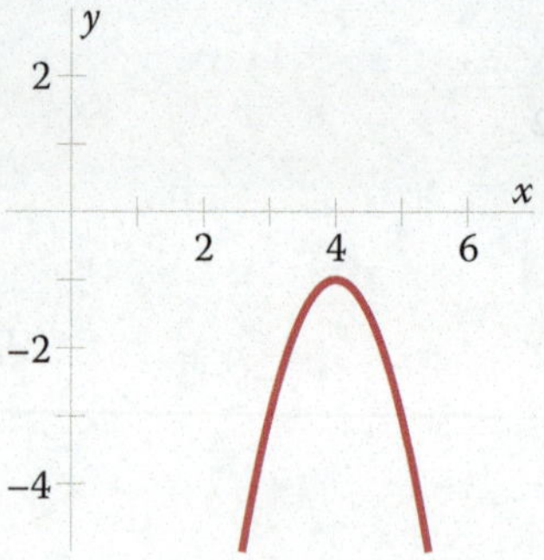

In Exercises 13–22, determine whether each pair of functions are inverses.

13. $g(x) = 4x + 1$ and $f(x) = -4x - 1$

14. $f(x) = \dfrac{2x+5}{3}$ and $g(x) = \dfrac{3x-5}{2}$

15. $g(x) = x$ and $f(x) = -x$

16. $g(x) = x + 2$ and $f(x) = -x - 2$

17. $f(x) = \dfrac{9x-13}{3}$ and $g(x) = \dfrac{3x+13}{9}$

18. $g(x) = \dfrac{5x+2}{9}$ and $f(x) = \dfrac{9x+2}{5}$

19. $g(x) = \dfrac{1}{x} + 2$ and $f(x) = \dfrac{1}{x-2}$

20. $f(x) = \dfrac{1}{2(x-5)}$ and $g(x) = \dfrac{1}{2x} + 5$

21. $g(x) = \dfrac{x^3}{8}$ and $f(x) = \sqrt[3]{2x}$

22. $f(x) = \sqrt[3]{x+1}$ and $g(x) = x^3 - 1$

In Exercises 23–24, graph the inverse function of each function, if it exists.

23.

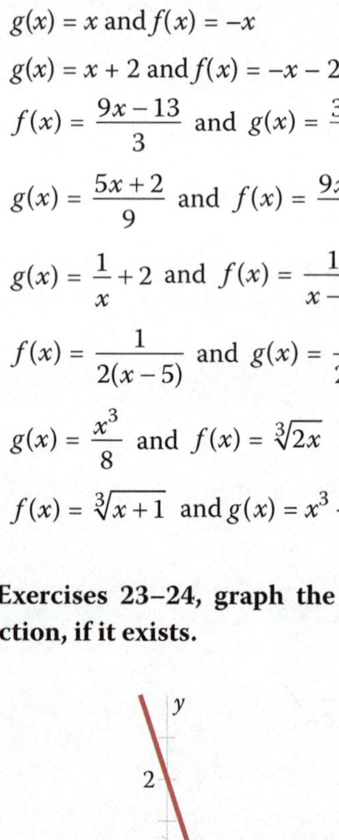

24.

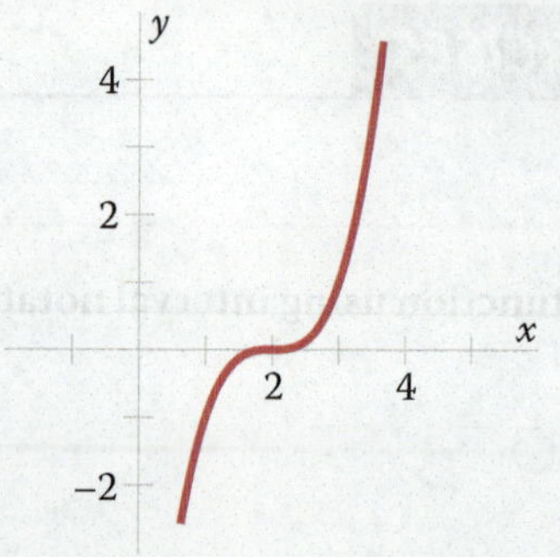

In Exercises 25–42, find the inverse function of each function, if it exists.

25. $g(x) = -x + 3$

26. $f(x) = \dfrac{5+7x}{2}$

27. $p(x) = 6x - 1$

28. $q(x) = \dfrac{4}{5}x - 3$

29. $t(x) = \dfrac{11+9x}{2}$

30. $r(x) = \dfrac{-3x+6}{13}$

31. $g(x) = \sqrt{x+4}$

32. $f(x) = \sqrt{3x-6}$

33. $h(x) = \sqrt{2x-1}$

34. $j(x) = \sqrt{8-5x}$

35. $w(x) = (x+3)^2$

36. $d(x) = x^2 - 7$

37. $k(x) = \dfrac{1}{x+6}$

38. $c(x) = \dfrac{1}{x} - 8$

39. $m(x) = \dfrac{1}{x-4} + 2$

40. $v(x) = -\dfrac{1}{x+1} - 3$

41. $s(x) = \sqrt[3]{2x+7}$

42. $n(x) = x^3 - 4$

Extensions

43. $f(x) = x^2 - 9$ is not a one-to-one function and therefore the inverse function of f does not exist. However, if the domain of f is restricted in a particular way (so that f passes the Horizontal Line Test), then $f^{-1}(x)$ exists. State the domain values over which $f^{-1}(x)$ exists.

In Exercises 44–50, use the following information.

Find the indicated value or expression, if it exists, given that $f(x) = x + 5$ and $g(x) = 2x - 1$.

44. $(f^{-1} \circ g)(4)$

45. $(f^{-1} \circ g^{-1})(0)$

46. $(g^{-1} \circ g^{-1})(2)$

47. $(f^{-1} \circ f^{-1})(9)$

48. $(f^{-1} \circ g)(x)$

49. $(g^{-1} \circ f^{-1})(x)$

50. $(g^{-1} \circ g^{-1})(x)$

51. Find the value of k such that $f(x) = kx^3 + kx + 6k$ has an inverse function and $f^{-1}(8) = -1$.

52. Always, sometimes, or never? If f is an odd function, then f^{-1} exists.

53. Find the inverse function of $g(x) = (x - 3)(x^2 + 3x + 9)$, if it exists.

CHAPTER 1 REVIEW EXERCISES

Section 1.1 Coordinates and Graphs

1. Given points $A(1, 3)$, $B(-1, -3)$, and $C(3, 9)$, find the length of the segments AB, AC, and BC and determine whether the points are collinear.

2. Write the equation $9x^2 + 9y^2 + 36x - 90y + 117 = 0$ in the standard form of a circle. Then identify the circle's center and radius, and graph the circle.

3. Use intercepts to sketch the graph of $y = x^2 + 5x + 6$.

4. Write the equation of the circle where $A(-2, -3)$ and $B(6, -3)$ are the endpoints of a diameter.

5. Use symmetry and intercepts to sketch the graph of $y = -|x| + 3$.

Section 1.2 Lines

6. Write the slope-intercept form equation of the line with slope 3 that passes through $(-2, 4)$. Graph the line.

7. Write the slope-intercept form equation of the line that passes through $(-1, 7)$ and $(-2, -15)$.

8. Find the slope and y-intercept of the graph of $4 + 2y = -5x$.

9. Line l passes through $(7, -2)$ and line p is the graph of $y - 1 = \frac{3}{4}(x + 5)$. If $l \parallel p$, what is the point-slope form equation of l?

10. Line l passes through $(4, -2)$ and line p is the graph of $y = \frac{4}{3}x - 1$. If $l \perp p$, what is the slope-intercept form equation of l?

Section 1.3 Variation

11. The pressure P exerted by a force varies inversely as the area A on which the force is applied. If 16 newtons per square feet of pressure are exerted by a force on an area of 3 square feet, what is the pressure exerted by the same force on an area of 6 square feet?

12. The number of chocolate bunnies that can be eaten varies directly with the number of people eating them and the time taken to eat them. If 39 chocolate bunnies can be eaten in 10 minutes by 3 people, how many chocolate bunnies can be eaten in 30 minutes by 4 people?

13. A satellite orbiting Earth has a period of 21 hours when its average distance is 3.8×10^7 meters. Find the orbital period of a satellite when the average distance is 5.1×10^7 meters. Use the law that the square of the time of orbit varies directly with the average radius cubed. Round to the nearest tenth, if necessary.

14. If a pendulum with length 81 centimeters has a frequency of 1/10 hertz, find the frequency of a 36-centimeter pendulum, given that frequency varies inversely with the square root of length.

15. The gravitational force F in newtons between two objects, having mass m_1 and m_2 respectively, varies jointly with respect to m_1 and m_2 and inversely with respect to the square of the distance d between the two objects. If the gravitational force between Earth (mass 5.98×1024 kilograms, radius 6.38×10^6 meters) and a math student (mass 65 kilograms) is 637.227 newtons, find the gravitational force F between Earth and a football player who weighs 90 kilograms. Round to the nearest thousandth.

Section 1.4 Functions

16. Use the Vertical Line Test to determine whether the graph represents a function.

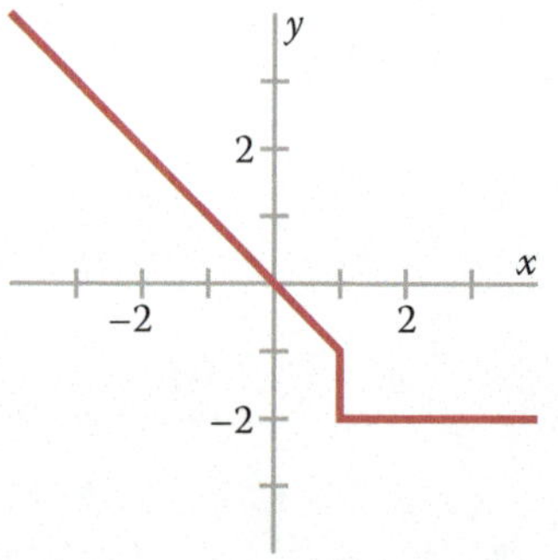

17. If $g(x) = x^2 - 5x + 1$, find $g(2)$, $g(-5)$, and $g(a - 2)$.

18. Given an equilateral triangle with side s, write a function for the area of the triangle in terms of s.

19. If $g(x) = 2x^2 + 3x - 1$ and $h(x) = -x + 5$, find all real x-values such the $g(x) = h(x)$.

20. Find the domain of $f(x) = \dfrac{x^2}{x^2 - 2x - 3}$.

Section 1.5 Graphs of Functions

21. Given the graph of function $h(x)$, use the graph to identify the domain and range of h, then evaluate $h(-1)$, $h(2)$, and find all x such that $h(x) = 0$.

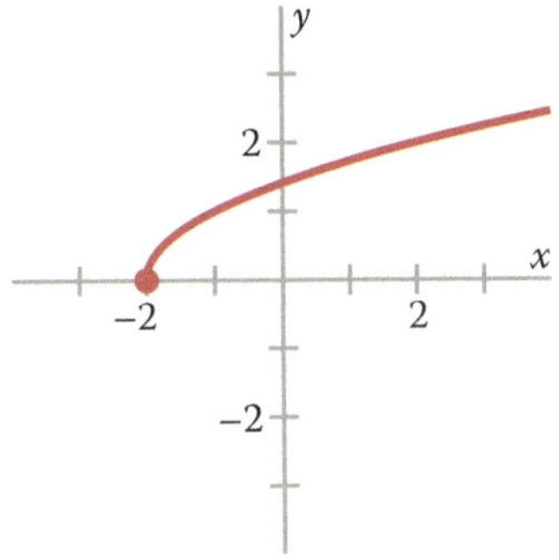

22. Graph. $k(x) = \begin{cases} -x-4 & \text{if } x \le -1 \\ 2x+1 & \text{if } -1 < x \le 2 \\ -4 & \text{if } x > 2 \end{cases}$

23. Graph. $r(x) = 4[[x - 1]]$

24. Find the real zeros of the function $m(x) = x^3 + 2x^2 - 36x - 72$.

25. Identify the intervals over which the function is strictly increasing, strictly decreasing, or constant.

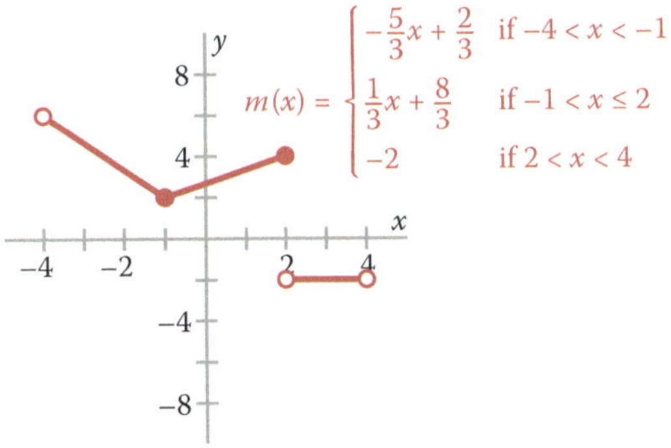

Section 1.6 Transformations of Functions

In Exercises 26–28, sketch the graph of each function.

26. $g(x) = (x - 6)^2 - 4$

27. $f(x) = 3|x + 1|$

28. $h(x) = -\sqrt{x-1}$

In Exercises 29–30, determine whether each function is even, odd, or neither.

29. $m(x) = 5|x| + 2$

30. $p(x) = 7(x + 6)^3$

Section 1.7 Combining Functions

31. Given $f(x) = 4x^2 - 3x + 3$ and $g(x) = 5 - 2x + x^2$, find $(f - g)(x)$ and $(f + g)(4)$.

32. Given $m(x) = x - 8$ and $n(x) = x + 1$, find $(mn)(x)$ and $\left(\frac{m}{n}\right)(x)$.

33. Simplify the difference quotient for $f(x) = -3x^2 + 5x - 6$.

34. Given $p(x) = x^2 - x - 5$ and $r(x) = 2x - 1$, find $(p \circ r)(x)$ and $(p \circ r)(3)$.

35. Find the function $g(x)$ that forms the composite $(f \circ g)(x) = (7x - 4)^2$ with $f(x) = x^2$.

Section 1.8 Inverse Functions

36. Suppose that p and g are inverse functions, and that $p(-5) = 8$, $g(6) = 0$, and $p(9) = 11$. Use the given information to evaluate $g(8)$, $p(0)$, and $p(11)$, if possible.

37. Use the Horizontal Line Test to determine whether the graph is a one-to-one function with an inverse function.

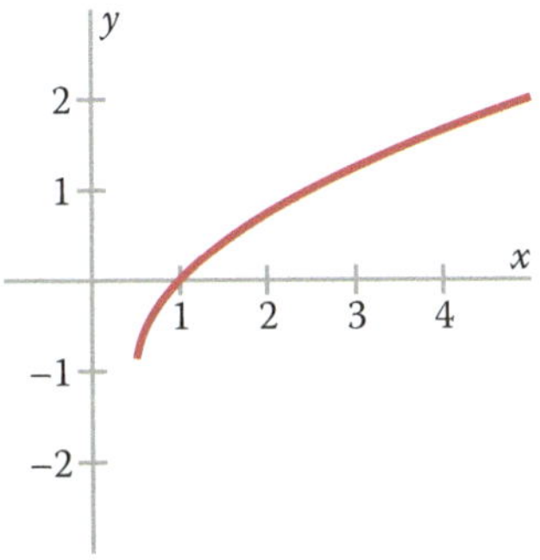

38. Determine whether $f(x) = 4x + \frac{1}{4}$ and $g(x) = \frac{1}{4}x - \frac{1}{16}$ are inverses of each other.

39. Graph the inverse function, if it exists.

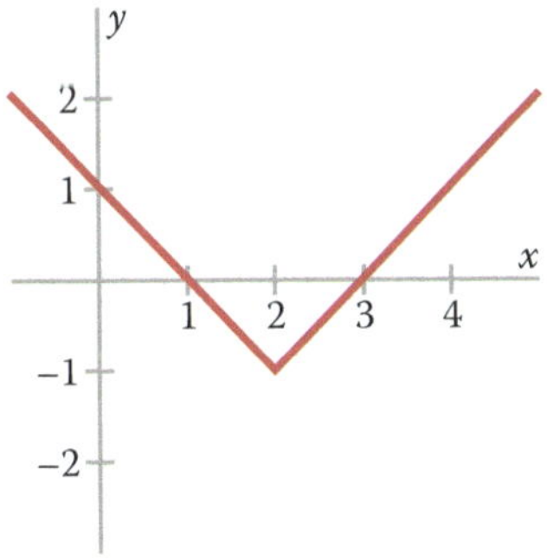

40. Find the inverse of $k(x) = 3\sqrt{x-5}$, if it exists.

POLYNOMIAL FUNCTIONS

Chapter

2

2.1 QUADRATIC FUNCTIONS AND MODELS

OBJECTIVES

- Graph quadratic functions using patterns.
- Find the vertex and axis of symmetry of graphs of quadratic functions.
- Understand the relationship between the discriminant of a quadratic function and the x-intercepts of its graph.
- Graph quadratic functions using the vertex and axis of symmetry.
- Write the equation of a quadratic function.
- Find the minimum or maximum of a quadratic function.
- Model with quadratic functions.

PREREQUISITE VOCABULARY TERMS

completing the square
function
leading coefficient
parabola
quadratic function
symmetry

Recall that a polynomial is an expression of the form $a_n x^n + a_{n-1}x^{n-1} + \ldots + a_1 x + a_0$, where $a_0, a_1, a_2, \ldots, a_n$ are real numbers, and n is a nonnegative integer. The expressions $2x + 4$ and $4y^5 - y^2 + y$ are examples of polynomial expressions. **Polynomial functions** will be discussed in this chapter.

DEFINITION

A **polynomial function** in terms of x with degree n is a function of the form

$$f(x) = a_n x^n + a_{n-1}x^{n-1} + \ldots + a_1 x + a_0,$$

where $a_0, a_1, a_2, \ldots, a_n$ are real numbers and n is a nonnegative integer.

Polynomial functions can be classified by their degree. For example, a polynomial function with degree 1 is a linear function, such as $f(x) = x + 1$ or $f(x) = -2x$. A polynomial function with degree 2 is a quadratic function, such as $f(x) = -x^2$ or $f(x) = x^2 + 3x$. A polynomial function with degree 3 is a cubic function, and so on. Quadratic functions will be discussed in this section.

2.1.1 Reflecting, Stretching, and Compressing Quadratic Functions

Recall that the graph of a **quadratic function** is a U-shaped curve called a **parabola**.

DEFINITION

A **quadratic function** in terms of x is a function of the form

$$f(x) = ax^2 + bx + c,$$

where a, b, and c are real numbers, and $a \neq 0$.

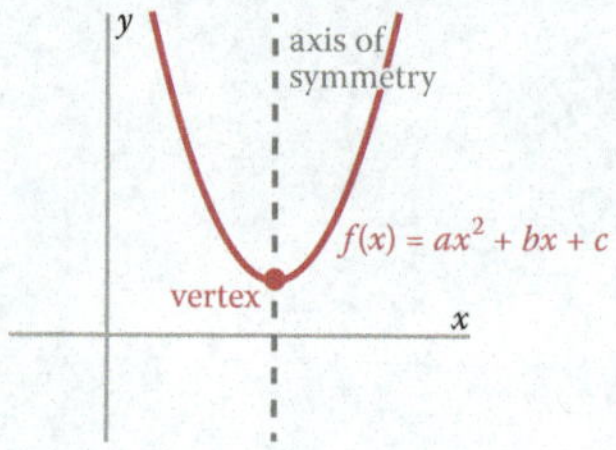

Figure 2.1a

All parabolas have one line of symmetry. This line of symmetry is called the **axis of symmetry** (or just the axis) of the parabola. The point on the parabola where the axis of symmetry intersects the parabola is called the parabola's **vertex**. The axis of symmetry and vertex of a parabola are shown in Figure 2.1a.

When the leading coefficient in $f(x) = ax^2 + bx + c$ is positive ($a > 0$), the parabola opens upward, as shown in Figure 2.1b. When the leading coefficient in $f(x) = ax^2 + bx + c$ is negative ($a < 0$), the parabola opens downward, as shown in Figure 2.1c. Notice that the vertex is the lowest point in a parabola that opens upward and the highest point in a parabola that opens downward.

Positive Leading Coefficient

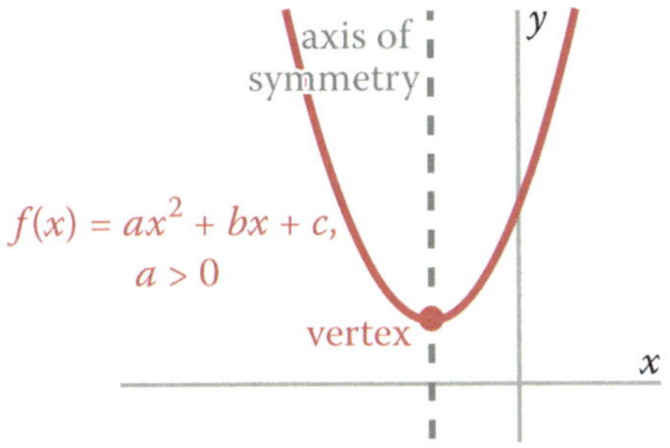

Parabola opens upward.
Vertex is parabola's lowest point.

Figure 2.1b

Negative Leading Coefficient

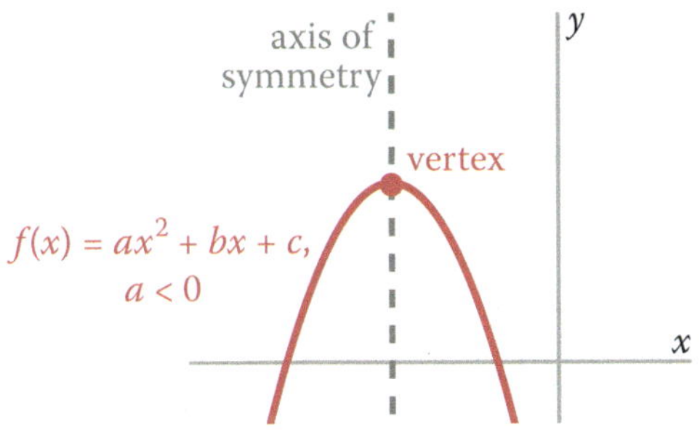

Parabola opens downward.
Vertex is parabola's highest point.

Figure 2.1c

Graphing $f(x) = ax^2$ where $a > 0$ or $a < 0$

Recall that the most basic quadratic function is $f(x) = x^2$, which passes through (−2, 4), (−1, 1), (0, 0), (1, 1), and (2, 4), as shown in Figure 2.1d. The graph is symmetric with respect to the y-axis, so the y-axis is the parabola's axis of symmetry. Notice that the parabola opens upward as expected, since $a > 0$ (the value of a is 1 in $f(x) = x^2$). The lowest point on the graph of $f(x) = x^2$ is (0, 0). This point is also the vertex of $f(x) = x^2$, and the point with the **minimum** y-value.

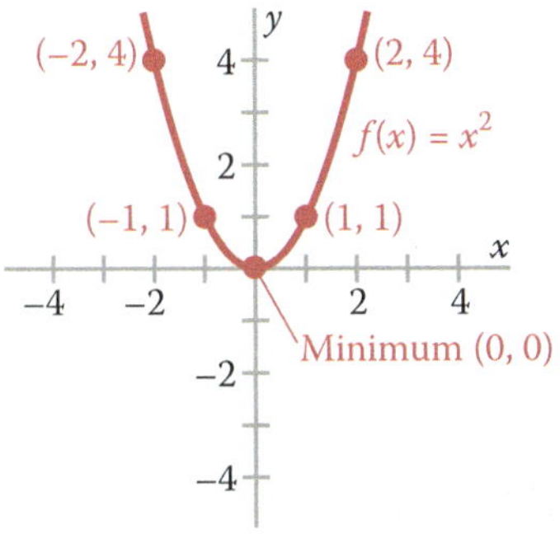

Parabola opens upward.
Vertex (0, 0) is parabola's lowest point.

Figure 2.1d

The graph of $f(x) = -x^2$ passes through (−2, −4), (−1, −1), (0, 0), (1, −1), and (2, −4), as shown in Figure 2.1e. This graph has the same axis of symmetry, the same y-intercept, and the same vertex (0, 0), as the graph of $f(x) = x^2$. However, the vertex of $f(x) = -x^2$ is the graph's highest point, and therefore the point with the **maximum** y-value. This graph is a reflection of the graph of $f(x) = x^2$ over the x-axis.

When graphing any quadratic function, it is often helpful to use the graph of $f(x) = x^2$ as a reference. Just as the graph of $f(x) = -x^2$ is a reflection of the graph of $f(x) = x^2$ over the x-axis, the graph of any quadratic function of the form $f(x) = -ax^2$ is a reflection of the graph of $f(x) = ax^2$ over the x-axis. Any quadratic function of the form $f(x) = ax^2$ will have a vertex at (0, 0). This vertex is either a minimum or a maximum, depending on the sign of a.

- If $a > 0$, then the parabola's vertex is a minimum.
- If $a < 0$, then the parabola's vertex is a maximum.

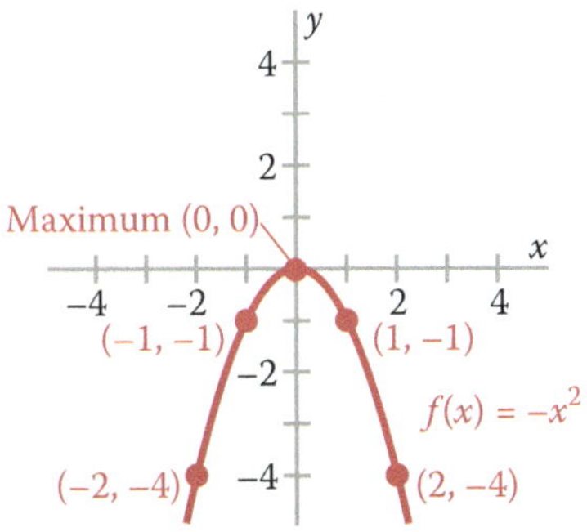

Parabola opens downward.
Vertex (0, 0) is parabola's highest point.

Figure 2.1e

Besides opening upward or downward, graphs of quadratic functions of the form $f(x) = ax^2$ also vary in terms of width. The width is described in comparison to the width of $f(x) = x^2$.

- If $|a| > 1$, then the parabola is more narrow than $f(x) = x^2$.
  The graph appears stretched in comparison to $f(x) = x^2$.
- If $0 < |a| < 1$, then the parabola is wider than $f(x) = x^2$.
  The graph appears compressed in comparison to $f(x) = x^2$.

The degree to which the parabola is stretched (made wider) or compressed (made more narrow) can be determined by multiplying the y-coordinates in the graph of $f(x) = x^2$ by $|a|$. This change to the graph of $f(x) = x^2$ is called a vertical stretch, even when the result is a graph that is wider than the original.

EXAMPLE 1 Graphing $f(x) = ax^2$

For each function, describe the graph by comparing it to the graph of $f(x) = x^2$. Then sketch the function's graph.

A. $g(x) = 3.5x^2$

B. $h(x) = -\frac{1}{4}x^2$

SOLUTION

A. Since g is of the form $f(x) = ax^2$ where $a = 3.5$, the graph of g

- is a parabola with a vertex at (0, 0),
- opens in the same direction (upward) as the graph of $f(x) = x^2$ (because $a > 0$),
- is more narrow than the graph of $f(x) = x^2$ (because $|a| > 1$), and
- is a vertical stretch of f by a factor of 3.5 (because $|a| = 3.5$).

Make a table to identify points on the graph of g. Note that multiplying the y-coordinate of a point on f by 3.5 gives the y-coordinate of a point on g (where both points have the same x-coordinate).

x	$f(x)$	$g(x)$
−2	4	14
−1	1	3.5
0	0	0
1	1	3.5
2	4	14

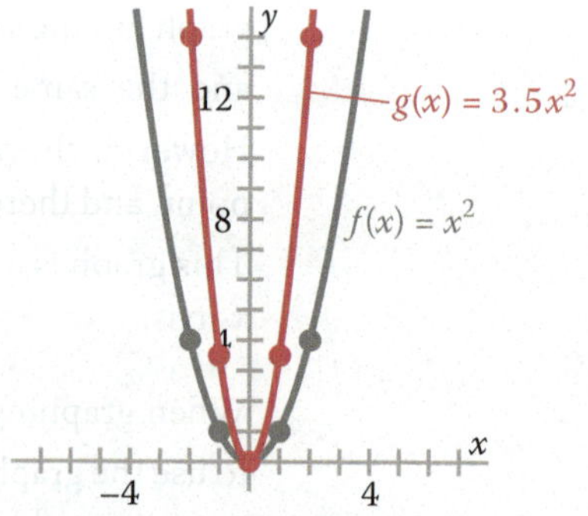

B. Since h is of the form $f(x) = ax^2$ where $a = -\frac{1}{4}$, the graph of h

- is a parabola with a vertex at (0, 0),
- opens downward (because $a < 0$),
- is wider than the graph of $f(x) = x^2$ (because $0 < |a| < 1$), and
- is a vertical stretch (compression) of f by a factor of $\frac{1}{4}$ (because $|a| = \frac{1}{4}$).

Make a table to identify points on the graph of h. Note that multiplying the y-coordinate of a point on f by $-1/4$ gives the y-coordinate of a point on h (where both points have the same x-coordinate).

x	$f(x)$	$h(x)$
-2	4	-1
-1	1	$-\frac{1}{4}$
0	0	0
1	1	$-\frac{1}{4}$
2	4	-1

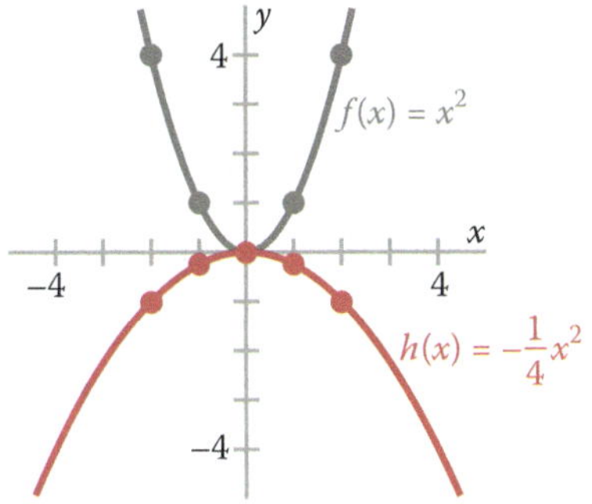

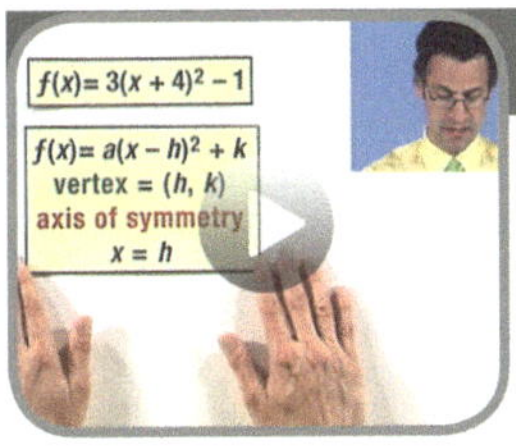

2.1.2 Identifying the Vertex and Axis of Symmetry

When the equation of a quadratic function is written in the **standard form of a quadratic function**, the parabola's vertex and axis of symmetry can be identified from values in the equation.

Standard Form of a Quadratic Function

The equation of a parabola with vertex at (h, k), where the axis of symmetry is the vertical line $x = h$, can be written as $f(x) = a(x - h)^2 + k$, where $a \neq 0$.

If $a > 0$, then the parabola opens upward and its vertex is a minimum.

If $a < 0$, then the parabola opens downward and its vertex is a maximum.

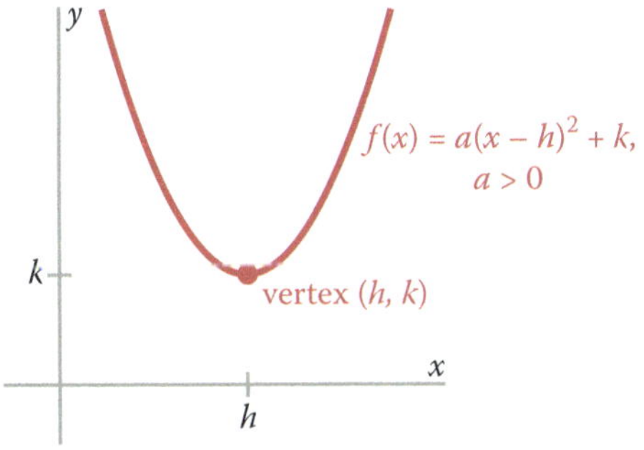

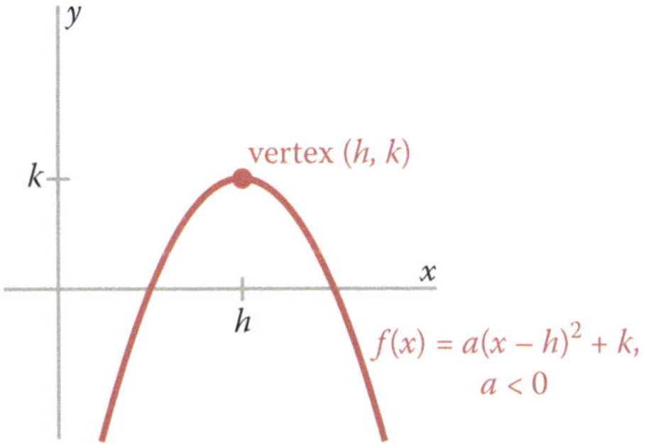

TIP

The standard form of a quadratic function is also called "vertex form."

The equation $f(x) = ax^2$ is a standard form quadratic equation where $h = 0$ and $k = 0$ (i.e., where the vertex is at the origin). As with the graph of $f(x) = ax^2$, the direction in which the graph of $f(x) = a(x - h)^2 + k$ opens, and whether the graph has a minimum or maximum y-value, depends on the sign of a.

EXAMPLE 2 Identifying the Vertex and Axis of Symmetry of a Parabola from Its Equation

For each function, identify the vertex and the axis of symmetry, and state whether the vertex is a minimum or maximum.

A. $p(x) = -(x - 6)^2 - 4$ **B.** $r(x) = 2(x + 1)^2$

SOLUTION

Each equation is given in the standard form of a quadratic function, $f(x) = a(x - h)^2 + k$. Identify the h- and k-values from the equation and then use these values to identify the vertex (h, k) and the axis of symmetry $x = h$.

A. $p(x) = -(x - 6)^2 - 4 = -(x - 6)^2 + (-4) \Rightarrow h = 6$ and $k = -4$

Vertex: $(6, -4)$ Axis of symmetry: $x = 6$

The value of a in $p(x) = -(x - 6)^2 - 4$ is -1. Since $a < 0$, the parabola opens downward and its vertex is a maximum.

B. $r(x) = 2(x + 1)^2 = 2(x - (-1))^2 + 0 \Rightarrow h = -1$ and $k = 0$

Vertex: $(-1, 0)$ Axis of symmetry: $x = -1$

The value of a in $r(x) = 2(x + 1)^2$ is 2. Since $a > 0$, the parabola opens upward and its vertex is a minimum.

The vertices and axes of symmetry for the parabolas from **Example 2** can be checked by graphing the functions using a graphing calculator, as shown in Figures 2.1f and 2.1g.

TIP

Adjust the graphing calculator's window settings so that the parabola's vertex can be viewed.

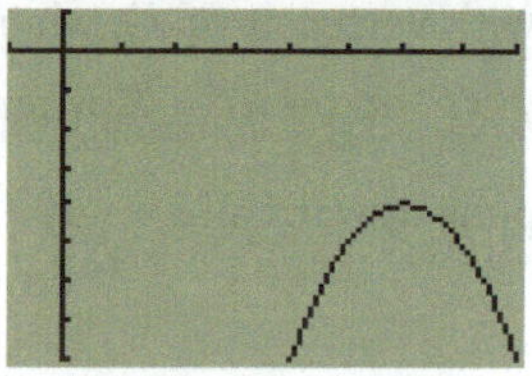

Figure 2.1f

The vertex is at $(6, -4)$ and the axis of symmetry is the vertical line $x = 6$.

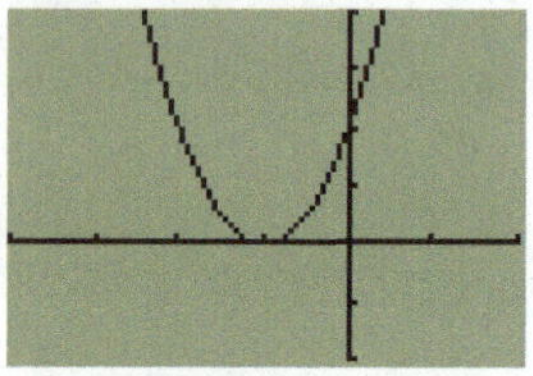

Figure 2.1g

The vertex is at $(-1, 0)$ and the axis of symmetry is the vertical line $x = -1$. The vertex is on the x-axis, so the vertex is also the parabola's x-intercept.

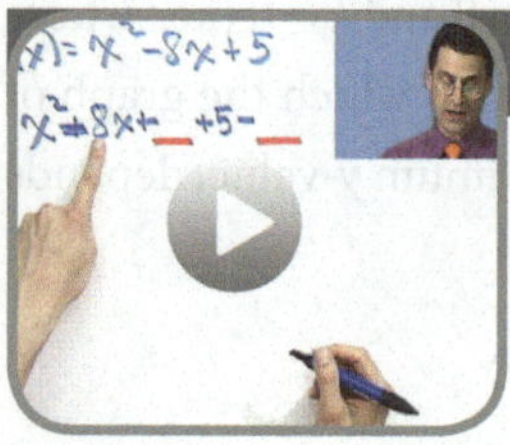

2.1.3 Finding the Vertex by Completing the Square

Review of Completing the Square for x in $x^2 + bx + c = 0$

Recall that a quadratic equation in the form $x^2 + bx + c = 0$ can be solved by completing the square. In this case, the square of half the coefficient of x is added to each side of the equation to complete the square.

$$x^2 + bx + c = 0$$

$$x^2 + bx = -c$$

$$x^2 + bx + \left(\frac{b}{2}\right)^2 = -c + \left(\frac{b}{2}\right)^2 \quad \textit{Add } \left(\frac{b}{2}\right)^2 \textit{ to both sides.}$$

$$\left(x + \frac{b}{2}\right)^2 = -c + \left(\frac{b}{2}\right)^2 \quad \textit{Factor to write the left side as the square of a binomial.}$$

REMEMBER

The graph of $f(x) = a(x - h)^2 + k$ (a quadratic function in standard form) is a parabola with vertex at (h, k).

Finding the Vertex of $f(x) = ax^2 + bx + c$ by Completing the Square

The graph of a quadratic function in general form, $f(x) = ax^2 + bx + c$, cannot be easily visualized because the coordinates of the vertex are not shown in the equation. So, before sketching the graph of a quadratic function in general form, it is often helpful to write the function in standard form, $f(x) = a(x - h)^2 + k$. This can be accomplished by completing the square.

To complete the square for the general form quadratic equation $f(x) = ax^2 + bx + c$ when $a = 1$, *add and subtract* half the square of the coefficient of x on the same side of the equation. The process can be more complicated when $a \neq 1$ than it is when $a = 1$. **Example 3B** demonstrates two methods for writing a quadratic function in the standard form when $a \neq 1$.

EXAMPLE 3 Writing a Quadratic Function in Standard Form by Completing the Square

Write each function in the standard form of a quadratic function, then identify the vertex of the parabola.

A. $g(x) = x^2 + 16x + 63$ **B.** $p(x) = -3x^2 + 36x - 106$ **C.** $f(x) = -5x^2 + 4$

SOLUTION

Complete the square to write each function in the standard form of a quadratic function, $f(x) = a(x - h)^2 + k$. Then identify the h- and k-values from the equation and use these values to identify the vertex (h, k).

A. Since $a = 1$, start by placing parentheses around $x^2 + 16x$. Then, since $\left(\frac{16}{2}\right)^2 = 64$, add 64 inside the parentheses and subtract 64 outside the parentheses.

$g(x) = (x^2 + 16x) + 63$

$= (x^2 + 16x + 64) + 63 - 64$ *Complete the square. Add 64 inside the parentheses and subtract 64 outside the parentheses.*

$= (x + 8)^2 - 1$ *Factor and simplify.*

NOTICE THAT

Adding 64 inside the parentheses and subtracting 64 outside the parentheses is equivalent to adding 0.

The function $g(x) = x^2 + 16x + 63$ can be expressed equivalently in standard form as $g(x) = (x + 8)^2 - 1$. And from the standard form equation, $h = -8$ and $k = -1$. Therefore, the vertex of the parabola formed by $g(x) = x^2 + 16x + 63$ is at $(-8, -1)$, as shown in Figure 2.1h.

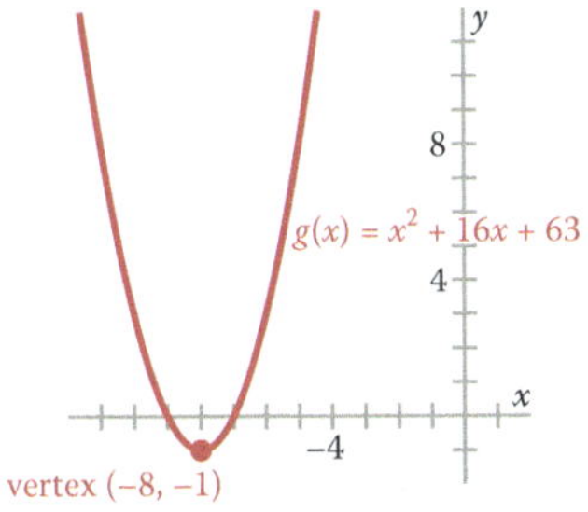

Figure 2.1h

B. Since $a \neq 1$, a must be factored out before completing the square. There are two methods for completing the square in order to write a quadratic function in standard form when $a \neq 1$.

Method 1: Factor *a* from all three terms.

$p(x) = -3x^2 + 36x - 106$

$= -3\left(x^2 - 12x + \frac{106}{3}\right)$ *Factor −3 from each term.*

$= -3\left((x^2 - 12x) + \frac{106}{3}\right)$ *Group $x^2 - 12x$ within parentheses.*

$= -3\left((x^2 - 12x + 36) + \frac{106}{3} - 36\right)$ *Complete the square. Add 36 inside the parentheses and subtract 36 outside the parentheses.*

$= -3\left((x - 6)^2 - \frac{2}{3}\right)$ *Factor and simplify.*

$= -3(x - 6)^2 - (-3)\left(\frac{2}{3}\right)$ *Distribute −3 to each of the two terms.*

$= -3(x - 6)^2 + 2$ *Multiply.*

IMPORTANT

Be sure to multiply both terms within parentheses by −3 (the a-value).

Method 2: Factor a from the x^2-term and the x-term, but not from the constant term.

$p(x) = -3x^2 + 36x - 106$

$= -3(x^2 - 12x) - 106$ *Factor -3 from $-3x^2 + 36x$.*

$= -3(x^2 - 12x + 36) - 106 - (-3)(36)$ *Complete the square. Add 36 inside and subtract $(-3)(36)$ outside the parentheses.*

$= -3(x - 6)^2 + 2$ *Factor and simplify.* *$106 - (-3)(36) = -106 + 108 = 2$*

Notice that 36 was added inside the parentheses (since $(-12/2)^2 = 36$) to complete the square, but then $(-3)(36)$ was subtracted outside the parentheses. This is because the expression within parentheses is being multiplied by -3, so adding 36 within parentheses has the effect of adding $(-3)(36)$ to the expression. So, $(-3)(36)$ must be subtracted outside the parentheses to undo the $(-3)(36)$ that was added to the expression.

Both methods show that $p(x) = -3x^2 + 36x - 106$ can be expressed equivalently in standard form as $p(x) = -3(x - 6)^2 + 2$. And from the standard form equation, $h = 6$ and $k = 2$. Therefore, the vertex of the parabola formed by $p(x) = -3x^2 + 36x - 106$ is at $(6, 2)$, as shown in Figure 2.1i.

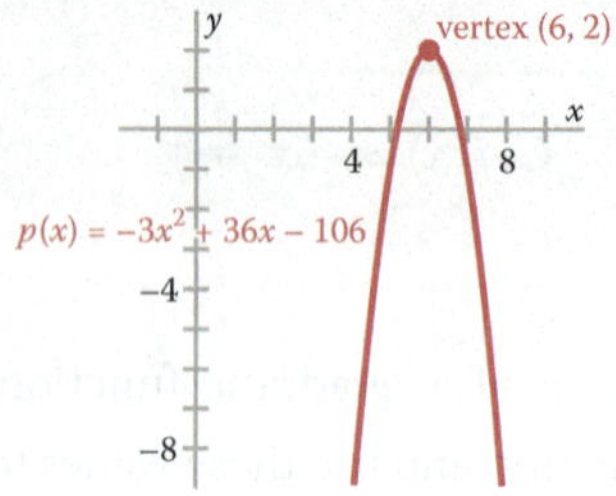

Figure 2.1i

C. Although the a-value is not 1 in the equation $f(x) = -5x^2 + 4$, there is no need to factor out the a-value because there is no x-term. It may not be obvious because there is no x-term, but this equation is already in standard form.

$$f(x) = -5x^2 + 4 \Rightarrow f(x) = -5(x - 0)^2 + 4 \Rightarrow h = 0 \text{ and } k = 4$$

Thus, the vertex of the parabola formed by $f(x) = -5x^2 + 4$ is at $(0, 4)$.

The Vertex of $f(x) = ax^2 + c$

TIP

The y-intercept of $f(x) = ax^2 + bx + c$ is always $(0, c)$.

All quadratic functions of the form $f(x) = ax^2 + c$ can be written as $f(x) = a(x - 0)^2 + c$. So, for $f(x) = ax^2 + c$, $h = 0$ and $k = c$. It follows that the graphs of all quadratic functions of the form $f(x) = ax^2 + c$ have a vertex at $(0, c)$, as shown in Figure 2.1j. Additionally, note that the vertex of the graph of $f(x) = ax^2 + c$ is always on the y-axis, and c is always the parabola's y-intercept.

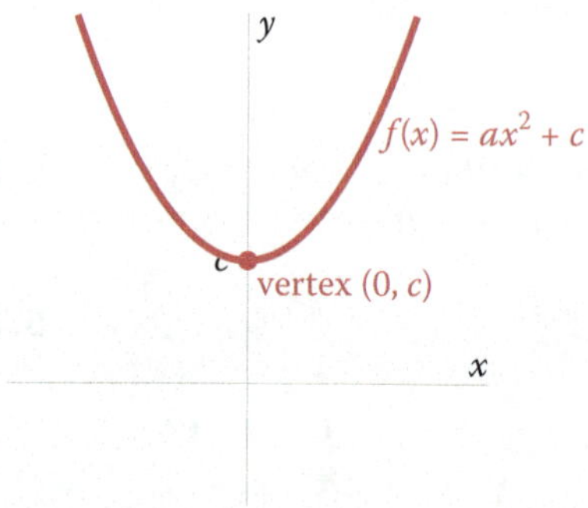

Figure 2.1j

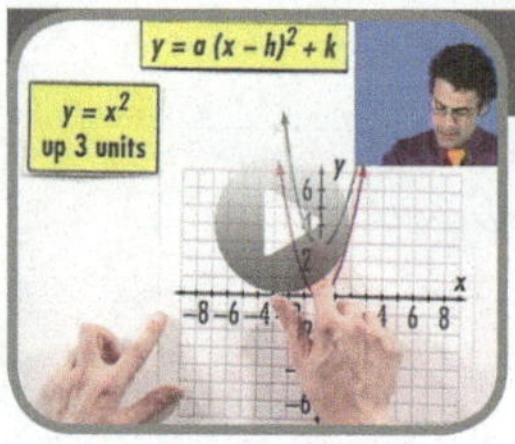

2.1.4 Translations of Quadratic Functions

Recall that a quadratic function can be graphed by translating the graph of the basic quadratic function $f(x) = x^2$. Translations can also be used to write the equation of a quadratic function when the equation of the pre-image is known, as demonstrated in the following example. Recall that translations change the position, but not the shape, of a graph. Applying a translation to the graph of $f(x) = a(x - h)^2 + k$ does not change the value of a, but it may change the value of h or k.

EXAMPLE 4 Writing the Equation of Translated Quadratic Function

Write the equation of g, and describe its graph.

A. g is the image of $y = x^2$ after a translation left 2 units and up 3 units.

B. g is the image of $y = -3(x + 4)^2 - 2$ after a translation right 1 unit and down 2 units.

C. g is the image of $y = 0.5x^2 - 6x + 19$ after a translation left 5 units.

SOLUTION

For each function, identify the vertex (h, k) and the a-value of g, then substitute these values into the standard form of a quadratic function, $f(x) = a(x - h)^2 + k$, to write the equation of g.

A. Translating $y = x^2$ left 2 units and up 3 units has the effect of translating each point (x, y) on $y = x^2$ to $(x - 2, y + 3)$. Since the vertex of $y = x^2$ is at $(0, 0)$, the vertex of g is at $(-2, 3)$. Since translations do not affect the shape of a graph, the a-value of g is also 1. Therefore, for g, $h = -2$, $k = 3$, and $a = 1$. Substituting these into the standard form equation gives $g(x) = (x + 2)^2 + 3$.

The graph of g is a parabola that opens upward (since $a > 0$) with vertex at $(-2, 3)$, as shown in Figure 2.1k.

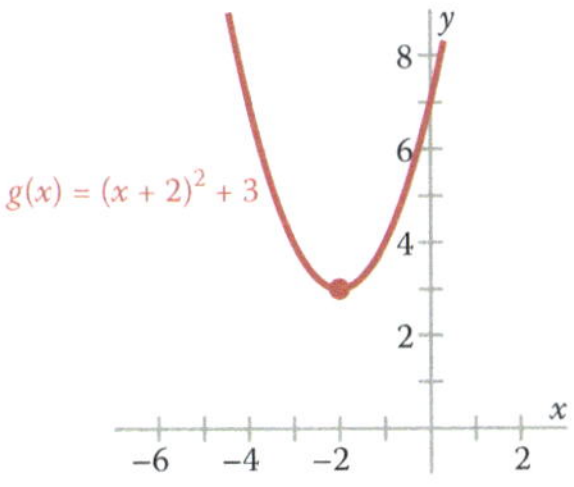

Figure 2.1k

B. Translating $y = -3(x + 4)^2 - 2$ right 1 unit and down 2 units has the effect of translating each point (x, y) on the graph to $(x + 1, y - 2)$. Since the vertex of $y = -3(x + 4)^2 - 2$ is at $(-4, -2)$, the vertex of g is at $(-3, -4)$. Since translations do not affect the shape of a graph, the a-value of g is also -3. Therefore, for g, $h = -3$, $k = -4$, and $a = -3$. Substituting into the standard form equation gives $g(x) = -3(x + 3)^2 - 4$.

The graph of g is a parabola that opens downward (since $a < 0$) with vertex at $(-3, -4)$, as shown in Figure 2.1l. Since $|a| > 1$, this parabola is more narrow than the parabola formed by $y = x^2$.

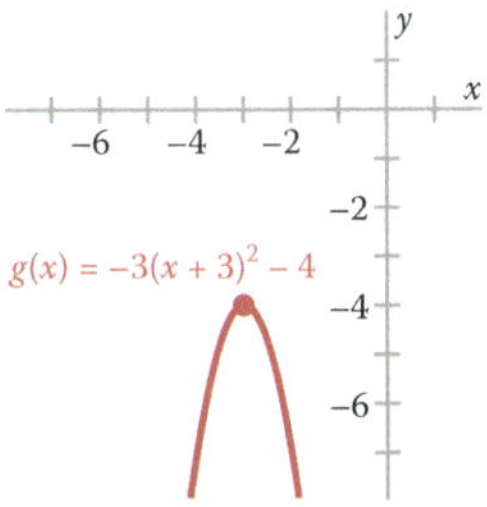

Figure 2.1l

C. Translating $y = 0.5x^2 - 6x + 19$ left 5 units has the effect of translating each point (x, y) on $y = 0.5x^2 - 6x + 19$ to $(x - 5, y)$. This equation is not in standard form, so the vertex cannot be identified by looking at the equation. The equation must then be written in standard form by completing the square.

$$\begin{aligned} y &= 0.5x^2 - 6x + 19 \\ &= 0.5(x^2 - 12x) + 19 && \textit{Factor 0.5 from } 0.5x^2 - 6x. \\ &= 0.5(x^2 - 12x + 36) + 19 - (0.5)(36) && \textit{Complete the square.} \\ &= 0.5(x - 6)^2 + 1 && \textit{Factor and simplify.} \end{aligned}$$

IMPORTANT

Subtract (0.5)(36) outside of the parentheses.

Since the vertex of $y = 0.5x^2 - 6x + 19$ is at $(6, 1)$, the vertex of g is at $(1, 1)$. Since translations do not affect the shape of a graph, the a-value of g is also 0.5. Therefore, for g, $h = 1$, $k = 1$, and $a = 0.5$. Substituting into the standard form equation gives $g(x) = 0.5(x - 1)^2 + 1$.

The graph of g is a parabola that opens upward (since $a > 0$) with vertex at $(1, 1)$, as shown in Figure 2.1m. Since $0 < |a| < 1$, this parabola is wider than the parabola formed by $y = x^2$.

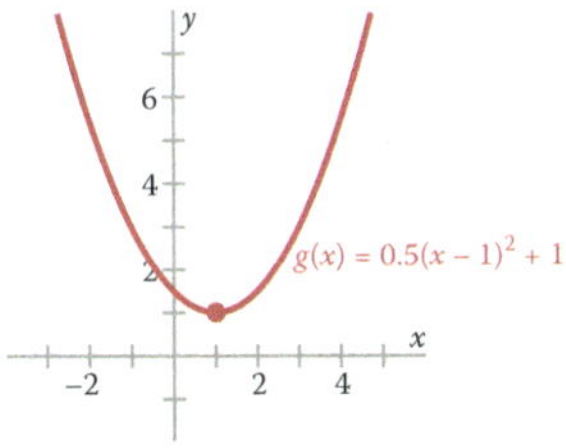

Figure 2.1m

2.1.5 Relating the Discriminant to the Graph of a Quadratic Function

The **discriminant** D of a quadratic equation $ax^2 + bx + c = 0$ is $b^2 - 4ac$, which is the radicand from the Quadratic Formula. The number and types of solutions of a quadratic equation can be determined by comparing the value of its discriminant to 0.

- If $D > 0$, then $ax^2 + bx + c = 0$ has two distinct real solutions.
- If $D = 0$, then $ax^2 + bx + c = 0$ has exactly one real solution.
- If $D < 0$, then $ax^2 + bx + c = 0$ has no real solutions, but two complex solutions.

Recall that the x-intercepts of a quadratic function $f(x) = ax^2 + bx + c$ are given by the solutions of the quadratic equation $ax^2 + bx + c = 0$. For example, the x-intercepts of the quadratic function $f(x) = x^2 - 5x + 4$ are at (4, 0) and (1, 0) because the solutions of the quadratic equation $x^2 - 5x + 4 = 0$ are $x = 4$ and $x = 1$.

REMEMBER

To find the x-intercept(s) of $f(x) = ax^2 + bx + c$, substitute 0 for $f(x)$ (i.e., for y) and solve for x.

Therefore, the discriminant gives information regarding the graph of a quadratic function.

The Discriminant and Graph of $f(x) = ax^2 + bx + c$

The discriminant D of $f(x) = ax^2 + bx + c$ is given by $D = b^2 - 4ac$.

- If $D > 0$, then $f(x)$ has two x-intercepts.
- If $D = 0$, then $f(x)$ has one x-intercept.
- If $D < 0$, then $f(x)$ has no x-intercepts.

Consider the graph of a quadratic function with one x-intercept. The graph must be a parabola with its vertex on the x-axis (so the vertex is the x-intercept), as shown in Figure 2.1n. So, when $D = 0$, the parabola's vertex is on the x-axis.

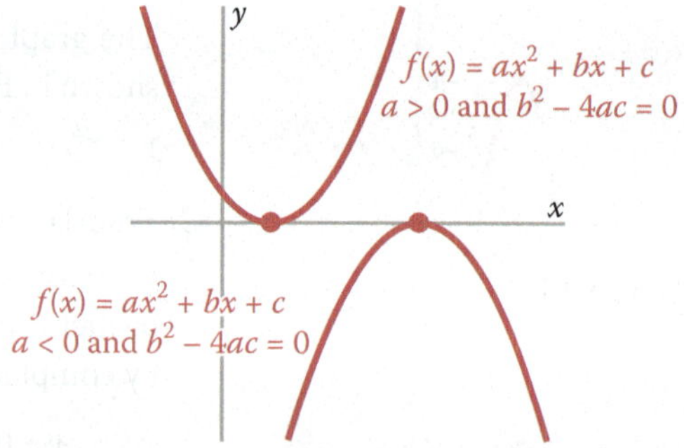

Figure 2.1n

EXAMPLE 5 Finding the Number of x-Intercepts of a Quadratic Function

Use the discriminant to determine the number of x-intercepts of each parabola.

A. $f(x) = -2x^2 - 3x + 12$ **B.** $g(x) = x^2 + 2$

SOLUTION

Find the discriminant, $D = b^2 - 4ac$, for each function.

A. $D = (-3)^2 - 4(-2)(12) = 9 + 96 = 105$

Since $D > 0$, the graph of f is a parabola with two x-intercepts.

B. $D = (0)^2 - 4(1)(2) = 0 - 8 = -8$

Since $D < 0$, the graph of g is a parabola with no x-intercepts.

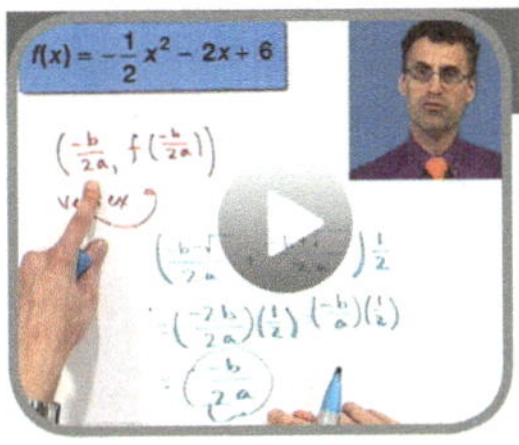

2.1.6 Graphing Quadratic Functions

The Vertex of $f(x) = ax^2 + bx + c$

One method for finding the vertex from a general form quadratic function, $f(x) = ax^2 + bx + c$, is to complete the square and write the function in the standard form, $f(x) = a(x - h)^2 + k$. A second method for finding the vertex from a general form quadratic function involves completing the square for $f(x) = ax^2 + bx + c$.

$$\begin{aligned} f(x) &= ax^2 + bx + c \\ &= a\left(x^2 + \frac{b}{a}x\right) + c && \textit{Factor } a \textit{ from } ax^2 + bx. \\ &= a\left(x^2 + \frac{b}{a}x + \left(\frac{b}{2a}\right)^2\right) + c - a\left(\frac{b}{2a}\right)^2 && \textit{Complete the square. Add } \left(\frac{b}{2a}\right)^2 \textit{ inside and subtract } a\left(\frac{b}{2a}\right)^2 \textit{ outside the parentheses.} \\ &= a\left(x + \frac{b}{2a}\right)^2 + c - a\left(\frac{b}{2a}\right)^2 && \textit{Factor.} \end{aligned}$$

This quadratic function in standard form, $f(x) = a\left(x + \frac{b}{2a}\right)^2 + c - a\left(\frac{b}{2a}\right)^2$, gives a formula for h in terms of a and b: $h = -\frac{b}{2a}$. The corresponding k-value (from the vertex) can be found by substituting the h-value into the original function. So, if $h = -\frac{b}{2a}$, then $k = f\left(-\frac{b}{2a}\right)$.

TIP

Substitute the h-value into the original function to find the k-value of the vertex (h, k).

The Vertex and Axis of Symmetry of $f(x) = ax^2 + bx + c$

The vertex of a quadratic function $f(x) = ax^2 + bx + c$ is given by $\left(-\frac{b}{2a}, f\left(-\frac{b}{2a}\right)\right)$, and the axis of symmetry is the vertical line $x = -\frac{b}{2a}$.

The Intercepts of $f(x) = ax^2 + bx + c$

The graph of every quadratic function $f(x) = ax^2 + bx + c$ has one y-intercept. This y-intercept is at $(0, c)$ because $f(0) = a(0)^2 + b(0) + c = c$.

The x-intercepts of the graph of $f(x) = ax^2 + bx + c$ are the real zeros of the related equation, $ax^2 + bx + c = 0$. So, solving $ax^2 + bx + c = 0$ gives the x-intercepts of the graph of $f(x) = ax^2 + bx + c$.

However, recall that some quadratic equations do not have real zeros, and therefore the graphs of some quadratic functions do not intersect the x-axis. You can use the discriminant to determine whether $ax^2 + bx + c = 0$ has one real solution, two real solutions, or no real solution (i.e., one x-intercept, two x-intercepts, or no x-intercepts) before actually solving the equation.

If $D < 0$, then the graph has no x-intercepts, so there is no need to solve $ax^2 + bx + c = 0$. If $D = 0$, then the graph has one x-intercept, and that x-intercept is also the parabola's vertex. If $D > 0$, then the graph has two x-intercepts.

Graphing $f(x) = ax^2 + bx + c$

Steps for Graphing a Quadratic Function $f(x) = ax^2 + bx + c$

❶ Find the vertex, $\left(-\frac{b}{2a}, f\left(-\frac{b}{2a}\right)\right)$.

❷ Find the intercepts.

❸ Plot the vertex, any x-intercepts, and the y-intercept.

❹ If the vertex is not also the y-intercept, then plot the reflection of the y-intercept over the axis of symmetry, $x = -\frac{b}{2a}$.

If the vertex is also the y-intercept, *and* if the parabola has no x-intercepts, then find another point on the graph by substituting. (An x-value near the vertex may be used.) Plot this point and its reflection over the axis of symmetry.

❺ Sketch the parabola through these points.

EXAMPLE 6 Graphing a Quadratic Function

Graph $f(x) = 0.5x^2 - 2x - 6$.

SOLUTION

The equation is quadratic and $a = 0.5$, so the graph is a parabola that opens upward (since $a > 0$), and is wider than the graph of $y = x^2$ (since $0 < |a| < 1$).

❶ Find the coordinates of the vertex.

$$h = -\frac{b}{2a} = -\frac{-2}{2(0.5)} = 2 \qquad k = f\left(-\frac{b}{2a}\right) = f(2) = 0.5(2)^2 - 2(2) - 6 = -8$$

So, the parabola's vertex is at $(2, -8)$.

❷ Find the intercepts. Since $D = (-2)^2 - 4(0.5)(-6) > 0$, the parabola has two x-intercepts. Find the zeros of f by factoring.

$$0.5x^2 - 2x - 6 = 0 \Rightarrow 0.5(x^2 - 4x - 12) = 0 \Rightarrow 0.5(x - 6)(x + 2) = 0 \Rightarrow x = 6 \text{ or } x = -2$$

So, the parabola has x-intercepts at $(6, 0)$ and $(-2, 0)$.

Since $c = -6$, the parabola has a y-intercept at $(0, -6)$.

❸ Plot the points $(2, -8)$, $(6, 0)$, $(-2, 0)$, and $(0, -6)$.

❹ The parabola's axis of symmetry is the line $x = 2$. So, the parabola also passes through the point $(4, -6)$, which is the reflection of the y-intercept over the line $x = 2$.

❺ Sketch the parabola.

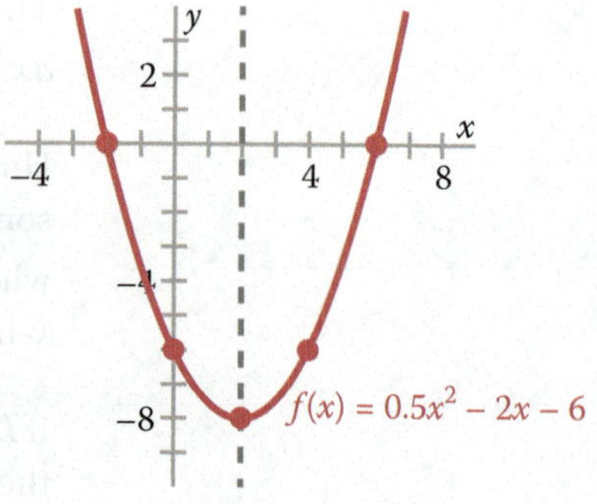

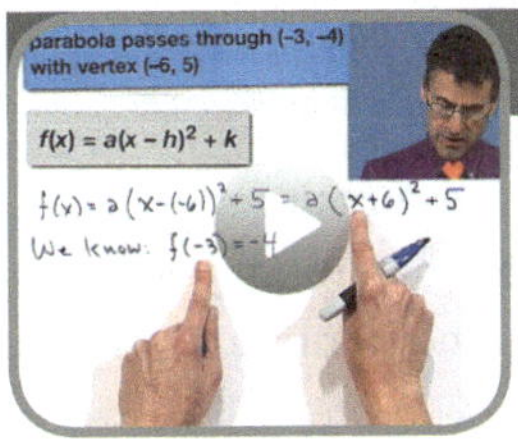

2.1.7 Writing the Equation of a Quadratic Function

Recall that the equation of a parabola can be written when the parabola is a translation of some known function by substituting the h- and k-values from the vertex and the known a-value into the standard form equation $f(x) = a(x - h)^2 + k$. In this case, the parabola's vertex and the function's a-value are known.

The equation of a parabola can also be written when only the vertex and one other point on the parabola are known. In this case, first use the coordinates of the vertex and the other point to find the value of a, then substitute the h- and k-values from the vertex and the a-value into the standard form equation $f(x) = a(x - h)^2 + k$.

Steps for Writing the General Form Equation of a Quadratic Function Using a Point on the Parabola (x, y) and the Parabola's Vertex (h, k)

❶ Substitute the x-value and the y-value (i.e., the $f(x)$-value) from the point and the h- and k-values from the vertex into $f(x) = a(x - h)^2 + k$, and then solve for a.

❷ Substitute the a-value (from step ❶) and the h- and k-values (from the vertex) into $f(x) = a(x - h)^2 + k$.

❸ Expand and simplify the equation from step ❷ to write the equation of the parabola in general form.

IMPORTANT

Do not substitute the x- and y-values from the point when writing the equation of the quadratic function in step ❷.

EXAMPLE 7 Writing the Equation of a Quadratic Function

Write the general form equation of the parabola that passes through $(3, -13)$ and has a vertex at $(5, -1)$.

SOLUTION

First, find the parabola's a-value.

The parabola's vertex is at $(5, -1)$, so $h = 5$ and $k = -1$.

The parabola passes through $(3, -13)$, so $x = 3$ and $f(x) = -13$.

Substitute these coordinates into the standard form equation, $f(x) = a(x - h)^2 + k$, and then solve for a.

$-13 = a(3 - 5)^2 + (-1)$	*Substitute.*
$-13 = a(-2)^2 + (-1)$	*Simplify within parentheses.*
$-13 = 4a - 1$	*Simplify.*
$-3 = a$	*Solve for a.*

Substitute $a = -3$, $h = 5$, and $k = -1$ into $f(x) = a(x - h)^2 + k$, and expand to write the equation of the parabola in the general form.

$$f(x) = -3(x - 5)^2 + (-1) = -3(x^2 - 10x + 25) - 1 = -3x^2 + 30x - 76$$

Therefore, the general form equation of the parabola that passes through $(3, -13)$ with vertex at $(5, -1)$ is $f(x) = -3x^2 + 30x - 76$.

TIP

The vertex is at $(5, -1)$ and the parabola passes through $(3, -13)$. So, the parabola opens downward and $a < 0$.

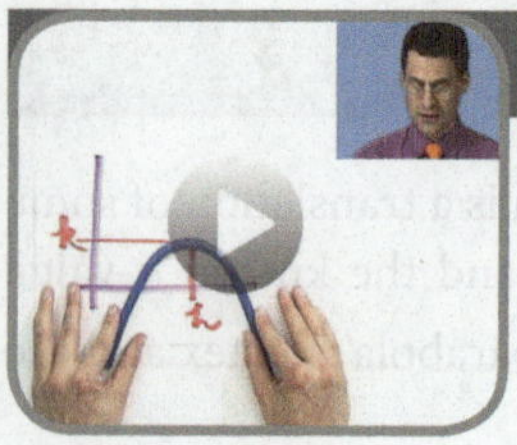

2.1.8 Finding the Maximum or Minimum of a Quadratic Function

As shown previously, a parabola's minimum or maximum value always occurs at its vertex. A parabola that opens upward (where $a > 0$) has a minimum y-value, but no maximum y-value. A parabola that opens downward (where $a < 0$) has a maximum y-value, but no minimum y-value.

When the equation of a parabola is given in the general form, $f(x) = ax^2 + bx + c$, then the coordinates of the parabola's vertex (h, k) are given by $\left(-\frac{b}{2a}, f\left(-\frac{b}{2a}\right)\right)$. Note that, since the h-coordinate is $h = -\frac{b}{2a}$, the k-coordinate can also be written as $k = f(h)$.

EXAMPLE 8 Finding the Maximum or Minimum of a Quadratic Function

Find the maximum or minimum value of $f(x) = -x^2 + 2x + 8$.

SOLUTION

The value of a is -1.

Since $a < 0$, the parabola opens downward, and thus has a maximum.

Find the x-coordinate of the maximum (i.e., the h-coordinate of the vertex).

$$h = -\frac{b}{2a} = -\frac{2}{2(-1)} = 1$$

Find the y-coordinate of the maximum (i.e., the k-coordinate of the vertex).

$$k = f(1) = -(1)^2 + 2(1) + 8 = -1 + 2 + 8 = 9$$

Therefore, the parabola's maximum value is at $f(1) = 9$ (i.e., $y = 9$).

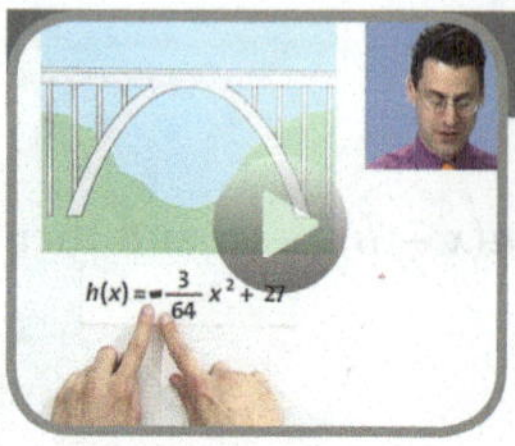

2.1.9 Modeling with Quadratic Functions

EXAMPLE 9 Using a Quadratic Model to Find Maximum Height

The height, in feet, of a baseball after x seconds is given by the function $f(x) = -16x^2 + 64x$. What is the maximum height of the baseball, and after how many seconds does it reach this height?

SOLUTION

The y-coordinate of the vertex gives the maximum height of the ball, and the x-coordinate of the vertex gives the number of seconds it takes to reach this height.

$$h = -\frac{b}{2a} = -\frac{64}{2(-16)} = 2 \qquad k = f(2) = -16(2)^2 + 64(2) = 64$$

Therefore, the ball reaches the maximum height of 64 feet after 2 seconds.

EXAMPLE 10 Modeling Area with a Quadratic Function

What is the maximum possible area of a rectangular garden enclosed by exactly 240 feet of fence? What is the area of the garden if the length of the garden is 30 feet? What are the dimensions of the garden if the area is 3500 square feet?

SOLUTION

Let x be the length of the rectangle.

> **TIP**
>
> *The formula for the perimeter of a rectangle is $P = 2l + 2w$. This rectangular garden is enclosed by exactly 240 feet of fence, so the garden's perimeter is 240 feet, or $240 = 2l + 2w$. Dividing both sides by 2 gives $120 = l + w$. Therefore, the sum of the garden's length and width must be 120 feet.*

The perimeter of the garden must be 240 feet, so 120 feet (half of 240 feet) must be the sum of the length and the width. Therefore, the width of the garden can be represented by $120 - x$.

Write the function for the area of the garden with length x and width $120 - x$.

$$A = x(120 - x) = 120x - x^2$$

The function is quadratic, so the y-coordinate of the vertex (i.e., the h-value) gives the maximum area of the garden.

$$h = -\frac{b}{2a} = -\frac{120}{2(-1)} = 60 \qquad A = 120(60) - (60)^2 = 3600$$

Thus, the maximum area of the garden is 3600 square feet.

Substitute 30 for x in the area function $A = 120x - x^2$ and simplify to find the area of the garden when the length is 30 feet.

$$A = 120(30) - (30)^2 = 2700$$

Thus, the area of the garden is 2700 square feet when the length is 30 feet.

Substitute 3500 into the function for A and solve for x to find the length of the garden when the area is 3500 square feet.

$$3500 = 120x - x^2$$
$$x^2 - 120x + 3500 = 0$$
$$(x - 70)(x - 50) = 0$$

The length of the garden is either 70 feet or 50 feet.

The dimensions of the garden are 70 feet by 50 feet when the area is 3500 square feet.

SECTION 2.1 EXERCISES

Warm Up

Solve by completing the square.

1. $x^2 - 12x + 36 = 81$
2. $p^2 + 6p = 27$
3. $m^2 - m - 12 = 0$

Just the Facts

Fill in the blanks.

4. Whether the equation is in standard form or general form, the _____ in a quadratic function determines whether the parabola opens _____ or _____ and whether it is _____ or more narrow when compared to the basic function.
5. The _____ of a parabola is the point at which the function has a(n) _____ or _____ value, depending upon whether the parabola opens upward or downward.
6. To find the axis of symmetry in general form, use _____.
7. When the equation of a quadratic function is written in standard form, the vertex, (h, _____) and the axis of symmetry, x = _____, can be identified easily from the equation.
8. If the discriminant $b^2 - 4ac$ _____ 0, then the graph has one x-intercept; if $b^2 - 4ac$ _____ 0, then the graph has two x-intercepts; and if $b^2 - 4ac$ _____ 0, then the graph has no x-intercepts.
9. In graphing a quadratic, once you have plotted the vertex and y-intercept, use the property of _____ to plot another point.
10. The graph of $h(x) = (x - 4)^2 + 2$ is a translation of the basic quadratic function _____ 4 units _____ and 2 units _____.

Essential Skills

In Exercises 1–2, graph each function.

1. $f(x) = 3x^2$
2. $g(x) = -\frac{4}{5}x^2$

In Exercises 3–8, identify each parabola's vertex and axis of symmetry, then state whether the vertex is a minimum or maximum.

3. $h(x) = (x - 6)^2 - 2$
4. $r(x) = -\frac{3}{2}(x + 10)^2$
5. $f(x) = -2(x + 1)^2$
6. $g(x) = \frac{9}{7}(x - 3)^2 + 1$
7. $m(x) = \frac{1}{3}(x - 5)^2 + 4$
8. $p(x) = -\frac{6}{7}(x + 4)^2 - 3$

In Exercises 9–22, identify the vertex of each parabola by writing the equation in the standard form of a quadratic function.

9. $p(x) = x^2 + 4x - 6$
10. $f(x) = x^2 - 22x + 116$
11. $h(x) = x^2 + 10x + 26$
12. $g(x) = x^2 + 6x + 8$
13. $m(x) = x^2 - 10x + 30$
14. $p(x) = x^2 + 4x - 6$
15. $h(x) = 2x^2 + 16x + 4$
16. $g(x) = -2x^2 - 40x - 193$
17. $m(x) = 4x^2 + 6$
18. $p(x) = -\frac{6}{5}x^2 - 1$
19. $m(x) = -\frac{5}{6}x^2 - 2$
20. $g(x) = -3x^2 - 24x - 49$
21. $f(x) = -2x^2 - 8x - 11$
22. $g(x) = \frac{3}{8}x^2 - 3$

In Exercises 23–24, write each equation in the standard form of a quadratic function. Then graph each function.

23. A function f is the image of $y = x^2 - 4x + 11$ after a translation right 3 units and up 1 unit.
24. A function g is the image of $y = 2x^2 + 44x + 249$ after a translation left 2 units and down 3 units.

In Exercises 25–28, write each equation in the standard form of a quadratic function.

25. A function f is the image of $y = 3x^2 + 36x + 2$ after a translation left 5 units and down 2 units.
26. A function g is the image of $y = -3x^2 - 18x + 2$ after a translation left 3 units and up 5 units.
27. A function f is the image of $y = -2x^2 - 12x + 2$ after a translation left 2 units and up 5 units.
28. A function g is the image of $y = -2x^2 - 16x + 2$ after a translation right 2 units and down 2 units.

In Exercises 29–34, use the discriminant to determine the number of x-intercepts of each parabola.

29. $f(x) = x^2 + 2x - 5$

30. $g(x) = 2x^2 + 5$

31. $f(x) = -4x^2 + 16x - 16$

32. $f(x) = 5x^2 + 50x + 125$

33. $g(x) = -2x^2 + 3x - 3$

34. $f(x) = -3x^2 - 4x - 4$

In Exercises 35–36, graph each function.

35. $f(x) = x^2 - 2x - 1$

36. $g(x) = -2x^2 - 8x - 9$

In Exercises 37–42, write the general form equation of each parabola that passes through the given point and has the given vertex.

37. (0, 3) with vertex at (–1, 0)

38. (2, –9) with vertex at (4, –8)

39. (4, –3) with vertex at (3, –1)

40. (3, 5) with vertex at (2, 2)

41. (3, 34) with vertex at (–1, 2)

42. (4, –53) with vertex at (–1, –3)

In Exercises 43–48, find the maximum or minimum value of each function.

43. $f(x) = 5x^2 - 20x$

44. $g(x) = -3x^2 + x + 9$

45. $h(x) = 3x^2 - 2x + 3$

46. $f(x) = 2x^2 - 3x + 8$

47. $m(x) = 2x^2 - 2x - 4$

48. $f(x) = -2x^2 + 2x + 6$

49. A fence of 120 feet is used to enclose a rectangular garden. Find the maximum area possible for the garden.

50. Write a function to model the volume of a rectangular prism if the length is 10 centimeters and the sum of the width and height is 18 centimeters. What is the maximum possible volume of the prism?

51. A fence of 650 feet is used to enclose a rectangular garden. Find the maximum area possible for the garden.

52. Write a function to model the volume of a rectangular prism if the length is 18 centimeters and the sum of the width and height is 28 centimeters. What is the maximum possible volume of the prism?

Extensions

53. Write the general form equation of the quadratic function that has x-intercepts of 5 and –5, and a y-intercept of 3.

54. True or False? If the graph of $f(x) = 2x^2 + 4x - 1$ is reflected across the x-axis, then translated right 4 units and up 4 units, the resulting graph will have the equation $g(x) = -2x^2 + 12x - 17$.

55. Write two quadratic equations in standard form, one where $a > 0$ and another where $a < 0$, which have x-intercepts of 2 and –2.

56. Write the equation of the graph in general form.

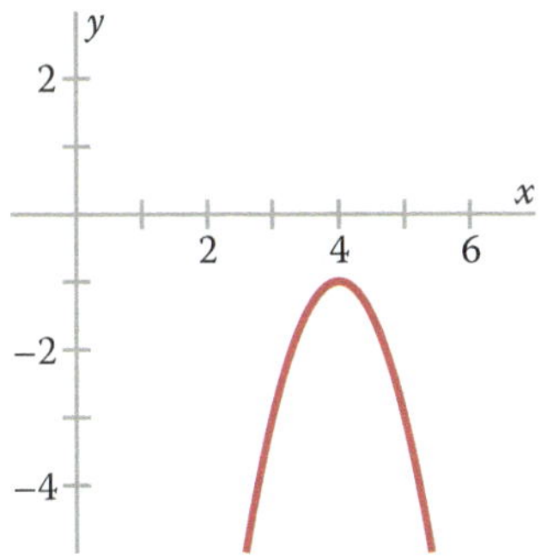

57. True or False? The graphs of the functions $g(x) = 3x^2 - 12x + 13$ and $f(x) = -0.5x^2 + 2x - 1$ share the same vertex.

58. A theater has 500 membership subscribers who each pay \$300 for a season pass. The theater wants to increase the price of the season pass by \$30. For each \$30 increase, they lose 10 subscribers. At what price can the theater maximize its revenue? Find the maximum revenue, and graph the function (over a reasonable domain).

2.2 POLYNOMIAL FUNCTIONS AND THEIR GRAPHS

OBJECTIVES

- Describe the end behavior of the graphs of polynomial functions.
- Graph polynomial functions.
- Find the real roots of polynomial functions.
- Find the multiplicity of the zeros of polynomial functions.
- Understand the Intermediate Value Theorem.
- Find a local maximum or minimum of a polynomial function.

PREREQUISITE VOCABULARY TERMS

degree (of a polynomial)
leading coefficient
maximum
minimum
polynomial
zero (of a function)

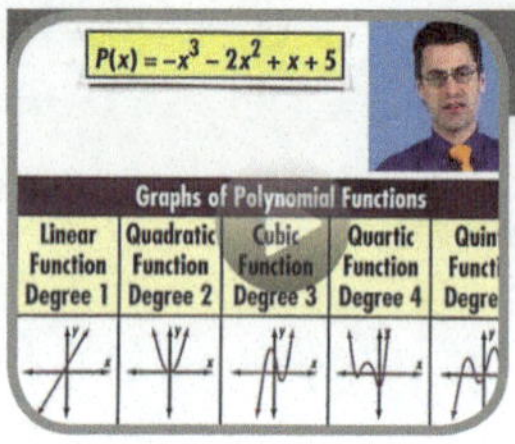

2.2.1 End Behavior of Graphs of Polynomial Functions

Monomial Functions

The graphs of polynomial functions of degree 2 (quadratic functions) and of degree 1 (linear functions) have been discussed in detail. We know that the graph of a quadratic function is a parabola and the graph of a linear function is a line. We have also seen graphs of polynomial functions of degree 3, such as the basic cubic function $f(x) = x^3$.

Recall that the graph of a linear, quadratic, or cubic function is always a smooth continuous curve with no corners or breaks. This is also the case for the graph of any polynomial function.

A polynomial function with one term is a **monomial function**. The simplest monomial functions are of the form $f(x) = x^n$, where n is a positive integer. The basic polynomial functions $f(x) = x$, $f(x) = x^2$, $f(x) = x^3$, $f(x) = x^4$, $f(x) = x^5$, and $f(x) = x^6$ are examples of monomial functions. The graphs of these functions are shown in the following table.

IMPORTANT

The graph of each basic monomial function passes through the origin. Each graph has no other x- or y-intercepts.

Basic Monomial Functions

Degree 1 (linear)	Degree 2 (quadratic)	Degree 3 (cubic)
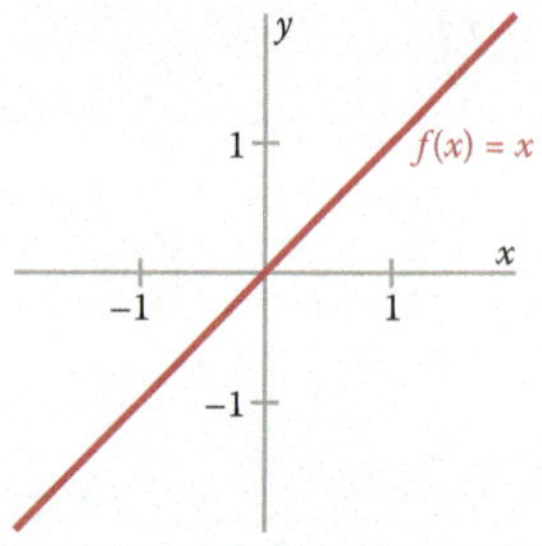	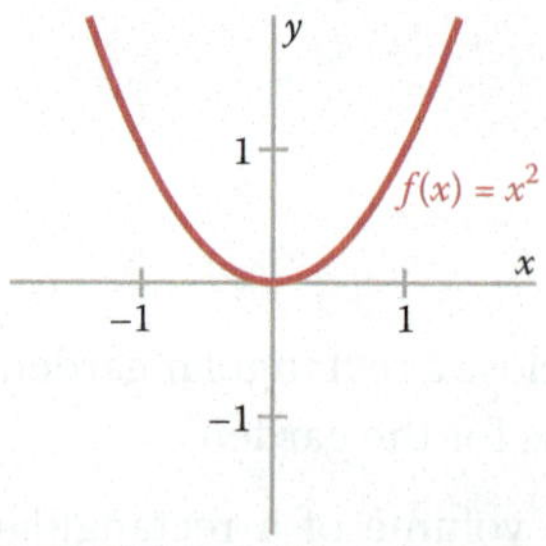	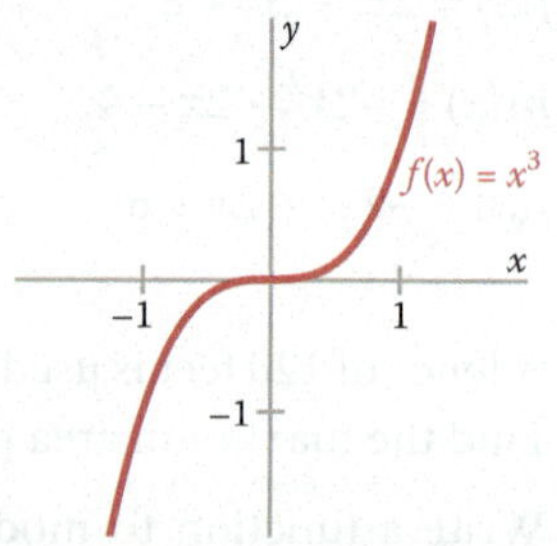
Degree 4	**Degree 5**	**Degree 6**
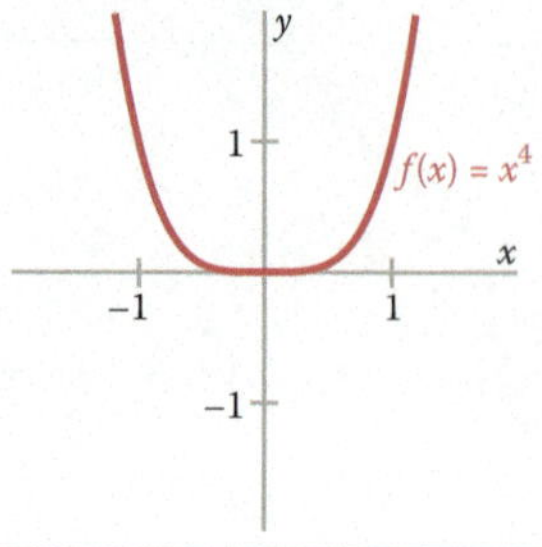	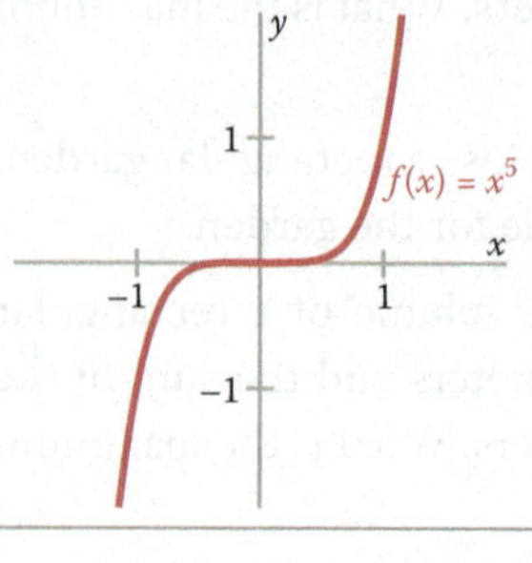	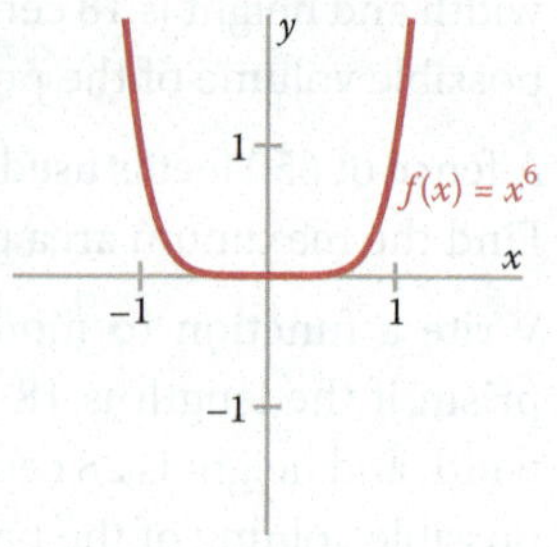

The End Behavior of the Graph of a Polynomial Function

Notice that the graphs of the monomial functions of the form $f(x) = x^n$ where n is odd are very similar, and the graphs of the monomial functions of the form $f(x) = x^n$ where n is even are also very similar. Specifically, the ends of each graph of $f(x) = x^n$ where n is odd are headed in opposite directions (one end is pointed down and the other end is pointed up), whereas the ends of each graph of $f(x) = x^n$ where n is even are both headed in the same direction. The direction that the graph of a function is pointing at its left and right ends is referred to as its **end behavior**.

The end behavior of a polynomial function's graph can be predicted by the polynomial's leading coefficient and the polynomial's degree, as shown in the following table. The middle part of each curve is dashed to indicate that this part of the graph is not determined by the polynomial's degree and leading coefficient.

REMEMBER

A polynomial's leading coefficient is the coefficient of the term with the greatest degree.

IMPORTANT

The notation "$x \to \infty$" is read as "x goes to positive infinity," which means that x becomes large in a positive direction. The notation "$x \to -\infty$" is read as "x goes to negative infinity," which means that x becomes large in a negative direction.

End Behavior of the Graphs of Polynomial Functions

The end behavior of the polynomial function $f(x) = a_n x^n + a_{n-1}x^{n-1} + \ldots + a_1 x + a_0$ is determined by the degree n (whether the degree is even or odd) and the sign of the leading coefficient a_n (whether the sign is positive or negative).

	$a_n > 0$	$a_n < 0$
n is even	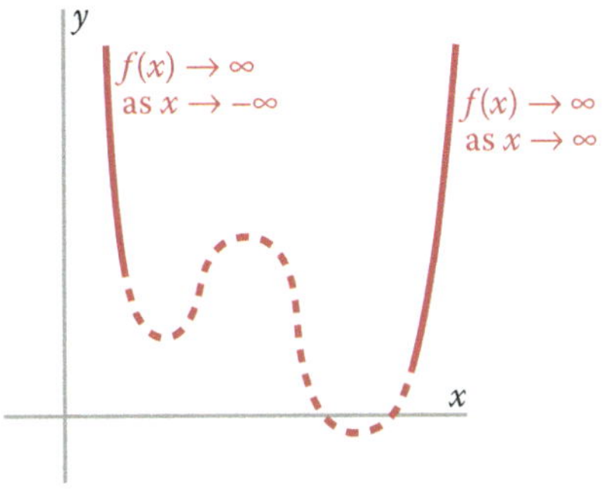 *If the degree is even and the leading coefficient is positive, then $f(x) \to \infty$ at the left and right ends.*	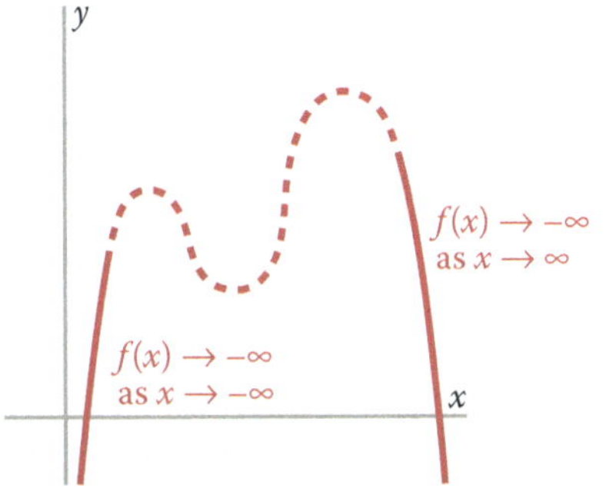 *If the degree is even and the leading coefficient is negative, then $f(x) \to -\infty$ at the left and right ends.*
n is odd	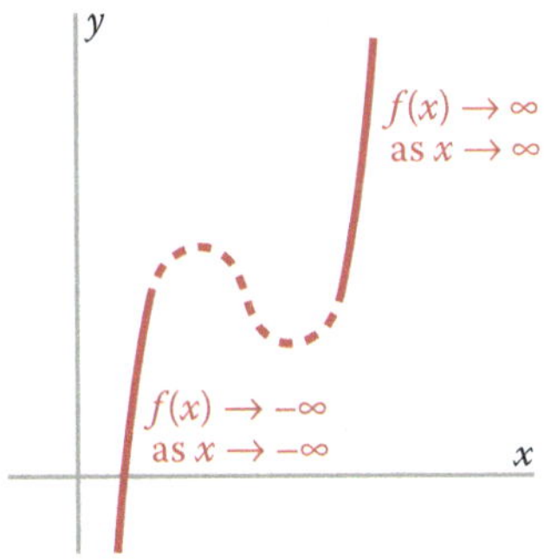 *If the degree is odd and the leading coefficient is positive, then $f(x) \to -\infty$ at the left end and $f(x) \to \infty$ at the right end.*	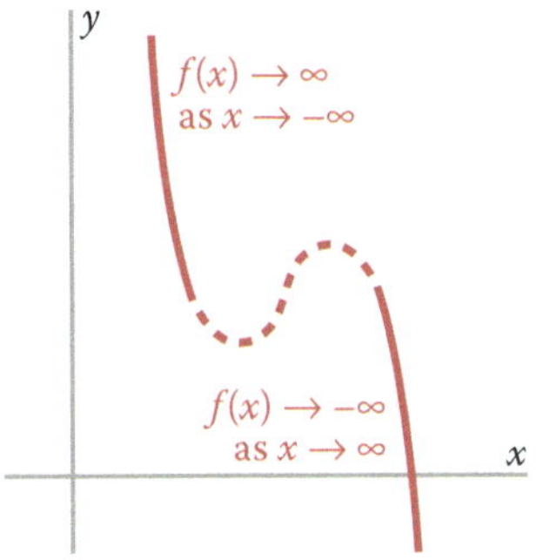 *If the degree is odd and the leading coefficient is negative, then $f(x) \to \infty$ at the left end and $f(x) \to -\infty$ at the right end.*

For example, consider the graph of a quadratic function $f(x) = ax^2$. Recall that when $a > 0$, the graph is a parabola that opens upward. So, $f(x)$ goes to ∞ as x goes to $-\infty$ *and* to ∞ (on the left and right ends), which is the expected end behavior of a polynomial function with even degree and a positive leading coefficient. Similarly, when $a < 0$, the graph is a parabola that opens downward. So, $f(x)$ goes to $-\infty$ as x goes to $-\infty$ and to ∞, which is the expected end behavior of a polynomial function with even degree and a negative leading coefficient.

EXAMPLE 1 Determining the End Behavior of a Polynomial Function

Describe the end behavior of each function.

A. $f(x) = 5x - x^3$ **B.** $f(x) = 5x^4 + x^2 + 1$

SOLUTION

A. The term with the greatest degree is $-x^3$, so the polynomial's degree is 3 and the leading coefficient is −1. Since the degree is odd and the leading coefficient is negative, $f(x) \to \infty$ at the left end and $f(x) \to -\infty$ at the right end.

End behavior: as $x \to -\infty, f(x) \to \infty$, and as $x \to \infty, f(x) \to -\infty$

B. The term with the greatest degree is $5x^4$, so the polynomial's degree is 4 and the leading coefficient is 5. Since the degree is even and the leading coefficient is positive, $f(x) \to \infty$ at the left and right ends.

End behavior: as $x \to -\infty, f(x) \to \infty$, and as $x \to \infty, f(x) \to \infty$

The end behavior described in **Example 1** can be confirmed by graphing each function using a graphing calculator, as shown in Figures 2.2a and 2.2b.

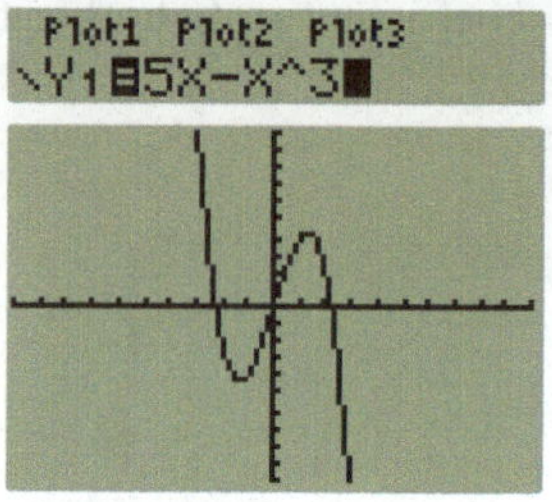

Figure 2.2a

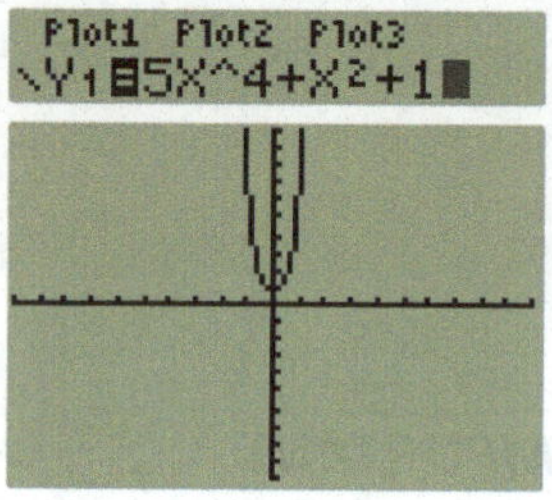

Figure 2.2b

2.2.2 Reflecting, Stretching, and Translating Polynomial Functions

To graph a monomial function g, start with the corresponding basic monomial function f (i.e., the basic function with the same degree as g), and then reflect, translate, or stretch (compress) the basic function as indicated by the equation of g.

TIP

The degree of g is even and the leading coefficient is negative, so $g(x) \to -\infty$ at the left and right ends.

For example, to graph $g(x) = -\frac{1}{8}x^4$, reflect the graph of $f(x) = x^4$ over the x-axis (since the leading coefficient is negative) and stretch by a factor of 1/8 (since the absolute value of the leading coefficient is 1/8). So, for each point (x, y) on $f(x) = x^4$, there is a point $\left(x, -\frac{1}{8}y\right)$ on $g(x) = -\frac{1}{8}x^4$. The graphs of $f(x) = x^4$ and $g(x) = -\frac{1}{8}x^4$ are shown in Figure 2.2c, along with a corresponding table. (Note that not all points from the table for f are shown on the graph of f.)

x	$f(x)$	$g(x)$
−2	16	−2
−1	1	$-\frac{1}{8}$
0	0	0
1	1	$-\frac{1}{8}$
2	16	−2

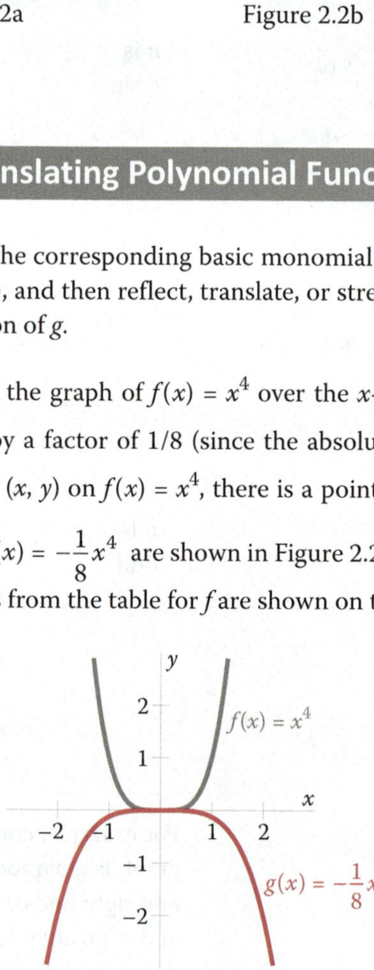

Figure 2.2c

EXAMPLE 2 Graphing Transformations of $f(x) = x^n$

Sketch the graph of each function.

A. $g(x) = x^4 + 2$ **B.** $h(x) = -(x + 1)^3$

SOLUTION

A. Translate the graph of $f(x) = x^4$ up 2 units to sketch the graph of $g(x) = x^4 + 2$.

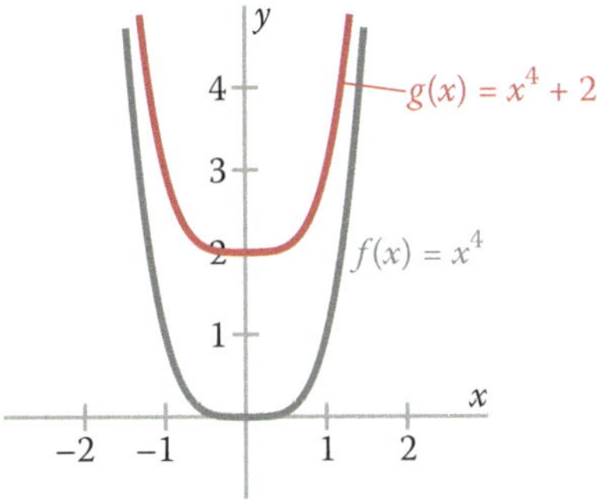

B. Reflect the graph of $f(x) = x^3$ over the x-axis, then translate the graph 1 unit to the left to sketch the graph of $h(x) = -(x + 1)^3$.

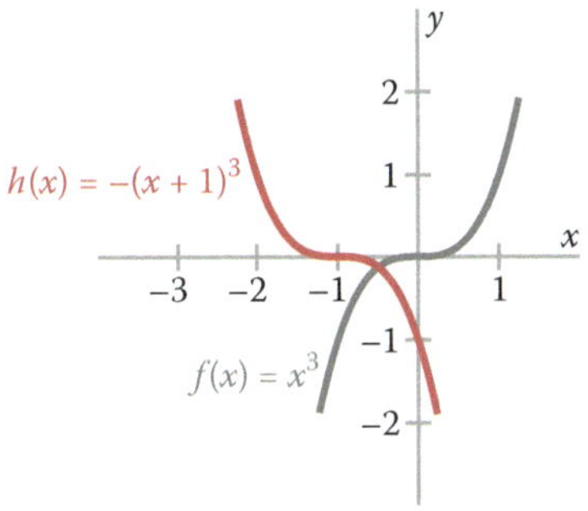

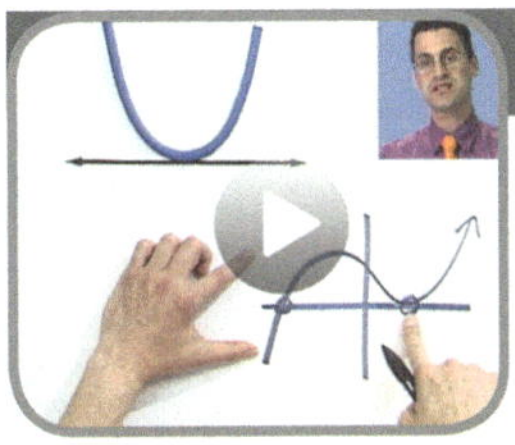

2.2.3 Finding Zeros and Their Multiplicities for a Polynomial

For a function $f(x)$, the **real zeros** of f are the real x-values that satisfy $f(x) = 0$. These real x-values are the x-intercepts of the function's graph. Knowing a function's zeros can be very helpful when sketching a graph of the function.

TIP

The graphs of some functions do not cross the x-axis, and therefore some functions do not have any real zeros.

Real Zeros of Polynomial Functions

If f is a polynomial function and c is a real number, then the following statements are equivalent.

1. c is a *zero* of f.

2. $x = c$ is a *solution* of the polynomial equation $f(x) = 0$.

3. $(c, 0)$ is an *x-intercept* of the graph of f.

4. $(x - c)$ is a *factor* of the polynomial $f(x)$.

To find a function's real zeros (if any exist), replace $f(x)$ with 0, and then solve for x. For example, to find the real zeros of $g(x) = 2x + 6$, substitute 0 for $g(x)$ and solve for x.

$$0 = 2x + 6 \Rightarrow 2x = -6 \Rightarrow x = -3$$

So, $g(x) = 2x + 6$ has one real zero, −3, meaning that the graph of $g(x) = 2x + 6$ crosses the x-axis only at the point $(-3, 0)$. Note also that since −3 is a real zero of g, $(x + 3)$ must be a factor of $g(x)$. Factor the expression to confirm: $g(x) = 2x + 6 = 2(x + 3)$.

TIP

The function $f(x) = x^2 - 9$ has degree 2 and two zeros.

Notice that the degree of the function $g(x) = 2x + 6$ is 1, and that the function has one zero. Functions with higher degrees can have more than one zero. For example, consider the function $f(x) = x^2 - 9$. Substituting 0 for $f(x)$ and solving gives $x = \pm 3$. Therefore, the function has two zeros, 3 and −3, and the graph's x-intercepts are at $(3, 0)$ and $(-3, 0)$. Furthermore, $(x - 3)$ and $(x + 3)$ must be factors of $f(x)$, which can be confirmed by factoring the expression $x^2 - 9$.

A function can also have repeated zeros. For example, the function $f(x) = (x - 1)^2$ has a zero at 1 that occurs twice. In this case, the zero at 1 is said to have **multiplicity** 2.

TIP

When a function has a zero that is repeated k times, the zero has multiplicity k.

Real Zeros of Polynomial Functions: Multiplicity

If f is a polynomial function, and c is a real number such that $(x - c)$ is a factor of $f(x)$ exactly k times (where $k \geq 1$), then c is a zero of f with multiplicity k.

- If k is even, then the graph of f touches, but does not cross, the x-axis at c.

The following graphs show the shape of the graph of f near the x-intercept c when k is even.

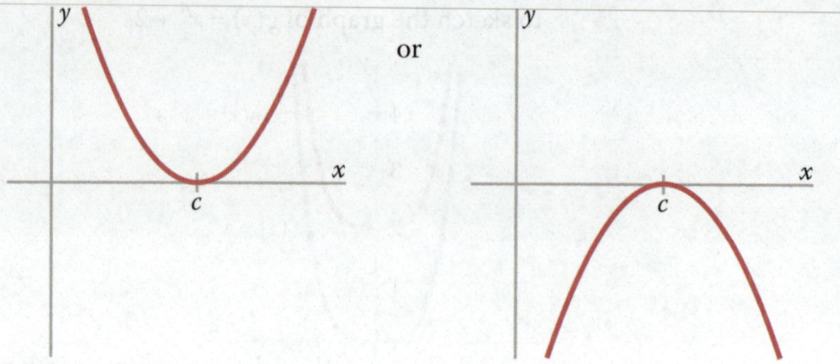

- If k is odd, then the graph of f crosses the x-axis at c.

The following graphs show the shape of the graph of f near the x-intercept c when k is odd.

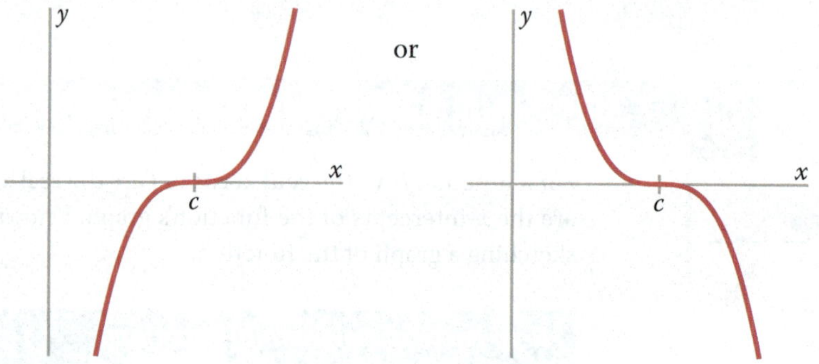

EXAMPLE 3 Finding All Zeros and Multiplicities

Find all real zeros and their multiplicity for each function.

A. $f(x) = x^2(x - 2)(x + 5)^3$ **B.** $f(x) = 4x^5 + 8x^4 + 4x^3$

SOLUTION

Set the polynomial equal to zero, and solve for x to find the zeros. The power of each factor gives the zero's multiplicity.

A. The polynomial is factored. Set each factor equal to 0 and solve to find the zeros.

$$x^2(x - 2)(x + 5)^3 = 0 \Rightarrow x = 0, x = 2, \text{ or } x = -5$$

Use the power of each factor to identify each zero's multiplicity.

$$f(x) = x^2(x - 2)(x + 5)^3$$

0 is a zero with multiplicity 2 — *The factor's power is 2, so the zero's multiplicity is 2.*

2 is a zero with multiplicity 1 — *The factor's power is 1, so the zero's multiplicity is 1.*

−5 is a zero with multiplicity 3 — *The factor's power is 3, so the zero's multiplicity is 3.*

B. Solve the polynomial equation $4x^5 + 8x^4 + 4x^3 = 0$ by factoring.

$4x^3(x^2 + 2x + 1) = 0$ *Factor out the GCF.*

$4x^3(x + 1)(x + 1) = 0$ *Factor the trinomial.*

The equation can also be written as $4x^3(x + 1)^2 = 0$ since the factor $(x + 1)$ is repeated twice.

$$f(x) = 4x^3(x + 1)^2$$

0 is a zero with multiplicity 3

The factor's power is 3, so the zero's multiplicity is 3.

−1 is a zero with multiplicity 2

The factor's power is 2, so the zero's multiplicity is 2.

The graphs of the polynomial functions from **Example 3** are shown in Figures 2.2d and 2.2e.

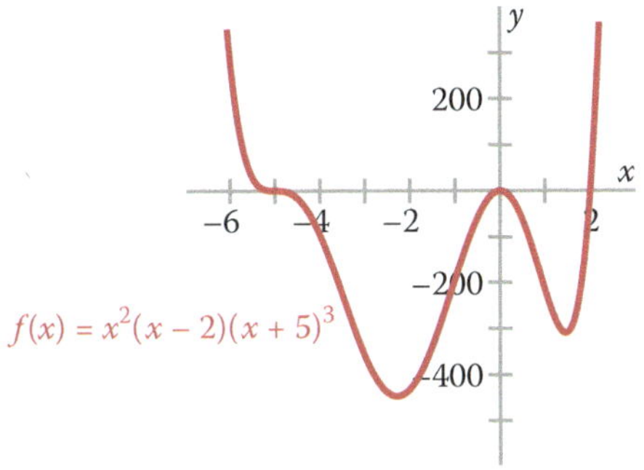

Figure 2.2d

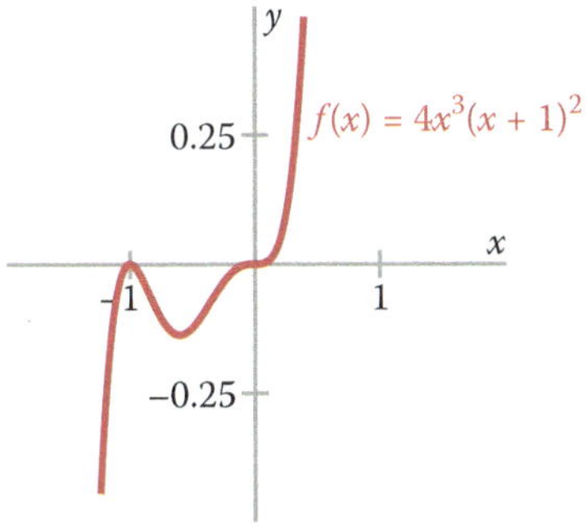

Figure 2.2e

Notice that the graph of $f(x) = x^2(x - 2)(x + 5)^3$ has x-intercepts at −5, 0, and 2, as expected, since −5, 0, and 2 are the equation's zeros. The zeros −5 and 2 both have odd multiplicities (the multiplicity of −5 is 3, and the multiplicity of 2 is 1), and the graph crosses the x-axis at these zeros. The zero 0 has even multiplicity, and the graph touches, but does not cross, the x-axis at this zero.

Similarly, the graph of $f(x) = 4x^3(x + 1)^2$ has x-intercepts at its zeros, −1 and 0. The graph crosses the x-axis at −1, which is a zero with odd multiplicity. The graph touches, but does not cross, the x-axis at 0, which is a zero with even multiplicity.

The zeros of a polynomial function and their multiplicities will be used later in this section to sketch the graph of a polynomial function.

2.2.4 Graphing Polynomial Functions

The graph of a polynomial function can be sketched by putting together all of the information covered in this section so far.

Steps for Sketching the Graph of a Polynomial Function

1. Determine the end behavior.
2. Find the y-intercept.
3. Find the zeros (x-intercepts), if any, and their multiplicities.
4. Find several additional points on the graph, choosing x-values near the x-intercepts.
5. Sketch the graph.

EXAMPLE 4 Graphing a Polynomial Function

Sketch the graph of the function. $f(x) = (x + 2)(x - 1)^2$

SOLUTION

1. Expand the polynomial to identify the degree and the leading coefficient.

$$f(x) = (x + 2)(x - 1)^2 = x^3 - 3x + 2$$

So, the degree is 3, and the leading coefficient is 1. Since the degree is odd and the leading coefficient is positive, $f(x) \to -\infty$ at the left end and $f(x) \to \infty$ at the right end.

TIP

Substitute 0 for x into either the original equation, or into the expanded form of the equation, to find the y-intercept.

2. Find the y-intercept. $f(0) = (0 + 2)(0 - 1)^2 = (2)(-1)^2 = 2$ *Substitute 0 for x.*

So, the graph crosses the y-axis at $(0, 2)$.

3. Find the zeros.

$$f(x) = (x + 2)(x - 1)^2$$

-2 is a zero with multiplicity 1
The factor's power is 1, so the zero's multiplicity is 1.

1 is a zero with multiplicity 2
The factor's power is 2, so the zero's multiplicity is 2.

So, the graph has x-intercepts at $(-2, 0)$ and $(1, 0)$. Since the multiplicity of the zero at 1 is even, the graph will not cross the x-axis at $(1, 0)$.

4. Find several additional points on the graph. Choose x-values for the table near the zeros, $x = 1$ and $x = -2$, such as $x = -1$ and $x = 2$.

$f(-1) = (-1 + 2)(-1 - 1)^2 = 4$ *The graph passes through (−1, 4).*

$f(2) = (2 + 2)(2 - 1)^2 = 4$ *The graph passes through (2, 4).*

5. Sketch the graph by plotting the points and extending the graph in a negative direction at its left end and a positive direction at its right end. Remember that the graph will not cross the x-axis at the x-intercept at $(1, 0)$.

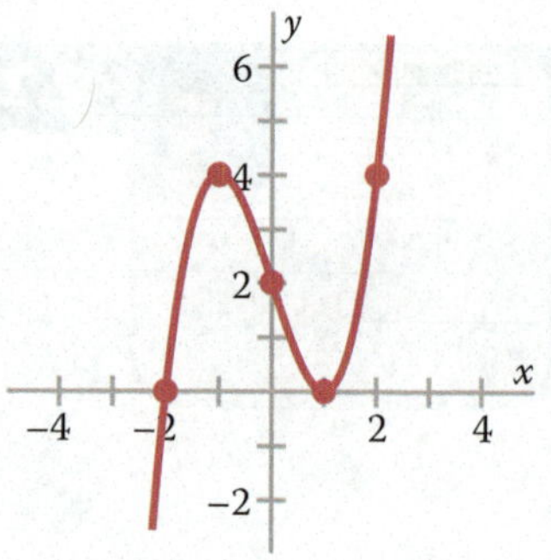

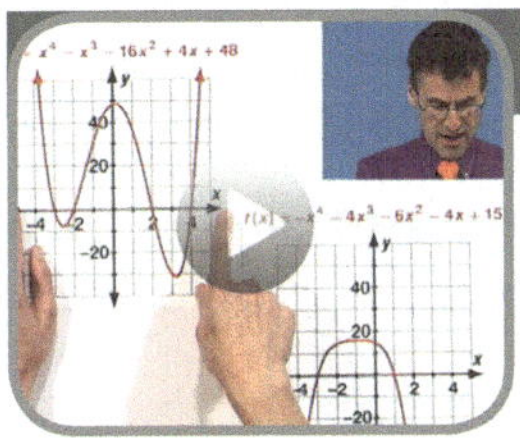

2.2.5 Intermediate Value Theorem and Local Extrema

The Intermediate Value Theorem

The **Intermediate Value Theorem** states that if a and b are real numbers such that $a < b$ and $(a, f(a))$ and $(b, f(b))$ are two points on the graph of the polynomial function f such that $f(a) < f(b)$, then for any real number c between $f(a)$ and $f(b)$, there must be at least one real number x between a and b such that $f(x) = c$.

Intermediate Value Theorem

Let a and b be real numbers such that $a < b$, and let f be a polynomial function such that $f(a) < f(b)$. Then for every c such that $f(a) < c < f(b)$, there is at least one number x in the interval $[a, b]$ such that $f(x) = c$.

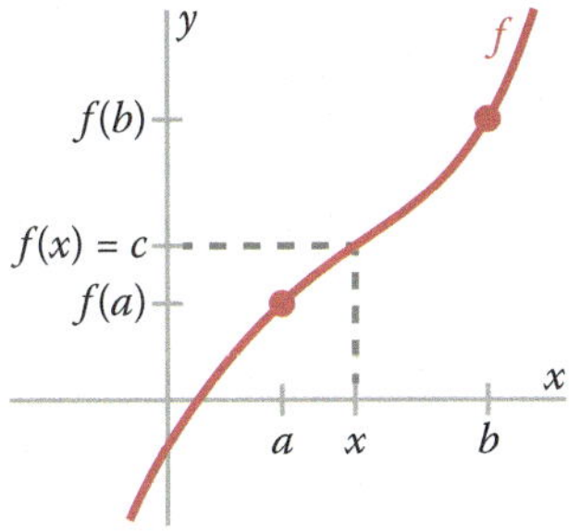

This theorem can be used to identify the existence of a real zero for a function that has points above and below the x-axis. For example, if there exists a point $(a, f(a))$ on the function f below the x-axis (where $f(a) < 0$), and there exists a point $(b, f(b))$ on the function above the x-axis (where $f(b) > 0$), then by the Intermediate Value Theorem, there must be some point on f where the y-value is 0. So, the function f must have a real zero.

Local Extrema

If the point $(x, f(x))$ is the lowest point on some interval $[a, b]$, then $f(x)$ is a **local minimum** within that interval. If the point $(x, f(x))$ is the highest point on some interval $[a, b]$, then $f(x)$ is a **local maximum** within that interval.

The graph of a polynomial function with degree n can have up to $n - 1$ local minimums and maximums, called **local extrema**.

Figure 2.2f shows the graph of a polynomial function with one local minimum (between $x = b$ and $x = c$) and one local maximum (between $x = a$ and $x = b$). Since there is one local maximum and one local minimum, there are two local extrema.

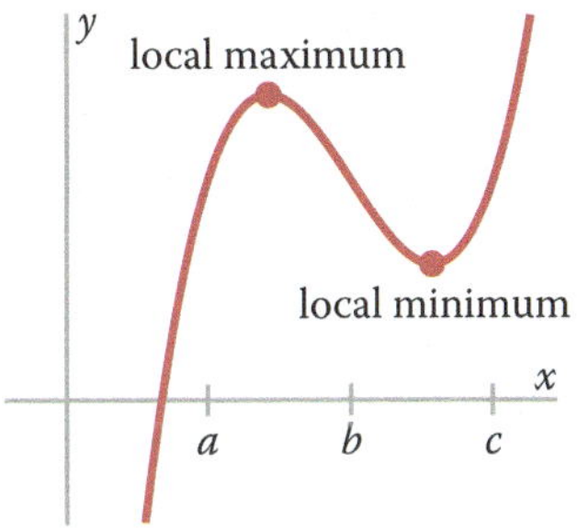

Figure 2.2f

EXAMPLE 5 **Identifying Local Extrema**

Identify the type and number of local extrema.

SOLUTION

The graph has two local maximums, one between $x = -2$ and $x = -1$, and the other between $x = 1$ and $x = 2$. The graph has one local minimum between $x = -1$ and $x = 0$. So, the graph has three local extrema.

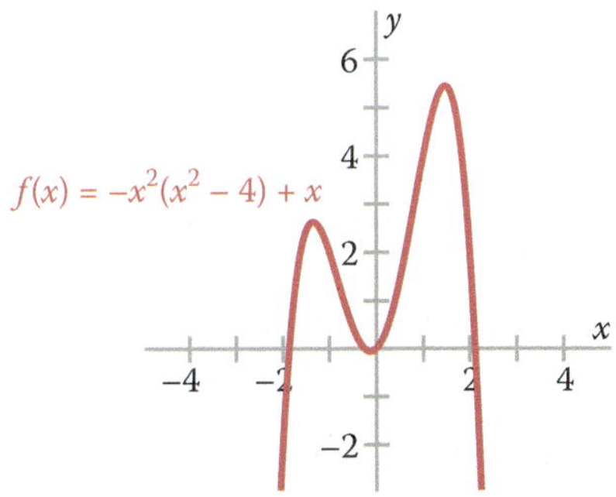

SECTION 2.2 EXERCISES

Warm Up

Write each polynomial in standard form, and identify its degree and leading coefficient.

1. $4x^2 + 9x^4 - 12x + 5x^3$
2. $-3x^7 - 10 - 1.4x^7 + 2x^5 - 11x$
3. $30 + 7x + x^2 - 4x^3 - x$

Just the Facts

Fill in the blanks.

4. A function's _____ and _____ determine the end behavior of its graph.
5. If the degree of a polynomial is _____ and the leading coefficient is _____, as x goes to negative infinity, y goes to negative infinity, and when x goes to positive infinity, y goes to positive infinity.
6. The _____ of a zero is the number of times that zero is a factor of the polynomial.
7. If the degree of a function is 5, then that function can have at most _____ local extrema.
8. By the _____, if the sign of $f(x)$ changes between two values, say, $f(c)$ and $f(d)$, then there exists a(n) _____ between c and d.
9. When a zero touches but does not cross the x-axis, that zero occurs a(n) _____ number of times.
10. $f(x) = (x + 2)^3 + 4$ has been translated 4 units _____ and 2 units _____ from the basic function _____.

Essential Skills

In Exercises 1–6, describe the end behavior of the graph of each function.

1. $h(x) = 3x^4 + x - 4$
2. $f(x) = 10 + x^3 - 2x^2$
3. $p(x) = 2x^2 + 3 - 4x^5$
4. $g(x) = 4 - 5x^6 - 8x^3$
5. $k(x) = -5x^6 + x^7 + 1$
6. $m(x) = 7x^8 + x^3$

In Exercises 7–10, graph each function.

7. $f(x) = (x + 1)^4 - 1$
8. $g(x) = (x - 2)^5 + 3$
9. $m(x) = -x^5 + 2$
10. $p(x) = -x^6 - 3$

In Exercises 11–22, find all real zeros (if any) of each function, and state their multiplicity.

11. $g(x) = x(x + 4)^4(x - 1)^2$
12. $f(x) = x^5(x - 2)^3(x + 7)$
13. $d(x) = x^2(x + 5)^6(x + 8)^5$
14. $h(x) = -x^4(x - 4)^4(x + 9)$
15. $n(x) = -(x + 1)^2(x + 9)(x - 7)^3$
16. $m(x) = x^6 (x + 6) (x + 3)^4$
17. $p(x) = x^3 + 4x^2 + 4x$
18. $f(x) = 5x^3 - 30x^2 + 45x$
19. $m(x) = 6x^6 + 12x^5 - 144x^4$
20. $h(x) = 2x^4 - 8x^3 - 24x^2$
21. $k(x) = 5x^4 - 10x^3 - 75x^2$
22. $g(x) = 3x^6 + 6x^5 - 72x^4$

In Exercises 23–26, graph each function.

23. $g(x) = x(x + 4)^2$
24. $f(x) = (x - 5)^2(x + 3)$
25. $h(x) = -(x - 2)(x + 1)^3$
26. $f(x) = (x - 3)^2(x^2 + 9)$

In Exercises 27–30, identify the type and number of local extrema for each function.

27.

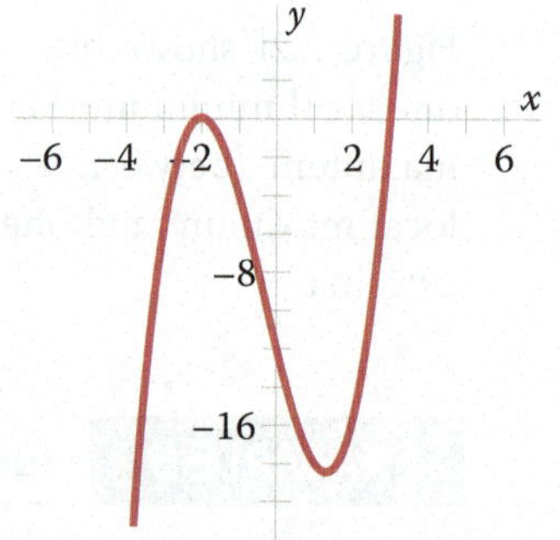

28.

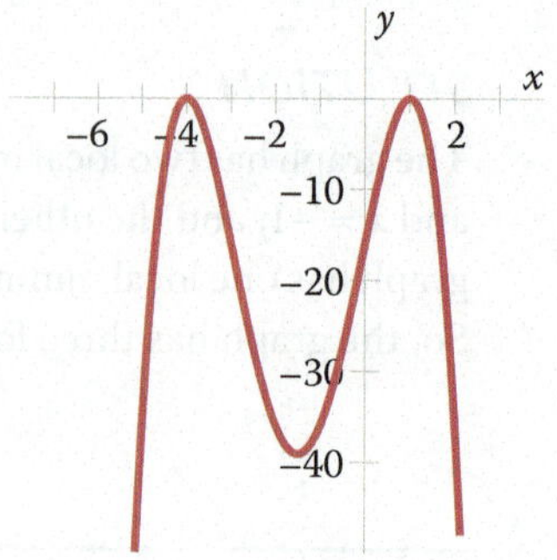

29.

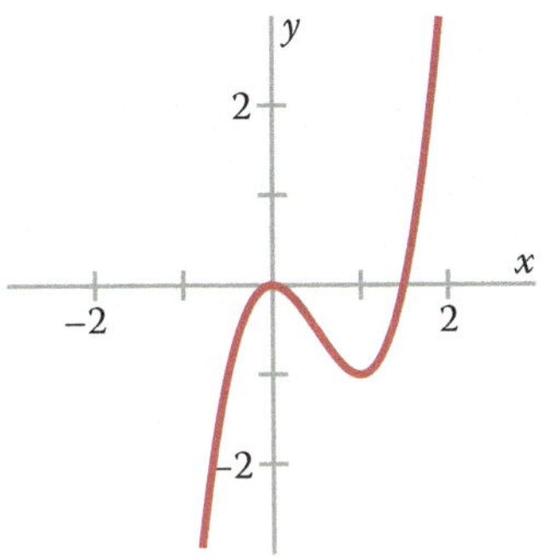

30.

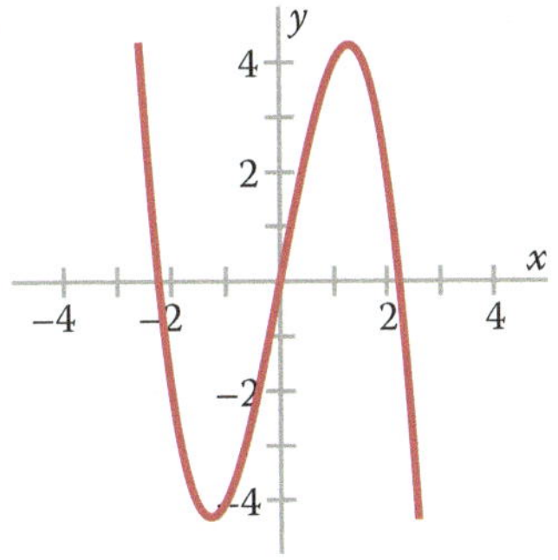

Extensions

In Exercises 31–33, identify the type and number of local extrema of each function. Indicate the integer values between which the extrema lie.

31. $h(x) = 0.2x^5 + x^4 - 3x^2 - 1$

32. $g(x) = x^4 + 5.4x^3 + 5.49x^2 - 4.86x + 0.81$

33. $f(x) = 0.1x^6 - 0.1x^5 - 1.5x^4 + 0.5x^3 + 7x^2 + 1.2x - 7.2$

34. Write the equation in factored form and then sketch the graph of the polynomial function that has the following characteristics.

- The function has zeros at 3 (multiplicity 2), −2 (multiplicity 1), and 0 (multiplicity 3).
- The graph passes through (−1, 8).
- The equation has a negative leading coefficient.

35. Write the equation in factored form of the polynomial function graphed here, given that the polynomial's degree is 7 and the y-intercept is 7.2.

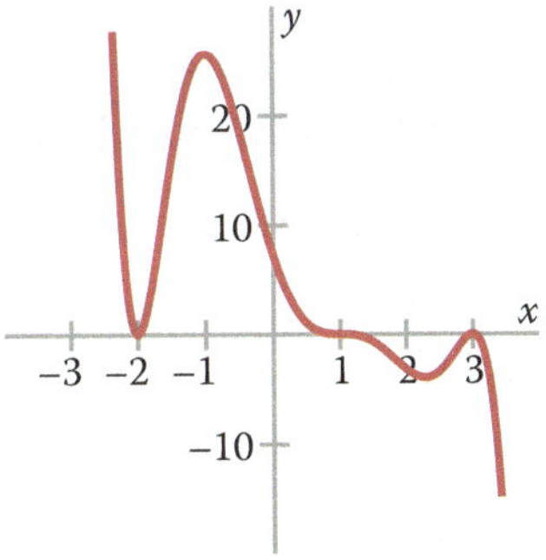

36. Write the general form equation of the function $p(x)$ that is the image of $f(x) = x^3$ after a reflection over the x-axis, a vertical stretch by a factor of 2, and translations 3 units left and 4 units down. Graph the function.

37. Write the equations of two functions of degree 3 that are reflections of each other over the x-axis, where both functions have zeros of −1, 2, and 6.

38. True or False? If a function's end behavior is as $x \to \infty, f(x) \to \infty$ and as $x \to -\infty, f(x) \to \infty$, then this function must be of even degree and have a negative leading coefficient.

39. A student graphed $g(x) = -x(x + 4)^4(x - 1)^2$ and plotted the x-intercepts at (−1, 0), (0, 0), and (4, 0). What error did the student make?

40. A rectangular box has a length that is 3 centimeters less than the width, and a height that is 8 centimeters less than the width. Write a function for the volume of the box. What do the zeros of this function represent?

In Exercises 41–44, write the equation in factored form of each polynomial function that has the given characteristics.

41.
- The function has zeros at 1 (multiplicity 3), 3 (multiplicity 2), and −1 (multiplicity 4).
- The graph passes through (0, −18).
- The equation has a positive leading coefficient.

42.
- The function has zeros at −4 (multiplicity 2), 0 (multiplicity 1), and 2 (multiplicity 1).
- The graph passes through (−2, −16).
- The equation has a negative leading coefficient.

43.
- The function has zeros at 5 (multiplicity 1), −0.5 (multiplicity 3), and 6 (multiplicity 1).
- The graph passes through (−1, 21).
- The equation has a negative leading coefficient.

44.
- The function has zeros at −6 (multiplicity 3), −1 (multiplicity 2), and 3 (multiplicity 2).
- The graph passes through (−2, 800).
- The equation has a negative leading coefficient.

45. Write the equation in general form of the polynomial function graphed here, given that the polynomial's degree is 3 and the graph passes through (1, 1).

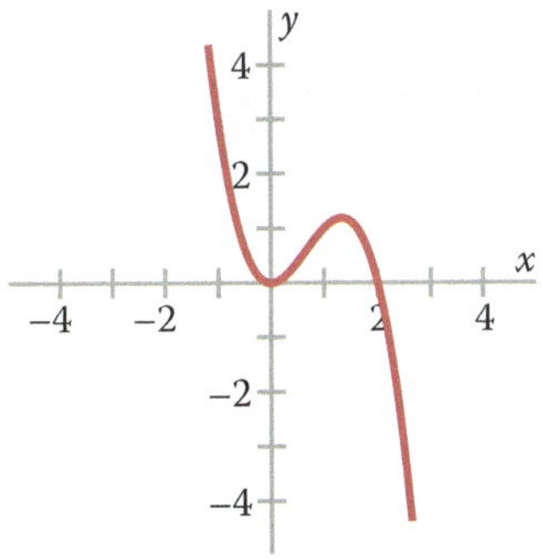

46. Write the general form equation of the function $g(x)$ that is the image of $f(x) = x^4$ after a reflection over the x-axis, a vertical stretch by a factor of 3, and translations 2 units right and 3 units up. Graph the function.

2.3 DIVIDING POLYNOMIALS

OBJECTIVES

- Divide polynomials using long division.
- Divide polynomials using synthetic division.
- Use the Remainder Theorem to evaluate polynomials.
- Use the Factor Theorem to determine whether binomials are factors of polynomials.

PREREQUISITE VOCABULARY TERMS

binomial
degree (of a polynomial)
factor (of a polynomial)
polynomial

2.3.1 Using Long Division with Polynomials

Division with polynomials is very similar to division with numbers. One method for dividing with numbers is to write the division expression as a fraction, then remove any common factors from the numerator and denominator. For example, 12 divided by 3 can be simplified as follows.

$$\frac{12}{3} = \frac{3(4)}{3} = \frac{\cancel{3}(4)}{\cancel{3}} = 4$$ *12 is the dividend, 3 is the divisor, and 4 is the quotient.*

Similarly, a quotient of polynomials can be written as a rational expression. Recall that a rational expression can be simplified (i.e., the polynomials can be divided) by factoring the numerator and denominator and removing any common factors.

Long division can be used to divide with numbers, as well as with polynomials. In long division with numbers, the result of dividing the dividend by the divisor is the quotient plus any remainder. For example, the result of dividing 13 by 4 is 3 with a remainder of 1.

$13 \div 4$

$$\begin{array}{r} 3 \\ 4\overline{)\,13} \\ -12 \\ \hline 1 \end{array}$$

3 –quotient; divisor–4; 13 –dividend; 1 –remainder

$$\frac{13}{4} = 3 + \frac{1}{4}$$

dividend–13; quotient–3; 1 –remainder; divisor–4

Also recall that long division can be checked by multiplying the quotient by the divisor and adding any remainder.

dividend = (quotient)(divisor) + remainder

Division of Polynomials

If $f(x)$ and $g(x)$ are polynomials, and $g(x) \neq 0$, then $\dfrac{f(x)}{g(x)} = q(x) + \dfrac{r(x)}{g(x)}$, where $q(x)$ is the quotient and $r(x)$ is the remainder.

The degree of $r(x)$ is less than the degree of $q(x)$, and $r(x)$ can equal 0.

Multiplying $\dfrac{f(x)}{g(x)} = q(x) + \dfrac{r(x)}{g(x)}$ by $g(x)$ gives $f(x) = g(x)q(x) + r(x)$.

$$f(x) = g(x)\,q(x) + r(x)$$

dividend: $f(x)$; divisor: $g(x)$; quotient: $q(x)$; remainder: $r(x)$

$$\frac{f(x)}{g(x)} = q(x) + \frac{r(x)}{g(x)}$$

dividend: $f(x)$; divisor: $g(x)$; quotient: $q(x)$; remainder: $r(x)$

EXAMPLE 1 Long Division of Polynomials

Use long division to divide $3x^3 + x^2 + 3$ by $x + 1$.

SOLUTION

The dividend is $3x^3 + x^2 + 3$, and the divisor is $x + 1$.

$$x+1\overline{)3x^3 + x^2 + 0x + 3}$$

Since the dividend does not contain an x-term, use 0x as a placeholder in the dividend.

To use long division, first divide the leading term in the dividend, $3x^3$, by the leading term in the divisor, x, to get the leading term of the quotient. So, $\frac{3x^3}{x} = 3x^2$ is the first term of the quotient. Next, multiply $3x^2$ (the first term of the quotient) by each term in the divisor, and then subtract this product from the dividend. Repeat this process until the degree of the difference is less than the degree of the divisor.

$$\begin{array}{r} 3x^2 \\ x+1\overline{)\;3x^3 + x^2 + 0x + 3} \\ -(3x^3 + 3x^2) \\ \hline -2x^2 + 0x \end{array}$$

The quotient's leading term is $\frac{3x^3}{x} = 3x^2$.

Multiply: $3x^2(x + 1) = 3x^3 + 3x^2$.
Subtract $3x^3 + 3x^2$ from the dividend, and bring down 0x.

The degree of the difference, $-2x^2 + 0x$, is 2. Continue dividing since the degree of the difference is not less than the degree of the divisor.

$$\begin{array}{r} 3x^2 - 2x \\ x+1\overline{)\;3x^3 + x^2 + 0x + 3} \\ -(3x^3 + 3x^2) \\ \hline -2x^2 + 0x \\ -(-2x^2 - 2x) \\ \hline 2x + 3 \end{array}$$

The quotient's 2nd term is $\frac{-2x^2}{x} = -2x$.

Multiply: $-2x(x + 1) = -2x^2 - 2x$.
Subtract $-2x^2 - 2x$, and bring down 3.

The degree of the difference, $2x + 3$, is 1. Continue dividing since the degree of the difference is not less than the degree of the divisor.

$$\begin{array}{r} 3x^2 - 2x + 2 \\ x+1\overline{)\;3x^3 + x^2 + 0x + 3} \\ -(3x^3 + 3x^2) \\ \hline -2x^2 + 0x \\ -(-2x^2 - 2x) \\ \hline 2x + 3 \\ -(2x + 2) \\ \hline 1 \end{array}$$

The quotient's 3rd term is $\frac{2x}{x} = 2$.

Multiply: $2(x + 1) = 2x + 2$.
Subtract $2x + 2$.

The degree of the difference, 1, is 0. The division is complete since the degree of the difference is less than the degree of the divisor.

The quotient is $3x^2 - 2x + 2$ with a remainder of 1.

Therefore, $\frac{3x^3 + x^2 + 3}{x+1} = 3x^2 - 2x + 2 + \frac{1}{x+1}$.

The quotient and remainder in **Example 1** can be checked using $f(x) = g(x)q(x) + r(x)$ (i.e., the dividend must be equal to the product of the divisor and the quotient, plus the remainder). Specifically, $3x^3 + x^2 + 3$ (the dividend) must be equal to the product of $x + 1$ (the divisor) and $3x^2 - 2x + 2$ (the quotient) plus 1 (the remainder).

$$(x + 1)(3x^2 - 2x + 2) + 1 = (3x^3 - 2x^2 + 2x + 3x^2 - 2x + 2) + 1 = 3x^3 + x^2 + 3 \checkmark$$

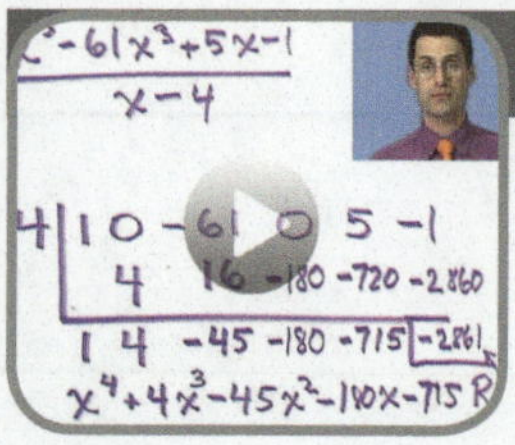

2.3.2 Using Synthetic Division with Polynomials

Synthetic division is a quick way of dividing polynomials, but it can be used only when the divisor can be written in the form $x - c$, where c is a real number not equal to 0. Only the coefficients are written when using synthetic division.

EXAMPLE 2 Using Synthetic Division

> ***NOTICE THAT***
>
> *Examples 1 and 2 show the same polynomials divided using the two different processes (long division and synthetic division) for comparison.*

Use synthetic division to divide $3x^3 + x^2 + 3$ by $x + 1$.

SOLUTION

The divisor is $x + 1$, or $x - (-1)$, so $c = -1$.

The dividend is $3x^3 + x^2 + 3$, or $3x^3 + x^2 + 0x + 3$. So, the coefficients are 3, 1, 0, and 3.

Write –1 outside of the box, and 3, 1, 0, and 3 inside of the box.

$$\begin{array}{r|rrrr} -1 & 3 & 1 & 0 & 3 \\ & & & & \\ \hline & & & & \end{array}$$

Bring down the first coefficient, 3. Then multiply –1 by 3. Write this product below the second coefficient (i.e., below the 1) inside the box.

$$\begin{array}{r|rrrr} -1 & 3 & 1 & 0 & 3 \\ & \downarrow & -3 & & \\ \hline & 3 & & & \end{array}$$

The coefficient of the quotient's squared term is 3.

Add 1 and –3.
Write this sum below the box.

$$\begin{array}{r|rrrr} -1 & 3 & 1 & 0 & 3 \\ & & -3 & & \\ \hline & 3 & -2 & & \end{array}$$

The coefficient of the quotient's linear term is –2.

Multiply –1 by –2. Write this product below the third coefficient (i.e., below the 0) inside the box.

$$\begin{array}{r|rrrr} -1 & 3 & 1 & 0 & 3 \\ & & -3 & 2 & \\ \hline & 3 & -2 & & \end{array}$$

Add 0 and 2.
Write this sum below the box.

$$\begin{array}{r|rrrr} -1 & 3 & 1 & 0 & 3 \\ & & -3 & 2 & \\ \hline & 3 & -2 & 2 & \end{array}$$

The coefficient of the quotient's constant term is 2.

Multiply –1 by 2. Write this product below the fourth coefficient (i.e., below the 3) inside the box.

$$\begin{array}{r|rrrr} -1 & 3 & 1 & 0 & 3 \\ & & -3 & 2 & -2 \\ \hline & 3 & -2 & 2 & \end{array}$$

> ***IMPORTANT***
>
> *The last number in the row below the box is the remainder. The rest of the numbers are the coefficients of the terms in the quotient. The degree of the first term in the quotient is always 1 less than the degree of the dividend. The degree of each consecutive term in the quotient decreases by 1.*

Add 3 and –2.
Write this sum below the box.

$$\begin{array}{r|rrrr} -1 & 3 & 1 & 0 & 3 \\ & & -3 & 2 & -2 \\ \hline & 3 & -2 & 2 & 1 \end{array}$$

The remainder is 1.

Use the numbers in the row below the box to write the quotient and remainder.

The degree of the quotient's first term is 2 because the degree of the dividend is 3.

$$\begin{array}{cccc} 3 & -2 & 2 & 1 \\ \downarrow & \downarrow & \downarrow & \downarrow \\ 3x^2 & -2x & 2 & \text{remainder} = 1 \end{array}$$

As in **Example 1**, the quotient is $3x^2 - 2x + 2$ with a remainder of 1.

So, $\dfrac{3x^3 + x^2 + 3}{x + 1} = 3x^2 - 2x + 2 + \dfrac{1}{x + 1}$.

Steps for Using Synthetic Division

1. Write the divisor in the form $x - c$ (c will be a positive or negative number).
2. Draw the bottom left corner of a box. Write the c-value on the left side of the box, and write the coefficients from the dividend inside the box, as shown below.

   c | *write the coefficients inside the box*

3. Bring down the first coefficient.
4. Multiply c by the first coefficient, and write this product below the second coefficient (inside the box).
5. Add the product to the coefficient above, and write this sum under the box.
6. Repeat step 4 until a product is added to the last coefficient.
7. Write the quotient using the sums as the coefficients, where the final sum in the list is the remainder.

IMPORTANT

Insert a "0" as a placeholder for the coefficient of any skipped term. For example, if the dividend is $5x^2 + 2$, write the coefficients 5, 0, and 2, since $5x^2 + 2 = 5x^2 + 0x + 2$.

2.3.3 The Remainder Theorem

The **Remainder Theorem** relates the process of division to evaluating polynomials.

Remainder Theorem

If the polynomial $f(x)$ is divided by $x - c$, then the remainder is the value $f(c)$.

For example, $f(5)$ is equal to the remainder when $f(x)$ is divided by $x - 5$. So, by the Remainder Theorem, a polynomial can be evaluated for some number c by dividing the polynomial (using long or synthetic division) by $x - c$.

EXAMPLE 3 Using the Remainder Theorem to Evaluate a Polynomial

Suppose $p(x)$ is a polynomial function such that $p(-2) = -14$, and the remainder is 1 when $p(x)$ is divided by $x - 3$. Use the Remainder Theorem to draw two conclusions about $p(x)$.

SOLUTION

If $p(-2) = -14$, then the remainder must be -14 when $p(x)$ is divided by $x + 2$.

If the remainder is 1 when $p(x)$ is divided by $x - 3$, then $p(3) = 1$.

2.3.4 The Factor Theorem

Recall that when f is a polynomial function and $f(c) = 0$, then c is a zero of the polynomial. Furthermore, if c is a zero of $f(x)$, then $x - c$ must be a factor of $f(x)$. For example, if 3 is a zero of $f(x)$, then $x - 3$ must be a factor of $f(x)$. The **Factor Theorem** relates a zero of a polynomial to a factor of the polynomial.

Factor Theorem

The binomial $(x - k)$ is a factor of a polynomial $f(x)$ if and only if $f(k) = 0$.

The Factor Theorem can be used to identify binomial factors of a polynomial. For example, to determine whether $(x + 7)$ is a factor of the polynomial function $f(x)$, find $f(-7)$. If $f(-7) = 0$, then $(x + 7)$ is a factor of $f(x)$. There are two methods for determining the value of $f(-7)$. One method is to evaluate $f(x)$ for $x = -7$ (i.e., substitute -7 into the function for x and simplify). The second method is to divide $f(x)$ by $(x + 7)$. By the Remainder Theorem, the remainder is equal to $f(-7)$. If the remainder is 0 when $f(x)$ is divided by $(x + 7)$, then $f(-7) = 0$, and therefore $(x + 7)$ is a factor of $f(x)$.

EXAMPLE 4 Using the Factor Theorem

Use synthetic division to show that $(x + 5)$ is a factor of $p(x) = x^3 + 5x^2 - 9x - 45$, and then factor the polynomial completely.

SOLUTION

By the Remainder Theorem, the remainder when $p(x)$ is divided by $(x + 5)$ is equal to $p(-5)$.

By the Factor Theorem, if $p(-5) = 0$, then $(x + 5)$ is a factor of $p(x)$.

So, if the remainder is 0 when $p(x)$ is divided by $(x + 5)$, then $p(-5) = 0$, and so $(x + 5)$ is a factor of $p(x)$.

Use synthetic division to find the remainder when $p(x)$ is divided by $(x + 5)$.

−5	1	5	−9	−45
		−5	0	45
	1	0	−9	0

Since the remainder is 0, $p(-5) = 0$, and therefore $(x + 5)$ is a factor of $p(x)$.

By the synthetic division result, we know that $p(x)$ divided by $(x + 5)$ is equal to $x^2 - 9$, or $\dfrac{p(x)}{x+5} = x^2 - 9$. Therefore, $p(x) = (x^2 - 9)(x + 5)$.

Factor $x^2 - 9$ to complete the factorization of $p(x)$.

$$p(x) = x^3 + 5x^2 - 9x - 45 = (x^2 - 9)(x + 5) = (x + 3)(x - 3)(x + 5)$$

REMEMBER

The factors can be written in any order because multiplication is commutative.

SECTION 2.3 EXERCISES

Warm Up

Simplify.

1. $\dfrac{x^2-25}{x+5}$

2. $\dfrac{2x^2+28x+98}{x+7}$

3. $\dfrac{2x^2-18}{x^2+6x+9}$

Just the Facts

Fill in the blanks.

4. When dividing polynomials, use either _____ division or _____ division.

5. When using synthetic division, only the _____ are written in the box. If a term is missing, _____ is written as a placeholder.

6. To determine whether a value is a zero of a function by synthetic division, check if the remainder is _____.

7. To evaluate a polynomial $f(x)$ at a value b, synthetically divide by _____. The remainder is the same as if you evaluated _____.

8. In $\dfrac{f(x)}{g(x)}=q(x)+\dfrac{r(x)}{g(x)}$, $f(x)$ is the _____, $g(x)$ is the _____, _____ is the quotient, and _____ is the remainder.

9. Polynomial long division can be checked by multiplying the _____ by the _____.

10. To synthetically divide $f(x)$ by $x+5$, write _____ in the left corner of the box.

Essential Skills

In Exercises 1–8, divide using long division.

1. $\dfrac{2x^3+3x^2+4x+4}{x+1}$

2. $\dfrac{5x^4+3x^3+2x-5}{x-11}$

3. $(x^3+4x-7)\div(x-2)$

4. $(2x^4-3x^2-12x-7)\div(x-2)$

5. $(2x^3+6x^2+5x+3)\div(x+2)$

6. $(4x^4+2x^3+2x-4)\div(x-3)$

7. $(3x^4+2x^3+9x-4)\div(x+2)$

8. $(-7x^5+5x^2+x+3)\div(x+3)$

In Exercises 9–16, divide using synthetic division.

9. $\dfrac{6x^3+5x^2-7x+12}{x-4}$

10. $\dfrac{x^4+6x^3-19x^2-144x-91}{x-5}$

11. $(3x^4+2x^3-2x^2+14x+14)\div(x+2)$

12. $(x^3-7x+6)\div(x+6)$

13. $(4x^4-46x^2-5x-105)\div(x-3)$

14. $(x^4+7x^3-19x^2-142x-89)\div(x-2)$

15. $(4x^4-3x^2+7x)\div(x+5)$

16. $(x^3-4x+3)\div(x+4)$

In Exercises 17–32, use the Remainder Theorem to draw a conclusion about each polynomial function.

17. $h(x)$ if $h(6)=-11$

18. $p(x)$ if $p(-1)=5$

19. $f(x)$ if $f(3)=10$

20. $s(x)$ if $s(-6)=4$

21. $m(x)$ if $m(-2)=-7$

22. $z(x)$ if $z(2)=8$

23. $q(x)$ if $q(-9)=0$

24. $w(x)$ if $w(-5)=3$

25. $g(x)$ if the remainder is 4 when $g(x)$ is divided by $x+1$

26. $f(x)$ if the remainder is -10 when $f(x)$ is divided by $x-4$

27. $h(x)$ if the remainder is 0 when $h(x)$ is divided by $x-8$

28. $s(x)$ if the remainder is 7 when $s(x)$ is divided by $x-3$

29. $k(x)$ if the remainder is 12 when $k(x)$ is divided by $x+9$

30. $p(x)$ if the remainder is 0 when $p(x)$ is divided by $x-2$

31. $j(x)$ if the remainder is -23 when $j(x)$ is divided by $x-10$

32. $w(x)$ if the remainder is 1 when $w(x)$ is divided by $x+7$

In Exercises 33–40, use the Factor Theorem to draw a conclusion about each polynomial function.

33. $g(x)$ if $g(-1)=-7$

34. $r(x)$ if $r(-9)=0$

35. $t(x)$ if $t(6)=0$

36. $h(x)$ if $h(8)=12$

37. $p(x)$ if $p(-2)=0$

38. $m(x)$ if $m(3)=0$

39. $r(x)$ if $r(4)=-9$

40. $s(x)$ if $s(-5)=-4$

Extensions

41. Use synthetic division to determine the value of k if $f(2) = -47$ and $f(x) = -3x^4 - 2x^2 + kx - 1$.

42. Divide $p(x) = 5x^5 + 1$ by $x + 6$. Express your answer as a quotient of a polynomial with a remainder.

43. Evaluate $g(-4)$ for $g(x) = x^6 - 3x^2 + 0.5x$.

44. Use synthetic division to determine if 3 is a solution to $h(x) = -4x^4 - 27x^3 + 10x^2 + 261x + 180$.

45. Is $2x - 3$ a factor of $h(x) = 2x^7 - 9x^6 + x^5 + 12x^4$? Hint: Synthetic division can be used only when the divisor is written in the form $x - c$.

46. True or False? $\sqrt{7}$ is a zero of $g(x) = x^4 - 3x^3 - 11x^2 + 21x + 28$.

47. Simplify the rational expression using either synthetic or long division, and state the domain.

$$\frac{x^4 + 4x^3 - 18x^2 - 59x + 12}{x^2 - x - 12}$$

In Exercises 48–52, divide using synthetic division.

48. $(9x^4 + 15x^3 - 3x^2 + 12) \div (3x + 3)$

49. $(x^4 + 5x^3 + 3x^2 - x - 5) \div (x^2 + 4)$

50. $(x^4 + 5x^3 - 4x^2 - x) \div (x^2 + 3x)$

51. $(x^5 + x^3 + x) \div (x^2 + x)$

52. $$\frac{\frac{1}{2}x^3 - \frac{1}{8}x^2 + \frac{3}{4}x + 1}{x - \frac{1}{2}}$$

In Exercises 53–56, use the Factor Theorem to determine which of the given binomials are factors of the polynomial.

53. $f(x) = 4x^4 - 21x^3 - 22x^2 + 117x + 90$; $(x - 3)$, $(x - 5)$, $(x - 3/4)$, $(x - 2)$, $(x + 2)$, $(x + 5)$, $(x + 3/4)$

54. $g(x) = 3x^4 - 19x^3 - 8x^2 + 124x + 8$; $(x + 4)$, $(x - 5)$, $(x - 2/3)$, $(x - 4)$, $(x + 2)$, $(x + 5)$, $(x - 8/3)$

55. $h(x) = 2x^4 - 5x^3 - 46x^2 + 69x + 180$; $(x + 4)$, $(x - 5)$, $(x - 3/2)$, $(x - 4)$, $(x + 3/2)$, $(x + 3)$, $(x + 9/2)$

56. $k(x) = 3x^4 - 10x^3 - 41x^2 + 68x + 60$; $(x + 4/3)$, $(x - 5)$, $(x - 2/3)$, $(x - 4/3)$, $(x + 2/3)$, $(x + 5)$, $(x - 3)$

2.4 REAL ZEROS OF POLYNOMIALS

OBJECTIVES

- Factor and find the real zeros of polynomials given a zero.
- Use the Rational Zero Theorem to find the real zeros of polynomials.
- Write the equation of a polynomial function given its zeros.
- Write the equation of a polynomial function given a zero, the degree, and a point on the graph.
- Use Descartes' Rule of Signs to determine the number of positive and negative real zeros of polynomials.
- Use the rules for upper and lower bounds.

PREREQUISITE VOCABULARY TERMS

degree (of a polynomial)
Factor Theorem
polynomial function
Remainder Theorem
synthetic division
zero (of a function)

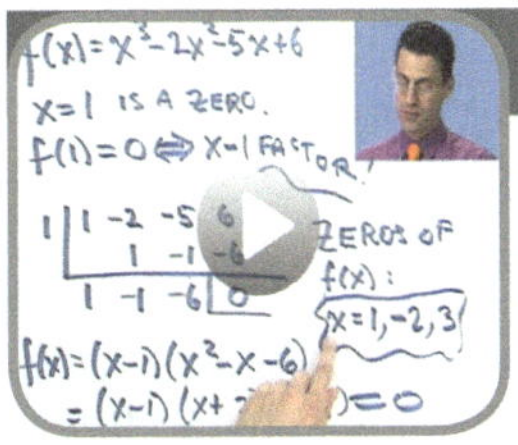

2.4.1 Factoring a Polynomial Given a Zero

When a polynomial function is written as a product of linear factors, its zeros can be read from the function. For example, consider the function $f(x) = (x + 1)(x - 3)^2(x - 5)$. This function has zeros at −1, 3 (with multiplicity 2), and 5. However, the zeros of a polynomial function cannot be immediately identified by looking at a polynomial in expanded form, such as $f(x) = x^3 + 4x^2 - 7x - 10$. So, factor an expanded polynomial function if possible to find its zeros.

The following example demonstrates a process for factoring an expanded polynomial function to find its zeros. This process utilizes the Factor Theorem and synthetic division.

EXAMPLE 1 Using a Zero to Factor a Polynomial and Find Its Zeros

Factor the polynomial $f(x) = x^4 + 3x^3 - 11x^2 - 3x + 10$ given that 2 and −5 are zeros. Then identify all of the zeros of $f(x)$.

SOLUTION

IMPORTANT

By the Remainder Theorem, since 2 is a zero of $f(x)$, the remainder when $f(x)$ is divided by $(x - 2)$ must be 0.

Since 2 and −5 are zeros, $(x - 2)$ and $(x + 5)$ must be factors of $f(x)$.

Use synthetic division to divide $f(x)$ by either one of its factors. Here, $f(x)$ is divided by $(x - 2)$ first.

$$\begin{array}{r|rrrrr} 2 & 1 & 3 & -11 & -3 & 10 \\ & & 2 & 10 & -2 & -10 \\ \hline & 1 & 5 & -1 & -5 & 0 \end{array}$$

So, $\dfrac{x^4 + 3x^3 - 11x^2 - 3x + 10}{x - 2} = x^3 + 5x^2 - x - 5$. It follows that $f(x) = (x - 2)(x^3 + 5x^2 - x - 5)$.

Now use synthetic division to divide $x^3 + 5x^2 - x - 5$ by $(x + 5)$.

$$\begin{array}{r|rrrr} -5 & 1 & 5 & -1 & -5 \\ & & -5 & 0 & 5 \\ \hline & 1 & 0 & -1 & 0 \end{array}$$

So, $\dfrac{x^3 + 5x^2 - x - 5}{x + 5} = x^2 - 1$.

Write all of the factors.

$$f(x) = (x - 2)(x + 5)(x^2 - 1)$$

The function is not fully factored. The final nonlinear factor, $(x^2 - 1)$, is the difference of two squares. So, $f(x) = (x - 2)(x + 5)(x - 1)(x + 1)$.

Therefore, the zeros of $f(x)$ are 2, −5, 1, and −1, each with multiplicity 1.

2.4.2 Using Zeros to Write a Polynomial Function

Recall that by the Factor Theorem, if c is a zero of $f(x)$, then $(x - c)$ must be a factor of $f(x)$. Use this fact, along with all of the zeros of a polynomial function, to write the equation of a polynomial function with those zeros. For example, if 2 and 3 are the only zeros of $f(x)$, and each has multiplicity 1, then $(x - 2)$ and $(x - 3)$ must be the only factors of $f(x)$. So, a polynomial function with zeros 2 and 3 is $f(x) = (x - 2)(x - 3)$, or $f(x) = x^2 - 5x + 6$.

EXAMPLE 2 Writing the Equation of a Polynomial Function

IMPORTANT

When the multiplicity of a zero is not given, assume the multiplicity is 1.

Write the simplest equation of a polynomial function $f(x)$ in expanded form where the only zeros are –7, –3, and 6.

SOLUTION

By the Factor Theorem, if –7, –3, and 6 are the only zeros of $f(x)$, then $(x + 7)$, $(x + 3)$, and $(x - 6)$ must be the only factors of $f(x)$. Multiply these factors to write the equation of the polynomial function in expanded form.

$$f(x) = (x + 7)(x + 3)(x - 6) = (x^2 + 10x + 21)(x - 6) = x^3 + 4x^2 - 39x - 126$$

So, $f(x) = x^3 + 4x^2 - 39x - 126$ is a function with zeros at –7, –3, and 6.

There are actually an infinite number of functions that have zeros at –7, –3, and 6. For example, multiplying the polynomial $f(x) = (x + 7)(x + 3)(x - 6)$ by 2 gives $f(x) = 2(x + 7)(x + 3)(x - 6)$, which does not change the values of the zeros, though the function is now different. Generally, any function of the form $f(x) = a(x + 7)(x + 3)(x - 6)$, where a is any nonzero real number, will have zeros at –7, –3, and 6.

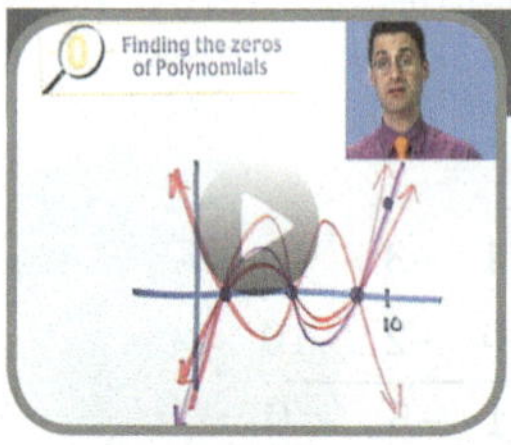

2.4.3 Using Zeros, Degree, and a Point to Write a Polynomial Function

When all of a function's zeros are known, along with the degree of the function and a point on the function's graph, then the unique function can be written, as shown in **Example 3**.

EXAMPLE 3 Writing the Equation of a Polynomial Function

Write the cubic polynomial function $f(x)$ in expanded form, where the only zeros are at –1, 2, and 0, given that $f(-2) = 8$.

SOLUTION

IMPORTANT

$f(x) = (x + 1)(x - 2)(x)$ is not necessarily the function, because the leading coefficient may not be 1.

The function's only zeros are –1, 2, and 0, so its only factors must be $(x + 1)$, $(x - 2)$, and x. If we let a represent the leading coefficient, then $f(x) = a(x + 1)(x - 2)(x)$. Use the fact that $f(-2) = 8$ to find the value of a.

$8 = a(-2 + 1)(-2 - 2)(-2)$	*Substitute $x = -2$ and $f(x) = 8$ into $f(x) = a(x + 1)(x - 2)(x)$.*
$8 = a(-1)(-4)(-2)$	*Simplify.*
$a = -1$	*Solve for a.*

The function is $f(x) = -(x + 1)(x - 2)(x)$. Multiply to write the function in expanded form.

$$f(x) = -(x + 1)(x - 2)(x) = -x^3 + x^2 + 2x$$

So, $f(x) = -x^3 + x^2 + 2x$ is the cubic function with zeros at –1, 2, and 0, where $f(-2) = 8$.

The answer in **Example 3** can be confirmed by graphing the function $f(x) = -x^3 + x^2 + 2x$ using a graphing calculator.

The graph of f crosses the x-axis at -1, 2, and 0, and passes through the point $(-2, 8)$.

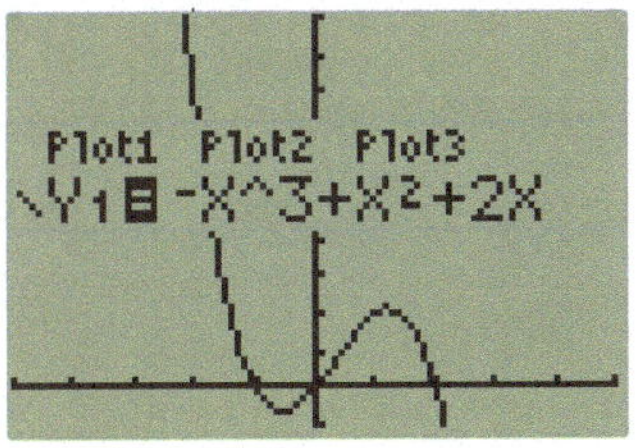

2.4.4 The Rational Zero Theorem

In Topic 2.4.1, we used given zeros to factor a polynomial (using synthetic division). In this topic, methods for finding the first zero (when no zeros are given) will be discussed.

Consider the relationship between the zeros and the expanded form of the following polynomial function.

Factored Form	**Zeros of $f(x)$**	**Expanded Form**
$f(x) = (x-2)(x-3)(x-5)$	2, 3, and 5	$f(x) = x^3 - 10x^2 + 31x - 30$

Notice that each of the function's zeros, 2, 3, and 5, is also a factor of the constant term of the expanded form, 30. This is not a coincidence. The zeros are all factors of the expanded form's constant term, because the constant term is produced by multiplying (2)(3)(5) when the factors are multiplied out.

From this example, we see that there is a relationship between a polynomial's constant term and its zeros. The leading coefficient can also play a role in the relationship, but this role is minimized when the value of the leading coefficient is 1, as is the case with $f(x) = x^3 - 10x^2 + 31x - 30$. The relationship between a polynomial's leading coefficient, its constant term, and its rational zeros is summarized in the **Rational Zero Theorem** (also named the Rational Root Theorem).

TIP

In other words, every rational zero of $f(x)$ will be a number of the form $\frac{p}{q}$ where p is a factor of the polynomial's constant term, and q is a factor of the polynomial's leading coefficient.

Rational Zero Theorem

If a polynomial function $f(x) = a_n x^n + a_{n-1}x^{n-1} + \ldots + a_1 x + a_0$ has integer coefficients, and $\frac{p}{q}$ is a rational zero of $f(x)$, then p must be a factor of a_0 (the polynomial's constant term), and q must be a factor of a_n (the polynomial's leading coefficient).

The Rational Zero Theorem gives a way of producing a finite list of rational numbers where each is a candidate to be a zero of the polynomial. The candidates in the list are all p/q where p is a factor of the polynomial's constant term and q is a factor of the polynomial's leading coefficient. Not all candidates will necessarily be zeros of the polynomial, and it is even possible that none of the candidates will be zeros. However, if $f(x)$ does have a rational zero, then that zero will be in the list of candidates. Each candidate c from the list can be tested to determine whether it is a zero of the function, using one of the two methods discussed previously.

TIP

Though synthetic division is typically easier, long division could be used for Method 2.

Determining Whether a Candidate is a Zero

Method 1: Evaluate the Function

Substitute the candidate c into the function and simplify. If $f(c) = 0$, then c is a zero of $f(x)$.

Method 2: Use Synthetic Division

Divide $f(x)$ by $(x - c)$ using synthetic division. If the remainder is 0, then c is a zero of $f(x)$.

EXAMPLE 4 Finding All Rational Zeros of a Polynomial Function

Make a list of all possible rational zeros for $f(x) = 2x^3 - 7x^2 + 4x + 4$ by the Rational Zero Theorem, and then use this list to find all of the function's rational zeros, if any exist.

SOLUTION

Start by identifying the constant term and the leading coefficient of $f(x)$, and then make a list of all factors of each.

Constant term = 4	**Leading coefficient = 2**
Factors: ±1, ±2, ±4	Factors: ±1, ±2
The p-values are ±1, ±2, or ±4.	*The q-values are ±1 or ±2.*

Make a list of all $\frac{p}{q}$.

$$\text{Possible rational zeros of } f(x)\text{: } \pm\frac{1}{1}, \pm\frac{1}{2}, \pm\frac{2}{1}, \pm\frac{2}{2}, \pm\frac{4}{1}, \pm\frac{4}{2} \qquad \frac{\textit{factor of 4}}{\textit{factor of 2}}$$

Simplify and remove the duplicates.

$$\text{Possible rational zeros of } f(x)\text{: } \pm 1, \pm\frac{1}{2}, \pm 2, \pm 4$$

IMPORTANT

If $f(c) = 0$, then c is a zero of $f(x)$.

TIP

By the Intermediate Value Theorem, since $f(1) = 3$ and $f(-1) = -9$, the function must have a zero between 1 and –1.

Use **Method 1** to determine which of the candidates (if any) are zeros of $f(x)$.

$f(1) = 2(1)^3 - 7(1)^2 + 4(1) + 4 = 3$ — *Since $f(1) \neq 0$, 1 is not a zero of $f(x)$.*

$f(-1) = 2(-1)^3 - 7(-1)^2 + 4(-1) + 4 = -9$ — *Since $f(-1) \neq 0$, –1 is not a zero of $f(x)$.*

$f\left(\frac{1}{2}\right) = 2\left(\frac{1}{2}\right)^3 - 7\left(\frac{1}{2}\right)^2 + 4\left(\frac{1}{2}\right) + 4 = \frac{9}{2}$ — *Since $f(1/2) \neq 0$, 1/2 is not a zero of $f(x)$.*

$f\left(-\frac{1}{2}\right) = 2\left(-\frac{1}{2}\right)^3 - 7\left(-\frac{1}{2}\right)^2 + 4\left(-\frac{1}{2}\right) + 4 = 0$ ✓ — *Since $f(-1/2) = 0$, –1/2 is a zero of $f(x)$.*

$f(2) = 2(2)^3 - 7(2)^2 + 4(2) + 4 = 0$ ✓ — *Since $f(2) = 0$, 2 is a zero of $f(x)$.*

$f(-2) = 2(-2)^3 - 7(-2)^2 + 4(-2) + 4 = -48$ — *Since $f(-2) \neq 0$, –2 is not a zero of $f(x)$.*

$f(4) = 2(4)^3 - 7(4)^2 + 4(4) + 4 = 36$ — *Since $f(4) \neq 0$, 4 is not a zero of $f(x)$.*

$f(-4) = 2(-4)^3 - 7(-4)^2 + 4(-4) + 4 = -252$ — *Since $f(-4) \neq 0$, –4 is not a zero of $f(x)$.*

So, the rational zeros of $f(x)$ are 2 and $-\frac{1}{2}$.

It is important to note that it *cannot* be assumed that all of the function's zeros have been found after checking the list of possible rational zero candidates. A polynomial function may have zeros that are not rational (i.e., are irrational or complex) in addition to (or instead of) the rational zeros. So, while 2 and –1/2 are the only *rational* zeros of $f(x)$ in **Example 4**, $f(x)$ may have additional nonrational zeros. Furthermore, it cannot be assumed that $(x - 2)$ and $\left(x + \frac{1}{2}\right)$ are the only factors of $f(x)$. These two known rational zeros can be used to factor $f(x)$ by using synthetic division, as demonstrated previously.

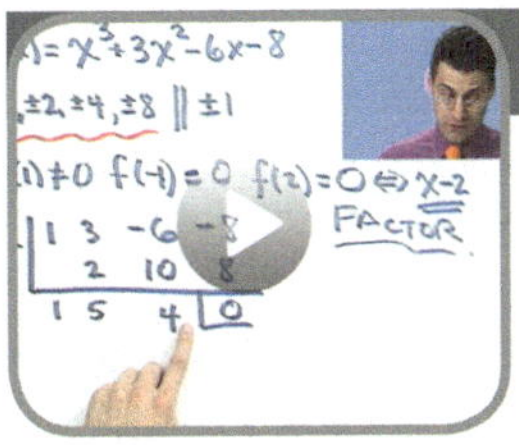

2.4.5 Finding the Zeros of a Polynomial from Start to Finish

In the preceding topic, we saw that the Rational Zero Theorem can be used to write a list of candidates for the possible rational zeros of a polynomial function. These candidates can then be tested (by evaluating the function or by using synthetic division) to determine whether or not each candidate is a zero of the function.

This process of testing the candidates is somewhat tedious and becomes even more tedious as the list of candidates grows.

For example, consider the function $f(x) = 6x^4 - 29x^3 - 53x^2 + 280x + 48$. The constant term is 48, and the leading coefficient is 6, so the list of possible rational zeros given by the Rational Zero Theorem is extensive.

Possible rational zeros of $f(x)$:

$$\pm 1, \pm\frac{1}{2}, \pm\frac{1}{3}, \pm\frac{1}{6}, \pm 2, \pm\frac{2}{3}, \pm 3, \pm\frac{3}{2}, \pm 4, \pm\frac{4}{3}, \ldots \qquad \frac{\textit{factor of } 48}{\textit{factor of } 6}$$

Testing all of the possible zeros is not a reasonable method for finding the zeros in this case.

The following steps can be used to factor a polynomial function and identify its rational zeros, instead of individually testing the candidates given by the Rational Zero Theorem.

Steps for Factoring a Polynomial Function $f(x)$ and Finding its Zeros

❶ Use the Rational Zero Theorem to list candidates for possible rational zeros, and test the candidates (by evaluating or using synthetic division) until a zero is identified. Use the zero to note the corresponding factor of $f(x)$.

❷ Divide (using synthetic division) the polynomial by the factor from step ❶.* Write the quotient, which is another factor of $f(x)$.

❸ Repeat steps ❶ and ❷ using the quotient from step ❷ as needed until the quotient is linear, quadratic, or an easily factored polynomial.

❹ Identify the zeros of $f(x)$ from the factored form of $f(x)$. (Use the Quadratic Formula as needed.)

**The division from step ❷ does not need to be completed if synthetic division was used in step ❶ to identify the zero.*

Note that in step ❶, only a few of the rational zero candidates need to be listed to start. Add additional candidates to the list as needed.

EXAMPLE 5 Finding All Zeros of a Polynomial Function

Factor $f(x) = x^4 - 13x^3 + 57x^2 - 99x + 54$, and find all of the function's rational zeros, if any exist.

SOLUTION

❶ Use the Rational Zero Theorem to list a few rational zero candidates. Test the candidates until a zero is identified.

TIP

The constant term is 54, so the p-values are ±1, ±2, ±3, ±6, ±9, ±18, ±27, and ±54. The leading coefficient is 1, so the q-values are ±1.

Possible rational zeros of $f(x)$: $\pm 1, \pm 2, \pm 3, \ldots$ *p/q*

$f(1) = (1)^4 - 13(1)^3 + 57(1)^2 - 99(1) + 54 = 0$ *Since f(1) = 0, 1 is a zero of f(x).*

The binomial $(x - 1)$ is a factor of $f(x)$, because 1 is a zero of $f(x)$.

❷ Divide $f(x)$ by $(x - 1)$ using synthetic division.

1	1	−13	57	−99	54
		1	−12	45	−54
	1	−12	45	−54	0

The quotient is $x^3 - 12x^2 + 45x - 54$.

So, $f(x) = (x - 1)(x^3 - 12x^2 + 45x - 54)$.

❸ Repeat steps ❶ and ❷ using the quotient from step ❷.

TIP

If none of those candidates are zeros, then more candidates would need to be found.

Start the testing with −1, and then test ±2 and ±3 as needed.

$(-1)^3 - 12(-1)^2 + 45(-1) - 54 = -112$ *−1 is not a zero of f(x).*

$(2)^3 - 12(2)^2 + 45(2) - 54 = -4$ *2 is not a zero of f(x).*

$(-2)^3 - 12(-2)^2 + 45(-2) - 54 = -200$ *−2 is not a zero of f(x).*

$(3)^3 - 12(3)^2 + 45(3) - 54 = 0$ *3 is a zero of f(x).*

The binomial $(x - 3)$ is a factor of $f(x)$, because 3 is a zero of $f(x)$.

Divide $x^3 - 12x^2 + 45x - 54$ by $(x - 3)$ using synthetic division.

3	1	−12	45	−54
		3	−27	54
	1	−9	18	0

The quotient is $x^2 - 9x + 18$.

So, $f(x) = (x - 1)(x - 3)(x^2 - 9x + 18)$.

The quotient, $x^2 - 9x + 18$, is quadratic, so move on to step ❹.

TIP

If a quadratic quotient is not easily factored, then the Quadratic Formula can be used to find the zeros.

❹ Identify the zeros from the factored form of $f(x)$.

$$f(x) = (x - 1)(x - 3)(x^2 - 9x + 18) = (x - 1)(x - 3)(x - 6)(x - 3)$$

So, the zeros of $f(x)$ are 1, 3 (multiplicity 2), and 6.

Note that all of the function's zeros were identified in **Example 5**. We know this because the zeros were identified from the fully factored polynomial.

2.4.6 Using Descartes' Rule of Signs

Consider again the polynomial function $f(x) = 6x^4 - 29x^3 - 53x^2 + 280x + 48$. We saw that the list of possible rational zeros is very long.

Possible rational zeros of $f(x)$: $\pm 1, \pm\frac{1}{2}, \pm\frac{1}{3}, \pm\frac{1}{6}, \pm 2, \pm\frac{2}{3}, \pm 3, \pm\frac{3}{2}, \pm 4, \pm\frac{4}{3}, \ldots$

Suppose you wanted to test the candidates. It would be helpful if the process of testing candidates could be somehow limited. **Descartes' Rule of Signs** provides a way to limit this process.

IMPORTANT

When using Descartes' Rule of Signs, a zero of multiplicity k counts as k zeros.

Descartes' Rule of Signs

Let $f(x)$ be a polynomial function with real coefficients.

Positive Real Zeros
The number of positive real zeros of $f(x)$ is equal to either the number of sign changes in $f(x)$ or the number of sign changes minus an even whole number.

Negative Real Zeros
The number of negative real zeros of $f(x)$ is equal to either the number of sign changes in $f(-x)$ or the number of sign changes in $f(-x)$ minus an even whole number.

To use Descartes' Rule of Signs, the polynomial function must first be written in general form (with its terms in descending order by degree). Then the number of changes in sign between the terms can be used to determine information regarding the number of real zeros. A *change in sign* occurs when consecutively positioned coefficients (ignoring any terms with a 0 coefficient) have opposite signs. For example, the polynomial function $f(x) = 6x^4 - 29x^3 - 53x^2 + 280x + 48$ has two changes in sign.

sign change #1 sign change #2

$$f(x) = \underset{\downarrow}{6x^4} \underset{\downarrow}{-29x^3} \underset{\downarrow}{-53x^2} \underset{\downarrow}{+280x} + 48$$

So, by Descartes' Rule of Signs, $f(x) = 6x^4 - 29x^3 - 53x^2 + 280x + 48$ must have either two positive real zeros (or one positive real zero with multiplicity 2), or no positive real zeros (since 2 minus an even whole number is 0). Evaluate $f(-x)$ to determine the possible number of negative real zeros.

$$f(-x) = 6(-x)^4 - 29(-x)^3 - 53(-x)^2 + 280(-x) + 48 = 6x^4 + 29x^3 - 53x^2 - 280x + 48$$

Since $f(-x)$ also has two sign changes, $f(x)$ must have either two negative real zeros (or one negative real zero with multiplicity 2), or no negative real zeros.

EXAMPLE 6 Using Descartes' Rule of Signs

Use Descartes' Rule of Signs to determine the possible number of positive and negative real zeros of $f(x) = x^4 + 14x^3 + 56x^2 + 34x - 105$.

SOLUTION

There is one sign change in $f(x)$, so $f(x)$ must have one positive real zero.

Evaluate $f(-x)$ to determine the possible number of negative real zeros.

$$f(-x) = (-x)^4 + 14(-x)^3 + 56(-x)^2 + 34(-x) - 105 = x^4 - 14x^3 + 56x^2 - 34x - 105$$

There are three sign changes in $f(-x)$.
Therefore, $f(x)$ has either three negative real zeros or one negative real zero.

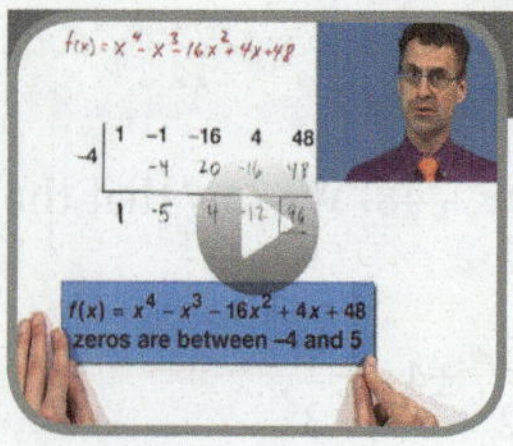

2.4.7 Upper and Lower Bounds

The rules for **upper and lower bounds** can also be used to limit the process of testing rational zero candidates.

DEFINITION

A number c is an **upper bound** for the real zeros of $f(x)$ if there are no real zeros greater than c.

A number c is a **lower bound** for the real zeros of $f(x)$ if there are no real zeros less than c.

The rules for upper and lower bounds are related to the sign pattern seen in the quotient row from synthetic division.

IMPORTANT

The rules for lower bounds are not held for all lower bounds. If a number c satisfies the lower bound rule, then it is a lower bound. However, not all lower bounds will satisfy the rules. In other words, a number that fails the test might still be a lower bound, but a number that passes the test definitely is a lower bound.

Rules for Upper and Lower Bounds

Let $f(x)$ be a polynomial function with real coefficients.

Upper Bound

If using synthetic division to divide $f(x)$ by $x - c$, where $c > 0$, results in a row with only nonnegative entries, then c is an upper bound for the real zeros of $f(x)$.

Lower Bound

If using synthetic division to divide $f(x)$ by $x - c$, where $c < 0$, results in a row with alternating positive and negative entries (where 0 counts as either positive or negative), then c is a lower bound for the real zeros of $f(x)$.

EXAMPLE 7 Using the Rules for Upper and Lower Bounds

Confirm that 2 is an upper bound and –15 is a lower bound for the real zeros of $f(x) = x^4 + 14x^3 + 56x^2 + 34x - 105$.

SOLUTION

If using synthetic division to divide $f(x)$ by $(x - 2)$ results in a row with only nonnegative entries, then 2 is an upper bound for $f(x)$.

Use synthetic division to divide $f(x)$ by $(x - 2)$.

2	1	14	56	34	–105
		2	32	176	420
	1	16	88	210	315

The entries are all nonnegative, so 2 must be an upper bound for the real zeros of $f(x)$.

If using synthetic division to divide $f(x)$ by $(x + 15)$ results in a row with alternating positive and negative entries, then –15 is a lower bound for $f(x)$.

Use synthetic division to divide $f(x)$ by $(x + 15)$.

–15	1	14	56	34	–105
		–15	15	–1065	15,465
	1	–1	71	–1031	15,360

The entries have alternating signs, so –15 must be a lower bound for the real zeros of $f(x)$.

SECTION 2.4 EXERCISES

Warm Up

True or False?

1. If $p(-5) = 0$, then $(x + 5)$ is a factor of $p(x)$.
2. The expression $(x^2 + 9)(x + 5)(x^2 - 4)$ is completely factored.
3. A polynomial function does not always have a y-intercept.

Just the Facts

Fill in the blanks.

4. According to the Rational Zero Theorem, the function $g(x) = x^4 + 8x^2 + 11x - 2$ has _____ possible rational zero candidates.
5. _____ can be used to determine the possible number of positive and negative real zeros that a function may have.
6. A polynomial with zeros at −4, 7, and 12 would have _____, _____, and _____ as factors.
7. The _____ is a way to limit the number of possible candidates for the real zeros of a function. The list of candidates comes from the function's _____ coefficient and its _____ term.
8. A real number c is a(n) _____ bound for the real zeros of f if there are no values less than c that are real zeros. A real number c is a(n) _____ bound if there are no values greater than c that are real zeros.
9. Suppose a polynomial function $f(x)$ is divided by $(x - b)$. If the final row of values obtained by synthetic division contains only positive values, then b is a(n) _____ bound for the real zeros of $f(x)$, whereas a row of alternating positive and negative values indicates that b is a(n) _____ bound.
10. The function $p(x) = x^3 + 4x^2 + 4x + 9$ has _____ possible positive real zeros by the Descartes' Rule of Signs.

Essential Skills

In Exercises 1–8, factor each polynomial with the given zero(s). Then identify all real zeros.

1. $g(x) = x^3 + 8x^2 + 11x - 20$; 1 is a zero
2. $f(x) = x^4 - 9x^3 + 12x^2 + 80x - 192$; 4 and −3 are zeros
3. $h(x) = x^4 - 2x^3 - 12x^2 + 40x - 32$; 2 and −4 are zeros
4. $f(x) = x^4 - 10x^3 + 20x^2 + 40x - 96$; 6 and −2 are zeros
5. $m(x) = x^4 - 12x^3 + 30x^2 + 100x - 375$; 5 and −3 are zeros
6. $f(x) = x^4 - 3x^3 - 7x^2 + 27x - 18$; 3 and −3 are zeros
7. $k(x) = x^4 - 9x^3 + 14x^2 + 36x - 72$; 6 and −2 are zeros
8. $f(x) = x^4 - 7x^3 + 9x^2 + 27x - 54$; 3 and −2 are zeros

In Exercises 9–16, write the equation of a polynomial function in expanded form given the only zeros of each function.

9. 0, 2, and −5
10. −1, 5, and 8
11. 1, 2, and 5
12. −4, 3, and 5
13. −6, −3, and −1
14. −2, 2, and 4
15. 1, 2, and 4
16. −4, 4, and 8

In Exercises 17–24, write the equation of the polynomial function in expanded form given the zeros, degree, and a point on the graph of the function.

17. a quadratic polynomial with zeros 3 and 9; $p(2) = -14$
18. a cubic polynomial with zeros −5, −2, and −1; $f(-3) = 4$
19. a cubic polynomial with zeros −5, −4, and 1; $g(-3) = -8$
20. a cubic polynomial with zeros −6, −5, and −1; $f(0) = 60$
21. a cubic polynomial with zeros −3, −2, and 1; $h(-5) = -36$
22. a cubic polynomial with zeros −4, −3, and 1; $f(-1) = -12$
23. a cubic polynomial with zeros −8, −1, and 8; $m(1) = -126$
24. a cubic polynomial with zeros −4, −2, and 3; $f(-3) = 18$

In Exercises 25–32, use the Rational Zero Theorem to make a list of all possible rational zeros for each polynomial function. Then use the list to find all rational zeros, if any exist.

25. $g(x) = x^3 - 8x^2 + 17x - 10$
26. $f(x) = 5x^3 - 9x^2 - 17x - 3$
27. $h(x) = 6x^3 + 11x^2 - 3x - 2$
28. $f(x) = 3x^3 - 11x^2 - 35x + 75$
29. $m(x) = 2x^3 - 10x^2 + 3x - 15$
30. $p(x) = 4x^3 + 12x^2 - x + 15$
31. $k(x) = 2x^3 - 7x^2 + 7x - 2$
32. $f(x) = 5x^3 + 7x^2 - 81x + 45$

In Exercises 33–40, find all real zeros of each function.

33. $h(x) = 6x^3 - 21x^2 - 12x$

34. $f(x) = 2x^4 - 21x^3 + 78x^2 - 119x + 60$

35. $f(x) = 6x^4 - 35x^3 + 13x^2 + 154x - 120$

36. $f(x) = 5x^4 - 57x^3 + 177x^2 - 167x + 42$

37. $m(x) = -4x^4 - 27x^3 + 10x^2 + 261x + 180$

38. $f(x) = 4x^4 + 5x^3 - 22x^2 - 20x + 24$

39. $f(x) = -2x^4 - 11x^3 + 20x^2 + 176x + 192$

40. $f(x) = 7x^4 - 59x^3 + 157x^2 - 141x + 36$

In Exercises 41–48, use Descartes' Rule of Signs to determine the possible number of positive and negative real zeros of each function.

41. $p(x) = 7x^4 - 2x^3 + 8x^2 - 11x + 6$

42. $f(x) = -5x^5 + 6x^4 + 3x^3 - 4x^2 + 5x + 9$

43. $h(x) = 2x^4 - 3x^3 + x^2 + 3x - 1$

44. $f(x) = -3x^5 - 6x^4 - 7x^3 + 7x^2 + 8x + 8$

45. $f(x) = 4x^5 - 5x^4 - 5x^3 + 5x^2 - 5x - 3$

46. $f(x) = -6x^5 + 8x^4 + 7x^3 - 8x^2 + 6x + 4$

47. $m(x) = -5x^5 - x^4 + x^3 - 4x^2 - 5x + 5$

48. $f(x) = 8x^5 + 6x^4 + 7x^3 + 5x^2 + 7x + 3$

In Exercises 49–56, what can be deduced about each function with the given synthetic division result?

49. $h(x) = x^3 + 10x^2 + 11x - 70;$

−11	1	10	11	−70
		−11	11	−242
	1	−1	22	−312

50. $f(x) = x^4 + 6x^3 - 24x^2 + 26x - 9;$

10	1	6	−24	26	−9
		10	160	1,360	13,860
	1	16	136	1,386	13,851

51. $g(x) = x^4 - 5x^3 - 8x^2 - 22x - 140;$

7	1	−5	−8	−22	−140
		7	14	42	140
	1	2	6	20	0

52. $f(x) = x^4 - 10x^3 - 20x^2 - 2x - 16;$

−2	1	−10	−20	−2	−16
		−2	24	−8	20
	1	−12	4	−10	4

53. $n(x) = x^4 + 8x^3 + 8x^2 + 298x - 10;$

8	1	8	8	298	−10
		8	128	1,088	11,088
	1	16	136	1,386	11,078

54. $f(x) = x^4 + 13x^3 + 40x^2 + 54x + 72;$

−3	1	13	40	54
		−3	−30	−30
	1	10	10	24

55. $f(x) = x^4 - 10x^3 - 12x^2 + 2x - 36;$

−2	1	−10	−12	2
		−2	24	−24
	1	−12	12	−22

56. $f(x) = x^4 + 2x^3 + 84x^2 + 588x - 7;$

6	1	2	84	588	−7
		6	48	792	8,280
	1	8	132	1,380	8,273

Extensions

57. A polynomial f was divided by $x + 9$, and the result was $x^4 - 9x^3 + 81x^2 - 726x + 6534 - \dfrac{58,807}{x+9}$. What was the original polynomial?

58. Use Descartes' Rule of Signs to determine the possible number of positive and negative real zeros of the function $f(x) = x(x - 2)^3(x + 7)$.

59. Suppose the given synthetic division applies to $p(x)$. Factor $p(x)$ completely.

9	2	−14	−94	462	540
		18	36	−522	−540
	2	4	−58	−60	0

60. Use the Rational Zero Theorem to list the possible rational zero candidates, and use this list to determine the real zeros of $f(x) = 6x^6 + x^5 - 92x^4 + 45x^3 + 184x^2 + 4x - 48$.

61. True or False? There exists only one function with a given set of real zeros.

62. Write a function f that passes through (4, 7), with exactly one negative real zero and exactly two positive real zeros.

63. Find all the real zeros of $h(x) = 2x^3 - 8.6x^2 - 2.1x + 1.35$, given that $x + 0.5$ is a factor.

2.5 COMPLEX ZEROS AND THE FUNDAMENTAL THEOREM OF ALGEBRA

OBJECTIVES

- Simplify powers of *i*.
- Add, subtract, and multiply complex numbers.
- Write quotients of complex numbers in the standard form of a complex number.
- Understand the Fundamental Theorem of Algebra.
- Find all real and complex solutions of polynomial equations.
- Understand the Conjugate Pair Theorem.
- Use the Conjugate Pair Theorem to find zeros of polynomial functions.

PREREQUISITE VOCABULARY TERMS

complex number
conjugates
imaginary unit
polynomial function
Quadratic Formula
synthetic division
zero (of a function)

Review of Imaginary and Complex Numbers

REMEMBER

The imaginary unit i is not a variable. Instead, it is a letter used to represent $\sqrt{-1}$.

Recall that the imaginary unit, represented by *i*, is defined to be $\sqrt{-1}$. The imaginary unit is used to write the square root of a negative number. For example, $\sqrt{-25}$ is equal to $5i$, which is called an imaginary number.

A complex number is an expression that can be written in the standard form $a + bi$, where a and b are real numbers and i is the imaginary unit. Complex numbers were introduced in Chapter P, where we saw that some quadratic equations have complex solutions. For example, the solutions of the quadratic equation $(x + 2)^2 = -9$ are complex numbers.

$$(x+2)^2 = -9$$

$$x+2 = \pm\sqrt{-9} \qquad \textit{Take } \pm \textit{ the square root of each side.}$$

$$x+2 = \pm 3i \qquad \textit{Simplify the square root.}$$

$$x = -2 \pm 3i \qquad \textit{Subtract 2 from each side.}$$

REMEMBER

The standard form of a complex number is $a + bi$.

Operations with complex numbers will be discussed in this section, then complex solutions of polynomial equations will be found.

2.5.1 Rewriting Powers of *i*

A "power of *i*", such as i^4, is a power where the base is the imaginary unit *i*. Consider the powers of *i* with consecutive integer exponents. Notice that simplification of these powers results in a pattern with four numbers.

$$i^2 = \left(\sqrt{-1}\right)^2 = -1 \qquad i^3 = i^2 \cdot i = (-1)i = -i$$

$$i^4 = i^2 \cdot i^2 = (-1)(-1) = 1 \qquad i^5 = i^4 \cdot i = (1)i = i$$

$$i^6 = i^4 \cdot i^2 = (1)(-1) = -1 \qquad i^7 = i^4 \cdot i^3 = (1)(-i) = -i$$

$$i^8 = i^4 \cdot i^4 = (1)(1) = 1 \qquad i^9 = i^4 \cdot i^4 \cdot i = (1)(1)(i) = i$$

So, $i^2 = -1$, $i^3 = -i$, $i^4 = 1$, $i^5 = i$, $i^6 = -1$, $i^7 = -i$, $i^8 = 1$, and $i^9 = i$.

It follows that all powers of *i* where the exponent is a whole number are equal to either -1, 1, i, or $-i$. No matter how large the exponent is, as long as the exponent is a whole number, the power of *i* can be simplified to -1, 1, i, or $-i$ by using the properties of exponents.

EXAMPLE 1 Simplifying Powers of *i*

Simplify. i^{38}

SOLUTION

Use the properties of exponents to simplify the power of *i*.

$$\begin{aligned} i^{38} &= i^{36} \cdot i^2 && \textit{Product of Powers Property} \\ &= (i^4)^9 \cdot i^2 && \textit{Power of a Power Property} \\ &= (1)^9(-1) && \textit{Simplify: } i^4 = 1 \textit{ and } i^2 = -1. \\ &= -1 && \textit{Multiply.} \end{aligned}$$

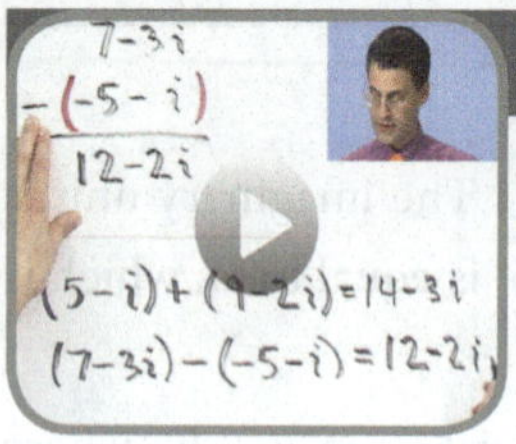

2.5.2 Adding and Subtracting Complex Numbers

Many of the properties of real numbers can be applied to complex numbers as well, such as the Associative Property (of Addition or Multiplication), Commutative Property (of Addition or Multiplication), and the Distributive Property. To add two complex numbers, add the real terms and add the imaginary terms separately. Similarly, to subtract two complex numbers, subtract the real terms and the imaginary terms separately. Basically, the process is the same as for adding and subtracting polynomials: combine the like terms. Once the expression is simplified, be sure to write the complex number in standard form, $a + bi$.

EXAMPLE 2 Adding and Subtracting Complex Numbers

Simplify. $(6 + 5i) - (2 - i) + (9 - i)$

SOLUTION

$$\begin{aligned} &(6 + 5i) - (2 - i) + (9 - i) \\ &= 6 + 5i - 2 + i + 9 - i && \textit{Distributive Property} \\ &= (6 - 2 + 9) + (5i + i - i) && \textit{Group the real terms, and group the imaginary terms.} \\ &= 13 + 5i && \textit{Combine the real terms, and combine the imaginary terms.} \end{aligned}$$

IMPORTANT

Be sure to distribute the negative to both terms within the subtracted complex number.
$-(2 - i) = -2 + i$

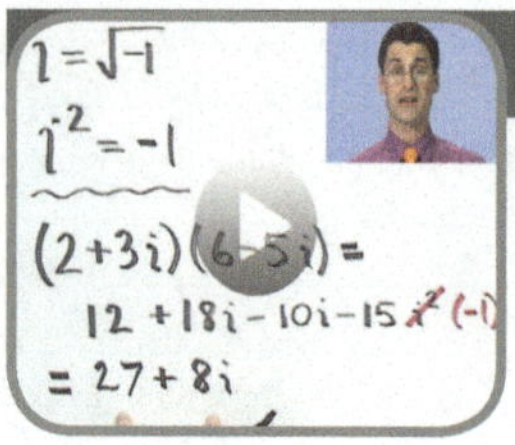

2.5.3 Multiplying Complex Numbers

Recall from Chapter P that two binomials can be multiplied by "FOILing," which is a way of applying the distributive steps in a particular order that makes the steps easier to remember.

$$(a + b)(c + d) = ac + ad + bc + bd$$
F O I L

The process of FOILing can also be used to multiply two complex numbers. When two complex numbers in standard form are FOILed, the final term will contain a factor of i^2. Note that this term can be simplified because $i^2 = -1$.

EXAMPLE 3 Multiplying Complex Numbers

Multiply. $(7-3i)(1-4i)$

SOLUTION

$$\begin{aligned}(7-3i)(1-4i) &= 7-28i-3i+12i^2 && \textit{FOIL}\\ &= 7-28i-3i+12(-1) && \textit{Simplify: } i^2=-1.\\ &= 7-28i-3i-12 && \textit{Multiply.}\\ &= -5-31i && \textit{Combine the like terms.}\end{aligned}$$

2.5.4 Dividing Complex Numbers

Writing a Quotient of Complex Numbers in Standard Form

A quotient of complex numbers, such as $\frac{c+di}{g+hi}$, is simplified when it is written in the standard form of a complex number, $a+bi$. Notice that a quotient of complex numbers is a fractional expression that contains a radical in the denominator because $i=\sqrt{-1}$.

$$\frac{c+di}{g+hi}=\frac{c+d\sqrt{-1}}{g+h\sqrt{-1}}$$

In Chapter P, we saw that a fractional expression containing a radical in the denominator can be rationalized by multiplying the fractional expression by some number equivalent to 1 that will remove the radical from the denominator. Recall that a denominator that is a sum or difference of a real number and a radical term, such as $\frac{c}{a+b\sqrt{n}}$ or $\frac{c}{a-b\sqrt{n}}$, is rationalized by multiplying the expression by the denominator's conjugate over itself. The conjugate of $a+b\sqrt{n}$ is $a-b\sqrt{n}$.

$$\frac{c}{a+b\sqrt{n}}\cdot\frac{a-b\sqrt{n}}{a-b\sqrt{n}}=\frac{c\left(a-b\sqrt{n}\right)}{\left(a+b\sqrt{n}\right)\left(a-b\sqrt{n}\right)}=\frac{c\left(a-b\sqrt{n}\right)}{a^2-b^2n} \quad \textit{The denominator is now rationalized.}$$

It follows that multiplying a quotient of complex numbers by the denominator's **complex conjugate** over itself will remove the radical (i.e., the imaginary number) from the denominator. In this way, a quotient of two complex numbers can be written in the standard form of a complex number, and thus simplified.

DEFINITION

Two complex numbers containing the same terms, where one is a sum of terms and the other is a difference of terms, are **complex conjugates**.

Complex Conjugates

$a+bi$ and $a-bi$

So, to simplify a quotient of complex numbers, multiply the quotient by the complex conjugate of the denominator, and write the expression in the standard form of a complex number, $a+bi$.

EXAMPLE 4 Dividing Complex Numbers

Simplify. $\dfrac{4+i}{3-2i}$

SOLUTION

The expression is a quotient of complex numbers, so multiply by the complex conjugate of the denominator over itself to simplify, and write the expression in the standard form of a complex number, $a + bi$. The complex conjugate of the denominator is $3 + 2i$.

$$\frac{4+i}{3-2i}\cdot\frac{3+2i}{3+2i}$$ *Multiply by the complex conjugate of the denominator over itself.*

$$=\frac{(4+i)(3+2i)}{(3-2i)(3+2i)}$$ *Multiply.*

$$=\frac{12+8i+3i+2i^2}{9+6i-6i-4i^2}$$ *FOIL in the numerator and in the denominator.*

$$=\frac{12+8i+3i+2(-1)}{9+6i-6i-4(-1)}$$ *Simplify: $i^2 = -1$.*

$$=\frac{12+8i+3i-2}{9+6i-6i+4}$$ *Multiply.*

$$=\frac{10+11i}{13}$$ *Combine the real terms, and combine the imaginary terms.*

$$=\frac{10}{13}+\frac{11}{13}i$$ *Write in the standard form of a complex number.*

2.5.5 The Fundamental Theorem of Algebra

So far in this course, the discussion of a polynomial function's zeros has focused on its real zeros. For the rest of this section, the discussion of a polynomial function's zeros will include complex zeros as well. We have seen polynomial functions that have one real zero, two real zeros, more than two real zeros, or even no real zeros. Polynomial functions that have no real zeros actually do have zeros, but they are complex zeros. This fact is stated in the **Fundamental Theorem of Algebra.**

Fundamental Theorem of Algebra

If P is a polynomial function of degree $n \geq 1$ with complex coefficients, then P has at least one complex zero.

It is important to point out that a complex number $a + bi$ may have a b-value equal to 0. In this case, the complex number is also a real number, since $a + bi = a + (0)i = a$. It follows that the real numbers are a subset of the complex numbers. In other words, any real number is also a complex number (but not all complex numbers are real numbers). So, the Fundamental Theorem of Algebra applies to polynomial functions with real coefficients as well as complex coefficients.

IMPORTANT

A complex coefficient may be a real coefficient, and a complex zero may be a real zero.

Therefore, by the Fundamental Theorem of Algebra, as long as a polynomial function has

- degree greater than or equal to 1, and
- coefficients that are all complex numbers (which includes real numbers),

then that polynomial function must have at least one complex zero (which may be a real zero).

The Relationship between a Polynomial's Degree and Number of Solutions

Previously in Chapter 2, we saw that every quadratic equation of the form $ax^2 + bx + c = 0$ has either one real solution (with multiplicity 2), two real solutions, or no real solution. Furthermore, when the value of the discriminant of $ax^2 + bx + c = 0$ is negative (i.e., $b^2 - 4ac < 0$), the quadratic equation has no real solution, but it does have two complex solutions. So, if complex solutions are included and a solution with multiplicity k is counted as k solutions, then we can say that every quadratic equation has exactly two complex solutions.

Recall that the degree of a quadratic equation is 2; a quadratic equation's degree is equal to its number of solutions. This is not a coincidence. When complex solutions are included, and the solutions with multiplicity k are counted as k solutions, then every polynomial equation with degree n also has n solutions.

> When complex solutions are included, and the zeros with multiplicity k are counted as k solutions, then every polynomial equation with degree n has n solutions.

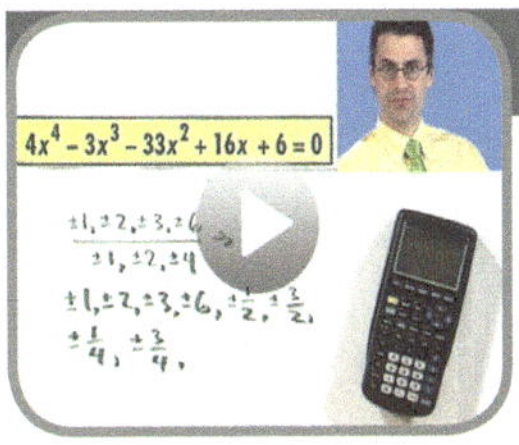

2.5.6 Finding All Solutions of a Polynomial Equation

Using Synthetic Division vs. Evaluating to Test Rational Zero Candidates

In all of the preceding examples, the rational zero candidates were tested by evaluating the function. Recall that synthetic division (or long division) is an alternative method for testing candidates (the candidate is a zero when the remainder is 0). When you are fairly sure that the candidate is a zero, synthetic division is a preferable method for testing.

Furthermore, to increase the chances that the candidate chosen for testing is a valid zero, graph the polynomial using a graphing calculator, and choose candidate(s) equal to the graph's x-intercept(s).

Solutions of a Polynomial Equation

A process similar to that used to find the real zeros of a polynomial function will be used to find the solutions of a polynomial equation in one variable.

Steps for Factoring a General Form Polynomial Equation and Finding Its Solutions

1. Use the Rational Zero Theorem to list candidates for possible rational zeros of the related function.
2. Graph the related function using a graphing calculator, and identify the x-intercepts.
3. Use synthetic division to test a zero that is equal to an x-intercept. If the remainder is 0, note the zero and use the quotient to test another zero. Repeat until the polynomial is written as a product of linear and quadratic factors.
4. Identify the solutions from the factored form of the polynomial, using the Quadratic Formula as needed.

IMPORTANT

The related polynomial function for a general form polynomial equation is found by replacing 0 with $f(x)$. For example, the related function for $ax^2 + bx + c = 0$ is $f(x) = ax^2 + bx + c$.

EXAMPLE 5 Finding All Solutions of a Polynomial Equation

Factor and find all of the solutions. $2x^4 + 7x^3 - 3x^2 - 5x - 1 = 0$

SOLUTION

The polynomial's degree is 4, so there are four solutions.

❶ Possible rational zeros of the related function $f(x) = 2x^4 + 7x^3 - 3x^2 - 5x - 1$:

$\pm 1, \pm \frac{1}{2}, \ldots$ $\quad \frac{\textit{factor of 1}}{\textit{factor of 2}}$ *There are only four rational zero candidates.*

CAUTION

The polynomial equation has four solutions, and there are four rational candidates, but all four candidates are not necessarily solutions, since the equation may have nonrational, nonreal, or repeated solutions.

❷ Using a graphing calculator, identify the x-intercepts of the related function.

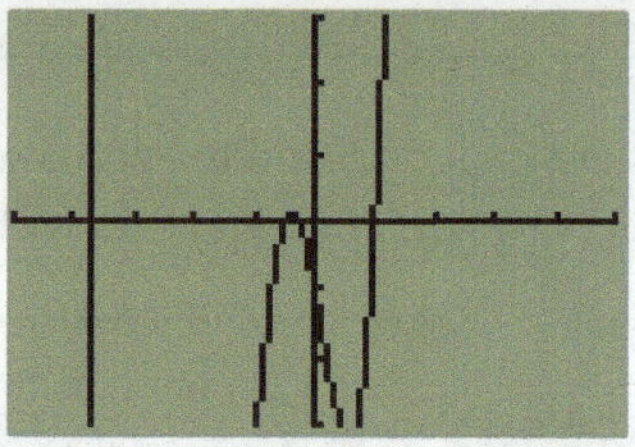

The graph has four x-intercepts, so all four of the equation's solutions must be real solutions. However, these real solutions may or may not be rational.

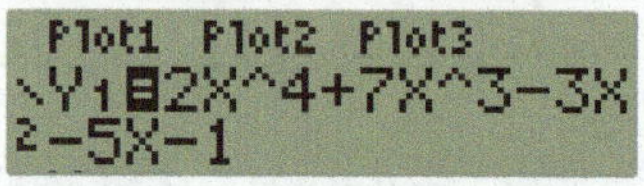

The graph appears to have an x-intercept at 1 and at $-\frac{1}{2}$.

❸ Test the zeros identified in step ❷ using synthetic division.

Here, we start with the x-intercept at 1.

$$\begin{array}{r|rrrrr} 1 & 2 & 7 & -3 & -5 & -1 \\ & & 2 & 9 & 6 & 1 \\ \hline & 2 & 9 & 6 & 1 & 0 \end{array}$$

The remainder is 0, so 1 is a solution, and (x – 1) is a factor.

The polynomial factors as $(x - 1)(2x^3 + 9x^2 + 6x + 1) = 0$.

Test the other zero from step ❷ (–1/2) in the quotient.

$$\begin{array}{r|rrrr} -\frac{1}{2} & 2 & 9 & 6 & 1 \\ & & -1 & -4 & -1 \\ \hline & 2 & 8 & 2 & 0 \end{array}$$

The remainder is 0, so –1/2 is a solution, and $\left(x + \frac{1}{2}\right)$ *is a factor.*

The polynomial factors as $(x-1)\left(x + \frac{1}{2}\right)(2x^2 + 8x + 2) = 0$.

❹ Solve $2x^2 + 8x + 2 = 0$ to find the remaining two solutions.

$$2x^2 + 8x + 2 = 0 \Rightarrow 2(x^2 + 4x + 1) = 0 \quad \textit{Factor out 2.}$$

Use the Quadratic Formula to find the remaining two solutions.

Quadratic Formula ▶

$$x = \frac{-b \pm \sqrt{b^2 - 4ac}}{2a}$$

$$x = \frac{-4 \pm \sqrt{16 - 4(1)(1)}}{2(1)} = \frac{-4 \pm \sqrt{12}}{2} = \frac{-4 \pm 2\sqrt{3}}{2} = -2 \pm \sqrt{3}$$

Write the factored equation. $\quad 2(x-1)\left(x + \frac{1}{2}\right)\left(x - \left(-2 + \sqrt{3}\right)\right)\left(x - \left(-2 - \sqrt{3}\right)\right) = 0$

The solutions are $1, -\frac{1}{2}, -2 + \sqrt{3}$, and $-2 - \sqrt{3}$.

2.5.7 Finding All Solutions of a Polynomial Equation: Another Example

The polynomial equation in the previous example had four solutions: two rational solutions and two irrational solutions. Polynomial equations can also have complex solutions, as seen in the following example.

EXAMPLE 6 Finding All Solutions of a Polynomial Equation

Factor and find all of the solutions. $x^4 - 3x^3 - x^2 - 27x - 90 = 0$

SOLUTION

The polynomial's degree is 4, so there are four solutions.

❶ Possible rational zeros of the related function $f(x) = x^4 - 3x^3 - x^2 - 27x - 90$:

$\pm1, \pm2, \pm3, \pm5, \pm6, \ldots$ *factor of 90/factor of 1*

❷ Using a graphing calculator, identify the x-intercepts of the related function.

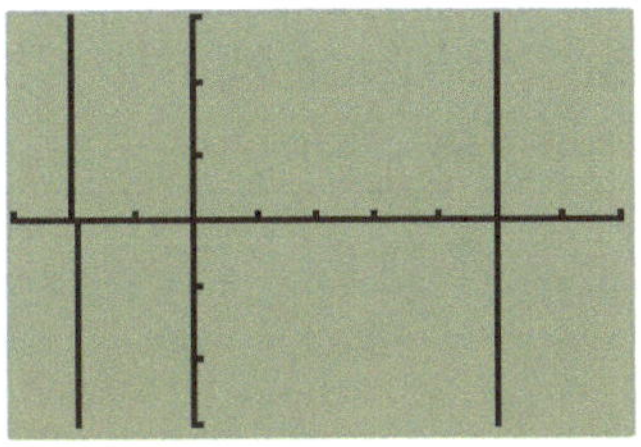

The graph has two x-intercepts, so two of the equation's four solutions must be real solutions.

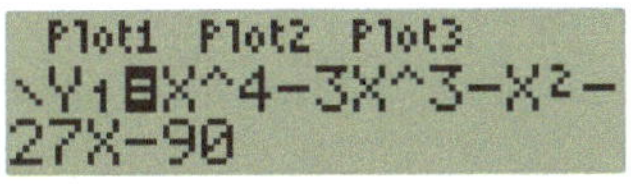

The graph appears to have x-intercepts at −2 and at 5.

❸ Test the zeros identified in step ❷ using synthetic division. If the remainder is 0, then 1 is a solution, and the quotient is a factor of the polynomial.

Test −2.

−2	1	−3	−1	−27	−90
		−2	10	−18	90
	1	−5	9	−45	0

The remainder is 0, so −2 is a solution, and (x + 2) is a factor.

The polynomial factors as $(x + 2)(x^3 - 5x^2 + 9x - 45) = 0$.

Test 5 in the quotient.

5	1	−5	9	−45
		5	0	45
	1	0	9	0

The remainder is 0, so 5 is a solution, and (x − 5) is a factor.

The polynomial factors as $(x + 2)(x - 5)(x^2 + 9) = 0$.

❹ Solve $x^2 + 9 = 0$ to find the remaining two solutions.

$$x^2 + 9 = 0 \Rightarrow x^2 = -9 \Rightarrow x = \pm\sqrt{-9} = \pm 3i$$

Write the fully factored equation. $(x + 2)(x - 5)(x + 3i)(x - 3i) = 0$

The solutions are −2, 5, −3*i*, and 3*i*.

REMEMBER

The quadratic equation $x^2 + 9 = 0$ can be solved using the square roots method because there is no b-term. Multiply the factors to check the answer.

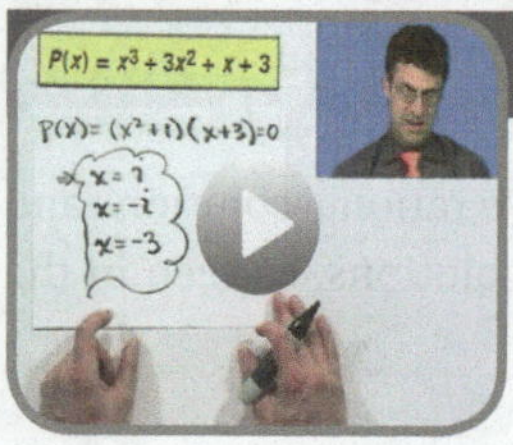

2.5.8 The Conjugate Pair Theorem

The solutions for the polynomial equation from the preceding example were -2, 5, $-3i$, and $3i$. Notice that the polynomial has two complex (nonreal) solutions, $-3i$ and $3i$, and that those two solutions are conjugates. You may have noticed that whenever a polynomial equation has a complex solution, that complex solution's conjugate is also a solution of the polynomial equation. In other words, complex solutions often come in pairs, as stated in the **Conjugate Pair Theorem**.

> ***Conjugate Pair Theorem***
>
> If $a + bi$ (with $b \neq 0$) is a complex zero of a polynomial function with real coefficients, then the conjugate $a - bi$ is also a complex zero of the polynomial function.

Note that the Conjugate Pair Theorem applies to polynomials where the coefficients are real numbers.

The Conjugate Pair Theorem is used in the following example to find all zeros of a polynomial function given one complex zero.

EXAMPLE 7 Using the Conjugate Pair Theorem to Find All Zeros of a Polynomial Function

Factor and find all of the zeros of $f(x) = x^3 - 4x^2 + 4x - 16$ given that $-2i$ is a zero.

SOLUTION

Use the Conjugate Pair Theorem to identify another zero of $f(x)$. The conjugate of $-2i$ is $2i$, so $2i$ is also a zero.

The polynomial's degree is 3, so there are three zeros. Therefore, there is only one remaining zero to find.

At this point, you could use the Rational Zero Theorem to identify rational zero candidates, and then test candidates by evaluating or using synthetic division, as was done in the previous examples. However, an alternative method is to divide the polynomial by the product of the factors resulting from the known zeros $2i$ and $-2i$.

Find the factors related to the zeros $2i$ and $-2i$, and then multiply those factors.

$$2i \text{ is a zero of } f(x) \Rightarrow (x - 2i) \text{ is a factor of } f(x)$$

$$-2i \text{ is a zero of } f(x) \Rightarrow (x + 2i) \text{ is a factor of } f(x)$$

The product of those factors is $(x - 2i)(x + 2i) = x^2 + 4$, so $x^2 + 4$ is a factor of $f(x)$.

Divide the polynomial $x^3 - 4x^2 + 4x - 16$ by $(x^2 + 4)$ to find another factor of $f(x)$.

Synthetic division cannot be used, because the divisor $(x^2 + 4)$ is not a linear binomial.

So, use long division.

$$\begin{array}{r} x - 4 \\ x^2 + 0x + 4 \overline{\big)\; x^3 - 4x^2 + 4x - 16} \\ \underline{-(x^3 + 0x + 4x)} \\ -4x^2 + 0x - 16 \\ \underline{-(-4x^2 + 0x - 16)} \\ 0 \end{array}$$

The quotient, $x - 4$, is linear, so the process of factoring is complete.

Write the factored polynomial.

$$f(x) = x^3 - 4x^2 + 4x - 16 = (x^2 + 4)(x - 4) = (x - 2i)(x + 2i)(x - 4)$$

The zeros are $-2i$, $2i$, and 4.

TIP

The given zero $-2i$ can be confirmed by evaluating $f(-2i)$.

$$\begin{aligned} f(-2i) &= (-2i)^3 - 4(-2i)^2 + 4(-2i) - 16 \\ &= -8i^3 - 4(4i^2) + 4(-2i) - 16 \\ &= -8(-i) - 4(4)(-1) + 4(-2i) - 16 \\ &= 8i + 16 - 8i - 16 \\ &= 0 \end{aligned}$$

Since $f(-2i) = 0$, $-2i$ is a zero of $f(x)$.

REMEMBER

$$\begin{aligned} &(x - 2i)(x + 2i) \\ &= x^2 + 2ix - 2ix - 4i^2 \\ &= x^2 - 4(-1) \\ &= x^2 + 4 \end{aligned}$$

TIP

The fact that the remainder is 0 confirms that $(x^2 + 4)$ is a factor of $x^3 - 4x^2 + 4x - 16$.

SECTION 2.5 EXERCISES

Warm Up

Simplify. Rationalize the denominator as needed.

1. $\dfrac{4}{4-\sqrt{5}}$
2. $\dfrac{7}{5+\sqrt{2}}$
3. $\left(4+\sqrt{3}\right)\left(4-\sqrt{3}\right)$

Just the Facts

Fill in the blanks.

4. The $\sqrt{-1}$ is the _____ unit, represented by _____.
5. $a + bi$ is a(n) _____ number, where a and b are _____ numbers and _____ is the imaginary unit.
6. If $a + bi$ is a complex zero of a polynomial function with real coefficients, then so is its conjugate, _____.
7. Any power of i where the exponent is a whole number is equal to _____, _____, _____, or _____.
8. It is important to remember when multiplying two complex numbers that _____ equals -1.
9. To simplify when dividing complex numbers, multiply by the denominator's complex _____ over itself.
10. The Fundamental Theorem of Algebra states that a polynomial of degree 4 with _____ coefficients will have at least _____ complex zero(s).

Essential Skills

In Exercises 1–22, simplify.

1. i^{14}
2. i^{31}
3. i^{37}
4. i^{30}
5. i^{23}
6. i^{26}
7. $(-12 - 14i) + (11 + 10i)$
8. $-2 + 3(2 - 5i) - 4(6 + i)$
9. $(4 - 2i) - 6(1 + 2i) + (7 + 8i)$
10. $-3(3 - 7i) - (5 - 9i) - 5(-7 + 2i)$
11. $5i + 4(11 - 12i) - 8(8 - 6i)$
12. $16 + 9(4 + 6i) - (3 + 10i) + (9 - 2i)$
13. $-22i + 10(4 + 6i) - 3(-6 - 4i)$
14. $4(9 + 6i) - 7(1 - 3i) - 2(-3 + 5i)$
15. $(3 + 4i)(1 + 2i)$
16. $i(9 + 8i)(3 - 2i)$
17. $i(2 - 5i)(1 - i)$
18. $(7 - 5i)(6 - 4i)$
19. $(4 - 7i)(-5 + 7i)$
20. $-3i(10 - 3i)(-1 - 6i)$
21. $-i(-9 + i)(4 - 4i)$
22. $5i(-6 - i)(7 - 5i)$

In Exercises 23–30, write each complex number in standard form.

23. $\dfrac{i}{6+i}$
24. $\dfrac{-6+3i}{-5-5i}$
25. $\dfrac{1-i}{-2-3i}$
26. $\dfrac{3-4i}{5+2i}$
27. $\dfrac{-4+6i}{-3+4i}$
28. $\dfrac{-2+7i}{8+i}$
29. $\dfrac{-9-2i}{4-7i}$
30. $\dfrac{5-8i}{-1-7i}$

In Exercises 31–46, find all the solutions of each equation.

31. $x^4 - 3x^3 - 5x^2 + 13x + 6 = 0$
32. $3x^4 - 8x^3 - 16x^2 + 24x + 32 = 0$
33. $3x^4 - 11x^3 - 11x^2 + 15x + 4 = 0$
34. $2x^4 - 9x^3 - 27x^2 + 80x + 150 = 0$
35. $2x^4 - 15x^3 + 17x^2 + 42x - 36 = 0$
36. $4x^4 - 11x^3 - 12x^2 + 26x + 20 = 0$
37. $2x^4 - 21x^3 + 37x^2 + 145x - 275 = 0$
38. $3x^4 - 28x^3 + 44x^2 + 84x - 135 = 0$
39. $x^4 - x^3 + 3x^2 - 9x - 54 = 0$
40. $x^4 + x^3 + 23x^2 + 25x - 50 = 0$
41. $x^4 - 2x^3 + x^2 - 8x - 12 = 0$
42. $x^4 - x^3 + 2x^2 - 4x - 8 = 0$
43. $x^4 - 2x^3 + 6x^2 - 18x - 27 = 0$

44. $x^4 - 2x^3 + 13x^2 - 32x - 48 = 0$

45. $x^4 - x^3 + 10x^2 - 16x - 96 = 0$

46. $x^4 + x^3 + 7x^2 + 9x - 18 = 0$

In Exercises 47–54, find all the zeros of each function with the given zero.

47. $p(x) = x^3 - x^2 + 25x - 25$, zero: $5i$

48. $f(x) = x^3 + 6x^2 - 11x + 40$, zero: $1 - 2i$

49. $g(x) = x^3 + 5x^2 + 12x + 182$, zero: $1 + 5i$

50. $f(x) = x^3 + 2x^2 - 14x + 272$, zero: $3 - 5i$

51. $m(x) = x^3 - 11x^2 + 44x - 60$, zero: $4 - 2i$

52. $f(x) = x^3 + 4x^2 - 19x + 104$, zero: $2 - 3i$

53. $d(x) = x^3 - x^2 - 6x + 72$, zero: $3 - 3i$

54. $f(x) = x^3 - 6x^2 + 28x - 40$, zero: $2 - 4i$

Extensions

55. Suppose $f(x)$ is a polynomial function with real coefficients and zeros $-i$, $2 + 3i$, $\sqrt{5}$, and $\frac{1-i}{3}$. Find three additional zeros of $f(x)$.

56. Simplify. i^{401}

In Exercises 57–63, perform the indicated operation. Write the result in standard form.

57. $\frac{1+i}{3-i} + \frac{5}{i-4}$

58. $\frac{-2-3i}{i} - \frac{6+i}{5-i}$

59. $(-5 - 8i^9)(3 + 7i^{19})$

60. $-3 + \frac{5i}{2-i} + \frac{i}{1+4i}$

61. $\frac{6-7i}{9+2i} \div \frac{-5+3i}{-9-2i}$

62. $(4 + 7i^8)(-2 - 3i^{15})$

63. $12 - \frac{2+i}{3+2i} + \frac{3i}{5-i}$

64. Find all the zeros of $p(x) = x^4 + x^3 - 29x^2 + 71x - 140$, and write the polynomial as a product of linear factors.

65. Write a polynomial function with real coefficients that has i, $2 + 3i$, $3 - 2i$, and 4 as zeros.

66. Let $p(x)$ be a cubic function with a leading coefficient of 3, where $p(5) = 0 = p(1 + i)$. Write an equation for $p(x)$.

CHAPTER 2 REVIEW EXERCISES

Section 2.1 Quadratic Functions and Models

1. Graph $g(x) = -2x^2 + 1$. Identify the parabola's vertex and axis of symmetry, then state whether the vertex is a minimum or maximum.
2. Identify the vertex of $f(x) = 5x^2 + 10x + 8$ by writing the equation in the standard form of a quadratic function.
3. Write $p(x) = -2x^2 + 12x - 19$ in the standard form of a quadratic function after a translation left 1 unit and down 4 units. Then graph.
4. Write the general form equation of the parabola that passes through $(-1, -3)$ with vertex at $(-2, 1)$.
5. Find the maximum or minimum value of $h(x) = -5x^2 + 3x - 7$.

Section 2.2 Polynomial Functions and Their Graphs

6. Describe the end behavior of the graph of the function. $d(x) = -2x^3 + 7x - 3x^5 + 8$
7. Graph. $m(x) = 2(x - 2)^3 + 1$
8. Find all real zeros of the function $f(x) = -\frac{1}{4}(x+5)(x+4)^2(4x+1)$, and state their multiplicity.
9. Graph. $w(x) = -(x + 2)^2 (x - 1)^2$
10. Identify the type and number of local extrema for the function.

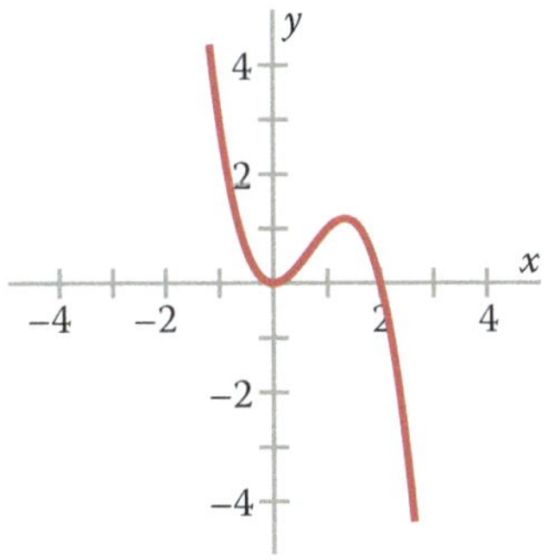

Section 2.3 Dividing Polynomials

11. Divide using long division. $\dfrac{5x^3 - 6x^2 + 8}{x - 4}$

In Exercises 12–13, divide using synthetic division.

12. $\dfrac{x^4 - 4x^3 - 5x^2 + 36x - 36}{x - 3}$
13. $\dfrac{3x^3 - 4x + 10}{x - 2}$
14. By the Remainder Theorem, what can be said about the polynomial function $f(x)$ if the remainder is 12 when $f(x)$ is divided by $x + 9$?
15. By the Factor Theorem, what can be said about the polynomial function $g(x)$ if $g(3) = 0$?

Section 2.4 Real Zeros of Polynomials

16. Factor $h(x) = x^3 + 4x^2 - 17x - 60$ given that one real zero is -3. Then identify all real zeros.
17. Write the cubic polynomial function $m(x)$ in expanded form, where the only zeros are at -2, -3, and -1, given that $m(4) = 210$.
18. Make a list of all possible rational zeros for $f(x) = 2x^3 - x^2 - 38x - 35$, and then use this list to find all of the function's rational zeros, if any exist.
19. Factor $n(x) = x^3 - 6x^2 - x + 30$, and find all of the function's rational zeros, if any exist.
20. Use Descartes' Rule of Signs to determine the possible number of positive and negative real zeros of $g(x) = 3x^5 + 3x^4 - 5x^3 + 4x^2 - 4x + 1$.

Section 2.5 Complex Zeros and the Fundamental Theorem of Algebra

In Exercises 21–22, simplify and write each complex number in standard form.

21. $i(3 - 7i)(2 + i)$
22. $\dfrac{8 - 5i}{8 + 9i}$

In Exercises 23–24, find all solutions of each equation.

23. $x^4 + x^3 - 18x^2 - 10x + 8 = 0$
24. $x^4 - 2x^3 + 13x^2 - 32x - 48 = 0$
25. Find all the zeros of $p(x) = x^4 - 4x^3 - x^2 - 16x - 20$ given that $2i$ is a zero.

RATIONAL FUNCTIONS AND CONICS

Chapter

3

3.1 GRAPHING RATIONAL FUNCTIONS

OBJECTIVES

- Use transformations to graph rational functions.
- Find the domain and range of rational functions.
- Graph rational functions having vertical and horizontal asymptotes.
- Graph rational functions containing a hole.
- Graph rational functions having oblique asymptotes.

PREREQUISITE VOCABULARY TERMS

function

rational expression

***x*-intercept**

***y*-intercept**

zero (of a function)

A function that can be expressed as a polynomial divided by a nonzero polynomial is called a **rational function.**

DEFINITION

A **rational function** $r(x)$ is a quotient of polynomials, $r(x) = \dfrac{f(x)}{g(x)}$, where f and g are polynomials such that $g(x) \neq 0$.

A rational function is a quotient of polynomials. Recall that the domain of any polynomial function (such as a linear or quadratic function) is all real numbers. However, the domain of a rational function may not be all real numbers when the denominator contains a variable because any x-value that results in a denominator equal to 0 must be excluded from the function's domain.

3.1.1 Graphing Basic Rational Functions

Recall that the graph of a polynomial function is a continuous line or smooth curve. The graph of a rational function may not be one continuous curve. When the graph of a rational function is not one continuous curve, the graph has a **discontinuity**, which is a point on the plane at which the function is undefined. Three examples of graphs of rational functions with one or more discontinuities are shown in the following table.

Examples of Graphs of Rational Functions

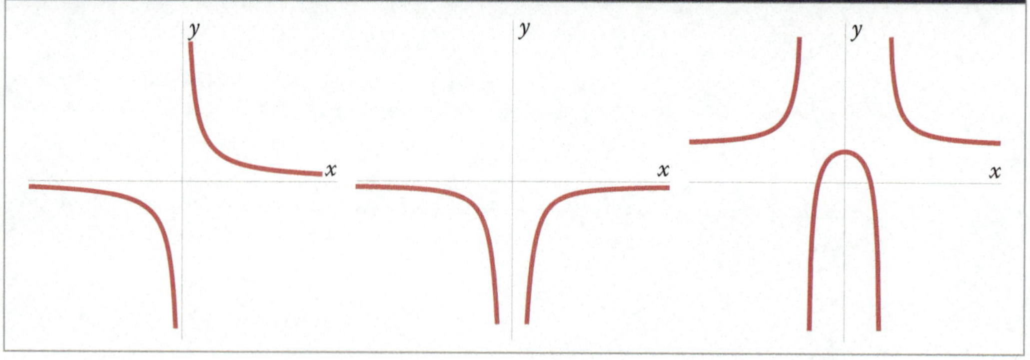

One method for graphing a rational function is to make a table of ordered pairs for the function. This method is demonstrated in **Example 1**. In this example, pay close attention to x-values that are excluded from the domain and to the behavior of the graph near these excluded values.

EXAMPLE 1 Graphing a Rational Function Using a Table

Graph the function $f(x) = \frac{1}{x}$.

SOLUTION

❶ Choose several x-values and make a table of ordered pairs.

x	-5	-4	-3	-2	-1	0	1	2	3	4	5
$f(x)$	$-\frac{1}{5}$	$-\frac{1}{4}$	$-\frac{1}{3}$	$-\frac{1}{2}$	-1	UND	1	$\frac{1}{2}$	$\frac{1}{3}$	$\frac{1}{4}$	$\frac{1}{5}$

❷ Plot the ordered pairs from the table on a coordinate plane. Sketch the graph through these ordered pairs.

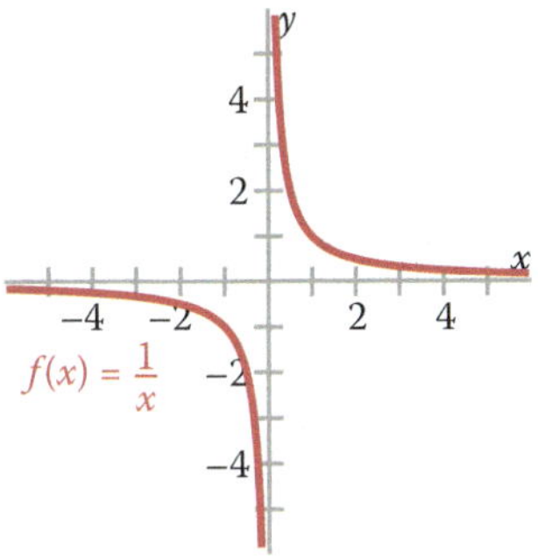

REMEMBER

Complete the table of ordered pairs by substituting each x-value into the given function to find the corresponding y-value.

The left and right ends of the curves in the graph in **Example 1** get closer and closer to the x-axis the further the graph is followed left or right, but those ends never intersect the x-axis (i.e., the line $y = 0$). The same phenomenon occurs at the top and bottom of the graph as the curves approach the y-axis (i.e., the line $x = 0$). The domain and range of the graph in **Example 1** and its behavior near $x = 0$ are examined in the following table.

Examining the Graph of $f(x) = 1/x$

	Description	Notation
Domain	All real numbers except $x = 0$	$\{x \mid x \neq 0\}$
Range	All real numbers except $y = 0$	$\{y \mid y \neq 0\}$
Behavior Near $x = 0$	As the x-values approach 0 from the left, $f(x)$ is heading down toward negative infinity.	as $x \to 0^-, f(x) \to -\infty$
	As the x-values approach 0 from the right, $f(x)$ is heading up toward positive infinity.	as $x \to 0^+, f(x) \to \infty$

Introducing Asymptotes

When the graph of a function approaches a line, this line is known as an **asymptote**. There are several types of asymptotes including **horizontal asymptotes** and **vertical asymptotes.**

DEFINITION

An **asymptote** is a line that a graph is heading toward and approaches more and more closely. A **horizontal asymptote** is an asymptote that is a horizontal line. A **vertical asymptote** is an asymptote that is a vertical line.

The graph of $f(x) = \frac{1}{x}$ has one vertical asymptote at $x = 0$ and one horizontal asymptote at $y = 0$ (Figure 3.1a).

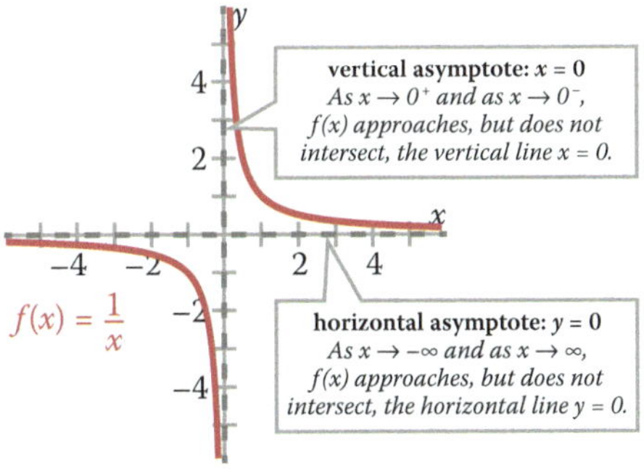

Figure 3.1a

WHAT'S THE BIG IDEA?

Why might the graph of a rational function have one or more asymptotes, but the graph of a polynomial function is guaranteed to have none?

Graphing a Rational Function by Translating $f(x) = \frac{1}{x}$

REMEMBER

Vertical translation k units: $(x, y) \rightarrow (x, y + k)$

Horizontal translation h units: $(x, y) \rightarrow (x + h, y)$

If a rational function can be expressed as $r(x) = \frac{1}{x-h} + k$ where h and k are any real numbers, then the graph of $r(x)$ is a translation of $f(x) = \frac{1}{x}$.

Vertical Translations	Horizontal Translations
$k > 0 \Rightarrow$ the graph of $r(x)$ is the graph of $f(x)$ translated $\lvert k \rvert$ units up	$h > 0 \Rightarrow$ the graph of $r(x)$ is the graph of $f(x)$ translated $\lvert h \rvert$ units right
$k < 0 \Rightarrow$ the graph of $r(x)$ is the graph of $f(x)$ translated $\lvert k \rvert$ units down	$h < 0 \Rightarrow$ the graph of $r(x)$ is the graph of $f(x)$ translated $\lvert h \rvert$ units left

EXAMPLE 2 Using a Translation to Graph a Rational Function

Graph each function by transforming the graph of $f(x) = \frac{1}{x}$.

A. $y = \frac{1}{x} - 1$ **B.** $y = \frac{1}{x+1}$ **C.** $y = 2 + \frac{1}{x-1}$

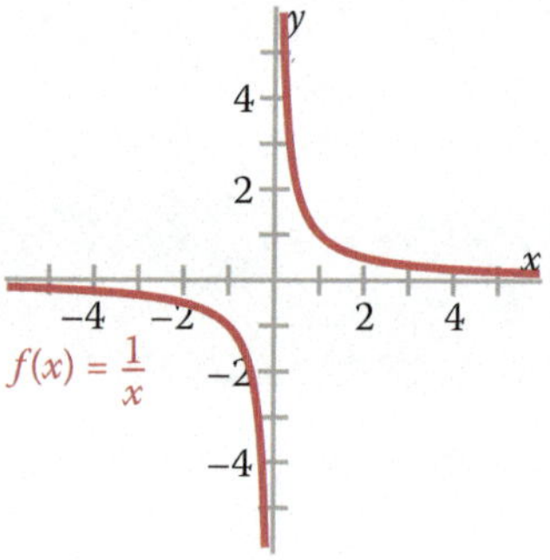

SOLUTION

NOTICE THAT

Vertical asymptotes do not change when a graph is horizontally translated. Horizontal asymptotes do not change when a graph is vertically translated.

IMPORTANT

In Part B, $h = -1$ because $x + 1 = x - (-1)$.

Apply any translations or reflections (identified from the equation) to the graph of $f(x) = \frac{1}{x}$.

A. $k = -1 \Rightarrow$ translate $f(x)$ down 1 unit

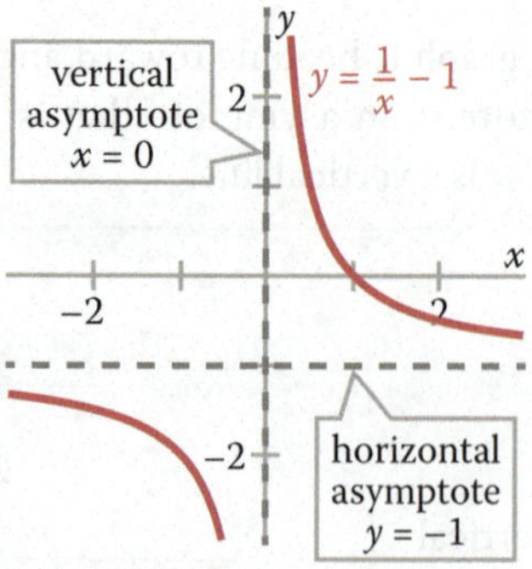

The horizontal asymptote is shifted down 1 unit.

B. $h = -1 \Rightarrow$ translate $f(x)$ left 1 unit

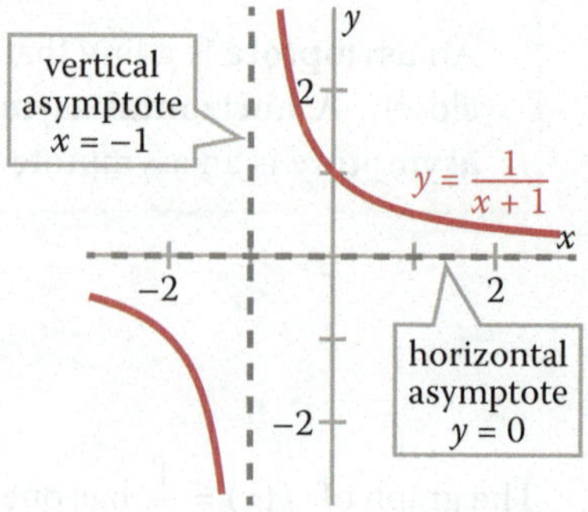

The vertical asymptote is shifted left 1 unit.

C. $h = 1$ and $k = 2 \Rightarrow$ translate $f(x)$ right 1 unit, up 2 units

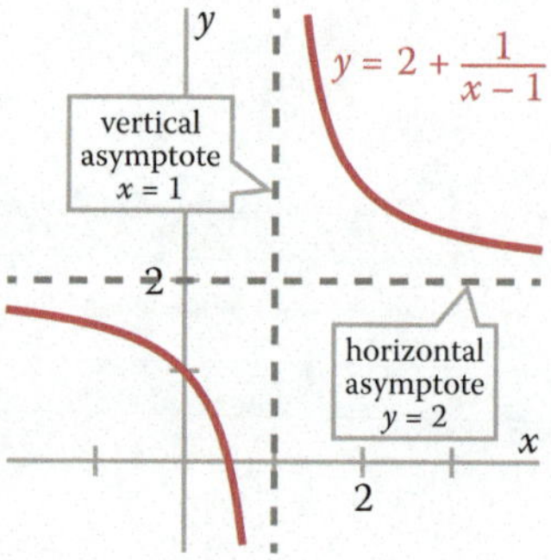

The horizontal asymptote is shifted up 2 units and the vertical asymptote is shifted right 1 unit.

EXAMPLE 3 Identifying the Domain and Range of a Rational Function

Write the domain and range of the function from **Example 2A** in interval notation.

SOLUTION

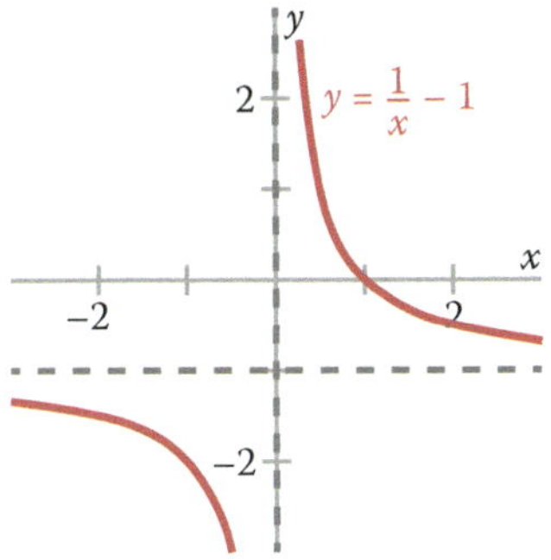

Domain: $(-\infty, 0) \cup (0, \infty)$

The only value excluded from the domain is at the vertical asymptote, $x = 0$.

Range: $(-\infty, -1) \cup (-1, \infty)$

The only value excluded from the range is at the horizontal asymptote, $y = -1$.

3.1.2 Finding the Vertical Asymptotes of a Rational Function

A rational function $r(x) = \dfrac{f(x)}{g(x)}$ is in **simplest form** (or lowest terms) when $f(x)$ and $g(x)$ have no common factors. When $r(x)$ is in simplest form, vertical asymptotes occur at the x-values where $g(x) = 0$. To find the equation of a function's vertical asymptote(s), set the expression in the function's denominator equal to 0 and solve. Any resulting x-value gives the equation of a vertical asymptote.

Steps for Writing the Equation of a Rational Function's Vertical Asymptote(s)

❶ Write the function in simplest form.

❷ Set the polynomial in the denominator equal to 0 and solve.

EXAMPLE 4 Writing the Equation of a Vertical Asymptote

Find each function's vertical asymptote(s), if any.

A. $f(x) = \dfrac{x}{2x-8}$ **B.** $f(x) = \dfrac{x-1}{x^2+3x-4}$

SOLUTION

A. The function is in simplest form. Set $2x - 8$ (from the denominator) equal to 0 and solve for x.

$2x - 8 = 0 \Rightarrow x = 4$ The function has a vertical asymptote at $x = 4$.

B. The function is not in simplest form.

❶ Simplify the function. Factor the denominator, then remove any common factors.

$$f(x) = \frac{x-1}{x^2+3x-4} = \frac{x-1}{(x+4)(x-1)} = \frac{1}{x+4},\ x \neq 1$$

Remove the common factor $x - 1$ from the numerator and the denominator.

❷ Set $x + 4$ equal to 0 and solve for x.

$x + 4 = 0 \Rightarrow x = -4$ The function has a vertical asymptote at $x = -4$.

TIP

$f(x)$ is in simplest form because x and $2x - 8$ have no common factors.

REMEMBER

The simplified function is equal to the original function only when $x = 1$ is excluded from the function's domain.

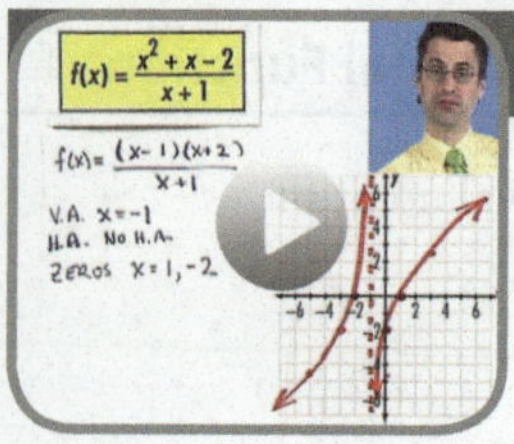

3.1.3 Graphing Rational Functions with Vertical Asymptotes

A vertical asymptote divides the plane into two regions (the right side and the left side of the asymptote). When there is one vertical asymptote, the graph of a rational function will be two curves, where one curve is on the left side of the vertical asymptote and the other curve is on the right side of the vertical asymptote. Remember, the graph of a rational function will never cross a vertical asymptote.

Steps for Graphing a Rational Function in Simplest Form

❶ Find any and all asymptote(s).

❷ Find the function's zeros, if any exist.

❸ Make a table to find additional points on the graph. Choose x-values near any zeros and on each side of any vertical asymptote(s).

❹ Sketch the asymptote(s), plot the zeros and the points, and then sketch the graph.

EXAMPLE 5 Graphing a Rational Function Having a Vertical Asymptote

Sketch the graph of $f(x) = \dfrac{x^2 - x - 6}{1 - x}$.

SOLUTION

❶ The function is in simplest form. Set the denominator equal to 0 and solve for x.

$1 - x = 0 \Rightarrow x = 1$ The function has a vertical asymptote at $x = 1$.

❷ Set the numerator equal to 0 and solve for x.

$$x^2 - x - 6 = 0$$
$$(x - 3)(x + 2) = 0$$
$x = 3$ or $x = -2$ The function has zeros at $x = 3$ and $x = -2$.

❸ Choose x-values near the zeros and vertical asymptote (i.e., near −2, 1, and 3) for the table.

x	−3	−1	0	2	4
$f(x)$	$\frac{3}{2}$	−2	−6	4	−2

❹ Sketch the vertical asymptote (the vertical line $x = 1$). There is one vertical asymptote, so the graph will be two curves (one on each side of the vertical asymptote).

Plot the zeros at (3, 0) and (−2, 0), and plot the five points from the table. Sketch the graph (the two curves) through the zeros and the other points.

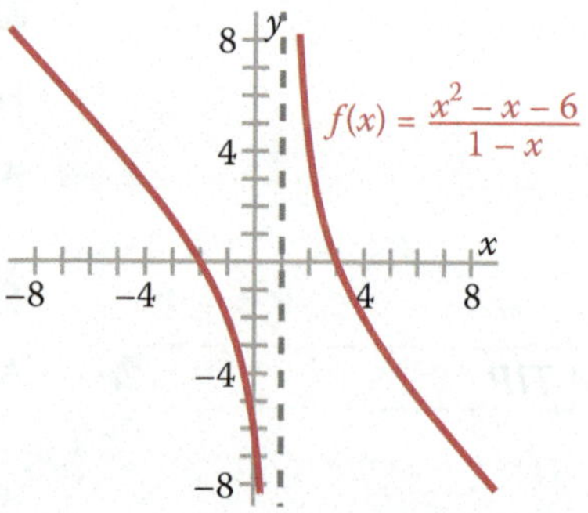

WHAT'S THE BIG IDEA?

What do you know about the equation of a rational function if the graph has an asymptote at $x = c$, where c is some particular value?

3.1.4 Graphing Rational Functions with Vertical and Horizontal Asymptotes

A horizontal asymptote is a line that the graph of a function approaches as $x \to \infty$ or as $x \to -\infty$. The existence and location of a horizontal asymptote depends upon the relationship between the degrees of the rational function's numerator and denominator. Note that while a graph of a rational function will never intersect a vertical asymptote, it may cross a horizontal asymptote.

Horizontal Asymptotes

If $r(x) = \dfrac{f(x)}{g(x)}$ is a rational function in simplest form where m is the degree of $f(x)$ and n is the degree of $g(x)$, then

- $m < n \to r(x)$ has a horizontal asymptote at $y = 0$,
- $m = n \to r(x)$ has a horizontal asymptote at $y = \dfrac{\text{leading coefficient of } f(x)}{\text{leading coefficient of } g(x)}$, and
- $m > n \to r(x)$ has no horizontal asymptote.

EXAMPLE 6 Graphing a Rational Function Having a Horizontal Asymptote

Sketch the graph of $f(x) = \dfrac{x^2 - x - 2}{x^2}$.

SOLUTION

The function is in simplest form.

❶ Find the asymptotes.

vertical asymptote: $x = 0$ *The denominator is 0 when $x = 0$.*

horizontal asymptote: $y = 1$ *The degrees of the numerator and denominator are equal,*
and the ratio of the leading coefficients is 1/1.

❷ Set the numerator equal to 0 and solve for x.

zeros: $x = -1$ and $x = 2$ *Solve $x^2 - x - 2 = 0$.*

❸ Choose x-values near $x = -1$, 0, and 2 for the table.

x	-3	-2	1	3	4
$f(x)$	$\frac{10}{9}$	1	-2	$\frac{4}{9}$	$\frac{5}{8}$

❹ Sketch the graph.

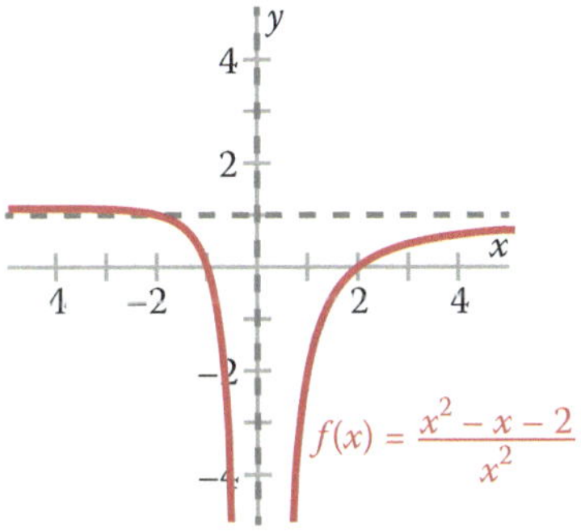

GRAPHING CALCULATOR

Enter the equation as $Y1=(X^2–X–2)/X^2$.

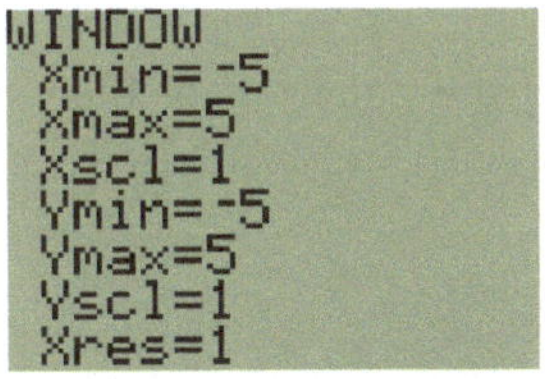

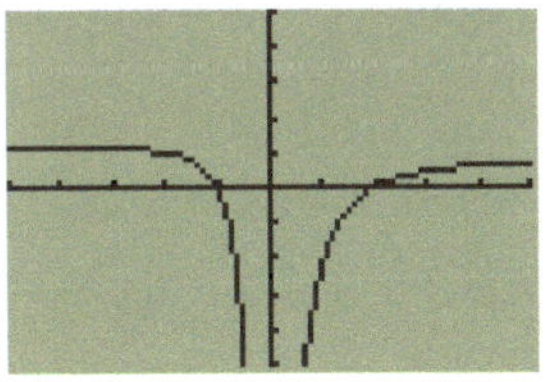

In **Example 6,** the left branch of the graph of $f(x)$ intersects the horizontal asymptote at the point $(-2, 1)$. The graph then appears to level out as x goes to $-\infty$ (Figure 3.1b). However, closer examination of this region reveals that the graph approaches the horizontal asymptote as x goes to $-\infty$ (Figure 3.1c).

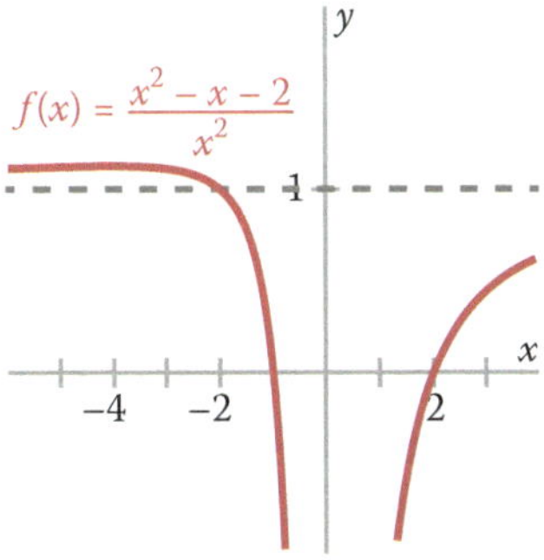

Figure 3.1b

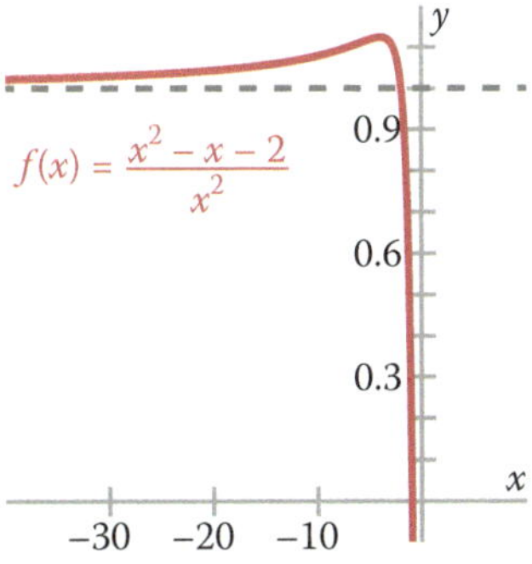

Figure 3.1c

WHAT'S THE BIG IDEA?

What do you know about the equation of a rational function if its graph has asymptotes at $x = c$ (where c is some particular value) and $y = 0$?

3.1.5 Oblique Asymptotes

As stated earlier, when the degree of a rational function's numerator is greater than the degree of its denominator, the function has no horizontal asymptote. However, if the degree of the numerator is *one greater* than the degree of the denominator, and the rational function is in simplest form, then the rational function has an **oblique asymptote**.

DEFINITION

An **oblique asymptote** is an asymptote that is neither vertical nor horizontal; it occurs when the degree of the numerator is one more than the degree of the denominator.

The equation of an oblique asymptote can be found using long division to write the function as the sum of the quotient and the remainder.

Oblique Asymptotes

If $r(x) = \dfrac{f(x)}{g(x)}$ is a rational function in simplest form where the degree of $f(x)$ is equal to 1 + the degree of $g(x)$, then using long division to write $r(x)$ as $r(x) = ax + b + \dfrac{R(x)}{g(x)}$ (where $ax + b$ is the quotient and $R(x)$ is the remainder) gives the equation of the oblique asymptote, $y = ax + b$.

EXAMPLE 7 Writing the Equation of an Oblique Asymptote

NOTICE THAT

$f(x)$ has an oblique asymptote because the degree of the numerator, 2, is one more than the degree of the denominator, 1.

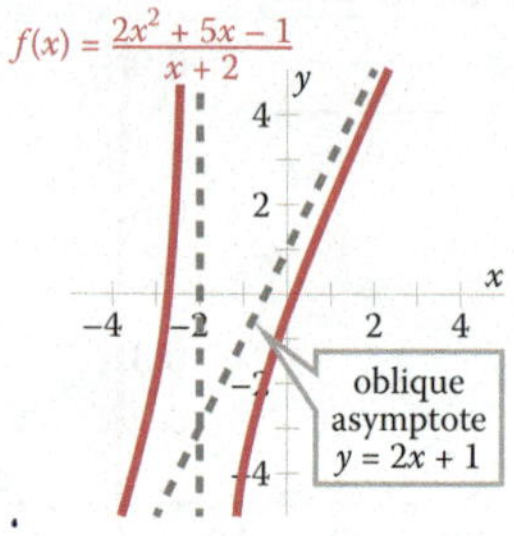

Write the equation of the oblique asymptote for the function $f(x) = \dfrac{2x^2 + 5x - 1}{x + 2}$.

SOLUTION

The function is in simplest form and the degree of the numerator is equal to 1 + the degree of the denominator. Therefore, the function has an oblique asymptote.

Use long division to find the quotient and the remainder.

$$
\begin{array}{r l}
 & 2x + 1 \quad \leftarrow \textit{quotient} \\
x + 2 \,\big)\!\! & 2x^2 + 5x - 1 \\
 & -(2x^2 + 4x) \\
 & x - 1 \\
 & -(x + 2) \\
 & -3 \quad \leftarrow \textit{remainder}
\end{array}
$$

Write $f(x)$ as the sum of the quotient and the remainder.

$$f(x) = 2x + 1 - \frac{3}{x + 2}$$

oblique asymptote: $y = 2x + 1$

EXAMPLE 8 Graphing a Rational Function Having an Oblique Asymptote

Sketch the graph of $f(x) = \dfrac{x^2 + x + 2}{x + 1}$.

SOLUTION

❶ The function is in simplest form. Find the asymptotes.

vertical asymptote: $x = -1$ *The denominator is 0 when $x = -1$.*

oblique asymptote: $y = x$ *Use long division to write the equation of the oblique asymptote (numerator's degree is equal to 1 + the denominator's degree).*

$$\begin{array}{r} x \\ x+1 \overline{) x^2 + x + 2} \\ \underline{-(x^2 + x)} \\ 2 \end{array}$$

❷ The function has no real zeros because the numerator is a quadratic polynomial where the discriminant is less than 0. *$D = 1^2 - 4(1)(2) = -7$*

❸ Choose x-values on each side of the vertical asymptote for the table.

x	−3	−2	0	1
$f(x)$	−4	−4	2	2

❹ Sketch the graph.
Note that $f(x)$ has no real zeros, so it has no x-intercepts.

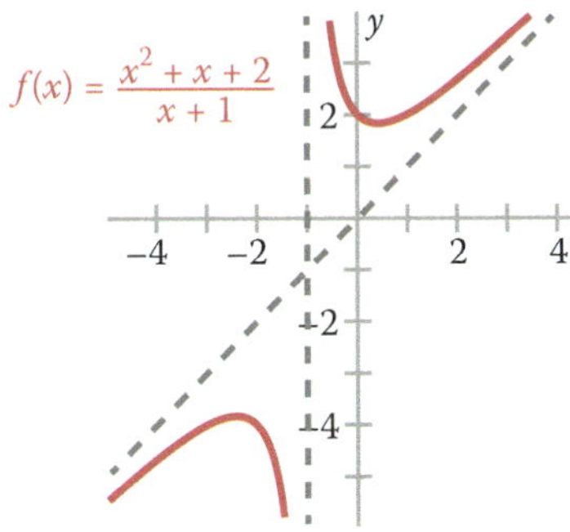

Graphing a Rational Function Not in Simplest Form

Additional steps will be necessary to graph a rational function that is not in simplest form (when the factor common to the numerator and denominator is not a constant). This function's graph may contain one or more **holes**. A hole is a discontinuity, represented with a open point on the graph. To find the x-coordinate of a hole, set the common factor equal to 0 and solve for x. Then substitute the x-coordinate into the simplified function to find the corresponding y-coordinate of the hole. Draw a hole (open point) at that point on the graph.

REMEMBER

The numerator and denominator of a rational function not in simplest form contain some common factor, which must be removed before completing the steps for graphing.

EXAMPLE 9 Graphing a Rational Function Containing a Hole

Sketch the graph of $f(x) = \dfrac{x^2 + 3x + 2}{x + 2}$.

SOLUTION

Simplify. $f(x) = \dfrac{x^2 + 3x + 2}{x + 2} = \dfrac{\cancel{(x+2)}(x+1)}{\cancel{x+2}} = x + 1, x \neq -2$

IMPORTANT

$x = -2$ is not in the function's domain.

Therefore, the function simplifies to the linear function $f(x) = x + 1$, for all $x \neq -2$.

The graph of $f(x) = x + 1$ for all x such that all $x \neq -2$ is a line with

- slope 1,
- y-intercept 1, and
- a hole at the point where $x = -2$.

Substitute −2 into $x + 1$ to find the y-coordinate of the hole. Since $-2 + 1 = -1$, the line has a hole at the point $(-2, -1)$.

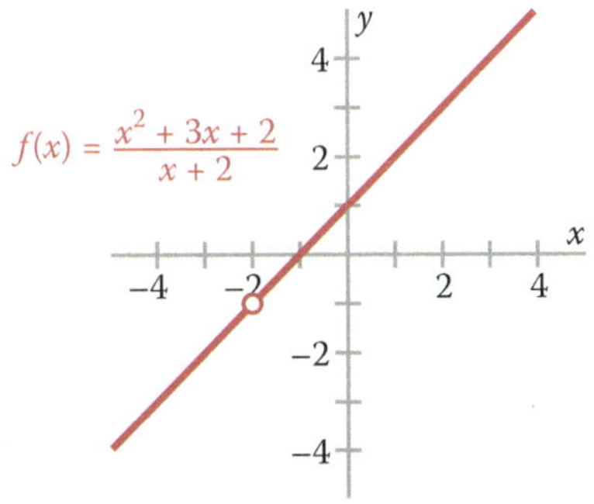

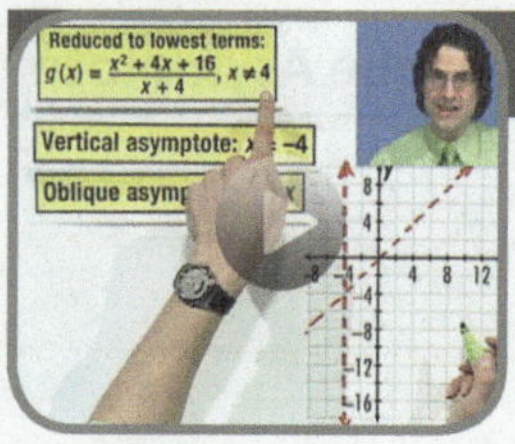

3.1.6 Oblique Asymptotes: Another Example

The graph of a rational function that is not in simplest form may have asymptotes and contain a hole, as demonstrated in **Example 10**.

> **Steps for Graphing a Rational Function Not in Simplest Form**
>
> ❶ Simplify the function. Then find any and all asymptote(s).
>
> ❷ Find the function's zeros, if any exist.
>
> ❸ Make a table to find additional points on the graph. Choose x-values near any zeros and on each side of any vertical asymptote(s).
>
> ❹ Find the coordinates of any holes.
>
> ❺ Sketch the asymptote(s), plot the zeros and the points, and then sketch the graph, including any holes.

EXAMPLE 10 Graphing a Rational Function

Sketch the graph of $f(x) = \dfrac{x^3 - 4x}{x^2 + x}$.

SOLUTION

❶ Simplify. $f(x) = \dfrac{x^3 - 4x}{x^2 + x} = \dfrac{x(x^2 - 4)}{x(x+1)} = \dfrac{x^2 - 4}{x+1}, x \neq 0$

Find the asymptotes.

vertical asymptote: $x = -1$ *The denominator is 0 when $x = -1$.*

oblique asymptote: $y = x - 1$ *Numerator's degree = 1 + denominator's degree. Use long division.*

$$\begin{array}{r} x-1 \\ x+1\overline{)\,x^2+0x-4} \\ -(x^2+x) \\ \hline -x-4 \\ -(-x-1) \\ \hline -3 \end{array}$$

❷ Set the numerator equal to 0 and solve for x.

zeros: $x = 2$ and $x = -2$ *Solve $x^2 - 4 = 0$.*

❸ Choose x-values near the zeros and vertical asymptote (i.e., near −2, −1, and 2) for the table.

x	−4	−3	1	3
$f(x)$	−4	$-\frac{5}{2}$	$-\frac{3}{2}$	$\frac{5}{4}$

Do not use $x = 0$ for the table because 0 is not in the function's domain.

❹ Find the y-coordinate of the hole by substituting $x = 0$ into the simplified rational expression.

$\dfrac{0^2 - 4}{0 + 1} = -4$ hole: $(0, -4)$

❺ Sketch the asymptotes. Plot the x-intercepts and the points from the table. Plot an open point at the hole, (0, −4). Then sketch the curves through these points, approaching the asymptotes.

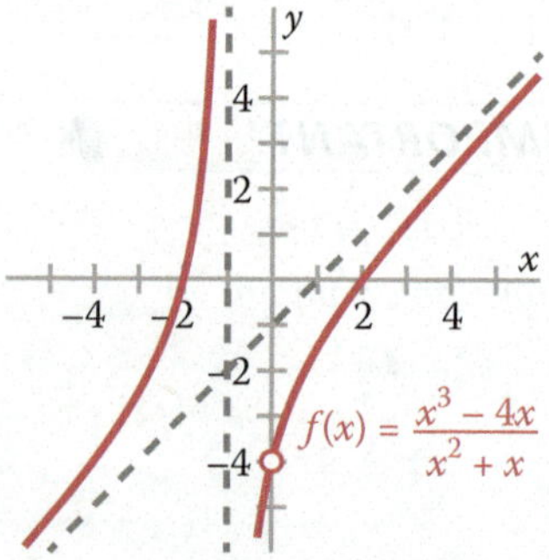

TIP

The asymptotes divide the plane into 4 regions. The two curves must be in the regions that include the zeros, $x = 2$, and $x = -2$.

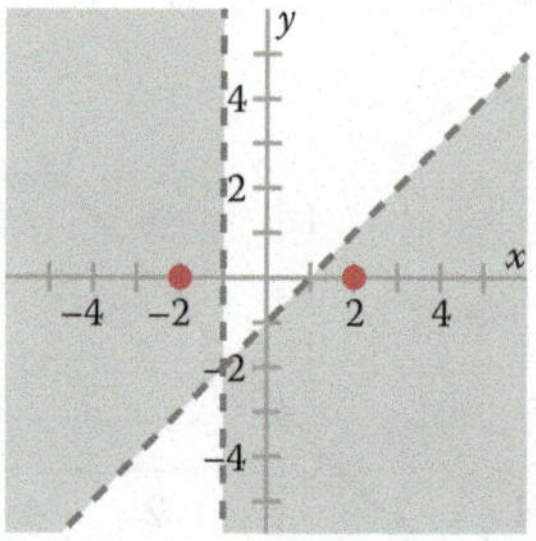

WHAT'S THE BIG IDEA?

If a rational function is simplified and graphed, and the graph contains a hole at $(0, c)$ where c is some particular value, what must be true about the original function?

SECTION 3.1 EXERCISES

Warm Up

1. Simplify. $\dfrac{x^2+4x+3}{x^3+2x^2-3x}$
2. Solve. $3x^3 - 3x = 0$
3. Graph each equation. $x = -2$, $y = 0$, and $y = -x + 2$

Just the Facts

Fill in the blanks.

4. Rational functions are ____ of polynomials.
5. The graph of $f(x) = \dfrac{1}{x}$ has a horizontal asymptote at ____ and a vertical asymptote at ____.
6. The graph of a rational function may have ____ or ____ asymptotes, but not both.
7. Graphs of rational functions may cross ____ asymptotes.
8. If the degree of a rational function's numerator is less than the degree of its denominator, then the graph has a(n) ____ asymptote at ____.
9. ____ is used to find the equation of a rational function's oblique asymptote when the degree of the numerator is equal to ____ the degree of the denominator.
10. If the quotient of a rational function's numerator and denominator is $2x + 5$ and the remainder is 10, then the graph has a(n) ____ asymptote at ____.

Essential Skills

In Exercises 1–2, write the equations of each graph's vertical and horizontal asymptotes.

1.

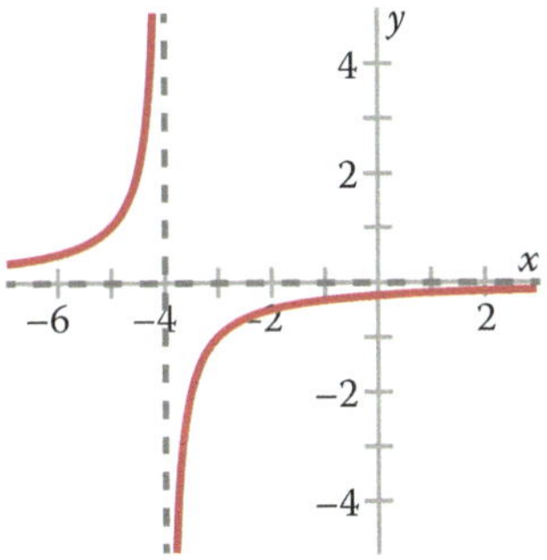

2.

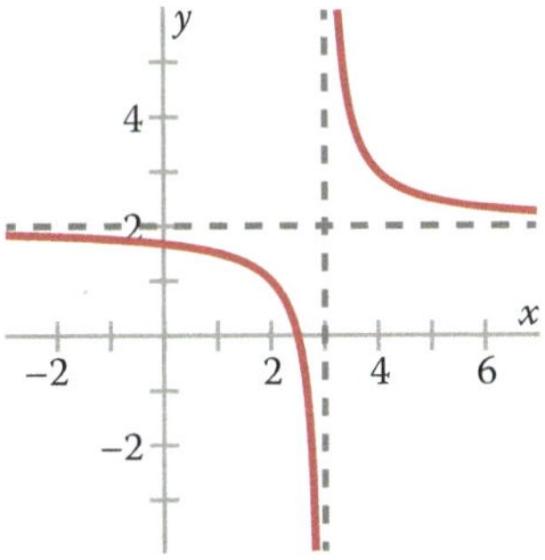

In Exercises 3–4, identify the domain and range of each function.

3.

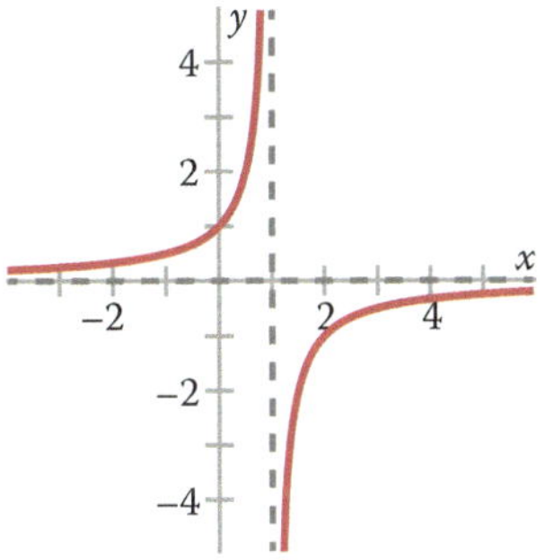

4.

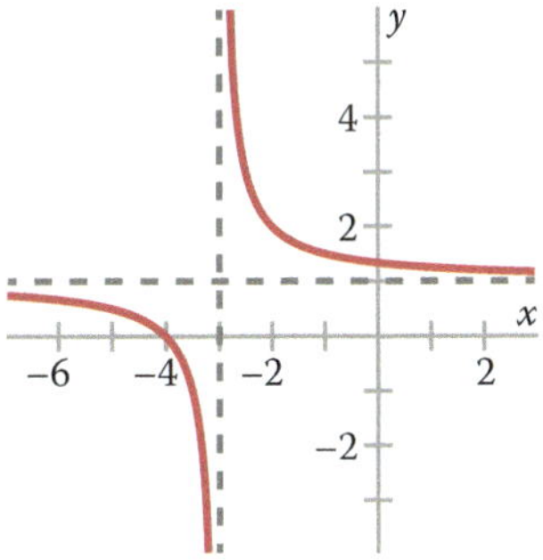

In Exercises 5–10, graph each function.

5. $f(x)=\dfrac{1}{x}+5$
6. $f(x)=-\dfrac{1}{x}+2$
7. $f(x)=\dfrac{1}{x-3}$
8. $f(x)=\dfrac{1}{x+3}$
9. $f(x)=\dfrac{1}{x-3}+1$
10. $f(x)=-\dfrac{1}{x+3}+2$

In Exercises 11–18, write the equation of the vertical asymptote of each function, if any.

11. $f(x)=-\dfrac{x}{x-2}$
12. $f(x)=\dfrac{x+2}{x+7}$
13. $f(x)=\dfrac{3-x}{x^2-36}$
14. $f(x)=\dfrac{4x+9}{x^2+5x-6}$
15. $f(x)=\dfrac{x}{6x^2+3x}$
16. $f(x)=\dfrac{x+5}{x^2+x-20}$
17. $f(x)=\dfrac{2x^2-18}{2x^3-5x^2+2x}$
18. $f(x)=\dfrac{x+1}{2x^3+x^2-3x}$

In Exercises 19–20, use each given function to complete the table.

19. $f(x)=\dfrac{x^2-1}{x+2}$

x	−3	−2	−1	0	1
$f(x)$					

20. $f(x)=\dfrac{x^2+3x-4}{x+1}$

x	−4	−3	−2	−1	0
$f(x)$					

In Exercises 21–22, find the zeros and vertical asymptote(s), if any, of each function.

21. $f(x)=\dfrac{6x^2+7x+2}{4x-4}$ 22. $f(x)=\dfrac{3x^3+5x^2-2x}{1-x^2}$

In Exercises 23–24, use the zeros and vertical asymptote(s) to sketch the graph of each function.

23. $f(x)=\dfrac{x^2-3x-10}{5+x}$ 24. $f(x)=\dfrac{x^2-5x-6}{x-1}$

In Exercises 25–26, choose one of the following phrases to complete each statement.

A. $r(x)$ has a horizontal asymptote at $y=0$.

B. $r(x)$ has a horizontal asymptote at $y=\dfrac{\text{numerator's leading coefficient}}{\text{denominator's leading coefficient}}$.

C. $r(x)$ has no horizontal asymptote.

25. If $r(x)$ is a rational function in simplest form where the degree of the numerator is 3 and the degree of the denominator is 3, then ________ .

26. If $r(x)$ is a rational function in simplest form where the degree of the numerator is 2 and the degree of the denominator is 3, then ________ .

In Exercises 27–32, find the horizontal asymptote(s), if any, of each function.

27. $f(x)=\dfrac{1-x^3}{2x^2+5x}$ 28. $f(x)=\dfrac{2x+5}{x+1}$

29. $f(x)=-\dfrac{1}{x^2}$ 30. $f(x)=\dfrac{x+3}{2x^3}$

31. $f(x)=\dfrac{8x^3}{2x^3+x+5}$ 32. $f(x)=\dfrac{3x^2+1}{x}$

In Exercises 33–36, use the zeros, vertical asymptote(s), and horizontal asymptote to sketch the graph of each function.

33. $f(x)=-\dfrac{2x}{x-1}$ 34. $f(x)=\dfrac{x}{x^2-16}$

35. $f(x)=\dfrac{3}{2-x}$ 36. $f(x)=\dfrac{2x^2-2}{x^2-4}$

In Exercises 37–38, identify the vertical, horizontal, and oblique asymptote(s), if any, of each function.

37. $f(x)=\dfrac{5-x^2}{x+3}$ 38. $f(x)=\dfrac{4x^2+6x+1}{2x-3}$

In Exercises 39–40, choose the equation of each graph.

39.

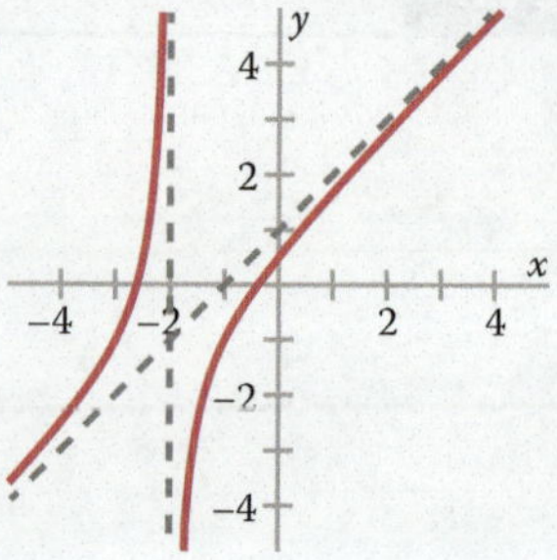

A. $f(x)=\dfrac{x^2+3x+1}{x-2}$ B. $f(x)=\dfrac{x+2}{x^2+3x+1}$

C. $f(x)=\dfrac{x-2}{x^2+3x+1}$ D. $f(x)=\dfrac{x^2+3x+1}{x+2}$

40.

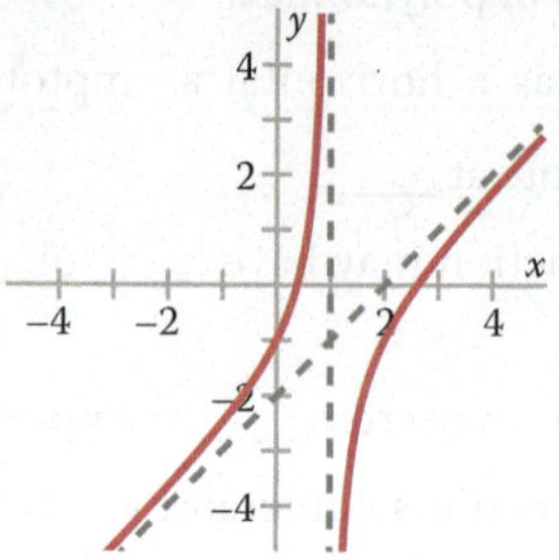

A. $f(x)=\dfrac{x^2+1}{x-1}$ B. $f(x)=\dfrac{x+1}{x-1}$

C. $f(x)=\dfrac{x^2-3x+1}{x-1}$ D. $f(x)=\dfrac{x^2+1}{x+1}$

In Exercises 41–42, find the coordinates of the hole, if it exists, in the graph of each function.

41. $h(x)=\dfrac{x^2+4}{x+2}$ 42. $g(x)=\dfrac{2x^2-5x-3}{x-3}$

In Exercises 43–44, sketch the graph of each function.

43. $y=\dfrac{3x^2+4}{x-2}$ 44. $y=\dfrac{5x^2-3x-2}{x-1}$

In Exercises 45–46, identify the vertical, horizontal, and oblique asymptote(s), if any, of each function.

45. $g(x)=\dfrac{x^3+3x^2-4x-12}{x^2+8x+15}$ 46. $f(x)=\dfrac{x^3-1}{x^2+3x-4}$

In Exercises 47–48, find the coordinates of the hole, if it exists, in the graph of each function.

47. $h(x)=\dfrac{x^3-125}{x^2-25}$ 48. $f(x)=\dfrac{x^3-2x^2+6x-12}{x^2+6x-16}$

In Exercises 49–50, sketch the graph of each function.

49. $f(x)=\dfrac{x^3-6x^2+6x-36}{x^2-5x-6}$ 50. $f(x)=\dfrac{x^3-8}{x^2-x-2}$

Extensions

In Exercises 51–53, correct any errors in each statement.

51. The graph of $f(x)=\dfrac{x^2}{x-2}$ has an x-intercept at $x = 2$ and one vertical asymptote at the line $x = 0$. The graph passes through the points $(1, -1)$ and $(4, 8)$.

52. The graph of a function with denominator $x^2 + x - 12$ must have two vertical asymptotes.

53. The graph of $f(x)=\dfrac{x^2}{(x+3)(x-1)}$ has an x-intercept at $x = 0$ and two vertical asymptotes at the lines $x = 1$ and $x = -3$. The graph has a horizontal asymptote at $y = 1$ and passes through the points $(2, 0.8)$ and $(-4, 3.2)$.

54. Describe the three types of asymptotes and explain how to find each one.

55. What do you know about the graph of a rational function in simplest form when the degree of the expression in the numerator is 4 and the degree of the expression in the denominator is 3?

In Exercises 56–59, write each rational function $g(x)$ if $f(x)=\dfrac{2}{x^2}$. Check each answer by graphing with a graphing calculator.

56. $g(x)$ is a reflection of $f(x)$ across the x-axis

57. $g(x)$ is a translation of $f(x)$ 6 units up

58. $g(x)$ is a translation of $f(x)$ 1 unit to the left

59. $g(x)$ is a translation of $f(x)$ 3 units down and 2 units to the right

In Exercises 60–65, write a rational function that satisfies each of the given constraints. Check each answer by graphing with a graphing calculator.

60. one vertical asymptote at $x = 7$

61. vertical asymptotes at $x = 3$ and $x = -1$

62. horizontal asymptote at $y = 0$ and two vertical asymptotes where one is at $x = 1/2$

63. hole at $(3, 4)$ and no vertical, horizontal, or oblique asymptotes

64. horizontal asymptote at $y = 2$, zeros at $x = 5$ and $x = -2$, two vertical asymptotes where one is at $x = 3$

65. oblique asymptote at $y = x + 2$ and vertical asymptote at $x = 1$

3.2 PARABOLAS

OBJECTIVES

- Understand the relationship between the four types of conics (circle, ellipse, parabola, and hyperbola) and a cone.
- Find the vertex, focus, and directrix of a parabola (vertex at the origin).
- Graph a parabola (vertex at the origin).
- Write the equation of a parabola (vertex at the origin).

PREREQUISITE VOCABULARY TERMS

axis of symmetry
circle
parabola
vertex (of a parabola)

3.2.1 Introduction to Conic Sections

When a plane intersects a cone (specifically, a right double cone with a circular base) without intersecting the cone's vertex, the figure formed is known as a **conic** or **conic section**. For example, a circle, one type of conic, is formed when the intersecting plane is parallel to the base of the cone. The four basic conics are the circle, ellipse, parabola, and hyperbola, as shown in Figure 3.2a.

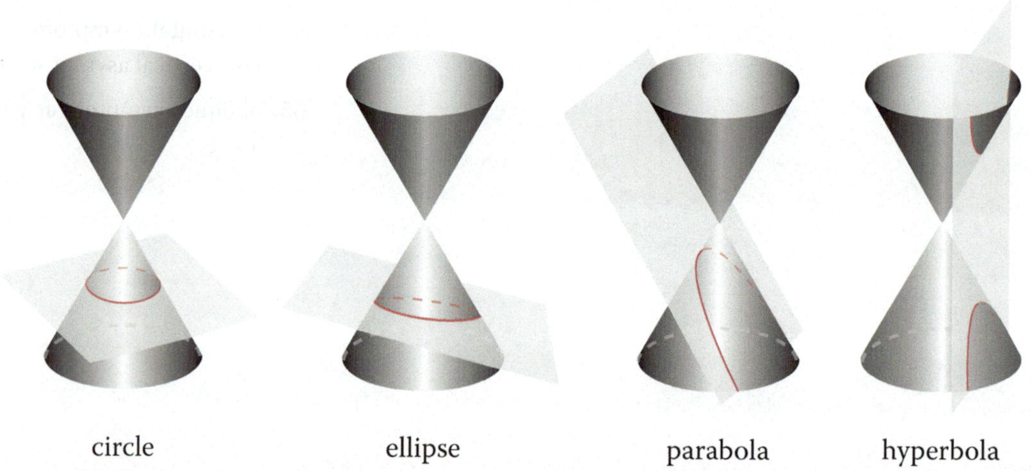

Figure 3.2a

An ellipse is formed by the intersection of a cone and a plane, where the plane does not intersect a base of the cone. (A special case of an ellipse is a circle, where the plane is parallel to the base.) A parabola is formed when a plane intersects only one part of the double cone through its base, and a hyperbola is formed when the plane intersects both parts of the double cone, including both bases.

When a plane intersects a double cone at the vertex of the cone, the resulting figure is called a **degenerate conic**, which can look like a single point, a single line, or a pair of lines.

The equations for two of the conics, the parabola and the circle, have been discussed previously. Recall that the standard form equation of a circle centered at the origin is $x^2 + y^2 = r^2$, where r is the radius of the circle, and the standard form equation of a parabola centered at the origin is $y = ax^2$, where the value of a describes the width and direction of the parabola. The equation of a parabola will be discussed further in this section, and the equations of the ellipse and hyperbola will be discussed in the following sections.

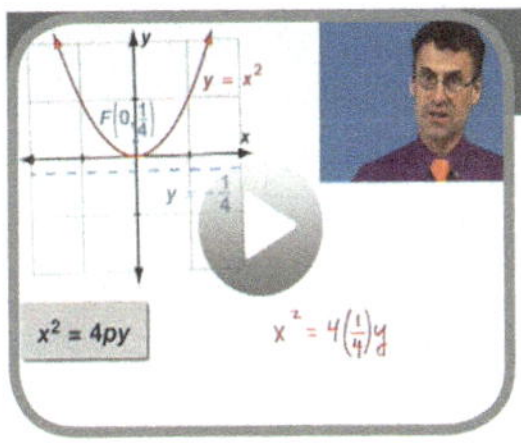

3.2.2 Graphing Parabolas

Recall that the graph of a quadratic function of the form $y = ax^2 + bx + c$ is a parabola with a vertical axis of symmetry that opens upward or downward, depending upon the value of a. The discussion of parabolas in this chapter will include parabolas that are not functions. These nonfunction parabolas, as shown in Figure 3.2b, will have a horizontal axis of symmetry and will open to the left or right.

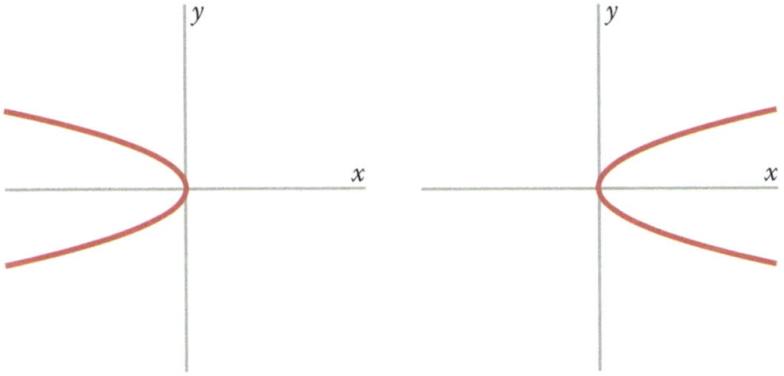

Figure 3.2b Parabolas with a Horizontal Axis of Symmetry

The Definition of a Parabola

Recall that, on a coordinate plane, a circle is the set of all points that are equidistant from a specific point called the circle's center. The distance from the center to any point on the circle is the circle's radius. Similarly, a **parabola**, on a coordinate plane, is the set of all points that are equidistant from a specific point and a specific line (i.e., the shortest distance from any given point on the parabola to a specified point in the plane is the same as the shortest distance from that same point on the parabola to the specified line). The point is called the parabola's **focus** F, and the line is called the parabola's **directrix**, as shown in Figure 3.2c.

DEFINITION

A **parabola** is the set of all points that are equidistant from a specific point, the **focus** F, and a specific line, the **directrix**.

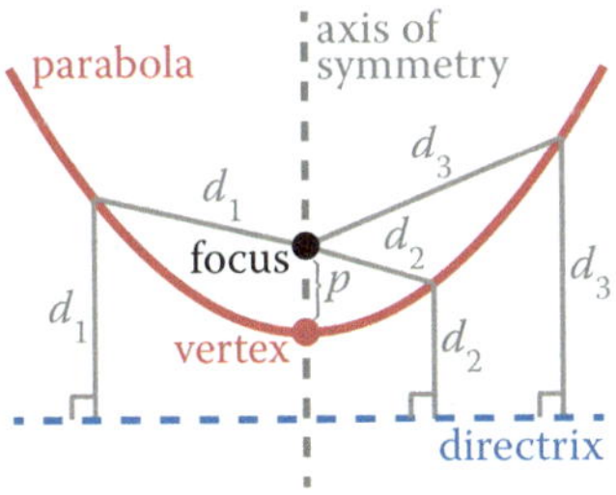

Figure 3.2c

The parabola's axis of symmetry passes through the focus and the vertex, and is perpendicular to the directrix. Since each point on the parabola is equidistant from the focus and the directrix, the parabola's vertex is the midpoint between the focus and the point at which the axis of symmetry intersects the directrix. The distance from the vertex to the focus is defined by p, where the sign of p depends on the direction of the parabola.

WHAT'S THE BIG IDEA?

Given the coordinates of a parabola's focus and the equation of the parabola's directrix, can the coordinates of the parabola's vertex be found? Explain.

Standard Form Equations of Parabolas: Vertex at the Origin

	Vertical Axis of Symmetry	Horizontal Axis of Symmetry
Equation	$x^2 = 4py$	$y^2 = 4px$
Focus	$(0, p)$	$(p, 0)$
Directrix	$y = -p$	$x = -p$
$p > 0$	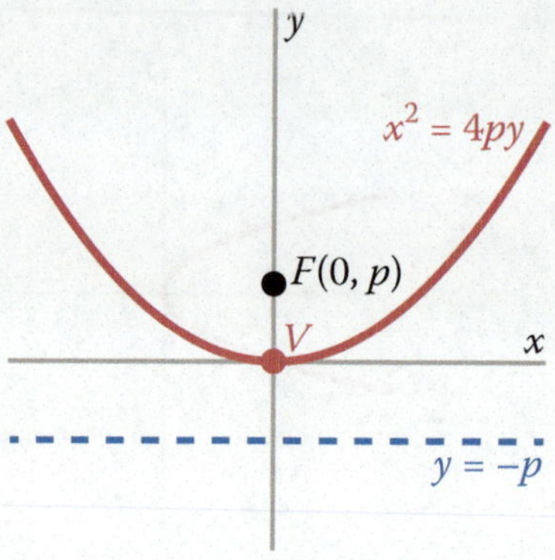	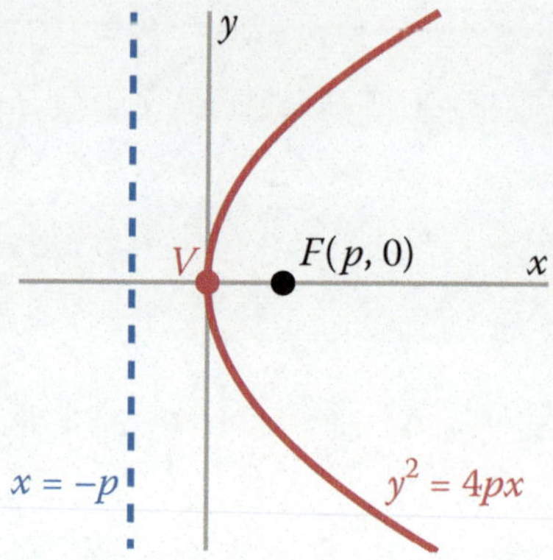
$p < 0$	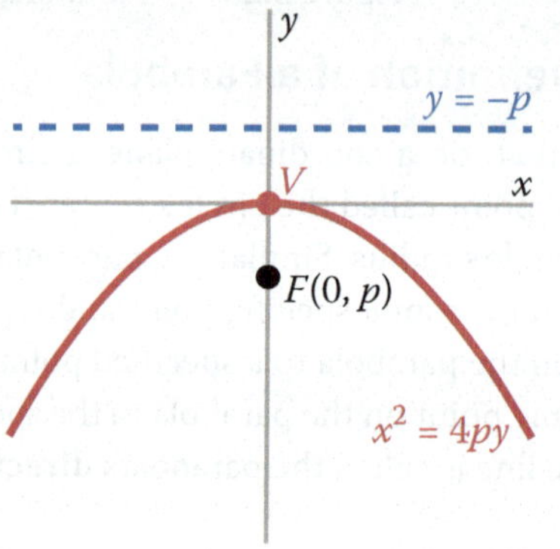	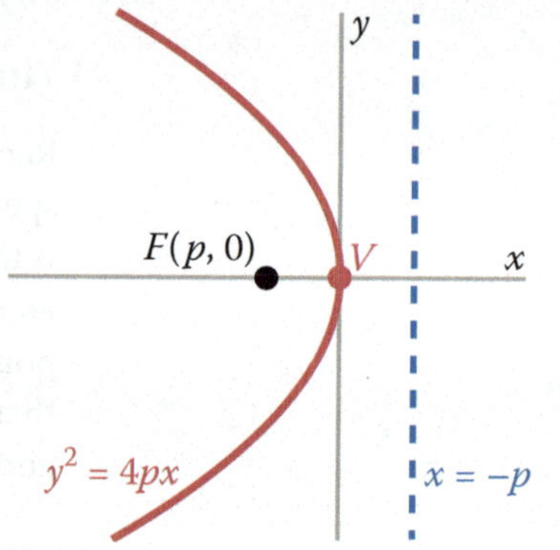

EXAMPLE 1 **Finding the Focus and the Directrix of a Parabola from Its Equation**

Find the focus and directrix of each parabola and describe its graph.

A. $y^2 + x = 0$ **B.** $y = \frac{1}{2}x^2$

GRAPHING CALCULATOR

Solve an equation with a y^2-term for y, and then enter the positive square root and the negative square root.

$y^2 = -x \Rightarrow y = \pm\sqrt{-x}$

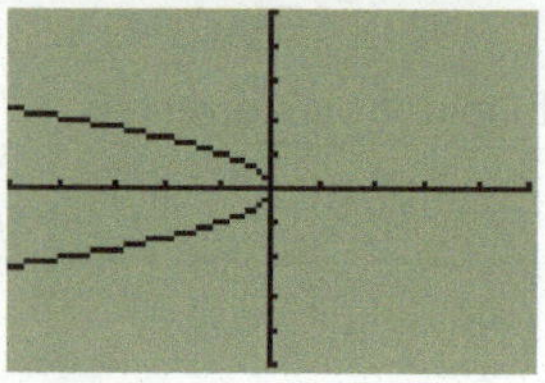

SOLUTION

Write the equation in the standard form of a parabola and then identify p.

Notice that the variable y is squared. Therefore, the standard form equation is $y^2 = 4px$, and the parabola has a horizontal axis of symmetry.

$$y^2 + x = 0$$
$$y^2 = -x$$
$$y^2 = 4\left(-\frac{1}{4}\right)x \qquad \text{Therefore, } p = -\frac{1}{4}.$$

Use the value of p to find the focus $(p, 0)$ and the directrix $x = -p$.

focus: $\left(-\frac{1}{4}, 0\right)$ directrix: $x = \frac{1}{4}$

Graph Description: Since the equation is of the form $y^2 = 4px$, the graph is a parabola with vertex at (0, 0) and a horizontal axis of symmetry. Additionally, since p is less than 0, the horizontal parabola opens to the left.

B. In this equation, the variable x is squared. Therefore, the standard form equation is $x^2 = 4py$, and the parabola has a vertical axis of symmetry.

$$y = \frac{1}{2}x^2$$
$$x^2 = 2y$$
$$x^2 = 4\left(\frac{1}{2}\right)y \qquad \text{Therefore, } p = \frac{1}{2}.$$

Use the value of p to find the focus $(0, p)$ and the directrix $y = -p$.

focus: $\left(0, \frac{1}{2}\right)$ directrix: $y = -\frac{1}{2}$

Graph Description: Since the equation is of the form $x^2 = 4py$, the graph is a parabola with vertex at (0, 0) and a vertical axis of symmetry. Additionally, since p is greater than 0, the vertical parabola opens upward. (The graph of the parabola is shown in Figure 3.2d.)

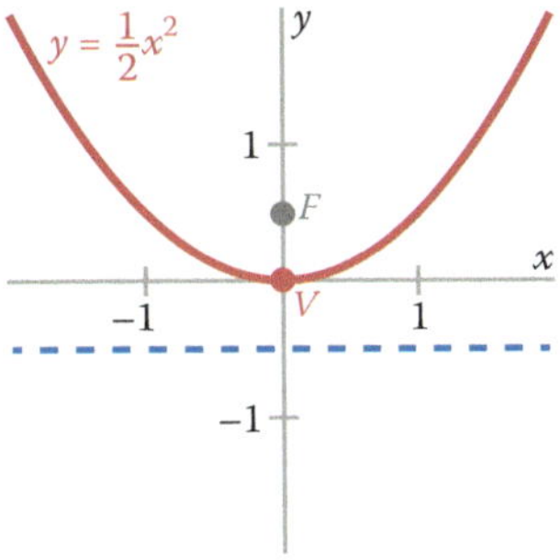

Figure 3.2d

EXAMPLE 2 Graphing a Parabola

Graph each parabola.

A. $x^2 = -2y$ B. $\frac{y^2}{3} - \frac{x}{6} = 0$

SOLUTION

A. The standard form equation is $x^2 = 4py$. Therefore, the graph is a parabola with vertex at the origin and a vertical axis of symmetry. Find the value of p to determine whether the parabola opens up or down.

$$x^2 = -2y$$
$$x^2 = 4\left(-\frac{1}{2}\right)y \qquad \text{Therefore, } p = -1/2. \text{ Since } p < 0, \text{ the parabola opens downward.}$$

Make a table to identify a few additional points on the parabola.

x	−2	−1	1	2
y	−2	$-\frac{1}{2}$	$-\frac{1}{2}$	−2

Plot the points from the table along with the vertex (0, 0), and sketch the graph of the parabola.

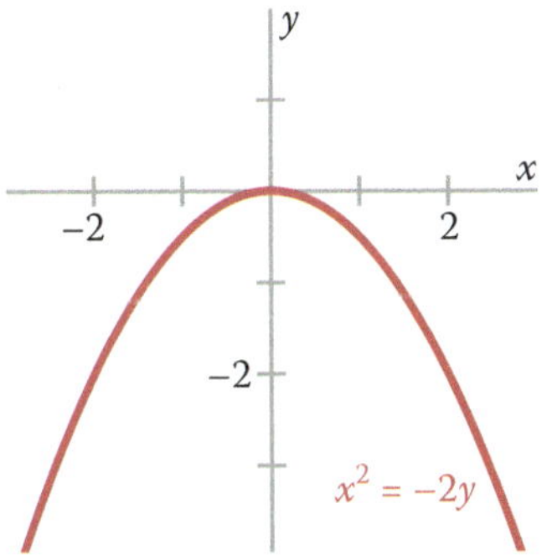

B. The standard form equation is $y^2 = 4px$. Therefore, the graph is a parabola with vertex at the origin and a horizontal axis of symmetry. Find the value of p to determine whether the parabola opens to the left or right.

$$\frac{y^2}{3} - \frac{x}{6} = 0$$
$$\frac{y^2}{3} = \frac{x}{6}$$
$$y^2 = \frac{1}{2}x$$
$$y^2 = 4\left(\frac{1}{8}\right)x \qquad \text{Therefore, } p = 1/8. \text{ Since } p > 0, \text{ the parabola opens to the right.}$$

TIP

When y is the squared variable, choose y-values for the table (instead of choosing x-values). Substitute each chosen y-value into the equation to find the corresponding x-value to complete the table.

Make a table to identify a few additional points on the parabola.

x	8	2	2	8
y	−2	−1	1	2

Plot the points from the table along with the vertex (0, 0), and sketch the graph of the parabola.

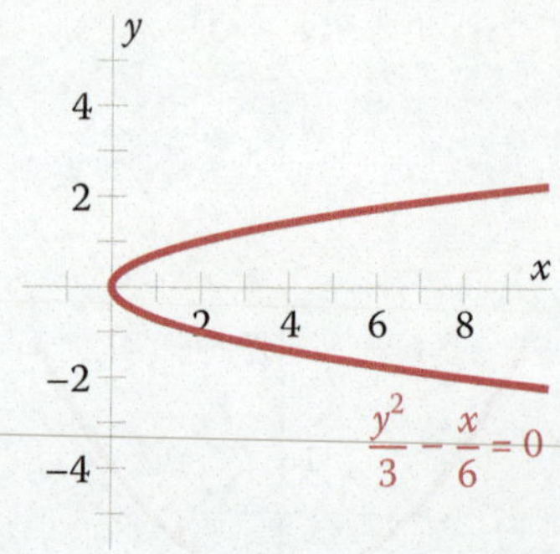

Focal Diameter

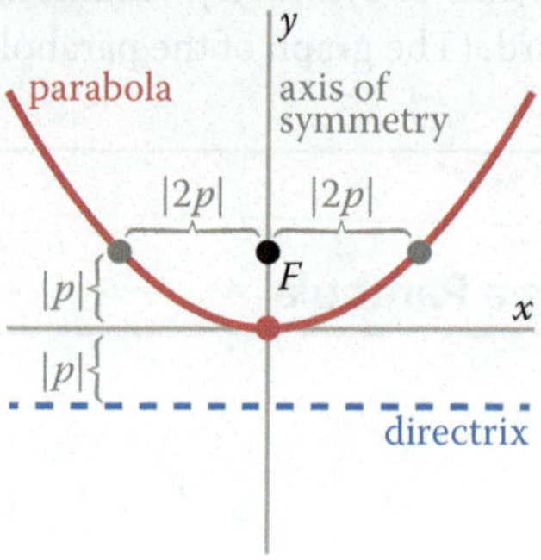

Figure 3.2e

Figure 3.2e shows that the length of each line segment perpendicular to the parabola's axis of symmetry with endpoints on the parabola and at the focus is equal to $|2p|$. The entire line segment (with endpoints on the parabola) is known as the **latus rectum**. The length of the latus rectum, $|4p|$, is the parabola's **focal diameter**. The focal diameter provides a measure of the "width" of a parabola through the focus.

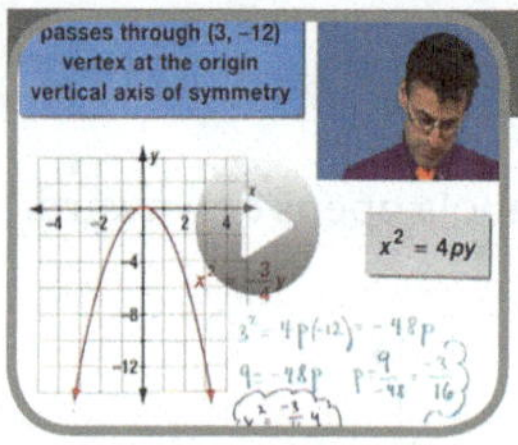

3.2.3 Writing the Equation of a Parabola

EXAMPLE 3 **Writing the Equation of a Parabola Given the Vertex and Focus**

Write the standard form equation of the parabola with focus at (−4, 0) and vertex at the origin.

SOLUTION

Since the focus is to the left of the vertex, on the x-axis, the parabola opens to the left and the axis of symmetry is horizontal. Therefore, the standard form equation is $y^2 = 4px$ where p is negative.

Since the focus, (−4, 0), is exactly 4 units to the left of the vertex, (0, 0), it follows that $p = -4$. Substitute the p-value into $y^2 = 4px$ to write the equation of the parabola.

$$y^2 = 4(-4)x$$

$$y^2 = -16x$$ (The graph of the parabola is shown in Figure 3.2f.)

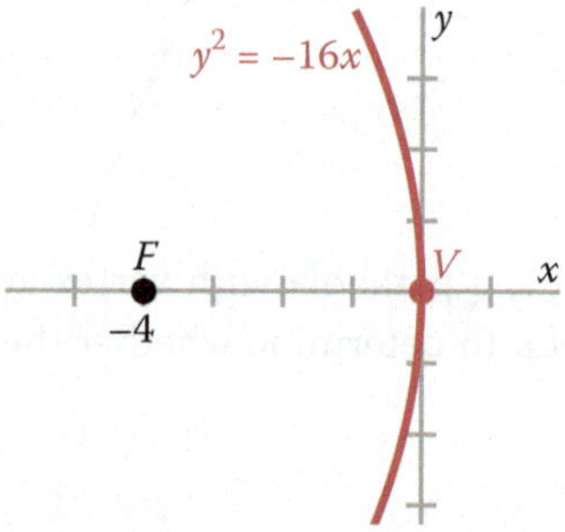

Figure 3.2f

EXAMPLE 4 Writing the Equation of a Parabola Given the Vertex and Directrix

Write the standard form equation of the parabola with directrix $y = \frac{1}{4}$ and vertex at the origin.

SOLUTION

Since the directrix is a horizontal line above the vertex, the parabola opens downward and the axis of symmetry is vertical. Therefore, the standard form equation is $x^2 = 4py$, where p is negative.

For a parabola with a vertical axis, the directrix is $y = -p$. From the given equation of the directrix $y = \frac{1}{4}$, it follows that $-p = \frac{1}{4}$, and so $p = -\frac{1}{4}$. Substitute the p-value into $x^2 = 4py$ to write the equation of the parabola.

$$x^2 = 4\left(-\frac{1}{4}\right)y$$

$$x^2 = -y$$ (The graph of the parabola is shown in Figure 3.2g.)

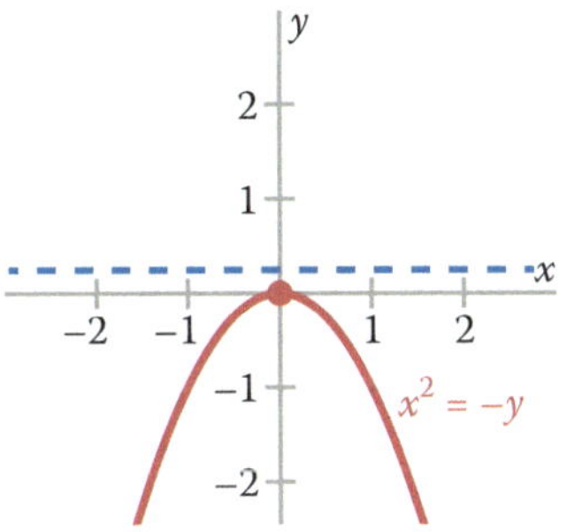

Figure 3.2g

EXAMPLE 5 Writing the Equation of a Parabola Given the Direction of the Axis of Symmetry, Vertex, and a Point on the Parabola

Write the standard form equation of the parabola that passes through (4, –1), if the vertex is at the origin and the axis of symmetry is horizontal.

SOLUTION

Since the vertex is at the origin and the axis of symmetry is horizontal, the standard form equation of the parabola is $y^2 = 4px$.

Use the x- and y-coordinates of the given point to find the value of p.

$$y^2 = 4px$$

$$(-1)^2 = 4p(4)$$

$$1 = 16p$$

$$p = \frac{1}{16}$$

Then use the p-value to write the standard form equation of the parabola.

$$y^2 = 4\left(\frac{1}{16}\right)x$$

$$y^2 = \frac{1}{4}x$$

(The graph of the parabola is shown in Figure 3.2h.)

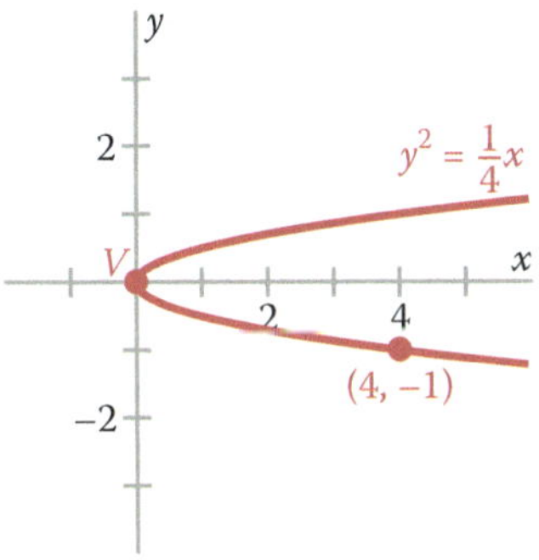

Figure 3.2h

WHAT'S THE BIG IDEA?

In each case, explain why the equation of a parabola cannot be found.

A. only the equation of the directrix and a point on the parabola are known

B. only a focus and the axis of symmetry's equation are known

C. only the vertex and a point on the parabola are known

SECTION 3.2 EXERCISES

Warm Up

Sketch the graph of each parabola and identify the vertex and axis of symmetry.

1. $f(x) = x^2$
2. $f(x) = -(x - 3)^2 + 1$
3. $f(x) = x^2 + 4x + 4$

Just the Facts

Fill in the blanks.

4. A conic is formed by the intersection of a(n) ____ and a(n) ____.
5. A parabola is the set of all points that are equidistant from a specific ____ called the focus and a specific ____ called the ____.
6. Equations of the form $x^2 = 4py$ where $p < 0$ describe ____ that open ____ with a vertical ____ on the ____, ____ at $(0, p)$, and directrix at ____.
7. Equations of the form $y^2 = 4px$ where $p > 0$ describe ____ that open ____ with a horizontal ____ on the ____, ____ at (0, 0), and directrix at ____.
8. If a parabola's ____ is to the right of its ____, then the parabola opens to the left.
9. The parabolas described by $x^2 = 4py$ and $y^2 = 4px$ have the same ____, located at ____.
10. If a parabola's directrix is a vertical line, then the parabola opens ____ or ____.

Essential Skills

In Exercises 1–2, find the focus and directrix of each parabola.

1. $y = \frac{2}{5}x^2$
2. $x = -\frac{1}{8}y^2$

In Exercises 3–4, determine the direction of each parabola.

3. $-3x + 4y^2 = 0$
4. $-y = \frac{x^2}{5}$

In Exercises 5–6, sketch the graph of each parabola.

5. $y^2 = \frac{x}{2}$
6. $-y^2 + 2x = 0$

In Exercises 7–12, write the standard form equation of each parabola with vertex at the origin and the following properties.

7. focus at $\left(\frac{1}{8}, 0\right)$
8. focus at $(0, -6)$
9. directrix at $y = -5/6$
10. directrix at $x = -5$
11. parabola passes through $(-3, -9)$ and the axis of symmetry is vertical
12. parabola passes through $(2, -2)$ and the axis of symmetry is horizontal

Extensions

In Exercises 13–15, correct any errors in the statements.

13. The graph of $4x - 20y^2 = 0$ is a parabola with a horizontal axis of symmetry. The value of p is positive, so the parabola opens to the right. The vertex is at the origin, the focus is at (20, 0), and the equation of the directrix is $y = -20$.
14. If the equation of a parabola is of the form $y^2 = 4px$ where $p < 0$, then the parabola's vertex is at $(p, 0)$, the axis of symmetry is at $x = 0$, and the parabola opens to the left.
15. The graph of $-x^2 = 6y$ is a parabola. The axis of symmetry is at $y = 0$, the vertex is at the origin, and the parabola opens to the left. Since $p = -3/2$, the focus is at $(-3/2, 0)$, and the equation of the directrix is $x = 3/2$.
16. Name the four types of conics and explain how each is formed by intersecting a plane and a cone.
17. Explain why $x^2 = 4py$ is a function and $y^2 = 4px$ is not a function.
18. If a parabola's vertex is at the origin, can the equation of the parabola's directrix be used to find the focus? Explain.
19. Can a parabola described by an equation of the form $x^2 = 4py$ have a directrix at $y = x$? Explain.
20. Can a parabola described by an equation of the form $x^2 = 4py$ have a directrix at $y = 0$? Explain.
21. Write the equations of two parabolas that pass through (2, 4), where the vertex is at the origin.
22. Write the equations of two parabolas that pass through (−1, 8), where the vertex is at the origin.

In Exercises 23–25, find the focal diameter of each parabola.

23. $x^2 = -\frac{1}{2}y$
24. $x = y^2$
25. $-x = -\frac{y^2}{12}$

3.3 ELLIPSES

OBJECTIVES

- Connect the geometric attributes of an ellipse (centered at the origin) to its standard form equation.
- Write the equation of an ellipse (centered at the origin).
- Graph an ellipse (centered at the origin).
- Find the vertices, co-vertices, and foci of an ellipse (centered at the origin).

PREREQUISITE VOCABULARY TERMS

circle
conic
Distance Formula
***x*-intercept**
***y*-intercept**

The intersection of a cone and a plane parallel to the cone's base forms a circle.

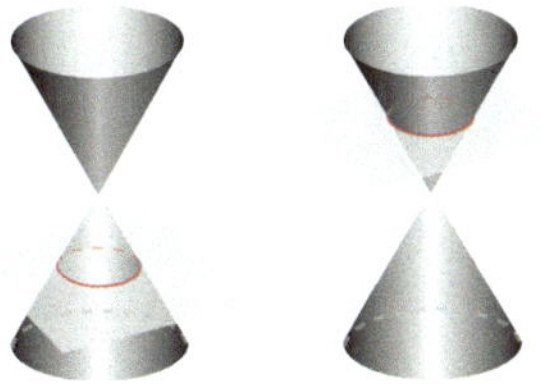

Figure 3.3a

Recall that an **ellipse** is the conic formed by the intersection of a cone and a plane that does not intersect the base of the cone. If the plane is parallel to the cone's base, then the ellipse formed is a circle, which is a special case of the ellipse, as shown in Figure 3.3a. If the plane is not parallel to the cone's base, then the ellipse formed is not circular, as shown in Figure 3.3b. The equations and graphs (on a coordinate plane) of these noncircular ellipses will be discussed in this section.

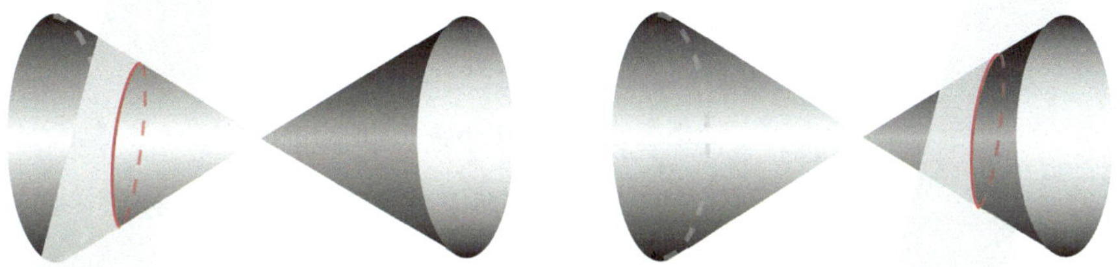

Figure 3.3b

Introduction to Ellipses

TIP

The plural of focus is foci.

Suppose there are two fixed points F_1 and F_2 on a coordinate plane such that the distance between some point (x, y) and F_1 is d_1, and the distance between that same point (x, y) and F_2 is d_2. An ellipse is the set of all points (x, y) such that $d_1 + d_2$ is constant. Each fixed point F_1 and F_2 is called a **focus** of the ellipse.

IMPORTANT

*The definition of an ellipse does **not** imply that the distance between a point on the ellipse and each focus must be constant (i.e., equal to each other).*

$$d_1 \neq d_2$$

*Instead, it is the **sum** of those distances that is **constant**.*

$$d_1 + d_2 = d_3 + d_4$$

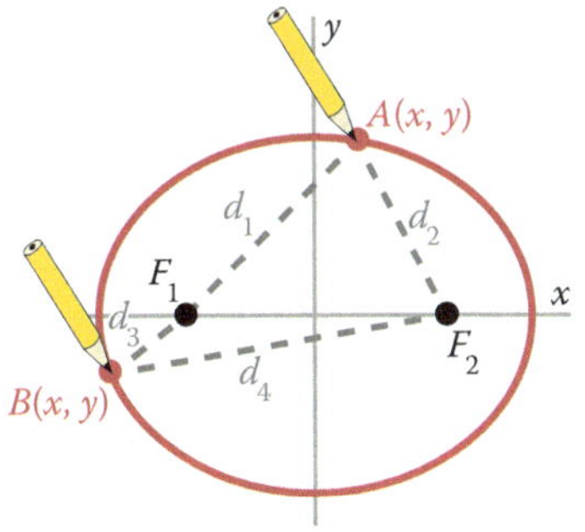

Figure 3.3c

In Figure 3.3c, A and B are points on the ellipse. For each point on the ellipse, the sum of the distances from the point to each focus is constant, n.

$$d_1 + d_2 = n = d_3 + d_4$$

In other words, the distance from a point on the ellipse to one focus plus the distance from that point to the other focus is the same for every point on the ellipse.

DEFINITION

An **ellipse** is the set of all points where the sum of the distances from two fixed points is constant. The fixed points are the **foci** of the ellipse.

Attributes of an Ellipse

Ellipses are symmetric figures with two perpendicular lines of symmetry called the **axes of symmetry**, which intersect at the **center of the ellipse**.

The **major** and **minor axes** of an ellipse are the segments on the axes of symmetry with endpoints on the ellipse. The major axis contains the foci, and its length is greater than or equal to that of the minor axis. The endpoints of the major axis are called the ellipse's **vertices,** and the endpoints of the minor axis are called the ellipse's **co-vertices** (Figure 3.3d).

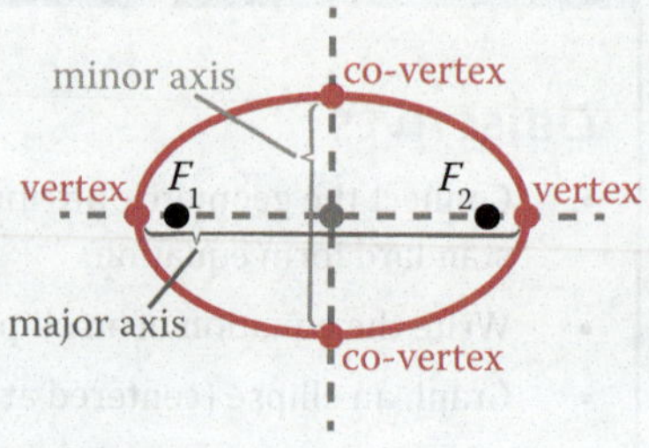

Figure 3.3d

3.3.1 Writing the Equation of an Ellipse

Horizontal and Vertical Ellipses Centered at the Origin

All of the ellipses discussed in this section will be centered at the origin. Each will be either a **horizontal ellipse** or a **vertical ellipse**.

DEFINITION

A **horizontal ellipse** is an ellipse whose major axis is a horizontal line.

A **vertical ellipse** is an ellipse whose major axis is a vertical line.

Horizontal & Vertical Ellipse Centered at the Origin

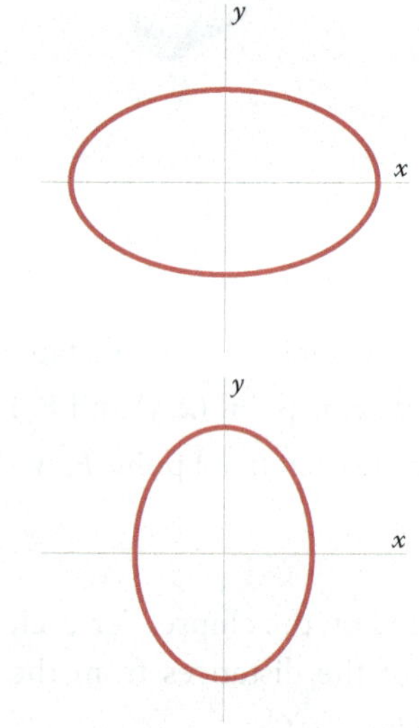

vertices & co-vertices = x- & y-intercepts

Figure 3.3e

Figure 3.3e shows a horizontal and a vertical ellipse. Notice that since both are centered *at the origin* their axes of symmetry are the x- and y-axes. Specifically, the major axis of the horizontal ellipse is on the x-axis, and the major axis of the vertical ellipse is on the y-axis. Since the vertices and co-vertices are the points at which the ellipse intersects its axes of symmetry, the vertices and co-vertices are also the ellipse's x- and y-intercepts.

Horizontal and Vertical Ellipses: Centered at the Origin

	Horizontal	Vertical
Major Axis	on the x-axis	on the y-axis
Minor Axis	on the y-axis	on the x-axis
Vertices	x-intercepts	y-intercepts
Co-Vertices	y-intercepts	x-intercepts
Foci	on the x-axis	on the y-axis

The Standard Form Equation of a Horizontal Ellipse Centered at the Origin

The standard form equation of a horizontal ellipse centered at the origin can be found using the distances from the origin to the foci, vertices, and co-vertices. Suppose there is a horizontal ellipse centered at the origin such that

- c is the distance between a focus and the origin,
- a is the distance between a vertex and the origin, and
- b is the distance between a co-vertex and the origin.

IMPORTANT

The distance between an ellipse's vertices is greater than or equal to the distance between its co-vertices (they are equal when the ellipse is a circle), so $a \geq b > 0$.

This ellipse's foci and vertices are on the x-axis and its co-vertices are on the y-axis (because it is horizontal and centered at the origin), so its foci occur at $(\pm c, 0)$, vertices at $(\pm a, 0)$, and co-vertices at $(0, \pm b)$.

By the definition of an ellipse, the sum of the distances from any point on the ellipse to the foci is a constant. In other words, for any point on an ellipse, the sum of the distance from the point to a focus and the distance from the point to the other focus is always the same. Use the coordinates of the vertex $(a, 0)$ and the foci $(\pm c, 0)$ to find an expression for that constant.

REMEMBER

The foci and vertices are on the x-axis, so the distance between a focus and a vertex is given by subtracting the x-coordinates.

$$\underset{(c,\,0)\text{ to }(a,\,0)}{\text{distance from}} + \underset{(-c,\,0)\text{ to }(a,\,0)}{\text{distance from}} = (a - c) + (a - (-c)) = a - c + a + c = 2a$$

Thus, the sum of the distances from any point on the ellipse to the foci is $2a$.

This fact can be used to find a relationship between a, b, and c, and to write a general equation for the ellipse. Since the co-vertex $(0, b)$ is a point on the ellipse, the sum of the distances from $(0, b)$ to the foci must be $2a$. Use the Distance Formula to write an expression for this sum.

$$\underset{(c,\,0)\text{ to }(0,\,b)}{\text{distance from}} + \underset{(-c,\,0)\text{ to }(0,\,b)}{\text{distance from}} = \sqrt{(0-c)^2+(b-0)^2}+\sqrt{(0-(-c))^2+(b-0)^2}$$
$$= \sqrt{c^2+b^2}+\sqrt{c^2+b^2}$$
$$= 2\sqrt{c^2+b^2}$$

IMPORTANT

If $a = b$, the lengths of the ellipse's major and minor axes are equal and the ellipse is a circle.

Therefore, $2\sqrt{c^2+b^2} = 2a$. Simplifying and then squaring both sides gives $c^2 + b^2 = a^2$. So, the relationship between a, b, and c is given by the equation $c^2 = a^2 - b^2$, where $a \ge b > 0$.

Now we can use the sum of the distances from a point on the ellipse (x, y) to the foci (which is equal to $2a$) to write an equation for the ellipse in terms of x and y. Again, use the Distance Formula to write an expression for this sum.

$$\underset{(c,\,0)\text{ to }(x,\,y)}{\text{distance from}} + \underset{(-c,\,0)\text{ to }(x,\,y)}{\text{distance from}} = \sqrt{(x-c)^2+(y-0)^2}+\sqrt{(x-(-c))^2+(y-0)^2}$$

Thus, $\sqrt{(x-c)^2+y^2}+\sqrt{(x+c)^2+y^2} = 2a$. Using quite a bit of algebra gives $\frac{x^2}{a^2}+\frac{y^2}{b^2}=1$, which is the standard form equation of a horizontal ellipse centered at the origin.

The following table summarizes the standard form equations for a horizontal and vertical ellipses and the relationships between these equations and the attributes of ellipses.

IMPORTANT

When writing the equation of an ellipse, first determine whether the ellipse is horizontal or vertical, then choose the appropriate standard form equation.

Standard Form Equations of Ellipses: Centered at the Origin

	Horizontal Major Axis	**Vertical Major Axis**
Equation	$\frac{x^2}{a^2}+\frac{y^2}{b^2}=1 \quad a \ge b$	$\frac{y^2}{a^2}+\frac{x^2}{b^2}=1 \quad a \ge b$
Vertices	$(\pm a, 0)$	$(0, \pm a)$
Foci	$(\pm c, 0)$	$(0, \pm c)$
Co-Vertices	$(0, \pm b)$	$(\pm b, 0)$
	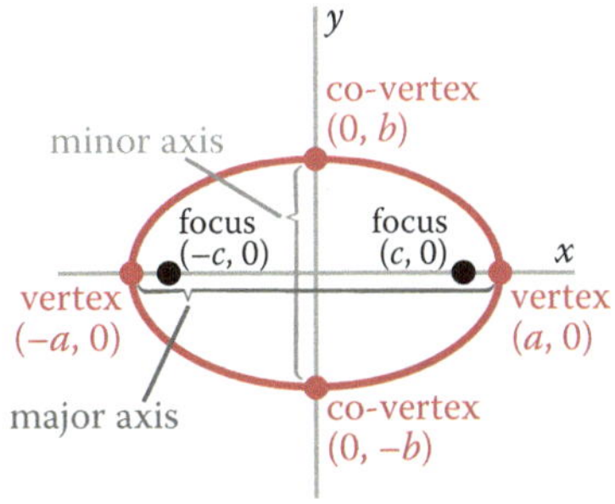	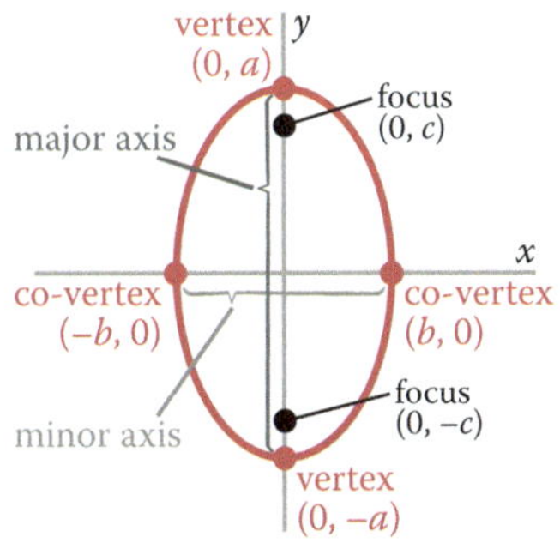

Writing the Equation of an Ellipse

The distance from the origin to a vertex a and the distance from the origin to a co-vertex b can be used to write the equation of a horizontal or vertical ellipse centered at the origin.

Steps for Writing the Equation of a Horizontal or Vertical Ellipse Centered at (0, 0)

❶ Determine the type of ellipse (horizontal or vertical).

❷ Find the values of a and b.

❸ Substitute the a- and b-values into the appropriate standard form equation.

EXAMPLE 1 Writing the Equation of an Ellipse Given Its Graph

Write the equation of each ellipse.

A.

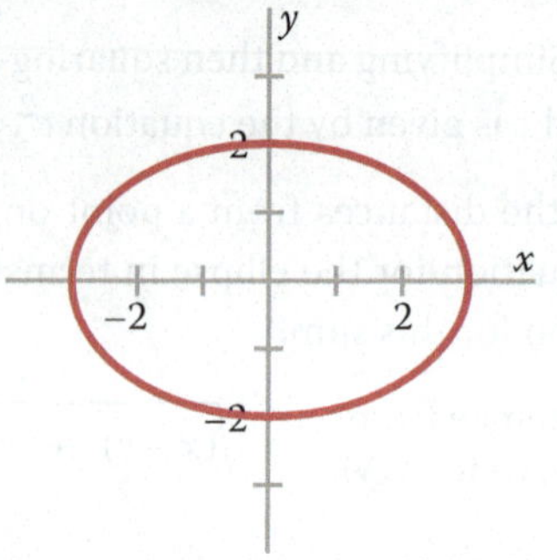

B.

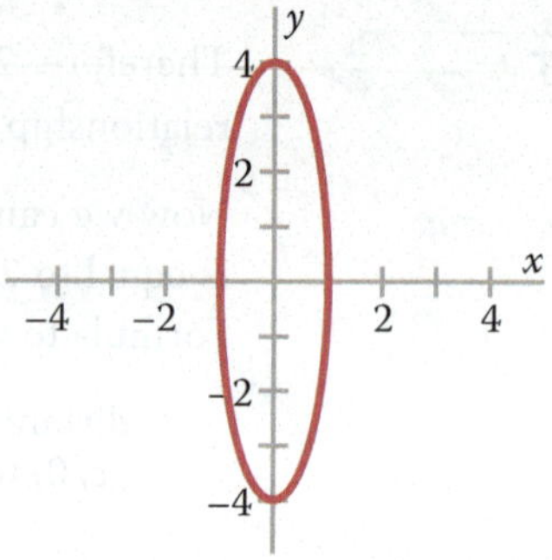

SOLUTION

A. From the graph, the x- and y-intercepts are at (±3, 0) and (0, ±2), respectively, and the center is at (0, 0).

❶ Since the major axis is horizontal, the ellipse is horizontal.

❷ The distance from the center to either vertex is 3, so $a = 3$.
The distance from the center to either co-vertex is 2, so $b = 2$.

❸ Substitute the values of a and b into the standard form equation of a horizontal ellipse.

$$\frac{x^2}{a^2}+\frac{y^2}{b^2}=1 \Rightarrow \frac{x^2}{3^2}+\frac{y^2}{2^2}=1 \Rightarrow \frac{x^2}{9}+\frac{y^2}{4}=1$$

B. From the graph, the x- and y-intercepts are at (±1, 0) and (0, ±4), respectively, and the center is at (0, 0).

❶ Since the major axis is vertical, the ellipse is vertical.

❷ The distance from the center to either vertex is 4, so $a = 4$.
The distance from the center to either co-vertex is 1, so $b = 1$.

❸ Substitute the values of a and b into the standard form equation of a horizontal ellipse.

$$\frac{y^2}{a^2}+\frac{x^2}{b^2}=1 \Rightarrow \frac{y^2}{4^2}+\frac{x^2}{1^2}=1 \Rightarrow \frac{y^2}{16}+\frac{x^2}{1}=1 \Rightarrow \frac{y^2}{16}+x^2=1$$

EXAMPLE 2 Writing the Equation of an Ellipse Given a Vertex and a Co-Vertex

Write the equation of an ellipse with center at (0, 0), a vertex at (0, –5), and a co-vertex at (4, 0).

SOLUTION

❶ The ellipse is centered at the origin, and a vertex is at (0, –5). So, the other vertex is at (0, 5). The vertices are on the y-axis, so the major axis must be vertical, and so the ellipse is vertical.

❷ The distance from the center to a vertex is 5, so $a = 5$.
The distance from the center to a co-vertex is 4, so $b = 4$.

❸ Substitute $a = 5$ and $b = 4$ into the standard form equation of a vertical ellipse.

$$\frac{y^2}{a^2} + \frac{x^2}{b^2} = 1 \Rightarrow \frac{y^2}{5^2} + \frac{x^2}{4^2} = 1 \Rightarrow \frac{y^2}{25} + \frac{x^2}{16} = 1$$

EXAMPLE 3 Writing the Equation of an Ellipse Given the Length of the Major and Minor Axes

Write the equation of an ellipse centered at the origin with a vertical minor axis of length 2 units and a major axis of length 16 units.

SOLUTION

❶ The ellipse has a vertical minor axis and is centered at the origin. Therefore, the major axis must be horizontal, and so the ellipse is horizontal.

IMPORTANT

The length of the major axis of an ellipse is 2a. The length of the minor axis of an ellipse is 2b.

❷ The lengths of the major and minor axes are equal to $2a$ and $2b$, respectively. The major axis is 16 units, so $2a = 16$, giving $a = 8$. The minor axis is 2 units, so $2b = 2$, giving $b = 1$.

❸ Substitute $a = 8$ and $b = 1$ into the standard form equation of a horizontal ellipse.

$$\frac{x^2}{a^2} + \frac{y^2}{b^2} = 1 \Rightarrow \frac{x^2}{8^2} + \frac{y^2}{1^2} = 1 \Rightarrow \frac{x^2}{64} + \frac{y^2}{1} = 1 \Rightarrow \frac{x^2}{64} + y^2 = 1$$

EXAMPLE 4 Writing the Equation of an Ellipse Given Intercepts and Foci

IMPORTANT

The foci of an ellipse are always on the ellipse's major axis.

Write the equation of an ellipse with center at (0, 0), y-intercepts at (0, ±8), and foci at (±6, 0).

SOLUTION

❶ The ellipse has foci on the x-axis at (±6, 0). Since foci are always located on the major axis, the major axis must be horizontal. So, the ellipse is horizontal.

❷ Since the major axis is horizontal, the minor axis must be vertical and the y-intercepts of the ellipse are therefore the co-vertices. Since the y-intercepts are at (0, ±8), $b = 8$.

IMPORTANT

$c^2 = a^2 - b^2$ can be used to find a when b and c are known.

Use $c^2 = a^2 - b^2$ to find a. The foci are at (±6, 0), so $c = 6$.

$$6^2 = a^2 - 8^2 \Rightarrow 36 = a^2 - 64 \Rightarrow a^2 = 100 \Rightarrow a = 10$$

❸ Substitute $a = 10$ and $b = 8$ into the standard form equation of a horizontal ellipse.

$$\frac{x^2}{a^2} + \frac{y^2}{b^2} = 1 \Rightarrow \frac{x^2}{10^2} + \frac{y^2}{8^2} = 1 \Rightarrow \frac{x^2}{100} + \frac{y^2}{64} = 1$$

EXAMPLE 5 Writing the Equation of an Ellipse Given a Focus and the Length of the Major Axis

Write the equation of an ellipse with center at (0, 0), a focus at (0, 5), and a 26 unit major axis.

SOLUTION

❶ The ellipse is centered at the origin and has a focus on the y-axis at (0, 5). Therefore, the other focus must be at (0, –5). Since the foci are on the y-axis, the ellipse's major axis must be vertical, and so the ellipse is vertical.

❷ Since the major axis is 26 units, $2a = 26$, so $a = 13$.

Use the equation $c^2 = a^2 - b^2$ to find b. The foci are at (±6, 0), so $c = 6$.

$$5^2 = 13^2 - b^2 \Rightarrow 25 = 169 - b^2 \Rightarrow b^2 = 144 \Rightarrow b = 12$$

❸ Substitute $a = 13$ and $b = 12$ into the standard form equation of a vertical ellipse.

$$\frac{y^2}{a^2} + \frac{x^2}{b^2} = 1 \Rightarrow \frac{y^2}{13^2} + \frac{x^2}{12^2} = 1 \Rightarrow \frac{y^2}{169} + \frac{x^2}{144} = 1$$

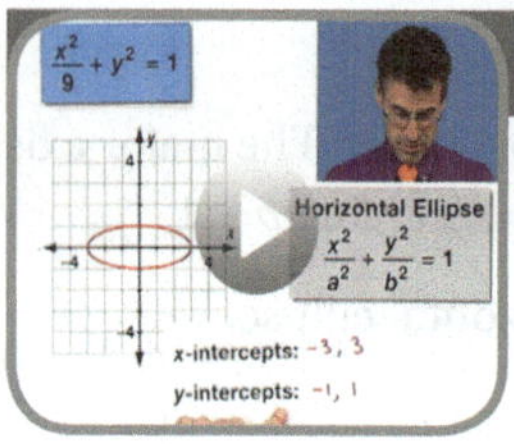

3.3.2 Graphing Ellipses

The values of a and b can be determined from the equation of an ellipse, as demonstrated in **Example 6**. The values of a and b can then be used to identify the ellipse's attributes, such as its vertices, co-vertices, and intercepts. As in Topic 3.3.1, all of the ellipses in Topic 3.3.2 will be centered at the origin.

EXAMPLE 6 Finding the Intercepts of an Ellipse

Find the x- and y-intercepts of each ellipse.

A. $\frac{x^2}{25} + \frac{y^2}{16} = 1$ **B.** $\frac{y^2}{49} + \frac{x^2}{16} = 1$

SOLUTION

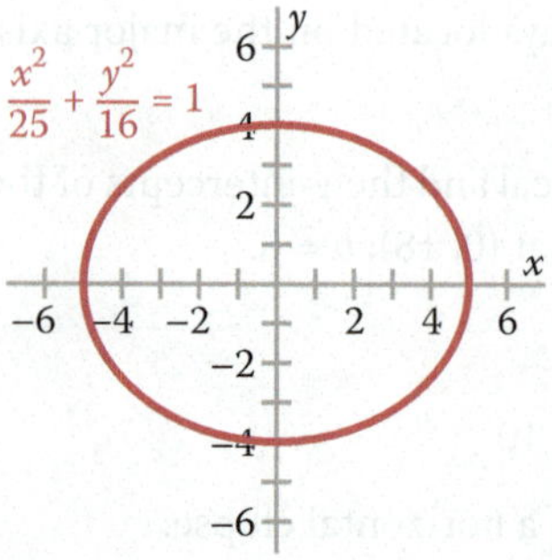

Figure 3.3f

A. The equation is of the form $\frac{x^2}{a^2} + \frac{y^2}{b^2} = 1$, where $a \geq b$, so the ellipse is horizontal and centered at the origin, with intercepts at $(\pm a, 0)$ and $(0, \pm b)$. Use the equation to find a and b.

$$a^2 = 25 \Rightarrow a = 5 \qquad b^2 = 16 \Rightarrow b = 4$$

The intercepts are at (±5, 0) and (0, ±4). (Figure 3.3f shows the graph of the ellipse.)

B. The equation is of the form $\frac{y^2}{a^2} + \frac{x^2}{b^2} = 1$, where $a \geq b$, so the ellipse is vertical and centered at the origin, with intercepts at $(\pm b, 0)$ and $(0, \pm a)$. Use the equation to find a and b.

$$a^2 = 49 \Rightarrow a = 7 \qquad b^2 = 16 \Rightarrow b = 4$$

The intercepts are at (±4, 0) and (0, ±7).

EXAMPLE 7 Finding the Length of the Major and Minor Axes of an Ellipse

Find the lengths of the major and minor axes of each ellipse.

A. $\frac{x^2}{12} + y^2 = 1$

B. $5y^2 + 9x^2 = 45$

SOLUTION

Find a and b, then use these values to find the lengths of the major and minor axes, which are $2a$ and $2b$, respectively.

TIP

The denominator of the y^2-term is 1.

A. From the equation, $a^2 = 12$ and $b^2 = 1$, since $a \geq b$.

$$a^2 = 12 \Rightarrow a = \sqrt{12} = \sqrt{4 \cdot 3} = 2\sqrt{3} \qquad b^2 = 1 \Rightarrow b = 1$$

major axis: $2(2\sqrt{3}) = 4\sqrt{3}$ units   minor axis: $2(1) = 2$ units

IMPORTANT

Write the equation of an ellipse in standard form in order to find a and b.

B. The right side of the equation is 45, not 1, so the equation is not in the standard form. The equation must be written in the standard form before a and b can be found. Therefore, divide each side by 45 to get 1 on the right side.

$$5y^2 + 9x^2 = 45 \Rightarrow \frac{5y^2}{45} + \frac{9x^2}{45} = \frac{45}{45} \Rightarrow \frac{y^2}{9} + \frac{x^2}{5} = 1$$

From the standard form equation, $a^2 = 9$ and $b^2 = 5$, since $a \geq b$.

$$a^2 = 9 \Rightarrow a = 3 \qquad b^2 = 5 \Rightarrow b = \sqrt{5}$$

major axis: $2(3) = 6$ units   minor axis: $2(\sqrt{5}) = 2\sqrt{5}$ units

Graphing a Horizontal or Vertical Ellipse Centered at the Origin

The following table summarizes three methods for graphing a horizontal or vertical ellipse centered at the origin.

Graphing a Horizontal or Vertical Ellipse Centered at the Origin Given Its Equation

Method 1: Use the Intercepts
Find the intercepts, then sketch the ellipse through the intercepts.

Method 2: Use the Major and Minor Axes
Find the lengths of the major and minor axes (line segments, each with midpoint at the center). Next, plot the endpoints of the major and minor axes on the x- and y-axes (as appropriate), and then sketch the ellipse through the four points.

Method 3: Use a Graphing Calculator
Solve the equation for y, and then enter the two equations in a graphing calculator.

EXAMPLE 8 Graphing an Ellipse

Graph the equation from **Example 6B**, $\frac{y^2}{49} + \frac{x^2}{16} = 1$.

SOLUTION

Method 1: Use the Intercepts
From **Example 6B**, the intercepts are at $(\pm 4, 0)$ and $(0, \pm 7)$. First plot these four points, and then sketch the ellipse (centered at the origin, and with vertical and horizontal axes of symmetry) through the points. The graph of the ellipse is shown in Figure 3.3g.

Method 2: Use the Major and Minor Axes

The equation is of the form $\frac{y^2}{a^2}+\frac{x^2}{b^2}=1$, where $a \geq b$, so

- the ellipse is vertical and centered at the origin,
- the major axis is vertical (on the y-axis), and
- the minor axis is horizontal (on the x-axis).

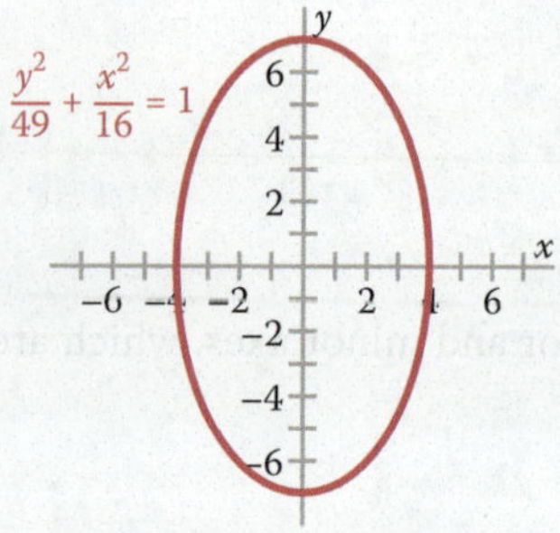

Figure 3.3g

Use a and b to find the lengths of the major and minor axes. From **Example 6B**, $a = 7$ and $b = 4$.

major axis: $2a = 2(7) = 14$ units   minor axis: $2b = 2(4) = 8$ units

Each axis is a line segment with midpoint at the origin. So, the 14 unit major axis extends from −7 to 7 on the y-axis and the 8 unit minor axis extends from −4 to 4 on the x-axis. Therefore, the endpoints of the axes are at (0, −7), (0, 7), (−4, 0), and (4, 0). The ellipse passes through the endpoints of the major and minor axes, so sketch the ellipse through these points (Figure 3.3g).

Method 3: Use a Graphing Calculator

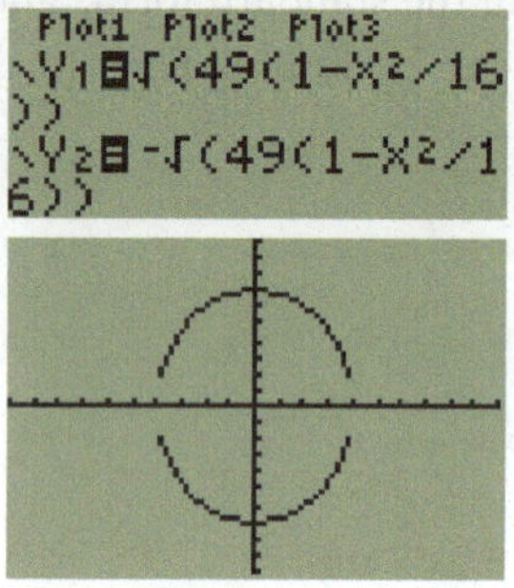

Figure 3.3h

Solve the equation for y.

$$\frac{y^2}{49}+\frac{x^2}{16}=1$$

$$\frac{y^2}{49}=1-\frac{x^2}{16}$$

$$y^2=49\left(1-\frac{x^2}{16}\right)$$

$$y=\pm\sqrt{49\left(1-\frac{x^2}{16}\right)}$$

The equation must be entered in two parts because it contains a ± symbol.

$$y=\sqrt{49\left(1-\frac{x^2}{16}\right)} \text{ and } y=-\sqrt{49\left(1-\frac{x^2}{16}\right)}$$

The calculator notation follows.

$$Y1=\sqrt{\ }(49(1-X^2/16))$$

$$Y2=-\sqrt{\ }(49(1-X^2/16))$$

The graph of the ellipse is shown in Figure 3.3h.

The Foci of an Ellipse

REMEMBER

For any ellipse, the distance between its center and a vertex is a, the distance between its center and a co-vertex is b, and a ≥ b > 0.

Recall that each focus of an ellipse is a point c units from the center on the ellipse's major axis. This distance c can be found using the relationship between the distances a and b, $c^2 = a^2 - b^2$.

EXAMPLE 9 **Finding the Foci of an Ellipse**

Find the foci of each ellipse.

A. $\frac{y^2}{9}=1-x^2$ **B.** $100x^2+50y^2-200=0$

SOLUTION

IMPORTANT

For the equation of an ellipse to be in standard form, one side of the equation must be equal to 1.

First, write each equation in standard form. Then, using the standard form equation, determine a^2, b^2, and whether the ellipse is horizontal with foci at ($\pm c$, 0) or vertical with foci at (0, $\pm c$).

A. Add x^2 to each side to write the equation in standard form. $\frac{y^2}{9}=1-x^2 \Rightarrow \frac{y^2}{9}+x^2=1$

From the equation, $a^2 = 9$ and $b^2 = 1$, because $a \geq b$. Find c.

$$c^2=a^2-b^2=9-1=8 \Rightarrow c=2\sqrt{2}$$

Since the equation is of the form $\frac{y^2}{a^2}+\frac{x^2}{b^2}=1$ where $a \geq b$, the ellipse is vertical with foci at (0, $\pm c$). Thus, the foci of the ellipse are at $(0, \pm 2\sqrt{2})$.

B. Add 200 to each side, then divide each side by 200 to write the equation in standard form.

$$100y^2+50x^2-200=0 \Rightarrow 100y^2+50x^2=200 \Rightarrow \frac{100y^2}{200}+\frac{50x^2}{200}=\frac{200}{200} \Rightarrow \frac{x^2}{4}+\frac{y^2}{2}=1$$

From the equation, $a^2=4$ and $b^2=2$, because $a \geq b$. Find c.

$$c^2=a^2-b^2=4-2=2 \quad \Rightarrow \quad c=\sqrt{2}$$

Since the equation is of the form $\frac{x^2}{a^2}+\frac{y^2}{b^2}=1$ where $a \geq b$, the ellipse is horizontal with foci at $(\pm c, 0)$. Thus, the foci of the ellipse are at $(\pm\sqrt{2}, 0)$.

TIP

The value of a^2 in $\frac{y^2}{2}+\frac{x^2}{4}=1$ is not 2. Since $a \geq b$ in the standard form equation of an ellipse, $a^2=4$.

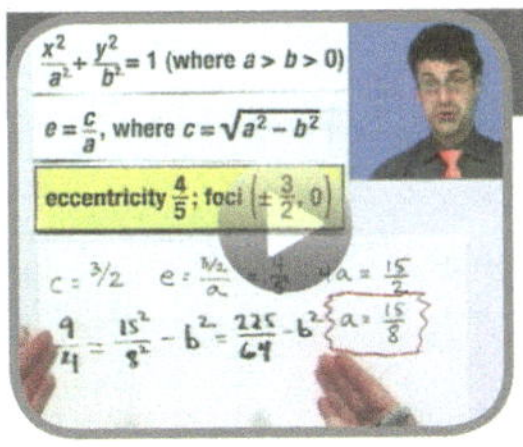

3.3.3 The Eccentricity of an Ellipse

One ellipse can be compared with another by comparing the degree to which each is stretched (elongated vertically or horizontally). For example, the first ellipse in Figure 3.3i can be described as "less stretched" than the second ellipse in Figure 3.3i.

less stretched *more stretched*

The first ellipse can also be described as "more circular" than the second ellipse.

Figure 3.3i

IMPORTANT

A circle is the special case of an ellipse where both foci are located at the center. So, for any circle, the distance between the foci and the center is 0 (i.e., $c=0$). Therefore, the eccentricity of any circle is $0/a$, or 0.

The degree to which an ellipse is stretched is called the it's **eccentricity**. Eccentricity is measured using the ratio of c to a (i.e., the ratio of the distance between its center and a focus and the distance between its center and a vertex). Because $c \leq a$, the ratio c/a is always a number between 0 and 1.

The Eccentricity of an Ellipse

The eccentricity e of an ellipse is the ratio of c to a, $e=\frac{c}{a}$, where c is the distance from the center to a focus and a is the distance from the center to a vertex.

The eccentricity of every ellipse is a number between 0 and 1. Ellipses with eccentricity close to 0 are more circular, and ellipses with eccentricity close to 1 are more stretched.

Examples of Eccentricity

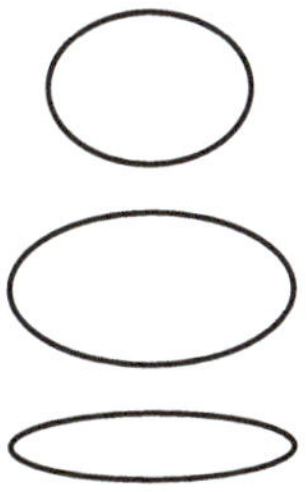

$e=0.24$

$e=0.6$

$e=0.87$

Figure 3.3j

Three examples of ellipses and their eccentricity are given in Figure 3.3j. These examples illustrate that an ellipse with an eccentricity close to 0 will be more circular, and an ellipse with an eccentricity close to 1 will be more stretched.

EXAMPLE 10 Finding the Eccentricity of an Ellipse

Find the eccentricity of each ellipse.

A. $x^2+4y^2=4$

B. $18x^2+9y^2=162$

SOLUTION

The eccentricity of an ellipse is given by c/a. Write each equation in standard form, find a, and then use $c^2=a^2-b^2$ to find c.

A. Divide each side by 4 to write the equation in standard form.

$$x^2 + 4y^2 = 4 \Rightarrow \frac{x^2}{4} + \frac{4y^2}{4} = \frac{4}{4} \Rightarrow \frac{x^2}{4} + y^2 = 1$$

From the equation, $a^2 = 4$ and $b^2 = 1$, since $a \geq b$. Therefore, $a = 2$. Find c.

$$c^2 = a^2 - b^2 = 4 - 1 = 3 \quad \Rightarrow \quad c = \sqrt{3}$$

Substitute the values of a and c into the eccentricity formula. $e = \dfrac{c}{a} = \dfrac{\sqrt{3}}{2}$

B. Divide each side by 162 to write the equation in standard form.

$$18x^2 + 9y^2 = 162 \Rightarrow \frac{18x^2}{162} + \frac{9y^2}{162} = \frac{162}{162} \Rightarrow \frac{y^2}{18} + \frac{x^2}{9} = 1$$

From the equation, $a^2 = 18$ and $b^2 = 9$, since $a \geq b$. Therefore, $a = 3\sqrt{2}$. Find c.

$$c^2 = a^2 - b^2 = 18 - 9 = 9 \quad \Rightarrow \quad c = 3$$

Substitute the values of a and c into the eccentricity formula.

$$e = \frac{c}{a} = \frac{3}{3\sqrt{2}} = \frac{1}{\sqrt{2}} = \frac{1}{\sqrt{2}} \cdot \frac{\sqrt{2}}{\sqrt{2}} = \frac{\sqrt{2}}{2}$$ *Rationalize the denominator.*

EXAMPLE 11 Using the Eccentricity of an Ellipse to Find Its Vertices

Find the vertices of an ellipse centered at the origin with eccentricity 3/4 and foci at $(0, \pm 6)$.

SOLUTION

The foci are at $(0, \pm 6)$. Therefore, $c = 6$. Use the eccentricity formula to solve for a.

$$e = \frac{c}{a} \Rightarrow \frac{3}{4} = \frac{6}{a} \Rightarrow 3a = 24 \Rightarrow a = 8$$

The foci are on the y-axis, and the ellipse is centered at the origin. Therefore, the vertices must be located at $(0, \pm a)$, that is, $(0, \pm 8)$.

EXAMPLE 12 Using the Eccentricity of an Ellipse to Write the Equation of the Ellipse

Write the equation of the ellipse centered at (0, 0) with eccentricity 1/2 and vertices at $(\pm 4, 0)$.

SOLUTION

The vertices are at $(\pm 4, 0)$. So, $a = 4$ and $a^2 = 16$. Use the eccentricity formula to solve for c.

$$e = \frac{c}{a} \Rightarrow \frac{1}{2} = \frac{c}{4} \Rightarrow 2c = 4 \Rightarrow c = 2$$

Use the equation $c^2 = a^2 - b^2$ to find b^2. $\quad 2^2 = 16 - b^2 \Rightarrow b^2 = 12$

The vertices are on the x-axis. Therefore, the ellipse is of the form $\dfrac{x^2}{a^2} + \dfrac{y^2}{b^2} = 1$.

$$\frac{x^2}{a^2} + \frac{y^2}{b^2} = 1 \Rightarrow \frac{x^2}{16} + \frac{y^2}{12} = 1$$

SECTION 3.3 EXERCISES

Warm Up

Solve each equation.

1. $c^2 = a^2 - b^2$; when $a = 5$ and $b = 1$
2. $c^2 = a^2 - b^2$; when $a^2 = 70$ and $c^2 = 20$
3. Rationalize the denominator. $\dfrac{2}{5\sqrt{3}}$

Just the Facts

Fill in the blanks.

4. An ellipse is the set of all points such that the ____ of the distances from two fixed points, called the ____, is constant.
5. A horizontal ellipse centered at the origin has its major axis on the ____-axis and its minor axis on the ____-axis.
6. A vertical ellipse centered at the origin has its major axis on the ____-axis and its minor axis on the ____-axis.
7. The ____, ____, and ____ of a vertical ellipse centered at the origin are on the y-axis.
8. The ____ and ____ of a vertical ellipse centered at the origin are on the x-axis.
9. An ellipse's co-vertices are the endpoints of its ____ .
10. An ellipse's vertices are the endpoints of its ____ .
11. The equation $\dfrac{x^2}{a^2} + \dfrac{y^2}{b^2} = 1$ where $a \geq b$ describes a ____ ellipse centered at the ____ with vertices at ____ and co-vertices at ____.
12. The equation $\dfrac{y^2}{a^2} + \dfrac{x^2}{b^2} = 1$ where $a \geq b$ describes a ____ ellipse centered at the ____ with vertices at ____ and co-vertices at ____.

Essential Skills

In Exercises 1–2, use the graph to write the equation of each ellipse.

1.

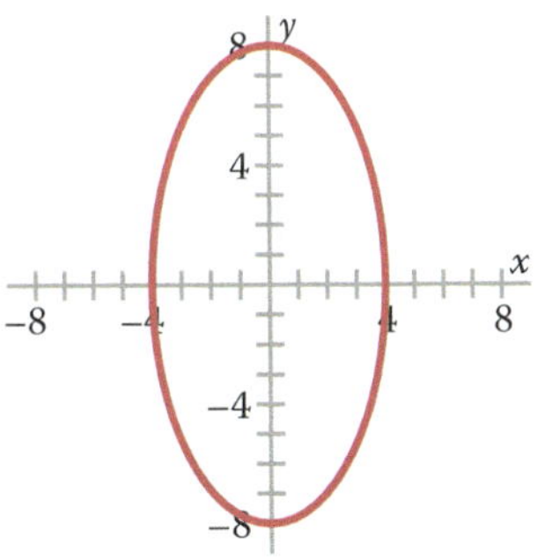

2.

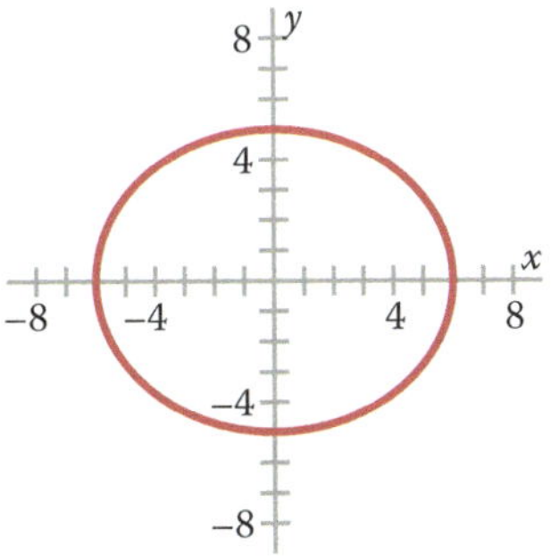

In Exercises 3–10, use the given information to write the equation of each ellipse in standard form. Assume that each ellipse is centered at the origin.

3. a vertex at (−3, 0) and a co-vertex at (0, −1)
4. a vertex at (0, 8) and a co-vertex at (2, 0)
5. a vertical major axis of length 8 and a horizontal minor axis of length 2
6. a vertical minor axis of length 10 and a horizontal major axis of length 14
7. x-intercepts at ±12 and foci located at (0, 5) and (0, −5)
8. x-intercepts at 5 and −5 and foci located at (4, 0) and (−4, 0)
9. minor axis of length 12 and foci located at (±8, 0)
10. major axis of length 8 and foci located at (0, 2) and (0, −2)

In Exercises 11–12, find the intercepts of each ellipse.

11. $\dfrac{y^2}{400} + \dfrac{x^2}{121} = 1$
12. $\dfrac{x^2}{16} + \dfrac{y^2}{9} = 1$

In Exercises 13–14, find the lengths of the major and minor axes of each ellipse.

13. $2x^2 = 36 - 6y^2$
14. $8x^2 + y^2 = 8$

In Exercises 15–18, graph each ellipse.

15. $\dfrac{y^2}{4} + x^2 = 1$
16. $\dfrac{x^2}{64} + \dfrac{y^2}{25} = 1$
17. $1 - \dfrac{y^2}{16} = x^2$
18. $4x^2 + 25y^2 = 100$

In Exercises 19–20, find the foci of each ellipse.

19. $\dfrac{x^2}{12} + \dfrac{y^2}{32} = 1$
20. $\dfrac{x^2}{100} + \dfrac{y^2}{36} = 1$

In Exercises 21–24, find the eccentricity of each ellipse.

21. $\dfrac{y^2}{9} + x^2 = 1$
22. $\dfrac{x^2}{20} + \dfrac{y^2}{4} = 1$
23. $\dfrac{14y^2}{25} + 2x^2 = 28$
24. $6x^2 + 12y^2 = 24$
25. Find the foci of the ellipse centered at the origin with eccentricity 2/5 and vertices at (0, 3) and (0, −3).
26. Find the vertices of the ellipse centered at the origin with eccentricity 1/3 and foci at (2, 0) and (−2, 0).

In Exercises 27–28, use the given information to write the equation of each ellipse. Assume that each ellipse is centered at the origin.

27. $e = \frac{1}{10}$ and vertices at $(0, \pm 20)$

28. $e = \frac{2}{3}$ and foci at $\left(\pm\frac{3}{2}, 0\right)$

Extensions

In Exercises 29–30, find the center, foci, major axis length, minor axis length, vertices, co-vertices, and eccentricity for each ellipse. Then sketch the graph.

29. $\frac{y^2}{2} + 2x^2 = 18$ **30.** $\frac{3x^2}{4} + 3y^2 = 12$

In Exercises 31–32, write each equation as it would be entered into a graphing calculator (i.e., solve for y).

31. $\frac{y^2}{18} + \frac{x^2}{7} = 1$ **32.** $\frac{14x^2}{25} + 2y^2 = 28$

In Exercises 33–45, write the equation of two ellipses that satisfy the given conditions. Assume that each ellipse is centered at the origin. Check the equations using a graphing calculator.

33. a major axis of length 12

34. a minor axis of length 12

35. intercepts at ±4 and ±7

36. co-vertices at (0, ±1)

37. co-vertices at (±1, 0)

38. vertices at (0, ±4)

39. vertices at (±4, 0)

40. foci at ±6

41. foci at ±3

42. a major axis of length 10 and a minor axis of length 4

43. x-intercepts at ±5 and a vertical major axis

44. y-intercepts at ±6 and a horizontal major axis

45. eccentricity of 1/3

46. If an ellipse is centered at the origin, do the lengths of the horizontal and vertical axes provide enough information to write the equation of the ellipse?

47. Given that ellipse A has eccentricity 1/2, ellipse B has eccentricity 9/10, and ellipse C has eccentricity 2/9, list the ellipses in order from most circular to most stretched.

48. Write the standard form equation of a circle centered at the origin and the standard form equation of a circle centered at (h, k) and describe the difference between the two equations. How might the standard form equation of an ellipse centered at the origin change if the ellipse is centered at (h, k) instead of at the origin?

3.4 HYPERBOLAS

OBJECTIVES

- Connect the geometric attributes of a hyperbola (centered at the origin) to its standard form equation.
- Write the equation of a hyperbola (centered at the origin).
- Graph a hyperbola (centered at the origin).
- Find the vertices, co-vertices, foci, and asymptotes of a hyperbola (centered at the origin).

PREREQUISITE VOCABULARY TERMS

conic
Distance Formula
ellipse
***x*-intercept**
***y*-intercept**

Recall that a **hyperbola** is the conic formed by the intersection of a double cone and a plane that intersects both parts of the cone, through the cone's bases. Notice that each hyperbola in Figure 3.4a is made up of two disconnected curves, called its **branches**. The equations and graphs of hyperbolas (on a coordinate plane) will be discussed in this section.

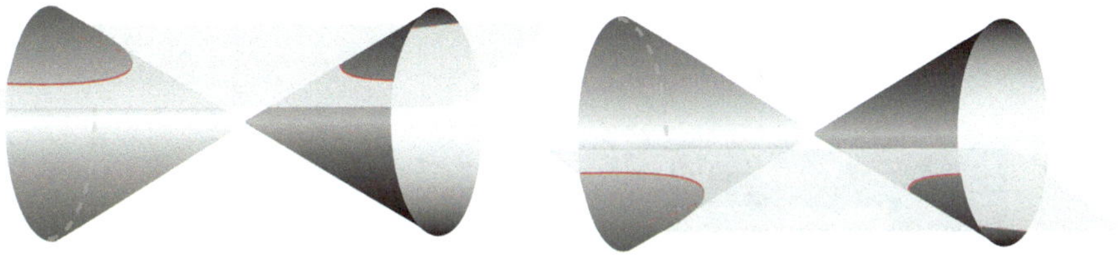

Figure 3.4a

Introduction to Hyperbolas

Like an ellipse, a hyperbola is defined on a coordinate plane using a relationship between two fixed points F_1 and F_2 called the **foci**. However, instead of the graph being all points such that the *sum* of the distances from the foci is a constant (as is the case with an ellipse), a hyperbola is the set of all points such that the *difference* of the distances from the foci is a constant.

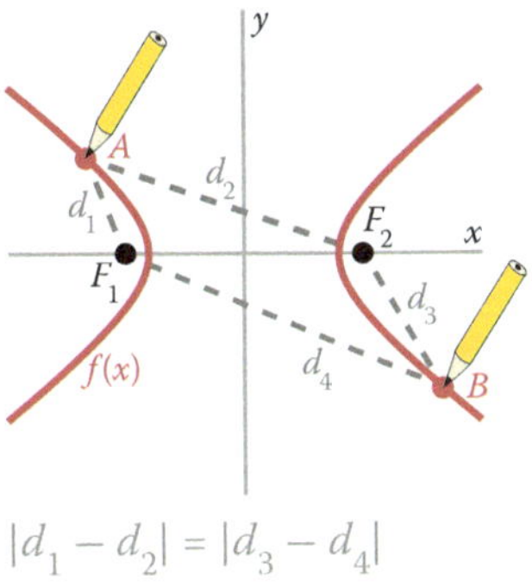

Figure 3.4b

DEFINITION

A **hyperbola** is the set of all points such that the difference of the distances from two fixed points is constant. The fixed points are the **foci** of the hyperbola.

The graph of $f(x)$ shown in Figure 3.4b is a hyperbola. The points F_1 and F_2 are its foci, and the points A and B are on the hyperbola. The distances from A to F_1 and F_2 are represented by d_1 and d_2 (as shown in the figure), and the distances from B to F_1 and F_2 are represented by d_3 and d_4 (as shown in the figure). Since a hyperbola is the set of all points such that the difference of the distances from the foci is a constant, the differences of the distances from the foci to any two points on the hyperbola $f(x)$ must be equal. Therefore, the difference of the distances from A to the foci, $|d_1 - d_2|$, must be equal to the difference of the distances from B to the foci, $|d_3 - d_4|$.

Attributes of a Hyperbola

TIP

A hyperbola's center is the midpoint between its vertices, which is also the midpoint of the transverse axis.

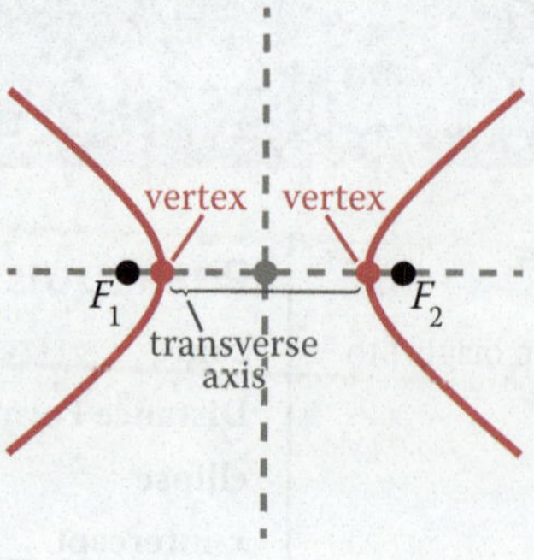

Figure 3.4c

As with an ellipse, a hyperbola is a symmetric figure with two perpendicular lines of symmetry called the axes of symmetry, intersecting at the **center of the hyperbola**. One axis of symmetry is between the branches, and the other passes through the branches and the foci, as shown in Figure 3.4c. The line segment on the axis of symmetry containing the foci, with endpoints on the hyperbola, is called the **transverse axis**. The endpoints of the transverse axis are the hyperbola's **vertices**.

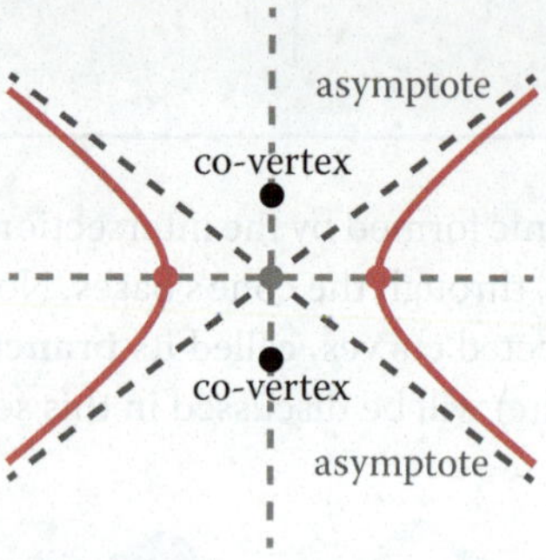

Figure 3.4d

A hyperbola has two asymptotes. Each asymptote passes through the hyperbola's center, as shown in Figure 3.4d. As with an ellipse, a hyperbola also has co-vertices on an axis of symmetry, as shown in Figure 3.4d. However, a hyperbola's co-vertices are not points on the graph. The relationship between the coordinates of the vertices and co-vertices will be discussed later in this section.

3.4.1 Writing the Equation of a Hyperbola

All of the hyperbolas discussed in this section will be centered at the origin, and will have either a horizontal or vertical transverse axis, as shown in Figure 3.4e.

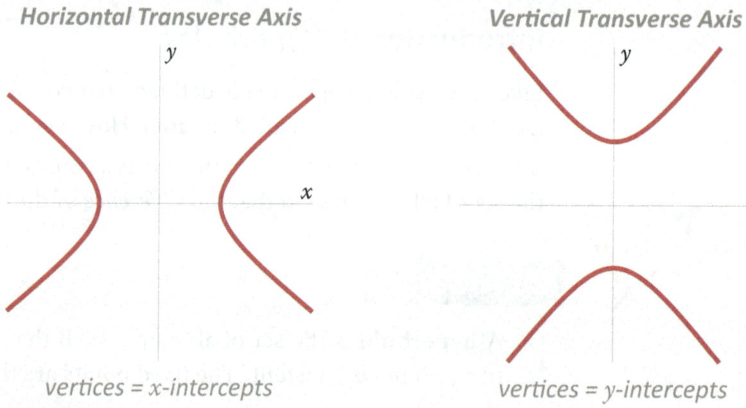

Figure 3.4e

Notice that any hyperbola *centered at the origin* with a horizontal or vertical transverse axis will be symmetric about the x- and y-axes. Therefore, a hyperbola centered at the origin with a horizontal transverse axis has vertices that are also its x-intercepts. Similarly, a hyperbola centered at the origin with a vertical transverse axis has vertices that are also its y-intercepts.

The Standard Form Equations

As with an ellipse, the standard form equations of horizontal and vertical hyperbolas centered at the origin can be derived using the Distance Formula, where a is the distance between the center and a vertex, b is the distance between the center and a co-vertex, and c is the distance between the center and a focus. For a hyperbola, the relationship between a, b, and c is defined by $c^2 = a^2 + b^2$, where $a \geq b > 0$.

REMEMBER

For an ellipse, $c^2 = a^2 - b^2$.

The following table contains the standard form equations for hyperbolas centered at the origin with a horizontal or vertical transverse axis. Additionally, the table summarizes the relationship between these equations and the attributes of the hyperbolas.

Standard Form Equations of Hyperbolas: Centered at the Origin

	Horizontal Transverse Axis (open left/right)	**Vertical Transverse Axis (open up/down)**
Equation	$\frac{x^2}{a^2} - \frac{y^2}{b^2} = 1$	$\frac{y^2}{a^2} - \frac{x^2}{b^2} = 1$
Vertices	$(\pm a, 0)$	$(0, \pm a)$
Foci	$(\pm c, 0)$	$(0, \pm c)$
Co-vertices	$(0, \pm b)$	$(\pm b, 0)$
Asymptotes	$y = \pm\frac{b}{a}x$	$y = \pm\frac{a}{b}x$

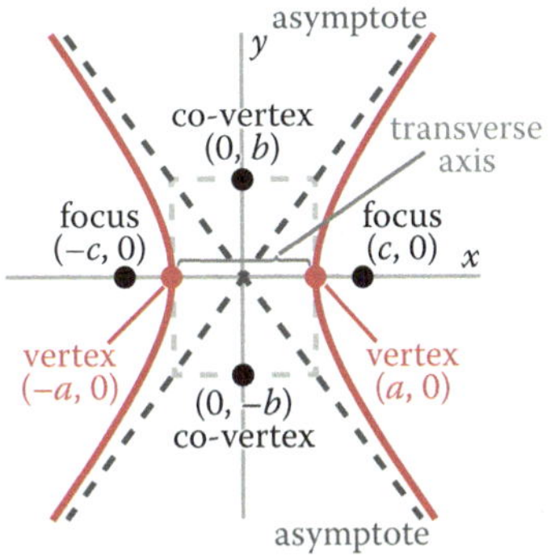

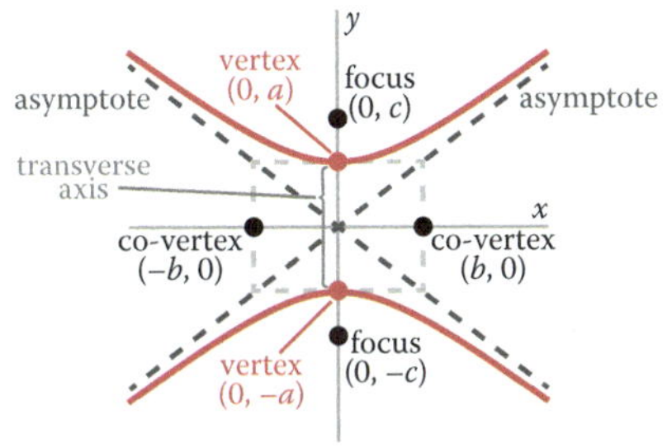

IMPORTANT

The relationship between a, b, and c is defined by the equation $c^2 = a^2 + b^2$.

IMPORTANT

The distance from the origin to a vertex is a, so the length of the transverse axis is $2a$.

The dashed-line box drawn between the hyperbola's branches is called the *central box.* Notice that the sides of the central box are horizontal and vertical lines that pass through the hyperbola's vertices and co-vertices, intersecting at the hyperbola's asymptotes.

A hyperbola's central box can be used to find its co-vertices when its vertices and asymptotes are known.

Steps for Using the Central Box to Find the Co-Vertices of a Hyperbola
Horizontal Hyperbola (centered at the origin)

1. Through each vertex, draw a vertical line segment with endpoints on the asymptotes.
2. Draw a horizontal line segment connecting those two endpoints that are above the hyperbola's center. Repeat for the endpoints below the center.
3. The y-intercept of each horizontal line is a co-vertex.

Vertical Hyperbola (centered at the origin)

1. Through each vertex, draw a horizontal line segment with endpoints on the asymptotes.
2. Draw a vertical line segment connecting those two endpoints that are to the left of the hyperbola's center. Repeat for the endpoints to the right of the center.
3. The x-intercept of each vertical line is a co-vertex.

Similarly, a hyperbola's vertices can be found by drawing the central box using the co-vertices and asymptotes.

Writing the Equation of a Hyperbola

The distance from the origin to a vertex (a) and the distance from the origin to a co-vertex (b) can be used to write the equation of a horizontal or vertical hyperbola centered at the origin.

Steps for Writing the Equation of a Horizontal or Vertical Hyperbola

❶ Determine the type of hyperbola (horizontal or vertical).

❷ Find the values of a and b.

❸ Substitute the values of a and b into the standard form equation of the hyperbola.

EXAMPLE 1 Writing the Equation of a Hyperbola Given Its Graph

Write the equation of each hyperbola.

A.

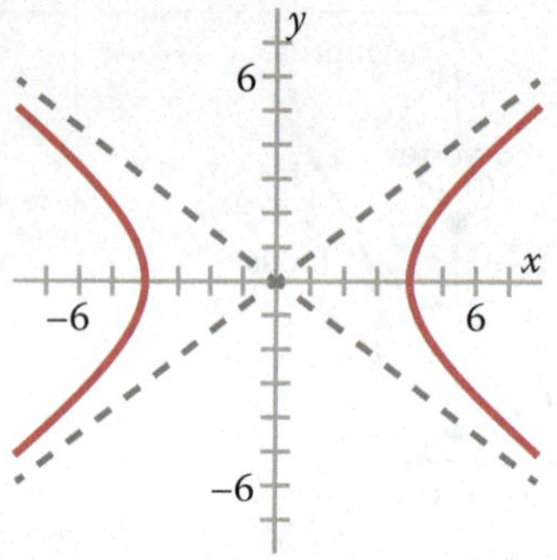

B.

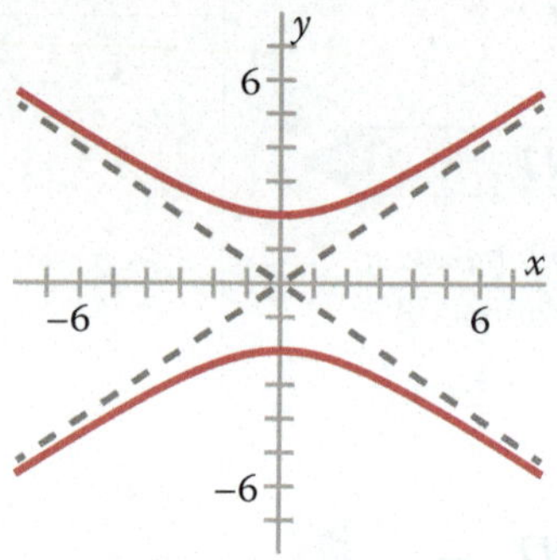

SOLUTION

A. From the graph, the vertices (x-intercepts) are at $(\pm4, 0)$, and the center is at $(0, 0)$.

❶ The hyperbola is horizontal and centered at the origin.

❷ The distance from the center to a vertex is 4, so $a = 4$.
Draw the central box to find the co-vertices (as shown in Figure 3.4f).
The distance from the center to a co-vertex is 3, so $b = 3$.

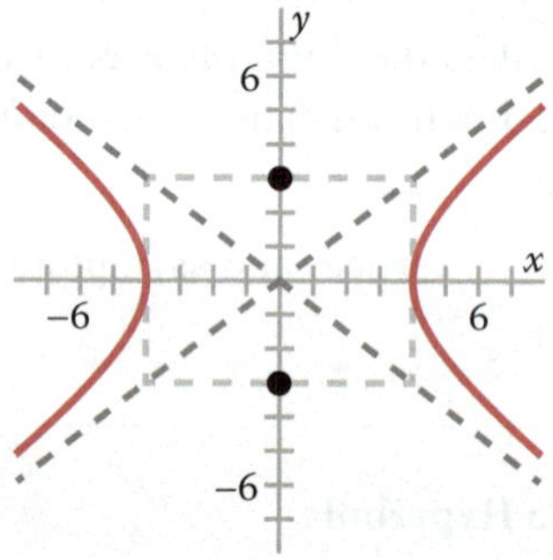

Figure 3.4f

❸ Substitute $a = 4$ and $b = 3$ into the standard form equation of a horizontal hyperbola.

$$\frac{x^2}{a^2} - \frac{y^2}{b^2} = 1 \Rightarrow \frac{x^2}{4^2} - \frac{y^2}{3^2} = 1 \Rightarrow \frac{x^2}{16} - \frac{y^2}{9} = 1$$

B. From the graph, the y-intercepts are at $(0, \pm2)$, and the center is at $(0, 0)$.

❶ The hyperbola is vertical and centered at the origin.

❷ The distance from the center to a vertex is 2, so $a = 2$.
Draw the central box to find the co-vertices (as shown in Figure 3.4g).
The distance from the center to a co-vertex is 3, so $b = 3$.

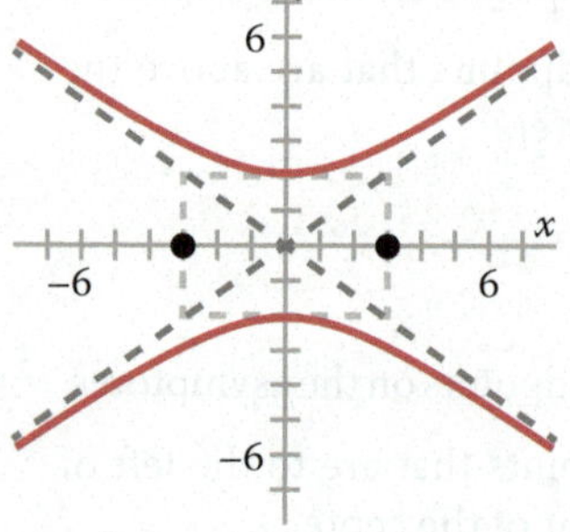

Figure 3.4g

❸ Substitute $a = 2$ and $b = 3$ into the standard form equation of a vertical hyperbola.

$$\frac{y^2}{a^2} - \frac{x^2}{b^2} = 1 \Rightarrow \frac{y^2}{2^2} - \frac{x^2}{3^2} = 1 \Rightarrow \frac{y^2}{4} - \frac{x^2}{9} = 1$$

EXAMPLE 2 Writing the Equation of a Hyperbola Given a Vertex and a Focus

Write the equation of a hyperbola centered at the origin with a vertex at (6, 0) and a focus at (8, 0).

SOLUTION

❶ The hyperbola is centered at the origin, so the other vertex must be at (–6, 0) and the other focus at (–8, 0). Since the vertices are on the x-axis, the hyperbola must be horizontal.

❷ The distance from the center to a vertex is 6, so $a = 6$. Therefore, $a^2 = 36$.

The distance from the center to a focus is 8, so $c = 8$.

Use $c^2 = a^2 + b^2$ to find b^2. $8^2 = 36 + b^2 \Rightarrow b^2 = 28$

TIP

$c^2 = a^2 + b^2$ can be used to find b when a and c are known.

❸ Substitute $a^2 = 36$ and $b^2 = 28$ into the standard form equation of a horizontal hyperbola.

$$\frac{x^2}{a^2} - \frac{y^2}{b^2} = 1 \Rightarrow \frac{x^2}{36} - \frac{y^2}{28} = 1$$

EXAMPLE 3 Writing the Equation of a Hyperbola Given the Foci and the Length of the Transverse Axis

Write the equation of a hyperbola centered at the origin with a transverse axis of length 10 units and foci at (0, ±9).

SOLUTION

❶ The hyperbola has foci at (0, ±9) and is centered at the origin. Since the foci are on the y-axis, the transverse axis must be vertical. Therefore, the hyperbola is vertical.

❷ The length of the transverse axis is equal to $2a$. Since the transverse axis is 10 units, $2a = 10$. Therefore, $a = 5$ and $a^2 = 25$.

The distance from the center to a focus is 8, so $c = 9$.

Use $c^2 = a^2 + b^2$ to find b^2. $9^2 = 25 + b^2 \Rightarrow b^2 = 56$

REMEMBER

A hyperbola's transverse axis always contains the vertices.

❸ Substitute $a^2 = 25$ and $b^2 = 56$ into the standard form equation of a vertical hyperbola.

$$\frac{y^2}{a^2} - \frac{x^2}{b^2} = 1 \Rightarrow \frac{y^2}{25} - \frac{x^2}{56} = 1$$

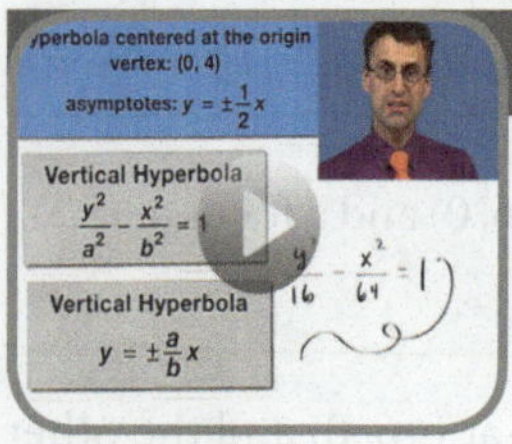

3.4.2 Writing the Equation of a Hyperbola: Another Example

The standard form equation of a hyperbola can be found by using the equations of its asymptotes, along with additional information about the hyperbola. Recall that a horizontal hyperbola has asymptotes at $y = \pm\frac{b}{a}x$, and a vertical hyperbola has asymptotes at $y = \pm\frac{a}{b}x$.

EXAMPLE 4 Writing the Equation of a Hyperbola Given the Vertices and Asymptotes

Write the equation of a hyperbola centered at the origin with vertices at (±6, 0) and asymptotes at $y = \pm\frac{4}{3}x$.

SOLUTION

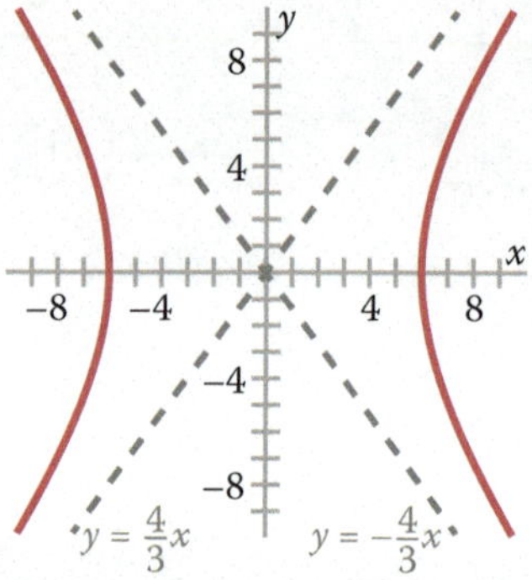

❶ Since the vertices (±6, 0) are on the x-axis, the hyperbola is horizontal.

❷ The distance from the center to a vertex is 6, so $a = 6$. Therefore, $a^2 = 36$.

The hyperbola's asymptotes are at $y = \pm\frac{4}{3}x$. The asymptotes of a horizontal hyperbola centered at the origin are given by $y = \pm\frac{b}{a}x$. Write an equation and solve for b.

$$\pm\frac{b}{a}x = \pm\frac{4}{3}x$$

$$\frac{b}{6} = \frac{4}{3} \qquad \textit{Substitute 6 for a and simplify.}$$

$$b = 8 \qquad \textit{Solve for b.}$$

Therefore, $b^2 = 64$.

❸ Substitute the values of a^2 and b^2 into the standard form equation of a horizontal hyperbola.

$$\frac{x^2}{a^2} - \frac{y^2}{b^2} = 1 \Rightarrow \frac{x^2}{36} - \frac{y^2}{64} = 1$$

TIP

Write the equation of the asymptotes to check that $a = 6$ and $b = 8$. The hyperbola is horizontal, so the hyperbola's asymptotes are $y = \pm\frac{b}{a}x = \pm\frac{8}{6}x = \pm\frac{4}{3}x$, as given in the question.

WHAT'S THE BIG IDEA?

For a horizontal or vertical hyperbola centered at the origin, explain the relationship between the equations of the asymptotes and the coordinates of the vertices and co-vertices.

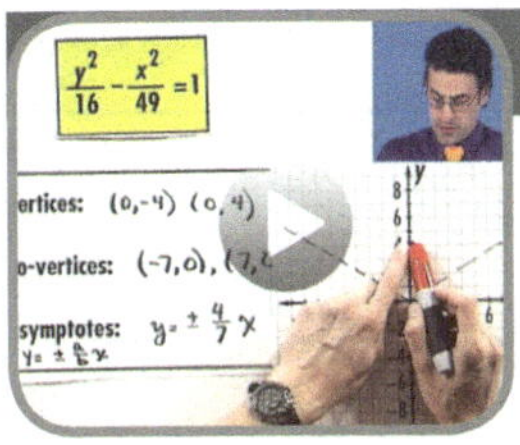

3.4.3 Graphing Hyperbolas

In this topic, the equation of a hyperbola will be used to identify the hyperbola's attributes, which will then be used to sketch the graph.

EXAMPLE 5 Finding the Vertices, Co-Vertices, Foci, and Asymptotes of a Hyperbola Given Its Equation

Find the vertices, co-vertices, foci, and asymptotes of each hyperbola.

A. $\dfrac{x^2}{8} - \dfrac{y^2}{18} = 1$ **B.** $y^2 - \dfrac{x^2}{4} = 1$

SOLUTION

A. The equation is of the form $\dfrac{x^2}{a^2} - \dfrac{y^2}{b^2} = 1$, so

- the hyperbola is horizontal, centered at the origin,
- the vertices are at $(\pm a, 0)$,
- the co-vertices are at $(0, \pm b)$,
- the foci are at $(\pm c, 0)$, and
- the asymptotes are at $y = \pm\dfrac{b}{a}x$.

Use the equation to find a and b.

$a^2 = 8 \Rightarrow a = \sqrt{8} = \sqrt{4 \cdot 2} = 2\sqrt{2}$ $b^2 = 18 \Rightarrow b = \sqrt{18} = \sqrt{9 \cdot 2} = 3\sqrt{2}$

vertices: $(\pm 2\sqrt{2}, 0)$ co-vertices: $(0, \pm 3\sqrt{2})$

Use $c^2 = a^2 + b^2$ to find c.

$c^2 = a^2 + b^2 = 8 + 18 = 26 \Rightarrow c = \sqrt{26}$ foci: $(\pm\sqrt{26}, 0)$

Use a and b to write the equation of the asymptotes.

$y = \pm\dfrac{b}{a}x = \pm\dfrac{3\sqrt{2}}{2\sqrt{2}}x = \pm\dfrac{3}{2}x$ asymptotes: $y = \pm\dfrac{3}{2}x$

B. The equation is of the form $\dfrac{y^2}{a^2} - \dfrac{x^2}{b^2} = 1$, so

- the hyperbola is vertical, centered at the origin,
- the vertices are at $(0, \pm a)$,
- the co-vertices are at $(\pm b, 0)$,
- the foci are at $(0, \pm c)$, and
- the asymptotes are at $y = \pm\dfrac{a}{b}x$.

Use the equation to find a and b. Use $c^2 = a^2 + b^2$ to find c.

$a^2 = 1 \Rightarrow a = 1$ $b^2 = 4 \Rightarrow b = 2$ $c^2 = a^2 + b^2 = 1 + 4 = 5 \Rightarrow c = \sqrt{5}$

vertices: $(0, \pm 1)$ co-vertices: $(\pm 2, 0)$ foci: $(0, \pm\sqrt{5})$

Use a and b to write the equation of the asymptotes.

$y = \pm\dfrac{a}{b}x = \pm\dfrac{1}{2}x$ asymptotes: $y = \pm\dfrac{1}{2}x$

EXAMPLE 6 Graphing a Hyperbola Given Its Equation

Graph each hyperbola.

A. $\frac{y^2}{9} - x^2 = 1$ **B.** $0 = 2 + \frac{y^2}{8} - 2x^2$

SOLUTION

To sketch the graph of a hyperbola, first plot the vertices and sketch the asymptotes. Then, starting at a vertex, sketch a branch of the hyperbola curving toward each asymptote. Repeat for the other branch using the other vertex.

A. The equation is in the form $\frac{y^2}{a^2} - \frac{x^2}{b^2} = 1$, so

- the hyperbola is vertical, centered at the origin,
- the vertices are at $(0, \pm a)$, and
- the asymptotes are at $y = \pm\frac{a}{b}x$.

Use the given equation of the hyperbola to find a^2.

$a^2 = 9 \Rightarrow a = 3$ vertices: $(0, \pm 3)$ *Plot the vertices.*

Use the given equation of the hyperbola to find b^2.

$b^2 = 1 \Rightarrow b = 1$

Write the equation of the asymptotes.

$y = \pm\frac{a}{b}x = \pm\frac{3}{1}x = \pm 3x$ ▶ asymptotes: $y = \pm 3x$ *Sketch the asymptotes.*

Starting at (0, 3), sketch the top branch of the hyperbola, curving toward each asymptote. Then sketch the bottom branch starting at (0, –3), curving toward each asymptote.

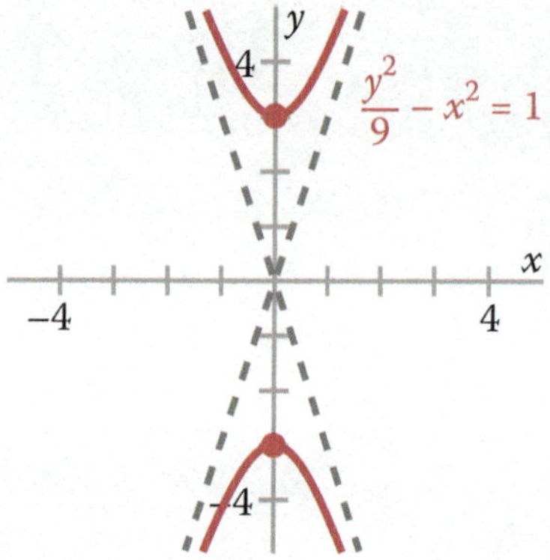

B. The equation is not in standard form. Write the equation in the standard form (where one side is equal to 1).

$$0 = 2 + \frac{y^2}{8} - 2x^2 \Rightarrow 2x^2 - \frac{y^2}{8} = 2 \Rightarrow x^2 - \frac{y^2}{16} = 1$$

Now the equation is of the form $\frac{x^2}{a^2} - \frac{y^2}{b^2} = 1$, so

- the hyperbola is horizontal, centered at the origin,
- the vertices are at $(\pm a, 0)$, and
- the asymptotes are at $y = \pm\frac{b}{a}x$.

Use the given equation of the hyperbola to find a^2.

$a^2 = 1 \Rightarrow a = 1$ vertices: $(\pm 1, 0)$ *Plot the vertices.*

Use the given equation of the hyperbola to find b^2.

$b^2 = 16 \Rightarrow b = 4$

Write the equation of the asymptotes.

$y = \pm\frac{b}{a}x = \pm\frac{4}{1}x = \pm 4x$ ▶ asymptotes: $y = \pm 4x$ *Sketch the asymptotes.*

Starting at (–1, 0), sketch the left branch of the hyperbola, curving toward each asymptote. Then sketch the right branch starting at (1, 0), curving toward each asymptote.

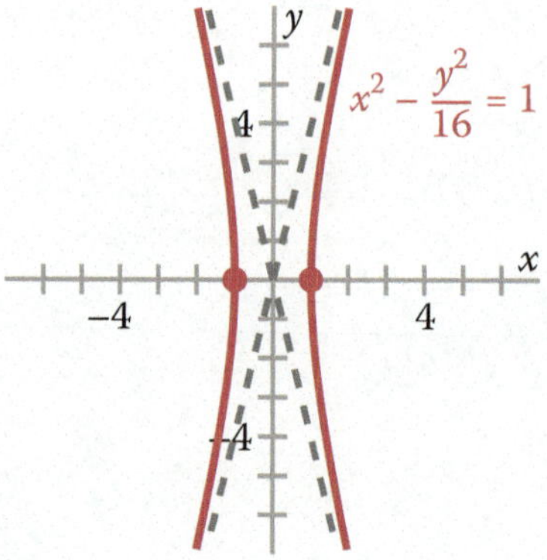

3.4.4 Applying Hyperbolas: Navigation

EXAMPLE 7 Finding the Location of a Ship Following a Hyperbolic Model

The path of a ship can be described by the hyperbolic model $\frac{x^2}{2501} - \frac{y^2}{3000} = 1$, where the ship travels along one branch of the hyperbola, relative to two stations on the shore line that are located at the foci. If the ship is 100 miles east of the hyperbola's vertical axis, what is the distance to the nearest mile between the ship and the shore (i.e., the horizontal axis)?

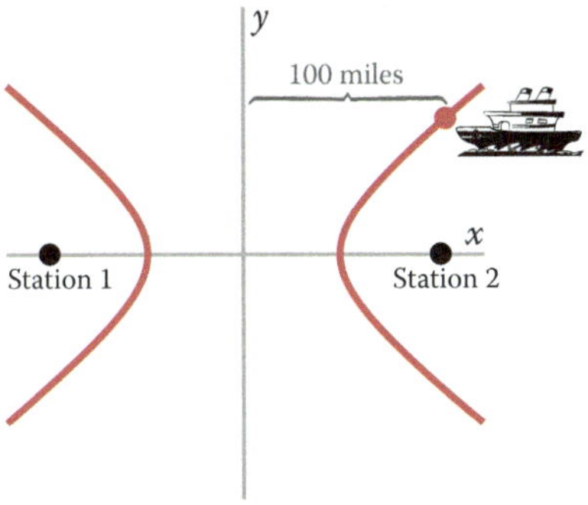

SOLUTION

The ship is following a hyperbolic model, where the hyperbola is horizontal and centered at the origin. Since the ship is located 100 miles east of the hyperbola's vertical axis, the ship is located on the hyperbola at the point $(100, y)$. Use the equation of the hyperbola to find the corresponding y-value, which is the distance between the ship and the shore.

$$\frac{x^2}{2501} - \frac{y^2}{3000} = 1$$

$$\frac{(100)^2}{2501} - \frac{y^2}{3000} = 1$$ *Substitute 100 for x.*

$$y^2 = 3000\left(\frac{10000}{2501} - 1\right)$$ *Simplify and solve for y^2.*

$$y \approx 95$$ *Take the square root of each side and round to the nearest mile.*

The distance between the ship and the shore is approximately 95 miles.

TIP

Wait until the final step in the calculations to round.

EXAMPLE 8 Finding the Equation of a Hyperbolic Model

The path of a ship can be described by a hyperbolic model centered at the origin, relative to two stations on the shore 160 miles apart that are located at the foci. If the ship is at a vertex 30 miles west of the hyperbola's center, find the equation of the hyperbola.

SOLUTION

The hyperbolic model that the ship follows is centered at the origin. Since the ship is located at a vertex, and that vertex is west of the hyperbola's center, the hyperbola is horizontal.

So, the standard form equation is $\frac{x^2}{a^2} - \frac{y^2}{b^2} = 1$, where the vertices are located at $(\pm a, 0)$.

Since the ship is at a vertex 30 miles west of the hyperbola's center, $a = 30$. So, $a^2 = 900$.

The stations are 160 miles apart from each other, so they are each 80 miles from the center. Since the stations are at the foci, $c = 80$. So, $c^2 = 6400$.

Use $c^2 = a^2 + b^2$ to find b^2. $\qquad 6400^2 = 900 + b^2 \Rightarrow b^2 = 5500$

Substitute the values of a^2 and b^2 into the standard form equation of a horizontal hyperbola.

$$\frac{x^2}{a^2} - \frac{y^2}{b^2} = 1 \Rightarrow \frac{x^2}{900} - \frac{y^2}{5500} = 1$$

SECTION 3.4 EXERCISES

Warm Up

1. Solve. $\dfrac{a}{45} = \dfrac{4}{9}$

2. Solve for y. $\dfrac{x^2}{10} - \dfrac{y^2}{12} = 1$

3. Simplify. $\sqrt{54}$

Just the Facts

Fill in the blanks.

4. A hyperbola is the set of all points whose distances from two fixed points, called the ____, are a constant ____.

5. The graph of a hyperbola consists of two disconnected curves called its ____.

6. A(n) ____ hyperbola opens up and down.

7. A(n) ____ hyperbola opens left and right.

8. The ____, ____, ____, and ____ of a horizontal hyperbola centered at the origin are on the ____-axis.

9. The ____, ____, ____, and ____ of a vertical hyperbola centered at the origin are on the ____-axis.

10. The endpoints of a hyperbola's transverse axis are its ____.

11. If the distance between a hyperbola's center and a vertex is x, then the length of its transverse axis is ____.

12. True or False? A hyperbola's co-vertices are located on the hyperbola.

Essential Skills

In Exercises 1–2, find the vertices and co-vertices of each hyperbola.

1.

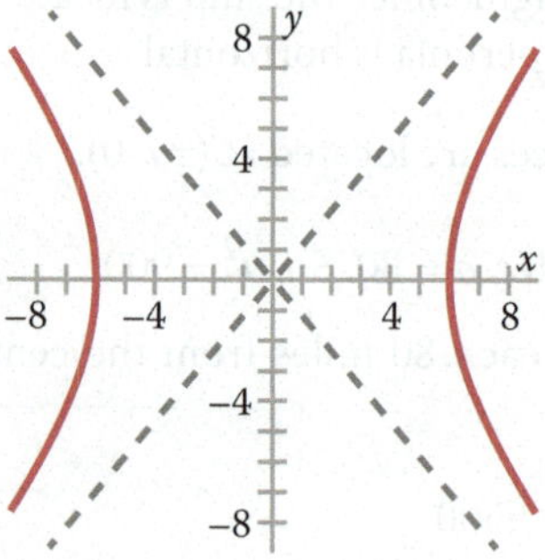

2.

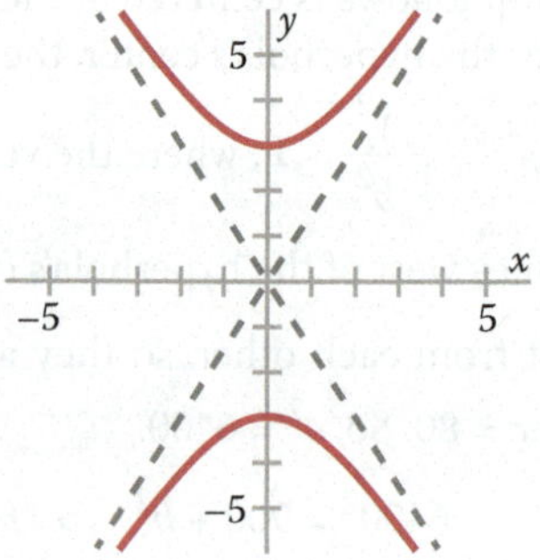

In Exercises 3–4, write the equation of each hyperbola.

3.

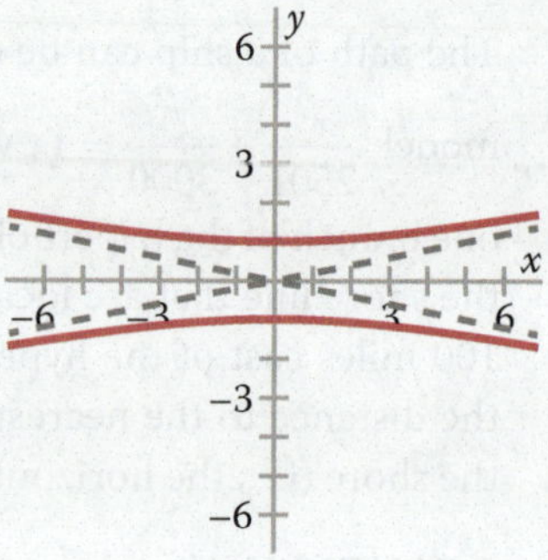

4.

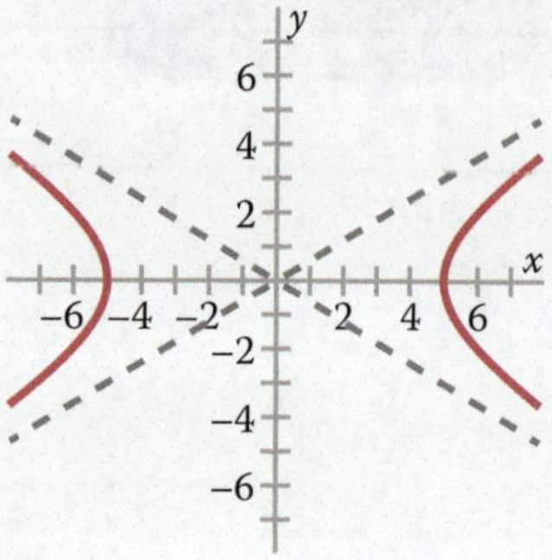

In Exercises 5–10, write the equation of each hyperbola.

5. vertices at (0, ±15) and foci at (0, ±20)

6. vertices at (±2, 0) and foci at (±9, 0)

7. centered at the origin with a focus at (0, 11) and a transverse axis of length 20

8. centered at the origin with a focus at (5, 0) and a transverse axis of length 8

9. vertices at (±9, 0) and asymptotes $y = \pm 5x$

10. vertices at (0, ±2) and asymptotes $y = \pm\dfrac{1}{3}x$

In Exercises 11–12, write the equation of each hyperbola using an algebraic method (i.e., without drawing the central box).

11.

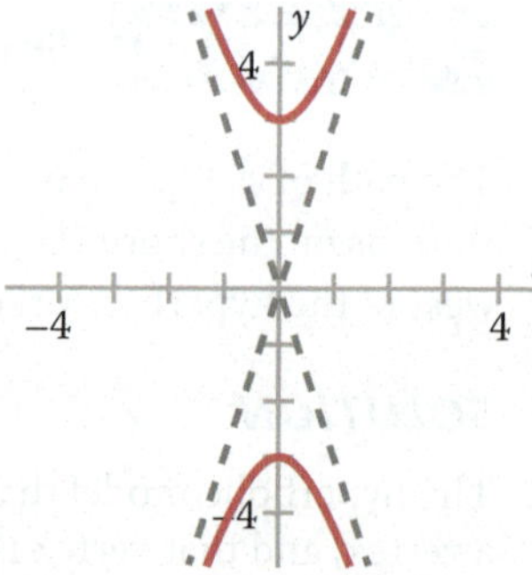

12.

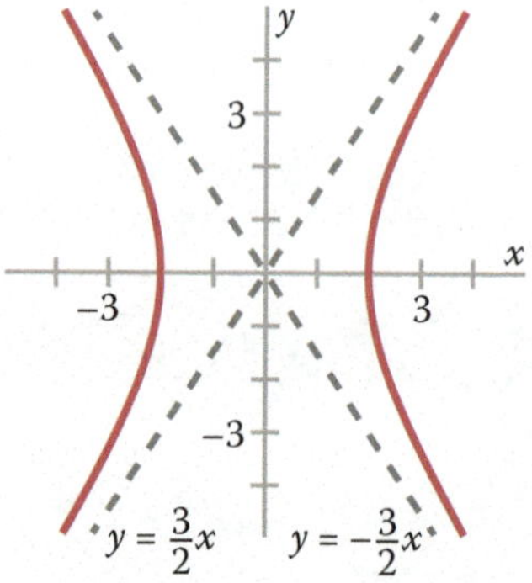

In Exercises 13–14, find the vertices, asymptotes, and foci of each hyperbola.

13. $y^2 - \dfrac{x^2}{49} = 1$

14. $\dfrac{x^2}{64} - \dfrac{y^2}{9} = 1$

In Exercises 15–18, graph each hyperbola.

15. $\dfrac{y^2}{9} - \dfrac{x^2}{4} = 1$

16. $\dfrac{x^2}{9} - \dfrac{y^2}{4} = 1$

17. $\dfrac{x^2}{2} = \dfrac{1+y^2}{2}$

18. $16y^2 - x^2 = 16$

19. The path of a ship can be described by the hyperbolic model $\frac{y^2}{983} - \frac{x^2}{1047} = 1$, where the ship travels along one branch of the hyperbola, relative to two stations on the shore that are located at the foci. If the ship is 115 miles north of the hyperbola's horizontal axis, what is the distance to the nearest mile between the ship and the shore (i.e., the vertical axis)?

20. The path of a ship can be described by the hyperbolic model $\frac{x^2}{784} - \frac{y^2}{2025} = 1$, where the ship travels along one branch of the hyperbola, relative to two stations on the shore that are located at the foci. If the ship is 125 miles east of the hyperbola's vertical axis, what is the distance to the nearest mile between the ship and the shore (i.e., the horizontal axis)?

21. The path of a ship can be described by a hyperbolic model centered at the origin, relative to two stations on the shore 168 miles apart that are located at the foci. If the ship is 60 miles south of the center of the hyperbola at a vertex, find the equation of the hyperbola.

22. The path of a ship can be described by a hyperbolic model centered at the origin, relative to two stations on the shore 212 miles apart that are located at the foci. If the ship is 90 miles east of the center of the hyperbola at a vertex, find the equation of the hyperbola.

Extensions

In Exercises 23–29, write the equations of two hyperbolas that satisfy the given conditions. If there are not two hyperbolas that satisfy the given conditions, state the reason.

23. transverse axis of length 12

24. vertices at $(0, \pm 3)$

25. co-vertices at $(0, \pm 10)$

26. foci at $(\pm 5, 0)$

27. foci and vertices such that $|c| = |a| + 1$

28. a central box with vertices $(\pm 5, 4)$ and $(\pm 5, -4)$

29. asymptotes at $y = \pm 2$

30. Explain how the co-vertices and asymptotes of a hyperbola could be used to sketch the graph of a hyperbola.

31. When $a = b$ in the standard form equation of an ellipse, the ellipse formed is a circle. What is true about a hyperbola when $a = b$?

32. Predict how the standard form equation of a horizontal parabola will change to describe a hyperbola with a horizontal transverse axis whose center is on the y-axis, n units above the origin.

33. Correct any errors in the following statement.

The graph of $\frac{y^2}{m^2} - \frac{x^2}{n^2} = 1$ is a hyperbola centered at the origin. The hyperbola has a vertical transverse axis of length $2n$ when $n > m$, and of length $2m$ when $m > n$. The hyperbola's vertices are at $(0, \pm m)$, its co-vertices are at $(0, \pm n)$, and its foci are at $(\pm(m^2 + n^2), 0)$.

3.5 TRANSLATIONS OF CONICS

OBJECTIVES

- Connect the geometric attributes of a translated conic (parabola, ellipse, and hyperbola) to its standard form equation.
- Write the equation of a translated conic (parabola, ellipse, hyperbola).
- Graph a translated conic (parabola, ellipse, hyperbola).
- Identify a conic given its equation in the general form.

PREREQUISITE VOCABULARY TERMS

conic
discriminant
ellipse
hyperbola
parabola

All of the conics (parabolas, ellipses, and hyperbolas) discussed in the preceding sections in Chapter 3 have had a vertex or center at the origin (vertex for the parabolas, center for the ellipses and hyperbolas). In this section, the discussion focuses on conics that are horizontally or vertically translated so their vertex or center is *not* at the origin. The following examples of translated conics are translated h units horizontally and k units vertically.

Examples of Translated Conics

Parabola | **Ellipse** | **Hyperbola**

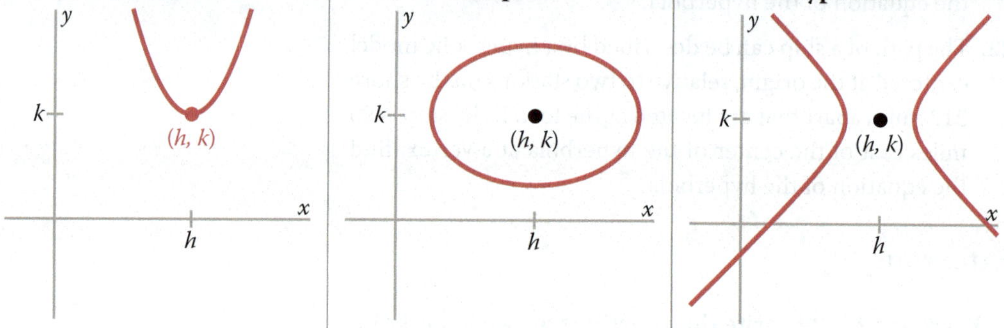

pre-image: $x^2 = 4py$ | pre-image: $\dfrac{x^2}{a^2} + \dfrac{y^2}{b^2} = 1$ | pre-image: $\dfrac{x^2}{a^2} - \dfrac{y^2}{b^2} = 1$

REMEMBER

When a tranformation is applied to the graph of an equation, the original graph is called the pre-image and the transformed graph is called the image.

The standard form equations of translated conics and conics with a center or vertex at the origin are shown in the following table.

Standard Form Equations of Conics

		Horizontal	Vertical
Parabola	vertex at (0, 0)	$y^2 = 4px$	$x^2 = 4py$
	vertex at (h, k)	$(y - k)^2 = 4p(x - h)$	$(x - h)^2 = 4p(y - k)$
Ellipse $(a \geq b)$	center at (0, 0)	$\dfrac{x^2}{a^2} + \dfrac{y^2}{b^2} = 1$	$\dfrac{y^2}{a^2} + \dfrac{x^2}{b^2} = 1$
	center at (h, k)	$\dfrac{(x-h)^2}{a^2} + \dfrac{(y-k)^2}{b^2} = 1$	$\dfrac{(y-k)^2}{a^2} + \dfrac{(x-h)^2}{b^2} = 1$
Hyperbola	center at (0, 0)	$\dfrac{x^2}{a^2} - \dfrac{y^2}{b^2} = 1$	$\dfrac{y^2}{a^2} - \dfrac{x^2}{b^2} = 1$
	center at (h, k)	$\dfrac{(x-h)^2}{a^2} - \dfrac{(y-k)^2}{b^2} = 1$	$\dfrac{(y-k)^2}{a^2} - \dfrac{(x-h)^2}{b^2} = 1$

In each standard form equation of a translated conic, the h-value gives the horizontal translation, and the k-value gives the vertical translation. A horizontal translation is either left or right, depending on the sign of h. Similarly, a vertical translation is either up or down, depending on the sign of k, as summarized in the following table.

Translations by h- and k-Units			
Horizontal Translation		**Vertical Translation**	
$h > 0$	right	$k > 0$	up
$h < 0$	left	$k < 0$	down

3.5.1 Translations of Parabolas

The Standard Form Equations

The standard form equations for translated horizontal and vertical parabolas are given in the following table. The vertex of each parabola is at (h, k). Recall that $\pm p$ is the distance between the vertex and the focus, where the sign of p depends on the direction of the parabola.

Standard Form Equations of Parabolas: Vertex (h, k) and Distance to Focus ±p			
	Equation	$p > 0$	$p < 0$
Vertical	$(x - h)^2 = 4p(y - k)$	 *opens upward*	 *opens downward*
Horizontal	$(y - k)^2 = 4p(x - h)$	 *opens to the right*	 *opens to the left*

As the vertex of a parabola is translated from (0, 0) to (h, k), the parabola's focus and axis of symmetry are also translated h units horizontally and k units vertically, as summarized in the following table.

	Vertical Parabola		***Horizontal Parabola***	
	vertex at (0, 0)	**vertex at (*h*, *k*)**	**vertex at (0, 0)**	**vertex at (*h*, *k*)**
axis of symmetry	$x = 0$	$x = h$	$y = 0$	$y = k$
focus (*F*)	$(0, p)$	$(h, k + p)$	$(p, 0)$	$(h + p, k)$

Writing the Equation of a Translated Parabola

The process for writing the equation of a translated parabola (where the vertex is not at the origin) is very similar to the process for writing the equation of a parabola with vertex at the origin.

Steps for Writing the Equation of a Translated Parabola

❶ Determine the type of parabola (horizontal or vertical) and its direction (left or right for a horizontal parabola, or up or down for a vertical parabola).

❷ Find the coordinates of the vertex (h, k) and the p-value.

❸ Substitute the values of h, k, and p into the appropriate standard form equation, then simplify as needed.

EXAMPLE 1 Writing the Equation of a Translated Parabola Given the Vertex and Focus

Write the equation of a parabola with vertex at (–7, 2) and focus at (–7, –1).

SOLUTION

❶ The vertex and focus are on the vertical line $x = -7$, and the focus is below the vertex, so
- the parabola has a vertical axis of symmetry,
- the equation is of the form $(x - h)^2 = 4p(y - k)$, and
- the parabola opens downward.

❷ The vertex is (–7, 2), so $h = -7$ and $k = 2$.
The distance between the vertex and the focus is 3 units. Since $p < 0$ for a parabola that opens downward, $p = -3$.

❸ Substitute $h = -7$, $k = 2$, and $p = -3$ into the standard form equation, then simplify.

$$(x - h)^2 = 4p(y - k) \;\Rightarrow\; (x - (-7))^2 = 4(-3)(y - 2) \;\Rightarrow\; (x + 7)^2 = -12(y - 2)$$

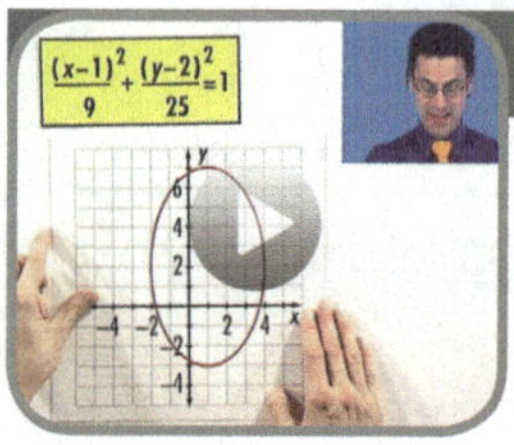

3.5.2 Translations of Ellipses

The Standard Form Equations

The standard form equations for translated horizontal and vertical ellipses centered at (h, k) are given in the following table. The table also summarizes the relationship between these equations and the attributes of a translated ellipse.

IMPORTANT

The relationship between a, b, and c is defined by the equation $c^2 = a^2 - b^2$, where $a \geq b > 0$.

Standard Form Equations of Ellipses: Centered at (h, k)

	Horizontal Major Axis	Vertical Major Axis
Equation	$\frac{(x-h)^2}{a^2} + \frac{(y-k)^2}{b^2} = 1 \quad a \geq b$	$\frac{(y-k)^2}{a^2} + \frac{(x-h)^2}{b^2} = 1 \quad a \geq b$
Vertices	$(h \pm a, k)$	$(h, k \pm a)$
Foci	$(h \pm c, k)$	$(h, k \pm c)$
Co-vertices	$(h, k \pm b)$	$(h \pm b, k)$
	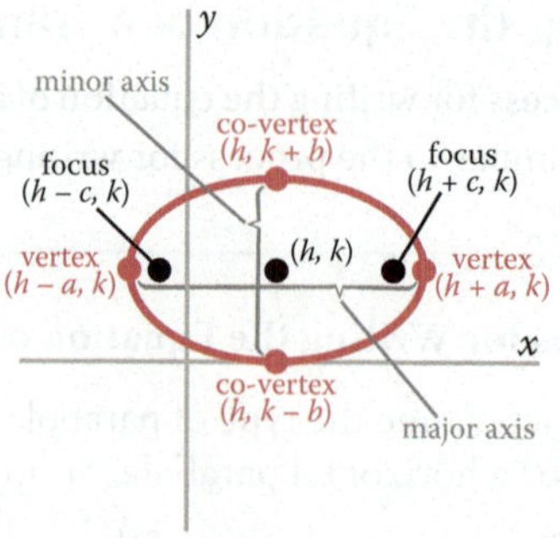	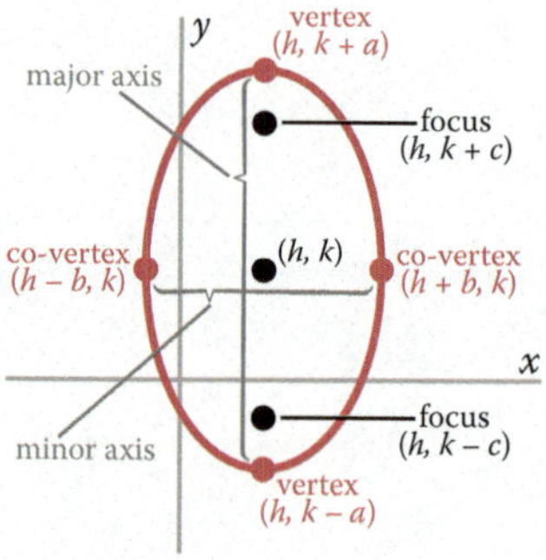

Major and Minor Axes

The major axis of an ellipse is the line segment through the center of the ellipse, whose endpoints are the ellipse's vertices. Recall that the length of the major axis for an ellipse centered at (0, 0) is $2a$. The major axis of a horizontal ellipse with center at (h, k) extends from the point $(h - a, k)$ to $(h + a, k)$, so the length of the major axis is equal to the distance between these points. Since the y-coordinate of $(h - a, k)$ and $(h + a, k)$ is k, these points form a horizontal line segment with length equal to the difference between the x-coordinates.

length of the major axis of a horizontal ellipse centered at (h, k):

$$(h + a) - (h - a) = h + a - h + a = 2a$$

Therefore, the length of the major axis for a horizontal ellipse centered at (h, k) is $2a$. A similar process can be used to show that the length of the minor axis for a horizontal ellipse centered at (h, k) is $2b$. Likewise, the lengths of the major and minor axes for a vertical ellipse centered at (h, k) are also $2a$ and $2b$, respectively.

TIP

The center of an ellipse is also the
- *midpoint of the major axis,*
- *midpoint of the minor axis, and*
- *the point at which the major and minor axes intersect.*

TIP

The fact that the lengths of the major and minor axes are $2a$ and $2b$, respectively, for ellipses centered at the origin, as well as for ellipses centered at (h, k), should not be surprising. Recall that a translation affects the position of the ellipse, but not the size or shape of the ellipse.

Graphing a Translated Ellipse

To graph any ellipse, plot the vertices and co-vertices, then sketch the ellipse through these points. To find the vertices and co-vertices of a translated ellipse, identify the values of a, b, h, and k from the standard form equation. Recall that the vertices and co-vertices, respectively, are given by $(h \pm a, k)$ and $(h, k \pm b)$ for a horizontal ellipse, and $(h, k \pm a)$ and $(h \pm b, k)$ for a vertical ellipse.

Note that for a horizontal ellipse, the vertices $(h \pm a, k)$ are a units to the left and right of the center (h, k), while the co-vertices $(h, k \pm b)$ are b units above and below the center (h, k). Similarly, for a vertical ellipse, the vertices $(h, k \pm a)$ are a units above and below the center (h, k), while the co-vertices $(h \pm b, k)$ are b units to the left and right of the center (h, k).

EXAMPLE 2 Graphing a Translated Ellipse

Find the center and graph each ellipse.

A. $\dfrac{(x-1)^2}{9} + \dfrac{(y+3)^2}{4} = 1$ **B.** $8(x + 5)^2 = 8 - 2y^2$

SOLUTION

A. The equation is in the standard form of a horizontal ellipse with center (h, k),

$$\frac{(x-h)^2}{a^2} + \frac{(y-k)^2}{b^2} = 1 \text{, where } a \geq b.$$

Identify the values of a, b, h, and k from the standard form equation.

$$\frac{(x-1)^2}{9} + \frac{(y+3)^2}{4} = 1 \Rightarrow \frac{(x-1)^2}{3^2} + \frac{(y-(-3))^2}{2^2} = 1$$

From the equation, $a = 3$, $b = 2$, $h = 1$, and $k = -3$. So, the center, (h, k), is at $(1, -3)$.

The ellipse is horizontal and centered at $(1, -3)$. So, the major axis and the vertices are on the horizontal line $y = -3$, and the minor axis and the co-vertices are on the vertical line $x = 1$.

vertices: $(h \pm a, k) = (1 \pm 3, -3)$, that is, $(4, -3)$ and $(-2, -3)$

co-vertices: $(h, k \pm b) = (1, -3 \pm 2)$, that is, $(1, -1)$ and $(1, -5)$

ALTERNATIVE METHOD

Vertices and Co-vertices

The center is at $(1, -3)$, the major axis is horizontal, and the a- and b-values are 3 and 2. So, the vertices are 3 units to the left and right of the center, at $(-2, -3)$ and $(4, -3)$. The co-vertices are 2 units above and below the center, at $(1, -1)$ and $(1, -5)$.

Plot the vertices and co-vertices, then sketch the ellipse through these points.

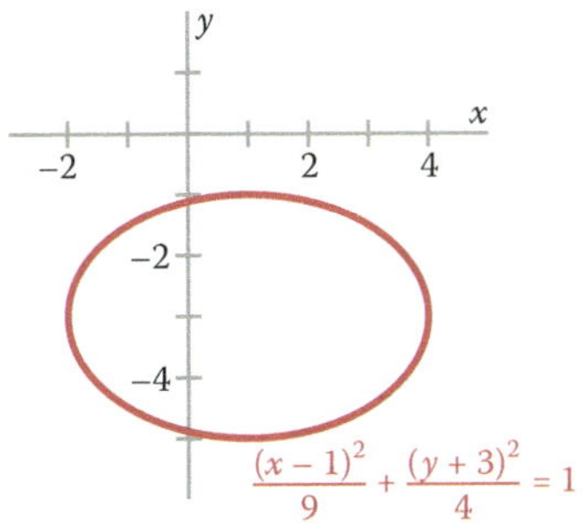

ALTERNATIVE METHOD

Graph Using Translations

Since $h = 1$ and $k = -3$, the graph of $\dfrac{(x-1)^2}{9} + \dfrac{(y+3)^2}{4} = 1$ is the image of $\dfrac{x^2}{9} + \dfrac{y^2}{4} = 1$ after a translation 1 unit to the right and 3 units down.

B. The equation is not in the standard form of an ellipse. Write the equation so that the right side is equal to 1.

$$8(x+5)^2 = 8 - 2y^2$$

$$8(x+5)^2 + 2y^2 = 8$$ *Add $2y^2$ to each side to isolate the constant term on the right side.*

$$\frac{8(x+5)^2}{8} + \frac{2y^2}{8} = \frac{8}{8}$$ *Divide each side by 8 so that the right side will equal 1.*

$$(x+5)^2 + \frac{y^2}{4} = 1$$ *Simplify.*

$$\frac{y^2}{4} + (x+5)^2 = 1$$ *For an ellipse, a must be greater than or equal to b.*

Now the equation is in the standard form of a vertical ellipse with center (h, k):

$$\frac{(y-k)^2}{a^2} + \frac{(x-h)^2}{b^2} = 1, \text{ where } a \geq b.$$

Identify the values of a, b, h, and k from the standard form equation.

$$\frac{y^2}{4} + (x+5)^2 = 1 \Rightarrow \frac{(y-0)^2}{2^2} + \frac{(x-(-5))^2}{1^2} = 1$$

From the equation, $a = 2$, $b = 1$, $h = -5$, and $k = 0$. So, the center, (h, k), is at $(-5, 0)$.

The ellipse is vertical and centered at $(-5, 0)$, so the vertical line $x = -5$ contains the major axis and the vertices, while the horizontal line $y = 0$ (the x-axis) contains the minor axis and the co-vertices.

vertices: $(h, k \pm a) = (-5, 0 \pm 2)$, that is, $(-5, 2)$ and $(-5, -2)$

co-vertices: $(h \pm b, k) = (-5 \pm 1, 0)$, that is, $(-4, 0)$ and $(-6, 0)$

Plot the vertices and co-vertices, then sketch the ellipse through these points.

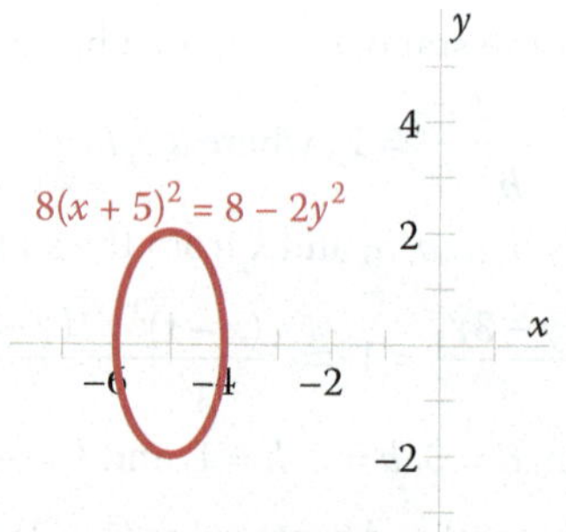

ALTERNATIVE METHOD

Graph Using Translations

Since $h = -5$ and $k = 0$, the graph of $\frac{y^2}{4} + (x+5)^2 = 1$ is the image of $\frac{y^2}{4} + x^2 = 1$ after a translation 5 units to the left.

ALTERNATIVE METHOD

Vertices and Co-vertices

The center is at $(-5, 0)$, the major axis is vertical, and the a- and b-values are 2 and 1. So, the vertices are 2 units above and below the center, at $(-5, 2)$ and $(-5, -2)$, while the co-vertices are 1 unit to the left and right of the center, at $(-6, 0)$ and $(-4, 0)$.

NOTICE THAT

Length of Major Axis

$2a = 2(2) = 4$ units

Length of Minor Axis

$2b = 2(1) = 2$ units

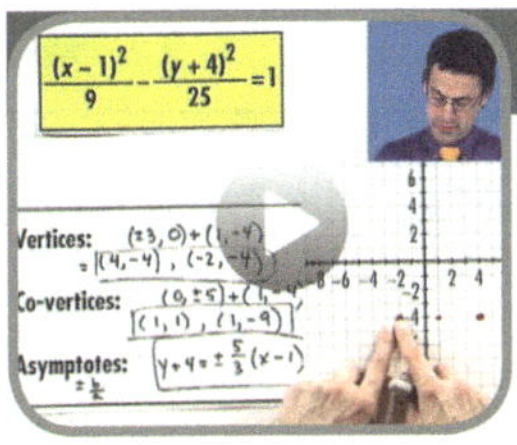

3.5.3 Translations of Hyperbolas

The Standard Form Equations

The standard form equations for translated horizontal and vertical hyperbolas centered at (h, k) are given in the following table. The table also summarizes the relationship between these equations and the attributes of a translated hyperbola.

Standard Form Equations of Hyperbolas: Centered at (h, k)

	Horizontal Transverse Axis (opens left or right)	Vertical Transverse Axis (opens up or down)
Equation	$\frac{(x-h)^2}{a^2} - \frac{(y-k)^2}{b^2} = 1$	$\frac{(y-k)^2}{a^2} - \frac{(x-h)^2}{b^2} = 1$
Vertices	$(h \pm a, k)$	$(h, k \pm a)$
Foci	$(h \pm c, k)$	$(h, k \pm c)$
Co-vertices	$(h, k \pm b)$	$(h \pm b, k)$
Asymptotes	$y - k = \pm\frac{b}{a}(x-h)$	$y - k = \pm\frac{a}{b}(x-h)$
	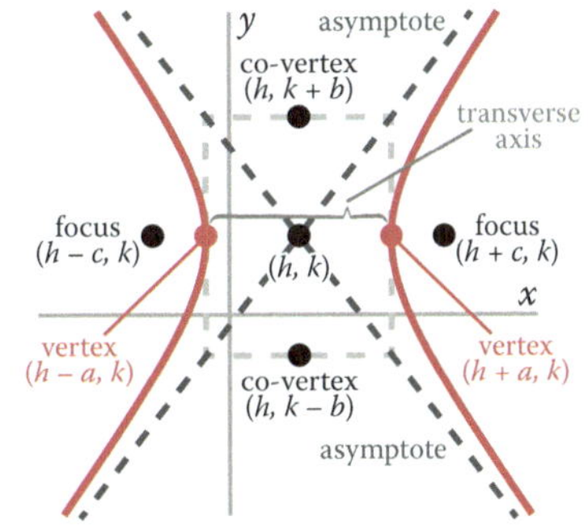	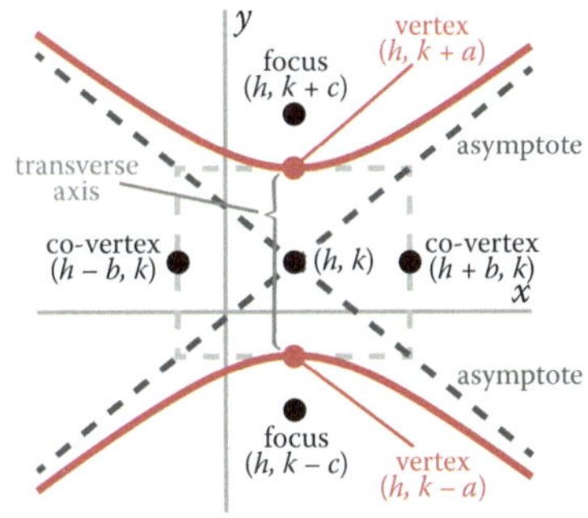

TIP

The midpoint of a hyperbola's transverse axis is always the center of the hyperbola.

Transverse Axis

As with the major axis of an ellipse, the length of a horizontal hyperbola's transverse axis can be found by subtracting the x-coordinates of its vertices.

length of the transverse axis of a horizontal hyperbola centered at (h, k):

$$(h + a) - (h - a) = h + a - h + a = 2a$$

The same is true for a vertical hyperbola centered at (h, k). Again, note that the length of the transverse axis is the same as for a hyperbola centered at the origin because a translation does not affect the size or shape of the hyperbola.

Graphing a Translated Hyperbola

To graph any hyperbola, plot the vertices and asymptotes, then sketch the two branches of the hyperbola, where each passes through a vertex and curves outward, toward the asymptotes. Use the values of a, b, h, and k (which can be identified from the standard form equation) to find the vertices and asymptotes of a translated hyperbola. Recall that the vertices and asymptotes, respectively, are given by $(h \pm a, k)$ and $y - k = \pm\frac{b}{a}(x-h)$ for a horizontal hyperbola, and by $(h, k \pm a)$ and $y - k = \pm\frac{a}{b}(x-h)$ for a vertical hyperbola.

As with an ellipse, note that for a horizontal hyperbola, the vertices $(h \pm a, k)$ are a units to the left and right of the center (h, k). Similarly, for a vertical hyperbola, the vertices $(h, k \pm a)$ are a units above and below the center (h, k).

EXAMPLE 3 Graphing a Translated Hyperbola

Find the center, vertices, co-vertices, and asymptotes of each hyperbola. Then sketch the graph.

A. $\dfrac{(x+1)^2}{4} - \dfrac{(y-5)^2}{9} = 1$ **B.** $\dfrac{(x+1)^2}{2} = 2y^2 - 2$

SOLUTION

A. The equation is in the standard form of a horizontal hyperbola with center (h, k).

$$\frac{(x-h)^2}{a^2} - \frac{(y-k)^2}{b^2} = 1$$

Identify the values of a, b, h, and k from the standard form equation.

$$\frac{(x+1)^2}{4} - \frac{(y-5)^2}{9} = 1 \Rightarrow \frac{(x-(-1))^2}{2^2} - \frac{(y-5)^2}{3^2} = 1$$

From the equation, $a = 2$, $b = 3$, $h = -1$, and $k = 5$.

center: $(h, k) = (-1, 5)$

vertices: $(h \pm a, k) = (-1 \pm 2, 5)$, that is, $(1, 5)$ and $(-3, 5)$

co-vertices: $(h, k \pm b) = (-1, 5 \pm 3)$, that is, $(-1, 8)$ and $(-1, 2)$

asymptotes: $y - k = \pm\dfrac{b}{a}(x-h) \Rightarrow y - 5 = \pm\dfrac{3}{2}(x+1)$

Plot the vertices and sketch the asymptotes. Then sketch the hyperbola's two branches.

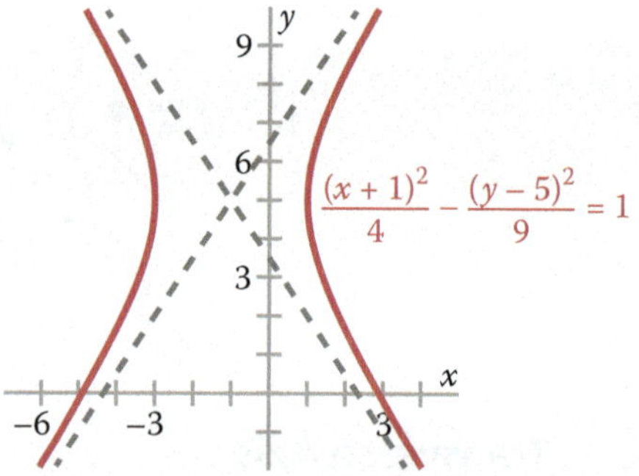

ALTERNATIVE METHOD

Vertices and Co-vertices

The center is at (−1, 5), the transverse axis is horizontal, and the a- and b-values are 2 and 3. So, the vertices are 2 units to the left and right of the center, at (−3, 5) and (1, 5), while the co-vertices are 3 units above and below the center, at (−1, 8) and (−1, 2).

NOTICE THAT

Since $h = -1$ and $k = 5$, the graph of $\frac{(x+1)^2}{4} - \frac{(y-5)^2}{9} = 1$ is the image of $\frac{x^2}{4} - \frac{y^2}{9} = 1$ after a translation 1 unit to the left and 5 units up.

B. The equation is not in the standard form of a hyperbola. Write the equation so that the right side is equal to 1.

$$\frac{(x+1)^2}{2} = 2y^2 - 2$$

$$\frac{(x+1)^2}{2} - 2y^2 = -2$$ *Subtract $2y^2$ from each side to isolate the constant term on the right side.*

$$\frac{(x+1)^2}{2(-2)} - \frac{2y^2}{-2} = \frac{-2}{-2}$$ *Divide each side by −2 so that the right side will be equal to 1.*

$$\frac{(x+1)^2}{-4} + y^2 = 1$$ *Simplify.*

$$y^2 - \frac{(x+1)^2}{4} = 1$$ *Write the equation using subtraction.*

Now the equation is in the standard form of a vertical hyperbola with center (h, k):

$$y^2 - \frac{(x+1)^2}{4} = 1 \Rightarrow \frac{(y-0)^2}{1^2} - \frac{(x-(-1))^2}{2^2} = 1$$

ALTERNATIVE METHOD

Vertices and Co-vertices

The center is at $(-1, 0)$, the transverse axis is vertical, and the a- and b-values are 1 and 2. So, the vertices are 1 unit above and below the center, at $(-1, 1)$ and $(-1, -1)$, while the co-vertices are 2 units to the left and right of the center, at $(-3, 0)$ and $(1, 0)$.

NOTICE THAT

Since $h = -1$ and $k = 0$, the graph of $\frac{(x+1)^2}{2} = 2y^2 - 2$ is the image of $y^2 - \frac{x^2}{4} = 1$ after a translation 1 unit to the left.

From the equation, $a = 1$, $b = 2$, $h = -1$, and $k = 0$.

center: $(h, k) = (-1, 0)$

vertices: $(h, k \pm a) = (-1, 0 \pm 1)$, that is, $(-1, 1)$ and $(-1, -1)$

co-vertices: $(h \pm b, k) = (-1 \pm 2, 0)$, that is, $(1, 0)$ and $(-3, 0)$

asymptotes: $y - k = \pm\frac{a}{b}(x - h) \;\Rightarrow\; y = \pm\frac{1}{2}(x+1)$

Plot the vertices and sketch the asymptotes. Then sketch the hyperbola's two branches.

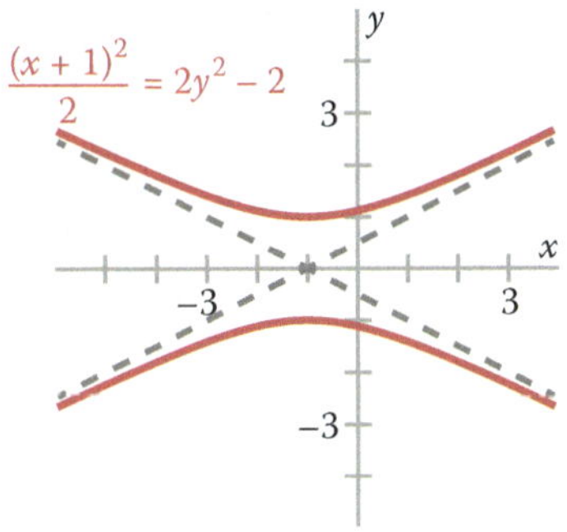

3.5.4 Identifying a Conic

When the equation of a conic is not given in standard form, the equation can be algebraically manipulated into standard form, and then the type of conic can be identified from the standard form equation.

When the equation of a conic is in the expanded form (where all parentheses have been removed by multiplying) and the right side is 0, the equation is said to be in the **general form.**

DEFINITION

The **general form of a conic** is $Ax^2 + Bxy + Cy^2 + Dx + Ey + F = 0$, where A, B, and C are not all equal to 0.

An equation written in general form can be algebraically manipulated into the standard form of a conic, often by completing the square.

EXAMPLE 4 Identifying a Conic

Identify each type of conic, if any, from its equation.

A. $y - x = \left(\frac{1}{2}\right)^2$ **B.** $\frac{(x+1)^2}{4} = 1 - \frac{(y-2)^2}{4}$ **C.** $2x + y^2 + 7y = x^2 + y - 4$

SOLUTION

A. Notice that the equation does not contain any terms where a variable is squared. However, the constant is squared. Simplify the equation.

$$y - x = \left(\frac{1}{2}\right)^2 \Rightarrow y - x = \frac{1}{4}$$

This equation can be written in the standard form of a linear equation, $Ax + By = C$.

$$-x + y = \frac{1}{4}$$

Therefore, the graph is a linear equation, not a conic.

B. Notice that the equation contains terms with variables that are squared. Therefore, the equation is not linear. Manipulate the equation to write it in the standard form of a conic. Begin by isolating the constant term on the right side.

$$\frac{(x+1)^2}{4} = 1 - \frac{(y-2)^2}{4} \Rightarrow \frac{(x+1)^2}{4} + \frac{(y-2)^2}{4} = 1$$

The equation is similar to the standard form of a horizontal ellipse centered at (h, k), $\frac{(x-h)^2}{a^2} + \frac{(y-k)^2}{b^2} = 1$. However, remember that when $a = b$, the ellipse is a circle. So, multiply the equation by 4.

$$\frac{(x+1)^2}{4} + \frac{(y-2)^2}{4} = 1 \Rightarrow (x+1)^2 + (y-2)^2 = 4$$

The equation is in the standard form of a circle with center (h, k) and radius r: $(x - h)^2 + (y - k)^2 = r^2$. Therefore, the conic is a circle centered at $(-1, 2)$ with radius 2.

C. Notice that the equation is in expanded form since it does not contain any parentheses. Algebraically manipulate the equation so that the constant term is isolated on the right side.

$$2x + y^2 + 7y = x^2 + y - 4 \Rightarrow -x^2 + 2x + y^2 + 6y = -4$$

Now complete the square for x and y.

$$-x^2 + 2x + y^2 + 6y = -4$$

$$-(x^2 - 2x) + (y^2 + 6y) = -4 \qquad \textit{Group the x-terms and group the y-terms.}$$

$$-(x^2 - 2x + 1) + (y^2 + 6y + 9) = -4 - 1 + 9 \qquad \textit{Complete the square twice.}$$

$$-(x - 1)^2 + (y + 3)^2 = 4 \qquad \textit{Factor each perfect-square trinomial.}$$

Divide each side by 4 to get 1 on the right side.

$$-(x-1)^2 + (y+3)^2 = 4 \Rightarrow -\frac{(x-1)^2}{4} + \frac{(y+3)^2}{4} = \frac{4}{4} \Rightarrow \frac{(y+3)^2}{4} - \frac{(x-1)^2}{4} = 1$$

Therefore, since the equation can be written in the standard form of a vertical hyperbola centered at (h, k), $\frac{(y-k)^2}{a^2} - \frac{(x-h)^2}{b^2} = 1$, the conic is a hyperbola centered at $(1, -3)$.

TIP

The equation is not in the standard form of an ellipse because the x^2-term is negative.

3.5.5 Using the Discriminant and Coefficients to Identify a Conic

All conic sections can be written in the general form, $Ax^2 + Bxy + Cy^2 + Dx + Ey + F = 0$, where A, B, and C are not all equal to 0. When the equation of a conic is in the general form, the coefficients of the terms and the discriminant, $B^2 - 4AC$, can be used to identify the conic. The relationship between the coefficients, the discriminant, and the type of conic is summarized in the following table.

REMEMBER

The discriminant of a quadratic equation in the form $ax^2 + bx + c = 0$ is $b^2 - 4ac$.

Classifying Conics (Equations in the General Form)

Conic	Discriminant / Coefficients
Circle	$B^2 - 4AC < 0$, $B = 0$, and $A = C$
Ellipse (non-circular)	$B^2 - 4AC < 0$, and $B \neq 0$ or $A \neq C$
Hyperbola	$B^2 - 4AC > 0$
Parabola	$B^2 - 4AC = 0$

Therefore, using the discriminant and the coefficients, the type of conic can be determined from an equation in the general form.

EXAMPLE 5 Identifying a Conic Using the Coefficients and Discriminant

Identify the type of conic, if any, from the equation $9x^2 + 108x + y^2 - 10y + 348 = 0$.

SOLUTION

Identify the coefficients and find the discriminant.

$$A = 9, B = 0, C = 1, D = 108, E = -10, \text{ and } F = 348$$

$$B^2 - 4AC = 0^2 - 4(9)(1) = -36 < 0$$

Since the discriminant is less than 0, the conic must be a circle or an ellipse. In order to be a circle, $B = 0$ and $A = C$ must both be true. Here, $B = 0$, but $A \neq C$. Therefore, the conic is not a circle. The conic is an ellipse if either $B \neq 0$ or $A \neq C$ is true. Here, $B \neq 0$ is false, but $A \neq C$ is true. Therefore, the conic is an ellipse.

SECTION 3.5 EXERCISES

Warm Up

1. Complete the square to write the equation $x^2 + 10x - 8 = 1$ as a squared binomial equal to a constant.
2. Complete the square twice to write the equation $2x^2 + 4x - 3y^2 + 24y + 10 = 0$ as the difference of two squared binomials equal to a constant.
3. Find the discriminant of $7 + 5x^2 - x = 3x$.

Just the Facts

Fill in the blanks.

4. Translating an ellipse left, right, ____, or ____ does not affect the length of its ____ or ____ axes.
5. Translating a hyperbola in any direction does not affect the length of its ____ axis.
6. The vertex of the graph of $y - 2 = -4(x + 1)^2$ is at ____.
7. The graph of $(y - k)^2 = 4p(x - h)$ is a(n) ____ that opens ____ when $p < 0$.
8. If the center of an ellipse is at (7, 9) and a vertex is at (7, 12), then the other vertex is at ____ and the ellipse has a(n) ____ major axis on the line ____.
9. If the center of a hyperbola is at ____ and $a =$ ____, then the vertices are at (–6, 0) and (0, 0) and the transverse axis is ____ on the line ____.
10. In the general form of a conic, the values of A, B, and ____ cannot be ____.

Essential Skills

In Exercises 1–2, for each parabola, state the direction, determine whether the standard form equation is $(y - k)^2 = 4p(x - h)$ or $(x - h)^2 = 4p(y - k)$, and determine whether $p > 0$ or $p < 0$.

1. vertex (–2, 0) and focus (6, 0)
2. vertex (5, –3) and focus (5, –1)

In Exercises 3–4, write the equation of each parabola.

3. vertex at (–2, 10) and focus at (–2, 6)
4. vertex at (1, –4) and focus at (–11, –4)

In Exercises 5–6, find the center of each ellipse.

5. $\dfrac{(x+4)^2}{20} + \dfrac{(y-3)^2}{6} = 1$
6. $\dfrac{(y+1)^2}{64} + x^2 = 1$

In Exercises 7–8, determine whether the major and minor axes of each ellipse are horizontal or vertical, and find the length of each axis.

7. $\dfrac{x^2}{40} + (y+3)^2 = 1$
8. $\dfrac{(x-4)^2}{9} + \dfrac{(y+3)^2}{32} = 1$

In Exercises 9–10, sketch the graph of each ellipse.

9. $\dfrac{y^2}{4} + (x-5)^2 = 1$
10. $\dfrac{(x+3)^2}{16} + \dfrac{(y-2)^2}{9} = 1$

In Exercises 11–12, find the center of each hyperbola.

11. $(x-10)^2 - \dfrac{(y+6)^2}{3} = 1$
12. $\dfrac{y^2}{12} - \dfrac{(x+8)^2}{20} = 1$

In Exercises 13–14, find the vertices of each hyperbola.

13. $\dfrac{(y-20)^2}{4} - \dfrac{(x+2)^2}{36} = 1$
14. $\dfrac{(x+3)^2}{25} - (y-2)^2 = 1$

In Exercises 15–16, write the equations of each hyperbola's asymptotes in point-slope form.

15. $\dfrac{(x+16)^2}{100} - \dfrac{(y-9)^2}{400} = 1$
16. $\dfrac{y^2}{8} - \dfrac{(x+5)^2}{18} = 1$

In Exercises 17–18, sketch the graph of each hyperbola.

17. $(y+4)^2 - \dfrac{x^2}{25} = 1$
18. $\dfrac{(x-1)^2}{36} - \dfrac{(y-2)^2}{16} = 1$

In Exercises 19–30, determine whether each equation is a circle, ellipse (not a circle), hyperbola, parabola, or not a conic by writing the equation in the standard form of a conic.

19. $(x - 1)^2 + y + 5 = 5$
20. $x + y = 5^2$
21. $(y + 10)^3 = x$

22. $x = y^2 + 7$

23. $18(y - 2)^2 + 2x^2 = 18$

24. $(x + 1)^2 - 24 = -(y - 9)^2$

25. $\frac{(x-2)^2}{2} + \frac{(y+3)^2}{2} = \frac{1}{5}$

26. $(y-2)^2 = 16 - \frac{16(x+4)^2}{9}$

27. $2y^2 - 8x^3 + 5 = 0$

28. $-x^2 + 4y^2 - 6x - 8y - 21 = 0$

29. $x^2 + y + 133 = 22x$

30. $-4x^2 - 12y^2 + 16x - 36y + 5 = 0$

In Exercises 31–34, write each equation in the standard form of a conic.

31. $x^2 - y^2 + 224 - 30x = 0$

32. $4x^2 + 3y^2 + 8x - 6y - 29 = 0$

33. $9y^2 + x^2 + 2x - 72y + 136 = 0$

34. $4y^2 - x^2 - 24y - 4x + 16 = 0$

In Exercises 35–36, write each equation in the general form of a conic, then identify the coefficients.

35. $x(x + 2) + 2(y + 10) = -1$

36. $3x(x - 2y + 5) + y = y(1 - y) - 2$

In Exercises 37–40, use the discriminant and the coefficients to determine the type of conic represented by each equation. Explain.

37. $9x^2 - 6xy + y^2 + 2x - 10 = 0$

38. $2x^2 - 5xy + 2y^2 - 11x - 7y - 4 = 0$

39. $-x^2 - y^2 - x + y = 0$

40. $9x^2 + 4y^2 - 36x + 32y + 64 = 0$

Extensions

In Exercises 41–43, use the following information. When a figure is transformed, the original figure is called the pre-image and the transformed figure is called the image. Write the equation of each image given the translation(s) and the pre-image $x^2 = -6y$.

41. 4 units up

42. 7 units to the left

43. 3 units to the right and 1 unit down

Exercises 44–49, write the equation in the standard form of a conic of each image given the translation(s) and the pre-image.

44. pre-image: $\frac{x^2}{9} + \frac{y^2}{4} = 1$,

translations: 10 units left and 6 units up

45. pre-image: $\frac{y^2}{6} + x^2 = 1$,

translations: 7 units right and 2 units up

46. pre-image: $y^2 - x^2 = 1$,

translations: 1 unit right and 20 units down

47. pre-image: $\frac{x^2}{a^2} + \frac{y^2}{b^2} = 1$,

translations: m units left and n units up

48. pre-image: $4x^2 - 3y^2 = 24$,

translations: 7 units left and 6 units down

49. pre-image: $\frac{(x-6)^2}{12} - \frac{(y+1)^2}{8} = 1$,

translation: 4 units left

In Exercises 50–53, find all *B*-values such that the graph of $3x^2 + Bxy + 3y^2 + x + 2y - 1 = 0$ will be the given type of conic. Write sets of values using interval notation.

50. circle

51. ellipse (not a circle)

52. hyperbola

53. parabola

CHAPTER 3 REVIEW EXERCISES

Section 3.1 Graphing Rational Functions

1. Graph $g(x) = -\frac{1}{x-4} - 2$. Write the domain and range in interval notation.

2. Write the equation of each vertical asymptote, if any, for the function $h(x) = \frac{x+1}{x^3 - x^2 - 9x + 9}$.

3. Use the zeros and vertical asymptote(s) to sketch the graph of $m(x) = \frac{x^2 - x - 12}{x+1}$.

4. Write the equation of the oblique asymptote for the function $p(x) = \frac{3x^2 - 2x + 5}{x - 4}$.

5. Graph the function $j(x) = \frac{x^3 - 27}{x^2 + 4x - 21}$ and state the coordinates of any holes.

Section 3.2 Parabolas

6. Find the focus and directrix of the graph of $x^2 + 3y = 0$.

7. Determine the direction of the graph of $-8y - 2x^2 = 0$.

8. Graph. $4x = -16y^2$

9. Write the standard form equation of the parabola with focus at (0, 2) and vertex at the origin.

10. Write the standard form equation of the parabola that passes through (–4, 8) with vertex at the origin and a vertical axis of symmetry.

Section 3.3 Ellipses

11. Use the graph to write the equation of the ellipse in standard form.

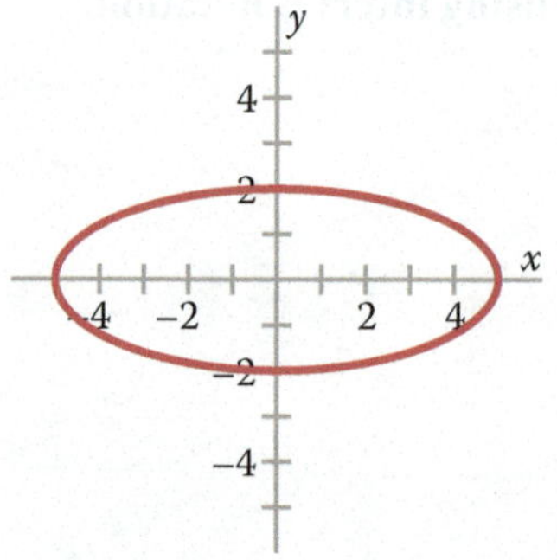

12. Write the standard form equation of an ellipse centered at the origin with a horizontal minor axis of length 4 units and a vertical major axis of length 8 units.

13. Find the intercepts and foci of $36x^2 + 100y^2 - 3600 = 0$.

14. Find the eccentricity and sketch the graph of $20x^2 + 36y^2 = 720$.

15. Write the equation of the ellipse centered at the origin with eccentricity 1/3 and foci at (±3, 0).

Section 3.4 Hyperbolas

In Exercises 16–17, write the standard form equation of each hyperbola.

16.

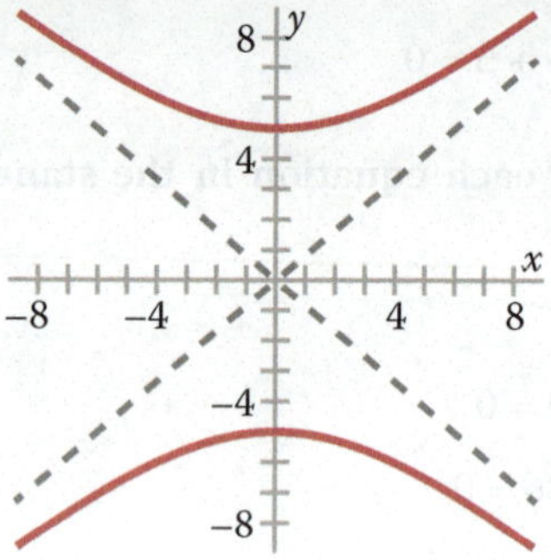

17. the hyperbola centered at the origin with a transverse axis of length 14 units and foci at $(\pm 6\sqrt{2}, 0)$

18. Find the vertices, co-vertices, foci, and asymptotes of $9x^2 - 4y^2 = 36$.

19. Graph. $\frac{y^2}{4} - \frac{x^2}{49} = 1$

20. From the LORAN system's use of hyperbolas, the equation that a ship can use to navigate its course, in nautical miles, using two stations as foci, is $\frac{y^2}{3025} - \frac{x^2}{2304} = 1$. Find the distance between the two stations.

Section 3.5 Translations of Conics

21. Write the standard form equation of the parabola with vertex (–3, –5) and focus (–7, –5).

In Exercises 22–23, find the center of each conic and graph the function.

22. $\frac{(x+4)^2}{9} + \frac{(y-2)^2}{16} = 1$

23. $\frac{(x-2)^2}{20} - \frac{(y-3)^2}{5} = 1$

24. Write $-x(x + 6) + 4y(-2 + y) = 21$ in the general form of a conic, then identify the coefficients.

25. Identify the type of conic, if any, from the equation $4x^2 - 6x + 9y^2 + 15y - 1 = 0$.

Precalculus

EXPONENTIAL AND LOGARITHMIC FUNCTIONS

Chapter

4

4.1 EXPONENTIAL FUNCTIONS

OBJECTIVES

- Evaluate exponential functions.
- Graph exponential functions.
- Model compound interest with exponential equations.
- Evaluate natural exponential functions.
- Graph natural exponential functions.
- Model population growth with natural exponential equations.

PREREQUISITE VOCABULARY TERMS

asymptote
function
interest
reflection
translation
***y*-intercept**

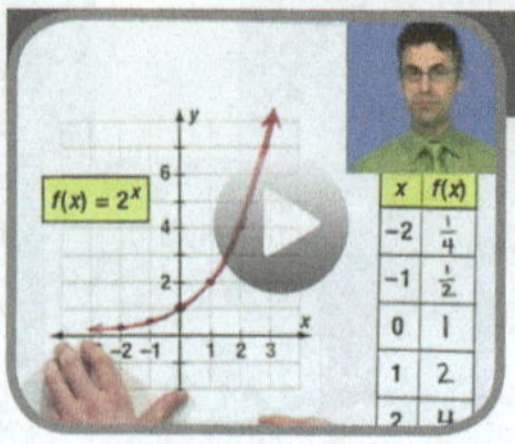

4.1.1 An Introduction to Exponential Functions

In previous sections, we have seen functions that contain powers. For example, the basic quadratic function $f(x) = x^2$ contains a power where the base is a variable and the exponent is 2. An **exponential function** is another type of function that contains a power, but the base is a number (a positive real number not equal to 1) and the exponent contains the variable. Several types of exponential functions will be discussed in this section.

DEFINITION

A function of the form $f(x) = b^x$ where b is a real number such that $b > 0$ and $b \neq 1$ is an **exponential function**.

The following are examples of exponential functions.

$$f(x) = 2^x \qquad f(x) = 1 + 5^x \qquad f(x) = 3(0.1^{x+5}) \qquad f(x) = \left(\frac{1}{2}\right)^{-x}$$

EXAMPLE 1 **Evaluating an Exponential Function**

Evaluate each exponential function.

A. $f(x) = 2^x$ for $x = 3$

B. $f(x) = 5^{-x}$ for $x = 2$

C. $f(x) = 16^x$ for $x = \frac{3}{4}$

D. $f(x) = \left(\frac{3}{4}\right)^{2x}$ for $x = -1$

SOLUTION

A. $f(3) = 2^3 = 8$

$a^{-m} = \frac{1}{a^m}$ ▶ **B.** $f(2) = 5^{-2} = \frac{1}{5^2} = \frac{1}{25}$

$a^{\frac{m}{n}} = \left(\sqrt[n]{a}\right)^m = \sqrt[n]{a^m}$ ▶ **C.** $f\left(\frac{3}{4}\right) = 16^{\frac{3}{4}} = \left(\sqrt[4]{16}\right)^3 = 2^3 = 8$

$\left(\frac{a}{b}\right)^{-n} = \left(\frac{b}{a}\right)^n$ ▶ **D.** $f(-1) = \left(\frac{3}{4}\right)^{2(-1)} = \left(\frac{3}{4}\right)^{-2} = \left(\frac{4}{3}\right)^2 = \frac{4^2}{3^2} = \frac{16}{9}$

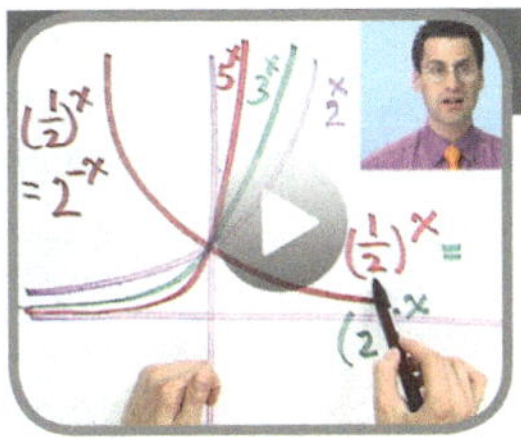

4.1.2 An Introduction to Graphing Exponential Functions

Exponential functions can be graphed by plotting points. Consider the graph of an exponential function of the form $f(x) = b^x$. Regardless of what the b-value is, the y-value will be 1 when $x = 0$, since any real number to the 0 power is 1 ($b^0 = 1$). So, the graph of $f(x) = b^x$ will always pass through (0, 1). Additionally, since b must be a positive number, the value of b^x will always be positive, so the graph of $f(x) = b^x$ will have a horizontal asymptote at $y = 0$ and no x-intercepts.

> The graph of the exponential function $f(x) = b^x$ has a horizontal asymptote at $y = 0$ (i.e., the x-axis), and passes through (0, 1). The domain of the function is all real numbers, and the range is $(0, \infty)$.

EXAMPLE 2 Graphing an Exponential Function $y = b^x$

Sketch the graphs of $f(x) = 2^x$ and $g(x) = \left(\frac{1}{2}\right)^x$ by plotting points.

TIP

$g(x)$ can be written equivalently as $g(x) = 2^{-x}$ by applying the Negative Exponent Property and the Power of a Power Property.

$$g(x) = \left(\frac{1}{2}\right)^x = \left(2^{-1}\right)^x = 2^{-x}$$

SOLUTION

Make a table of values for each function. Then plot the points and sketch the curves.

x	$f(x)$	$g(x)$
-2	$\frac{1}{4}$	4
-1	$\frac{1}{2}$	2
0	1	1
1	2	$\frac{1}{2}$
2	4	$\frac{1}{4}$

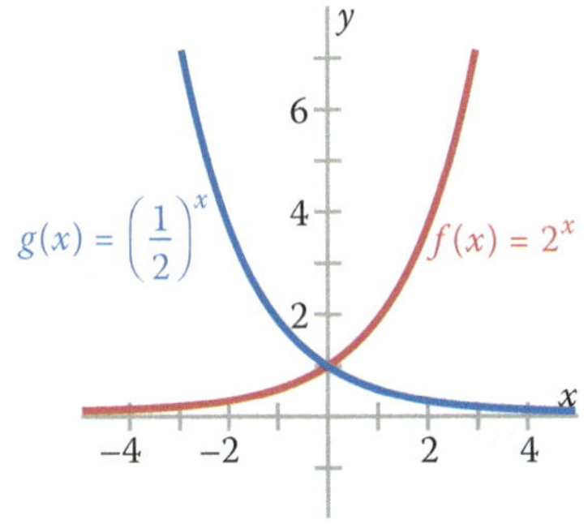

Comparing the Graphs of $f(x) = b^x$ and $f(x) = b^{-x}$

Recall that the graphs of $f(x)$ and $f(-x)$ are reflections over the y-axis. It follows that the graphs of the exponential functions $f(x) = b^x$ and $f(x) = b^{-x}$ are reflections over the y-axis. Note that since $b^{-x} = \left(\frac{1}{b}\right)^x$, we can also say that the graphs of $f(x) = b^x$ and $f(x) = \left(\frac{1}{b}\right)^x$ are reflections over the y-axis. In other words, if two functions of the form $f(x) = b^x$ have reciprocal bases, then their graphs are reflections over the y-axis. For example, the functions in **Example 2** have reciprocal bases, and their graphs are reflections over the y-axis.

Exponential Functions: Reflections Over the y-Axis

$$f(x) = b^x \text{ and } f(-x) = b^{-x} = \left(\frac{1}{b}\right)^x$$

$$f(x) = \left(\frac{a}{b}\right)^x \text{ and } f(-x) = \left(\frac{a}{b}\right)^{-x} = \left(\frac{b}{a}\right)^x$$

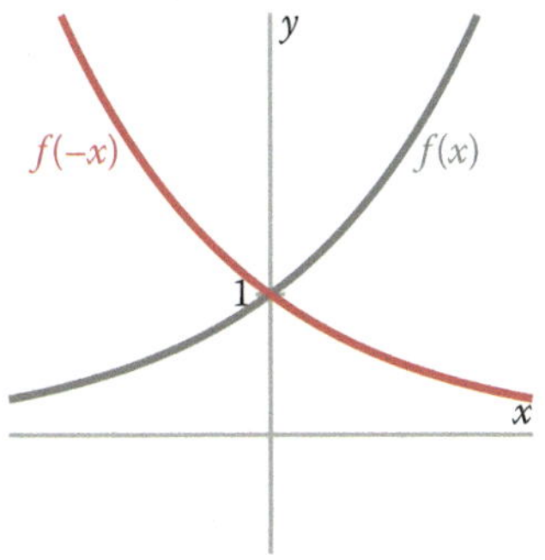

Examining the Graph of $f(x) = b^x$ for Different Values of b

When $b > 1$, the graph of $f(x) = b^x$ increases from left to right (i.e., as x increases, $f(x)$ increases). However, when $0 < b < 1$, the graph of $f(x) = b^x$ decreases from left to right (i.e., as x increases, $f(x)$ decreases).

Another aspect of the graph of $f(x) = b^x$ that can be deduced from the value of b is the steepness of the curve's incline (or decline). Consider the functions $f(x) = a^x$ and $f(x) = b^x$ when $b > a > 1$. For all $x > 1$, it is true that $b^x > a^x$. It follows that when $x > 1$, the y-values for $f(x) = b^x$ will increase faster than those of $f(x) = a^x$. So, the curve $f(x) = b^x$ is steeper than the curve $f(x) = a^x$ in the first quadrant. Furthermore, for all $x < 1$, $b^x < a^x$. So, the curve $f(x) = b^x$ approaches the y-axis more quickly than the curve $f(x) = a^x$ in the second quadrant. Several examples of this are shown in Figure 4.1a.

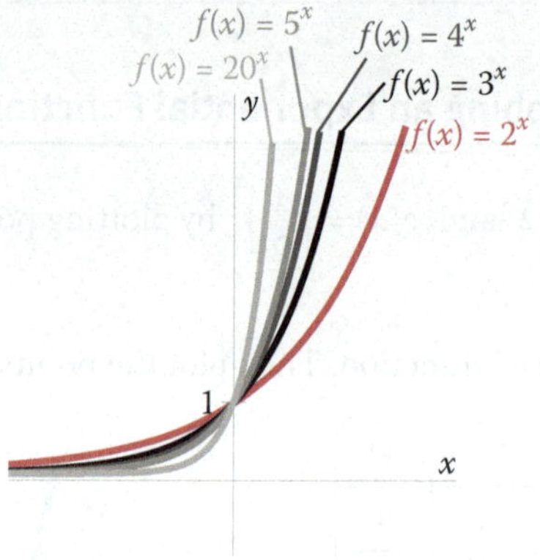

Figure 4.1a

NOTICE THAT

The graph of each equation passes through (0, 1).

In Figure 4.1a, we see that in the first quadrant, the graph of $f(x) = 3^x$ is steeper than the graph of $f(x) = 2^x$, and the graph of $f(x) = 4^x$ is steeper than the graph of $f(x) = 3^x$. However, note that this steepness does not appear to increase at a constant rate. The difference in the steepness between the graph of $f(x) = 3^x$ and $f(x) = 2^x$ is quite obvious, but the difference in the steepness between the graph of $f(x) = 4^x$ and $f(x) = 5^x$ is much less.

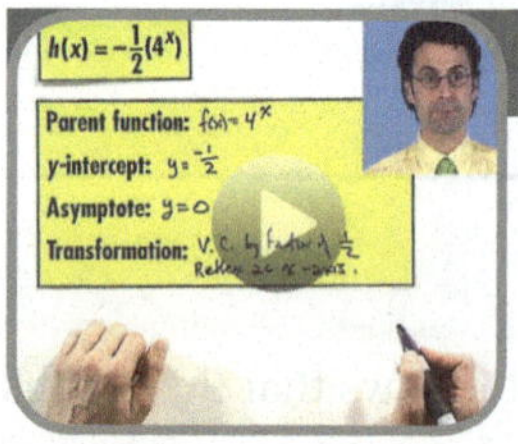

4.1.3 Transformations of Exponential Functions

Exponential functions can be graphed by using transformations, as we saw with other types of functions. The following table summarizes the transformations of exponential functions.

Transformations of Functions	
Vertical translation	$f(x) + k$ or $f(x) - k$
Horizontal translation	$f(x + h)$ or $f(x - h)$
Vertical stretch or compression	$cf(x)$
Reflection over the x-axis	$-f(x)$
Reflection over the y-axis	$f(-x)$

Remember, for a vertical translation $f(x) + k$, the graph of $f(x)$ is translated k units *up*, and for $f(x) - k$, the graph of $f(x)$ is translated k units *down*. For a horizontal translation $f(x - h)$, the graph of $f(x)$ is translated h units to the *right*, and for $f(x + h)$, the graph of $f(x)$ is translated h units to the *left*.

Reflections are used to graph the exponential functions in **Example 3**. Consider the function $g(x) = -2^x$. Notice that $-2^x = -(2^x)$. Therefore, $g(x) = -f(x)$ when $f(x) = 2^x$. So, the graph of $g(x) = -2^x$ is the graph of $f(x) = 2^x$ reflected over the x-axis. Specifically, this means that for every point (x, y) on the graph of $f(x) = 2^x$, the point $(x, -y)$ is on the graph of $g(x) = -2^x$.

EXAMPLE 3 Using Reflections and Stretches to Graph an Exponential Function

Sketch the graph of each function.

A. $g(x) = -3^x$ **B.** $h(x) = (0.5)2^{-x}$

> ***IMPORTANT***
>
> *The base of -3^x is 3, not -3. A base is negative only when enclosed within parentheses. For example, the base of $(-3)^x$ is -3. However, the base of an exponential function cannot be negative.*

SOLUTION

A. Sketch the graph of the basic function $f(x) = 3^x$.

Then reflect that graph over the x-axis, since $g(x) = -3^x = -f(x)$.

x	$f(x)$	$g(x)$
-2	$\frac{1}{9}$	$-\frac{1}{9}$
-1	$\frac{1}{3}$	$-\frac{1}{3}$
0	1	-1
1	3	-3
2	9	-9

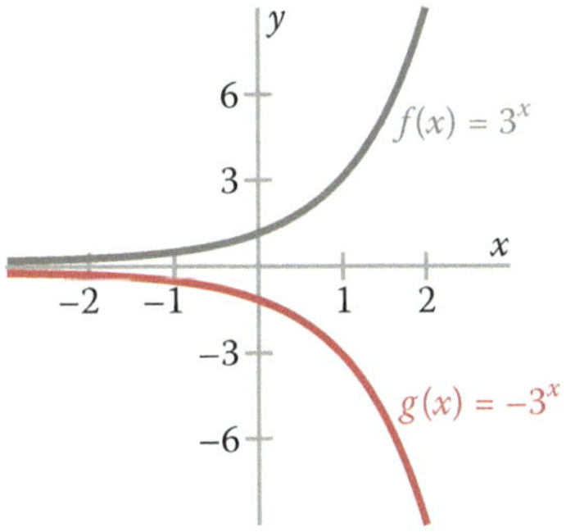

For every point (x, y) on the graph of $f(x) = 3^x$, the point $(x, -y)$ is on the graph of $g(x) = -3^x$.

B. Sketch the graph of the basic function $f(x) = 2^x$.

Then reflect that graph over the y-axis and vertically compress the curve by a factor of 1/2 (since $g(x) = (0.5)2^{-x} = (0.5)f(-x)$ and $0.5 = 1/2$).

x	$f(x)$
-2	$\frac{1}{4}$
-1	$\frac{1}{2}$
0	1
1	2
2	4

x	$h(x)$
2	$\frac{1}{8}$
1	$\frac{1}{4}$
0	$\frac{1}{2}$
-1	1
-2	2

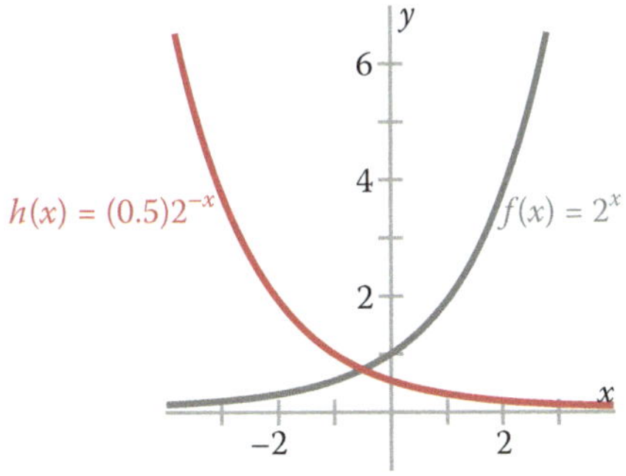

For every point (x, y) on the graph of $f(x) = 2^x$, the point $(-x, 0.5y)$ is on the graph of $g(x) = (0.5)2^{-x}$.

4.1.4 Graphing Exponential Functions: Another Example

Vertical and Horizontal Translations of $f(x) = b^x$

Translations of the exponential function $f(x) = b^x$ occur when a number is added to, or subtracted from, x or b^x. The following functions are all vertical or horizontal translations of $f(x) = b^x$.

$g(x) = b^x + 1$ *The graph of $f(x)$ is translated up 1 unit to form the graph of $g(x)$.*

$g(x) = b^x - 1$ *The graph of $f(x)$ is translated down 1 unit to form the graph of $g(x)$.*

$g(x) = b^{x+1}$ *The graph of $f(x)$ is translated to the left 1 unit to form the graph of $g(x)$.*

$g(x) = b^{x-1}$ *The graph of $f(x)$ is translated to the right 1 unit to form the graph of $g(x)$.*

For example, the graph of $g(x) = 2^x + 3$ is the graph of $f(x) = 2^x$ translated up 3 units (Figure 4.1b). Additionally, the graph of $g(x) = -3^{x-1}$ is the graph of $f(x) = 3^x$ reflected across the x-axis and translated to the right 1 unit (Figure 4.1c).

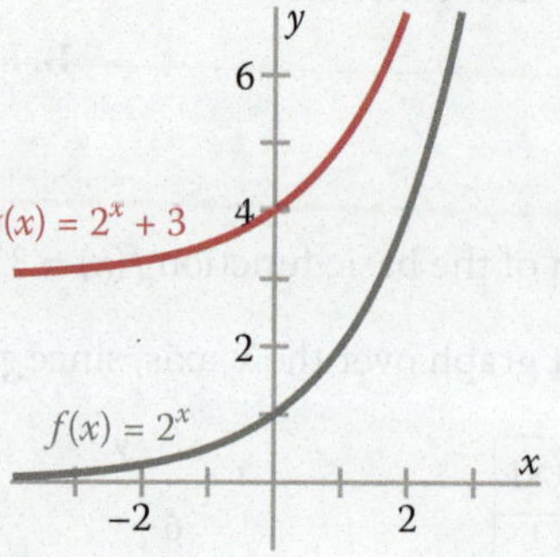

Figure 4.1b

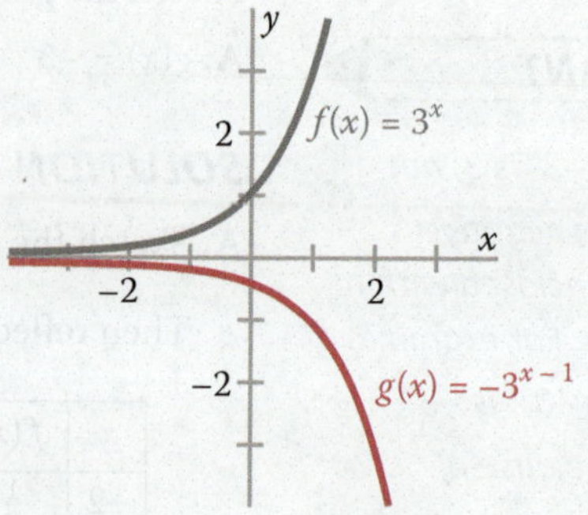

Figure 4.1c

EXAMPLE 4 Using Translations to Graph an Exponential Function

Sketch the graph of each function.

A. $g(x) = 2^{x-1}$ **B.** $h(x) = 5 + 2^x$

SOLUTION

A. Sketch the graph of the basic function $f(x) = 2^x$.

Then translate that graph 1 unit to the right since $g(x) = 2^{x-1} = f(x-1)$.

x	$f(x)$
−2	$\frac{1}{4}$
−1	$\frac{1}{2}$
0	1
1	2
2	4

x	$g(x)$
−1	$\frac{1}{4}$
0	$\frac{1}{2}$
1	1
2	2
3	4

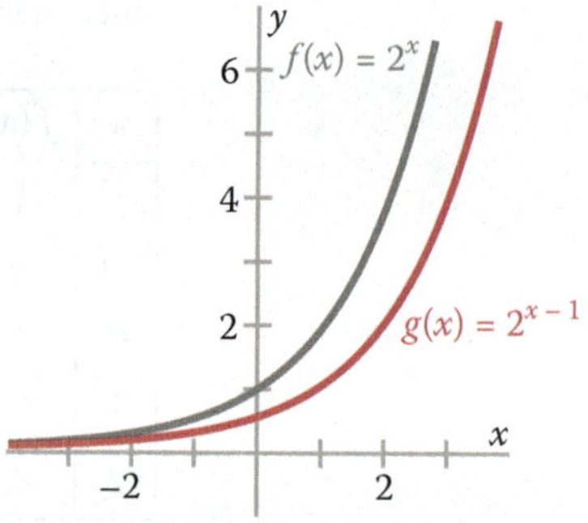

For every point (x, y) on the graph of $f(x) = 2^x$, the point (x + 1, y) is on the graph of $g(x) = 2^{x-1}$.

B. Sketch the graph of the basic function $f(x) = 2^x$.

Then translate that graph 5 units up, since $h(x) = 5 + 2^x = f(x) + 5$.

x	$f(x)$
−2	$\frac{1}{4}$
−1	$\frac{1}{2}$
0	1
1	2
2	4

x	$h(x)$
−2	$5\frac{1}{4}$
−1	$5\frac{1}{2}$
0	6
1	7
2	9

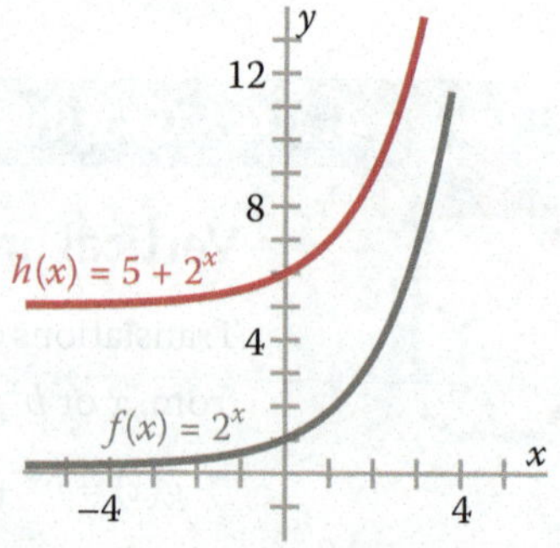

For every point (x, y) on the graph of $f(x) = 2^x$, the point (x, y + 5) is on the graph of $h(x) = 5 + 2^x$.

4.1.5 Finding Present Value and Future Value

When a bank or other financial institution loans a customer sum of money, called the **principal**, the amount that is eventually repaid by the customer is greater than the principal. The difference between the principal and the total amount repaid is called **interest**. Interest can also be earned and added to a principal that is invested, instead of loaned.

There are several ways to make interest calculations. The most basic type of interest is called **simple interest**. When a principal amount P is borrowed (or invested) for t years at an **interest rate** of r, the amount of simple interest I paid (or earned) is given by the formula $I = Prt$.

When the interest on an account (or loan) is compounded (calculated and added to the principal) m times per year, the interest is called **compound interest**. Compound interest is calculated using the compound interest formula, and is an example of exponential growth.

TIP

If the interest rate r is given as a percent, convert r to a fraction or decimal for use in the formula.

Compound Interest Formula

$$A = P\left(1 + \frac{r}{m}\right)^{mt}$$

Compound interest is compounded m times per year.

P = the principal amount (original amount) in the account

t = the length of time that the money is invested, in years

A = the balance in the account after t years

r = the annual interest rate (as a decimal)

m = the number of times per year that the interest is compounded

The compound interest formula can also be applied to debt. In that case, P is the original amount borrowed, and A is the amount owed after t years.

EXAMPLE 5 Using the Compound Interest Formula to Find a Balance

The original price of an appliance is \$2200. A customer financed the appliance at 6% interest, compounded monthly, and paid off the loan after 7 years. How much did the customer pay for the appliance, including interest?

SOLUTION

Identify the values of the quantities to be substituted into the compound interest formula from the given information.

The original amount borrowed is \$2200, so $P = 2200$.

The interest rate is 6%, so $r = 0.06$.

The interest is compounded monthly, so $m = 12$ (because the interest is compounded 12 times per year).

The loan is paid off after 7 years, so $t = 7$.

Substitute these values into the formula, and simplify to find the amount paid after 7 years.

IMPORTANT

Do not round until the final step in the calculations.

$$A = 2{,}200\left(1 + \frac{0.06}{12}\right)^{12(7)} = 2{,}200(1.005)^{84} \approx 3{,}344.81 \quad \textit{Round to the nearest cent.}$$

The customer paid a total of \$3344.81 for the appliance that originally cost \$2200.

EXAMPLE 6 Using the Compound Interest Formula to Find the Principal

How much needs to be deposited into an account offering 7% interest, compounded quarterly, in order for the account to be worth $250,000 after 30 years? Assume that no money is deposited after the initial deposit.

SOLUTION

Identify the values of the quantities to be substituted into the compound interest formula from the given information.

The original amount invested, P, is the unknown.

The interest rate is 7%, so $r = 0.07$. The interest is compounded quarterly, so $m = 4$ (because the interest is compounded 4 times per year).

The interest accumulates for 30 years, so $t = 30$.

The amount in the account after 30 years should be $250,000, so $A = 250{,}000$.

Substituting these values into the formula yields $250{,}000 = P\left(1+\frac{0.07}{4}\right)^{4(30)}$.

IMPORTANT

The unknown value P is multiplied by $\left(1+\frac{0.07}{4}\right)^{4(30)}$. So, divide both sides by this amount to solve the equation for P.

Simplify, and then solve the equation for P.

Do not round until the final step in the calculations.

$$250{,}000 = P\left(1+\frac{0.07}{4}\right)^{4(30)} \qquad \text{Substitute. Use the compound interest formula.}$$

$$250{,}000 = P(1.0175)^{120} \qquad \text{Simplify.}$$

$$\frac{250{,}000}{(1.0175)^{120}} = P \qquad \text{Solve for } P.$$

$$31{,}175.24 \approx P \qquad \text{Round to the nearest cent.}$$

$31,175.24 should be deposited initially so that the value of the account will be $250,000 after 30 years.

The answer for **Example 6** can be checked by substituting the given values of r, m, and t, along with the found value of P, into the compound interest formula and simplifying to confirm that the balance after 30 years will be approximately $250,000 when $31,175.24 is deposited into a quarterly compounding account with a 7% interest rate.

$$A = 31{,}175.24\left(1+\frac{0.07}{4}\right)^{4(30)} \approx 250{,}000 \checkmark$$

4.1.6 Finding an Interest Rate to Match Given Goals

In **Examples 5** and **6**, the compound interest formula was used to find A and P, respectively. The compound interest formula can also be solved for the interest rate r, as demonstrated in **Example 7**. (In Section 4.4, the compound interest formula will be solved for each exponent, m and t, using a slightly more complicated process.)

EXAMPLE 7 Using the Compound Interest Formula to Find the Interest Rate

An initial deposit of \$8050 was made into an account that compounds interest semi-annually. No other deposits were made. At the end of 18 years, the balance in the account had doubled. Find the interest rate on this account to the nearest hundredth of a percent.

TIP

The rate r in the compound interest formula is a decimal value. To write the rate as a percent rounded to the nearest hundredth of a percent, round the r-value to the nearest ten thousandth and then move the decimal point two places to the right.

SOLUTION

Identify the values of the variables in the compound interest formula.

The original amount deposited is \$8050, so $P = 8050$.

The interest rate is the unknown.

The interest is compounded semi-annually, so $m = 2$.

The interest accumulates for 18 years, so $t = 18$.

The amount in the account after 18 years is double the original amount deposited, so $A = 2(8050)$.

Substituting these values into the formula yields $2(8050) = 8050\left(1+\frac{r}{2}\right)^{2(18)}$.

Simplify, and then solve the equation for r. Do not round until the final step in the calculations.

$2(8050) = 8050\left(1+\frac{r}{2}\right)^{2(18)}$	*Substitute. Use the compound interest formula.*
$2 = \left(1+\frac{r}{2}\right)^{36}$	*Divide each side by 8050 and simplify.*
$\sqrt[36]{2} = 1+\frac{r}{2}$	*Take the 36th root of each side.*
$2\left(\sqrt[36]{2}-1\right) = r$	*Solve for r.*
$0.0389 \approx r$	*Simplify and round to the nearest ten thousandth.*

The interest rate is approximately 3.89%.

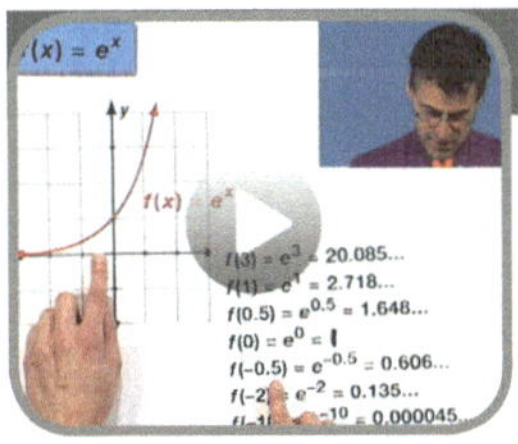

4.1.7 Evaluating and Graphing a Natural Exponential Function

The Natural Base e

The base of an exponential function can be any positive number not equal to 1. A base value that is commonly seen in applications of exponential functions is the irrational **number e**. Since e is an irrational number, its decimal expansion never terminates or repeats.

$$e = 2.7182818284\ldots$$

Any exponential function with base e is called a **natural exponential function**.

DEFINITION

The **natural exponential function** is the exponential function $f(x) = e^x$.

A natural exponential function can be evaluated by using a calculator or by substituting some approximation of e into the function.

EXAMPLE 8 Evaluating a Natural Exponential Function

Use a calculator to evaluate the natural exponential function $f(x) = e^x$ for $x = -2, 0, 1$, and 5. Round the answer to the nearest hundredth, if needed.

SOLUTION

Most calculators have an exponential function key, e^x. Enter the value of the exponent after pressing the exponential function key.

$$f(-2) = e^{-2} \approx 0.14$$
$$f(0) = e^0 = 1$$
$$f(1) = e^1 = e \approx 2.72$$
$$f(5) = e^5 \approx 148.41$$

TIP

Any number to the 0 power is equal to 1, even when the base is an irrational number.

Graphing Natural Exponential Functions

IMPORTANT

The natural exponential function is a type of exponential function, so all properties of exponential functions apply to natural exponential functions.

The graph of the natural exponential function $f(x) = e^x$ and its corresponding table are shown in Figure 4.1d. Like the graph of the exponential function $f(x) = b^x$, the graph of $f(x) = e^x$ has no x-intercepts, the y-intercept is at (0, 1), and the graph of $f(x) = e^x$ approaches, but does not intersect, the x-axis as x goes to $-\infty$ (i.e., the graph has a horizontal asymptote at $y = 0$). Moreover, all properties of the exponential function $f(x) = b^x$ apply to the natural exponential function $f(x) = e^x$ since the natural exponential function is an exponential function.

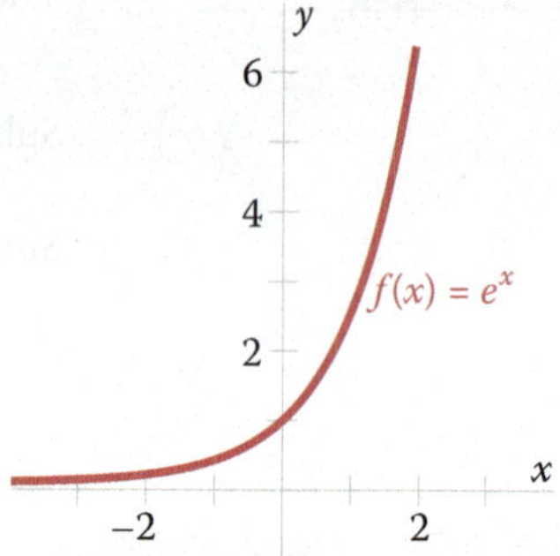

x	−2	−1	0	1	2
$f(x)$	0.14	0.37	1	2.72	7.39

Figure 4.1d

Natural exponential functions can be graphed using translations, reflections, and stretches, just like any other type of function.

EXAMPLE 9 Using Translations to Graph a Natural Exponential Function

Sketch the graph of each function.

A. $g(x) = e^{x+2}$ **B.** $h(x) = 3 + e^x$

SOLUTION

A. Sketch the graph of the basic function $f(x) = e^x$. Then translate that graph 2 units to the left, since $g(x) = e^{x+2} = f(x+2)$.

x	−2	−1	0	1	2
$f(x)$	0.14	0.37	1	2.72	7.39

x	−4	−3	−2	−1	0
$g(x)$	0.14	0.37	1	2.72	7.39

For every point (x, y) on the graph of $f(x) = e^x$, the point $(x - 2, y)$ is on the graph of $g(x) = e^{x+2}$.

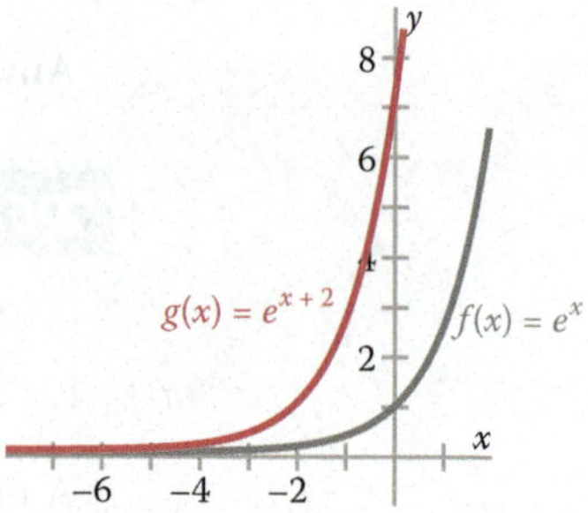

B. Sketch the graph of the basic function $f(x) = e^x$.

Then translate that graph 3 units up, since $h(x) = 3 + e^x = f(x) + 3$.

x	-2	-1	0	1	2
$f(x)$	0.14	0.37	1	2.72	7.39

x	-2	-1	0	1	2
$h(x)$	3.14	3.37	4	5.72	10.39

For every point (x, y) on the graph of $f(x) = e^x$, the point $(x, y + 3)$ is on the graph of $h(x) = 3 + e^x$.

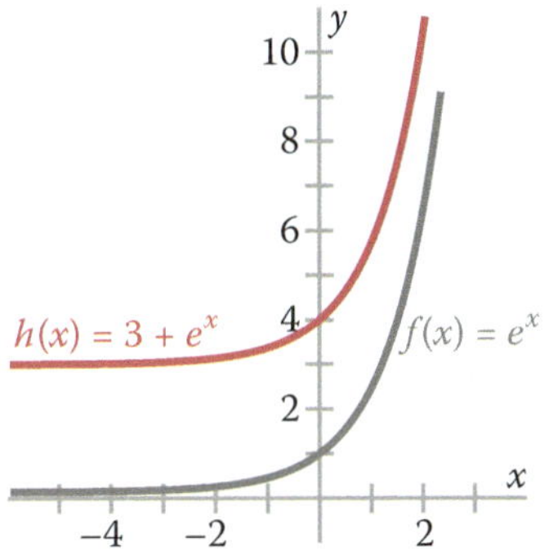

4.1.8 Applying Natural Exponential Functions

The natural base e is commonly used with **exponential models** involving a quantity that increases exponentially over time, such as population growth.

EXAMPLE 10 Using a Natural Exponential Model to Estimate Population

In 1990, the population of the United States was approximately 250 million people and was expected to grow according to the function $P(x) = 250{,}000{,}000e^{0.009x}$, where x is the number of years after 1990. According to the U.S. Census Bureau, the population of the United States in 2010 was approximately 310 million. Compare the actual population of the United States in 2010 to the expected population of the United States in 2010, according to this model.

SOLUTION

In the population model $P(x) = 250{,}000{,}000e^{0.009x}$, x represents the number of years after 1990. So, the population in 1990 corresponds to $x = 0$.

The expected population in 2010 corresponds to $x = 20$, because 2010 is 20 years after 1990.

Evaluate the population model for $x = 20$ to find the expected population in 2010.

$$P(20) = 250{,}000{,}000e^{0.009(20)} = 250{,}000{,}000e^{0.18} \approx 299{,}304{,}341 \approx 299{,}000{,}000$$

Thus, according to the model, the expected population in 2010 was approximately 299 million.

The model's estimate is approximately 11 million less than the actual population in 2010.

SECTION 4.1 EXERCISES

Warm Up

Describe the graph of each quadratic function as a transformation of $f(x) = x^2$.

1. $g(x) = x^2 - 4$
2. $h(x) = -5x^2$
3. $p(x) = (x + 3)^2 + 1$

Just the Facts

Fill in the blanks.

4. For a function of the general form $f(x) = b^x$ to be considered exponential, the base must be _____, _____, and not equal to _____.
5. Given the function $f(x) = b^x$, $f(x)$ will be equal to _____ for any positive real value of b if x is equal to _____.
6. If $f(x) = b^x$, then $g(x) =$ _____ is the reflection of $f(x)$ across the y-axis, and $g(x) =$ _____ is the reflection of $f(x)$ across the x-axis.
7. The graph of $f(x) = 5^x$ is _____ than the graph of $g(x) = 2^x$ for all $x > 0$.
8. A principal of $1000 was invested at 5% over 2 years. If the principal was compounded yearly using the compound interest formula, the interest gained would be _____. If the interest was calculated using the simple interest formula, the interest gained would be _____.
9. For the equation $7500 = \left(1 + \frac{r}{4}\right)^{40}$, the first step in isolating r is _____.
10. An exponential function with a base of e is called a(n) _____ exponential function.

Essential Skills

In Exercises 1–4, identify the exponential functions in each list.

1. $g(x) = 5^x$
   $h(x) = 2^{3x}$
   $p(x) = -4$
   $f(x) = 1^{3x+4}$
   $m(x) = 510 + x$
   $q(x) = 8^x - 3$
   $r(x) = x^2$
   $n(x) = 5x$
2. $p(x) = x^3$
   $g(x) = 7^{2x+1}$
   $h(x) = (-3)^x$
   $f(x) = \left(\frac{4}{7}\right)^x$
   $q(x) = -(3)^x$
   $r(x) = \pi^{2x}$
   $m(x) = 1^x$
   $n(x) = 0.8^{x+1}$
3. $g(x) = 2^{x+3}$
   $h(x) = 16x^2 + 3$
   $p(x) = \frac{x^4}{3}$
   $r(x) = (x - 1)^5$
   $m(x) = 7\pi^x$
   $q(x) = -5^x - 9$
   $f(x) = 8^{\frac{1}{2}x}$
   $n(x) = 1^{6+2x}$
4. $d(x) = 1^{3x}$
   $k(x) = (-4)^{6x}$
   $q(x) = 3x^3 + 4x - 5$
   $f(x) = \left(\frac{3}{2}\right)^{-x}$
   $t(x) = -(0.9)^x$
   $w(x) = 3^{5x+2}$
   $n(x) = \left(-\frac{5}{9}\right)^{4x-3}$
   $h(x) = (8x)^{1/3}$

In Exercises 5–24, evaluate each exponential function.

5. If $f(x) = 2^x$, find $f(4)$, $f(0)$, and $f(7)$.
6. If $f(x) = 8^x$, find $f(3)$, $f(-2)$, and $f\left(\frac{1}{2}\right)$.
7. If $g(x) = -(12)^x$, find $g(3)$, $g(-1)$, and $g\left(\frac{1}{2}\right)$.
8. If $c(x) = -5^x$, find $c(4)$, $c(-2)$, and $c\left(\frac{1}{2}\right)$.
9. If $d(x) = 3^{-x}$, find $d(5)$, $d(-3)$, and $d\left(\frac{3}{2}\right)$.
10. If $h(x) = 18^{-x}$, find $h(-3)$, $h(2)$, and $h\left(\frac{1}{2}\right)$.
11. If $r(x) = 49^x$, find $r(-2)$, $r\left(\frac{1}{2}\right)$, and $r\left(-\frac{3}{2}\right)$.
12. If $b(x) = 27^x$, find $b\left(\frac{1}{2}\right)$, $b\left(-\frac{1}{3}\right)$, and $b\left(\frac{2}{3}\right)$.
13. If $p(x) = \left(\frac{1}{27}\right)^x$, find $p(-2)$, $p\left(-\frac{1}{2}\right)$, and $p\left(\frac{2}{3}\right)$.
14. If $q(x) = \left(\frac{1}{32}\right)^x$, find $q(-1)$, $q(0)$, and $q\left(-\frac{1}{2}\right)$.
15. If $m(x) = \left(\frac{1}{20}\right)^x$, find $m(-3)$, $m(2)$, and $m\left(-\frac{3}{2}\right)$.
16. If $v(x) = \left(\frac{8}{5}\right)^x$, find $v(3)$, $v(-2)$, and $v\left(-\frac{3}{2}\right)$.
17. If $t(x) = \left(\frac{1}{12}\right)^{-x}$, find $t(2)$, $t(-1)$, and $t\left(\frac{1}{2}\right)$.
18. If $u(x) = \left(\frac{1}{45}\right)^{-x}$, find $u(-1)$, $u\left(-\frac{1}{2}\right)$, and $u\left(\frac{1}{2}\right)$.
19. If $n(x) = \left(\frac{3}{4}\right)^{-x}$, find $n(-5)$, $n(4)$, and $n\left(\frac{1}{2}\right)$.
20. If $w(x) = \left(\frac{25}{9}\right)^{-x}$, find $w(1)$, $w(-2)$, and $w\left(-\frac{3}{2}\right)$.
21. If $f(x) = 8^{-2x}$, find $f(2)$, $f\left(-\frac{3}{2}\right)$, and $f\left(\frac{5}{6}\right)$.

22. If $h(x) = 18^{3x}$, find $h(-1)$, $h\left(-\frac{2}{3}\right)$, and $h\left(\frac{1}{2}\right)$.

23. If $p(x) = \left(\frac{2}{3}\right)^{2x}$, find $p(0)$, $p(-1)$, and $p\left(-\frac{5}{2}\right)$.

24. If $q(x) = \left(\frac{5}{4}\right)^{2x}$, find $q(-1)$, $q(2)$, and $q\left(-\frac{3}{2}\right)$.

In Exercises 25–30, identify the equivalent exponential functions in each list.

25. $f(x) = 5^x$
$h(x) = (-5)^x$
$k(x) = -5^{-x}$
$g(x) = \left(\frac{1}{5}\right)^{-x}$
$m(x) = x^{-5}$
$b(x) = 5x^{-5}$
$g(x) = \left(\frac{1}{5}\right)^{x}$

26. $f(x) = \left(\frac{1}{3}\right)^{-x}$
$g(x) = 3^x$
$a(x) = -3^x$
$d(x) = -x^3$
$k(x) = -3^{-x}$
$m(x) = x^{-3}$
$h(x) = 3^{-x}$

27. $g(x) = -8^{-x}$
$d(x) = 8^{-x}$
$p(x) = \left(\frac{1}{8}\right)^{-x}$
$h(x) = 8^x$
$f(x) = -8^x$
$c(x) = x^{-8}$
$m(x) = \left(\frac{1}{8}\right)^{x}$

28. $p(x) = \left(\frac{4}{7}\right)^{-x}$
$q(x) = -\left(\frac{7}{4}\right)^{x}$
$m(x) = -\left(\frac{4}{7}\right)^{x}$
$f(x) = \left(\frac{7}{4}\right)^{x}$
$n(x) = -\left(\frac{7}{4}\right)^{-x}$
$c(x) = -\left(\frac{4}{7}\right)^{-x}$
$r(x) = \frac{7^{-x}}{4}$

29. $w(x) = -\left(\frac{3}{5}\right)^{-x}$
$g(x) = \left(\frac{5}{3}\right)^{x}$
$r(x) = -\left(\frac{5}{3}\right)^{x}$
$n(x) = \frac{5^{-x}}{3}$
$p(x) = -\left(\frac{5}{3}\right)^{-x}$
$f(x) = -\left(\frac{3}{5}\right)^{x}$
$h(x) = \left(\frac{3}{5}\right)^{-x}$

30. $h(x) = \left(\frac{11}{9}\right)^{-x}$
$p(x) = \frac{11^{-x}}{9}$
$w(x) = -\left(\frac{9}{11}\right)^{x}$
$r(x) = -\left(\frac{9}{11}\right)^{-x}$
$f(x) = \left(\frac{11}{9}\right)^{-x}$
$b(x) = \left(\frac{11}{9}\right)^{x}$
$g(x) = -\left(\frac{11}{9}\right)^{-x}$

In Exercises 31–36, graph each exponential function.

31. $g(x) = 4^x$

32. $f(x) = \left(\frac{1}{5}\right)^{x}$

33. $h(x) = 3^{-x}$

34. $g(x) = \left(\frac{1}{4}\right)2^{-x}$

35. $f(x) = 4^x - 2$

36. $g(x) = 3^{x-1} + 4$

37. An $1800 appliance is paid off over 6 years at 7% interest, compounded monthly. Find the total amount paid for the appliance including interest.

In Exercises 38–40, use the following information.

A credit union client deposits $900 in an account earning 9.5% interest, compounded quarterly. Find the balance of the account at the end of the given time period.

38. 19 years

39. 25 years

40. 40 years

In Exercises 41–43, use the following information.

A painter wants to have $100,000 at retirement in 25 years by making a one-time deposit in a bank account. How much money does the painter need to deposit, given the following account information?

41. compounds quarterly at 6% interest

42. compounds monthly at 5.5% interest

43. compounds daily at 5% interest

In Exercises 44–46, use the following information.

How much money must be deposited in an account that pays 4.75% interest, compounded semi-annually, in order to accrue a balance of $7,000 after the given time period?

44. 5 years and 6 months

45. 8 years and 3 months

46. 12 years and 9 months

47. Suppose $34,000 is invested in an account where interest is compounded monthly. After 18 years, the balance is $90,481. What was the interest rate? Round the answer to the nearest hundredth of a percent.

48. Suppose $80,000 is invested in an account where interest is compounded monthly. After 30 years, the balance is $637,803. What was the interest rate to the nearest hundredth of a percent?

49. An initial deposit of \$5525 was made into an account that compounds interest quarterly. No other deposits were made. At the end of 10 years, the balance in the account had doubled. Find the interest rate on this account. Round the answer to the nearest hundredth of a percent.

50. Suppose \$600 is compounded yearly for 20 years. If no other deposits are made, what rate is needed for the balance to triple in that time? Round the answer to the nearest hundredth of a percent.

In Exercises 51–58, evaluate each exponential function for $x = -2, -1, 0, 1,$ and 2. Round each value to the nearest hundredth.

51. $g(x) = 3e^x$

52. $f(x) = 5e^{-x}$

53. $p(x) = -2e^x$

54. $q(x) = 0.25e^x$

55. $w(x) = 1.4e^{-x}$

56. $r(x) = 2.6e^{-x}$

57. $d(x) = 0.325e^x$

58. $c(x) = -0.286e^x$

In Exercises 59–60, graph the exponential function.

59. $h(x) = e^{x+3}$

60. $g(x) = 5 + e^{x-2}$

In Exercises 61–63, use the following information.

The average atmospheric pressure P in pounds per square inch is $P = 14.7e^{-0.21x}$, where x is the altitude in miles above sea level. Find the average atmospheric pressure for the given altitude. Round to the nearest tenth.

61. 5.2 miles

62. 8.6 miles

63. 3.7 miles

In Exercises 64–66, use the following information.

A certain type of bacteria grows according to the function $P(t) = 1000e^{0.1912t}$, where P is the number of bacteria present after t hours. Find the number of bacteria present after the given time period. Round to the nearest whole number.

64. 6 hours

65. 8 hours

66. 12 hours

Extensions

67. A certain type of bacteria grows according to the function $P = 1000e^{0.1912t}$, where P is the number of bacteria present after t hours. Find the number of bacteria present after 5 hours and 15 minutes, rounded to the nearest whole number.

68. Graph the exponential function $h(x) = 2e^{-x-2} + 1$.

69. The average atmospheric pressure P in pounds per square inch is $P = 14.7e^{-0.21x}$, where x is the altitude in miles above sea level. Find the average atmospheric pressure for an altitude of 12,762 feet, rounded to the nearest tenth.

In Exercises 70–74, write the equation of an exponential function $g(x)$ that has the given characteristics.

70. base of 5, translated down 3 units, left 4 units, and reflected across the x-axis

71. base of 2, translated up 6 units, and right 3 units

72. base of 4, translated down 7 units, and right 1 unit

73. base of 1/2, translated down 5 units, and reflected across the y-axis

74. base of 9/4, translated up 2 units, left 8 units, and reflected across the x-axis

In Exercises 75–77, determine whether each statement is true or false.

75. The graph of $h(x) = e^{2x+3}$ has a y-intercept of –3/2.

76. The graph of $p(x) = -2e^{-x}$ is a reflection of the graph of $q(x) = 2e^{-x}$ across the y-axis.

77. The graph of $d(x) = 5e^{-(4x-3)}$ is a horizontal translation of $c(x) = 5e^{-4x}$ by 3/4 units to the right.

78. A student assigned to graph $g(x) = 4 - e^{x+2}$ transformed the coordinates of $f(x) = e^x$ by subtracting 2 from the x-coordinates and subtracting 4 from the y-coordinates. State the error(s), if any, that the student made.

4.2 LOGARITHMIC FUNCTIONS

OBJECTIVES

- Convert between exponential and logarithmic forms.
- Evaluate logarithmic functions, common logs, and natural logs.
- Use properties of logarithmic functions.
- Graph logarithmic functions.
- Find the domain of natural logarithmic functions.

PREREQUISITE VOCABULARY TERMS

domain
exponential function
Horizontal Line Test
inverse function
natural exponential function

4.2.1 An Introduction to Logarithmic Functions

Recall that when the graph of a function passes the Horizontal Line Test (i.e., no horizontal line intersects the graph of the function at more than one point), then that function is one-to-one and has an inverse function. Figure 4.2a shows the graph of the general exponential function $f(x) = b^x$, where $b > 1$. This function passes the Horizontal Line Test, is one-to-one, and must therefore have an inverse function. The inverse of an exponential function $f(x) = b^x$ is called the **logarithmic function with base *b***.

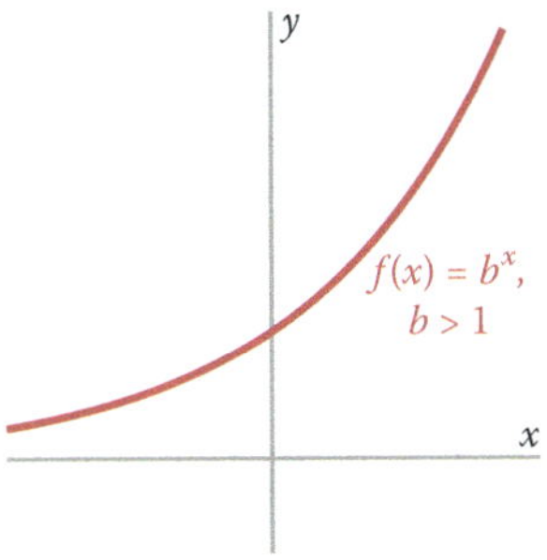

Figure 4.2a

REMEMBER

The x- and y-variables are switched for inverse functions.

DEFINITION

The function $f(x) = \log_b x$ is called the **logarithmic function with base *b***.
If b is a positive number such that $b \neq 1$, then $y = \log_b x \iff b^y = x$.

TIP

The symbol ⇔ is read as "if and only if." The expression $\log_b x$ is read as "log base b of x."

The equation $3^2 = 9$ can be written as a logarithm (or as a "log") where the base of the power is the base of the log, and the exponent from the power is the value of the log.

$$3^2 = 9 \iff 2 = \log_3 9 \qquad \textit{The exponent is the value of the log.}$$

The equation $y = \log_b x$ states that the expression $\log_b x$ is equal to the value y. Notice that the value of y is the exponent in the equation $b^y = x$. Therefore, the value of $\log_b x$ is the exponent to which the base b must be raised to get x. So, a log is an exponent.

EXAMPLE 1 Simplifying Log Expressions

Find the value of each log.

A. $\log_5 25$ **B.** $\log_2 16$ **C.** $\log_5 125$

SOLUTION

To simplify a log expression $\log_b x$, find the power of b that gives x.

A. The log's base is 5. What power of 5 is 25? $\log_5 25 = 2$ $\quad 5^2 = 25$

B. The log's base is 2. What power of 2 is 16? $\log_2 16 = 4$ $\quad 2^4 = 16$

C. The log's base is 5. What power of 5 is 125? $\log_5 125 = 3$ $\quad 5^3 = 125$

TIP

Keep in mind that the value of a log is an exponent.

4.2.2 Converting between Exponential and Logarithmic Functions

In the previous topic, you saw that $y = \log_b x$ if and only if $b^y = x$. So, any exponential equation of the form $b^y = x$ can be written as a logarithmic equation $y = \log_b x$, and vice versa.

Exponential Form Equation	Logarithmic Form Equation
$b^y = x$	$y = \log_b x$

EXAMPLE 2 Writing an Exponential Equation in Logarithmic Form

Write each exponential equation as a log statement.

A. $5^4 = 625$ **B.** $\left(\frac{3}{2}\right)^{-2} = \frac{4}{9}$ **C.** $4^{\frac{3}{2}} = 8$

SOLUTION

A. The power's base is 5, and the exponent is 4, so $\log_5 625 = 4$.

B. The power's base is $\frac{3}{2}$, and the exponent is −2, so $\log_{\frac{3}{2}} \frac{4}{9} = -2$.

C. The power's base is 4, and the exponent is $\frac{3}{2}$, so $\log_4 8 = \frac{3}{2}$.

An equation in logarithmic form can be written as an equivalent equation in exponential form where the log's base is the base of the power and the log's value is the exponent.

EXAMPLE 3 Writing a Logarithmic Equation in Exponential Form

Write each logarithmic equation as an exponential statement.

A. $\log_3 2187 = 7$ **B.** $\log_{81} 3 = \frac{1}{4}$ **C.** $\log_6 \frac{1}{216} = -3$

SOLUTION

A. The log's base is 3, and the log's value is 7, so $3^7 = 2187$.

B. The log's base is 81, and the log's value is $\frac{1}{4}$, so $81^{\frac{1}{4}} = 3$.

C. The log's base is 6, and the log's value is −3, so $6^{-3} = \frac{1}{216}$.

REMEMBER

The 1/4th power of 81 is equivalent to the 4th root of 81.

$81^{\frac{1}{4}} = \sqrt[4]{81} = \sqrt[4]{(3)^4} = 3$

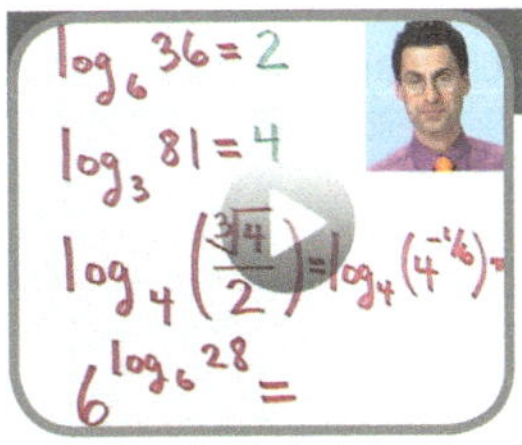

4.2.3 Evaluating Logarithms

The properties of exponents will often be used when simplifying a logarithmic expression.

EXAMPLE 4 **Simplifying a Logarithmic Expression**

Find the value of each log.

A. $\log_3 \frac{1}{81}$ **B.** $\log_{25} 5$ **C.** $\log_{16} 8$

SOLUTION

A. The log's base is 3. What power of 3 is $\frac{1}{81}$?

$a^{-m} = \frac{1}{a^m}$ ▶ $\log_3 \frac{1}{81} = -4$ $\quad 3^{-4} = \frac{1}{3^4} = \frac{1}{81}$

B. The log's base is 25. What power of 25 is 5?

$a^{\frac{1}{2}} = \sqrt{a}$ ▶ $\log_{25} 5 = \frac{1}{2}$ $\quad 25^{\frac{1}{2}} = \sqrt{25} = 5$

C. The log's base is 16. What power of 16 is 8?

$a^{\frac{m}{n}} = \left(\sqrt[n]{a}\right)^m$ ▶ $\log_{16} 8 = \frac{3}{4}$ $\quad 16^{\frac{3}{4}} = \left(\sqrt[4]{16}\right)^3 = 2^3 = 8$

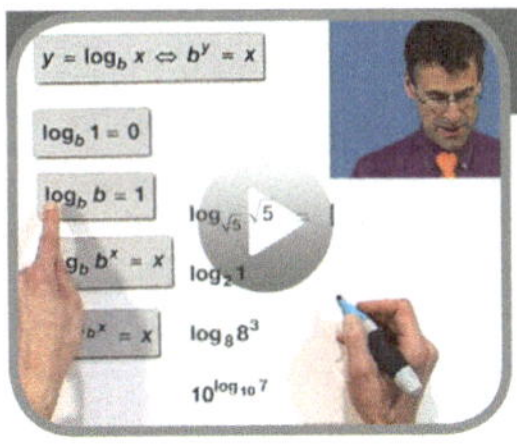

4.2.4 Using Properties to Evaluate Logarithms

There are several properties of logarithms that can be used when evaluating logarithmic expressions. These log properties follow directly from the properties of exponents and the definition of a logarithmic function.

By the Zero Exponent Property, any real number to the 0 power is equal to 1 (i.e., $b^0 = 1$). Writing this exponential equation in logarithmic form yields $\log_b 1 = 0$.

Consider the power of some number b where the exponent is equal to 1 (i.e., b^1). Any number to the power of 1 is equal to that number, so $b^1 = b$. Writing this exponential equation in logarithmic form yields $\log_b b = 1$. A third property of logs follows from the fact that a power is equal to itself: $b^x = b^x$. Writing this exponential equation in logarithmic form yields $\log_b b^x = x$.

A fourth property of logs is derived by substituting the expression for y from $y = \log_b x$ into the equivalent exponential equation $b^y = x$. Since $y = \log_b x$ and $b^y = x$, it follows that $b^{\log_b x} = x$.

Properties of Logarithms with Base b

Property	Description	Exponential Form
$\log_b 1 = 0$	*The power of b that equals 1 is 0.*	$b^0 = 1$
$\log_b b = 1$	*The power of b that equals b is 1.*	$b^1 = b$
$\log_b b^x = x$	*The power of b that equals b^x is x.*	$b^x = b^x$
$b^{\log_b x} = x$	*$\log_b x$ is the power to which b must be raised to get x.*	$y = \log_b x$ and $b^y = x$

EXAMPLE 5 Using Properties to Evaluate a Logarithmic Expression

Evaluate each logarithm.

A. $\log_{12} 12$ **B.** $\log_7 7^{\sqrt{2}}$ **C.** $\sqrt{3}^{\log_{\sqrt{3}} \frac{2}{3}}$ **D.** $\log_{0.2} 1$

SOLUTION

Identify the property that applies to each log. Then use that property to evaluate the log.

A. $\log_{12} 12 = 1$ $\quad$ $\log_b b = 1$

B. $\log_7 7^{\sqrt{2}} = \sqrt{2}$ $\quad$ $\log_b b^x = x$

C. $\sqrt{3}^{\log_{\sqrt{3}} \frac{2}{3}} = \dfrac{2}{3}$ $\quad$ $b^{\log_b x} = x$

D. $\log_{0.2} 1 = 0$ $\quad$ $\log_b 1 = 0$

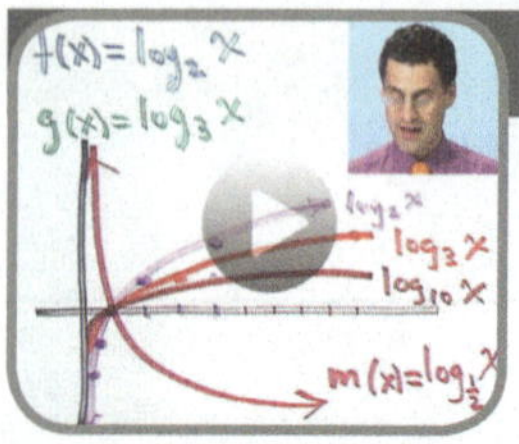

4.2.5 Graphing Logarithmic Functions

A logarithmic function of the form $f(x) = \log_b x$ can be graphed by making a table and plotting points. To find the ordered pairs for the table, first write the logarithmic function in exponential form: $x = b^{f(x)}$. Then choose y-values ($f(x)$-values) for the table, and find the corresponding x-values by evaluating $b^{f(x)}$.

Logarithmic Function		Exponential Function
$f(x) = \log_b x$	$\Leftrightarrow$	$x = b^{f(x)}$

EXAMPLE 6 Graphing a Logarithmic Function $y = \log_b x$

Sketch the graph of each logarithmic function.

A. $g(x) = \log_2 x$ **B.** $h(x) = \log_{\frac{1}{3}} x$

SOLUTION

A. In exponential form, the function is $x = 2^{g(x)}$. Make a table of values. Choose several values for $g(x)$, such as −2, −1, 0, 1, and 2, and then evaluate the power $2^{g(x)}$ to find each corresponding x-value. Plot the points from the table on a coordinate plane, and sketch a curve through the points.

x	$g(x)$	
$\frac{1}{4}$	−2	$2^{-2} = \frac{1}{2^2} = \frac{1}{4}$
$\frac{1}{2}$	−1	$2^{-1} = \frac{1}{2^1} = \frac{1}{2}$
1	0	$2^0 = 1$
2	1	$2^1 = 2$
4	2	$2^2 = 4$

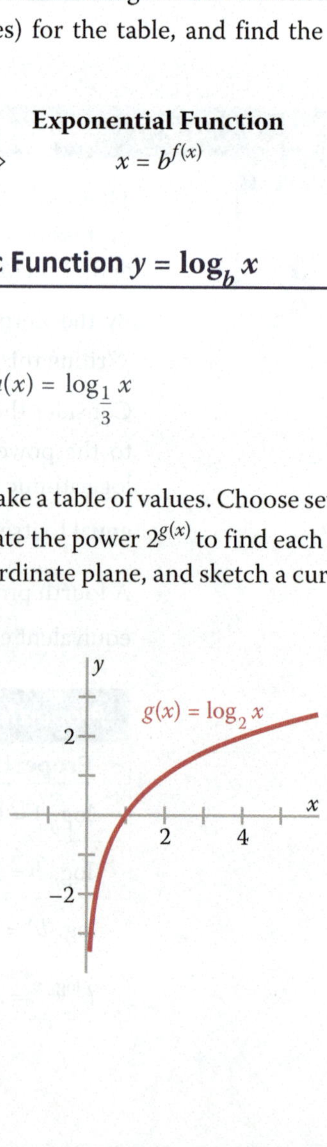

B. In exponential form, the function is $x = \left(\frac{1}{3}\right)^{h(x)}$. Make a table of values. Plot the points from the table, and sketch a curve through the points.

x	$h(x)$	
9	−2	$\left(\frac{1}{3}\right)^{-2} = 3^2 = 9$
3	−1	$\left(\frac{1}{3}\right)^{-1} = 3^1 = 3$
1	0	$\left(\frac{1}{3}\right)^{0} = 1$
$\frac{1}{3}$	1	$\left(\frac{1}{3}\right)^{1} = \frac{1}{3}$
$\frac{1}{9}$	2	$\left(\frac{1}{3}\right)^{2} = \frac{1}{9}$

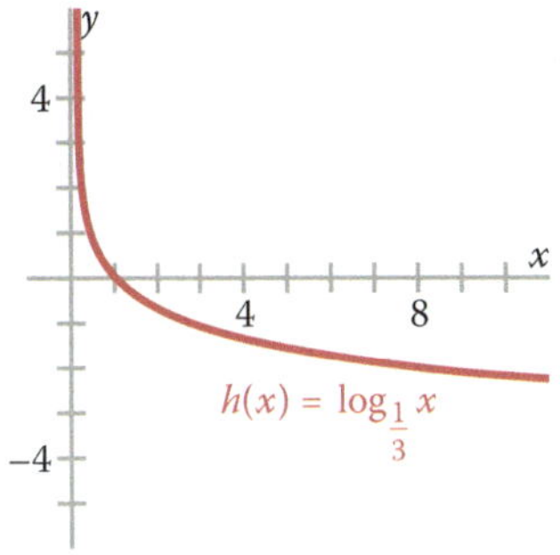

Using the Inverse to Graph a Logarithmic Function

Recall that if two functions are inverses, then their graphs are reflections over the line $y = x$ (i.e., for each point (x, y) on the graph of one function, the point (y, x) is on the graph of the other function). It follows that since $y = b^x$ and $y = \log_b x$ are inverses, their graphs are reflections over the line $y = x$, as shown in Figure 4.2b.

So, to graph a logarithmic function of the form $y = \log_b x$, graph the inverse function (the exponential function with the same base as the logarithmic function, $y = b^x$), and then reflect that graph over the line $y = x$.

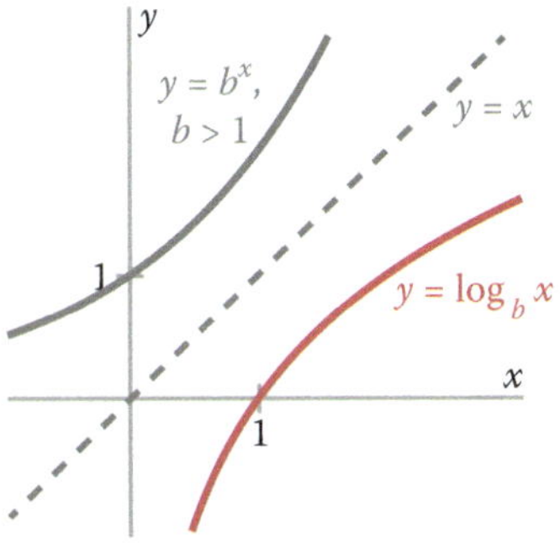

Figure 4.2b

Consider the functions graphed in **Example 6**, $g(x) = \log_2 x$ and $h(x) = \log_{\frac{1}{3}} x$.

The graphs of $g(x) = \log_2 x$ and $f(x) = 2^x$ are reflections over the line $y = x$ since f and g are inverse functions (Figure 4.2c).

The graphs of $h(x) = \log_{\frac{1}{3}} x$ and $f(x) = \left(\frac{1}{3}\right)^x$ are reflections over the line $y = x$ since f and h are inverse functions (Figure 4.2d).

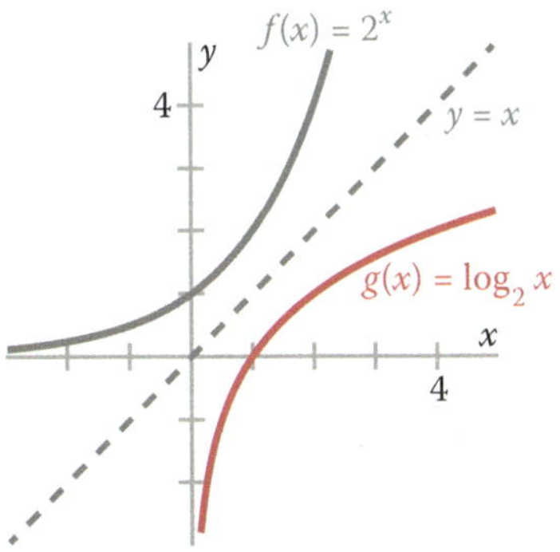

Figure 4.2c

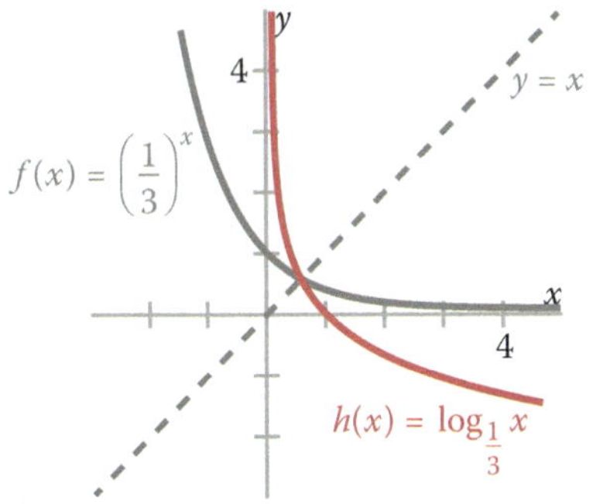

Figure 4.2d

Note that the functions $f(x) = \log_b x$ and $x = b^{f(x)}$ form the same graph. Additionally, the graph of $f(x) = \log_b x$ is the reflection of the graph of its inverse, $f(x) = b^x$, over the line $y = x$.

4.2.6 Matching Logarithmic Functions with Their Graphs

You are now familiar with the general shape of the graph of a logarithmic function of the form $f(x) = \log_b x$. By referring to two of the properties of logs, $\log_b b = 1$ and $\log_b 1 = 0$, we can generalize that all graphs of functions of the form $f(x) = \log_b x$ will pass through the points $(b, 1)$ and $(1, 0)$, which is the graph's x-intercept. You have also seen that all graphs of functions of the form $f(x) = \log_b x$ will have a vertical asymptote at $x = 0$ (i.e., at the y-axis).

Other forms of logarithmic functions can be graphed by using the property $\log_b 1 = 0$ to find the graph's x-intercept and then using $\log_b b = 1$ to find an additional point on the graph.

EXAMPLE 7 Graphing a Logarithmic Function

Sketch the graph of each logarithmic function.

A. $f(x) = \log_4 x$ **B.** $g(x) = \log_4 -2x$ **C.** $h(x) = \log_2 (x - 1)$

SOLUTION

Use the property $\log_b 1 = 0$ to find the graph's x-intercept, and use the property $\log_b b = 1$ to find another point on the graph. Then sketch the curve through those two points.

A. Since $\log_4 1 = 0$, the graph's x-intercept must be at $(1, 0)$. Since $\log_4 4 = 1$, another point on the graph is $(4, 1)$. Sketch the curve through these points.

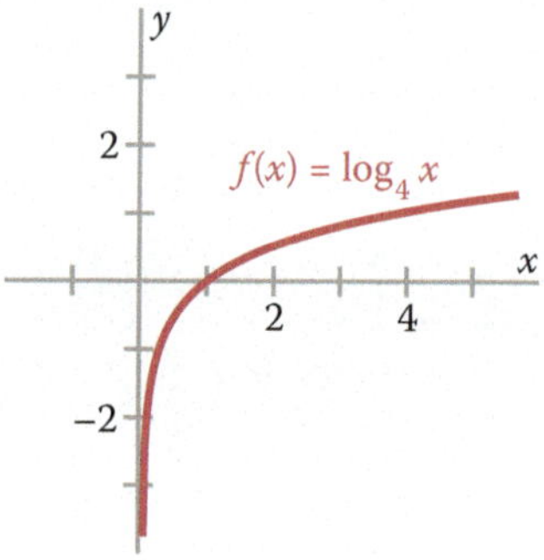

B. Because −2 is being multiplied by x in $g(x) = \log_4 -2x$, using the properties $\log_b 1 = 0$ and $\log_b b = 1$ to find two points on the graph is more complicated for $g(x) = \log_4 -2x$ than it was for $f(x) = \log_4 x$. So, find the values of x that make $-2x$ equal to 1 (to use $\log_b 1 = 0$) and equal to 4 (to use $\log_b b = 1$).

TIP

The expression $\log_4 -2x$ is equal to $\log_4 1$ when $x = -\frac{1}{2}$ because $\log_4\left(-2 \cdot -\frac{1}{2}\right) = \log_4 1$. So, when $x = -\frac{1}{2}$, $y = 0$.

Since $\log_4 1 = 0$ and $\log_4 -2x = \log_4 1$ when $x = -\frac{1}{2}$, the graph's x-intercept must be at $\left(-\frac{1}{2}, 0\right)$.

Since $\log_4 4 = 1$ and $\log_4 -2x = \log_4 4$ when $x = -2$, another point on the graph is $(-2, 1)$. Sketch the curve through these points.

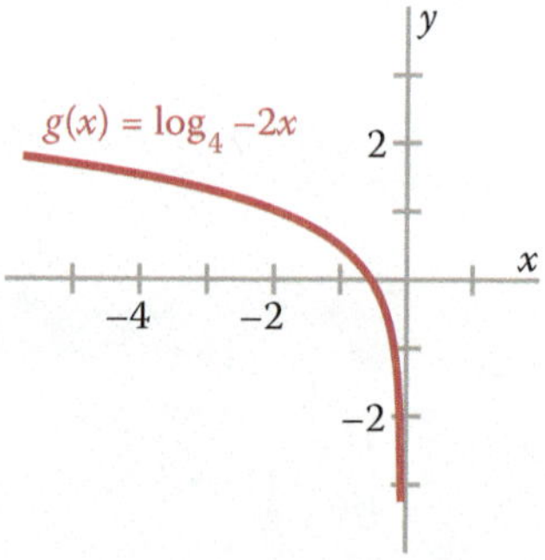

ALTERNATIVE METHOD

The graph of $h(x) = \log_2 (x - 1)$ is the graph of $f(x) = \log_2 x$ translated 1 unit to the right, since $h(x) = f(x - 1)$.

C. Since $\log_2 1 = 0$ and $\log_2 (x - 1) = \log_2 1$ when $x = 2$, the graph's x-intercept must be at $(2, 0)$.

Since $\log_2 2 = 1$ and $\log_2 (x - 1) = \log_2 2$ when $x = 3$, another point on the graph is $(3, 1)$. Sketch the curve through these points.

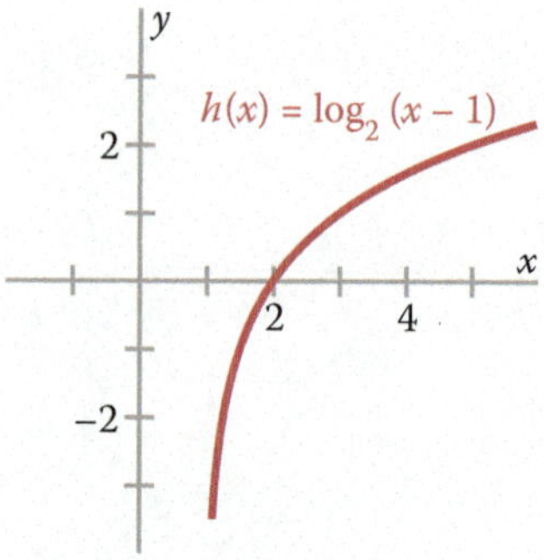

CAUTION

The expressions $\log_b (x + 1)$ and $\log_b x + 1$ are not equivalent.

$\log_b (x + 1)$ — *Log base b of (x + 1)*

$\log_b x + 1$ — *The sum of log base b of x and 1; equivalent to $1 + \log_b x$*

So, the graph of $y = \log_b (x + 1)$ is the graph of $y = \log_b x$ translated 1 unit to the left. The graph of $y = \log_b x + 1$ is the graph of $y = \log_b x$ translated 1 unit up.

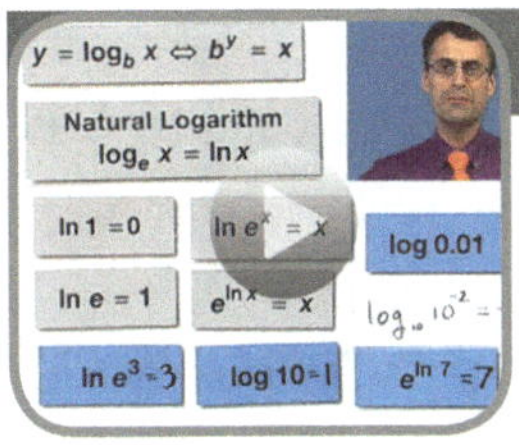

4.2.7 Common Logs and Natural Logs

Common Logs

A logarithm with base 10 is called a **common log**. The base of a common log is not written. So, when a log has no base written, the base is understood to be 10. The graph of the basic common log function $f(x) = \log x$ is shown in Figure 4.2e.

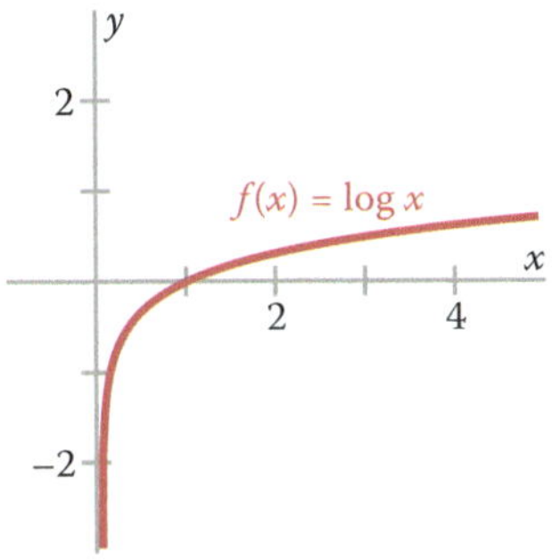

Figure 4.2e

DEFINITION

A **common log** is a logarithm where the base is 10.

$$\log x = \log_{10} x$$

The common log function $y = \log x$ is the inverse function of $y = 10^x$.

Natural Logs

Recall that the natural exponential function is the exponential function with base e, $f(x) = e^x$. There is also a **natural logarithmic function**. A natural logarithm has base e and is denoted $\ln x$. The graph of the basic natural log function $f(x) = \ln x$ is shown in Figure 4.2f.

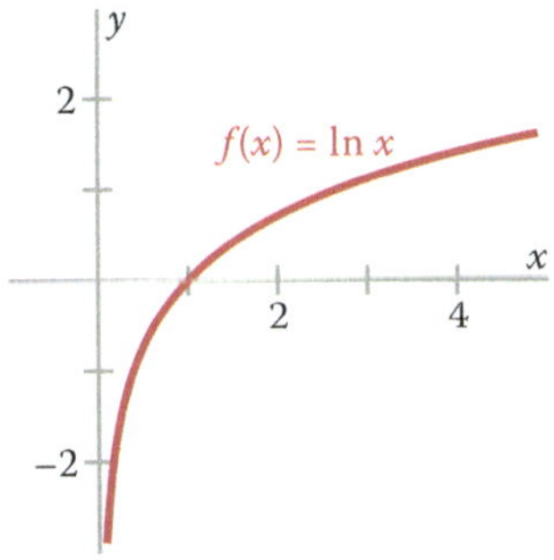

Figure 4.2f

DEFINITION

A **natural log** is a logarithm where the base is the number e.

$$\ln x = \log_e x \quad (\text{where } x > 0)$$

The natural log function $y = \ln x$ is the inverse function of $y = e^x$.

Properties of Common Logs and Natural Logs

Recall that the exponential form of the logarithmic function $y = \log_b x$ is $b^y = x$. A common log and a natural log can also be written in exponential form. All properties of logs can also be applied to common logs and natural logs.

Logarithmic Form and Exponential Form

	Logarithmic Form	Exponential Form
Logarithm with Base *b*	$y = \log_b x$	$b^y = x$
Common Logarithm	$y = \log x$	$10^y = x$
Natural Logarithm	$y = \ln x$	$e^y = x$

REMEMBER

Properties of Logarithms with Base b

$\log_b 1 = 0$

$\log_b b = 1$

$\log_b b^x = x$

$b^{\log_b x} = x$

Properties of Common Logarithms and Natural Logarithms

Property	Description	Exponential Form
$\log 1 = 0$ $\ln 1 = 0$	*The power of 10 that equals 1 is 0.* *The power of e that equals 1 is 0.*	$10^0 = 1$ $e^0 = 1$
$\log 10 = 1$ $\ln e = 1$	*The power of 10 that equals 10 is 1.* *The power of e that equals e is 1.*	$10^1 = 10$ $e^1 = e$
$\log 10^x = x$ $\ln e^x = x$	*The power of 10 that equals 10^x is x.* *The power of e that equals e^x is x.*	$10^x = 10^x$ $e^x = e^x$
$10^{\log x} = x$ $e^{\ln x} = x$	*log x is the power to which 10 must be raised to get x.* *ln x is the power to which e must be raised to get x.*	$y = \log x$ and $10^y = x$ $y = \ln x$ and $e^y = x$

These properties can be used to simplify common log expressions and natural log expressions.

EXAMPLE 8 Using Properties to Evaluate Common Logs and Natural Logs

Evaluate each log without using a calculator.

A. $\log 10^7$ **B.** $(\ln e)(\ln 1)$ **C.** $10^{\log\sqrt{5}}$ **D.** $\ln e^{\log 10}$

SOLUTION

A. Since $\log 10^x = x$, $\log 10^7 = 7$.

Alternatively, find the power of 10 (the base) that equals 10^7.

Since the power of 10 that equals 10^7 is 7, $\log 10^7 = 7$.

B. By the properties, $\ln e = 1$ and $\ln 1 = 0$. So, $(\ln e)(\ln 1) = (1)(0) = 0$.

C. Since $10^{\log x} = x$, $10^{\log\sqrt{5}} = \sqrt{5}$.

Alternatively, notice that $\log\sqrt{5}$ is the power to which 10 must be raised to get $\sqrt{5}$.

So, $10^{\log\sqrt{5}} = \sqrt{5}$.

D. Since $\ln e^x = x$, $\ln e^{\log 10} = \log 10$. By the property $\log 10 = 1$, $\log 10$ simplifies to 1.

So, $\ln e^{\log 10} = 1$.

4.2.8 Evaluating Common Logs and Natural Logs Using a Calculator

Most calculators have a LOG key and a LN key. The LOG key is for the common log, and the LN key is for the natural log. For example, to simplify log 5 with a calculator, use the keystrokes LOG 5 ENTER. Similarly, use the keystrokes LN 5 ENTER to simplify ln 5 with a calculator.

EXAMPLE 9 Using a Calculator to Evaluate Common Logs and Natural Logs

Evaluate log 80 and ln 80 using a calculator. Round to the nearest tenth.

SOLUTION

> **TIP**
>
> *The log's number must be entered before the log key for some scientific calculators.*

The value of log 80 is the power of 10 that equals 80. Since $10^1 < 80 < 10^2$, the value of log 80 must be between 1 and 2. Furthermore, since 80 is much closer to 100 than it is to 10, the value of log 80 must be close to 2.

$\log 80 \approx 1.9$ Calculator Keystrokes: LOG 80 ENTER

The value of ln 80 is the power of e that equals 80. Since $2.7^4 < 80 < 2.7^5$ (2.7 is being used as the approximation for the value of e), the value of ln 80 must be between 4 and 5.

$\ln 80 \approx 4.4$ Calculator Keystrokes: LN 80 ENTER

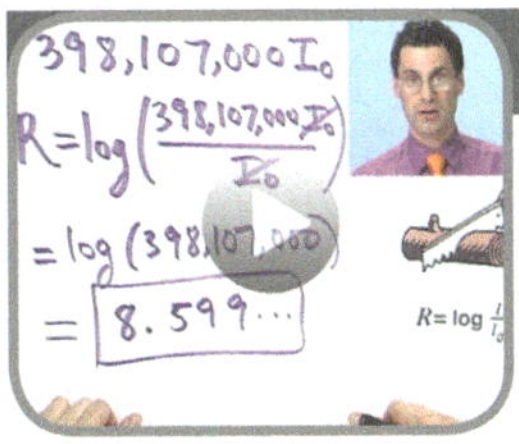

4.2.9 Evaluating Logarithmic Models

Logarithms can be used to measure quantities in real-world situations.

EXAMPLE 10 Measuring Loudness

The loudness L (in decibels) of a sound of intensity I is given by $L = 10\log\left(\frac{I}{I_0}\right)$, where I_0 is a small threshold intensity. Find the approximate loudness of an average coffee shop that has a loudness of $3{,}000{,}000I_0$. Round to the nearest whole decibel.

SOLUTION

Substitute $3{,}000{,}000I_0$ for I and simplify.

$$L = 10\log\left(\frac{3{,}000{,}000I_0}{I_0}\right) = 10\log(3{,}000{,}000) \approx 65$$

The loudness of the coffee shop is approximately 65 decibels.

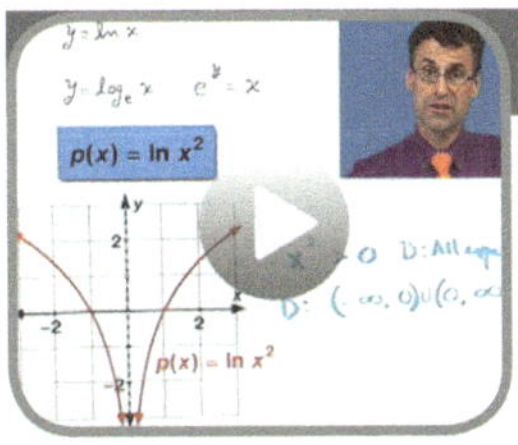

4.2.10 Domain of a Natural Log Function

The domain of any logarithmic function, $y = \log_b x$, $y = \log x$, or $y = \ln x$, is all positive real numbers.

EXAMPLE 11 Finding the Domain of a Logarithmic Function

Find the domain of $f(x) = \ln(x - 4)$.

SOLUTION

The natural logarithm $\ln(x - 4)$ is defined only when the expression $x - 4$ is positive. So, solve the inequality $x - 4 > 0$ to find the domain of $f(x)$.

$$x - 4 > 0 \Rightarrow x > 4$$

Therefore, the domain of $f(x) = \ln(x - 4)$ is $(4, \infty)$.

SECTION 4.2 EXERCISES

Warm Up

Simplify.

1. $4^0 + 5^1 - 6$
2. $27^{-\frac{2}{3}}$
3. $32^{\frac{3}{5}}$

Just the Facts

Fill in the blanks.

4. The inverse of an exponential function of base b will be a(n) _____ function with base _____.
5. $\log_b 1 =$ _____ by the _____ property.
6. One cannot take the logarithm of a(n) _____ number.
7. A logarithmic function can be graphed by converting the logarithm to _____ form and making a(n) _____ .
8. A common log is written without a(n) _____ and has a base of _____.
9. The natural log is written as _____ and has a base of _____.
10. The expression $\log_b b^x$ can be rewritten as _____.

Essential Skills

In Exercises 1–8, find each value.

1. $\log_3 9$
2. $\log_{10} 10{,}000$
3. $\log_7 343$
4. $\log_{11} 121$
5. $\log_5 625$
6. $\log_2 128$
7. $\log_3 243$
8. $\log_8 512$

In Exercises 9–22, write the exponential equation in logarithmic form.

9. $3^4 = 81$
10. $6^{-2} = \frac{1}{36}$
11. $\frac{2}{5}^{-2} = \frac{25}{4}$
12. $8^{\frac{2}{3}} = 4$
13. $\frac{1}{81} = 9^{-2}$
14. $\left(\frac{4}{5}\right)^4 = \frac{256}{625}$
15. $\left(\frac{4}{11}\right)^{-2} = \frac{121}{16}$
16. $2^{-9} = \frac{1}{512}$
17. $27^{\frac{5}{3}} = 243$
18. $\left(\frac{9}{5}\right)^3 = \frac{729}{125}$
19. $7^{-4} = \frac{1}{2401}$
20. $81^{\frac{3}{4}} = 27$
21. $\left(\frac{2}{7}\right)^2 = \frac{4}{49}$
22. $\left(\frac{8}{9}\right)^{-3} = \frac{729}{512}$

In Exercises 23–36, write the logarithmic equation in exponential form.

23. $\log_6 216 = 3$
24. $\log_{\frac{1}{2}} 8 = -3$
25. $\log_{\frac{1}{5}} 625 = -4$
26. $\log_{81} \frac{1}{3} = -\frac{1}{4}$
27. $\log_7 16{,}807 = 5$
28. $\log_{32} 8 = \frac{3}{5}$
29. $\log_{\frac{4}{5}} \frac{64}{125} = 3$
30. $\log_{14} \frac{1}{196} = -2$
31. $\frac{4}{3} = \log_{27} 81$
32. $\log_{13} 2197 = 3$
33. $\log_{\frac{4}{3}} \frac{16}{9} = 2$
34. $-4 = \log_6 \frac{1}{1296}$
35. $\log_8 128 = \frac{7}{3}$
36. $4 = \log_{\frac{7}{5}} \frac{2401}{625}$

In Exercises 37–56, simplify the expression.

37. $\log_{\frac{3}{5}} \frac{9}{25}$

38. $\log_{10} \frac{1}{1000}$

39. $\log_2 512$

40. $\log_5 \frac{1}{3125}$

41. $\log_8 4096$

42. $\log_{17} \frac{1}{289}$

43. $\log_{27} 3$

44. $\log_{25} 125$

45. $\log_{81} 27$

46. $\log_{512} 8$

47. $\log_{343} 49$

48. $\log_{216} 36$

49. $\log_{78} 78$

50. $\log_{\sqrt{7}} 1$

51. $\log_4 4^5$

52. $0.4^{\log_{0.4} 10}$

53. $\log_{2.8} 1$

54. $\log_{\sqrt{23}} \sqrt{23}$

55. $9^{\log_9 64}$

56. $\log_{\frac{2}{3}} \left(\frac{2}{3}\right)^{-5}$

In Exercises 57–60, graph the function.

57. $h(x) = \log_5 x$

58. $g(x) = \log_{\frac{1}{5}} x$

59. $m(x) = \log_5(-x)$

60. $g(x) = \log_2(x+5)$

In Exercises 61–68, evaluate without using a calculator.

61. $\log 10^e$

62. $\ln e^6$

63. $\ln e^{\frac{4}{7}}$

64. $\log 10^\pi$

65. $e^{\ln 9} + \ln e$

66. $\log 1 + 10^{\log 2}$

67. $4 \cdot 10^{\log 7} + 8 \log 1$

68. $\ln e + 7e^{\ln 5}$

In Exercises 69–74, evaluate with a calculator. Round to the nearest hundredth.

69. $\ln \frac{3}{8}$

70. $(\log 5)(\ln 12)$

71. $\log 25 - \log 16$

72. $\ln 48 + \log 63$

73. $(\log 67)(\ln 8)$

74. $(\ln 93)/(\ln 7)$

In Exercises 75–77, use the following information.

The loudness L (in decibels) of a sound of intensity I is given by $L = 10 \log\left(\frac{I}{I_0}\right)$, where I_0 is a small threshold intensity.

75. Find the approximate loudness to the nearest hundredth on a calm day on Crater Lake, which has an intensity of $48{,}000I_0$.

76. Find the approximate loudness to the nearest hundredth of a vacuum cleaner, which has an intensity of $90{,}000{,}000I_0$.

77. Find the approximate loudness to the nearest hundredth of a normal conversation, which has an intensity of $800{,}000I_0$.

In Exercises 78–80, use the following information.

The Richter scale magnitude R of an earthquake of intensity I is defined as $R = \log\left(\frac{I}{I_0}\right)$, where I_0 is a small threshold intensity. Find the magnitude of an earthquake to the nearest hundredth with the given intensity.

78. $52{,}000{,}000I_0$

79. $76{,}000I_0$

80. $8{,}500{,}000I_0$

In Exercises 81–86, express the domain of the function in interval notation.

81. $g(x) = \ln(x - 7)$

82. $f(x) = \ln(6 - 2x)$

83. $p(x) = \log(4x + 3)$

84. $w(x) = \ln(9 - 5x)$

85. $t(x) = \log(3x - 1)$

86. $n(x) = \ln(22x + 19)$

Extensions

87. Simplify. $\log_6 6^5 + 7\ln e^3 - \log 10^{-4} + 6^0$

88. Determine the x-intercept of the graph of $m(x) = \log_4 (2x)$.

89. Which are valid logarithmic equation?
$y = \log_{-5} 6$, $y = \log_1 4$, $y = \log_0 7$,
$y = \log_b 0$, $y = \log_6 -10$, $y = \log 1$

In Exercises 90–93, use the following information.

The equation $\log_b x = y$ has three variables: *b*, *x*, and *y*:

90. Which variable can never be negative?

91. Which variable can never be 1?

92. Which variable may at times be zero?

93. Which variable may at times be negative?

94. Express the domain of $p(x) = \log\sqrt{2x-5}$ in interval notation.

95. Which of these two functions, $g(x) = \log_{\frac{4}{7}} x$ or $h(x) = \log_{\frac{9}{10}} x$, increases as x increases?

4.3 PROPERTIES OF LOGARITHMS

OBJECTIVES

- Understand the Product Property, Quotient Property, and Power Property of logarithms.
- Use the properties of logarithms to expand logarithmic expressions.
- Use the properties of logarithms to simplify logarithmic expressions.
- Use the Change of Base Formula to simplify logarithms.

PREREQUISITE VOCABULARY TERMS

common logarithm
exponential function
logarithmic function
natural logarithm

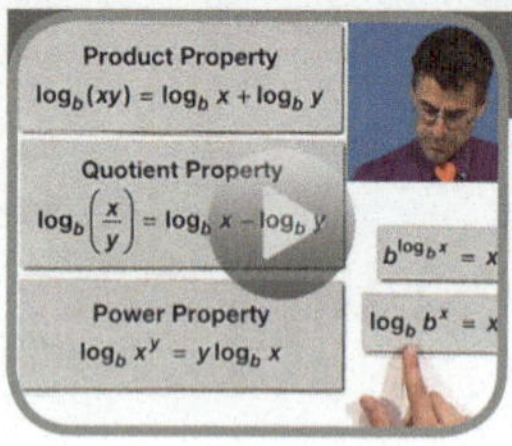

4.3.1 Properties of Logarithms

Three types of logarithms were discussed in Section 4.2: logs with base b (where b is a positive number not equal to 1), common logs (where the base is 10), and natural logs (where the base is e). Recall that each type of log can be written in logarithmic form or exponential form.

Logarithmic Form and Exponential Form		
	Logarithmic Form	**Exponential Form**
Logarithm with Base b	$y = \log_b x$	$b^y = x$
Common Logarithm	$y = \log x$	$10^y = x$
Natural Logarithm	$y = \ln x$	$e^y = x$

The following table shows the four basic properties of logarithms, which were introduced in Section 4.2.

Properties of Logarithms				
Logarithm with Base b	$\log_b 1 = 0$	$\log_b b = 1$	$\log_b b^x = x$	$b^{\log_b x} = x$
Common Logarithm	$\log 1 = 0$	$\log 10 = 1$	$\log 10^x = x$	$10^{\log x} = x$
Natural Logarithm	$\ln 1 = 0$	$\ln e = 1$	$\ln e^x = x$	$e^{\ln x} = x$

The following table lists three additional properties of logarithms. These additional properties provide alternative ways to express the log of a product, the log of a quotient, and the log of a power.

NOTICE THAT

The Product and Quotient Properties of logarithms are very similar to the Product and Quotient Properties of Powers.

Product of Powers

$a^m a^n = a^{m+n}$

Quotient of Powers

$\frac{a^m}{a^n} = a^{m-n}$

Properties of Logarithms

	Product Property	Quotient Property	Power Property
Logarithm with Base *b*	$\log_b(xy) = \log_b x + \log_b y$	$\log_b\left(\frac{x}{y}\right) = \log_b x - \log_b y$	$\log_b x^y = y\log_b x$
Common Logarithm	$\log(xy) = \log x + \log y$	$\log\left(\frac{x}{y}\right) = \log x - \log y$	$\log x^y = y\log x$
Natural Logarithm	$\ln(xy) = \ln x + \ln y$	$\ln\left(\frac{x}{y}\right) = \ln x - \ln y$	$\ln x^y = y\ln x$

These properties can be used to reduce a log of a product, quotient, or power into logs that can be simplified.

EXAMPLE 1 Using Properties to Evaluate Logarithmic Expressions

Evaluate each expression without using a calculator.

A. $\log_3\sqrt{27}$ **B.** $\log 20 + \log 5$ **C.** $\log_2 6 - \log_2 15 + \log_2 20$

SOLUTION

A. Write the radical as a rational exponent, then apply the Power Property to the log to simplify.

$\sqrt{a} = a^{\frac{1}{2}}$ ▶

$$\log_3\sqrt{27} = \log_3 27^{\frac{1}{2}}$$

$$= \frac{1}{2}\log_3 27 \quad \textit{Write the radical as a rational exponent.}$$

$$= \frac{1}{2}(3) \quad \textit{Power Property}$$

$$= \frac{3}{2} \quad \textit{Simplify: } 3^3 = 27.$$

B. The expression is the sum of two common logs. Since the base is 10 for both of the logs in the sum, use the Product Property to simplify.

$$\log 20 + \log 5 = \log(20)(5) \quad \textit{Product Property}$$

$$= \log 100 \quad \textit{Multiply.}$$

$$= 2 \quad \textit{Simplify: } 10^2 = 100.$$

C. The expression is the difference and sum of three logs with base 2. Use the Product Property and the Quotient Property to simplify.

TIP

Follow the order of operations. Since addition and subtraction are completed from left to right, subtract first and then add.

$$\log_2 6 - \log_2 15 + \log_2 20 = \log_2\frac{6}{15} + \log_2 20 \quad \textit{Quotient Property}$$

$$= \log_2\left(\frac{6}{15}\right)(20) \quad \textit{Product Property}$$

$$= \log_2 8 \quad \textit{Multiply and divide.}$$

$$= 3 \quad \textit{Simplify: } 2^3 = 8.$$

4.3.2 Expanding Logarithmic Expressions

The properties of logs can be used to expand logarithmic expressions. For example, the log of a quotient is expanded when it is written as the difference of the log of the dividend and the log of the divisor.

EXAMPLE 2 Using Properties to Expand Logarithmic Expressions

Use the properties of logarithms to expand each expression.

A. $\ln\sqrt{ab}$ **B.** $\log\dfrac{\sqrt[3]{x^2}}{y}$ **C.** $\log_4\left(\dfrac{4(p+q)}{\sqrt[3]{r}}\right)^2$

SOLUTION

A. Write the radical as a rational exponent. Then apply the Power Property, followed by the Product Property, to expand the natural log.

$$\begin{aligned}
\ln\sqrt{ab} &= \ln(ab)^{\frac{1}{2}} && \textit{Write the radical as a rational exponent.}\\
&= \frac{1}{2}\ln(ab) && \textit{Power Property}\\
&= \frac{1}{2}(\ln a+\ln b) && \textit{Product Property}\\
&= \frac{1}{2}\ln a+\frac{1}{2}\ln b && \textit{Distribute.}
\end{aligned}$$

B. Write the radical as a rational exponent. Then apply the Quotient Property, and then the Power Property, to expand the common log.

$$\begin{aligned}
\log\frac{\sqrt[3]{x^2}}{y} &= \log\frac{x^{\frac{2}{3}}}{y} && \textit{Write the radical as a rational exponent.}\\
&= \log x^{\frac{2}{3}}-\log y && \textit{Quotient Property}\\
&= \frac{2}{3}\log x-\log y && \textit{Power Property}
\end{aligned}$$

TIP

The Quotient Property must be applied before the Power Property, because the exponent applies to the numerator only.

C. Apply the Power Property first. Then apply the Quotient Property to expand the log. Be sure to place the difference of the logs within parentheses because the multiple 2 applies to both logs in the difference.

$$\begin{aligned}
\log_4\left(\frac{4(p+q)}{\sqrt[3]{r}}\right)^2 &= 2\log_4\frac{4(p+q)}{\sqrt[3]{r}} && \textit{Power Property}\\
&= 2\left(\log_4 4(p+q)-\log_4\sqrt[3]{r}\right) && \textit{Quotient Property}\\
&= 2\left(\log_4 4+\log_4(p+q)-\log_4 r^{\frac{1}{3}}\right) && \textit{Product Property; Write the radical as a rational exponent.}\\
&= 2\left(1+\log_4(p+q)-\frac{1}{3}\log_4 r\right) && \textit{Simplify: } 4^1=4\textit{. Power Property}\\
&= 2+2\log_4(p+q)-\frac{2}{3}\log_4 r && \textit{Distribute.}
\end{aligned}$$

CAUTION

When expanding the log in **Example 2C**, you may be tempted to write the term $2\log_4 (p + q)$ as $2(\log_4 p + \log_4 q)$. However, the Product Property of logs states that $\log_b (xy) = \log_b x + \log_b y$, not that $\log_b (x + y) = \log_b x + \log_b y$.

4.3.3 Combining Logarithmic Expressions

The properties of logs can also be used to write a sum or difference of logs as a single log. This process is called combining logs, and it is the opposite of expanding logs. When combining logs, it is important to remember that the Product and Quotient Properties of logs can be applied only when the logs have the same base.

EXAMPLE 3 Using Properties to Combine Logarithmic Expressions

Use the properties to write each expression as a single log, if possible.

A. $5 \ln c + \ln (d + 1)$ **B.** $\log_2 x + \frac{1}{2}\log_2 (x + y) - 3\log_2 (x - y)$

SOLUTION

A. The expression is the sum of two natural logs: $5 \ln c$ and $\ln (d + 1)$. Since both logs are natural logs, the bases are the same. Use the Product Property to write the sum as the product of a single log, but first use the Power Property to write 5 as an exponent.

$$5 \ln c + \ln (d + 1) = \ln c^5 + \ln (d + 1) \qquad \textit{Power Property}$$
$$= \ln c^5(d + 1) \qquad \textit{Product Property}$$

B. Use the Power Property to write the coefficients as exponents first. Then, since all three log terms have base 2, the Product and Quotient Properties can be used to write the terms as a single log.

$$\log_2 x + \frac{1}{2}\log_2 (x + y) - 3\log_2 (x - y)$$
$$= \log_2 x + \log_2 (x + y)^{\frac{1}{2}} - \log_2 (x - y)^3 \qquad \textit{Power Property}$$
$$= \log_2 x + \log_2 \sqrt{x + y} - \log_2 (x - y)^3 \qquad \textit{Write the rational exponent as a radical.}$$
$$= \log_2 \frac{x\sqrt{x + y}}{(x - y)^3} \qquad \textit{Product and Quotient Properties}$$

4.3.4 Using the Change of Base Formula

The **Change of Base Formula** provides a method for rewriting a logarithm as a quotient of logs with like bases.

Change of Base Formula

If a, b, and x are positive real numbers such that $a \neq 1$ and $b \neq 1$, then $\log_b x$ can be written as a quotient of logs with any base a by using the Change of Base Formula, $\log_b x = \dfrac{\log_a x}{\log_a b}$.

The Change of Base Formula: Evaluating a Logarithm Using a Calculator

Most calculators cannot evaluate a log unless the base is 10 (common log) or e (natural log). So, a logarithm such as $\log_2 7$ (where the base of the log is 2), for example, cannot be entered directly into a typical calculator.

However, by using the Change of Base Formula, any logarithm can be written as a quotient of common logs or as a quotient of natural logs: $\dfrac{\log x}{\log b} = \log_b x = \dfrac{\ln x}{\ln b}$. Then, a calculator can be used to evaluate the expression.

IMPORTANT

Since a (in the Change of Base Formula) can be any positive real number except 1, we can either let $a = 10$ or $a = e$.

For example, if the Change of Base Formula is used to write $\log_2 7$ as a quotient of logs with base 10 or base e, the expression can be entered into a calculator using the LOG key or the LN key, respectively, and then evaluated.

$$\log_2 7 = \frac{\log 7}{\log 2} = \frac{\ln 7}{\ln 2} = 2.807\ldots$$

TIP

The value of $\log_2 7$ is between 2 and 3, because $2^2 = 4$ and $2^3 = 8$.

EXAMPLE 4 **Using the Change of Base Formula**

Use the Change of Base Formula to approximate the value of each expression to the nearest hundredth using a calculator.

A. $\log_8 20$ **B.** $\log_9 \dfrac{3}{2}$

SOLUTION

A. $\log_8 20 = \dfrac{\log 20}{\log 8} \approx 1.44$ **B.** $\log_9 \dfrac{3}{2} = \dfrac{\ln \frac{3}{2}}{\ln 9} \approx 0.18$

Graphing a Logarithmic Function with Base b Using a Graphing Calculator

We have now seen that the Change of Base Formula can be used to write a logarithmic expression that is not a common or natural log as a quotient of common or natural logs, thus providing a method for evaluating a log with any base using a calculator. The Change of Base Formula can also be used to write a logarithmic *function* that is not a common or natural log as a quotient of common or natural logs. Then the logarithmic function can be entered into a graphing calculator in order to graph the function.

$$f(x) = \log_b x = \frac{\log x}{\log b} = \frac{\ln x}{\ln b} \quad \text{(where } b \text{ is a number such that } b > 0 \text{ and } b \neq 1\text{)}$$

SECTION 4.3 EXERCISES

Warm Up

State the property used to rewrite each expression.

1. $4 \ln w + \ln (d+3) = \ln w^4 + \ln (d+3)$
2. $\log 23 + \log 6 = \log (23 \cdot 6)$
3. $\log_6(5x+2) - \log_6(3x) = \log_6 \frac{5x+2}{3x}$

Just the Facts

Fill in the blanks.

4. The Product Property of logarithms states that the log of a product xy can be rewritten as the _____ of the logarithms of _____ and _____.
5. The Quotient Property of logarithms says that the log of a quotient x/y can be rewritten as the _____ of the logarithms of _____ and _____.
6. By the Power Property of logarithms, the $\log\sqrt{20}$ can be rewritten as _____.
7. The properties of logs can be used to _____ or _____ logarithmic expressions.
8. The Product and Quotient Properties of logarithms cannot be applied to two logs if they do not share the same _____.
9. To evaluate a log that is neither common nor natural using a simple calculator, one must use the _____.
10. The Change of Base Formula allows a logarithm to be written as a quotient of logarithms with _____ bases.

Essential Skills

In Exercises 1–20, evaluate each logarithm without using a calculator.

1. $\frac{5}{2}\log_4 \sqrt[3]{4}$
2. $\log_5 \sqrt[4]{\frac{1}{125}}$
3. $\frac{1}{2}\log_3 \sqrt{243}$
4. $2\log_5 \sqrt{125}$
5. $\frac{3}{4}\log_6 \sqrt{6}$
6. $5\log_2 \sqrt[4]{\frac{1}{32}}$
7. $\log_5 3 - \log_5 75$
8. $\log_9 \frac{1}{27} + \log_9 3$
9. $\log_4 2 + \log_4 \frac{1}{32}$
10. $\log_7 8 - \log_7 56$
11. $\log_8 \frac{2}{3} + \log_8 48$
12. $\log_9 \frac{3}{5} + \log_9 45$
13. $\log_{16} \frac{5}{4} - \log_{16} 10$
14. $\log_4 \frac{3}{8} - \log_4 12$
15. $\log_4 24 - \log_4 6 + \log_4 16$
16. $\log_3 2 + \log_3 4 - \log_3 72$
17. $\log_3 4 - \log_3 648 + \log_3 2$
18. $\log_2 3 + \log_2 4 - \log_2 48$
19. $\log_6 18 + \log_6 24 - \log_6 12$
20. $\log_8 40 - \log_8 5 - \log_8 64$

In Exercises 21–34, use the properties of logarithms to expand each expression.

21. $\log_4 x^5 \sqrt[9]{y}$
22. $\log_2 \sqrt{\frac{8x}{x+1}}$
23. $\log_6 \frac{a\sqrt{c}}{b^2}$
24. $\log_5 \frac{\sqrt{xy}}{z^4}$
25. $\ln x^2 y \sqrt[3]{z}$
26. $\log a^2 \sqrt[3]{bc}$
27. $\ln \frac{p(q+4)^5}{r}$
28. $\log \frac{g}{h(k-2)^3}$
29. $\log \frac{r^2}{w\sqrt{x+1}}$
30. $\log \frac{\sqrt{d-5}}{b^4 c}$
31. $\ln \sqrt[5]{\frac{6+x}{y^4}}$
32. $\log_5 \sqrt[4]{\frac{5+m}{625k}}$
33. $\log_2 \sqrt[3]{\frac{2+t}{16s}}$
34. $\ln \sqrt[4]{\frac{2-g}{h^3}}$

In Exercises 35–48, use the properties of logarithms to write each expression as a single log, if possible.

35. $4\log_6 x - \frac{1}{5}\log_6 y$

36. $2\log_a 5x^3 - \frac{1}{2}\log_a(2x+3)$

37. $3\log_7 2m^4 - \frac{1}{6}\log_7(2x+1)$

38. $4\log 3p^2 - \frac{1}{5}\log(8-3x)$

39. $3\log_5 x - 2\log_5 y + 4\log_5 z$

40. $6\log_2 a - 5\log_2 b - 3\log_2 c$

41. $3\log_b 2x + 5\log_b y - 3\log_b z^2$

42. $5\log_3 4w - 3\log_3 5y + 2\log_3 x$

43. $\frac{1}{2}\log_8(3w-2) + 4\log_8 2r - 2\log_8(x+8)$

44. $\frac{1}{3}\log_5(2t+4) - 5\log_5 u^2 - 3\log_5(v-1)$

45. $\frac{1}{4}\log_2(6-a)^3 - 6\log_2(b-4)^2 - 3\log_2 c$

46. $2\log_6(x+2)^4 - 8\log_6\sqrt{y+3} + 2\log_6(w-z)$

47. $\frac{1}{3}\left(\log_5(5-m)^2 - 4\log_5 2nk^2\right)$

48. $\frac{1}{2}\left(3\log_3 2b^4 - 7\log_3(2-5a)^2\right)$

In Exercises 49–54, use the Change of Base Formula and common logs or natural logs to approximate the value of each expression to the nearest hundredth using a calculator.

49. $\log_{\frac{1}{2}} 5$

50. $\log_{12}\left(\frac{7}{8}\right)$

51. $\log_8\left(\frac{24}{9}\right)$

52. $\log_8\left(\frac{25}{7}\right)$

53. $\log_2\left(\frac{43}{19}\right)$

54. $\log_5\left(\frac{38}{3}\right)$

Extensions

In Exercises 55–57, use the properties of logarithms to expand each expression.

55. $\log_2 \frac{(1-x)(3+x)}{2-x}$

56. $\log_8\left(\frac{x^6 y^5}{11z^2}\right)$

57. $\log_a\left[\frac{5x^2\left(\sqrt[3]{1-4x}\right)}{7(x+1)^2}\right]^{\frac{3}{2}}$

In Exercises 58–60, use the properties of logarithms to write each expression as a single log, if possible.

58. $\frac{1}{7}\log_3 6 + 2\log_3 x + \frac{2}{3}\log_3(2+x) - \log_3 7 - \log_3(x+3)$

59. $\log_4(10+x) + \log_5(3-2x) - \log_4(4-x)$

60. $\frac{1}{2}\log_4(x+3) - 5\log_4 x + 2\log_4(7-x) - 3\log_4(x+1)$

61. Two students simplified the expression

$\log_a\left(\frac{x}{x-1}\right) + \log_a\left(\frac{x+1}{x}\right) - \log_a\left(x^2-1\right)$.

One student's answer was $\log_a\left(\frac{1}{(x-1)^2}\right)$ and the other student's answer was $\log_a\left(\frac{1}{x^2-2x+1}\right)$.

Which answer is correct?

62. True or False? In order to graph the function $g(x) = \log_3 x$ with a graphing calculator, $\frac{\log x}{\ln 3}$ would be entered into Y= .

In Exercises 63–67, use the Change of Base Formula and common logs or natural logs to evaluate each expression to the nearest hundredth.

63. $\log_8 9$

64. $5\log_3 1$

65. $-3\log_7 102$

66. $\log_{-3} 27$

67. $\log_{\frac{2}{5}} \frac{7}{8} - \log_{\frac{1}{9}} \frac{3}{8}$

4.4 EXPONENTIAL AND LOGARITHMIC EQUATIONS

OBJECTIVES

- Solve exponential equations.
- Solve logarithmic equations.
- Use the Distance Modulus Formula.
- Use the Compound Interest Formula to find the length of investments.

PREREQUISITE VOCABULARY TERMS

common logarithm
exponential equation
logarithm
natural logarithm

4.4.1 Using the One-to-One Property to Solve Exponential Equations

Consider the exponential equation (i.e., an equation where the variable is in the exponent of a power) $5^x = 125$. We know that the solution is $x = 3$ because 125 is a perfect cube ($125 = 5^3$). Notice that the right side of the equation $5^x = 125$ can also be written as a power with base 5.

$$5^x = 125 \quad \Rightarrow \quad 5^x = 5^3$$

The powers are equivalent, so $x = 3$. Concluding that $x = 3$ because $5^x = 5^3$ is an example of the **One-to-One Property** for exponential equations.

One-to-One Property (Exponential Equations)

If $b > 0$ and $b \neq 1$, then

$$b^x = b^y \text{ if and only if } x = y.$$

So, if two equal powers are written with like bases, then the exponents can be equated. This property can be used to solve exponential equations.

IMPORTANT

Each side of the exponential equation must be written as a single power, and the powers' bases must be the same, before the exponents can be set equal to each other.

Steps for Solving an Exponential Equation by Using the One-to-One Property

❶ Write the equation in the form $b^x = b^y$ (i.e., the expressions on each side of the equation are powers with like bases).

❷ Set the exponents equal to each other.

❸ Solve the equation made from the exponents.

EXAMPLE 1 Solving an Exponential Equation

Solve each equation.

A. $2^x = \dfrac{1}{32}$

B. $125^x = 5$

C. $16^{x+2} = 64^{1-2x}$

SOLUTION

A. $2^x = \dfrac{1}{32} \Rightarrow 2^x = 2^{-5} \Rightarrow x = -5$

B.

$125^x = 5$	
$(5^3)^x = 5$	*Write 125 as a power with base 5.*
$5^{3x} = 5^1$	*Power of a Power Property of Exponents:* $(a^m)^n = a^{mn}$
$3x = 1$	*Set the exponents equal to each other.*
$x = \dfrac{1}{3}$	*Solve for x.*

C.

$16^{x+2} = 64^{1-2x}$	
$(4^2)^{x+2} = (4^3)^{1-2x}$	*Write 16 and 64 as powers with base 4.*
$4^{2(x+2)} = 4^{3(1-2x)}$	*Power of a Power Property of Exponents:* $(a^m)^n = a^{mn}$
$4^{2x+4} = 4^{3-6x}$	*Simplify the exponents.*
$2x + 4 = 3 - 6x$	*Set the exponents equal to each other.*
$8x = -1$	*Set the exponents equal to each other.*
$x = -\dfrac{1}{8}$	*Solve for x.*

4.4.2 Solving Exponential Equations Using Logs

The One-to-One Property also applies to logarithmic equations with the same base.

One-to-One Property (Logarithmic Equations)

If $b > 0$ and $b \neq 1$, then for all x and y where $\log_b x$ and $\log_b y$ are defined,

$$\log_b x = \log_b y \text{ if and only if } x = y.$$

In other words, if two values are equal, then their logs (of the same base) are also equal. Conversely, if the logs of two values are equal, then those values are equal.

When an exponential equation cannot be written in the form $b^x = b^y$ (i.e., two equivalent powers with like bases), then you can take the log of each side and algebraically manipulate the equation so that the One-to-One Property of logs can be applied.

Steps for Solving an Exponential Equation Using Common Logs

❶ Write the equation in the form $a^x = b^y$ (i.e., the expressions on each side of the equation are powers).

❷ Take the common log of each side.

❸ Use the Power Property of logs to write the exponent(s) as coefficients of the logs.

❹ Use the Product or Quotient Properties of logs to expand the logs as needed.

❺ Solve for x.

TIP

Alternatively, you could take the log base b of both sides, or the natural log of both sides.

EXAMPLE 2 Using Common Logs to Solve an Exponential Equation

Solve each equation. Find the exact value of x and the approximate value to the nearest hundredth.

A. $2(6^x) = 22$ **B.** $8^x - 5^{x+9} = 0$

SOLUTION

A. Divide each side by 2 to write the equation in the form $a^x = b^y$. $2(6^x) = 22 \Rightarrow 6^x = 11$

Try solving the equation by taking the log of each side, because 6 and 11 cannot be written as powers with like bases. The equation is now in the form $a^x = b^y$, so begin by taking the common log of each side.

$$6^x = 11$$

$$\log 6^x = \log 11 \quad \textit{Take the common log of each side.}$$

$$x \log 6 = \log 11 \quad \textit{Power Property of Logs}$$

$$x = \frac{\log 11}{\log 6} \quad \textit{Divide each side by log 6 to solve for x.}$$

$$x \approx 1.34 \quad \textit{Simplify using a calculator.}$$

The solution is $x = \dfrac{\log 11}{\log 6} \approx 1.34$.

B. Write the equation in the form $a^x = b^y$. $8^x - 5^{x+9} = 0 \Rightarrow 8^x = 5^{x+9}$

The powers cannot be written with like bases, so take the common log of each side first.

$$8^x = 5^{x+9}$$

$$\log 8^x = \log 5^{x+9} \quad \textit{Take the common log of each side.}$$

$$x \log 8 = (x+9)\log 5 \quad \textit{Power Property of Logs}$$

$$x \log 8 = x \log 5 + 9 \log 5 \quad \textit{Distribute.}$$

$$x \log 8 - x \log 5 = 9 \log 5 \quad \textit{Group the x-terms on the left side.}$$

$$x(\log 8 - \log 5) = 9 \log 5 \quad \textit{Factor x from each log.}$$

$$x \log \frac{8}{5} = 9 \log 5 \quad \textit{Quotient Property of Logs}$$

$$x = \frac{9 \log 5}{\log \frac{8}{5}} \quad \textit{Divide each side by } \log\tfrac{8}{5} \textit{ to solve for x.}$$

$$x \approx 30.82 \quad \textit{Simplify using a calculator.}$$

The solution is $x = \dfrac{9 \log 5}{\log \frac{8}{5}} \approx 30.82$.

4.4.3 Solving Natural Exponential Equations

Natural exponential equations (where the base of a power is e) can also be solved by using the One-to-One Properties. When a natural exponential equation can be written in the form $e^x = e^y$, then you can set x equal to y and solve that equation.

When a natural exponential equation cannot be written in the form $e^x = e^y$, then you can take the natural log of each side and algebraically manipulate the equation so that the One-to-One Property of logs can be applied.

Steps for Solving a Natural Exponential Equation Using Natural Logs

❶ Isolate e^x on one side of the equation.

❷ Take the natural log of each side.

❸ Use the Power Property of logs to write the exponent(s) as a coefficient of the natural log.

❹ Use the fact that $\ln e = 1$ to simplify.

❺ Solve for x.

EXAMPLE 3 Using Natural Logs to Solve a Natural Exponential Equation

Solve $10 - 2e^{5-3x} = 4$. Find the exact value of x and the approximate value to the nearest tenth.

SOLUTION

Manipulate the equation to write it in the form $e^x = c$ (i.e., solve for the power with base e). So, subtract 10 from each side and then divide each side by -2.

$$10 - 2e^{5-3x} = 4 \;\Rightarrow\; -2e^{5-3x} = -6 \;\Rightarrow\; e^{5-3x} = 3$$

Take the natural log of each side, then solve for x.

$$\begin{aligned} e^{5-3x} &= 3 & & \\ \ln e^{5-3x} &= \ln 3 & & \textit{Take the natural log of each side.} \\ (5-3x)\ln e &= \ln 3 & & \textit{Power Property of Logs} \\ 5-3x &= \ln 3 & & \textit{Simplify: } \ln e = 1. \\ -3x &= -5+\ln 3 & & \textit{Subtract 5 from each side.} \\ x &= \frac{-5+\ln 3}{-3} & & \textit{Divide each side by } -3 \textit{ to solve for } x. \\ x &\approx 1.3 & & \textit{Simplify using a calculator.} \end{aligned}$$

The solution is $x = -\dfrac{-5+\ln 3}{3} \approx 1.3$.

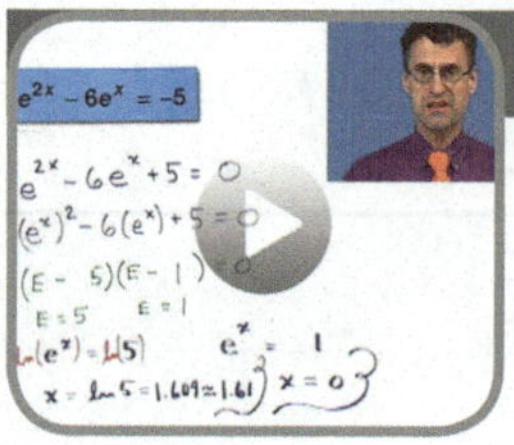

4.4.4 Solving Exponential Equations of Quadratic Type

Recall that equations of quadratic type can be solved by making a special substitution that results in a quadratic equation. One of the quadratic solving techniques (such as using the Quadratic Formula or factoring) can then be used to solve the new equation. This process is illustrated in the next example with a natural exponential equation.

EXAMPLE 4 Using Natural Logs to Solve an Equation of Quadratic Type

Solve $e^{6x} - e^{3x} = 6$. Find the exact value of x and the approximate value to the nearest hundredth.

SOLUTION

Write the equation as a quadratic by substituting some variable, for example, z, for e^{3x}. Then solve the equation in z by using one of the techniques for solving a quadratic equation. After the equation is solved for z, substitute e^{3x} for z and solve the resulting equations for x.

TIP

If the quadratic expression cannot be factored easily, then use completing the square or the Quadratic Formula to solve the quadratic equation.

$e^{6x} - e^{3x} = 6$	
$(e^{3x})^2 - (e^{3x}) = 6$	*Write e^{6x} as a power where the exponent is 2.*
$z^2 - z = 6$	*Substitute $e^{3x} = z$.*
$z^2 - z - 6 = 0$	*Write the equation in general form.*
$(z - 3)(z + 2) = 0$	*Factor.*
$z = 3$ or $z = -2$	*Solve each equation for z.*
$e^{3x} = 3 \quad e^{3x} = -2$	*Substitute $z = e^{3x}$.*

The equation $e^{3x} = -2$ has no solution, because $e^x > 0$ for all x.

The first equation $e^{3x} = 3$, can be solved for x.

$\ln e^{3x} = \ln 3$	*Take the natural log of each side.*
$3x(\ln e) = \ln 3$	*Power Property of Logs*
$3x = \ln 3$	*Simplify: ln e = 1.*
$x = \dfrac{\ln 3}{3}$	*Solve for x.*
$x \approx 0.37$	*Simplify using a calculator.*

The solution is $x = \dfrac{\ln 3}{3} \approx 0.37$.

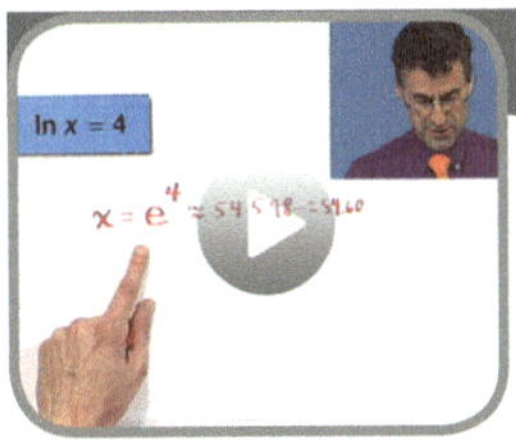

4.4.5 Using Exponential Form to Solve Logarithmic Equations

In a logarithmic equation, the log is of a variable. The following are examples of logarithmic equations.

$$1 + \log_2 x = 6 \qquad \log(x + 4) = 2 \qquad \ln x = e^2$$

REMEMBER

$y = \log_b x \Leftrightarrow b^y = x$

$y = \log x \Leftrightarrow 10^y = x$

$y = \ln x \Leftrightarrow e^y = x$

A logarithmic equation where the log is isolated on one side of the equation can be solved for the variable by writing the log in exponential form.

For example, to solve the logarithmic equation $\log(x + 4) = 2$, write the common log in exponential form: $10^2 = x + 4$. Then simplify and solve for x.

$\log(x + 4) = 2$	
$10^2 = x + 4$	*Write the common log in exponential form.*
$100 = x + 4$	*Simplify the power.*
$x = 96$	*Solve for x.*

Alternatively, $\log(x + 4) = 2$ can be solved by making each side a power where the base is 10 (the base of the equation's log).

$$\log(x + 4) = 2 \Rightarrow 10^{\log(x+4)} = 10^2$$

This equation may appear to be more complicated, but the log property $b^{\log_b x} = x$ can be used to simplify the equation.

$\log(x + 4) = 2$	
$10^{\log(x+4)} = 10^2$	*Raise 10 to each side.*
$x + 4 = 10^2$	*Use a property of logarithms to simplify: $b^{\log_b x} = x$.*
$x + 4 = 100$	*Simplify the power.*
$x = 96$	*Solve for x.*

Either method can be used when solving logarithmic equations.

Steps for Solving a Logarithmic Equation

1. Isolate the log on one side of the equation.
2. Write the equation in exponential form (or raise the base of the log to each side of the equation).
3. Solve the resulting equation.

EXAMPLE 5 Using Exponential Form to Solve a Logarithmic Equation

Solve each equation. Round to the nearest hundredth, if needed.

A. $\ln x^3 = 15$ **B.** $5\log_2(8x) + 1 = 16$

SOLUTION

A. Use the Power Property to write the exponent as a coefficient. Then isolate the log and write the equation in exponential form.

$\ln x^3 = 15$

$3\ln x = 15$ *Power Property*

$\ln x = 5$ *Divide each side by 3 to isolate the natural log.*

$x = e^5$ *Write the natural log in exponential form.*

The exact solution is $x = e^5$. Using a calculator to simplify gives $x \approx 148.41$.

B. Isolate the log, and then write the equation in exponential form.

$5\log_2(8x) + 1 = 16$

$5\log_2(8x) = 15$ *Subtract 1 from each side.*

$\log_2(8x) = 3$ *Divide each side by 5 to isolate the log.*

$8x = 2^3$ *Write the log in exponential form.*

$8x = 8$ *Simplify the power.*

$x = 1$ *Solve for x.*

CHECK

Substitute x = 1 into the original equation to check the solution.

$5\log_2(8 \cdot 1) + 1$
$= 5\log_2(8) + 1$
$= 5(3) + 1$
$= 16$ ✓

4.4.6 The Distance Modulus Formula

The distance modulus is a way of expressing a distance between astronomical objects. The **Distance Modulus Formula** is a logarithmic equation that relates distance modulus to distance in parsecs (1 parsec ≈ 3.3 light years).

Distance Modulus Formula

$$M = 5\log r - 5,$$

where M is the distance modulus and r is the distance in parsecs between two astronomical objects.

EXAMPLE 6 Using the Distance Modulus Formula

If a star's distance modulus from Earth is 3.6, what is the distance between the star and Earth in parsecs and in light years? Round each answer to the nearest hundredth.

SOLUTION

Substituting $M = 3.6$ into the Distance Modulus Formula yields $3.6 = 5\log r - 5$. This is a logarithmic equation. So, isolate the log and then write the log in exponential form to solve for r.

$3.6 = 5\log r - 5$	
$8.6 = 5\log r$	*Add 5 to each side.*
$1.72 = \log r$	*Divide each side by 5.*
$r = 10^{1.72}$	*Write the common log in exponential form.*
$r \approx 52.48$	*Simplify using a calculator.*

The distance between the star and Earth is approximately 52.48 parsecs.

Since 1 parsec ≈ 3.3 light years, the distance between the star and Earth is approximately $3.3(10^{1.72}) \approx 173.19$ light years.

IMPORTANT

Do not round until the final step in the calculations. If an interim step yields a nonterminating decimal value, use the fraction form of the number, not a rounded version.

4.4.7 Solving Logarithmic Equations

When a logarithmic equation contains more than one logarithmic term where the logs have the same base, use the properties of logs to combine the logs into a single log, then follow the steps for solving a logarithmic equation to solve. All possible solutions should be checked in the original equation, because it is possible to get extraneous solutions.

EXAMPLE 7 Solving Logarithmic Equations

Solve. $\log(x-1) + \log(x+2) = 1$

SOLUTION

The equation contains the sum of two common logs, $\log(x-1)$ and $\log(x+2)$. The bases are the same (both 10), so the Product Property of logs can be used to combine the logarithmic terms into a single log.

Product Property $\log(xy) = \log x + \log y$

$\log(x-1) + \log(x+2) = 1$	
$\log(x-1)(x+2) = 1$	*Product Property of Logs*
$(x-1)(x+2) = 10^1$	*Write the common log in exponential form.*
$(x-1)(x+2) = 10$	*Simplify the power.*

The resulting equation is quadratic, so expand the binomials and subtract 10 from each side to get 0 on one side, then solve by using one of the techniques for solving a quadratic equation.

$x^2 + x - 2 = 10$	*FOIL*
$x^2 + x - 12 = 0$	*Write the equation in general form.*
$(x-3)(x+4) = 0$	*Factor.*
$x = 3$ or $x = -4$	*Solve each equation for x.*

Substitute each solution into the original equation to check for extraneous solutions.

$x = 3$: $\log(3-1) + \log(3+2) = \log(2) + \log(5) = \log(2)(5) = \log 10 = 1$ ✔

The value of log 10 is 1 since $10^1 = 10$. So, the solution $x = 3$ is a solution.

$x = -4$: $\log(-4-1) + \log(-4+2) = \log(-5) + \log(-2) \neq 1$

The log of a negative number is undefined. So, $x = -4$ is a not a solution.

4.4.8 Compound Interest

In Section 4.1, the Compound Interest Formula was used to find a total balance A, an interest rate r, and an initial amount deposited or borrowed P (the principal).

$$\text{Compound Interest Formula: } A = P\left(1+\frac{r}{m}\right)^{mt}$$

P = the principal amount (original amount) in the account
t = the length of time (in years) that the money is invested
A = the balance in the account after t years
r = the annual interest rate (as a decimal)
m = the number of times per year that the interest is compounded

A total balance A was found by evaluating the expression $P\left(1+\frac{r}{m}\right)^{mt}$ using known values for P, r, m, and t. A principal P was found by evaluating the expression $\left(1+\frac{r}{m}\right)^{mt}$ using known values for r, m, and t, and then dividing the known balance A by that number. The most complicated variable solved for so far was the interest rate r, which was found by simplifying $P\left(1+\frac{r}{m}\right)^{mt}$ using known values for P, m, and t, and then taking the mth root of both sides of the equation.

If t is the unknown, then $A = P\left(1+\frac{r}{m}\right)^{mt}$ is an exponential equation. The Compound Interest Formula is solved for an unknown time t in the following example.

EXAMPLE 8 Using the Compound Interest Formula

Suppose \$14,900 is invested in a quarterly compounded account at 6%. In how many years will the balance reach \$20,000? Round to the nearest year.

SOLUTION

Substitute the known values into the Compound Interest Formula and solve for t.

$$A = P\left(1+\frac{r}{m}\right)^{mt}$$

$$20{,}000 = 14{,}900\left(1+\frac{0.06}{4}\right)^{4t} \quad \textit{Substitute the known values into the formula.}$$

$$\frac{200}{149} = (1.015)^{4t} \quad \textit{Simplify.}$$

$$\ln\frac{200}{149} = \ln(1.015)^{4t} \quad \textit{Take the natural log of each side. (Alternatively, take the common log of each side.)}$$

$$\ln\frac{200}{149} = 4t\ln 1.015 \quad \textit{Power Property of Logs}$$

$$t = \frac{\ln\frac{200}{149}}{4\ln 1.015} \quad \textit{Solve for t.}$$

$$t \approx 4.94 \quad \textit{Simplify using a calculator.}$$

The balance will reach \$20,000 after approximately 5 years.

IMPORTANT

Since $20{,}000/14{,}900 = 1.34228\ldots$, leave the quotient as a fraction in simplest form so that you will not have to round until the final step in the calculations.

SECTION 4.4 EXERCISES

Warm Up

Rewrite each value as a power with the given base.

1. 625 with a base of 5
2. 64 with a base of 2
3. 243 with base of 3

Just the Facts

Fill in the blanks.

4. The _____ Property can be used to solve exponential equations. It states that if _____ and _____, $b^x = b^y$ if and only if $x = y$.
5. _____ are used to solve exponential equations when the powers cannot be written with like bases.
6. _____ logarithms are used to solve exponential equations of the form $e^x = c$.
7. The first step to solving a logarithmic equation is to _____ the log. Then the equation can be rewritten in _____ form.
8. If Betelgeuse has a distance modulus from Earth of 6.0, the Distance Modulus Formula can be used to calculate the distance from Earth to Betelgeuse in _____.
9. It is possible to get _____ solutions when combining multiple _____ terms in a logarithmic equation.
10. The _____ Formula can be used to solve exponential equations with two exponential terms, the exponent for one of which is twice the exponent for the other.

Essential Skills

In Exercises 1–48, solve each exponential equation. Round to the nearest hundredth, if needed.

1. $64^{7x-8} = 16$
2. $\frac{1}{125} = 25^{5x-3}$
3. $\frac{1}{81} = 27^{4m-5}$
4. $\frac{1}{256} = 32^{3m-4}$
5. $625 = 125^{3x-2}$
6. $27^{6x-2} = 81$
7. $243^{5m-5} = 729$
8. $\frac{1}{729} = 81^{6t-2}$
9. $7^{2x+3} = 49^{8-x}$
10. $9^{x+2} = \left(\frac{1}{27}\right)^{1-x}$
11. $27^{3x-4} = 81^{1-2x}$
12. $16^{4x-7} = 64^{3+x}$
13. $4^{2p+1} = 32^{3-p}$
14. $\left(\frac{1}{32{,}768}\right)^{3-2x} = 64^{x+3}$
15. $125^{1-2m} = 5^{3m+2}$
16. $4^{x+2} = \left(\frac{1}{32}\right)^{-2x+3}$
17. $3^x = 12$
18. $2(5^x) - 1 = 15$
19. $4(8^x) - 4 = 16$
20. $3\left(\frac{1}{2}\right)^x - 1 = 17$
21. $16 = 4\left(\frac{1}{3}\right)^m - 4$
22. $4(10^x) - 1 = 23$
23. $35 = 4\left(\frac{1}{5}\right)^m - 1$
24. $13 = 2\left(\frac{1}{10}\right)^x - 5$
25. $3^x = 4^{x+8}$
26. $\left(\frac{3}{5}\right)^x - 7^{1-x} = 0$
27. $\left(\frac{1}{8}\right)^x - 6^{5-x} = 0$
28. $5^x - 7^{6-x} = 0$
29. $\left(\frac{8}{5}\right)^x - 8^{4-x} = 0$
30. $\left(\frac{5}{4}\right)^x - 5^{3-x} = 0$
31. $0 = 10^x - 5^{4-x}$
32. $8^x - 4^{5-x} = 0$
33. $0 = -12 + 5e^{2x}$
34. $\frac{5 + 13e^{4-x}}{7} + 1 = 11$
35. $6e^{12x-4} - 6 = 3$
36. $3e^{6-x} + 10 = 19$
37. $47 = 11e^{4-x} + 3$
38. $\frac{24 + 6e^{3-x}}{6} + 1 = 10$
39. $5 + \frac{6 + 3e^{2-x}}{3} = 14$

40. $5 + 4e^{7x-8} = 8$

41. $e^{2x} + e^x - 12 = 0$

42. $2(5^{8x}) = 5^{4x} + 3$

43. $4(8^{6x}) - 20 = -11(8^{3x})$

44. $-13(4^{4x}) = -6 - 5(4^{8x})$

45. $3(4^{6x}) = -10 + 17(4^{3x})$

46. $4(10^{6x}) = -3(10^{3x}) + 10$

47. $-15(8^{4x}) + 18 = -2(8^{8x})$

48. $-15(3^{4x}) = -25 - 2(3^{8x})$

In Exercises 49–64, solve each logarithmic equation. Round to the nearest hundredth, if needed.

49. $8 - \ln x^5 = 0$

50. $\ln (x + 5)^3 + 12 = 21$

51. $\ln (x + 4)^5 = 20$

52. $34 = \ln (x + 3)^5 + 9$

53. $30 - \ln (x - 1)^4 = 10$

54. $\ln (x - 2)^2 - 12 = 15$

55. $18 = \ln (x - 6)^2 + 6$

56. $\ln (x + 7)^7 + 13 = 34$

57. $\log (4x + 7) = 2$

58. $-2 + \log_3(1 - x)^5 = 8$

59. $\log_4(3 + 2x)^4 + 3 = 27$

60. $-11 = 7 - \log_3(3 - x)^9$

61. $-1 = 5 - \log_5(x + 6)^3$

62. $0 = 6 - \log_7(2 - x)^3$

63. $\log_6(x + 6)^5 + 4 = 19$

64. $-19 + \log_2(4 - x)^7 = 9$

In Exercises 65–68, use the Distance Modulus Formula.

65. Find the distance from Earth to a star if its distance modulus from Earth is –0.26. Round the answer to the nearest hundredth of a parsec.

66. Find the distance in light years (1 parsec ≈ 3.3 light years) from Earth to a star if its distance modulus from Earth is 0.67. Round the answer to the nearest hundredth of a light year.

67. Find the distance from Earth to a star if its distance modulus from Earth is 2.38. Round the answer to the nearest hundredth of a parsec.

68. Find the distance in light years (1 parsec ≈ 3.3 light years) from Earth to a star if its distance modulus from Earth is 0.79. Round the answer to the nearest hundredth of a light year.

In Exercises 69–76, solve.

69. $\log_2 x + \log_2(x - 6) = 4$

70. $2\log_3 x - \log_3(x + 4) = 2$

71. $\log_4(x + 62) - \log_4(x - 1) = 3$

72. $\log_2(x + 2) + \log_2(x + 4) = 3$

73. $\log_3(2x + 1) - \log_3(x - 1) = 1$

74. $\log_3 x + \log_3(x - 6) = 3$

75. $\log_4(x + 16) + \log_4(x + 4) = 3$

76. $2\log_2 x - \log_2(x + 5) = 4$

77. Suppose \$2000 is invested in an annually compounding account at 12%. Find approximately how long it will take for the balance to reach \$5,000.

78. Suppose \$11,400 is invested in a monthly compounded account at 4.98%. Approximately how long will it take for the balance to reach \$25,650? Round the answer to the nearest tenth of a year.

79. Suppose \$18,700 is invested in a quarterly compounded account at 3.39%. Approximately how long will it take for the balance to reach \$50,490? Round the answer to the nearest tenth of a year.

80. Suppose \$23,300 is invested in a daily compounded account at 5.28%. Approximately how long will it take for the balance to reach \$46,600? Round the answer to the nearest tenth of a year.

Extensions

81. Solve $\ln (10x - 5) - \ln 5 + 4 = 3$. Round to the nearest thousandth.

82. Solve $A = P\left(1 + \frac{r}{m}\right)^{mt}$ for t using common logs.

83. Find the extraneous solution, if any, for $\log_3\left(x + \sqrt{19}\right) + \log_3\left(x - \sqrt{19}\right) = 4$.

84. Solve $e^{8x} - 6e^{4x} = 7$. Find the exact value of x and the approximate value to the nearest hundredth.

85. With a distance modulus from Earth of –2.88, Sirius is the brightest star in the night sky, almost twice the brightness of the second brightest star, Canopus. How far is Sirius from Earth in parsecs and in light years (1 parsec ≈ 3.3 light years), each rounded to the nearest hundredth?

86. Solve $3^x = 8^{x-8}$. Find the exact value of x and the approximate value to the nearest hundredth.

4.5 EXPONENTIAL AND LOGARITHMIC MODELS

OBJECTIVES

- Find exponential models of growth and decay.
- Use Newton's Law of Cooling.
- Use the formula for continuously compounded interest.
- Solve logarithmic models.
- Use logistic and Gaussian models.

PREREQUISITE VOCABULARY TERMS

exponential function
interest
logarithmic function
natural exponential function
natural logarithmic function

4.5.1 Predicting Change with Logarithmic Models

The techniques used for solving a logarithmic equation can be used to solve a logarithmic model for one of its variables, or to predict some value given by the equation.

EXAMPLE 1 Using a Logarithmic Model

The power gain P for an amplifier is given by $P = 10\log\frac{x}{y}$, where x is the power output and y is the power input, both in watts. Solve the equation for y to find an equation for the power input in terms of P and x. If an amplifier has a power gain of 20 watts and the output is 12 watts, find the power input in watts.

SOLUTION

REMEMBER

To solve a logarithmic equation, isolate the log on one side of the equation, then write the equation in exponential form.

Solve the logarithmic equation for y, the power input.

$$P = 10\log\frac{x}{y}$$

$$\frac{P}{10} = \log\frac{x}{y} \qquad \textit{Divide each side by 10.}$$

$$10^{\frac{P}{10}} = \frac{x}{y} \qquad \textit{Write the logarithmic equation in exponential form.}$$

$$y = \frac{x}{10^{\frac{P}{10}}} \qquad \textit{Solve for } y.$$

Use the equation for y to find the power input when the power gain is 20 ($P = 20$) and the output is 12 ($x = 12$).

$$y = \frac{12}{10^{\frac{20}{10}}} = \frac{12}{10^2} = \frac{12}{100} = 0.12$$

The power input is 0.12 watts.

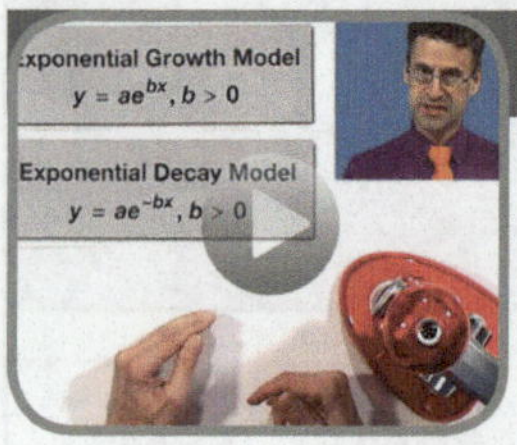

4.5.2 Exponential Growth and Decay

Two of the most commonly used exponential models are the **exponential growth model** and the **exponential decay model**. These models are used to make predictions concerning the time it takes for some quantity (e.g., a population) to grow (increase) or decay (decrease) to some number.

IMPORTANT

x represents the elapsed time of the growth (or decay), y represents the quantity that is growing (or decaying), and a is the quantity y at the initial time x.

Exponential Growth and Decay Models

Exponential Growth Model	Exponential Decay Model
$y = ae^{bx}$, $a > 0$ and $b > 0$	$y = ae^{-bx}$, $a > 0$ and $b > 0$

EXAMPLE 2 Finding and Using an Exponential Model

The number of frogs in a wildlife preserve is increasing according to an exponential model. After the preserve has been open for 2 years, the population of frogs is 100. After 4 years, the population is 300. Use an exponential growth model to predict the preserve's frog population after 5, 6, and 10 years. According to the model, how long (to the nearest tenth of a year) will it take for the preserve's frog population to increase to 20,000 frogs?

SOLUTION

Let y be the number of frogs after x years.

Find the values of a and b, then use those values to write the exponential growth model, $y = ae^{bx}$.

Use the known values to write two equations using $y = ae^{bx}$.

$x = 2$ and $y = 100 \Rightarrow 100 = ae^{2b}$ *After 2 years, there were 100 frogs.*

$x = 4$ and $y = 300 \Rightarrow 300 = ae^{4b}$ *After 4 years, there were 300 frogs.*

TIP

The ordered pairs (2, 100) and (4, 300) are solutions to the exponential growth model $y = ae^{bx}$.

Solve one of the two equations for either a or b. Since b is in the exponent and a is a coefficient, it will be easier to solve for a. Here, $100 = ae^{2b}$ is solved for a.

$100 = ae^{2b} \Rightarrow a = \dfrac{100}{e^{2b}}$ *Divide both sides by e^{2b}.*

TIP

If one of the equations had been solved for b (instead of for a), then the next step would be to substitute the expression for b into the other equation and solve for a.

Substitute the expression for a into $300 = ae^{4b}$, and then solve for b.

$300 = \dfrac{100}{e^{2b}}\left(e^{4b}\right)$ *Substitute $a = \dfrac{100}{e^{2b}}$ into $300 = ae^{4b}$.*

$300 = 100e^{2b}$ *Simplify.*

$3 = e^{2b}$ *Divide each side by 100.*

$\ln 3 = \ln e^{2b}$ *Take the natural log of each side.*

$\ln 3 = 2b$ *Use the Power Property of logs and simplify ($\ln e = 1$).*

$b = \dfrac{\ln 3}{2}$ *Divide each side by 2 to solve for b.*

Substitute this b-value into $a = \dfrac{100}{e^{2b}}$ to find the a-value.

$$a = \frac{100}{e^{2\left(\frac{\ln 3}{2}\right)}} = \frac{100}{e^{\ln 3}} = \frac{100}{3}$$

Use the a- and b-values to write the exponential growth model.

$$y = ae^{bx} \Rightarrow y = \frac{100}{3}e^{\left(\frac{\ln 3}{2}\right)x}$$

TIP

The model does not have to be written without the logarithmic exponent, but doing so will typically make the calculations easier.

Write the model in exponential form without the logarithmic exponent by simplifying.

$$y = \frac{100}{3}e^{\left(\frac{\ln 3}{2}\right)x} = \frac{100}{3}e^{\ln 3\left(\frac{x}{2}\right)} \qquad \textit{Factor out ln 3 in the exponent.}$$

$$= \frac{100}{3}\left(e^{\ln 3}\right)^{\frac{x}{2}} \qquad \textit{Power of a Power Property (of Exponents)}$$

$$= \frac{100}{3}\cdot 3^{\frac{x}{2}} \qquad \textit{Simplify using the fact that } e^{\ln x} = x.$$

Substitute $x = 5$, 6, and 10 into the model to predict the frog population after 5, 6, and 10 years.

5 years	**6 years**	**10 years**
$y = \frac{100}{3}\cdot 3^{\frac{5}{2}} \approx 520$	$y = \frac{100}{3}\cdot 3^{\frac{6}{2}} = 900$	$y = \frac{100}{3}\cdot 3^{\frac{10}{2}} = 8100$

According to the model, the preserve's frog population will be approximately 520 after 5 years, 900 after 6 years, and 8100 after 10 years.

Substitute $y = 20{,}000$ into the model and solve for x to find the length of time it will take for the population to increase to 20,000.

$$20{,}000 = \frac{100}{3}\cdot 3^{\frac{x}{2}} \qquad \textit{Substitute } y = 20{,}000 \textit{ into the model.}$$

$$600 = 3^{\frac{x}{2}} \qquad \textit{Multiply each side by 3/100.}$$

$$\ln 600 = \ln 3^{\frac{x}{2}} \qquad \textit{Take the natural log of each side.}$$

$$\ln 600 = \left(\frac{x}{2}\right)\ln 3 \qquad \textit{Power Property of Logs}$$

$$x = \frac{2(\ln 600)}{\ln 3} \qquad \textit{Multiply each side by 2 and divide each side by ln 3 to solve for } x.$$

$$x \approx 11.6 \qquad \textit{Use a calculator to simplify.}$$

According to the model, it will take approximately 11.6 years for the preserve's frog population to reach 20,000.

4.5.3 Half-Life

The **half-life** of a decaying substance is the amount of time required for the substance to decrease by half. In other words, it is the period of time over which one-half of the substance decomposes. For example, consider a 6 milligram sample of some exponentially decaying substance. This substance's half-life is the amount of time it takes for the substance to decay so that there are only 3 milligrams remaining. This substance's half-life is also the amount of time it takes for the substance to decay from 3 milligrams to 1.5 milligrams, or from 1.5 milligrams to 0.75 milligrams, or from 0.75 milligrams to 0.375 milligrams, and so on.

Suppose the substance's half-life is 3 hours. Then a 10-milligram sample of the substance will decay to 5 milligrams after 3 hours. After 6 total hours, the substance will measure 2.5 milligrams, and after 9 total hours, the substance will measure 1.25 milligrams.

The amount of radioactive material and the amount of a medication in a person's body are commonly described by their half-lives.

EXAMPLE 3 Finding the Half-Life of a Substance

The amount remaining from a 90-milligram sample of a radioactive substance after t years is given by $A = 90e^{-0.052t}$. Find the substance's half-life to the nearest year.

SOLUTION

Let $A = 45$ (half of 90 milligrams) and solve for t to find the substance's half-life.

$$45 = 90e^{-0.052t} \qquad \textit{Substitute A = 45 into the model.}$$

$$\frac{1}{2} = e^{-0.052t} \qquad \textit{Divide each side by 90.}$$

$$\ln\frac{1}{2} = \ln e^{-0.052t} \qquad \textit{Take the natural log of each side.}$$

$$\ln\frac{1}{2} = -0.052t \qquad \textit{Use the Power Property of logs and simplify (ln e = 1).}$$

$$t = \frac{\ln\frac{1}{2}}{-0.052} \qquad \textit{Divide each side by –0.052 to solve for t.}$$

$$t \approx 13 \qquad \textit{Use a calculator to simplify.}$$

The half-life of the substance is approximately 13 years.

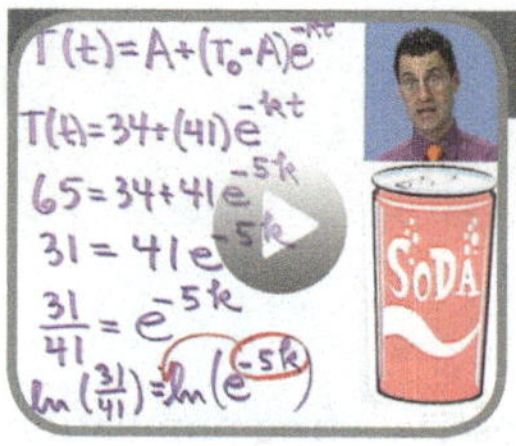

4.5.4 Newton's Law of Cooling

Newton's Law of Cooling states that the rate of change of the temperature of some object (or substance) is proportional to the difference between its own temperature and the temperature of its surrounding environment.

Newton's Law of Cooling

$$T = A + (T_0 - A)e^{-kt}$$

T = object's temperature at time t

A = temperature of the object's surroundings

T_0 = object's initial temperature

k = constant (specific to the object)

t = time

The k-value in the Newton's Law of Cooling equation depends upon the object or substance. This value may be given, or you may solve for the constant k using a given initial temperature, final temperature (i.e., temperature at time t), time, and temperature of the object's surroundings.

EXAMPLE 4 Using Newton's Law of Cooling

Suppose the temperature of a liquid decreases from 100° C to 60° C after being placed in a 5° C freezer for 10 minutes. How long (to the nearest tenth of a minute) would it take for the same liquid to cool from 90° C to 10° C in the 5° C freezer?

SOLUTION

Find the constant k.

$60 = 5 + (100 - 5)e^{-k(10)}$	*Substitute.*
$55 = 95e^{-10k}$	*Simplify.*
$\frac{11}{19} = e^{-10k}$	*Divide each side by 95.*
$\ln\frac{11}{19} = \ln e^{-10k}$	*Take the natural log of each side.*
$\ln\frac{11}{19} = -10k$	*Use the Power Property of logs and simplify (ln e = 1).*
$k = \frac{\ln\frac{11}{19}}{-10}$	*Divide each side by –10.*
$k \approx 0.055$	*Use a calculator to simplify.*

Using this k-value, find the time needed to cool the liquid from 90° C to 10° C in the 5° C freezer.

$10 = 5 + (90 - 5)e^{-0.055t}$	*Substitute.*
$5 = 85e^{-0.055t}$	*Simplify.*
$\frac{1}{17} = e^{-0.055t}$	*Divide each side by 85.*
$\ln\frac{1}{17} = \ln e^{-0.055t}$	*Take the natural log of each side.*
$\ln\frac{1}{17} = -0.055t$	*Use the Power Property of logs and simplify (ln e = 1).*
$t = \frac{\ln\frac{1}{17}}{-0.055}$	*Divide each side by –0.055.*
$t \approx 51.5$	*Use a calculator to simplify.*

According to this model, the liquid's temperature will decrease from 90° C to 10° C in the 5° C freezer after approximately 51.5 minutes.

4.5.5 Continuously Compounded Interest

The formulas for simple and compound interest were discussed previously. The formula for **continuously compounded interest** will be discussed in this topic, which is used when the interest is compounded continuously, instead of some set number of times per year.

Continuously Compounded Interest Formula

$$A = Pe^{rt}$$

A = balance of the account

P = the principal (initial) amount in the account

r = annual interest rate (as a decimal)

t = the length of time that the money is invested, in years

When the continuously compounded interest formula is solved for the interest rate r or the length of time the money is invested (or borrowed) t, then the equation solved is exponential.

EXAMPLE 5 Using the Continuously Compounded Interest Formula

What was the interest rate (to the nearest tenth of a percent) on an account with continuously compounded interest if the initial amount deposited tripled after 14 years, assuming there were no other deposits?

SOLUTION

Since the interest is compounded continuously, the formula $A = Pe^{rt}$ can be used. The value of t is 14, and no other values are given. However, it is given that the account's balance after 14 years was triple the initial amount. Therefore, $A = 3P$. Substitute into the formula and solve for r.

$3P = Pe^{r(14)}$	*Substitute.*
$3 = e^{14r}$	*Divide each side by P.*
$\ln 3 = \ln e^{14r}$	*Take the natural log of each side.*
$\ln 3 = 14r$	*Use the Power Property of logs and simplify (ln e = 1).*
$r = \dfrac{\ln 3}{14}$	*Divide each side by 14 to solve for r.*
$r \approx 0.078$	*Use a calculator to simplify.*

Therefore, the interest rate was approximately 7.8%.

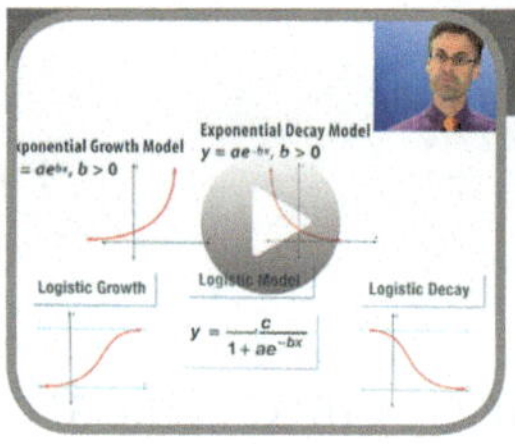

4.5.6 Logistic Models

A **logistic model** is an exponential equation that describes the growth (or decay) of a population over time. A logistic model is characterized by initial rapid growth (or decay), and then a slower rate of growth (or decay) over time.

Logistic Model

If a population's growth (or decay) can be described by a logistic model, then the population's total y after time x is given by $y = \dfrac{c}{1+ae^{-bx}}$, where a, b, and c are constants such that $a > 0$, $b \neq 0$, and $c > 0$.

IMPORTANT

If $b > 0$, then the logistic model describes population growth. If $b < 0$, then the logistic model describes population decay.

EXAMPLE 6 Using a Logistic Growth Model

In a small town with a population of 4000 people, the total number of people y infected by a virus x days after it is introduced into the population is modeled by $y = \dfrac{4000}{1+2500e^{-0.9x}}$ where $x \geq 0$. To the nearest tenth of a percent, find the percent of the town's total population that will be infected after 5 days. Find the approximate number of days before 50% of the town's population will be infected.

SOLUTION

Substitute $x = 5$ into the model and simplify to find the number of people infected after 5 days.

$$y = \frac{4000}{1+2500e^{-0.9(5)}} \approx 139 \text{ people}$$

Since $139/4000 \approx 0.03475$, approximately 3.5% of the population will be infected after 5 days.

Substitute $y = 2000$ into the model and solve for x to find the number of days before 2000 people (i.e., 50% of the population) are infected.

$2000 = \dfrac{4000}{1+2500e^{-0.9x}}$	*Substitute.*
$1+2500e^{-0.9x} = 2$	*Multiply each side by $1 + 2500e^{-0.9x}$ and divide each side by 2000.*
$2500e^{-0.9x} = 1$	*Subtract 1 from each side.*
$e^{-0.9x} = 0.0004$	*Divide each side by 2500.*
$-0.9x = \ln 0.0004$	*Take the natural log of each side.*
$x = \dfrac{\ln 0.0004}{-0.9}$	*Divide each side by −0.9.*
$x \approx 9$	*Use a calculator to simplify.*

After approximately 9 days, 50% of the population will be infected.

4.5.7 Gaussian Models

A **Gaussian model** is an exponential equation of the form $y = ae^{-(x-b)^2/c}$, where a, b, and c are constants. This model describes a **normally distributed** population (represented by the independent variable, typically x), which is characterized by a graph that forms a bell-shaped curve. Moreover, the maximum y-value (i.e., the top of the bell) occurs at the mean of the x-values, and all other points on the graph are symmetric about the mean, as shown in **Example 7**.

EXAMPLE 7 Using a Gaussian Model

The weight of the dogs seen by a veterinarian during one year roughly followed the normal distribution given by $y = 0.03577e^{-(x-50)^2/325}$ where x is the weight of a dog. Use the given graph to estimate the mean weight of a dog seen by this veterinarian during that year.

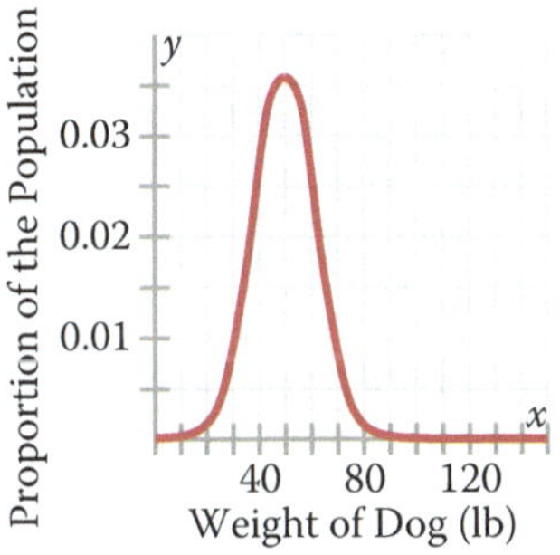

SOLUTION

From the graph, the maximum y-value corresponds to $x \approx 50$. Therefore, the mean weight of a dog seen by this veterinarian during that year was approximately 50 pounds.

SECTION 4.5 EXERCISES

Warm Up

Use a calculator to evaluate each expression to the nearest thousandth.

1. $\ln\frac{3}{7} - \ln 5$

2. $\log\left(\frac{\frac{9}{6}}{10}\right) - 4$

3. $\frac{\ln\left(\frac{3}{5}\right)}{-9}$

Just the Facts

Fill in the blanks.

4. $P = 10\log\frac{x}{y}$ is a(n) _____ equation in _____ variables.

5. The equation $y = ae^{-bx}$ is the exponential _____ model with the condition that b is _____ than 0. This model is used to predict the time it will take for some quantity to _____.

6. If a sample contains 8 milligrams of ununtrium, which has a half-life of 20 minutes, after _____ hour(s) the sample will contain only 1 milligram of ununtrium.

7. In addition to modeling radioactive material, half-life is often used to measure the amount of _____ in a person's body.

8. If it takes _____ hour(s) for a sample of meitnerium to decay from 48 grams to 12 grams, it will take 1.5 hours to decay from 12 grams to 1.5 grams. The half-life of meitnerium is _____.

9. To find the cooling constant of a sample of some substance, the object's _____ temperature, the temperature of the object's _____, and the temperature of the substance at some time _____ are needed.

10. An investment made in an account that compounds interest continuously will earn _____ interest than an investment made in an account at the same interest rate that is compounded monthly.

Essential Skills

1. In chemistry, pH is given by the formula: $\text{pH} = -\log\left[H_3O^+\right]$, where H_3O^+ is the hydronium ion concentration in moles per liter. Given that the pH of the drinking water of Yukon, Oklahoma is equal to 5.3, what is the concentration of the hydronium ion in moles per liter?

2. The mathematical model for learning an assembly-line procedure needed for assembling one component of a manufactured item is $n = -5\ln\left(\frac{0.94}{P} - 1\right)$, where P is the proportion of correctly assembled components after n practice sessions. Solve the equation for P to find an equation for the proportion of correctly assembled components in terms of the number of practice sessions n. Use the model to predict the percent of components (to the nearest tenth of a percent) that will be correctly assembled after 5 practice sessions.

3. The population of a city has been rising according to the model $A = pe^{0.036t}$, where p is the population in 2000 and A is the population t years later. If the city's population was 468,000 in 2000, in what year is the city expected to reach 1 million people?

4. The number of bacteria in a sample is increasing according to an exponential model. After 2 hours, the sample contained 200 bacteria; and after 6 hours, the sample contained 800 bacteria. Write an exponential growth model for the number of bacteria in the sample after x hours.

5. Sodium-24 is a radioactive substance used for medical purposes. Suppose that a patient is injected with sodium-24 and the amount of sodium-24 in the patient's body in milligrams after t hours is given by $S(t) = 5e^{-0.35t}$. What is the half-life of sodium-24?

6. Find the half-life to the nearest tenth of a year of a radioactive substance that decays from 80 milligrams to 15 milligrams in 25 years according to the exponential decay model $y = ae^{-bx}$, where a is the initial amount and y is the amount remaining after x years. Hint: Find the b-value, then use this value to write the exponential decay model for this substance with initial amount 80 milligrams.

7. At time $t = 0$, a cup of coffee is placed in a room with a constant temperature of 81° F. The temperature of the coffee after t minutes in the room is described by the function $C(t) = 81 + 101e^{-0.039t}$. Based on this model, how long does it take the coffee to cool to a temperature of 92° F?

8. Suppose the temperature of a liquid decreases from 70° F to 55° F after being placed in a 33° F freezer for 60 minutes. Use Newton's Law of Cooling, $T = A + (T_0 - A)e^{-kt}$, to find the number of minutes (to the nearest minute) needed for the liquid to cool from 82° F to 60° F in a 29° F freezer.

9. Suppose a savings account offers 7.5% interest compounded continuously. How long does it take for a \$3500 deposit in this account to reach \$20,000? Round the number of years to the nearest tenth if needed.

10. An investor wants to analyze the earnings of a mutual fund account. Four years ago, the value of the account was \$16,000 and it is now worth \$24,000 (no additional deposits were made). If the account is compared to a bank account paying interest that is compounded continuously, what interest rate (rounded to the nearest hundredth of a percent) would the bank account have to pay to match the mutual fund account's earnings?

In Exercises 11–12, use the following information.

In a small town with a population of 1720, the total number of people y infected by a virus x days after it is introduced into the population is modeled by $y = \dfrac{1720}{1+550e^{-0.88x}}$ where $x \geq 0$.

11. Find the number of people that will be infected after 2 days.

12. Find the percent of the town's total population that will be infected after 4 days to the nearest whole number.

In Exercises 13–14, use the following information.

The amount remaining of a 45 mL sample of a liquid substance, y (in mg) after x hours is modeled by $y = \dfrac{45}{1+0.07e^{0.09x}}$, where $x \geq 0$.

13. To the nearest tenth of an hour, how long will it take for the substance to decay to 30 mL?

14. To the nearest tenth of an hour, how long will it take for the substance to decay from 40 mL to 30 mL?

15. The total amount spent on books by the students at a college during one year roughly followed the normal distribution given by $y = 0.00313e^{-(x-500)^2/32{,}500}$, where x is the total amount spent on books in dollars. Use the graph of that equation to estimate the mean amount spent on books per student during that year.

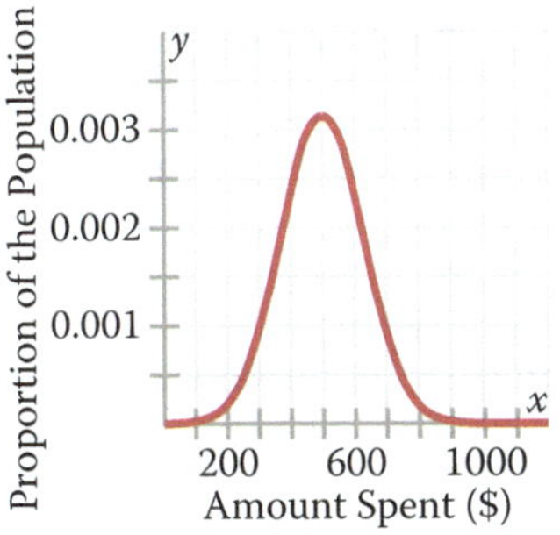

16. The total number of tacos eaten in a month by individuals in Austin, Texas last month roughly followed the normal distribution given by $y = 0.356825e^{-(x-7)^2/25}$, where x is the total number of tacos eaten by a given individual. Use the graph of that equation to estimate the mean number of tacos eaten per Austinite during that month.

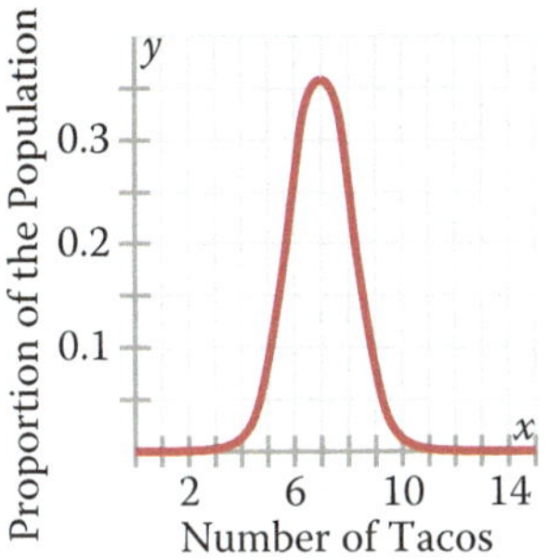

Extensions

17. The voltage across a resistor in a circuit is 12 volts at time $t = 1.2$. The formula for voltage is $V(t) = V_0\left(1-e^{-4.5t}\right)$, where t is the time in seconds, and V_0 is the voltage at $t = 0$. Find the voltage to the nearest hundredth at $t = 0$.

18. The growth of a city is described by the population function $P(t) = P_0e^{kt}$, where P_0 is the initial population of the city, t is the time in years, and k is a constant. If the population of the city at $t = 0$ is 15,000 and the population of the city at $t = 3$ is 19,000, what is the approximation to the population of the city at $t = 6$ rounded to the nearest hundredth?

19. If 80 milligrams of a radioactive substance decays to 15 milligrams in 25 years, find its half-life. Use $A = pe^{-kt}$, where p is the initial amount, k is a constant, and A is the amount remaining after t years. Round to the nearest thousandth.

20. The loudness of sound is measured by the formula $d = 10\log\dfrac{I}{I_0}$, where d is the loudness of sound measured in decibels, I is the intensity of sound, and I_0 is the softest audible sound. Solve the formula for I_0.

21. The corpse of a famous math professor had a temperature of 80° F when it was discovered at midnight in his mansion. Two hours later, the temperature of the corpse had dropped to 75° F. The temperature of the room was kept constant at 60° F all night. Use Newton's Law of Cooling to find his time of death to the nearest minute if his body temperature was 98.6° F at the time of death.

22. Under ideal conditions, the population of a certain bacterial colony will double in 65 minutes. How long will it take for the population to increase to 16 times the initial population?

CHAPTER 4 REVIEW EXERCISES

Section 4.1 Exponential Functions

In Exercises 1–3, graph each exponential function.

1. $p(x) = 2^{x+1} - 1$
2. $h(x) = (2)3^x + 1$
3. $g(x) = -\left(\frac{1}{3}\right)e^x - 2$

4. How much should be deposited in an account paying 5.75% interest compounded monthly in order to have a balance of $14,000 after 5 years?

5. The number of streptococcus bacteria in a Petri dish, kept in a medical laboratory, is $A = pe^{0.047t}$, where p is the initial number (when the culture was first brought in) and A is the final number of bacteria after t hours. If a culture containing 40,000 bacteria was brought in, how many bacteria will there be after 10 hours? after a week?

Section 4.2 Logarithmic Functions

In Exercises 6–7, simplify each expression without using a calculator.

6. $\log 1000 + \log_{64} 4 - \log_{29} 1$
7. $\ln e^5 - e^{\ln 7} + \log 10^{\ln e}$

In Exercises 8–9, graph each function.

8. $f(x) = \log_2(-x) + 1$
9. $n(x) = \log_{\frac{1}{2}} x - 2$

10. Express the domain of $k(x) = \log_4(-x + 8) + 1$ in interval notation.

Section 4.3 Properties of Logarithms

In Exercises 11–12, simplify each expression without using a calculator.

11. $\log_6 144 - \log_6 12 + \log_6 3$
12. $\log_3 15 + \log_3 6 - \log_3 810$

13. Use the properties of logarithms to expand
$$\log_8\left(\frac{8y(x-5)}{\sqrt[4]{x}}\right)^3$$

14. Use the properties of logarithms to write
$$4\log_b(x+2) - \frac{1}{3}\log_b 27 + \frac{2}{3}\log_b y$$
as a single log, if possible.

15. Use the Change of Base Formula and common logs or natural logs to approximate the value of $\log_{\frac{2}{3}}\left(\frac{7}{9}\right)$ to the nearest hundredth, using a calculator.

Section 4.4 Exponential and Logarithmic Equations

In Exercises 16–17, solve each exponential equation. Round to the nearest hundredth, if needed.

16. $5^{3x+2} = 125^{1-2x}$
17. $3e^{4x} = 5e^{2x} + 2$

In Exercises 18–19, solve each logarithmic equation. Round to the nearest hundredth, if needed.

18. $\log_4(x + 4) + \log_4(x + 16) = 3$
19. $\log_6(x + 4) - \log_6(x - 4) = \log_6 5$

20. Suppose a total of $25,000 is invested in a semi-annually compounded account at 4.8%. Approximately how long will it take for the balance to reach $36,000? Round to the nearest tenth of a year.

Section 4.5 Exponential and Logarithmic Models

21. In an amplifier, the power gain P is given by the function $P = 10\log\frac{P_{OUT}}{P_{IN}}$, where P_{OUT} and P_{IN} are the power output and input in watts. If an amplifier has a power gain of 24 watts and the output power was 16 watts, find the power input in watts. Round to the nearest thousandth.

22. The number of fruit flies present in a population at time t (in days) is given by the function $P(t) = \frac{240}{1 + 56.5e^{-0.37t}}$. How long will it take for the population to reach 200? Round to the nearest tenth.

23. If 40 milligrams of strontium-90 radioactively decays to 12 milligrams in 30 years, find its half-life. Use the formula $A = pe^{-kt}$, where p is the initial amount and A the final amount. Round to the nearest tenth.

24. When an object is removed from a furnace and placed in an environment with a constant temperature of 31° C, its core temperature is 730° C. The temperature of the object t hours after it has been removed from the furnace can be described by the function $F(t) = 31 + 699e^{-0.154t}$. How long will it take the object to cool to 510° C? Round to the nearest tenth.

25. Suppose an account offers 8.5% interest compounded continuously. How many years does it take for an investment in this account to triple? Round to the nearest tenth of a year.

Precalculus

TRIGONOMETRIC FUNCTIONS

Chapter

5

5.1 ANGLES AND THEIR MEASURE

OBJECTIVES

- Draw an angle in standard position in a coordinate plane.
- Identify coterminal angles.
- Find the supplement and complement of an angle.
- Measure angles in degrees and radians.
- Use the arc length formula and the formula for the area of a sector of a circle.
- Find linear speed and angular speed.

PREREQUISITE VOCABULARY TERMS

circle
circumference
coordinate plane
degrees
quadrants
radius

5.1.1 Measuring Angles Using Degrees

A **ray** is a line with one endpoint. Two rays with a common endpoint define an **angle**, which is formed by rotating one ray about its endpoint. The ray in the starting position is called the **initial side**, and the ray in the ending position (i.e., the position after the rotation) is called the **terminal side**. The endpoint is called the angle's **vertex** (Figure 5.1a).

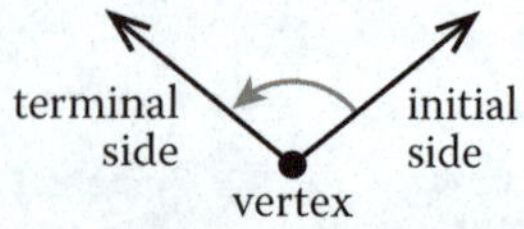

Figure 5.1a

We have seen angles measured using degrees. For example, a right angle (i.e., the angle formed by two perpendicular rays) measures 90°. Note that a fraction of 1° can be expressed using a decimal (e.g., $\frac{1}{2}(1°) = 0.5°$). Alternatively, a fraction of 1° can be expressed in **minutes** and **seconds**, where 1 minute is defined as 1/60 of 1°, and 1 second is defined as 1/60 of 1 minute.

A Fraction of a Degree

$1 \text{ minute} = 1' = \dfrac{1}{60}(1°) \approx 0.01667°$ *Use ′ (prime) to denote minutes.*

$1 \text{ second} = 1'' = \dfrac{1}{60}(1') = \dfrac{1}{3600}(1°) \approx 0.0002778°$ *Use ″ (double prime) to denote seconds.*

For example, 10 degrees, 5 minutes, and 80 seconds is denoted 10° 5′ 80″. To find the approximate degree value of 10° 5′ 80″, convert the minutes and the seconds to degrees, then add the three degree values. Since $10 + \dfrac{5}{60} + \dfrac{80}{3600} \approx 10.10556$, 10° 5′ 80″ is approximately 10.10556°.

Angles in a Coordinate Plane

Consider a coordinate plane, as shown in Figure 5.1b. Recall that the x- and y-axes are perpendicular lines that divide the plane into four quadrants. Since the axes intersect at right angles (which measure 90°), the sum of the measures of the four angles formed by the intersections of the axes is 4(90°) = 360°.

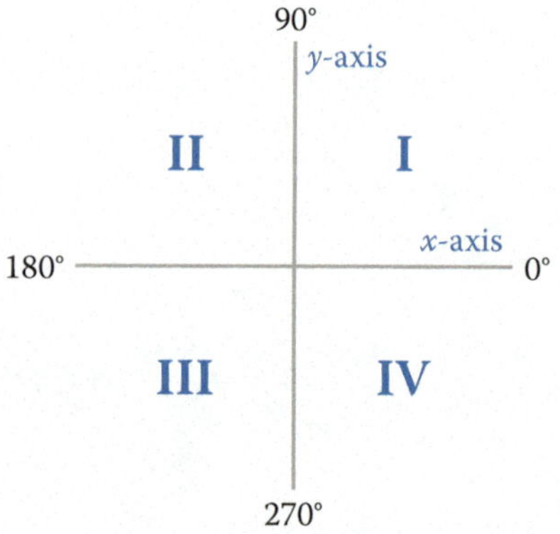

Figure 5.1b

An angle in a coordinate plane with its vertex at the origin and its initial side on the positive x-axis is said to be in **standard position**. The terminal side of an angle in standard position may lie in one of the four quadrants or on either the positive or negative x- or y-axis.

Angles with Positive Measure

Angles with positive measure are formed by rotating the terminal side in a counterclockwise direction, as shown in Figures 5.1c and 5.1d. Angle θ in Figure 5.1c is called an **acute angle** because $0° < \theta < 90°$. Angle θ in Figure 5.1d is called an **obtuse angle** because $90° < \theta < 180°$.

IMPORTANT

Greek letters, such as θ (theta), are often used to represent the measure of an angle.

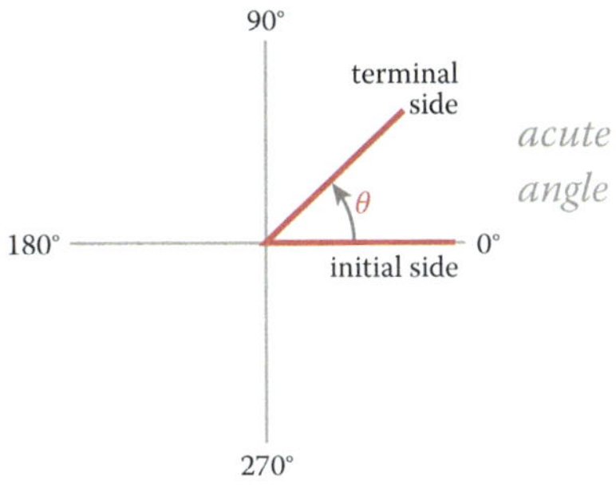

Figure 5.1c

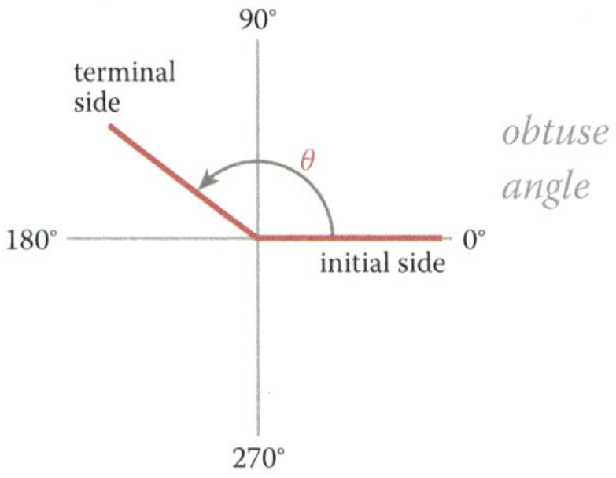

Figure 5.1d

Consider an angle in standard position. If the angle measures 90°, then the terminal side is on the positive y-axis (Figure 5.1e). As we have seen, angles measuring 90° are called right angles. If the angle measures 180°, then the terminal side is on the negative x-axis (Figure 5.1f). Angles measuring 180° are called **straight angles**. If the angle measures 360°, then the terminal side is on the positive x-axis (Figure 5.1g). In this case, the terminal side is in the same position as the initial side.

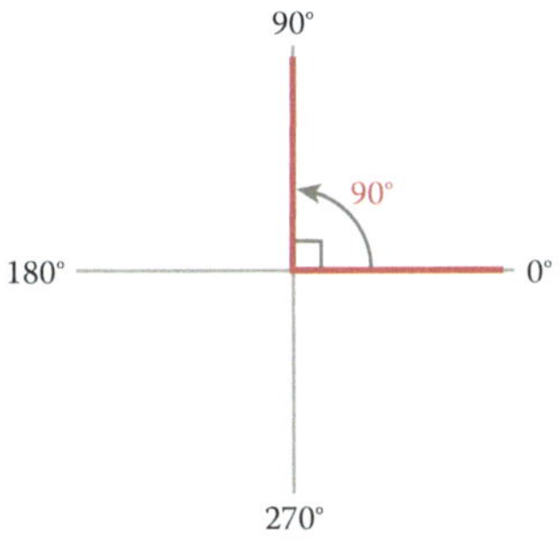

Figure 5.1e

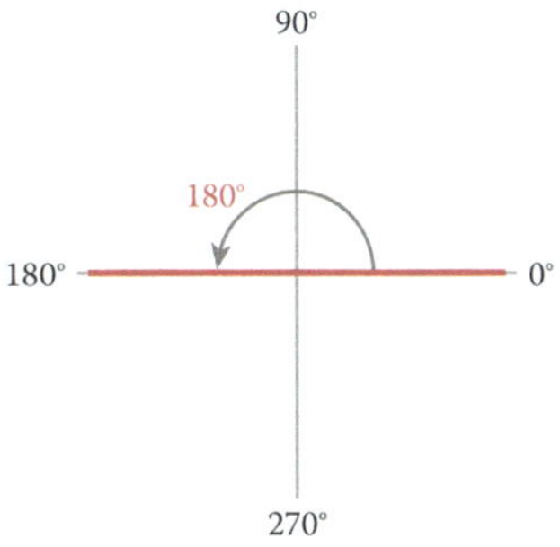

Figure 5.1f

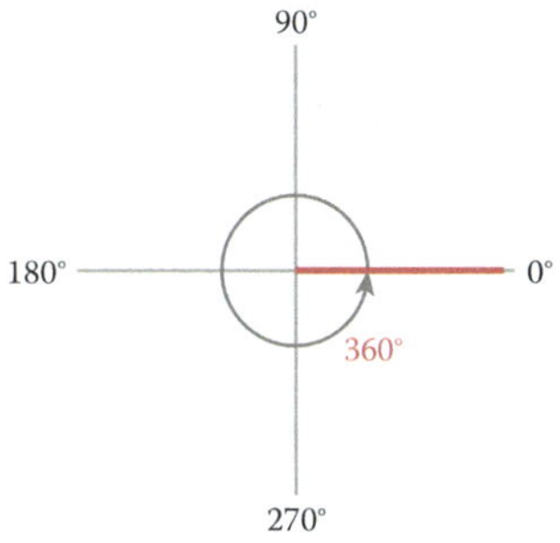

Figure 5.1g

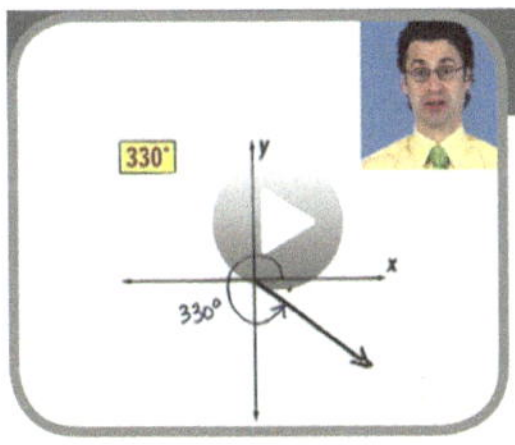

5.1.2 Drawing an Angle in Standard Position

To draw an angle in standard position, first draw the initial side on the positive x-axis, then rotate the terminal side from this position. For angles with positive measure, rotate the terminal side in a *counterclockwise* direction, using the angle's measure at each axis as a guide.

For angles with negative measure, rotate the terminal side in a *clockwise* direction (Figure 5.1h), again using the angle's measure at each axis as a guide. The terminal side of the angle in Figure 5.1h is rotated clockwise from the positive x-axis, through the negative y-axis, to a position between the negative y-axis and the negative x-axis (i.e., in quadrant III). So, its measure must be between $-90°$ and $-180°$.

Angles with measure greater than 360° (or less than $-360°$) are formed by rotating the terminal side more than one full rotation around the axes. For example, a 380° angle is formed by rotating the terminal side 20° beyond the positive x-axis (i.e., beyond 360°). Note that the terminal side of a 20° angle is in the same position as the terminal side of a 380° angle, and this is true for any two angles which differ by a multiple of 360°.

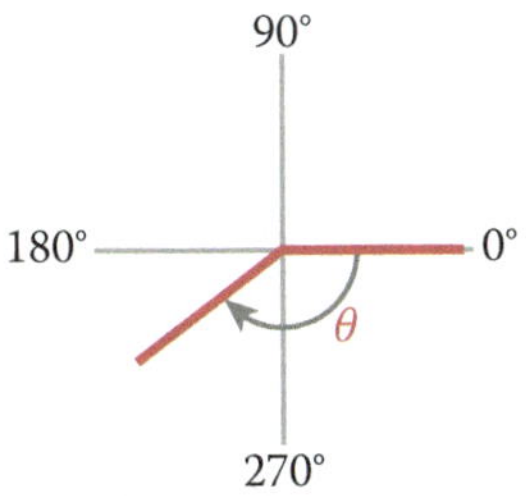

θ is a negative angle.

Figure 5.1h

EXAMPLE 1 **Drawing an Angle in Standard Position**

Draw each angle in standard position.

A. 160° B. –45° C. 400°

SOLUTION

A. The angle measure is positive, so rotate the terminal side in a counterclockwise direction from the positive x-axis.

Note that rotating 90° counterclockwise from the positive x-axis brings the terminal side to the positive y-axis.

To draw a 160° angle, rotate the terminal side counterclockwise 70° past the positive y-axis because 90° + 70° = 160°.

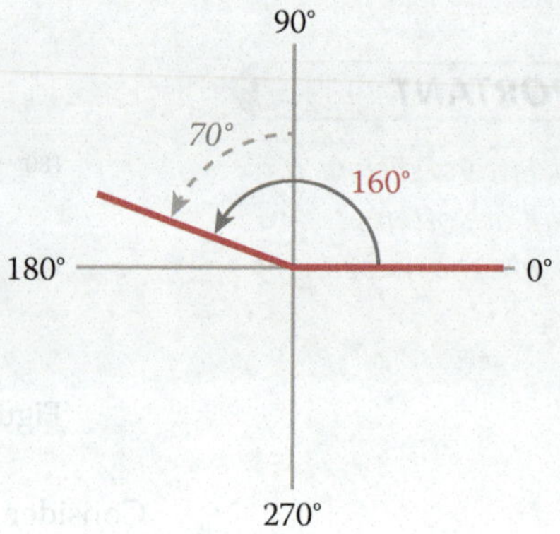

TIP

Since 160° is between 90° and 180°, the terminal side of a 160° angle is in quadrant II.

B. The angle measure is negative, so rotate the terminal side in a clockwise direction from the positive x-axis.

To draw a –45° angle, rotate the terminal side 45° clockwise.

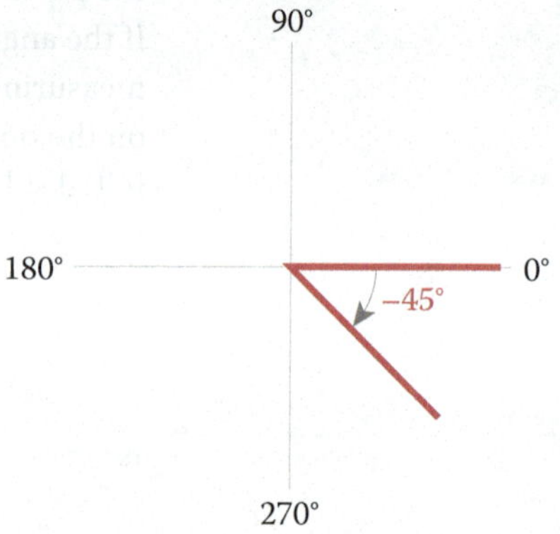

TIP

Since –45° is between 0° and –90°, the terminal side of a –45° angle is in quadrant IV.

C. The angle measure is positive, so rotate the terminal side in a counterclockwise direction from the positive x-axis.

Note that rotating 360° counterclockwise from the positive x-axis returns the terminal side to the positive x-axis.

To draw a 400° angle, make a full counterclockwise rotation of the terminal side (i.e., rotate 360°) and then rotate 40° past the positive x-axis because 360° + 40° = 400°.

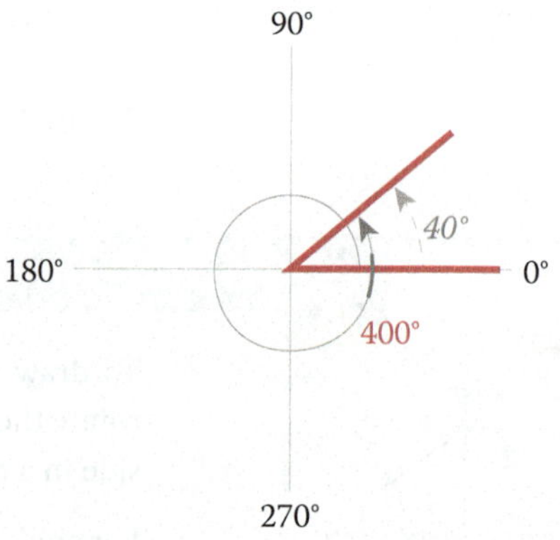

NOTICE THAT

The terminal side of a 40° angle is in the same position as the terminal side of a 400° angle.

5.1.3 Finding the Quadrant in Which an Angle Lies

From here forward, assume that all angles in a coordinate plane are in standard position.

An angle in standard position where the terminal side is on the x- or y-axis is called a **quadrantal angle**. If an angle is not a quadrantal angle, then its terminal side lies in a quadrant. An angle is said to lie in a particular quadrant when its terminal side lies in that quadrant. For example, a 100° angle lies in quadrant II because its terminal side lies in quadrant II.

EXAMPLE 2 Finding the Quadrant in Which an Angle Lies

IMPORTANT

A quadrantal angle does not lie in a quadrant because its terminal side lies on an axis.

Identify the quadrant in which each angle lies, if any.

A. 190° **B.** –120° **C.** –800°

SOLUTION

A. A 190° angle is formed by rotating the terminal side counterclockwise because its measure is positive. The terminal side of a 190° angle is rotated 10° beyond the negative x-axis because 180° + 10° = 190°. Since the angle's terminal side is between the negative x-axis and the negative y-axis, a 190° angle lies in quadrant III.

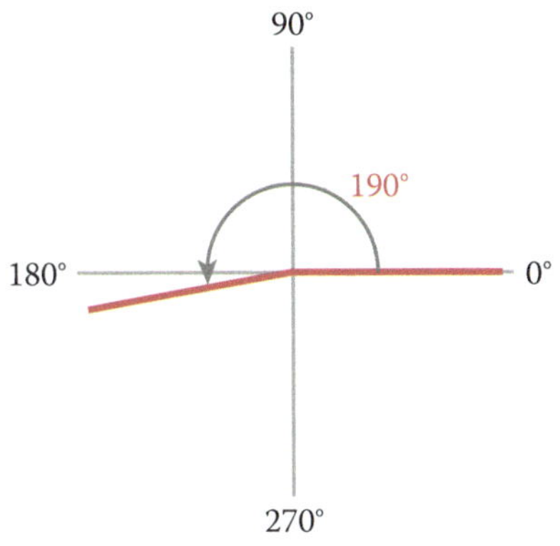

NOTICE THAT

The terminal side of a –120° angle is between –90° and –180°, so a –120° angle lies in quadrant III.

B. A –120° angle is formed by rotating the terminal side clockwise because its measure is negative. The terminal side of a –120° angle is rotated –30° beyond the negative y-axis because –90° + (–30°) = –120°. Since the angle's terminal side is between the negative x-axis and the negative y-axis, a –120° angle lies in quadrant III.

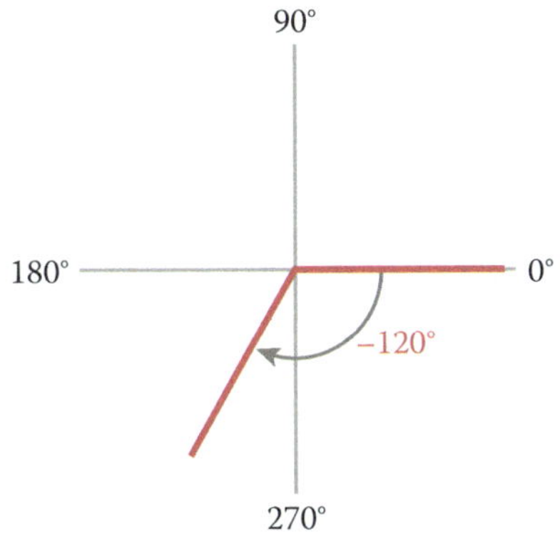

NOTICE THAT

Angles measuring –80° and –800° have their terminal side in the same position.

C. A –800° angle is formed by rotating the terminal side clockwise because its measure is negative. The terminal side of a –800° angle is rotated –80° beyond the positive x-axis because –360° + (–360°) + (–80°) = –800°. Since the angle's terminal side is between the positive x-axis and the negative y-axis, a –800° angle lies in quadrant IV.

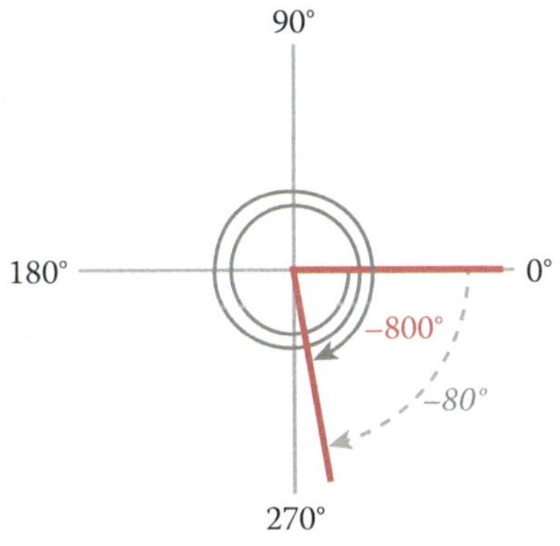

5.1.4 Finding Coterminal Angles

Since rotating the terminal side of an angle 360° clockwise or counterclockwise is a full rotation around the plane, adding (or subtracting) a multiple of 360° to the measure of any angle results in an angle whose terminal side is in the same position as the original angle.

Angles in standard position with common terminal sides are called **coterminal angles**. For example, angles measuring –80° and 280° are coterminal (Figure 5.1i).

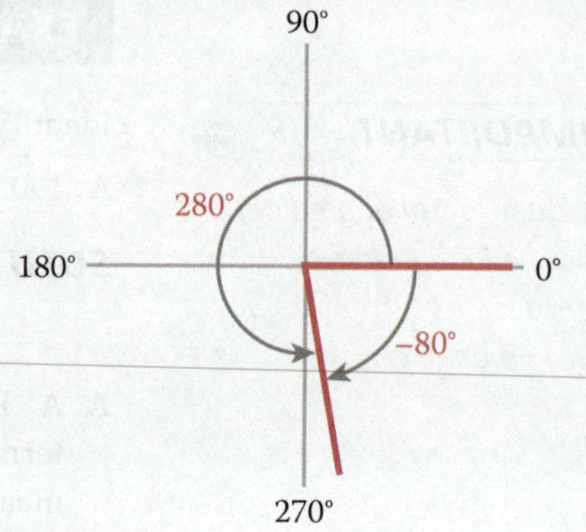

Figure 5.1i

All angles that are coterminal with a given angle can be identified by adding (or subtracting) multiples of 360° to the angle's measure. The difference of two coterminal angles is always a multiple of 360°.

TIP

Multiples of 360°: . . . –1080°, –720°, –360°, 0°, 360°, 720°, 1080°, . . .

EXAMPLE 3 Identifying Coterminal Angles

Determine whether angles measuring –1180° and 20° are coterminal with θ if $\theta = -460°$. Then find the smallest positive angle that is coterminal with θ, and find the negative angle closest to 0° that is coterminal with θ.

SOLUTION

ALTERNATIVE METHOD

Sketch the two angles in a coordinate plane to determine whether or not the angles are coterminal.

Angles measuring –1180° and 20° are coterminal with θ if the difference between each measure and θ is a multiple of 360°.

$-1180° - (-460°) = -720°$ ✓ *The difference is a multiple of 360°, so the angles are coterminal.*

$20° - (-460°) = 480°$ ☒ *The difference is not a multiple of 360°, so the angles are not coterminal.*

Adding (or subtracting) multiples of 360° to the measure of θ gives the measures of the angles coterminal with θ. The measure of the smallest positive angle that is coterminal with θ is between 0° and 360°. The measure of the negative angle closest to 0° that is coterminal with θ is between 0° and –360°.

$-460° + 360° = -100°$ The negative angle closest to 0° that is coterminal with θ is –100°.

$-100° + 360° = 260°$ The smallest positive angle that is coterminal with θ is 260°.

5.1.5 Finding the Complement and Supplement of an Angle

Two positive angles are **complementary** if the sum of their measures is 90° (Figure 5.1j), or **supplementary** if the sum of their measures is 180° (Figure 5.1k). When two angles are complementary, each angle is the **complement** of the other. When two angles are supplementary, each angle is the **supplement** of the other.

TIP

The Greek letters α (alpha) and β (beta) represent the measures of the angles.

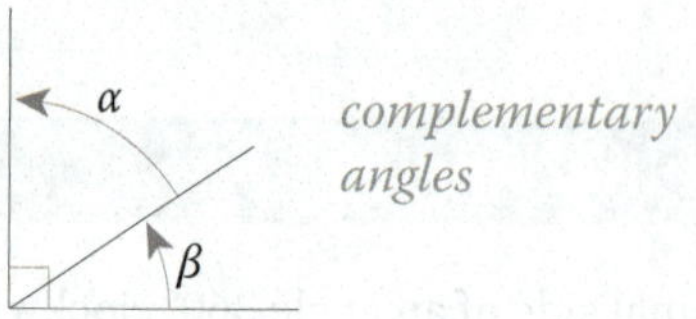

$\alpha + \beta = 90°$

Figure 5.1j

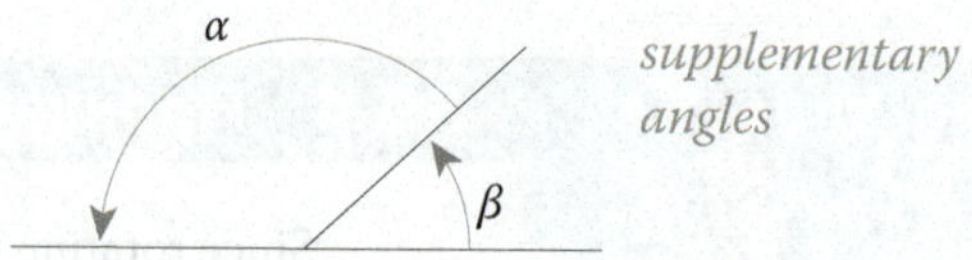

$\alpha + \beta = 180°$

Figure 5.1k

Since the sum of the measures of complementary angles is 90°, only acute angles have complements, and the complement of the angle is also acute. Since the sum of supplementary angles is 180°, an angle and its supplement may both be right angles, or one angle may be acute and the other obtuse.

IMPORTANT

Complementary and supplementary angles must be positive angles.

EXAMPLE 4 Finding Complementary and Supplementary Angles

For each angle, identify the complement and supplement, if it exists.

A. 15° **B.** 129° **C.** 200°

SOLUTION

Let θ be the measure of the given angle.

- If $0 < \theta < 90$, then θ has a complement given by x in the equation $\theta + x = 90$.
- If $0 < \theta < 180$, then θ has a supplement given by y in the equation $\theta + y = 180$.

IMPORTANT

If $\theta > 90°$, then θ has no complement.
If $\theta > 180°$, then θ has no complement or supplement.

A. Since 15° < 90° and 15° < 180°, 15° has a complement and a supplement.

$15° + x = 90° \Rightarrow x = 75°$ The complement of 15° is 75°.

$15° + y = 180° \Rightarrow y = 165°$ The supplement of 15° is 165°.

B. Since 129° < 180°, 129° has a supplement. However, since 129° > 90°, 129° has no complement.

$129° + y = 180° \Rightarrow y = 51°$ The supplement of 129° is 51°.

C. Since 200° > 180°, 200° has neither a complement nor a supplement.

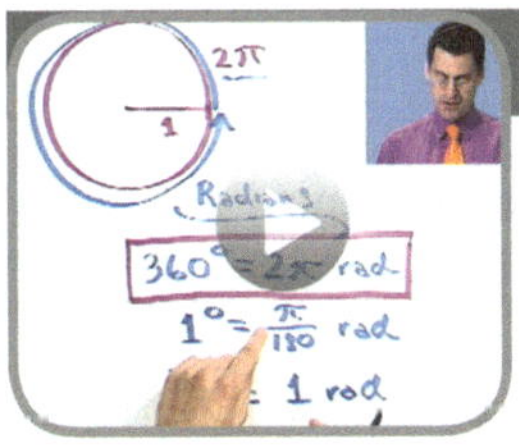

5.1.6 Converting between Degrees and Radians

Introduction to Radians

We have seen that an angle can be measured in degrees, and that this degree measure indicates the amount of rotation from the angle's initial side to its terminal side. An amount of rotation can also be measured in **radians**.

DEFINITION

Radians are a unit of angular measure defined such that 2π radians = 360°.

When the amount of rotation from the initial side to the terminal side forms a complete circle, the angle's measure is 360°. If the radius of this complete circle is 1 unit, then the circumference of this circle (i.e., the length around the circle) is 2π. It follows that radians relate the length around a circle to the angle forming the circle (Figure 5.11).

REMEMBER

The formula for the circumference of a circle is $C = 2\pi r$.

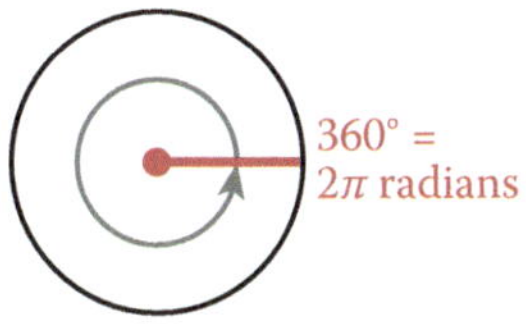

Figure 5.11

IMPORTANT

$360° = 2\pi$ radians
$180° = \pi$ radians

When the amount of rotation from the initial side to the terminal side forms a semicircle, the angle's measure is 180° because 360°/2 = 180°. This amount of rotation is equal to π radians (Figure 5.1m) because $2\pi/2 = \pi$. Therefore, π radians = 180°.

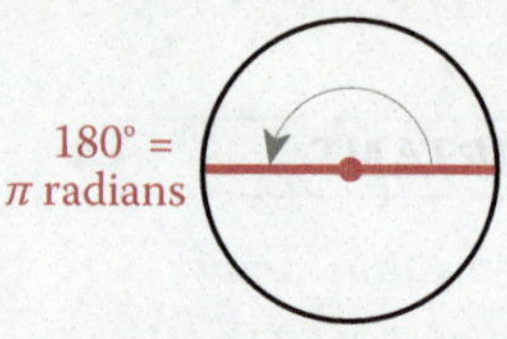

Figure 5.1m

The radian measures of additional common angles are shown in the following table. Notice that a 90° angle (a right angle) is 1/4 of a complete rotation, so the radian measure of a right angle is $2\pi/4$ radians, that is, $\pi/2$ radians. Alternatively, a 90° angle is 1/2 of π radians, that is, $\pi/2$ radians. Additionally, a 45° angle is 1/2 of $\pi/2$ radians, that is, $\pi/4$ radians.

Common Angles Measured in Degrees and Radians

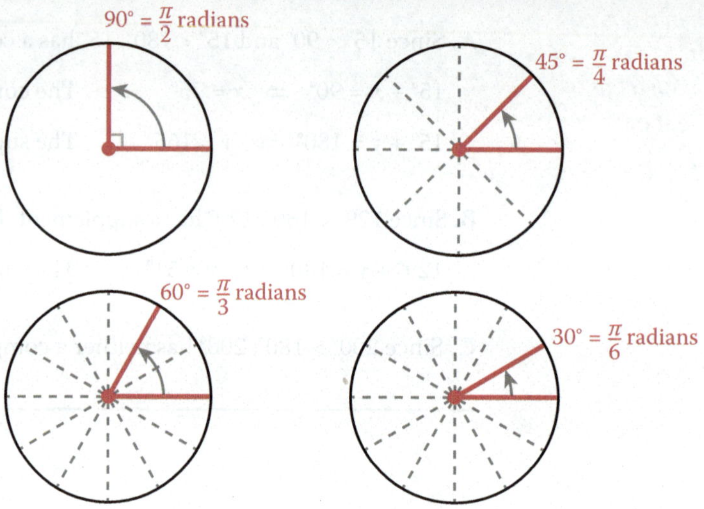

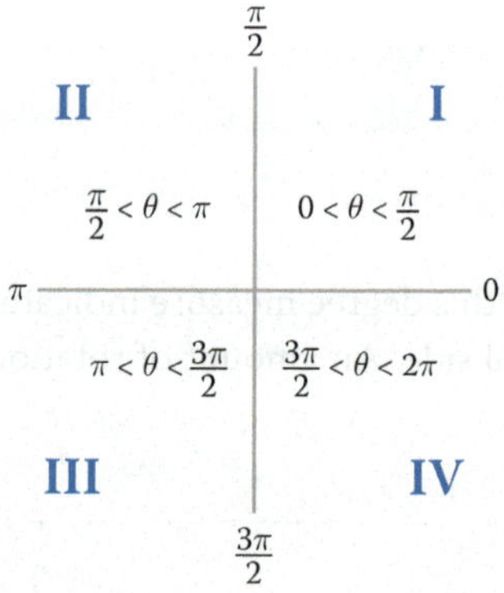

Figure 5.1n

Figure 5.1n shows the range of radian measures for an angle θ within each quadrant (when $0 < \theta < 2\pi$).

Converting between Degree and Radian Measures

Since π radians is equal to 180°, it follows that $\frac{\pi \text{ radians}}{180°} = 1$ and $\frac{180°}{\pi \text{ radians}} = 1$. Therefore, multiplying the degree measure of an angle by $\frac{\pi \text{ radians}}{180°}$ will convert the degree measure into radians. Similarly, multiplying the radian measure of an angle by $\frac{180°}{\pi \text{ radians}}$ will convert the radian measure into degrees.

Converting between Degrees and Radians

Degrees to Radians	Radians to Degrees
Multiply the degrees by $\frac{\pi \text{ radians}}{180°}$.	Multiply the radians by $\frac{180°}{\pi \text{ radians}}$.

EXAMPLE 5 Converting from Degrees to Radians

Convert each degree measure to radian measure.

A. 225° **B.** –420°

SOLUTION

Multiply each degree measure by $\frac{\pi \text{ radians}}{180°}$.

A. $225° = (225°)\left(\frac{\pi \text{ radians}}{180°}\right) = \frac{225\pi}{180} \text{ radians} = \frac{5\pi}{4} \text{ radians}$

B. $-420° = (-420°)\left(\frac{\pi \text{ radians}}{180°}\right) = \frac{-420\pi}{180} \text{ radians} = -\frac{7\pi}{3} \text{ radians}$

ALTERNATIVE METHOD

IMPORTANT

Since 180° = π radians, 1° = π/180 radians.

Multiply each side of the equation $1° = \frac{\pi}{180}$ radians by the given degree measure.

A. Multiply each side by 225.

$$(225)1° = \frac{\pi}{180} \text{ radians}(225)$$
$$225° = \frac{225\pi}{180} \text{ radians}$$
$$225° = \frac{5\pi}{4} \text{ radians}$$

B. Multiply each side by –420.

$$(-420)1° = \frac{\pi}{180} \text{ radians}(-420)$$
$$-420° = \frac{-420\pi}{180} \text{ radians}$$
$$-420° = -\frac{7\pi}{3} \text{ radians}$$

EXAMPLE 6 Converting from Radians to Degrees

Convert each radian measure to degree measure.

A. 3π radians **B.** –5π/6 radians

SOLUTION

Multiply each radian measure by $\frac{180°}{\pi \text{ radians}}$.

A. $3\pi \text{ radians} = (3\pi \text{ radians})\left(\frac{180°}{\pi \text{ radians}}\right) = \frac{3\pi(180°)}{\pi} = 540°$

B. $-\frac{5\pi}{6} \text{ radians} = (-\frac{5\pi}{6} \text{ radians})\left(\frac{180°}{\pi \text{ radians}}\right) = \frac{-5\pi(180°)}{6\pi} = -150°$

ALTERNATIVE METHOD

IMPORTANT

Since 180° = π radians, 180°/π = 1 radian.

Multiply each side of the equation $\frac{180°}{\pi} = 1$ radian by the given radian measure.

A. Multiply each side by 3π.

$$(3\pi)\frac{180°}{\pi} = 1 \text{ radian}(3\pi)$$
$$(3)(180°) = 3\pi \text{ radians}$$
$$540° = 3\pi \text{ radians}$$

B. Multiply each side by $-\frac{5\pi}{6}$.

$$\left(-\frac{5\pi}{6}\right)\frac{180°}{\pi} = 1 \text{ radian}\left(-\frac{5\pi}{6}\right)$$
$$-\frac{5(180°)}{6} = -\frac{5\pi}{6} \text{ radians}$$
$$-150° = -\frac{5\pi}{6} \text{ radians}$$

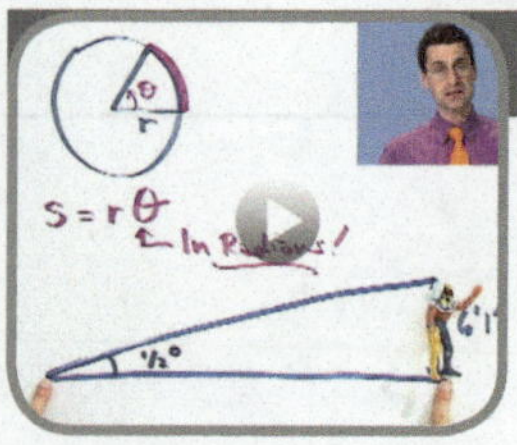

5.1.7 Using the Arc Length Formula

Figure 5.1o shows a circle where **central angle** θ defines **arc** s.

DEFINITION

An **arc** is a portion of the circumference of a circle.

A **central angle** is an angle within a circle where the angle's vertex is at the circle's center.

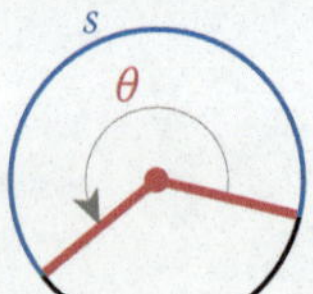

Figure 5.1o

The **arc length formula** gives the measure of an arc as a length.

The Arc Length Formula

If a central angle θ (measured in radians) intercepts an arc of length s in a circle with radius r, then $s = r\theta$.

NOTICE THAT

If $r = 1$, then the radian measure of θ is equal to the length of the arc.
$s = (1)\theta = \theta$

EXAMPLE 7 Finding the Length of an Arc

Suppose an arc on a circle with a radius of 5 meters is intercepted by a 90° angle. Find the length of the arc to the nearest hundredth of a meter.

SOLUTION

Substitute the measure of the central angle (in radians) and the radius of the circle into the arc length formula $s = r\theta$, and then round the answer to the nearest hundredth of a meter.

The measure of the central angle is 90°, so substitute $\pi/2$ for θ (since $90° = \pi/2$ radians).

$$s = r\theta = (5)\left(\frac{\pi}{2}\right) = \frac{5\pi}{2} \approx 7.85 \text{ meters}$$

Therefore, the length of the arc is approximately 7.85 meters.

NOTICE THAT

A 90° central angle corresponds to 1/4 of a circle, so the length of its intercepted arc is 1/4 of the circumference of the circle ($C = 2\pi r$).
$s = C/4 = 2\pi(5)/4 = 5\pi/2$

Using the Arc Length Formula to Approximate a Distance

If a central angle is very small, then the length of the intercepted arc can be used to approximate the subtended distance perpendicular to the angle's initial side, as shown in **Example 8**.

EXAMPLE 8 Using the Arc Length Formula to Approximate a Distance

A tower 2500 feet from an observer subtends a 2.5° angle, as shown in the picture. Use the arc length formula to approximate the height of the tower to the nearest foot.

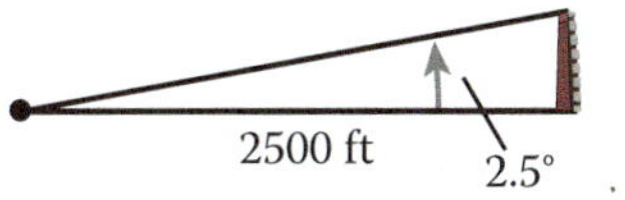

SOLUTION

The height of the tower is approximately equal to the length of the intercepted arc of a circle, where the central angle is 2.5° and the radius is 2500 feet.

IMPORTANT

The measure of an angle used in the arc length formula must be in radians, not degrees.

Convert the angle measure from degrees to radians.

$$2.5° = (2.5°)\left(\frac{\pi \text{ radians}}{180°}\right) = \frac{2.5\pi}{180} \text{ radians} = \frac{\pi}{72} \text{ radians}$$

Substitute the measure of the central angle and the radius of the circle into the arc length formula and round to the nearest foot.

$$s = r\theta = (2500)\left(\frac{\pi}{72}\right) \approx 109 \text{ feet}$$

The height of the tower is approximately 109 feet.

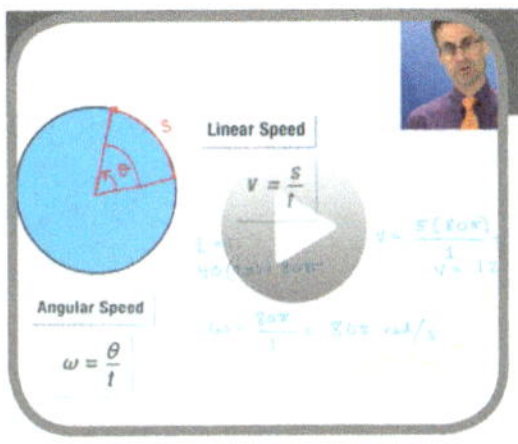

5.1.8 Finding Linear and Angular Speeds

As an object travels along a circular path (i.e., along an arc) at a constant speed, the central angle changes uniformly. The rate at which the central angle is changing is the object's **angular speed** $\boldsymbol{\omega}$. The object's **linear speed** $\boldsymbol{v}$ is the rate at which the object is moving along the arc.

Formulas for Linear and Angular Speed

Suppose an object travels at a constant speed along an arc on a circle with radius r. If s is the length of the arc traveled in time t and the corresponding central angle is θ (in radians), then

$$\pi = \frac{\theta}{t} \text{ and } v = \frac{s}{t}.$$

EXAMPLE 9 Finding Linear and Angular Speed

The blades on a ceiling fan are 3 feet long. When set to medium speed, the blades rotate at 12 revolutions per second. Find the linear speed of the end of a blade in feet per second and the angular speed of a blade in radians per second, each to the nearest whole number.

IMPORTANT

One complete revolution is 2π radians.

SOLUTION

In 1 second, a blade completes 12 revolutions. So, when $t = 1$ second, $\theta = 12(2\pi) = 24\pi$ radians.

Find the linear speed.

$$v = \frac{s}{t} = \frac{r\theta}{t} = \frac{3(24\pi)}{1} = 72\pi \text{ ft/s} \approx 226 \text{ ft/s}$$

Find the angular speed.

$$\pi = \frac{\theta}{t} = \frac{24\pi}{1} = 24\pi \text{ rad/s} \approx 75 \text{ rad/s}$$

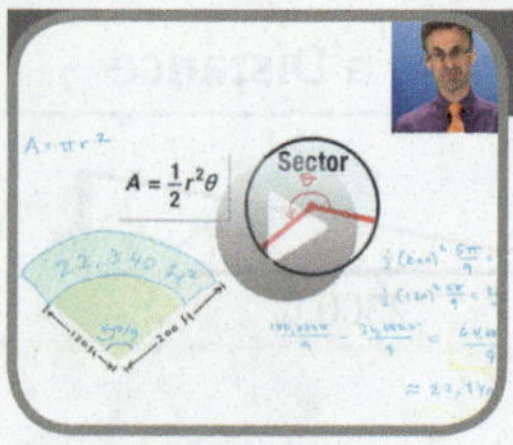

5.1.9 Finding the Area of a Sector of a Circle

A **sector** of a circle is the interior portion of a circle defined by a central angle θ (i.e., the portion of the circle enclosed by two radii and their intercepted arc) where θ is between 0 and 2π radians, as shown in Figure 5.1p.

The **area of a sector** of a circle is a fraction of the circle's area. This fraction of the circle's area is given by $\theta/2\pi$. For example, the area of a sector defined by a central angle θ of π radians is 1/2 the area of the circle because $\pi/2\pi = 1/2$. Therefore, since the area of a circle with radius r is πr^2, the area of a sector of a circle with radius r and angle θ is $\frac{\theta}{2\pi} \cdot \pi r^2$, which simplifies to $\frac{1}{2}r^2\theta$.

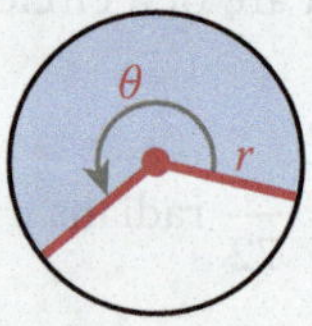

Figure 5.1p

The Area of a Sector of a Circle

The area of a sector of a circle defined by central angle θ is given by

$$A = \frac{1}{2}r^2\theta$$

where r is the circle's radius and θ is measured in radians.

NOTICE THAT

If $\theta = 2\pi$, then the sector is the entire circle. In this case, the formula gives $A = \frac{1}{2}r^2(2\pi) = \pi r^2$, which is the area of an entire circle.

EXAMPLE 10 Finding the Radius of a Circle

Find the radius of a circle to the nearest tenth of a centimeter if the area of a sector defined by a 120° central angle is 54.8 square centimeters.

SOLUTION

Convert the angle measure to radians.

$$120° = (120°)\left(\frac{\pi \text{ radians}}{180°}\right) = \frac{120\pi}{180} \text{ radians} = \frac{2\pi}{3} \text{ radians}$$

Substitute the given angle and area into the formula, then solve for r.

$$54.8 = \frac{1}{2}r^2\left(\frac{2\pi}{3}\right) \Rightarrow 54.8 = r^2\left(\frac{\pi}{3}\right) \Rightarrow \frac{(3)(54.8)}{\pi} = r^2 \Rightarrow r = \sqrt{\frac{(3)(54.8)}{\pi}} \approx 7.2 \text{ cm}$$

REMEMBER

Since a radius is a length, r must be positive.

SECTION 5.1 EXERCISES

Warm Up

1. Find the circumference and area of a circle with a radius of 15 centimeters. Round each answer to the nearest tenth.
2. Identify the quadrant or axis in which each point lies: (–2, 3), (5, 9), (7, 0), (5, –4), and (–10, –9).
3. Convert 158 inches to feet.

Just the Facts

Fill in the blanks.

4. An angle in standard position has its initial side on the positive _____ and its _____ at the origin. To graph an angle in standard position that has a measure of –81°, rotate 81 degrees in the _____ direction, starting from the _____ x-axis.
5. A straight angle has a measure of _____, a(n) _____ angle has a measure of 90°, and a(n) _____ angle has a measure between 0° and 90°.
6. Angles whose difference is a multiple of 360° are _____.
7. An angle whose measure is 250° is located in quadrant _____ and is _____ with –110°. The measure of 250° in radians is _____.
8. A nail moving at a rate of 145 feet per second in a wheel of a car is an example of the nail's _____.
9. An angle that is its own complement has a measure of _____. An angle that is its own supplement has a measure of _____.
10. To convert $-\frac{10\pi}{3}$ radians to degrees, multiply by _____, and the result is _____. This angle is located in quadrant _____.

Essential Skills

In Exercises 1–12, draw an angle with each given measure in standard position.

1. 100°
2. 240°
3. 59°
4. 319°
5. –80°
6. –210°
7. –155°
8. –320°
9. 420°
10. –840°
11. –630°
12. 910°

In Exercises 13–16, choose which angle(s) lies in the indicated quadrant.

13. quadrant III
A. 45° B. 195° C. –20°

14. quadrant IV
A. 280° B. 175° C. –70° D. –120° E. 10°

15. quadrant II
A. 455° B. 175° C. –181° D. –275° E. 1100°

16. quadrant I
A. 1105° B. 775° C. –710° D. –320° E. 91°

In Exercises 17–24, identify the quadrant in which each angle lies.

17. 380°
18. 1150°
19. 1349°
20. 650°
21. –445°
22. –1100°
23. –1650°
24. –938°

In Exercises 25–30, choose all angles that are coterminal with each given angle θ.

25. $\theta = 70°$
A. –70° B. 430° C. –290° D. –430° E. 790°

26. $\theta = 200°$
A. –160° B. 160° C. 560° D. –200° E. –520°

27. $\theta = -300°$
A. 60° B. –60° C. 560° D. –1020° E. 420°

28. $\theta = -140°$
A. 260° B. –400° C. 220° D. –500° E. 580°

29. $\theta = 150°$
A. –110° B. 1230° C. 410° D. –210° E. –570°

30. $\theta = 340°$
A. –20° B. –700° C. 700° D. –380° E. –1100°

In Exercises 31–42, find the smallest positive angle and the negative angle closest to 0° that are coterminal with each given angle (other than itself).

31. 240°
32. 655°
33. 1020°
34. 981°
35. 158°
36. 556°
37. –170°
38. –795°
39. –642°
40. –287°
41. –1033°
42. –714°

In Exercises 43–46, find the measure of the complementary angle, if it exists, for the angle with each given measure.

43. 7°
44. 103°
45. 90.5°
46. 22.7°

In Exercises 47–50, find the measure of the supplementary angle, if it exists, for the angle with each given measure.

47. 75°
48. 58°
49. 144°
50. 280°

In Exercises 51–56, convert each degree measure to radian measure.

51. 45°
52. 132°
53. –201°
54. –94°
55. –355°
56. 314°

In Exercises 57–62, convert each radian measure to degree measure. Round to the nearest degree.

57. $\frac{\pi}{6}$ radians
58. $\frac{5\pi}{12}$ radians
59. $-\frac{11\pi}{3}$ radians
60. –2.4 radians
61. 4.7 radians
62. $\frac{3\pi}{8}$ radians

63. What is the length of the arc intercepted by an angle of 10° on a circle with a radius of 10 meters? Round the answer to the nearest hundredth of a meter.

64. A tree 1200 yards from an observer subtends an angle of 1.5°. Use the arc length formula to estimate the height of the tree to the nearest yard.

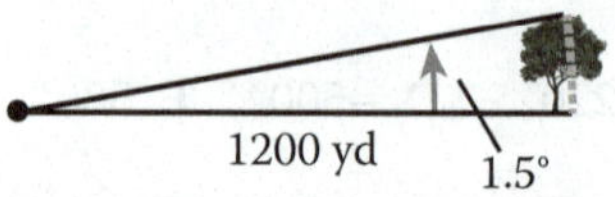

65. What is the length of the arc intercepted by an angle of 324° on a circle with a radius of 47.2 feet? Round the answer to the nearest tenth of a foot.

66. A flagpole 1600 yards from an observer subtends an angle of 3°. Use the arc length formula to estimate the height of the flagpole to the nearest yard.

67. What is the length of the arc intercepted by an angle of 115° on a circle with a radius of 7 inches? Round the answer to the nearest hundredth of an inch.

68. A telephone mast 1300 yards from an observer subtends an angle of 1.5°. Use the arc length formula to estimate the height of the telephone mast to the nearest yard.

69. Find the radius of the circle to the nearest unit.

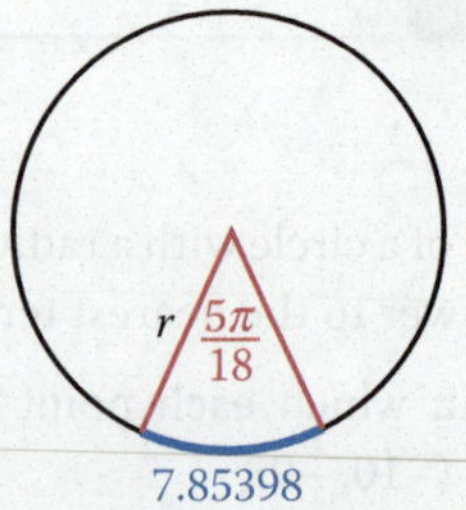

70. A 50-foot radio tower r feet from an observer subtends an angle of 0.75°. Use the arc length formula to estimate r (the distance between the observer and the radio tower) to the nearest foot.

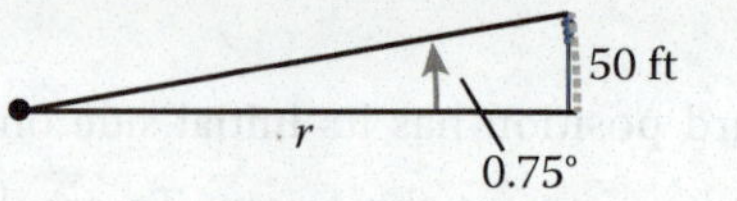

71. Find the radius of the circle to the nearest tenth of an inch.

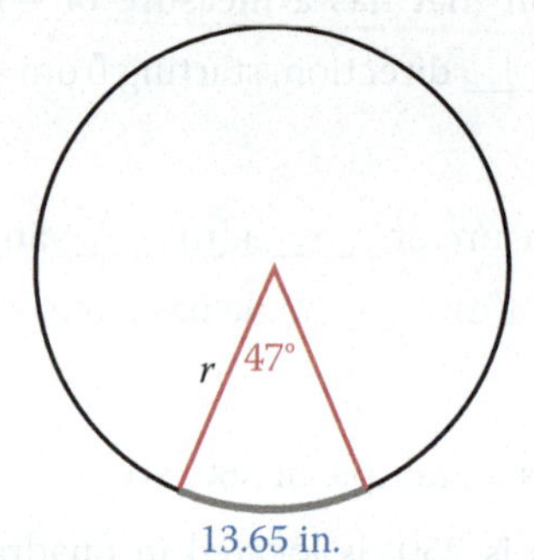

72. A 65-foot flagpole r feet from an observer subtends an angle of 3.5°. Use the arc length formula to estimate r (the distance between the observer and the flagpole) to the nearest foot.

73. Find the radius of the circle to the nearest tenth of a meter.

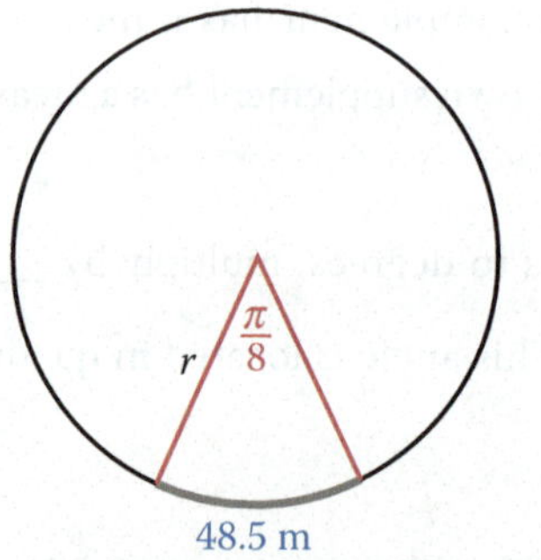

74. A 50-foot building r feet from an observer subtends an angle of 3.25°. Use the arc length formula to estimate r (the distance between the observer and the building) to the nearest foot.

75. An object is traveling around a circle of radius 13 centimeters at a constant speed. If in 10 seconds the object travels 27 centimeters, what is its linear speed in centimeters per second? Round to the nearest whole number.

76. A wheel with diameter 24 inches completes 2 revolutions in 0.5 seconds. Find the linear speed of the edge of the wheel in inches per second. Round to the nearest whole number.

77. An object is traveling around a circle of radius 74 millimeters at a constant speed. If in 36 seconds the object travels 137 millimeters, what is its linear speed in millimeters per second? Round to the nearest tenth.

78. A wheel with diameter 32 inches completes 4 revolutions in 0.5 seconds. Find the linear speed of the edge of the wheel in inches per second. Round to the nearest whole number.

79. An object is traveling around a circle at a constant speed. If in 4 seconds the object travels 29 inches, what is its linear speed in inches per second? Round to the nearest whole number.

80. A wheel with diameter 9 inches completes 10 revolutions in 2.5 seconds. Find the linear speed of the edge of the wheel in inches per second. Round to the nearest whole number.

81. The original Ferris wheel, built in Chicago in 1893, completed 4 revolutions in 40 minutes. Find the angular speed of the Ferris wheel in radians per minute. Leave the answer in terms of π.

82. If a 5-foot blade on a propeller completes 20 revolutions per second, find the angular speed of the blade in radians per second. Round to the nearest hundredth.

83. A small pulley is turning at 145 revolutions per minute. Find the angular speed of the pulley in radians per second. Round to the nearest tenth.

84. If a 9-foot blade on a propeller completes 38 revolutions in 16 seconds, find the angular speed of the blade in radians per second. Round to the nearest hundredth.

85. The average speed in professional bike racing is 30 miles per hour. The bicycles' wheels have a diameter of 24.5 inches. Find the angular speed of the wheels in radians per hour. Round to the nearest tenth. Hint: 1 mile = 5280 feet.

86. If a 3-foot blade on a propeller completes 10 revolutions in 42 seconds, find the angular speed of the blade in radians per second. Round to the nearest hundredth.

87. Find the area of a sector of a circle with a radius of 7 millimeters and a central angle of $\frac{2\pi}{3}$ radians. Round to the nearest square millimeter.

88. When a sprinkler is set to full rotation, the water covers a circular section of a garden with a radius of 22 feet. If the sprinkler is set to rotate back and forth through an angle of 100°, find the area of the garden that is covered by the sprinkler to the nearest square foot.

89. Find the area of a sector of a circle with a radius of 22 millimeters and a central angle of 210°. Round to the nearest square millimeter.

90. When a sprinkler is set to full rotation, the water covers a circular section of a front yard with a radius of 30 feet. If the sprinkler is set to rotate back and forth through an angle of $\frac{9\pi}{5}$ radians, find the area of the front yard that is covered by the sprinkler to the nearest square foot.

91. If an 18-inch windshield wiper rotates at an angle of 115°, find the area covered by the wiper to the nearest tenth of an inch.

92. Find the area of a sector of a circle with a radius of 65 meters and a central angle of 10°. Round to the nearest square meter.

93. A speedometer needle covers an area of approximately 40 square centimeters when it rotates at an angle of $\frac{\pi}{5}$ radians. Find the length of the speedometer needle to the nearest centimeter.

94. If a windshield wiper covers an area of approximately 200 square inches when it rotates at an angle of 98°, find the length of the wiper to the nearest tenth of an inch.

95. Find the radius of a circle to the nearest hundredth of an inch if the area of a sector defined by a 302° central angle is 214.5 square inches.

96. An area of approximately 180 square inches is covered when a rear windshield wiper rotates at an angle of $\frac{11\pi}{8}$ radians. Find the length of the wiper to the nearest tenth of an inch.

97. Find the radius of a circle to the nearest tenth of a meter if the area of a sector defined by a central angle of $\frac{\pi}{3}$ radians is 456.8 square meters.

98. The second hand of a clock covers an area of approximately 50 square centimeters when it rotates at an angle of 215°. Find the length to the nearest hundredth of a centimeter of the second hand.

Extensions

99. Find the measure of an angle y, in degrees, if its complement and supplement can be represented by $6x$ and $12x + 4$, respectively.

100. A wall clock reads exactly four o'clock. What is the radian measure of the acute angle formed by the hour and minute hands of the clock?

101. The cities of Cape Town and Stockholm are located at the same longitude (the angular distance of a place on Earth measured east or west of a north-south line called the Prime Meridian), but at different latitudes (the angular distance of a place on Earth measured north or south of the equator). Lines of longitude are circles on Earth's surface that go through both poles. Cape Town is at latitude 34° S, while Stockholm is at 59° N. Ignoring topographic features such as mountains and valleys, what is the distance (along Earth's surface) between the two cities? Assume that Earth is perfectly round and that its radius is 3959 miles. Round to the nearest mile.

102. An arc that is 3/8 the circumference of a circle subtends what central angle in radians? in degrees?

In Exercises 103–105, use the following information.

A van is moving at a rate of 62 miles per hour, and the circumference of its wheels is approximately 7.5 feet. Round each answer to the nearest tenth.

103. Find the linear speed of the wheels in feet per second.

104. Find the angular speed of the wheels in radians per second.

105. Find the number of revolutions that the wheels are rotating per second.

106. Find the area of the yellow region to the nearest square centimeter if the diameter of the larger circle is 12 centimeters and the radius of the red circle is 2/3 of the larger circle's radius.

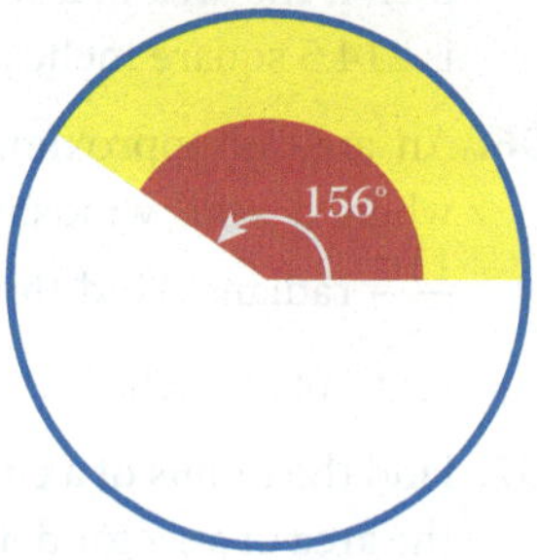

107. Which is the larger angle, 1.2 radians or 68° 45′ 13″? Explain.

108. Suppose a sector comprises 32% of the area of a circle. If the circle has a diameter of 34.6 centimeters, what is the area of the sector to the nearest square centimeter?

109. The wheels on a child's bicycle have a radius of 50 centimeters. The child rides her bicycle so that its wheels rotate at 1.2 revolutions per second. How fast is she riding in meters per second? Round to the nearest tenth.

110. At a birthday party, a circular cake with a diameter of 14 inches is cut into 16 equal pieces. What is the length to the nearest hundredth of an inch of the curved end of each piece of cake?

111. The clock in the tower of City Hall has a circular face that measures 15 feet in diameter. What is the arc length to the nearest thousandth of a foot of the smaller arc between the number 3 and the number 8?

In Exercises 112–115, use the following information.

A pepper shaker is sitting 4.5 inches from the center of a rotating tray. A woman spins the tray through an angle of 132°. Round each answer to the nearest tenth.

112. Through how many radians does the shaker turn?

113. What distance in inches does the shaker travel?

114. If the woman turned the rotating tray 42° in a half second, what was the angular speed of the shaker in radians per second?

115. What was the linear speed of the shaker in inches per second?

5.2 RIGHT ANGLE TRIGONOMETRY

OBJECTIVES

- Write trigonometric ratios.
- Evaluate trigonometric functions of acute angles.
- Use trigonometric functions to find the length of a side of a right triangle.
- Use trigonometric functions to model situations with angles of elevation and depression.

PREREQUISITE VOCABULARY TERMS

acute angle
equilateral triangle
hypotenuse
isosceles triangle
perpendicular bisector
Pythagorean Theorem
reciprocal
right triangle
similar triangles
vertex (of a polygon)

Trigonometry is the study of angles and the relationships between angles and sides in triangles. The discussion of trigonometry in this course begins with the relationships between angles and sides in right triangles.

IMPORTANT

A length of a side of a triangle may be referenced by naming the side (e.g., "the length of the hypotenuse" will often be referred to as "the hypotenuse"). Similarly, the measure of an angle may be referenced by naming the angle.

Review of Right Triangles

A right triangle has one right angle and two acute angles. The side opposite the right angle is called the hypotenuse. The other two sides (opposite of the acute angles) are called the legs. The hypotenuse is always the longest side of a right triangle because it is opposite the triangle's largest angle. The Pythagorean Theorem, $a^2 + b^2 = c^2$, relates the lengths of the sides of a right triangle, where c is the length of the hypotenuse, and a and b are the lengths of the legs.

There are two **special right triangles**: the 45°-45°-90° triangle (a right triangle with two equal legs, called an isosceles right triangle) and the 30°-60°-90° triangle. These two triangles are referenced often in trigonometry, so the relationship between the lengths of their sides should be memorized. For example, in a 30°-60°-90° triangle, the ratio of the short side (i.e., the side opposite the 30° angle) to the hypotenuse is always equal to 1/2, and in a 45°-45°-90° triangle, the ratio of either leg to the hypotenuse is always equal to $\frac{1}{\sqrt{2}}$.

IMPORTANT

If $a = 1$, then the side lengths of a 45°-45°-90° triangle are 1, 1, and $\sqrt{2}$, and the side lengths of a 30°-60°-90° triangle are 1, $\sqrt{3}$, and 2. Special right triangles with these specific side lengths will be referenced often in this chapter.

Special Right Triangles: 45°-45°-90° and 30°-60°-90°

The sides of each special right triangle are expressed in terms of a (a positive real number).

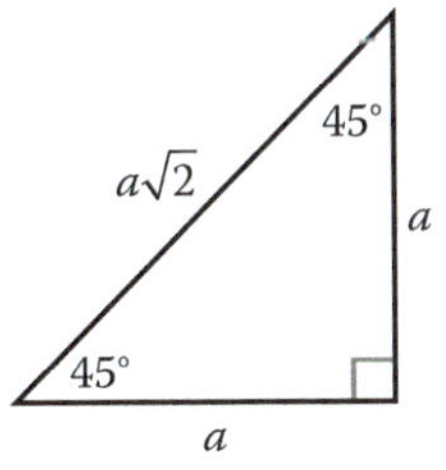

45°-45°-90° Triangle

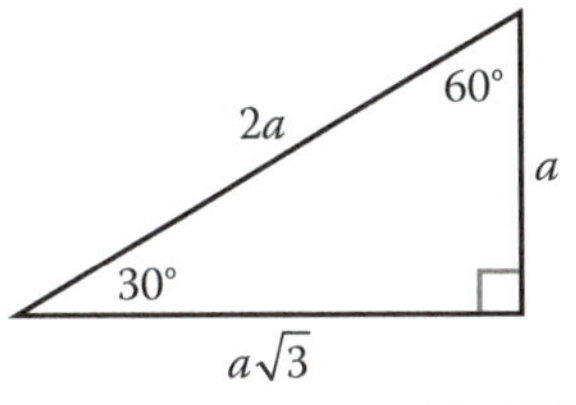

30°-60°-90° Triangle

Note that an equilateral triangle is divided into two 30°-60°-90° triangles by a perpendicular bisector (as shown by the dashed line in Figure 5.2a), which bisects both the angle and the opposite side. In Figure 5.2a, the equilateral triangle's sides have length 2. Therefore, each 30°-60°-90° triangle has side lengths 1, $\sqrt{3}$, and 2.

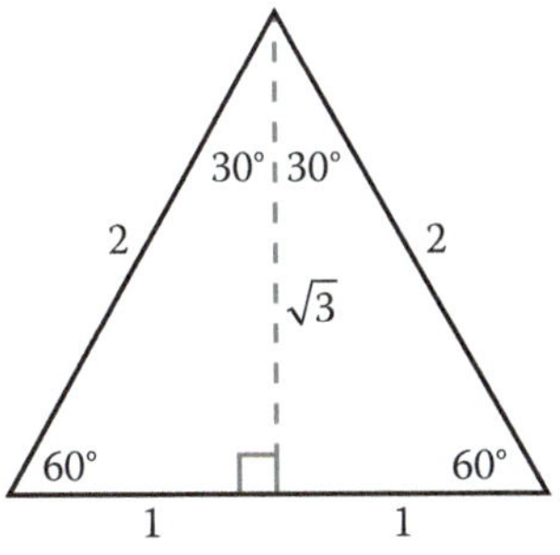

Figure 5.2a

An equilateral triangle is divided into two 30°-60°-90° triangles by a perpendicular bisector.

5.2.1 Introduction to the Trigonometric Functions

Consider a right triangle where one of the acute angles is identified as θ. Then the leg opposite θ is called the **opposite side** and the leg next to θ is called the **adjacent side**. Note that the opposite and the adjacent sides of a right triangle are identified as such only in relation to angle θ (Figures 2.2b and 2.2c).

TIP

The opposite and adjacent sides of acute angle θ are always legs (i.e., they are never the hypotenuse).

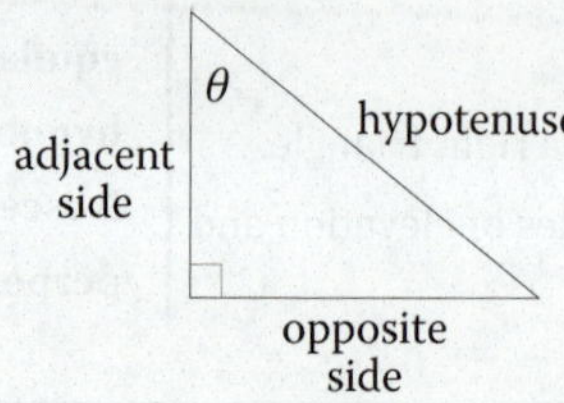

Figure 2.2b

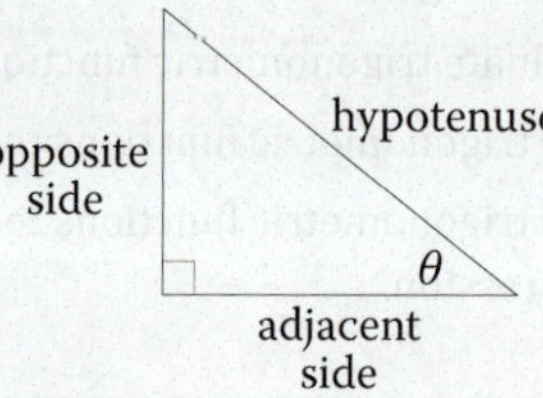

Figure 2.2c

The six **trigonometric functions** of θ are **sine θ**, **cosine θ**, **tangent θ**, **cosecant θ**, **secant θ**, and **cotangent θ**, abbreviated as sin θ, cos θ, tan θ, csc θ, sec θ, and cot θ, respectively. If θ is an acute angle, then each of the trigonometric functions of θ is a ratio of the lengths of two sides of a right triangle, as defined below. Again, θ is an acute angle, so $0° < \theta < 90°$.

The Trigonometric Functions of Acute Angle θ

$$\text{sine } \theta = \sin\theta = \frac{\text{opp}}{\text{hyp}} \qquad \text{cosecant } \theta = \csc\theta = \frac{\text{hyp}}{\text{opp}}$$

$$\text{cosine } \theta = \cos\theta = \frac{\text{adj}}{\text{hyp}} \qquad \text{secant } \theta = \sec\theta = \frac{\text{hyp}}{\text{adj}}$$

$$\text{tangent } \theta = \tan\theta = \frac{\text{opp}}{\text{adj}} \qquad \text{cotangent } \theta = \cot\theta = \frac{\text{adj}}{\text{opp}}$$

IMPORTANT

The abbreviations opp, adj, and hyp will be used to represent the lengths of the three sides of a right triangle.
opp = side opposite θ
adj = side adjacent to θ
hyp = hypotenuse

The ratio for the cosecant of θ is the reciprocal of the ratio for the sine of θ. Furthermore, the ratios for cosine and secant of θ are reciprocals, and the ratios for the tangent and cotangent of θ are reciprocals. The cosecant, secant, and cotangent functions are called the **reciprocal trigonometric functions**.

The reciprocal relationship can be used to evaluate a trigonometric function. For example, if $\sin\theta = 1/2$, then $\csc\theta = 2$, because sine and cosecant are reciprocal functions, and 2 is the reciprocal of 1/2.

The Value of the Sine or Cosine of an Acute Angle θ

Since the length of the hypotenuse is always greater than the length of the side opposite of or adjacent to an acute angle θ in a right triangle, the ratio opp/hyp must always be a value between 0 and 1. For the same reason, the ratio adj/hyp must also be a value between 0 and 1. Therefore, the value of the sine or cosine of an acute angle θ will always be between 0 and 1.

Additionally, since cosecant is the reciprocal of sine, and $0 < \sin\theta < 1$ (where θ is an acute angle), the value of the cosecant of an acute angle θ will always be greater than 1. The same is true for the value of the secant of an acute angle θ.

EXAMPLE 1 Evaluating the Trigonometric Functions of an Unknown Angle

Find the sine, cosine, tangent, cosecant, secant, and cotangent of θ in each triangle.

A.

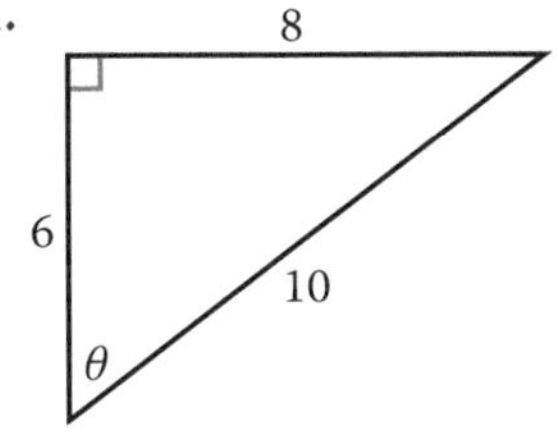

B.

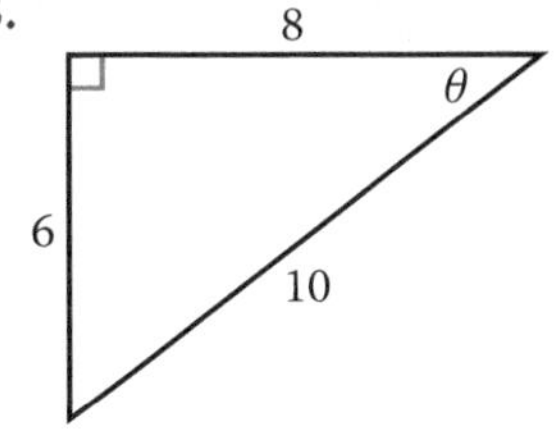

C.

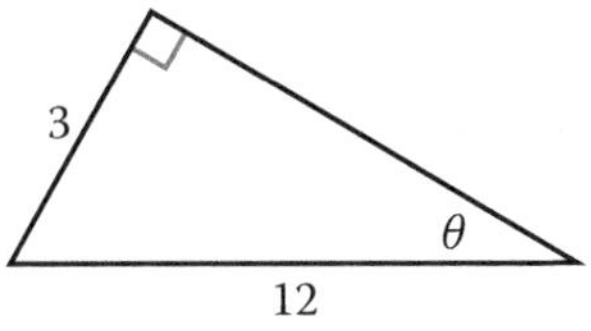

SOLUTION

A. The hypotenuse is 10, the side opposite θ is 8, and the side adjacent to θ is 6.

$$\sin\theta = \frac{\text{opp}}{\text{hyp}} = \frac{8}{10} = \frac{4}{5} \qquad \cos\theta = \frac{\text{adj}}{\text{hyp}} = \frac{6}{10} = \frac{3}{5} \qquad \tan\theta = \frac{\text{opp}}{\text{adj}} = \frac{8}{6} = \frac{4}{3}$$

$$\csc\theta = \frac{\text{hyp}}{\text{opp}} = \frac{10}{8} = \frac{5}{4} \qquad \sec\theta = \frac{\text{hyp}}{\text{adj}} = \frac{10}{6} = \frac{5}{3} \qquad \cot\theta = \frac{\text{adj}}{\text{opp}} = \frac{6}{8} = \frac{3}{4}$$

NOTICE THAT

Since cosecant is the reciprocal of sine, $\csc\theta$ can be determined by just finding the reciprocal of $\sin\theta$. Similarly, $\cos\theta$ and $\sec\theta$ are reciprocals, as well as $\tan\theta$ and $\cot\theta$.

B. The triangles given in **A** and **B** are the same, but the angle identified as θ is different. Therefore, the hypotenuse is again 10, but here the side opposite θ is 6, and the side adjacent to θ is 8.

$$\sin\theta = \frac{\text{opp}}{\text{hyp}} = \frac{6}{10} = \frac{3}{5} \qquad \cos\theta = \frac{\text{adj}}{\text{hyp}} = \frac{8}{10} = \frac{4}{5} \qquad \tan\theta = \frac{\text{opp}}{\text{adj}} = \frac{6}{8} = \frac{3}{4}$$

$$\csc\theta = \frac{\text{hyp}}{\text{opp}} = \frac{10}{6} = \frac{5}{3} \qquad \sec\theta = \frac{\text{hyp}}{\text{adj}} = \frac{10}{8} = \frac{5}{4} \qquad \cot\theta = \frac{\text{adj}}{\text{opp}} = \frac{8}{6} = \frac{4}{3}$$

Pythagorean Theorem $a^2 + b^2 = c^2$

C. Let b represent the side adjacent to θ. Use the Pythagorean Theorem to find b.

$$b = \sqrt{12^2 - 3^2} = \sqrt{135} = 3\sqrt{15}$$

Thus, the side adjacent to θ is $3\sqrt{15}$. From the triangle, the hypotenuse is 12 and the side opposite θ is 3.

$$\sin\theta = \frac{\text{opp}}{\text{hyp}} = \frac{3}{12} = \frac{1}{4} \qquad \cos\theta = \frac{\text{adj}}{\text{hyp}} = \frac{3\sqrt{15}}{12} = \frac{\sqrt{15}}{4} \qquad \tan\theta = \frac{\text{opp}}{\text{adj}} = \frac{3}{3\sqrt{15}} = \frac{1}{\sqrt{15}}$$

Use the reciprocals of $\sin\theta$, $\cos\theta$, and $\tan\theta$ to find $\csc\theta$, $\sec\theta$, and $\cot\theta$.

$$\csc\theta = 4 \qquad \sec\theta = \frac{4}{\sqrt{15}} \qquad \cot\theta = \sqrt{15}$$

Rationalizing the Denominator

Recall from Section P.3 that when a fractional expression contains a radical in the denominator, we can rationalize the denominator, and thus write the fractional expression as an equivalent expression in which the denominator does not contain a radical. For example, the value of $\tan\theta$ from **Example 1C** can be rationalized and written equivalently as $\tan\theta = \frac{\sqrt{15}}{15}$ because $\tan\theta = \frac{1}{\sqrt{15}} = \frac{1}{\sqrt{15}} \cdot \frac{\sqrt{15}}{\sqrt{15}} = \frac{1 \cdot \sqrt{15}}{\sqrt{15} \cdot \sqrt{15}} = \frac{\sqrt{15}}{15}$. Since $\frac{1}{\sqrt{15}} = \frac{\sqrt{15}}{15}$, either answer is acceptable.

*From here forward, many denominators in **Examples** will remain irrational, but all answers to Section Exercises and Chapter Review Exercises will be given with a rationalized denominator.*

Using Similar Triangles to Evaluate Trigonometric Functions

If the measures of the angles in a triangle are equal to the measures of the corresponding angles in another triangle, then the two triangles are similar. When triangles are similar, the corresponding side lengths are proportional, and therefore, the ratio of any two sides from one triangle will always be equal to the ratio of the two corresponding sides from the other triangle.

TIP

$30° = \pi/6$ radians
$45° = \pi/4$ radians
$60° = \pi/3$ radians

The fact that corresponding sides of similar triangles are proportional can be very helpful when evaluating trigonometric functions. Since all 45°-45°-90° triangles are similar, a 45°-45°-90° triangle with side lengths 1, 1, and $\sqrt{2}$ can be used to evaluate the trigonometric functions of any 45° angle. Similarly, since all 30°-60°-90° triangles are similar, a 30°-60°-90° triangle with side lengths 1, $\sqrt{3}$, and 2 can be used to evaluate the trigonometric functions of any 30° angle or any 60° angle.

EXAMPLE 2 Evaluating the Trigonometric Functions of an Angle in a Special Right Triangle

TIP

The tangent of $\pi/3$ radians can be written as "tan ($\pi/3$)."

Find the sine, cosine, and tangent of $\pi/3$ radians.

SOLUTION

Recall that $\pi/3$ radians = 60° and a 60° angle is the angle opposite the longer leg in a 30°-60°-90° triangle. So, use the 60° angle in a 30°-60°-90° triangle to evaluate the trigonometric functions of $\pi/3$ radians.

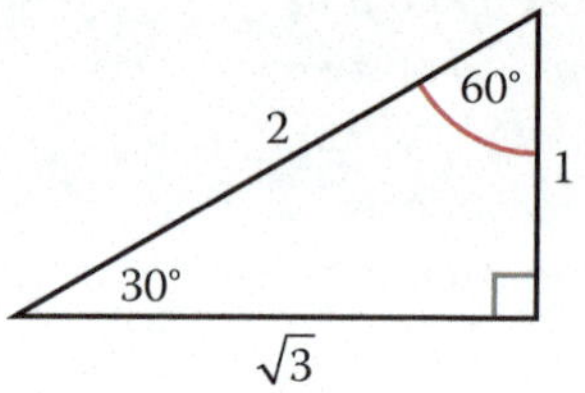

IMPORTANT

The 60° angle in any 30°-60°-90° triangle can be used to evaluate the trigonometric functions of $\pi/3$ radians.

From the 30°-60°-90° triangle, the hypotenuse is 2, the side opposite the 60° angle is $\sqrt{3}$, and the side adjacent to the 60° angle is 1.

$$\sin\frac{\pi}{3} = \frac{\text{opp}}{\text{hyp}} = \frac{\sqrt{3}}{2} \qquad \cos\frac{\pi}{3} = \frac{\text{adj}}{\text{hyp}} = \frac{1}{2} \qquad \tan\frac{\pi}{3} = \frac{\text{opp}}{\text{adj}} = \frac{\sqrt{3}}{1} = \sqrt{3}$$

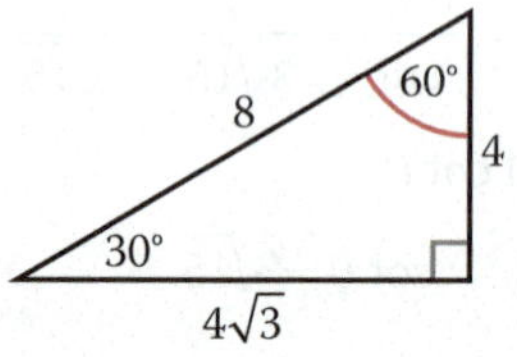

Figure 5.2d

To confirm that the 60° angle in any 30°-60°-90° triangle can be used to evaluate the trigonometric functions of $\pi/3$ radians, evaluate the functions again using a 30°-60°-90° triangle where the side lengths are something other than 1, 2, and $\sqrt{3}$. For example, use a triangle with side lengths of 4, 8, and $4\sqrt{3}$, as shown in Figure 5.2d. In this case, the hypotenuse is 8, the side opposite the 60° angle is $4\sqrt{3}$, and the side adjacent to the 60° angle is 4.

$$\sin\frac{\pi}{3} = \frac{\text{opp}}{\text{hyp}} = \frac{4\sqrt{3}}{8} = \frac{\sqrt{3}}{2} \checkmark \qquad \cos\frac{\pi}{3} = \frac{\text{adj}}{\text{hyp}} = \frac{4}{8} = \frac{1}{2} \checkmark \qquad \tan\frac{\pi}{3} = \frac{\text{opp}}{\text{adj}} = \frac{4\sqrt{3}}{4} = \sqrt{3} \checkmark$$

5.2.2 Evaluating Trigonometric Functions

Knowing the value of one trigonometric function of an acute angle θ is sufficient for evaluating the rest of the trigonometric functions. For example, suppose θ is an acute angle such that $\cos\theta = 5/13$. Then θ can be drawn as an acute angle in a right triangle where the ratio of the side adjacent to θ to the hypotenuse is 5/13 (since $\cos\theta$ = adj/hyp). If the Pythagorean Theorem is used to find the length of the side opposite θ in this right triangle, then the rest of the trigonometric functions can be evaluated, as demonstrated in **Example 3**.

EXAMPLE 3 Using the Value of One Trigonometric Function to Evaluate the Rest of the Trigonometric Functions

Suppose that θ is an acute angle such that $\tan\theta = 1/3$. Use a right triangle to find $\sin\theta$ and $\sec\theta$.

SOLUTION

Draw a right triangle where θ is an acute angle, the side opposite θ is 1, and the side adjacent to θ is 3 ($\tan\theta = 1/3 = \text{opp/adj}$).

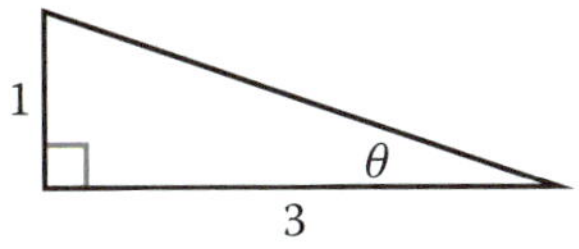

Use the Pythagorean Theorem to find the hypotenuse c of this right triangle.

$$c = \sqrt{1^2 + 3^2} = \sqrt{10} \qquad \sin\theta = \frac{\text{opp}}{\text{hyp}} = \frac{1}{\sqrt{10}} \qquad \sec\theta = \frac{\text{hyp}}{\text{adj}} = \frac{\sqrt{10}}{3}$$

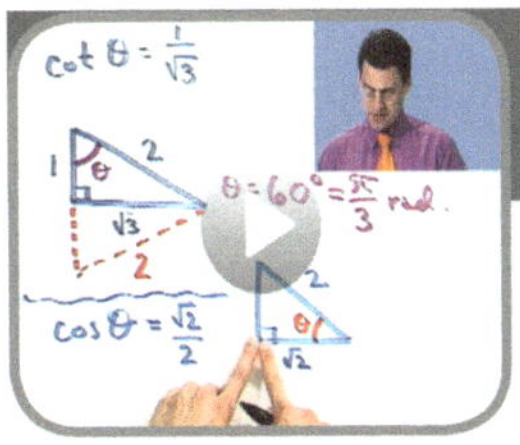

5.2.3 Finding an Angle Measure in a Special Right Triangle Using the Value of a Trigonometric Function

In **Example 3**, the value of a given trigonometric function of an acute angle θ and a right triangle were used to find the values of other trigonometric functions of θ. In that example, the value of θ could not be determined using only the given information. However, if the value of a given trigonometric function of an acute angle θ can be used to form a *special* right triangle, then the value of the angle θ can be determined. In this case, we must recognize the right triangle formed from the given information as a special right triangle.

EXAMPLE 4 Using a Special Right Triangle to Find the Measure of an Acute Angle

Suppose that θ is an acute angle such that $\csc\theta = \dfrac{8\sqrt{128}}{64}$.

Use a right triangle to find the measure of θ in radians and in degrees.

SOLUTION

Simplify. $\csc\theta = \dfrac{8\sqrt{128}}{64} = \dfrac{64\sqrt{2}}{64} = \sqrt{2}$

Draw a right triangle where θ is an acute angle. Since $\csc\theta = \text{hyp/opp}$, the hypotenuse is $\sqrt{2}$ and the side opposite θ is 1.

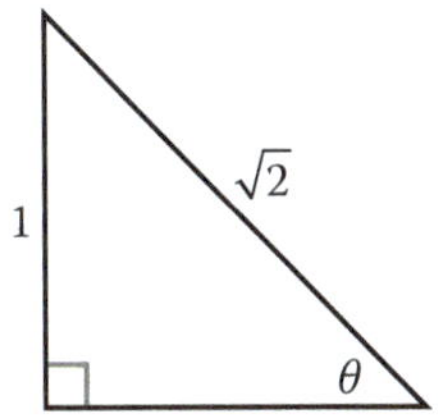

You should recognize that this is a 45°-45°-90° triangle. Use the Pythagorean Theorem to confirm that this is an isosceles right triangle. Let b represent the unknown leg. $b = \sqrt{\sqrt{2}^2 - 1^2} = \sqrt{2-1} = 1$

Since both legs have length 1, the triangle is a 45°-45°-90° triangle.

Therefore, when θ is an acute angle such that $\csc\theta = \sqrt{2}$, $\theta = 45° = \pi/4$ radians.

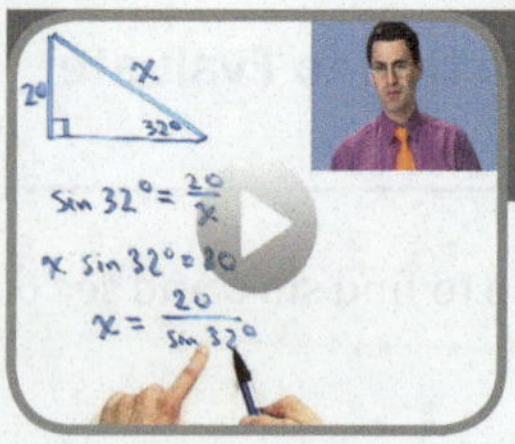

5.2.4 Using Trigonometric Functions to Find Unknown Sides of Right Triangles

The value of a trigonometric function of a given angle (in degrees or radians) can be determined using a calculator, as demonstrated in **Example 5**. The calculator must be set to degree mode or radian mode, depending on whether the given angle is measured in degrees or radians.

EXAMPLE 5 Using a Calculator to Evaluate a Trigonometric Function of a Given Angle Measured in Degrees or Radians

> **TIP**
>
> *The default setting of a calculator is typically radian mode.*

Use a calculator to evaluate each trigonometric function.

A. tan 50° **B.** cos (π/7)

SOLUTION

The expressions can be entered directly as "tan 50" and "cos (π/7)", followed by **ENTER** when using a graphing calculator. However, when using some scientific calculators, the angle measure must be entered first. The following directions are for this type of scientific calculator.

A. To calculate tan 50°, set your calculator to degree mode, enter 50, then press **TAN**.

$$\tan 50° = 1.1917535\ldots$$

> **IMPORTANT**
>
> *When an angle measure does not include a degree symbol, assume that the measure is in radians.*

B. To calculate cos (π/7), set your calculator to radian mode, then use the keystrokes **π**, **/**, **7**, **ENTER**, and then press **COS**.

$$\cos(\pi/7) = 0.9009688\ldots$$

When the length of one side and the measure of one acute angle in a right triangle are known, the length of either of the other two sides can be found by solving an equation that involves a trigonometric function.

EXAMPLE 6 Finding the Length of the Sides of a Right Triangle Given One Acute Angle Measure and One Side Length

> **IMPORTANT**
>
> *Be sure to set your calculator to radian mode or degree mode as needed.*

Use the given information to find the length of the indicated unknown side of each right triangle. Round to the nearest hundredth as needed.

A. Find the length of the side opposite a 52° angle if its adjacent side has length 14.

B. Find the length of the side adjacent to an angle measuring $2\pi/5$ radians if the hypotenuse is 3.7.

SOLUTION

A. Draw the right triangle.

Write an equation using one of the trigonometric functions. Since the equation should include the sides opposite and adjacent to the known angle, use $\tan\theta = \dfrac{\text{opp}}{\text{adj}}$.

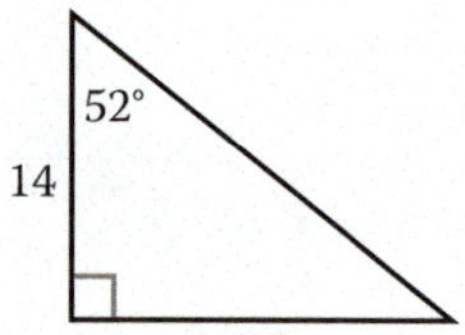

$$\tan 52° = \frac{\text{opp}}{14} \quad \Rightarrow \quad \text{opp} = 14 \tan 52° \approx 17.92$$

Multiply both sides by 14.
Use a calculator to evaluate tan 52°.
Round to the nearest hundredth.

B. Draw the right triangle.

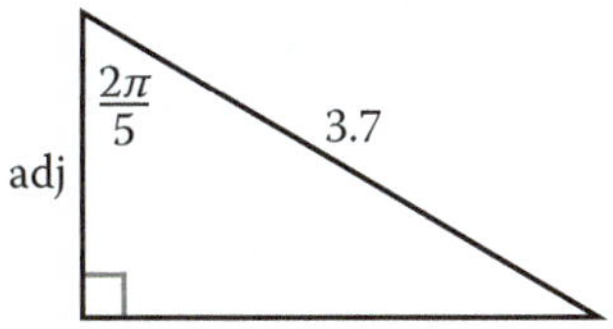

Write an equation using one of the trigonometric functions. Since the equation should include the side adjacent to the known angle and the hypotenuse, use $\cos\theta = \frac{\text{adj}}{\text{hyp}}$.

$$\cos\frac{2\pi}{5} = \frac{\text{adj}}{3.7} \quad\Rightarrow\quad \text{adj} = 3.7\cos\frac{2\pi}{5} \approx 1.14$$

Multiply both sides by 3.7.
Use a calculator to evaluate cos 2π/5.
Round to the nearest hundredth.

ALTERNATIVE METHOD

In **Example 6A**, the cotangent function $\cot\theta = \text{adj}/\text{opp}$ can be used to write the equation because, like tangent, cotangent relates the sides opposite of and adjacent to an angle.

Similarly, in **Example 6B**, the secant function $\sec\theta = \text{hyp}/\text{adj}$ can used to write the equation because, like cosine, secant relates the hypotenuse and the side adjacent to an angle.

5.2.5 Using Trigonometric Functions to Determine a Length

EXAMPLE 7 Using a Trigonometric Function to Determine a Length

A 27-foot support wire attached to the top of a pole makes a 65° angle with the ground. Find the height of the pole to the nearest hundredth of a foot.

SOLUTION

The pole forms a right angle with the ground. The wire and the pole are therefore two sides of a right triangle, where the length of the wire is the hypotenuse and the height of the pole is the side opposite the 65° angle.

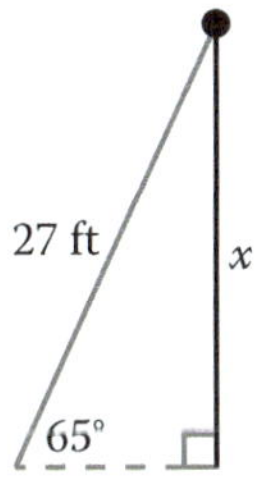

Let x be the height of the pole.

Since the hypotenuse is known and the side opposite the known angle is x, use $\sin\theta = \frac{\text{opp}}{\text{hyp}}$ to write an equation.

$$\sin 65° = \frac{x}{27} \quad\Rightarrow\quad x = 27\sin 65°$$

Multiply both sides by 27 to solve for x.

Use a calculator to evaluate sin 65°.

$$x = 27\sin 65° \approx 24.47$$

Round to the nearest hundredth.

Therefore, the height of the pole is approximately 24.47 feet.

5.2.6 Using Trigonometric Functions with an Angle of Elevation or Depression

Angles of **elevation** or **depression** are formed with a horizontal line. The horizontal line is often the ground or a horizontal line at the eye level of an observer. An angle of elevation is measured *up* from this horizontal line and an angle of depression is measured *down* from this horizontal line (Figure 5.2e).

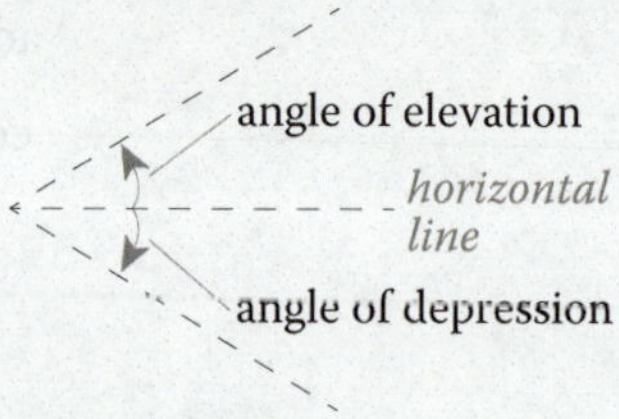

Figure 5.2e

EXAMPLE 8 Using a Trigonometric Function to Determine a Length

An observer stands on a platform that is 100 feet from a building. The angle of depression formed by the observer's line of sight to the base of the building is 13°, and the angle of elevation formed by the observer's line of sight to the top of the building is 51°. Find the height of the building to the nearest hundredth of a foot.

SOLUTION

The height of the building is divided into two parts. Let x be the upper part and let y be the lower part, as shown in the drawing.

Two right triangles can be formed. In each triangle, the side adjacent to the known acute angle is 100 feet, and the side opposite the known acute angle is part of the height of the building (i.e., x or y).

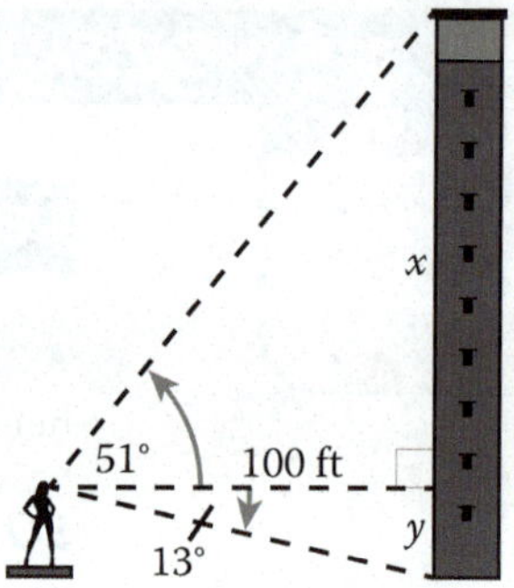

Since tangent relates the side opposite an acute angle to the side adjacent to an acute angle, use tangent to write two equations.

$$\tan 51° = \frac{x}{100} \quad \Rightarrow \quad x = 100 \tan 51°$$ *Multiply each side by 100 to solve for x.*

$$\tan 13° = \frac{y}{100} \quad \Rightarrow \quad y = 100 \tan 13°$$ *Multiply each side by 100 to solve for y.*

Therefore, the height of the building is approximately $100 \tan 51° + 100 \tan 13° \approx 146.58$ feet.

IMPORTANT

Set your calculator to degree mode. Alternatively, convert the degree measures to radians and use radian mode.

SECTION 5.2 EXERCISES

Warm Up

1. Classify each angle according to its measure: 150° 15′ 22″, 32°, 90°, 180°, and 47.325°.
2. Find the length of the hypotenuse of an isosceles right triangle if each leg has length 10 feet.
3. Find the height of an equilateral triangle if a side has length 24 centimeters.

Just the Facts

Fill in the blanks.

4. In any triangle, the longest side is opposite the _____ angle. In a right triangle, this side is called the _____, and it is opposite the _____ angle. The other two sides of a right triangle are called the _____.
5. In the two special right triangles, the 30°-60°-90° triangle has sides of length _____, _____, and _____. The 45°-45°-90° triangle has sides of length _____, _____, and _____.
6. When the lengths of the opposite and adjacent sides to angle θ are known, the trigonometric functions _____ or _____ may be used to find θ.
7. The values of sine and _____ of acute angle θ will always be between _____ and 1.
8. The reciprocal function of cosine is _____, of cotangent is _____, and of cosecant is _____.
9. When a radical is eliminated from the denominator of a fractional expression, that is called _____ the denominator.
10. The angle of _____ is always measured up from a horizontal line, and the angle of _____ is always measured _____ from a horizontal line.

Essential Skills

In Exercises 1–6, find the sine, cosine, tangent, cosecant, secant, and cotangent of θ.

1.
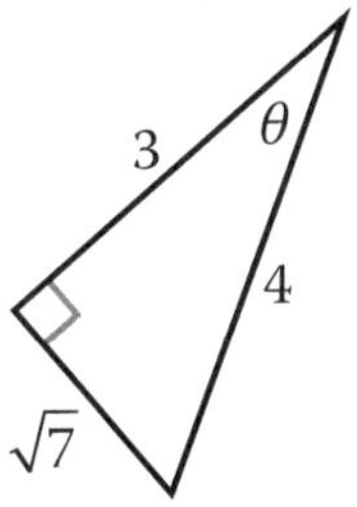

2.
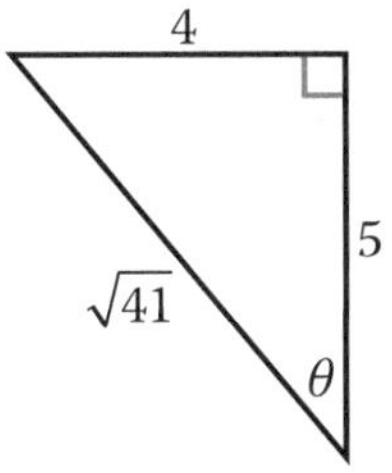

3.
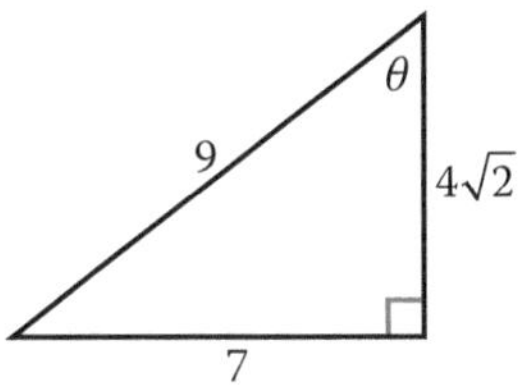

4.
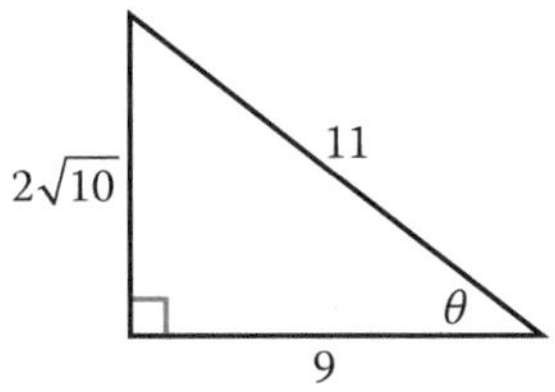

5.
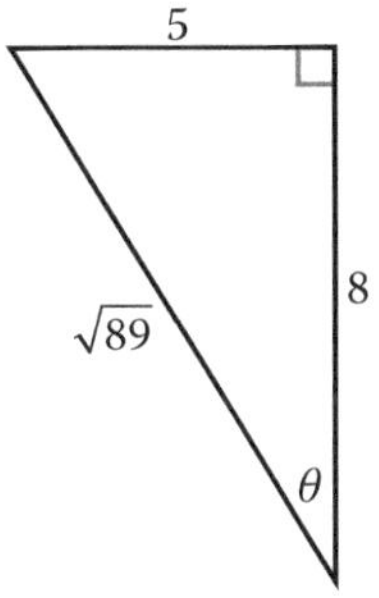

6.
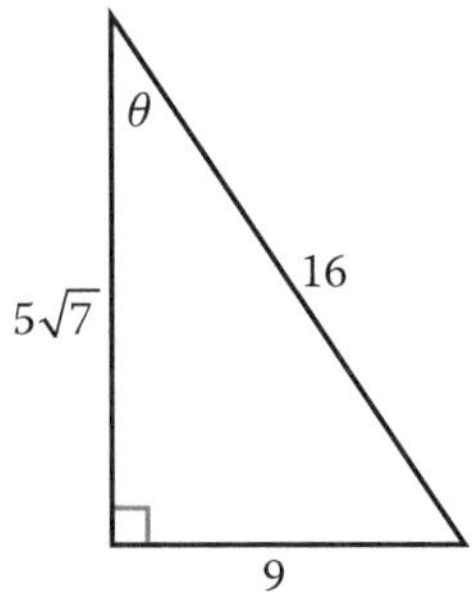

In Exercises 7–12, find the sine, cosine, and tangent of θ.

7.
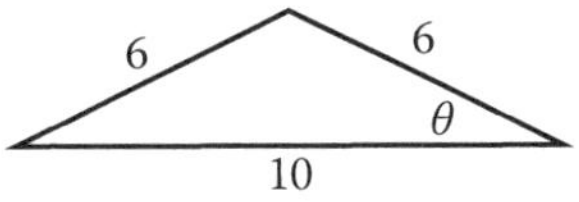

8.
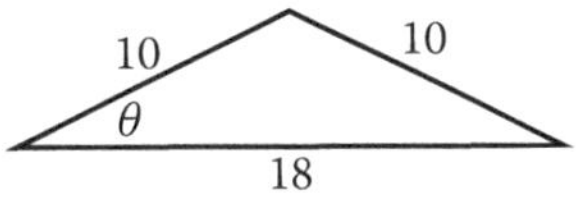

9.
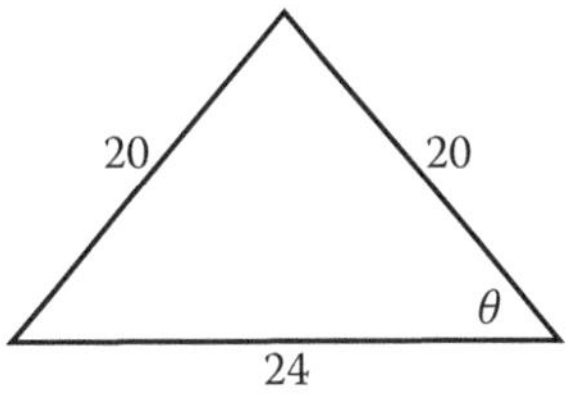

10.
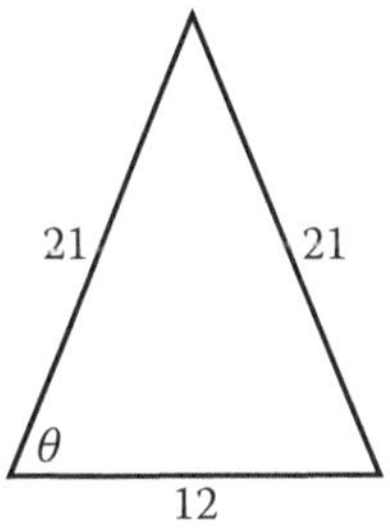

11.
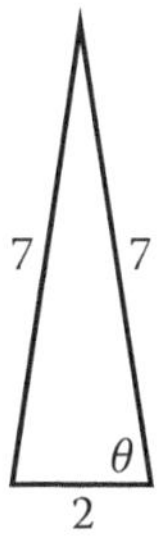

12.
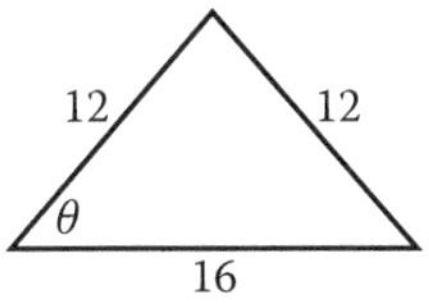

In Exercises 13–18, find the cosecant, secant, and cotangent of θ.

13.

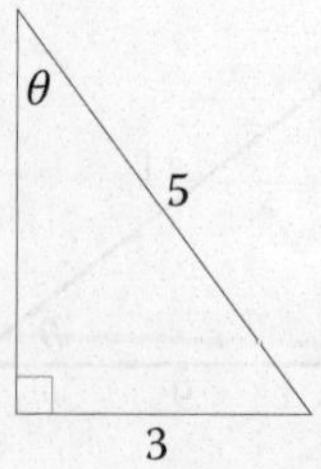

14.

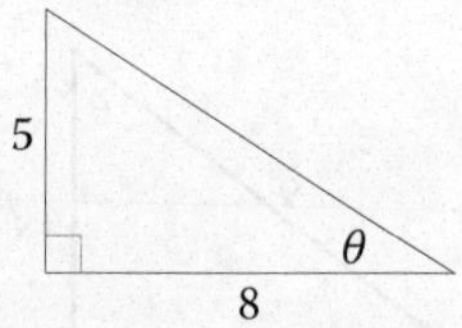

15.

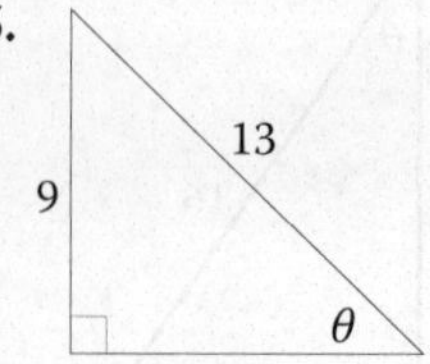

16.

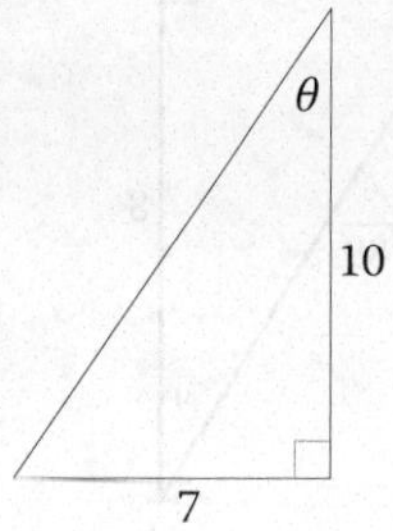

17.

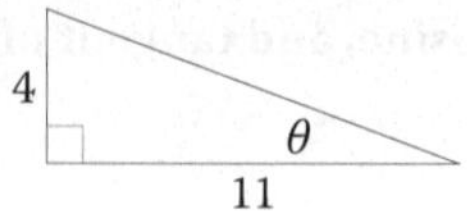

18.

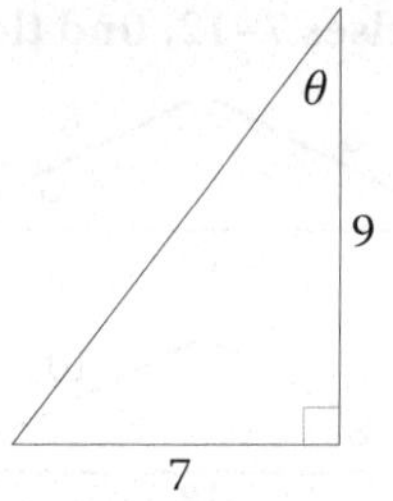

In Exercises 19–20, find the sine, cosine, tangent, cosecant, secant, and cotangent of each given angle.

19. 45°

20. $\pi/6$ radians

In Exercises 21–26, use a right triangle and the value of each trigonometric function to find all other trigonometric functions of acute angle θ.

21. $\cos\theta = 5/12$

22. $\sin\theta = 4/9$

23. $\cos\theta = \dfrac{9}{\sqrt{85}}$

24. $\tan\theta = \dfrac{4\sqrt{65}}{65}$

25. $\sec\theta = \dfrac{\sqrt{51}}{7}$

26. $\csc\theta = \dfrac{2\sqrt{55}}{11}$

In Exercises 27–34, use a right triangle to find the exact measure of each acute angle θ in radians and degrees.

27. $\cos\theta = \dfrac{\sqrt{3}}{2}$

28. $\sin\theta = \dfrac{\sqrt{12}}{4}$

29. $\csc\theta = \sqrt{2}$

30. $\cot\theta = \sqrt{3}$

31. $\cos\theta = 19/38$

32. $\sec\theta = \dfrac{12}{\sqrt{72}}$

33. $\csc\theta = \dfrac{8}{\sqrt{32}}$

34. $\tan\theta = \dfrac{\sqrt{108}}{18}$

In Exercises 35–40, use a calculator to evaluate each trigonometric function. Round to the nearest thousandth as needed.

35. $\sin(\pi/13)$ radians

36. tan 75°

37. csc 259°

38. tan 0.42 radians

39. sec 312.6°

40. cos 1.26 radians

In Exercises 41–52, round the unknown side of each right triangle to the nearest hundredth as needed.

41. Find the length of the side opposite a 20° angle if its hypotenuse has length 25.

42. Find the length of the side adjacent to a 38° angle if its opposite side has length 7.

43. Find the length of the side adjacent to a 75° angle if its hypotenuse has length 18.

44. Find the length of the side opposite a 35° angle if its hypotenuse has length 10.

45. Find the length of the side opposite a 52° angle if its adjacent side has length 32.

46. Find the length of the side adjacent to a 46° angle if its opposite side has length 13.

47. Find the length of the side opposite an angle measuring $\pi/5$ radians if its hypotenuse has length 14.

48. Find the length of the side opposite an angle measuring $2\pi/5$ radians if its hypotenuse has length 12.

49. Find the length of the side adjacent to an angle measuring $\pi/11$ radians if its opposite side has length 25.2.

50. Find the length of the side adjacent to an angle measuring $2\pi/21$ radians if its hypotenuse has length 13.6.

51. Find the length of the side opposite an angle measuring $\pi/25$ radians if its adjacent side has length 148.

52. Find the length of the side opposite an angle measuring $\pi/7$ radians if its adjacent side has length 15.74.

53. The angle of elevation from the foot of an observer to the basket of a hot air balloon is 30 degrees. Find the height of the balloon. Round to the nearest foot.

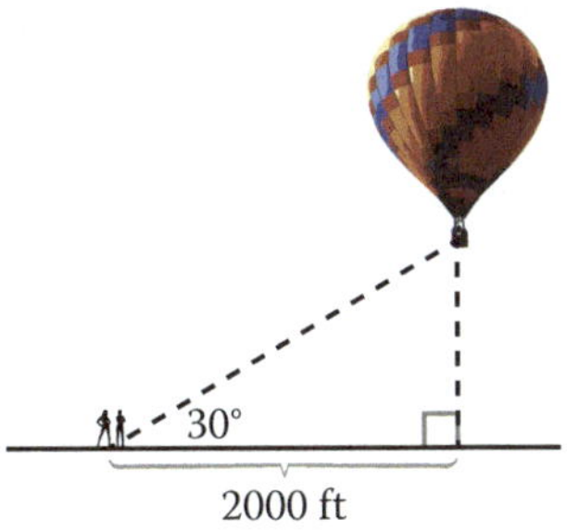

54. For safety reasons, the angle a ladder makes with the ground should be no greater than 70°. Find the length to the nearest foot of the shortest ladder that will safely reach a roof that is 25 feet high.

55. If the distance between a kicking tee and a goalpost is 63 yards and a football is kicked from this tee at an angle of 40°, how far to the nearest tenth of a yard will the football have traveled when it passes through the goalpost? Assume the effects of gravity are negligible and that the ball travels in a straight line.

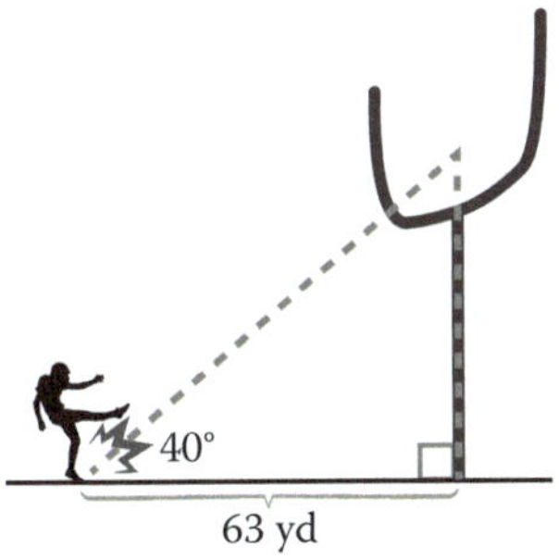

56. A 50-foot ladder is set against a wall so that the top of the ladder makes a 20° angle with the wall. How high up the wall to the nearest tenth of a foot does the ladder reach?

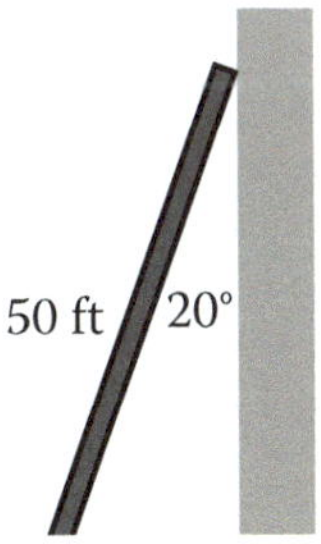

57. Find the distance to the nearest foot from the lighthouse to the boat.

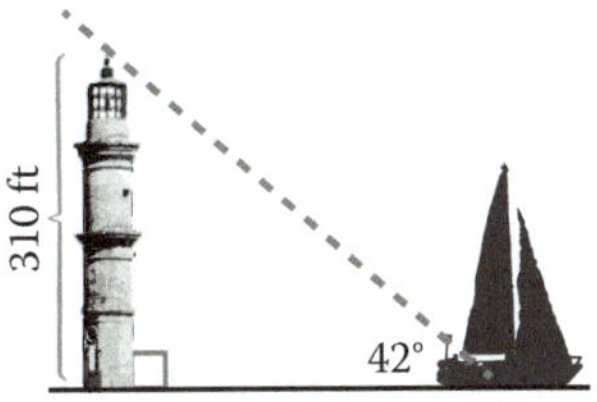

58. A hiker is hiking up a 14° slope. If she hikes at a constant rate of 2 miles per hour, how much altitude to the nearest hundredth of a mile does she gain in 4 hours of hiking?

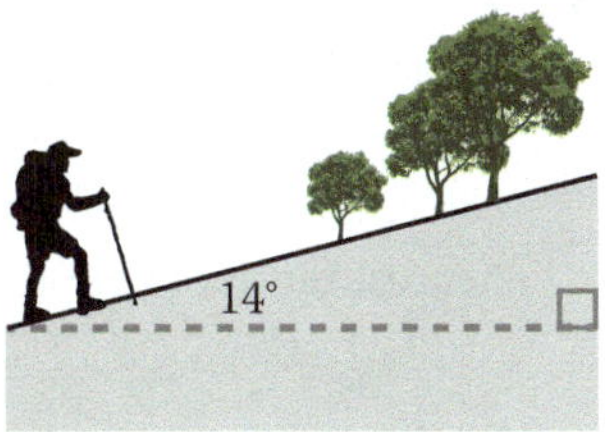

59. A 60-foot ladder is set against a wall so that the top of the ladder makes a 15° angle with the wall. How high up the wall does the ladder reach? Round to the nearest ten-thousandth of a foot.

60. A holiday tree is formed by extending single strands of lights from the top of a pole to an anchor point on the ground. Each strand's anchor point is on the circumference of a circle with radius 60 feet, where the bottom of the pole is the center of the circle. If the angle formed between the ground and a strand of lights is $7\pi/18$ radians, find the length of a single strand of lights to the nearest hundredth of a foot.

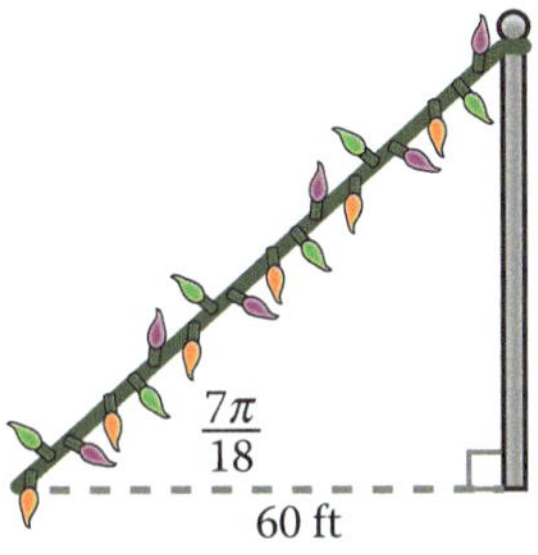

61. Find the height x of the indicated mountain to the nearest hundredth of a mile.

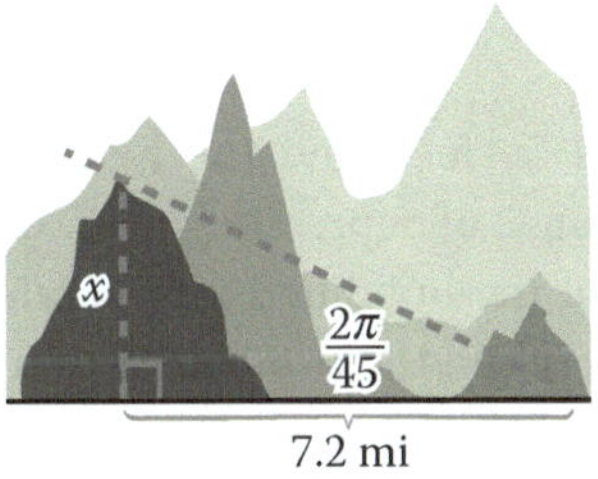

62. Find the height of the building to the nearest hundredth of a foot.

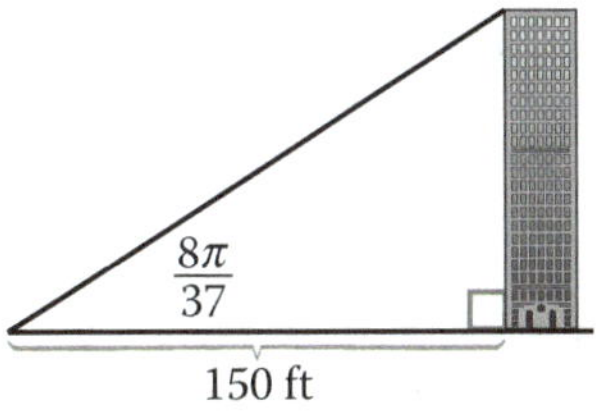

63. A moat surrounds a castle. A knight wants to extend a ladder from one side of the moat to a window 45 feet above ground level on the other side of the moat so he can rescue a princess. How long must the ladder be to the nearest tenth of a foot?

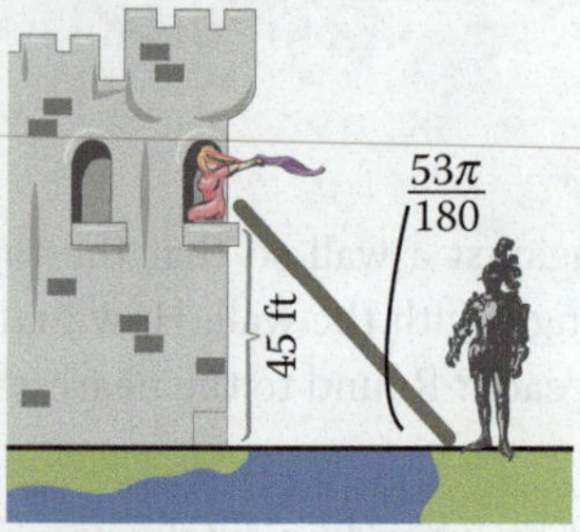

64. Find the height of the hot air balloon to the nearest tenth of a foot.

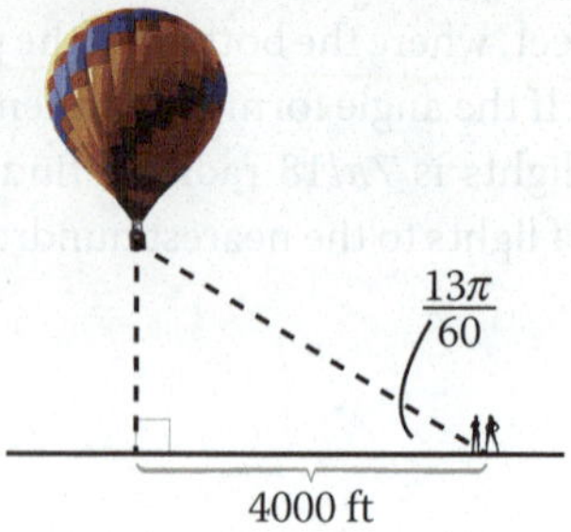

65. A telephone pole casts an 18-foot-long shadow. If the angle of elevation from the ground to the sun is 68°, what is the height of the pole? Round to the nearest hundredth of a foot.

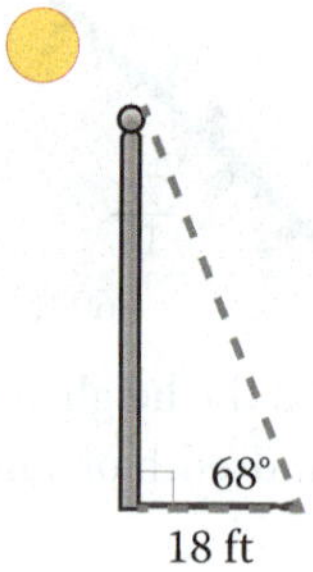

66. The height of a tower is 50 feet, and the height of a building is 20 feet. If the angle of depression from the top of the tower to the top of the building is 32.8°, what is the horizontal distance between the two buildings to the nearest foot?

67. A blimp is flying at an elevation of 6000 feet. The camera under the blimp has to swivel up 58° from the vertical to focus on the football stadium. What is the distance x from the blimp to the stadium to the nearest tenth of a foot?

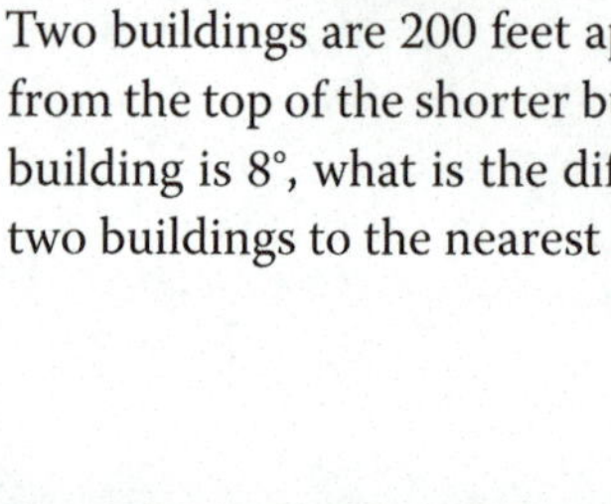

68. Two buildings are 200 feet apart. If the angle of elevation from the top of the shorter building to the top of the taller building is 8°, what is the difference in the heights of the two buildings to the nearest foot?

69. The height of a building is 40 feet, and the height of a tree is 20 feet. If the angle of depression from the top of the building to the top of the tree is 12.3°, what is the horizontal distance between the two to the nearest tenth of a foot?

70. A vertical pole is 15 feet high and the angle of elevation of the sun is 36.9°. What is the length of the pole's shadow rounded to the nearest tenth of a foot?

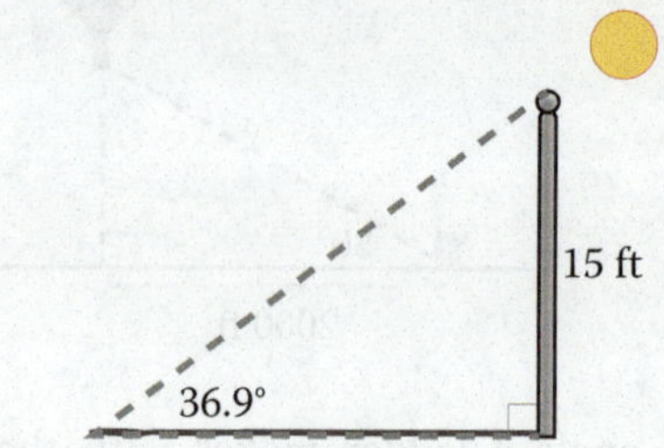

71. When observed from the top of a 250-foot-tall lighthouse, the angle of depression of an approaching ship is 50°. Identify the horizontal distance from the lighthouse to the ship rounded to the nearest foot.

72. A man standing on a lighthouse of height 162 feet sights two boats directly in front of him. One is at an angle of depression of 50°, and the other is at an angle of depression of 45°. Identify the distance between the two boats rounded to the nearest foot.

Extensions

73. Suppose that $\tan\theta = 4$. If the area of the square is 25 square inches, what is the area of the triangle shown in the picture?

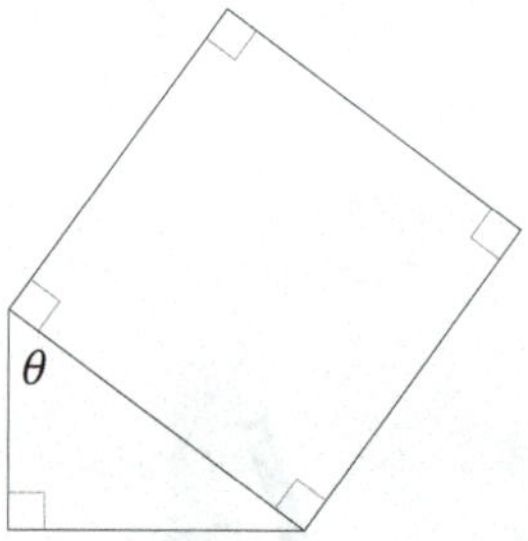

74. Suppose C is the center of the circle. What is the value of $\cos\theta$?

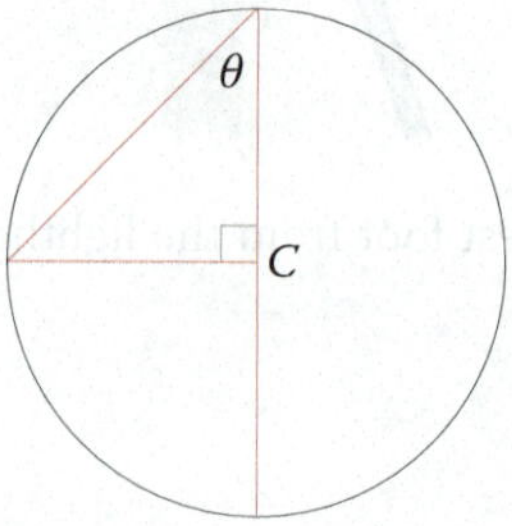

75. If $\tan \pi = \frac{2}{7}$, what is the area of the triangle?

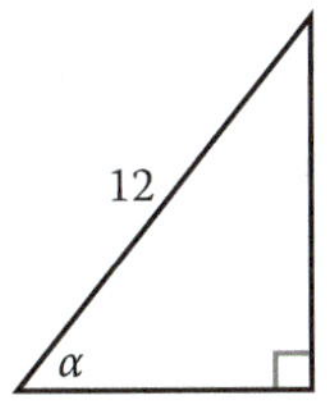

76. Suppose the square has a side length 2 and the hexagon is regular, having all equal sides and angles, what is the value of $\sin \theta$?

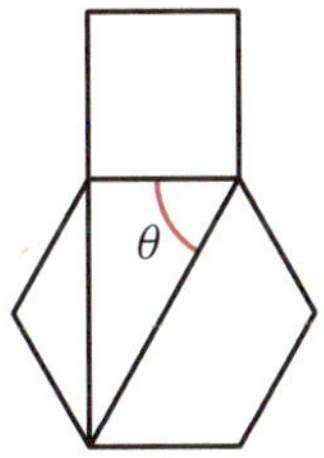

77. From a point on the North Rim of the Grand Canyon to a point on the South Rim, a surveyor measures an angle of depression of $\pi/179$ radians. From an aerial photograph, she determines that the horizontal distance is 20 miles. How many feet is the South Rim below the North Rim? Round to the nearest tenth of a foot.

78. Suppose θ is the angle between AF and the rectangular base $BCDF$. What is the value of $\cos \theta$?

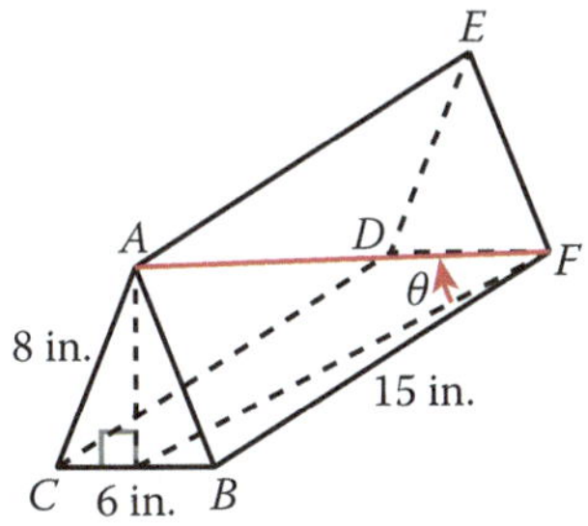

79. Suppose that $\tan \theta = \frac{3421}{\sqrt{11703241}}$. What is the value of θ in degrees? Assume that θ is an acute angle in a right triangle.

80. Given that $\sin 72° \approx 0.95$, what is the value of x?

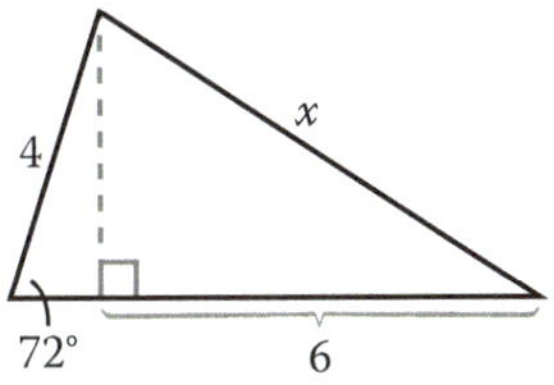

81. An isosceles triangle has a base of length 5. The angle opposite the base measures $\pi/5$ radians. What are the lengths of the two equal sides? Round to the nearest tenth.

82. A person is walking towards a billboard sign directly in front of him. The angle of elevation from his position to the top of the billboard is 10.7°. After he walks 10 more feet, the angle of elevation increases by 5°. Approximate the height of the billboard sign to the nearest tenth of a foot.

83. A surveillance camera is mounted in a pawn shop on a ceiling that is 12 feet high. Is an angle of depression of 12.3° sufficient if the camera needs to focus on the cash register, which is 17 feet from the camera and 4 feet above the floor?

84. A bird is fluttering 30 feet above the ground. At an angle of depression of 35°, it notices a cockroach. At that moment, what is the shortest distance to the nearest tenth of a foot the bird must travel to reach the cockroach?

85. A neap tide occurs when the positions of the sun, moon, and Earth form a right triangle in which Earth is at the vertex of the right angle and the moon is at the vertex of an angle that measures approximately 89.8528 degrees. Find the distance to the nearest kilometer from Earth to the sun at neap tide, given that the distance from the moon to Earth at this time is 381,550 kilometers.

86. Use the triangle to prove that $(\tan \theta)^2 + 1 = (\sec \theta)^2$.

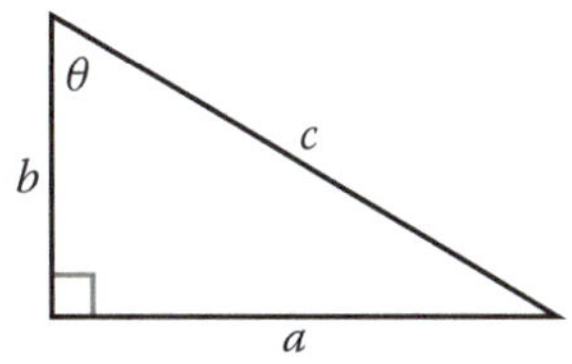

87. A submarine at the surface of the Atlantic Ocean makes an emergency dive at an angle of 39° 10′ 25″ with the surface of the water. How deep will it be if it goes for 700 meters along its path? What is the submarine's horizontal distance from its starting point? How much farther does it need to dive along its path to reach a depth of 1200 meters? Round all answers to the nearest tenth of a meter.

88. An observer 6 kilometers away from the launching pad observes a space shuttle ascend at an angle of elevation of 24° 5′. After 6 seconds, the angle has increased by 12° 10′. How far did the space shuttle travel to the nearest hundredth of a kilometer during the 6-second interval? What was its average speed during that interval?

89. Commercial airliners typically fly at an altitude of 10.9728 kilometers. They start descending toward the airport when they are far away so that they will not have to dive at a steep angle. If the pilot wants the plane's path to make an angle of 2° 15′ 39″ with the ground, what horizontal distance to the nearest tenth of a kilometer from the airport must she start descending?

90. A beam of gamma rays is used to treat a tumor that is 4.6 centimeters beneath the patient's skin. To avoid damaging a vital organ, the radiologist aims the gamma ray source at an angle of depression of 39° 1′ 2″. What is the distance to the nearest hundredth of a centimeter that the gamma ray has to travel to reach the tumor, if the machine is 11 centimeters above the patient?

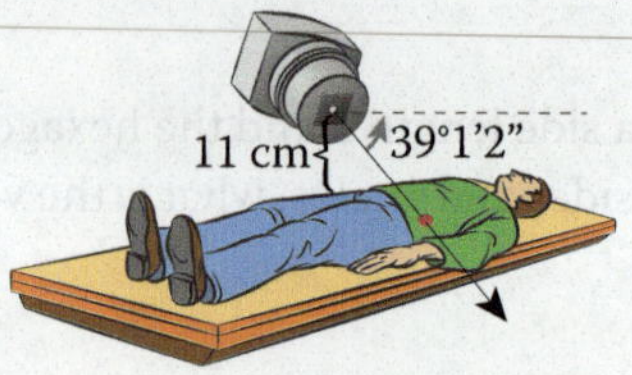

5.3 THE TRIGONOMETRIC FUNCTIONS IN THE COORDINATE PLANE

OBJECTIVES

- Evaluate the trigonometric functions of an angle in standard position in a coordinate plane.
- Use a reference angle to evaluate the trigonometric functions.
- Use the value of a given trigonometric function and the sign of another to evaluate other trigonometric functions.
- Evaluate and memorize the trigonometric functions of important angles.

PREREQUISITE VOCABULARY TERMS

coordinate plane
degrees
radians
special right triangles
standard position
trigonometric functions

5.3.1 Evaluating Trigonometric Functions of an Angle in the Coordinate Plane

In Section 5.2, the trigonometric functions of an *acute* angle θ ($0° < \theta < 90°$) were evaluated using a right triangle (Figure 5.3a). In this section, the trigonometric functions of angles of *any measure* will be evaluated by using a related right triangle graphed on a coordinate plane (Figure 5.3b).

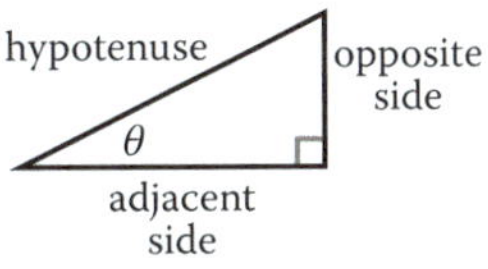

Figures 5.3a

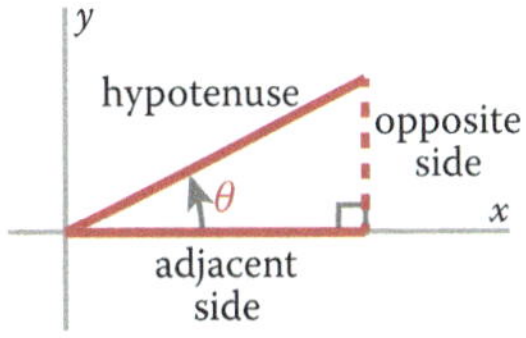

Figure 5.3b

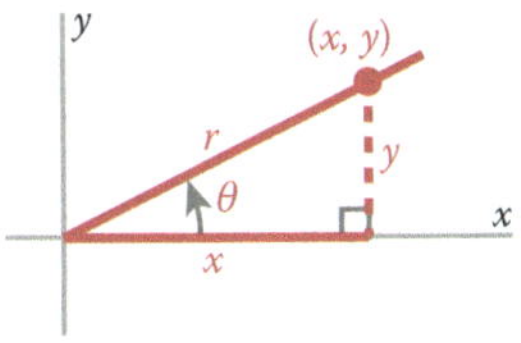

Figure 5.3c

Suppose (x, y) is a point on the terminal side of an angle θ in standard position. A right triangle can be formed with the x-axis where the legs have lengths $|x|$ and $|y|$. The hypotenuse r is given by the Pythagorean Theorem, $r = \sqrt{x^2 + y^2}$ such that $r > 0$ (Figure 5.3c). The signed side lengths of this triangle (x, y, and r) can be used to evaluate the trigonometric functions of θ. For example, since $\sin\theta = \text{opp/hyp}$, using the right triangle in Figure 5.3c gives since $\sin\theta = \text{opp/hyp} = y/r$. The definitions of the trigonometric functions in terms of x, y, and r are given in the following table.

IMPORTANT

*These definitions of the trigonometric functions can be used for any angle θ. However, θ is an angle in the triangle with sides x, y, and r **only** when $0° < \theta < 90°$ (i.e., when (x, y) is in quadrant I).*

The Trigonometric Functions of Angle θ

If (x, y) is a point on the terminal side of an angle θ in standard position and r is the distance from the origin to (x, y) such that $r = \sqrt{x^2 + y^2} \neq 0$, then the trigonometric functions of θ can be defined in terms of x, y, and r.

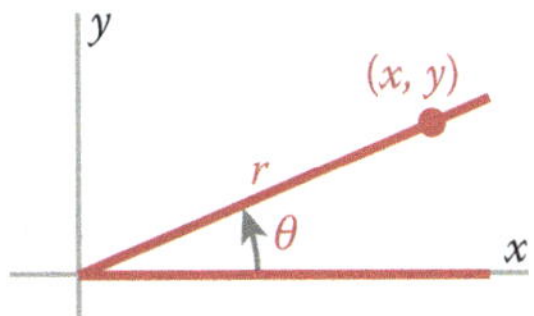

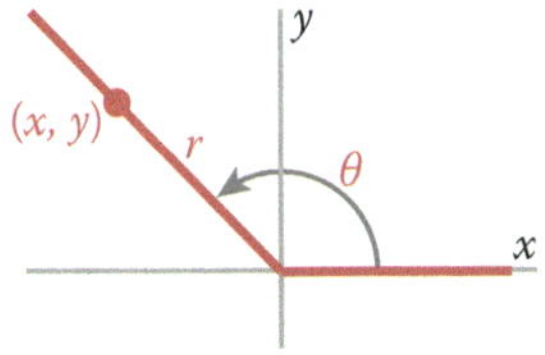

$\sin\theta = \dfrac{y}{r}$ $\qquad \cos\theta = \dfrac{x}{r}$ $\qquad \tan\theta = \dfrac{y}{x},\ x \neq 0$

$\csc\theta = \dfrac{r}{y},\ y \neq 0$ $\qquad \sec\theta = \dfrac{r}{x},\ x \neq 0$ $\qquad \cot\theta = \dfrac{x}{y},\ y \neq 0$

EXAMPLE 1 Evaluating Trigonometric Functions Given a Point on the Terminal Side of θ in Quadrant I

Find the sine, cosine, tangent, cosecant, secant, and cotangent of θ, given that θ is an angle in standard position and (6, 8) is a point on the terminal side of θ.

SOLUTION

Sketch angle θ in standard position with its terminal side passing through the point (6, 8). Then draw a line perpendicular to the x-axis through the point to form a right triangle where r is the hypotenuse.

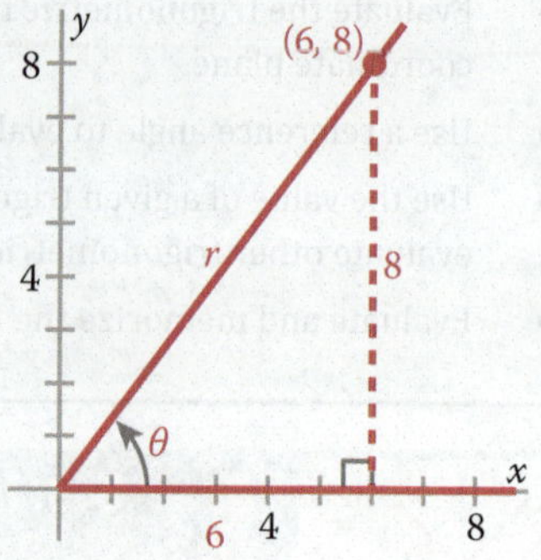

Since the terminal side of θ passes through (6, 8), $x = 6$, $y = 8$, and $r = \sqrt{6^2 + 8^2} = \sqrt{100} = 10$.

Use the definitions of the trigonometric functions of θ where $x = 6$, $y = 8$, and $r = 10$ to evaluate each function.

$$\sin\theta = \frac{y}{r} = \frac{8}{10} = \frac{4}{5} \qquad \cos\theta = \frac{x}{r} = \frac{6}{10} = \frac{3}{5} \qquad \tan\theta = \frac{y}{x} = \frac{8}{6} = \frac{4}{3}$$

$$\csc\theta = \frac{r}{y} = \frac{10}{8} = \frac{5}{4} \qquad \sec\theta = \frac{r}{x} = \frac{10}{6} = \frac{5}{3} \qquad \cot\theta = \frac{x}{y} = \frac{6}{8} = \frac{3}{4}$$

NOTICE THAT

The point on the terminal side of θ is in quadrant I, and the value of each trigonometric function of θ is positive.

When the Point on the Terminal Side of θ Is Not in Quadrant I

We have seen that if a point (x, y) on the terminal side of angle θ is in quadrant I, then the values of both x and y are positive, which means that the values of all six trigonometric functions of θ are positive. However, if a point (x, y) on the terminal side of angle θ is in quadrant II, III, or IV, then the values of some of the trigonometric functions of θ are negative because at least one of x and y is negative (depending on the quadrant of the terminal side). For example, if (x, y) is in quadrant II, then $\tan\theta < 0$ because $x < 0$ and $y > 0$ in quadrant II and $\tan\theta = y/x$.

NOTICE THAT

If a point (x, y) on the terminal side of θ is in quadrant II, then $x < 0$, $y > 0$, and $r > 0$ (r is always positive). Thus, $\sin\theta = y/r > 0$, $\cos\theta = x/r < 0$, and $\tan\theta = y/x < 0$.

EXAMPLE 2 Evaluating Trigonometric Functions Given a Point on the Terminal Side of θ in Quadrant II

Find the sine, cosine, tangent, cosecant, secant, and cotangent of θ, given that θ is an angle in standard position and (−4, 2) is a point on the terminal side of θ.

SOLUTION

Since the terminal side of θ passes through (−4, 2), $x = -4$, $y = 2$, and $r = \sqrt{(-4)^2 + 2^2} = \sqrt{20} = 2\sqrt{5}$.

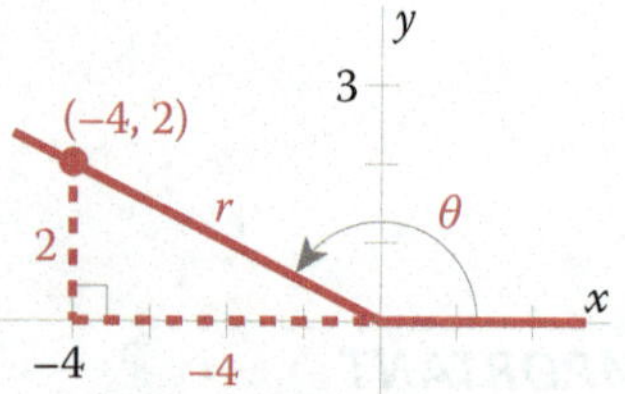

Since (x, y) is not in quadrant I, θ is not an angle in the triangle with sides x, y, and r. However, the definitions in terms of x, y, and r can still be used to evaluate the trigonometric functions of θ.

Evaluate the trigonometric functions of θ using $x = -4$, $y = 2$, and $r = 2\sqrt{5}$.

$$\sin\theta = \frac{y}{r} = \frac{2}{2\sqrt{5}} = \frac{1}{\sqrt{5}} \qquad \cos\theta = \frac{x}{r} = \frac{-4}{2\sqrt{5}} = -\frac{2}{\sqrt{5}} \qquad \tan\theta = \frac{y}{x} = \frac{2}{-4} = -\frac{1}{2}$$

$$\csc\theta = \frac{r}{y} = \frac{2\sqrt{5}}{2} = \sqrt{5} \qquad \sec\theta = \frac{r}{x} = \frac{2\sqrt{5}}{-4} = -\frac{\sqrt{5}}{2} \qquad \cot\theta = \frac{x}{y} = \frac{-4}{2} = -2$$

5.3.2 Evaluating Trigonometric Functions of an Angle in the Coordinate Plane: Another Example

The sign of each trigonometric function of θ depends on the quadrant in which θ lies. For example, if an angle θ is in quadrant III and (x, y) is on the terminal side of θ, then $x < 0$ and $y < 0$. Therefore, $y/r < 0$ and $x/r < 0$, but $y/x > 0$. It also follows that $\sin\theta < 0$ and $\cos\theta < 0$, but $\tan\theta > 0$. As a result, only $\tan\theta$ and its reciprocal, $\cot\theta$, are positive when θ lies in quadrant III.

Similarly, only $\sin\theta$, and its reciprocal $\csc\theta$, are positive when θ lies in quadrant II, and only $\cos\theta$, and its reciprocal, $\sec\theta$, are positive when θ lies in quadrant IV.

Use the phrase "all students take calculus" to remember the correspondence between the quadrant containing θ and the sign of the trigonometric function of θ.

- "All" (quadrant I): if θ is in quadrant I, then *all* trigonometric functions of θ are positive.
- "Students" (quadrant II): if θ is in quadrant II, then only $\sin\theta$ and $\csc\theta$ are positive.
- "Take" (quadrant III): if θ is in quadrant III, then only $\tan\theta$ and $\cot\theta$ are positive.
- "Calculus" (quadrant IV): if θ is in quadrant IV, then only $\cos\theta$ and $\sec\theta$ are positive.

REMEMBER

$r > 0$, regardless of the quadrant in which θ is located.

TIP

students (sine)
take (tangent)
calculus (cosine)

IMPORTANT

Reciprocals have the same sign, so cosecant is positive in the same quadrants as sine, cosine is positive in the same quadrants as secant, and tangent is positive in the same quadrants as cotangent.

Signs of the Trigonometric Functions of Angle θ

II $\sin\theta$: + $\cos\theta$: – $\tan\theta$: –	I $\sin\theta$: + $\cos\theta$: + $\tan\theta$: +
III $\sin\theta$: – $\cos\theta$: – $\tan\theta$: +	IV $\sin\theta$: – $\cos\theta$: + $\tan\theta$: –

"students" *Sine and cosecant are positive.*	"all" *All trigonometric functions are positive.*
"take" *Tangent and cotangent are positive.*	"calculus" *Cosine and secant are positive.*

EXAMPLE 3 Evaluating Trigonometric Functions Given a Point on the Terminal Side of θ in Quadrant III

Suppose θ is an angle in standard position such that its terminal side passes through $(-2, -5)$. Determine the signs of $\sin\theta$, $\cos\theta$, and $\tan\theta$, and then evaluate each function.

SOLUTION

Since θ lies in quadrant III, only tangent and cotangent will be positive. Therefore, $\sin\theta < 0$, $\cos\theta < 0$, and $\tan\theta > 0$.

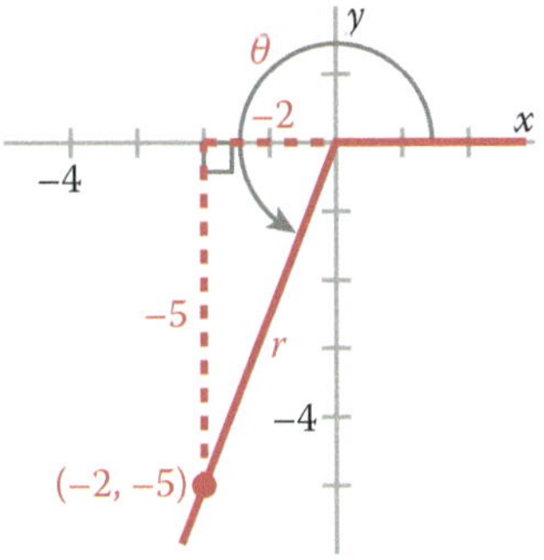

Use the definitions of the trigonometric functions of θ to evaluate $\sin\theta$, $\cos\theta$, and $\tan\theta$.

Find x, y, and r. Since the terminal side of θ passes through $(-2, -5)$, $x = -2$, $y = -5$, and $r = \sqrt{(-2)^2 + (-5)^2} = \sqrt{29}$.

$$\sin\theta = \frac{y}{r} = \frac{-5}{\sqrt{29}} = -\frac{5}{\sqrt{29}} \qquad \cos\theta = \frac{x}{r} = \frac{-2}{\sqrt{29}} = -\frac{2}{\sqrt{29}} \qquad \tan\theta = \frac{y}{x} = \frac{-5}{-2} = \frac{5}{2}$$

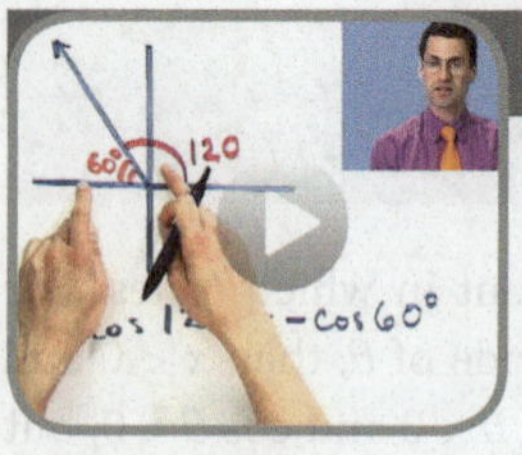

5.3.3 Evaluating Trigonometric Functions Using the Reference Angle

If θ is a standard position angle that is *not a quadrantal angle*, then the positive acute angle formed by the terminal side of θ and the x-axis is called the **reference angle** of θ, denoted θ'. In Figure 5.3d we see $\theta = 150°$ and its reference angle $\theta' = 30°$. Note that θ' is not in standard position. Other examples of θ and the corresponding reference angle θ' are shown below.

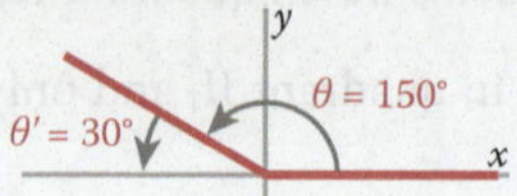

Figure 5.3d

NOTICE THAT

If $0 < \theta < \pi/2$, then θ is its own reference angle ($\theta = \theta'$). Angle θ' is a standard position angle only when θ is in quadrant I.

Reference Angles

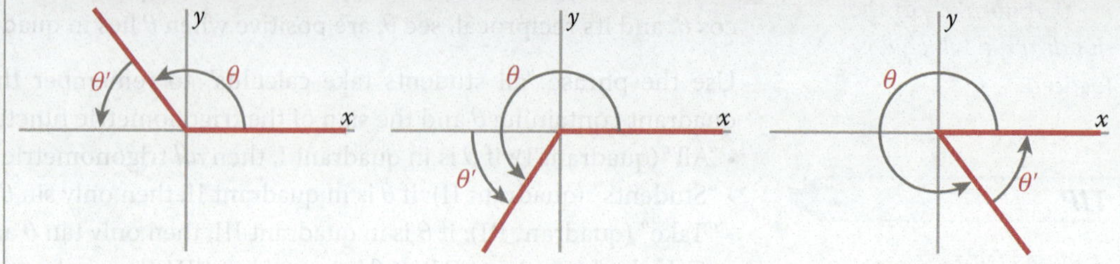

If $\pi/2 < \theta < \pi$, then $\theta' = \pi - \theta$.	*If $\pi < \theta < 3\pi/2$, then $\theta' = \theta - \pi$.*	*If $3\pi/2 < \theta < 2\pi$, then $\theta' = 2\pi - \theta$.*

The value of a trigonometric function of θ' is the absolute value of the same trigonometric function of θ (e.g., $\sin\theta' = |\sin\theta|$). It follows that the value of a trigonometric function of θ can be determined by applying the appropriate sign (as given by the quadrant containing θ) to the trigonometric function of θ', as demonstrated in **Example 4**.

IMPORTANT

A reference angle θ' is always an acute positive angle. Therefore, all six trigonometric functions of θ' are positive.

EXAMPLE 4 Evaluating a Trigonometric Function by Using a Reference Angle and a Special Right Triangle

Use a reference angle to evaluate each trigonometric function.

A. sec $(3\pi/4)$ **B.** cos $(-120°)$

SOLUTION

A. If $\theta = 3\pi/4$, then $\theta' = \pi/4$.

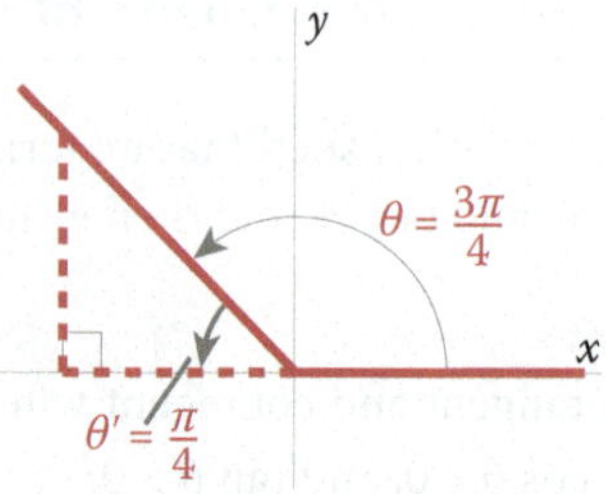

Evaluate sec θ'. Since $\pi/4$ is an angle in a special right triangle, use a 45°-45°-90° triangle to evaluate sec $(\pi/4)$.

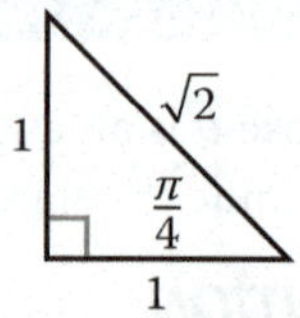

$$\sec\frac{\pi}{4} = \frac{\text{hyp}}{\text{adj}} = \frac{\sqrt{2}}{1} = \sqrt{2}$$

Apply the appropriate sign. Since θ is in quadrant III, $\sec\theta < 0$. So, $\sec\frac{3\pi}{4} = -\left(\sec\frac{\pi}{4}\right) = -\sqrt{2}$.

NOTICE THAT

Angles θ and θ' are supplementary angles (i.e., $\theta' = \pi - \theta$) only when $\pi/2 < \theta < \pi$.

B. θ is in quadrant III, but $\theta' \neq \theta - \pi$ because θ is not between π and $3\pi/2$. However, θ' is equal to the reference angle for the smallest positive angle coterminal with θ, which is 240°.

So, $\theta' = 240° - \pi = 240° - 180° = 60°$. Now evaluate cos θ' using a 30°-60°-90° triangle.

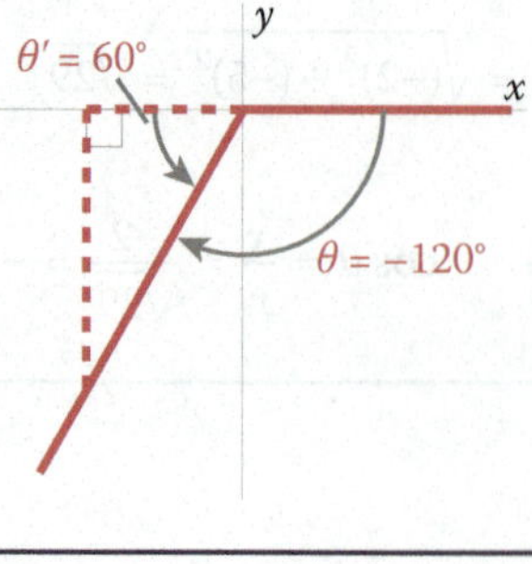

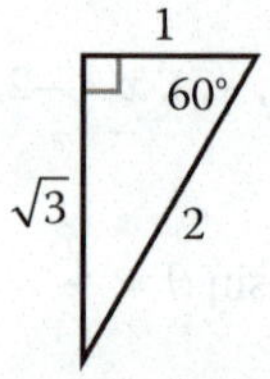

$$\cos 60° = \frac{\text{adj}}{\text{hyp}} = \frac{1}{2}$$

Apply the appropriate sign. Since θ is in quadrant III, $\cos\theta < 0$. So, $\cos(-120°) = -(\cos 60°) = -1/2$.

IMPORTANT

When $\theta < 0$ or $\theta > 2\pi$, θ' is equal to the reference angle for the angle between 0 and 2π that is coterminal with θ.

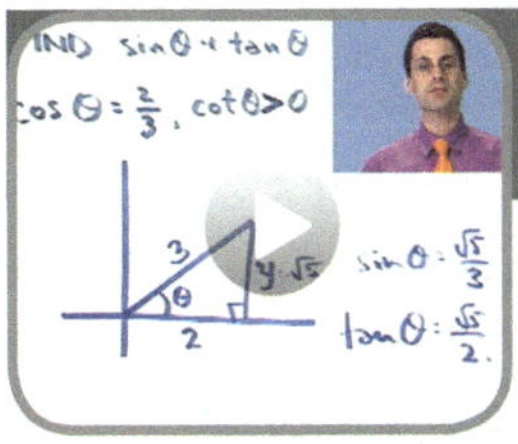

5.3.4 Using Values of Trigonometric Functions to Find the Value of Another Trigonometric Function

If θ is a non-quadrantal angle, then the value of one trigonometric function of θ and the sign of a second trigonometric function of θ (that is not the reciprocal of the first) can be used to find the values of all other trigonometric functions of θ. Two methods for doing this are demonstrated in **Example 5**.

EXAMPLE 5 Evaluating Trigonometric Functions by Using the Value of One Trigonometric Function and the Sign of Another

Find $\cos\theta$ and $\cot\theta$, given that $\sin\theta = \dfrac{4}{\sqrt{65}}$ and $\tan\theta < 0$.

SOLUTION

REMEMBER

To evaluate a trigonometric function of θ using its reference angle θ', first evaluate the trigonometric function of θ'. Then apply the appropriate sign, which depends on the quadrant that contains θ.

Method 1: Use the Reference Angle

Use the signs of both trigonometric functions to identify the quadrant containing θ. Then, use the x-axis to draw a right triangle in the quadrant corresponding to θ, where one angle is θ' and two sides are given by the value of the known trigonometric function.

Since $\sin\theta > 0$ and $\tan\theta < 0$, θ is in quadrant II.

Draw the right triangle in quadrant II.

Since $\sin\theta = \dfrac{4}{\sqrt{65}} = \dfrac{\text{opp}}{\text{hyp}}$, the side opposite θ' is 4 and the hypotenuse is $\sqrt{65}$.

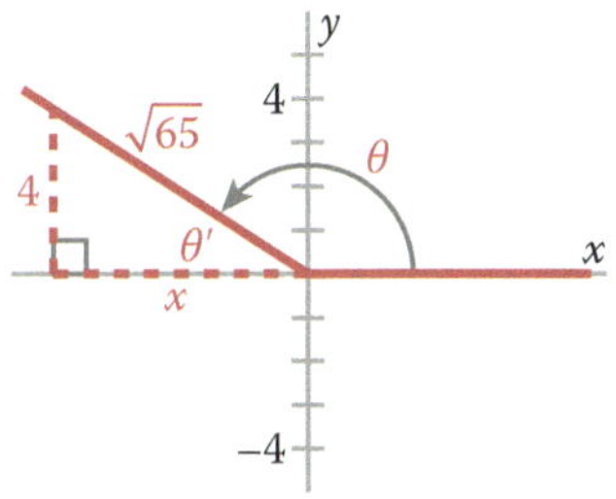

REMEMBER

x is a length, so only the positive square root is needed.

Use the Pythagorean Theorem to find the unknown side x. $x^2 + 16 = 65 \Rightarrow x^2 = 49 \Rightarrow x = 7$

Use the triangle to evaluate $\cos\theta'$ and $\cot\theta'$.

$$\cos\theta' = \text{adj/hyp} = 7/\sqrt{65} \qquad \cot\theta' = \text{adj/opp} = 7/4$$

Since cosine and cotangent are both negative in quadrant II, $\cos\theta = -7/\sqrt{65}$ and $\cot\theta = -7/4$.

Method 2: Use the Definitions of the Trigonometric Functions of θ

Let (x, y) be a point on the terminal side of θ. Since $\sin\theta > 0$ and $\tan\theta < 0$, θ is in quadrant II. Therefore, $x < 0$ and $y > 0$.

REMEMBER

$r > 0$, regardless of the signs of x and y.

Since $\sin\theta = \dfrac{4}{\sqrt{65}} = \dfrac{y}{r}$, $y = 4$ and $r = \sqrt{65}$.

Use $r = \sqrt{x^2 + y^2}$ to find x. $\sqrt{65} = \sqrt{x^2 + 4^2} \Rightarrow 65 = x^2 + 16 \Rightarrow x^2 = 49 \Rightarrow x = \pm 7$

However, since $x < 0$ in quadrant II, $x = -7$.

Use the definitions of the trigonometric functions of θ where $x = -7$, $y = 4$, and $r = \sqrt{65}$ to evaluate $\cos\theta$ and $\cot\theta$.

$$\cos\theta = \frac{x}{r} = \frac{-7}{\sqrt{65}} = -\frac{7}{\sqrt{65}} \qquad \cot\theta = \frac{x}{y} = \frac{-7}{4} = -\frac{7}{4}$$

5.3.5 Trigonometric Functions of Important Angles

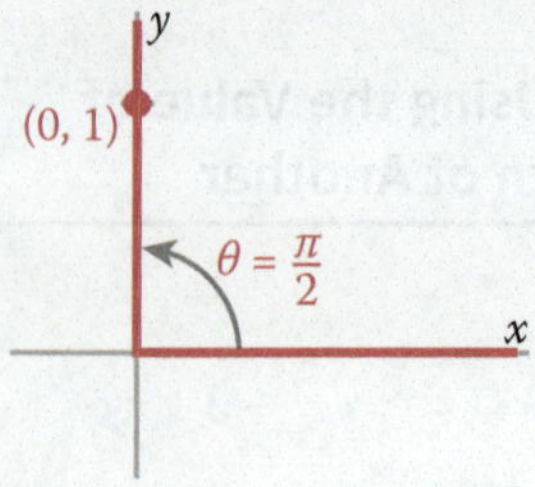

Figure 5.3e

If $\theta = \pi/2$ radians (i.e, $\theta = 90°$) and $r = 1$, then $x = 0$ and $y = 1$ (Figure 5.3e). If $\theta = 0$ radians (i.e, $\theta = 0°$) and $r = 1$, then $x = 1$ and $y = 0$ (Figure 5.3f). These values can be used to evaluate the trigonometric functions of the quadrantal angles $\pi/2$ radians and 0 radians.

The Trigonometric Functions of π/2 and 0 Radians (90° and 0°)

$\sin\frac{\pi}{2} = 1$ $\quad sin\frac{\pi}{2} = \frac{y}{r} = \frac{1}{1} = 1$	$\sin 0 = 0$ $\quad sin\, 0 = \frac{y}{r} = \frac{0}{1} = 0$
$\cos\frac{\pi}{2} = 0$ $\quad cos\frac{\pi}{2} = \frac{x}{r} = \frac{0}{1} = 0$	$\cos 0 = 1$ $\quad cos\, 0 = \frac{x}{r} = \frac{1}{1} = 1$
$\tan\frac{\pi}{2}$ is undefined $\quad tan\frac{\pi}{2} = \frac{y}{x} = \frac{1}{0}$	$\tan 0 = 0$ $\quad tan\, 0 = \frac{y}{x} = \frac{0}{1} = 0$

From the special right triangles, we know the values of $\sin\theta$, $\cos\theta$, and $\tan\theta$ when $\theta = \pi/6$ radians, $\theta = \pi/4$ radians, and $\theta = \pi/3$ radians (i.e, when $\theta = 30°$, $\theta = 45°$, and $\theta = 60°$). The values of $\sin\theta$, $\cos\theta$, and $\tan\theta$ for these angle values, along with $\theta = \pi/2$ radians and $\theta = 0$ radians, are listed in the following chart for reference.

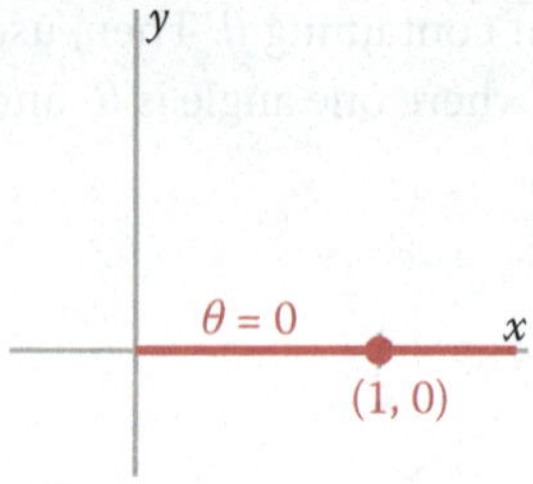

Figure 5.3f

The Trigonometric Functions of Important Angles

θ (deg)	θ (rad)	$\sin\theta$	$\cos\theta$	$\tan\theta$
0°	0	0	1	0
30°	$\frac{\pi}{6}$	$\frac{1}{2}$	$\frac{\sqrt{3}}{2}$	$\frac{1}{\sqrt{3}}$
45°	$\frac{\pi}{4}$	$\frac{\sqrt{2}}{2}$	$\frac{\sqrt{2}}{2}$	1
60°	$\frac{\pi}{3}$	$\frac{\sqrt{3}}{2}$	$\frac{1}{2}$	$\sqrt{3}$
90°	$\frac{\pi}{2}$	1	0	und

TIP

Memorize each of these values.

EXAMPLE 6 **Evaluating Trigonometric Functions of Important Angles without Using a Calculator**

Evaluate each trigonometric function without using a calculator.

A. $\cos 90°$ **B.** $\cot\frac{\pi}{3}$ radians

SOLUTION

A. From the chart, $\cos 90° = 0$.

B. From the chart, $\tan\frac{\pi}{3} = \sqrt{3}$.

Cotangent is the reciprocal of tangent, so $\cot\frac{\pi}{3} = \frac{1}{\tan\frac{\pi}{3}} = \frac{1}{\sqrt{3}}$.

SECTION 5.3 EXERCISES

Warm Up

1. Convert 135°, 240°, 90°, and 330° to radian measure.
2. Convert $2\pi/3$, $5\pi/4$, $\pi/5$, and $3\pi/2$ radians to degree measure.
3. Draw 135°, 240°, 90°, and 180° in standard position.

Just the Facts

Fill in the blanks.

4. When the lengths of any two sides in a right triangle are given, the length of the third side can be found using the _____ Theorem.
5. If (x, y) is a point on the terminal side of an angle θ in standard position, and r is the distance from the origin to (x, y) such that $r = \sqrt{x^2 + y^2} \neq 0$, then the trigonometric functions of θ can be defined in terms of x, y, and r. The trigonometric function defined as x/r is _____, y/x is _____, and r/y is _____. All the values of the trigonometric functions are _____ in quadrant I. Only the values of _____ and _____ are positive in quadrant III.
6. If $\tan \theta < 0$ and $\csc \theta > 0$, θ must lie in quadrant _____.
7. A reference angle, usually denoted by _____, is the positive _____ angle formed by the _____ side of θ and the _____.
8. If $\theta = 325°$, then the reference angle is _____.
9. The phrase "All Students Take Calculus" is useful for recalling in which quadrants the trigonometric functions are _____. When $x < 0$ and $y > 0$, _____ and its reciprocal function _____ are positive. The word _____ in the phrase indicates this fact.
10. If $\cos \theta = 1/2$, θ is either in the first or _____ quadrant.

Essential Skills

In Exercises 1–16, find the sine, cosine, tangent, cosecant, secant, and cotangent of *θ*, given that *θ* is an angle in standard position and the given point is on the terminal side of *θ*.

1. (4, 7)
2. (5, 3)
3. (1, 2)
4. (9, 7)
5. (−3, 5)
6. (−1, 4)
7. (−6, 3)
8. (−4, 9)
9. (−1, −2)
10. (−4, −5)
11. (−2, −6)
12. (−9, −7)
13. (5, −12)
14. (7, −3)
15. (3.4, −5.1)
16. (5, −7)

In Exercises 17–28, evaluate each trigonometric function without using a calculator.

17. $\sin 150°$
18. $\sec 240°$
19. $\cot 315°$
20. $\csc 300°$
21. $\tan(-120°)$
22. $\cos 135°$
23. $\cot \frac{5\pi}{6}$
24. $\sin\left(-\frac{\pi}{3}\right)$
25. $\sec \frac{4\pi}{3}$
26. $\csc \frac{3\pi}{4}$
27. $\cos\left(-\frac{5\pi}{6}\right)$
28. $\tan \frac{11\pi}{6}$

In Exercises 29–46, evaluate the trigonometric functions.

29. Find $\sec \theta$ and $\cot \theta$, given that $\sin \theta = -5/13$ and $\sec \theta < 0$.
30. Find $\sin \theta$ and $\tan \theta$, given that $\cos \theta = -3/5$ and $\sin \theta > 0$.
31. Find $\sin \theta$ and $\cot \theta$, given that $\cos \theta = -24/25$ and $\tan \theta < 0$.
32. Find $\sec \theta$ and $\csc \theta$, given that $\tan \theta = 12/5$ and $\cos \theta > 0$.
33. Find $\cos \theta$ and $\csc \theta$, given that $\sin \theta = \frac{\sqrt{5}}{7}$ and $\sec \theta > 0$.
34. Find $\sec \theta$ and $\csc \theta$, given that $\tan \theta = -24/7$ and $\cos \theta > 0$.
35. Find $\sin \theta$ and $\tan \theta$, given that $\cos \theta = -\frac{4\sqrt{3}}{9}$ and $\csc \theta > 0$.
36. Find $\sin \theta$ and $\cos \theta$, given that $\tan \theta = -24/7$ and $\sin \theta < 0$.
37. Find $\cos \theta$ and $\cot \theta$, given that $\sin \theta = -\frac{\sqrt{6}}{3}$ and $\tan \theta < 0$.
38. Find $\sec \theta$ and $\csc \theta$, given that $\cos \theta = -24/25$ and $\sin \theta > 0$.
39. Find $\cos \theta$ and $\cot \theta$, given that $\tan \theta = -2/3$ and $\csc \theta < 0$.
40. Find $\sin \theta$ and $\tan \theta$, given that $\cos \theta = 3/5$ and $\sin \theta < 0$.
41. Find $\sec \theta$ and $\csc \theta$, given that $\cot \theta = 7/5$ and $\sin \theta < 0$.
42. Find $\cos \theta$ and $\tan \theta$, given that $\csc \theta = -8/5$ and $\sec \theta > 0$.
43. Find $\cos \theta$ and $\sin \theta$, given that $\cot \theta = -9/4$ and $\sin \theta < 0$.
44. Find $\csc \theta$ and $\cot \theta$, given that $\sec \theta = -25/7$ and $\csc \theta < 0$.

45. Find $\tan\theta$ and $\csc\theta$, given that $\sec\theta = -\frac{5\sqrt{7}}{4}$ and $\tan\theta > 0$.

46. Find $\cos\theta$ and $\tan\theta$, given that $\csc\theta = 12/7$ and $\sec\theta < 0$.

In Exercises 47–64, evaluate each trigonometric function without using a calculator.

47. $\tan 60°$

48. $\cos 45°$

49. $\sin 30°$

50. $\cos 60°$

51. $\tan 0°$

52. $\sin 90°$

53. $\cos(\pi/4)$

54. $\sin(\pi/3)$

55. $\tan(\pi/6)$

56. $\sin(\pi/4)$

57. $\cos(\pi/6)$

58. $\tan(\pi/4)$

59. $\csc(\pi/6)$

60. $\cot(\pi/6)$

61. $\sec(\pi/4)$

62. $\csc(\pi/3)$

63. $\cot(\pi/2)$

64. $\sec(\pi/6)$

Extensions

In Exercises 65–68, simplify the expression without using a calculator.

65. $\left(\sin\frac{\pi}{6}\right)^2 + \left(\cos\frac{\pi}{4}\right)^2 + \left(\tan\left(-\frac{5\pi}{3}\right)\right)^2$

66. $\cos\frac{5\pi}{6} - \sin\left(-\frac{\pi}{3}\right) + \cot\frac{\pi}{2}$

67. $\left(\cos\pi\right)^2\left(\left(\csc\frac{5\pi}{4}\right)^2 + \cos\left(-\frac{\pi}{2}\right) + \left(\sin\frac{\pi}{4}\right)^2\right)$

68. $\dfrac{\left(\sec\frac{\pi}{4}\right)\left(\cot\frac{\pi}{6}\right)^2}{\left(\cos\frac{5\pi}{6}\right)^2 + \left(\tan\frac{5\pi}{3}\right)^2 + \left(\sin\frac{2\pi}{3}\right)^2}$

69. Suppose that $\sin\theta < 0$ and $\cos\theta = a$, where $0 < a < 1$. Find the value of $\csc\theta$ in terms of a.

70. Suppose that $\tan\theta < 0$ and $\sin\theta = a$, where $0 < a < 1$. Find the value of $\cot\theta$ in terms of a.

In Exercises 71–73, use the following information.

The angular position x and velocity x' with respect to a crank angle for a piston engine are given by

$$x = r\cos A + \sqrt{l^2 - r^2(\sin A)^2} \quad \text{and}$$

$$x' = -r\sin A - \frac{r^2\sin A\cos A}{\sqrt{l^2 - r^2(\sin A)^2}},$$ **respectively, where r is the crank radius, A is the crank angle, l is the rod length, x is the position of the piston pin from the crank center, and x' is the velocity of the piston pin. Determine the position and velocity of the piston pin in each of the following situations. Leave answers in simplified radical form.**

71. $r = 4$ inches, $A = 30°$, $l = 15$ inches

72. $r = 6$ inches, $A = 60°$, $l = 11$ inches

73. $r = 2$ inches, $A = 45°$, $l = 5$ inches

In Exercises 74–76, use the following information.

The acceleration a of a block sliding down a plane is represented by the function $a = g\cos\theta(\tan\theta - \mu_k)$, where g is the acceleration due to gravity, θ is the angle of elevation of the plane, and μ_k is the kinetic friction coefficient. Use $g \approx 32$ feet per second squared to determine the acceleration in each of the following situations. Leave answers in simplified radical form.

74. $\theta = 30°$, $\mu_k = 0.4$

75. $\theta = 45°$, $\mu_k = 0.5$

76. $\theta = 60°$, $\mu_k = 0.3$

In Exercises 77–80, determine the smallest possible positive value in radians to make each statement true.

77. $\sin x = \cos\frac{\pi}{3}$

78. $\sin\frac{\pi}{2} = \cos x$

79. $\tan\frac{\pi}{6} = \cot x$

80. $\cos 0 = \sin x$

81. Explain why a reference angle cannot be used to evaluate the trigonometric functions of θ when θ is a quadrantal angle.

5.4 UNIT CIRCLE TRIGONOMETRY

OBJECTIVES

- Find points on a unit circle.
- Evaluate the trigonometric functions using the unit circle.
- Define the period and domain of sine and cosine functions.
- Use the domain and period to evaluate sine and cosine functions.
- Use the even and odd identities to evaluate the trigonometric functions.

PREREQUISITE VOCABULARY TERMS

central angle
domain
radians
radius
trigonometric functions

5.4.1 The Unit Circle

The **unit circle** is the circle on a coordinate plane with center at (0, 0) and radius 1, as shown in Figure 5.4a. Therefore, the equation of the unit circle is $x^2 + y^2 = 1$.

Since the equation of a unit circle is $x^2 + y^2 = 1$, any point (x, y) in the plane such that $x^2 + y^2 = 1$ is on the unit circle. And for any such point (x, y), a corresponding right triangle can be formed with legs of length $|x|$ and $|y|$ and hypotenuse of length 1 (Figure 5.4b).

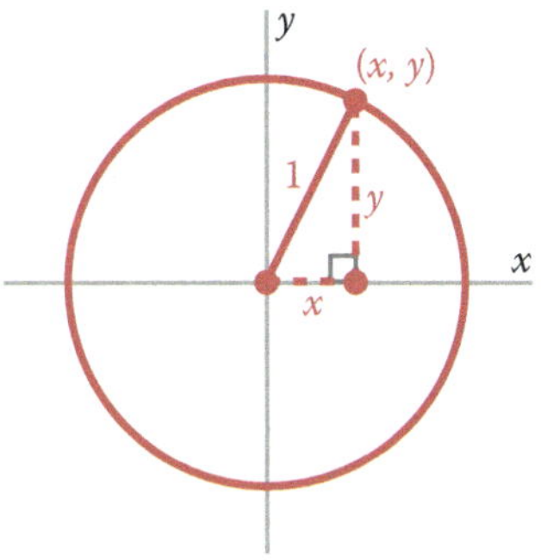

Figure 5.4b

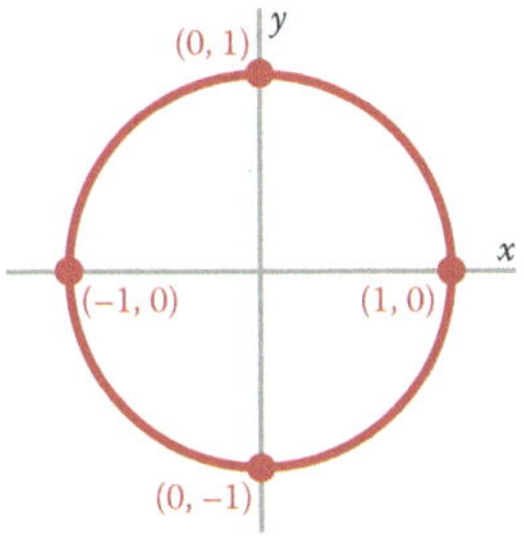

Figure 5.4a

Recall that the length of a circular arc s intercepted by a central angle θ is given by $s = r\theta$. Since the radius of a unit circle is 1, the length of any arc on the unit circle is equal to the measure of the intercepting central angle. Therefore, $s = \theta$ for the unit circle.

Let t represent the radian measure of a central angle in standard position of the unit circle. Then t is also equal to the length of the intercepted arc on the unit circle (i.e., $s = t$). Every t-value has a corresponding point (x, y) on the unit circle that satisfies the equation $x^2 + y^2 = 1$.

If t is a multiple of $\pi/6$ or $\pi/4$, then the coordinates of the corresponding point (x, y) on the unit circle can be determined using special right triangles. The unit circle in Figure 5.4c shows the point (x, y) corresponding to each t that is a multiple of $\pi/6$ or $\pi/4$ such that $0 \le t < 2\pi$.

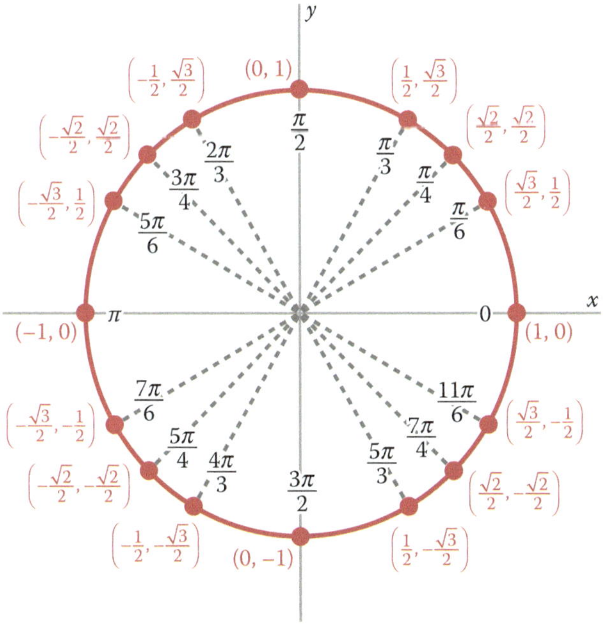

Figure 5.4c

NOTICE THAT

The important angles discussed in Topic 5.3.5 (0, π/6, π/4, π/3, and π/2) are t-values in this unit circle.

Determining the x- and y-Coordinates of Points on the Unit Circle

Each point on the unit circle in Figure 5.4c corresponds to a t-value that is a multiple of either $\pi/6$ or $\pi/4$ (i.e., 30° and 45°). It follows that the coordinates of each point (x, y) can be determined by using a special right triangle.

For example, consider the point (x, y) corresponding to $t - 5\pi/3$. Since the measure of the reference angle for $5\pi/3$ is 60°, the x- and y-coordinates of this point are the signed lengths of the legs of a 30°-60°-90° triangle formed with the x-axis, as shown in Figure 5.4d. The triangle's hypotenuse is a radius of the unit circle, so the length of the hypotenuse must be 1.

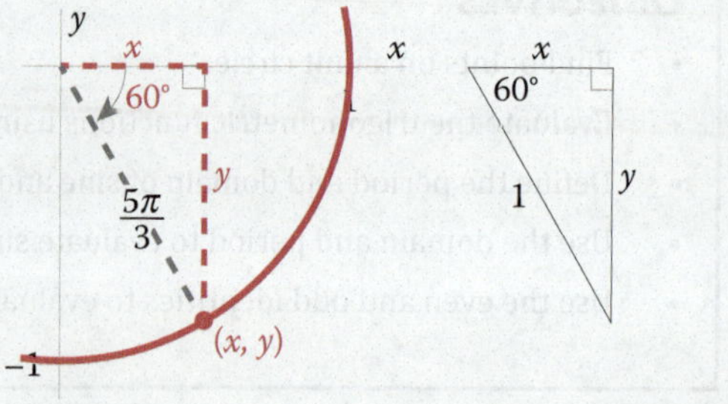

Figure 5.4d

NOTICE THAT

In (x, y), x is the signed length of the side opposite the 30° angle, and y is the signed length of the side opposite the 60° angle.

Recall that the sides of a 30°-60°-90° triangle can be expressed as a, $a\sqrt{3}$, and $2a$, where, a is the side opposite the 30° angle, $a\sqrt{3}$ is the side opposite the 60° angle, and $2a$ is the hypotenuse. The hypotenuse of this 30°-60°-90° triangle is 1, so $2a = 1$. Thus, $a = 1/2$. It follows that the side opposite the 30° angle is 1/2, and the side opposite the 60° angle is $\sqrt{3}/2$.

TIP

If $a = 1/2$, then

$$a\sqrt{3} = \left(\frac{1}{2}\right)\left(\sqrt{3}\right) = \frac{\sqrt{3}}{2}.$$

The point (x, y) is in quadrant IV, so $x > 0$ and $y < 0$.

Therefore, as shown in Figure 5.4e, the point on the unit circle that corresponds to $t = \frac{5\pi}{3}$ is $\left(\frac{1}{2}, -\frac{\sqrt{3}}{2}\right)$.

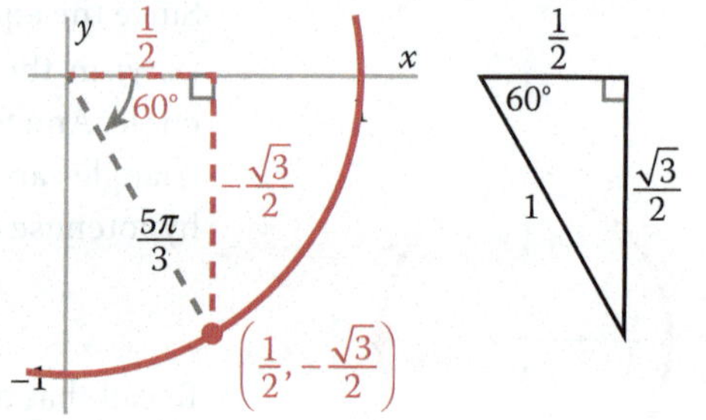

Figure 5.4e

Coterminal Angles and the Unit Circle

Each point (x, y) on the unit circle corresponds not only to the given t-value, but also to $t + 2\pi n$ (i.e., to any angle coterminal with t), where n is an integer. For example, (1, 0) corresponds to $t = 0$, as well as to $t = 2\pi$, $t = 4\pi$, $t = 6\pi$, . . . , and to $t = -2\pi$, $t = -4\pi$, ...

Additional examples are provided in the following table.

(x, y)	t (radians)
(0, 1)	$\pi/2, 5\pi/2, 9\pi/2, \ldots$ $-3\pi/2, -7\pi/2, -11\pi/2, \ldots$
(−1, 0)	$\pi, 3\pi, 5\pi, \ldots$ $-\pi, -3\pi, -5\pi, \ldots$
$\left(\frac{1}{2}, \frac{\sqrt{3}}{2}\right)$	$\pi/3, 7\pi/3, 13\pi/3, \ldots$ $-5\pi/3, -11\pi/3, -17\pi/3, \ldots$

TIP

$\pi/2 + 2\pi = \pi/2 + 4\pi/2 = 5\pi/2$

$\pi/2 + 4\pi = \pi/2 + 8\pi/2 = 9\pi/2$

$\pi/2 - 2\pi = \pi/2 - 4\pi/2 = -3\pi/2$

$\pi/2 - 4\pi = \pi/2 - 8\pi/2 = -7\pi/2$

$\pi/2 - 6\pi = \pi/2 - 12\pi/2 = -11\pi/2$

EXAMPLE 1 Identifying a Point on the Unit Circle

Identify the point (x, y) on the unit circle that corresponds to each given real number t.

A. $3\pi/2$ **B.** $-2\pi/3$ **C.** $19\pi/4$

SOLUTION

Use the unit circle in Figure 5.4c to find the point that corresponds to each t-value.

TIP

All angles coterminal to $3\pi/2$ also correspond to $(0, -1)$.

A. From the unit circle, $t = 3\pi/2$ corresponds to $(0, -1)$.

B. $t = -2\pi/3$ is not shown on the unit circle. However, since $-\frac{2\pi}{3} + 2\pi = \frac{4\pi}{3}$, the angles measuring $-2\pi/3$ and $4\pi/3$ are coterminal. So, the point corresponding to $t = -2\pi/3$ also corresponds to $t = 4\pi/3$, which is $\left(-\frac{1}{2}, -\frac{\sqrt{3}}{2}\right)$.

IMPORTANT

If $t > 2\pi$, subtract a multiple of 2π from t to produce a value between 0 and 2π, then find its corresponding point on the unit circle.

C. $t = 19\pi/4$ is not shown on the unit circle. However, since $\frac{19\pi}{4} - 4\pi = \frac{3\pi}{4}$, the angles measuring $19\pi/4$ and $3\pi/4$ are coterminal. So, the point corresponding to $t = 19\pi/4$ also corresponds to $t = 3\pi/4$, which is $\left(-\frac{\sqrt{2}}{2}, \frac{\sqrt{2}}{2}\right)$.

5.4.2 Using the Unit Circle to Evaluate Trigonometric Functions

Recall from Section 5.3 that $\sin\theta = y/r$ and $\cos\theta = x/r$.

Since $r = 1$ on a unit circle, if (x, y) is a point on a unit circle, then $\sin\theta = y$ and $\cos\theta = x$. Therefore, any point (x, y) on a unit circle can be expressed as $(\cos\theta, \sin\theta)$ (Figure 5.4f).

The following definitions of the trigonometric functions follow from the definitions given in Section 5.3, here using $r = 1$.

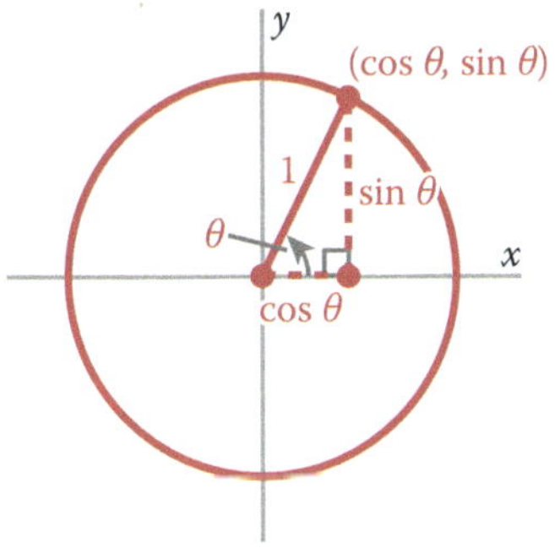

Figure 5.4f

The Trigonometric Functions of Real Number t on the Unit Circle

Let t be a real number and let (x, y) be a point on the unit circle corresponding to t.

$\sin t = y$ $\csc t = \frac{1}{y}, y \neq 0$

$\cos t = x$ $\sec t = \frac{1}{x}, x \neq 0$

$\tan t = \frac{y}{x}, x \neq 0$ $\cot t = \frac{x}{y}, y \neq 0$

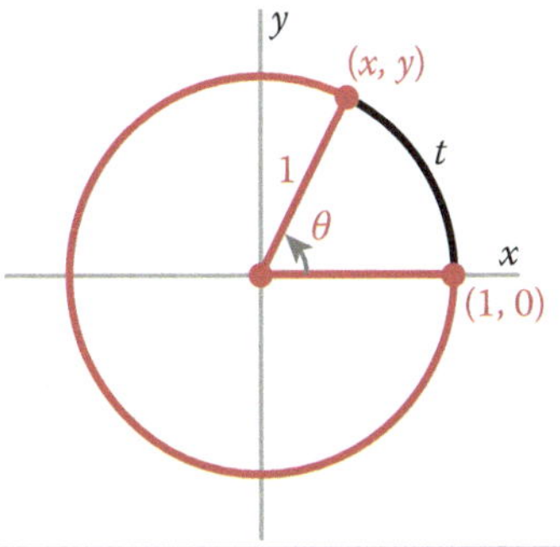

EXAMPLE 2 Using the Unit Circle to Evaluate the Trigonometric Functions

Evaluate the sine, cosine, tangent, cosecant, secant, and cotangent of each real number.

A. $t = 7\pi/6$ **B.** $t = -7\pi/6$

SOLUTION

A. On the unit circle, $t = \dfrac{7\pi}{6}$ corresponds to the point $\left(-\dfrac{\sqrt{3}}{2}, -\dfrac{1}{2}\right)$. So, $x = -\dfrac{\sqrt{3}}{2}$ and $y = -\dfrac{1}{2}$.

$$\sin\frac{7\pi}{6} = y = -\frac{1}{2} \qquad \cos\frac{7\pi}{6} = x = -\frac{\sqrt{3}}{2} \qquad \tan\frac{7\pi}{6} = \frac{y}{x} = \frac{-\frac{1}{2}}{-\frac{\sqrt{3}}{2}} = \frac{1}{\sqrt{3}}$$

$$\csc\frac{7\pi}{6} = \frac{1}{y} = \frac{1}{-\frac{1}{2}} = -2 \qquad \sec\frac{7\pi}{6} = \frac{1}{x} = \frac{1}{-\frac{\sqrt{3}}{2}} = -\frac{2}{\sqrt{3}} \qquad \cot\frac{7\pi}{6} = \frac{x}{y} = \frac{-\frac{\sqrt{3}}{2}}{-\frac{1}{2}} = \sqrt{3}$$

B. On the unit circle, $t = -\dfrac{7\pi}{6}$ corresponds to the point $\left(-\dfrac{\sqrt{3}}{2}, \dfrac{1}{2}\right)$ because $-\dfrac{7\pi}{6} + 2\pi = \dfrac{5\pi}{6}$.

So, $x = -\dfrac{\sqrt{3}}{2}$ and $y = \dfrac{1}{2}$.

$$\sin\left(-\frac{7\pi}{6}\right) = y = \frac{1}{2} \qquad \cos\left(-\frac{7\pi}{6}\right) = x = -\frac{\sqrt{3}}{2} \qquad \tan\left(-\frac{7\pi}{6}\right) = \frac{y}{x} = \frac{\frac{1}{2}}{-\frac{\sqrt{3}}{2}} = -\frac{1}{\sqrt{3}}$$

Using reciprocals gives $\csc\left(-\dfrac{7\pi}{6}\right) = 2$, $\sec\left(-\dfrac{7\pi}{6}\right) = -\dfrac{2}{\sqrt{3}}$, and $\cot\left(-\dfrac{7\pi}{6}\right) = -\sqrt{3}$.

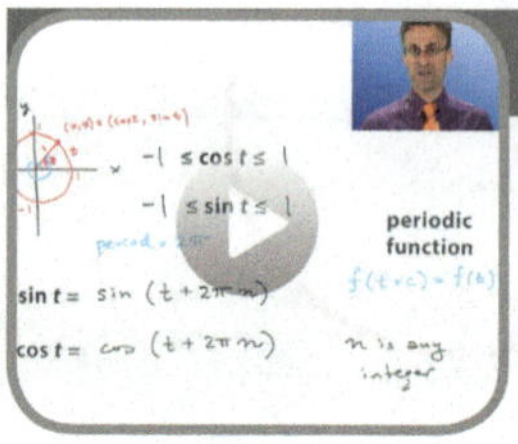

5.4.3 The Domain and Period of Sine and Cosine

Since the input, t, can be any angle, the domains of $\sin t$ and $\cos t$ consist of all of real numbers. To determine the ranges of $\sin t$ and $\cos t$, consider the unit circle. Since $-1 \le x \le 1$ and $-1 \le y \le 1$ on the unit circle, and since $\sin t = y$ and $\cos t = x$ by definition, it follows that $-1 \le \sin t \le 1$, and $-1 \le \cos t \le 1$. Therefore, the ranges of $\sin t$ and $\cos t$ are the interval $[-1, 1]$.

Periodic Functions

Recall that a complete rotation around a circle measures 2π radians. So, $\sin 0 = \sin 2\pi = \sin 4\pi \ldots$, since adding or subtracting 2π from any t-value completes another rotation around the unit circle. This is true not only for $\sin 0$, but also for the sine of any angle. Thus, for any integer n, $\sin t = \sin(t + 2\pi n)$, and $\cos t = \cos(t + 2\pi n)$. When the value of a function is cyclic (i.e., repeated regularly), as is the case for sine and cosine, the function is called a **periodic function**.

DEFINITION

A function f is a **periodic function** if there exists a positive real number c such that $f(t) = f(t + c)$ for all t in the domain of f. The smallest value of c for which f is periodic is called the **period** of f.

Since the period of a function is the smallest number for which the function is periodic, the periods of $\sin t$ and $\cos t$ are 2π.

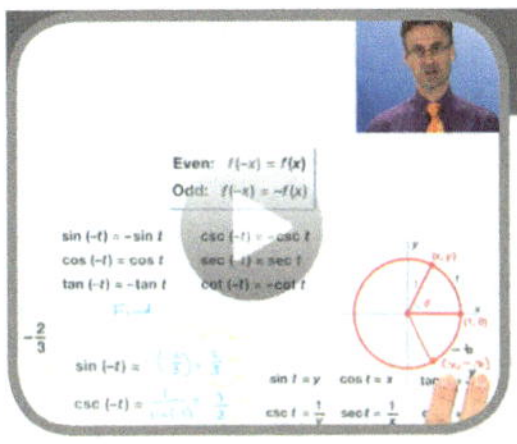

5.4.4 Even and Odd Trigonometric Functions

Recall that if f is an even function (for all x in the domain of f), then $f(-x) = f(x)$, and if f is an odd function, then $f(-x) = -f(x)$. These definitions apply to the trigonometric functions as well. Cosine and its reciprocal function, secant, are even functions. For example, recall that $\cos(\pi/3) = 1/2$, and $\cos(-\pi/3) = 1/2$. Therefore, $\cos(-\pi/3) = \cos(\pi/3)$, since cosine is an even function. The remaining trigonometric functions are all odd functions.

Even and Odd Trigonometric Functions		
$\sin(-t) = -\sin t$	$\cos(-t) = \cos t$	$\tan(-t) = -\tan t$
$\csc(-t) = -\csc t$	$\sec(-t) = \sec t$	$\cot(-t) = -\cot t$

EXAMPLE 3 Using Properties of Even and Odd Functions to Evaluate Trigonometric Functions

Evaluate the sine and cosine of $t = -3\pi/4$.

SOLUTION

Since sine is an odd function, $\sin(-t) = -\sin t$. Therefore, $\sin\left(-\frac{3\pi}{4}\right) = -\sin\frac{3\pi}{4}$. And since cosine is an even function, $\cos(-t) = \cos t$. Therefore, $\cos\left(-\frac{3\pi}{4}\right) = \cos\frac{3\pi}{4}$.

So, $t = \frac{3\pi}{4}$ can be used to evaluate $\sin\left(-\frac{3\pi}{4}\right)$ and $\cos\left(-\frac{3\pi}{4}\right)$.

From the unit circle, $t = \frac{3\pi}{4}$ corresponds to the point $\left(-\frac{\sqrt{2}}{2}, \frac{\sqrt{2}}{2}\right)$. So, $x = -\frac{\sqrt{2}}{2}$ and $y = \frac{\sqrt{2}}{2}$.

$$\sin\left(-\frac{3\pi}{4}\right) = -\sin\frac{3\pi}{4} = -\left(\frac{\sqrt{2}}{2}\right) = -\frac{\sqrt{2}}{2} \qquad \cos\left(-\frac{3\pi}{4}\right) = \cos\frac{3\pi}{4} = -\frac{\sqrt{2}}{2}$$

REMEMBER

If (x, y) is a point on the unit circle corresponding to t, then $\cos t = x$, and $\sin t = y$.

EXAMPLE 4 Using Properties of Even and Odd Functions to Evaluate Trigonometric Functions

If $\tan t = -1/5$, evaluate $\tan(-t)$ and $\cot(-t)$.

SOLUTION

Use the fact that $\tan(-t) = -\tan t$ (i.e., tangent is an odd function) to evaluate $\tan(-t)$.

$$\tan(-t) = -\tan t = -\left(-\frac{1}{5}\right) = \frac{1}{5}$$

Cotangent is the reciprocal of tangent, so $\cot(-t) = \frac{1}{\tan(-t)} = \frac{1}{\frac{1}{5}} = 5$.

SECTION 5.4 EXERCISES

Warm Up

1. Find the values of tan 60°, cos 45°, sin 60°, sec 30°, and cot 45° without using a calculator.

2. Identify the quadrant in which $\cot\theta < 0$ and $\csc\theta > 0$.

3. If $\csc\theta = 7/5$, evaluate $\sin\theta$.

Just the Facts

Fill in the blanks.

4. The equation of the unit circle is _____ with center _____ and radius _____.

5. On the unit circle, the coordinates _____ correspond to $t = \frac{7\pi}{4}$, the coordinates $\left(-\frac{1}{2}, -\frac{\sqrt{3}}{2}\right)$ correspond to $t =$ _____, and the coordinates of the point on the negative x-axis (i.e., $(-1, 0)$) correspond to $t =$ _____.

6. The range of sine and cosine is _____, since the least value that is possible is _____, and the greatest value possible is _____.

7. A function that repeats its values in regular intervals is called a(n) _____ function. The _____ of sine and cosine is 2π.

8. For a(n) even function, $f(-x) =$ _____, and for an odd function, $f(-x) =$ _____. Cosine and secant are examples of _____ functions.

9. Since the radius on the unit circle is _____, the length of any arc on the unit circle is _____ to the measure of its corresponding central angle.

10. A full rotation around a circle is _____ radians, and every additional full rotation in either a(n) _____ or a(n) _____ direction is a multiple of _____.

Essential Skills

In Exercises 1–18, identify the point (x, y) on the unit circle that corresponds to each given real number t.

1. $\frac{3\pi}{4}$

2. $\frac{7\pi}{6}$

3. $\frac{2\pi}{3}$

4. $\frac{3\pi}{2}$

5. $\frac{5\pi}{6}$

6. $\frac{7\pi}{4}$

7. $-\pi$

8. $-\frac{4\pi}{3}$

9. $-\frac{11\pi}{6}$

10. $-\frac{3\pi}{2}$

11. $-\frac{2\pi}{3}$

12. $-\frac{3\pi}{4}$

13. $\frac{21\pi}{2}$

14. $-\frac{13\pi}{6}$

15. $\frac{15\pi}{4}$

16. $\frac{17\pi}{3}$

17. $-\frac{29\pi}{6}$

18. $-\frac{19\pi}{2}$

In Exercises 19–42, evaluate the sine, cosine, tangent, cosecant, secant, and cotangent of each real number.

19. $t = \frac{4\pi}{3}$

20. $t = \frac{3\pi}{4}$

21. $t = \frac{5\pi}{6}$

22. $t = \frac{3\pi}{2}$

23. $t = \frac{11\pi}{4}$

24. $t = \frac{5\pi}{3}$

25. $t = \pi$

26. $t = \frac{11\pi}{6}$

27. $t = \frac{2\pi}{3}$

28. $t = \frac{5\pi}{4}$

29. $t = \frac{\pi}{2}$

30. $t = \frac{7\pi}{6}$

31. $t = -\frac{\pi}{6}$

32. $t = -\frac{5\pi}{3}$

33. $t = -\frac{5\pi}{4}$

34. $t = -\frac{3\pi}{2}$

35. $t = -\frac{2\pi}{3}$

36. $t = -\frac{7\pi}{4}$

37. $t = -\frac{\pi}{2}$

38. $t = -\frac{11\pi}{6}$

39. $t = -\frac{3\pi}{4}$

40. $t = -\frac{\pi}{3}$

41. $t = -\frac{5\pi}{6}$

42. $t = -\frac{\pi}{4}$

43. Evaluate the cotangent and tangent of $t = -\frac{5\pi}{3}$.

44. Evaluate the cosecant and secant of $t = -\frac{7\pi}{6}$.

45. Evaluate the sine and cosine of $t = -\frac{5\pi}{6}$.

46. Evaluate the cotangent and tangent of $t = -\frac{7\pi}{4}$.

47. Evaluate the cosecant and secant of $t = -\frac{3\pi}{4}$.

48. Evaluate the sine and cosine of $t = -\frac{2\pi}{3}$.

49. If $\sin t = -4/9$, evaluate $\sin(-t)$ and $\csc(-t)$.

50. If $\cos t = -2/7$, evaluate $\cos(-t)$ and $\sec(-t)$.

51. If $\tan t = 3/5$, evaluate $\tan(-t)$ and $\cot(-t)$.

52. If $\sec t = 10/3$, evaluate $\sec(-t)$ and $\cos(-t)$.

53. If $\csc t = -5/4$, evaluate $\csc(-t)$ and $\sin(-t)$.

54. If $\cot t = -12/11$, evaluate $\cot(-t)$ and $\tan(-t)$.

Extensions

55. If $\tan t = \dfrac{\sqrt{3}}{3}$ and $\sin t = 1/2$, what is the smallest positive value of t?

56. True or False? If $\sin x = 0$, then $\cos x =$ undefined.

57. If $\cos t = -3/5$ and t is in quadrant II, evaluate $\cos(-t) + \sec(-t) - \sin(-t) + \csc(-t) - \tan(-t) + \cot(-t)$.

In Exercises 58–60, evaluate without using a calculator.

58. $\cos \dfrac{4001\pi}{2}$

59. $\sin 15{,}000\pi$

60. $\left(\sin\left(-\dfrac{\pi}{2}\right)\right)\left(\cos\dfrac{29\pi}{6}\right)\left(\sec\left(-\dfrac{17\pi}{3}\right)\right)\left(\tan\dfrac{76\pi}{4}\right)$

In Exercises 61–64, use the given function $f(x)$ and the given value of $f(t)$ to find the exact value of $f(-t)$, $f(-t) + f(t + 2\pi)$, and $f(t) + f(t + 4\pi) + f(-t)$ without using a calculator.

61. $f(x) = \sin x, f(t) = -2/3$

62. $f(x) = \cos x, f(t) = -1/4$

63. $f(x) = \sec x, f(t) = -11/3$

64. $f(x) = \csc x, f(t) = -7/4$

In Exercises 65–66, use the given function $f(x)$ and the given value of $f(t)$ to find the exact value of $f(-t)$, $f(-t) + f(t + \pi)$, and $f(t) + f(t + 2\pi) + f(-t)$ without using a calculator.

65. $f(x) = \cot x, f(t) = -2$

66. $f(x) = \tan x, f(t) = -5/8$

In Exercises 67–72, determine whether each statement is true or false.

67. $\left(\sin\dfrac{2\pi}{3}\right)^2 + \left(\cos\dfrac{2\pi}{3}\right)^2 = 1$

68. $\left(\tan\dfrac{5\pi}{4}\right)^2 = 1 - \left(\sec\dfrac{5\pi}{4}\right)^2$

69. $\left(\tan\dfrac{\pi}{3}\right)^2 = \dfrac{1-\cos\left(2\cdot\dfrac{\pi}{3}\right)}{1+\cos\left(2\cdot\dfrac{\pi}{3}\right)}$

70. $\sin\dfrac{\pi}{2} + \sin\pi = 2\sin\dfrac{1}{2}\left(\dfrac{\pi}{2}+\pi\right)\cos\dfrac{1}{2}\left(\dfrac{\pi}{2}+\pi\right)$

71. $\cos\dfrac{5\pi}{6}\sin\dfrac{\pi}{3} = \dfrac{1}{2}\left[\sin\left(\dfrac{5\pi}{6}+\dfrac{\pi}{3}\right) - \cos\left(\dfrac{5\pi}{6}-\dfrac{\pi}{3}\right)\right]$

72. $\tan\left(2\cdot\dfrac{5\pi}{3}\right) = \dfrac{2\tan\dfrac{5\pi}{3}}{1+\left(\tan\dfrac{5\pi}{3}\right)^2}$

5.5 GRAPHING SINE AND COSINE FUNCTIONS

OBJECTIVES

- Graph the basic sine function and the basic cosine function.
- Use the amplitude and period to graph sine and cosine functions.
- Use translations to graph sine and cosine functions.
- Use a graph to find minimum and maximum values of sine and cosine functions.
- Use a graph to find the zeros of sine and cosine functions.

PREREQUISITE VOCABULARY TERMS

cosine

maximum and minimum

period

sine

translation

zero (of a function)

5.5.1 Introduction to the Graphs of the Sine and Cosine Functions

In Chapter 1, we used various methods to graph functions such as $y = x + 1$ and $f(x) = x^2$. In one method, we made a table of ordered pairs representing points on the graph of the function, plotted the points on a coordinate plane, and then connected them using a smooth line or curve.

This procedure can also be used to graph trigonometric functions, such as the basic sine and cosine functions, $y = \sin x$ and $y = \cos x$, where x represents an angle (or a real number). First, choose several x-values (multiples of $\pi/4$ or $\pi/6$) for the table, then find each corresponding y-value (i.e., the values of $\sin x$ and $\cos x$). To make the points easier to plot, choose multiples of $\pi/2$ as x-values so that each corresponding y-value is -1, 0, or 1.

IMPORTANT

Choosing multiples of $\pi/4$ or $\pi/6$ as x-values (in radians) for the table ensures that exact values of $\sin x$ and $\cos x$ (i.e., the y-values) can be found using the unit circle, a special right triangle, or a reference angle.

$y = \sin x$	
x	y
0	0
$\pi/2$	1
π	0
$3\pi/2$	-1
2π	0
$5\pi/2$	1

$y = \cos x$	
x	y
0	1
$\pi/2$	0
π	-1
$3\pi/2$	0
2π	1
$5\pi/2$	0

From the tables, we see that the graph of $y = \sin x$ passes through the points $(0, 0)$, $(\pi/2, 1)$, $(\pi, 0)$, $(3\pi/2, -1)$, $(2\pi, 0)$, and $(5\pi/2, 1)$, and the graph of $y = \cos x$ passes through the points $(0, 1)$, $(\pi/2, 0)$, $(\pi, -1)$, $(3\pi/2, 0)$, $(2\pi, 1)$, and $(5\pi/2, 0)$. Plot the points on a coordinate plane where the scale on the x-axis is $\pi/2$ and the scale on the y-axis is 1.

Since the sine and cosine functions are periodic, the graphs continue to follow the patterns established by these points. The change in the y-coordinate between these points is gradual, resulting in the smooth curves shown here.

NOTICE THAT

The graph of $y = \sin x$ is symmetric about the origin and the graph of $y = \cos x$ is symmetric about the y-axis (confirming that $\sin x$ is an odd function and $\cos x$ is an even function). The range of both functions is the interval $[-1, 1]$.

The Graphs of the Basic Sine and Cosine Functions: $y = \sin x$ and $y = \cos x$

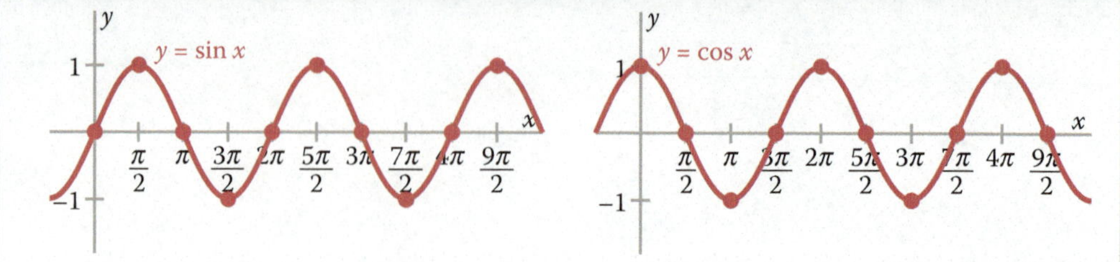

Domain: $(-\infty, \infty)$ Range: $[-1, 1]$ Domain: $(-\infty, \infty)$ Range: $[-1, 1]$

Recall from Section 5.4 that the period of $\sin\theta$ and $\cos\theta$ is 2π. When a function's period is 2π, the graph of the function will repeat itself every 2π radians. Figure 2.5a shows the graph of $y = \sin x$ over four periods.

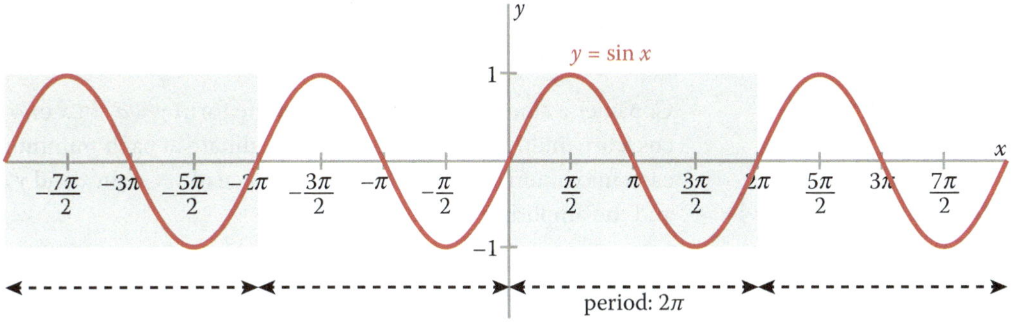

Figure 2.5a

The Amplitude, Zeros, Minimums, and Maximums of $y = \sin x$ and $y = \cos x$

The **amplitude** of sine and cosine functions is half the distance between the minimum y-value and the maximum y-value. For the graph of $y = \sin x$, the minimum y-value is −1 and the maximum y-value 1. Therefore, the amplitude of the graph of $y = \sin x$ is 1 because $|-1-1|/2 = 1$. Similarly, the amplitude of the graph of $y = \cos x$ is also 1.

REMEMBER

A zero of a function is an x-value where the function intersects the x-axis.

Within each period of $y = \sin x$ and $y = \cos x$ there are two zeros, one minimum, and one maximum. Since $y = \sin x$ and $y = \cos x$ are periodic functions, each function has an infinite number of zeros, minimums, and maximums, each occurring at regular intervals.

The zeros of $y = \sin x$ and $y = \cos x$ occur every π units. The minimums (points where $y = -1$) and the maximums (points where $y = 1$) of $y = \sin x$ and $y = \cos x$ occur every 2π units

IMPORTANT

There are many ways to write a general equation using n (where n always represents any integer). For example, $x = \pi + 2\pi n$ can also be written as $x = 3\pi + 2\pi n$ or as $x = -\pi + 2\pi n$. However, the equation typically includes the positive zero (minimum or maximum) that is closest to 0.

Specifically, the zeros of $y = \sin x$ occur when $x = \ldots -\pi, 0, \pi, 2\pi, \ldots$, that is, at multiples (positive and negative) of π. If we let n be any integer, then the *general equation* of the zeros of $y = \sin x$ is $x = \pi n$. Additionally, if n is an integer, then the general equations for the minimums and maximums of $y = \sin x$ are $x = \frac{3\pi}{2} + 2\pi n$ and $x = \frac{\pi}{2} + 2\pi n$, respectively.

The Attributes of the Graphs of $y = \sin x$ and $y = \cos x$

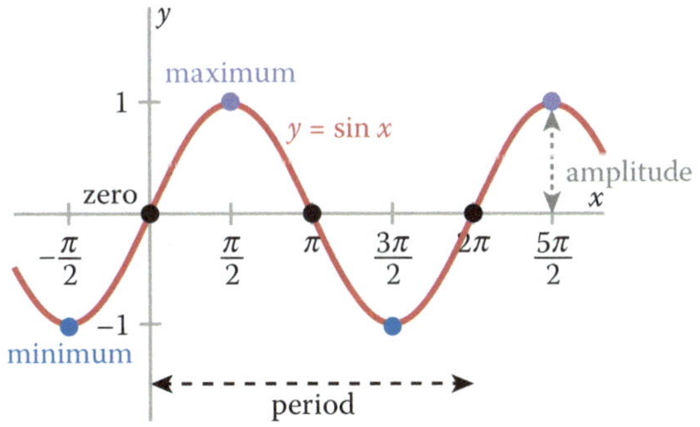

Amplitude: 1

Zeros: $x = \pi n$

Minimums: $\left(\frac{3\pi}{2} + 2\pi n, -1\right)$

Maximums: $\left(\frac{\pi}{2} + 2\pi n, 1\right)$

Period: 2π

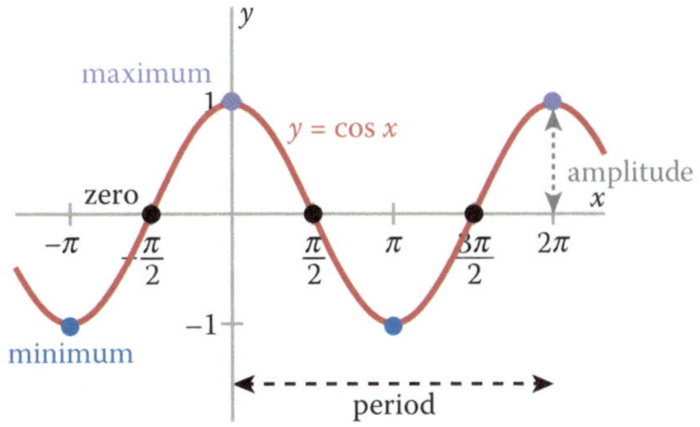

Amplitude: 1

Zeros: $x = \frac{\pi}{2} + \pi n$

Minimums: $(\pi + 2\pi n, -1)$

Maximums: $(2\pi n, 1)$

Period: 2π

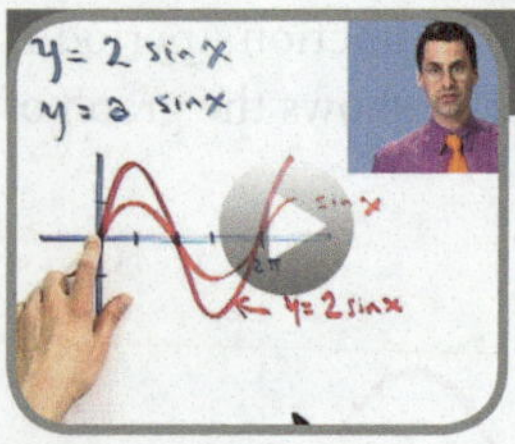

5.5.2 Changing the Amplitude of Sine and Cosine Functions

Recall that for $y = \sin x$ and $y = \cos x$, the y-coordinate at each minimum is -1, and the y-coordinate at each maximum is 1. Therefore, the range of $y = \sin x$ and $y = \cos x$ is the interval $[-1, 1]$, and the amplitude is 1.

Consider a sine or cosine function of the form $y = a \sin x$ or $y = a \cos x$. The values of $\sin x$ and $\cos x$ are multiplied by a, so the y-coordinate at each minimum is $-|a|$, and the y-coordinate at each maximum is $|a|$. Therefore, the range of $y = a \sin x$ and $y = a \cos x$ is the interval $[-|a|, |a|]$, and the amplitude is $|a|$.

TIP

Since the amplitude is half the distance between the minimum and maximum y-values, the amplitude of $y = a \sin x$ and $y = a \cos x$ is $\frac{\left||a|-(-|a|)\right|}{2} = \frac{\left|2|a|\right|}{2}$, that is, $|a|$.

EXAMPLE 1 Finding the Amplitude of Sine and Cosine Functions

Find the amplitude of each function.

A. $y = \frac{1}{3} \sin x$ **B.** $y = -4 \cos x$

SOLUTION

The amplitude of $y = a \sin x$ and $y = a \cos x$ is $|a|$.

A. The amplitude of $y = \frac{1}{3} \sin x$ is $\frac{1}{3}$ because $a = \frac{1}{3}$ and $\left|\frac{1}{3}\right| = \frac{1}{3}$.

B. The amplitude of $y = -4 \cos x$ is 4 because $a = -4$ and $|-4| = 4$.

The Graphs of $y = a \sin x$ and $y = a \cos x$ When $a > 0$

The coefficient a of the trigonometric functions $y = a \sin x$ and $y = a \cos x$ is a scale factor that applies either a vertical stretch or a vertical compression to the graph of the corresponding basic trigonometric function. If $|a| > 1$, then the function's graph is stretched vertically (i.e., becomes taller) relative to the graph of the corresponding basic trigonometric function. This is because, relative to the corresponding basic function, the y-coordinate at each maximum is increased, and the y-coordinate at each minimum is decreased.

If $0 < |a| < 1$, then the function's graph is compressed vertically (i.e., becomes shorter) relative to the graph of the corresponding basic trigonometric function. This is because the y-coordinate at each maximum is *decreased* and the y-coordinate at each minimum is *increased*, relative to the corresponding basic function.

The graphs of $y = \sin x$, $y = 2 \sin x$, and $y = \frac{1}{2} \sin x$ are shown in Figure 5.5b for comparison.

NOTICE THAT

Compared to the graph of $y = \sin x$, the graph of $y = 2 \sin x$ appears stretched, and the graph of $y = \frac{1}{2} \sin x$ appears compressed.

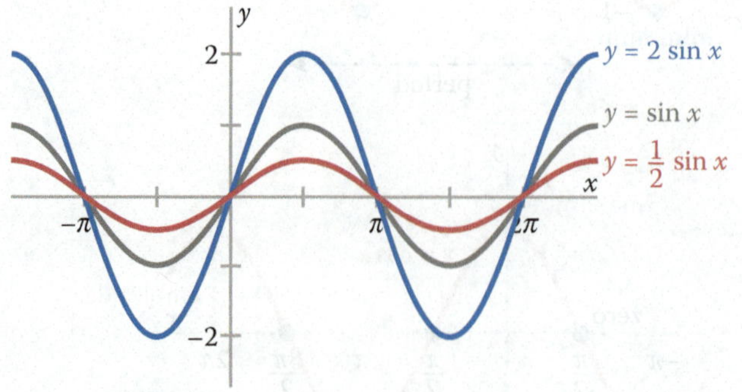

	maximum	minimum
$y = \sin x$	1	-1
$y = 2 \sin x$	2	-2
$y = \frac{1}{2} \sin x$	$\frac{1}{2}$	$-\frac{1}{2}$

Figure 5.5b

Notice that the period, zeros, and x-values of the minimums and maximums are the same for $y = a \sin x$ as they are for $y = \sin x$. The same is true for $y = a \cos x$ and $y = \cos x$.

IMPORTANT

When $a > 0$, the graphs of $y = a \sin x$ and $y = \sin x$ have maximums and minimums at the same x-coordinates, the same zeros, and the same period. However (unless $a = \pm 1$), the two graphs have different amplitudes, and so the maximums and minimums have different y-coordinates.

The Graphs of $y = a \sin x$ and $y = a \cos x$, $a > 0$

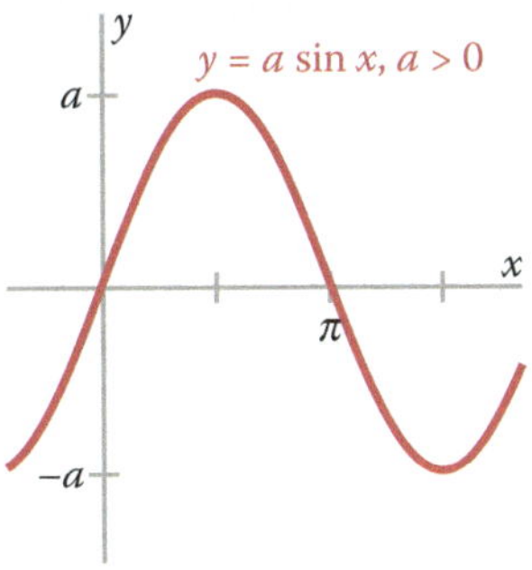

Domain: $(-\infty, \infty)$

Range: $[-a, a]$

Amplitude: a

Zeros: $x = \pi n$

Period: 2π

Maximums: $\left(\frac{\pi}{2} + 2\pi n, a\right)$

Minimums: $\left(\frac{3\pi}{2} + 2\pi n, -a\right)$

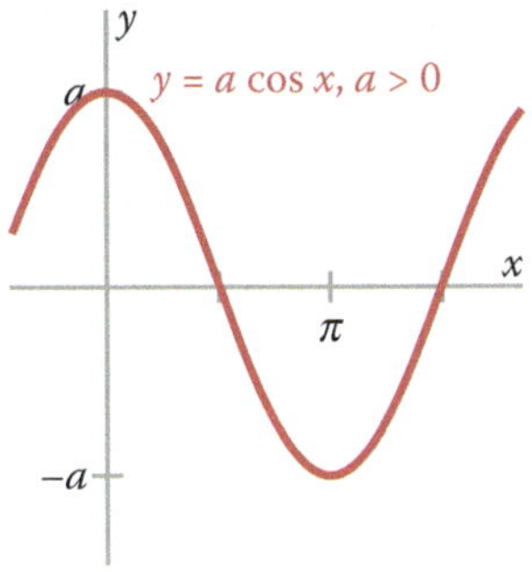

Domain: $(-\infty, \infty)$

Range: $[-a, a]$

Amplitude: a

Zeros: $x = \frac{\pi}{2} + \pi n$

Period: 2π

Maximums: $(2\pi n, a)$

Minimums: $(\pi + 2\pi n, -a)$

The Graphs of $y = a \sin x$ and $y = a \cos x$ ($a < 0$ and $a > 0$)

If $a < 0$, then the graph of $y = a \sin x$ is a reflection of the graph of $y = |a| \sin x$ over the x-axis. Similarly, if $a < 0$, then the graph of $y = a \cos x$ is a reflection of the graph of $y = |a| \cos x$ over the x-axis. For example, the graphs of $y = \sin x$ and $y = -\sin x$ are reflections over the x-axis, as shown in Figure 5.5c. Notice that these graphs have the same amplitude, because $|-1| = |1|$. Furthermore, notice that the y-values for the minimums and maximums have switched.

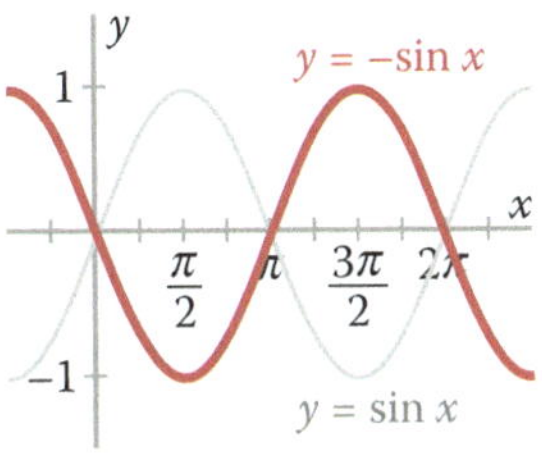

Figure 5.5c

To graph $y = a \sin x$ when $a > 0$, use the graph of $y = \sin x$ to identify the zeros. Then change the y-coordinates of the minimums and maximums of $y = \sin x$ from -1 and 1 to $-a$ and a, respectively. When $a < 0$, graph $y = |a| \sin x$ first, then reflect that graph over the x-axis to graph $y = a \sin x$. Follow the same general procedure to graph $y = a \cos x$.

EXAMPLE 2 Graphing Sine and Cosine Functions: Amplitude Is Not 1

Graph at least two periods of each function.

A. $y = \frac{1}{3} \sin x$

B. $y = -4 \cos x$

SOLUTION

A. To graph $y = \frac{1}{3} \sin x$, first sketch the graph of the basic sine function $y = \sin x$, then compare the period, zeros, maximums, and minimums of the two functions. Since the equation is of the form $y = a \sin x$, the period and the zeros are not changed (i.e., the period is 2π and the zeros occur at $x = \pi n$ for both graphs). Furthermore, the minimums and maximums of both graphs occur at the same x-values, but the y-values at those points are not the same. Instead, since the amplitude of $y = \frac{1}{3} \sin x$ is 1/3, the range is $\left[-\frac{1}{3}, \frac{1}{3}\right]$ and the y-coordinates at the maximums and minimums are 1/3 and −1/3, respectively.

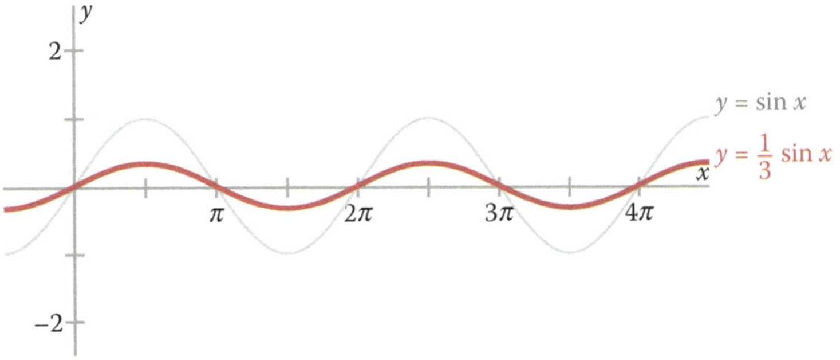

NOTICE THAT

Since $0 < a < 1$, the graph is vertically compressed as compared to the graph of $y = \sin x$.

Confirm several points on the graph by making a table, as shown here. Note that multiplying each value of sin x by 1/3 gives the corresponding y-coordinate for the point on the graph of $y = \frac{1}{3}\sin x$.

From the table, we see that the graph passes through (0, 0), (π/2, 1/3), (π, 0), (3π/2, –1/3), (2π, 0), and (5π/2, 1/3).

x	sin x	$\frac{1}{3}\sin x$
0	0	0
$\pi/2$	1	1/3
π	0	0
$3\pi/2$	–1	–1/3
2π	0	0
$5\pi/2$	1	1/3

IMPORTANT

If $a < 0$, then the graph of $y = a\cos x$ is a reflection of the graph of $y = |a|\cos x$ over the x-axis.

B. Since $a < 0$, sketch the graph of $y = 4\cos x$ first, then reflect that graph over the x-axis.

Since $y = 4\cos x$ is of the form $y = a\cos x$, the period is 2π and the zeros occur at $x = \frac{\pi}{2} + \pi n$ (like $y = \cos x$).

Furthermore, the minimums and maximums of $y = 4\cos x$ and $y = \cos x$ occur at the same x-values, but the y-values at those points are not the same. Instead, since the amplitude of $y = 4\cos x$ is 4, the range is [–4, 4] and the y-coordinates at the maximums and minimums are 4 and –4, respectively.

Reflect the graph of $y = 4\cos x$ over the x-axis to graph $y = -4\cos x$.

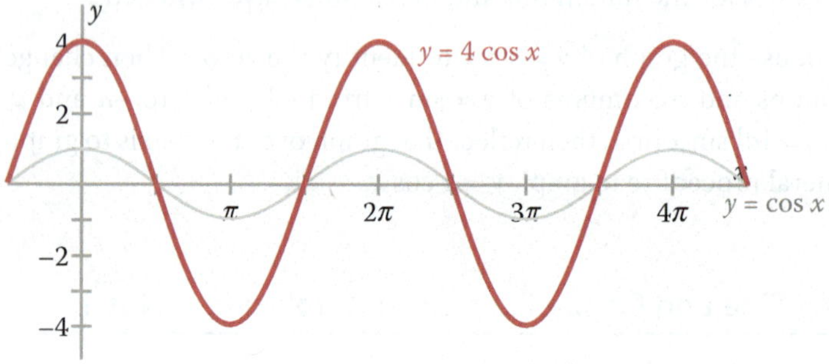

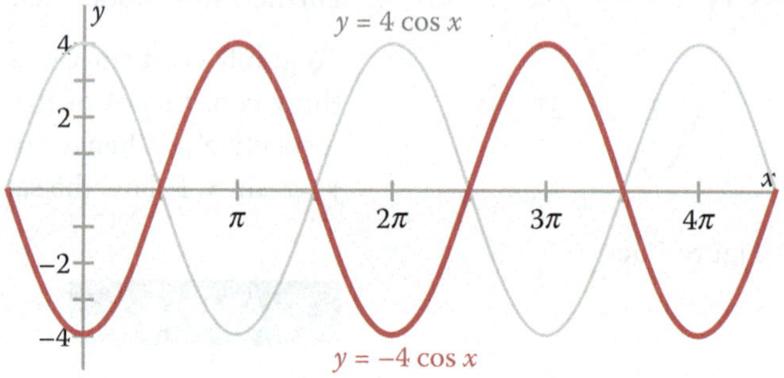

Notice that the point where $x = 0$ is a minimum on the graph of $y = -4\cos x$, but a maximum on the graph of $y = 4\cos x$.

Also, notice that the point where $x = \pi$ is a maximum on the graph of $y = -4\cos x$, but a minimum on the graph of $y = 4\cos x$.

In general, any x-value that corresponds to a minimum on the graph of $y = 4\cos x$ also corresponds to a maximum on the graph of $y = -4\cos x$, and any x-value that corresponds to a maximum on the graph of $y = 4\cos x$ also corresponds to a minimum on the graph of $y = -4\cos x$.

Confirm several points on the graph by making a table, as shown here. Note that multiplying each value of cos x by –4 gives the corresponding y-coordinate for the point on the graph of $y = -4\cos x$.

From the table, we see that the graph passes through (0, –4), (π/2, 0), (π, 4), (3π/2, 0), (2π, –4), and (5π/2, 0).

x	cos x	–4 cos x
0	1	–4
$\pi/2$	0	0
π	–1	4
$3\pi/2$	0	0
2π	1	–4
$5\pi/2$	0	0

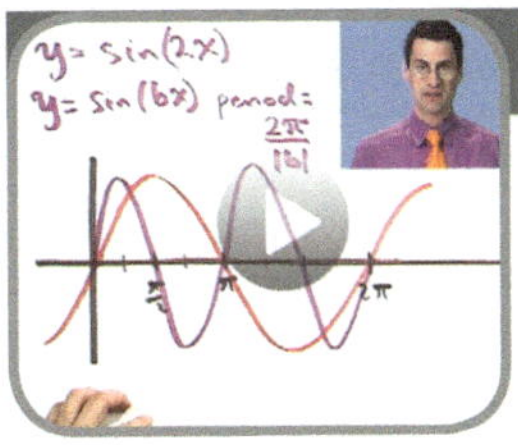

5.5.3 Changing the Period of Sine and Cosine Functions

Recall that the period of $y = \sin x$ and $y = \cos x$ is 2π. Note that the coefficient of x in each function is 1. If the coefficient of x is not 1, then the period is not 2π, and the function's graph will appear either horizontally stretched or horizontally compressed because the periods of sine and cosine functions are affected by changes to the coefficient of the independent variable (in this case x). In general, the period of $y = a \sin bx$ and $y = a \cos bx$ is $\dfrac{2\pi}{|b|}$.

EXAMPLE 3 Finding the Period of Sine and Cosine Functions

Find the period of each function.

A. $y = 2 \sin \frac{1}{2}x$

B. $y = -\cos 2x$

SOLUTION

The period of $y = a \sin bx$ and $y = a \cos bx$ is $\dfrac{2\pi}{|b|}$.

A. The period of $y = 2 \sin \frac{1}{2}x$ is 4π because $b = \frac{1}{2}$ and $\dfrac{2\pi}{\left|\frac{1}{2}\right|} = 2\pi(2) = 4\pi$.

B. The period of $y = -\cos 2x$ is π because $b = 2$ and $\dfrac{2\pi}{|2|} = \pi$.

The Graphs of $y = a \sin bx$ and $y = a \cos bx$

If $b \neq 1$, then the graphs of $y = a \sin bx$ and $y = a \sin x$ have the same amplitude (and thus the same range), but differ in their periods and zeros. The same is true for $y = a \cos bx$ and $y = a \cos x$.

REMEMBER

n is an integer.

The zeros of $y = a \sin bx$ and $y = a \cos bx$ occur at $x = \dfrac{\pi}{|b|}n$ and $x = \dfrac{\pi}{2|b|} + \dfrac{\pi}{|b|}n$, respectively.

To graph $y = a \sin bx$, make a table of ordered pairs to identify several points on the graph. Then sketch the curve with amplitude $|a|$ through those points (and through the function's zeros), following the general curve of a $\sin x$ graph.

Follow the same general process to graph $y = a \cos bx$.

Note that since the period of $y = a \sin bx$ and $y = a \cos bx$ is $2\pi/|b|$, each function's graph completes one cycle between 0 and $2\pi/|b|$.

The Graph of $y = a \sin bx$, $a > 0$

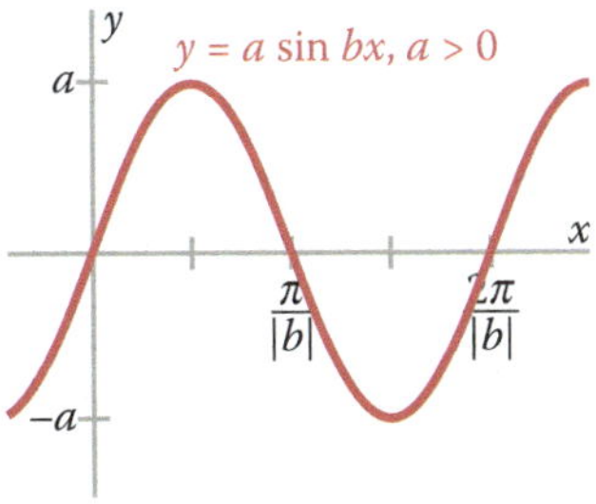

Domain: $(-\infty, \infty)$

Range: $[-a, a]$

Amplitude: a

Zeros: $x = \dfrac{\pi}{|b|}n$

Period: $\dfrac{2\pi}{|b|}$

The Graph of $y = a \cos bx,\ a > 0$

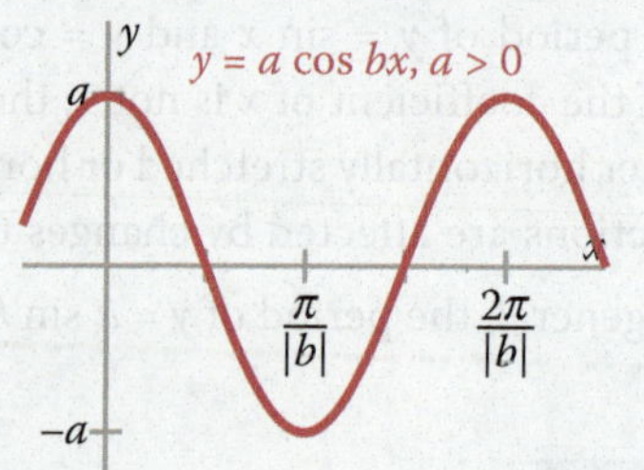

Domain: $(-\infty, \infty)$

Range: $[-a, a]$

Amplitude: a

Zeros: $x = \dfrac{\pi}{2\,|\,b\,|} + \dfrac{\pi}{|\,b\,|}n$

Period: $\dfrac{2\pi}{|\,b\,|}$

EXAMPLE 4 Graphing Sine and Cosine Functions When the Period Is Not 2π

Graph at least one period of each function.

A. $y = \sin \frac{1}{2}x$ **B.** $y = \cos 2x$

SOLUTION

A. The function is of the form $y = a \sin bx$, so the amplitude a is 1 and the range is the interval $[-1, 1]$. Since the period is $\dfrac{2\pi}{|1/2|} = 2\pi(2) = 4\pi$, the graph completes one cycle in 4π radians.

When $x = 0$, $y = \sin 0 = 0$. Thus, the graph passes through the origin $(0, 0)$. Since the period is 4π, the graph also passes through $(4\pi, 0)$.

The period of $y = \sin \frac{1}{2}x$ is twice the period of $y = \sin x$, so its graph is the graph of $y = \sin x$ stretched horizontally by a factor of 2.

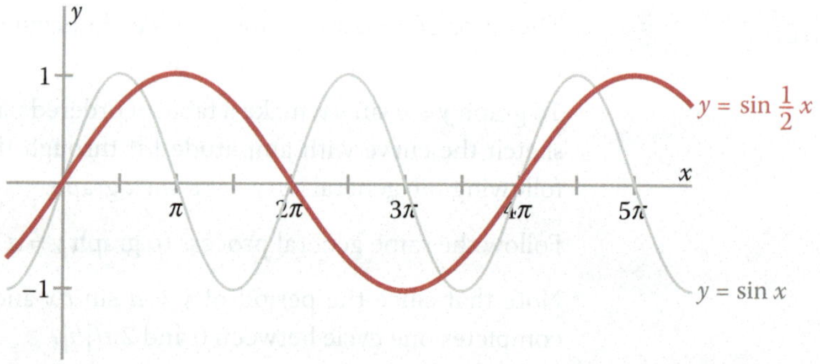

NOTICE THAT

If each $\frac{1}{2}x$-value is a multiple of $\pi/2$, then each corresponding y-value (i.e., each value of $\sin \frac{1}{2}x$) is either −1, 0, or 1.

Confirm several points on the graph by making a table, as shown here. Note that since the coefficient of x is 1/2, choosing x-values such that $\frac{1}{2}x$ is a multiple of $\pi/2$ will result in points where the y-value is −1, 0, or 1.

From the table, we see that the graph passes through $(0, 0)$, $(\pi, 1)$, $(2\pi, 0)$, $(3\pi, -1)$, $(4\pi, 0)$, and $(5\pi, 1)$.

x	$\frac{1}{2}x$	$\sin \frac{1}{2}x$
0	0	0
π	$\pi/2$	1
2π	π	0
3π	$3\pi/2$	−1
4π	2π	0
5π	$5\pi/2$	1

B. The function is of the form $y = a \cos bx$ and the amplitude a is 1, so the range is the interval $[-1, 1]$. The period is $2\pi/|2| = \pi$, so the function completes one cycle in π radians.

The period of $y = \cos 2x$ is 1/2 the period of $y = \cos x$, so its graph is the graph of $y = \cos x$ stretched by a factor of 1/2 (resulting in a horizontal compression).

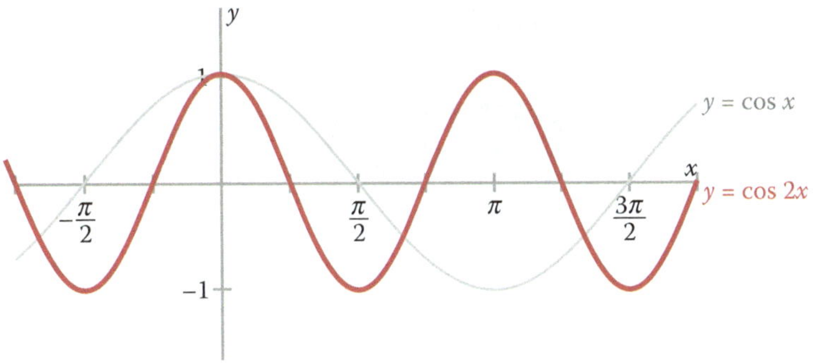

Confirm several points on the graph by making a table, as shown here. Note that since the coefficient of x is 2, choosing x-values such that $2x$ is a multiple of $\pi/2$ will result in points where the y-value is –1, 0, or 1.

x	$2x$	$\cos 2x$
0	0	1
$\pi/4$	$\pi/2$	0
$\pi/2$	π	–1
$3\pi/4$	$3\pi/2$	0
π	2π	1
$5\pi/4$	$5\pi/2$	0

NOTICE THAT

If each $2x$-value is a multiple of $\pi/2$, then each corresponding y-value (i.e., each value of $\cos 2x$) is either –1, 0, or 1. This makes the graph easier to sketch.

So, the graph passes through $(0, 1)$, $\left(\frac{\pi}{4}, 0\right)$, $\left(\frac{\pi}{2}, -1\right)$, $\left(\frac{3\pi}{4}, 0\right)$, $(\pi, 1)$, and $\left(\frac{5\pi}{4}, 0\right)$.

The Zeros of $y = \cos 2x$

The zeros of $y = \cos 2x$ can be found using the b-value 2.

$$x = \frac{\pi}{2|2|} + \frac{\pi}{|2|}n = \frac{\pi}{4} + \frac{\pi}{2}n$$

This general equation of the zeros of $y = \cos 2x$ can be confirmed using the graph of $y = \cos 2x$ from **Example 4B**, where we see that the x-intercepts of the graph of $y = \cos 2x$ occur at $x = -\pi/4$, $x = \pi/4$, $x = 3\pi/4$, and $x = 5\pi/4$, or generally at $x = \frac{\pi}{4} + \frac{\pi}{2}n$, as expected.

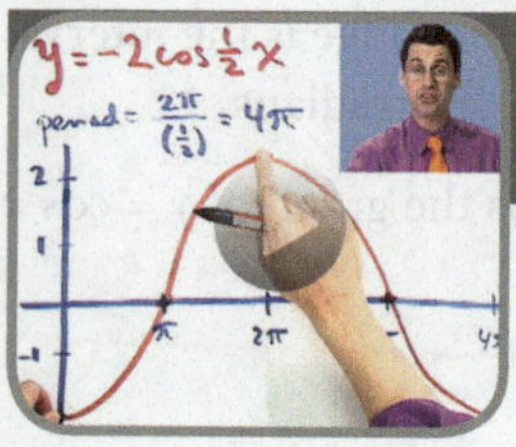

5.5.4 Changing the Amplitude and Period of Sine and Cosine Functions

Each sine and cosine function graphed in the preceding topic had a period other than 2π, but had amplitude 1. The procedure for graphing a sine (or cosine) function when the period is not 2π and the amplitude is not 1 is demonstrated in **Example 5**.

EXAMPLE 5 Graphing Sine and Cosine Functions When the Period Is Not 2π and the Amplitude Is Not 1

Graph at least one period of $y = -5 \sin \frac{1}{3}x$.

SOLUTION

The function is of the form $y = a \sin bx$ and the amplitude is 5, so the range is the interval $[-5, 5]$.

The period is 6π, so the function will complete one cycle in 6π radians.

Since $a = -5$, the graphs of $y = -5 \sin \frac{1}{3}x$ and $y = 5 \sin \frac{1}{3}x$ are reflections over the x-axis.

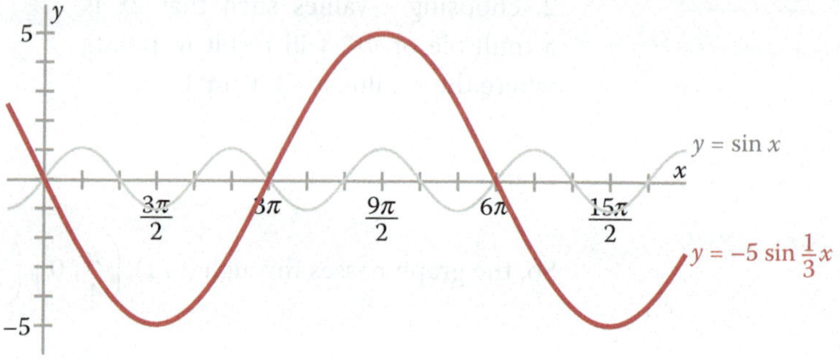

Make a table to confirm several points on the graph.

Choose x-values such that $\frac{1}{3}x$ is a multiple of $\pi/2$ because the coefficient of x is $1/3$.

Multiply each value of $\sin \frac{1}{3}$ by -5 to find the corresponding y-value.

x	$\frac{1}{3}x$	$\sin \frac{1}{3}x$	$-5 \sin \frac{1}{3}x$
0	0	0	0
$3\pi/2$	$\pi/2$	1	-5
3π	π	0	0
$9\pi/2$	$3\pi/2$	-1	5
6π	2π	0	0
$15\pi/2$	$5\pi/2$	1	-5

So, the graph passes through $(0, 0)$, $\left(\frac{3\pi}{2}, -5\right)$, $(3\pi, 0)$, $\left(\frac{9\pi}{2}, 5\right)$, $(6\pi, 0)$, and $\left(\frac{15\pi}{2}, -5\right)$.

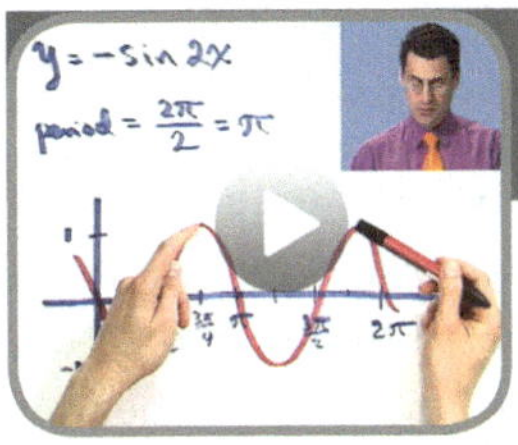

5.5.5 Finding Maximum and Minimum Values and Zeros of Sine and Cosine

We have seen that functions of the form $y = a \sin bx$ and $y = a \cos bx$ are periodic, each having an infinite number of minimums, maximums, and zeros, all of which occur at regular intervals. More specifically, the x-coordinates of the minimums, maximums, or zeros of the functions of these forms occur at $x = r + pn$, where p is the function's period, r is the x-coordinate of any minimum, maximum, or zero, and n is any integer.

EXAMPLE 6 Finding the Minimums, Maximums, and Zeros of Sine and Cosine Functions

Write the general equations for the x-values of all minimums, maximums, and zeros of $y = -5\sin\frac{1}{3}x$. (This function was graphed in **Example 5**).

TIP

The function's graph can be used to identify the zero, the minimum, or the maximum with the smallest positive x-coordinate. This value is typically used in the general equation.

SOLUTION

Use the graph of from **Example 5** to identify the smallest positive x-value corresponding to a minimum, a maximum, and a zero. Note that the function's period is 6π.

The smallest positive minimum occurs at $x = \frac{3\pi}{2}$, so the minimums occur at $x = \frac{3\pi}{2} + 6\pi n$.

The smallest positive maximum occurs at $x = \frac{9\pi}{2}$, so the maximums occur at $x = \frac{9\pi}{2} + 6\pi n$.

The smallest positive zero occurs at $x = 0$, so the zeros occur at $x = 0 + \frac{6\pi}{2}n = 0 + 3\pi n = 3\pi n$.

Substitute the n-values and compare the x-values to the graph to confirm the general equations.

Zeros

$x = -6\pi$	$3\pi(-2)$	$x = 0$	$3\pi(0)$	$x = 6\pi$	$3\pi(2)$
$x = -3\pi$	$3\pi(-1)$	$x = 3\pi$	$3\pi(1)$	$x = 9\pi$	$3\pi(3)$

Minimums

$x = -\frac{9\pi}{2}$	$\frac{3\pi}{2} + 6\pi(-1) = \frac{3\pi}{2} - 6\pi = -\frac{9\pi}{2}$
$x = \frac{3\pi}{2}$	$\frac{3\pi}{2} + 6\pi(0) = \frac{3\pi}{2} + 0 = \frac{3\pi}{2}$
$x = \frac{15\pi}{2}$	$\frac{3\pi}{2} + 6\pi(1) = \frac{3\pi}{2} + 6\pi = \frac{15\pi}{2}$
$x = \frac{27\pi}{2}$	$\frac{3\pi}{2} + 6\pi(2) = \frac{3\pi}{2} + 12\pi = \frac{27\pi}{2}$

Maximums

$x = -\frac{3\pi}{2}$	$\frac{9\pi}{2} + 6\pi(-1) = \frac{9\pi}{2} - 6\pi = -\frac{3\pi}{2}$
$x = \frac{9\pi}{2}$	$\frac{9\pi}{2} + 6\pi(0) = \frac{9\pi}{2} + 0 = \frac{9\pi}{2}$
$x = \frac{21\pi}{2}$	$\frac{9\pi}{2} + 6\pi(1) = \frac{9\pi}{2} + 6\pi = \frac{21\pi}{2}$
$x = \frac{33\pi}{2}$	$\frac{9\pi}{2} + 6\pi(2) = \frac{9\pi}{2} + 12\pi = \frac{33\pi}{2}$

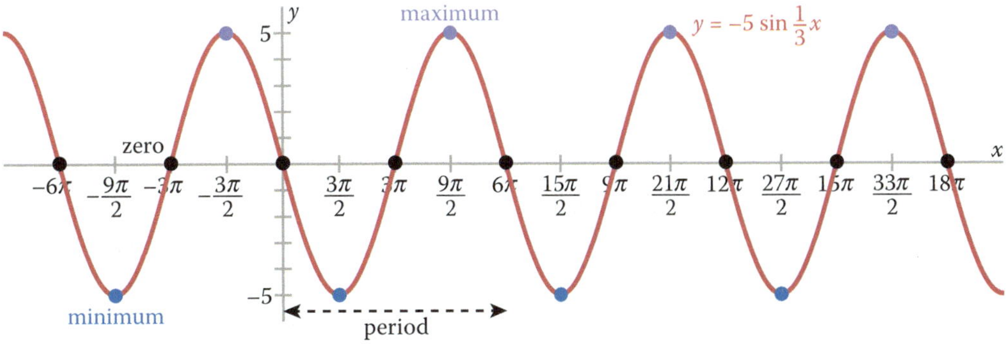

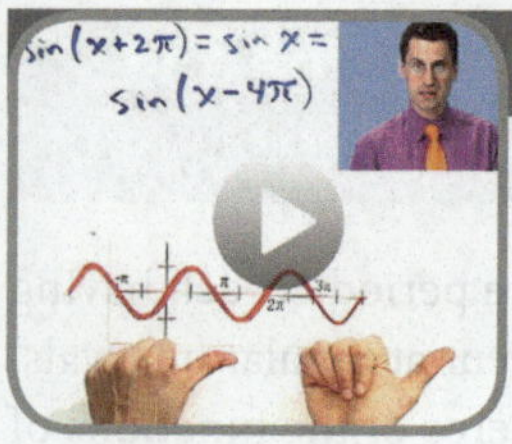

5.5.6 Graphing Sine and Cosine Functions with Phase Shifts

Functions of the form $y = \sin(x - c)$ and $y = \cos(x - c)$, where c is any real number, can be graphed by horizontally translating the graph of the corresponding basic trigonometric function. Specifically, the graph of $y = \sin(x - c)$ is the graph of $y = \sin x$ shifted c units to the right if $c > 0$ or c units to the left if $c < 0$. The graphs of cosine functions behave in the same way.

The horizontal translation applied to the graph of $y = \sin x$ when graphing a function of the form $y = \sin(x - c)$ is also called a **phase shift**. If the graph of $y = \sin x$ is horizontally translated c units to the right, then the phase shift is c. If the graph of $y = \sin x$ is horizontally translated c units to the left, then the phase shift is $-c$.

When the Phase Shift Is $2\pi n$

Since the period of $y = \sin x$ and $y = \cos x$ is 2π, shifting the graph of $y = \sin x$ or $y = \cos x$ to the left or right 2π units does not change the appearance of the graph of the function. Furthermore, shifting the graph of $y = \sin x$ or $y = \cos x$ to the left or right of any multiple of 2π units does not change the resulting graph of the function. For example, $y = \sin x$, $y = \sin(x + 2\pi)$, $y = \sin(x + 4\pi)$, and $y = \sin(x - 4\pi)$ all result in the same graph.

When the Phase Shift Is $\pi/2$

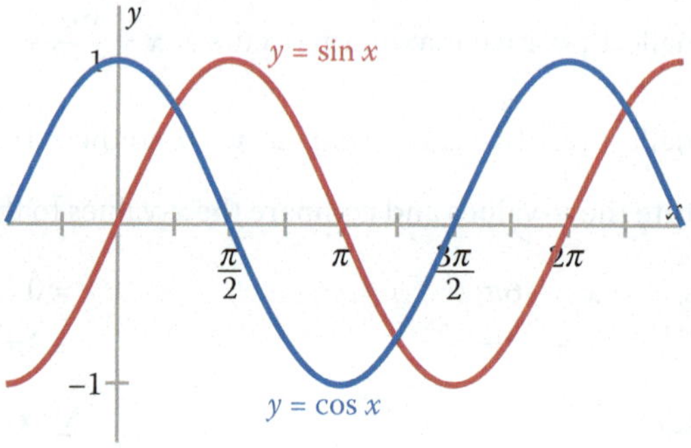

Figure 5.5d

Figure 5.5d shows the graphs of $y = \sin x$ and $y = \cos x$. Notice that each graph is a horizontal translation of the other. Specifically, if the graph of $y = \cos x$ is shifted $\pi/2$ units to the right, then the resulting graph is the same as the graph of $y = \sin x$, as shown in Figure 5.5e. Therefore, the graphs of $y = \cos\left(x - \frac{\pi}{2}\right)$ and $y = \sin x$ are the same.

Furthermore, shifting the graph of $y = \cos x$ to the right $\frac{\pi}{2} + 2\pi n$ units (where n is an integer) will also result in the graph of $y = \sin x$.

Therefore, the graph of any equation of the form $y = \cos\left(x - \left(\frac{\pi}{2} + 2\pi n\right)\right)$ is the same as the graph of $y = \sin x$.

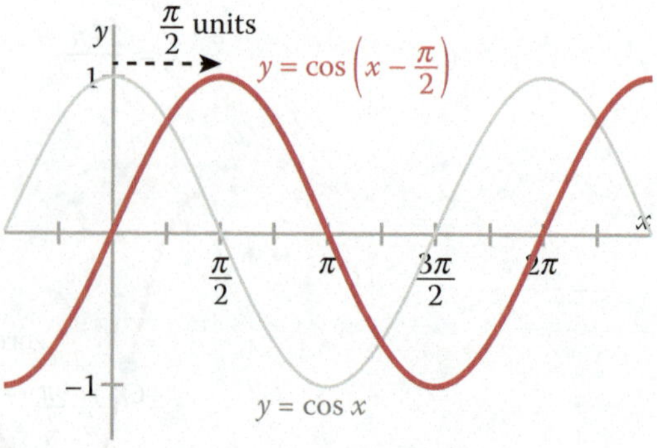

Figure 5.5e

> **NOTICE THAT**
>
> *Shifting the graph of $y = \sin x$ to the left $\pi/2 + 2\pi n$ units (where n is an integer) will also result in the graph of $y = \cos x$.*

If the graph of $y = \sin x$ is shifted $\pi/2$ units to the left, the resulting graph is the same as the graph of $y = \cos x$, as shown in Figure 5.5f. So, $y = \cos x$ and $y = \sin\left(x + \dfrac{\pi}{2}\right)$ have the same graph.

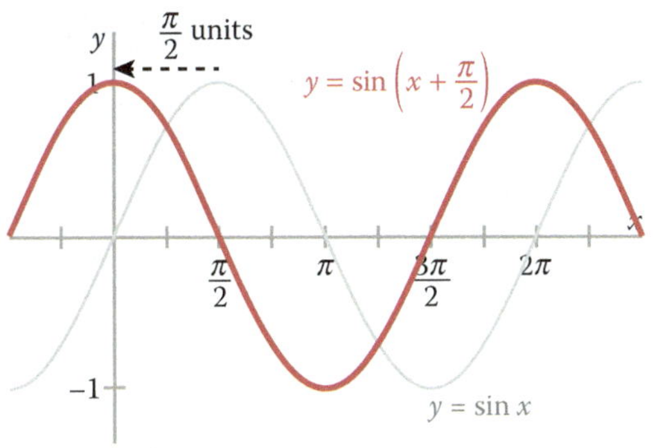

Figure 5.5f

EXAMPLE 7 Graphing Sine and Cosine Functions with a Phase Shift

Graph at least two periods of $y = \cos(x + \pi)$ and identify all zeros, if any.

SOLUTION

Since $y = \cos(x + \pi)$ is of the form $y = \cos(x - c)$ where $c = -\pi$ (i.e., π is added to x), the graph of $y = \cos(x + \pi)$ is the graph of $y = \cos x$ shifted π units to the left. From the graph, we see that $y = \cos(x + \pi)$ and $y = \cos x$ have the same zeros.

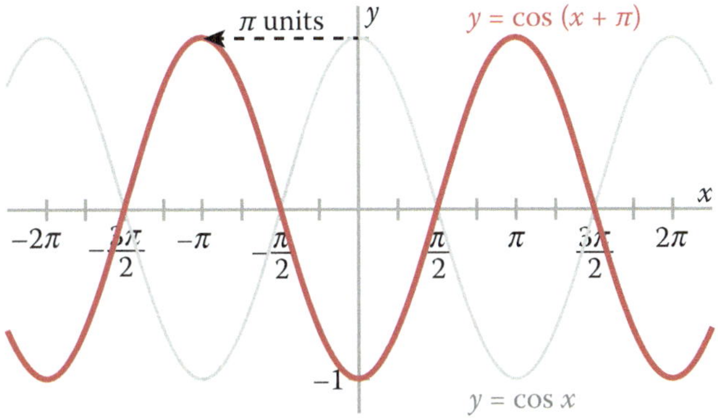

Therefore, the zeros of $y = \cos(x + \pi)$ occur at $x = \dfrac{\pi}{2} + \pi n$.

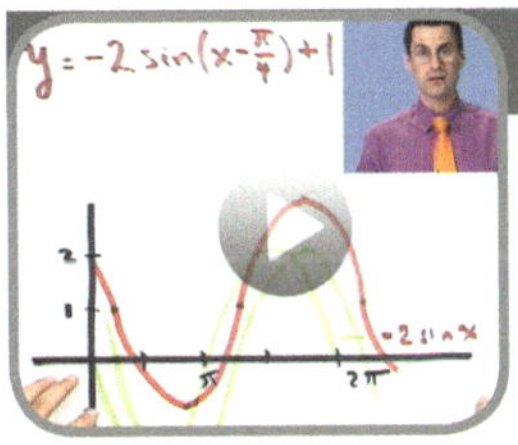

5.5.7 Graphing All Types of Sine and Cosine Functions

Functions of the form $y = \sin x + d$ and $y = \cos x + d$ (i.e., where a constant d is added to the trigonometric function $\sin x$ or $\cos x$) can be graphed by translating the graph of $y = \sin x$ or $y = \cos x$ vertically d units (upward if $d > 0$ or downward if $d < 0$).

Note that a vertical shift of d units affects a function's range. The amplitude of $y = a \sin x + d$ is still $|a|$, but the range is not $[-|a|, |a|]$ unless $d = 0$. For example, consider the function $y = 5 \sin x + 1$. The amplitude is 5 and the vertical shift is 1 unit up. Therefore, the function's range is $[-4, 6]$, as shown in Figure 5.5g.

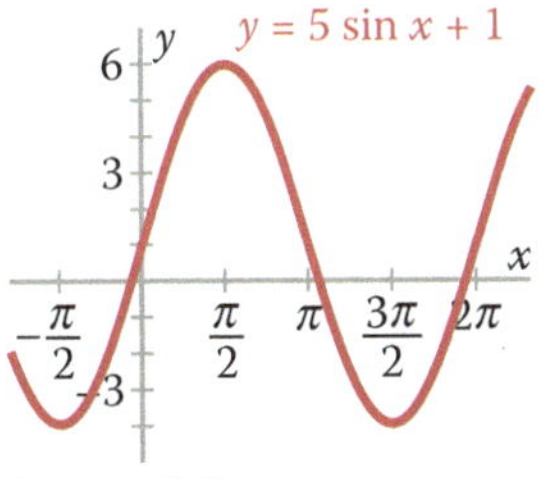

Figure 5.5g

The Graphs of $y = a \sin b(x - c) + d$ and $y = a \cos b(x - c) + d$

- The amplitude is $|a|$.
- The phase shift is c units right if $c > 0$, or left if $c < 0$.
- The period is $\dfrac{2\pi}{|b|}$.
- The vertical shift is d units upward if $d > 0$, or downward if $d < 0$.

EXAMPLE 8 Describing the Attributes of Sine and Cosine Functions

Find the amplitude and period of each function. Then describe any phase shifts or vertical shifts, as compared to the corresponding basic trigonometric function.

A. $y = 2\cos 4(x - \pi) + 3$ **B.** $y = 0.5\sin 2x - 1$

SOLUTION

A. The amplitude of $y = a\cos b(x - c) + d$ is $|a|$ and the period is $2\pi/|b|$. Therefore, since $a = 2$ and $b = 4$, the amplitude of the graph of $y = 2\cos 4(x - \pi) + 3$ is 2 and its period is $\pi/2$.

When compared to $y = \cos x$, the graph has a phase shift of π units to the right (because π is subtracted from x), and a vertical shift 3 units upward (because 3 is added to y).

B. The amplitude of $y = a\sin b(x - c) + d$ is $|a|$ and the period is $2\pi/|b|$. Therefore, since $a = 0.5$ and $b = 2$, the amplitude of the graph of $y = 0.5\sin 2x - 1$ is 0.5 and its period is π.

When compared to $y = \sin x$, the graph has no phase shift (because nothing is added to or subtracted from x). However, the graph of $y = 0.5\sin 2x - 1$ does have a vertical shift 1 unit downward (because 1 is subtracted from y).

CAUTION

$y = 0.5\sin 2x - 1$ is equivalent to $y = -1 + 0.5\sin 2x$. but is not equivalent to $y = 0.5\sin(2x - 1)$, because $\sin 2x - 1 \neq \sin(2x - 1)$.

EXAMPLE 9 Graphing Sine and Cosine Functions

Graph at least two periods of each function.

A. $y = 2\cos 4(x - \pi) + 3$ **B.** $y = 0.5\sin 2x - 1$

SOLUTION

A. As shown in **Example 8**, the amplitude of the graph $y = 2\cos 4(x - \pi) + 3$ is 2, its period is $\pi/2$, and, when compared to $y = \cos x$, the graph has a phase shift π units to the right and a vertical shift 3 units upward.

First, graph $y = 2\cos 4x$ (i.e., $y = \cos x$ but with amplitude 2 and period $\pi/2$.)

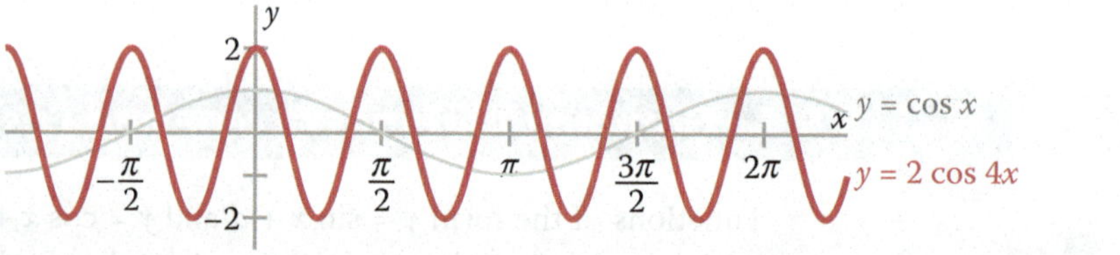

Apply the phase shift, π units to the right.

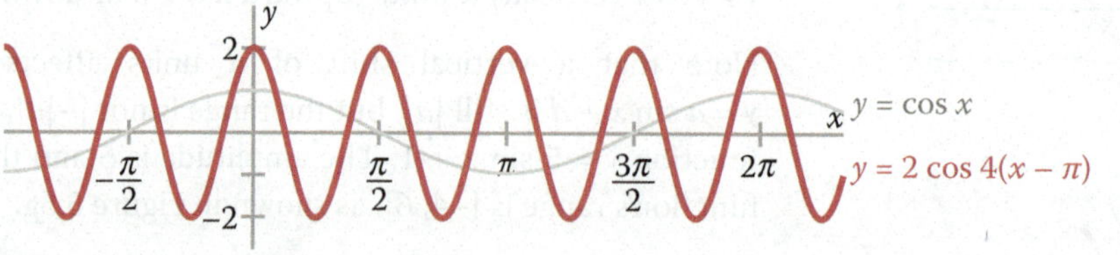

NOTICE THAT

The phase shift does not change the appearance of the graph, because the phase shift is a multiple of the period.

Apply the vertical shift, 3 units upward.

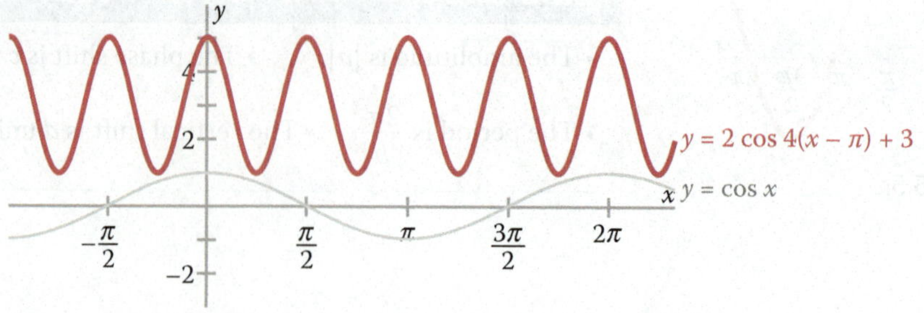

NOTICE THAT

The function $y = 2\cos 4(x - \pi) + 3$ has no zeros.

B. From **Example 8**, the amplitude of the graph of $y = 0.5 \sin 2x - 1$ is 0.5, its period is π, and as compared to $y = \sin x$, the graph has a vertical shift 1 unit downward (but no phase shift).

First, graph $y = 0.5 \sin 2x$ (i.e., $y = \sin x$ but with amplitude 0.5 and period π.)

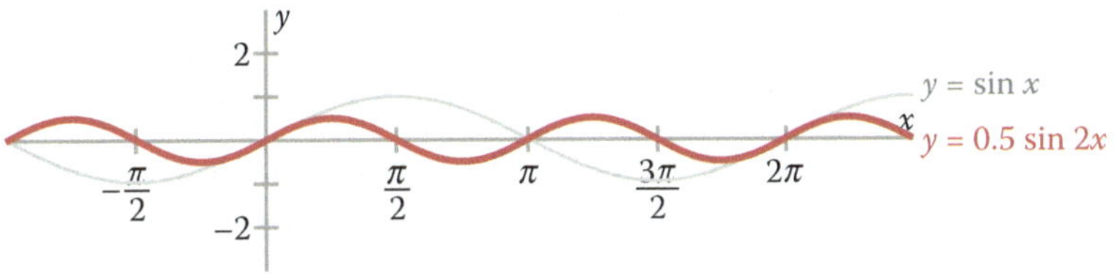

Apply the vertical shift, 1 unit downward.

NOTICE THAT

The function $y = 0.5 \sin 2x - 1$ has no zeros.

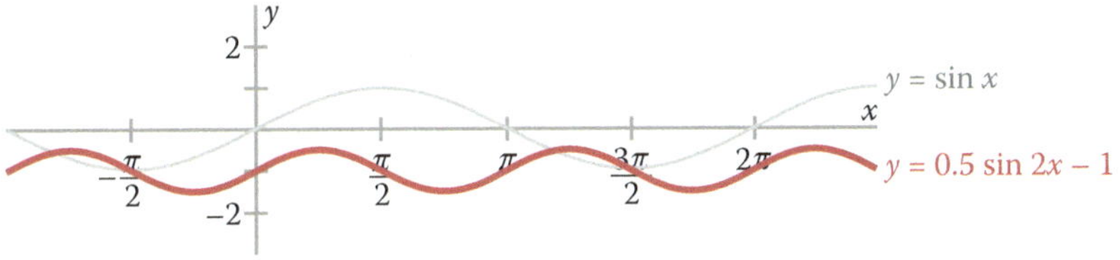

5.5.8 Modeling with Sine and Cosine Functions

If a trigonometric function of a binomial contains a coefficient of the independent variable that is not equal to 1, then the coefficient must be factored from the terms in the binomial before determining the period or any horizontal shifts. This is demonstrated in **Example 10**.

EXAMPLE 10 Graphing Sine and Cosine Functions

Graph at least one period of each function.

A. $y = \sin(2x - \pi)$ **B.** $y = 5\cos\left(\frac{1}{2}x + \frac{3\pi}{2}\right) + 1$

SOLUTION

A. First, factor 2 from $2x - \pi$, then graph the resulting function.

$$y = \sin(2x - \pi) = \sin 2\left(x - \frac{\pi}{2}\right)$$

IMPORTANT

The phase shift is not π units. The coefficient 2 must be factored from $2x - \pi$ before the phase shift is identified.

Since $y = \sin 2\left(x - \frac{\pi}{2}\right)$ is of the form $y = a \sin b(x - c)$, where $a = 1$, $b = 2$, and $c = \pi/2$, the amplitude is 1, the period is $2\pi/|2| = \pi$, and the phase shift is $\pi/2$ units to the right.

Graph $y = \sin 2x$ (i.e., $y = \sin x$ but with period π).

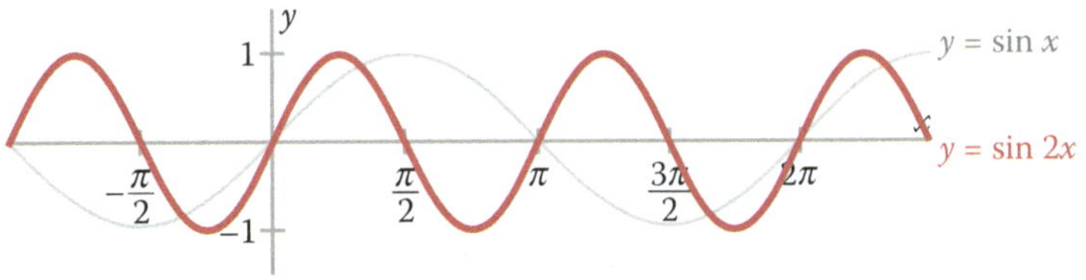

NOTICE THAT

The zeros of $y = \sin(2x - \pi)$ occur at $x = \frac{\pi}{2}n$.

Apply the phase shift, $\pi/2$ units to the right.

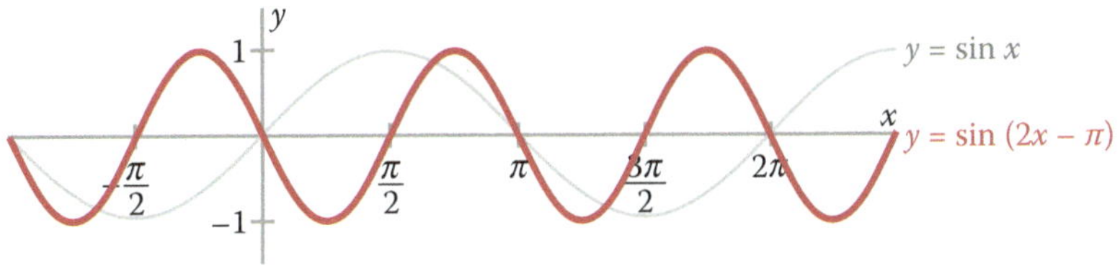

B. First, factor $\frac{1}{2}$ from $\frac{1}{2}x + \frac{3\pi}{2}$, then graph the resulting function.

$$y = 5\cos\left(\frac{1}{2}x + \frac{3\pi}{2}\right) + 1 = 5\cos\frac{1}{2}(x + 3\pi) + 1$$

Since $y = 5\cos\frac{1}{2}(x + 3\pi) + 1$ is of the form $y = a\cos b(x - c) + d$, where $a = 5$, $b = \frac{1}{2}$, $c = -3\pi$, and $d = 1$, the amplitude is 5, the period is $2\pi/|1/2| = 4\pi$, the phase shift is 3π units to the left, and the vertical shift is 1 unit upward.

Graph $y = 5\cos\frac{1}{2}x$ (i.e., $y = \cos x$ but with amplitude 5 and period 4π).

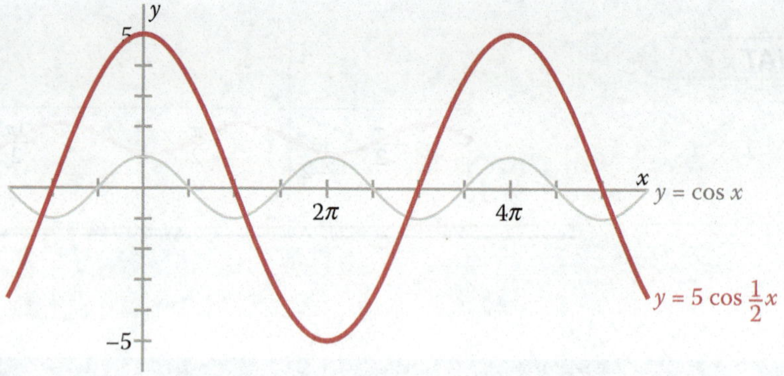

Apply the phase shift, 3π units to the left, and the vertical shift, 1 unit upward.

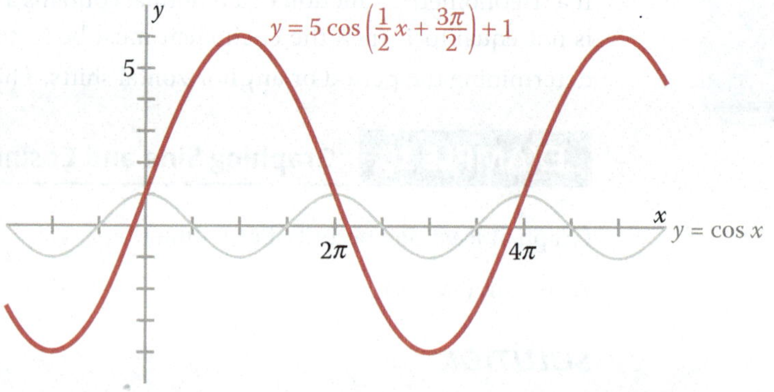

Graphs Where the Scale on the x-Axis Is Not in Terms of π

Thus far in Section 5.5, all of the sine and cosine functions have been graphed on a coordinate plane in which the scale on the x-axis is in terms of π. This type of graph is useful when the period or phase shift of the function is in terms of π.

However, sine and cosine functions can be graphed on a coordinate plane in which the x-axis is not in terms of π. If the scale on the x-axis is not in terms of π, but the function's period or phase shift is in terms of π, then the zeros (x-intercepts) will need to be approximated.

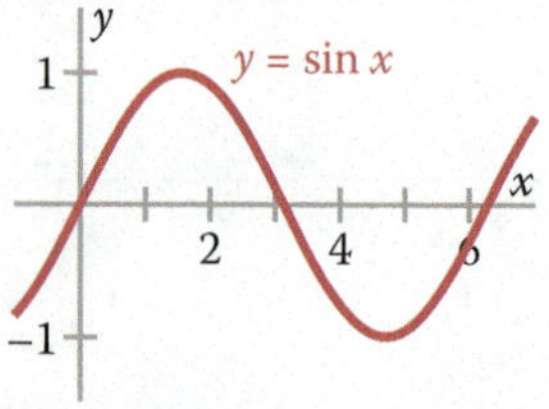

Figure 5.5h

For example, Figure 5.5h shows the graph of $y = \sin x$ on a coordinate plane in which the scales on the x- and y-axes are both 1. Since the graph $y = \sin x$ has zeros at multiples of π, the graph in Figure 5.5.h crosses the x-axis when $x = 0$, $x = \pi \approx 3.1$, $x = 2\pi \approx 6.3$, and so on.

EXAMPLE 11 Modeling with a Sine Function

A windmill takes 6 seconds to make one complete rotation. The height y in feet above the ground of the tip of one blade can be modeled by $y = 40 \sin\left(\frac{\pi}{3}x - \frac{1.5\pi}{3}\right) + 60$, where x is the time in seconds. Graph at least two periods of the function and find the maximum height of the tip of the blade.

SOLUTION

First, factor $\frac{\pi}{3}$ from $\frac{\pi}{3}x - \frac{1.5\pi}{3}$, then graph the resulting function.

$$y = 40 \sin\left(\frac{\pi}{3}x - \frac{1.5\pi}{3}\right) + 60 = 40 \sin\frac{\pi}{3}(x - 1.5) + 60$$

The equation $y = 40 \sin\frac{\pi}{3}(x - 1.5) + 60$ is of the form $y = a \sin b(x - c) + d$, where $a = 40$, $b = \pi/3$, $c = 1.5$, and $d = 60$. So, the phase shift is 1.5 units to the right, the vertical shift is 60 units upward, the amplitude is 40, and the period is 6 (because $\frac{2\pi}{\left|\frac{\pi}{3}\right|} = 2\pi\left(\frac{3}{\pi}\right) = 6$).

Graph $y = 40 \sin\frac{\pi}{3}x$ (i.e., $y = \sin x$, but with amplitude 40 and period 6.) Since the period is not in terms of π, make the scale used on the x-axis some whole number (i.e., not in terms of π).

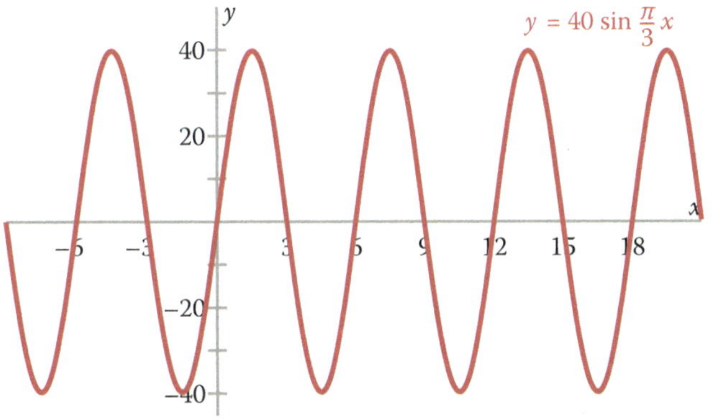

Apply the phase shift, 1.5 units to the right, and the vertical shift, 60 units upward.

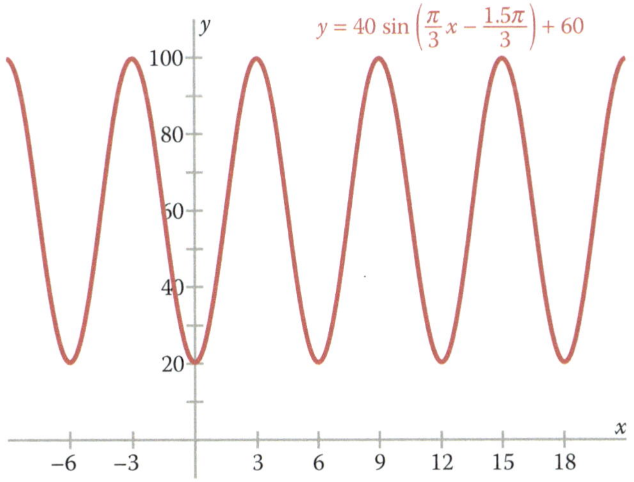

The curve represents the height of the tip of the blade of the windmill after x seconds, and so the graph's maximum and minimum represent the high and low positions of the tip of the blade. Therefore, from the graph, the maximum height of a tip of the blade is 100 feet.

NOTICE THAT

The period of the function corresponds to the number of seconds needed for the blade to make one complete rotation. The vertical shift relates to the height of the center axis that turns the blades, and the phase shift relates to the starting position of the blade, which happens to be at the lowest point of the rotation. The amplitude represents the radius of the windmill.

SECTION 5.5 EXERCISES

Warm Up

Identify the real number(s) t that satisfy each equation.

1. $\cos t = 0$
2. $\sin t = 1$
3. $\cos t = -1$

Just the Facts

Fill in the blanks.

4. The distance from the minimum y-value to the maximum y-value divided by _____ is called the _____ of the sine and cosine functions.
5. Due to the periodic nature of the sine and _____ functions, the x-intercepts, or _____, occur at regular intervals, as do the _____ and minimums.
6. The range of $y = 10 \sin x$ is _____.
7. One major difference between $y = 2 \cos x$ and $y = \cos x$ is the _____ of the cosine functions, which affects the range of the functions.
8. When the graph of $y = a \sin x$ is shifted $\pi/2$ units left, the resulting graph is the same as _____.
9. The period of $y = a \cos bx$ is _____. Therefore, the graph of $y = -4 \cos \frac{1}{4}x$ has period _____ and amplitude _____.
10. A sine function of the form $y = a \sin b(x - c) + d$ has a phase shift to the right when _____, and to the left when _____.

Essential Skills

In Exercises 1–4, find the amplitude of each function.

1. $y = \frac{1}{2} \sin x$
2. $y = 3 \cos x$
3. $y = -5 \cos x$
4. $y = -11 \sin x$

In Exercises 5–12, graph at least two periods of each function.

5. $y = 4 \cos x$
6. $y = \frac{1}{2} \sin x$
7. $y = \frac{3}{4} \cos x$
8. $y = 3 \sin x$
9. $y = -5 \sin x$
10. $y = -3 \cos x$
11. $y = -\frac{1}{4} \sin x$
12. $y = -\frac{1}{3} \cos x$

In Exercises 13–16, find the period of each function.

13. $y = -6 \sin \pi x$
14. $y = 2 \cos 4x$
15. $y = \frac{1}{3} \cos \frac{3}{4}x$
16. $y = -\frac{1}{2} \sin \frac{1}{4}x$

In Exercises 17–24, graph at least two periods of each function.

17. $y = \cos 4x$
18. $y = \sin 8x$
19. $y = \sin 3x$
20. $y = \cos 2x$
21. $y = \sin \frac{1}{3}x$
22. $y = \cos \frac{1}{2}x$
23. $y = \cos \frac{2}{3}x$
24. $y = \sin \frac{3}{4}x$

In Exercises 25–28, find the period and amplitude of each function.

25. $y = -4 \cos 3x$
26. $y = 3 \sin 5x$
27. $y = -\frac{1}{2} \sin \frac{1}{4}\pi x$
28. $y = -\frac{2}{3} \cos \frac{3}{5}x$

In Exercises 29–36, graph at least two periods of each function.

29. $y = 2 \cos 2x$
30. $y = \frac{1}{2} \sin 3x$
31. $y = -\frac{1}{4} \sin \frac{1}{3}x$
32. $y = -3 \cos \frac{1}{2}x$
33. $y = 4 \sin \frac{1}{4}x$

34. $y = -2 \sin 2x$

35. $y = -\frac{3}{4} \cos \frac{3}{4}x$

36. $y = \frac{1}{3} \cos 4x$

In Exercises 37–44, find all x-values for the maximums, minimums, and zeros of each function.

37. $y = \frac{1}{4} \sin \frac{1}{2}x$

38. $y = -5 \sin 3x$

39. $y = -2 \sin 4x$

40. $y = \frac{1}{3} \sin \frac{2}{3}x$

41. $y = \frac{1}{2} \cos 2x$

42. $y = 3 \cos \frac{1}{2}x$

43. $y = -4 \cos \frac{1}{3}x$

44. $y = -\frac{1}{3} \cos 3x$

In Exercises 45–48, graph at least two periods of each function and identify all zeros, if any.

45. $y = \cos\left(x + \frac{\pi}{2}\right)$

46. $y = \sin\left(x - \frac{\pi}{4}\right)$

47. $y = \sin\left(x + \frac{2\pi}{3}\right)$

48. $y = \cos\left(x - \frac{\pi}{6}\right)$

In Exercises 49–52, find the amplitude and period of each function. Then describe any phase shifts or vertical shifts, as compared to the corresponding basic trigonometric function.

49. $y = 2 \sin\left(x - \frac{\pi}{3}\right) + 1$

50. $y = -3 \cos\left(x + \frac{\pi}{6}\right) + 2$

51. $y = -\frac{3}{4} \cos(x + \pi) - 5$

52. $y = \frac{2}{5} \sin\left(x - \frac{\pi}{4}\right) - 3$

In Exercises 53–60, graph at least two periods of each function.

53. $y = 2 \sin\left(x - \frac{\pi}{6}\right)$

54. $y = 3 \cos\left(x + \frac{\pi}{6}\right)$

55. $y = -\frac{1}{2} \cos\left(x + \frac{\pi}{4}\right)$

56. $y = -\frac{3}{4} \sin(x + \pi)$

57. $y = 2 \cos\left(x + \frac{\pi}{3}\right) - 1$

58. $y = 3 \sin\left(x - \frac{\pi}{6}\right) + 2$

59. $y = -\frac{4}{5} \sin(x - \pi) + 3$

60. $y = -\frac{1}{3} \cos\left(x + \frac{\pi}{4}\right) - 2$

In Exercises 61–64, find the amplitude and period of each function. Then describe any phase shifts or vertical shifts, as compared to the corresponding basic trigonometric function.

61. $y = 2 \sin\left(x - \frac{\pi}{3}\right) + 1$

62. $y = -2 \cos\left(2x + \frac{\pi}{3}\right) + 3$

63. $y = -\frac{1}{4} \cos\left(\frac{1}{2}x + \frac{\pi}{2}\right) - 1$

64. $y = -\frac{4}{7} \sin\left(\frac{1}{3}x - \frac{\pi}{2}\right) - 2$

In Exercises 65–76, graph at least two periods of each function.

65. $y = -\sin \frac{1}{2}x + 3$

66. $y = -2 \cos 2\left(x + \frac{\pi}{4}\right) - 1$

67. $y = \frac{1}{3} \cos\left(x - \frac{\pi}{6}\right) - 1$

68. $y = \frac{1}{2} \sin\left(x - \frac{\pi}{3}\right) + 2$

69. $y = \frac{1}{2} \cos(4x + \pi)$

70. $y = 4 \sin\left(2x - \frac{\pi}{2}\right)$

71. $y = -2 \sin(3x - 2\pi)$

72. $y = -\frac{2}{3} \cos\left(2x - \frac{\pi}{3}\right)$

73. $y = 4 \sin\left(\frac{1}{2}x - \frac{\pi}{4}\right)$

74. $y = -2 \cos\left(\frac{1}{3}x + \frac{\pi}{6}\right)$

75. $y = 3 \cos\left(\frac{1}{4}x - \frac{\pi}{12}\right)$

76. $y = \frac{1}{2} \sin\left(\frac{2}{3}x - \frac{\pi}{4}\right)$

77. The height h of a seat above ground on a Ferris wheel at time t is given by $h(t) = 30 \sin \frac{\pi}{10}t + 35$, where $t \geq 0$ seconds and h is in feet. Graph the function for two periods and find the maximum height of the seat.

78. The current I, in amperes, flowing through an alternating circuit at time t, in seconds, is given by $I(t) = 65 \sin\left(20\pi t - \frac{\pi}{6}\right)$, where $t \geq 0$. Graph the function for two periods and find the maximum current flowing through the circuit.

79. A person's blood pressure, $P(t)$, in millimeters of mercury, is modeled by the function $P(t) = 20 \cos \frac{7\pi}{3}t + 85$, where t is the time in seconds. Graph the function for two periods and find the maximum blood pressure.

80. The height h of a seat above ground on a Ferris wheel at time t is given by $h(t) = 45 \cos \frac{\pi}{6}t + 50$, where $t \geq 0$ seconds and h is in feet. Graph the function for two periods and find the maximum height of the seat.

81. The current I, in amperes, flowing through an alternating circuit at time t, in seconds, is given by $I(t) = 200 \sin\left(45\pi t - \frac{\pi}{3}\right)$, where $t \geq 0$. Graph the function for two periods and find the maximum current flowing through the circuit.

82. A person's blood pressure, $P(t)$, in millimeters of mercury, is modeled by the function $P(t) = -20 \cos \frac{5\pi}{3}t + 90$, where t is the time in seconds. Graph the function for two periods and find the maximum blood pressure.

Extensions

83. In a particular wildlife refuge the variation in the total number of white spotted owls can be described by the formula $P(t) = 15 \sin \frac{\pi}{6}t + 20$, where P is the population and t is the time in months since January 1, 1991. What is the maximum population of white spotted owls in the wildlife refuge?

In Exercises 84–87, determine whether the statements are true or false. If false, explain why.

84. $\sin 9\pi = \sin \pi$

85. $\cos \frac{210\pi}{2} = \sin(-52.5\pi)$

86. The graph of $y = \cos\left(x - \frac{\pi}{2}\right)$ is identical to the graph of $y = \sin\left(x + \frac{\pi}{2}\right)$.

87. The function $y = \sin \frac{1}{3}x$ has a period that is one-third that of its corresponding basic trigonometric function.

88. A group of scientists conducting a statistical survey found that traffic in a local mall over the course of a day could be estimated by the equation $P(t) = -3500 \cos \frac{\pi}{5}t + 3500$, where P is the population and t is the time in hours after the mall opens. After the mall opens, how many hours pass before the number of people in the mall reaches its maximum? What is the maximum number of people?

89. The temperature of a patient ranges from 96° Fahrenheit to 100° Fahrenheit as modeled by $T(t) = 2 \sin \frac{\pi}{14}t + C$. What is the value of C?

In Exercises 90–93, use the given amplitude, period, phase shift, and vertical shift to write the equation for the function $y = a \sin b(x - c) + d$.

	Amplitude	Period	Phase Shift	Vertical Shift
90.	3	8π	$\pi/6$ left	6 down
91.	4	6	3 right	7 up
92.	1/2	$3\pi/2$	none	1 up
93.	15	$\pi/3$	$4\pi/5$ left	30 down

94. A Ferris wheel ride at the local carnival lasts 4 minutes, during which time the Ferris wheel makes 8 revolutions. Patrons get on the ride from a platform that is 5 feet above the ground, and after 15 seconds they reach a maximum height of 60 feet. Write the trigonometric function that models this situation.

In Exercises 95–98, graph at least two periods of each function.

95. $y = -15 \sin\left(4\pi x - \frac{2}{3}\right) + 5$

96. $y = -4 \cos\left(\frac{3\pi}{4}x + \frac{4}{5}\right) - 2$

97. $y = 3 + 8 \cos\left(\frac{\pi}{2}x + \frac{\pi}{8}\right)$

98. $y = 7 + 2 \sin\left(2\pi x - \frac{7\pi}{4}\right)$

5.6 GRAPHING OTHER TRIGONOMETRIC FUNCTIONS

OBJECTIVES

- Graph the basic tangent and cotangent functions.
- Graph the basic cosecant and secant functions.
- Use the period to graph cosecant and secant functions.
- Use the period and stretches to graph tangent and cotangent functions.
- Find the vertical asymptotes of cosecant, secant, tangent, and cotangent functions.
- Use a graph to find the zeros of tangent and cotangent functions.
- Use translations to graph cosecant, secant, tangent, and cotangent functions.
- Match graphs of trigonometric functions to their equations.

PREREQUISITE VOCABULARY TERMS

cosecant
cotangent
maximum
minimum
period
phase shift
secant
tangent
zero (of a function)

5.6.1 Introduction to the Graph of the Tangent Function

Recall from Section 5.4 that tangent is an odd function. Therefore, since all odd functions are symmetric with respect to the origin, the graph of $y = \tan x$ must also be symmetric with respect to the origin.

Points on the graph of $y = \tan x$ can be determined by considering the relationship between the tangent function and the sine and cosine functions. Recall the definitions of sine, cosine, and tangent for any angle θ from Section 5.4.

$$\sin\theta = \frac{y}{r} \qquad \cos\theta = \frac{x}{r} \qquad \tan\theta = \frac{y}{x}$$

Notice that the result of dividing $\sin\theta$ by $\cos\theta$ is equal to $\tan\theta$. $\dfrac{\sin\theta}{\cos\theta} = \dfrac{\frac{y}{r}}{\frac{x}{r}} = \dfrac{y}{r}\cdot\dfrac{r}{x} = \dfrac{y}{x} = \tan\theta$

So, $\tan\theta = \dfrac{\sin\theta}{\cos\theta}$ and similarly, $\cot\theta = \dfrac{\cos\theta}{\sin\theta}$.

Therefore, the graph of $y = \tan x$ is the same as the graph of $y = \dfrac{\sin x}{\cos x}$. It follows that for all (x, y) on the graph of $y = \tan x$, the y-coordinate is equal to $\sin x$ divided by $\cos x$.

IMPORTANT

Unless otherwise stated, assume that n represents any integer.

From this definition of tangent (where $\cos x$ is in the denominator), we see that $\tan x$ is undefined for all x such that $\cos x = 0$. Therefore, the graph of $y = \tan x$ has a vertical asymptote at each zero of $y = \cos x$. From Section 5.5, we know that the zeros of $y = \cos x$ occur when $x = \pi/2 + \pi n$ (where n is an integer). Thus, the graph of $y = \tan x$ has a vertical asymptote at each $x = \pi/2 + \pi n$.

Similarly, $y = \tan x = \dfrac{\sin x}{\cos x} = 0$ for all x such that $\sin x = 0$. Since the zeros of $y = \sin x$ occur where $x = \pi n$ (where n is an integer), the equation $x = \pi n$ gives the zeros of $y = \tan x$.

To graph the basic tangent function $y = \tan x$, it is convenient to use a coordinate plane in which the scale on the x-axis is $\pi/4$ and the scale on the y-axis is 1 (since $\tan x$ is equal to -1, 0, or 1 when x is a multiple of $\pi/4$). As shown in Figure 5.6a, use a table to list the values of $\tan x$, where x is a multiple of $\pi/4$.

x	$\tan x$
$-5\pi/4$	-1
$-\pi$	0
$-3\pi/4$	1
$-\pi/2$	undefined
$-\pi/4$	-1
0	0
$\pi/4$	1
$\pi/2$	undefined
$3\pi/4$	-1
π	0
$5\pi/4$	1

Figure 5.6a

We have now seen that tangent is undefined at $x = \pi/2 + \pi n$, and equal to 0 at $x = \pi n$. Therefore, the graph of $y = \tan x$ has a vertical asymptote at $x = -3\pi/2$, $x = -\pi/2$, $x = \pi/2$, $x = 3\pi/2$, $x = 5\pi/2$, and so on (to the left and to the right), and x-intercepts at $x = -\pi$, $x = 0$, $x = \pi$, $x = 2\pi$, and so on (to the left and to the right).

Consider the shape and behavior of the graph of $y = \tan x$ between pairs of consecutive vertical asymptotes. From the table, we see that between the asymptotes at $x = -3\pi/2$ and $x = -\pi/2$, the graph of $y = \tan x$ passes through the points $(-5\pi/4, -1)$, $(-\pi, 0)$, and $(-3\pi/4, 1)$. And between the asymptotes at $x = -\pi/2$ and $x = \pi/2$, the graph of $y = \tan x$ passes through the points $(-\pi/4, -1)$, $(0, 0)$, and $(\pi/4, 1)$. These y-coordinates are repeated between the interval made by each pair of asymptotes. Therefore, the tangent function is periodic and the graph continues to the left and right following this pattern.

NOTICE THAT

The graph of $y = \tan x$ is symmetric about the origin. The graph has no minimums or maximums.

The Basic Tangent Function: $y = \tan x$

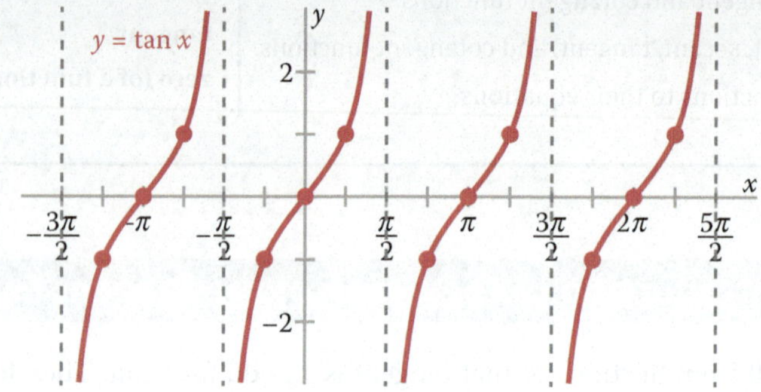

Domain: $\{x : x \neq \frac{\pi}{2} + \pi n\}$

Range: $(-\infty, \infty)$

Vertical Asymptotes: $x = \frac{\pi}{2} + \pi n$

Zeros: $x = \pi n$

Period: π

The same curve is repeated between each pair of consecutive vertical asymptotes. Therefore, the period of $y = \tan x$ is π.

5.6.2 Introduction to the Graphs of the Cosecant and Secant Functions

Recall that the cosecant function is the reciprocal of the sine function, and the secant function is the reciprocal of the cosine function. This reciprocal relationship can be used to graph the basic cosecant function $y = \csc x$ and the basic secant function $y = \sec x$ by first considering the graphs of $y = \sin x$ and $y = \cos x$.

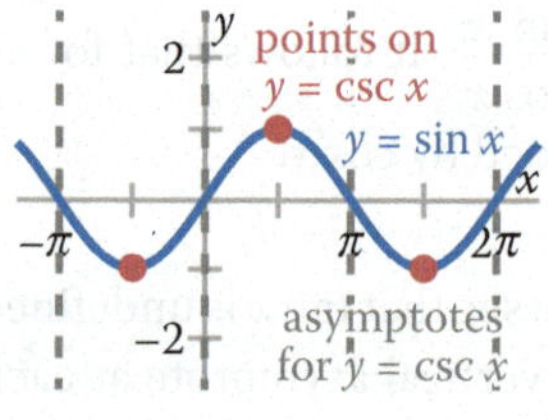

Figure 5.6b

Generally, if (x, y) is a point on the graph of $y = \sin x$ such that $y \neq 0$, then $(x, 1/y)$ is a point on the graph of $y = \csc x$ because the functions are reciprocals of one another. The same relationship exists between the points on the graphs of $y = \cos x$ and $y = \sec x$ as these functions are reciprocals as well.

Since the reciprocal of 0 is undefined, $y = \csc x$ is undefined at each x-value where $y = \sin x$ has a zero. Recall that the zeros of $y = \sin x$ are at $x = \pi n$ (i.e., when x is a multiple of π). Therefore, $y = \csc x$ is undefined at $x = \pi n$, so the graph of $y = \csc x$ has vertical asymptotes when $x = \pi n$.

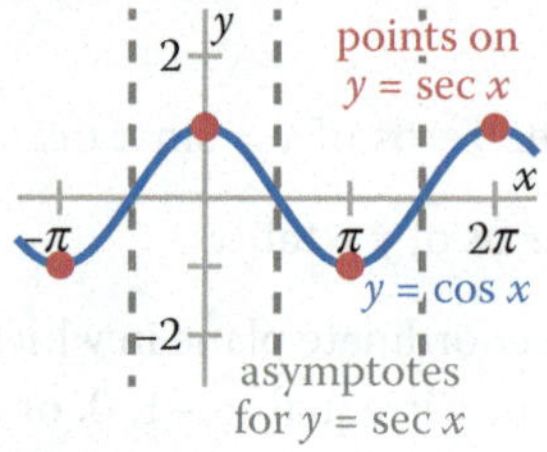

Figure 5.6c

Similarly, since the zeros of $y = \cos x$ are at $x = \pi/2 + \pi n$, these x-values give the equations of the vertical asymptotes on the graph of $y = \sec x$. The vertical asymptotes at the zeros are shown in Figures 5.6b and 5.6c.

Since the reciprocal of 1 is 1, and the reciprocal of -1 is -1, the points on the graph of $y = \sin x$, where $y = 1$ or $y = -1$, are also on the graph of $y = \csc x$. Similarly, the points on the graph of $y = \cos x$, where $y = 1$ or $y = -1$, are also on the graph of $y = \sec x$. These common points are shown in red in Figures 5.6b and 5.6c.

The behavior of the graphs of $y = \csc x$ and $y = \sec x$ between the asymptotes can be determined by generalizing the relationship between reciprocals. If a value is between 0 and 1, then its reciprocal is greater than 1.

Moreover, the closer the value is to 0, the greater its reciprocal is. Therefore, for each (x, y) on the graph of $y = \sin x$ such that $0 < y < 1$, there is a point on the graph of $y = \csc x$ with the same x-coordinate, and a y-coordinate that is greater than 1. In addition, as the value of the y-coordinate for $y = \sin x$ decreases toward 0, the value of the y-coordinate on the graph of $y = \csc x$ increases.

This also describes the relationship between the points on the graphs of $y = \cos x$ and $y = \sec x$.

The Basic Cosecant Function: $y = \csc x$

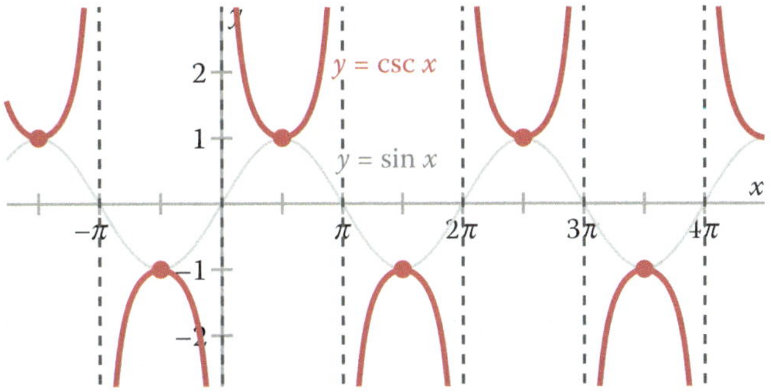

Domain: $\{x : x \neq \pi n\}$

Range: $(-\infty, -1] \cup [1, \infty)$

Period: 2π

Vertical Asymptotes: $x = \pi n$

Local Minimums: $\left(\frac{\pi}{2} + 2\pi n, 1\right)$

Local Maximums: $\left(\frac{3\pi}{2} + 2\pi n, -1\right)$

The Basic Secant Function: $y = \sec x$

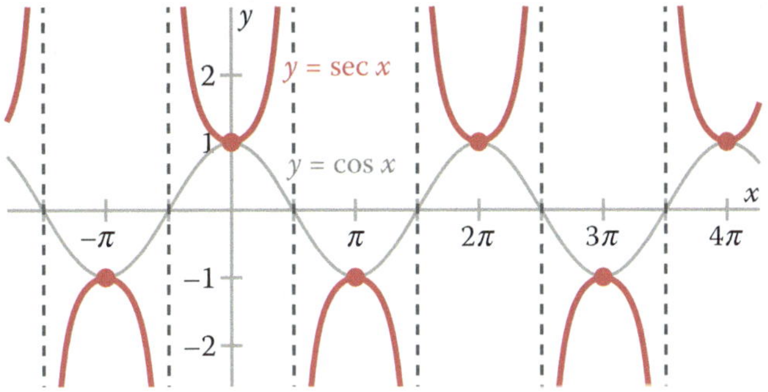

Domain: $\{x : x \neq \frac{\pi}{2} + \pi n\}$

Range: $(-\infty, -1] \cup [1, \infty)$

Period: 2π

Vertical Asymptotes: $x = \frac{\pi}{2} + \pi n$

Local Minimums: $(2\pi n, 1)$

Local Maximums: $(\pi + 2\pi n, -1)$

Notice that each local maximum of $y = \sin x$ is a local minimum of $y = \csc x$, and each local minimum of $y = \sin x$ is a local maximum of $y = \csc x$. The relationship is the same for $y = \cos x$ and $y = \sec x$. From the graph, the period of $y = \csc x$ and $y = \sec x$ is 2π.

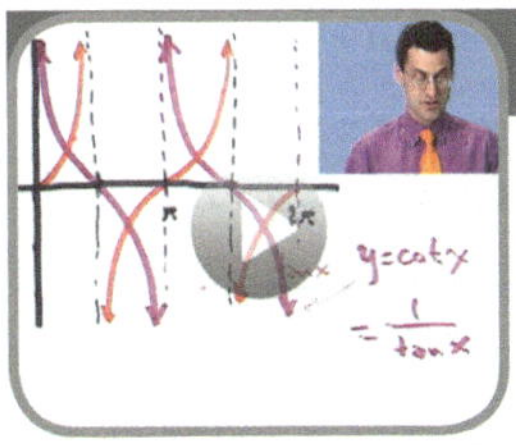

5.6.3 Introduction to the Graph of the Cotangent Function

The basic cotangent function $y = \cot x$ can be graphed by again using the relationship between reciprocals and the graph of $y = \tan x$ because the cotangent function is the reciprocal of the tangent function. Generally, if (x, y) is a point on the graph of $y = \tan x$ such that $y \neq 0$, then $(x, 1/y)$ is a point on the graph of $y = \cot x$.

Again, where the graph of $y = \tan x$ has zeros, the graph of $y = \cot x$ will have asymptotes. And, all points on the graph of $y = \tan x$ where $y = 1$ and where $y = -1$ will also be points on the graph of $y = \cot x$, because the reciprocal of 1 is 1 and the reciprocal of -1 is -1.

The Basic Cotangent Function: $y = \cot x$

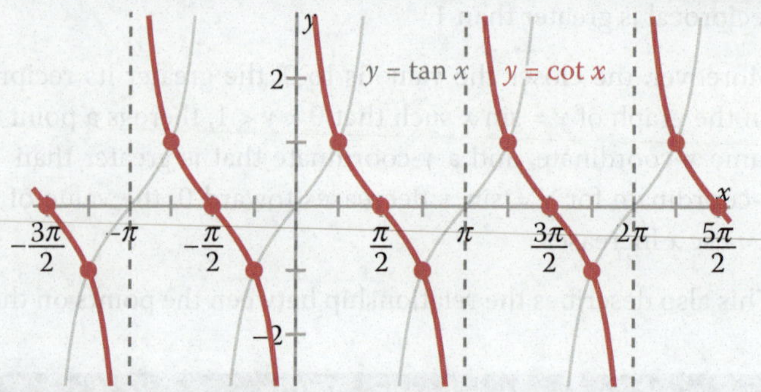

Domain: $\{x : x \neq \pi n\}$

Range: $(-\infty, \infty)$

Vertical Asymptotes: $x = \pi n$

Zeros: $x = \dfrac{\pi}{2} + \pi n$

Period: π

NOTICE THAT

Like the graph of $y = \tan x$, the graph of $y = \cot x$ is symmetric about the origin, and has no minimums or maximums. Where the graph of $y = \tan x$ has zeros, the graph of $y = \cot x$ has vertical asymptotes.

The same curve is repeated between each pair of consecutive vertical asymptotes. Since the distance between any two asymptotes is π, the period of $y = \cot x$ is π.

5.6.4 Graphing Tangent Functions: Stretches

The graph of $y = a \tan x$ $(a > 0)$ is the graph of $y = \tan x$ stretched vertically by a factor of a. So, for each point (x, y) on the graph of $y = \tan x$, there is a point (x, ay) on the graph of $y = a \tan x$.

The Graph of $y = a \tan x$, $a > 0$

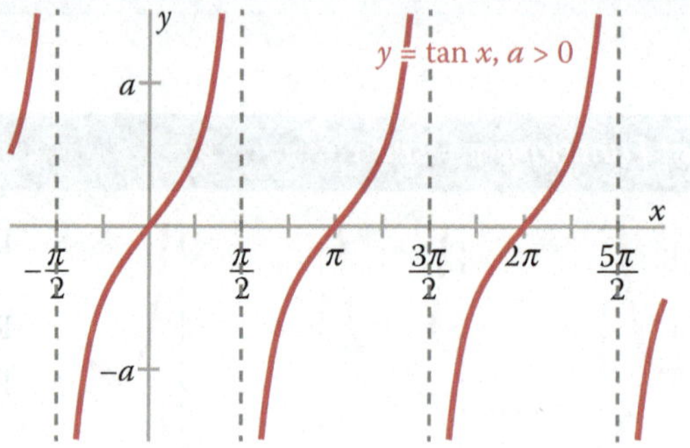

Period: π

Vertical Stretch Factor: $|a|$

Vertical Asymptotes: $x = \dfrac{\pi}{2} + \pi n$

Zeros: $x = \pi n$

If $a < 0$, then the graph of $y = \tan x$ is not only stretched by a factor of $|a|$, but also reflected across the x-axis.

EXAMPLE 1 Graphing a Tangent Function with a Stretch

IMPORTANT

All functions of the form $y = a \tan x$ will have the same zeros, vertical asymptotes, and period as $y = \tan x$.

Graph at least two periods of $y = 2 \tan x$, $y = \dfrac{1}{2} \tan x$, and $y = -2 \tan x$. Write the general equation of all vertical asymptotes and zeros of each function, if any.

SOLUTION

Find several points on the graph by making a table. Begin with the table for $y = \tan x$, then multiply each value in the $\tan x$ column by 2, 1/2, and −2 to find the corresponding y-values for the points on the graphs of $y = 2 \tan x$, $y = \dfrac{1}{2} \tan x$, and $y = -2 \tan x$, respectively.

x	$\tan x$	$2 \tan x$	$\frac{1}{2} \tan x$	$-2 \tan x$
$-\pi$	0	0	0	0
$-3\pi/4$	1	2	1/2	−2
$-\pi/4$	−1	−2	−1/2	2
0	0	0	0	0
$\pi/4$	1	2	1/2	−2
$3\pi/4$	−1	−2	−1/2	2
π	0	0	0	0

From the table, the graphs of all three functions pass through $(-\pi, 0)$, $(0, 0)$, and $(\pi, 0)$ as expected, since graphs of functions of the form $y = a \tan x$ have the same zeros.

Since $y = 2\tan x$, $y = \frac{1}{2}\tan x$, and $y = -2\tan x$ are of the form $y = a \tan x$, each graph has vertical asymptotes at $x = \frac{\pi}{2} + \pi n$, zeros at $x = \pi n$, and a period of π, like $y = \tan x$.

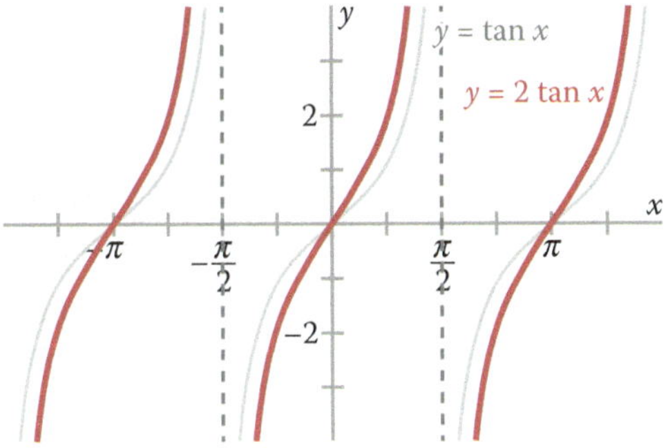

Because each equation has a coefficient of $\tan x$ that is not equal to 1 (or −1), each graph appears stretched or shrunk vertically as compared to the graph of $y = \tan x$.

NOTICE THAT

The graphs of $y = 2 \tan x$ and $y = -2 \tan x$ are reflections of each other over the x-axis.

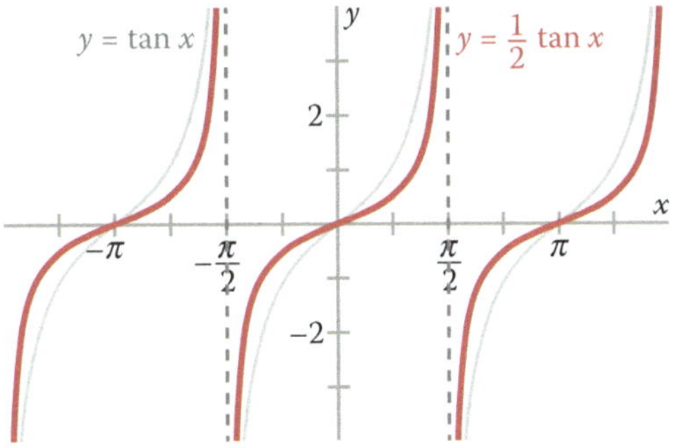

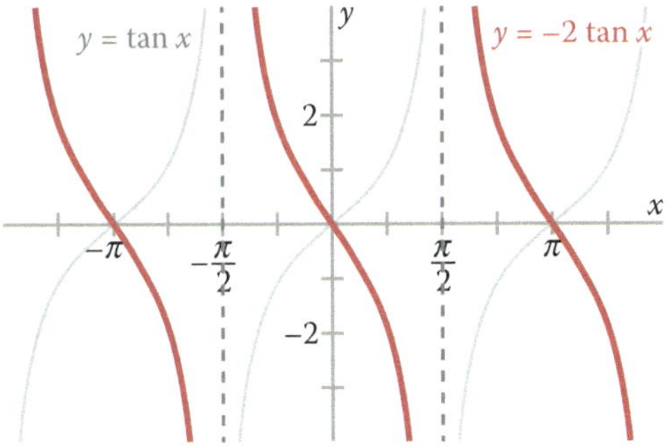

5.6.5 Graphing Tangent and Cotangent Functions: Stretches and Changes to the Period

In Section 5.5, we saw that the period of $y = \sin x$ and $y = \cos x$ is 2π, and that the period of $y = a \sin bx$ and $y = a \cos bx$ can be found by dividing 2π by $|b|$. The period of tangent and cotangent functions of the form $y = a \tan bx$ and $y = a \cot bx$ can be found in a similar way. Since the period of $y = \tan x$ and $y = \cot x$ is π, the period of $y = a \tan bx$ and $y = a \cot bx$ is $\frac{\pi}{|b|}$.

EXAMPLE 2 Graphing a Tangent Function When the Period Is Not π

Graph at least two periods of $y = \tan 2x$. Write the general equation of all vertical asymptotes and zeros, if any.

SOLUTION

Since $y = \tan 2x$ is of the form $y = \tan bx$ where $b = 2$, the function's period is $\pi/2$, and the graph has vertical asymptotes at $x = \frac{\pi}{4}$ and $x = -\frac{\pi}{4}$, and a zero at 0.

Sketch the tangent curve between those asymptotes.

The tangent function is periodic, so the graph repeats itself between each pair of consecutive vertical asymptotes, which occur every $\pi/2$ radians.

From the graph, the vertical asymptotes are at $x = \frac{\pi}{4} + \frac{\pi}{2}n$ and the zeros are at $x = \frac{\pi}{2}n$.

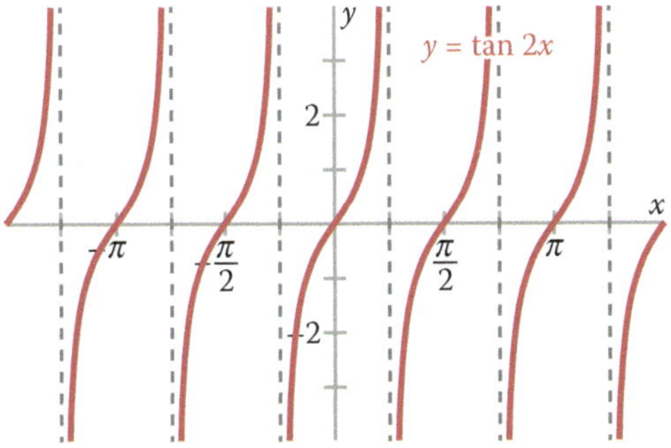

Confirm several points on the graph by making a table. Since $\tan x$ is equal to -1, 0, 1, or undefined at multiples of $\pi/4$, and since the coefficient of x in $y = \tan 2x$ is 2, choose x-values such that $2x$ is a multiple of $\pi/4$ (thus making the points easier to plot).

x	$2x$	$\tan 2x$
$-3\pi/8$	$-3\pi/4$	1
$-\pi/4$	$-\pi/2$	undefined
$-\pi/8$	$-\pi/4$	-1
0	0	0
$\pi/8$	$\pi/4$	1
$\pi/4$	$\pi/2$	undefined
$3\pi/8$	$3\pi/4$	-1

From the table, the graph of $y = \tan 2x$ has vertical asymptotes at $x = \dfrac{\pi}{4}$ and $x = -\dfrac{\pi}{4}$, a zero at 0, and passes through the points $(-3\pi/8, 1)$, $(-\pi/8, -1)$, $(\pi/8, 1)$, and $(3\pi/8, -1)$.

EXAMPLE 3 Graphing a Tangent Function with a Stretch and a Reflection When the Period Is Not π

Graph at least two periods of $y = -2\tan\dfrac{1}{2}x$. Write the general equation of all vertical asymptotes and zeros, if any.

SOLUTION

Since $y = -2\tan\dfrac{1}{2}x$ is of the form $y = a\tan bx$ where $b = 1/2$, the graph of the function is horizontally stretched by a factor of 2 and the period is $\pi/|1/2| = 2\pi$.

Therefore, the graph has vertical asymptotes at $x = \pi$ and $x = -\pi$, and a zero at 0.

Sketch the tangent curve between the asymptotes. Note that the tangent curve is

- stretched by a factor of 2 because $|a| = 2$ and
- reflected over the x-axis because $a < 0$.

Since the tangent function is periodic, the graph repeats itself between each pair of consecutive vertical asymptotes, which occur every 2π radians.

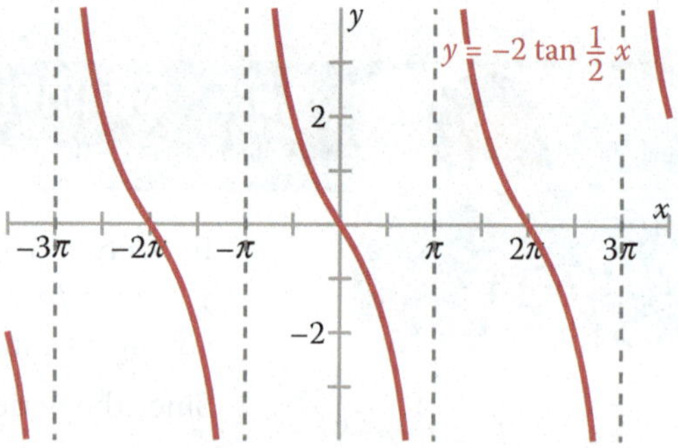

From the graph, the vertical asymptotes are at $x = \pi + 2\pi n$, and the zeros are at $x = 2\pi n$.

Confirm several points on the graph by making a table. Since the coefficient of x is 1/2, choose x-values such that $\dfrac{1}{2}x$ is a multiple of $\pi/4$ (thus making the points easier to plot).

x	$\frac{1}{2}x$	$\tan\frac{1}{2}x$	$-2\tan\frac{1}{2}x$
$-3\pi/2$	$-3\pi/4$	1	-2
$-\pi$	$-\pi/2$	undefined	undefined
$-\pi/2$	$-\pi/4$	-1	2
0	0	0	0
$\pi/2$	$\pi/4$	1	-2
π	$\pi/2$	undefined	undefined
$3\pi/2$	$3\pi/4$	-1	2

From the table, the graph of $y = -2\tan\dfrac{1}{2}x$ has vertical asymptotes at $x = \pi$ and $x = -\pi$, a zero at 0, and passes through the points $(-3\pi/2, -2)$, $(-\pi/2, 2)$, $(\pi/2, -2)$, and $(3\pi/2, 2)$.

Graphing $y = a \cot bx$

The strategies used for graphing tangent functions can also be applied to graphing cotangent functions.

EXAMPLE 4 Graphing a Cotangent Function with a Stretch and a Reflection When the Period Is Not π

Graph at least two periods of $y = -3\cot\frac{1}{4}x$. Write the general equation of all vertical asymptotes and zeros, if any.

SOLUTION

Since $y = -3\cot\frac{1}{4}x$ is of the form $y = a\cot bx$ where $b = 1/4$, the graph of the function is horizontally stretched by a factor of 4 and the period is $\pi/|1/4| = 4\pi$. Therefore, the graph has vertical asymptotes at $x = -4\pi$, $x = 0$, and $x = 4\pi$, and zeros at $x = -2\pi$ and $x = 2\pi$ (i.e., the midpoints on the x-axis between the vertical asymptotes).

Draw the vertical asymptotes, then sketch the cotangent curve between each pair of consecutive asymptotes.

Note that the cotangent curve is
- stretched vertically by a factor of 3 because $|a| = 3$ and
- reflected over the x-axis because $a < 0$.

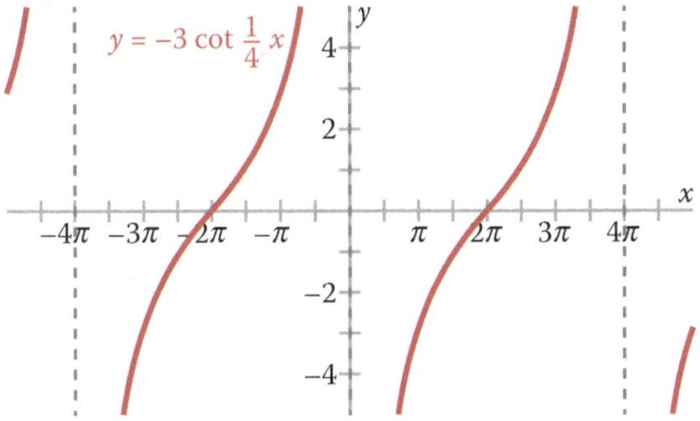

From the graph, the vertical asymptotes are at $x = 4\pi n$, and the zeros are at $x = 2\pi + 4\pi n$.

Confirm several points on the graph by making a table. Since the coefficient of x is 1/4, choose x-values such that $\frac{1}{4}x$ is a multiple of $\pi/4$.

Multiply each value of $\cot\frac{1}{4}x$ by −3 to find the corresponding y-values.

x	$\frac{1}{4}x$	$\cot\frac{1}{4}x$	$-3\cot\frac{1}{4}x$
-3π	$-3\pi/4$	1	−3
-2π	$-\pi/2$	0	0
$-\pi$	$-\pi/4$	−1	3
0	0	undefined	undefined
π	$\pi/4$	1	−3
2π	$\pi/2$	0	0
3π	$3\pi/4$	−1	3

TIP

Since $\cot x$ is equal to −1, 0, 1, or undefined at multiples of $\pi/4$, choosing x-values such that $\frac{1}{4}x$ is a multiple of $\pi/4$ makes the graph easier to sketch.

From the table, the graph of $y = -3\cot\frac{1}{4}x$ has a vertical asymptotes at $x = 0$, zeros at $x = -2\pi$ and $x = 2\pi$, and passes through the points $(-3\pi, -3)$, $(-\pi/2, 3)$, $(\pi, -3)$, and $(3\pi, 3)$.

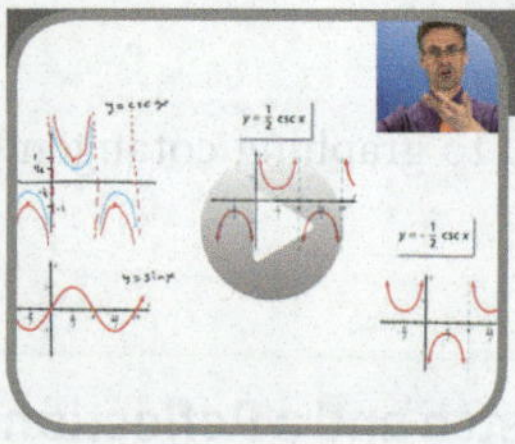

5.6.6 Graphing Cosecant and Secant Functions

Recall that the amplitudes of $y = a \sin x$ and $y = a \cos x$ is $|a|$, so the vertical distance between the local minimums and maximums is $2|a|$. Similarly, the vertical distance between the local minimums and maximums of a graph of a function of the form $y = a \csc x$ or $y = a \sec x$ is $2|a|$.

The Graph of $y = a \csc x,\ a > 0$

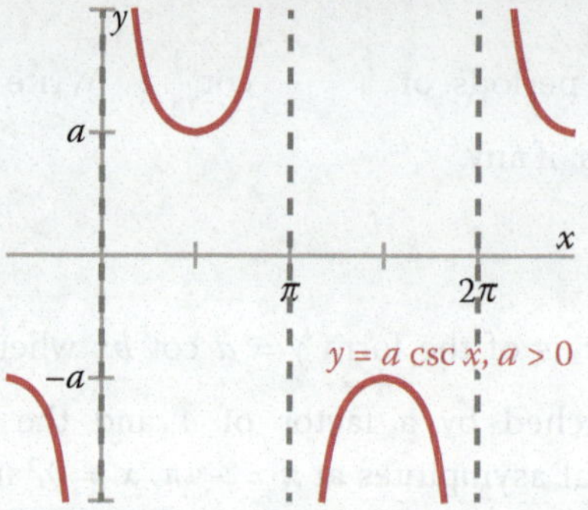

Period: 2π

Vertical Asymptotes: $x = \pi n$

Local Maximums: $\left(\frac{3\pi}{2} + 2\pi n,\ -a\right)$

Local Minimums: $\left(\frac{\pi}{2} + 2\pi n,\ a\right)$

The Graph of $y = a \sec x,\ a > 0$

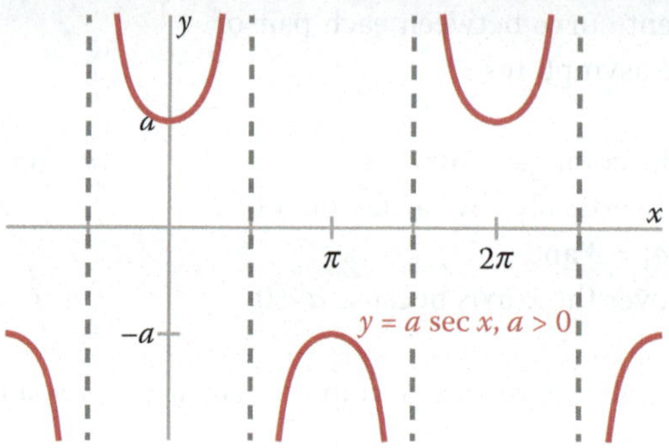

Period: 2π

Vertical Asymptotes: $x = \frac{\pi}{2} + \pi n$

Local Maximums: $(\pi + 2\pi n,\ -a)$

Local Minimums: $(2\pi n,\ a)$

NOTICE THAT

The graphs of $y = a \csc x$ and $y = a \sec x$ have no zeros.

Note that the graph of $y = a \csc x$ is very similar to the graph of $y = \csc x$. They have the same vertical asymptotes, the same period, and when $a > 0$, the same x-values corresponding to the local minimums and local maximums (though not the same y-values). The same is true for the graphs of $y = a \sec x$ as compared to $y = \sec x$.

The Graphs of $y = a \csc x$ and $y = a \sec x$ When $a < 0$

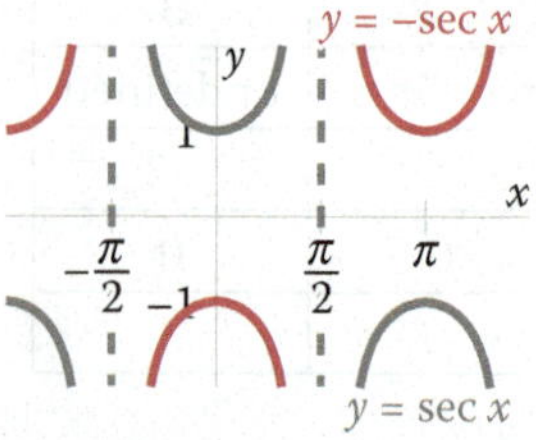

Figure 5.6d

Figure 5.6d shows the graphs of $y = \sec x$ and $y = -\sec x$. Notice that the graphs are reflections over the x-axis. Generally (as with $y = a \sin x$ and $y = a \cos x$), if $a < 0$, then the graph of $y = a \sec x$ is a reflection of the graph of $y = |a| \sec x$ across the x-axis. Similarly, if $a < 0$, then the graph of $y = a \csc x$ is a reflection of the graph of $y = |a| \csc x$ across the x-axis.

Graphing $y = a \csc x$ and $y = a \sec x$

Functions of the form $y = a \csc x$ and $y = a \sec x$ can be graphed by transforming the corresponding basic trigonometric function.

Alternatively, to graph $y = a \csc x$ when $a \neq 1$, first graph the corresponding reciprocal function $y = \sin x$ and sketch vertical asymptotes through the zeros of $y = \sin x$. Then multiply the y-coordinates at the local minimums and maximums of $y = \sin x$ by $|a|$. The resulting points are the local maximums and minimums of $y = |a| \csc x$. Sketch the graph of $y = |a| \csc bx$ using these local minimum and local maximum points between the vertical asymptotes. If $a < 0$, reflect the graph of $y = |a| \csc x$ over the x-axis, resulting in the graph of $y = a \csc x$.

The process is the same for graphing $y = a \sec x$ when $a \neq 1$, but begin by graphing the corresponding reciprocal function $y = \cos x$ (instead of $y = \sin x$).

EXAMPLE 5 Graphing Cosecant and Secant Functions

Graph at least one period of each function and write the general equation of all vertical asymptotes and zeros, if any.

A. $y = \frac{1}{2}\sec x$ **B.** $y = -2\csc x$

SOLUTION

A. Graph the corresponding reciprocal function $y = \cos x$. Draw vertical asymptotes through the zeros. For each maximum on the graph of $y = \cos x$, there is a local minimum with 1/2 of the y-coordinate on the graph of $y = \frac{1}{2}\sec x$. For each minimum on the graph of $y = \cos x$, there is a local maximum with 1/2 of the y-coordinate on the graph of $y = \frac{1}{2}\sec x$.

Graph $y = \frac{1}{2}\sec x$ using these points and the vertical asymptotes.

The zeros of $y = \cos x$ occur at $x = \frac{\pi}{2} + \pi n$, so the vertical asymptotes of $y = \frac{1}{2}\sec x$ occur at $x = \frac{\pi}{2} + \pi n$.

Like $y = \sec x$, $y = \frac{1}{2}\sec x$ has no zeros.

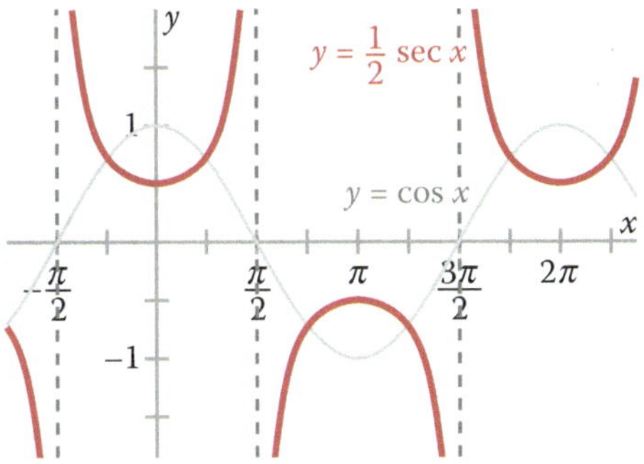

NOTICE THAT

The function is of the form $y = a\sec x$ where $a = 1/2$, so:

- *the period is 2π,*
- *the vertical asymptotes are at $x = \frac{\pi}{2} + \pi n$,*
- *$y = 1/2$ at the local minimums, and*
- *$y = -1/2$ at the local maximums.*

B. Graph the corresponding reciprocal function $y = \sin x$. Draw vertical asymptotes at the zeros.

For each maximum on the graph of $y = \sin x$, there is a local minimum with 2 times the y-coordinate on the graph of $y = 2\csc x$. For each minimum on the graph of $y = \sin x$, there is a local maximum with 2 times the y-coordinate on the graph of $y = 2\sin x$.

Graph $y = 2\csc x$ using these points and the vertical asymptotes.

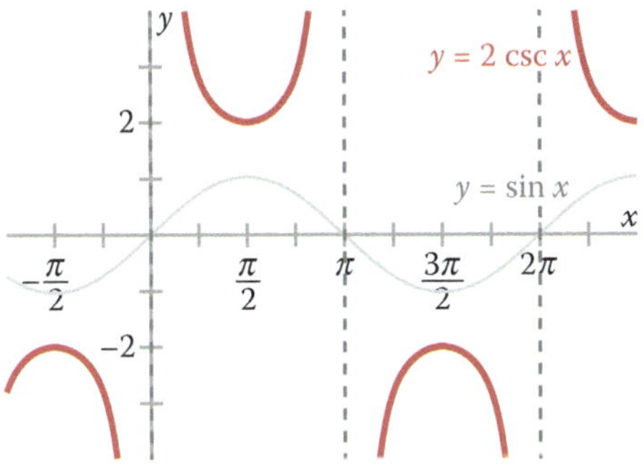

Reflect the graph of $y = 2\csc x$ over the x-axis to graph $y = -2\csc x$.

The zeros of $y = \sin x$ occur at $x = \pi n$, so the vertical asymptotes of $y = -2\csc x$ occur at $x = \pi n$.

Like $y = \csc x$, $y = -2\csc x$ has no zeros.

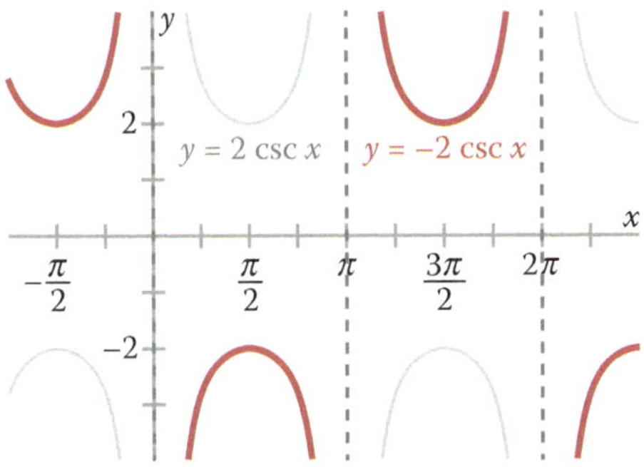

NOTICE THAT

The function is of the form $y = a\csc x$ where $a = -2$, so:

- *the period is 2π,*
- *the vertical asymptotes are at $x = \pi n$,*
- *$y = 2$ at the local minimums, and*
- *$y = -2$ at the local maximums.*

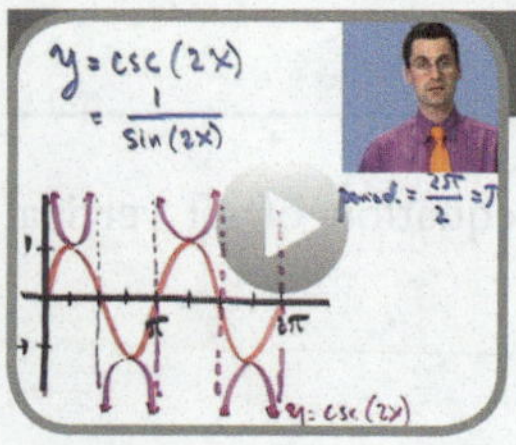

5.6.7 Graphing Cosecant and Secant Functions: Changes to the Period

Recall that the period of $y = a \sin bx$ and $y = a \cos bx$ is $2\pi/|b|$. Since the period of a function and its reciprocal is the same, the period of functions of the form $y = a \csc bx$ and $y = a \sec bx$ is also $2\pi/|b|$. Then, if $b \neq 1$, the period of $y = a \csc bx$ and $y = a \sec bx$ is not 2π.

Graphing $y = \csc bx$ and $y = \sec bx$

To graph $y = \csc bx$, first graph the corresponding reciprocal function $y = \sin bx$ and sketch vertical asymptotes through the zeros of $y = \sin bx$. Each maximum on the graph of $y = \sin bx$ is a local minimum on the graph of $y = \csc bx$. Each minimum on the graph of $y = \sin bx$ is a local maximum on the graph of $y = \csc bx$. Sketch the graph of $y = \csc bx$ using these local minimum and local maximum points between the vertical asymptotes.

The process is the same for graphing $y = \sec bx$, but begin by graphing the corresponding reciprocal function $y = \cos bx$ (instead of $y = \sin bx$).

EXAMPLE 6 **Graphing Cosecant and Secant Functions When the Period Is Not 2π**

Graph at least one period of each function and write the general equation of all vertical asymptotes and zeros, if any.

A. $y = \csc \frac{1}{2}x$ **B.** $y = \sec 3x$

SOLUTION

NOTICE THAT

Since the function is of the form $y = a \csc bx$ where $b = 1/2$, the period is 4π.

A. Graph the corresponding reciprocal function $y = \sin \frac{1}{2}x$ with a period of 4π. Draw vertical asymptotes at the zeros.

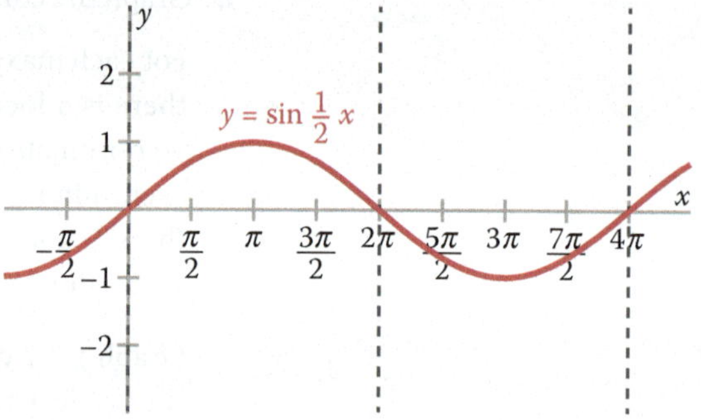

Each maximum and minimum on the graph of $y = \sin \frac{1}{2}x$ is a local minimum or local maximum, respectively, on the graph of $y = \csc \frac{1}{2}x$.

Graph $y = \csc \frac{1}{2}x$ using these points and the vertical asymptotes.

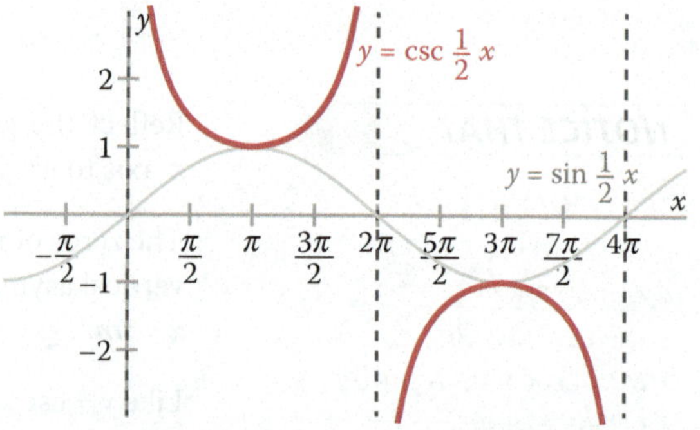

The zeros of $y = \sin \frac{1}{2}x$ occur at $x = 2\pi n$, so the vertical asymptotes of $y = \csc \frac{1}{2}x$ occur at $x = 2\pi n$.

Like $y = \csc x$, $y = \csc \frac{1}{2}x$ has no zeros.

NOTICE THAT

Since the function is of the form $y = a \sec bx$ where $b = 3$, the period is $2\pi/3$.

B. Graph the corresponding reciprocal function $y = \cos 3x$, which has a period of $2\pi/3$. Draw vertical asymptotes through the zeros.

Each maximum and minimum on the graph of $y = \cos 3x$ is a local minimum or local maximum, respectively, on the graph of $y = \sec 3x$.

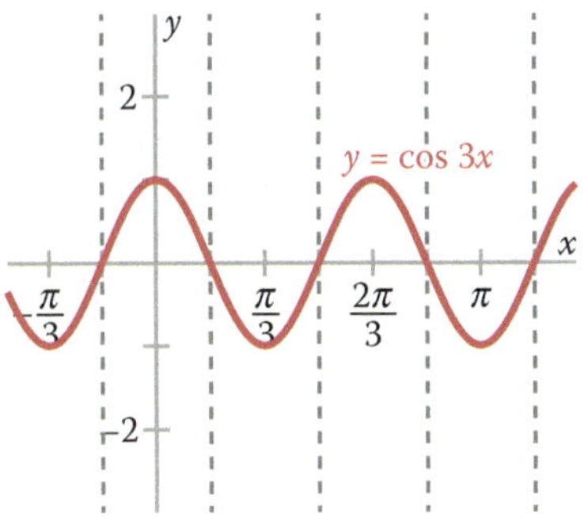

Graph $y = \sec 3x$ using these points and the vertical asymptotes.

The zeros of $y = \cos 3x$ occur at $x = \frac{\pi}{6} + \frac{\pi}{3}n$, as do the vertical asymptotes of $y = \sec 3x$.

Like $y = \sec x$, $y = \sec 3x$ has no zeros.

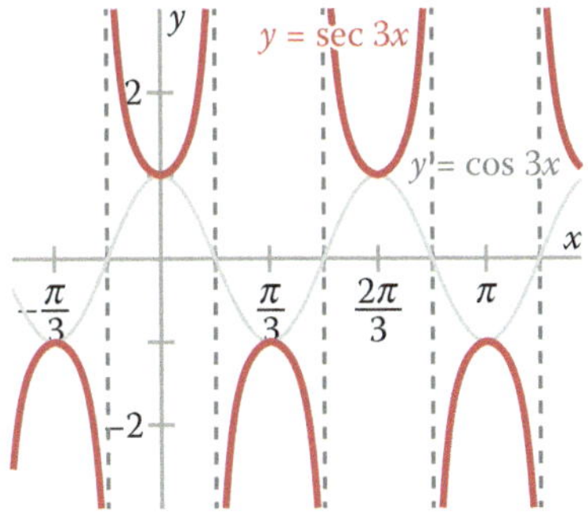

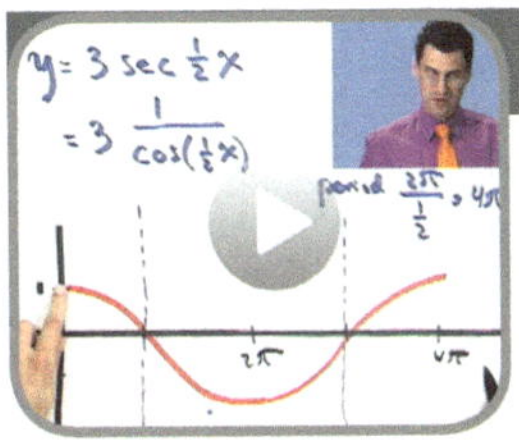

5.6.8 Graphing Cosecant and Secant Functions: Another Example

Graphing $y = a \csc bx$ and $y = a \sec bx$

To graph $y = a \csc bx$ when $a \neq 1$ and $b \neq 1$, first graph the corresponding reciprocal function $y = \sin bx$ and sketch vertical asymptotes through the zeros of $y = \sin bx$. Then multiply the y-coordinates at the minimums and maximums of $y = \sin bx$ by $|a|$. The resulting points are the local maximums and local minimums of $y = |a| \csc bx$. Sketch the graph of $y = |a| \csc bx$ using these local minimum and local maximum points between the vertical asymptotes. If $a < 0$, reflect the graph of $y = |a| \csc bx$ over the x-axis, resulting in the graph of $y = a \csc bx$.

The process is the same for graphing $y = a \sec bx$ when $b \neq 1$, but begin by graphing the corresponding reciprocal function $y = \cos bx$ (instead of $y = \sin bx$).

IMPORTANT

The graphs of $y = a \csc bx$ and $y = a \sec bx$ have the same vertical asymptotes and x-values of the zeros, minimums, and maximums as the graphs of $y = \csc bx$ and $y = \sec bx$, respectively.

EXAMPLE 7 **Graphing Cosecant and Secant Functions When the Period Is Not 2π**

Graph at least two periods of $y = -\frac{1}{2}\csc 3x$. Write the general equation of all vertical asymptotes and zeros, if any.

SOLUTION

Graph the corresponding reciprocal function $y = \sin 3x$. Draw vertical asymptotes at the zeros.

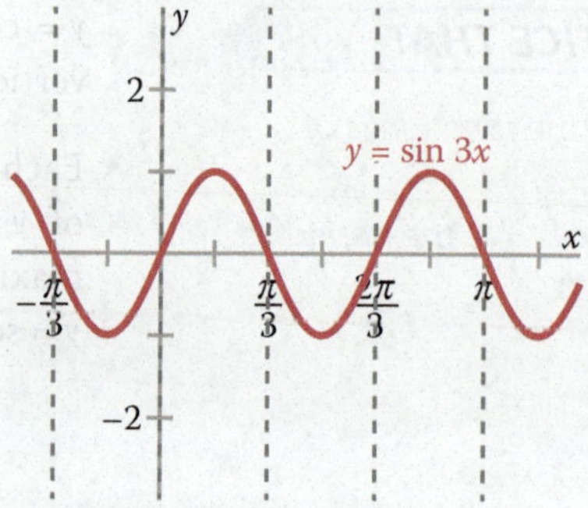

NOTICE THAT

Since the function is of the form $y = a \csc bx$ where $b = 3$, the period is $2\pi/3$.

For each maximum on the graph of $y = \sin 3x$, there is a local minimum with 1/2 of the y-coordinate on the graph of $y = \frac{1}{2}\csc 3x$.

For each minimum on the graph of $y = \sin 3x$, there is a local maximum with 1/2 of the y-coordinate on the graph of $y = \frac{1}{2}\csc 3x$.

ALTERNATIVE METHOD

If the y-coordinates of the minimums and maximums on the graph of $y = \sin 3x$ are multiplied by $-1/2$ to find the points on the graph of $y = -\frac{1}{2}\csc 3x$, there is no need to first graph $y = \frac{1}{2}\csc 3x$ and reflect it over the x-axis.

Graph $y = \frac{1}{2}\csc 3x$ using these points and the vertical asymptotes.

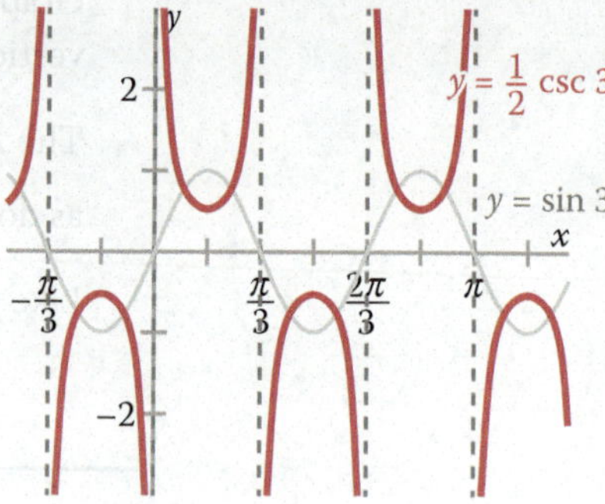

Reflect the graph of $y = \frac{1}{2}\csc 3x$ over the x-axis to graph $y = -\frac{1}{2}\csc 3x$.

The zeros of $y = \sin 3x$ occur at $x = \frac{\pi}{3}n$, as do the vertical asymptotes of $y = -\frac{1}{2}\csc 3x$.

Like $y = \csc x$, $y = -\frac{1}{2}\csc 3x$ has no zeros.

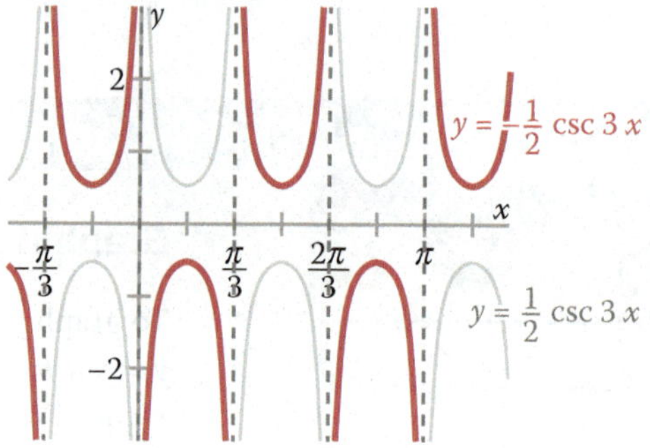

5.6.9 Shifting the Graphs of Other Trigonometric Functions

Graphing $y = \tan(x - c)$ and $y = \cot(x - c)$

Recall from Section 5.5 that a phase shift is a horizontal shift of a trigonometric function.

Functions of the form $y = \tan(x - c)$ and $y = \cot(x - c)$, where c is any real number, can be graphed by horizontally translating the graph of the corresponding basic trigonometric function. Specifically, the graph of $y = \tan(x - c)$ is the graph of $y = \tan x$ shifted c units to the right if $c > 0$, or c units to the left if $c < 0$.

TIP

The graphs of $y = \tan x$ and $y = \tan(x - c)$ are the same when the phase shift c is a multiple of the period π. The same is true for $y = \cot x$ and $y = \cot(x - c)$.

The graphs of cotangent functions behave in the same way.

EXAMPLE 8 **Graphing Tangent and Cotangent Functions with a Phase Shift**

Graph at least two periods of $y = \cot\left(x - \frac{\pi}{3}\right)$. Write the general equation of all vertical asymptotes and zeros, if any.

SOLUTION

Graph the corresponding basic trigonometric function $y = \cot x$.

Recall that the zeros occur at $x = \dfrac{\pi}{2} + \pi n$ and the vertical asymptotes at $x = \pi n$.

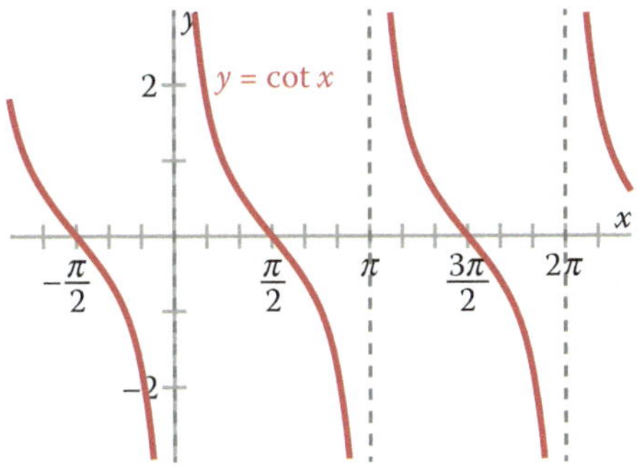

TIP

Since the phase shift is $\pi/3$, and the zeros of $y = \cot x$ occur at $x = \dfrac{\pi}{2} + \pi n$, graph $y = \cot x$ using an interval of $\pi/6$ on the x-axis.

The graph of $y = \cot\left(x - \dfrac{\pi}{3}\right)$ is the graph of $y = \cot x$ shifted $\pi/3$ units to the right.

Thus, the zeros occur at

$$x = \left(\frac{\pi}{2} + \pi n\right) + \frac{\pi}{3} = \frac{5\pi}{6} + \pi n,$$

and the vertical asymptotes occur at

$$x = (\pi n) + \frac{\pi}{3} = \frac{\pi}{3} + \pi n.$$

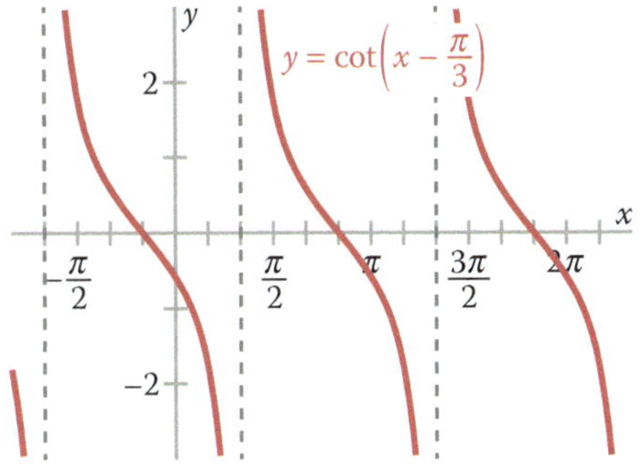

Graphing $y = \csc(x - c)$ and $y = \sec(x - c)$

Trigonometric functions of the form $y = \csc(x - c)$ and $y = \sec(x - c)$ can also be graphed by horizontally translating the graph of the corresponding basic trigonometric function. However, an alternative method is demonstrated in **Example 9**, where the corresponding reciprocal function is graphed first.

EXAMPLE 9 **Graphing Cosecant and Secant Functions with a Phase Shift**

Graph at least two periods of $y = \sec\left(x + \dfrac{3\pi}{2}\right)$. Write the general equation of all vertical asymptotes and zeros, if any.

SOLUTION

REMEMBER

Horizontally shifting the graph of $y = \cos x$ $\dfrac{\pi}{2} + 2\pi n$ units will result in the graph of $y = \sin x$. Therefore, the graph of $y = \cos\left(x + \dfrac{3\pi}{2}\right)$ is the same as the graph of $y = \sin x$.

Graph the corresponding reciprocal function $y = \cos\left(x + \dfrac{3\pi}{2}\right)$ by shifting the graph of $y = \cos x$ left $3\pi/2$ units.

Draw vertical asymptotes at the zeros.

Since $a = 1$, each maximum and minimum on the graph of $y = \cos\left(x + \dfrac{3\pi}{2}\right)$ is a local minimum or local maximum, respectively, on the graph of $y = \sec\left(x + \dfrac{3\pi}{2}\right)$.

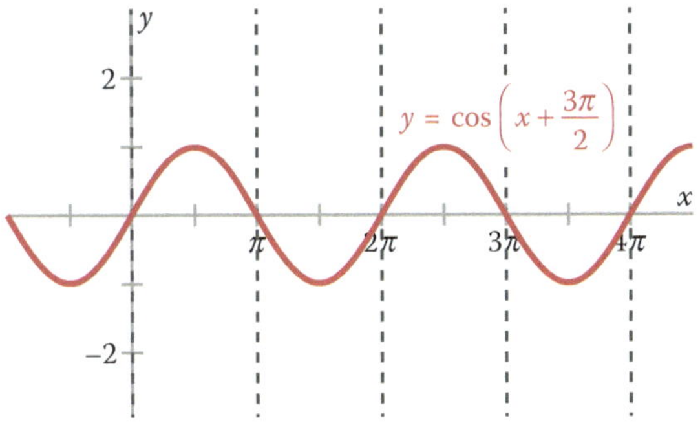

Graph $y = \sec\left(x + \dfrac{3\pi}{2}\right)$ using these points and the vertical asymptotes.

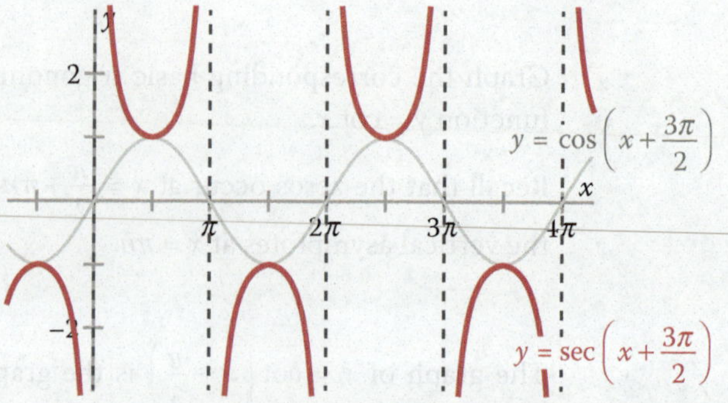

Since the zeros of $y = \cos\left(x + \dfrac{3\pi}{2}\right)$ occur at $x = \pi n$, the vertical asymptotes of $y = \sec\left(x + \dfrac{3\pi}{2}\right)$ occur at $x = \pi n$. Like $y = \sec x$, $y = \sec\left(x + \dfrac{3\pi}{2}\right)$ has no zeros.

Graphing $y = a \tan (bx - c)$ and $y = a \cot (bx - c)$

To graph a function of the form $y = a \tan (bx - c)$ or $y = a \cot (bx - c)$, first factor the b-value from $bx - c$, then graph the resulting function, as demonstrated in **Example 10**.

EXAMPLE 10 Graphing Tangent and Cotangent Functions with a Phase Shift When the Period Is Not π

Graph at least two periods of $y = -\tan (3x + 2\pi)$. Write the general equation of all vertical asymptotes and zeros, if any.

SOLUTION

First, factor 3 from $3x + 2\pi$.

$$y = -\tan(3x + 2\pi) = -\tan 3\left(x + \frac{2\pi}{3}\right)$$

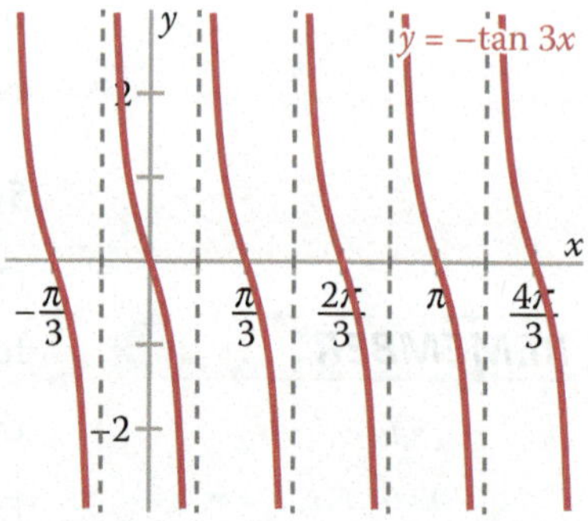

First, graph $y = -\tan 3x$, which is the reflection of $y = \tan 3x$ over the x-axis.

Note that since $b = 3$, the period is $\pi/3$ and the asymptotes are located at $x = \dfrac{\pi}{6} + \dfrac{\pi}{3}n$.

Now graph $y = -\tan 3\left(x + \dfrac{2\pi}{3}\right)$, which is the graph of $y = -\tan 3x$ shifted $2\pi/3$ units to the left.

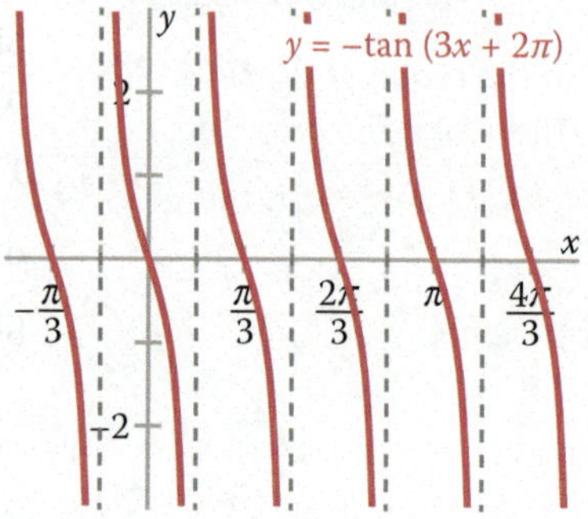

Note that the phase shift $2\pi/3$ is a multiple of the period of $y = -\tan 3x$, $\pi/3$. So the graph of $y = -\tan (3x + 2\pi)$ is the same graph as $y = -\tan 3x$.

The zeros occur at $x = \dfrac{\pi}{3}n$, and the vertical asymptotes occur at $x = \dfrac{\pi}{6} + \dfrac{\pi}{3}n$.

Graphing $y = a \csc(bx - c)$ and $y = a \sec(bx - c)$

To graph $y = a \csc(bx - c)$ or $y = a \sec(bx - c)$, first factor the b-value from $bx - c$, then graph the resulting function, as demonstrated in **Example 11**.

EXAMPLE 11 Graphing Cosecant and Secant Functions with a Phase Shift When the Period Is Not 2π

Graph at least two periods of $y = 3\sec\left(\frac{1}{2}x - \frac{\pi}{4}\right)$. Write the general equation of all vertical asymptotes and zeros, if any.

SOLUTION

First, factor $\frac{1}{2}$ from $\frac{1}{2}x - \frac{\pi}{4}$. $\quad y = 3\sec\left(\frac{1}{2}x - \frac{\pi}{4}\right) = 3\sec\frac{1}{2}\left(x - \frac{\pi}{2}\right)$

Graph the corresponding reciprocal function $y = \cos\frac{1}{2}\left(x - \frac{\pi}{2}\right)$. Since $b = 1/2$, the period is 4π. Draw vertical asymptotes through the zeros.

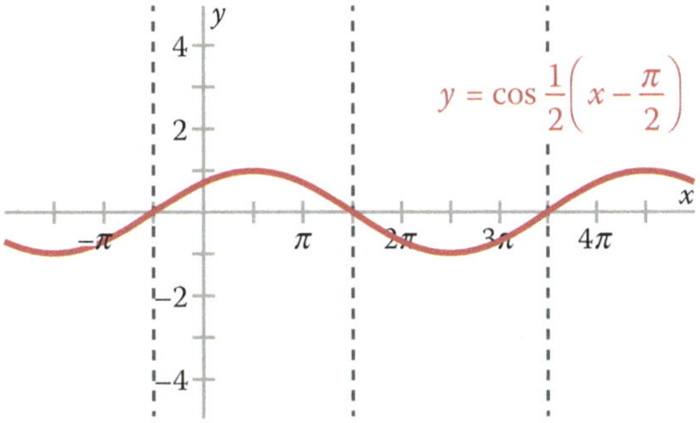

For each maximum on the graph of $y = \cos\frac{1}{2}\left(x - \frac{\pi}{2}\right)$, there is a local minimum with 3 times the y-coordinate on the graph of $y = 3\sec\frac{1}{2}\left(x - \frac{\pi}{2}\right)$. For each minimum on the graph of $y = \cos\frac{1}{2}\left(x - \frac{\pi}{2}\right)$, there is a local maximum with 3 times the y-coordinate on the graph of $y = 3\sec\frac{1}{2}\left(x - \frac{\pi}{2}\right)$. Graph $y = 3\sec\frac{1}{2}\left(x - \frac{\pi}{2}\right)$ (i.e., graph $y = 3\sec\left(\frac{1}{2}x - \frac{\pi}{4}\right)$) using these points and the vertical asymptotes.

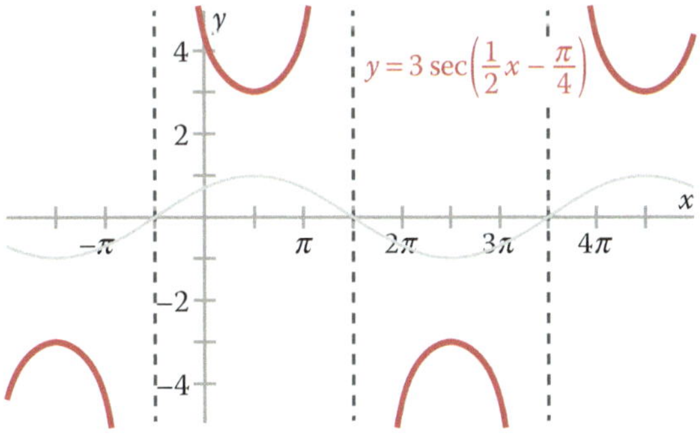

Since the zeros of $y = \cos\frac{1}{2}\left(x - \frac{\pi}{2}\right)$ occur at $x = \frac{3\pi}{2} + 2\pi n$, the vertical asymptotes of $y = 3\sec\left(\frac{1}{2}x - \frac{\pi}{4}\right)$ occur at $x = \frac{3\pi}{2} + 2\pi n$.

Like $y = \sec x$, $y = 3\sec\left(\frac{1}{2}x - \frac{\pi}{4}\right)$ has no zeros.

5.6.10 Identifying the Equation of a Trigonometric Function from Its Graph

EXAMPLE 12 Using the Attributes of the Graph of a Trigonometric Function to Identify Its Equation

Choose the equation that corresponds to the graph.

A. $y = 3\sec\left(x - \frac{\pi}{2}\right)$ **B.** $y = 2\sin\left(2x - \frac{\pi}{2}\right)$

C. $y = -3\sin\left(2x - \frac{\pi}{2}\right)$ **D.** $y = -3\cos\left(x - \frac{\pi}{4}\right)$

E. $y = -3\tan\left(2x - \frac{\pi}{4}\right)$ **F.** $y = 3\csc\left(x - \frac{\pi}{2}\right)$

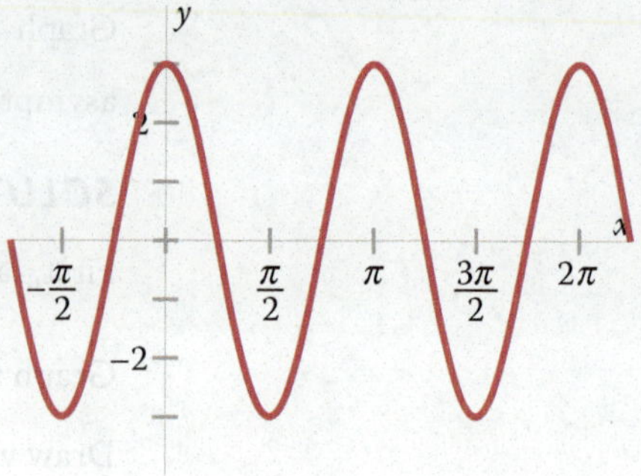

SOLUTION

The given graph has no vertical asymptotes, so the graph cannot be a cosecant, secant, tangent, or cotangent function. Therefore, the graph must be a sine or cosine function.

The amplitude of the given graph is 3, so the equation of the graph cannot be $y = 2\sin\left(2x - \frac{\pi}{2}\right)$ (as the amplitude of that function is 2).

The period of the given graph is π. So, the coefficient of x in the sine or cosine function must be 2, and thus cannot be $y = -3\cos\left(x - \frac{\pi}{4}\right)$.

So, the equation of the given graph must be $y = -3\sin\left(2x - \frac{\pi}{2}\right)$.

Verify that the equation corresponds to the given graph.

First, factor 2 from $2x - \frac{\pi}{2}$. $y = -3\sin\left(2x - \frac{\pi}{2}\right) = -3\sin 2\left(x - \frac{\pi}{4}\right)$

The graph of this equation is a sine function with an amplitude of 3, a period of π, and a phase shift of $\pi/4$, all of which match the given graph.

SECTION 5.6 EXERCISES

Warm Up

1. Identify the domain and range of $y = \cos x$.
2. Identify the zeros of $y = \sin x$.
3. Identify the values for which the function $y = \tan x$ is undefined.

Just the Facts

Fill in the blanks.

4. Unlike the graphs of $y = \sin x$ and $y = \cos x$, the graph of _____ has _____ asymptotes at $\frac{\pi}{2} + \pi n$ radians, where _____ is an integer.
5. The reciprocal of the sine function is the _____ function, and the reciprocal of the _____ function is the secant function.
6. The reciprocal of the _____ function is the cotangent function. Both functions have a period of _____. When $y = a \tan bx$ or $y = a \cot bx$, the _____ stretch is _____ and the period is _____. When _____ is negative, the graph is reflected across the x-axis.
7. The graph of $y = \tan x$ is identical to the graph of $y = \cot x$ if the cotangent graph is shifted _____ units to the _____ or _____, and is reflected across the _____.
8. In the graphs of the cosecant and _____ functions, there are _____ asymptotes, since where sine and cosine are 0 the cosecant and _____ functions are undefined.
9. The graph of $y = \csc (x + c)$ has a phase shift left c units if _____. When c is a multiple of _____, the graph of $y = \csc (x + c)$ is identical to $y = \csc x$.
10. To graph an equation when of the form $y = a \csc (bx + c)$, first factor out the _____.

Essential Skills

In Exercises 1–32, graph each function and identify all vertical asymptotes and zeros, if any.

1. $y = 3 \tan x$
2. $y = \frac{1}{4} \tan x$
3. $y = -\frac{2}{3} \tan x$
4. $y = -4 \tan x$
5. $y = \tan 3x$
6. $y = \tan \frac{1}{2}x$
7. $y = \tan \frac{1}{4}x$
8. $y = \tan 5x$
9. $y = \frac{1}{4} \tan 2x$
10. $y = 3 \tan 4x$
11. $y = 5 \tan \frac{1}{3}x$
12. $y = -\frac{1}{2} \tan 3x$
13. $y = -\frac{3}{4} \tan 4x$
14. $y = 2 \tan \frac{2}{3}x$
15. $y = \cot \frac{1}{2}x$
16. $y = \cot 3x$
17. $y = \cot 2x$
18. $y = \cot \frac{1}{3}x$
19. $y = -\frac{1}{2} \cot 4x$
20. $y = 2 \cot \frac{1}{4}x$
21. $y = \frac{1}{4} \cot 3x$
22. $y = -\frac{1}{3} \cot 2x$
23. $y = 3 \cot \frac{1}{2}x$
24. $y = -\frac{3}{4} \cot \frac{1}{3}x$
25. $y = \frac{1}{2} \csc x$
26. $y = -3 \csc x$
27. $y = -4 \csc x$
28. $y = \frac{1}{4} \csc x$
29. $y = -2 \sec x$
30. $y = \frac{3}{4} \sec x$
31. $y = \frac{1}{3} \sec x$
32. $y = -3 \sec x$

In Exercises 33–36, find the period of each function.

33. $y = \csc \frac{1}{2}x$
34. $y = \sec 3x$
35. $y = \sec \frac{2}{5}x$
36. $y = \csc 6x$

In Exercises 37–66, graph each function and identify all vertical asymptotes and zeros, if any.

37. $y = \csc 3x$
38. $y = \sec 4x$
39. $y = \sec 2x$
40. $y = \csc 2x$
41. $y = \sec \frac{1}{3}x$
42. $y = \csc \frac{1}{4}x$
43. $y = \csc \frac{3}{4}x$
44. $y = \sec \frac{1}{2}x$
45. $y = \frac{1}{2} \csc 3x$
46. $y = -2 \sec \frac{1}{3}x$
47. $y = -3 \sec 2x$
48. $y = \frac{1}{4} \csc 4x$
49. $y = -\frac{3}{4} \csc \frac{1}{2}x$
50. $y = -\frac{2}{3} \sec \frac{2}{3}x$
51. $y = \cot \left(x + \frac{\pi}{6}\right)$
52. $y = \tan \left(x - \frac{\pi}{8}\right)$
53. $y = \tan \left(x + \frac{\pi}{4}\right)$
54. $y = \cot \left(x - \frac{\pi}{3}\right)$
55. $y = \csc \left(x - \frac{\pi}{4}\right)$
56. $y = \sec \left(x + \frac{3\pi}{2}\right)$
57. $y = \sec \left(x - \frac{\pi}{6}\right)$
58. $y = \csc \left(x + \frac{2\pi}{3}\right)$
59. $y = -2 \tan (3x - \pi)$
60. $y = 3 \cot \left(\frac{1}{2}x + \frac{\pi}{3}\right)$

61. $y = \frac{1}{2}\cot(4x - \pi)$

62. $y = -\frac{3}{4}\tan\left(\frac{1}{3}x + \frac{\pi}{6}\right)$

63. $y = 2\sec\left(2x + \frac{\pi}{3}\right)$

64. $y = -4\csc\left(\frac{1}{4}x - \frac{\pi}{6}\right)$

65. $y = \frac{1}{3}\csc(3x + \pi)$

66. $y = -\frac{1}{2}\sec\left(\frac{1}{2}x - \frac{\pi}{6}\right)$

In Exercises 67–70, choose the equation from the following list that corresponds to each graph.

A. $y = \sin\frac{1}{2}\left(x - \frac{\pi}{4}\right)$ **B.** $y = -\cos\frac{1}{3}\left(x - \frac{\pi}{3}\right)$

C. $y = -\sin 3\left(x - \frac{\pi}{3}\right)$ **D.** $y = -\sin\frac{1}{2}\left(x - \frac{\pi}{4}\right)$

E. $y = \cos 2\left(x + \frac{\pi}{6}\right)$ **F.** $y = \sin 4\left(x - \frac{\pi}{6}\right)$

67.

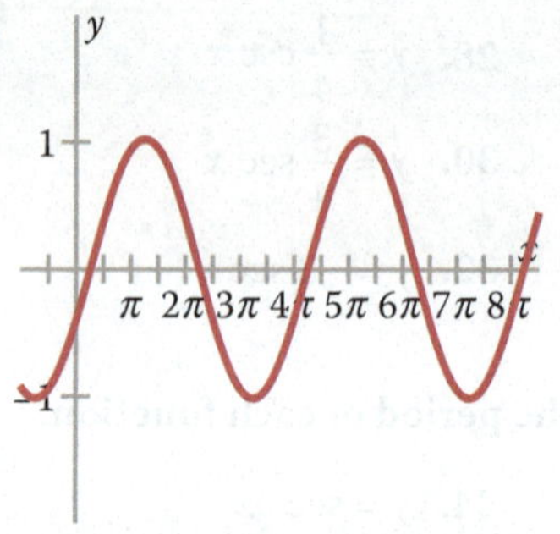

68.

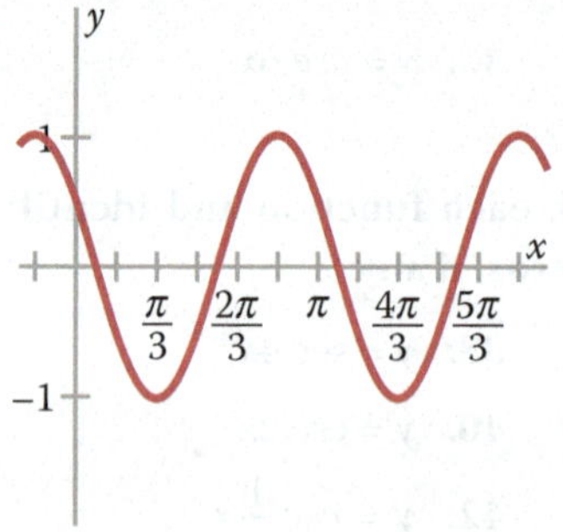

69.

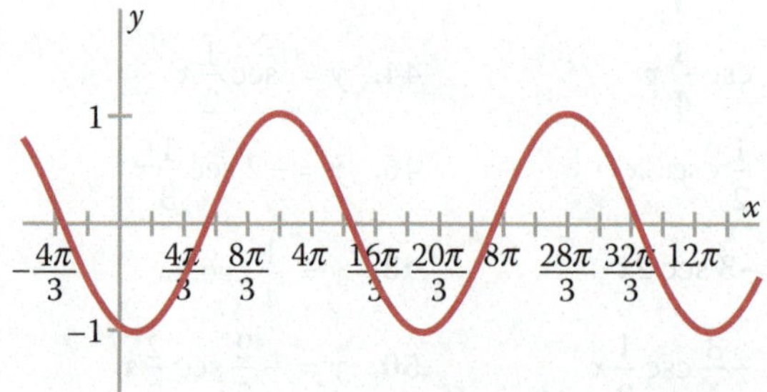

70.

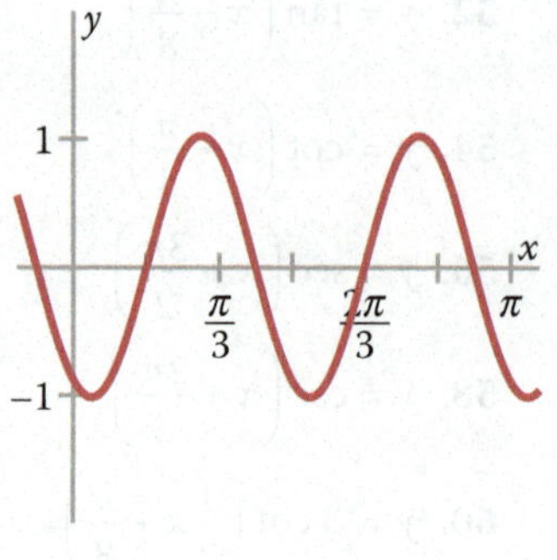

In Exercises 71–74, choose the equation from the following list that corresponds to each graph.

A. $y = \cot\frac{1}{4}\left(x + \frac{\pi}{3}\right)$ **B.** $y = -\cot 3\left(x + \frac{\pi}{4}\right)$

C. $y = -\tan\frac{1}{2}\left(x + \frac{\pi}{6}\right)$ **D.** $y = \cot\frac{1}{4}\left(x - \frac{\pi}{3}\right)$

E. $y = \tan\frac{1}{2}\left(x - \frac{\pi}{3}\right)$ **F.** $y = \tan 2\left(x - \frac{\pi}{3}\right)$

71.

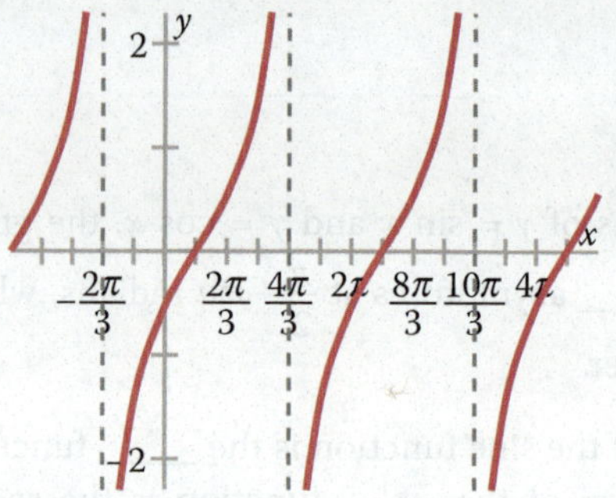

72.

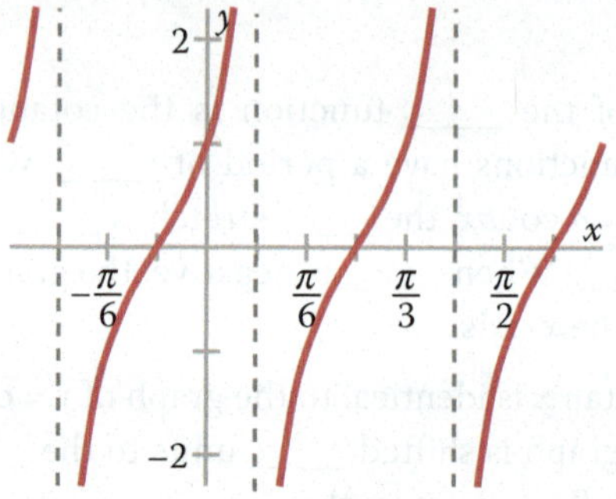

73.

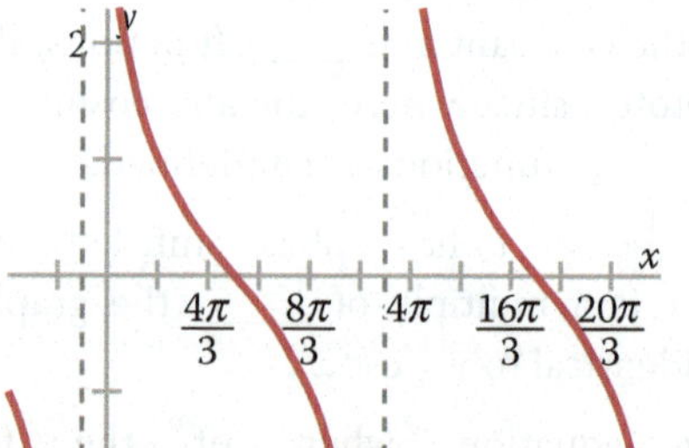

74.

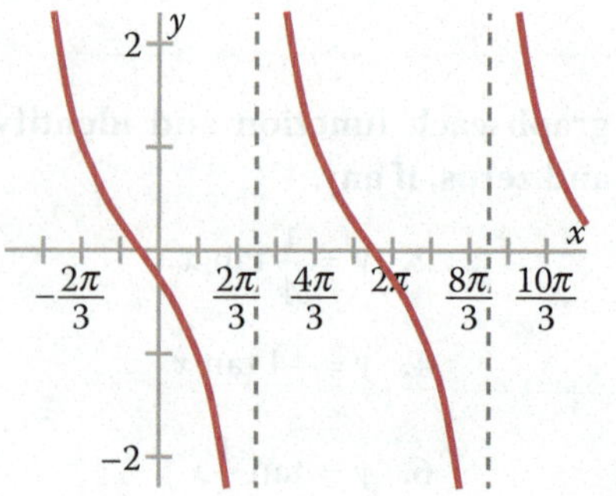

In Exercises 75–78, choose the equation from the following list that corresponds to each graph.

A. $y = \sec 2\left(x - \frac{\pi}{4}\right)$

B. $y = -\csc 4\left(x + \frac{\pi}{2}\right)$

C. $y = -\csc 3\left(x + \frac{\pi}{6}\right)$

D. $y = \sec 2\left(x + \frac{\pi}{3}\right)$

E. $y = \csc 2\left(x + \frac{\pi}{4}\right)$

F. $y = \sec \frac{1}{2}\left(x - \frac{\pi}{3}\right)$

75.

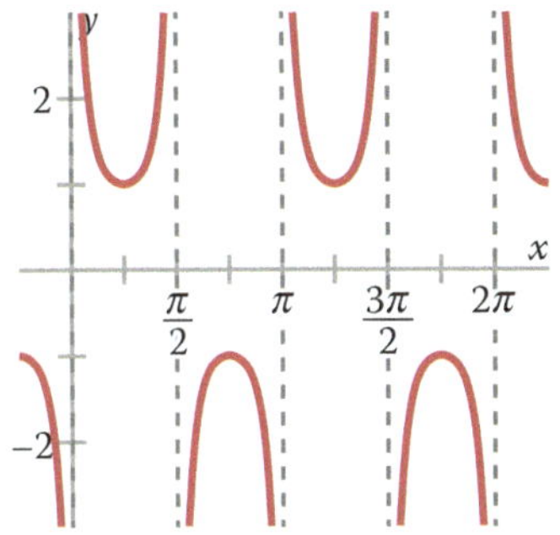

76.

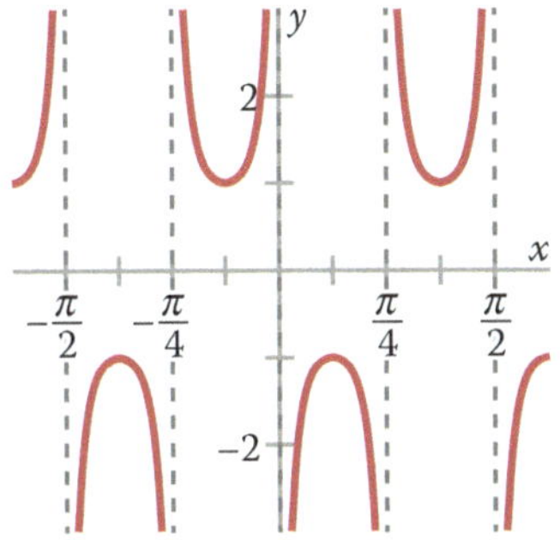

77.

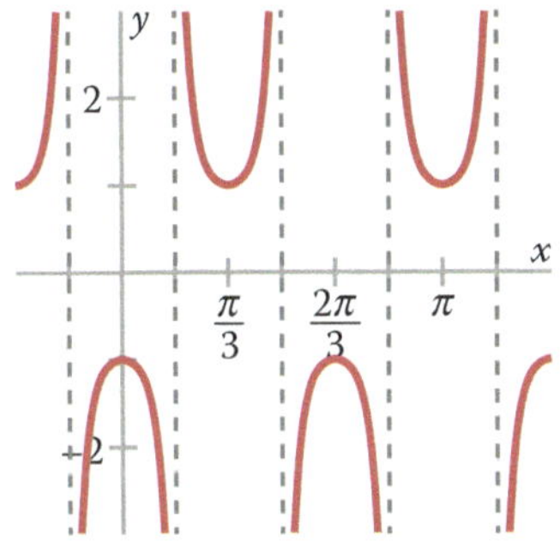

78.

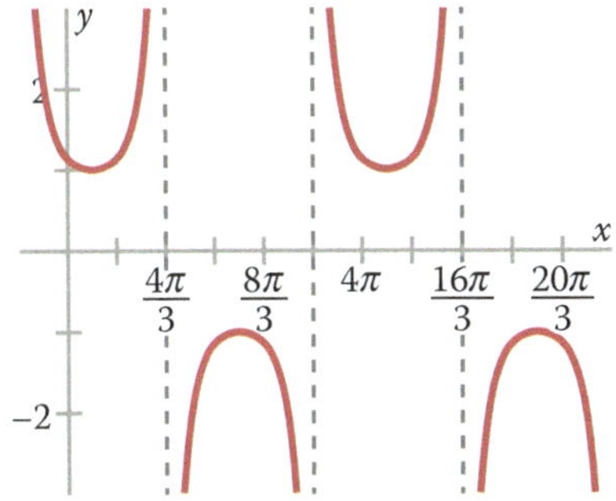

Extensions

79. How does the graph of $y = 2 \cot \pi x$ differ from the graph of $y = \cot x$?

80. Identify the vertical stretch, period, phase shift, and reflection, if any, of the graph of $y = -5 \sec\left(\frac{2}{3}x + \frac{\pi}{4}\right)$, as compared to the corresponding basic trigonometric function. Graph.

81. True or False? When the phase shift of any trigonometric function is a multiple of its period, the graph is identical to the graph of the same function without the phase shift.

82. Graph. $y = 2 \csc\left(\frac{\pi}{3}x - \frac{\pi}{6}\right)$

83. Write the equation of the tangent function that has period $\pi/4$, vertical stretch 5, phase shift $\pi/5$ radians to the left, and a reflection across the x-axis.

84. Identify any similarities between the graphs of

$y = -2 \csc(2x - 2\pi)$ and $y = -2 \sec 2\left(x - \frac{\pi}{4}\right)$.

In Exercises 85–86, write the equation for the function $y = a \tan b(x - c)$ with the given characteristics.

85. vertical asymptotes: $x = \frac{\pi}{6} + \frac{\pi}{3}n$

zeros: $x = \frac{\pi}{3}n$

vertical stretch: 4

phase shift: none

reflection over the x-axis: none

86. period: 6π

vertical stretch: 2/3

phase shift: $\pi/4$ units left

reflection over the x-axis: yes

In Exercises 87–88, write the equation for the function $y = a \cot b(x - c)$ with the given characteristics.

87. period: $3\pi/2$

vertical stretch: 5

phase shift: $\pi/6$ units right

reflection over the x-axis: yes

88. vertical asymptotes: $x = -\frac{\pi}{4} + 3\pi n$

zeros: $x = \frac{5\pi}{4} + 3\pi n$

vertical stretch: 4

phase shift: none

reflection over the x-axis: none

In Exercises 89–90, write the equation for the function $y = a \csc b(x - c)$ with the given characteristics.

89. vertical asymptotes: $x = -\dfrac{\pi}{6} + \dfrac{3\pi}{2}n$
zeros: none
vertical stretch: 1/2
phase shift: $\pi/6$ units left
reflection over the x-axis: yes

90. period: $\pi/2$
vertical stretch: 2
phase shift: none
reflection over the x-axis: yes

In Exercises 91–92, write the equation for the function $y = a \sec b(x - c)$ with the given characteristics.

91. period: $\pi/4$
vertical stretch: 8
phase shift: $2\pi/5$ units right
reflection over the x-axis: yes

92. vertical asymptotes: $x = \dfrac{\pi}{8} + \dfrac{\pi}{4}n$
zeros: none
vertical stretch: 4
phase shift: none
reflection over the x-axis: none

5.7 INVERSE TRIGONOMETRIC FUNCTIONS

OBJECTIVES

- Define the inverse of the sine, cosine, tangent, cosecant, secant, and cotangent functions.
- Find exact values of inverse trigonometric functions.
- Use a calculator to evaluate inverse trigonometric functions.
- Evaluate the composition of a trigonometric function and its inverse.
- Evaluate the composition of a trigonometric function and the inverse of another trigonometric function.

PREREQUISITE VOCABULARY TERMS

composition (of functions)
domain
Horizontal Line Test
inverse function
one-to-one function
range
trigonometric functions

Review of Inverse Functions

REMEMBER

If any horizontal line would intersect the graph of a function f at most once, then the inverse function of f exists.

Recall from Section 1.8 that if a function f passes the Horizontal Line Test, then f is a one-to-one function and f has an inverse function, denoted f^{-1} or $f^{-1}(x)$, with the following properties.

1. For all x in the domain of f and f^{-1}: $f^{-1}(f(x)) = x = f(f^{-1}(x))$.
2. The graphs of f and f^{-1} are reflections of each other over the line $y = x$.
3. The domain of f is the range of f^{-1}, and the domain of f^{-1} is the range of f.

5.7.1 An Introduction to Inverse Trigonometric Functions

The Inverse Sine Function

In Figure 5.7a, we see that the graph of $y = \sin x$ does not pass the Horizontal Line Test, and so its inverse function does not exist. However, when the domain of $y = \sin x$ is restricted to the interval $\left[-\frac{\pi}{2}, \frac{\pi}{2}\right]$ (Figure 5.7b), the graph of $y = \sin x$ does pass the Horizontal Line Test. On this restricted domain (among others), its inverse function does exist. Note that the range of $y = \sin x$ is still $[-1, 1]$ within this interval.

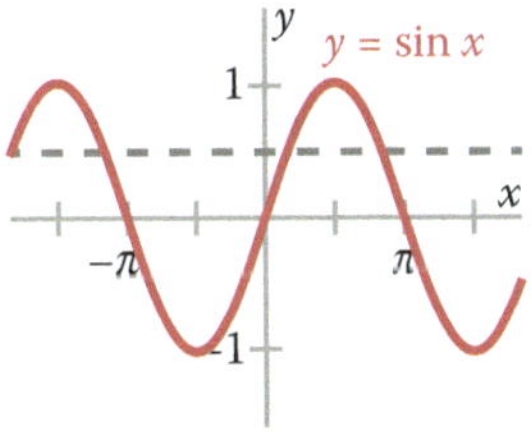

Figure 5.7a

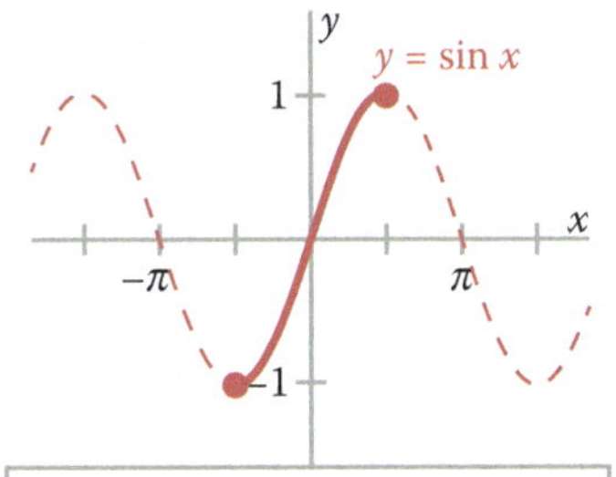

$y = \sin x$ has an inverse function on the interval $-\frac{\pi}{2} \le x \le \frac{\pi}{2}$.

Figure 5.7b

Therefore, $y = \sin x$ with domain $-\frac{\pi}{2} \le x \le \frac{\pi}{2}$ is a one-to-one function which has an inverse function. Though there are other intervals of x-values where $y = \sin x$ is a one-to-one function, such as $\frac{\pi}{2} \le x \le \frac{3\pi}{2}$, we will use $-\frac{\pi}{2} \le x \le \frac{\pi}{2}$ to define the **inverse sine function**.

TIP

Think of $\sin^{-1} x$ as "the angle (or real number) in the interval $[-\pi/2, \pi/2]$ whose sine is x."

DEFINITION

The **inverse sine function**, denoted $y = \sin^{-1} x$ or $y = \arcsin x$, is the inverse function of $y = \sin x$ with restricted domain $\left[-\frac{\pi}{2}, \frac{\pi}{2}\right]$.

$$y = \sin^{-1} x \iff \sin y = x \text{ where } -1 \le x \le 1 \text{ and } -\frac{\pi}{2} \le y \le \frac{\pi}{2}$$

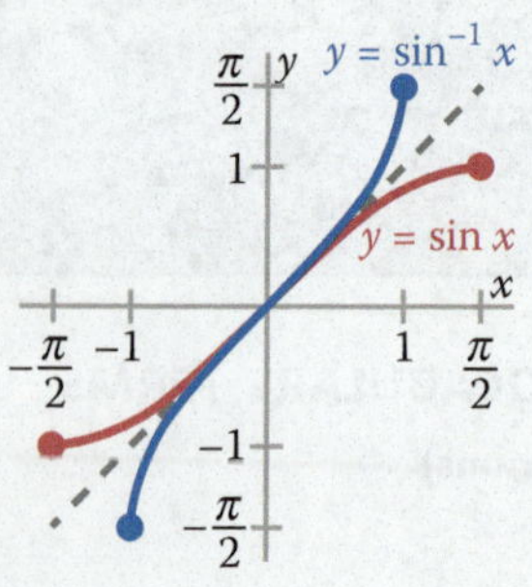

Figure 5.7c

Since $y = \sin x$ where $-\frac{\pi}{2} \le x \le \frac{\pi}{2}$ and $y = \sin^{-1} x$ are inverse functions, their graphs must be reflections over the line $y = x$, as shown in Figure 5.7c.

The Inverse Cosine Function

Like the graph of $y = \sin x$, the graph of $y = \cos x$ does not pass the Horizontal Line Test when the function is graphed over its entire domain (all real numbers), so its inverse function does not exist. However, if the domain of $y = \cos x$ is restricted to an interval where the function is either only increasing or only decreasing, such as $[0, \pi]$, then the graph does pass the Horizontal Line Test (Figure 5.7d), and then its inverse function does exist.

So, $y = \cos x$ with domain $0 \le x \le \pi$ is a one-to-one function which has an inverse function.

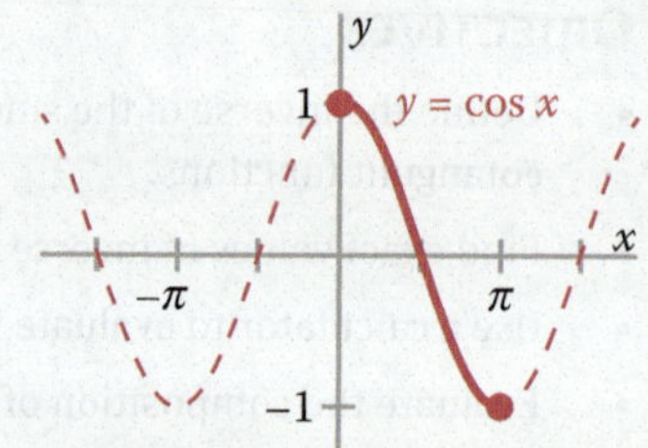

$y = \cos x$ has an inverse function on the interval $0 \le x \le \pi$.

Figure 5.7d

NOTICE THAT

The range of $y = \cos x$ does not change when its domain is restricted to $[0, \pi]$.

DEFINITION

The **inverse cosine function**, denoted $y = \cos^{-1} x$ or $y = \arccos x$, is the inverse function of $y = \cos x$ with restricted domain $[0, \pi]$.

$$y = \cos^{-1} x \quad \Leftrightarrow \quad \cos y = x \text{ where } -1 \le x \le 1 \text{ and } 0 \le y \le \pi$$

TIP

Think of $\cos^{-1} x$ as "the angle (or real number) in the interval $[0, \pi]$ whose cosine is x."

The Inverse Tangent Function

The graph $y = \tan x$ passes the Horizontal Line Test when the domain is restricted to $\left(-\frac{\pi}{2}, \frac{\pi}{2}\right)$, as shown in Figure 5.7e, and then its inverse function exists.

Note that within this restricted domain, the range of $y = \tan x$ is still $(-\infty, \infty)$.

So, $y = \tan x$ with domain $-\frac{\pi}{2} < x < \frac{\pi}{2}$ is a one-to-one function which has an inverse function.

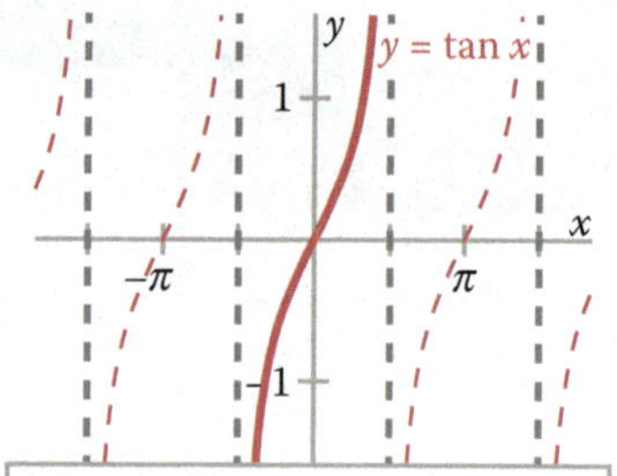

$y = \tan x$ has an inverse function on the interval $-\frac{\pi}{2} < x < \frac{\pi}{2}$.

Figure 5.7e

NOTICE THAT

The graph of $y = \tan x$ has vertical asymptotes at $x = -\pi/2$ and $x = \pi/2$, so those values are not in the restricted domain.

DEFINITION

The **inverse tangent function**, denoted $y = \tan^{-1} x$ or $y = \arctan x$, is the inverse function of $y = \tan x$ with restricted domain $\left(-\frac{\pi}{2}, \frac{\pi}{2}\right)$.

$$y = \tan^{-1} x \quad \Leftrightarrow \quad \tan y = x \text{ where } -\infty < x < \infty \text{ and } -\frac{\pi}{2} < y < \frac{\pi}{2}$$

TIP

Think of $\tan^{-1} x$ as "the angle (or real number) in the interval $(-\pi/2, \pi/2)$ whose tangent is x."

The Graphs of the Basic Inverse Trigonometric Functions

Recall that the domain of a function is the range of its inverse function, and the range of a function is the domain of its inverse function. So, the restricted domains of $y = \sin x$, $y = \cos x$, and $y = \tan x$ are the ranges of $y = \sin^{-1} x$, $y = \cos^{-1} x$, and $y = \tan^{-1} x$, respectively. Furthermore, the ranges $y = \sin x$, $y = \cos x$, and $y = \tan x$ are the domains of $y = \sin^{-1} x$, $y = \cos^{-1} x$, and $y = \tan^{-1} x$, respectively.

The following table shows the graphs of $y = \sin^{-1} x$, $y = \cos^{-1} x$, and $y = \tan^{-1} x$, along with the domain and range of each.

The Graphs of the Basic Inverse Trigonometric Functions

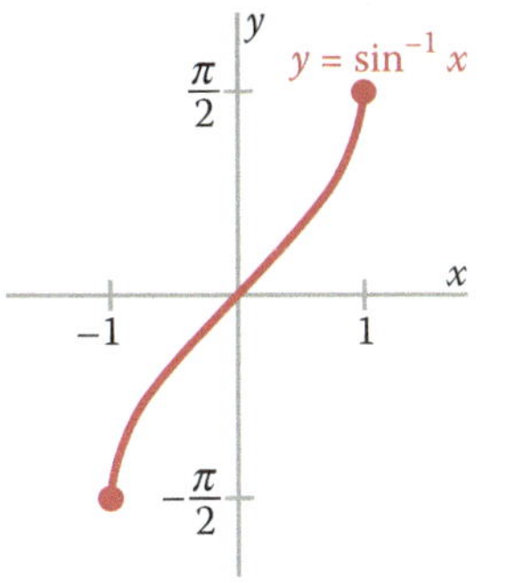

Domain: $[-1, 1]$

Range: $\left[-\frac{\pi}{2}, \frac{\pi}{2}\right]$

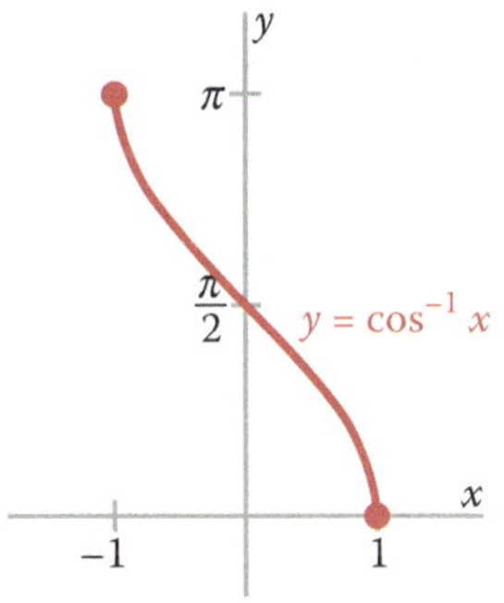

Domain: $[-1, 1]$

Range: $[0, \pi]$

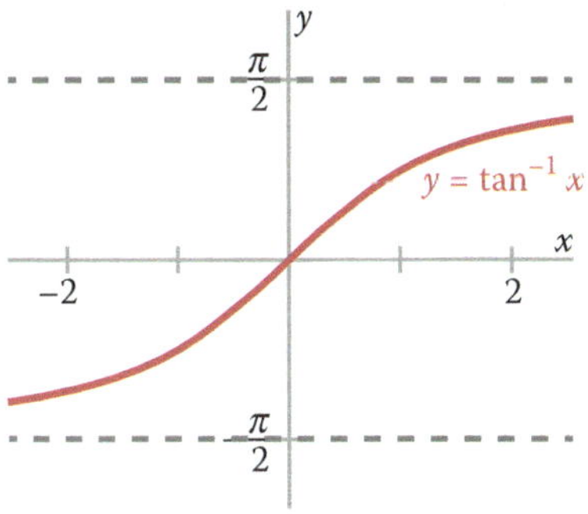

Domain: $(-\infty, \infty)$

Range: $\left(-\frac{\pi}{2}, \frac{\pi}{2}\right)$

NOTICE THAT

The graph of $y = \tan x$ has vertical asymptotes at $x = \pm\pi/2$, so the graph of $y = \tan^{-1} x$ has horizontal asymptotes at $y = \pm\pi/2$.

CAUTION

Inverse function notation can easily be confused with exponential notation.

Recall that x^{-1} means the reciprocal of x (i.e., $x^{-1} = \frac{1}{x}$). Similarly,

- $\sin x^{-1}$ means the sine of the reciprocal of x (i.e., $\sin x^{-1} = \sin \frac{1}{x}$), and
- $(\sin x)^{-1}$ means the reciprocal of the sin of x (i.e., $(\sin x)^{-1} = \frac{1}{\sin x} = \csc x$).

However, $\sin^{-1} x$ means the inverse sine function of x (i.e., $\sin^{-1} x = \arcsin x$).

5.7.2 Evaluating Inverse Sine, Cosine, and Tangent Functions

When evaluating an inverse trigonometric function, that inverse function's domain and range must be considered. Recall that the domain of $y = \tan^{-1} x$ is all real numbers x, but the domains of $y = \sin^{-1} x$ and $y = \cos^{-1} x$ are limited to only those x-values in the interval $[-1, 1]$. Therefore, while the value of $\tan^{-1} x$ exists for any real number x, the values of $\sin^{-1} x$ and $\cos^{-1} x$ do not exist when $|x| > 1$.

REMEMBER

$|x| > 1 \Rightarrow x < -1$ or $x > 1$

$|x| \le 1 \Rightarrow -1 \le x \le 1$

Since an inverse function is a one-to-one function, each value in its domain corresponds to exactly one value in its range. For example, if $|x| \le 1$ (which indicates that the value of $\cos^{-1} x$ exists), then $\cos^{-1} x$ is equal to exactly one value, and that value is in the interval $[0, \pi]$. In other words, when the value of $\cos^{-1} x$ exists for some given x-value, $\cos^{-1} x$ is equal to exactly one angle which is located either in quadrant I, quadrant II, on the positive y-axis, or on the x-axis.

EXAMPLE 1 Finding the Exact Value of an Inverse Sine, Inverse Cosine, or Inverse Tangent Function

REMEMBER

If an angle θ is a multiple of $\pi/6$ or $\pi/4$ (i.e., 30° or 45°), then the exact value of a trigonometric function of θ can be determined. Otherwise, the approximate value of the trigonometric function of θ can be found using a calculator.

Find the exact value of each inverse trigonometric function, if possible.

A. $\cos^{-1} 1$ **B.** $\sin^{-1}\left(-\frac{\sqrt{3}}{2}\right)$ **C.** $\tan^{-1}\left(-\sqrt{3}\right)$

SOLUTION

A. Since 1 is in the domain of the inverse cosine function, $\cos^{-1} 1$ exists.

The value of $\cos^{-1} 1$ is the angle θ in the interval $[0, \pi]$, whose cosine is 1. In other words, if we let $\cos^{-1} 1 = \theta$, then $\cos\theta = 1$ when $0 \le \theta \le \pi$.

Find θ. Since $\cos 0 = 1$ and 0 is in the interval $[0, \pi]$, $\theta = 0$. Therefore $\cos^{-1} 1 = 0$.

B. Since $-\frac{\sqrt{3}}{2}$ is in the domain of the inverse sine function, $\sin^{-1}\left(-\frac{\sqrt{3}}{2}\right)$ exists.

The value of $\sin^{-1}\left(-\frac{\sqrt{3}}{2}\right)$ is the angle θ in the interval $\left[-\frac{\pi}{2}, \frac{\pi}{2}\right]$, whose sine is $-\frac{\sqrt{3}}{2}$.

In other words, if we let $\sin^{-1}\left(-\frac{\sqrt{3}}{2}\right) = \theta$, then $\sin\theta = -\frac{\sqrt{3}}{2}$ where $-\frac{\pi}{2} \le \theta \le \frac{\pi}{2}$.

IMPORTANT

The inverse sine function's range does not include quadrant III, so θ cannot be coterminal with $4\pi/3$.

Find θ. From the unit circle, $\sin\frac{4\pi}{3} = -\frac{\sqrt{3}}{2}$ and $\sin\frac{5\pi}{3} = -\frac{\sqrt{3}}{2}$, but neither angle is in the interval $\left[-\frac{\pi}{2}, \frac{\pi}{2}\right]$. Therefore, θ must be coterminal with one of these angles. Specifically, since the interval includes quadrant IV, but not quadrant III, θ must be coterminal with $5\pi/3$, which is in quadrant IV. The angle in quadrant IV that is coterminal with $5\pi/3$ is $-\pi/3$. Thus, $\theta = -\pi/3$.

Therefore, $\sin^{-1}\left(-\frac{\sqrt{3}}{2}\right) = -\frac{\pi}{3}$.

REMEMBER

The domain of the inverse tangent function is all real numbers.

C. Since $-\sqrt{3}$ is in the domain of the inverse tangent function, $\tan^{-1}\left(-\sqrt{3}\right)$ exists.

The value of $\tan^{-1}\left(-\sqrt{3}\right)$ is the angle θ in the interval $\left(-\frac{\pi}{2}, \frac{\pi}{2}\right)$, whose tangent is $-\sqrt{3}$.

In other words, if we let $\tan^{-1}\left(-\sqrt{3}\right) = \theta$, then $\tan\theta = -\sqrt{3}$ where $-\frac{\pi}{2} < \theta < \frac{\pi}{2}$.

IMPORTANT

The inverse tangent function's range does not include quadrant II, so θ cannot be coterminal with $2\pi/3$.

Find θ. From the unit circle, $\tan\frac{2\pi}{3} = -\sqrt{3}$ and $\tan\frac{5\pi}{3} = -\sqrt{3}$, but neither angle is in the interval $\left(-\frac{\pi}{2}, \frac{\pi}{2}\right)$. Therefore, θ must be coterminal with one of these angles. Specifically, since the interval includes quadrant IV, but not quadrant II, θ must be coterminal with $5\pi/3$, which is in quadrant IV. The angle in quadrant IV coterminal with $5\pi/3$ is $-\pi/3$. Thus, $\theta = -\pi/3$.

Therefore, $\tan^{-1}\left(-\sqrt{3}\right) = -\frac{\pi}{3}$.

EXAMPLE 2 Using a Calculator to Evaluate Inverse Sine, Cosine, and Tangent Functions

Use a calculator to evaluate $\sin^{-1} x$, $\cos^{-1} x$, and $\tan^{-1} x$ for each given x-value, if possible. Round to the nearest thousandth.

A. 4 **B.** $-\frac{\sqrt{3}}{2}$

SOLUTION

The expressions can be entered directly (e.g., "$\tan^{-1} 4$") followed by ENTER when using a graphing calculator. However, when using a scientific calculator, first enter the value, and then press the button for the inverse trigonometric function. Be sure in either case to have the calculator set to radian mode.

A. Since 4 is not in the domain of the inverse sine function or the inverse cosine function, $\sin^{-1} 4$ and $\cos^{-1} 4$ do not exist. Entering either value into a calculator results in an error.

IMPORTANT

Unless otherwise stated, calculate the angle measure given by an inverse trigonometric function in radians.

To calculate $\tan^{-1} 4$ with a scientific calculator, press 4, then press TAN⁻¹.

If using a graphing calculator, press TAN⁻¹, 4 , then ENTER.

$$\tan^{-1} 4 \approx 1.326$$

B. Since $-1 \le -\frac{\sqrt{3}}{2} \le 1$, all three inverse trigonometric functions of $-\frac{\sqrt{3}}{2}$ exist.

For a scientific calculator, always enter $-\frac{\sqrt{3}}{2}$ first.

3, √, −, /, 2, ENTER.

Then press the button for the appropriate inverse trigonometric function.

$$\sin^{-1}\left(-\frac{\sqrt{3}}{2}\right) \approx -1.047 \qquad \cos^{-1}\left(-\frac{\sqrt{3}}{2}\right) \approx 2.618 \qquad \tan^{-1}\left(-\frac{\sqrt{3}}{2}\right) \approx -0.714$$

TIP

While the value of any inverse trigonometric function that exists can be approximated using a calculator, the ability to find an exact value (i.e., without using a calculator) is important because recognizing exact values will simplify calculations involving more complicated inverse trigonometric functions.

The exact value of $\sin^{-1}\left(-\frac{\sqrt{3}}{2}\right)$ is $-\pi/3$ and its approximate value is −1.047 (as shown in **Example 1B** and **Example 2B**, respectively). The approximate value can be confirmed by finding the decimal approximation of $-\pi/3$.

$$\sin^{-1}\left(-\frac{\sqrt{3}}{2}\right) = -\frac{\pi}{3} \approx -1.047 \checkmark$$ *Divide $-\pi$ by 3 and round to the nearest thousandth.*

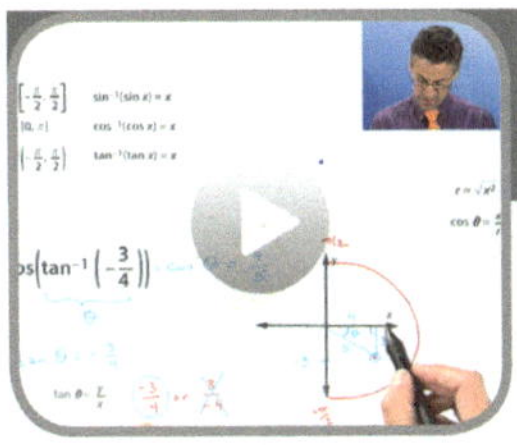

5.7.3 Evaluating a Composition Involving Inverse Trigonometric Functions

Recall from Section 1.7 that the composition of a function f with a function g, $f(g(x))$, is the result of evaluating f at g. In this case, the domain of $f(g(x))$ is the set of all x in the domain of g such that $g(x)$ is in the domain of f.

Compositions can involve trigonometric functions. For example, the composition $\cos(\tan \pi)$ is the result of evaluating the cosine function for the value of $\tan \pi$. Since $\tan \pi = 0$, $\cos(\tan \pi) = \cos 0 = 1$.

Compositions can also involve inverse trigonometric functions, as shown in **Example 3** and **Example 4**. Note that if f is a trigonometric function and g is its inverse trigonometric function, then $f(g(x)) = x$ for all x in the domain of inverse function g. However, $g(f(x)) = x$ *only* when x is in the range of inverse function g, as summarized in the following properties.

TIP

For all real numbers x, $\tan(\tan^{-1} x) = x$.
For all x in the interval $[-1, 1]$, $\sin(\sin^{-1} x) = x$ and $\cos(\cos^{-1} x) = x$.
*For all x **not** in the interval $[-1, 1]$, $\sin(\sin^{-1} x)$ and $\cos(\cos^{-1} x)$ do not exist.*

Properties of Inverse Sine, Cosine, and Tangent Functions

For all x in the interval $\left[-\frac{\pi}{2}, \frac{\pi}{2}\right]$, $\sin^{-1}(\sin x) = x$.

For all x in the interval $[0, \pi]$, $\cos^{-1}(\cos x) = x$.

For all x in the interval $\left(-\frac{\pi}{2}, \frac{\pi}{2}\right)$, $\tan^{-1}(\tan x) = x$.

EXAMPLE 3 Evaluating a Composition of a Trigonometric Function and Its Inverse Function

Find the exact value of each composition, if possible.

A. $\sin^{-1}\left(\sin\left(-\frac{\pi}{6}\right)\right)$ **B.** $\cos^{-1}\left(\cos\frac{7\pi}{6}\right)$ **C.** $\tan^{-1}\left(\tan\left(-\frac{2\pi}{3}\right)\right)$

SOLUTION

IMPORTANT

$\sin^{-1}(\sin x) = x$ only when x is in the range of the inverse sine function.

A. Since $-\pi/6$ is in the range of the inverse sine function, $\sin^{-1}\left(\sin\left(-\frac{\pi}{6}\right)\right) = -\frac{\pi}{6}$ by the property of the inverse sine function.

ALTERNATIVE METHOD

First evaluate $\sin\left(-\frac{\pi}{6}\right)$. $\sin\left(-\frac{\pi}{6}\right) = -\frac{1}{2}$

So, $\sin^{-1}\left(\sin\left(-\frac{\pi}{6}\right)\right) = \sin^{-1}\left(-\frac{1}{2}\right)$. Now evaluate the inverse sine function.

Since $-\frac{\pi}{6}$ is in the interval $\left[-\frac{\pi}{2}, \frac{\pi}{2}\right]$ and is an angle whose sine is $-\frac{1}{2}$, $\sin^{-1}\left(-\frac{1}{2}\right) = -\frac{\pi}{6}$.

Therefore, $\sin^{-1}\left(\sin\left(-\frac{\pi}{6}\right)\right) = \sin^{-1}\left(-\frac{1}{2}\right) = -\frac{\pi}{6}$.

B. Since $\frac{7\pi}{6}$ is not in the range of the inverse cosine function, $\cos^{-1}\left(\cos\frac{7\pi}{6}\right) \neq \frac{7\pi}{6}$.

First evaluate $\cos\frac{7\pi}{6}$. $\cos\frac{7\pi}{6} = -\frac{\sqrt{3}}{2}$

So, $\cos^{-1}\left(\cos\frac{7\pi}{6}\right) = \cos^{-1}\left(-\frac{\sqrt{3}}{2}\right)$. Now evaluate the inverse cosine function.

From the unit circle, the angle in the interval $[0, \pi]$ whose cosine is equal to $-\frac{\sqrt{3}}{2}$ is $\frac{5\pi}{6}$.

Therefore, $\cos^{-1}\left(\cos\frac{7\pi}{6}\right) = \cos^{-1}\left(-\frac{\sqrt{3}}{2}\right) = \frac{5\pi}{6}$.

C. Since $-\frac{2\pi}{3}$ is not in the range of the inverse tangent function, $\tan^{-1}\left(\tan\left(-\frac{2\pi}{3}\right)\right) \neq -\frac{2\pi}{3}$.

First evaluate $\tan\left(-\frac{2\pi}{3}\right)$. $\tan\left(-\frac{2\pi}{3}\right) = \sqrt{3}$

So, $\tan^{-1}\left(\tan\left(-\frac{2\pi}{3}\right)\right) = \tan^{-1}\sqrt{3}$. Now evaluate the inverse tangent function.

From the unit circle, the angle in the interval $\left(-\frac{\pi}{2}, \frac{\pi}{2}\right)$ whose tangent is equal to $\sqrt{3}$ is $\frac{\pi}{3}$.

Therefore, $\tan^{-1}\left(\tan\left(-\frac{2\pi}{3}\right)\right) = \tan^{-1}\sqrt{3} = \frac{\pi}{3}$.

EXAMPLE 4 Evaluating a Composition of a Trigonometric Function and the Inverse of Another Trigonometric Function

Find the exact value of each composition, if possible.

A. tan (arcsin 0.6) **B.** cos (arctan (–0.6))

REMEMBER

$\arcsin x = \sin^{-1} x$
$\arctan x = \tan^{-1} x$

SOLUTION

Use $\sin\theta = y/r$, $\cos\theta = x/r$, $\tan\theta = y/x$, and $r = \sqrt{x^2+y^2}$ to evaluate the compositions.

A. Let arcsin 0.6 = θ. It follows that $\sin\theta = 0.6$ and tan (arcsin 0.6) = $\tan\theta = y/x$.

We can use the value of $\sin\theta$ to find $\tan\theta$ (i.e., y/x), without finding the value of θ.

Since $\sin\theta = 0.6 = 3/5$ and $\sin\theta = y/r$ where $r > 0$, it follows that $y = 3$ and $r = 5$.

Use $r = \sqrt{x^2+y^2}$ to find x. $5 = \sqrt{x^2+3^2} \Rightarrow 25 = x^2 + 9 \Rightarrow x^2 = 16 \Rightarrow x = \pm 4$

Find the quadrant containing θ to determine whether $x = 4$ or $x = -4$.

Since $\sin\theta = 0.6 > 0$, θ must be in quadrant I or II (where sine is positive). However, the range of arcsin is $\left[-\frac{\pi}{2}, \frac{\pi}{2}\right]$ (i.e., the range includes quadrants I and IV), so θ is in quadrant I.

Since $x > 0$ in quadrant I, $x = 4$. Therefore, tan (arcsin 0.6) = $\tan\theta = y/x = 3/4$.

REMEMBER

We let arcsin 0.6 = θ, so tan (arcsin 0.6) = tan θ.

B. Let arctan (–0.6) = θ. It follows that $\tan\theta = -0.6$ and cos (arctan (–0.6)) = $\cos\theta = x/r$.

Use the value of $\tan\theta$ to find $\cos\theta$ (i.e., x/r).

Since $\tan\theta = -0.6 = -3/5$ and $\tan\theta = y/x$, it follows that $x = 5$ and $y = -3$ *or* $x = -5$ and $y = 3$.

Find the quadrant containing θ to determine the signs of x and y.

Because $\tan\theta = -0.6 < 0$ and the range of arctan is $\left(-\frac{\pi}{2}, \frac{\pi}{2}\right)$, θ is in quadrant IV.

Since $x > 0$ and $y < 0$ in quadrant IV, $x = 5$ and $y = -3$.

Use $r = \sqrt{x^2+y^2}$ to find r. $r = \sqrt{5^2+(-3)^2} = \sqrt{25+9} = \sqrt{34}$

Therefore, cos (arctan (–0.6)) = $\cos\theta = x/r = \frac{5}{\sqrt{34}}$.

REMEMBER

We let arctan (–0.6) = θ, so cos(arctan(–0.6)) = cos θ.

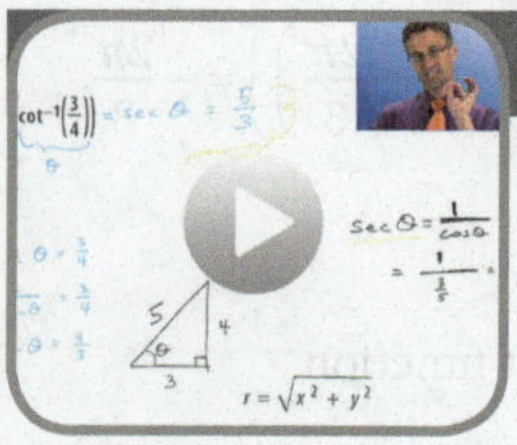

5.7.4 Inverse Cosecant, Secant, and Cotangent Functions

Like sine, cosine, and tangent, the cosecant, secant, and cotangent functions also have inverse functions when their domains are restricted.

The restricted domain of $y = \csc x$ is similar to that of $y = \sin x$ (i.e., $[-\pi/2, \pi/2]$), except 0 is excluded because csc 0 is undefined. The restricted domain of $y = \csc x$ (and thus the range of $y = \csc^{-1} x$) is the union of the intervals $\left[-\frac{\pi}{2}, 0\right)$ and $\left(0, \frac{\pi}{2}\right]$.

The restricted domain of $y = \sec x$ is similar to that of $y = \cos x$ (i.e., $[0, \pi]$), except $\pi/2$ is excluded because sec $\pi/2$ is undefined. The restricted domain of $y = \sec x$ (and thus the range of $y = \sec^{-1} x$) is the union of the intervals $\left[0, \frac{\pi}{2}\right)$ and $\left(\frac{\pi}{2}, \pi\right]$.

To define the inverse cotangent function, we restrict the domain of $y = \cot x$ to $(0, \pi)$. Thus, the range of $y = \cot^{-1} x$ is the interval $(0, \pi)$.

The graphs of the inverse cosecant, inverse secant, and inverse cotangent functions are reflections over the line $y = x$ of the graphs of the basic cosecant, secant, and cotangent functions, each over its respective restricted domain.

Inverse Cosecant, Secant, and Cotangent Functions

$y = \csc^{-1} x$ means that $\csc y = x$	$y = \sec^{-1} x$ means that $\sec y = x$	$y = \cot^{-1} x$ means that $\cot y = x$
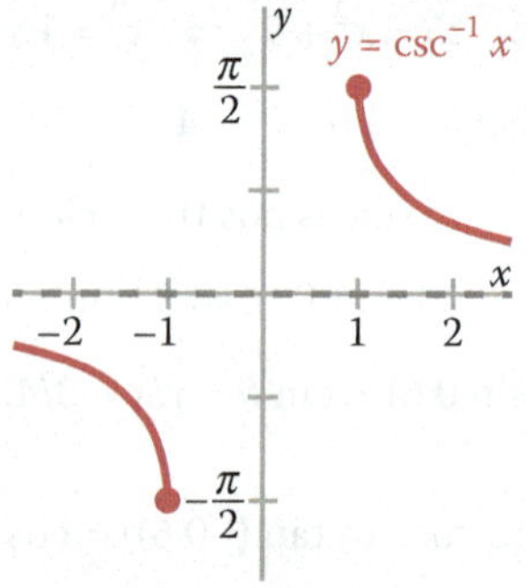	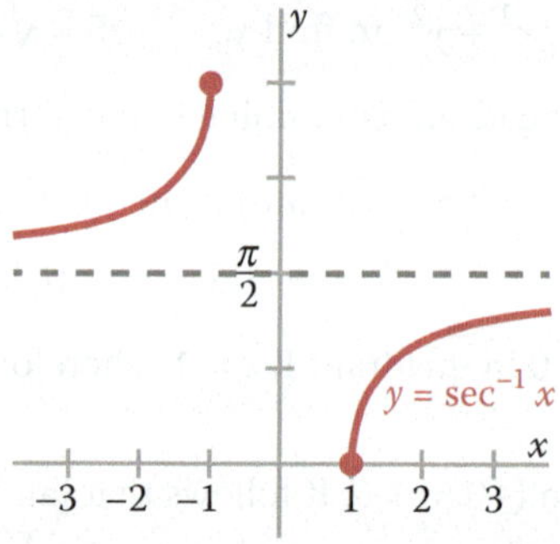	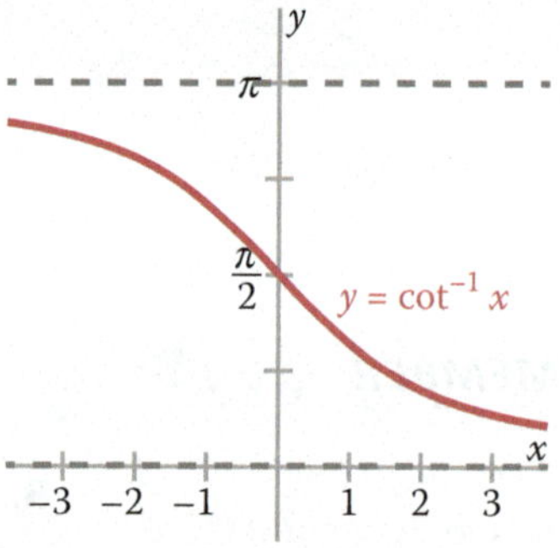
Domain: $(-\infty, -1] \cup [1, \infty)$	Domain: $(-\infty, -1] \cup [1, \infty)$	Domain: $(-\infty, \infty)$
Range: $\left[-\frac{\pi}{2}, 0\right) \cup \left(0, \frac{\pi}{2}\right]$	Range: $\left[0, \frac{\pi}{2}\right) \cup \left(\frac{\pi}{2}, \pi\right]$	Range: $(0, \pi)$

EXAMPLE 5 Finding the Exact Value of an Inverse Cosecant, Inverse Secant, or Inverse Cotangent Function

REMEMBER

The domain of $\cot^{-1} x$ is all real numbers. The domain of both $\csc^{-1} x$ and $\sec^{-1} x$ is all real numbers x such that $|x| \geq 1$.

Find the exact value of each inverse trigonometric function, if possible.

A. $\csc^{-1}(-1)$ **B.** $\sec^{-1}(-2)$ **C.** $\operatorname{arccot}\left(-\frac{1}{\sqrt{3}}\right)$

SOLUTION

A. The value of $\csc^{-1}(-1)$ is the angle θ in the interval $\left[-\frac{\pi}{2}, 0\right) \cup \left(0, \frac{\pi}{2}\right]$, whose cosecant is -1.

In other words, if we let $\csc^{-1}(-1) = \theta$, then $\csc\theta = -1$ where $-\frac{\pi}{2} \leq \theta < 0$ or $0 < \theta \leq \frac{\pi}{2}$.

Find θ. From the unit circle, we see that $\csc\frac{3\pi}{2} = -1$. However, since $\frac{3\pi}{2}$ is not in the interval $\left[-\frac{\pi}{2}, 0\right) \cup \left(0, \frac{\pi}{2}\right]$, θ must be an angle in this interval that is coterminal with $\frac{3\pi}{2}$. Thus, $\theta = -\frac{\pi}{2}$, and therefore $\csc^{-1}(-1) = -\frac{\pi}{2}$.

B. The value of $\sec^{-1}(-2)$ is the angle θ in the interval $\left[0, \frac{\pi}{2}\right) \cup \left(\frac{\pi}{2}, \pi\right]$, whose secant is -2.

In other words, if we let $\sec^{-1}(-2) = \theta$, then $\sec\theta = -2$ where $0 \leq \theta < \frac{\pi}{2}$ or $\frac{\pi}{2} < \theta \leq \pi$.

NOTICE THAT

$\sec 4\pi/3 = -2$, but $\sec^{-1}(-2) \neq 4\pi/3$ because $4\pi/3$ is in quadrant III, which is not in the range of the inverse secant function.

Find θ. From the unit circle, $\sec\frac{2\pi}{3} = -2$. Since $\frac{2\pi}{3}$ is in the interval $\left[0, \frac{\pi}{2}\right) \cup \left(\frac{\pi}{2}, \pi\right]$, $\theta = \frac{2\pi}{3}$. Therefore, $\sec^{-1}(-2) = \frac{2\pi}{3}$.

C. The value of $\operatorname{arccot}\left(-\frac{1}{\sqrt{3}}\right)$ is the angle in the interval $(0, \pi)$, whose cotangent is $-\frac{1}{\sqrt{3}}$.

From the unit circle, $\cot\frac{2\pi}{3} = -\frac{1}{\sqrt{3}}$. Since $0 < 2\pi/3 < \pi$, $\operatorname{arccot}\left(-\frac{1}{\sqrt{3}}\right) = \frac{2\pi}{3}$.

SECTION 5.7 EXERCISES

Warm Up

Determine whether each function has an inverse function. If so, write the equation of the inverse function.

1. $f(x) = 3x - 1$
2. $y = x^2$
3. $h(x) = \sqrt{2x+1}$

Just the Facts

Fill in the blanks.

4. A function f has an inverse function if any possible _____ line drawn through the graph touches the graph at most _____.
5. The graphs of a function and its inverse are _____ across the line _____.
6. The domains of the _____ trigonometric functions are restricted in order for the graphs to pass the _____ and thus have _____ that are functions.
7. The domain of $y = \sin x$ is restricted to _____ so it is a(n) _____ function with an inverse function. The range of the inverse sine function is _____ and its _____ is $[-1, 1]$.
8. The two ways to denote the inverse cosine function are $y = \cos^{-1} x$ and _____.
9. In order to define the inverse cosine function, the domain of the cosine function is restricted to _____.
10. The secant and _____ functions have undefined values at $\pi/2$ and _____. Therefore, the ranges of the inverse secant and inverse _____ functions are _____.

Essential Skills

In Exercises 1–8, find the exact value of $\sin^{-1} x$, $\cos^{-1} x$, and $\tan^{-1} x$ for each given x-value without using a calculator, if possible.

1. $\frac{\sqrt{2}}{2}$
2. $-\frac{\sqrt{2}}{2}$
3. 0
4. $-\frac{\sqrt{3}}{2}$
5. -1
6. $\frac{\sqrt{3}}{3}$
7. $\frac{\sqrt{3}}{2}$
8. $\sqrt{3}$

In Exercises 9–20, use a calculator to evaluate each inverse trigonometric function, if possible. Round to the nearest thousandth.

9. $\tan^{-1} 4.72$
10. $\sin^{-1} (-0.715)$
11. $\sin^{-1} 0.176$
12. $\cos^{-1} (-0.262)$
13. $\cos^{-1} 1.4411$
14. $\tan^{-1} (-10.619)$
15. $\cos^{-1} \frac{\sqrt{11}}{5}$
16. $\tan^{-1} 24.1$
17. $\sin^{-1} \frac{6\sqrt{3}}{17}$
18. $\cos^{-1} (-0.9)$
19. $\sin^{-1} (-0.068)$
20. $\tan^{-1} (-2.759)$

In Exercises 21–46, find the exact value of each composition, if possible.

21. $\sin^{-1}\left(\sin \frac{7\pi}{4}\right)$
22. $\sin^{-1}\left(\sin \frac{11\pi}{6}\right)$
23. $\sin^{-1}\left(\sin\left(-\frac{2\pi}{3}\right)\right)$
24. $\arcsin (\sin \pi)$
25. $\cos^{-1}\left(\cos \frac{5\pi}{6}\right)$
26. $\cos^{-1}\left(\cos \frac{5\pi}{4}\right)$
27. $\cos^{-1}\left(\cos\left(-\frac{\pi}{7}\right)\right)$
28. $\cos^{-1}\left(\cos\left(-\frac{7\pi}{3}\right)\right)$
29. $\tan^{-1}\left(\tan \frac{7\pi}{6}\right)$
30. $\tan^{-1}\left(\tan \frac{2\pi}{3}\right)$
31. $\tan^{-1}\left(\tan\left(-\frac{5\pi}{4}\right)\right)$
32. $\arctan (\tan \pi)$
33. $\tan\left(\cos^{-1} \frac{2}{5}\right)$
34. $\sin\left(\arctan \frac{2}{7}\right)$
35. $\cos\left(\sin^{-1} \frac{\sqrt{3}}{5}\right)$
36. $\cos\left(\tan^{-1} \frac{14}{5}\right)$
37. $\tan\left(\arcsin \frac{\sqrt{3}}{5}\right)$
38. $\sin\left(\cos^{-1} \frac{5}{7}\right)$
39. $\sin\left(\tan^{-1}\left(-\frac{15}{8}\right)\right)$
40. $\cos\left(\arcsin\left(-\frac{3}{8}\right)\right)$
41. $\sin\left(\arccos\left(-\frac{4}{5}\right)\right)$
42. $\tan\left(\cos^{-1}\left(-\frac{7}{9}\right)\right)$
43. $\cos\left(\tan^{-1}\left(-\frac{2}{5}\right)\right)$
44. $\tan\left(\sin^{-1}\left(-\frac{\sqrt{5}}{4}\right)\right)$
45. $\sin (\arctan (-5))$
46. $\sin\left(\cos^{-1}\left(-\frac{1}{3}\right)\right)$

In Exercises 47–50, find the exact value of $\csc^{-1} x$, $\sec^{-1} x$, and $\cot^{-1} x$ for each given x-value, if possible.

47. $\frac{2}{\sqrt{3}}$

48. $-\frac{2}{\sqrt{3}}$

49. 2

50. 1

In Exercises 51–56, find the exact value of each composition, if possible.

51. $\csc\left(\sec^{-1}\frac{5}{2}\right)$

52. $\cot\left(\sec^{-1}\left(-\frac{15}{2}\right)\right)$

53. $\sec\left(\cot^{-1}\left(-\frac{4}{7}\right)\right)$

54. $\sec\left(\csc^{-1}\frac{11}{5}\right)$

55. $\csc\left(\cot^{-1}\frac{12}{5}\right)$

56. $\cot\left(\csc^{-1}\left(-\frac{12}{11}\right)\right)$

Extensions

57. What is the domain of the function $f(t) = 3\sin^{-1}(2t)$?

58. Use a calculator to compute the smallest angle measure in a right triangle with side lengths 3, 4, and 5 in degrees to the nearest hundredth.

In Exercises 59–60, use the following information. A fighter jet cruises at an altitude of 10.32 miles toward a ship's landing deck. Consider θ and x as shown below.

59. Write θ as a function of x.

60. Find θ to the nearest tenth of a degree when $x = 4$ miles and when $x = 400$ feet.

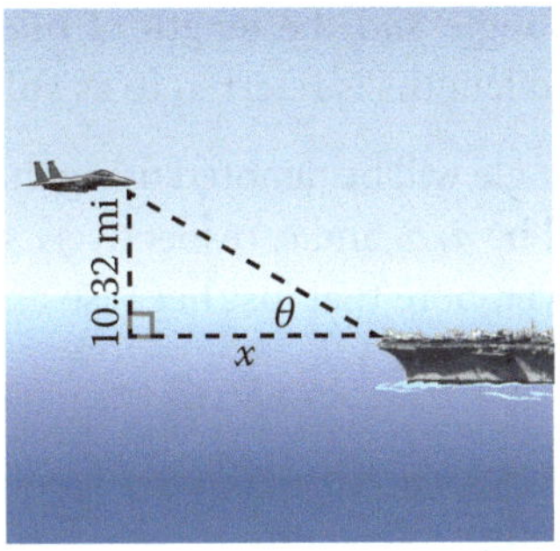

61. True or False? $\frac{5\pi}{4} = \arcsin\left(-\frac{\sqrt{2}}{2}\right)$

In Exercises 62–66, evaluate.

62. $\sin\left(\cos^{-1}\frac{\sqrt{3}}{2} - \sin^{-1}\left(-\frac{1}{2}\right)\right)$

63. $\cos(\arctan(-1) + \arccos(-1) + \arcsin 0)$

64. $\cot\left(\csc^{-1}(-2) - \sec^{-1}(-\sqrt{2}) + \arctan(-1)\right)$

65. $\tan^{-1}\left(\frac{\cos^{-1}\left(-\frac{1}{2}\right) + \sin^{-1}(-1)}{\cot^{-1}\sqrt{3}}\right)$

66. $\cot\left(\csc^{-1}\left(-\frac{7}{5}\right)\right) + \sec\left(\sin^{-1}\frac{6}{11}\right)$

5.8 APPLICATIONS OF TRIGONOMETRIC FUNCTIONS

OBJECTIVES

- Use the Triangle Sum Theorem to find the measure of an unknown angle in a right triangle.
- Use trigonometric functions to find the lengths of two unknown sides in a right triangle.
- Use trigonometric functions to find bearings and distance.
- Use trigonometric functions to find the frequency of harmonic motion.
- Introduction to damped harmonic motion.

PREREQUISITE VOCABULARY TERMS

acute angle
right angle
right triangle
trigonometric functions

5.8.1 Finding Unknown Sides and Angles in a Right Triangle

By the **Triangle Sum Theorem**, the sum of the measures of the three angles in a triangle must equal 180°. Therefore, if two angle measures are known, the measure of the third angle can be found by subtracting the sum of the two known angles' measure from 180°. It follows that if the measure of one acute angle in a right triangle is known, then the measure of the third angle can be found by subtracting the acute angle's measure from 90°.

These facts regarding the angles of a right triangle will be used, along with the trigonometric functions, to find the measures of all unknown sides and all unknown angles in a right triangle when the measure of one acute angle and the length of one side are given. This process of finding all unknown measures and lengths is referred to as **solving a triangle.**

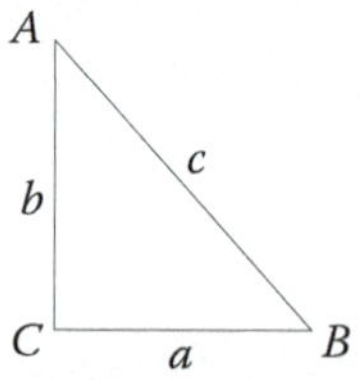

Figure 5.8a

In this section, the angles of a triangle will be denoted using the letters A, B, and C, and then the sides opposite of these angles will be a, b, and c, respectively, as shown in Figure 5.8a, where C represents the right angle. However, note that just because an angle is labeled C, this does not necessarily mean the angle is a right angle.

EXAMPLE 1 Solving a Right Triangle

Suppose triangle ABC is a right triangle such that $B = 42°$, and $a = 12.5$. Solve the triangle for all unknown sides and angles. Round to the nearest tenth, as needed.

SOLUTION

Draw the right triangle where

- C is the right angle,
- B (one of the acute angles) is 42°, and
- a (the side opposite of A) is 12.5.

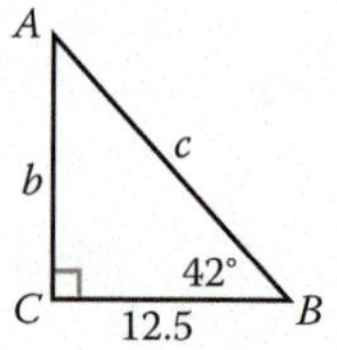

Since C is a right angle, the sum of angles A and B must be 90°, so $A = 90° - 42° = 48°$.

$$\tan 42° = \frac{b}{12.5} \quad \Rightarrow \quad b = 12.5(\tan 42°) \approx 11.3 \qquad \sin 48° = \frac{12.5}{c} \quad \Rightarrow \quad c = \frac{12.5}{\sin 48°} \approx 16.8$$

ALTERNATIVE METHOD

There are many alternative procedures for solving triangle ABC from **Example 1**.

For example, c can be found using $\cos\theta = \dfrac{\text{adj}}{\text{hyp}}$ where $\theta = B$.

$$\cos 42° = \frac{12.5}{c} \quad \Rightarrow \quad c = \frac{12.5}{\cos 42°} \approx 16.8$$

IMPORTANT

Whenever possible, use exact values for procedures.

And instead of using another trigonometric function to find the third unknown side length (in this case c), the Pythagorean Theorem can be used.

$$c = \sqrt{12.5^2 + 11.3^2} = \sqrt{283.94} = 16.85\ldots \approx 16.9$$

However, the value of c obtained using a trigonometric function involving the given side length (12.5) is more precise than the value of c obtained using the Pythagorean Theorem because the Pythagorean Theorem used an approximate value for b, 11.3.

A more precise value of c can be obtained using the Pythagorean Theorem by using the exact value of b, 12.5(tan 42°), instead of the approximate value 11.3.

$$c = \sqrt{12.5^2 + (12.5(\tan 42°))^2} = 16.82\ldots \approx 16.8$$

5.8.2 Trigonometry and Bearings

A **bearing** measures the acute angle that a direct path (straight line) makes with a fixed, vertical north-south line. The graph of a bearing is the acute angle formed by rotating an initial side (where the bearing begins) clockwise or counterclockwise from either the north or south end of the north-south line, depending on the measure and direction of the bearing.

The notation for a bearing begins with either the letters N or S (representing north and south), followed by a degree value, and then by the letters E or W (representing east and west). N or S indicates whether the rotation begins from the north or south end of the north-south line, the degree value gives the amount of rotation, and E or W indicates whether the rotation is toward the east or west end of the east-west line.

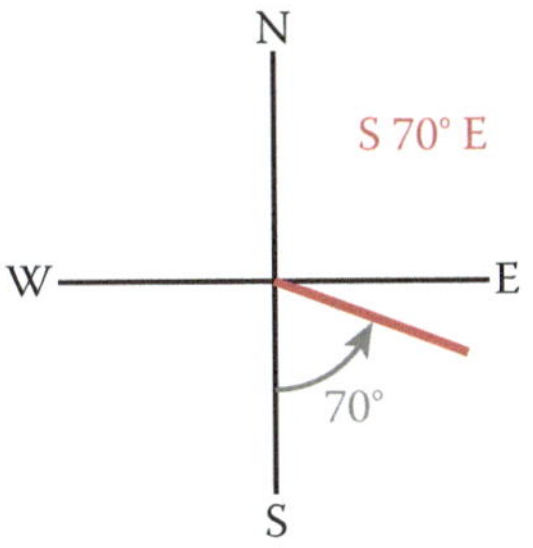

Figure 5.8b

For example, Figure 5.8b shows the graph of the bearing S 70° E. Notice that the terminal side is rotated counterclockwise from the south end of the north-south line a total of 70° toward the east end of the east-west line.

In **Example 2**, a bearing is used to find a distance between two points.

EXAMPLE 2 Using a Bearing to Find a Distance

A submarine departed from a naval base, traveled due west at 10 nautical miles per hour for 2 hours, and then traveled S 35° W at 20 nautical miles per hour for 1.5 hours. Find the distance (to the nearest tenth of a nautical mile) between the submarine and the base and find the submarine's bearing to the base at the end of the trip.

SOLUTION

Since the submarine traveled due west for 2 hours at 10 nautical miles per hour, it traveled 20 NM due west.

It then traveled along the S 35° W bearing for 1.5 hours at 20 nautical miles per hour, that is, a distance of 30 NM.

Draw a picture representing the submarine's path.

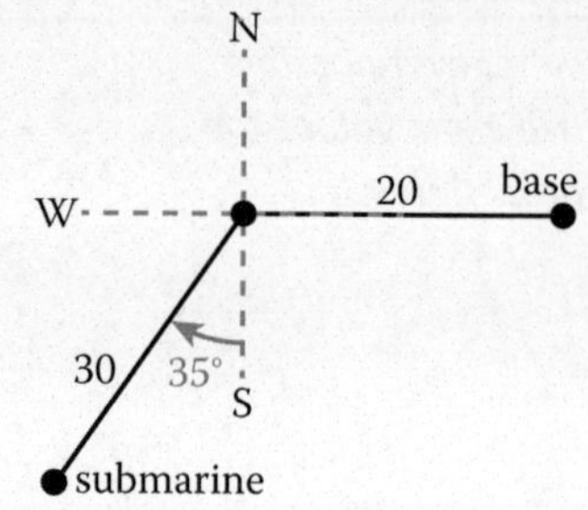

Now draw a right triangle where the distance between the submarine and the naval base is the hypotenuse c, as shown in the picture.

Let $a + 20$ and b represent the right triangle's legs.

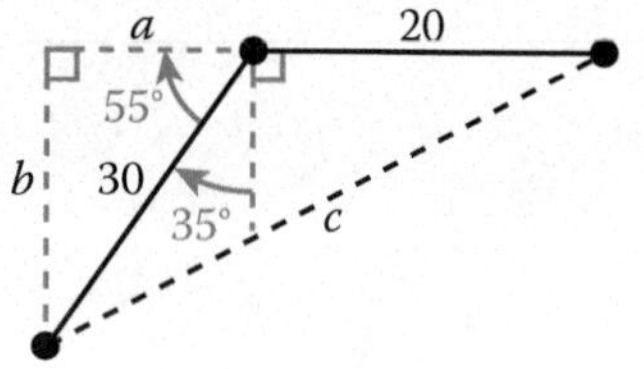

Consider the smaller right triangle with hypotenuse 30 and legs a and b. Since angle B (the angle opposite of side b) is complementary with the 35° angle, it measures $90° - 35° = 55°$.

Use cos 55° to find a and sin 55° to find b.

$$\cos 55° = a/30 \Rightarrow a = 30(\cos 55°) \qquad \sin 55° = b/30 \Rightarrow b = 30(\sin 55°)$$

So, the legs of the larger right triangle (with hypotenuse c) are $30(\sin 55°)$ and $30(\cos 55°) + 20$.

> ***TIP***
>
> *The approximate values of a and b (17.2 and 24.6) can be used to find θ (instead of the exact values as seen here), but then the value of θ will not be as precise.*

Let θ be the angle opposite the naval base.

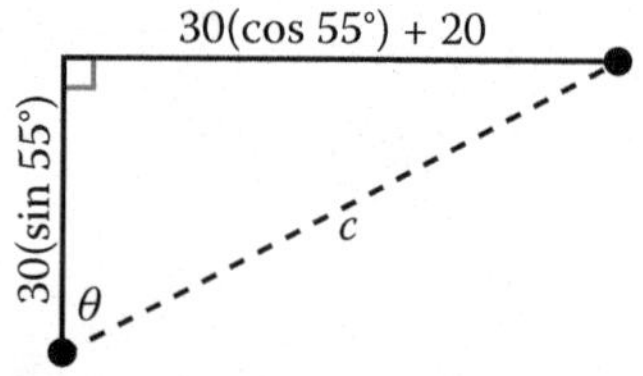

Since tan θ = opp/adj, $\tan\theta = \dfrac{30(\cos 55°) + 20}{30(\sin 55°)}$.

Use the inverse tangent function (and a calculator) to find θ.

$$\theta = \tan^{-1}\left(\frac{30(\cos 55°) + 20}{30(\sin 55°)}\right) \approx 56.6°$$

Now use a trigonometric function to find c.

$$\sin 56.6° = \frac{30(\cos 55°) + 20}{c} \Rightarrow c = \frac{30(\cos 55°) + 20}{\sin 56.6°} \approx 44.6 \text{ NM}$$

> ***IMPORTANT***
>
> *A bearing measures an acute angle, so the bearing's degree measure must be less than 90°.*

Thus, the distance between the submarine and the base is approximately 44.6 nautical miles. The submarine's bearing with the naval base (the acute angle made with the north-south line and the line of sight between the submarine and the naval base) is approximately N 56.6° E.

ALTERNATIVE METHOD

As with **Example 1**, there are alternative ways to find the distance between the submarine and the naval base in **Example 2**. For example, the Pythagorean Theorem can be used to find c.

$$c = \sqrt{(30(\cos 55°) + 20)^2 + (30(\sin 55°))^2} \approx 44.6$$

5.8.3 Harmonic Motion

The motion of a point on an object that vibrates or rotates at a regular interval can often be modeled by a trigonometric function because the trigonometric functions are periodic, as is the motion of the point. For example, consider a point on an oscillating fan where the distance between the initial point, the **equilibrium point**, and the position at time t is d. At $t = 0$, $d = 0$. The distance then increases to some maximum displacement, decreases again back to 0, and then to some minimum displacement. Since d repeats at regular intervals of t, a function modeling this motion is therefore periodic. The total time needed for some point to complete a full cycle is the period, and the **frequency** is the number of cycles completed per unit of time.

Motion that can be modeled by a sine or cosine function of the form $d = a \sin \omega t$ or $d = a \cos \omega t$ is called **simple harmonic motion.**

IMPORTANT

If $d = 0$ at $t = 0$, then the simple harmonic motion can be modeled by $d = a \sin \omega t$.

If the initial motion is in the positive direction, then $a > 0$, but if the initial motion is in the negative direction, then $a < 0$.

Simple Harmonic Motion

If the distance d of a point from the origin at time t can be modeled by either

$$d = a \sin \omega t \qquad \text{or} \qquad d = a \cos \omega t$$

(where a and ω are real numbers with $\omega > 0$), then the motion of the point is called simple harmonic motion. In this case, the motion can be characterized by the following.

amplitude: $|a|$ period: $2\pi/\omega$ frequency: $\omega/2\pi$

EXAMPLE 3 Modeling Simple Harmonic Motion

A weight is attached to the bottom of a spring, and the top of the spring is attached to a surface such that the maximum distance d the weight can move up or down from its point of equilibrium (the point at which the spring is relaxed) is 8 inches, as shown in the picture. Suppose the weight begins from its equilibrium point at $t = 0$ seconds, moves to its maximum displacement above 0, to its minimum displacement below 0, and then back to equilibrium every 2 seconds (assume there is no air resistance, or friction). Model the weight's motion with a simple harmonic equation, find its frequency, and graph the model over at least two periods.

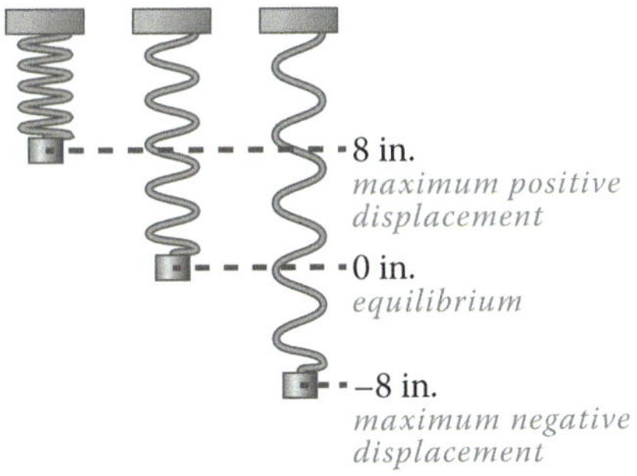

SOLUTION

Because the motion is harmonic and because $d = 0$ when $t = 0$ (i.e., the weight begins its motion at the point of equilibrium), use the sine equation $d = a \sin \omega t$. Since the weight completes 1 full cycle of its motion in 2 seconds, the period is 2 seconds.

Use the period to find ω. $2 = 2\pi/\omega \Rightarrow \omega = 2\pi/2 = \pi$

The displacement from the equilibrium point is 8, and the initial motion is up, so $a = 8$.

Using these values to write the equation gives $d = 8 \sin \pi t$.

The frequency is $\pi/2\pi$, that is, 1/2 cycle per second.

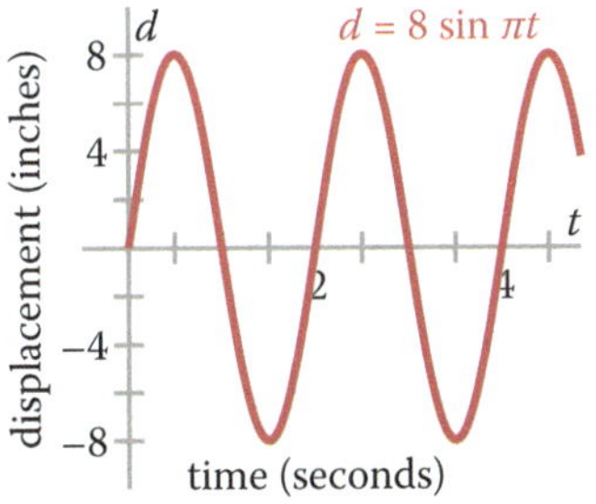

5.8.4 Damped Harmonic Motion

The motion of the spring in **Example 3** continued regularly because the motion was not affected by gravity or friction. However, this is not a realistic situation as motion is typically affected by some sort of friction, and eventually slows until it stops. When the rate of harmonic motion decreases over time, the motion is called **damped harmonic motion.**

IMPORTANT

Damped harmonic motion is not strictly periodic because its cycles are not the same, but the term is still used to describe the total time needed to cycle from the point of equilibrium and back.

Damped Harmonic Motion

If the distance d of a point from the origin at time t can be modeled by either

$$d = ke^{-ct}\sin\omega t \quad \text{or} \quad d = ke^{-ct}\cos\omega t$$

(where k, ω, and c are real numbers with $\omega > 0$ and $c > 0$), then the motion of the point is called damped harmonic motion. In this case, the motion can be characterized by its:

initial amplitude: $|k|$ period: $2\pi/\omega$ damping constant: c

The simple and damped harmonic models are the same except for the coefficients of sine and cosine (a and ke^{-ct}). Recall that local minimums and local maximums of functions of the form $y = a\sin bx$ and $y = a\sin bx$ occur when its graph intersects the lines $y = -a$ and $y = a$ (i.e., the function's amplitude is $|a|$), as is the case for a simple harmonic model (Figure 5.8c). However, the relative minimums and maximums of a damped harmonic model occur when its graph intersects the curves $d = -ke^{-ct}$ and $d = ke^{-ct}$ (Figure 5.8d).

NOTICE THAT

The graph of a damped harmonic model is easily recognized because the amplitude decreases as the curve is followed to the right.

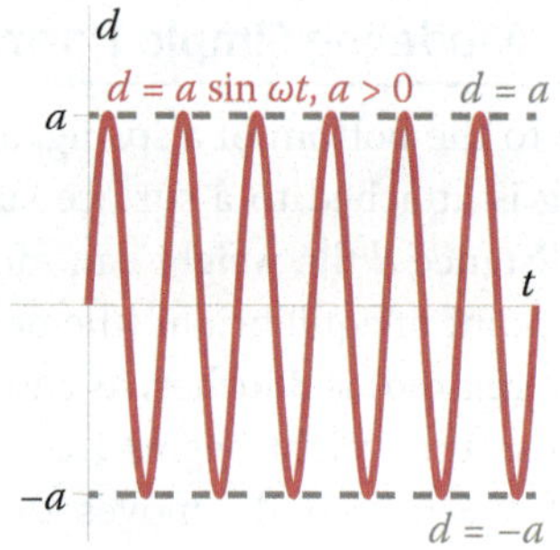

Figure 5.8c

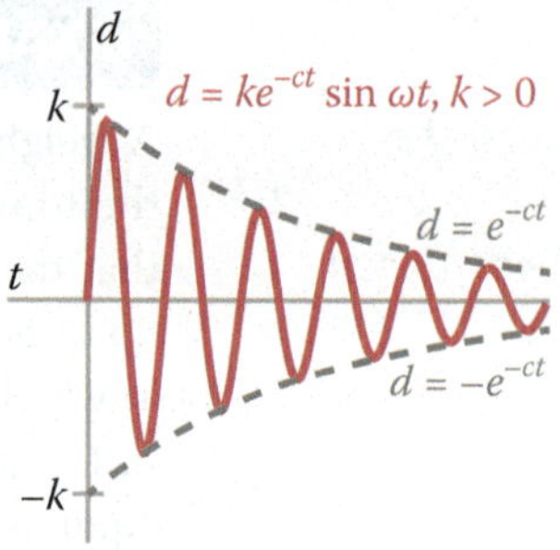

Figure 5.8d

EXAMPLE 4 Modeling Damped Harmonic Motion

Suppose a gust of wind causes a tree to sway (oscillate) back and forth with frequency 0.25 cycles per second. Assume that at the instant the wind stops, $t = 0$ seconds, the tree has reached its maximum displacement of 2 feet. Model the motion of the tree with a damped harmonic equation if the damping constant is 0.8.

SOLUTION

Use the cosine equation $d = ke^{-ct}\cos\omega t$ because in this equation $d = 2$ when $t = 0$. The tree's frequency is 0.25 (i.e., the tree completes one-quarter cycle of its motion every 1 second).

Use the frequency to find ω. $\quad 0.25 = \dfrac{\omega}{2\pi} \;\Rightarrow\; \omega = 2\pi(0.25) = 0.5\pi$

The initial displacement is 2, so $k = 2$. Using these values, along with the given damping constant $c = 0.8$, to write the equation gives $d = 2e^{-0.8t}\cos 0.5\pi t$.

SECTION 5.8 EXERCISES

Warm Up

In Exercises 1–3, translate the standard position angle into a bearing.

1. 150°
2. 40°
3. 290°

Just the Facts

Fill in the blanks.

4. When given the measures of two angles of a triangle, the third angle can be found by _____ the sum of the two angles from _____.
5. When the measure of an acute angle in a right triangle is known, the missing angle can be found by _____ the acute angle from _____.
6. When one angle and the hypotenuse of a triangle are known, either the _____ or _____ function can be used to find a missing side in a right triangle.
7. A bearing measures a(n) _____ angle. The first letter is always _____ or _____ to indicate whether the rotation begins from the _____ or _____ end of the north-south line.
8. The position where a spring is relaxed is called the _____ point.
9. The function $d = 6 \cos 9\pi t$ is an example of _____ harmonic motion. The frequency of the function is _____, and the period is _____.
10. Simple and damped harmonic motion models are the same except for the _____ of the sine and cosine models. With the simple harmonic model, the coefficient is _____. In the damped harmonic motion model, the coefficient is _____ because the motion is decreasing over time.

Essential Skills

In Exercises 1–6, use the given information for right triangle *ABC* to solve for all unknown sides and angles. Assume that *C* is the right angle. Round to the nearest tenth as needed.

1. $c = 12$, $B = 35°$
2. $b = 4$, $A = 20°$
3. $b = 7.2$, $A = 43°$
4. $a = 8$, $B = 16.2°$
5. $a = 24.3$, $B = 82°$
6. $c = 16.9$, $A = 54°$

In Exercises 7–10, use the following information.

A cellphone tower with height h feet must be secured with a guy-wire with length l feet such that θ is the angle of elevation of the wire as measured from the ground. Use the given information to find each indicated measure. Round to the nearest tenth as needed.

7. Find h if $l = 222.5$ feet and $\theta = 37°$.
8. Find l if $h = 150$ feet and $\theta = 22°$.
9. Find l if $h = 173$ feet and $\theta = 18°$.
10. Find h if $l = 89.2$ feet and $\theta = 46°$.

In Exercises 11–16, use the following information.

A firefighter is using a ladder to reach a person in a second-story window of a burning building. The window is h feet high, the ladder has length l feet, the firefighter is d feet away from the base of the building, and θ is the angle of elevation of the ladder. Use the given information to find each indicated measure. Round to the nearest tenth as needed.

11. Find h if $l = 46$ feet and $\theta = 15°$.
12. Find l if $d = 24$ feet and $\theta = 28°$.
13. Find d if $h = 28$ feet and $\theta = 32°$.
14. Find h if $d = 14$ feet and $\theta = 18°$.
15. Find l if $h = 50$ feet and $\theta = 36°$.
16. Find d if $l = 35$ feet and $\theta = 27°$.

In Exercises 17–22, use the following information.

A U.S. Coast Guard Response Boat leaves Charleston, South Carolina at time t heading due east. At time h, the boat changes course to bearing b. Find the boat's bearing and distance from Charleston, South Carolina (to the nearest hundredth of a unit) at the given time if it travels at an average of x knots (nautical miles per hour) for the entire trip.

17. at 6 p.m., if $t = 1$ p.m., $h = 4$ p.m., $b =$ N 24° W, $x = 35$ knots
18. at 8 p.m., if $t = 2$ p.m., $h = 5{:}30$ p.m., $b =$ N 16° E, $x = 42$ knots
19. at 9:30 a.m., if $t = 3{:}45$ a.m., $h = 6$ a.m., $b =$ S 18° E, $x = 28$ knots
20. at 11 a.m., if $t = 4$ a.m., $h = 8{:}30$ a.m., $b =$ S 11° W, $x = 24$ knots
21. at 7:45 p.m., if $t = 2{:}15$ p.m., $h = 3{:}30$ p.m., $b =$ N 32° E, $x = 20$ knots
22. at 10:30 a.m., if $t = 1{:}30$ a.m., $h = 7{:}30$ a.m., $b =$ S 26° E, $x = 38$ knots

In Exercises 23–28, find the frequency of each simple harmonic motion function.

23. $d = \frac{1}{3} \sin \frac{\pi}{4} t$

24. $d = 6 \cos 9\pi t$

25. $d = -2 \cos \frac{3\pi}{7} t$

26. $d = -\frac{4}{5} \sin \frac{6\pi}{11} t$

27. $d = 4 \sin \frac{8\pi}{17} t$

28. $d = 7 \cos \frac{4\pi}{5} t$

In Exercises 29–36, find a model for the displacement of a spring that exhibits simple harmonic motion that has displacement d_0 when $t = 0$, period p, and a maximum displacement m.

29. $d_0 = -12$ centimeters, $p = 16$ seconds, $m = 12$

30. $d_0 = 15$ inches, $p = 8$ seconds, $m = 15$

31. $d_0 = 0$, $p = 11$ seconds, $m = 23$

32. $d_0 = 0$, $p = 21$ seconds, $m = 8$

33. $d_0 = 0$, $p = 6$ seconds, $m = 22$

34. $d_0 = 37$ centimeters, $p = 25$ seconds, $m = 37$

35. $d_0 = -26$ centimeters, $p = 42$ seconds, $m = 26$

36. $d_0 = 0$, $p = 15$ seconds, $m = 43$

In Exercises 37–40, use the following information.

A string on a guitar is pulled a distance k above its rest position, then released and allowed to vibrate. Suppose that the note produced has a frequency of x cycles per second with a damping constant c. Find an equation that describes the motion of the point at which the string was pulled.

37. $k = 0.3$ millimeters, $c = 2$, $x = 261.63$

38. $k = 0.35$ millimeters, $c = 1.2$, $x = 293.665$

39. $k = 0.38$ millimeters, $c = 2$, $x = 293.665$

40. $k = 0.28$ millimeters, $c = 1.8$, $x = 329.628$

Extensions

41. A person sitting in a courtyard 50 feet from a building can see a flowerpot in a window as he looks up at an angle of 30° 24′ 10″. Looking directly above the flowerpot from the same position, he sees a dog in another window at an angle of 65° 55′ 8″. Find the distance between the dog and the flowerpot to the nearest hundredth of a foot.

42. A pilot spots a lake at an angle of depression from her line of sight of 20° 14′ 53″ while flying at an altitude of 15,000 feet. After two minutes, the angle of depression is 46° 5′ 18″. Find the horizontal distance the plane has traveled.

In Exercises 43–48, use the following information.

A point on the tip of a tuning fork vibrates in simple harmonic motion. Because the movement is short in duration, it is described by the equation $D = a \sin \omega t$. The velocity and acceleration are described by $V = \omega a \cos \omega t$ and $A = -\omega^2 a \sin \omega t$, respectively. Round each answer to the nearest tenth unless instructed otherwise.

43. Find the time t to the nearest millionth of a second at which the end of the prong has a displacement of 0.217 millimeters, frequency 1200 hertz (cycles per second), and amplitude 0.4944 millimeters.

44. Find the velocity at time t in Exercise 43.

45. Find the acceleration at time t in Exercise 43.

46. Find the displacement of the tuning fork that has a frequency of 800 hertz and an amplitude of 0.14 millimeters at 0.003 seconds. Round to the nearest thousandth of a millimeter.

47. Find the velocity at time t in Exercise 46.

48. Find the acceleration at time t in Exercise 46.

49. Find four possible models that relate the displacement d of an object that has 1.5 oscillations per second and a maximum displacement of 15 inches.

50. At the local carnival, a water gun is aimed at the mouth of a plastic clown in order to inflate a balloon until it bursts. Assuming the water is projected with enough pressure to reach the clown's mouth along a linear path, at what range of values for the angle of elevation can the water gun be positioned to ensure that the water goes through the clown's mouth? The clown's mouth is a circle of radius 1.75 inches, the distance between the water gun and the clown is 5 feet, and the clown's mouth is 7 inches higher than the water gun.

51. Solve for c in $d = ke^{-ct} \sin \omega t$.

52. A yacht and a sailboat leave the same port at the same time. The yacht travels at an average of 6 knots heading due east. The sailboat travels at an average of 5 knots heading due south. After 3 hours, what is the distance between the two boats, and what is the bearing of the yacht to the boat?

53. A buoy oscillates in simple harmonic motion as waves go past. At a given time, it is noted that the buoy moves a total of 4.5 feet from its low point to its high point, and that it completes a cycle every 2.4 seconds. Write an equation that describes the motion of the buoy if, at $t = 0$, it is at its high point.

54. A helicopter is hovering 300 feet over the finish line of a boat race. As Boats A and B approach the finish line (each traveling at a constant speed), the angle of elevation from Boat A to the helicopter is 3° at the same moment that the angle of elevation from Boat B to the helicopter is 3.5°. After 2 minutes, the angle of elevation from Boat A to the helicopter is 32°, and the angle of elevation from Boat B to the helicopter is 32.2°. If each boat maintains its constant speed, which boat will win the race and by how many seconds?

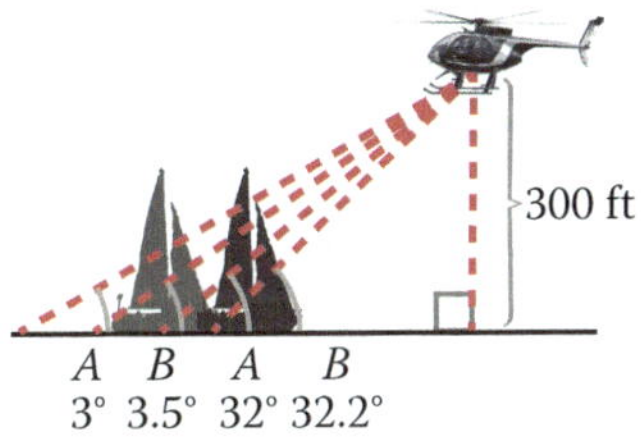

CHAPTER 5 REVIEW EXERCISES

Section 5.1 Angles and Their Measure

In Exercises 1–4, use each given angle θ.

- **Identify the quadrant in which θ lies.**
- **Find the smallest positive angle and the negative angle closest to 0° that are coterminal with θ (other than itself).**
- **Find the complement of θ, if it exists.**
- **Find the supplement of θ, if it exists.**
- **Find the radian measure of θ.**

1. $\theta = 45°$

2. $\theta = 160°$

3. $\theta = 715°$

4. $\theta = -400°$

5. When a sprinkler is set to full rotation, the water covers a circular section of a garden with a radius of 18.5 feet. The sprinkler is set to rotate back and forth through an angle of $11\pi/12$ radians. Find the area of the garden that is covered by the sprinkler, find the length of the outer edge of the area that the water covers, and convert the radian measure to degrees. Round to the nearest tenth, as necessary.

Section 5.2 Right Angle Trigonometry

6. Find the six trigonometric functions of θ, if θ is an angle in a right triangle where the hypotenuse is 10 and the side opposite θ is 7.

7. Find the length of the side adjacent to a 22.5° angle in a right triangle if its opposite side has length 19.2. Round to the nearest tenth.

8. Find the length of the side opposite an angle measuring $\pi/8$ radians in a right triangle if the hypotenuse has length 16. Round to the nearest tenth.

9. Find the length of the hypotenuse of a right triangle if the side adjacent to an angle measuring $\pi/5$ radians has length 4.5. Round to the nearest tenth.

10. When observed from the top of a 175-foot-tall lighthouse, the angle of depression of an approaching ship is 32°.Find the straight line distance from the lighthouse to the ship. Round to the nearest tenth of a foot.

Section 5.3 The Trigonometric Functions in the Coordinate Plane

In Exercises 11–12, find the sine, cosine, tangent, cosecant, secant, and cotangent of θ, given that θ is an angle in standard position and the given point is on the terminal side of θ.

11. $(-1, 5)$

12. $(-7, -3)$

In Exercises 13–14, find the six trigonometric functions of θ, if θ is a standard position angle.

13. $\theta = -225°$

14. $\theta = 7\pi/3$

15. Find $\sin\theta$ and $\cos\theta$, given that $\cot\theta = -5/2$ and $\sin\theta < 0$.

Section 5.4 Unit Circle Trigonometry

In Exercises 16–18, evaluate the six trigonometric functions of each real number.

16. $t = 7\pi/6$

17. $t = -\pi$

18. $t = 9\pi/2$

19. If $\csc t = -6/5$, evaluate $\csc(-t)$ and $\sin(-t)$.

20. If $\tan(-t) = 7/6$, evaluate $\tan t$ and $\cot t$.

Section 5.5 Graphing Sine and Cosine Functions

In Exercises 21–23, find each function's amplitude and period, describe any phase shifts or vertical shifts (as compared to the corresponding basic trigonometric function), and then graph at least two periods of the function. Identify the x-values of all maximums, minimums, and zeros.

21. $y = -2\cos \pi x + 3$

22. $y = -\dfrac{1}{2}\sin\left(x + \dfrac{2\pi}{3}\right) + 1$

23. $y = \dfrac{2}{3}\cos(3x - 2\pi)$

24. Find the function's amplitude and period, describe any phase shift or vertical shift (as compared to $y = \sin x$), and then graph at least two periods of the function. Identify the x-values of all maximums and minimums.

$y = 3\sin\frac{\pi}{2}x - 1$

25. The height h of a seat above ground on a ferris wheel at time t is given by $h(t) = 25\sin\left(\frac{\pi}{6}t - \frac{5\pi}{6}\right) + 55$, where $t \geq 0$ seconds and h is in feet. Graph the function for two periods and find the maximum height of the seat.

Section 5.6 Graphing Other Trigonometric Functions

In Exercises 26–27, find each function's period, describe any phase shifts or vertical shifts (as compared to the corresponding basic trigonometric function), and then graph at least two periods of the function. Identify all vertical asymptotes and zeros, if any.

26. $y = \frac{1}{2}\tan\left(\frac{2}{3}x - \frac{\pi}{4}\right)$

27. $y = 4\cot\left(\frac{1}{2}x + \frac{\pi}{2}\right)$

In Exercises 28–30, find each function's period, describe any phase shifts or vertical shifts (as compared to the corresponding basic trigonometric function), and then graph at least two periods of the function. Identify all vertical asymptotes.

28. $y = -\frac{1}{3}\csc(\pi x + 2\pi) - 2$

29. $y = 2\sec\left(2x - \frac{\pi}{2}\right) + 1$

30. $y = -\frac{3}{4}\cot 2(x + \pi) - 1$

Section 5.7 Inverse Trigonometric Functions

In Exercises 31–35, find the exact value of each expression, if possible. If not possible, use a calculator to find the value of the expression to the nearest hundredth.

31. $\sin^{-1}(1/2)$

32. $\cos^{-1}(-0.882)$

33. $\arccos\left(\cos\left(-\frac{2\pi}{3}\right)\right)$

34. $\cos\left(\tan^{-1}\left(-\frac{\sqrt{6}}{7}\right)\right)$

35. $\csc\left(\cot^{-1}\left(-\frac{\sqrt{7}}{3}\right)\right)$

Section 5.8 Applications of Trigonometric Functions

36. In right triangle ABC with $a = 11.9$, $B = 36.2°$, and $C = 90°$, solve for the unknown sides and angle. Round to the nearest tenth, as needed.

37. A cat is trapped on a tree branch 12.3 feet above the ground. A nearby ladder is 13 feet long. If the ladder's tip is placed on the branch, what angle will the ladder make with the ground? Round to the nearest tenth of a degree.

38. An observer 75 feet above the surface of the water measures an angle of depression of 12° to a distant ship. How many feet to the nearest hundredth is the ship from the observer?

39. A ship leaves the port of Galveston with a bearing of S 68° E and a speed of 12 knots (nautical miles per hour). After 2 hours, the ship turns directly south. If the ship maintains its speed, what is the bearing of the ship relative to the port after 3 hours of travel? Round to the nearest hundredth of a degree.

40. A weight is attached to the bottom of a spring, and the top of the spring is attached to a surface so that the maximum distance d the weight can move vertically up or down from its point of equilibrium (the point at which the spring is relaxed) is 20 centimeters. Model the motion of the weight with a simple harmonic equation and find its frequency. Assume that the weight begins from its equilibrium point at $t = 0$ seconds and moves to its maximum displacement above 0, to its minimum displacement below 0, and then back to equilibrium every 1.5 seconds. Assume also that the weight is not affected by friction.

Precalculus

ANALYTIC TRIGONOMETRY

Chapter

6

6.1 FUNDAMENTAL TRIGONOMETRIC IDENTITIES

OBJECTIVES

- Use fundamental trigonometric identities to evaluate a trigonometric function.
- Use fundamental trigonometric identities to simplify or rewrite a trigonometric expression.
- Use fundamental trigonometric identities to factor a trigonometric expression.
- Prove a trigonometric identity.

PREREQUISITE VOCABULARY TERMS

binomial
Distributive Property
factors (of a polynomial)
Pythagorean Theorem
reciprocal
trigonometric functions

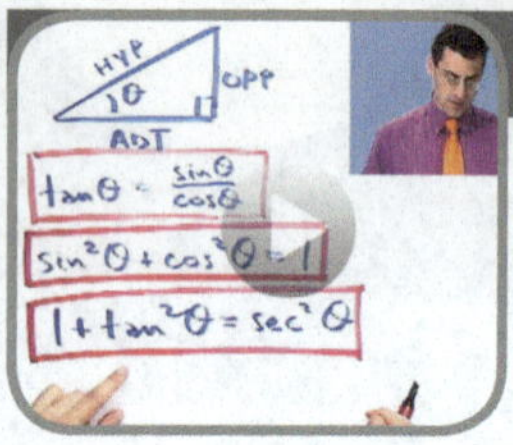

6.1.1 Fundamental Trigonometric Identities

An **identity** is an equation that is always true. For example, 1 + 2 = 3 is a simple identity. If an identity contains variables, then the equation must be true for all values of the variables (within the equation's domain). For example, $2x + 6 = 2(x + 3)$ is an identity because any real number substituted for x results in a true statement.

Several groups of **fundamental trigonometric identities** will be presented in this topic. These identities can be used to simplify or rewrite a **trigonometric expression** (i.e., an expression involving at least one trigonometric function).

The **even-odd identities** were presented in Chapter 5.

Even-Odd Identities		
$\sin(-\theta) = -\sin\theta$	$\cos(-\theta) = \cos\theta$	$\tan(-\theta) = -\tan\theta$
$\csc(-\theta) = -\csc\theta$	$\sec(-\theta) = \sec\theta$	$\cot(-\theta) = -\cot\theta$

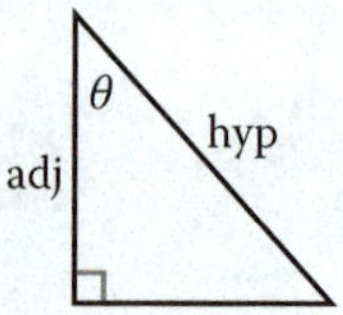

Figure 6.1a

The rest of the fundamental trigonometric identities introduced in this topic will follow from the definitions of the trigonometric functions given in Chapter 5. Recall that if θ is an acute angle in a right triangle (Figure 6.1a), then

$$\sin\theta = \frac{\text{opp}}{\text{hyp}},\ \cos\theta = \frac{\text{adj}}{\text{hyp}},\ \tan\theta = \frac{\text{opp}}{\text{adj}},\ \csc\theta = \frac{\text{hyp}}{\text{opp}},\ \sec\theta = \frac{\text{hyp}}{\text{adj}},\ \text{and}\ \cot\theta = \frac{\text{adj}}{\text{opp}}.$$

Recall that $\sin\theta$ and $\csc\theta$ are reciprocals. The same is true for $\cos\theta$ and $\sec\theta$, as well as for $\tan\theta$ and $\cot\theta$. The **reciprocal identities** follow from these reciprocal relationships.

Reciprocal Identities		
$\sin\theta = \dfrac{1}{\csc\theta}$	$\cos\theta = \dfrac{1}{\sec\theta}$	$\tan\theta = \dfrac{1}{\cot\theta}$
$\csc\theta = \dfrac{1}{\sin\theta}$	$\sec\theta = \dfrac{1}{\cos\theta}$	$\cot\theta = \dfrac{1}{\tan\theta}$

The next fundamental trigonometric identities are the **quotient identities**. Note that the quotient of sin θ and cos θ can be simplified, as shown here.

$$\frac{\sin\theta}{\cos\theta} = \frac{\frac{\text{opp}}{\text{hyp}}}{\frac{\text{adj}}{\text{hyp}}} = \frac{\text{opp}}{\text{hyp}} \cdot \frac{\text{hyp}}{\text{adj}} = \frac{\text{opp}}{\text{adj}} = \tan\theta$$

NOTICE THAT

The same result is found when the definitions of the sine and cosine of any angle θ are used (i.e., $\sin\theta = y/r$ and $\cos\theta = x/r$).

$$\frac{\sin\theta}{\cos\theta} = \frac{\frac{y}{r}}{\frac{x}{r}} = \frac{y}{r} \cdot \frac{r}{x} = \frac{y}{x} = \tan\theta$$

Therefore, $\tan\theta = \dfrac{\sin\theta}{\cos\theta}$. Furthermore, since tan θ and cot θ are reciprocals, $\cot\theta = \dfrac{\cos\theta}{\sin\theta}$.

Quotient Identities

$$\tan\theta = \frac{\sin\theta}{\cos\theta} \qquad \cot\theta = \frac{\cos\theta}{\sin\theta}$$

The **Pythagorean identities** are fundamental trigonometric identities following from the Pythagorean Theorem. To derive the first of the Pythagorean identities, consider the three sides of the right triangle in Figure 6.1a, which can be represented by opp, adj, and hyp. So, by the Pythagorean Theorem, $(\text{opp})^2 + (\text{adj})^2 = (\text{hyp})^2$. Dividing each side of the equation by $(\text{hyp})^2$ and simplifying gives the first Pythagorean identity.

$$(\text{opp})^2 + (\text{adj})^2 = (\text{hyp})^2$$

$$\frac{(\text{opp})^2}{(\text{hyp})^2} + \frac{(\text{adj})^2}{(\text{hyp})^2} = \frac{(\text{hyp})^2}{(\text{hyp})^2} \qquad \textit{Divide each side by } (\text{hyp})^2.$$

$$\left(\frac{\text{opp}}{\text{hyp}}\right)^2 + \left(\frac{\text{adj}}{\text{hyp}}\right)^2 = 1 \qquad \textit{Simplify and use the Power of a Quotient Property.}$$

$$(\sin\theta)^2 + (\cos\theta)^2 = 1 \qquad \textit{Substitute } \sin\theta \textit{ for opp/hyp and } \cos\theta \textit{ for adj/hyp.}$$

$$\sin^2\theta + \cos^2\theta = 1 \qquad \textit{Write the square of } \sin\theta \textit{ and the square of } \cos\theta.$$

IMPORTANT

$(\sin\theta)(\sin\theta) = (\sin\theta)^2 = \sin^2\theta$

$(\sin\theta)(\sin\theta) \neq \sin\theta^2$

Pythagorean Identities

$$\sin^2\theta + \cos^2\theta = 1 \qquad \tan^2\theta + 1 = \sec^2\theta \qquad 1 + \cot^2\theta = \csc^2\theta$$

IMPORTANT

Pythagorean identities can be written in many forms. For example, $\sin^2\theta + \cos^2\theta = 1$ can be written as $\cos^2\theta + \sin^2\theta = 1$, $\cos^2\theta = 1 - \sin^2\theta$, or $\sin^2\theta = 1 - \cos^2\theta$.

The Pythagorean identities $1 + \tan^2\theta = \sec^2\theta$ and $1 + \cot^2\theta = \csc^2\theta$ can be derived using a number of methods. For example, dividing $\sin^2\theta + \cos^2\theta = 1$ by $\cos^2\theta$ yields $\tan^2\theta + 1 = \sec^2\theta$.

$$\frac{\sin^2\theta}{\cos^2\theta} + \frac{\cos^2\theta}{\cos^2\theta} = \frac{1}{\cos^2\theta} \quad \Rightarrow \quad \tan^2\theta + 1 = \sec^2\theta$$

The **cofunction identities** follow from the periodic nature of trigonometric functions.

Cofunction Identities

$\sin\left(\frac{\pi}{2} - \theta\right) = \cos\theta$	$\cos\left(\frac{\pi}{2} - \theta\right) = \sin\theta$	$\tan\left(\frac{\pi}{2} - \theta\right) = \cot\theta$
$\csc\left(\frac{\pi}{2} - \theta\right) = \sec\theta$	$\sec\left(\frac{\pi}{2} - \theta\right) = \csc\theta$	$\cot\left(\frac{\pi}{2} - \theta\right) = \tan\theta$

6.1.2 Using Trigonometric Identities

In Section 5.3, the value of one trigonometric function of θ and the sign of another trigonometric function of θ (not the reciprocal of the first function) were used to evaluate all other trigonometric functions of θ. The first step in this procedure was to use the signs of the two trigonometric functions to determine the quadrant containing θ. Then a right triangle was drawn using the reference angle, and the Pythagorean Theorem was used to find this triangle's unknown side length. Finally, the lengths of the triangle's sides were used to evaluate the other trigonometric functions of θ.

Fundamental trigonometric identities can be used to evaluate all trigonometric functions of θ when the value of one trigonometric function of θ and the quadrant containing θ are known, as demonstrated in **Example 1**.

EXAMPLE 1 Using Trigonometric Identities to Evaluate the Trigonometric Functions of θ

IMPORTANT

If $\pi < \theta < 3\pi/2$, then θ is in quadrant III. In quadrant III, $\tan\theta > 0$ and $\cot\theta > 0$, but the rest of the trigonometric functions of θ are negative.

Evaluate $\sin\theta$, $\cos\theta$, $\tan\theta$, $\sec\theta$, and $\cot\theta$ if $\csc\theta = -2$ and $\pi < \theta < 3\pi/2$.

SOLUTION

Since $\csc\theta = -2$ and $\sin\theta = \dfrac{1}{\csc\theta}$ (reciprocal identity), $\sin\theta = \dfrac{1}{-2} = -\dfrac{1}{2}$.

Use the Pythagorean identity $1 + \cot^2\theta = \csc^2\theta$ to find $\cot\theta$.

$1 + \cot^2\theta = \csc^2\theta$ *Pythagorean Identity*

$1 + \cot^2\theta = (-2)^2$ *Substitute –2 for csc θ.*

$1 + \cot^2\theta = 4$ *Simplify.*

$\cot^2\theta = 3$ *Subtract 1 from each side.*

$\cot\theta = \pm\sqrt{3}$ *Take ± the square root of each side.*

Since $\pi < \theta < 3\pi/2$, θ is in quadrant III, where cotangent is positive. Therefore, $\cot\theta = \sqrt{3}$.

Since $\cot\theta = \sqrt{3}$ and $\tan\theta = \dfrac{1}{\cot\theta}$ (reciprocal identity), $\tan\theta = \dfrac{1}{\sqrt{3}}$.

Use the quotient identity $\cot\theta = \dfrac{\cos\theta}{\sin\theta}$ to find $\cos\theta$.

$\cot\theta = \dfrac{\cos\theta}{\sin\theta}$ *Quotient Identity*

$\sqrt{3} = \dfrac{\cos\theta}{-\frac{1}{2}}$ *Substitute $\sqrt{3}$ for cot θ and –1/2 for sin θ.*

$\cos\theta = -\dfrac{1}{2}\cdot\sqrt{3} = -\dfrac{\sqrt{3}}{2}$ *Multiply both sides by –1/2 and simplify.*

Since $\cos\theta = -\dfrac{\sqrt{3}}{2}$ and $\sec\theta = \dfrac{1}{\cos\theta}$ (reciprocal identity), $\sec\theta = \dfrac{1}{-\frac{\sqrt{3}}{2}} = -\dfrac{2}{\sqrt{3}}$.

ALTERNATIVE METHOD

There are many different ways to use the fundamental trigonometric identities to evaluate the trigonometric functions of θ. For example, the Pythagorean identity $\sin^2\theta + \cos^2\theta = 1$ can be used along with $\sin\theta = -1/2$ to find $\cos\theta$.

$\sin^2\theta + \cos^2\theta = 1$	*Pythagorean Identity*
$\left(-\frac{1}{2}\right)^2 + \cos^2\theta = 1$	*Substitute −1/2 for sin θ.*
$\frac{1}{4} + \cos^2\theta = 1$	*Simplify.*
$\cos^2\theta = \frac{3}{4}$	*Subtract 1/4 from each side.*
$\cos\theta = \pm\sqrt{\frac{3}{4}} = \pm\frac{\sqrt{3}}{2}$	*Take ± the square root of each side and simplify the radical.*

Again, since $\pi < \theta < 3\pi/2$, θ is in quadrant III. Cosine is negative in quadrant III, so $\cos\theta = -\frac{\sqrt{3}}{2}$.

6.1.3 Simplifying a Trigonometric Expression Using Trigonometric Identities

The fundamental trigonometric identities can be used to simplify trigonometric expressions. Note that you may need to manipulate the expression by applying a property of real numbers before substituting an expression from one of the fundamental trigonometric identities into the given trigonometric expression.

There will often be several possible procedures for simplifying a trigonometric expression, each resulting in an equivalent form of the expression when completed correctly.

EXAMPLE 2 Using Trigonometric Identities to Simplify an Expression

Simplify each expression.

A. $\sin^2 x + \tan^2 x + \cos^2 x$ **B.** $\cos^2\theta \cdot \tan^2\theta \cdot \csc\theta$

SOLUTION

> **REMEMBER**
>
> *Commutative Property*
>
> $a + c + b = a + b + c$
>
> $a \cdot c \cdot b = a \cdot b \cdot c$

A. Notice that the expression contains the terms $\sin^2 x$ and $\cos^2 x$. By the Commutative Property of Addition, $\sin^2 x + \tan^2 x + \cos^2 x$ can be written as $\sin^2 x + \cos^2 x + \tan^2 x$. The sin-cos Pythagorean identity states that $\sin^2\theta + \cos^2\theta = 1$. So, substitute 1 for $\sin^2 x + \cos^2 x$.

$\sin^2 x + \tan^2 x + \cos^2 x = \sin^2 x + \cos^2 x + \tan^2 x$	*Commutative Property of Addition*
$= 1 + \tan^2 x$	*Substitute 1 for $\sin^2 x + \cos^2 x$.*
$= \sec^2 x$	*Substitute $\sec^2 x$ for $1 + \tan^2 x$.*

> *Pythagorean Identity* ▶
>
> $1 + \tan^2\theta = \sec^2\theta$

So, $\sin^2 x + \tan^2 x + \cos^2 x = \sec^2 x$.

TIP

When simplifying, if an identity cannot be used immediately, it may be useful to first write all of the trigonometric functions in terms of sin θ and cos θ.

B. Since no Pythagorean identities involve both $\cos^2 x$ and $\tan^2 x$, use identities to write the expression in terms of sine and cosine.

$$\cos^2\theta\cdot\tan^2\theta\cdot\csc\theta = \cos^2\theta\cdot(\tan\theta)^2\cdot\csc\theta$$ *$\tan^2\theta = (\tan\theta)^2$*

$$= \cos^2\theta\cdot\left(\frac{\sin\theta}{\cos\theta}\right)^2\cdot\csc\theta$$ *Substitute $\sin\theta/\cos\theta$ for $\tan\theta$.*

$$= \cos^2\theta\cdot\frac{\sin^2\theta}{\cos^2\theta}\cdot\frac{1}{\sin\theta}$$ *Power of a Quotient Property*

$$= \sin\theta$$ *Simplify (remove the common factors from the numerator and denominator).*

So, $\cos^2\theta\cdot\tan^2\theta\cdot\csc\theta = \sin\theta$.

6.1.4 Simplifying a Trigonometric Expression Using Trigonometric Identities: Another Example

EXAMPLE 3 Using Trigonometric Identities to Simplify an Expression

Simplify. $\dfrac{1}{\tan x+1}+\dfrac{\cot x}{\tan x+1}$

SOLUTION

Quotient Identities ▶

$$\tan\theta = \frac{\sin\theta}{\cos\theta}$$

$$\cot\theta = \frac{\cos\theta}{\sin\theta}$$

$$\frac{1}{\tan x+1}+\frac{\cot x}{\tan x+1} = \frac{1+\cot x}{\tan x+1}$$ *Add the terms.*

$$= \frac{1+\dfrac{\cos x}{\sin x}}{\dfrac{\sin x}{\cos x}+1}$$ *Substitute $\sin x/\cos x$ for $\tan x$ and $\cos x/\sin x$ for $\cot x$.*

$$= \frac{\dfrac{\sin x}{\sin x}+\dfrac{\cos x}{\sin x}}{\dfrac{\sin x}{\cos x}+\dfrac{\cos x}{\cos x}}$$ *Write the terms with a common denominator.*

$$= \frac{\sin x+\cos x}{\sin x}\cdot\frac{\cos x}{\sin x+\cos x}$$ *Add the terms and multiply by the reciprocal.*

$$= \frac{(\sin x+\cos x)(\cos x)}{(\sin x+\cos x)(\sin x)}$$ *Multiply the numerators and the denominators.*

$$= \frac{\cos x}{\sin x}$$ *Simplify by removing common factor $(\sin x + \cos x)$.*

$$= \cot x$$ *Substitute $\cot x$ for $\cos x/\sin x$.*

So, $\dfrac{1}{\tan x+1}+\dfrac{\cot x}{\tan x+1} = \cot x$.

6.1.5 Simplifying Products of Binomials Involving Trigonometric Functions

Recall from Section P.4 that the FOIL method is a procedure for multiplying two binomials.

$$(a + b)(c + d) = ac + ad + bc + bd$$

Each letter in the word "FOIL" represents a multiplication step used in the procedure.

- F: multiply the *first* terms from each binomial
- O: multiply the *outer* terms from each binomial
- I: multiply the *inner* terms from each binomial
- L: multiply the *last* terms from each binomial

The FOIL method can also be used to multiply two binomials involving trigonometric functions. For example, the expansion of $(3 + \cos x)^2$ is shown here.

$$(3 + \cos x)^2 = (3 + \cos x)(3 + \cos x) \qquad \textit{Write the power as a product of two factors.}$$
$$= 9 + 3\cos x + 3\cos x + \cos^2 x \qquad \textit{FOIL}$$
$$= 9 + 6\cos x + \cos^2 x \qquad \textit{Combine the like terms.}$$

If a trigonometric expression contains a product of two binomials, FOIL first and then use trigonometric identities to simplify, if possible.

EXAMPLE 4 Using Trigonometric Identities and FOIL to Simplify an Expression

Expand and simplify $(\sec x - 1)(\sec x + 1)$, if possible.

SOLUTION

FOIL the two binomials.

$$(\sec x - 1)(\sec x + 1) = \sec^2 x + \sec x - \sec x - 1 \qquad \textit{FOIL}$$
$$= \sec^2 x - 1 \qquad \textit{Simplify.}$$
$$= \tan^2 x \qquad \textit{Pythagorean Identity}$$

So, $(\sec x - 1)(\sec x + 1) = \tan^2 x$.

TIP

Subtracting 1 from both sides of the Pythagorean Identity $1 + \tan^2 \theta = \sec^2 \theta$ *gives* $\tan^2 \theta = \sec^2 \theta - 1$.

6.1.6 Simplifying Products of Binomials Involving Trigonometric Functions: Another Example

EXAMPLE 5 Using Trigonometric Identities and FOIL to Simplify an Expression

Expand and simplify $(\tan^2 x + 1)(1 - \cos^2 x)$, if possible.

SOLUTION

FOIL the two binomials.

$$\begin{aligned}(\tan^2 x + 1)(1 - \cos^2 x) &= \tan^2 x + (\tan^2 x)(-\cos^2 x) + 1 - \cos^2 x && \textit{FOIL}\\ &= \tan^2 x - \tan^2 x \cos^2 x + 1 - \cos^2 x && \textit{Simplify.}\\ &= \tan^2 x - \tan^2 x \cos^2 x + (1 - \cos^2 x) && \textit{Associative Property}\\ &= \tan^2 x - \tan^2 x \cos^2 x + \sin^2 x && \textit{Pythagorean Identity}\\ &= \tan^2 x - \frac{\sin^2 x}{\cos^2 x} \cdot \cos^2 x + \sin^2 x && \textit{Quotient Identity}\\ &= \tan^2 x - \sin^2 x + \sin^2 x && \textit{Simplify.}\\ &= \tan^2 x && \textit{Subtract the like terms.}\end{aligned}$$

TIP

Subtracting $\cos^2 \theta$ from both sides of the Pythagorean identity $\sin^2 \theta + \cos^2 \theta = 1$ gives $\sin^2 \theta = 1 - \cos^2 \theta$.

So, $(\tan^2 x + 1)(1 - \cos^2 x) = \tan^2 x$.

ALTERNATIVE METHOD

Use an identity to simplify the expression in each binomial first.

$$\begin{aligned}(\tan^2 x + 1)(1 - \cos^2 x) &= (\sec^2 x)(\sin^2 x) && \textit{Pythagorean Identities}\\ &= \left(\frac{1}{\cos^2 x}\right)(\sin^2 x) && \textit{Reciprocal Identity}\\ &= \frac{\sin^2 x}{\cos^2 x} && \textit{Multiply.}\\ &= \tan^2 x && \textit{Quotient Identity}\end{aligned}$$

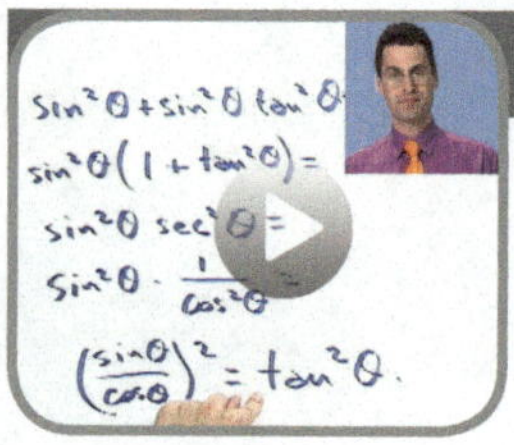

6.1.7 Factoring Trigonometric Expressions

Factoring is another form of algebraic manipulation that may be helpful when simplifying trigonometric expressions. Factoring is used in the following example to demonstrate a third procedure for simplifying $(\tan^2 x + 1)(1 - \cos^2 x)$ (from **Example 5**).

EXAMPLE 6 Using Trigonometric Identities and Factoring to Simplify an Expression

Simplify. $(\tan^2 x + 1)(1 - \cos^2 x)$

SOLUTION

Suppose only the second binomial is simplified using a Pythagorean identity.

$$(\tan^2 x + 1)(1 - \cos^2 x) = (\tan^2 x + 1)(\sin^2 x) \qquad \textit{Pythagorean Identity}$$

The Distributive Property can now be used to expand the expression.

$= \tan^2 x \cdot \sin^2 x + \sin^2 x$	*Distributive Property*
$= \dfrac{\sin^2 x}{\cos^2 x} \cdot \sin^2 x + \sin^2 x$	*Use a quotient identity to write the expression in terms of sine and cosine.*
$= \dfrac{\sin^4 x}{\cos^2 x} + \sin^2 x$	*Multiply.*
$= \dfrac{\sin^4 x}{\cos^2 x} + \dfrac{\sin^2 x \cos^2 x}{\cos^2 x}$	*Write the terms with a common denominator.*
$= \dfrac{\sin^4 x + \sin^2 x \cos^2 x}{\cos^2 x}$	*Add.*
$= \dfrac{\sin^2 x(\sin^2 x + \cos^2 x)}{\cos^2 x}$	*Factor $\sin^2 x$ from the numerator.*
$= \dfrac{\sin^2 x}{\cos^2 x}$	*Pythagorean Identity*
$= \tan^2 x$	*Quotient Identity*

Pythagorean Identity ▶
$\sin^2 \theta + \cos^2 \theta = 1$

In addition to factoring a common factor from an expression (as seen in **Example 6**), simplification of trigonometric expressions may involve factoring a trinomial into two binomials. Factoring a trigonometric trinomial into two binomials is demonstrated in **Example 7**.

EXAMPLE 7 Factoring a Trigonometric Trinomial

IMPORTANT

Some trigonometric trinomials cannot be factored into two binomials, just like with polynomials.

Factor. $\cot^2 x - 7 \csc x + 11$

SOLUTION

Substitute $\csc^2 x - 1$ for $\cot^2 x$ (the Pythagorean identity) to write the expression in terms of cosecant.

$\cot^2 x - 7 \csc x + 11 = (\csc^2 x - 1) - 7 \csc x + 11$	*Pythagorean Identity*
$= \csc^2 x - 7 \csc x + 10$	*Simplify.*
$= (\csc x - 5)(\csc x - 2)$	*Factor.*

ALTERNATIVE METHOD

If factoring $\csc^2 x - 7 \csc x + 10$ into $(\csc x - 5)(\csc x - 2)$ is difficult because of the trigonometric terms, substitute S (or any variable) for the trigonometric function first and factor as usual. Then substitute the trigonometric function back into the expression.

$\cot^2 x - 7 \csc x + 11 = (\csc^2 x - 1) - 7 \csc x + 11$	*Pythagorean Identity*
$= \csc^2 x - 7 \csc x + 10$	*Simplify.*
$= S^2 - 7S + 10$	*Let $S = \csc x$.*
$= (S - 5)(S - 2)$	*Factor.*
$= (\csc x - 5)(\csc x - 2)$	*Substitute $\csc x$ for S.*

Factoring a 4-term Expression

Recall that a polynomial expression containing four terms, such as $1 - a^2 + b^2 - a^2b^2$, can sometimes be factored into two binomials using the grouping method.

$$\begin{aligned} &1 - a^2 + b^2 - a^2b^2 \\ &= (1 - a^2) + b^2(1 - a^2) && \textit{Group the first two terms and factor } b^2 \textit{ from the last two terms.} \\ &= (1 - a^2)(1 + b^2) && \textit{Factor } 1 - a^2 \textit{ from each term.} \end{aligned}$$

In **Example 8**, this grouping method is used to factor a 4-term trigonometric expression, and then the expression is simplified using trigonometric identities.

EXAMPLE 8 **Using Trigonometric Identities and Factoring to Simplify an Expression**

Simplify. $1 - \cos^2 x + \cot^2 x - \cos^2 x \cdot \cot^2 x$

SOLUTION

To make the possible factors easier to find, let a be $\cos^2 x$ and let b be $\cot^2 x$.

$$\begin{aligned} 1 - \cos^2 x + \cot^2 x - \cos^2 x \cdot \cot^2 x &= 1 - a^2 + b^2 - a^2b^2 && \textit{Substitute.} \\ &= (1 - a^2)(1 + b^2) && \textit{Factor.} \\ &= (1 - \cos^2 x)(1 + \cot^2 x) && \textit{Substitute.} \\ &= (\sin^2 x)(\csc^2 x) && \textit{Pythagorean Identities} \\ &= (\sin^2 x)\left(\frac{1}{\sin^2 x}\right) && \textit{Quotient Identity} \\ &= \frac{\sin^2 x}{\sin^2 x} && \textit{Multiply.} \\ &= 1 && \textit{Simplify.} \end{aligned}$$

So, $1 - \cos^2 x + \cot^2 x - \cos^2 x \cdot \cot^2 x = 1$.

6.1.8 Proving an Identity

Recall that an identity is an equation that is true for all values of the variable within the equation's domain. So, an identity is an equality between two expressions (the two sides of the equation).

To prove that an equation is an identity, we can show that these two expressions are equivalent by simplifying each until both expressions are the same. If the two expressions cannot be written as the same expression, then the equation is not an identity.

Proving an identity may appear to be similar to solving an equation. However, solving an equation involves applying an operation to both sides of the equation (e.g., both sides of an equation might be divided by the coefficient of the variable), which is a strategy that cannot be used when proving an identity. To prove an identity, each side of the equation must be simplified separately (i.e., operations are not applied to both sides of the equation).

EXAMPLE 9 Proving a Trigonometric Identity

Determine whether $\dfrac{1+\csc^2 t+\cot^2 t}{(1+\sin t)(1-\sin t)} = \dfrac{2}{\sin^2 t \cdot \cos^2 t}$ is an identity or not.

SOLUTION

The expression on the right side appears to be simplified because it is written in terms of sine and cosine. So, simplify the left side. Keep in mind that the purpose of simplifying is to manipulate one side until it looks the same as the other side.

$$\frac{1+\csc^2 t+\cot^2 t}{(1+\sin t)(1-\sin t)} = \frac{\csc^2 t+(1+\cot^2 t)}{1-\sin^2 t}$$

Use the Associative and Commutative Properties to group the terms in the numerator and use FOIL in the denominator.

$$= \frac{\csc^2 t+\csc^2 t}{\cos^2 t}$$

Pythagorean Identities

$$= \frac{2\csc^2 t}{\cos^2 t}$$

Combine the like terms.

$$= \frac{2\left(\dfrac{1}{\sin^2 t}\right)}{\cos^2 t}$$

Reciprocal Identity

$$= 2\left(\frac{1}{\sin^2 t}\right)\cdot\left(\frac{1}{\cos^2 t}\right)$$

Multiply by the reciprocal to divide.

$$= \frac{2}{\sin^2 t \cdot \cos^2 t}$$

Multiply.

So, $\dfrac{1+\csc^2 t+\cot^2 t}{(1+\sin t)(1-\sin t)} = \dfrac{2}{\sin^2 t \cdot \cos^2 t}$ is an identity.

6.1.9 Proving an Identity: Another Example

If there are any x-values in an equation's domain that yield a false statement when substituted into the equation, then the equation is not an identity. In this case, any one of those x-values is a **counterexample**. One counterexample is sufficient evidence for concluding that an equation is not an identity.

For example, consider the equation $\sin x = \cos x$. We know that this equation is not an identity because there are many x-values for which $\sin x$ does not equal $\cos x$, such as $x = 0$.

$$\sin 0 \neq \cos 0$$
$$0 \neq 1$$

Therefore, $x = 0$ is a counterexample proving that $\sin x = \cos x$ is not an identity.

NOTICE THAT

The equation $\sin x = \cos x$ is true for some x-values, which are the solutions to the equation, but to be an identity the equation must be true for all values.

EXAMPLE 10 Proving a Trigonometric Identity

Determine whether $\dfrac{(\tan^2 t)^2}{\sec^4 t - 2\sec^2 t + 1} = \dfrac{\sin^2 t + \sin^2 t \cdot \cot^2 t}{\sin^2 t}$ is an identity or not.

SOLUTION

IMPORTANT

There are many ways to proceed, but do not apply any operation to both sides of this equation. For example, do not multiply both sides of the equation by $\sin^2 t$, as this is not allowed when proving an identity.

Simplify the left side of the equation.

$$\begin{aligned} \frac{(\tan^2 t)^2}{\sec^4 t - 2\sec^2 t + 1} &= \frac{\tan^4 t}{(\sec^2 t - 1)(\sec^2 t - 1)} && \textit{Power of a Power; Factor the denominator.} \\ &= \frac{\tan^4 t}{(\tan^2 t)(\tan^2 t)} && \textit{Pythagorean Identity} \\ &= \frac{\tan^4 t}{\tan^4 t} && \textit{Simplify.} \\ &= 1 && \textit{Pythagorean Identity} \end{aligned}$$

Now, simplify the right side of the equation.

$$\begin{aligned} \frac{\sin^2 t + \sin^2 t \cdot \cot^2 t}{\sin^2 t} &= \frac{\sin^2 t(1 + \cot^2 t)}{\sin^2 t} && \textit{Factor the numerator.} \\ &= 1 + \cot^2 t && \textit{Simplify.} \\ &= \csc^2 t && \textit{Pythagorean Identity} \end{aligned}$$

Since $\csc^2 t \neq 1$ for all values of t, $\dfrac{(\tan^2 t)^2}{\sec^4 t - 2\sec^2 t + 1} = \dfrac{\sin^2 t + \sin^2 t \cdot \cot^2 t}{\sin^2 t}$ is not an identity.

While the equation given in **Example 10** is not an identity, it is however an equation that can be solved for t (i.e., there are t-values for which $\csc^2 t = 1$). Solving a trigonometric equation for a variable is covered in the next section.

A counterexample can also be used to prove that the equation is not an identity. Consider the equation when $t = \pi/4$.

$$\frac{\tan^2 \frac{\pi}{4}^2}{\sec^4 \frac{\pi}{4} - 2\sec^2 \frac{\pi}{4} + 1} \neq \frac{\sin^2 \frac{\pi}{4} + \sin^2 \frac{\pi}{4}\cot^2 \frac{\pi}{4}}{\sin^2 \frac{\pi}{4}} \qquad \textit{Substitute } \frac{\pi}{4} \textit{ for } t.$$

$$\frac{1}{4 - 4 + 1} \neq \frac{\frac{1}{2} + \frac{1}{2}}{\frac{1}{2}} \qquad \textit{Evaluate.}$$

$$1 \neq 2 \qquad \textit{Simplify.}$$

Therefore, $t = \pi/4$ is a counterexample proving that the equation is not an identity.

SECTION 6.1 EXERCISES

Warm Up

True or False?

1. $\cot 45° = 1$
2. $\sec 240° + 2 = \sec 120° + 2$
3. $2 \sin 150° = -1$

Just the Facts

Fill in the blanks.

4. An equation that is always true is an _____.
5. Writing $\sin(-50°)$ as $-\sin 50°$ and $\cos(-50°)$ as $\cos 50°$ is an application of the _____ identities.
6. $\sec^2\theta - 1 = \tan^2\theta$ is an example of a(n) _____ trigonometric identity.
7. The fundamental identities that follow from the periodic nature of trigonometric functions are the _____ identities.
8. In order for an equation to be an identity, _____ sides must be _____ for all values of the _____ in the equation's domain.
9. Replacing $\tan\theta$ with its quotient identity of _____ would simplify $\tan\theta \cdot \cos\theta$ to _____.
10. To show that an equation is not an identity, it is sufficient to provide a(n) _____.

Essential Skills

In Exercises 1–18, use the value of each given trigonometric function and quadrant to evaluate all other trigonometric functions.

1. $\sin\theta = -5/13$ and $3\pi/2 < \theta < 2\pi$
2. $\cos\theta = 1/4$ and $3\pi/2 < \theta < 2\pi$
3. $\cos\theta = 7/9$ and $0 < \theta < \pi/2$
4. $\sin\theta = 4/5$ and $\pi/2 < \theta < \pi$
5. $\sin\theta = -2/3$ and $\pi < \theta < 3\pi/2$
6. $\cos\theta = -3/8$ and $\pi < \theta < 3\pi/2$
7. $\tan\theta = 7/4$ and $0 < \theta < \pi/2$
8. $\cot\theta = -2$ and $\pi/2 < \theta < \pi$
9. $\cot\theta = -3$ and $3\pi/2 < \theta < 2\pi$
10. $\tan\theta = -5/6$ and $3\pi/2 < \theta < 2\pi$
11. $\tan\theta = -8$ and $\pi/2 < \theta < \pi$
12. $\cot\theta = 7/3$ and $\pi < \theta < 3\pi/2$
13. $\sec\theta = -3/2$ and $\pi < \theta < 3\pi/2$
14. $\csc\theta = 6$ and $\pi/2 < \theta < \pi$
15. $\csc\theta = -25/24$ and $\pi < \theta < 3\pi/2$
16. $\sec\theta = \dfrac{\sqrt{73}}{3}$ and $3\pi/2 < \theta < 2\pi$
17. $\sec\theta = 11/4$ and $3\pi/2 < \theta < 2\pi$
18. $\csc\theta = -\dfrac{\sqrt{290}}{13}$ and $\pi < \theta < 3\pi/2$

In Exercises 19–36, simplify.

19. $\tan^3 x \cdot \csc^3 x$
20. $\cot x \cdot \sec x \cdot \csc x \cdot \sin^2 x$
21. $\cos x \cdot \sin^3 x \cot^3 x \cdot \sec^2 x$
22. $\tan^3 x \cdot \cos^5 x \cdot \csc x \cdot \sec x$
23. $\sin x \cdot \tan x \cdot \csc x \cdot \cot x \cdot \cos x$
24. $\tan^6 x \cdot \csc^2 x \cdot \sin^2 x \cdot \cot^3 x$
25. $\sec x \cdot \cos x - \cos^2 x$
26. $\cot^2 x + \sin^2 x + \cos^2 x$
27. $\sec^2 x - \sin^2 x - \tan^2 x$
28. $\csc^2 x - \cot^2 x - \cos^2 x$
29. $\dfrac{\sin x \cdot \sec x}{\tan x}$
30. $\dfrac{\sec\theta \cdot \tan\theta}{\cot\theta \cdot \csc\theta}$
31. $\dfrac{\csc\theta \cdot \cot\theta}{\tan\theta \cdot \sec\theta}$
32. $\dfrac{\cos^4 x \cdot \tan^2 x}{\sin^3 x}$
33. $\dfrac{\cos x - 1}{\sin x} - \dfrac{\sin x}{\cos x - 1}$
34. $\dfrac{\sec^2\theta - 1}{\cot^2\theta} - \dfrac{1 + \tan^2\theta}{\csc^2\theta - 1}$
35. $\dfrac{\cos x}{1 + \sin x} + \dfrac{1 + \sin x}{\cos x}$
36. $\dfrac{1}{1 - \sec x} + \dfrac{1}{1 + \sec x}$

In Exercises 37–42, expand and simplify, if possible.

37. $(\sec x + \tan x)(\sec x - \tan x)$
38. $(1 + \sin^2\theta)(1 + \cot^2\theta)$
39. $(\tan x + \sin x)(\tan x - \sin x)$
40. $(\cos^2 x + \sec^2 x)(\cos^2 x - 2)$
41. $(3\csc x + 3)(3\csc x - 3)$
42. $(1 + \cos^2\theta)(1 + \tan^2\theta)$

In Exercises 43–48, expand and simplify, if possible.

43. $(\tan x + \cot x)^2$

44. $(\cot x - 1)^2$

45. $(\sin x + \cos x)^2$

46. $(\sec^2 \theta - 1)(\csc^2 \theta - 1)$

47. $(1 - \sin^2 \theta)(\cot^2 \theta + 1)$

48. $(\sin x - \csc x)^2$

In Exercises 49–52, factor and simplify, if possible.

49. $\sin^2 \theta + \cot^2 \theta \cdot \sin^2 \theta$

50. $\cot x \cdot \csc^2 x - \cot x$

51. $\tan^2 \theta - \tan^2 \theta \cdot \sin^2 \theta$

52. $\csc^5 x \cdot \cot^2 x + \csc^5 x$

In Exercises 53–58, factor.

53. $\sec^2 x + 6 \tan x + 4$

54. $\cot^2 x - 4 \csc x + 5$

55. $-2 \sin^2 \theta + 5 \cos \theta - 10$

56. $36 + 48 \cos t \cdot \sin t - 20 \sin^2 t$

57. $9 \sin^2 t + 42 \cos t \cdot \sin t + 49 \cos^2 t$

58. $\tan^2 x - 8 \sec x + 17$

In Exercises 59–62, factor and simplify, if possible.

59. $\cos^2 t \cdot \sin t - 2 \cos^4 t \cdot \sin t + \cos^6 t \cdot \sin t$

60. $1 - \sin^2 t + \tan^2 t - \sin^2 t \cdot \tan^2 t$

61. $\cos^4 t - \cos^4 t \cdot \sin^2 t + \cos^4 t \cdot \tan^2 t - \cos^4 t \cdot \tan^2 t \cdot \sin^2 t$

62. $\sin t - \sin t \cdot \cos^2 t + \sin t \cdot \cot^2 t - \sin t \cdot \cot^2 t \cdot \cos^2 t$

In Exercises 63–70, determine whether each equation is an identity or not.

63. $(\cos x - \sec x)^2 = \tan^2 x - \sin^2 x$

64. $\dfrac{1 - \cos t}{\sin t} + \dfrac{\sin t}{1 - \cos t} = 2 \csc t$

65. $\dfrac{1 - \sin t}{\cos t} + \dfrac{\cos t}{1 - \sin t} = 2 \sec t$

66. $\dfrac{1 + \cos t}{\sin t} + \dfrac{\sin t}{1 + \cos t} = 2 \sec t$

67. $\dfrac{1 + 2 \cot^2 x + \cot^4 x}{1 - \cot^2 x} = \dfrac{\csc^4 x}{1 - \cot^2 x}$

68. $\dfrac{\csc^4 x - 2 \csc^2 x + 1}{\cot^2 x} = \cot^2 x$

69. $\csc^2 x - \cos^2 x \cdot \csc^2 x = \cot x$

70. $\dfrac{1 - 2 \sin^2 x + \sin^4 x}{\cos^2 x} = \sin^2 x$

Extensions

In Exercises 71–74, evaluate without using a calculator.

71. $(\cos^2 (33°) + 4 + \sin^2 (33°))^3$

72. $\cot^2 (159°) - \csc^2 (159°)$

73. $5(\sec^2 (-221°) - \tan^2 (-221°)) + 7$

74. $-\sin^2 (317°) - \cos^2 (317°)$

In Exercises 75–80, determine whether each equation is an identity or not.

75. $\dfrac{1 + \sin x + \cos x}{1 - \sin x + \cos x} = \dfrac{1 + \sin x}{\cos x}$

76. $\dfrac{1 + \sin x + \cos x}{1 - \sin x - \cos x} = \dfrac{1 + \cos x}{\sin x}$

77. $\dfrac{\sec^3 \theta - \cos^3 \theta}{\sec \theta - \cos \theta} = \sec^2 \theta + \cos^2 \theta + 1$

78. $\csc^6 x - \cot^6 x = 3 \csc^2 x \cdot \cot^2 x + 1$

79. $(2 \sin x + 3 \cos x)^2 + (3 \sin x - 2 \cos x)^2 = 13$

80. $\dfrac{\sec^2 \theta - 6 \tan \pi + 7}{\sec^2 \theta - 5} = \dfrac{\tan \theta - 4}{\tan \theta + 2}$

81. Simplify. $(\tan x + \cos^2 x + \sin^2 x)(\tan x - \cos^2 x - \sin^2 x)$

82. Factor and simplify, if possible. $2 - \dfrac{\cos^2 x}{1 - \sin x}$

6.2 SOLVING TRIGONOMETRIC EQUATIONS

OBJECTIVES

- Use algebraic techniques to solve trigonometric equations.
- Use inverse trigonometric functions to solve trigonometric equations.
- Solve trigonometric equations of quadratic type.
- Solve trigonometric equations involving a multiple of an angle.
- Use fundamental identities to solve trigonometric equations.

PREREQUISITE VOCABULARY TERMS

fundamental trigonometric identities
inverse trigonometric functions
quadratic equation
Quadratic Formula
Zero-Product Property

6.2.1 Solving Equations with a Single Trigonometric Function

The objective of solving a trigonometric equation is to find the values of the variable that make the equation true. These values are called the equation's solutions. Several procedures for solving trigonometric equations will be demonstrated in this section. Because trigonometric functions are periodic in nature, a trigonometric equation can have infinitely many solutions. These solutions can be expressed using general form equations, where each equation is the sum of a solution and all multiples of the function's period, as demonstrated in **Example 1**.

In some cases, all solutions within a given interval, such as $[0, 2\pi)$, are found. In these cases, the equation will either have no solution or a finite number of solutions.

Like the solutions of a polynomial equation, the solutions of a trigonometric equation can be determined (or checked) using a graph. For example, the equation $x + 1 = 4$ has solution $x = 3$, so graphically, the solution of $x + 1 = 4$ is the x-coordinate of the point of intersection of the graph of $y = x + 1$ and the graph of $y = 4$ (Figure 6.2a). The same relationship exists for the solutions of a trigonometric equation, as shown in **Example 1**.

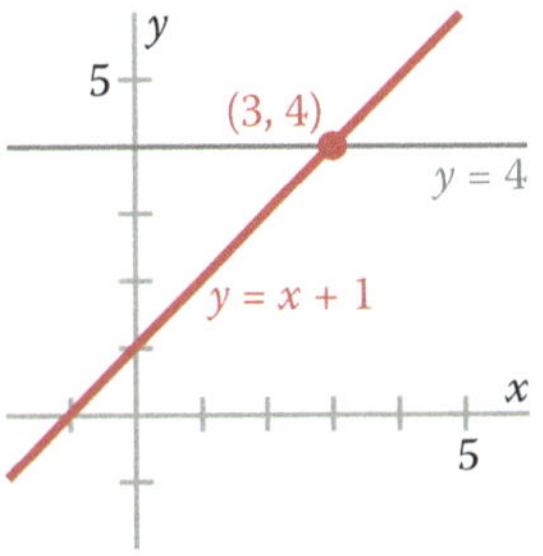

Figure 6.2a

EXAMPLE 1 Solving a Basic Trigonometric Equation

IMPORTANT

No interval is given, so find all solutions to each equation.

Solve each equation.

A. $\sin x = 1$ **B.** $\tan x = 0$ **C.** $\sec x = -\sqrt{2}$

SOLUTION

Identify the solutions (if any) within one period. Then, if any solutions exist, add multiples of the function's period to each solution to write each solution in general form.

A. The period of $y = \sin x$ is 2π, so first identify all solutions within the interval $[0, 2\pi)$.

We know that $\sin \pi/2 = 1$ because $\pi/2$ is an important angle. From the graph, we see that $\pi/2$ is the only solution within the interval $[0, 2\pi)$. Since the period is 2π, add $2\pi n$ (where n is any integer) to $\pi/2$ to write the general form equation of all solutions.

$$x = \frac{\pi}{2} + 2\pi n$$

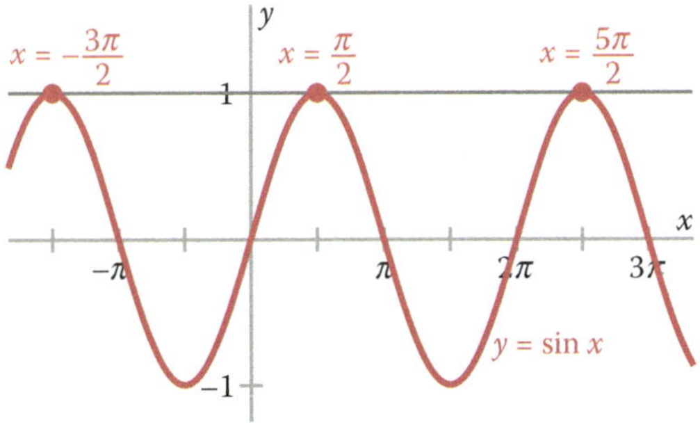

IMPORTANT

Solutions should be written as exact values, if possible.

B. The period of $y = \tan x$ is π, so identify all solutions on the interval $[0, \pi)$.

We know that $\tan 0 = 0$ because 0 is an important angle. Notice that this is the only solution within the interval $[0, \pi)$.

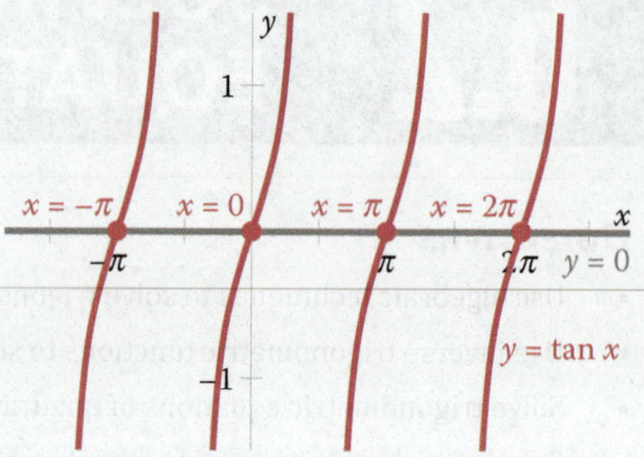

To write the general form equation of all solutions, add multiples of π (the period of $\tan x$) to 0.

$$x = 0 + \pi n = \pi n$$

Notice that the graph of $y = \tan x$ intersects the line $y = 0$ at $x = -\pi,\ 0,\ \pi$, and 2π.

IMPORTANT

Unless otherwise specified, n will always represent any integer in a general form equation.

C. The period of $y = \sec x$ is 2π, so identify all solutions on the interval $[0, 2\pi)$.

If $\sec x = -\sqrt{2}$, then $\cos x = -\frac{1}{\sqrt{2}} = -\frac{\sqrt{2}}{2}$ because secant and cosine are reciprocal functions.

Since $\cos x < 0$, all solutions must be in quadrants II and III (i.e., in quadrants where cosine is negative).

Find all t-values in quadrants II and III on the unit circle such that $\cos x = -\frac{\sqrt{2}}{2}$.

$$\cos\frac{3\pi}{4} = -\frac{\sqrt{2}}{2} \text{ and } \cos\frac{5\pi}{4} = -\frac{\sqrt{2}}{2}$$

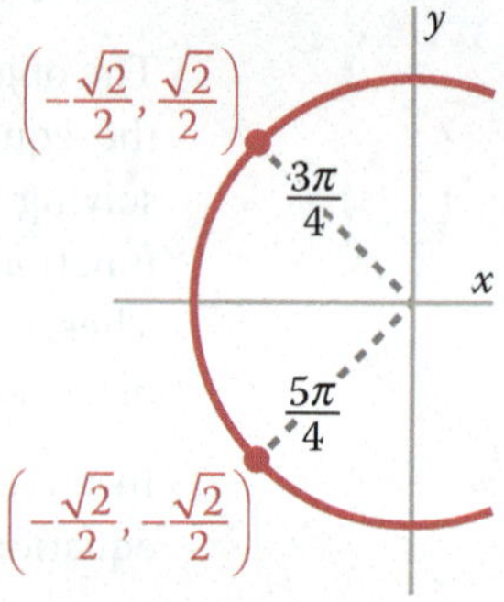

REMEMBER

$\cos t = x$, where (x, y) is the point on the unit circle that corresponds to t.

To write the general form equations of all solutions, add multiples of 2π to each solution.

$$x = \frac{3\pi}{4} + 2\pi n \text{ and } x = \frac{5\pi}{4} + 2\pi n$$

6.2.2 Solving Equations with a Single Trigonometric Function: Another Example

If a trigonometric equation is not solved for the trigonometric function, use the standard algebraic methods to isolate the trigonometric function on one side of the equation, and then find the solutions in one period.

Steps for Finding All Solutions for a Trigonometric Equation

❶ Isolate the trigonometric function on one side of the equation.

❷ Identify all solutions within one period.

❸ Add multiples of the period to each solution to write the general form of all solutions.

IMPORTANT

Use algebraic methods (such as dividing both sides by a number, or taking ± the square root of both sides) to isolate the trigonometric function on one side of the equation.

EXAMPLE 2 Solving a Trigonometric Equation

Solve each equation.

A. $3 = -2 \sin x + 1$ **B.** $3 \cot^2 x - 1 = 0$

SOLUTION

Use algebraic methods to isolate the trigonometric function in each equation.

TIP

Substitute a variable for the trigonometric function, then solve. Let S represent sin x.

$3 = -2S + 1$

$2 = -2S$

$-1 = S$

So, $-1 = \sin x$.

A. Subtract 1 from each side and then divide each side by –2.

$$3 = -2 \sin x + 1 \;\Rightarrow\; 2 = -2 \sin x \;\Rightarrow\; -1 = \sin x$$

The period of $y = \sin x$ is 2π, so identify all solutions of $\sin x = -1$ on the interval $[0, 2\pi)$.

We know that $\sin 3\pi/2 = -1$ because $3\pi/2$ is an important angle. Notice that this is the only solution within the interval $[0, 2\pi)$.

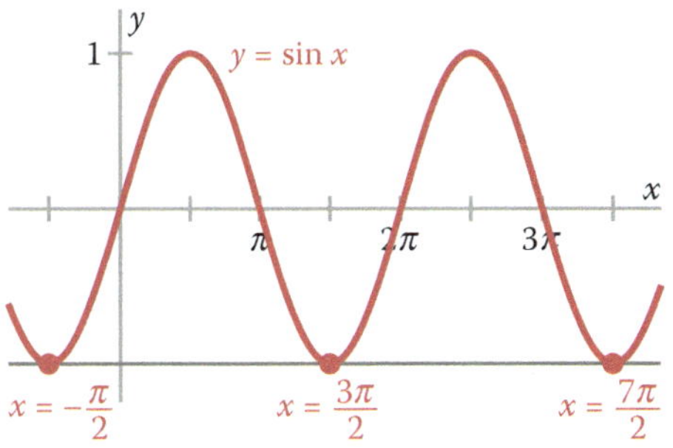

To write the general form equation of all solutions, add multiples of 2π (the period of sin x) to $3\pi/2$.

$$x = \frac{3\pi}{2} + 2\pi n$$

The graph of $y = \sin x$ confirms the solution.

TIP

Let S represent cot x.

$3S^2 - 1 = 0$

$3S^2 = 1$

$S^2 = 1/3$

$S = \pm\sqrt{\frac{1}{3}}$

So, $\cot x = \pm\sqrt{\frac{1}{3}}$.

B. Add 1 to each side and then divide each side by 3.

$$3 \cot^2 x - 1 = 0 \;\Rightarrow\; 3 \cot^2 x = 1 \;\Rightarrow\; \cot^2 x = 1/3$$

Take the ± square root of each side. $\cot^2 x = 1/3 \;\Rightarrow\; \cot x = \pm\sqrt{\frac{1}{3}} = \pm\frac{1}{\sqrt{3}}$

The period of $y = \cot x$ is π, so identify all solutions of $\cot x = \pm\frac{1}{\sqrt{3}}$ on the interval $[0, \pi)$.

The equation $\cot x = \pm\frac{1}{\sqrt{3}}$ is equivalent to $\tan x = \pm\sqrt{3}$ because cotangent and tangent are reciprocal functions.

Since the value of tan x is both positive and negative, solutions are in all four quadrants.

Find all t-values on the unit circle within $[0, \pi)$ such that

$\tan t = \sqrt{3}$ or $\tan t = -\sqrt{3}$.

$$\tan\frac{\pi}{3} = \sqrt{3} \text{ and } \tan\frac{2\pi}{3} = -\sqrt{3}$$

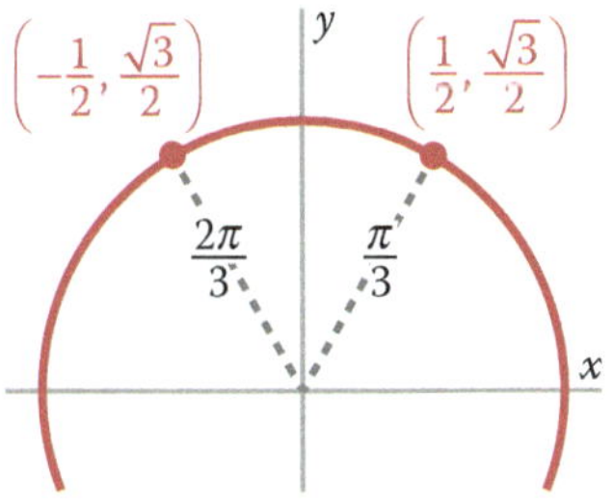

Since the period of cot x is π, identify the solutions within the interval [0, π) (i.e., in quadrants I and II).

To write the general form equations of all solutions, add multiples of π (the period of cot x) to each solution.

$$x = \frac{\pi}{3} + \pi n \text{ and } x = \frac{2\pi}{3} + \pi n$$

REMEMBER

tan t = sin t/cos t = y/x, where (x, y) is the point on the unit circle that corresponds to t.

$$\tan\frac{\pi}{3} = \frac{y}{x} = \frac{\frac{\sqrt{3}}{2}}{\frac{1}{2}} = \sqrt{3}$$

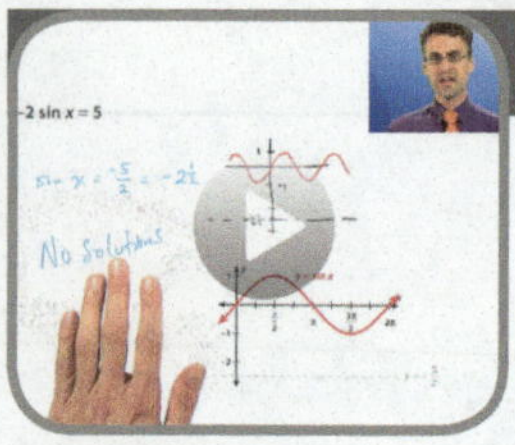

6.2.3 Solving Trigonometric Equations Using Inverse Functions

Recall from Section 5.7 that an inverse trigonometric function can be used to find the value of an angle in a trigonometric function. For example, the inverse sine function, $y = \sin^{-1} x$, means that $\sin y = x$. It follows that a trigonometric equation can be solved by using an inverse trigonometric function. For example, if $\sin x = 1$, then $x = \sin^{-1} 1$.

Inverse trigonometric functions can be evaluated using a calculator. However, if the angle is a multiple or combination of $\pi/6$ or $\pi/4$, then the solution should be expressed as an exact value instead of a decimal approximation.

IMPORTANT

The solutions obtained from using an inverse trigonometric function will be within the domain of the inverse function. Check the original function's graph to identify any additional solutions.

EXAMPLE 3 Solving a Trigonometric Equation Using an Inverse Function

Use an inverse trigonometric function to solve each equation. Round to the nearest tenth as needed.

A. $5 = 4 - \cos x$ **B.** $7 \csc x - 5 = 3 \csc x$ **C.** $6(\sec x + 1) = 2$

SOLUTION

A. Subtract 4 from each side and then divide each side by −1.

$$5 = 4 - \cos x \;\Rightarrow\; 1 = -\cos x \;\Rightarrow\; -1 = \cos x$$

If $\cos x = -1$, then $x = \cos^{-1}(-1)$. We know that $\cos \pi = -1$ because π is an important angle. Add multiples of 2π (the period of $\cos x$) to π to write the general form equation.

$$x = \pi + 2\pi n$$

IMPORTANT

The graph of $y = \cos x$ and $y = -1$ confirms that $x = \pi$ is the only solution of $-1 = \cos x$ within the interval $[0, 2\pi)$.

B. There is a cosecant term on each side of the equation, so subtract $3 \csc x$ from each side.

$$7 \csc x - 5 = 3 \csc x \;\Rightarrow\; 4 \csc x - 5 = 0 \;\Rightarrow\; 4 \csc x = 5 \;\Rightarrow\; \csc x = 5/4$$

If $\csc x = 5/4$, then $\sin x = 4/5$, because cosecant and sine are reciprocal functions. It follows that $x = \sin^{-1}(4/5)$.

$$x = \sin^{-1}(4/5) = 0.927\ldots \approx 0.9$$ *Use a calculator to evaluate the inverse function.*

NOTICE THAT

The graph of $y = 4/5$ intersects the graph of $y = \sin x$ at two points between 0 and 2π. Thus, the equation $4/5 = \sin x$ has two solutions in the interval $[0, 2\pi)$.

Check the graph of $y = \sin x$ to identify any additional solutions within $[0, 2\pi)$. Because $\sin x = 4/5 > 0$, and because sine is positive only in quadrants I and II, the interval can be limited to $[0, \pi]$.

The graph of $y = \sin x$ intersects the graph of $y = 4/5$ when $x \approx 0.9$ and at one additional point within the interval $[0, 2\pi)$.

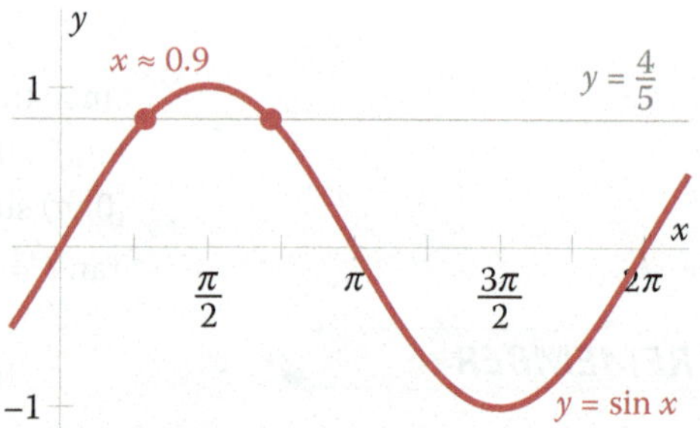

Let the x-coordinate of the second solution be x_2. Because of the symmetry of the graph of $y = \sin x$, the distance between 0 and $0.927\ldots$ (approximately 0.9) is equal to the distance between x_2 and π.

Therefore, $x_2 = \pi - 0.927\ldots \approx 2.2$.

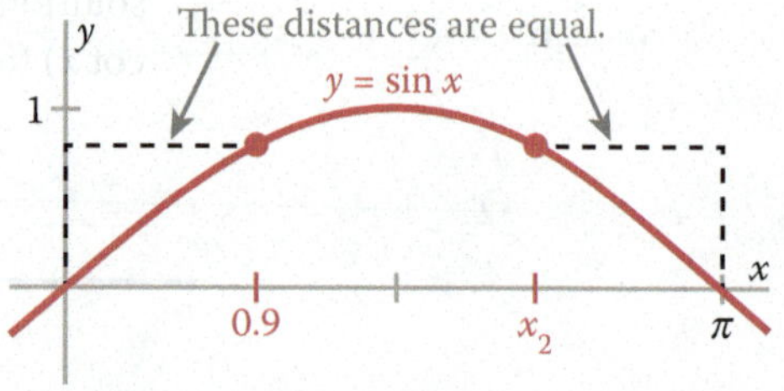

IMPORTANT

The graph of $y = \sin x$ is symmetric about $x = \pi/2$ within the interval $[0, \pi]$, so if (x_1, y) and (x_2, y) are points on $y = \sin x$ such that $x_2 > x_1$, then $x_2 = \pi - x_1$.

So, within $[0, 2\pi)$, $x \approx 0.9$ and $x \approx 2.2$. Add multiples of 2π (the period of $\csc x$) to each solution to write the general form equations of all solutions.

$$x \approx 0.9 + 2\pi n \text{ and } x \approx 2.2 + 2\pi n$$

C. Divide each side by 6 and then subtract 1 from each side.

$$6(\sec x + 1) = 2 \;\Rightarrow\; \sec x + 1 = 1/3 \;\Rightarrow\; \sec x = -2/3$$

IMPORTANT

If the y-value is not in the range of the trigonometric function, then the equation has no real solution.

If $\sec x = -2/3$, then $\cos x = -3/2$. Recall that the range of cosine is $-1 \le y \le 1$. Since $y = -3/2$ is not in this interval, the equation has no real solution.

CHECK

Confirm that the equation has no solution by graphing $y = \cos x$.

The graph of $y = \cos x$ does not intersect the graph of $y = -3/2$, so $6(\sec x + 1) = 2$ has no real solution.

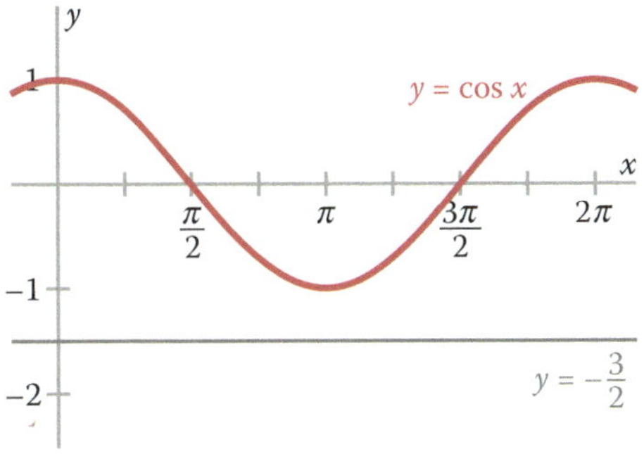

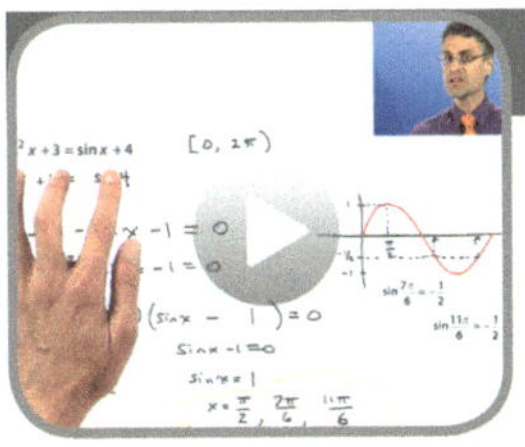

6.2.4 Solving Trigonometric Equations by Factoring

Recall from Section P.7 that the Zero-Product Property states that if $A \cdot B = 0$, then either $A = 0$ or $B = 0$. This property is applied to solve an equation in which a product of factors is equal to 0. For example, if $(x + 1)(x + 2) = 0$, then the equation's solutions are given by $x + 1 = 0$ or $x + 2 = 0$. The Zero-Product Property can be applied to trigonometric equations as well.

EXAMPLE 4 Solving a Trigonometric Equation by Factoring the GCF

IMPORTANT

When a solution interval is given, identify only specific solutions within that interval. If no solution interval is given, then all solutions should be expressed in the general form.

Find all solutions to $2\cos^2 x \cdot \sin x = -\cos^2 x$ within the interval $[0, 2\pi)$.

SOLUTION

Add $\cos^2 x$ to each side to get 0 on one side of the equation, then factor.

$$2\cos^2 x \cdot \sin x = -\cos^2 x$$

$$2\cos^2 x \cdot \sin x + \cos^2 x = 0 \qquad \textit{Add } \cos^2 x \textit{ to each side.}$$

$$\cos^2 x\,(2\sin x + 1) = 0 \qquad \textit{Factor } \cos^2 x \textit{ from each term.}$$

By the Zero-Product Property, $\cos^2 x = 0$ or $2\sin x + 1 = 0$. Solve each equation.

$$\cos^2 x = 0$$

$$\cos x = 0 \qquad \textit{Take } \pm \textit{ the square root of each side.}$$

$$2\sin x + 1 = 0$$

$$2\sin x = -1 \qquad \textit{Subtract 1 from each side.}$$

$$\sin x = -1/2 \qquad \textit{Divide each side by 2.}$$

Find all solutions to $\cos x = 0$ and $\sin x = -1/2$ within the interval $[0, 2\pi)$. We know that $\cos x = 0$ exactly when $x = \pi/2$ and $x = 3\pi/2$ because $\pi/2$ and $3\pi/2$ are important angles.

Use the unit circle to solve $\sin x = -1/2$.

REMEMBER

$\sin t = y$, where (x, y) is the point on the unit circle that corresponds to t.

Since $\sin x < 0$, all solutions must be in quadrants III and IV (i.e., in quadrants where sine is negative).

Find all t-values on the unit circle in quadrants III and IV such that $\sin t = -1/2$.

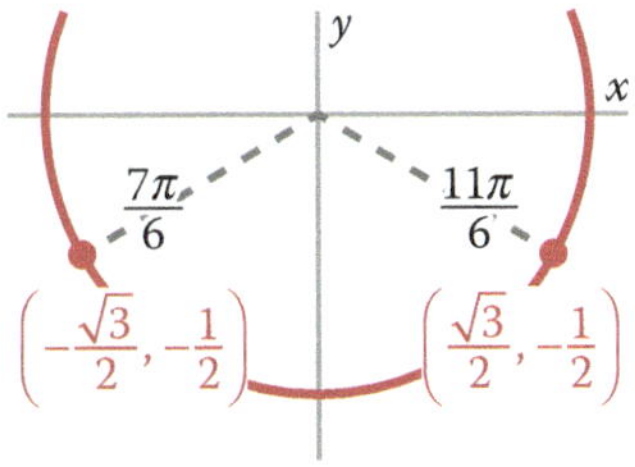

$$\sin\frac{7\pi}{6} = -\frac{1}{2} \text{ and } \sin\frac{11\pi}{6} = -\frac{1}{2}$$

Therefore, the solutions of $2\cos^2 x \cdot \sin x = -\cos^2 x$ in $[0, 2\pi)$ are $x = \pi/2, 7\pi/6, 3\pi/2$, and $11\pi/6$.

CHECK

To confirm the solutions, graph $y = \cos x$ and $y = 0$, and graph $y = \sin x$ and $y = -1/2$, both on the interval on $[0, 2\pi)$.

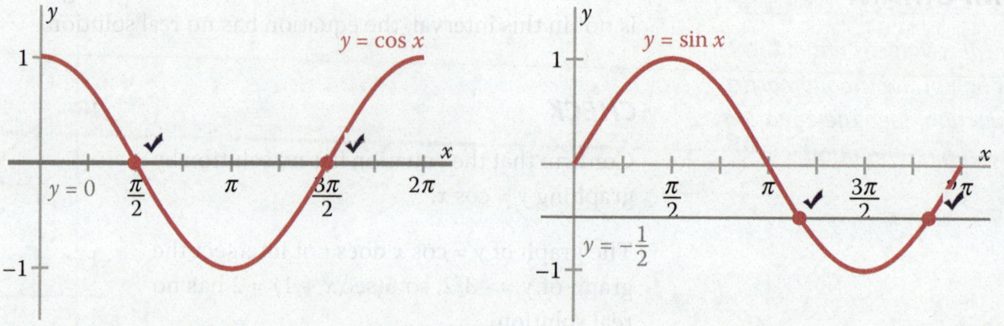

EXAMPLE 5 Solving a Trigonometric Equation by Factoring

Solve. $2 - 2\sin^2 x = 1 + \cos x$

SOLUTION

Write the equation in general form.

$$2 - 2\sin^2 x = 1 + \cos x$$

$$-2\sin^2 x - \cos x + 1 = 0 \qquad \textit{Subtract 1 and } \cos x \textit{ from each side.}$$

The expression $-2\sin^2 x - \cos x + 1$ cannot be factored. So, use the Pythagorean identity $\sin^2 x + \cos^2 x = 1$ to write the equation in terms of only cosine, and then factor (if possible).

$$-2(1 - \cos^2 x) - \cos x + 1 = 0 \qquad \textit{Substitute } 1 - \cos^2 x \textit{ for } \sin^2 x.$$

$$-2 + 2\cos^2 x - \cos x + 1 = 0 \qquad \textit{Distribute } -2.$$

$$2\cos^2 x - \cos x - 1 = 0 \qquad \textit{Simplify.}$$

$$(2\cos x + 1)(\cos x - 1) = 0 \qquad \textit{Factor.}$$

$$2\cos x + 1 = 0 \quad \text{or} \quad \cos x - 1 = 0 \qquad \textit{Set each factor equal to 0.}$$

$$\cos x = -1/2 \qquad \cos x = 1 \qquad \textit{Solve each equation for } \cos x.$$

TIP

Let S be cos x.

$2S^2 - S - 1 = 0$

$(2S + 1)(S - 1) = 0$

Substitute cos x for S.

$(2\cos x + 1)(\cos x - 1) = 0$

Now solve each equation. We know that $\cos x = 1$ when $x = 0$ because 0 is an important angle.

Use the unit circle to solve $\cos x = -1/2$.

Since $\cos x < 0$, all solutions must be in quadrants II and III (i.e., in quadrants where cosine is negative).

Find all t-values on the unit circle in quadrants II and III such that $\cos t = -1/2$.

$$\cos\frac{2\pi}{3} = -\frac{1}{2} \text{ and } \cos\frac{4\pi}{3} = -\frac{1}{2}$$

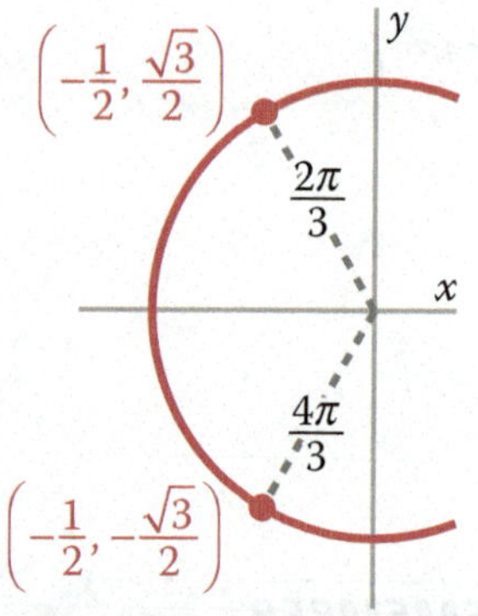

REMEMBER

$\cos t = x$, where (x, y) is the point on the unit circle that corresponds to t.

To write the general form equations of all solutions, add multiples of 2π (the period of $\cos x$) to each solution.

$$x = 2\pi n,\ x = \frac{2\pi}{3} + 2\pi n, \text{ and } x = \frac{4\pi}{3} + 2\pi n$$

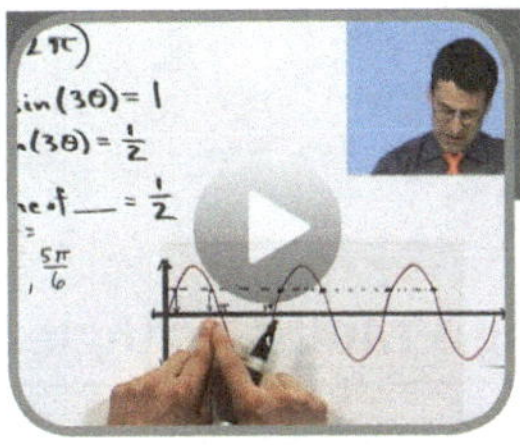

6.2.5 Solving Trigonometric Equations Involving a Multiple of an Angle

If the angle of a trigonometric function has a coefficient (e.g., sin $2x$), then the trigonometric function involves a multiple of the angle.

The process for solving a trigonometric equation involving a multiple of an angle is demonstrated in **Example 6**.

EXAMPLE 6 Solving a Trigonometric Equation Involving a Multiple Angle

Find all solutions to $2\sin 2x + \sqrt{3} = 0$ within the interval $[0, 2\pi)$.

SOLUTION

Isolate the trigonometric function.

$$\sin 2x = -\frac{\sqrt{3}}{2}$$ *Subtract $\sqrt{3}$ from each side and divide each side by 2.*

TIP

The solutions are all x-values within $[0, 2\pi)$ such that the sine of twice that value is equal to $-\sqrt{3}/2$.

The coefficient of x is 2 and the given interval for the x-values is $0 \le x < 2\pi$, so the interval for the $2x$-values is $2(0) \le 2(x) < 2(2\pi)$, that is, $0 \le 2x < 4\pi$.

Find all $2x$-values within $[0, 4\pi)$ for which $\sin 2x = -\frac{\sqrt{3}}{2}$.

From the unit circle we know that $\sin\frac{4\pi}{3} = -\frac{\sqrt{3}}{2}$ and $\sin\frac{5\pi}{3} = -\frac{\sqrt{3}}{2}$.

Add 2π (the period of $\sin x$) to each $2x$-value.

NOTICE THAT

Adding $2(2\pi)$ to either $4\pi/3$ or $5\pi/3$ results in a number greater than 4π, and thus not in the interval $[0, 4\pi)$.

$$\frac{4\pi}{3} + 2\pi = \frac{4\pi}{3} + \frac{6\pi}{3} = \frac{10\pi}{3} < 4\pi$$

$$\frac{5\pi}{3} + 2\pi = \frac{5\pi}{3} + \frac{6\pi}{3} = \frac{11\pi}{3} < 4\pi$$

Therefore, $2x = \frac{4\pi}{3}, \frac{5\pi}{3}, \frac{10\pi}{3}, \frac{11\pi}{3}$.

Divide each $2x$-value by 2 to find the x-values.

$$x = \frac{2\pi}{3}, \frac{5\pi}{6}, \frac{5\pi}{3}, \frac{11\pi}{6}$$

So, $x = 2\pi/3, 5\pi/6, 5\pi/3$, and $11\pi/6$ are all of the solutions of $2\sin 2x + \sqrt{3} = 0$ in $[0, 2\pi)$.

6.2.6 Solving Trigonometric Equations Involving a Multiple of an Angle: Another Example

EXAMPLE 7 Solving a Trigonometric Equation Involving a Multiple Angle

Find all solutions to $\left(\sin\frac{1}{2}x\right)-\left(\tan\frac{1}{2}x\right)\left(\sin\frac{1}{2}x\right)=0$ within the interval $[0, 10\pi)$.

SOLUTION

Factor $\sin\frac{1}{2}x$ from each term. $\quad \left(\sin\frac{1}{2}x\right)\left(1-\tan\frac{1}{2}x\right)=0$

Set each factor equal to 0 and solve each equation for the trigonometric function.

$$\sin\frac{1}{2}x=0 \quad \text{or} \quad 1-\tan\frac{1}{2}x=0 \quad \Rightarrow \quad \tan\frac{1}{2}x=1$$

Subtract 1 from each side.
Divide each side by −1.

The coefficient of x is 1/2 and the given interval for the x-values is $0 \le x < 10\pi$, so the interval for the $\frac{1}{2}x$-values is $\frac{1}{2}(0) \le \frac{1}{2}(x) < \frac{1}{2}(10\pi)$, that is, $0 \le \frac{1}{2}x < 5\pi$.

Find all $\frac{1}{2}x$-values within $[0, 5\pi)$ for which $\sin\frac{1}{2}x=0$ or $\tan\frac{1}{2}x=1$.

We know that $\sin 0 = 0$, $\sin\pi = 0$, and $\tan(\pi/4) = 1$ because 0, π, and $\pi/4$ are important angles. Furthermore, $x = 0$ and $x = \pi$ are the only solutions of $\sin\frac{1}{2}x=0$ within one period of sine, $[0, 2\pi)$. And, $x = \pi/4$ is the only solution of $\tan\frac{1}{2}x=1$ within the period of tangent, $[0, \pi)$.

IMPORTANT

To find all solutions in $[0, 5\pi)$, keep adding the period of $\sin x$ to 0 and π, and the period of $\tan x$ to $\pi/4$, until each sum is a value greater than or equal to 5π.

To find all $\frac{1}{2}x$-values within $[0, 5\pi)$ that are solutions for each equation, add positive multiples of each function's period to each solution. So, add 2π (the period of $\sin x$) to 0 and π to find all $\frac{1}{2}x$-values within $[0, 5\pi)$ for which $\sin\frac{1}{2}x=0$.

$0 + 2\pi = 2\pi < 5\pi \qquad 0 + 2(2\pi) = 4\pi < 5\pi \qquad 0 + 3(2\pi) = 6\pi > 5\pi$

$\pi + 2\pi = 3\pi < 5\pi \qquad \pi + 2(2\pi) = 5\pi = 5\pi \qquad \pi + 3(2\pi) = 7\pi > 5\pi$

Add positive multiples of π (the period of $\tan x$) to $\pi/4$ to find all $\frac{1}{2}x$-values within $[0, 5\pi)$ for which $\tan\frac{1}{2}x=1$.

$\frac{\pi}{4}+\pi=\frac{\pi}{4}+\frac{4\pi}{4}=\frac{5\pi}{4}<5\pi \qquad \frac{\pi}{4}+2\pi=\frac{\pi}{4}+\frac{8\pi}{4}=\frac{9\pi}{4}<5\pi$

$\frac{\pi}{4}+3\pi=\frac{\pi}{4}+\frac{12\pi}{4}=\frac{13\pi}{4}<5\pi \qquad \frac{\pi}{4}+4\pi=\frac{\pi}{4}+\frac{16\pi}{4}=\frac{17\pi}{4}<5\pi$

IMPORTANT

Because the sum of 5π and $\pi/4$ is greater than 5π, it is clear that $17\pi/4$ is the greatest solution.

Therefore, $\frac{1}{2}x = 0, \frac{\pi}{4}, \pi, \frac{5\pi}{4}, 2\pi, \frac{9\pi}{4}, 3\pi, \frac{13\pi}{4}, 4\pi, \frac{17\pi}{4}$.

Multiply each $\frac{1}{2}x$-value by 2. $\quad x = 0, \frac{\pi}{2}, 2\pi, \frac{5\pi}{2}, 4\pi, \frac{9\pi}{2}, 6\pi, \frac{13\pi}{2}, 8\pi, \frac{17\pi}{2}$

Therefore, $x = 0, \frac{\pi}{2}, 2\pi, \frac{5\pi}{2}, 4\pi, \frac{9\pi}{2}, 6\pi, \frac{13\pi}{2}, 8\pi$, and $\frac{17\pi}{2}$ are all of the solutions of $\left(\sin\frac{1}{2}x\right)-\left(\tan\frac{1}{2}x\right)\left(\sin\frac{1}{2}x\right)=0$ within $[0, 10\pi)$.

6.2.7 Solving Trigonometric Equations of Quadratic Type

Recall from Section P.8 that an equation of quadratic type can be solved using a special substitution which results in a quadratic equation. Then the equation can be solved using one of the techniques for solving a quadratic equation, which include factoring and using the Quadratic Formula.

EXAMPLE 8 Solving a Trigonometric Equation by Factoring

Solve. $4\cos^4 x + 3 = 7\cos^2 x$

SOLUTION

Subtract $7\cos^2 x$ from each side so that the right side is 0.

$$4\cos^4 x + 3 = 7\cos^2 x \;\Rightarrow\; 4\cos^4 x - 7\cos^2 x + 3 = 0$$

Notice that substituting S for $\cos^2 x$ results in a quadratic equation in terms of S. Solve the quadratic equation.

$4S^2 - 7S + 3 = 0$		*Let $S = \cos^2 x$. Substitute.*
$(4S - 3)(S - 1) = 0$		*Factor.*
$4S - 3 = 0$ or	$S - 1 = 0$	*Set each factor equal to 0.*
$S = 3/4$	$S = 1$	*Solve each equation for S.*
$\cos^2 x = 3/4$	$\cos^2 x = 1$	*Substitute $\cos^2 x$ into each equation for S.*
$\cos x = \pm\dfrac{\sqrt{3}}{2}$	$\cos x = \pm 1$	*Take ± the square root of each side.*

TIP

If the quadratic expression cannot be factored, solve using the Quadratic Formula.

Now solve each equation.

We know that $\cos x = 1$ when $x = 0$ because 0 is an important angle.

Additionally, we know that $\cos x = -1$ when $x = \pi$ because π is an important angle.

Use the unit circle to solve $\cos x = \pm\dfrac{\sqrt{3}}{2}$. From the unit circle we know that

$\cos\dfrac{\pi}{6} = \dfrac{\sqrt{3}}{2}$, $\cos\dfrac{11\pi}{6} = \dfrac{\sqrt{3}}{2}$, $\cos\dfrac{5\pi}{6} = -\dfrac{\sqrt{3}}{2}$, and $\cos\dfrac{7\pi}{6} = -\dfrac{\sqrt{3}}{2}$.

TIP

Since $\cos x > 0$ and $\cos x < 0$, the solutions are in all four quadrants.

To write the general form equations of all solutions, add multiples of 2π (the period of $\cos x$) to each solution.

$$x = 2\pi n,\; x = \pi + 2\pi n,\; x = \frac{\pi}{6} + 2\pi n,\; x = \frac{5\pi}{6} + 2\pi n,\; x = \frac{7\pi}{6} + 2\pi n, \text{ and } x = \frac{11\pi}{6} + 2\pi n$$

Note that all solutions given by $x = \dfrac{\pi}{6} + 2\pi n$ and $x = \dfrac{7\pi}{6} + 2\pi n$ can be expressed as $x = \dfrac{\pi}{6} + \pi n$. Similarly, all solutions given by $x = \dfrac{5\pi}{6} + 2\pi n$ and $x = \dfrac{11\pi}{6} + 2\pi n$ can be expressed as $x = \dfrac{5\pi}{6} + \pi n$.

Therefore, $x = \dfrac{\pi}{6} + \pi n$ and $x = \dfrac{5\pi}{6} + \pi n$ represent all the solutions of $4\cos^4 x + 3 = 7\cos^2 x$.

6.2.8 Solving Trigonometric Equations of Quadratic Type: Another Example

Recall that the three common methods for solving quadratic equations are factoring, completing the square, and using the Quadratic Formula, and that not all quadratic equations can be solved by factoring.

The Quadratic Formula will be used in **Example 9** to solve a trigonometric equation of quadratic type that cannot be factored.

EXAMPLE 9 Using the Quadratic Formula to Solve a Trigonometric Equation

Find all solutions to $\sin^2 x + \sin x = 1$. Round to the nearest thousandth as needed.

SOLUTION

Subtract 1 from each side so that the right side is 0.

$$\sin^2 x + \sin x = 1 \;\Rightarrow\; \sin^2 x + \sin x - 1 = 0$$

Notice that substituting S for $\sin x$ results in a quadratic equation in terms of S.

$$S^2 + S - 1 = 0 \qquad \textit{Substitute S for sin x.}$$

Solve the quadratic equation. The expression $S^2 + S - 1$ cannot be factored, so use the Quadratic Formula to solve the equation and then substitute $\sin x$ for S.

$$S = \frac{-1 \pm \sqrt{1^2 - 4(1)(-1)}}{2(1)} = \frac{-1 \pm \sqrt{5}}{2} \qquad \textit{Substitute a = 1, b = 1, and c = −1 into the Quadratic Formula to solve for S.}$$

$$\sin x = \frac{-1+\sqrt{5}}{2} \text{ or } \sin x = \frac{-1-\sqrt{5}}{2} \qquad \textit{Substitute sin x for S.}$$

Next, use the inverse sine function to solve each equation, if possible.

However, note that $\sin^{-1}\left(\frac{-1-\sqrt{5}}{2}\right)$ does not exist, because $\frac{-1-\sqrt{5}}{2} < -1$.

Therefore, use the inverse sine function to solve $\sin x = \frac{-1+\sqrt{5}}{2}$.

$$x = \sin^{-1}\left(\frac{-1+\sqrt{5}}{2}\right) = 0.6662\ldots \approx 0.666 \qquad \textit{Use a calculator. Round to the nearest thousandth.}$$

Since $\sin x = \frac{-1+\sqrt{5}}{2} > 0$ and $\sin x > 0$ in quadrants I and II, there is a solution in quadrant II.

$$x = \pi - 0.6662\ldots \approx 2.475 \qquad \textit{Subtract 0.6662... from } \pi \textit{ to find the solution in quadrant II.}$$

To write the general form equations of all solutions, add multiples of 2π (the period of $\sin x$) to each solution.

$$x \approx 0.666 + 2\pi n \text{ and } x \approx 2.475 + 2\pi n$$

6.2.9 Solving Trigonometric Equations Using Fundamental Identities

Any of the fundamental trigonometric identities from Section 6.1 can be used to rewrite a trigonometric equation in a form that is easier to solve.

EXAMPLE 10 Using Identities to Solve a Trigonometric Equation

Find all solutions to $\sec^2 x + \tan x - 2 = \tan(-x)$ within the interval $[0, 2\pi]$. Round to the nearest thousandth as needed.

SOLUTION

Use an even-odd identity to write $\tan(-x)$ as a function of x (instead of $-x$), and use a Pythagorean identity to write the equation in terms of tangent only.

$$\sec^2 x + \tan x - 2 = \tan(-x)$$

$$\sec^2 x + \tan x - 2 = -\tan x \quad \textit{Even-Odd Identity}$$

$$\sec^2 x + \tan x - 1 - 1 = -\tan x \quad \textit{Express } -2 \textit{ as } -1-1.$$

$$(\sec^2 x - 1) + \tan x - 1 = -\tan x \quad \textit{Associative and Commutative Properties}$$

$$\tan^2 x + \tan x - 1 = -\tan x \quad \textit{Pythagorean Identity}$$

Pythagorean Identity ▶ $1 + \tan^2\theta = \sec^2\theta$

$$\tan^2 x + 2\tan x - 1 = 0 \quad \textit{Add } \tan x \textit{ to each side.}$$

$$S^2 + 2S - 1 = 0 \quad \textit{Substitute } S \textit{ for } \tan x.$$

$$S = \frac{-2 \pm \sqrt{2^2 - 4(1)(-1)}}{2(1)} = \frac{-2 \pm 2\sqrt{2}}{2} = -1 \pm \sqrt{2}$$

Substitute $a = 1$, $b = 2$, and $c = -1$ into the Quadratic Formula to solve for S.

$$\tan x = -1 + \sqrt{2} \quad \text{or} \quad \tan x = -1 - \sqrt{2} \quad \textit{Substitute } \tan x \textit{ for } S.$$

Use the inverse tangent function to solve each equation.

$$x = \tan^{-1}\left(-1 + \sqrt{2}\right) = 0.3926\ldots \approx 0.393 \text{ or}$$

$$x = \tan^{-1}\left(-1 - \sqrt{2}\right) = -1.1780\ldots \approx -1.178$$

Use a calculator and round to the nearest thousandth.

Since $\tan x = -1 + \sqrt{2} > 0$ and $\tan x > 0$ in quadrants I and III, there is a solution in quadrant III.

$$x = \pi + 0.3926\ldots \approx 3.533 \quad \textit{Add } \pi \textit{ to } 0.3926\ldots \textit{ to find the solution in quadrant III.}$$

Note that $x \approx -1.178$ is not in the interval $[0, 2\pi]$.

However, since $\tan x = -1 - \sqrt{2} < 0$ and $\tan x < 0$ in quadrants II and IV, use $x = -1.1780\ldots$ to find the solutions in quadrants II and IV.

$$x = \pi + (-1.1780\ldots) \approx 1.962 \quad \textit{Add } \pi \textit{ to } -1.1780\ldots \textit{ to find the solution in quadrant II.}$$

$$x = 2\pi + (-1.1780\ldots) \approx 5.102 \quad \textit{Add } 2\pi \textit{ to } -1.1780\ldots \textit{ to find the solution in quadrant IV.}$$

Therefore, $x \approx 0.393, 1.962, 3.533$, and 5.102 are all of the solutions of $\sec^2 x + \tan x - 2 = \tan(-x)$ within the interval $[0, 2\pi]$.

6.2.10 Applications Involving Trigonometric Equations

EXAMPLE 11 Modeling with a Trigonometric Equation

Suppose the electromotive force E (in volts) in an electric circuit after t seconds is given by $E(t) = 2 \cos 50\pi t$. Find the smallest positive t-value for which the electromotive force is 2 volts.

SOLUTION

Substitute 2 for $E(t)$ and solve for t.

$2 = 2 \cos 50\pi t$	*Substitute 2 for $E(t)$.*
$1 = \cos 50\pi t$	*Divide each side by 2.*
$\cos^{-1}(1) = 50\pi t$	*Take the inverse cosine of each side.*

We know that $\cos^{-1}(1) = 0$. Solving $0 = 50\pi t$ for t yields $t = 0$, so at time 0 (after 0 seconds) the electromotive force is 2 volts. However, the question asks for the smallest *positive* t-value. Since 0 is not positive, evaluate $\cos^{-1}(1)$ as the smallest *positive* angle with cosine equal to 1, which is 2π.

Substitute 2π for $\cos^{-1}(1)$ and solve for t.

$2\pi = 50\pi t$	*The smallest positive angle with cosine equal to 1 is 2π.*
$t = 1/25$	*Divide each side by 50π.*

Thus, the electromotive force is 2 volts after 1/25 seconds.

SECTION 6.2 EXERCISES

Warm Up

Solve each equation by factoring.

1. $6x^2 + 7x = 3$
2. $20x = 4x^2$
3. $13x = -6 + 5x^2$

Just the Facts

Fill in the blanks.

4. Solutions to polynomial equations or trigonometric equations are the values that make the equations _____.
5. A trigonometric equation may have an infinite number of solutions due to its _____ nature.
6. To write the general form for all solutions to a trigonometric equation, _____ multiples of the _____ to the solutions found within one period.
7. Use _____ trigonometric functions to find the value of an angle.
8. If $y = f(x)$ is a trigonometric function and the y-value is not in the range of the trigonometric function, then the equation has _____.
9. Applying the even-odd identity, $\tan(-x) =$ _____.
10. Checking the _____ of the calculator is very important when solving for an angle.

Essential Skills

In Exercises 1–24, solve each equation.

1. $\sin x = \frac{\sqrt{3}}{2}$
2. $\sec x = -\frac{2}{\sqrt{3}}$
3. $\tan x = -1$
4. $\cos x = 1/2$
5. $\cot x = -\sqrt{3}$
6. $\csc x = -2$
7. $\sec x = 1$
8. $\csc x = -1$
9. $\cos x = 0$
10. $\cot x = 0$
11. $2\cos x - 3 = -4$
12. $2\sqrt{3} - \tan x = \sqrt{3}$
13. $2 = 4 + 2\cot x$
14. $4 + \sqrt{3}\sec x = 6$
15. $-3 = -5 + \sqrt{3}\csc x$
16. $7 + 2\sin x = 8$
17. $\tan^2 x = 3$
18. $2\csc^2 x = 4$
19. $-3 = -4\sin^2 x$
20. $10\cos^2 x = 5$
21. $\tan^2 x + 6 = 2\tan^2 x + 5$
22. $5\cos^2 x + 4 = 7 + \cos^2 x$
23. $2 + 7\csc^2 x = 8 + 4\csc^2 x$
24. $\sin^2 x + 6 = 5 + 5\sin^2 x$

In Exercises 25–28, using a calculator, find all solutions to each equation within the interval $[0, 2\pi)$ using inverse functions. Round each answer to the nearest thousandth.

25. $2\cos x + 1.3 = 2.06$
26. $3.6 - 1.6\sin x = 3.2$
27. $8.96 - 4.8\csc x = 3.4$
28. $0.96\cot x + 3.16 = 4.48$

In Exercises 29–68, find all solutions to each equation within the interval $[0, 2\pi)$, unless otherwise stated.

29. $\tan^2 x \cdot \csc x = \tan^2 x$
30. $2\sin x \cdot \cos x = \sin x$
31. $3\tan x = \tan^3 x$
32. $-5\sin^3 x = -5\sin x$
33. $\tan^2 x + \tan x = 0$
34. $\tan x \cdot \sec x = \tan x$
35. $2\sin^2 x + \sin x = 1$
36. $2\cos^2 x - 5\cos x = -2$
37. $4\csc^2 x + 4\csc x + 1 = 0$
38. $2\sec^2 x - 3\sec x = 2$
39. $2\cos^2 x + \cos x = 1$
40. $\sin^2 x + 5\sin x + 6 = 0$
41. $3 - 3\sin x = 2\cos^2 x$
42. $2\cos^2 x = 1 + \sin x$
43. $\tan^2 x - \sec x - 1 = 0$
44. $\sec^2 x = (\sqrt{3} + 1)\tan x - \sqrt{3} + 1$
45. $\tan^5 x - 9\tan x = 0$
46. $3 - \cos^2 x + 3\sin x + 1 = \cos^2 x$

47. $2 \cot \frac{x}{2} = 2\sqrt{3}$

48. $\sec \frac{3\pi}{2} = -2$

49. $2 \cos 2x = 1$

50. $\tan 2\pi = \sqrt{3}$; $[0, \pi)$

51. $3 \tan \frac{x}{2} + 3 = 0$

52. $\csc \frac{\pi}{2} = -\frac{2}{\sqrt{3}}$; $[0, 4\pi)$

53. $\sin^2 2x = 1/4$

54. $4 \cos^2 2x - 4 \cos 2x + 1 = 0$

55. $\cos^2 2\theta = 1/4$

56. $\tan^2 2x - 2 \tan 2x + 1 = 0$

57. $\cos^2 2x - \sin^2 2x = 0$

58. $4 \csc^2 2x - 5 \csc 2x + 1 = 0$

59. $\sin 2x = 2 \tan 2x$

60. $\sin 3x \cdot \cos 2x = 0$

61. $\tan 3x + 1 = \sec 3x$

62. $\sec 3x \cdot \cos 2x = 0$

63. $\cos 3x \cdot \sin 2x = 0$

64. $\csc 5x \cdot \cos 2x = 0$

65. $\tan^4 x - 4 \tan^2 x + 3 = 0$

66. $8 \sin^4 \theta + 3 = 10 \sin^2 \theta$

67. $3 \csc^4 x - 16 \csc^2 x + 16 = 0$

68. $4 \cot^4 \theta - 3 = \cot^2 \theta$

In Exercises 69–78, find all solutions to each equation within the interval $[0, 2\pi)$. Round each answer to the nearest thousandth.

69. $\cos^2 x - 3 \cos x + 1 = 0$

70. $5 \tan^2 x - 3 \tan x - 5 = 0$

71. $\csc^2 x - \csc x = 1$

72. $10 \sin^2 x - 3 \sin x - 2 = 0$

73. $\sin^2 2x - 3 \sin 2x + 1 = 0$

74. $2 \cos^2 3x + 6 \cos 3x = 3$

75. $\sin^2 \frac{x}{2} + 4 \sin \frac{x}{2} + 2 = 0$

76. $8 \sin^2 2x + 7 \sin 2x = 4$

77. $\cos x - \sin (-x) \cdot \tan x - 2 = 1$

78. $\frac{\cos (-x) + \cos x}{\sin x + 3 \sin (-x)} = \sqrt{3}$

In Exercises 79–88, find all solutions to each equation within the interval $[0, 2\pi)$.

79. $\cos x(\sec x - \cos x) - 1 = 0$

80. $\cos x + \tan x \cdot \sin x = \frac{2}{\sqrt{3}}$

81. $\sin x \cdot \csc (-x) - \sec (-x) \cdot \cot (-x) = 0$

82. $\cot (-x) \cdot \tan (-x) + \sin (-x) \cdot \sin x = 2$

83. $\sec^2 x = \frac{1}{1 - \sin x} - \frac{1}{1 + \sin x}$

84. $\sec^2 x = 2 \tan x$

85. $\frac{\cos^2 x}{\sin x - 1} = -1$

86. $1 + \cos x = 2 - 2 \sin^2 x$

87. $\sec x = \tan x + \cot x$

88. $\csc^2 x = -2 \cot x$

In Exercises 89–94, use the following information.

A child is playing with a yo-yo. When the child does a certain trick, the yo-yo's position in feet relative to the ground is $S(t)$, where t is the time in seconds. After how many seconds, to the nearest thousandth of a second, is the yo-yo first at the specified height?

89. $S(t) = -2 \sin 2\pi t + 2.2$; 0.25 feet

90. $S(t) = -2 \sin 2\pi t + 3$; 1.5 feet

91. $S(t) = -2.25 \sin 2\pi t + 2.26$; 0.75 feet

92. $S(t) = -1.75 \sin 2\pi t + 2$; 0.25 feet

93. $S(t) = -1.6 \sin 2\pi t + 2.9$; 4.35 feet

94. $S(t) = -2.25 \sin 2\pi t + 2.8$; 0.5 feet

Extensions

95. A weight is oscillating on the end of a spring. The position of the weight relative to the point of equilibrium is given by the equation $y = \frac{1}{12}(\sqrt{3} \cos 8t - 3 \sin 8t)$ where y is the displacement in meters and t is the time in seconds. Find the smallest positive value of t when the weight is at the point of equilibrium. Leave answer in terms of π.

96. An electric generator produces a 30-cycle alternating current described by the equation below.

$$I(t) = 40 \sin\left(60\pi\left(t - \frac{7}{72}\right)\right)$$

If $I(t)$ is the current measured in amperes at t seconds, find the smallest positive value of t for which the current is 20 amperes. Leave answer in terms of π.

97. In an electric circuit, the electromotive force is $E(t)$ volts where $E(t) = 2\cos 50\pi t$. Find the smallest positive value of t to the nearest thousandth of a second for which the electromotive force is 1.6 volts.

98. Verify $4\cos^2 2x - 4\cos 2x + 1 = 0$ when $x = 5\pi/6$.

99. Solve $\cos^2 \frac{x}{2} - \frac{1}{2}\cos \frac{x}{2} = \frac{1}{2}$ on the interval $[0, 10\pi)$.

100. Verify $3\csc^4 x - 16\csc^2 x + 16 = 0$ when $x = 7\pi/6$.

101. To the nearest hundredth of a degree, at what angle must a cannon ball be shot in order to hit a target 500 meters away, if the muzzle velocity V_0 is 150 meters per second and the horizontal distance r traveled is given by the equation $r = \frac{1}{9.8}V_0^2 \sin 2\pi$?

102. A baseball was thrown from first base at an angle θ with the horizontal and with an initial velocity V_0 of 80 feet per second. The ball was caught by the second-base player who was 100 feet away. Find θ if the horizontal distance covered by the ball, r, is given by the equation $r = \frac{1}{32}V_0^2 \sin 2\theta$.

103. Verify $\tan 3x - \sec 3x = -1$ when $x = 2\pi/3$, if possible.

104. Solve $10\sec^2 x - 10 = 7\tan x + 2$ on the interval $[0, 2\pi)$.

6.3 THE SUM AND DIFFERENCE FORMULAS

OBJECTIVES	PREREQUISITE VOCABULARY TERMS
• Use sum and difference formulas to evaluate trigonometric functions.	**cofunction identities**
• Prove the cofunction identities using sum and difference formulas.	**fundamental trigonometric identities**
• Use sum and difference formulas to verify an identity.	**trigonometric functions**

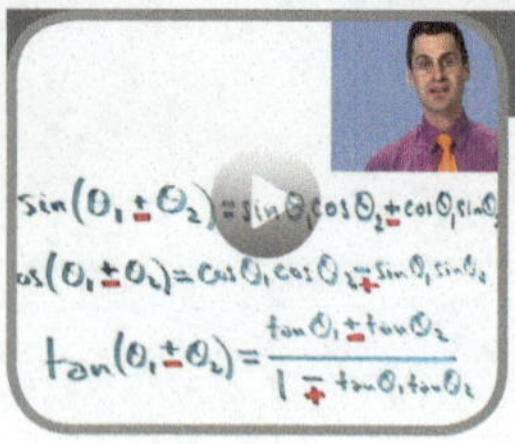

6.3.1 Introduction to the Sum and Difference Formulas

Consider the equation $\sin(\pi + \pi) = \sin\pi + \sin\pi$. This equation is a true statement because $\sin(\pi + \pi) = \sin 2\pi = 0$, $\sin\pi + \sin\pi = 0 + 0 = 0$, and, of course, $0 = 0$. Based on that result, we might wonder whether $\sin(x + y) = \sin x + \sin y$ is an identity (i.e., whether $\sin(x + y) = \sin x + \sin y$ is a true statement for all real numbers x and y). However, $x = \pi/2$ and $y = \pi$ provide a counterexample.

$$\sin(\pi/2 + \pi) = \sin(3\pi/2) = -1 \quad \text{and} \quad \sin(\pi/2) + \sin\pi = 1 + 0 = 1$$

Because $-1 \neq 1$, the equation $\sin(x + y) = \sin x + \sin y$ is not an identity.

However, identities for the sum of angles of trigonometric functions do exist. The following **sum and difference formulas** provide ways to express the sine, cosine, or tangent of a sum or a difference of angles θ_1 and θ_2. Each formula is an identity, so each is true for all values in the trigonometric function's domain.

Sum and Difference Formulas

Sum of Angles Formulas	Difference of Angles Formulas
$\sin(\theta_1 + \theta_2) = \sin\theta_1\cos\theta_2 + \cos\theta_1\sin\theta_2$	$\sin(\theta_1 - \theta_2) = \sin\theta_1\cos\theta_2 - \cos\theta_1\sin\theta_2$
$\cos(\theta_1 + \theta_2) = \cos\theta_1\cos\theta_2 - \sin\theta_1\sin\theta_2$	$\cos(\theta_1 - \theta_2) = \cos\theta_1\cos\theta_2 + \sin\theta_1\sin\theta_2$
$\tan(\theta_1 + \theta_2) = \dfrac{\tan\theta_1 + \tan\theta_2}{1 - \tan\theta_1\tan\theta_2}$	$\tan(\theta_1 - \theta_2) = \dfrac{\tan\theta_1 - \tan\theta_2}{1 + \tan\theta_1\tan\theta_2}$

6.3.2 Using the Sum and Difference Formulas to Find Exact Values

Recall that if an angle θ is a multiple of $\pi/6$ or $\pi/4$ (i.e., 30° or 45°), then the exact value of a trigonometric function of θ can be determined using the function's graph, the unit circle, or a reference angle on a coordinate plane (among other methods). Otherwise (i.e., if θ is not a multiple of $\pi/6$ or $\pi/4$), a calculator has been used to find a decimal approximation of the trigonometric function of θ.

The sum and difference formulas provide methods for finding the exact value of a trigonometric function for values of θ such that $\theta = \theta_1 + \theta_2$ or $\theta = \theta_1 - \theta_2$, where θ_1 and θ_2 are multiples of $\pi/6$ or $\pi/4$. In other words, a sum formula or a difference formula gives the exact value of a trigonometric function of θ when θ can be expressed as a sum or as a difference of multiples of $\pi/6$ or $\pi/4$.

EXAMPLE 1 Using the Sum and Difference Formulas to Find the Exact Value of a Trigonometric Function

Find the exact value of each trigonometric function, if possible.

A. cos 105° **B.** tan $5\pi/12$

SOLUTION

Find two angles θ_1 and θ_2 (multiples of $\pi/6$ or $\pi/4$) such that their sum or difference is equal to the measure of the angle in the given trigonometric function. Then substitute the values of θ_1 and θ_2 into the corresponding sum or difference formula and simplify.

A. Notice that 105° = 150° – 45°, 150° is a multiple of 30°, and 45° is a multiple of 45°. So, use the difference formula $\cos(\theta_1 - \theta_2) = \cos\theta_1 \cos\theta_2 + \sin\theta_1 \sin\theta_2$, where $\theta_1 = 150°$ and $\theta_2 = 45°$, to evaluate cos 105°.

ALTERNATIVE METHOD

60° is a multiple of 30°, 45° is a multiple of 45°, and 60° + 45° = 105°, so the cosine of a sum formula can be used to evaluate cos 105°, where $\theta_1 = 60°$ and $\theta_2 = 45°$ (or $\theta_1 = 45°$ and $\theta_2 = 60°$).

$$\cos 105° = \cos(150° - 45°)$$

$$= \cos 150° \cos 45° + \sin 150° \sin 45° \quad \textit{Use the cosine of a difference formula.}$$

$$= -\frac{\sqrt{3}}{2}\cdot\frac{\sqrt{2}}{2} + \frac{1}{2}\cdot\frac{\sqrt{2}}{2} \quad \textit{Evaluate each trigonometric function.}$$

$$= -\frac{\sqrt{6}}{4} + \frac{\sqrt{2}}{4} \quad \textit{Multiply.}$$

$$= \frac{-\sqrt{6}+\sqrt{2}}{4} \quad \textit{Add.}$$

B. Since $\frac{5\pi}{12} = \frac{3\pi}{12} + \frac{2\pi}{12} = \frac{\pi}{4} + \frac{\pi}{6}$, use the tangent of a sum formula, where $\theta_1 = \pi/4$ and $\theta_2 = \pi/6$, to evaluate tan $5\pi/12$.

ALTERNATIVE METHOD

The tangent of a difference formula can be used to evaluate tan $5\pi/12$, where $\theta_1 = 2\pi/3$ and $\theta_2 = \pi/4$.

$$\tan\frac{5\pi}{12} = \tan\left(\frac{\pi}{4} + \frac{\pi}{6}\right)$$

$$= \frac{\tan\frac{\pi}{4} + \tan\frac{\pi}{6}}{1 - \tan\frac{\ }{4}\cdot\tan\frac{\ }{6}} \quad \textit{Use the tangent of a sum formula.}$$

$$= \frac{1 + \frac{1}{\sqrt{3}}}{1 - \frac{1}{\sqrt{3}}} \quad \textit{Evaluate each trigonometric function.}$$

$$= \frac{\left(1 + \frac{1}{\sqrt{3}}\right)\sqrt{3}}{\left(1 - \frac{1}{\sqrt{3}}\right)\sqrt{3}} \quad \textit{Multiply the numerator and denominator by } \sqrt{3} \textit{ to simplify the complex fraction.}$$

$$= \frac{\sqrt{3}+1}{\sqrt{3}-1} \quad \textit{Simplify.}$$

TIP

Rationalizing the denominator results in $\tan\frac{5\pi}{12} = 2 + \sqrt{3}$.

CHECK

Use a calculator to check each answer.

$\cos 105° \approx -0.259$ and $\frac{-\sqrt{6}+\sqrt{2}}{4} \approx -0.259$ ✓ $\tan\frac{5\pi}{12} \approx 3.732$ and $\frac{\sqrt{3}+1}{\sqrt{3}-1} \approx 3.732$ ✓

6.3.3 Using the Sum and Difference Formulas to Find Exact Values: Another Example

If an expression is given in the expanded form of a sum or difference formula, then the expression can be simplified and written as the trigonometric function of the sum (or difference) of the angles, as demonstrated in **Example 2**. The exact value of the function can be determined if the sum (or difference) of those angles is a multiple of $\pi/6$ or $\pi/4$.

EXAMPLE 2 Using the Sum and Difference Formulas to Simplify a Trigonometric Expression

Find the exact value of $\cos\frac{\pi}{12}\cos\frac{5\pi}{12}+\sin\frac{\pi}{12}\sin\frac{5\pi}{12}$, if possible.

SOLUTION

The expression is in the expanded form of the cosine of a difference formula, $\cos\theta_1\cos\theta_2+\sin\theta_1\sin\theta_2$, where $\theta_1=\pi/12$ and $\theta_2=5\pi/12$. So, use this formula to simplify.

$$\cos\frac{\pi}{12}\cos\frac{5\pi}{12}+\sin\frac{\pi}{12}\sin\frac{5\pi}{12}=\cos\left(\frac{\pi}{12}-\frac{5\pi}{12}\right)$$ *Use the cosine of a difference formula.*

$$=\cos\left(-\frac{4\pi}{12}\right)$$ *Subtract.*

$$=\cos\left(-\frac{\pi}{3}\right)$$ *Simplify.*

$$=\cos\frac{\pi}{3}$$ *Even-Odd Identity*

$$=\frac{1}{2}$$ *Evaluate the trigonometric function.*

REMEMBER

Since cosine is an even function, $\cos(-x)=\cos x$.

6.3.4 Using the Sum and Difference Formulas to Verify a Cofunction Identity

Verifying the Cofunction Identities

In this section we will establish a more formal proof of the cofunction identities that were introduced in Section 6.1.

Cofunction Identities

$\sin\left(\frac{\pi}{2}-\theta\right)=\cos\theta$	$\cos\left(\frac{\pi}{2}-\theta\right)=\sin\theta$	$\tan\left(\frac{\pi}{2}-\theta\right)=\cot\theta$
$\csc\left(\frac{\pi}{2}-\theta\right)=\sec\theta$	$\sec\left(\frac{\pi}{2}-\theta\right)=\csc\theta$	$\cot\left(\frac{\pi}{2}-\theta\right)=\tan\theta$

Since each cofunction identity involves a trigonometric function of a difference of angles, the difference formulas can be used to expand the expression. Moreover, these identities can be proven using the difference formulas.

For example, expand $\sin\left(\frac{\pi}{2}-\theta\right)$ using the sine of a difference of angles formula, where $\theta_1 = \pi/2$ and $\theta_2 = \theta$, to verify that $\sin\left(\frac{\pi}{2}-\theta\right) = \cos\theta$ is an identity.

$$\begin{aligned}\sin\left(\frac{\pi}{2}-\theta\right) &= \sin\frac{\pi}{2}\cos\theta - \cos\frac{\pi}{2}\sin\theta \\ &= 1\cdot\cos\theta - 0\cdot\sin\theta \\ &= \cos\theta \ \checkmark\end{aligned}$$

Therefore, $\sin\left(\frac{\pi}{2}-\theta\right) = \cos\theta$ is an identity.

A similar process can be used to verify the other five cofunction identities.

Deriving Additional Cofunction Identities

Additional identities can be derived using the sum and difference formulas. For example, suppose the difference of angles is $\theta - \pi/2$ (instead of $\pi/2 - \theta$). Then an identity involving $\sin\left(\theta-\frac{\pi}{2}\right)$ can be derived by using the sine of a difference of angles formula, where $\theta_1 = \theta$ and $\theta_2 = \pi/2$, as demonstrated in **Example 3**.

EXAMPLE 3 Using the Sum and Difference Formulas to Derive a Cofunction Identity

Complete the identity $\sin\left(\theta-\frac{\pi}{2}\right) =$ ________.

SOLUTION

$$\begin{aligned}\sin\left(\theta-\frac{\pi}{2}\right) &= \sin\theta\cos\frac{\pi}{2} - \cos\theta\sin\frac{\pi}{2} && \textit{Use the sine of a difference formula.} \\ &= \sin\theta\cdot 0 - \cos\theta\cdot 1 && \textit{Evaluate } \cos(\pi/2) \textit{ and } \sin(\pi/2). \\ &= 0 - \cos\theta && \textit{Simplify.} \\ &= -\cos\theta && \textit{Subtract.}\end{aligned}$$

Therefore, $\sin\left(\theta-\frac{\pi}{2}\right) = -\cos\theta$ is an identity.

Identities for the other five trigonometric functions can be determined using a similar process.

Cofunction Identities ($\theta - \pi/2$)

$\sin\left(\theta-\frac{\pi}{2}\right) = -\cos\theta$	$\cos\left(\theta-\frac{\pi}{2}\right) = \sin\theta$	$\tan\left(\theta-\frac{\pi}{2}\right) = -\cot\theta$
$\csc\left(\theta-\frac{\pi}{2}\right) = -\sec\theta$	$\sec\left(\theta-\frac{\pi}{2}\right) = \csc\theta$	$\cot\left(\theta-\frac{\pi}{2}\right) = -\tan\theta$

6.3.5 Using the Sum and Difference Formulas to Verify an Identity

EXAMPLE 4 Proving a Trigonometric Identity

Determine whether $\cos(x + y)\cos y + \sin(x + y)\sin y = \cos x$ is an identity or not.

SOLUTION

The expression on the right side is simplified. So, simplify the left side to determine whether or not it is equal to the right side.

The expression on the left side contains $\cos(x + y)$ and $\sin(x + y)$, so use the cosine of a sum formula to expand $\cos(x + y)$, and use the sine of a sum of angles formula to expand $\sin(x + y)$.

$$\cos(x + y)\cos y + \sin(x + y)\sin y$$

$$= (\cos x \cos y - \sin x \sin y)\cos y + (\sin x \cos y + \cos x \sin y)\sin y \quad \textit{Expand.}$$

$$= \cos x \cos^2 y - \sin x \sin y \cos y + \sin x \cos y \sin y + \cos x \sin^2 y \quad \textit{Distribute.}$$

$$= \cos x \cos^2 y + \cos x \sin^2 y \quad \textit{Simplify.}$$

$$= \cos x(\cos^2 y + \sin^2 y) \quad \textit{Factor out } \cos x.$$

$$= \cos x(1) \quad \textit{Pythagorean Identity}$$

$$= \cos x \ \checkmark$$

NOTICE THAT

The sum of the second term ($-\sin x \sin y \cos y$) and the third term ($\sin x \cos y \sin y$) is 0.

So, $\cos(x + y)\cos y + \sin(x + y)\sin y = \cos x$ is an identity.

6.3.6 Using the Sum and Difference Formulas to Solve an Equation

Recall that an equation where the domain is not all real numbers may have an extraneous solution (i.e., a solution not in the equation's domain). So, if an equation's domain is not all real numbers, each possible solutions should be checked to confirm that it is in the equation's domain (and therefore not extraneous).

For example, consider the equation $\tan(x + \pi) = -2\sin(x + \pi)$, as seen in **Example 5**. This equation contains the tangent of an unknown angle and the sine of an unknown angle. We know that sine is defined for all real numbers (because the domain of $\sin x$ is all real numbers). However, tangent is not defined for all real numbers (because the domain of $\tan x$ is all real numbers except all x such that $x = \pi/2 + \pi n$, where n is an integer). Therefore, extraneous solutions may be found when the equation is solved.

EXAMPLE 5 Solving a Trigonometric Equation

Find all solutions to $\tan(x + \pi) = -2\sin(x + \pi)$ within the interval $[0, 2\pi)$.

SOLUTION

The equation contains the tangent of a sum and the sine of a sum, so use the tangent and sine of a sum formulas to expand.

$$\tan(x+\pi) = -2\sin(x+\pi)$$

$$\frac{\tan x + \tan \pi}{1 - \tan x \tan \pi} = -2(\sin x \cos \pi + \cos x \sin \pi) \quad \textit{Expand.}$$

$$\frac{\tan x + 0}{1 - \tan x \cdot 0} = -2(\sin x \cdot (-1) + \cos x \cdot 0) \quad \textit{Evaluate the trigonometric functions.}$$

$$\tan x = -2(-\sin x) \quad \textit{Simplify.}$$

$$\tan x = 2\sin x \quad \textit{Multiply.}$$

$$\tan x - 2\sin x = 0 \quad \textit{Subtract 2 sin x from each side.}$$

$$\frac{\sin x}{\cos x} - 2\sin x = 0 \quad \textit{Quotient Identity}$$

$$\frac{\sin x}{\cos x} - \frac{2\sin x \cos x}{\cos x} = 0 \quad \textit{Write the terms with a common denominator.}$$

$$\frac{\sin x - 2\sin x \cos x}{\cos x} = 0 \quad \textit{Subtract.}$$

Since the fraction is equal to 0, the numerator must also be equal to 0. Therefore, we can set the numerator equal to 0, and then solve the resulting equation. Note, however, that all values of x such that $\cos x = 0$ must be excluded from the equation's domain.

$$\sin x - 2\sin x \cos x = 0 \quad \textit{Set the numerator equal to 0.}$$

$$\sin x(1 - 2\cos x) = 0 \quad \textit{Factor out sin x.}$$

$$\sin x = 0 \quad \text{or} \quad 1 - 2\cos x = 0 \quad \textit{Set each factor equal to 0.}$$

$$\cos x = 1/2 \quad \textit{Solve for cos x.}$$

Find all solutions to $\sin x = 0$ and $\cos x = 1/2$ within the interval $[0, 2\pi)$.

- $\sin 0 = 0$, so $x = 0$.
- $\sin \pi = 0$, so $x = \pi$.
- $\cos(\pi/3) = 1/2$, so $x = \pi/3$.
- $\cos(5\pi/3) = 1/2$, so $x = 5\pi/3$.

Check for extraneous solutions.

The equation's domain excludes
- all x-values such that $\cos x = 0$ (i.e., all x such that $x = \pi/2 + \pi n$) and
- all x-values for which $\tan(x + \pi)$ is undefined (i.e., all x such that $x = \pi/2 + \pi n$).

The solutions $x = 0, \pi/3, \pi$, and $5\pi/3$ are not excluded from the domain, so they are not extraneous solutions.

Therefore, $x = 0, \pi/3, \pi$, and $5\pi/3$ are all of the solutions of $\tan(x + \pi) = -2\sin(x + \pi)$ within the interval $[0, 2\pi)$.

SECTION 6.3 EXERCISES

Warm Up

Write a sum of multiples of $\pi/6$ or $\pi/4$ equal to each given angle.

1. $7\pi/12$
2. $75°$
3. $-105°$

Just the Facts

Fill in the blanks.

4. A(n) _____ is an equation that is true for all values in the trigonometric function's domain.
5. True or False? $\cos(A + B) = \cos A + \cos B$ is an identity.
6. By the tangent of a _____ formula, the expansion of $\tan(60° + 45°)$ is _____.
7. By the cosine of a _____ formula, the expansion of _____ is $\cos A \cos B - \sin A \sin B$.
8. By the sine of a _____ formula, the expansion of _____ is $\sin A \cos B - \cos A \sin B$.
9. When solving an equation that involves a trigonometric function that is not defined for all values of x, the solutions must be checked to ensure that there are no _____ solutions.
10. A sum or difference formula can be used to find an exact value of a trigonometric function of θ when θ can be expressed as a sum or _____ of multiples of $\pi/6$ or _____.

Essential Skills

In Exercises 1–4, write the sine, cosine, and tangent of each angle using a difference formula.

1. $75°$
2. $165°$
3. $7\pi/12$
4. $195°$

In Exercises 5–16, find the exact value of each trigonometric function, if possible.

5. $\sin 255°$
6. $\sin(7\pi/12)$
7. $\sin 195°$
8. $\sin(\pi/12)$
9. $\cos 165°$
10. $\cos(5\pi/12)$
11. $\cos 15°$
12. $\cos(11\pi/12)$
13. $\tan 165°$
14. $\tan(11\pi/12)$
15. $\tan(-\pi/12)$
16. $\tan(7\pi/12)$

In Exercises 17–24, find the exact value of each trigonometric expression, if possible.

17. $\cos 229° \cos 4° + \sin 229° \sin 4°$
18. $\dfrac{\tan 290° + \tan 10°}{1 - \tan 290° \tan 10°}$
19. $\dfrac{\tan 75° - \tan 15°}{1 + \tan 75° \tan 15°}$
20. $\sin 70° \cos 10° - \cos 70° \sin 10°$
21. $\sin \dfrac{11\pi}{8} \cos \dfrac{3\pi}{8} + \cos \dfrac{11\pi}{8} \sin \dfrac{3\pi}{8}$
22. $\cos \dfrac{5\pi}{12} \cos \dfrac{7\pi}{12} - \sin \dfrac{5\pi}{12} \sin \dfrac{7\pi}{12}$
23. $\sin \dfrac{\pi}{4} \cos \dfrac{\pi}{12} - \cos \dfrac{\pi}{4} \sin \dfrac{\pi}{12}$
24. $\dfrac{\tan \frac{7\pi}{12} - \tan \frac{5\pi}{12}}{1 + \tan \frac{7\pi}{12} \tan \frac{5\pi}{12}}$

In Exercises 25–30, complete each identity.

25. $\cos(\theta - 3\pi/2) =$
26. $\tan(\theta - \pi/2) =$
27. $\tan(4\pi - x) =$
28. $\sin(\theta + \pi) =$
29. $\sin(3\pi/2 - x) =$
30. $\cos(\theta + \pi/2) =$

In Exercises 31–34, determine whether each equation is an identity or not.

31. $\cos(x + \pi/6) + \sin(x - \pi/3) = 1$
32. $\cos(a + b)\cos(a - b) = \cos^2 a - \sin^2 b$
33. $\sin\left(\dfrac{\pi}{6} + x\right) = \dfrac{1}{2}\left(\cos x + \sqrt{3} \sin x\right)$
34. $\sin(a + b)\sin(a - b) = \sin^2 b - \cos^2 a$

In Exercises 35–38, find the exact value solutions to each equation on the interval $[0, 2\pi)$.

35. $\sin(x + \pi/3) + \sin(x - \pi/3) - 1 = 0$

36. $\tan(x - \pi) + 2\sin(x - \pi) = 0$

37. $\cos(x + \pi/4) - \cos(x - \pi/4) = 1$

38. $\sin\left(x + \frac{\pi}{4}\right) + \sin\left(x - \frac{\pi}{4}\right) + \frac{\sqrt{6}}{2} = 0$

Extensions

In Exercises 39–44, find the exact value of each trigonometric function given that $\tan x = -5/12$, $\cos y = 4/5$, and both angles are in quadrant IV.

39. $\sin(x - y)$

40. $\sin(x + y)$

41. $\cos(x - y)$

42. $\cos(x + y)$

43. $\tan(x - y)$

44. $\tan(x + y)$

In Exercises 45–50, find the exact value of each trigonometric function given that $\tan x = 3/4$, $\sin y = -7/25$, and x and y are in quadrant III.

45. $\sin(x + y)$

46. $\sin(x - y)$

47. $\cos(x + y)$

48. $\cos(x - y)$

49. $\tan(x + y)$

50. $\tan(x - y)$

In Exercises 51–54, find the exact value of each trigonometric function, if possible.

51. $\sin(-7\pi/12)$

52. $\cos(-105°)$

53. $\tan(-13\pi/12)$

54. $\cot(19\pi/12)$

In Exercises 55–56, determine whether each equation is an identity or not.

55. $\tan\left(x + \frac{\pi}{4}\right) + 1 = 2\cos x \sec\left(x + \frac{\pi}{4}\right)$

56. $(\cos B \cos C - \sin B \sin C)^2 + (\sin B \cos C + \cos B \sin C)^2 = 1$

6.4 DOUBLE-ANGLE, HALF-ANGLE, AND PRODUCT-SUM FORMULAS

OBJECTIVES

- Use double-angle formulas to simplify trigonometric expressions, evaluate trigonometric functions, and solve trigonometric equations.
- Use power-reducing formulas to simplify trigonometric expressions.
- Use half-angle formulas to evaluate trigonometric functions.
- Use product-sum formulas to expand a product of trigonometric functions and to simplify a sum of trigonometric functions.

PREREQUISITE VOCABULARY TERMS

fundamental trigonometric identities

Pythagorean Theorem

sum and difference formulas

trigonometric functions

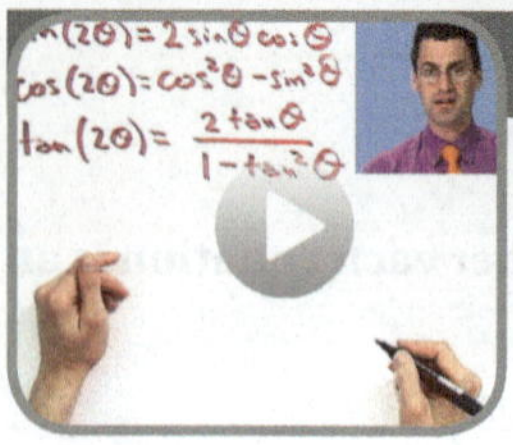

6.4.1 Double-Angle Formulas

The sine of a sum formula and the cosine of a sum formula (introduced in Section 6.3) can each be simplified if we assume the condition that $\theta = \theta_1 = \theta_2$.

$\sin(\theta_1 + \theta_2) = \sin\theta_1 \cos\theta_2 + \cos\theta_1 \sin\theta_2$	*Sine of a Sum Formula*
$\sin(\theta + \theta) = \sin\theta\cos\theta + \cos\theta\sin\theta$	*Substitute θ for θ_1 and θ_2.*
$\sin 2\theta = 2\sin\theta\cos\theta$	*Simplify.*

$\cos(\theta_1 + \theta_2) = \cos\theta_1 \cos\theta_2 - \sin\theta_1 \sin\theta_2$	*Cosine of a Sum Formula*
$\cos(\theta + \theta) = \cos\theta\cos\theta - \sin\theta\sin\theta$	*Substitute θ for θ_1 and θ_2.*
$\cos 2\theta = \cos^2\theta - \sin^2\theta$	*Simplify.*

Formulas for the sine, cosine, or tangent of twice an angle (such as $\sin 2\theta = 2\sin\theta\cos\theta$ and $\cos 2\theta = \cos^2\theta - \sin^2\theta$) are called **double-angle formulas**.

NOTICE THAT

The additional forms of the cosine of a double-angle formula are given by substituting using the Pythagorean identity $\cos^2\theta + \sin^2\theta = 1$.

Double-Angle Formulas	
$\sin 2\theta = 2\sin\theta\cos\theta$	$\cos 2\theta = \cos^2\theta - \sin^2\theta$
$\tan 2\theta = \dfrac{2\tan\theta}{1-\tan^2\theta}$	$= 1 - 2\sin^2\theta$
	$= 2\cos^2\theta - 1$

EXAMPLE 1 **Using a Double-Angle Formula to Evaluate a Trigonometric Function**

Use a double-angle formula to evaluate $\sin(11\pi/3)$.

SOLUTION

NOTICE THAT

$11\pi/3 = 2(1/2)(11\pi/3)$ because $2(1/2) = 1$.

Write $11\pi/3$ in the form 2θ. Note that $11\pi/3 = 2(1/2)(11\pi/3)$, and multiplying $1/2$ by $11\pi/3$ yields the double angle $2(11\pi/6)$. So, apply the sine of a double-angle formula where $\theta = 11\pi/6$.

$$\sin\frac{11\pi}{3} = \sin 2\left(\frac{1}{2}\right)\left(\frac{11\pi}{3}\right) = \sin 2\left(\frac{11\pi}{6}\right) = 2\sin\frac{11\pi}{6}\cos\frac{11\pi}{6} = 2\left(-\frac{1}{2}\right)\left(\frac{\sqrt{3}}{2}\right) = -\frac{\sqrt{3}}{2}$$

EXAMPLE 2 Using a Double-Angle Formula to Simplify an Expression

Use a double-angle formula to simplify each expression, if possible.

A. $3\cos^2 t - 3\sin^2 t$

B. $\dfrac{4\tan\frac{x}{2}}{3 - 3\tan^2\frac{x}{2}}$

SOLUTION

A. $3\cos^2 t - 3\sin^2 t = 3(\cos^2 t - \sin^2 t)$ *Factor 3 from each term.*

$= 3\cos 2t$ *Cosine of a Double-Angle Formula*

B. $\dfrac{4\tan\frac{x}{2}}{3 - 3\tan^2\frac{x}{2}} = \dfrac{4\tan\frac{x}{2}}{3\left(1 - \tan^2\frac{x}{2}\right)}$ *Factor 3 from each term in the denominator.*

$= \dfrac{2}{3}\cdot\dfrac{2\tan\frac{x}{2}}{1 - \tan^2\frac{x}{2}}$ *Factor 2 from the numerator.*

$= \dfrac{2}{3}\tan 2\left(\dfrac{x}{2}\right)$ *Tangent of a Double-Angle Formula*

$= \dfrac{2}{3}\tan x$ *Multiply.*

IMPORTANT

In order to substitute using the tangent of a double-angle formula, the coefficient of tangent must be 2.

6.4.2 Using Double-Angle Formulas

The double-angle formulas can be used to evaluate a trigonometric function of twice an angle (e.g., the value of $\tan 2\theta$), as demonstrated in **Example 3**.

EXAMPLE 3 Using the Double-Angle Formulas

Evaluate $\sin 2\theta$, $\cos 2\theta$, and $\tan 2\theta$.

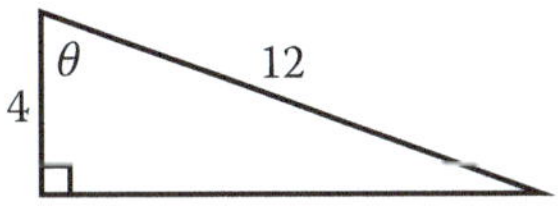

SOLUTION

Pythagorean Theorem $a^2 + b^2 = c^2$

Find the length of the side opposite of θ. $\text{opp} = \sqrt{12^2 - 4^2} = \sqrt{128} = 8\sqrt{2}$

So, $\sin\theta = \dfrac{8\sqrt{2}}{12} = \dfrac{2\sqrt{2}}{3}$, $\cos\theta = 4/12 = 1/3$, and $\tan\theta = \dfrac{8\sqrt{2}}{4} = 2\sqrt{2}$.

Substitute into the double angle formulas to evaluate $\sin 2\theta$, $\cos 2\theta$, and $\tan 2\theta$.

$$\sin 2\theta = 2\sin\theta\cos\theta = 2\left(\frac{2\sqrt{2}}{3}\right)\left(\frac{1}{3}\right) = \frac{4\sqrt{2}}{9}$$

$$\cos 2\theta = 2\cos^2\theta - 1 = 2\left(\frac{1}{3}\right)^2 - 1 = -\frac{7}{9}$$

$$\tan 2\theta = \frac{2\tan\theta}{1 - \tan^2\theta} = \frac{2\cdot 2\sqrt{2}}{1 - \left(2\sqrt{2}\right)^2} = \frac{4\sqrt{2}}{1 - 8} = -\frac{4\sqrt{2}}{7}$$

EXAMPLE 4 **Using the Double-Angle Formulas to Evaluate the Trigonometric Functions of a Double Angle**

Evaluate each trigonometric function of 2θ if $\csc\theta = -13/12$ and $\pi < \theta < 3\pi/2$.

SOLUTION

Let (x, y) be a point on the terminal side of θ.

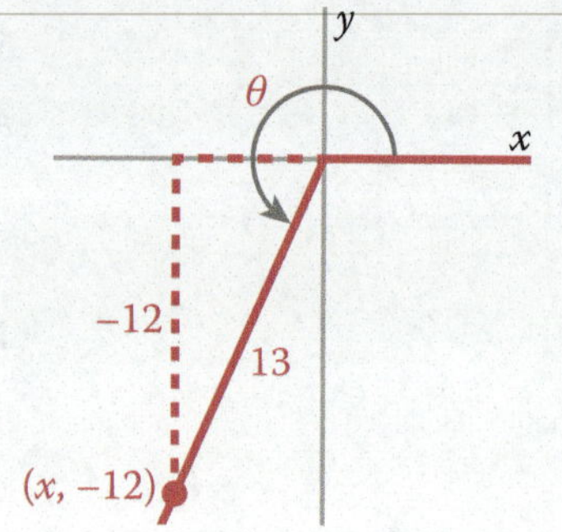

The angle θ is in quadrant III (because $\pi < \theta < 3\pi/2$), so $x < 0$ and $y < 0$.

REMEMBER

r is the distance from the origin to the point (x, y) on the terminal side of the angle, so r > 0.

Since $\csc\theta = -13/12 = r/y$, $r = 13$ and $y = -12$.

Use $r = \sqrt{x^2 + y^2}$ to find x.

$$13 = \sqrt{x^2 + (-12)^2} \;\Rightarrow\; 169 = x^2 + 144 \;\Rightarrow\; x^2 = 25 \;\Rightarrow\; x = \pm 5$$

Since $x < 0$, $x = -5$.

REMEMBER

θ is in quadrant III, so sin θ < 0, cos θ < 0, and tan θ > 0.

Use x, y, and r to evaluate $\sin\theta$, $\cos\theta$, and $\tan\theta$.

$\sin\theta = -12/13$ $\quad\cos\theta = -5/13$ $\quad\tan\theta = -12/(-5) = 12/5$

Use the double-angle formulas to evaluate the trigonometric functions of 2θ.

$$\sin 2\theta = 2\sin\theta\cos\theta = 2(-12/13)(-5/13) = 120/169$$

$$\cos 2\theta = 1 - 2\sin^2\theta = 1 - 2(-12/13)^2 = 1 - 2(144/169) = -119/169$$

$$\tan 2\theta = \frac{2\tan\theta}{1-\tan^2\theta} = \frac{2(12/5)}{1-(12/5)^2} = \frac{24/5}{1-(144/25)} = \frac{24/5}{-119/25} = -\frac{120}{119}$$

By the reciprocal identities, $\csc 2\theta = 169/120$, $\sec 2\theta = -169/119$, and $\cot 2\theta = -119/120$.

The Quadrant of a Double Angle

Multiplying the interval of an angle θ by 2 gives the interval for the double angle 2θ. For example, if θ is in quadrant II, then the interval containing θ is $\pi/2 < \theta < \pi$. Multiplying the interval of θ by 2 gives the interval for 2θ.

$$2(\pi/2) < 2(\theta) < 2(\pi) \;\Rightarrow\; \pi < 2\theta < 2\pi$$

Therefore, if θ is in quadrant II, then 2θ is in either quadrant III or quadrant IV.

NOTICE THAT

In Example 4, sin 2x > 0, as expected, since x is in (π, 3π/2) (quadrant III). Therefore, 2x is in (2π, 3π) (quadrants I or II), where sine must be positive.

If the interval of 2θ is limited to two quadrants, then the sign of one of the trigonometric functions of 2θ can be determined. For example, if θ is in quadrant II (and so 2θ is in quadrant III or IV), then $\sin 2\theta < 0$ because sine is negative in quadrants III and IV. However, knowing that θ is in quadrant II does not provide conclusive information regarding the signs of cosine or tangent of 2θ because these functions are positive in one quadrant, but negative in the other.

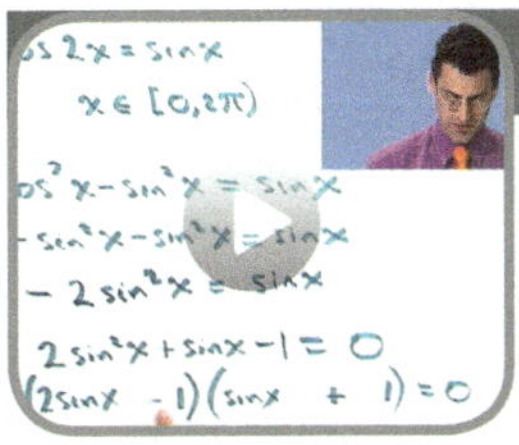

6.4.3 Using Double-Angle Formulas to Solve Equations

An equation involving trigonometric functions of a double angle (i.e., 2θ) can be solved by using algebraic techniques to isolate the trigonometric function and then using the inverse of the trigonometric function, as demonstrated in Section 6.2. An equation involving trigonometric functions of a single angle θ *and* of a double angle 2θ should first be rewritten using a double-angle formula, and then solved in a similar manner.

EXAMPLE 5 Solving a Trigonometric Equation Involving a Double Angle

REMEMBER

For $\sin x$ and $\cos x$, the domain is $(-\infty, \infty)$, and the range is $[-1, 1]$.

Find all solutions to $\sin 2t + \sin t + 4\cos t = -2$ within the interval $[0, 2\pi)$.

SOLUTION

The equation involves trigonometric functions of the single angle t and of the double angle $2t$. So, use the sine of a double-angle formula to write $\sin 2t$ as the sine of a single angle t, and then solve as usual.

$\sin 2t + \sin t + 4\cos t = -2$	
$2\sin t\cos t + \sin t + 4\cos t = -2$	*Sine of a Double-Angle Formula*
$2\sin t\cos t + 4\cos t + \sin t + 2 = 0$	*Add 2 to each side. Group the cosine terms.*
$(2\cos t)(\sin t + 2) + \sin t + 2 = 0$	*Factor 2 cos t from the first two terms.*
$(\sin t + 2)(2\cos t + 1) = 0$	*Factor out sin t + 2.*
$\sin t + 2 = 0$ or $2\cos t + 1 = 0$	*Set each factor equal to 0.*
$\sin t = -2$ $\quad$ $\cos t = -1/2$	*Solve for sin t and for cos t.*

$\sin t = -2$ has no real solution because -2 is not within the range of $\sin t$.

Solve the other equation $\cos t = -1/2$.

$\cos t = -1/2$ when $t = 2\pi/3$ or $t = 4\pi/3$.

Therefore, the solutions of $\sin 2t + \sin t + 4\cos t = -2$ within $[0, 2\pi)$ are $t = 2\pi/3$ and $t = 4\pi/3$.

The graph $y = \cos t$ and $y = -1/2$ (Figure 6.4a) confirms that there are no additional solutions within $[0, 2\pi)$.

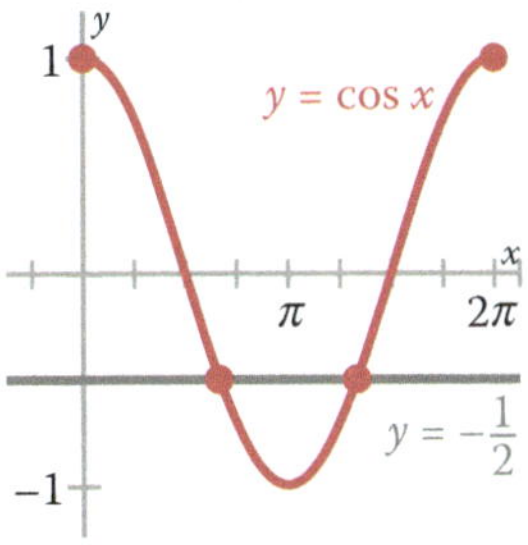

Figure 6.4a

6.4.4 Using a Power-Reducing Formula

We have seen that the double-angle formulas can be used to write a trigonometric function of a double angle 2θ as a trigonometric function of a single angle θ. Similarly, the **power-reducing formulas** (which are derived from the double-angle formulas) can be used to write an expression involving a power of a trigonometric function as an expression involving only powers of 1.

Power-Reducing Formulas

$$\sin^2\theta = \frac{1-\cos 2\theta}{2} \qquad \cos^2\theta = \frac{1+\cos 2\theta}{2} \qquad \tan^2\theta = \frac{1-\cos 2\theta}{1+\cos 2\theta}$$

EXAMPLE 6 Reducing a Power of a Trigonometric Function

Write $\cos^3 x - \sin^3 x$ as an expression with only first powers.

SOLUTION

Write each term as a product containing a power of 2, then use a power-reducing formula.

$$\begin{aligned}
\cos^3 x - \sin^3 x &= \cos x \cos^2 x - \sin x \sin^2 x \\
&= \cos x\,\frac{1+\cos 2x}{2} - \sin x\,\frac{1-\cos 2x}{2} && \textit{Power-Reducing Formulas} \\
&= \frac{\cos x + \cos x \cos 2x}{2} - \frac{\sin x - \sin x \cos 2x}{2} && \textit{Multiply.} \\
&= \frac{\cos x + \cos x \cos 2x - \sin x + \sin x \cos 2x}{2} && \textit{Add.}
\end{aligned}$$

6.4.5 Using Half-Angle Formulas to Find Exact Values

The **half-angle formulas** are derived from the power-reducing formulas by replacing θ with $\theta/2$. These formulas can be used to write an expression involving a trigonometric function of a half of an angle as an expression involving the single angle.

Half-Angle Formulas

$$\sin\frac{\theta}{2} = \pm\sqrt{\frac{1-\cos\theta}{2}} \qquad \cos\frac{\theta}{2} = \pm\sqrt{\frac{1+\cos\theta}{2}} \qquad \tan\frac{\theta}{2} = \frac{1-\cos\theta}{\sin\theta}$$

The signs of $\sin\theta/2$ and $\cos\theta/2$ depend on the quadrant containing the angle $\theta/2$.

EXAMPLE 7 Using a Half-Angle Formula

Find the exact value of $\cos(-22.5°)$.

SOLUTION

Write $-22.5°$ as a half angle (i.e., as a fraction with denominator 2). $-22.5° = -45°/2$

IMPORTANT

$-22.5°$ is in quadrant IV, so $\cos(-22.5°) > 0$.

Use the cosine of a half-angle formula to evaluate $\cos(-45°/2)$.

$$\begin{aligned}
\cos\left(-\frac{45°}{2}\right) &= \sqrt{\frac{1+\cos(-45°)}{2}} && \textit{Cosine of a Half-Angle Formula} \\
&= \sqrt{\frac{1+\frac{\sqrt{2}}{2}}{2}} && \textit{Evaluate } \cos(-45°). \\
&= \sqrt{\frac{2+\sqrt{2}}{4}} && \textit{Multiply by 2/2.} \\
&= \frac{\sqrt{2+\sqrt{2}}}{2} && \textit{Simplify the radical.}
\end{aligned}$$

CHECK

IMPORTANT

The calculator must be in degree mode.

Use a calculator to evaluate cos (−22.5°) and to simplify $\frac{\sqrt{2+\sqrt{2}}}{2}$.

$$\cos(-22.5°) = 0.9238\ldots \qquad \frac{\sqrt{2+\sqrt{2}}}{2} = 0.9238\ldots \quad ✔$$

EXAMPLE 8 Using a Half-Angle Formula and a Triangle to Evaluate the Trigonometric Functions of a Half Angle

Suppose θ is an acute angle in a right triangle with an opposite side of length 3 and an adjacent side of length 4. Evaluate the trigonometric functions of $\theta/2$.

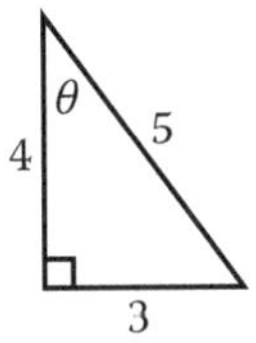

Figure 6.4b

SOLUTION

By the Pythagorean Theorem, if the legs of a right triangle have lengths 3 and 4, then the hypotenuse is 5, as shown in Figure 6.4b.

Evaluate sin θ and cos θ. $\quad \sin\theta = \text{opp/hyp} = 3/5 \qquad \cos\theta = \text{adj/hyp} = 4/5$

Substitute these values into the half-angle formulas.

TIP

Only the values of sin θ and cos θ are needed to evaluate the sine, cosine, and tangent of θ/2 using the half-angle formulas.

Since θ is an acute angle, $\theta/2$ is in quadrant I. Therefore, $\sin\theta > 0$ and $\cos\theta > 0$.

$$\sin\frac{\theta}{2} = \sqrt{\frac{1-\cos\theta}{2}} = \sqrt{\frac{1-4/5}{2}} = \sqrt{\frac{1/5}{2}} = \sqrt{\frac{1}{10}} = \frac{1}{\sqrt{10}}$$

$$\cos\frac{\theta}{2} = \sqrt{\frac{1+\cos\theta}{2}} = \sqrt{\frac{1+4/5}{2}} = \sqrt{\frac{9/5}{2}} = \sqrt{\frac{9}{10}} = \frac{3}{\sqrt{10}}$$

$$\tan\frac{\theta}{2} = \frac{1-\cos\theta}{\sin\theta} = \frac{1-4/5}{3/5} = \frac{1/5}{3/5} = \frac{1}{3}$$

By the reciprocal identities, $\csc\frac{\theta}{2} = \sqrt{10}$, $\sec\frac{\theta}{2} = \frac{\sqrt{10}}{3}$, and $\cot\frac{\theta}{2} = 3$.

EXAMPLE 9 Using a Half-Angle Formula

Use a half-angle formula to evaluate sin $t/2$, if $\sin t = -1/2$ and $-\frac{\square}{2} < t < 0$.

SOLUTION

Use the sine of a half-angle formula $\sin\frac{\theta}{2} = \pm\sqrt{\frac{1-\cos\theta}{2}}$ to find sin $t/2$.

First, evaluate cos t.

Since $\sin t = -1/2 = y/r$, and $r > 0$, $y = -1$ and $r = 2$.

Use $r = \sqrt{x^2+y^2}$ to find x. Note that $x > 0$ because $-\frac{\pi}{2} < t < 0$ (i.e., t is in quadrant IV).

$$2 = \sqrt{x^2+(-1)^2} \Rightarrow 4 = x^2+1 \Rightarrow x = \sqrt{3}$$

Therefore, $\cos t = \frac{x}{r} = \frac{\sqrt{3}}{2}$.

Determine the sign of sin $(t/2)$.

Dividing the interval for t by 2 gives the interval for $t/2$.

$$-\frac{\pi}{2} < t < 0 \;\Rightarrow\; -\frac{\pi}{4} < \frac{t}{2} < 0$$

Therefore, $t/2$ is in quadrant IV and $\sin(t/2) < 0$.

Now evaluate sin $(t/2)$.

Substitute the value of $\cos t$ into the sine of a half-angle formula, using only the negative square root, because $\sin(t/2) < 0$.

$$\sin\frac{t}{2} = -\sqrt{\frac{1-\frac{\sqrt{3}}{2}}{2}} = -\sqrt{\frac{\frac{2-\sqrt{3}}{2}}{2}} = -\sqrt{\frac{2-\sqrt{3}}{4}} = -\frac{\sqrt{2-\sqrt{3}}}{2}$$

ALTERNATIVE METHOD

The value of sin $(t/2)$ can be found using a difference of angles formula instead of a half-angle formula, first by finding the exact value of $t/2$ using the fact that $\sin t = -1/2$ and $-\frac{\pi}{2} < t < 0$.

It is given that t is in quadrant IV. And in quadrant IV, $\sin t = -1/2$ when $t = 11\pi/6 + 2\pi n$. More specifically, we know that t is between $-\pi/2$ and 0. Therefore, $t = -\pi/6$. By dividing each side of $t = -\pi/6$ by 2, we see that $t/2 = -\pi/12$. Therefore, $\sin(t/2) = \sin(-\pi/12)$.

TIP

The only value of $11\pi/6 + 2\pi n$ that is between $-\pi/2$ and 0 is $-\pi/6$.
$11\pi/6 - 2\pi(1) = -\pi/6$

The value of sin $(-\pi/12)$ can be found using the sine of a difference of angles formula.

$$\sin\left(-\frac{\pi}{12}\right) = \sin\left(\frac{\pi}{6}-\frac{\pi}{4}\right) = \sin\frac{\pi}{6}\cos\frac{\pi}{4} - \cos\frac{\pi}{6}\sin\frac{\pi}{4} = \frac{1}{2}\cdot\frac{\sqrt{2}}{2} - \frac{\sqrt{3}}{2}\cdot\frac{\sqrt{2}}{2} = \frac{\sqrt{2}-\sqrt{6}}{4}$$

Although the numbers $-\frac{\sqrt{2-\sqrt{3}}}{2}$ and $\frac{\sqrt{2}-\sqrt{6}}{4}$ are equal, this may not be obvious visually.

Determining the Equivalence of Numeric Expressions

Using a calculator to simplify each expression, we find that $-\frac{\sqrt{2-\sqrt{3}}}{2} = -0.25881\ldots$ and $\frac{\sqrt{2}-\sqrt{6}}{4} = -0.25881\ldots$, so we can infer that $-\frac{\sqrt{2-\sqrt{3}}}{2} = \frac{\sqrt{2}-\sqrt{6}}{4}$.

However, since it is possible that there could be some difference in decimal digits much further to the right, this does not prove that the two numbers are equal.

A more accurate numeric procedure for determining equivalence is to subtract the numbers (without using a calculator). If the difference of two numbers is 0, then the numbers are equal.

6.4.6 Using Product-Sum Formulas

Product-to-Sum Formulas

The **product-to-sum formulas** can be verified using the sum and difference formulas. These formulas can be used to simplify an expression involving a product of two trigonometric functions (sine or cosine).

Product-to-Sum Formulas

$$\sin\theta_1 \sin\theta_2 = \frac{1}{2}[\cos(\theta_1 - \theta_2) - \cos(\theta_1 + \theta_2)]$$

$$\cos\theta_1 \cos\theta_2 = \frac{1}{2}[\cos(\theta_1 - \theta_2) + \cos(\theta_1 + \theta_2)]$$

$$\sin\theta_1 \cos\theta_2 = \frac{1}{2}[\sin(\theta_1 + \theta_2) + \sin(\theta_1 - \theta_2)]$$

$$\cos\theta_1 \sin\theta_2 = \frac{1}{2}[\sin(\theta_1 + \theta_2) - \sin(\theta_1 - \theta_2)]$$

EXAMPLE 10 Using a Product-to-Sum Formula

Use a product-to-sum formula to rewrite each expression and simplify, if possible.

A. $\sin 6x \sin x$ **B.** $\sin 5t \cos 5t$

SOLUTION

A. Use the product-to-sum formula for $\sin\theta_1 \sin\theta_2$ where $\theta_1 = 6x$ and $\theta_2 = 6x$.

$$\sin 6x \sin x = \frac{1}{2}[\cos(6x - x) - \cos(6x + x)] \quad \textit{Product-to-Sum Formula}$$

$$= \frac{1}{2}[\cos 5x - \cos 7x] \quad \textit{Simplify (combine the like terms).}$$

$$= \frac{1}{2}\cos 5x - \frac{1}{2}\cos 7x \quad \textit{Distribute.}$$

B. Use the product-to-sum formula for $\sin\theta_1 \cos\theta_2$, where $\theta_1 = 5t$ and $\theta_2 = 5t$.

$$\sin 5t \cos 5t = \frac{1}{2}[\sin(5t + 5t) + \sin(5t - 5t)] \quad \textit{Product-to-Sum Formula}$$

$$= \frac{1}{2}[\sin 10t + \sin 0] \quad \textit{Simplify (combine the like terms).}$$

$$= \frac{1}{2}[\sin 10t + 0] \quad \textit{Evaluate sin 0.}$$

$$= \frac{1}{2}\sin 10t \quad \textit{Multiply.}$$

Sum-to-Product Formulas

The **sum-to-product formulas** reverse the process and allow for a sum or difference of sine functions or cosine functions to be written as a product of sine or cosine functions.

Sum-to-Product Formulas

$$\sin\theta_1 + \sin\theta_2 = 2\sin\left(\frac{\theta_1+\theta_2}{2}\right)\cos\left(\frac{\theta_1-\theta_2}{2}\right)$$

$$\sin\theta_1 - \sin\theta_2 = 2\cos\left(\frac{\theta_1+\theta_2}{2}\right)\sin\left(\frac{\theta_1-\theta_2}{2}\right)$$

$$\cos\theta_1 + \cos\theta_2 = 2\cos\left(\frac{\theta_1+\theta_2}{2}\right)\cos\left(\frac{\theta_1-\theta_2}{2}\right)$$

$$\cos\theta_1 - \cos\theta_2 = -2\sin\left(\frac{\theta_1+\theta_2}{2}\right)\sin\left(\frac{\theta_1-\theta_2}{2}\right)$$

EXAMPLE 11 Using a Sum-to-Product Formula

Use a sum-to-product formula to rewrite each expression and simplify, if possible.

A. $\sin 4t + \sin 6t$ **B.** $\cos x - \cos(-x)$

SOLUTION

A. Use the sum-to-product formula for $\sin\theta_1 + \sin\theta_2$, where $\theta_1 = 4t$ and $\theta_2 = 6t$.

$$\sin 4t + \sin 6t = 2\sin\left(\frac{4t+6t}{2}\right)\cos\left(\frac{4t-6t}{2}\right) \quad \textit{Sum-to-Product Formula}$$

$$= 2\sin\left(\frac{10t}{2}\right)\cos\left(\frac{-2t}{2}\right) \quad \textit{Simplify (combine the like terms).}$$

$$= 2\sin 5t\cos(-t) \quad \textit{Simplify (divide).}$$

$$= 2\sin 5t\cos t \quad \cos(-t) = \cos t$$

B. Use the product-to-sum formula for $\cos\theta_1 - \cos\theta_2$, where $\theta_1 = x$ and $\theta_2 = -x$.

$$\cos x - \cos(-x) = -2\sin\left(\frac{x+(-x)}{2}\right)\sin\left(\frac{x-(-x)}{2}\right) \quad \textit{Sum-to-Product Formula}$$

$$= -2\sin\left(\frac{0}{2}\right)\sin\left(\frac{2x}{2}\right) \quad \textit{Simplify (combine the like terms).}$$

$$= -2\sin 0\sin x \quad \textit{Simplify (divide).}$$

$$= 0 \quad \textit{Evaluate sin 0 and multiply.}$$

Note that the expression in **Example 11B** can be simplified using the fact that cosine is an even function, that is, $\cos(-x) = \cos x$.

$$\cos x - \cos(-x) = \cos x - \cos x = 0$$

SECTION 6.4 EXERCISES

Warm Up

Use Pythagorean identities to simplify each expression.

1. $3\cos^2 x - 3\sin^2 x - 3$
2. $\tan^2 x - \sec^2 x$
3. $5\csc^2 x - 5\cot^2 x$

Just the Facts

Fill in the blank to complete each formula.

4. $\sin 2\theta =$ _____
5. $1 - 2\sin^2\theta =$ _____
6. $\tan 2\theta =$ _____
7. $\dfrac{1-\cos 2\theta}{1+\cos 2\theta} =$ _____
8. $\sin(\theta/2) =$ _____
9. $\pm\sqrt{\dfrac{1+\cos\theta}{2}} =$ _____
10. $\dfrac{1-\cos 2\theta}{2} =$ _____

Essential Skills

In Exercises 1–4, use a double-angle formula to evaluate each expression.

1. $\tan(5\pi/3)$
2. $\cos(7\pi/2)$
3. $\cos(5\pi/2)$
4. $\sin(7\pi/3)$

In Exercises 5–10, use a double-angle formula to rewrite each expression.

5. $20\sin\theta\cos\theta$
6. $-8 + 16\sin^2\theta$
7. $10\sin^2\theta - 10\cos^2\theta$
8. $-7 + 7(\cos\theta - \sin\theta)^2$
9. $\dfrac{20\tan\theta}{1-\tan^2\theta}$
10. $\dfrac{1}{\tan x - 1} + \dfrac{1}{\tan x + 1}$

In Exercises 11–14, evaluate the six trigonometric functions of 2θ.

11.
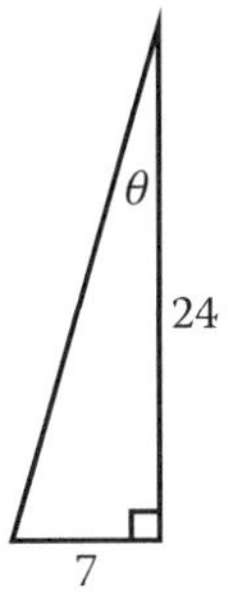

12.
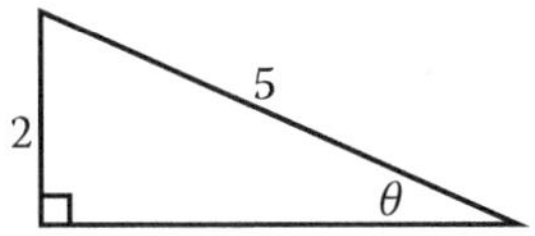

13.
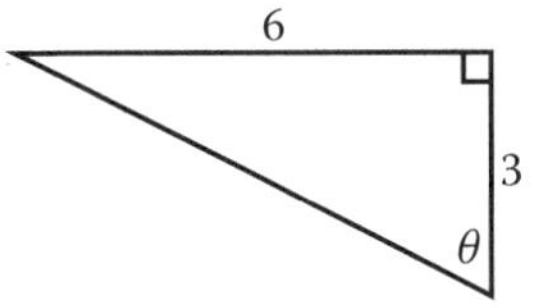

14.
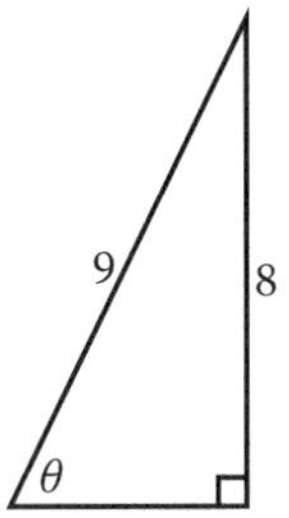

In Exercises 15–20, evaluate the six trigonometric functions of 2θ with the given information.

15. $\sec\theta = 5/3$ and $3\pi/2 < \theta < 2\pi$
16. $\tan\theta = -6/5$ and $\pi/2 < \theta < \pi$
17. $\cot\theta = -8/7$ and $3\pi/2 < \theta < 2\pi$
18. $\sec\theta = -\dfrac{\sqrt{61}}{5}$ and $\pi < \theta < 3\pi/2$
19. $\sin\theta = 5/13$ and $\pi/2 < \theta < \pi$
20. $\csc\theta = -\dfrac{\sqrt{13}}{3}$ and $\pi < \theta < 3\pi/2$

In Exercises 21–28, find all exact solutions to each equation within the interval [0, 2π).

21. $\sin 2x \sin x - \cos x = 0$
22. $\sin 2x = 2\sin x$
23. $\sin 2x = -\cos x$
24. $\sin 2x = 3/2\tan x$
25. $\cos 2x + 3\sin x + 1 = 0$
26. $\cos 2x = \cos x$
27. $\cos 2x + \cos 4x = 0$
28. $\cos 2x = \sin x$

In Exercises 29–32, write each trigonometric expression with only first powers.

29. $\dfrac{\cos^4 t}{16}$

30. $\dfrac{\tan^4 t}{4}$

31. $\cos^4 \theta - \sin^4 \theta$

32. $\sin^6 \theta$

In Exercises 33–38, find the exact value of each trigonometric expression.

33. tan 105°

34. cos 22.5°

35. sin (–15°)

36. sin 165°

37. tan (–157.5°)

38. cos (–67.5°)

In Exercises 39–42, evaluate the six trigonometric functions of $\theta/2$ for each given triangle.

39.

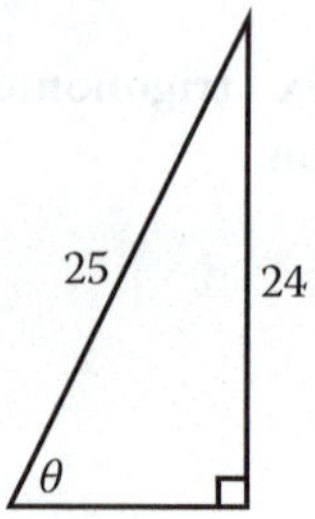

40.

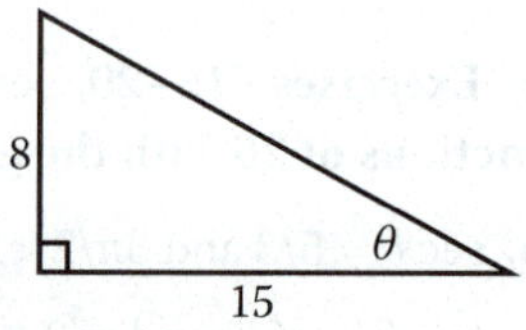

41.

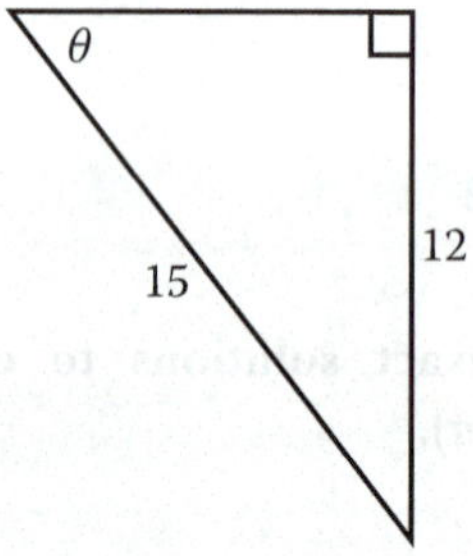

42.

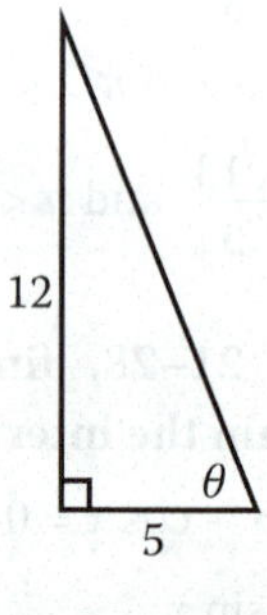

In Exercises 43–46, use a half-angle formula to evaluate the given trigonometric expression of $\theta/2$ using the given information.

43. $\sin(\theta/2)$, given $\tan\theta = -5/12$ and $\pi/2 < \theta < \pi$

44. $\cos(\theta/2)$, given $\cos\theta = -3/4$ and $\pi/2 < \theta < \pi$

45. $\tan(\theta/2)$, given $\sin\theta = -5/8$ and $3\pi/2 < \theta < 2\pi$

46. $\sin(\theta/2)$, given $\sec\theta = -7/4$ and $\pi < \theta < 3\pi/2$

In Exercises 47–56, use a product-to-sum formula to rewrite each expression and simplify, if possible.

47. $\sin 20x \cos 7x$

48. $\cos 3x \sin 5x$

49. $\cos 4x \cos 2x$

50. $\sin 11x \sin 6x$

51. $\cos 13x - \cos 3x$

52. $\sin 7x + \sin 3x$

53. $\cos 5x + \cos 3x$

54. $\cos 9x + \cos 3x$

55. $\sin x + \sin 5x$

56. $\sin 11x - \sin 3x$

Extensions

57. The area of a figure is given by the equation $A = \dfrac{1+\cos 2\theta}{2}$. In the interval $0 \le \theta \le \pi$, at what value of θ is the area of the figure equal to 0?

58. Suppose (x, y) is a point on the circle with equation $x^2 + y^2 = 9$, then the points (x, y), $(-x, y)$, $(-x, -y)$, and $(x, -y)$ determine a rectangle. Also, for each value of θ, $x = 3\cos\theta$ and $y = 3\sin\theta$. Since the area of the rectangle pictured below is $2x \cdot 2y$, the area can be expressed in terms of θ as $18 \sin 2\theta$. In the interval $0 \le \theta \le \pi$, what is the value of θ that maximizes the area of the rectangle?

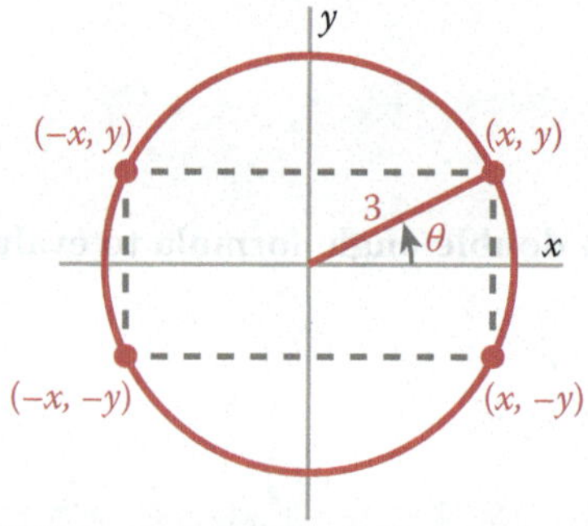

59. What is the area of the rectangle in the figure below in terms of r and θ?

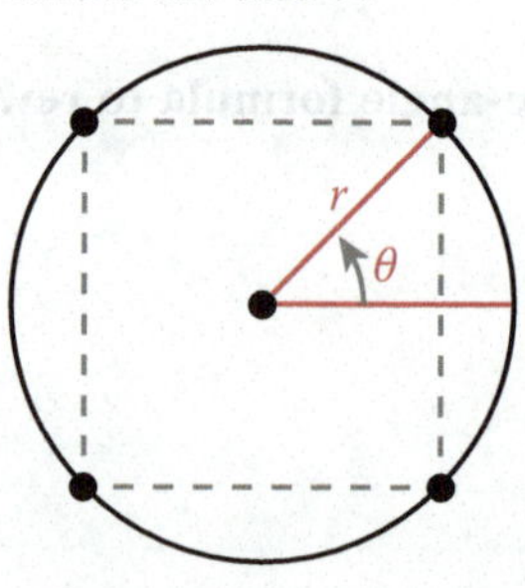

60. What is the difference in areas of squares $ABCD$ and $DEFG$ in the figure below in terms of h and θ?

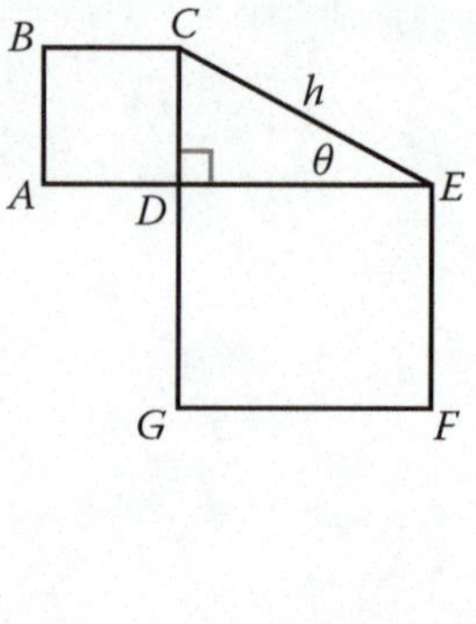

61. What is the area of an arbitrary right triangle in terms of the hypotenuse, c, and one acute angle, A?

In Exercises 62–64, write each trigonometric expression with only first powers.

62. $\tan^4 \theta$

63. $\tan^3 \theta$

64. $\sin^4 2\theta$

In Exercises 65–69, find all exact solutions to each equation within the interval $[0, 2\pi)$.

65. $\sin 2\theta + 4 \sin \theta + 2 \cos \theta = -4$

66. $8 \cos^4 \theta = \cos 4\theta + 4 \cos 2\theta + 3$

67. $\sec^4 2\theta = 4$

68. $\tan^2 2\theta - 1 = 0$

69. $\csc^4 2\theta - 4 = 0$

70. Suppose (x, y) is a point along the arc in the first quadrant of a circle represented by $x^2 + y^2 = 900$. At any location on the circle, a ray from the origin through (x, y) determines an angle θ with the x-axis. A right triangle is formed by a horizontal line through (x, y), a vertical line through (x, y), and a diameter of the circle. Find an expression for the area of the triangle in terms of $\cos \theta$ and $\sin \theta$.

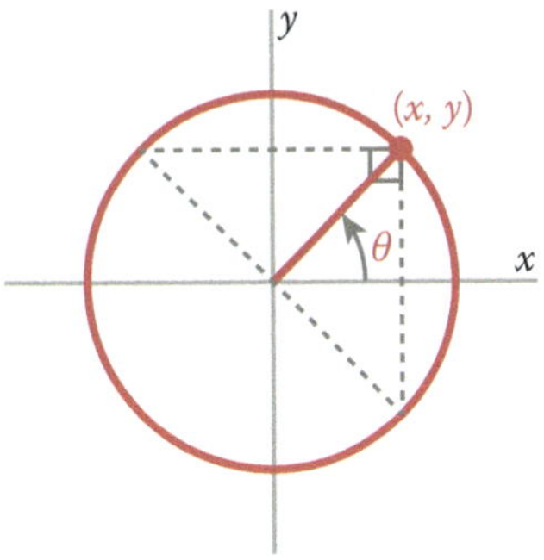

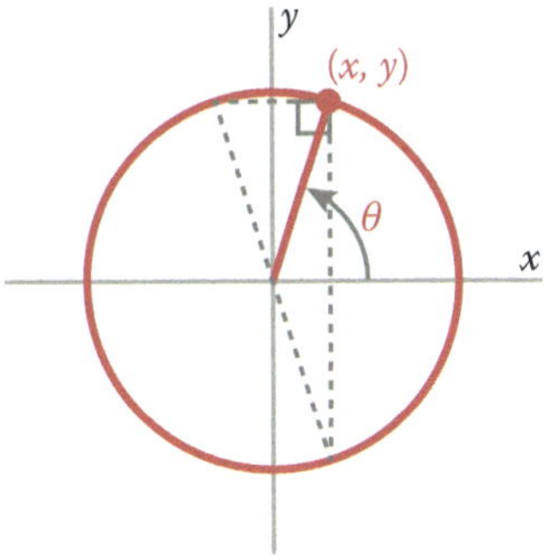

71. Suppose (x, y) is a point on the circle represented by $x^2 + y^2 = 100$. At any location on the circle, a ray from the origin through (x, y) determines an angle θ with the x-axis. The points (x, y), $(-x, y)$, $(-x, 0)$, and $(x, 0)$ determine a rectangle. Find an expression for the shaded portion of the circle in terms of θ.

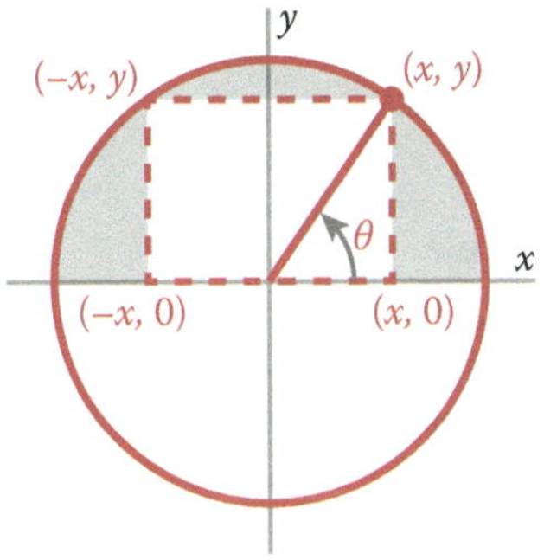

In Exercises 72–75, find all solutions to each equation within the interval $[0, 2\pi)$. Round each answer to the nearest hundredth.

72. $3 \sin 2\theta = 2 \cos \theta$

73. $5 \cos 2\theta + \sin^2 \theta = 0$

74. $6 \cos^2 \theta + 3 \cos 2\theta = 0$

75. $6 \sin 2\theta = -5 \sin \theta$

CHAPTER 6 REVIEW EXERCISES

Section 6.1 Fundamental Trigonometric Identities

1. Use $\csc \theta = -12/5$ and $\pi < \theta < 3\pi/2$ to evaluate all other trigonometric functions.
2. Expand and simplify, if possible.
$(\cot x - \csc x)(\cos x + 1)$
3. Factor and simplify, if possible.
$\tan x \sin x \cos x + \cot x \sin x \cos x$
4. Prove $\dfrac{\cos^2 x + \tan^2 x - 1}{\sin^2 x} = \tan^2 x$, if possible.
5. Prove $\dfrac{\sec x + \csc x}{\cot x + \tan x} = \cos x + \sin x$, if possible.

Section 6.2 Solving Trigonometric Equations

In Exercises 6–8, solve the equation.

6. $3 \tan (x/2) + 3 = 0$
7. $2 \sin^2 x - 5 \sin x + 2 = 0$
8. $\tan^3 x + \tan^2 x - 3 \tan x - 3 = 0$

In Exercises 9–10, find all solutions to the equation, within the interval $[0, 2\pi)$. Round to the nearest thousandth.

9. $3 \cos^2 x - 5 \cos x - 4 = 0$
10. $10 \sec^2 x - 10 = 7 \tan x + 2$

Section 6.3 The Sum and Difference Formulas

In Exercises 11–12, if possible, find the exact value of the trigonometric function.

11. $\tan 255°$
12. $\sin 11\pi/12$
13. Complete the identity.
$\cos (17\pi - x)$
14. Prove that $\sin (x + y) - \sin (x - y) = 2 \cos x \sin y$ is an identity, if possible.
15. Prove that $\dfrac{\sin (x + y) - \sin (x - y)}{\cos (x + y) + \cos(x - y)} = \tan y$ is an identity, if possible.

Section 6.4 Double-Angle, Half-Angle, and Product-Sum Formulas

16. Evaluate the six trigonometric functions of 2θ and $\theta/2$ using the given triangle.

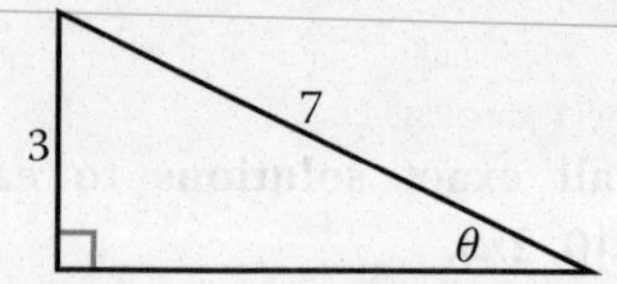

17. Evaluate the six trigonometric functions of 2θ and $\theta/2$, if $\cot \theta = -2/9$ and $\pi/2 < \theta < \pi$.
18. Find all exact solutions to $\sin 2x \cot x - 2 \sin^2 x = 1$ within the interval $[0, 2\pi)$.
19. Write $3 \cos^3 x \cos^2 x$ as an expression with only first powers.
20. Use a product-to-sum formula to rewrite $\cos 4x - \cos 6x$, and simplify, if possible.

Applications of Trigonometry

Chapter

7

7.1 THE LAW OF SINES

OBJECTIVES

- Use the Law of Sines to solve an oblique triangle.
- Determine whether zero, one, or two triangles exist with given measurements.
- Find the area of an oblique triangle.

PREREQUISITE VOCABULARY TERMS

acute angle
obtuse angle
supplementary angles
Triangle Sum Theorem
trigonometric functions

The six trigonometric functions were used in Chapter 5 to determine side lengths and angle measures of right triangles. In this chapter, we will determine side lengths and angle measures of **oblique triangles**, which are triangles that have no right angles.

IMPORTANT

An oblique triangle has no right angles, so the sides are not related by the Pythagorean Theorem (i.e., $a^2 + b^2 \neq c^2$), which holds only for right triangles.

The angles of an oblique triangle will typically be represented by the Greek letters α (alpha), β (beta), and γ (gamma). The *corresponding opposite sides* in the oblique triangle will be represented by a, b, and c, respectively (Figure 7.1a). For any triangle with sides a, b, and c, and angles α, β, and γ, we can assume that side a is opposite angle α, side b is opposite angle β, and side c is opposite angle γ.

Oblique Triangle

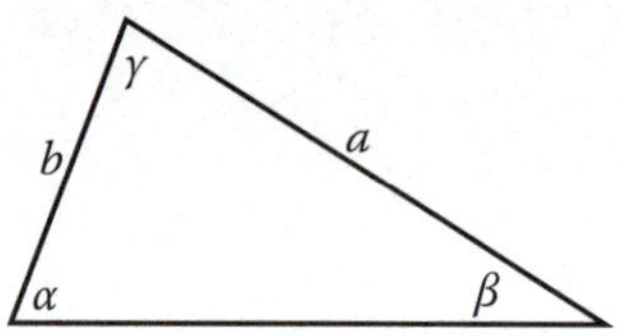

Figure 7.1a

Note that the angle between two specific sides is called the **included angle** (e.g., the included angle for sides b and c is angle α). Similarly, the side between two specific angles is called the **included side** (e.g., the included side for angles α and γ is side b).

7.1.1 Using the Law of Sines Given Two Angles and One Side

For any oblique (or right) triangle, the ratio of the sine of an angle to its opposite side is equal to the ratio of the sine of another angle to that angle's opposite side, as stated in the **Law of Sines**.

The Law of Sines

For any triangle with sides a, b, and c, and respective opposite angles α, β, and γ, $\dfrac{\sin\alpha}{a} = \dfrac{\sin\beta}{b} = \dfrac{\sin\gamma}{c}$ and $\dfrac{a}{\sin\alpha} = \dfrac{b}{\sin\beta} = \dfrac{c}{\sin\gamma}$.

NOTICE THAT

Each ratio is the sine of an angle to that angle's opposite side.

The Law of Sines can be used to solve an oblique triangle (i.e., to find the unknown side and angle measures of an oblique triangle), given three specific measures of the triangle (either two sides and one angle or two angles and one side), as stated in the following cases.

- ASA (angle-side-angle): given two angles and the included side
- AAS (angle-angle-side): given two angles and the side opposite either given angle
- SSA (side-side-angle): given two sides and the angle opposite either given side

REMEMBER

The Triangle Sum Theorem, $\alpha + \beta + \gamma = 180°$, gives the measure of a triangle's third angle when the two other angles are known.

In other words, the Law of Sines gives the unknown measures in an oblique triangle when either two angles and any side (ASA or AAS), or two sides and the non-included angle (SSA) are given.

EXAMPLE 1 Using the Law of Sines to Solve a Triangle

Use the Law of Sines to solve each triangle, if possible. Round the measures to the nearest hundredth, as needed.

A. $\beta = 18°$, $\gamma = 121°$, and $b = 60$ meters **B.** $a = 10$ feet, $c = 14$ feet, and $\beta = 55°$

SOLUTION

A. Sketch the triangle.

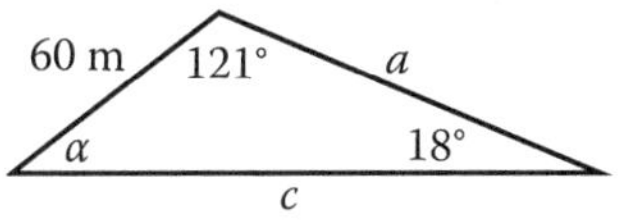

The measures of angle β and opposite side b, as well as angle γ, are given, so the AAS case applies to this triangle and the Law of Sines can be used to solve the triangle.

Use the Triangle Sum Theorem to find the unknown angle α. $\alpha = 180° - (121° + 18°) = 41°$

IMPORTANT

In each equation (proportion), one ratio must include an angle and its opposite side for which both measures are given (known).

Write two equations using the Law of Sines, then solve each to find a and c.

$$\frac{a}{\sin \alpha} = \frac{b}{\sin \beta} \Rightarrow \frac{a}{\sin 41°} = \frac{60}{\sin 18°} \Rightarrow a = \frac{60 \sin 41°}{\sin 18°} \approx 127.38 \text{ m}$$

$$\frac{c}{\sin \gamma} = \frac{b}{\sin \beta} \Rightarrow \frac{c}{\sin 121°} = \frac{60}{\sin 18°} \Rightarrow c = \frac{60 \sin 121°}{\sin 18°} \approx 166.43 \text{ m}$$

B. Sketch the triangle.

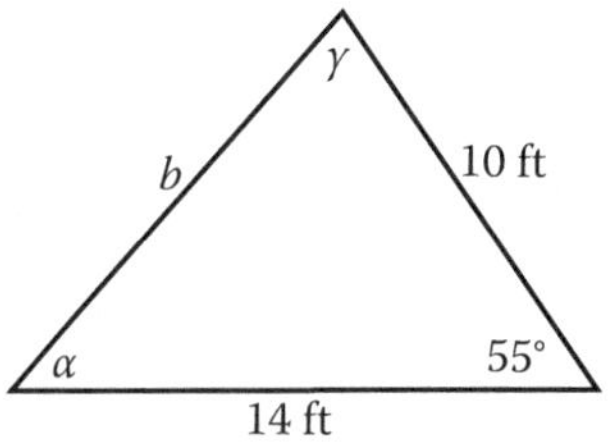

The measures of two sides, a and c, and the included angle β are given. However, since β is the included angle between the given sides (i.e., not the angle opposite one of the given sides), the Law of Sines cannot be used to solve this triangle.

Solving an oblique triangle where two sides and the included angle are given (the SAS case) will be discussed later in this chapter.

For any triangle, the smallest side must be opposite the smallest angle, and the largest side must be opposite the largest angle. Therefore, if a triangle's angles are related by the inequality $\beta < \gamma < \alpha$, then the triangle's sides must be related by the inequality $b < c < a$. These relationships should be confirmed after solving a triangle, as demonstrated below with the triangle solved in **Example 1A** (Figure 7.1b).

IMPORTANT

Consistency of the inequalities of the sides and their opposite angles does not confirm that the measures are correct. However, if the sides are not consistent with their opposite angles, then at least one measure (side or angle) must be incorrect.

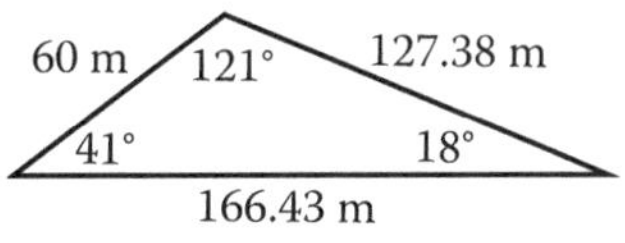

Figure 7.1b

From the triangle in Figure 7.1b:

- 60 is opposite the 18° angle,
- 127.38 is opposite the 41° angle, and
- 166.43 is opposite the 121° angle.

Sides from least to greatest: $60 < 127.38 < 166.43$

Angles from least to greatest: $18° < 41° < 121°$

Since the smallest side (60 m) is opposite the smallest angle (18°) and the largest side (166.43 m) is opposite the largest angle (121°), the lengths of the triangle's sides are consistent with their opposite angles.

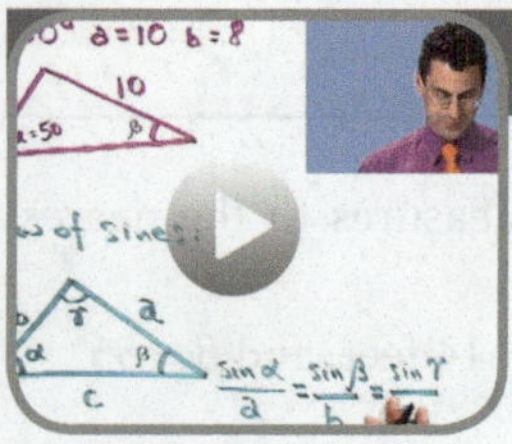

7.1.2 Using the Law of Sines Given One Angle and Two Sides

Three side lengths uniquely determine a triangle, but three angle measures do not. For example, there is exactly one triangle with sides 4, 5, and 8 units, but there are an infinite number of triangles with angle measures 60°, 40°, and 80°.

If the measures of two angles and one side of an oblique triangle are given (i.e., the ASA or AAS case), then there is exactly one triangle that fits these criteria. For example, consider the oblique triangle given in Figure 7.1c, where one angle is 60°, another is 40°, and the included side has length 10 units (ASA). The lengths of sides a and b are unknown, but the point at which these two sides meet is determined by the given angles, thus setting the length of each unknown side. Therefore, exactly one triangle exists with the given angles and side shown in Figure 7.1c.

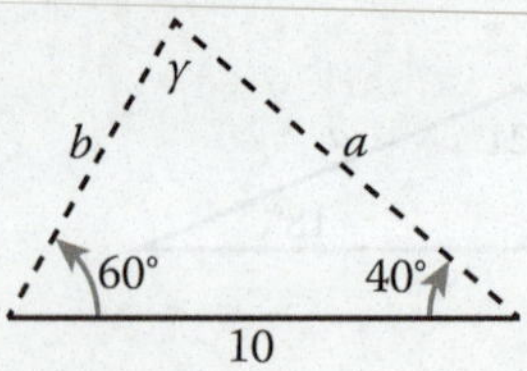

Figure 7.1c

If the measures of two sides and the non-included angle are given (i.e., the SSA case), there can be exactly one triangle, exactly two triangles, or zero triangles with the given measures. In both of the following SSA situations, exactly one triangle exists with the given measures.

The SSA cases where zero or two triangles exist will be discussed further in Topic 7.1.3.

1. The given angle is *obtuse* and its opposite side is greater than the other given side.
2. The given angle is *acute* and its opposite side is greater than or equal to the other given side.

EXAMPLE 2 Using the Law of Sines to Solve a Triangle

Use the Law of Sines to solve a triangle where $a = 10$ feet, $b = 14$ feet, and $\beta = 55°$. Round the measures to the nearest hundredth, as needed.

SOLUTION

Exactly one triangle with the given measures exists because the given angle is acute ($\beta = 55° < 90°$) and its opposite side is greater than the other given side ($b > a$).

IMPORTANT

For an SSA triangle, the measure of the unknown angle opposite the given side must be found first.

Use the Law of Sines to find the measure of the unknown angle, opposite the given side.

$$\frac{\sin \alpha}{a} = \frac{\sin \beta}{b} \quad \textit{Law of Sines}$$

$$\frac{\sin \alpha}{10} = \frac{\sin 55^\circ}{14} \quad \textit{Substitute the given measures.}$$

$$\sin \alpha = \frac{10 \sin 55^\circ}{14} \quad \textit{Solve for } \sin \alpha.$$

$$\alpha = \sin^{-1} \frac{10 \sin 55^\circ}{14} \quad \textit{Take the inverse sine of both sides to solve for } \alpha.$$

$$\alpha \approx 35.81^\circ \quad \textit{Use a calculator to evaluate the inverse sine function.}$$

Use the Triangle Sum Theorem to find the unknown angle γ. $\gamma \approx 180° - (55° + 35.81°) = 89.19°$

Use the Law of Sines find the measure of side c.

$$\frac{c}{\sin \gamma} = \frac{b}{\sin \beta}$$

$$\frac{c}{\sin 89.19^\circ} \approx \frac{14}{\sin 55^\circ}$$

$$c \approx \frac{14 \sin 89.19^\circ}{\sin 55^\circ}$$

$$c \approx 17.09 \text{ ft}$$

TIP

Use b and $\sin \beta$ in one ratio (as opposed to a and $\sin \alpha$) because the exact values of β and b are known.

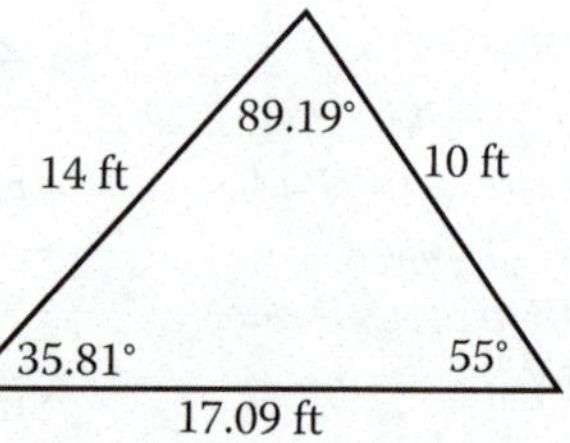

The side lengths in increasing order are consistent with their opposite angles in increasing order.

Therefore, $\alpha \approx 35.81°$, $\gamma \approx 89.19°$, and $c \approx 17.09$ feet.

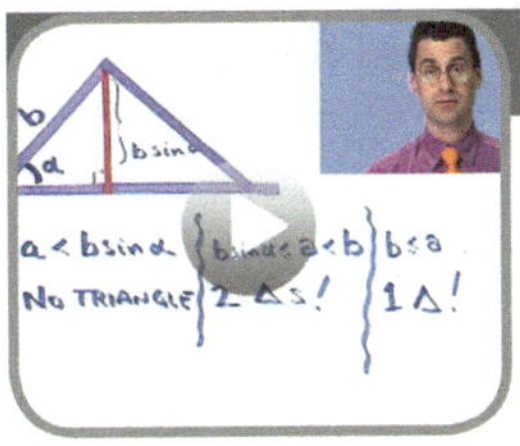

7.1.3 Using the Law of Sines: The Ambiguous Case

SSA is called the **ambiguous case** because the given measures (two side lengths and a non-included angle) might not correspond to exactly one triangle. Instead, zero, one, or even two triangles with the given measures can exist.

To determine whether zero or two triangles exist with the given measures, we compare the triangle's height (i.e., an altitude) to the side opposite the given angle. However, since the height of an oblique triangle is typically not given, we need an expression for the height in terms of a known side and a known angle to make this comparison.

REMEMBER

The base of a triangle can be any of its sides.

Recall that the height h of a triangle is the distance between a vertex and a line containing the opposite side (the triangle's base), as shown in Figure 7.1d.

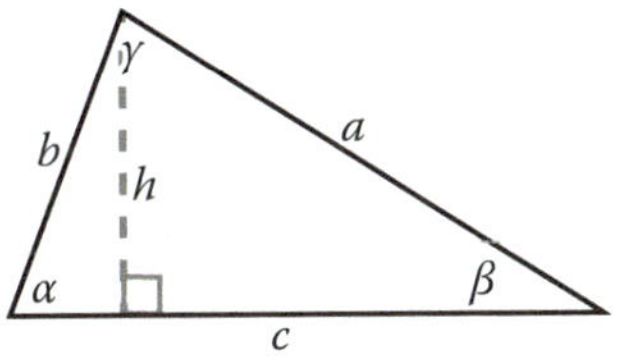

Figure 7.1d

Notice that h is a leg in the right triangle with hypotenuse b, where α is the angle opposite h. Therefore, $\sin \alpha = h/b$. Solving for h gives $h = b \sin \alpha$. Thus, the height of an oblique triangle is given by the expression $b \sin \alpha$.

The criteria for determining whether zero, one, or two triangles exists in an SSA case is summarized in the following tables. Note that α represents the given angle, and a and b represent the given sides, where a is the side opposite the given angle α.

NOTICE THAT

If $a = b \sin \alpha$, then a is the height of the triangle and the triangle is a right triangle.

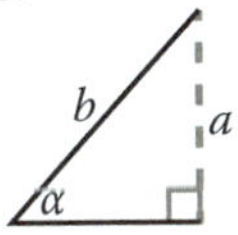

SSA: The Ambiguous Case Given an Acute Angle α

One Oblique Triangle	Two Oblique Triangles	Zero Triangles
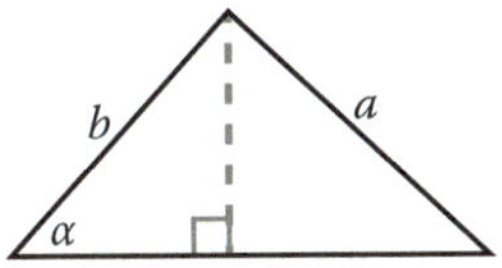	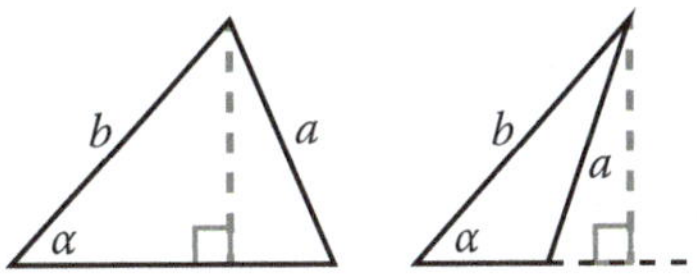	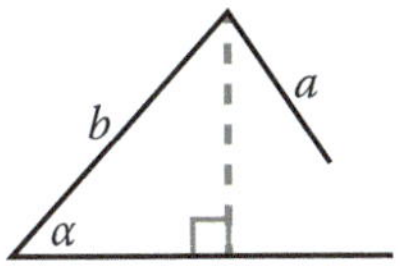
$a \geq b$	$b \sin \alpha < a < b$	$a < b \sin \alpha$
The side opposite the given acute angle is greater than or equal to the other given side.	*The height ($b \sin \alpha$) is less than the side opposite the given acute angle, which is less than the other side.*	*The side opposite the given acute angle is less than the height ($b \sin \alpha$).*

SSA: The Ambiguous Case Given an Obtuse Angle α

One Oblique Triangle	Zero Triangles
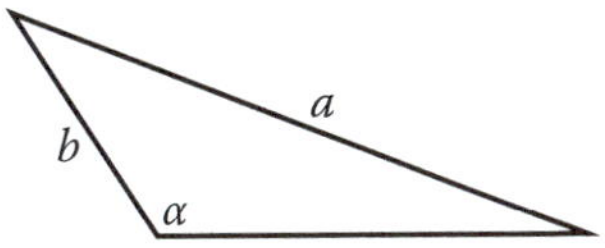	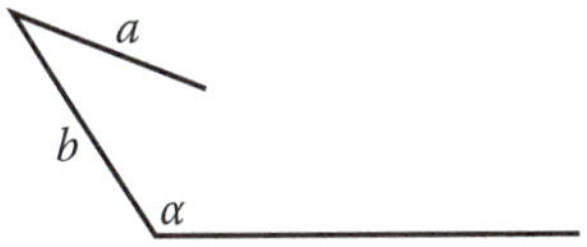
$a > b$	$a \leq b$
The side opposite the given acute angle is greater than the other given side.	*The side opposite the given acute angle is less than or equal to the other given side.*

EXAMPLE 3 Determining the Number of Triangles that Exist (SSA)

Determine whether zero, one, or two triangles exist with the given measures.

A. $a = 12, c = 8, \alpha = 100°$ **B.** $a = 5, c = 7, \gamma = 25°$ **C.** $b = 4, c = 6, \beta = 51°$

SOLUTION

A. The given angle is obtuse. Since $a > c$ (because $12 > 8$), the side opposite the given angle (side a) is greater than the other given side (side c). Thus, there is exactly one oblique triangle with $a = 12$, $c = 8$, and $\alpha = 100°$.

IMPORTANT

The given angle is γ, so c is the side opposite the given angle in this case.

B. The given angle is acute. Since $c > a$ (because $7 > 5$), the side opposite the given angle (side c) is greater than the other given side (side a). Thus, there is exactly one oblique triangle with $a = 5$, $c = 7$, and $\gamma = 25°$.

C. The given angle is acute. Since $b < c$ (because $4 < 6$), the side opposite the given angle (side b) is less than the other given side (side c). Therefore, use the height of the triangle to determine whether the given measures result in zero or two oblique triangles.

The height of the triangle is the product of the sine of the given angle and the other given side (side c): $h = 6 \sin 51° \approx 4.7$. Since $b < c \sin \beta$ (because $4 < 4.7$), zero triangles exist with $b = 4$, $c = 6$, and $\beta = 51°$.

EXAMPLE 4 Using the Law of Sines to Solve a Triangle

IMPORTANT

Since α is an acute angle and $b > a$, either zero or two triangles exist with the given measures. Therefore, begin by finding $b \sin \alpha$.

Use the Law of Sines to solve a triangle where $a = 7$, $b = 9$, and $\alpha = 40°$, if possible. Round the measures to the nearest hundredth, as needed. If two triangles exist, solve both triangles.

SOLUTION

Find $b \sin \alpha$. $b \sin \alpha = 9 \sin 40° \approx 5.79$

Since $5.79 < 7 < 9$ (i.e., $b \sin \alpha < a < b$), there are two oblique triangles with $a = 7$, $b = 9$, and $\alpha = 40°$. First, find the two possible measures of angle β (the angle opposite the given side). These angle measures are supplementary, so one possible measure of β is an acute angle and the other is an obtuse angle. Use the Law of Sines to find the acute measure of angle β.

$$\frac{\sin 40°}{7} = \frac{\sin \beta}{9} \Rightarrow \sin \beta = \frac{9 \sin 40°}{7} \Rightarrow \beta = \sin^{-1} \frac{9 \sin 40°}{7} \approx 55.73°$$

IMPORTANT

The two possible measures of β are supplementary angles. Therefore, subtract the acute measure of β (given by using the Law of Sines) from 180° to find the obtuse measure of β.

$\beta \approx 180° - 55.73° = 124.27°$

Subtracting 55.73° from 180° gives the approximate measure of obtuse angle β. 124.27°.

Use the Triangle Sum Theorem to find the unknown angle γ in each triangle. Then use the Law of Sines to find the measure of side c in each triangle.

Triangle with Acute Angle β

$$\gamma \approx 180° - (40° + 55.73°) = 84.27°$$

$$\frac{c}{\sin 84.27°} \approx \frac{7}{\sin 40°}$$

$$c \approx \frac{7 \sin 84.27°}{\sin 40°}$$

$$c \approx 10.84$$

Triangle with Obtuse Angle β

$$\gamma \approx 180° - (40° + 124.27°) = 15.73°$$

$$\frac{c}{\sin 15.73°} \approx \frac{7}{\sin 40°}$$

$$c \approx \frac{7 \sin 15.73°}{\sin 40°}$$

$$c \approx 2.95$$

Thus, $\beta \approx 55.73°$, $\gamma \approx 84.27°$, and $c \approx 10.84$ in one triangle where $a = 7$, $b = 9$, and $\alpha = 40°$, and $\beta \approx 124.27°$, $\gamma \approx 15.73°$, and $c \approx 2.95$ in the other triangle where $a = 7$, $b = 9$, and $\alpha = 40°$.

7.1.4 Finding the Area of an Oblique Triangle

Recall that the area of a triangle is equal to 1/2 of the product of the triangle's base and height. The oblique triangle shown in Figure 7.1e has height h and base c. So, the area A of this triangle is given by the formula $A = \frac{1}{2}ch$.

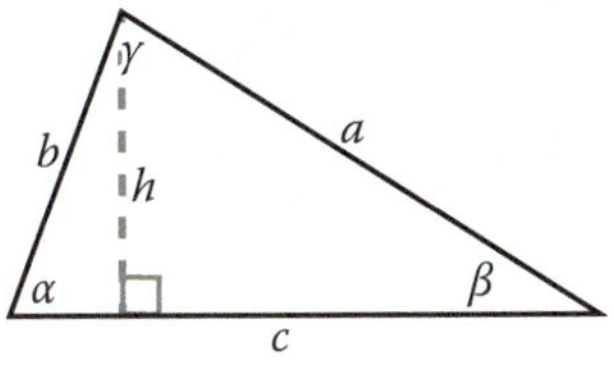

Figure 7.1e

The formula for the area of an oblique triangle can also be written in terms of two sides and an angle. We saw in Topic 7.1.3 that $h = b \sin \alpha$. It follows that the area of an oblique triangle is given by the formula $A = \frac{1}{2}ch = \frac{1}{2}cb \sin \alpha$. Since α is the included angle between sides b and c, the area of an oblique triangle is half the product of two sides and the sine of the included angle.

This formula can be further generalized to include any two sides and the included angle.

The Area of an Oblique Triangle

For any oblique triangle with side lengths a, b, and c, and angle measures α, β, and γ, the area of the triangle is given by $A = \frac{1}{2}cb \sin \alpha = \frac{1}{2}ca \sin \beta = \frac{1}{2}ab \sin \gamma$.

EXAMPLE 5 Finding the Area of an Oblique Triangle

Find the area of a triangle to the nearest hundredth of a square inch if $a = 56$ inches, $\alpha = 63°$, and $\gamma = 46°$.

SOLUTION

Sketch the triangle.

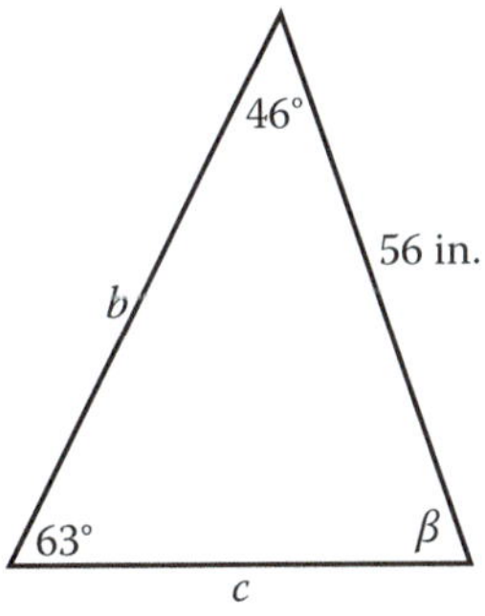

Two sides and the included angle must be known in order to use the formula for the area of an oblique triangle. Therefore, use the Triangle Sum Theorem to find β.

$$\beta \approx 180° - (63° + 46°) = 71°$$

Use the Law of Sines to find c (or b).

$$\frac{c}{\sin 46°} = \frac{56}{\sin 63°} \quad \Rightarrow \quad c = \frac{56 \sin 46°}{\sin 63°} \approx 45.21 \text{ in.}$$ *Law of Sines*

Use the area formula $A = \frac{1}{2}ca \sin \beta$.

$$A \approx \frac{1}{2}(45.21)(56)(\sin 71°) \approx 1196.91 \text{ in}^2$$

IMPORTANT

Always use two sides and the included angle to evaluate the formula for the area of an oblique triangle.

7.1.5 The Law of Sines: An Application

EXAMPLE 6 Using the Law of Sines

Suppose a pole is secured by two wires, where the longer wire forms a 20° angle with the ground, the shorter wire forms a 42° angle with the ground, and the distance between the point at which the two wires are attached to the ground is 28 meters, as shown in the picture.

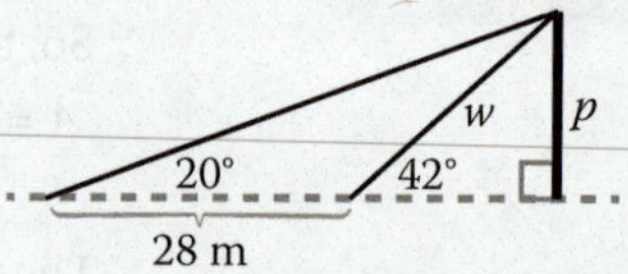

Assume that the pole forms a right angle with the ground. Find the length of the short wire w and the height of the pole p to the nearest tenth of a meter.

SOLUTION

Let α be the included angle between sides of lengths w and 28 meters, and let β be the third angle in the oblique triangle.

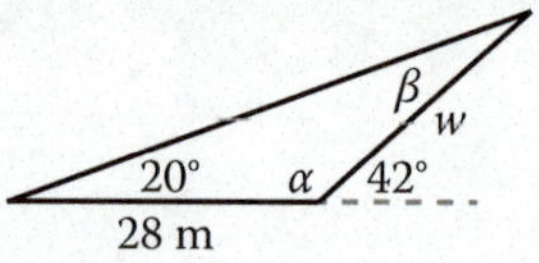

Since α and the 42° angle are supplementary angles, $\alpha = 180° - 42° = 138°$.

By the Triangle Sum Theorem, $\beta = 180° - (20° + 138°) = 22°$.

Use the Law of Sines to find w. $\dfrac{w}{\sin 20°} = \dfrac{28}{\sin 22°} \Rightarrow w = \dfrac{28 \sin 20°}{\sin 22°} \approx 25.6 \text{ m}$

NOTICE THAT

The side opposite the 22° angle (28 meters) is longer than the side opposite the 20° angle (25.6 meters).

Since the height of the pole p is a leg in a right triangle with hypotenuse approximately 25.6 meters, and the angle opposite p is 42°, we can use $\sin \theta$ = opp/hyp to find the approximate value of p.

$$\sin 42° \approx p/25.6 \Rightarrow p \approx 25.6 \sin 42° \approx 17.1 \text{ m}$$

Therefore, the height of the pole and the length of the short wire are approximately 17.1 meters and 25.6 meters, respectively.

SECTION 7.1 EXERCISES

Warm Up

Use the sine function, the given right triangle, and the given measures to find each indicated side or angle to the nearest hundredth, as needed.

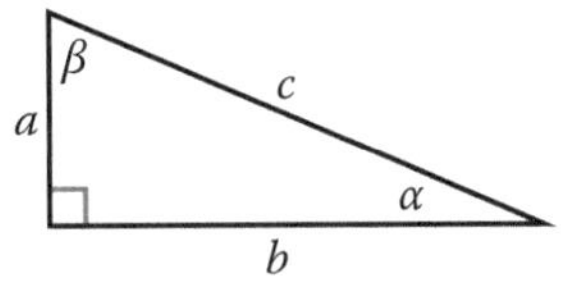

1. $\alpha = 15°$, $a = 25.86$; find c
2. $\beta = 21°$, $c = 22.55$; find b
3. $a = 73.5$, $c = 91$; find α

Just the Facts

Fill in the blanks.

4. A triangle where none of the angles is a right angle is called a(n) _____ triangle.
5. The Law of Sines states that for a triangle with sides a, b, c, and opposite angles α, β, γ, respectively, then _____ = _____ = _____.
6. The situation that applies to triangles for which two sides and the angle opposite one of them are known (SSA) is called the _____ case.
7. In an SSA triangle where α is given as an acute angle, if $a < b \sin \alpha$, then there are exactly _____ triangle(s) with the given measures.
8. In an SSA triangle where α is given as an acute angle, if $a = b \sin \alpha$, then exactly _____ triangle(s) exist with the given measures.
9. In an SSA triangle where α is given as an acute angle, if $a > b \sin \alpha$, and if $a < b$, then there are exactly _____ triangle(s) with the given measures.
10. Given a, c, and β, the area of an oblique triangle can be found using the formula $A =$ _____.

Essential Skills

In Exercises 1–16, use the Law of Sines to solve each triangle, if possible. Round each answer to the nearest hundredth, as needed.

1.

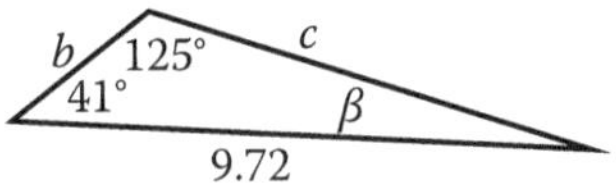

2.

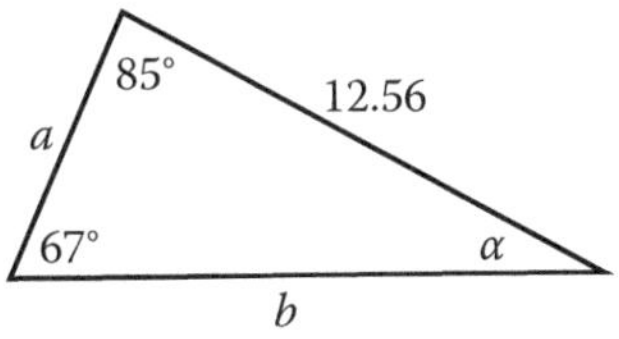

3.

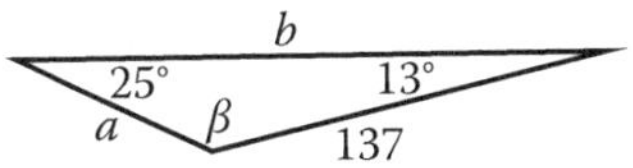

4.

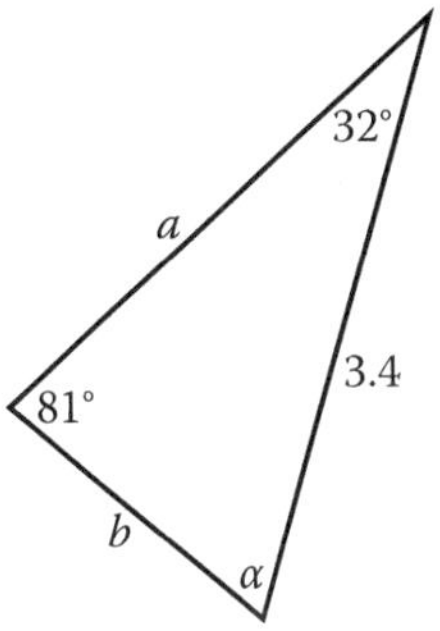

5.

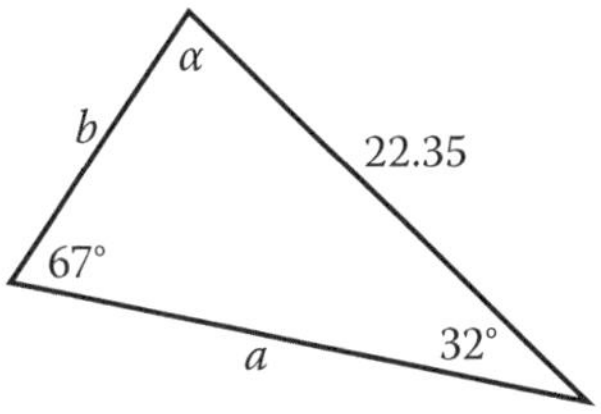

6.

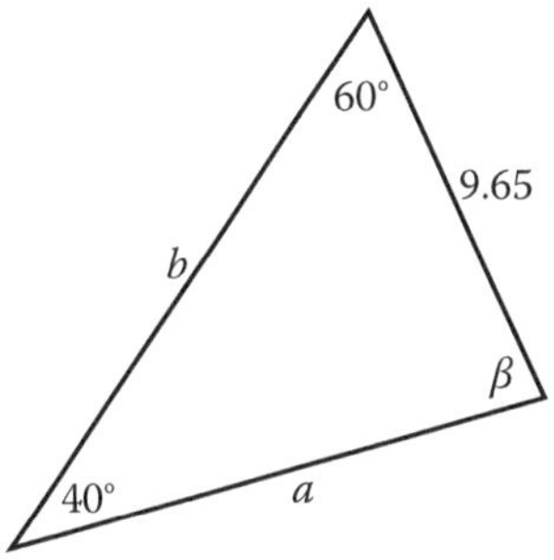

7.

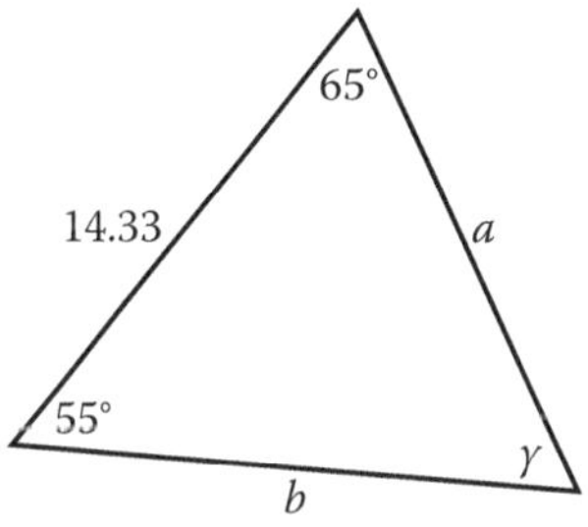

8.

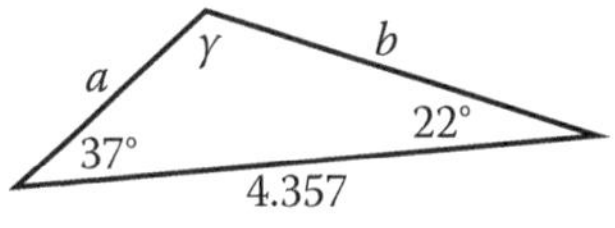

9. $\alpha = 15°$, $\gamma = 127°$, $b = 25.86$
10. $\alpha = 75°$, $\beta = 60°$, $c = 15$
11. $\beta = 45°$, $\gamma = 120°$, $a = 12$
12. $\alpha = 24°$, $\beta = 37°$, $c = 5.192$
13. $\gamma = 122°$, $b = 128$, $c = 453$
14. $\gamma = 97°$, $b = 17.21$, $c = 27.32$
15. $\alpha = 39°$, $a = 91$, $b = 73.5$
16. $\beta = 110°$, $b = 22.55$, $c = 12$

In Exercises 17–20, determine whether exactly zero, one, or two triangles exist with the given measures.

17. $\beta = 30°$, $a = 10$, and $b = 9$

18. $\alpha = 35°$, $a = 7$, and $b = 9$

19. $\alpha = 70°$, $a = 7$, $c = 8$

20. $\gamma = 35°$, $a = 8$, and $c = 6$

In Exercises 21–24, use the Law of Sines to solve each triangle, if possible. Round answers to the nearest hundredth, as needed. If two triangles exist, solve both triangles.

21. $\alpha = 35°$, $a = 6$, and $b = 8$

22. $\alpha = 57°$, $a = 10.8$, and $b = 12.8$

23. $\gamma = 25°$, $a = 2$, and $c = 1$

24. $\alpha = 41°$, $a = 9.9$, and $b = 10.2$

In Exercises 25–28, use the given information to determine all measures for b that will result in zero, one, and two oblique triangles. Round b-values to the nearest hundredth, as needed.

25. $\alpha = 40°$ and $a = 11$

26. $\alpha = 35°$ and $a = 10$

27. $\gamma = 45°$ and $c = 8$

28. $\gamma = 50°$ and $c = 9$

In Exercises 29–36, find the area to the nearest tenth of an inch for each triangle.

29. $\beta = 49°$, $a = 115$ inches, and $c = 170$ inches

30. $\gamma = 119°$, $a = 10$ inches, and $b = 15$ inches

31. $\gamma = 124°$, $a = 17$ inches, and $b = 31$ inches

32. $\alpha = 10°$, $b = 14$ inches, and $c = 11$ inches

33. $\beta = 64°$, $b = 73$ inches, and $c = 68$ inches

34. $\gamma = 97°$, $b = 7$ inches, and $c = 12$ inches

35. $\alpha = 103°$, $a = 11$ inches, and $b = 9$ inches

36. $\alpha = 30°$, $a = 11$ inches, and $b = 9$ inches

37. Two wires help support a pole. One wire forms an angle of 48° with the ground and the other wire forms an angle of 72° with the ground. The wires are 20 meters apart. How tall is the pole to the nearest meter?

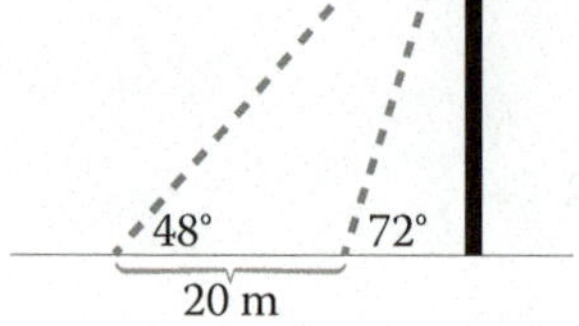

38. In a baseball game, a batter standing at home plate sees the second baseman standing on the baseline between 1st and 2nd base, as shown in the figure. How far is the second baseman from second base? Round to the nearest hundredth of a foot.

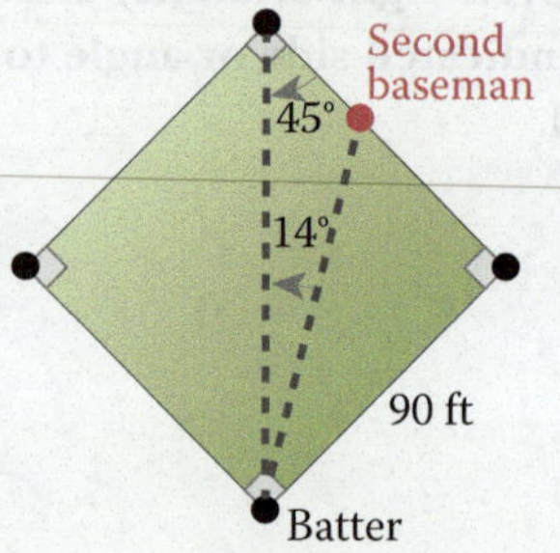

39. A person has a kite out on 1650 feet of string at an angle of 72° with the ground. An observer notes that the angle formed by the kite and the kite flier is 103°. Find the distance between the observer and the kite flier to the nearest foot.

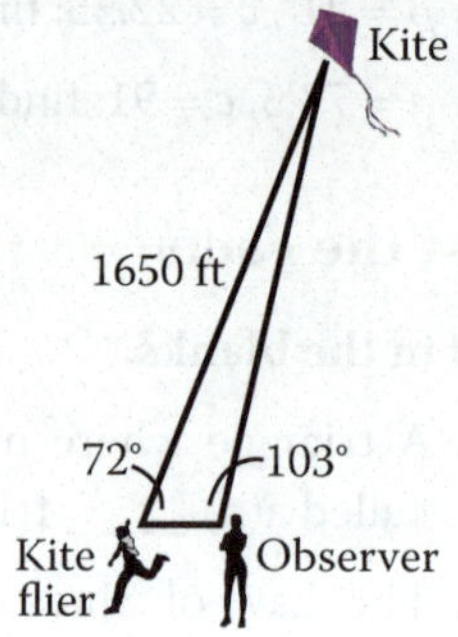

40. Two wires are part of the supports of a pole. One wire forms an angle of 53° with the ground and the other wire forms an angle of 78° with the ground. The distance between the wires on the ground is 25 meters. Find the height of the pole to the nearest tenth of a meter.

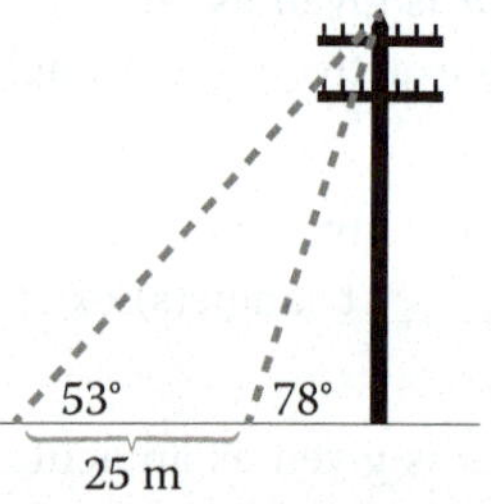

41. Troop A left base camp and hiked 6 hours at 4.4 mph. Troop B left the same camp and hiked 7 hours at 4.2 mph. That evening, Troop A noticed that the angle formed by beacons at the base camp and Troop B's camp was 87°. Find the distance to the nearest tenth of a mile between Troop A and Troop B.

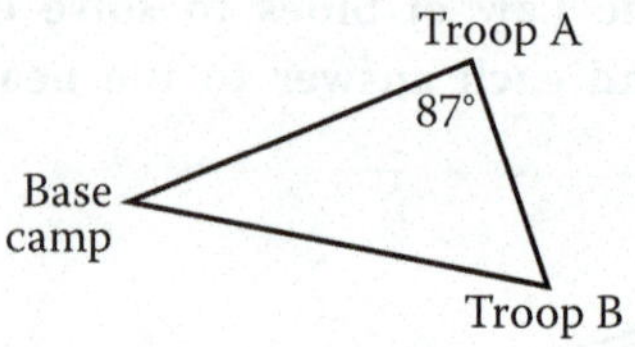

42. Two wireless telephone transmission centers are 35 km apart. A cell phone user is 19 km from one of the centers. At the cell phone user, the angle formed by the two centers is 101°. Find the distance to the nearest kilometer between the cell phone user and the other transmission center.

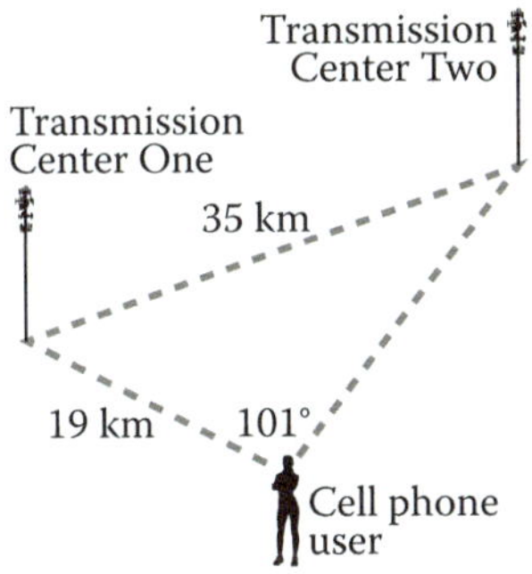

Extensions

43. In triangle *ABC*, $a = 700$ feet, $\beta = 73°\ 39'$, and $\gamma = 37°\ 21'$. If *M* is the midpoint of *c* (i.e., *M* is the midpoint of line segment *AB*), find the length of *AM* to the nearest hundredth of a foot.

44. Find the area of triangle *MNP* to the nearest tenth of a square unit.

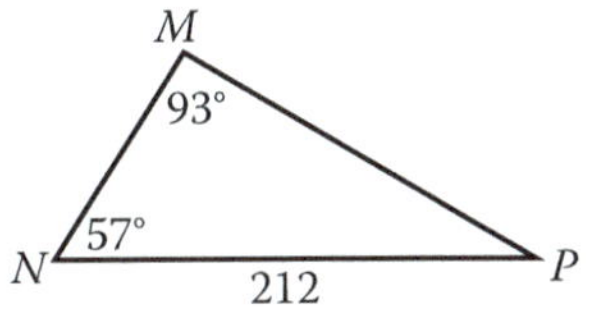

45. Suppose *AB* is a 652 foot line segment along a bank of a stream and *C* is a point on the opposite bank. If $A = 50°$ and $B = 48°$ in triangle *ABC* as shown in the picture, find the perpendicular distance from *C* to *AB* to the nearest hundredth of a foot.

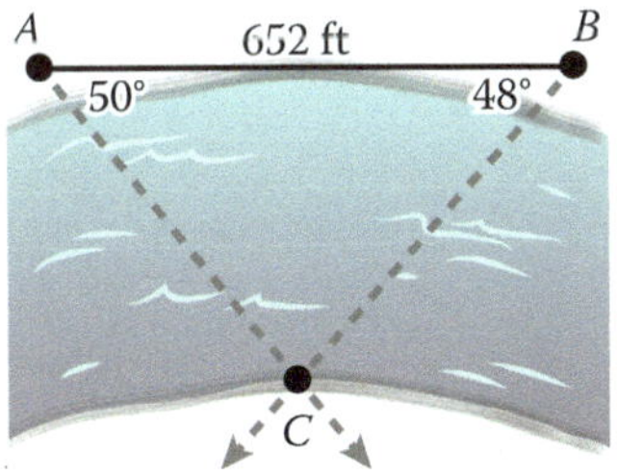

46. Suppose the ratio of the angles in a triangle are in a ratio of 4:5:8 and the shortest side of the triangle is 2.65 inches. Find the length of the longest side to the nearest hundredth of an inch.

47. A flagpole is leaning with the wind 8° from the vertical. 7 meters from the bottom of the flagpole, the angle of elevation to the top of the flagpole is 68°. How tall is the flagpole to the nearest hundredth of a meter?

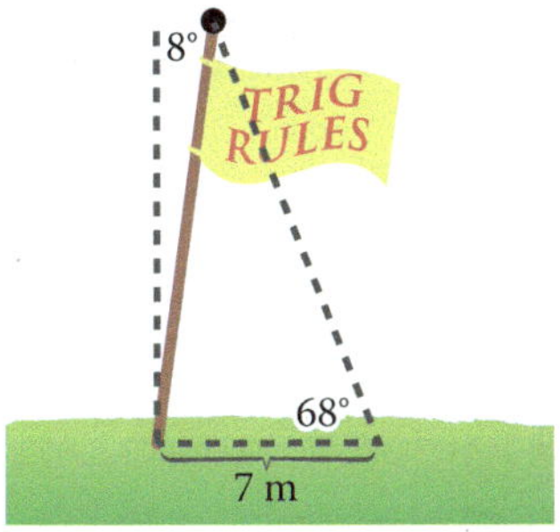

48. A person flying his jet airplane over Waikiki noticed a party on a yacht. His angle of depression to the yacht was 17.5°. Then he noticed the surfer on a huge pipeline wave. His angle of depression to the surfer was 24.167°. His navigation copilot told him he was 5120 feet from the yacht. How far to the nearest foot was he from the surfer?

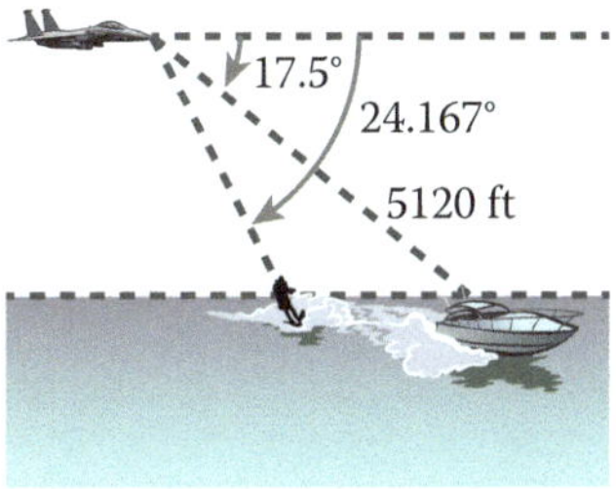

49. Two ships are 2 km apart. A submarine lies directly below the line passing through the two ships. From the submarine, the angle formed by the ships is 41°, while from ship A, the angle formed by ship B and the submarine is 97°. To the nearest hundredth of a kilometer, how much farther is ship B from the submarine than ship A?

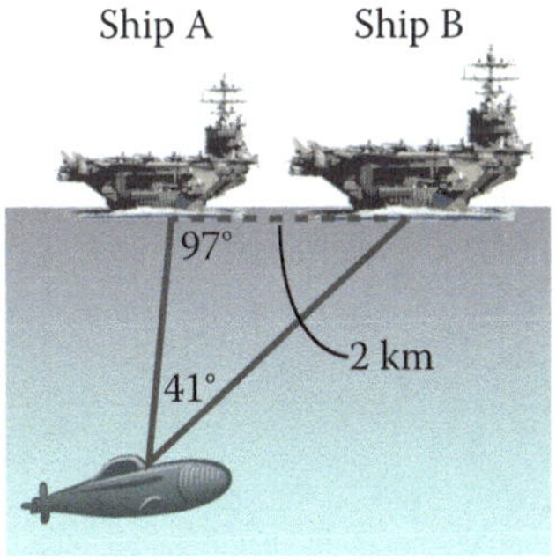

50. A park ranger standing on a hill can see two firewatch towers in the distance. She knows that the two towers are 8 miles from each other. She thinks she is approximately 4.8 miles from one tower. She roughly calculates the angle between her two lines of sight to the towers to be 34.6°. How far to the nearest tenth of a mile is she from the second tower?

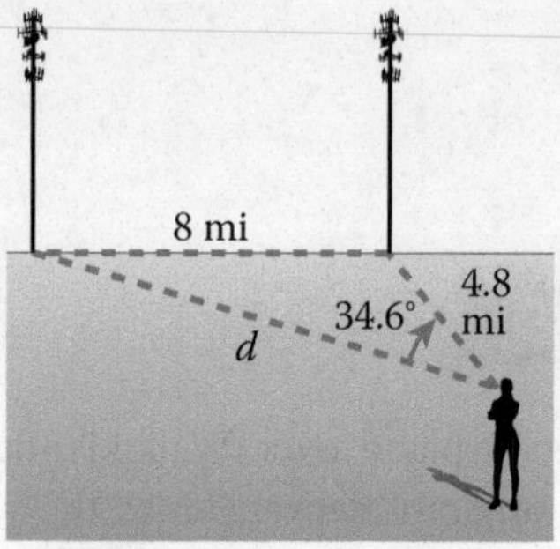

51. In the figure below, what is the measure of x to the nearest tenth of a unit?

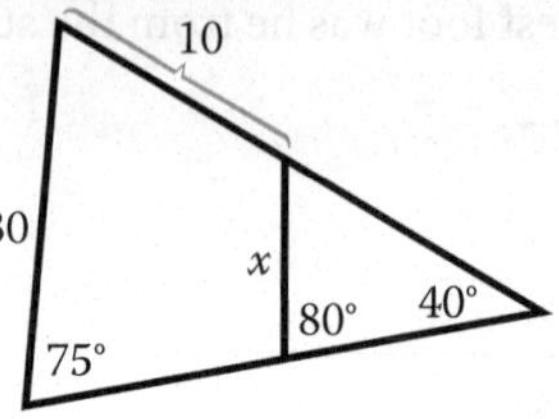

52. In the following triangles, $\alpha = 41°$, $b = 15.56$ mm, and $a = 13.05$ mm. Find the measures of γ_1 and γ_2 to the nearest degree.

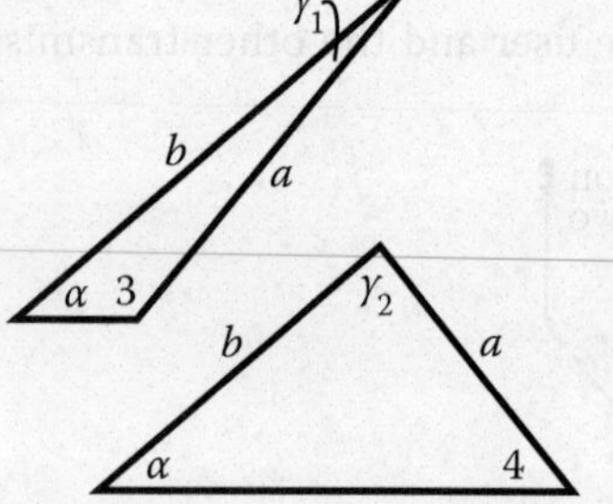

53. In the following triangles, $\alpha = 52°$, $b = 22.71$ cm, and $a = 18.95$ cm. Find the measures of β_1 and β_2 to the nearest degree.

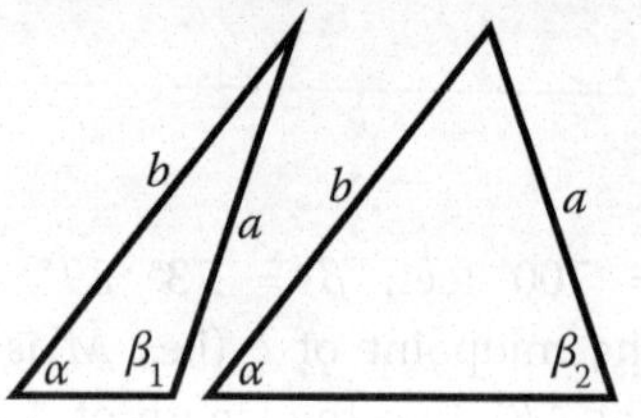

7.2 THE LAW OF COSINES

OBJECTIVES

- Use the Law of Cosines to find the angles of an oblique triangle given the lengths of three sides (SSS).
- Use the Law of Cosines to find the unknown angles and side of an oblique triangle given the lengths of two sides and the included angle (SAS).
- Use Heron's Formula to find the area and height of a triangle.

PREREQUISITE VOCABULARY TERMS

cosine
included angle
Law of Sines
oblique triangle
supplementary angles

7.2.1 The Law of Cosines

The **Law of Cosines** gives a relationship between the sides of an oblique triangle and one of its angles. Specifically, that the sum of the squares of two sides minus twice the product of the two sides and the cosine of the included angle is equal to the square of the third side.

NOTICE THAT

The included angle between b and c is α, the included angle between a and c is β, and the included angle between a and b is γ.

The Law of Cosines

For any triangle with sides a, b, and c,
and opposite angles α, β, and γ, respectively,

$a^2 = b^2 + c^2 - 2bc\cos\alpha$,

$b^2 = a^2 + c^2 - 2ac\cos\beta$, and

$c^2 = a^2 + b^2 - 2ab\cos\gamma$.

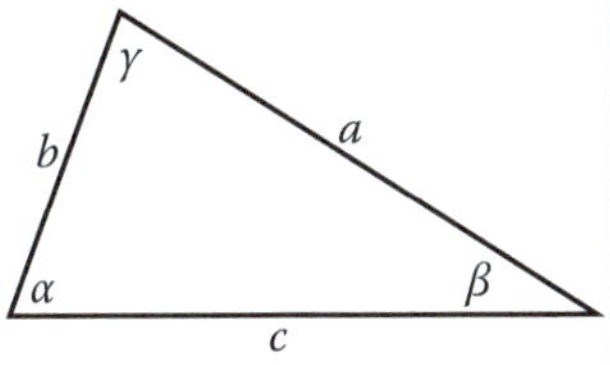

7.2.2 The Law of Cosines (SSS)

Use the Law of Cosines to solve an oblique triangle when three sides are given (SSS).

EXAMPLE 1 Using the Law of Cosines

Find the measures of the largest and smallest angles in a triangle where $a = 10$, $b = 14$, and $c = 6$.

SOLUTION

A triangle's largest and smallest angles are opposite its longest and shortest sides, respectively. So, the largest angle is β (opposite $b = 14$) and the smallest angle is γ (opposite $c = 6$). Use the Law of Cosines to find β and γ.

$$
\begin{aligned}
b^2 &= a^2 + c^2 - 2ac\cos\beta \\
14^2 &= 10^2 + 6^2 - 2(10)(6)\cos\beta \\
196 &= 136 - 120\cos\beta \\
-1/2 &= \cos\beta \\
\beta &= \cos^{-1}(-1/2) = 120^\circ
\end{aligned}
$$

$$
\begin{aligned}
c^2 &= a^2 + b^2 - 2ab\cos\gamma \\
6^2 &= 10^2 + 14^2 - 2(10)(14)\cos\gamma \\
36 &= 296 - 280\cos\gamma \\
13/14 &= \cos\gamma \\
\gamma &= \cos^{-1}(13/14) \approx 21.79^\circ
\end{aligned}
$$

CAUTION

By the order of operations, the product $2ac\cos\beta$ is subtracted from $a^2 + c^2$. So, do not subtract $2ac$ from $a^2 + c^2$.

ALTERNATIVE METHOD

After finding β, γ can be found using the Law of Sines.

7.2.3 The Law of Cosines (SAS): An Application

The Law of Cosines can be used to solve an oblique triangle when the lengths of two sides and the measure of their included angle are known (SAS), as demonstrated in **Example 2.**

Remember, use the Law of Sines to find an unknown side length or angle measure of an oblique triangle given the lengths of two sides and the measure of their *nonincluded* angle (SSA).

EXAMPLE 2 **Using the Law of Cosines**

The obtuse angle formed by the intersection of Highway 1 and Highway 2 is 148°, as shown in the picture, where the points *A*, *B*, and *C* represent exits from the highways.

The distance between exit A and exit B is 47 miles and the distance between exit B and exit C is 55 miles. Find the shortest distance between exit A and exit C to the nearest hundredth of a mile.

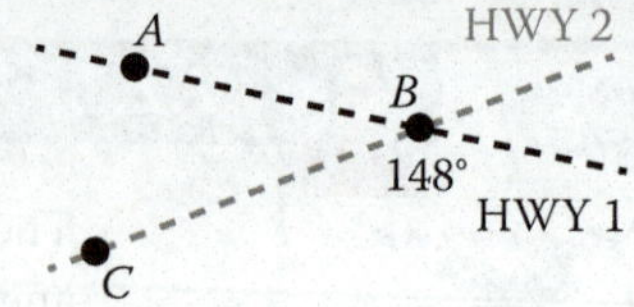

SOLUTION

Sketch a triangle.

Let b be the distance between A and C.

Then b is a side of triangle ABC, where the other two sides are 47 miles and 55 miles.

REMEMBER

The shortest distance between any two points is a straight line.

Angle B has supplement 148°, so
$B = 180° - 148° = 32°$.

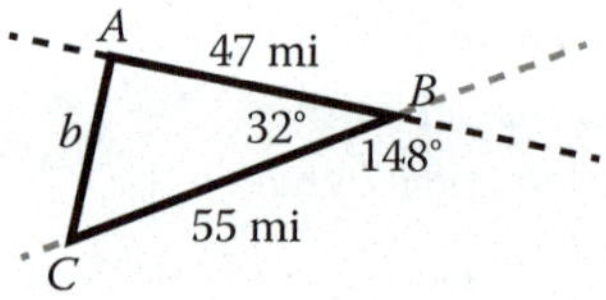

Since two sides and their included angle are known (SAS), the Law of Cosines can be used to find the length of the triangle's remaining side (side b).

Use the Law of Cosines where $a = 55$ miles and $c = 47$ miles.

$b^2 = a^2 + c^2 - 2ac\cos\beta$	*Law of Cosines*
$b^2 = 55^2 + 47^2 - 2(55)(47)\cos 32°$	*Substitute.*
$b^2 = 5234 - 5170\cos 32°$	*Simplify.*
$b = \sqrt{5234 - 5170\cos 32°}$	*Take the positive square root of each side.*
$b \approx 29.15$	*Use a calculator and round to the nearest hundredth.*

TIP

Since b represents a distance, the square root must be positive.

Therefore, the shortest distance between exit A and exit C is approximately 29.15 miles.

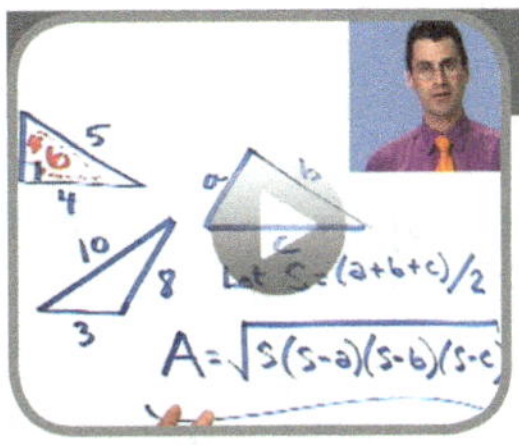

7.2.4 Heron's Formula

A formula for the area of an oblique triangle $A = \frac{1}{2}cb\sin\alpha$ was given in Section 7.1. Recall that in order to use this formula, the lengths of two sides and the measure of their included angle must be known.

The Law of Cosines can be used to derive a formula for the area of a triangle (oblique or right) in terms of only the triangle's side lengths in an expression known as **Heron's formula**. Note that the **semiperimeter** of a triangle is half the triangle's perimeter.

Heron's Formula for the Area of a Triangle

The area of any triangle with sides a, b, and c, is given by $A = \sqrt{s(s-a)(s-b)(s-c)}$, where s is the semiperimeter, $s = \frac{1}{2}(a+b+c)$.

REMEMBER

The perimeter P of a polygon is the sum of the lengths of its sides. So, for triangle with sides a, b, and c, $P = a + b + c$.

EXAMPLE 3 Using Heron's Formula to Find the Area of a Triangle

Find the area to the nearest tenth of a foot of the triangle with sides 10 feet, 12 feet, and 18 feet. Then, letting the triangle's base be 12 feet, find the triangle's height to the nearest tenth of a foot.

SOLUTION

Find the semiperimeter. $s = \dfrac{1}{2}(10+12+18) = \dfrac{1}{2}(40) = 20 \text{ ft}$

Use Heron's formula to find the triangle's area.

$$A = \sqrt{20(20-10)(20-12)(20-18)} = \sqrt{20(10)(8)(2)} = \sqrt{3200} = 40\sqrt{2} \approx 56.6 \text{ ft}^2$$

NOTICE THAT

The triangle's exact area is $40\sqrt{2}$ ft^2.

Use the formula for the area of a triangle in terms of its height and base, $A = \frac{1}{2}bh$, to find the height of the triangle.

$$40\sqrt{2} = \frac{1}{2}(12)h$$ *Substitute 12 for b and $40\sqrt{2}$ for A.*

$$h = \frac{40\sqrt{2}}{6} \approx 9.4 \text{ ft}$$ *Solve for h. Use a calculator to simplify.*

Therefore, the area of the triangle is approximately 56.6 square feet and the height of the triangle (when the base is 12 feet) is approximately 9.4 feet.

SECTION 7.2 EXERCISES

Warm Up

1. Solve $c^2 = a^2 + b^2 - 2ab \cos \gamma$ for γ.
2. In the triangle where $a = 12$, $b = 17$, and $c = 15$, is α, β, or γ the smallest angle?
3. In the triangle where $a = 20.1$, $b = 20.12$, and $c = 20.01$, is α, β, or γ the largest angle?

Just the Facts

Fill in the blanks.

4. If three sides of a triangle are given, the Law of _____ is used to solve the triangle.
5. The Law of Cosines states that for any triangle with sides a, b, c, and opposite angles α, β, γ, respectively, $a^2 =$ _____, $b^2 =$ _____, and $c^2 =$ _____.
6. If one side and two angles of a triangle are given, the Law of _____ is used to solve the triangle.
7. If two sides and the included angle of a triangle are given, the Law of _____ is used to solve the triangle.
8. If two sides of a triangle and an angle opposite one of them are given, the Law of _____ is used to solve the triangle.
9. To find the area of a triangle if the three sides are known, _____ Formula can be used.
10. The area A of a triangle with given sides a, b, and c is $A =$ _____, where $s =$ _____.

Essential Skills

In Exercises 1–4, find the largest angle to the nearest thousandth of a degree in each triangle with the given sides.

1. 11, 6, 9
2. 50, 70, 85
3. 3, 8, 7
4. 42, 64, 77

In Exercises 5–8, find the smallest angle to the nearest thousandth of a degree in each triangle with the given sides.

5. 6, 9, 4
6. 40, 68, 83
7. 12, 17, 15
8. 38, 54, 71

9. Suppose two people stand 400 meters apart on one side of a river and a house is on the opposite side of the river. The people are 700 meters and 1065 meters from the house, as shown in the picture below. In the triangle formed by the two people and the house, what is the measure of the angle with vertex at the person 700 meters from the house? Round to the nearest degree.

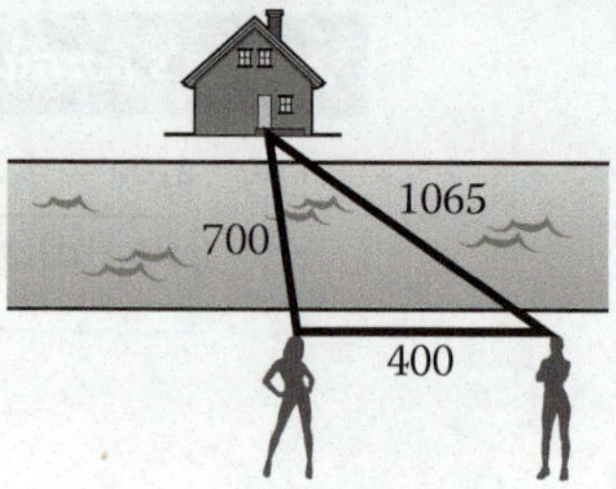

10. A Geiger counter is 4 stories (40 feet) above a sample of radium in a stairwell and 13 feet along the stairs away from directly above the substance. If based on the counter readings the radium is 42 feet away, find θ, as shown in the picture, to the nearest hundredth of a degree.

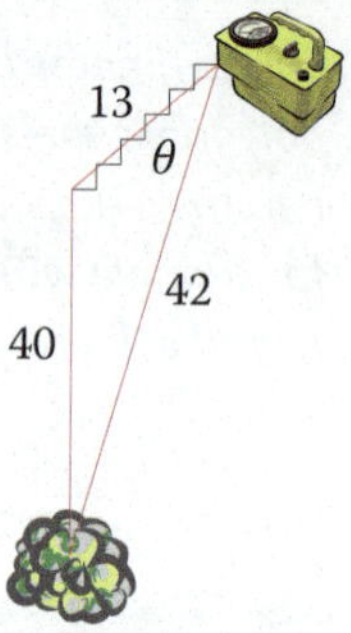

11. To straighten a tree, an 8.4 foot guy-wire is stretched from the top of the tree trunk (6.8 feet above the ground) to a stake in the ground that is 4.9 feet from the base of the tree. What is the angle to the nearest hundredth of a degree that the base of the tree makes with the ground?

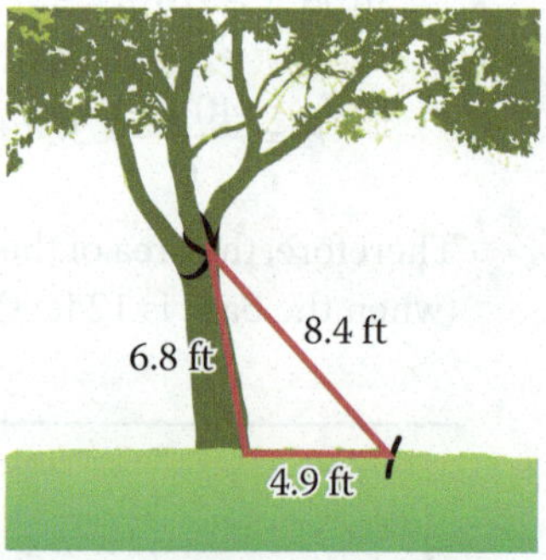

12. A 40-foot flag pole stands atop a hill. A woman stands at the base of the hill, 40 feet from the base of the flag pole, and 60 feet from the top. Find θ, as shown in the picture, to the nearest hundredth of a degree.

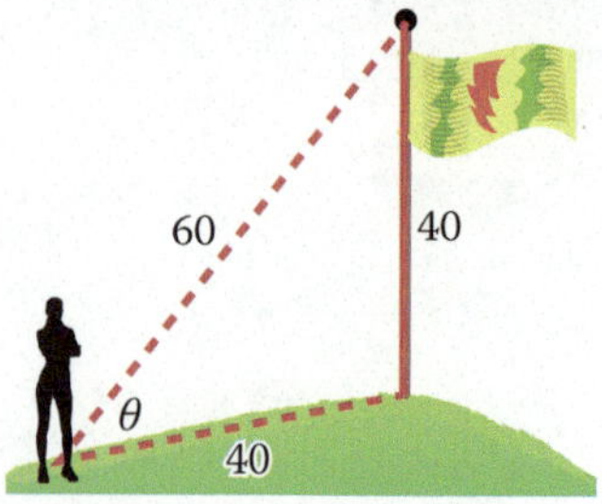

13. The sides of a parallelogram are 40 feet and 70 feet. The length of the longer diagonal is 105 feet. What is the measure of the smaller angle in the parallelogram? Round to the nearest tenth of a degree.

14. A Geiger counter is 3 stories (30 feet) above a sample of radium in a stairwell and 13 feet along the stairs away from being directly above the radium sample. If Geiger counter's readings indicate the radium is 37 feet away, θ, as shown in the picture, to the nearest hundredth of a degree.

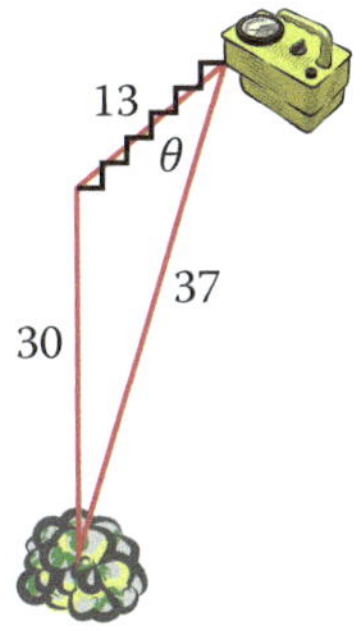

15. A blue boat and a red boat moored along the same side of a lake and are 18 miles apart. The blue boat is 30 miles from a lighthouse on the opposite side of the lake. The angle formed by the boats and the lighthouse, and whose vertex is the blue boat, measures 120°. Find the distance to the nearest mile from the red boat to the lighthouse.

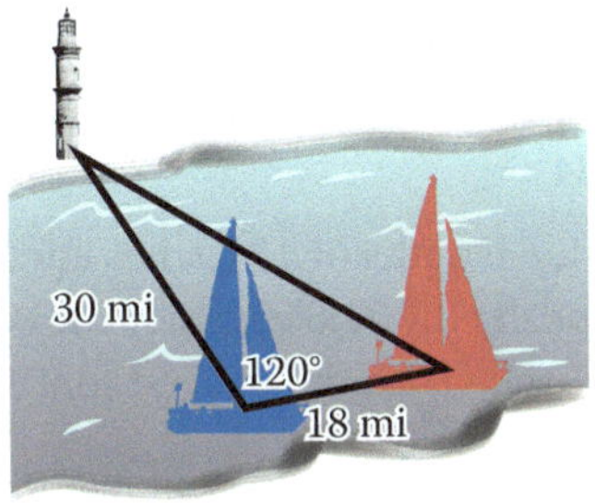

16. A man went off-roading in his Jeep. After driving 50 miles from the highway, he turned 80° to his left. After another 40 miles he reached the highway again. How much of the highway did he bypass? Round to the nearest hundredth of a mile.

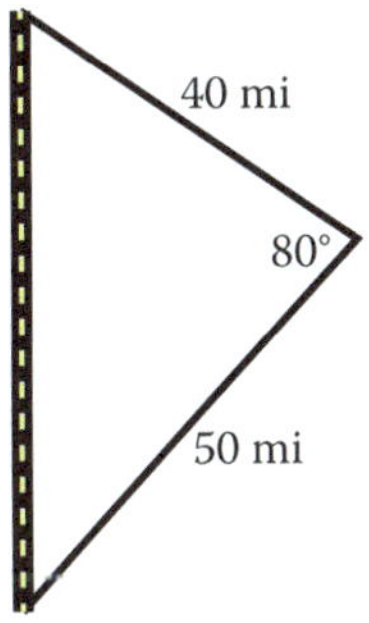

17. Two runners began running at the same time, but at a 25° angle from each other. One runner averages 7.5 mph, and the other averages 8 mph. How far apart are the runners after two hours? Round to the nearest tenth of a mile.

18. A piece of wire is bent into the shape of a triangle. Two sides have lengths of 18 inches and 22 inches. The angle between these two sides is 45°. What is the length of the third side to the nearest hundredth of an inch?

19. Determine the width of Lake Minnehaha to the nearest meter, given the following information. A surveyor standing at a point south of the lake is 2125 meters from the western end, and 3250 meters from the eastern end of the lake. The angle between the two lines-of-sight and either end of the lake is 77°.

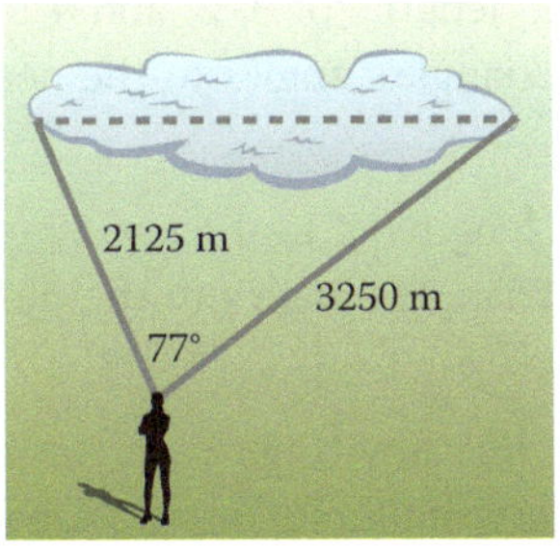

20. A man went off-roading in his Jeep. After driving 45 miles from the highway, he turned 52° to his left. After another 65 miles he reached the highway again. How much of the highway did he bypass? Round to the nearest hundredth of a mile.

In Exercises 21–26, find the area of each triangle with the given lengths. Round each answer to the nearest tenth, as needed.

21. 4 inches, 6 inches, and 9 inches

22. 5.9 centimeters, 6.7 centimeters, and 10.3 centimeters

23. 35 feet, 45 feet, and 50 feet

24. 7 centimeters, 7.2 centimeters, and 9 centimeters

25. 3 inches, 2 inches, and 4 inches

26. 6.4 centimeters, 8.9 centimeters, and 9.4 centimeters

In Exercises 27–32, find the height of each triangle with the given sides. Round each answer to the nearest hundredth, as needed.

27. base 14 feet; sides 9 feet and 7 feet

28. base 6 inches; sides 3 inches and 5 inches

29. base 8 feet; sides 6 feet and 4 feet

30. base 12 inches; sides 6 inches and 7 inches

31. base 4 inches; sides 7 inches and 6 inches

32. base 9 inches; sides 3 inches and 7 inches

Extensions

33. Suppose a triangle has two sides with lengths of 4 units, and their included angle measures 110°. Find the unknown side to the nearest tenth of a unit for the triangle.

34. Suppose a triangle has side lengths of 3, 2, and 4 units. Find the measures of the triangle's angles to the nearest tenth of a degree.

35. Suppose a triangle has side lengths of 14, 12, and 17 units. Find the measures of the triangle's angles to the nearest tenth of a degree.

36. Two boats leave a dock at the same time. Each travels in a straight line. The angle between their courses measures 54° 10′ 15″. One boat travels at 36.2 kilometers per hour and the other travels at 45.6 kilometers per hour. How far apart, to the nearest tenth of a kilometer, will they be after 3 hours?

37. The perimeter of a triangle is 42 centimeters. Two sides of the triangle have lengths of 15 centimeters and 20 centimeters. What is the area of the triangle?

38. Find the area of quadrilateral *ABCD* to the nearest tenth of a square foot.

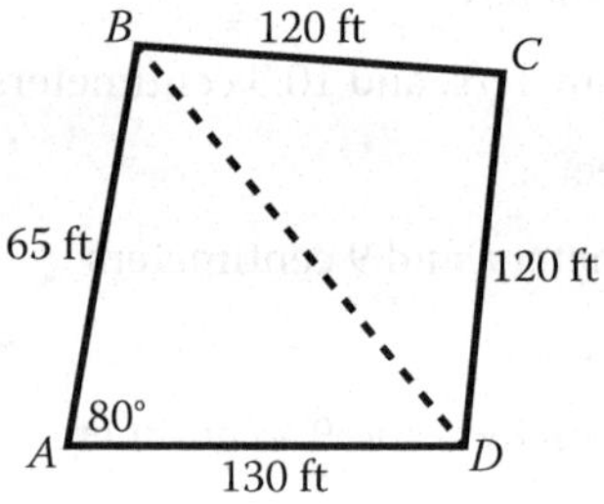

39. Find the area of the figure to the nearest tenth of a square foot.

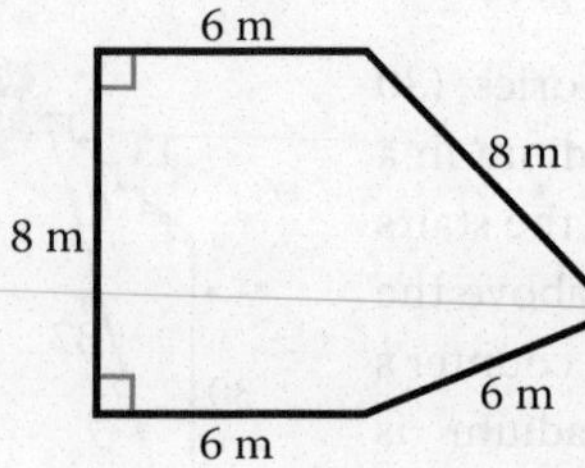

40. Find the area of the parallelogram to the nearest tenth of a square meter.

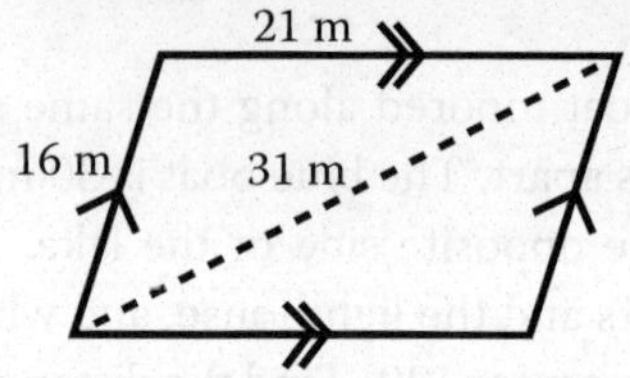

41. The lengths of the diagonals of a parallelogram are 8 feet and 12 feet. Find the lengths, to the nearest tenth of a foot, of the short sides of the parallelogram if the diagonals intersect at an angle of 24°.

42. Suppose a triangle has side lengths of 18 inches and 21.6 inches, and the angle opposite the 21.6 inch side measures 52°. Find the area of the triangle to the nearest tenth of a square inch.

7.3 VECTORS

OBJECTIVES

- Define a vector and the magnitude of a vector.
- Find the magnitude and direction of a vector.
- Multiply a vector by a scalar.
- Add and subtract vectors.
- Find the components of a vector.
- Find a unit vector.
- Write a vector as a linear combination of the standard unit vectors.
- Find the direction angle of a vector.

PREREQUISITE VOCABULARY TERMS

bearing
inverse trigonometric functions
parallelogram
Pythagorean Theorem
reference angle
trigonometric functions

7.3.1 Introduction to Vectors

A **vector** is a quantity that has both **magnitude** (length) and direction. It can be represented graphically by a directed line segment. **Equivalent vectors** have the same magnitude and the same direction. The following table shows several pairs of equivalent vectors.

Equivalent Vectors

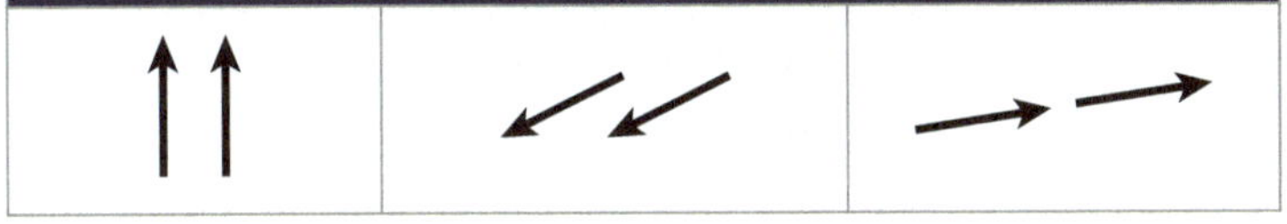

IMPORTANT

The equivalence of two vectors is determined by their direction and magnitude, not by their position. Equivalent vectors have the same slope.

Vectors are named using a single letter with a small arrow over the letter, or with a single boldface letter, such as $\vec{v}$ or $\boldsymbol{v}$ (both read as "vector v").

A vector is represented by a directed line segment on the coordinate plane, where the line segment's endpoints are called the vector's **initial point** and **terminal point**. An arrowhead on the segment indicates the direction of the vector. Figure 7.3a shows $\vec{v}$ with initial point at the origin and terminal point (x, y). A vector on the coordinate plane is said to be in **standard position** if its initial point is at the origin.

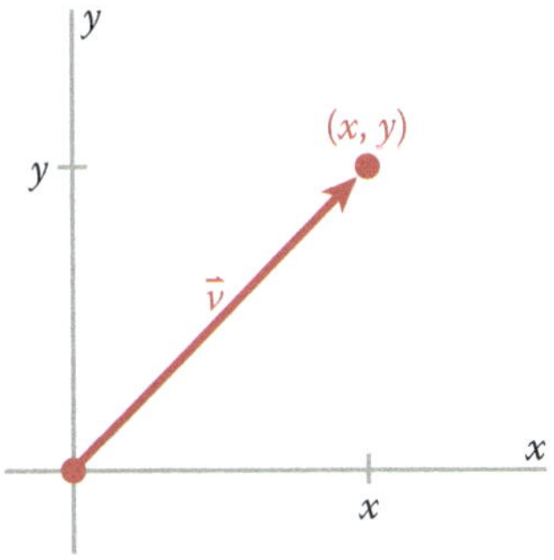

Figure 7.3a

IMPORTANT

Two vectors may be equivalent even if their initial and terminal points differ. However, if two equivalent vectors have the same initial point, then they must have the same terminal point.

If $\vec{v}$ is a vector on the coordinate plane with initial point at (x_1, y_1) and terminal point at (x_2, y_2), then the **component form** of $\vec{v}$ is $\vec{v} = \langle x_2 - x_1, y_2 - y_1 \rangle$. If a vector $\vec{v}$ is in standard position with terminal point at (x, y), then the component form of $\vec{v}$ is $\langle x, y \rangle$.

The two values in the component form of a vector are called the vector's **components**, specifically the **horizontal component** x and the **vertical component** y. The components indicate the change in x and the change in y from the initial point to the terminal point of a vector. The vector's components also indicate the coordinates of the terminal point of the vector in standard position.

EXAMPLE 1 Finding the Component Form of a Vector

Write the component form of each vector.

A. $\vec{v}$ with initial point (0, 0) and terminal point (–1, 3)

B. $\vec{u}$ with initial point (0, –4) and terminal point (–5, 2)

C. $\vec{w}$ with initial point (–3, 1) and terminal point (0, 0)

SOLUTION

IMPORTANT

The terminal point must be (x_2, y_2) and the initial point must be (x_1, y_1).

A. $\vec{v} = \langle x_2 - x_1, y_2 - y_1 \rangle = \langle -1 - 0, 3 - 0 \rangle = \langle -1, 3 \rangle$ $\quad x_1 = 0, y_1 = 0, x_2 = -1, y_2 = 3$

The component form of any vector with initial point at the origin and terminal point at (x, y) is $\langle x, y \rangle$. So, since $\vec{v}$ has initial point at (0, 0) and terminal point at (–1, 3), $\vec{v} = \langle -1, 3 \rangle$.

B. $\vec{u} = \langle x_2 - x_1, y_2 - y_1 \rangle = \langle -5 - 0, 2 - (-4) \rangle = \langle -5, 6 \rangle$ $\quad x_1 = 0, y_1 = -4, x_2 = -5, y_2 = 2$

C. $\vec{w} = \langle x_2 - x_1, y_2 - y_1 \rangle = \langle 0 - (-3), 0 - 1 \rangle = \langle 3, -1 \rangle$ $\quad x_1 = -3, y_1 = 1, x_2 = 0, y_2 = 0$

Equivalent Vectors

Two vectors are equivalent if and only if their respective components are equal. Specifically, $\vec{v} = \langle v_1, v_2 \rangle$ and $\vec{u} = \langle u_1, u_2 \rangle$ are equivalent if and only if $v_1 = u_1$ and $v_2 = u_2$.

Figure 7.3b shows the graphs of two vectors: one with initial point at (0, –4) and terminal point at (–5, 2), and the other with initial point at (0, 0) and terminal point at (–5, 6) (from **Example 1B**). Since the component form of each vector is $\langle -5, 6 \rangle$, they are equivalent vectors (i.e., both line segments are graphs of $\vec{u} = \langle -5, 6 \rangle$).

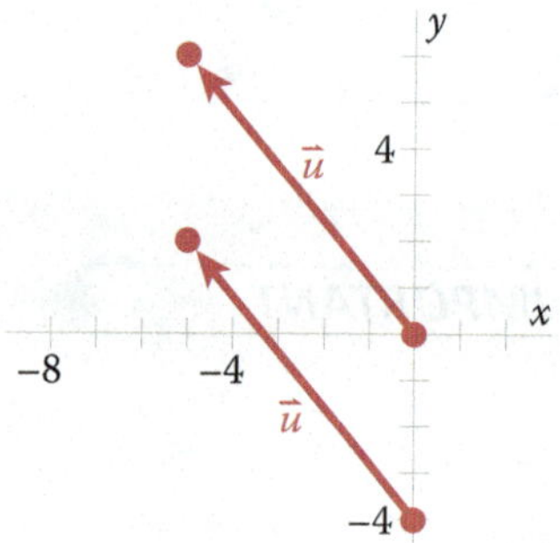

Figure 7.3b

Furthermore, any vector with initial point (x_1, y_1) and terminal point (x_2, y_2) such that $\langle x_2 - x_1, y_2 - y_1 \rangle = \langle -5, 6 \rangle$ is equivalent to $\vec{u}$. For example, a vector with initial point (5, –6) and terminal point (0, 0) is also equivalent to $\vec{u}$.

It follows that there are an infinite number of directed line segments representing a vector with component form $\langle x, y \rangle$. Typically, the most convenient of these is the standard position vector, with initial point at the origin and terminal point at (x, y).

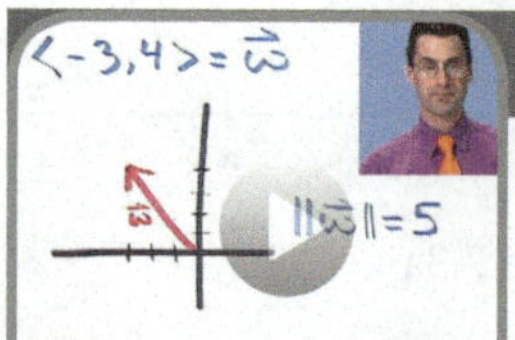

7.3.2 Finding the Magnitude and Direction of a Vector

The magnitude of a vector is equal to the length of its graphical representation (i.e., the length of the line segment from its initial point to its terminal point). Consider the graphs of $\vec{u}$ shown in Figure 7.3b. Each representation of $\vec{u}$ is the hypotenuse of a right triangle with legs 6 and 5 units (Figure 7.3c). By the Pythagorean Theorem, the length of the hypotenuse is $\sqrt{5^2 + 6^2}$, that is, $\sqrt{61}$ units. Therefore, the magnitude of $\vec{u}$ is $\sqrt{61}$.

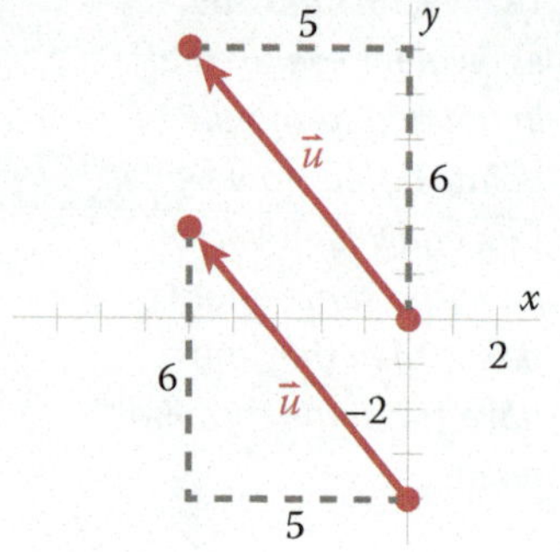

Figure 7.3c

The Magnitude of a Vector

For vector $\vec{v} = \langle x, y \rangle$, the magnitude, denoted $||\vec{v}||$, is $||\vec{v}|| = \sqrt{x^2 + y^2}$.

To find the magnitude of a vector when the coordinates of the initial and terminal points are given, first find the component form of the vector, then use the x- and y-components to find the magnitude. Alternatively, use the Distance Formula.

$$||\vec{v}|| = \sqrt{(x_2 - x_1)^2 + (y_2 - y_1)^2}$$

TIP

A vector's direction angle is the positive standard position angle with the vector on its terminal side.

The **direction angle** θ of $\vec{v} = \langle x, y \rangle$ is the angle formed by a counterclockwise rotation of a ray from the positive x-axis to the standard position vector (i.e., to the line segment representing the vector with endpoints at the origin and (x, y)), as shown in Figure 7.3d. It follows that (x, y) is a point on the terminal side of θ.

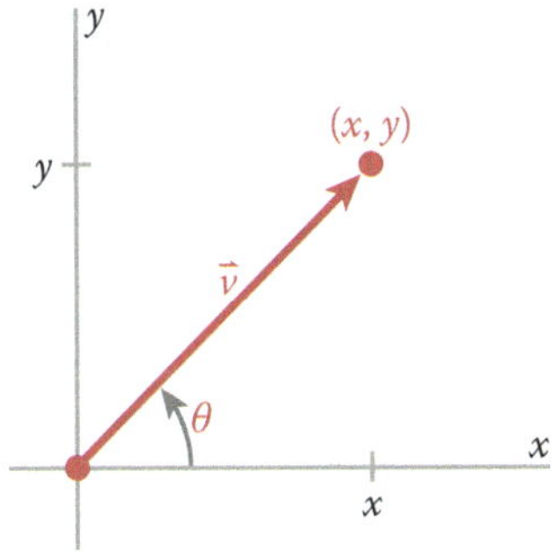

Figure 7.3d

In this case, we know that $\sin\theta = y/r$, $\cos\theta = x/r$, and $\tan\theta = y/x$, where x and y are the components of the vector and r is the vector's magnitude. So, a trigonometric function can be used to find θ, the vector's direction angle. This procedure is demonstrated in **Example 2**.

EXAMPLE 2 Finding the Magnitude and Direction Angle of a Vector

Find the magnitude and direction angle (to the nearest thousandth of a degree) of each vector.

A. $\vec{v} = \langle -6, 4 \rangle$ **B.** $\vec{u} = \langle -2, -2 \rangle$

SOLUTION

A. Sketch the graph of $\vec{v} = \langle -6, 4 \rangle$ in standard position.

Find the magnitude.

$$||\vec{v}|| = \sqrt{(-6)^2 + (4)^2} = \sqrt{52} = 2\sqrt{13}$$

Use $\cos\theta = x/r$ to find the direction angle.

$$\cos\theta = -6/\left(2\sqrt{13}\right) \Rightarrow \theta = \cos^{-1}\left(-3/\sqrt{13}\right) \approx 146.31°$$

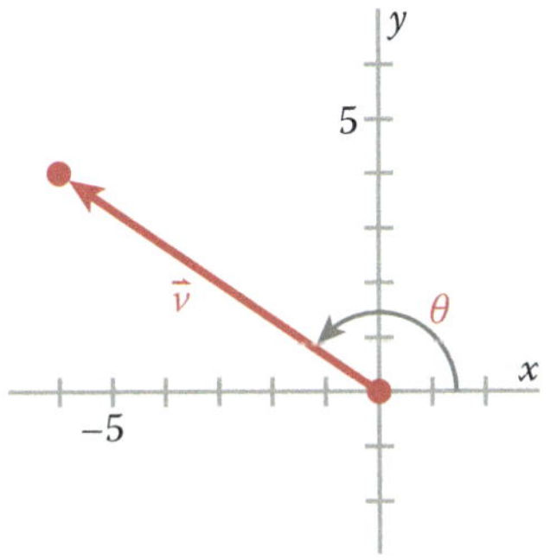

TIP

Use cosine to find the direction angle when the vector is in quadrant II. Otherwise, additional steps are necessary because the range of the inverse sine and tangent functions do not include quadrant II.

B. Sketch the graph of $\vec{u} = \langle -2, -2 \rangle$ in standard position and find the magnitude.

$$||\vec{u}|| = \sqrt{(-2)^2 + (-2)^2} = \sqrt{8} = 2\sqrt{2}$$

Use $\tan\theta = y/x$ to find the direction angle.

$$\tan\theta = -2/(-2) = 1$$

Since θ is in quadrant III, $\theta = \tan^{-1} 1 + \pi = 5\pi/4 = 225°$.

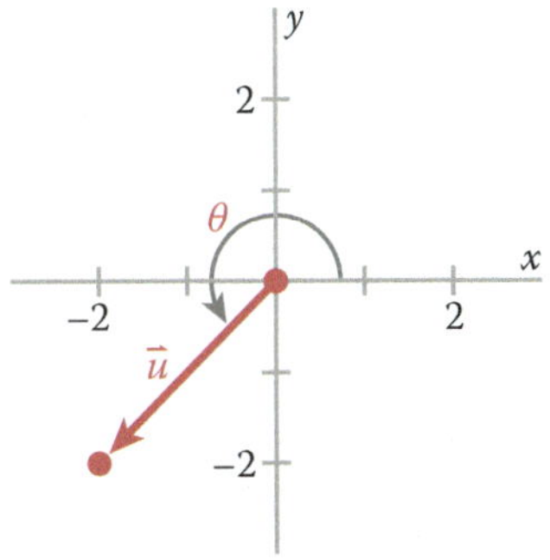

REMEMBER

The range of the inverse tangent function is $(-\pi/2, \pi/2)$.

TIP

In quadrant II,
$\theta = \cos^{-1}(x/r)$.
In quadrant III,
$\theta = 180° + \tan^{-1}(y/x)$.
In quadrant IV,
$\theta = 360° - \sin^{-1}(y/r)$.

Any trigonometric function can be used to find a vector's direction angle θ. However, because the range of the inverse trigonometric functions is not $[0, 360°)$, the direction angle is not always equal to the value given by the inverse trigonometric function, and depends on which quadrant contains the vector and which trigonometric function is used to find the direction angle.

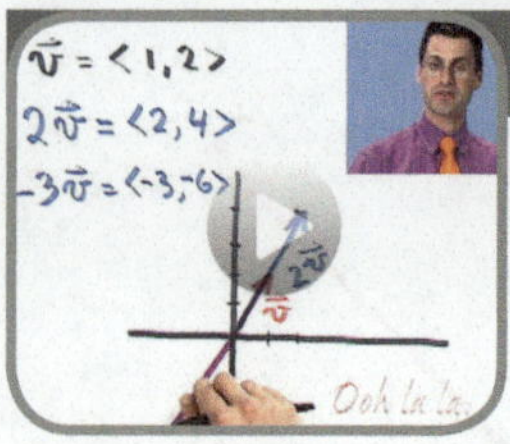

7.3.3 Scalar Multiplication

If $\bar{v} = \langle x, y \rangle$ is a vector and k is a **scalar** (a constant real number), then the components of the **scalar multiple** $k\bar{v}$ are found by multiplying each component of $\bar{v}$ by k.

Scalar Multiplication of a Vector

If $\bar{v} = \langle x, y \rangle$ and k is a scalar, then $k\bar{v} = \langle kx, ky \rangle$.

EXAMPLE 3 **Finding a Scalar Multiple of a Vector**

If $\bar{v} = \langle -3, 1 \rangle$, find each scalar multiple.

A. $2\bar{v}$

B. $-\bar{v}$

SOLUTION

A. Multiply each component by 2. $2\bar{v} = \langle 2(-3), 2(1) \rangle = \langle -6, 2 \rangle$

B. Multiply each component by -1. $-\bar{v} = \langle -(-3), -(1) \rangle = \langle 3, -1 \rangle$

Graphing a Scalar Multiple

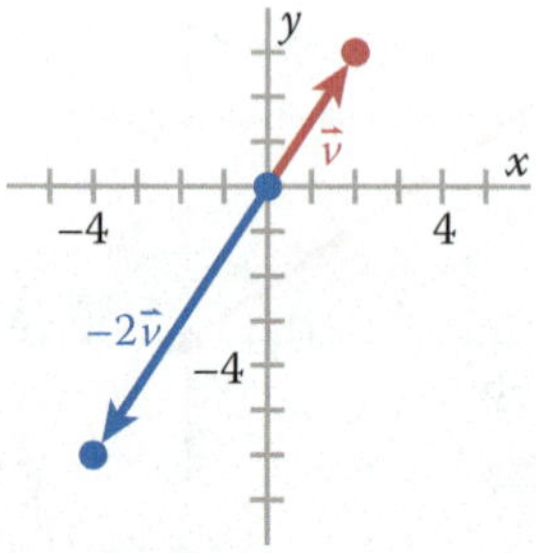

Figure 7.3e

If the magnitude of $\bar{v} = \langle x, y \rangle$ is $||\bar{v}||$, then the magnitude of the scalar multiple vector $k\bar{v} = \langle kx, ky \rangle$ is $|k| \cdot ||\bar{v}||$. If $k < 0$, then the direction of $k\bar{v} = \langle kx, ky \rangle$ is opposite the direction of $\bar{v} = \langle x, y \rangle$. Therefore, the vector $-2\bar{v}$ is twice the length of $\bar{v}$, and $-2\bar{v}$ points in the direction opposite $\bar{v}$, as shown in Figure 7.3e.

7.3.4 Vector Arithmetic

The sum or difference of two vectors is called a **resultant vector**.

Vector Addition and Subtraction

If $\bar{v} = \langle v_1, v_2 \rangle$ and $\bar{u} = \langle u_1, u_2 \rangle$, then $\bar{v} + \bar{u} = \langle v_1 + u_1, v_2 + u_2 \rangle$, and $\bar{v} - \bar{u} = \langle v_1 - u_1, v_2 - u_2 \rangle$.

EXAMPLE 4 Finding a Resultant Vector

IMPORTANT

The sum or difference of two vectors is called the resultant vector.

If $\vec{v} = \langle 5, 0 \rangle$ and $\vec{u} = \langle 2, -1 \rangle$, find each resultant vector.

A. $\vec{v} + \vec{u}$ **B.** $\vec{u} - \vec{v}$ **C.** $\vec{v} - 2\vec{u}$

SOLUTION

A. Add. $\vec{v} + \vec{u} = \langle 5, 0 \rangle + \langle 2, -1 \rangle = \langle 5 + 2, 0 + (-1) \rangle = \langle 7, -1 \rangle$

B. Subtract. $\vec{u} - \vec{v} = \langle 2, -1 \rangle - \langle 5, 0 \rangle = \langle 2 - 5, -1 - 0 \rangle = \langle -3, -1 \rangle$

C. Multiply each component of $\vec{u}$ by 2, then subtract.

$$\vec{v} - 2\vec{u} = \langle 5, 0 \rangle - 2\langle 2, -1 \rangle = \langle 5, 0 \rangle - \langle 4, -2 \rangle = \langle 1, 2 \rangle$$

Graphically, the resultant vector $\vec{v} + \vec{u}$ is the diagonal of the parallelogram formed with $\vec{v}$ and $\vec{u}$ as the adjacent sides. For example, the parallelogram formed by adjacent sides $\vec{v} = \langle 5, 0 \rangle$ and $\vec{u} = \langle 2, -1 \rangle$ is shown in Figure 4.3f.

Notice that the resultant vector $\vec{v} + \vec{u} = \langle 7, -1 \rangle$ (from **Example 4A**) is a diagonal of the parallelogram.

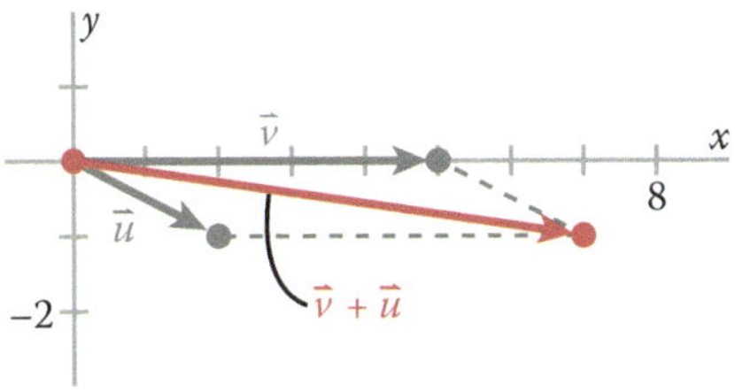

Figure 4.3f

7.3.5 Finding the Components of a Vector

The components of a vector can be found using the vector's magnitude and direction angle (or the reference angle formed with the standard position vector), as demonstrated in **Example 5**. Recall that if the terminal point of a standard position vector is (x, y), the vector's magnitude is r, and the direction angle is θ, then $\sin\theta = y/r$, $\cos\theta = x/r$, and the vector's components are $\langle x, y \rangle$.

EXAMPLE 5 Finding the Components of a Vector

Find the component form of $\vec{v}$ if $||\vec{v}|| = 10$, and the vector's direction angle is 300°.

SOLUTION

REMEMBER

Evaluate sin 300° and cos 300° by using the unit circle or by using the reference angle (be sure to apply the appropriate signs for quadrant IV).

Sketch $\vec{v}$ on the coordinate plane in standard position with length 10 units and direction angle 300°. Use the sine and cosine functions to find x and y (where the vector's terminal point is (x, y) and $r = 10$ is the vector's magnitude).

$$\cos 300° = x/10 \Rightarrow x = 10\cos 300° = 10(1/2) = 5$$

$$\sin 300° = y/10 \Rightarrow y = 10\sin 300° = 10\left(-\sqrt{3}/2\right) = -5\sqrt{3}$$

Thus, $\vec{v} = \left\langle 5, -5\sqrt{3} \right\rangle$.

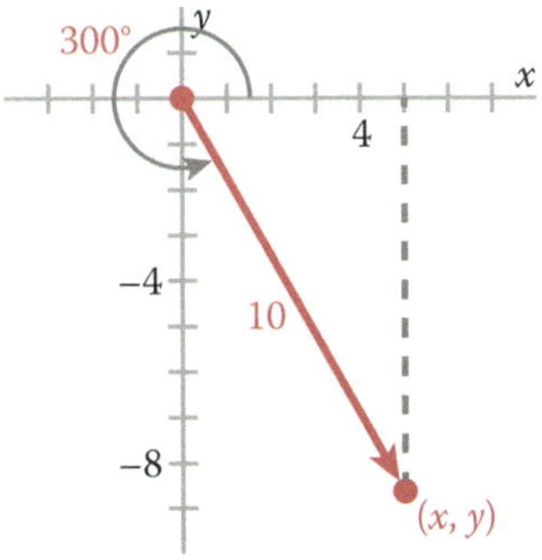

ALTERNATIVE METHOD

The reference angle of a 300° angle is 60°, so the vector is the hypotenuse of a 30°-60°-90° triangle with legs x and y.

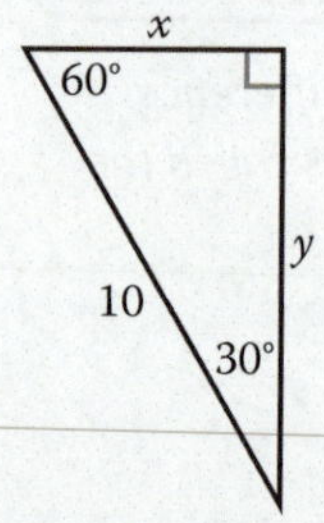

REMEMBER

The side lengths of a 30°-60°-90° triangle are a, $a\sqrt{3}$, and $2a$, where the hypotenuse is $2a$. In this case, $a = 5$.

The hypotenuse is 10, so the leg opposite the 30° angle is 5 and the leg opposite the 60° angle is $5\sqrt{3}$.

The vector is in quadrant IV, so $x > 0$ and $y < 0$. Thus, $x = 5$ and $y = -5\sqrt{3}$, and $\vec{v} = \langle 5, -5\sqrt{3} \rangle$.

7.3.6 Finding a Unit Vector

If both the initial and terminal points of a vector in standard position are at the origin, then that vector is called the **zero vector** $\vec{0}$. Thus, $\vec{0} = \langle 0, 0 \rangle$. The magnitude of the zero vector is 0.

A **unit vector**, denoted $\vec{u}$, is a vector with magnitude 1. When a nonzero vector's direction is important, but the magnitude is not, then the **unit vector in the same direction** can be considered. Multiply each of a nonzero vector's components by the reciprocal of its magnitude to find the components of the unit vector in the same direction.

NOTICE THAT

If $\vec{u}$ is the unit vector in the same direction as $\vec{v}$, then $\vec{u}$ is a scalar multiple of $\vec{v}$.

Unit Vectors

If $\vec{v} = \langle x, y \rangle$ and $||\vec{v}|| \neq 0$, then the unit vector in the same direction as $\vec{v}$ is $\vec{u} = \frac{1}{||\vec{v}||}\vec{v} = \left\langle \frac{x}{||\vec{v}||}, \frac{y}{||\vec{v}||} \right\rangle$.

EXAMPLE 6 Finding a Unit Vector

Find the component form of the unit vector in the same direction as $\vec{v} = \langle -3, -4 \rangle$.

SOLUTION

NOTICE THAT

The vector forms a 3:4:5 right triangle, so the vector's magnitude must be 5.

Find the magnitude of $\vec{v}$. $\quad ||\vec{v}|| = \sqrt{(-3)^2 + (-4)^2} = \sqrt{25} = 5$

Multiply $\vec{v}$ by the reciprocal of $||\vec{v}||$, 1/5, to find the unit vector in the same direction as $\vec{v}$.

$\vec{u} = (1/5)\langle -3, -4 \rangle = \langle -3/5, -4/5 \rangle$ *Multiply each component by the scalar 1/5.*

CHECK

Find the magnitude of the unit vector.

$||\vec{u}|| = \sqrt{(-3/5)^2 + (-4/5)^2} = \sqrt{25/25} = 1$ ✔

Sketch the graph of $\vec{u}$ and $\vec{v}$.

$\vec{u}$ and $\vec{v}$ have the same direction. ✔

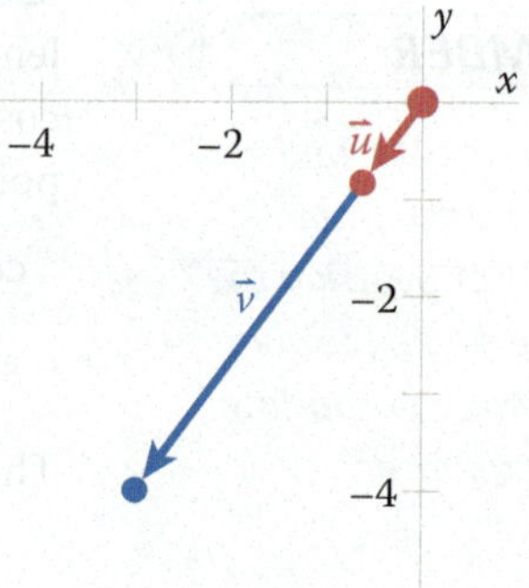

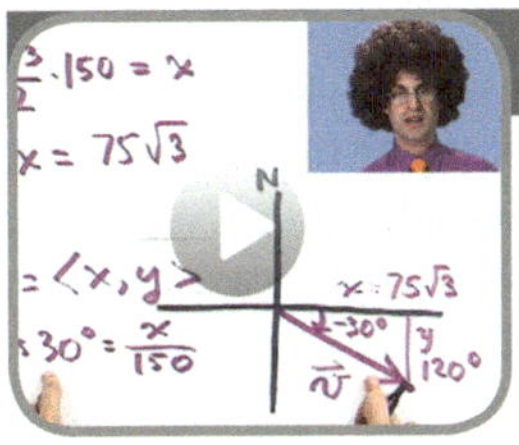

7.3.7 Applications of Vectors

A vector representing the velocity (a measure of speed and direction) of an object is called a **velocity vector**. A velocity vector's magnitude is the speed of the object and its direction is the direction the object is traveling.

EXAMPLE 7 Finding a Velocity Vector

REMEMBER

A bearing with no stated direction is measured clockwise from due north.

Find the velocity vector of a boat traveling at 40 kilometers per hour at a 210° bearing.

SOLUTION

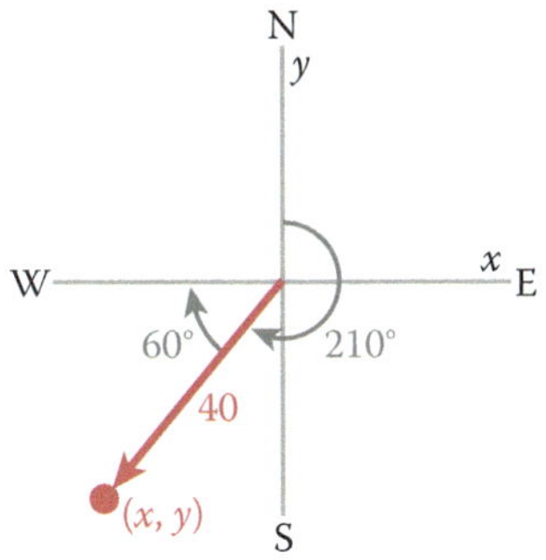

A bearing of 210° means the ray is rotated 210° clockwise from due north. So, the velocity vector is in quadrant III and its reference angle is 60°.

Use the sine and cosine functions to find x and y (where the vector's terminal point is (x, y) and $r = 40$ is the vector's magnitude).

$$\cos 60° = x/40 \quad \Rightarrow \quad x = 40 \cos 60° = 40(1/2) = 20$$

$$\sin 60° = y/40 \quad \Rightarrow \quad y = 40 \sin 60° = 40\left(\sqrt{3}/2\right) = 20\sqrt{3}$$

The vector is in quadrant III, so $x < 0$ and $y < 0$. Thus, the velocity vector is $\vec{v} = \left\langle -20, -20\sqrt{3} \right\rangle$.

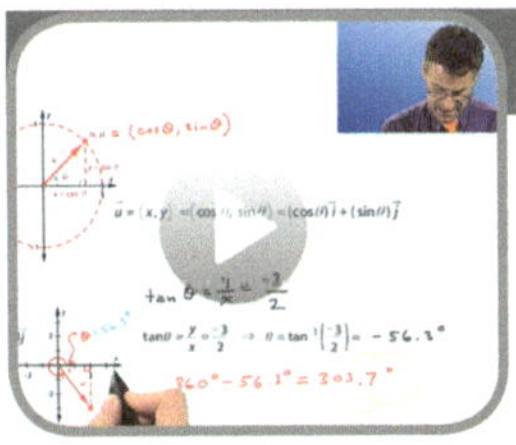

7.3.8 Linear Combinations and Direction Angles of Vectors

A sum of vectors is called a **linear combination** of vectors. For example, the sums $5\vec{u} + 2\vec{v}$, $-\vec{u} + 6\vec{v}$, and $\vec{u} + 3v$ are each a linear combination of vectors $\vec{u}$ and $\vec{v}$. In general, if r and s are scalars, then $r\vec{u} + s\vec{v}$ is a linear combination of the vectors $\vec{u}$ and $\vec{v}$.

A linear combination of **standard unit vectors** is a special type of linear combination. In the coordinate plane there are two standard unit vectors, typically denoted by $\vec{i}$ and $\vec{j}$.

Standard Unit Vectors

The standard unit vectors are $\vec{i} = \langle 1, 0 \rangle$ and $\vec{j} = \langle 0, 1 \rangle$.

The graphs of the standard unit vectors $\vec{i}$ and $\vec{j}$ in standard position are line segments on the x- and y-axis, respectively, and are thus perpendicular. Furthermore, any scalar multiples of the two standard unit vectors in standard position are also perpendicular to each other. Therefore, the resultant vector formed from a sum of scalar multiples of the two standard unit vectors is the diagonal of a rectangle (i.e., a parallelogram with perpendicular sides).

Any vector can be written as a linear combination of the standard unit vectors $\vec{i}$ and $\vec{j}$.

NOTICE THAT

The components of $\vec{v}$, x and y, are the scalars in the linear combination.

$$\vec{v} = \langle x, y \rangle = x\langle 1, 0 \rangle + y\langle 0, 1 \rangle = x\vec{i} + y\vec{j}$$

For example, suppose $\vec{v} = \langle 2, 1 \rangle$. Then $\vec{v} = \langle 2, 1 \rangle = 2\langle 1, 0 \rangle + 1\langle 0, 1 \rangle = 2\vec{i} + \vec{j}$. So, $\vec{v} = 2\vec{i} + \vec{j}$ is the linear combination of the standard unit vectors $\vec{i}$ and $\vec{j}$ for $\vec{v} = \langle 2, 1 \rangle$.

EXAMPLE 8 Writing a Vector as a Linear Combination of Standard Unit Vectors

REMEMBER

The magnitude of each standard unit vector is 1.

Suppose $\vec{v}$ is a vector with initial point (1, −2) and terminal point (−3, 6). Write $\vec{v}$ as a linear combination of the standard unit vectors.

SOLUTION

Find the component form of $\vec{v}$.

$$\vec{v} = \langle x_2 - x_1, y_2 - y_1 \rangle = \langle -3 - 1, 6 - (-2) \rangle = \langle -4, 8 \rangle \qquad x_1 = 1, y_1 = -2, x_2 = -3, y_2 = 6$$

Use the horizontal and vertical components of $\vec{v}$ to write $\vec{v}$ as a linear combination of the standard unit vectors $\vec{i}$ and $\vec{j}$.

$$\vec{v} = \langle -4, 8 \rangle = -4\langle 1, 0 \rangle + 8\langle 0, 1 \rangle = -4\vec{i} + 8\vec{j}$$

The component forms of the vectors $-4\vec{i}$, $8\vec{j}$, and $\vec{v}$ from **Example 8** are $\langle -4, 0 \rangle$, $\langle 0, 8 \rangle$, and $\langle -4, 8 \rangle$, respectively. So, the terminal points of vectors $-4\vec{i}$, $8\vec{j}$, and $\vec{v}$ in standard position are (−4, 0), (0, 8), and (−4, 8), respectively. The graph of these three vectors in standard position (Figure 7.3g) confirms that $\vec{v} = -4\vec{i} + 8\vec{j}$ is the resultant vector of $-4\vec{i}$ and $8\vec{j}$ because $\vec{v} = -4\vec{i} + 8\vec{j}$ is a diagonal of the parallelogram with adjacent sides $-4\vec{i}$ and $8\vec{j}$.

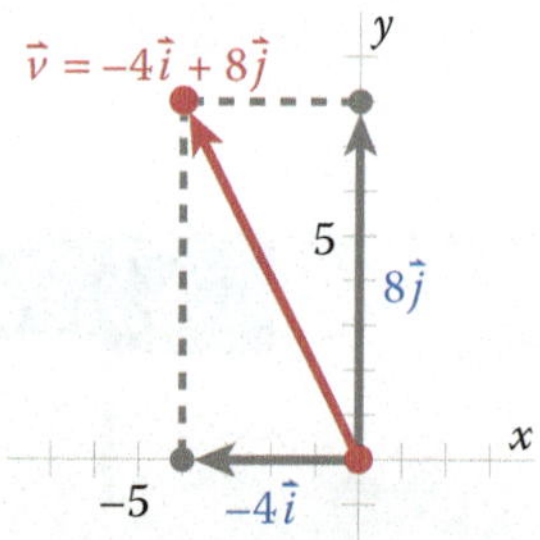

Figure 7.3g

Writing a Vector as a Linear Combination of the Standard Unit Vectors in Terms of the Vector's Direction Angle and Magnitude

A standard position vector $\vec{v} = \langle x, y \rangle$ has terminal point (x, y). So, the terminal side of the vector's direction angle θ contains the point (x, y). Therefore, by the definitions of the trigonometric functions of θ, $x = r\cos\theta$ and $y = r\sin\theta$, where r is the vector's magnitude. The components of $\vec{v}$ can thus be written in terms of the vector's magnitude r and direction angle θ.

$$\vec{v} = \langle x, y \rangle = \langle r\cos\theta, r\sin\theta \rangle = (r\cos\theta)\vec{i} + (r\sin\theta)\vec{j}$$

EXAMPLE 9 Finding the Direction Angle of a Vector Given as a Linear Combination of Standard Unit Vectors

Find the direction angle of each vector. Round to the nearest tenth of a degree as needed.

A. $\vec{w} = 2(\cos 30°)\vec{i} + 2(\sin 30°)\vec{j}$ **B.** $-3\vec{i} - \vec{j}$

SOLUTION

A. The vector is expressed in the form $r(\cos\theta)\vec{i} + r(\sin\theta)\vec{j}$, so its direction angle is 30°.

B. Since $-3\vec{i} - \vec{j} = (r\cos\theta)\vec{i} + (r\sin\theta)\vec{j}$, we can conclude that $-3 = (r\cos\theta)$ and $-1 = (r\sin\theta)$, where $r = \sqrt{10}$. Solving either equation gives θ.

Use $x = r\cos\theta$.

$$-3 = \sqrt{10}(\cos\theta) \quad \Rightarrow \quad \cos\theta = -\frac{3}{\sqrt{10}}$$

Since θ is in quadrant III, $q = 360° - \cos^{-1}\left(-\frac{3}{\sqrt{10}}\right) \approx 360° - 161.6° = 198.4°$.

ALTERNATIVE METHOD

Use $\tan\theta = y/x$.

$$\tan\theta = -1/(-3)$$

Since θ is in quadrant III, $\theta = \tan^{-1}(1/3) + \pi \approx 198.4°$.

So, the direction angle is approximately 198.4°.

SECTION 7.3 EXERCISES

Warm Up

Find the distance between each pair of points.

1. (2, –5), (–1, 4)
2. (–3, 6), (7, –2)
3. (–8, –3), (1, 9)

Just the Facts

Fill in the blanks.

4. A quantity that has both magnitude and direction can be represented by a _____.
5. The magnitude of a vector is the _____ of the vector and is found using the _____ Theorem.
6. The direction angle of a vector is the _____ formed by a counterclockwise rotation of a ray from the positive x-axis to the standard position vector.
7. The magnitude of vector $\vec{v}$ is denoted by _____.
8. A vector that has magnitude 1 is a(n) _____ vector.
9. To find the unit vector $\vec{u}$ for a given vector $\vec{v} = \langle x, y \rangle$, use the formula $\vec{u} =$ _____.
10. Given the magnitude r and direction angle θ of a vector $\vec{v}$, the component form can be found using the formulas _____ for x and _____ for y.

Essential Skills

In Exercises 1–6, write the component form of each vector with an initial point at the origin and the given terminal point.

1. (3, –5)
2. (–2, 4)
3. (–6, 7)
4. (–8, –8)
5. (–10, –15)
6. (4, –7)

In Exercises 7–12, find the component form of each vector.

7.

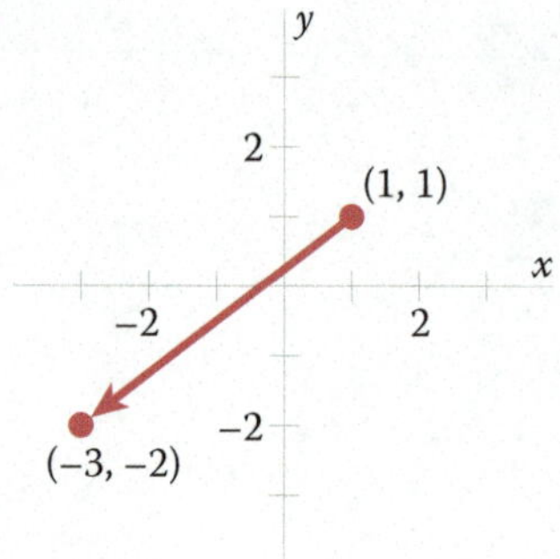

8.

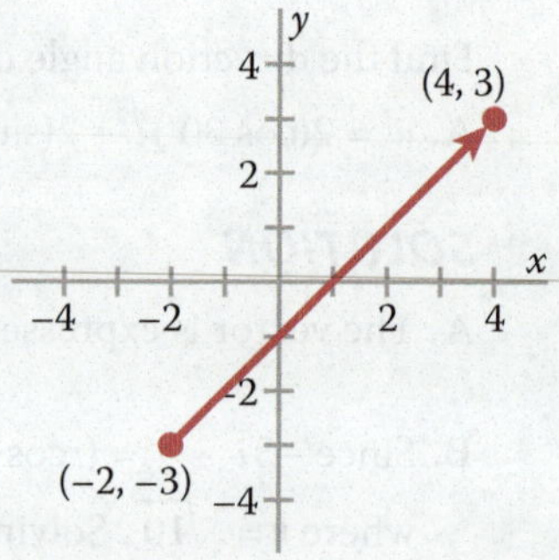

9.

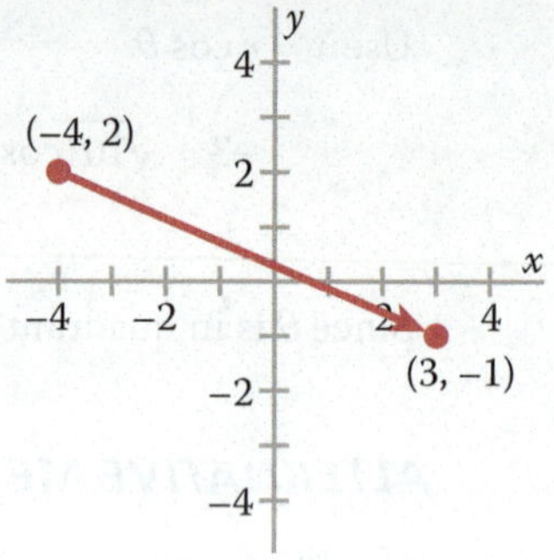

10.

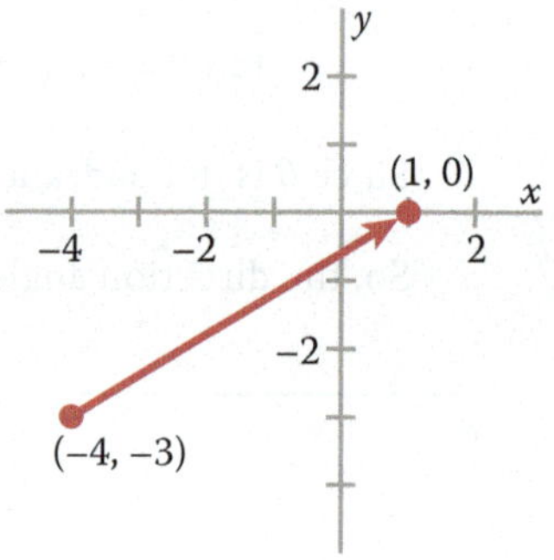

11.

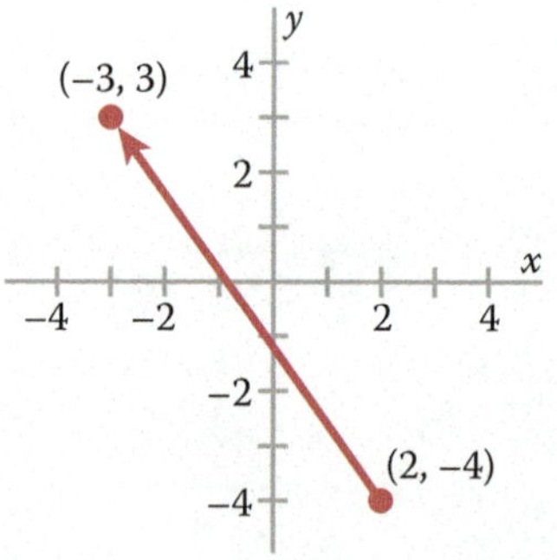

12.

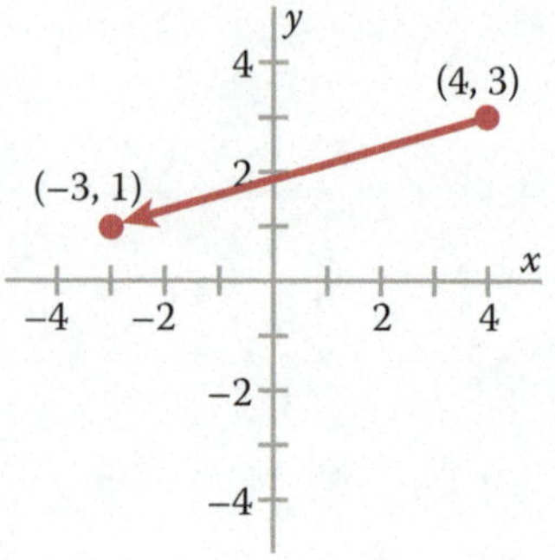

In Exercises 13–18, find the magnitude and direction angle of each vector.

13. $\langle -\sqrt{3}, 1 \rangle$

14. $\langle 3, -3 \rangle$

15. $\langle -4, -4 \rangle$

16. $\langle 3\sqrt{3}, 3 \rangle$

17. $\langle 2, -2\sqrt{3} \rangle$

18. $\langle -5, -5\sqrt{3} \rangle$

In Exercises 19–24, find the magnitude and direction angle of each vector. Round each direction angle to the nearest hundredth of a degree.

19. $\langle 3, 6 \rangle$

20. $\langle -2, 1 \rangle$

21. $\langle 7, -4 \rangle$

22. $\langle -7, -5 \rangle$

23. $\langle -5, 1 \rangle$

24. $\langle 2, 3 \rangle$

In Exercises 25–30, use $\vec{u} = \langle -2, 5 \rangle$, $\vec{v} = \langle 8.2, -1.3 \rangle$, and $\vec{w} = \langle -1, -1.5 \rangle$ to find each scalar multiple.

25. $2\vec{u}$

26. $4\vec{w}$

27. $5\vec{v}$

28. $-3\vec{u}$

29. $-2.5\vec{w}$

30. $-6\vec{v}$

In Exercises 31–36, use $\vec{u} = \langle -2, -2 \rangle$, $\vec{v} = \langle -6, 8 \rangle$, and $\vec{w} = \langle 3, 2 \rangle$ to graph each scalar multiple.

31. $3\vec{w}$

32. $-\frac{1}{2}\vec{v}$

33. $-1.5\vec{u}$

34. $2\vec{u}$

35. $0.5\vec{v}$

36. $-3\vec{w}$

In Exercises 37–40, use $\vec{w} = \langle 1.4, -3.6 \rangle$, $\vec{x} = \langle -3, -2 \rangle$, $\vec{y} = \langle -6, -11 \rangle$, and $\vec{z} = \langle 4, 1 \rangle$ to find each resultant vector.

37. $\vec{x} + \vec{z}$

38. $\vec{x} + \vec{y}$

39. $\vec{w} + \vec{y}$

40. $\vec{y} + \vec{z}$

In Exercises 41–44, use $\vec{u} = \langle -5, 1 \rangle$, $\vec{v} = \langle 1, 5 \rangle$, $\vec{w} = \langle 4, 3 \rangle$, and $\vec{x} = \langle -7, 6 \rangle$ to graph each resultant vector.

41. $\vec{u} + \vec{v}$

42. $\vec{v} + \vec{w}$

43. $\vec{w} + \vec{x}$

44. $\vec{v} + \vec{x}$

In Exercises 45–50, use $\vec{w} = \langle -2.8, -5.7 \rangle$, $\vec{x} = \langle -6.3, 6.3 \rangle$, $\vec{y} = \langle 8.9, 4.1 \rangle$, and $\vec{z} = \langle 3, 0 \rangle$ to find each resultant vector.

45. $\vec{x} - \vec{z}$

46. $\vec{y} - \vec{x}$

47. $\vec{w} - \vec{x}$

48. $\vec{z} - \vec{w}$

49. $\vec{y} - \vec{w}$

50. $\vec{x} - \vec{y}$

In Exercises 51–56, use $\vec{u} = \langle 5, -3 \rangle$, $\vec{v} = \langle 2, 6 \rangle$, $\vec{w} = \langle 3, 1 \rangle$, and $\vec{x} = \langle -4, -1 \rangle$ to graph each resultant vector.

51. $\vec{w} - \vec{x}$

52. $\vec{v} - \vec{w}$

53. $\vec{u} - \vec{v}$

54. $\vec{x} - \vec{u}$

55. $\vec{v} - \vec{u}$

56. $\vec{w} - \vec{v}$

In Exercises 57–66, find the component form of the vector with each given magnitude and direction angle θ.

57. $10, \theta = 30°$

58. $\sqrt{3}, \theta = 60°$

59. $9\sqrt{2}, \theta = 45°$

60. $6\sqrt{2}, \theta = 45°$

61. $6, \theta = 150°$

62. $2, \theta = 210°$

63. $3\sqrt{2}, \theta = 315°$

64. $6, \theta = 300°$

65. $10, \theta = 240°$

66. $5\sqrt{2}, \theta = 225°$

In Exercises 67–72, find the component form of each vector, to the nearest tenth, with the given magnitude and reference angle.

67.

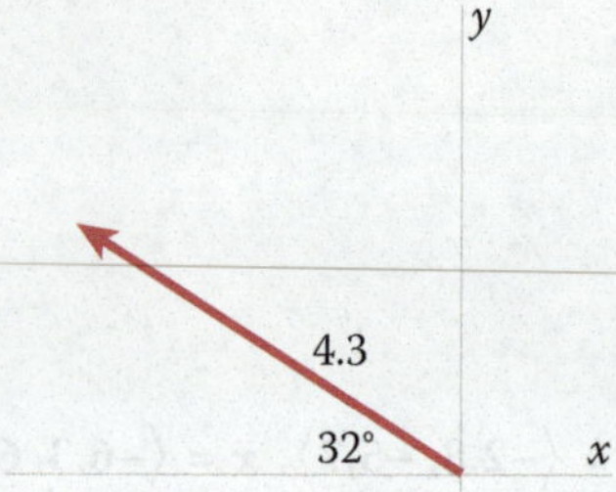

68.

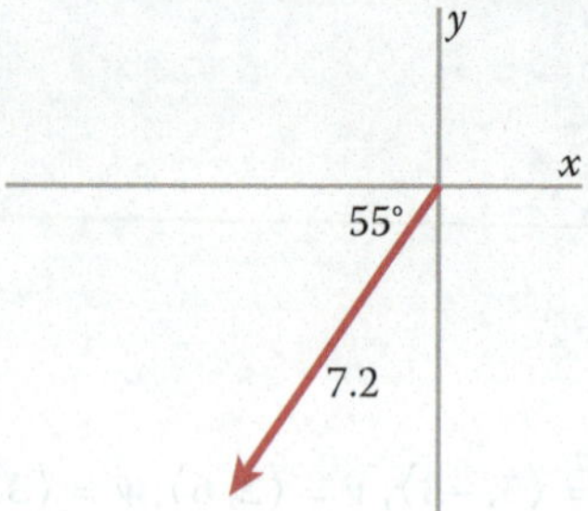

69.

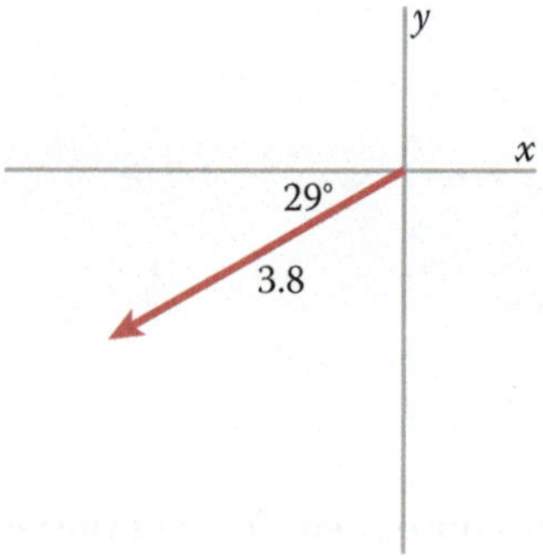

70.

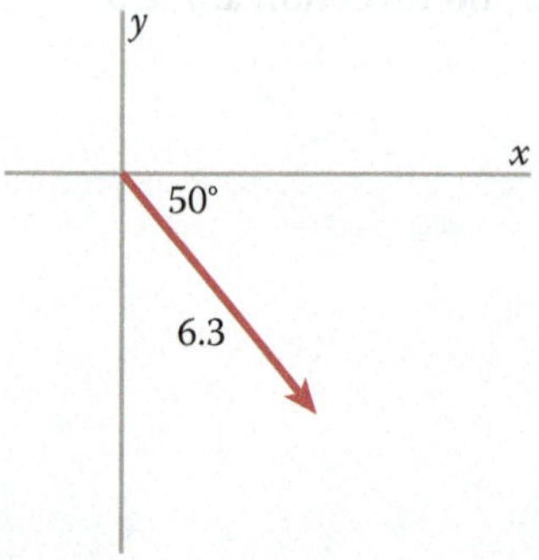

71.

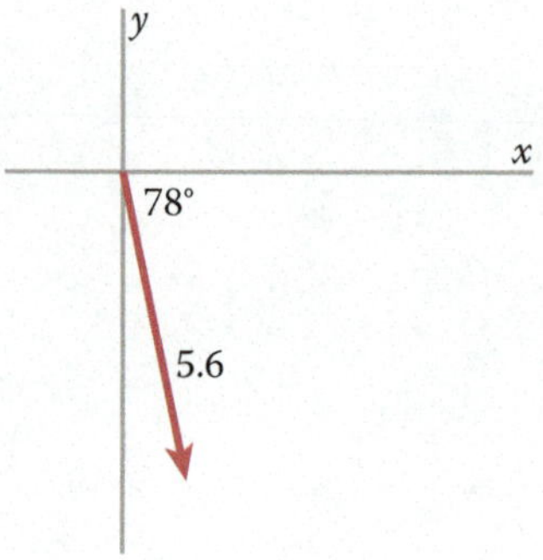

72.

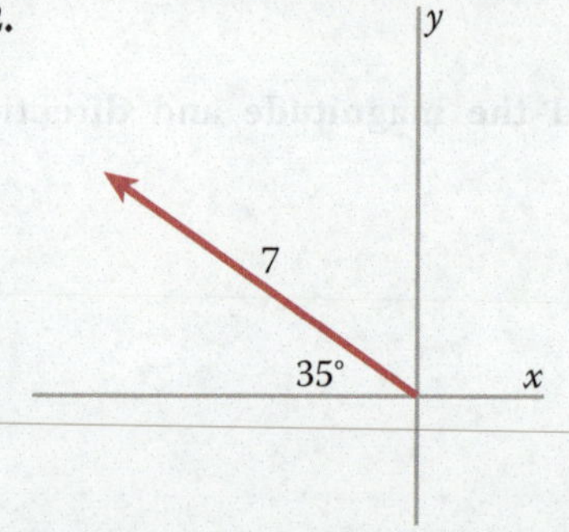

In Exercises 73–78, find a unit vector with the same direction as each given vector.

73. $\langle 3, 5\rangle$

74. $\langle 1/2, -2/5\rangle$

75. $\langle -1, -2\rangle$

76. $\langle -1/4, -5/3\rangle$

77. $\langle 1/3, -1/2\rangle$

78. $\langle -3/2, 4/3\rangle$

In Exercises 79–84, find the velocity vector in component form for a bus traveling at each given speed and bearing (on a compass).

79. 70 mph; 60°

80. 40 mph; 330°

81. 60 mph; 150°

82. 46 mph; 135°

83. 54 mph; 225°

84. 32 mph; 210°

In Exercises 85–88, suppose $\vec{v}$ is a vector with the given initial and terminal points. Write $\vec{v}$ as a linear combination of the standard unit vectors $\vec{i}$ and $\vec{j}$.

85. initial point (–3, –5), terminal point (1, 4)

86. initial point (1, –3), terminal point (–2, 5)

87. initial point (–2, 1), terminal point (3, –4)

88. initial point (3, 4), terminal point (–2, 0)

In Exercises 89–94, find the direction angle of each vector.

89. $\vec{w} = (3\cos 45°)\vec{i} + (3\sin 45°)\vec{j}$

90. $\vec{v} = (6\cos 240°)\vec{i} + (6\sin 240°)\vec{j}$

91. $\vec{z} = (2\cos 330°)\vec{i} + (2\sin 330°)\vec{j}$

92. $\vec{w} = \left(5\cos\dfrac{\pi}{3}\right)\vec{i} + \left(5\sin\dfrac{\pi}{3}\right)\vec{j}$

93. $\vec{y} = \left(9\cos\dfrac{2\pi}{3}\right)\vec{i} + \left(9\sin\dfrac{2\pi}{3}\right)\vec{j}$

94. $\vec{u} = \left(12\cos\dfrac{5\pi}{6}\right)\vec{i} + \left(12\sin\dfrac{5\pi}{6}\right)\vec{j}$

In Exercises 95–100, find the direction angle to the nearest tenth of a degree.

95. $\vec{z} = -2\vec{i} + \vec{j}$

96. $\vec{w} = -4\vec{i} - \vec{j}$

97. $\vec{v} = 6\vec{i} - 2\vec{j}$

98. $\vec{w} = 5\vec{i} - 3\vec{j}$

99. $\vec{x} = -7\vec{i} - 4\vec{j}$

100. $\vec{y} = -3\vec{i} + 2\vec{j}$

Extensions

In Exercises 101–104, use $\vec{w} = \langle -2.4, 1.8 \rangle$, $\vec{x} = \langle 4.5, -1.2 \rangle$, $\vec{y} = \langle -5, -2 \rangle$, and $\vec{z} = \langle 3, 4 \rangle$ to find each resultant vector.

101. $2\vec{z} - 3\vec{w}$

102. $6\vec{y} + 4\vec{x}$

103. $5\vec{x} + 2\vec{y}$

104. $3\vec{w} - 5\vec{z}$

In Exercises 105–108, write the initial velocity vector in component form for a projectile shot at the given angle and initial velocity. Round to the nearest hundredth.

105. 45°; 50 miles per hour

106. 62°; 37 miles per hour

107. 38°; 46 miles per hour

108. 29°; 56 miles per hour

109. Use $\vec{z} = \langle -5, -12 \rangle$ to find the unit vector in the same direction for $\vec{z}$, $2\vec{z}$, and $3\vec{z}$.

In Exercises 110–112, use the following information.

A boat is moving due east across a river with an average velocity of 15.2 feet per second. The river current is moving due south with an average velocity of 4.8 feet per second. Round each answer to the nearest tenth.

110. What is the resultant average velocity (both magnitude and direction) of the boat?

111. If the width of the river is 60 feet, then in how many seconds will the boat travel from shore to shore?

112. How far downstream has the boat traveled when it reaches the opposite shore?

7.4 THE DOT PRODUCT

OBJECTIVES

- Find the dot product of two vectors.
- Use properties of dot products.
- Use the dot product of a vector with itself to determine its magnitude.
- Find the angle between two vectors.
- Determine whether two vectors are orthogonal, parallel, or neither.
- Find the projection of one vector onto another.
- Decompose a vector into two orthogonal vectors.

PREREQUISITE VOCABULARY TERMS

Commutative Property
magnitude
resultant vector
scalar
standard position vector
trigonometric functions
vector
zero vector

7.4.1 The Dot Product of Vectors

In Section 7.3 we discussed operations with vectors, such as scalar multiplication of a vector and vector addition. Recall that in each case the operation produced a vector (specifically, a scalar multiple of a vector and a resultant vector). In this section another vector operation is introduced, the **dot product**. In this case, the operation produces a scalar instead of a vector.

IMPORTANT

The dot product of two vectors is a scalar (a real number), not a vector.

The Dot Product

If $\vec{u} = \langle u_1, u_2 \rangle$ and $\vec{v} = \langle v_1, v_2 \rangle$, then the dot product of $\vec{u}$ and $\vec{v}$ is $\vec{u} \cdot \vec{v} = u_1 v_1 + u_2 v_2$.

EXAMPLE 1 Finding a Dot Product

Find the dot product $\vec{u} \cdot \vec{v}$ if $\vec{v} = \langle 10, -4 \rangle$ and $\vec{u} = \langle 3, 6 \rangle$.

SOLUTION

Find the sum of the product of the horizontal components and the product of the vertical components.

$$\vec{u} \cdot \vec{v} = 3(10) + 6(-4) = 30 + (-24) = 6$$

Since multiplication is commutative for real numbers (i.e., $ab = ba$), changing the order of the components in each product of $u_1v_1 + u_2v_2$ does not change the value of the sum: $u_1v_1 + u_2v_2 = v_1u_1 + v_2u_2$. Therefore, the dot product of $\vec{u}$ and $\vec{v}$ ($u_1v_1 + u_2v_2$) is equal to the dot product of $\vec{v}$ and $\vec{u}$ ($v_1u_1 + v_2u_2$), that is, $\vec{u} \cdot \vec{v} = \vec{v} \cdot \vec{u}$. This property of the dot product is listed in the following table, along with several other properties of the dot product.

Properties of the Dot Product	
If $\vec{u}$, $\vec{v}$, and $\vec{w}$ are vectors and c is a scalar, then the following properties are true.	
$\vec{u} \cdot \vec{v} = \vec{v} \cdot \vec{u}$	*The dot product of $\vec{u}$ and $\vec{v}$ is equal to the dot product of $\vec{v}$ and $\vec{u}$.*
$\vec{0} \cdot \vec{v} = 0$	*The dot product of the zero vector and any other vector is 0.*
$\vec{u} \cdot (\vec{v} + \vec{w}) = \vec{u} \cdot \vec{v} + \vec{u} \cdot \vec{w}$	*The dot product of a vector and a resultant vector $\vec{v} + \vec{w}$ is equal to the sum of the dot product of the vector and $\vec{v}$ and the dot product of the vector and $\vec{w}$.*
$\vec{v} \cdot \vec{v} = \|\|\vec{v}\|\|^2$	*The dot product of a vector and itself is equal to the square of the vector's magnitude.*
$c(\vec{u} \cdot \vec{v}) = c\vec{u} \cdot \vec{v} = \vec{u} \cdot c\vec{v}$	*The product of a scalar and the dot product of two vectors is equal to the dot product of one of the vectors and the scalar multiple of the other vector.*

EXAMPLE 2 Using the Properties of the Dot Product

Find $-2(\vec{u} \cdot \vec{v}) + \vec{v} \cdot \vec{v}$ if $\vec{u} = \langle 3, 1 \rangle$ and $\vec{v} = \langle -3, 4 \rangle$.

SOLUTION

Use the applicable properties of the dot product: $\vec{v} \cdot \vec{v} = ||\vec{v}||^2$ and $c(\vec{u} \cdot \vec{v}) = c\vec{u} \cdot \vec{v} = \vec{u} \cdot c\vec{v}$.

$$-2(\vec{u} \cdot \vec{v}) + \vec{v} \cdot \vec{v} = -2\vec{u} \cdot \vec{v} + ||\vec{v}||^2$$

$$= -2\langle 3, 1 \rangle \cdot \langle -3, 4 \rangle + \left(\sqrt{(-3)^2 + 4^2}\right)^2 \quad \textit{Apply the properties.}$$

$$= \langle -6, -2 \rangle \cdot \langle -3, 4 \rangle + 25 \quad \textit{Find the scalar multiple and simplify.}$$

$$= -6(-3) + -2(4) + 25 \quad \textit{Find the dot product.}$$

$$= 35 \quad \textit{Simplify.}$$

ALTERNATIVE METHOD

Instead of using a property to evaluate $\vec{v} \cdot \vec{v}$, find the dot product of the vector and itself.

$$\vec{v} \cdot \vec{v} = \langle -3, 4 \rangle \cdot \langle -3, 4 \rangle = -3(-3) + 4(4) = 25$$

EXAMPLE 3 Using the Dot Product to Find the Magnitude of a Vector

If $\vec{v} \cdot \vec{v} = 20$, what is the magnitude of $\vec{v}$?

SOLUTION

By the properties of the dot product, $\vec{v} \cdot \vec{v} = ||\vec{v}||^2$. So, $20 = ||\vec{v}||^2$. Taking the square root of each side gives $||\vec{v}|| = \sqrt{20} = 2\sqrt{5}$. So, the magnitude of $\vec{v}$ is $2\sqrt{5}$.

REMEMBER

The magnitude of a vector is always positive, so only the positive square root is considered.

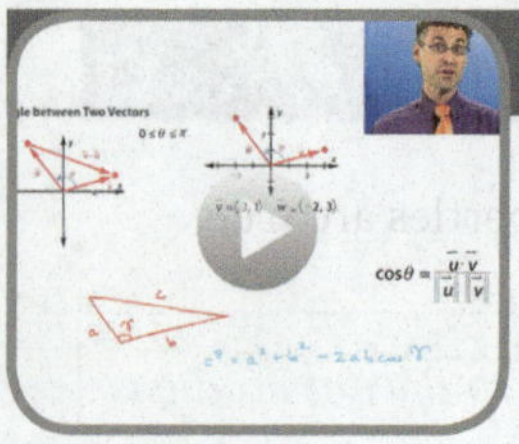

7.4.2 The Angle between Two Vectors

Consider two nonzero standard position vectors $\vec{u}$ and $\vec{v}$ with direction angles θ_u and θ_v, respectively. Regardless of their direction angles, there are two angles formed by these vectors (i.e., two angles with one vector as its initial side and the other as its terminal side, though these are not standard position angles). For example, if $\theta_u = 80°$ and $\theta_v = 20°$, then the two angles formed by $\vec{u}$ and $\vec{v}$ measure 60° (i.e., 80° – 20°) and 300° (i.e., 360° – (80° – 20°)). We will define the **angle between two vectors** as the angle θ between two standard position nonzero vectors such that $0 \le \theta \le \pi$ (that is, $0° \le \theta \le 180°$). In the case of $\vec{u}$ and $\vec{v}$, the angle between the two vectors is 60° because $0° \le 60° \le 180°$.

NOTICE THAT

The two angles formed by the vectors will have the same measure, 180°, if their directions are opposite.

So, if the direction angles of the two vectors are known, then the angle between the two vectors can be found by subtracting. If the result of subtracting the smaller angle from the larger angle is greater than 180°, then subtract that difference from 360° to find the angle between the vectors.

EXAMPLE 4 Finding the Angle between Two Vectors

Find the angle between $\vec{u} = \cos\frac{\pi}{4}\vec{i} + \sin\frac{\pi}{4}\vec{j}$ and $\vec{v} = \cos\frac{3\pi}{2}\vec{i} + \sin\frac{3\pi}{2}\vec{j}$.

SOLUTION

The direction angle of $\vec{u} = \cos\frac{\pi}{4}\vec{i} + \sin\frac{\pi}{4}\vec{j}$ is $\pi/4$.

The direction angle of $\vec{v} = \cos\frac{3\pi}{2}\vec{i} + \sin\frac{3\pi}{2}\vec{j}$ is $3\pi/2$.

Subtracting $\pi/4$ from $3\pi/2$ gives $5\pi/4$. Since $5\pi/4 > \pi$, subtract $5\pi/4$ from 2π to find the angle between the vectors. So, the angle between $\vec{u}$ and $\vec{v}$ is $\theta = 2\pi - (3\pi/2 - \pi/4) = 3\pi/4$.

If the direction angles of the two vectors $\vec{u}$ and $\vec{v}$ are not known, then the angle between the vectors θ can be found using the dot product of $\vec{u}$ and $\vec{v}$, and the magnitudes of $\vec{u}$ and $\vec{v}$.

The Cosine of the Angle Between Two Vectors

If θ is the angle between two standard position nonzero vectors $\vec{u}$ and $\vec{v}$, then $\cos\theta = \dfrac{\vec{u}\cdot\vec{v}}{\|\vec{u}\|\,\|\vec{v}\|}$.

EXAMPLE 5 Finding the Angle between Two Vectors

Find the angle between $\vec{u} = 3\vec{i} + \vec{j}$ and $\vec{v} = 2\vec{i} - 5\vec{j}$.

SOLUTION

Write the vectors in component form. $\vec{u} = \langle 3, 1\rangle$ $\vec{v} = \langle 2, -5\rangle$

Use the formula for the cosine of the angle between two vectors.

$$\cos\theta = \frac{3(2)+1(-5)}{\sqrt{3^2+1^2}\sqrt{2^2+(-5)^2}} = \frac{1}{\sqrt{10}\sqrt{29}} = \frac{1}{\sqrt{290}}$$

So, $\theta = \cos^{-1}\dfrac{1}{\sqrt{290}} \approx 86.6°$.

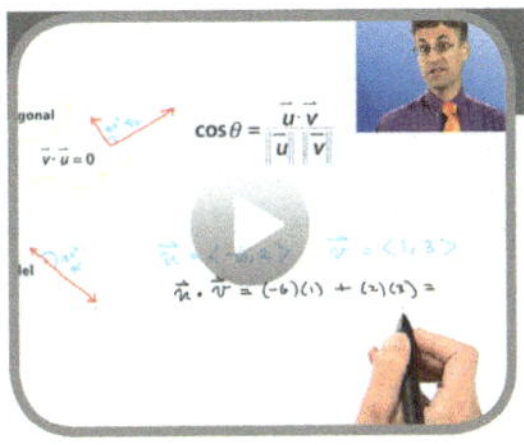

7.4.3 Determine Whether Two Vectors Are Parallel or Orthogonal

If the angle between two vectors is 0° or 180°, then we say the two vectors are **parallel vectors**.

If the angle between two vectors is 90° (i.e., the vectors form a right angle and are perpendicular), then we say the vectors are **orthogonal vectors**. The dot product of any two vectors is 0 if and only if they are orthogonal. It follows that since the dot product of any vector and the zero vector is 0, every vector is orthogonal with the zero vector.

Orthogonal Vectors and Parallel Vectors

$\vec{u}$ **and** $\vec{v}$ **are orthogonal**	$\vec{u}$ **and** $\vec{v}$ **are parallel**
The angle between $\vec{u}$ and $\vec{v}$ is $\pi/2$. $\vec{u}\cdot\vec{v}=0$	The angle between $\vec{u}$ and $\vec{v}$ is 0 or π.

IMPORTANT

Knowing that the angle between two vectors is $\pi/2$ is sufficient information for stating that the vectors are orthogonal. Similarly, knowing that their dot product is 0 is sufficient information for stating that the vectors are orthogonal.

EXAMPLE 6 Determining Whether Vectors are Orthogonal or Parallel

Determine whether $\vec{u}=5\cos\frac{\pi}{6}\vec{i}+5\sin\frac{\pi}{6}\vec{j}$ and $\vec{v}=7\cos\frac{2\pi}{3}\vec{i}+7\sin\frac{2\pi}{3}\vec{j}$ are orthogonal, parallel, or neither.

SOLUTION

The direction angle of $\vec{u}$ is $\pi/6$ and the direction angle of $\vec{v}$ is $2\pi/3$.

Subtracting $\pi/6$ from $2\pi/3$ gives $3\pi/6$, that is, $\pi/2$.

So, the angle between $\vec{u}$ and $\vec{v}$ is $\theta=\pi/2$. Therefore, the vectors are orthogonal.

7.4.4 Decompose a Vector into Two Orthogonal Vectors

If $\vec{u}$ is the nonzero resultant vector $\vec{u}=\vec{w}_1+\vec{w}_2$, where $\vec{w}_1$ and $\vec{w}_2$ are orthogonal (i.e., the angle between $\vec{w}_1$ and $\vec{w}_2$ is 90°), then $\vec{w}_1$ and $\vec{w}_2$ are called **vector components** of $\vec{u}$.

If $\vec{w}_1$ and $\vec{w}_2$ are vector components of $\vec{u}$, and $\vec{v}$ is any vector that is parallel to $\vec{w}_1$, then $\vec{w}_1$ is the **projection of $\vec{u}$ onto $\vec{v}$**, denoted $\vec{w}_1=\text{proj}_{\vec{v}}\vec{u}$. If $\vec{u}$ and $\vec{v}$ are nonzero vectors, then the component form of the projection of $\vec{u}$ onto $\vec{v}$ is given by the following formula.

$$\text{proj}_{\vec{v}}\vec{u}=\left(\frac{\vec{u}\cdot\vec{v}}{\|\vec{v}\|^2}\right)\vec{v}$$

Since the dot product of two vectors is a scalar and the magnitude of a vector is a scalar, $\frac{\vec{u}\cdot\vec{v}}{\|\vec{v}\|^2}$ is also a scalar. Therefore, the projection of $\vec{u}$ onto $\vec{v}$ is a scalar multiple of $\vec{v}$.

EXAMPLE 7 Finding the Projection of One Vector Onto Another

Find the projection of $\vec{v} = -\vec{i} + 4\vec{j}$ onto $\vec{z} = 2\vec{i} + \vec{j}$. Then decompose $\vec{v}$ as the sum of two orthogonal vectors, one of which is $\text{proj}_{\vec{z}}\vec{v}$.

SOLUTION

Use $\text{proj}_{\vec{z}}\vec{v} = \left(\frac{\vec{v} \cdot \vec{z}}{\|\vec{z}\|^2}\right)\vec{z}$ to find the projection of $\vec{v}$ onto $\vec{z}$ where $\vec{v} = \langle -1, 4 \rangle$ and $\vec{z} = \langle 2, 1 \rangle$.

$$\text{proj}_{\vec{z}}\vec{v} = \left(\frac{-1(2) + 4(1)}{\sqrt{2^2 + 1^2}^{\,2}}\right)\langle 2, 1 \rangle$$ *Substitute. Find the dot product and the magnitude.*

$$= \left(\frac{2}{5}\right)\langle 2, 1 \rangle$$ *Simplify the scalar.*

$$= \left\langle \frac{4}{5}, \frac{2}{5} \right\rangle$$ *Multiply the scalar by the components of $\vec{z}$.*

Since $\langle 4/5, 2/5 \rangle$ is the projection of $\vec{v}$ onto $\vec{z}$, we know that $\vec{w}_1 = \langle 4/5, 2/5 \rangle$ where $\vec{v} = \vec{w}_1 + \vec{w}_2$ and $\vec{w}_1$ and $\vec{w}_2$ are orthogonal.

Solve $\vec{v} = \vec{w}_1 + \vec{w}_2$ for $\vec{w}_2$. $\quad \vec{w}_2 = \vec{v} - \vec{w}_1 = \langle -1, 4 \rangle - \langle 4/5, 2/5 \rangle = \langle -9/5, 18/5 \rangle$

Therefore, $\vec{v} = \langle 4/5, 2/5 \rangle + \langle -9/5, 18/5 \rangle$.

CHECK

Find the sum of $\langle 4/5, 2/5 \rangle$ and $\langle -9/5, 18/5 \rangle$ to confirm that $\vec{v} = \langle 4/5, 2/5 \rangle + \langle -9/5, 18/5 \rangle$.

$$\langle 4/5, 2/5 \rangle + \langle -9/5, 18/5 \rangle = \langle -5/5, 20/5 \rangle = \langle -1, 4 \rangle = \vec{v} \checkmark$$

Find the dot product of $\langle 4/5, 2/5 \rangle$ and $\langle -9/5, 18/5 \rangle$ to confirm that the vectors are orthogonal.

$$\langle 4/5, 2/5 \rangle \cdot \langle -9/5, 18/5 \rangle = 4/5(-9/5) + 2/5(18/5) = -36/25 + 36/25 = 0$$

The dot product is 0, so $\langle 4/5, 2/5 \rangle$ and $\langle -9/5, 18/5 \rangle$ are orthogonal. ✓

SECTION 7.4 EXERCISES

Warm Up

Identify the property demonstrated in each equation.

1. $3(x + 2y) = 3x + 6y$

2. $y \cdot 5 = 5 \cdot y$

3. $0 \cdot 2x = 0$

Just the Facts

Fill in the blanks.

4. If $\vec{v} = \langle v_1, v_2 \rangle$ and $\vec{w} = \langle w_1, w_2 \rangle$, the dot product $\vec{v} \cdot \vec{w} =$ _____.

5. If $\vec{u}$ and $\vec{v}$ are two nonzero vectors, the angle θ, $0 \le \theta \le \pi$, between $\vec{u}$ and $\vec{v}$, is determined by the formula $\cos\theta =$ _____.

6. Vectors $\vec{u}$ and $\vec{v}$ are orthogonal if $\vec{u} \cdot \vec{v} =$ _____.

7. Vectors $\vec{u}$ and $\vec{v}$ are parallel if $\theta =$ _____.

8. If $\vec{u}$ and $\vec{v}$ are two nonzero vectors, the projection of $\vec{u}$ onto $\vec{v}$ is $\text{proj}_{\vec{v}}\vec{u} =$ _____

9. The dot product of two vectors is a(n) _____, not a vector.

10. The dot product $\vec{v} \cdot \vec{v}$ is equal to the square of the vector's _____.

Essential Skills

In Exercises 1–6, find each dot product.

1. $\langle 3, -5 \rangle \cdot \langle -6.1, 0 \rangle$

2. $\langle -2.6, 3.1 \rangle \cdot \langle -1.6, -4 \rangle$

3. $\langle -1, 2 \rangle \cdot \langle -4, 5.1 \rangle$

4. $\langle 2.3, 2 \rangle \cdot \langle -1.4, -2.8 \rangle$

5. $\langle 7, 8 \rangle \cdot \langle -0.1, -6 \rangle$

6. $\langle -1.6, 0.3 \rangle \cdot \langle 3.8, -1.4 \rangle$

In Exercises 7–12, find $\vec{u} \cdot (\vec{v} + \vec{w})$, $\vec{v} \cdot \vec{v}$, and $4(\vec{u} \cdot \vec{v})$.

7. $\vec{u} = \langle 4, -5 \rangle$, $\vec{v} = \langle 8, 2 \rangle$, $\vec{w} = \langle -3, 0 \rangle$

8. $\vec{u} = \langle -2, 3 \rangle$, $\vec{v} = \langle 7, -1.4 \rangle$, $\vec{w} = \langle -3.6, 5 \rangle$

9. $\vec{u} = \langle -1.2, -1 \rangle$, $\vec{v} = \langle -3.5, 4 \rangle$, $\vec{w} = \langle -4, -2 \rangle$

10. $\vec{u} = \langle -0.3, -1.9 \rangle$, $\vec{v} = \langle -0.1, 3.3 \rangle$, $\vec{w} = \langle 3, -5 \rangle$

11. $\vec{u} = \langle 3, 0.5 \rangle$, $\vec{v} = \langle -6, -2.3 \rangle$, $\vec{w} = \langle -1, 9 \rangle$

12. $\vec{u} = \langle -4.9, 0.8 \rangle$, $\vec{v} = \langle 0.1, 3.9 \rangle$, $\vec{w} = \langle 3.4, 0.9 \rangle$

In Exercises 13–18, find the magnitude of the vector given in each equation.

13. $\vec{x} \cdot \vec{x} = 40$

14. $7\vec{v} \cdot \vec{v} = 21$

15. $6\vec{u} \cdot \vec{u} = 876$

16. $2.8\vec{z} \cdot \vec{z} = 61.6$

17. $-3\vec{v} \cdot \vec{v} = -77.43$

18. $-5\vec{u} \cdot \vec{u} = -80$

In Exercises 19–22, find the angle between each pair of vectors.

19. $\vec{y} = 3\cos\frac{\pi}{7}\vec{i} + 3\sin\frac{\pi}{7}\vec{j}$ and $\vec{w} = \cos\frac{2\pi}{3}\vec{i} + \sin\frac{2\pi}{3}\vec{j}$

20. $\vec{u} = \cos\frac{5\pi}{6}\vec{i} + \sin\frac{5\pi}{6}\vec{j}$ and $\vec{v} = \cos\frac{7\pi}{4}\vec{i} + \sin\frac{7\pi}{4}\vec{j}$

21. $\vec{v} = \cos\frac{5\pi}{8}\vec{i} + \sin\frac{5\pi}{8}\vec{j}$ and $\vec{x} = \cos\frac{11\pi}{6}\vec{i} + \sin\frac{11\pi}{6}\vec{j}$

22. $\vec{z} = \cos\frac{7\pi}{4}\vec{i} + \sin\frac{7\pi}{4}\vec{j}$ and $\vec{u} = \cos\frac{\pi}{2}\vec{i} + \sin\frac{\pi}{2}\vec{j}$

In Exercises 23–38, find the angle to the nearest tenth of a degree between each pair of vectors.

23. $\vec{x} = \cos 102°\vec{i} + \sin 102°\vec{j}$ and $\vec{w} = 5\cos 36°\vec{i} + 5\sin 36°\vec{j}$

24. $\vec{u} = 8\cos 96°\vec{i} + 8\sin 96°\vec{j}$ and $\vec{w} = \cos 21°\vec{i} + \sin 21°\vec{j}$

25. $\vec{z} = 3\cos 206°\vec{i} + 3\sin 206°\vec{j}$ and $\vec{x} = \cos 65°\vec{i} + \sin 65°\vec{j}$

26. $\vec{v} = \cos 332°\vec{i} + \sin 332°\vec{j}$ and $\vec{z} = \cos 189°\vec{i} + \sin 189°\vec{j}$

27. $\vec{x} = \langle 3, 4 \rangle$ and $\vec{z} = \langle 5, 2 \rangle$

28. $\vec{w} = \langle -2, 5 \rangle$ and $\vec{v} = \langle 1, 3 \rangle$

29. $\vec{z} = \langle -4, 2.5 \rangle$ and $\vec{y} = \langle 0, -7 \rangle$

30. $\vec{u} = \langle 4.5, -2.5 \rangle$ and $\vec{v} = \langle -2, 2.8 \rangle$

31. $\vec{y} = \langle 5.2, 1 \rangle$ and $\vec{x} = \langle -2.1, 6 \rangle$

32. $\vec{u} = \langle 3.4, 4.4 \rangle$ and $\vec{v} = \langle 0.8, -2.6 \rangle$

33. $\vec{z} = 3\vec{i} + 3\vec{j}$ and $\vec{x} = 4\vec{i} - 5\vec{j}$

34. $\vec{u} = 4\vec{i} - 2\vec{j}$ and $\vec{w} = -3\vec{i} + 2\vec{j}$

35. $\vec{u} = -2\vec{i} + 4.1\vec{j}$ and $\vec{v} = -6\vec{i} - 3\vec{j}$

36. $\vec{v} = -2.3\vec{i} - 3.8\vec{j}$ and $\vec{w} = 4.6\vec{i} - 0.6\vec{j}$

37. $\vec{y} = 2.6\vec{i} - 1.3\vec{j}$ and $\vec{z} = 0.6\vec{i} + \vec{j}$

38. $\vec{u} = 3.1\vec{i} + 2.1\vec{j}$ and $\vec{v} = -2.2\vec{i} - 1.1\vec{j}$

In Exercises 39–42, using the three pairs of vectors given in each exercise, determine which pair of vectors is parallel, which pair is orthogonal, and which pair is neither.

39. $\vec{u} = -\dfrac{3}{10}\vec{i} + \dfrac{3}{4}\vec{j}$ and $\vec{w} = 3\vec{i} - \vec{j}$;

$\vec{v} = \langle 1/3, -1/3 \rangle$ and $\vec{w} = \langle -3, -3 \rangle$;

$\vec{u} = 3\vec{i} - 5\vec{j}$ and $\vec{v} = -9\vec{i} + 15\vec{j}$

40. $\vec{v} = \langle 1/2, -3/4 \rangle$ and $\vec{w} = \langle -2, -4 \rangle$;

$\vec{u} = 3\vec{i} - 4\vec{j}$ and $\vec{v} = -21\vec{i} + 28\vec{j}$;

$\vec{u} = \langle -3/5, 3/4 \rangle$ and $\vec{w} = \langle 4, -1 \rangle$

41. $\vec{v} = \langle 1/3, -1/2 \rangle$ and $\vec{w} = \langle -6, -4 \rangle$;

$\vec{u} = 4\vec{i} - 4\vec{j}$ and $\vec{v} = -16\vec{i} + 16\vec{j}$;

$\vec{u} = \langle -3/8, 3/4 \rangle$ and $\vec{w} = \langle 4, -1 \rangle$

42. $\vec{u} = 3\vec{i} - 4\vec{j}$ and $\vec{v} = -9\vec{i} + 12\vec{j}$;

$\vec{u} = \langle -3/7, 3/4 \rangle$ and $\vec{w} = \langle 4, -1 \rangle$;

$\vec{v} = \langle 1/3, -1/2 \rangle$ and $\vec{w} = \langle -9, -6 \rangle$

In Exercises 43–46, find each projection.

43. $\vec{u}$ onto $\vec{v}$ if $\vec{u} = \langle -2, 9 \rangle$ and $\vec{v} = \langle -1, 2 \rangle$

44. $\vec{v}$ onto $\vec{w}$ if $\vec{w} = \langle 2, -4 \rangle$ and $\vec{v} = \langle -1/2, 3/4 \rangle$

45. $\vec{v}$ onto $\vec{u}$ if $\vec{u} = \langle -1, 1 \rangle$ and $\vec{v} = \langle -2, 4 \rangle$

46. $\vec{v}$ onto $\vec{u}$ if $\vec{v} = \langle -1/2, 3/4 \rangle$ and $\vec{u} = \langle -1/5, -4 \rangle$

In Exercises 47–50, decompose $\vec{w}$ into the sum of two orthogonal vectors $\vec{w}_1$ and $\vec{w}_2$, where $\vec{w}_1 = \text{proj}_{\vec{v}}\vec{u}$.

47. $\vec{w} = \langle 2, 7 \rangle$ and $\vec{v} = \langle 1, -1 \rangle$

48. $\vec{w} = \langle 5, -12 \rangle$ and $\vec{v} = \langle -3, 4 \rangle$

49. $\vec{w} = \langle -6, 1 \rangle$ and $\vec{v} = \langle 2, 3 \rangle$

50. $\vec{w} = \langle 4, -12 \rangle$ and $\vec{v} = \langle -3, 2 \rangle$

Extensions

In Exercises 51–54, use vectors to find the interior angles to the nearest tenth of a degree for the triangle with the given vertices.

51. (9, 8), (3, 4), (7, 7)

52. (1, 1), (5, 2), (4, 6)

53. (0, 3), (6, 3), (5, 4)

54. (2, 5), (8, 7), (5, 0)

In Exercises 55–60, use $\vec{u} = \langle 1/2, -4 \rangle$, $\vec{v} = \langle -2, -1/3 \rangle$, and $\vec{w} = \langle -3, 2 \rangle$ to find the indicated quantity.

55. $\| \vec{u} \| - \| \vec{v} \|$

56. $(\vec{w} \cdot \vec{u}) \cdot \vec{v}$

57. $\| \vec{v} \| - 4$

58. $(\vec{w} \cdot \vec{v}) + (\vec{v} \cdot \vec{u})$

59. $(2\vec{u} + \vec{v}) \cdot \vec{w}$

60. $\| \vec{u} \|^2$

7.5 TRIGONOMETRIC FORM AND ROOTS OF COMPLEX NUMBERS

OBJECTIVES

- Represent a complex number as a point in the complex plane.
- Find the modulus of a complex number.
- Express a complex number in trigonometric form.
- Multiply and divide complex numbers in trigonometric form.
- Use DeMoivre's Theorem to find a power of a complex number.
- Find the *n*th roots of a complex number.

PREREQUISITE VOCABULARY TERMS

complex number
degree of a polynomial
double-angle formulas
imaginary number
sum and difference formulas
trigonometric functions

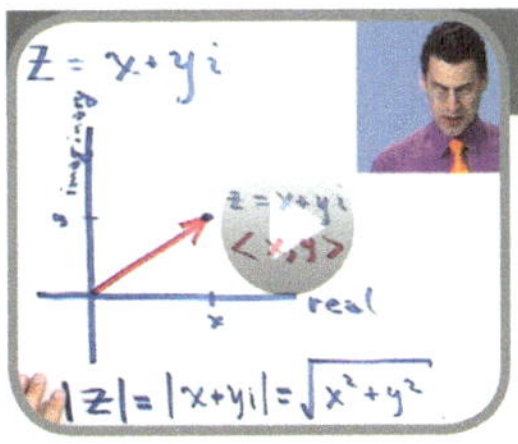

7.5.1 Graphing a Complex Number and Finding Its Modulus

Review of Complex Numbers

Complex numbers can be written in the standard form $a + bi$, where a and b are real numbers and i is the imaginary unit (i.e., $i = \sqrt{-1}$). The real numbers a and b in a complex number $a + bi$ are called the **real part** and the **imaginary part**, respectively.

Complex numbers are equal when their real parts are equal and their imaginary parts are equal. Therefore, $a + bi = c + di$ if and only if $a = c$ and $b = d$.

Complex numbers were added, subtracted, multiplied, and divided in preceding chapters.

Additionally, we saw that equations of the form $a^n = b$, where a is an algebraic expression, n is an *even* whole number, and b is a *negative* real number, have complex solutions with a nonzero imaginary part. For example, the equation $(x - 2)^2 = -9$ has two complex solutions: $2 + 3i$ and $2 - 3i$.

NOTICE THAT

The bi-term in $a + bi$ is an imaginary number.

REMEMBER

To solve $(x - 2)^2 = -9$, first take ± the square root of both sides, then add 2 to both sides. Simplifying $\pm\sqrt{-9}$ yields the imaginary term of each complex solution: $3i$ and $-3i$.

Graphing a Complex Number

Just as real numbers can be plotted on a real number line, and ordered pairs of real numbers can be graphed on a rectangular (Cartesian) plane (where the horizontal and vertical axes are real number lines), complex numbers can be plotted on a coordinate plane called a **complex plane**.

In a complex plane, the horizontal axis (**real axis**) is a real number line and the vertical axis (**imaginary axis**) is an imaginary number line.

If z is the complex number $x + yi$, then the graph of $z = x + yi$ is the point (x, y) on a complex plane, as shown in Figure 7.5a. The procedure for graphing a complex number in a complex plane is demonstrated in **Example 1**.

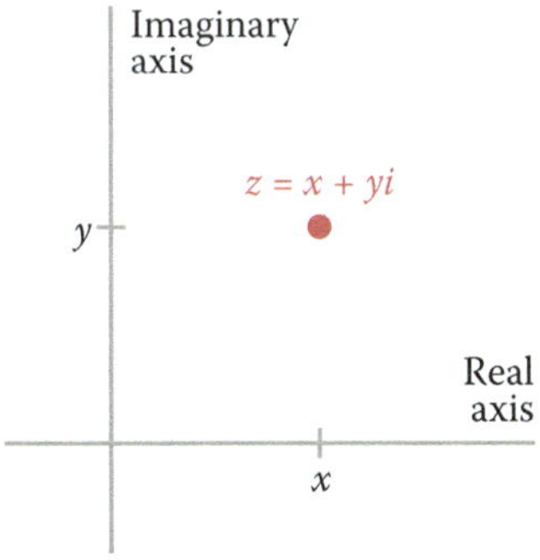

Figure 7.5a

EXAMPLE 1 Plotting a Complex Number on the Complex Plane

TIP

For $x + yi$:
$x > 0, y > 0 \Rightarrow$ quadrant I
$x < 0, y > 0 \Rightarrow$ quadrant II
$x < 0, y < 0 \Rightarrow$ quadrant III
$x > 0, y < 0 \Rightarrow$ quadrant IV
$x = 0 \Rightarrow$ on imaginary axis
$y = 0 \Rightarrow$ on real axis

Plot $z = 3 + 2i$, $w = 2 - i$, and $p = -2i$ on the complex plane.

SOLUTION

Use each complex number's real part and imaginary part to plot the point.

In $z = 3 + 2i$, the real part is 3 (so $x = 3$) and the imaginary part is 2 (so $y = 2$). To plot $z = 3 + 2i$, move 3 units to the right from the origin (since $x = 3 > 0$), then move 2 units up (since $y = 2 > 0$). Therefore, z is in quadrant I.

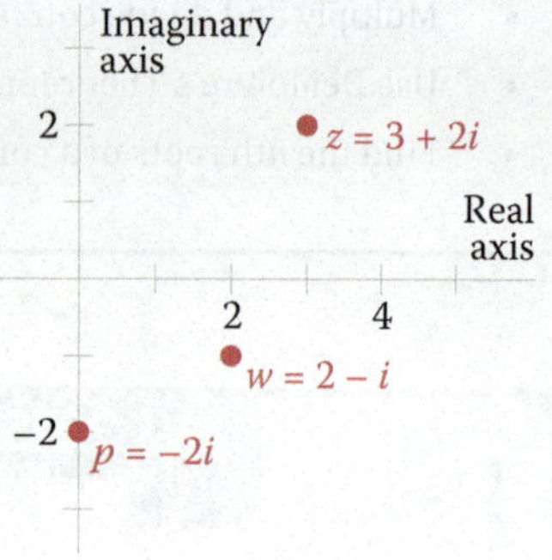

To plot $w = 2 - i$, move 2 units to the right from the origin (since $2 > 0$), then move 1 unit down (since $-1 < 0$), because the real part is 2 and the imaginary part is -1. Therefore, w is in quadrant IV.

To plot $p = -2i$, move 2 units down from the origin, because the real part is 0 and the imaginary part is -2. Therefore, p is on the imaginary axis.

The Modulus of a Complex Number

The **modulus** of a complex number (or **absolute value of a complex number**) is the distance from the origin to the point corresponding to the complex number on the complex plane. This measure is sometimes referred to as the *length* of the complex number.

The Modulus or the Absolute Value of a Complex Number

The modulus of a complex number $z = x + yi$, denoted $|z|$, is $|z| = |x + yi| = \sqrt{x^2 + y^2}$.

EXAMPLE 2 Finding the Modulus of a Complex Number

Find the modulus of $w = 2 - i$.

SOLUTION

Substitute $x = 2$ and $y = -1$ into the modulus formula and simplify.

$$|w| = |2 - i| = \sqrt{2^2 + (-1)^2} = \sqrt{5}$$

7.5.2 Expressing a Complex Number in Trigonometric or Polar Form

Let z be the complex number $x + yi$, let r be the modulus of z, and let θ be the angle between the positive real axis and the ray containing z and the origin (Figure 7.5b). Then θ is a standard position angle with the point (x, y) on its terminal side. Therefore, $\sin\theta = y/r$ and $\cos\theta = x/r$. Solving these equations for x and y, respectively, gives $y = r\sin\theta$ and $x = r\cos\theta$.

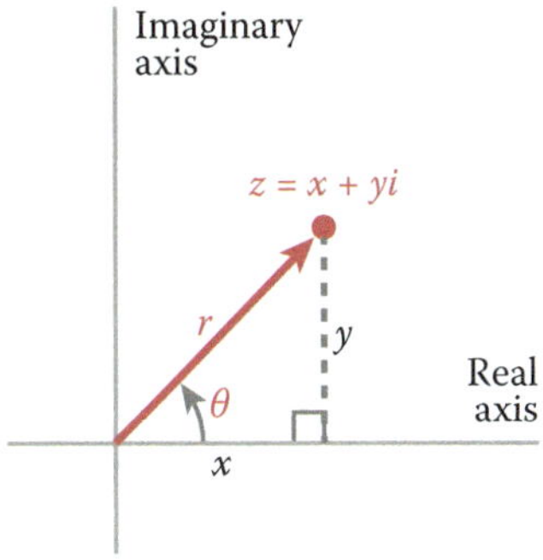

Figure 7.5b

The complex number $z = x + yi$ can be written in terms of $\sin\theta$ and $\cos\theta$ by substituting $r\cos\theta$ and $r\sin\theta$ for x and y, respectively, in $z = x + yi$, which gives $r\cos\theta + (r\sin\theta)i$. Factoring r from each term in the equation gives the complex number's **trigonometric form** (or **polar form**).

Trigonometric Form (or Polar Form) of a Complex Number

$$z = x + yi = r(\cos\theta + i\sin\theta)$$

Steps for Writing a Complex Number in Trigonometric Form

1. Find the complex number's modulus r.
2. Find the angle θ between the positive real axis and the ray containing the graph of the complex number and the origin.
3. Substitute the r- and θ-values into $r(\cos\theta + i\sin\theta)$.

EXAMPLE 3 Writing a Complex Number in Trigonometric Form

Write $w = 2 - i$ in trigonometric form. Round θ to the nearest degree if an exact value of θ cannot be found.

SOLUTION

1. From **Example 2**, the modulus of $w = 2 - i$ is $\sqrt{5}$, so $r = \sqrt{5}$.

2. Use a trigonometric function (e.g., $\cos\theta = x/r$) to find θ.

$$\cos\theta = 2/\sqrt{5}$$

Using a calculator, we see that $\cos^{-1}\left(2/\sqrt{5}\right) \approx 27°$. However, since $x = 2 > 0$, and $y = -1 < 0$, θ is in quadrant IV. Therefore, $\theta \approx 360° - 27° = 333°$.

3. Substitute $r = \sqrt{5}$ and $\theta \approx 333°$ into $r(\cos\theta + i\sin\theta)$ to write $w = 2 - i$ in trigonometric form.

$$w = 2 - i = r(\cos\theta + i\sin\theta) \approx \sqrt{5}(\cos 333° + i\sin 333°)$$

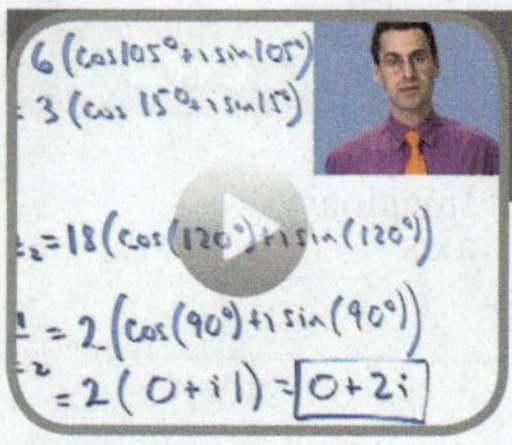

7.5.3 Multiplying and Dividing Complex Numbers in Trigonometric or Polar Form

Multiplying the complex numbers $z_1 = r_1(\cos\theta_1 + i\sin\theta_1)$ and $z_2 = r_2(\cos\theta_2 + i\sin\theta_2)$ and then applying the sine of a sum formula and the cosine of a sum formula to simplify yields a formula for the product of any two complex numbers in trigonometric form.

$$\begin{aligned} z_1 z_2 &= (r_1(\cos\theta_1 + i\sin\theta_1)(r_2(\cos\theta_2 + i\sin\theta_2) \\ &= r_1 r_2[\cos\theta_1\cos\theta_2 + i\cos\theta_1\sin\theta_2 + i\sin\theta_1\cos\theta_2 + i^2\sin\theta_1\sin\theta_2] && \textit{FOIL} \\ &= r_1 r_2[\cos\theta_1\cos\theta_2 + i\cos\theta_1\sin\theta_2 + i\sin\theta_1\cos\theta_2 - \sin\theta_1\sin\theta_2] && \textit{Simplify } (i^2 = -1). \\ &= r_1 r_2[\cos\theta_1\cos\theta_2 - \sin\theta_1\sin\theta_2 + i(\cos\theta_1\sin\theta_2 + \sin\theta_1\cos\theta_2)] && \textit{Factor out i.} \\ &= r_1 r_2[\cos(\theta_1 + \theta_2) + i\sin(\theta_1 + \theta_2)] && \textit{Sum Formulas} \end{aligned}$$

REMEMBER

Cosine of a Sum Formula

$\cos(\theta_1 + \theta_2) = \cos\theta_1\cos\theta_2 - \sin\theta_1\sin\theta_2$

Sine of a Sum Formula

$\sin(\theta_1 + \theta_2) = \sin\theta_1\cos\theta_2 + \cos\theta_1\sin\theta_2$

A formula for the quotient of two complex numbers can be found in a similar way.

The Product and Quotient of Two Complex Numbers in Trigonometric Form

If $z_1 = x_1 + y_1 i = r_1(\cos\theta_1 + i\sin\theta_1)$ and $z_2 = x_2 + y_2 i = r_2(\cos\theta_2 + i\sin\theta_2)$, then the product $z_1 z_2$ and quotient z_1/z_2 (where $z_2 \neq 0$) are given by the following formulas.

$$z_1 z_2 = (x_1 + y_1 i)(x_2 + y_2 i) = r_1 r_2[\cos(\theta_1 + \theta_2) + i\sin(\theta_1 + \theta_2)]$$

$$\frac{z_1}{z_2} = \frac{x_1 + y_1 i}{x_2 + y_2 i} = \frac{r_1}{r_2}[\cos(\theta_1 - \theta_2) + i\sin(\theta_1 - \theta_2)]$$

NOTICE THAT

In the product formula, the r-values are multiplied and the θ-values are added. In the quotient formula, the r-values are divided and the θ-values are subtracted.

EXAMPLE 4 Multiplying and Dividing Two Complex Numbers in Trigonometric Form

Write zw and z/w in the form $a + bi$, if $w = 2(\cos 25° + i\sin 25°)$ and $z = 5(\cos 125° + i\sin 125°)$.

SOLUTION

Use the formulas for the product and quotient of two complex numbers in trigonometric form.

$$\begin{aligned} zw &= 5(\cos 125° + i\sin 125°)\cdot 2(\cos 25° + i\sin 25°) && r_1 = 5,\ r_2 = 2,\ \theta_1 = 125,\ \theta_2 = 25 \\ &= 5\cdot 2[\cos(125° + 25°) + i\sin(125° + 25°)] && \textit{Substitute into the product formula.} \\ &= 10[\cos 150° + i\sin 150°] && \textit{Multiply and add.} \\ &= 10[-\sqrt{3}/2 + i(1/2)] && \textit{Evaluate the sine and cosine.} \\ &= -5\sqrt{3} + 5i && \textit{Simplify and write in standard form.} \end{aligned}$$

NOTICE THAT

The trigonometric form of the product zw is $10[\cos 150° + i\sin 150°]$.

$$\begin{aligned} z/w &= 5(\cos 125° + i\sin 125°)/[2(\cos 25° + i\sin 25°)] && r_1 = 5,\ r_2 = 2,\ \theta_1 = 125,\ \theta_2 = 25 \\ &= (5/2)[\cos(125° - 25°) + i\sin(125° - 25°)] && \textit{Substitute into the quotient formula.} \\ &= (5/2)[\cos 100° + i\sin 100°] && \textit{Subtract.} \\ &\approx (5/2)[-0.17 + i(0.98)] && \textit{Use a calculator and round.} \\ &\approx -0.43 + 2.45i && \textit{Simplify and write in standard form.} \end{aligned}$$

NOTICE THAT

The trigonometric form of the quotient z/w is $(5/2)[\cos 100° + i\sin 100°]$.

7.5.4 Using DeMoivre's Theorem to Raise a Complex Number to a Power

Just as FOIL can be used to expand a power of a binomial, FOIL can also be used to expand a power of a complex number in trigonometric form. For example, z^2 (where $z = r(\cos\theta + i\sin\theta)$) is expanded and simplified here.

$$z^2 = [r(\cos\theta + i\sin\theta)]^2$$

$$z^2 = r^2(\cos\theta + i\sin\theta)^2 \qquad \textit{Power of a Product Property}$$

$$= r^2(\cos\theta + i\sin\theta)(\cos\theta + i\sin\theta) \qquad \textit{Expand the power of the binomial.}$$

$$= r^2(\cos\theta\cos\theta + i\sin\theta\cos\theta + i\sin\theta\cos\theta + i\cdot i\sin\theta\sin\theta) \qquad \textit{FOIL}$$

$$= r^2(\cos^2\theta + 2i\sin\theta\cos\theta - \sin^2\theta) \qquad \textit{Simplify } (i^2 = -1).$$

$$= r^2(\cos^2\theta + i\sin 2\theta - \sin^2\theta) \qquad \textit{Double-Angle Formula: } \sin 2\theta = 2\sin\theta\cos\theta.$$

$$= r^2(\cos 2\theta + i\sin 2\theta) \qquad \textit{Double-Angle Formula: } \cos 2\theta = \cos^2\theta - \sin^2\theta.$$

TIP

To simplify, write $\cos\theta\cos\theta$ and $\sin\theta\sin\theta$ as powers, combine like terms ($i\sin\theta\cos\theta + i\sin\theta\cos\theta = 2i\sin\theta\cos\theta$), and write i^2 as -1.

Therefore, if $z = r(\cos\theta + i\sin\theta)$, then $z^2 = r^2(\cos 2\theta + i\sin 2\theta)$. Thus, the square of a complex number in trigonometric form (i.e., $z = r(\cos\theta + i\sin\theta)$) can be found by simply squaring its modulus r and doubling θ.

DeMoivre's Theorem states that this relationship between a complex number in trigonometric form z and its square can be extended to any positive integer power of z.

DeMoivre's Theorem: Powers of a Complex Number

If $z = x + yi = r(\cos\theta + i\sin\theta)$ and n is an integer such that $n > 0$, then $z^n = (x + yi)^n = r^n(\cos n\theta + i\sin n\theta)$.

EXAMPLE 5 Simplifying a Power of a Complex Number

If $w = 5\left(\cos\frac{\pi}{8} + i\sin\frac{\pi}{8}\right)$, write w^4 in trigonometric form and in standard form.

SOLUTION

Use DeMoivre's Theorem.

$$w^4 = \left(5\left(\cos\frac{\pi}{8} + i\sin\frac{\pi}{8}\right)\right)^4$$
$$= 5^4\left(\cos\frac{4\pi}{8} + i\sin\frac{4\pi}{8}\right)$$
$$= 625\left(\cos\frac{\pi}{2} + i\sin\frac{\pi}{2}\right)$$

Simplify and write in the standard form of a complex number.

$$w^4 = 625\left(\cos\frac{\pi}{2} + i\sin\frac{\pi}{2}\right) = 625(0 + i\cdot 1) = 625i$$

7.5.5 Roots of Complex Numbers

We have seen that all *real* solutions of the polynomial equation $x^4 = 81$ can be found by taking ± the fourth root of each side.

$$x^4 = 81 \quad \Rightarrow \quad x = \pm\sqrt[4]{81} = \pm 3$$

Therefore, the real solutions of $x^4 = 81$ are $x = 3$ and $x = -3$.

REMEMBER

By the conjugate pair theorem, if $a + bi$ (where a and b are real numbers such that $b \neq 0$) is a complex solution to a polynomial equation with only real coefficients, then its conjugate $a - bi$ is also a solution.

However, recall that (by the Fundamental Theorem of Algebra) every polynomial equation with degree n has n complex solutions. Therefore, since the degree of $x^4 = 81$ is 4, $x^4 = 81$ must have four complex solutions. All four complex solutions of $x^4 = 81$ can be found by factoring.

$x^4 - 81 = 0$	*Write the equation in general form.*
$(x + 3)(x - 3)(x^2 + 9) = 0$	*Factor.*
$x + 3 = 0$ or $x - 3 = 0$ or $x^2 + 9 = 0$	*Set each factor equal to 0.*
$x = -3$ $\quad$ $x = 3$ $\quad$ $x = \pm\sqrt{-9}$	*Solve each equation for x.*

Therefore, the four complex solutions of the fourth degree equation $x^4 = 81$ are $x = -3$, $x = 3$, $x = 3i$, and $x = -3i$. Each solution is called a "fourth root of 81" because the 4th power of each value is equal to 81.

REMEMBER

$i^4 = i^2\,i^2 = (-1)(-1) = 1$

$$3^4 = 81 \qquad (-3)^4 = 81 \qquad (3i)^4 = 3^4 i^4 = (81)(1) = 81 \qquad (-3i)^4 = (-3)^4 i^4 = (81)(1) = 81$$

In general, if z and $a + bi$ are complex numbers and n is a positive integer such that $z = (a + bi)^n$, then $a + bi$ is an ***n*th root of the complex number *z***. From the definition of an nth root of a complex number z and DeMoivre's Theorem for powers of a complex number, the following result can be concluded.

The *n*th Roots of a Complex Number

If n is an integer such that $n > 0$, and z is a complex number such that $z = x + yi = r(\cos\theta + i\sin\theta)$,
then there are exactly n distinct nth roots of z given by

$$\sqrt[n]{r}\left(\cos\frac{\theta + 2\pi k}{n} + i\sin\frac{\theta + 2\pi k}{n}\right),$$

where $k = 0, 1, 2, \ldots, n - 1$.

TIP

Each nth root of a complex number z is given by evaluating this expression. Evaluate the expression n times to find all n of the nth roots: once where $k = 0$, once where $k = 1, \ldots$, and then finally where $k = n - 1$. The value of n stays constant in each instance of the expression.

Steps for Finding the *n*th Roots of a Complex Number *z*

❶ Find r (the modulus of z) and θ (the angle between the positive real axis and a ray containing the graph of z and the origin).

❷ Substitute the values of n, r, and θ into $\sqrt[n]{r}\left(\cos\frac{\theta + 2\pi k}{n} + i\sin\frac{\theta + 2\pi k}{n}\right)$ and simplify.

❸ Evaluate the expression from step ❷ n times, each time using a different value of k (i.e., $k = 0, k = 1, k = 2, \ldots,$ and $k = n - 1$).

This procedure is used in **Example 6** to find all fourth roots of 81, which we know from the preceding example are -3, 3, $3i$, and $-3i$.

EXAMPLE 6 Finding the nth Roots of a Complex Number

Find all fourth roots of 81.

SOLUTION

❶ Since $81 = 81 + 0i$, 81 is the point $(81, 0)$ on a complex plane. *(81, 0) on the positive real axis.*

Substitute $x = 81$ and $y = 0$ into the modulus formula and simplify to find r.

$$r = \sqrt{81^2 + 0^2} = \sqrt{81^2} = 81$$

Use a trigonometric function (e.g., $\sin\theta = y/r$) to find θ.

$$\sin\theta = 0/81 \Rightarrow \sin\theta = 0 \quad \text{Since } \sin 0 = 0,\ \theta = 0.$$

❷ Substitute $n = 4$, $r = 81$, and $\theta = 0$ into $\sqrt[n]{r}\left(\cos\dfrac{\theta + 2\pi k}{n} + i\sin\dfrac{\theta + 2\pi k}{n}\right)$ and simplify.

$$\sqrt[4]{81}\left(\cos\frac{0 + 2\pi k}{4} + i\sin\frac{0 + 2\pi k}{4}\right) = 3\left(\cos\frac{\pi k}{2} + i\sin\frac{\pi k}{2}\right)$$

❸ Repeat the process of substituting a k-value into the expression and simplifying four times (once for each k-value: $k = 0$, $k = 1$, $k = 2$, and $k = 3$) to find the 4 roots.

$$k = 0:\ 3\left(\cos\frac{\pi(0)}{2} + i\sin\frac{\pi(0)}{2}\right) = 3(\cos 0 + i\sin 0) = 3(1 + i(0)) = 3$$

$$k = 1:\ 3\left(\cos\frac{\pi(1)}{2} + i\sin\frac{\pi(1)}{2}\right) = 3\left(\cos\frac{\pi}{2} + i\sin\frac{\pi}{2}\right) = 3(0 + i(1)) = 3i$$

$$k = 2:\ 3\left(\cos\frac{\pi(2)}{2} + i\sin\frac{\pi(2)}{2}\right) = 3(\cos\pi + i\sin\pi) = 3(-1 + i(0)) = -3$$

$$k = 3:\ 3\left(\cos\frac{\pi(3)}{2} + i\sin\frac{\pi(3)}{2}\right) = 3\left(\cos\frac{3\pi}{2} + i\sin\frac{3\pi}{2}\right) = 3(0 + i(-1)) = -3i$$

Therefore, the fourth roots of 81 are -3, 3, $3i$, and $-3i$.

TIP

Since the graph of $81 + 0i$ is a point 81 units to the right of the origin on the positive real axis, we know that the distance between the origin and the point is 81. Additionally, we know that the angle between the positive real axis and itself is 0 radians. Thus, $r = 81$ and $\theta = 0$.

The Solutions of an nth Degree Polynomial Equation and the nth Roots of a Complex Number

We have now seen that the complex solutions of the equation $x^4 = 81$ are the fourth roots of 81. In general, if n is a positive integer and z is a complex number, then the complex solutions of the polynomial equation $x^n = z$ are the nth roots of z. Therefore, if a polynomial equation can be written in the form $x^n = z$, then all n complex solutions of the polynomial equation can be found using the procedure for finding the nth roots of z. This procedure should be used to find the solutions of a polynomial equation when the equation cannot be solved easily by factoring.

IMPORTANT

The complex solutions of a polynomial equation in terms of x are the nth roots of z only when the equation can be written in the form $x^n = z$, where n is positive integer and z is a complex number.

EXAMPLE 7 Finding All Solutions of a Polynomial Equation

Find all solutions of $2x^4 + 32 = 0$.

SOLUTION

Subtracting 32 from each side and then dividing each side by 2 yields $x^4 = -16$. Since 4 is a positive integer and -16 is a complex number, the solutions of $2x^4 + 32 = 0$ are the fourth roots of -16. So, find the fourth roots of -16.

NOTICE THAT

The distance from (−16, 0) to the origin is 16 units. The angle between the positive real axis and the negative real axis is π radians.

❶ On a complex plane, −16 is the point (−16, 0) on the negative real axis, so $r = 16$ and $\theta = \pi$.

❷ Substitute $n = 4$, $r = 16$, and $\theta = \pi$ into the expression for the nth roots of z and simplify.

$$\sqrt[4]{16}\left(\cos\frac{\pi+2\pi k}{4}+i\sin\frac{\pi+2\pi k}{4}\right)=2\left(\cos\frac{\pi+2\pi k}{4}+i\sin\frac{\pi+2\pi k}{4}\right)$$

❸ Repeat the process of substituting a k-value into the expression and simplifying four times (once for each k-value: $k = 0$, $k = 1$, $k = 2$, and $k = 3$) to find the 4 roots.

$$k=0:\ 2\left(\cos\frac{\pi}{4}+i\sin\frac{\pi}{4}\right)=2\left(\frac{\sqrt{2}}{2}+i\frac{\sqrt{2}}{2}\right)=\sqrt{2}+i\sqrt{2} \qquad \frac{\pi+2\pi(0)}{4}=\frac{\pi}{4}$$

$$k=1:\ 2\left(\cos\frac{3\pi}{4}+i\sin\frac{3\pi}{4}\right)=2\left(-\frac{\sqrt{2}}{2}+i\frac{\sqrt{2}}{2}\right)=-\sqrt{2}+i\sqrt{2} \qquad \frac{\pi+2\pi(1)}{4}=\frac{3\pi}{4}$$

$$k=2:\ 2\left(\cos\frac{5\pi}{4}+i\sin\frac{5\pi}{4}\right)=2\left(-\frac{\sqrt{2}}{2}-i\frac{\sqrt{2}}{2}\right)=-\sqrt{2}-i\sqrt{2} \qquad \frac{\pi+2\pi(2)}{4}=\frac{5\pi}{4}$$

$$k=3:\ 2\left(\cos\frac{7\pi}{4}+i\sin\frac{7\pi}{4}\right)=2\left(\frac{\sqrt{2}}{2}-i\frac{\sqrt{2}}{2}\right)=\sqrt{2}-i\sqrt{2} \qquad \frac{\pi+2\pi(3)}{4}=\frac{7\pi}{4}$$

Therefore, the solutions of $2x^4 + 32 = 0$ (i.e., the fourth roots of −16) are $\sqrt{2}+i\sqrt{2}$, $-\sqrt{2}+i\sqrt{2}$, $-\sqrt{2}-i\sqrt{2}$, and $\sqrt{2}-i\sqrt{2}$.

7.5.6 Roots of Complex Numbers: Another Example

EXAMPLE 8 Finding the *n*th Roots of a Complex Number

Find all cube roots of $-8i$.

SOLUTION

❶ Since $-8i$ is the point (0, −8) on the negative imaginary axis, $r = 8$ and $\theta = 3\pi/2$.

❷ Substitute $n = 3$, $r = 8$, and $\theta = 3\pi/2$ into the expression for the nth roots z and simplify.

$$\sqrt[3]{8}\left(\cos\frac{3\pi/2+2\pi k}{3}+i\sin\frac{3\pi/2+2\pi k}{3}\right)=2\left(\cos\frac{3\pi/2+2\pi k}{3}+i\sin\frac{3\pi/2+2\pi k}{3}\right)$$

❸ Substitute the k-values (0, 1, and 2) to find the 3 roots.

$$k=0:\ 2\left(\cos\frac{3\pi/2}{3}+i\sin\frac{3\pi/2}{3}\right)=2\left(\cos\frac{\pi}{2}+i\sin\frac{\pi}{2}\right)=2(0+1i)=2i$$

$$k=1:\ 2\left(\cos\frac{7\pi}{6}+i\sin\frac{7\pi}{6}\right)=2\left(-\frac{\sqrt{3}}{2}+\left(-\frac{1}{2}\right)i\right)=-\sqrt{3}-i \qquad \frac{3\pi/2+2\pi(1)}{3}=\frac{7\pi}{6}$$

$$k=2:\ 2\left(\cos\frac{11\pi}{6}+i\sin\frac{11\pi}{6}\right)=2\left(\frac{\sqrt{3}}{2}+\left(-\frac{1}{2}\right)i\right)=\sqrt{3}-i \qquad \frac{3\pi/2+2\pi(2)}{3}=\frac{11\pi}{6}$$

Therefore, the 3 cube roots of $-8i$ are $2i$, $\sqrt{3}-i$, and $-\sqrt{3}-i$.

NOTICE THAT

By the conjugate pair theorem, we might expect that $-2i$ is also a cube root of $-8i$. However, the theorem does not apply because all cube roots of $-8i$ are solutions of the equation $x^3 = -8i$, which has a non-real coefficient ($-8i$). So, the complex conjugate of each root is not necessarily a root of the equation.

SECTION 7.5 EXERCISES

Warm Up

Find the complex conjugate for each of the following.

1. $1 + 3i$
2. $-5 + 4i$
3. $-2 - 6i$

Just the Facts

Fill in the blanks.

4. In the complex plane, the x-axis is referred to as the _____ axis and the y-axis is referred to as the _____ axis.
5. Let $z = x + yi$ be a complex number. The distance from the origin to the point (x, y) is called the _____, denoted by $|z|$.
6. If $r > 0$ and $0 \le \theta \le 2\pi$, the complex number $z = x + yi$ can be written in trigonometric form as _____.
7. If $z_1 = r_1(\cos\theta_1 + i\sin\theta_1)$ and $z_2 = r_2(\cos\theta_2 + i\sin\theta_2)$, then $z_1 z_2 =$ _____.
8. If $z_1 = r_1(\cos\theta_1 + i\sin\theta_1)$, $z_2 = r_2(\cos\theta_2 + i\sin\theta_2)$, and $z_2 \neq 0$, then $z_1/z_2 =$ _____.
9. If $z = r(\cos\theta + i\sin\theta)$ is a complex number, then $z^n =$ _____.
10. Let x be a complex number, and let $n \ge 2$ denote a positive integer. Any complex number z that satisfies the equation $z^n = x$ is called a complex _____ root of x.

Essential Skills

In Exercises 1–6, plot each complex number in the complex plane.

1. $z = 1 + 5i$
2. $z = -5 - 4i$
3. $z = -3 + i$
4. $z = 2 - 3i$
5. $z = -4 - 2i$
6. $z = 2 + 6i$

In Exercises 7–12, determine the modulus of each complex number.

7. $4 + 2i$
8. $12 - 5i$
9. $-3 + 6i$
10. $-7 + 24i$
11. $1 - 4i$
12. $-9 - 12i$

In Exercises 13–18, write each complex number in trigonometric form.

13. $2 + 2\sqrt{3}i$
14. $-5 + 5i$
15. $3 - 3i$
16. $-4\sqrt{3} - 4i$
17. $-6\sqrt{3} + 6i$
18. $4\sqrt{2} + 4\sqrt{2}i$

In Exercises 19–24, write each complex number in trigonometric form. Round θ to the nearest degree.

19. $3 + 6i$
20. $2 - 3i$
21. $-1 + 2i$
22. $-5 + 4i$
23. $-4 - i$
24. $-2 - 5i$

In Exercises 25–30, find each product $z_1 z_2$.

25. $z_1 = 2(\cos 50^\circ + i\sin 50^\circ)$, $z_2 = 3(\cos 70^\circ + i\sin 70^\circ)$
26. $z_1 = 5(\cos 140^\circ + i\sin 140^\circ)$, $z_2 = 12(\cos 20^\circ + i\sin 20^\circ)$
27. $z_1 = 8(\cos 65^\circ + i\sin 65^\circ)$, $z_2 = 4(\cos 95^\circ + i\sin 95^\circ)$
28. $z_1 = 6(\cos 160^\circ + i\sin 160^\circ)$, $z_2 = 7(\cos 130^\circ + i\sin 130^\circ)$
29. $z_1 = 9(\cos 170^\circ + i\sin 170^\circ)$, $z_2 = 2(\cos 120^\circ + i\sin 120^\circ)$
30. $z_1 = 4(\cos 80^\circ + i\sin 80^\circ)$, $z_2 = 5(\cos 100^\circ + i\sin 100^\circ)$

In Exercises 31–36, find each quotient z_1/z_2.

31. $z_1 = 36(\cos 150^\circ + i\sin 150^\circ)$, $z_2 = 18(\cos 80^\circ + i\sin 80^\circ)$
32. $z_1 = 82(\cos 207^\circ + i\sin 207^\circ)$, $z_2 = 2(\cos 15^\circ + i\sin 15^\circ)$
33. $z_1 = 45(\cos 310^\circ + i\sin 310^\circ)$, $z_2 = 5(\cos 290^\circ + i\sin 290^\circ)$
34. $z_1 = 12(\cos 230^\circ + i\sin 230^\circ)$, $z_2 = 4(\cos 140^\circ + i\sin 140^\circ)$
35. $z_1 = 48(\cos 140^\circ + i\sin 140^\circ)$, $z_2 = 12(\cos 20^\circ + i\sin 20^\circ)$
36. $z_1 = 81(\cos 136^\circ + i\sin 136^\circ)$, $z_2 = 9(\cos 92^\circ + i\sin 92^\circ)$

In Exercises 37–42, write each complex number in trigonometric form.

37. $(-7i)^3$
38. $(5 - 5\sqrt{3}i)^2$
39. $(3 - 3i)^4$
40. $(-6 + 6i)^3$
41. $(-4 - 4\sqrt{3}i)^2$
42. $(2\sqrt{3} - 2i)^5$

In Exercises 43–54, simplify and write in standard form.

43. $(2 + 2i)^3$

44. $(-4 + 4i)^5$

45. $(-3\sqrt{3} + 3i)^4$

46. $(-5 - 5\sqrt{3}i)^4$

47. $(\sqrt{3} - i)^5$

48. $(4\sqrt{3} - 4i)^3$

49. $\left(2\left(\cos\frac{\pi}{2} + i\sin\frac{\pi}{2}\right)\right)^3$

50. $\left(3\left(\cos\frac{2\pi}{3} + i\sin\frac{2\pi}{3}\right)\right)^4$

51. $\left(4\left(\cos\frac{3\pi}{4} + i\sin\frac{3\pi}{4}\right)\right)^4$

52. $\left(5\left(\cos\frac{11\pi}{6} + i\sin\frac{11\pi}{6}\right)\right)^3$

53. $\left(5\left(\cos\frac{5\pi}{6} + i\sin\frac{5\pi}{6}\right)\right)^5$

54. $\left(3\left(\cos\frac{7\pi}{4} + i\sin\frac{7\pi}{4}\right)\right)^5$

In Exercises 55–62, solve.

55. $v^4 = 9$

56. $4z^4 = 100$

57. $4u^6 = 16$

58. $2v^6 = -98$

59. $u^4 - 81 = 0$

60. $z^5 + 243 = 0$

61. $v^6 - 729 = 0$

62. $t^4 + 81 = 0$

In Exercises 63–70, find the indicated roots.

63. fourth roots of 16

64. fifth roots of 243

65. cube roots of 216

66. fourth roots of 100

67. cube roots of i

68. fourth roots of $2 - 2\sqrt{3}i$

69. cube roots of $-8 - 8i$

70. cube roots of $4\sqrt{3} + 4i$

Extensions

In Exercises 71–74, write each complex number in trigonometric form. Round θ to the nearest degree.

71. $(4 - 3i)^5$

72. $(-5 - 12i)^4$

73. $(-6 + 10i)^3$

74. $(8 + 9i)^5$

In Exercises 75–78, find the indicated roots. Round each answer to the nearest hundredth, as needed.

75. fifth roots of $32(\cos 300° + i\sin 300°)$

76. fourth roots of $81(\cos 150° + i\sin 150°)$

77. cube roots of $125\left(\cos\frac{5\pi}{4} + i\sin\frac{5\pi}{4}\right)$

78. fifth roots of $1024(\cos\pi + i\sin\pi)$

CHAPTER 7 REVIEW EXERCISES

Section 7.1 The Law of Sines

In Exercises 1–5, solve each triangle and find its area, if possible. If two triangles exist, solve and find the area of both triangles. Round to the nearest tenth, as needed.

1. $\alpha = 38°$, $a = 11$, $b = 14.5$

2. $\beta = 130°$, $a = 17$, and $b = 20$

3. $\gamma = 46°$, $a = 33$, and $c = 42$

4. $\alpha = 141°$, $a = 19.9$, and $b = 10.2$

5. $\beta = 58°$, $\gamma = 112°$, $a = 18$

Section 7.2 The Law of Cosines

In Exercises 6–10, solve each triangle and find its area and height (using the shortest side as the triangle's base). Round to the nearest tenth, as needed.

6. 41, 60, 72

7. 14, 11, 12

8. 16, 19, 91° (included angle)

9. 43, 92, 76° (included angle)

10. 24, 28, 132° (included angle)

Section 7.3 Vectors

In Exercises 11–13, find the component form of each vector. Then find the vector's magnitude, direction angle to the nearest tenth of a degree, the component form of the unit vector in the same direction, and write the vector as a linear combination of the standard unit vectors.

11. initial point (–4, 7), terminal point (2, 5)

12. initial point (1, –6), terminal point (–8, 3)

13. initial point (–3.2, –5.8), terminal point (4, –4)

In Exercises 14–15, use $\vec{w} = \langle 3.5, -2 \rangle$, $\vec{x} = \langle -6.2, 4 \rangle$, $\vec{y} = \langle -7, -4 \rangle$, and $\vec{z} = \langle -2.8, 5.4 \rangle$ to find each resultant vector.

14. $-\vec{z} - 4\vec{w}$

15. $2\vec{y} + 3\vec{x}$

Section 7.4 The Dot Product

In Exercises 16–18, find $\vec{u} \cdot (\vec{v} - \vec{w})$, $\vec{u} \cdot \vec{u}$, $-3(\vec{u} \cdot \vec{v})$, and $||\vec{w}||$.

16. $\vec{u} = \langle -3, -1.8 \rangle$, $\vec{v} = \langle 8, -2 \rangle$, $\vec{w} = \langle -3, 2 \rangle$

17. $\vec{u} = \langle 2.6, 5 \rangle$, $\vec{v} = \langle -7, 3 \rangle$, $\vec{w} = \langle -5, 6 \rangle$

18. $\vec{u} = \langle -1, 4.2 \rangle$, $\vec{v} = \langle 9, -1.5 \rangle$, $\vec{w} = \langle -6.1, -3.5 \rangle$

In Exercises 19–20, determine whether $\vec{u}$ and $\vec{v}$ are parallel, orthogonal, or neither. Then find the projection of $\vec{u}$ onto $\vec{v}$ and decompose $\vec{u}$ into the sum of two orthogonal vectors $\vec{w}_1$ and $\vec{w}_2$, where $\vec{w}_1 = \text{proj}_{\vec{v}}\vec{u}$.

19. $\vec{u} = \langle -2, -4 \rangle$, $\vec{v} = \langle -5, 3 \rangle$

20. $\vec{u} = \langle -1, 8 \rangle$, $\vec{v} = \langle 1/2, -4 \rangle$

Section 7.5 Trigonometric Form and Roots of Complex Numbers

21. Convert $-2 + 2\sqrt{3}i$ to trigonometric form.

22. Write both zw and z/w in the form $a + bi$, given that $w = 4(\cos 110° + i \sin 110°)$ and $z = 3(\cos 130° + i \sin 130°)$. Round to the nearest hundredth, as needed.

23. Write $(-5\sqrt{3} - 5i)^5$ in standard and trigonometric form.

24. Solve $-3v^6 = -69$.

25. Find all fourth roots of $6 - 6i$.

PRECALCULUS

SYSTEMS OF EQUATIONS AND INEQUALITIES

CHAPTER

8

8.1 SOLVING SYSTEMS OF TWO LINEAR EQUATIONS IN TWO VARIABLES

OBJECTIVES

- Find solutions to systems of linear equations in two variables by graphing.
- Find solutions to systems of linear equations in two variables algebraically by using substitution and elimination.
- Identify systems of linear equations in two variables that have no solution or infinitely many solutions.

PREREQUISITE VOCABULARY TERMS

linear equation
slope
slope-intercept form
***y*-intercept**

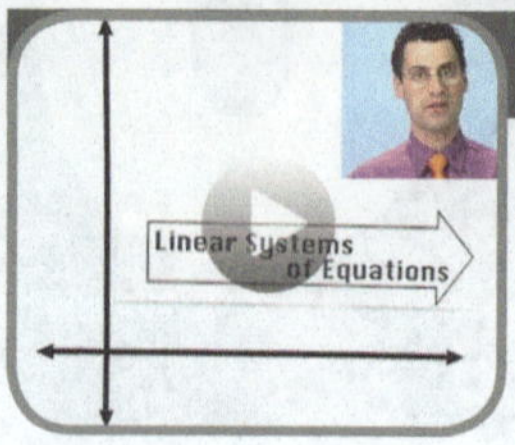

8.1.1 An Introduction to Linear Systems

A **system of equations** is a set of two or more equations. It is possible to find solutions that concurrently satisfy each of the equations that comprise a system of equations.

In a **linear system of equations**, each equation is linear, and so the graph of each equation is a line. A system of two linear equations has the following general form.

$$\begin{cases} Ax + By = C & \textit{equation 1} \\ Dx + Ey = F & \textit{equation 2} \end{cases}$$

IMPORTANT

The first equation in a system will be referred to as "equation 1," the second equation in a system will be referred to as "equation 2," and so on.

Any ordered pair (x, y) that is a solution for each of the equations in the system is a solution for the system of equations. Several methods for solving a system of equations will be discussed in this section.

Finding a solution to a system of two linear equations requires finding values for the two variables that satisfy both equations. Graphically, this means finding the point where the two lines intersect. For example, Figure 8.1a shows the graph of a linear system of two equations where the solution (i.e., the point of intersection) is (3, 2). A system of equations with at least one solution is called a **consistent system**.

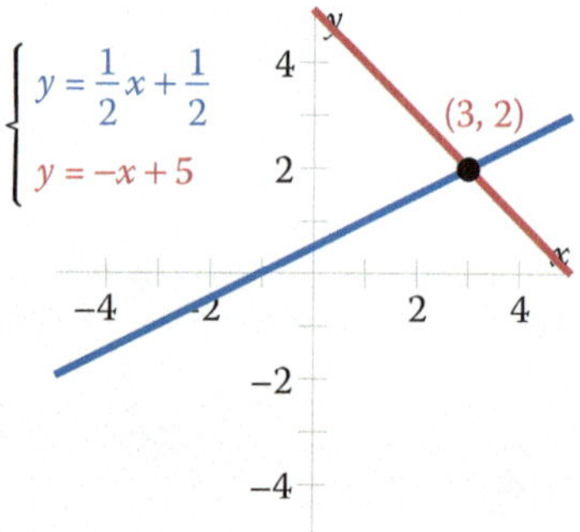

Figure 8.1a

The two equations in a system of linear equations may form lines that are parallel. Since parallel lines do not intersect, there is no solution for systems of linear equations where the lines are parallel. For example, the system of linear equations graphed in Figure 8.1b has no solution. This type of system is called an **inconsistent** system.

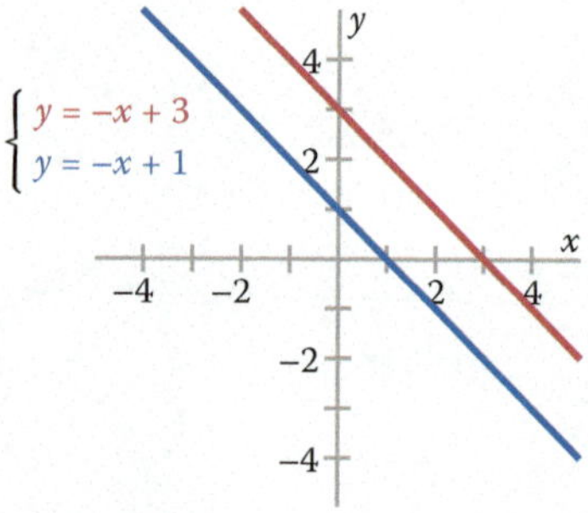

Figure 8.1b

8.1.2 Solving Systems by Graphing

One method used to find a solution to a system of linear equations is to graph each line and find a point of intersection, if one exists.

EXAMPLE 1 Solving a System of Linear Equations by Graphing

Solve the system by graphing. $\begin{cases} 4x - y = 1 \\ 3x + y = 6 \end{cases}$

SOLUTION

Write each equation in slope-intercept form, $y = mx + b$, and then identify the slope m and y-intercept b.

$4x - y = 1 \Rightarrow y = 4x - 1 \quad m = 4$ and $b = -1$

$3x + y = 6 \Rightarrow y = -3x + 6 \quad m = -3$ and $b = 6$

Graph the two lines using the respective slope and y-intercept.

The point of intersection is (1, 3).

Therefore, the solution to the system is (1, 3), or $x = 1$ and $y = 3$.

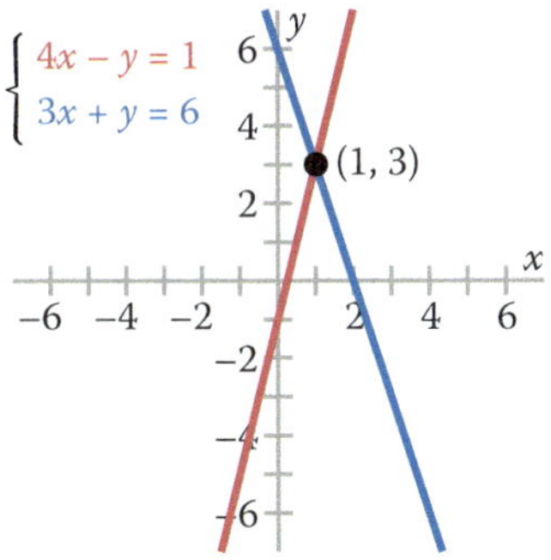

GRAPHING CALCULATOR

Enter the equations in $Y_1 =$ and $Y_2 =$.

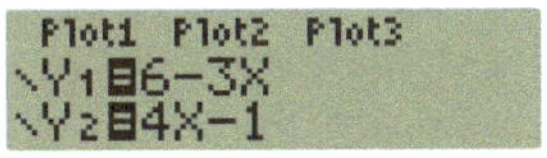

Adjust the viewing window as needed, graph the equations, and then select the "Intersection" function from the "CALC" menu. The solution is shown below.

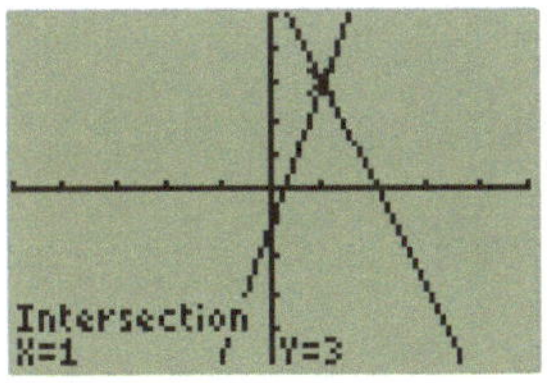

CHECK

To verify that (1, 3) is the solution to the system, substitute the point's coordinates into each of the original equations. If the result of substituting is a true statement, then the point is on the graph of that equation. If the point is on the graphs of both equations, then the point must be a solution to the system.

$4x - y = 1$	$3x + y = 6$
$4(1) - 3 \stackrel{?}{=} 1$	$3(1) + 3 \stackrel{?}{=} 6$
$1 = 1$ ✓	$6 = 6$ ✓

Therefore, the solution to the system of equations is (1, 3).

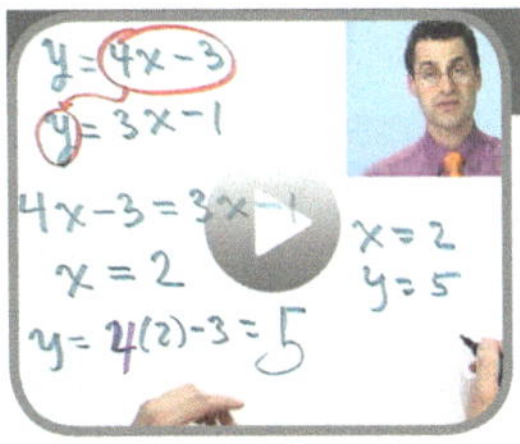

8.1.3 The Substitution Method

The **substitution method** is an algebraic method for finding the solution to a system of equations. The strategy of this method is to solve one of the equations for one of the variables, then substitute the expression for that variable into the other equation. This substitution results in an equation in one variable that can be solved to find the value of the first of the two variables. That value is then substituted into either of the equations, and solving yields the value of the second of the two variables.

Steps for Solving a System of Two Equations Using the Substitution Method

❶ Solve one of the equations for one of the variables.*

❷ To find the value of one of the variables, substitute the expression for the variable from step ❶ into the other equation and solve.

❸ To find the value of the other variable, substitute the value from step ❷ into either of the original equations (or into the equation from step ❶) and solve.

**Skip step ❶ when one of the given equations is already solved for one of the variables.*

EXAMPLE 2 Solving a System of Linear Equations Using Substitution

Use substitution to solve the system. $\begin{cases} y = 2x - 9 \\ y = -x + 6 \end{cases}$

SOLUTION

TIP

The slope-intercept form of a linear equation provides an expression for y.

❶ Both equations are solved for y.

❷ Since both equations are already solved for y, you can substitute the expression for y from either equation into the other equation, and then solve the resulting equation for x. Here, the expression for y from equation 1 is substituted for y in equation 2.

$2x - 9 = -x + 6$	*Substitute $2x - 9$ for y in equation 2.*
$2x + x = 9 + 6$	*Move the variables to one side and the constant terms to the other.*
$3x = 15$	*Simplify.*
$x = 5$	*Divide each side by 3.*

❸ Substitute $x = 5$ into either of the original equations and simplify to find the y-value. Here, equation 1 is used.

$$y = 2(5) - 9 = 10 - 9 = 1$$

Therefore, the solution to the system of equations is (5, 1).

TIP

Check the solution by substituting the x- and y-values into each of the equations.

EXAMPLE 3 Solving a System of Linear Equations Using Substitution

Use substitution to solve the system. $\begin{cases} 2x - 3y = -12 \\ 3x + 2y = -5 \end{cases}$

SOLUTION

❶ Solve either of the equations for either of the variables. Here, equation 1 is solved for x.

$2x - 3y = -12$	
$2x = 3y - 12$	*Add 3y to each side.*
$x = \dfrac{3y - 12}{2}$	*Divide by 2.*

TIP

The expression for x, $\frac{3y - 12}{2}$, cannot be substituted into equation 1 because equation 1 was solved for x to get this expression.

❷ Substitute the expression for x into equation 2 and solve for y.

$3x + 2y = -5$	
$3\left(\dfrac{3y - 12}{2}\right) + 2y = -5$	*Substitute the expression for x in equation 2.*
$\dfrac{9y - 36}{2} + 2y = -5$	*Distribute.*
$9y - 36 + 4y = -10$	*Multiply the equation by 2 to remove the denominator.*
$13y = 26$	*Add 36 to each side and combine the like terms.*
$y = 2$	*Solve for y.*

❸ Substitute $y = 2$ into either of the original equations, or into the expression for x from step ❶. Here, the equation from step ❶ is used.

$$x = \frac{3(2) - 12}{2} = \frac{6 - 12}{2} = \frac{-6}{2} = -3$$

Therefore, the solution to the system of equations is (–3, 2).

8.1.4 The Elimination Method: Adding

The substitution method will always work to find a solution for a system of linear equations, but the algebra can sometimes be complicated. An alternative, often simpler approach, is called the **elimination method**.

EXAMPLE 4 Solving a System of Linear Equations Using Elimination by Addition

Use elimination to solve the system. $\begin{cases} 2x+4y=-12 \\ -x-4y=14 \end{cases}$

SOLUTION

IMPORTANT

When adding two equations, be sure to combine only the like terms.

Notice that the y-terms have coefficients that are opposites. So, adding the two equations will result in an equation without a y-term (i.e., an equation in one variable), because $4y + (-4y) = 0$.

Add the equations.

$$\begin{aligned} 2x+4y &= -12 \\ +(-x-4y &= 14) \\ \hline x &= 2 \end{aligned} \qquad \textit{The x-value of the solution is 2.}$$

Substitute $x = 2$ into either one of the original equations and solve for y. Here, equation 1 is used.

$$\begin{aligned} 2x+4y &= -12 \\ 2(2)+4y &= -12 && \textit{Substitute 2 for x.} \\ 4y &= -16 && \textit{Subtract 4 from both sides.} \\ y &= -4 && \textit{Divide by 4.} \end{aligned}$$

Therefore, the solution to the system of equations is $(2, -4)$.

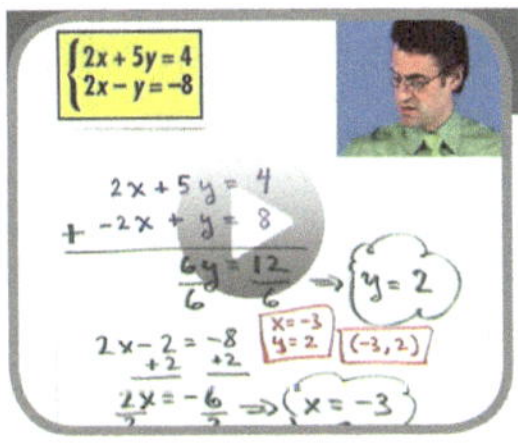

8.1.5 The Elimination Method: Subtracting

Recall that the coefficients of the y-terms were opposites in the system solved in **Example 4**. Because those coefficients were opposites, the y-terms canceled out when the equations were added. It follows that if a system contains a pair of like terms (whether the x-terms or the y-terms) with opposite coefficients, then the value of one variable can be found by simply adding the two equations together and simplifying.

But how could you solve a system where there are no like terms with coefficients that are opposites? For example, consider the system $\begin{cases} x+y=10 \\ 2x+y=16 \end{cases}$. Adding the two equations yields $3x + 2y = 26$, which cannot be solved for one of the variables because it still contains two variables. Simply adding the equations will not work unless there are opposite coefficients on like terms.

Remember that multiplying an equation by a number does not change the value of the equation as long as the entire equation is multiplied by that number. So, one (or both) of the equations could be multiplied by some number that will result in opposite coefficients on like terms. Again consider $\begin{cases} x+y=10 \\ 2x+y=16 \end{cases}$; multiplying equation 1 by -1 would make -1 the coefficient of y. Since the coefficient of y in equation 2 is 1, the y-terms would have opposite coefficients. Then the equations could be added, and the y-terms would cancel out.

$$\begin{array}{rcr} -1(x+y=10) & \Rightarrow & -x-y=-10 \\ 2x+y=16 & & +\ (2x+y=16) \\ \hline & & x=6 \end{array}$$

It is worth noting that there are many ways to manipulate the equation(s) so that a pair of like terms will cancel out.

For example, equation 2 could be multiplied by –1, instead of equation 1.

Recall that adding a negative is equivalent to subtracting.

$$\begin{array}{rcr} x+y=10 & & x+y=10 \\ -1(2x+y=16) & \Rightarrow & +\ (-2x-y=-16) \\ \hline & & -x=-6 \\ & & x=6 \end{array}$$

In this way, multiplying an equation by –1 and then adding the two equations is an equivalent process to subtracting one equation from the other.

Steps for Solving a System of Two Equations Using the Elimination Method

❶ Multiply the equation(s) through by an appropriate factor so that the coefficients of either the x-terms or the y-terms will be opposites.

❷ To find the value of one of the variables, add the equations and solve for the remaining variable.

❸ To find the value of the other variable, substitute the solution from step ❷ into either of the original equations and solve.

EXAMPLE 5 Solving a System of Linear Equations Using Elimination by Subtraction

Use elimination to solve the system. $\begin{cases} 3x+2y=-8 \\ 3x-3y=-3 \end{cases}$

SOLUTION

❶ Notice that the x-terms have the same coefficient. So, multiply either equation 1 or equation 2 by –1 (but not both equations) to make the x-terms have opposite coefficients. Here, equation 2 is multiplied by –1.

$$\begin{array}{rcr} 3x+2y=-8 & & 3x+2y=-8 \\ -1(3x-3y=-3) & \Rightarrow & -3x+3y=3 \end{array}$$

❷ Add the equations to eliminate x, and solve the resulting equation for y.

$$\begin{array}{r} 3x+2y=-8 \\ +\ (-3x+3y=3) \\ \hline 5y=-5 \\ y=-1 \end{array}$$

The y-value of the solution is –1.

❸ Substitute $y = -1$ into either one of the original equations and solve for x. Here, equation 1 is used.

$3x+2y=-8$

$3x+2(-1)=-8$ *Substitute –1 for y.*

$3x=-6$ *Add 2 to both sides.*

$x=-2$ *Divide by 3.*

Therefore, the solution to the system of equations is (–2, –1).

8.1.6 Solving Systems by Elimination

In some cases, it will not be sufficient to multiply one equation by −1. It may also be necessary to multiply one equation, or both equations, by some other number in order to make a pair of like terms have opposite coefficients.

EXAMPLE 6 Solving a System of Linear Equations Using Elimination

Use elimination to solve the system. $\begin{cases} 6x+3y=21 \\ 2x+4y=22 \end{cases}$

SOLUTION

❶ Notice that neither the x- nor y-terms have the same coefficient, so multiplying one of the equations by −1 will not be sufficient. If equation 2 is multiplied by −3, then the coefficients of the x-terms will be 6 and −6. So, multiply equation 2 by −3.

$$\begin{aligned} 6x+3y&=21 & & & 6x+3y&=21 \\ -3(2x+4y&=22) & &\Rightarrow & -6x-12y&=-66 \end{aligned}$$

❷ Add the equations to eliminate x, and solve the resulting equation for y.

$$\begin{aligned} 6x+3y&=21 & & \\ \underline{+\;(-6x-12y}&\underline{=-66)} & & \textit{Add the equations.} \\ -9y&=-45 & & \textit{Divide by } -9. \\ y&=5 & & \end{aligned}$$

❸ Substitute $y=5$ into either one of the original equations and solve for x. Here, equation 1 is used.

$$\begin{aligned} 6x+3y&=21 & & \\ 6x+3(5)&=21 & & \textit{Substitute 5 for y.} \\ 6x&=6 & & \textit{Subtract 15 from both sides.} \\ x&=1 & & \textit{Divide by 6.} \end{aligned}$$

Therefore, the solution to the system of equations is (1, 5).

TIP

There are many different ways to manipulate the equations to make a variable cancel out when the equations are added. Multiplying equation 2 by −3 is just one of these ways.

Sometimes the equations in a system can be multiplied by different numbers to create a pair of like terms with opposite coefficients.

EXAMPLE 7 Solving a System of Linear Equations Using Elimination

Use elimination to solve the system. $\begin{cases} 4x-3y=-5 \\ 3x+2y=-8 \end{cases}$

SOLUTION

❶ Notice that neither the x- nor y-terms have the same coefficient. If equation 1 is multiplied by 2 and equation 2 is multiplied by 3, then the coefficients of the y-terms will cancel out with addition. So, multiply equation 1 by 2 and equation 2 by 3.

$$\begin{aligned} 2(4x-3y&=-5) & &\Rightarrow & 8x-6y&=-10 \\ 3(3x+2y&=-8) & &\Rightarrow & 9x+6y&=-24 \end{aligned}$$

❷ Add the equations to eliminate y, and solve the resulting equation for x.

$$\begin{aligned} 8x-6y&=-10 & & \\ \underline{+\;(9x+6y}&\underline{=-24)} & & \\ 17x&=-34 & & \\ x&=-2 & & \textit{The x-value of the solution is } -2. \end{aligned}$$

TIP

The choice of whether to eliminate the x-terms or the y-terms is arbitrary when solving a system for both x and y. Multiplying equation 1 by −3 and equation 2 by 4 would eliminate the x-terms and be an equally valid route to a solution.

❸ Substitute $x = -2$ into either of the original equations and solve for y. Here, equation 1 is used.

$$4x - 3y = -5$$
$$4(-2) - 3y = -5 \quad \textit{Substitute } -2 \textit{ for } x.$$
$$-3y = 3 \quad \textit{Add 8 to both sides.}$$
$$y = -1 \quad \textit{Divide by } -3.$$

The solution to the system of equations is $(-2, -1)$.

8.1.7 Three Cases for Linear Systems

There are three possible cases for solutions to systems of two linear equations in two variables since there are three possible ways in which two lines can exist on a plane in relation to one another.

1. Two lines may intersect at one point.
2. Two lines may be parallel (i.e., they do not intersect).
3. Two lines may intersect at all points (when the two equations form the same line).

Therefore, a system of two linear equations in two variables may have one solution, no solution, or infinitely many solutions.

Recall that an inconsistent system has no solution. When a system of linear equations in two variables is inconsistent, the equations must form parallel lines. Therefore, the equations in these systems must form lines that have the same slope and different y-intercepts.

Recall also that a consistent system has at least one solution. When a system of linear equations in two variables is consistent, the equations must form lines that intersect. These lines may intersect at one point, or at all points. If the two lines intersect at one point, then the system is consistent and **independent**. The equations in these systems must form lines with different slopes.

The other possibility for a consistent system of linear equations in two variables is that the equations form lines that intersect at all points (i.e., the equations form the same line). In this case, all of the points on the line are a solution to the system, and the system is consistent and **dependent**. Since there are infinitely many points on a line, there are infinitely many solutions to the system. The equations in these systems must form lines that have the same slope and the same y-intercept. The cases and their terminology are summarized in the following table.

Three Cases for Linear Systems

	Number of Solutions	Type of System	Graph Description	Slopes
Case 1	Exactly one	Consistent and independent	Lines intersect at 1 point	Not equal
Case 2	Infinitely many	Consistent and dependent	Lines coincide (are identical)	Equal
Case 3	None	Inconsistent	Lines are parallel	Equal

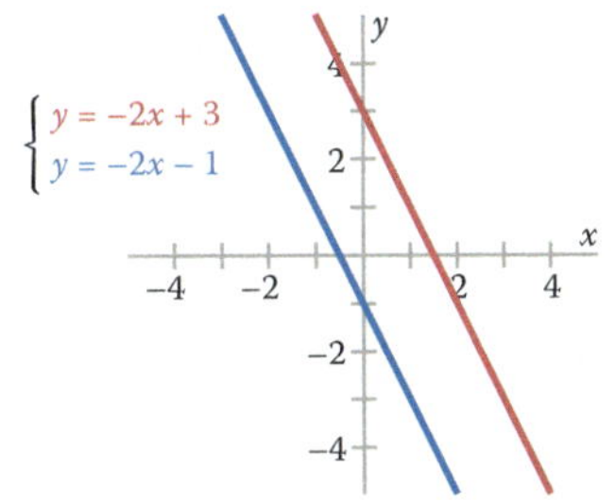

Figure 8.1c

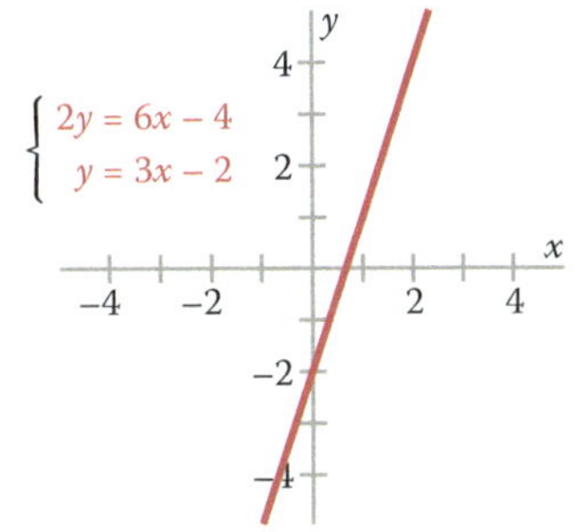

Figure 8.1d

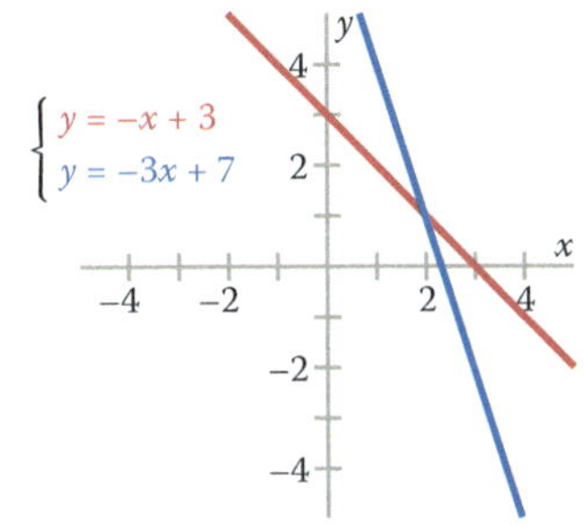

Figure 8.1e

EXAMPLE 8 Three Cases for Systems of Linear Equations

Without solving, identify each system as consistent and independent, consistent and dependent, or inconsistent.

A. $\begin{cases} y = -2x + 3 \\ y = -2x - 1 \end{cases}$ **B.** $\begin{cases} 2y = 6x - 4 \\ y = 3x - 2 \end{cases}$ **C.** $\begin{cases} y = -x + 3 \\ y = -3x + 7 \end{cases}$

SOLUTION

A. From the equations, the slope of both lines is -2, and the y-intercepts are 3 and -1. Therefore, the lines are parallel, but they do not coincide (Figure 8.1c). The system is inconsistent.

B. Divide equation 1 by 2 so it is in slope-intercept form.

$$2y = 6x - 4 \Rightarrow y = 3x - 2$$

From the equations, the slope of both lines is 3, and the y-intercept of both lines is -2. Therefore, the lines coincide (Figure 8.1d). The system is consistent and dependent.

C. From the equations, the slope of the first line is -1, and the slope of the second line is -3. Therefore, the lines are not parallel, and they intersect at exactly one point (Figure 8.1e). The system is consistent and independent.

The system from **Example 8A** has no solution. Note that solving the system algebraically (using the substitution method) yields a false statement.

$$\begin{aligned} -2x + 3 &= -2x - 1 \\ 3 &\neq -1 \end{aligned}$$

The procedure results in a false statement, indicating the system has no solution.

The system from **Example 8B** has infinitely many solutions. Note that solving the system algebraically yields a true statement.

$$\begin{aligned} 2y &= 6x - 4 \\ -2(y &= 3x - 2) \end{aligned} \quad \Rightarrow \quad \begin{array}{r} 2y = 6x - 4 \\ +\ (-2y = -6x + 4) \\ \hline 0 = 0 \end{array}$$

The procedure results in a true statement, indicating there are infinitely many solutions.

SECTION 8.1 EXERCISES

Warm Up

True or False?

1. (2, 4) is on the graph of $y = 7x - 10$.
2. $(-4, -1)$ is on the graph of $3x + 5y = -17$.
3. The x-intercept of $-3x + 1 = 2y$ is 1/2.

Just the Facts

Fill in the blanks.

4. A set of two or more equations is called a(n) _____ of equations.
5. A system of linear equations that consists of two parallel lines will have _____ solution(s), whereas a system of linear equations that consists of the same line will have _____ solution(s).
6. Three methods that are useful when solving a system of linear equations are _____, _____, and _____.
7. To use the _____ method, one equation must be solved for a variable.
8. A system with exactly one solution is classified as _____ and _____.
9. A system with two linear equations that have the same slope and different y-intercepts will have _____ solution(s) and be classified as _____.
10. The _____ method is best used when the coefficients of the same variable are opposite values.

Essential Skills

In Exercises 1–2, graph each system.

1. $\begin{cases} y = -2x + 7 \\ y = -2x - 5 \end{cases}$

2. $\begin{cases} 3x + y = 2 \\ y - x = -6 \end{cases}$

In Exercises 3–4, solve each system by graphing.

3. $\begin{cases} y = -x + 2 \\ y = 4x - 8 \end{cases}$

4. $\begin{cases} x - 2y = 3 \\ y + 3x = -5 \end{cases}$

In Exercises 5–40, solve each system using any method.

5. $\begin{cases} x = 4y - 19 \\ x = -3 \end{cases}$

6. $\begin{cases} y = 2x - 2 \\ y = -2x + 6 \end{cases}$

7. $\begin{cases} y = 3x + 3 \\ y = -5x - 13 \end{cases}$

8. $\begin{cases} y = 3x + 7 \\ y = -5x - 9 \end{cases}$

9. $\begin{cases} x = \dfrac{y - 8}{3} \\ x = -\dfrac{y + 7}{2} \end{cases}$

10. $\begin{cases} x = \dfrac{y + 6}{5} \\ x = \dfrac{y - 3}{4} \end{cases}$

11. $\begin{cases} 2x - y = -10 \\ x + y = -2 \end{cases}$

12. $\begin{cases} 5x + 2y = 10 \\ 3x + 4y = 6 \end{cases}$

13. $\begin{cases} 3x - 2y = 3 \\ 2x + 5y = 2 \end{cases}$

14. $\begin{cases} 5x + 2y = 12 \\ 2x + 4y = 8 \end{cases}$

15. $\begin{cases} 5x - 2y = 9 \\ 3x + 2y = -1 \end{cases}$

16. $\begin{cases} 5x + 3y = -11 \\ 3x + 4y = -11 \end{cases}$

17. $\begin{cases} x - 2y = 6 \\ -x + 6y = -6 \end{cases}$

18. $\begin{cases} -4x + 4y = -12 \\ 7x - 4y = 24 \end{cases}$

19. $\begin{cases} -4x + 7y = -35 \\ 6x - 7y = 49 \end{cases}$

20. $\begin{cases} 3x - 5y = -14 \\ -3x + 3y = 6 \end{cases}$

21. $\begin{cases} 2x - 5y = -6 \\ -2x + 2y = 0 \end{cases}$

22. $\begin{cases} -5x + 6y = -54 \\ 7x - 6y = 66 \end{cases}$

23. $\begin{cases} x - 4y = 8 \\ x + 3y = -6 \end{cases}$

24. $\begin{cases} -x-2y=2 \\ 4x-2y=-8 \end{cases}$

25. $\begin{cases} -x-5y=8 \\ 2x-5y=-1 \end{cases}$

26. $\begin{cases} 5x-6y=14 \\ 5x+4y=24 \end{cases}$

27. $\begin{cases} 6x-4y=-10 \\ -x-4y=-24 \end{cases}$

28. $\begin{cases} 2x-3y=6 \\ 2x+5y=22 \end{cases}$

29. $\begin{cases} 4x-y=2 \\ -12x+3y=-6 \end{cases}$

30. $\begin{cases} 6x+2y=-19 \\ 5x+4y=-17 \end{cases}$

31. $\begin{cases} 5x+8y=-9 \\ 3x+4y=19 \end{cases}$

32. $\begin{cases} 2x+3y=12 \\ 6x+4y=-9 \end{cases}$

33. $\begin{cases} 10x+5y=-4 \\ 5x+3y=-19 \end{cases}$

34. $\begin{cases} 3x+6y=-20 \\ 4x+12y=-14 \end{cases}$

35. $\begin{cases} 3x-2y=17 \\ -4x-5y=8 \end{cases}$

36. $\begin{cases} 5x+4y=-2 \\ 3x-3y=2 \end{cases}$

37. $\begin{cases} 5x+5y=-1 \\ 7x-2y=7 \end{cases}$

38. $\begin{cases} 6x+2y=-1 \\ 7x-3y=3 \end{cases}$

39. $\begin{cases} 2x+3y=2 \\ 5x-4y=2 \end{cases}$

40. $\begin{cases} 2x+2y=7 \\ 7x+5y=-1 \end{cases}$

In Exercises 41–48, describe each system as consistent and independent, consistent and dependent, or inconsistent.

41. $\begin{cases} y=5x-9 \\ y=5x-9 \end{cases}$

42. $\begin{cases} y=2x-8 \\ y=2x-3 \end{cases}$

43. $\begin{cases} 3x-7y=8 \\ 7x+y=8 \end{cases}$

44. $\begin{cases} 3x-y=10 \\ -15x+5y=-50 \end{cases}$

45. $\begin{cases} 4x-y=3 \\ y=4x-8 \end{cases}$

46. $\begin{cases} x=\frac{3}{2}x+7 \\ -10x+15y=-70 \end{cases}$

47. $\begin{cases} 8x-2y=10 \\ 3x-y=3 \end{cases}$

48. $\begin{cases} 3x-5y=2 \\ 6x-10y=9 \end{cases}$

Extensions

49. Find the value of k such that the system of linear equations is consistent and dependent.

$$\begin{cases} \frac{4}{3}x-ky=8 \\ -8x+14y=-48 \end{cases}$$

50. True or False? A system is consistent and independent if it has an infinite number of solutions.

51. Use any method to solve. $\begin{cases} 0.002x+0.005y=-2.45 \\ -0.004x-0.008y=8.6 \end{cases}$

52. A student solved the system $\begin{cases} -2x-5y=17 \\ 7x-6y=8 \end{cases}$ by multiplying equation 1 by 6 and equation 2 by 5 to eliminate the y-value. Is this a valid approach?

53. Use any method to solve. $\begin{cases} \frac{2}{3}x+\frac{1}{4}y=\frac{5}{4} \\ \frac{3}{2}x-\frac{4}{5}y=\frac{29}{20} \end{cases}$

In Exercises 54–56, write a system of linear equations with the indicated solutions.

54. consistent and dependent

55. consistent and independent

56. inconsistent

57. Solve the system using substitution. $\begin{cases} 23x-11y=1 \\ 5x-4y=-3 \end{cases}$

8.2 NONLINEAR SYSTEMS

OBJECTIVES

- Identify systems of nonlinear equations.
- Find solutions to systems of nonlinear equations in two variables by graphing.
- Find solutions to systems of nonlinear equations in two variables by substitution.
- Find solutions to systems of nonlinear equations in two variables by elimination.

PREREQUISITE VOCABULARY TERMS

axis of symmetry
circle
ellipse
parabola
Quadratic Formula
radius (of a circle)
vertex (of a parabola)

8.2.1 Solving Nonlinear Systems by Graphing

A **nonlinear system of equations** will contain at least one equation where the graph is not a line. The number of possible solutions depends on the equations themselves. As with linear systems of equations, the solution, or solutions, occur graphically where the graphs of the two equations intersect.

DEFINITION

A **nonlinear system of equations** is a set of two or more equations, at least one of which is a nonlinear equation.

A nonlinear system that includes a linear equation and a quadratic equation (i.e., where the graph is a line and a parabola) can have two solutions (Figure 8.2a), one solution (Figure 8.2b), or no solution (Figure 8.2c).

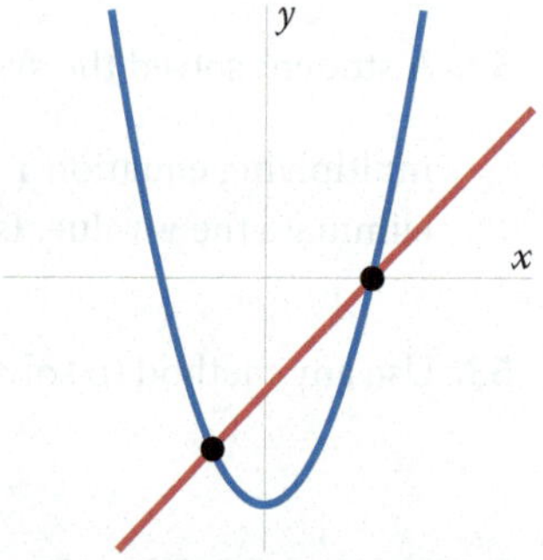

Figure 8.2a
The system has two solutions.

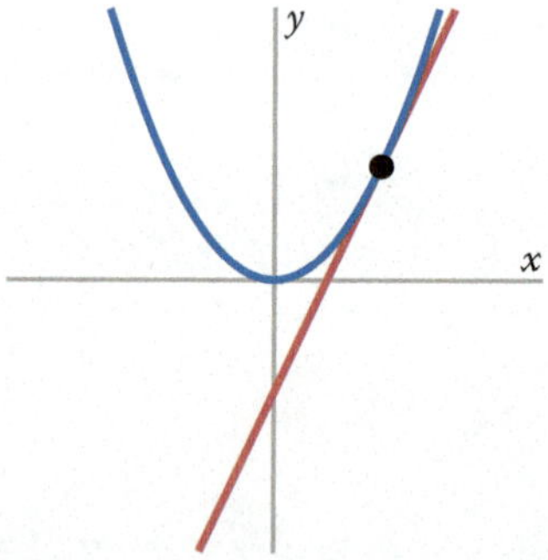

Figure 8.2b
The system has one solution.

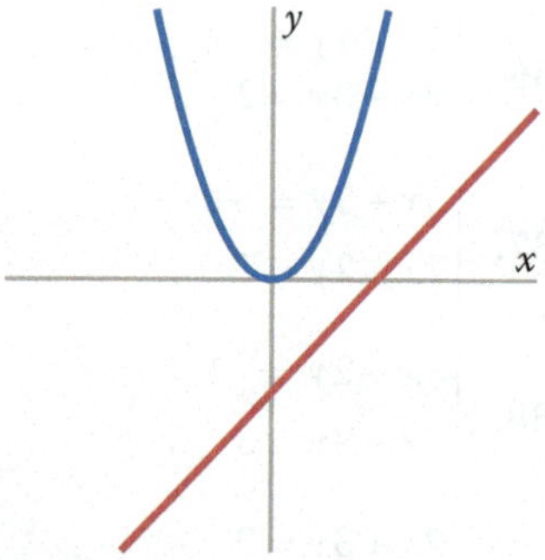

Figure 8.2c
The system has no solution.

A line that touches, but does not cross through a parabola, is said to be **tangent** to the parabola. For example, the line in Figure 8.2b is tangent to the parabola.

DEFINITION

A line is **tangent** to a curve when it touches but does not cross a curve.

EXAMPLE 1 Solving a Nonlinear System of Equations by Graphing

GRAPHING CALCULATOR ▶ *To solve the system of equations, enter the equations in $Y_1 =$ and $Y_2 =$, then use the "Intersection" function from the "CALC" menu.*

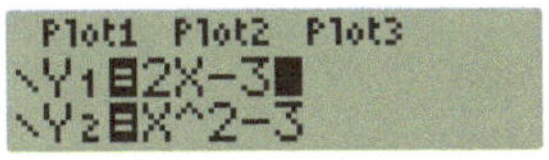

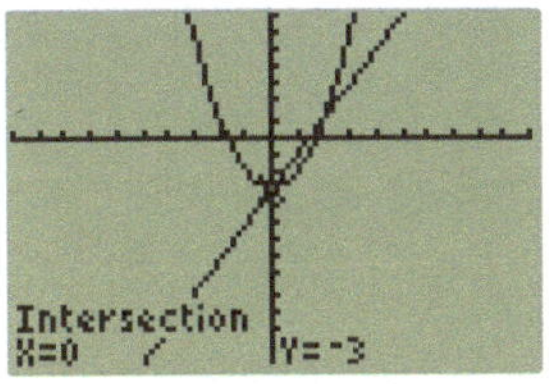

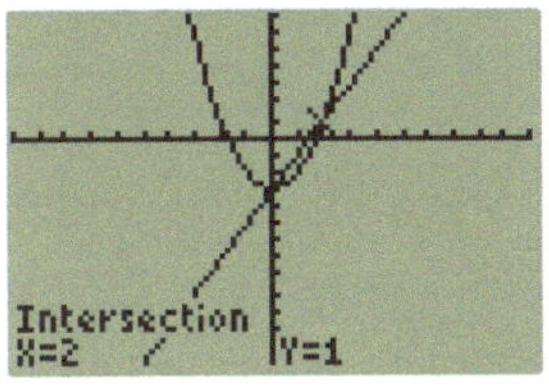

Solve the system by graphing. $\begin{cases} y = 2x - 3 \\ y = x^2 - 3 \end{cases}$

SOLUTION

Graph each equation, then identify the point(s) at which the graphs intersect, if any.

The graph of $y = 2x - 3$ is a line with slope of 2 and a y-intercept of -3.

The graph of $y = x^2 - 3$ is the graph of the parabola $y = x^2$ translated down 3 units.

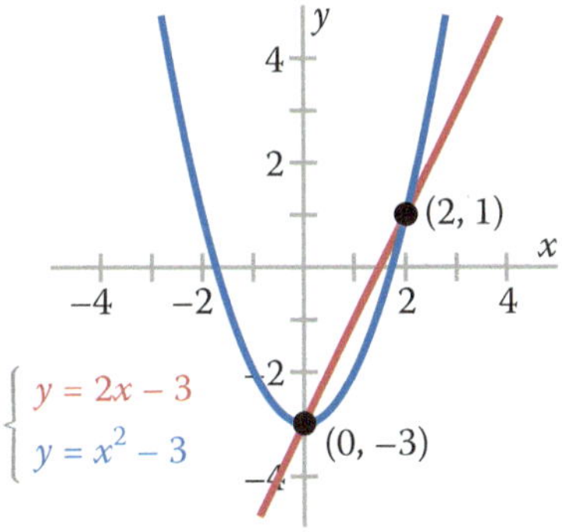

The line and the parabola intersect at the points (2, 1) and (0,–3). Therefore, the system has two solutions, (2, 1) and (0,–3).

8.2.2 Solving Nonlinear Systems with Substitution

As with systems of linear equations, nonlinear systems can be solved using the substitution method. (The steps for solving a system of equations by substitution are covered in Topic 8.1.3.)

EXAMPLE 2 Solve a Nonlinear System of Equations Using Substitution

Use substitution to solve the system. $\begin{cases} 2x - y = 4 \\ -x^2 + y = -x - 2 \end{cases}$

SOLUTION

❶ Solve one of the equations for one of the variables. Here, equation 1 is solved for y.

$$2x - y = 4 \;\Rightarrow\; -y = 4 - 2x \;\Rightarrow\; y = 2x - 4$$

❷ Substitute the expression for y into equation 2 and solve for x.

$$-x^2 + 2x - 4 = -x - 2 \qquad \textit{Substitute } (2x - 4) \textit{ for } y \textit{ in equation 2.}$$
$$0 = x^2 - 3x + 2 \qquad \textit{Write the equation in general form.}$$
$$0 = (x - 2)(x - 1) \qquad \textit{Factor.}$$

So, $x = 2$ and $x = 1$.

❸ Substitute $x = 2$ and $x = 1$ into either of the original equations (or an equivalent form) and simplify to find the corresponding y-values. Here, an equivalent form of equation 1 is used.

$x = 2$: $y = 2(2) - 4 = 0$ $\qquad$ $x = 1$: $y = 2(1) - 4 = -2$

The system has two solutions, (2, 0) and (1, –2) (Figure 8.2d).

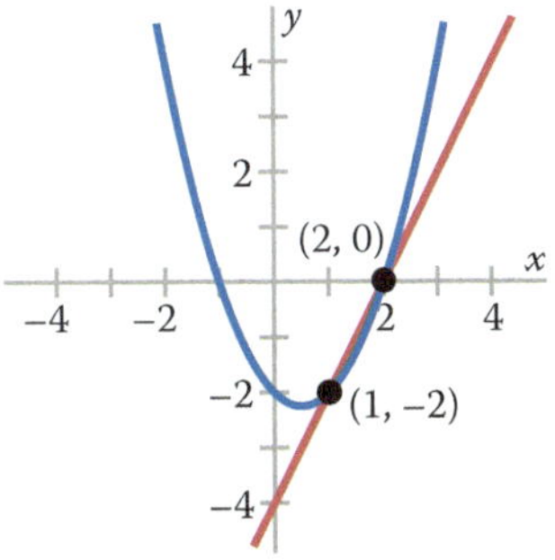

Figure 8.2d

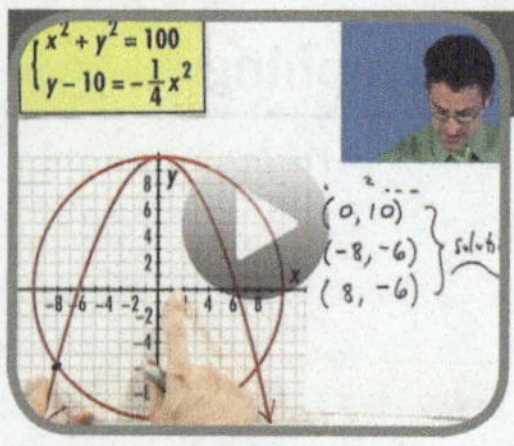

8.2.3 Solving Nonlinear Systems with Substitution: Another Example

When both equations in a nonlinear system of equations are nonlinear, the number of possibilities for solutions increases. For example, if the equations are for an ellipse (or circle) and a parabola, there may be up to four solutions, points at which the graphs intersect or are tangent with each other.

EXAMPLE 3 Solve a Nonlinear System of Equations Using Substitution

Use substitution to solve the system. $\begin{cases} x^2 + y^2 = 36 \\ y = -6 + \frac{1}{3}x^2 \end{cases}$

SOLUTION

TIP

Solving for x^2, rather than x, avoids the extra step of having to substitute a square root expression. Also, substituting the expression for y from equation 2 will yield a 4th power term.

❶ Solve one of the equations for one of the variables. Here, equation 2 is solved for x^2.

$$y = -6 + \frac{1}{3}x^2$$

$3(y+6) = x^2$ *Add 6 to both sides and multiply both sides by 3.*

$3y + 18 = x^2$ *Distribute.*

❷ Substitute the expression for x^2 into equation 1 and solve for y.

$$x^2 + y^2 = 36$$

$(3y + 18) + y^2 = 36$ *Substitute 3y + 18 for x^2 in equation 1.*

$y^2 + 3y - 18 = 0$ *Write the equation in general form.*

$(y + 6)(y - 3) = 0$ *Factor.*

So, $y = -6$ and $y = 3$.

❸ Substitute $y = -6$ and $y = 3$ into either of the original equations and solve to find the corresponding x-values. Here, equation 1 is used.

$x^2 + (-6)^2 = 36$ *Substitute −6 for y.*

$x^2 = 0$ *Subtract 36 from both sides.*

$x = 0$ *Solve for x.*

$x^2 + (3)^2 = 36$ *Substitute 3 for y.*

$x^2 = 27$ *Subtract 9 from both sides.*

$x = \pm\sqrt{27}$ *Solve for x.*

$x = \pm 3\sqrt{3}$ *Simplify the radical.*

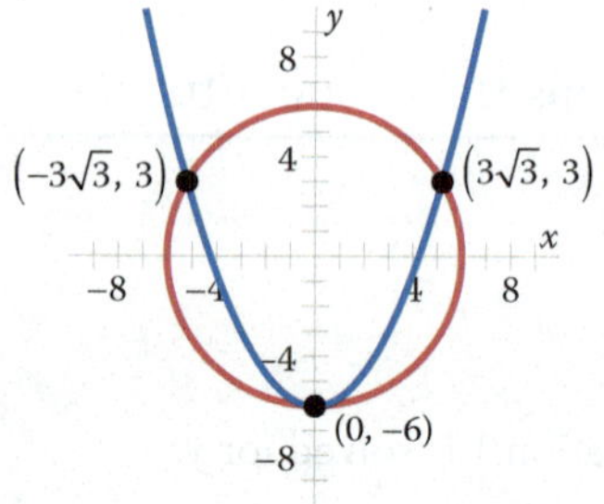

Figure 8.2e

The system has three solutions, $(0, -6)$, $(3\sqrt{3}, 3)$, and $(-3\sqrt{3}, 3)$ (Figure 8.2e).

WHAT'S THE BIG IDEA?

Write a nonlinear system of equations comprising a circle and a parabola with exactly one solution.

8.2.4 Solving Nonlinear Systems with Elimination

Elimination can also be used to solve nonlinear systems of equations. (The steps for solving a system of equations by elimination are covered in Section 8.1.)

EXAMPLE 4 Solve a Nonlinear System of Equations Using Elimination

Use elimination to solve the system. $\begin{cases} x - y = 3 \\ y = x^2 + 5x + 1 \end{cases}$

NOTICE THAT

Adding the equations will result in an equation in terms of only x, because the coefficients of the y-terms are opposites.

SOLUTION

❶ No multiplication is necessary because the coefficients of the y-terms are opposites.

❷ Add the equations to eliminate y, and solve the resulting equation for x.

$$\begin{aligned} x - y &= 3 \\ +(\quad y &= x^2 + 5x + 1) \\ \hline x &= x^2 + 5x + 4 && \text{Add the equations.} \\ 0 &= x^2 + 4x + 4 && \text{Write the equation in general form.} \\ 0 &= (x+2)(x+2) && \text{Factor.} \end{aligned}$$

So, $x = -2$.

❸ Substitute $x = -2$ into either of the original equations and solve to find the corresponding y-value. Here, equation 1 is used.

$$x - y = 3 \;\Rightarrow\; -2 - y = 3 \;\Rightarrow\; y = -5$$

The system has one solution, $(-2, -5)$ (Figure 8.2f).

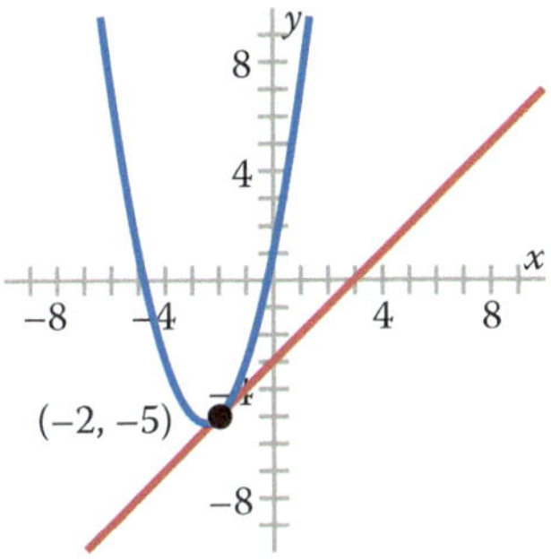

Figure 8.2f

EXAMPLE 5 Solve a Nonlinear System of Equations Using Elimination

Use elimination to solve the system. $\begin{cases} y = 2x^2 + 2x + 2 \\ x - 3y = 3 \end{cases}$

SOLUTION

❶ Multiply equation 1 by 3 so that the coefficients of the y-terms will be opposites.

$$\begin{aligned} 3(y &= 2x^2 + 2x + 2) \quad\Rightarrow & 3y &= 6x^2 + 6x + 6 \\ x - 3y &= 3 & x - 3y &= 3 \end{aligned}$$

❷ Add the equations to eliminate y, and solve the resulting equation for x.

$$\begin{aligned} 3y &= 6x^2 + 6x + 6 \\ +(x - 3y &= 3) \\ \hline x &= 6x^2 + 6x + 9 && \text{Add the equations.} \\ 0 &= 6x^2 + 5x + 9 && \text{Write the equation in general form.} \end{aligned}$$

Use the Quadratic Formula.

$$x = \frac{-5 \pm \sqrt{5^2 - 4(6)(9)}}{2(6)} = \frac{-5 \pm \sqrt{-191}}{12}$$

The discriminant (i.e., the radicand) is negative, so the equation has no real solutions.

The system has no real solution (Figure 8.2g).

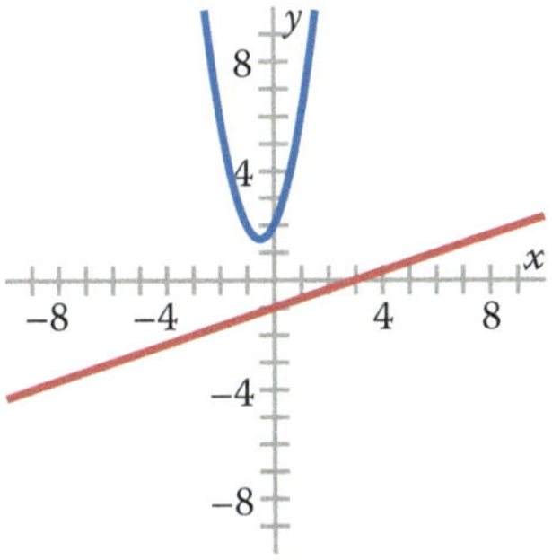

Figure 8.2g

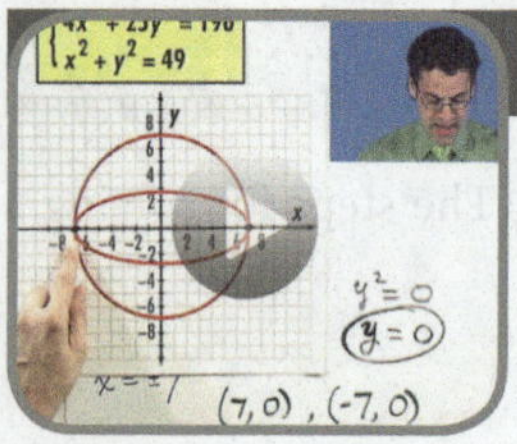

8.2.5 Solving Nonlinear Systems with Elimination: Another Example

The elimination method can be used to remove x^2- or y^2-terms in a nonlinear system of equations.

EXAMPLE 6 Solve a Nonlinear System of Equations Using Elimination

Use elimination to solve the system. $\begin{cases} 4x^2 + 16y^2 = 172 \\ x^2 + y^2 = 16 \end{cases}$

SOLUTION

❶ Multiply equation 2 by –4 so that the coefficients of the x^2-terms will be opposites.

$$\begin{aligned} 4x^2 + 16y^2 &= 172 \\ -4(x^2 + y^2 &= 16) \end{aligned} \quad \Rightarrow \quad \begin{aligned} 4x^2 + 16y^2 &= 172 \\ -4x^2 - 4y^2 &= -64 \end{aligned}$$

❷ Add the equations to eliminate the x^2-terms, and solve the resulting equation for y.

$$\begin{aligned} 4x^2 + 16y^2 &= 172 \\ + (-4x^2 - 4y^2 &= -64) \\ \hline 12y^2 &= 108 && \textit{Add the equations.} \\ y^2 &= 9 && \textit{Divide both sides by 12.} \\ y &= \pm 3 && \textit{Take } \pm \textit{ the square root of both sides.} \end{aligned}$$

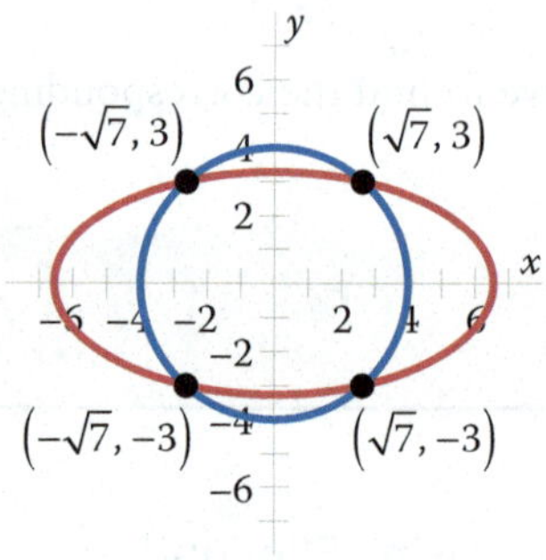

Figure 8.2h

❸ Substitute $y = -3$ and $y = 3$ into either of the original equations and solve to find the corresponding x-values. Here, equation 2 is used.

$$\begin{aligned} x^2 + (-3)^2 &= 16 && \textit{Substitute } -3 \textit{ for } y. \\ x^2 &= 7 && \textit{Subtract 9 from both sides.} \\ x &= \pm\sqrt{7} && \textit{Take } \pm \textit{ the square root of both sides.} \end{aligned}$$

$$\begin{aligned} x^2 + (3)^2 &= 16 && \textit{Substitute 3 for } y. \\ x^2 &= 7 && \textit{Subtract 9 from both sides.} \\ x &= \pm\sqrt{7} && \textit{Take } \pm \textit{ the square root of both sides.} \end{aligned}$$

The system has four solutions, $(\sqrt{7}, 3)$, $(\sqrt{7}, -3)$, $(-\sqrt{7}, 3)$, and $(-\sqrt{7}, -3)$ (Figure 8.2h).

SECTION 8.2 EXERCISES

Warm Up

Solve each equation for the indicated variable.

1. $5x - y = -14; x$
2. $6x^2 + y = 12; y$
3. $x^2 + 5y^2 = 20; x^2$

Just the Facts

Fill in the blanks.

4. A(n) _____ equation has a graph that is not a straight line.
5. A system of nonlinear equations has at least _____ equation that is nonlinear. If the nonlinear system consists of a line and a parabola, then the most solutions it may have is _____ and the least amount is _____. If the linear equation is _____ to the parabola, then it has exactly one solution.
6. As with linear systems, nonlinear systems may be solved using the methods of _____, _____, or _____.
7. When the graph of a system of equations has no points of intersection, the system has _____ solution(s).
8. If a system of nonlinear equations consists of a parabola and a circle or ellipse, there can be up to _____ solutions.
9. When using substitution to solve a system, one equation must be solved for a(n) _____.
10. As with linear systems, if solving by _____, then one variable's coefficient must be the opposite of the other.

Essential Skills

In Exercises 1–2, graph each system.

1. $\begin{cases} y = 3 \\ y = x^2 + 1 \end{cases}$

2. $\begin{cases} y = -2x - 2 \\ y = -x^2 - 2 \end{cases}$

In Exercises 3–4, solve each system by graphing.

3. $\begin{cases} y = 2 \\ y = -x^2 + 2 \end{cases}$

4. $\begin{cases} y = x^2 - 4 \\ y = -2x - 1 \end{cases}$

In Exercises 5–34, solve each system using any method.

5. $\begin{cases} y = 2x - 2 \\ y = -x^2 - 6x - 9 \end{cases}$

6. $\begin{cases} -2x + y = -6 \\ y = x^2 - 2x - 3 \end{cases}$

7. $\begin{cases} x^2 + y = -6x + 9 \\ y = -x^2 + 9 \end{cases}$

8. $\begin{cases} -x^2 + y = 4x + 4 \\ x^2 + y = 4 \end{cases}$

9. $\begin{cases} x^2 - y = 2x \\ -x^2 - y = x - 1 \end{cases}$

10. $\begin{cases} -4x + 2y = -26 \\ y = x^2 - 5x - 3 \end{cases}$

11. $\begin{cases} x^2 + y^2 = 16 \\ y = x^2 - 4 \end{cases}$

12. $\begin{cases} x^2 + y^2 = 25 \\ y = \frac{1}{2}x^2 - 5 \end{cases}$

13. $\begin{cases} x^2 - y^2 = 16 \\ -x^2 + y^2 = 9 \end{cases}$

14. $\begin{cases} x^2 + y^2 = 225 \\ 3y = x^2 - 45 \end{cases}$

15. $\begin{cases} y = -\frac{1}{6}x^2 + 3 \\ x^2 + y^2 = 25 \end{cases}$

16. $\begin{cases} x^2 + y^2 = 49 \\ y = -\frac{1}{28}x^2 + 7 \end{cases}$

17. $\begin{cases} -x + y = 4 \\ y = -x^2 + 4 \end{cases}$

18. $\begin{cases} x - y = -3 \\ y = x^2 + 4x - 1 \end{cases}$

19. $\begin{cases} y = x^2 + 5x - 3 \\ x - y = -2 \end{cases}$

20. $\begin{cases} x - y = -5 \\ y = x^2 + 6x - 1 \end{cases}$

21. $\begin{cases} 2x - y = -7 \\ y = x^2 + 4x - 1 \end{cases}$

22. $\begin{cases} 2x - y = -6 \\ y = x^2 + 4x - 2 \end{cases}$

23. $\begin{cases} y = x^2 - 4x + 4 \\ 2x + y = 7 \end{cases}$

24. $\begin{cases} y = x^2 + 3x - 4 \\ x - 2y = 1 \end{cases}$

25. $\begin{cases} x - 4y = 4 \\ y = x^2 + 3x - 6 \end{cases}$

26. $\begin{cases} 3x - 6y = 9 \\ y = x^2 + 3x - 5 \end{cases}$

27. $\begin{cases} y + 6 = x^2 + 3x \\ x - 2y = 5 \end{cases}$

28. $\begin{cases} y = x^2 + 3x - 5 \\ x - 4y = 5 \end{cases}$

29. $\begin{cases} 3x^2 + y^2 = 27 \\ x^2 + y^2 = 9 \end{cases}$

30. $\begin{cases} x^2 + y^2 = 25 \\ 25x^2 + 9y^2 = 225 \end{cases}$

31. $\begin{cases} 25x^2 + 16y^2 = 400 \\ x^2 + y^2 = 25 \end{cases}$

32. $\begin{cases} x^2 + y^2 = 81 \\ 100x^2 + 64y^2 = 6{,}400 \end{cases}$

33. $\begin{cases} 169x^2 + 25y^2 = 4{,}225 \\ x^2 + y^2 = 100 \end{cases}$

34. $\begin{cases} x^2 + y^2 = 16 \\ 100x^2 + 36y^2 = 3{,}600 \end{cases}$

Extensions

35. Solve by graphing. $\begin{cases} (x-1)^2 + y^2 = 36 \\ \frac{1}{6}x^2 - \frac{7}{3}x + \frac{49}{6} = y \end{cases}$

36. Write a system of nonlinear equations consisting of two circles and exactly one solution.

37. Using a nonlinear system, find two numbers whose sum is 16 and whose squares have a sum of 146.

38. Solve using any method. $\begin{cases} 2x + \frac{1}{10}y = -4 \\ -x^2 - 14x - 49 = y \end{cases}$

39. Solve by graphing. $\begin{cases} y = |x+4| \\ x^2 + y^2 = 16 \end{cases}$

40. Find the extraneous solution of the system.

$\begin{cases} -\sqrt{x-5} = y \\ x^2 + y^2 = 36 \end{cases}$

41. Solve by any method. $\begin{cases} (x-2)^2 + y^2 = 20 \\ x^2 + 4x + y^2 = 16 \end{cases}$

8.3 MODELING WITH SYSTEMS

OBJECTIVES

- Model a set of investments with a linear system of equations.
- Model rates of motion with a linear system of equations.
- Model mixtures with a linear system of equations.
- Model motion with acceleration with a nonlinear system of equations.
- Model trajectories with a nonlinear system of equations.

PREREQUISITE VOCABULARY TERMS

elimination
simple interest
system of linear equations
system of nonlinear equations
substitution

8.3.1 Applying Linear Systems: Investments

Recall from Section 4.1 that when a principal amount P is borrowed (or invested) for t years at an interest rate of r, the amount of simple interest paid (or earned) is given by $I = Prt$.

EXAMPLE 1 Modeling Investments with a System of Linear Equations

TIP

The money is invested for 1 year, so the formula for simple interest is $I = Pr$.

A bank customer invests a total of \$6681 in two savings accounts. One account yields 8% simple interest and the other 9.5% simple interest. The customer earned a total of \$552 interest for the year. How much was invested in the 9.5% account?

SOLUTION

Identify the unknown values. Let x be the principal in the 8% account and let y be the principal in the 9.5% account.

Write a system of equations using the given information.

$x + y = 6681$ *The total principal in the two accounts is \$6681.*

$0.08x + 0.095y = 552$ *The total interest earned from the two accounts is \$552.*

TIP

The interest from the 8% account is $x(8\%)(1)$, that is, $0.08x$. The interest from the 9.5% account is $y(9.5\%)(1)$, that is, $0.095y$.

Use the substitution method to solve this system.

Solve one of the equations for one of the variables. Here, equation 1 is solved for x.

$$x + y = 6681 \Rightarrow x = 6681 - y$$

Substitute the expression for x into equation 2, and solve for y.

$0.08(6681 - y) + 0.095y = 552$ *Substitute.*

$534.48 - 0.08y + 0.095y = 552$ *Distribute.*

$534.48 + 0.015y = 552$ *Combine like terms.*

$y = 1168$ *Solve for y.*

\$1168 was invested in the 9.5% account.

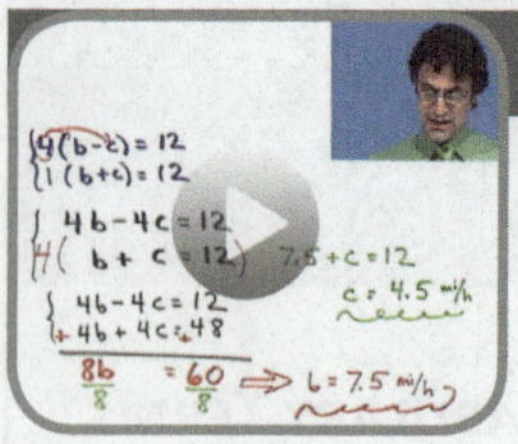

8.3.2 Applying Linear Systems: Distance, Rate, and Time

Recall that the relationship between distance, rate, and time can be modeled by the linear equation $d = rt$.

EXAMPLE 2 Modeling with a System of Linear Equations

An airplane travels 460 miles against the wind in 1 hour and 15 minutes. Another airplane travels the same distance, but in the opposite direction (with the wind) in 1 hour. Find the speed of the airplanes (with no wind) and find the speed of the wind. (Assume that both remain constant.)

SOLUTION

Identify the unknown values. Let a be the speed of the airplane and let w be the wind speed.

TIP

1 hour and 15 minutes is 1.25 hours (15 minutes is one-quarter of an hour).

Organize the given information in a table using $d = rt$.

	Rate	·	Time	=	Distance
Plane 1 (against the wind)	$a - w$		1.25 h		460 mi
Plane 2 (with the wind)	$a + w$		1 h		460 mi

Write a system of linear equations.

$$\begin{cases} (a-w)1.25 = 460 \\ (a+w)1 = 460 \end{cases} \Rightarrow \begin{cases} 1.25a - 1.25w = 460 \\ a + w = 460 \end{cases}$$

TIP

Alternatively, the substitution method can be used to solve the system.

Use the elimination method to solve this system.

Multiply equation 2 by 1.25 so that the coefficients of the w-terms are opposites.

$$\begin{aligned} 1.25a - 1.25w &= 460 \\ +\ \underline{(1.25a + 1.25w = 575)} & \quad \textit{Multiply equation 2 by 1.25.} \\ 2.5a &= 1035 \quad \textit{Add the equations.} \\ a &= 414 \quad \textit{Divide each side by 2.5.} \end{aligned}$$

Substitute $a = 414$ into either of the original equations and solve to find the corresponding w-value. Here, equation 2 is used.

$$414 + w = 460 \Rightarrow w = 46$$

The speed of each airplane is 414 miles per hour and the wind speed is 46 miles per hour.

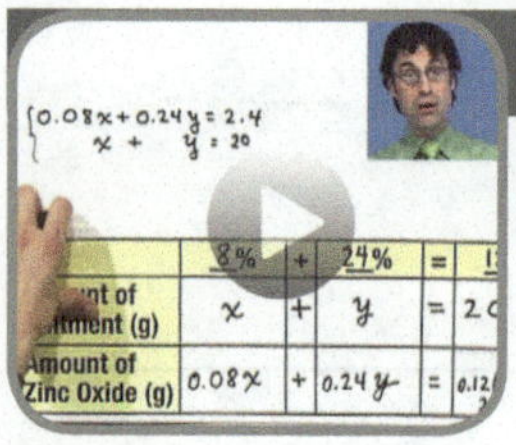

8.3.3 Applying Linear Systems: Mixtures

The amount of some substance in a mixture can be found by multiplying the substance's concentration C, which is typically given as a percent, by the mixture's total volume V. For a mixture that is divided into two parts, the amount of the substance in the total mixture is the sum of CV from one part and CV from the second part.

$$C_{\text{total}}V_{\text{total}} = C_1V_1 + C_2V_2$$

EXAMPLE 3 Modeling a Mixture with a System of Linear Equations

Suppose a paint that is 15% gloss is mixed with a paint that is 30% gloss to make 5 gallons of a paint that is 20% gloss. How many gallons of the 15% gloss paint were used in the mixture?

SOLUTION

Identify the unknown values. Let l be the number of gallons of the 15% gloss paint used in the mixture and let h be the number of gallons of the 30% gloss paint used in the mixture.

Organize the given information in a table.

	15% gloss	+	30% gloss	=	20% gloss
Amount of paint	l gal	+	h gal	=	5 gal
Percent of gloss	$0.15l$	+	$0.3h$	=	0.2(5)

TIP

The mixture is divided into two parts. Part 1 is the 15% gloss and part 2 is the 30% gloss.

Write a system of linear equations. $\begin{cases} l + h = 5 \\ 0.15l + 0.3h = 1 \end{cases}$

Use the substitution method to solve this system.
Solve one of the equations for one of the variables. Here, equation 1 is solved for h.

$$l + h = 5 \Rightarrow h = 5 - l$$

Substitute the expression for h into equation 2, and solve for l.

$0.15l + 0.3(5 - l) = 1$ *Substitute.*

$0.15l + 1.5 - 0.3l = 1$ *Distribute.*

$1.5 - 0.15l = 1$ *Combine like terms.*

$l = 3\frac{1}{3}$ *Solve for l.*

The mixture contained $3\frac{1}{3}$ gallons of the 15% gloss paint.

8.3.4 Applying Nonlinear Systems: Physics

EXAMPLE 4 Modeling Motion with a System of Nonlinear Equations

An elevator begins to ascend at a rate of 15 feet per second, and at the same time, a ball is thrown straight up into the air near the elevator. The following system of equations describes the height h in feet of the elevator (equation 1) and the ball (equation 2) after t seconds. After how many seconds (to the nearest hundredth) will the ball and the elevator reach the same height? $\begin{cases} h = 15t \\ h = -16t^2 + 40t \end{cases}$

SOLUTION

Substitute the expression for h from equation 1 into equation 2.

$15t = -16t^2 + 40t$ *Substitute.*

$0 = -16t^2 + 25t$ *Write the equation in general form.*

$0 = -t(16t - 25)$ *Factor.*

So, $t = 0$ and $t = \frac{25}{16} \approx 1.56$.

The ball and the elevator will reach the same height after approximately 1.56 seconds.

TIP

Both equations are solved for h, so use the substitution method to solve the system.

TIP

The elevator and the ball begin to move at t = 0.

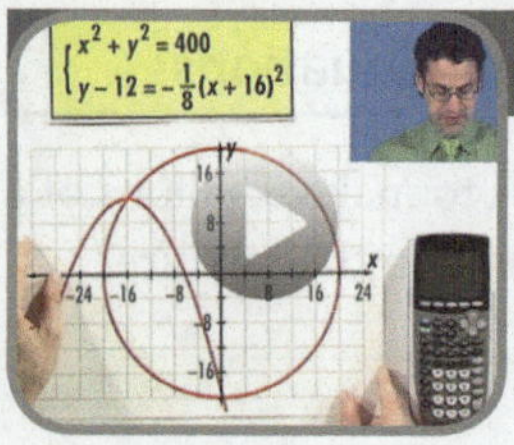

8.3.5 Applying Nonlinear Systems: Paths of Objects

EXAMPLE 5 Modeling the Paths of Objects with a System of Nonlinear Equations

A satellite is in a circular orbit around Earth. A space rock is observed being drawn toward Earth by Earth's gravitational field. The rock will travel a parabolic path around Earth and continue on its way. The rock and the satellite are traveling in the same plane. Their trajectories, in kilometers, are described by a system of nonlinear equations.

$$\begin{cases} x^2 + y^2 = 250{,}000 \\ y = \dfrac{1}{1600}x^2 - 400 \end{cases}$$

Do the paths of the objects intersect, thus raising the possibility of a collision? If so, how many times do the paths intersect within a single orbit?

SOLUTION

GRAPHING CALCULATOR ▶

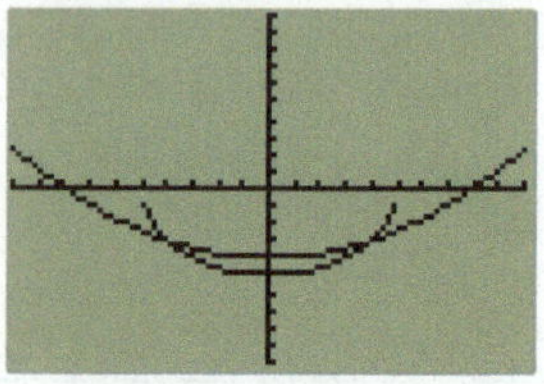

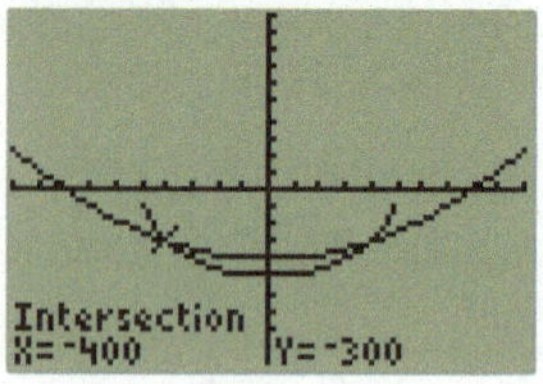

Use the substitution method to solve the system.

Solve one of the equations for one of the variables. Here, equation 2 is solved for x^2.

$$y = \frac{1}{1600}x^2 - 400$$

$$y + 400 = \frac{1}{1600}x^2 \qquad \textit{Add 400 to both sides.}$$

$$1600(y + 400) = x^2 \qquad \textit{Multiply both sides by 1600.}$$

Substitute the expression for x^2 into equation 1, and solve for y.

$$1600(y + 400) + y^2 = 250000 \qquad \textit{Substitute.}$$

$$1600y + 640000 + y^2 = 250000 \qquad \textit{Distribute.}$$

$$y^2 + 1600y + 390000 = 0 \qquad \textit{Write the equation in general form.}$$

$$(y + 1300)(y + 300) = 0 \qquad \textit{Factor.}$$

So, $y = -1300$ and $y = -300$.

Substitute $y = -1300$ and $y = -300$ into $x^2 = 1600(y + 400)$ (alternatively, the y-values could be substituted into either of the original equations), and solve to find each corresponding x-value.

$$x^2 = 1600(-1300 + 400)$$

$$x^2 = -1440000$$

The equation has no real solutions.

$$x^2 = 1600(-300 + 400)$$

$$x^2 = 160000$$

$$x = \pm\sqrt{160000}$$

$$x = \pm 400$$

The system has two real solutions, $(400, -300)$ and $(-400, -300)$.

The solutions are locations on a coordinate plane of the orbital system where Earth is at the center of the circular orbit and at the focus of the parabolic orbit. Since two solutions exist, the paths of the satellite and rock intersect twice, and so the potential for two collisions within a single orbit around Earth exists.

SECTION 8.3 EXERCISES

Warm Up

Write an equation that represents each situation.

1. The sum of two consecutive even integers is 18.
2. The simple interest earned at a 6% rate over 3 years was \$45. Use P for the principal.
3. A distance d was traveled in h hours by a van traveling at a rate of 50 miles per hour.

Just the Facts

Fill in the blanks.

4. Suppose a total of \$5000 is invested in two mutual fund accounts. If x is the amount in one account and y is the amount in the other account, then _____ = 5000.
5. When writing a linear system involving percents, the percent values are written as _____.
6. When modeling with systems, assign variables to the _____.
7. _____, _____, and _____ are three methods used to solve systems of equations.
8. In solving systems involving distance, rate, and time, it is helpful to organize the data into a(n) _____.
9. If x gallons of a 40% solution are combined with y gallons of a 15% solution to make 10 gallons of a 30% solution, then _____ = 3.
10. The easiest method to solve the system $\begin{cases} h = 7t \\ h = -16t^2 + 20t \end{cases}$ would be the _____ method, since both equations are solved for a(n) _____.

Essential Skills

1. A person invests a total of \$9681 in two savings accounts. One account yields 7% simple interest and the other 7.5% simple interest. A total of \$689.40 interest for the year was earned. How much was invested in the 7.5% account?
2. A man invests a total of \$9493 in two savings accounts. One account yields 9% simple interest and the other 10% simple interest. He earned a total of \$930.80 interest for the year. How much was invested in the 9% account?
3. A woman paddles a kayak 4 miles upstream in 2 hours. The return trip downstream takes her 30 minutes. What is the speed that she paddles in still water? What is the speed of the current? (Assume that both speeds remain constant.)
4. It took an airplane 2 hours to fly 600 miles against a headwind. The return trip (with the wind) took 1 and 2/3 hours. Find the speed of the plane with no wind and find the speed of the wind. (Assume that both speeds remain constant.)
5. A coffee house blended 12 pounds of espresso-flavored coffee beans with 11 pounds of vanilla-flavored coffee beans. The 23-pound mixture cost \$166.50. A second mixture included 17 pounds of espresso-flavored coffee beans and 18 pounds of vanilla-flavored coffee beans. The 35-pound mixture cost \$254. Find the cost per pound of the espresso- and vanilla-flavored coffee beans.
6. A chemist is mixing a 40% salt solution with a 20% salt solution to make 50 liters of a new solution that will contain 25% salt. How much of the 40% solution should the chemist use?
7. A crane begins to lift its load at a rate of 4 feet per second. At the same time, a ball is tossed into the air. The following system of equations describes the height h in feet of the crane's load (equation 1) and the ball (equation 2) after t seconds.

   $$\begin{cases} h = 4t \\ h = -16t^2 + 30t \end{cases}$$

   How long does it take until the ball and the crane's load reach the same height? (Round to the nearest hundredth.)
8. A freight elevator begins to ascend at a rate of 9 feet per second. At the same time, a bottle rocket is ignited. The following system of equations describes the height h in feet of the elevator (equation 1) and the bottle rocket (equation 2) after t seconds.

   $$\begin{cases} h = 9t \\ h = -16t^2 + 54t \end{cases}$$

   How long does it take until the bottle rocket and the elevator reach the same height? (Round to the nearest hundredth.)
9. An air traffic controller is monitoring two planes. Plane A is circling the airport, waiting for its turn to land, in a trajectory with the equation $x^2 + y^2 = 9$. Plane B has just taken off and is turning around to get on course. Its path can be described by the equation $y = -x^2 - 3$. If these two planes are flying at the same altitude, is there a danger of collision? If so, at what coordinates?
10. A local radio station has a broadcast range bounded by the circle with equation $x^2 + y^2 = 100$. There is a stretch of a highway nearby that is modeled by the equation $y + 6 = -\frac{1}{16}(x - 8)^2$. At what points does a car on this highway enter or exit the broadcast range of the radio station?

Extensions

11. At the traveling rodeo, adult tickets cost \$7.50 and children's tickets cost \$3.50. On Saturday, 13,000 tickets were sold for \$85,480. How many of each type of ticket were sold?

12. A corporate executive invests \$35,000 into two agency bonds that pay 2.4% and 3.8% simple interest. How much should the executive invest in the 3.8% bond if she wants \$910 in annual interest from the bonds?

13. A tour boat travels a path around a small island that can be modeled by $x^2 + y^2 = 225$. A fishing boat is approaching the island by a path that can be modeled by $y + 12 = -\frac{1}{27}(x + 9)^2$. Is there any danger of the two colliding? If so, at what locations?

14. A plane flying with a tailwind travels 1698 miles in 3 hours and 24 minutes. The return flight takes 3 hours and 48 minutes. Find the speed of the wind and find the speed of the plane with no wind. (Assume that both speeds remain constant.)

15. Solution A is 20% saline and solution B is 45% saline. How many ounces of each should be used to make 30 ounces of a 35% saline solution?

16. A radiator in an automobile contains 3 gallons of a mixture of antifreeze and water. The mixture in the radiator is 28% antifreeze. How much of this mixture should be drained and replaced with pure antifreeze to make the mixture 50% antifreeze?

8.4 MULTIVARIABLE LINEAR SYSTEMS

OBJECTIVES

- Describe geometric interpretations of systems of linear equations with three variables.
- Solve triangular systems of equations in three variables using back-substitution.
- Solve three-variable systems of equations using Gaussian elimination.
- Identify three-variable systems of linear equations with no solutions and those with infinitely many solutions.
- Solve nonsquare three-variable systems of equations.

PREREQUISITE VOCABULARY TERMS

consistent system
dependent system
elimination
inconsistent system
linear system of equations
substitution
system of equations

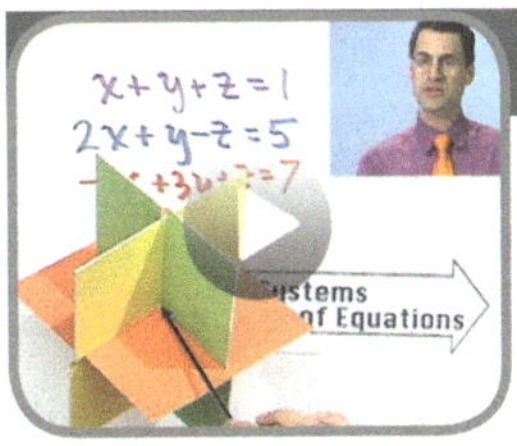

8.4.1 An Introduction to Linear Systems in Three Variables

Systems of equations can have more than two variables and more than two equations. Understanding the geometry of a three-variable system requires thinking in three dimensions rather than the two dimensions of the Cartesian plane. For example, consider the three-variable equation $x + y + z = 1$. This is a linear equation, since none of the variables are raised to a power higher than 1, but it is not the equation of a line. Just as a linear equation with two variables describes a line in two-dimensional space, a linear equation in three variables describes a plane in three-dimensional space (Figure 8.4a).

One Linear Equation, Three Variables

Figure 8.4a

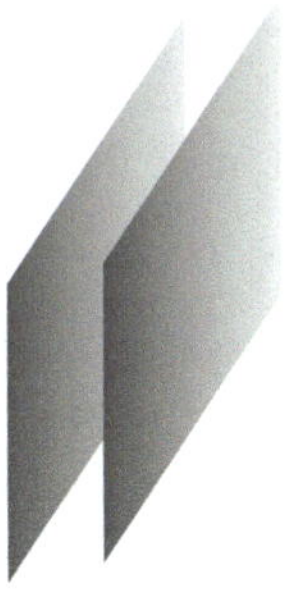

Two Linear Equations, No Solutions

Figure 8.4b

Two Linear Equations, Infinitely Many Solutions

Figure 8.4c

Consider a system of two linear equations in three variables. The graph of each equation is a plane, and those planes either do or do not intersect. If the two planes do not intersect (i.e., they are parallel, as shown in Figure 8.4b), then the system has no solutions.

If the two planes do intersect, and the two equations are not equivalent equations, then the intersection the of two planes is a line (Figure 8.4c) along which lie the infinitely many solutions to the two-equation system.

If the two equations form the same plane (meaning that the two equations in the system are equivalent), then the infinitely many solutions of the two-equation system form that plane.

Note that a system of two linear equations in three variables cannot have one unique solution. In order for the possibility of a unique solution to the system to exist, a third equation must be added to the system.

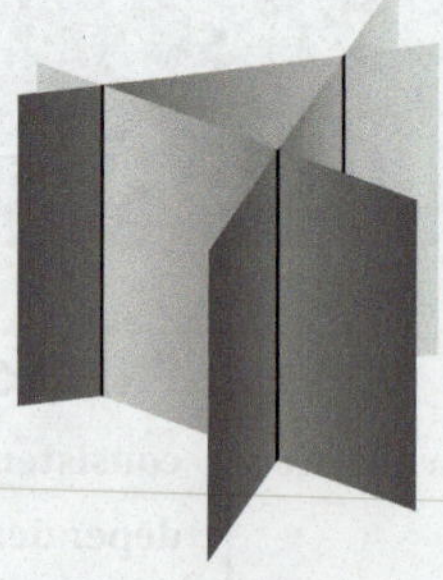

No Solutions

Figure 8.4d

Infinitely Many Solutions

Figure 8.4e

One Unique Solution

Figure 8.4f

NOTICE THAT

A system of two equations in three variables cannot have one unique solution because two planes never intersect at one point.

Adding a third linear equation to the system adds a third plane to the graph of the system. A solution to this system is a point at which all three planes intersect. If two of the planes, or all three of the planes, are parallel, then there is no solution. If each plane only intersects one other plane, then there is also no solution (Figure 8.4d). If the third plane intersects the other two planes along their line of intersection, then there are infinitely many solutions (Figure 8.4e). Finally, all three planes may intersect each other at a single point, then there is one unique solution (Figure 8.4f).

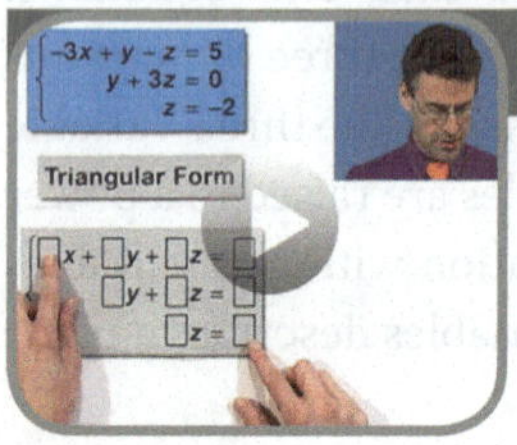

8.4.2 Solving a Triangular System Using Back-Substitution

A system of linear equations in three variables is represented similarly to a two-variable system. Note that a system of equations in three variables will not necessarily have all three variables represented in all three equations. Each of the equations in the following systems still represents a plane in three dimensions.

Examples of Systems of Linear Equations in Three Variables

$$\begin{cases} x+y+z=1 \\ x-y+2z=3 \\ -2x+4y-z=2 \end{cases} \qquad \begin{cases} z=6 \\ x-y=1 \\ y-4z=0 \end{cases}$$

TIP

The "triangular form" refers to the triangle shape made by the system of equations.

When a system of linear equations is in the following form, called **triangular form**, the system can easily be solved because one of the equations is already solved for one of the variables

$$\begin{cases} \_\_x+\_\_y+\_\_z=\_\_ \\ \_\_y+\_\_z=\_\_ \\ \_\_z=\_\_ \end{cases}$$

The advantage of the triangular form is that it allows **back-substitution** of the value of the variable (from the bottom equation) into the equation above it, which has only one other variable. Solving this equation gives a value for a second variable. Then the two known variables can be back-substituted again into the top equation (with three variables) to solve for the final variable.

DEFINITION

Triangular form describes a system of equations in which the top equation contains all variables and each subsequent equation contains one less variable until the final equation has only one variable.

Back-substitution is the method of taking the value of one variable in a system of equations and substituting it into the equation above to obtain the value of another variable.

EXAMPLE 1 Using Back-Substitution to Solve a System of Equations

Solve the linear system of equations. $\begin{cases} x+2y+z=9 \\ 3y-z=7 \\ 2z=4 \end{cases}$

SOLUTION

Solve equation 3 for z by dividing both sides by 2. $2z=4 \Rightarrow z=2$

Now use back-substitution to solve for y, and then x.

Substitute the value for z into equation 2 and solve for y.

$$\begin{aligned} 3y-2&=7 \\ 3y&=9 \\ y&=3 \end{aligned}$$

Substitute the values for z and y into equation 1 to solve for x.

$$\begin{aligned} x+2(3)+2&=9 \\ x+6+2&=9 \\ x&=1 \end{aligned}$$

The solution to the system of equations is (1, 3, 2).

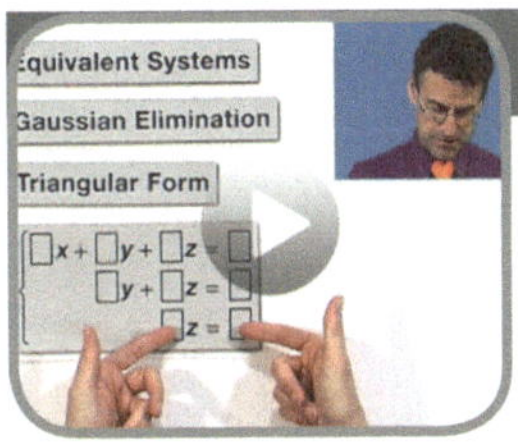

8.4.3 Using Gaussian Elimination to Solve a System

Certain operations can be performed on equations that transform them into equivalent equations. These transformations, such as dividing both sides of an equation by a constant, are the basis for solving equations and for solving systems of equations. When these operations are performed on a system of equations, the system that results is an **equivalent system of equations** (i.e., a system of equations that has the same solution as the original system).

IMPORTANT

Subtracting one equation from another is also allowed since subtraction is the same as multiplying an equation by −1 and then adding. Dividing an equation by a constant is allowed as well, since division by a constant is the same as multiplying by the reciprocal of that constant.

Operations That Yield Equivalent Systems of Linear Equations
1. Add a nonzero multiple of one equation to another.
2. Multiply an equation by a nonzero constant.
3. Interchange the position of two equations.

Gaussian elimination is a method for solving systems of linear equations where the system is first written as an equivalent system in triangular form by using these three allowable operations. There are many ways to reach the goal of a triangular system (i.e., many different ways to apply the allowable operations to the system resulting in triangular form) and no one way is correct. Once the system is in triangular form, back-substitution is used to solve the triangular system. Because the triangular system is equivalent to the original system, its solution is also the solution to the original system.

Steps for Solving a System of Linear Equations Using Gaussian Elimination

1. Use the allowable operations to write an equivalent system in triangular form.
2. Solve the triangular system using back-substitution.

EXAMPLE 2 Solving a System of Linear Equations Using Gaussian Elimination

Solve the system of equations. $\begin{cases} 2x+2y+z=16 \\ x-y+3z=14 \\ 2x+y-z=6 \end{cases}$

SOLUTION

Solve the system using Gaussian elimination.

IMPORTANT

The process shown here is just one of many different ways to manipulate the system into triangular form.

❶ Use the allowable operations to write an equivalent system in triangular form.

Switch equations 1 and 2 so that the coefficient of x in equation 1 will be 1.

$$\begin{cases} x-y+3z=14 \\ 2x+2y+z=16 & \textit{Switch equation 1 and equation 2.} \\ 2x+y-z=6 & \textit{Write equation 3.} \end{cases}$$

Eliminate the x-term from equation 3.

$$\begin{array}{r} -2x-2y-z=-16 \\ 2x+y-z=6 \\ \hline -y-2z=-10 \end{array}$$

$$\begin{cases} x-y+3z=14 & \textit{Write equation 1.} \\ 2x+2y+z=16 & \textit{Write equation 2.} \\ -y-2z=10 & \textit{new equation 3 = equation 3 + (–1)equation 2} \end{cases}$$

Eliminate the x-term from equation 2.

$$\begin{array}{r} -2x+2y-6z=-28 \\ 2x+2y+z=16 \\ \hline 4y-5z=-12 \end{array}$$

$$\begin{cases} x-y+3z=14 & \textit{Write equation 1.} \\ 4y-5z=-12 & \textit{new equation 2 = equation 2 + (–2)equation 1} \\ -y-2z=-10 & \textit{Write equation 3.} \end{cases}$$

Switch equations 2 and 3.

$$\begin{cases} x-y+3z=14 & \textit{Write equation 1.} \\ -y-2z=-10 \\ 4y-5z=-12 & \textit{Switch equation 2 and equation 3.} \end{cases}$$

Eliminate the y-term of equation 3, resulting in a system of equations in triangular form.

$$\begin{array}{r} -4y-8z=-40 \\ 4y-5z=-12 \\ \hline -13z=-52 \end{array}$$

$$\begin{cases} x-y+3z=14 & \textit{Write equation 1.} \\ -y-2z=-10 & \textit{Write equation 2.} \\ -13z=-52 & \textit{new equation 3 = equation 3 + (4)equation 2} \end{cases}$$

❷ Solve the triangular system using back-substitution.

Solve equation 3 for z by dividing both sides by –13. $-13z=-52 \Rightarrow z=4$

Now solve for y, and then x.

Substitute the value for z into equation 2 and solve for y.

$$\begin{aligned} -y-2(4) &= -10 \\ -y-8 &= -10 \\ -y &= -2 \\ y &= 2 \end{aligned}$$

Substitute the values for z and y into equation 1 to solve for x.

$$\begin{aligned} x-2+3(4) &= 14 \\ x-2+12 &= 14 \\ x+10 &= 14 \\ x &= 4 \end{aligned}$$

The solution for the system of equations is (4, 2, 4).

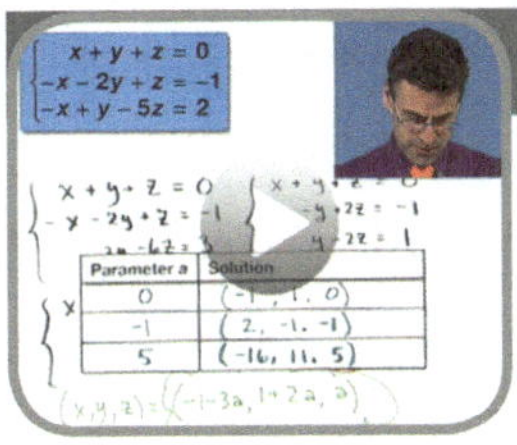

8.4.4 Gaussian Elimination: Special Cases

A system of linear equations with three equations and three variables may have one unique solution, infinitely many solutions, or no solutions at all. Gaussian elimination can be used to determine whether a system has one unique solution, infinitely many solutions, or no solutions, but back-substitution will not be used when the system has no solutions.

EXAMPLE 3 Determining That a System of Linear Equations Is Inconsistent

Solve the system of equations. $\begin{cases} 3x - 2y - z = 7 \\ 3x + y - 3z = 2 \\ 12x - 2y - 8z = -5 \end{cases}$

SOLUTION

Solve the system using Gaussian elimination.

Use the allowable operations to write an equivalent system in triangular form.

Eliminate the x-term from equation 2.

$$\begin{array}{r} -3x + 2y + z = -7 \\ 3x + y - 3z = 2 \\ \hline 3y - 2z = -5 \end{array}$$

$$\begin{cases} 3x - 2y - z = 7 & \textit{Write equation 1.} \\ 3y - 2z = -5 & \textit{new equation 2 = equation 2 + (-1)equation 1} \\ 12x - 2y - 8z = -5 & \textit{Write equation 3.} \end{cases}$$

Eliminate the x-term from equation 3.

$$\begin{array}{r} -12x + 8y + 4z = -28 \\ 12x - 2y - 8z = -5 \\ \hline 6y - 4z = -33 \end{array}$$

$$\begin{cases} 3x - 2y - z = 7 & \textit{Write equation 1.} \\ 3y - 2z = -5 & \textit{Write equation 2.} \\ 6y - 4z = -33 & \textit{new equation 3 = equation 3 + (-4)equation 1} \end{cases}$$

Eliminate the y-term of equation 3.

$$\begin{array}{r} -6y + 4z = 10 \\ 6y - 4z = -33 \\ \hline 0 = -23 \end{array}$$

$$\begin{cases} 3x - 2y - z = 7 & \textit{Write equation 1.} \\ 3y - 2z = -5 & \textit{Write equation 2.} \\ 0 = -23 & \textit{new equation 3 = equation 3 + (-2)equation 2} \end{cases}$$

Equation 3 is now a false statement. This result signals that there is no solution to the system of equations.

EXAMPLE 4 Solving a System of Linear Equations with Infinitely Many Solutions

Solve the system of equations. $\begin{cases} x+3y-2z=2 \\ x-2y+2z=-3 \\ x-7y+6z=-8 \end{cases}$

SOLUTION

Solve the system using Gaussian elimination.

❶ Use the allowable operations to write an equivalent system in triangular form.

Eliminate the x-term from equation 2.

$$\begin{array}{r} -x-3y+2z=-2 \\ x-2y+2z=-3 \\ \hline -5y+4z=-5 \end{array}$$

$\begin{cases} x+3y-2z=2 & \textit{Write equation 1.} \\ \quad -5y+4z=-5 & \textit{new equation 2 = equation 2 + (−1)equation 1} \\ x-7y+6z=-8 & \textit{Write equation 3.} \end{cases}$

Eliminate the x-term from equation 3.

$$\begin{array}{r} -x-3y+2z=-2 \\ x-7y+6z=-8 \\ \hline -10y+8z=-10 \end{array}$$

$\begin{cases} x+3y-2z=2 & \textit{Write equation 1.} \\ \quad -5y+4z=-5 & \textit{Write equation 2.} \\ \quad -10y+8z=-10 & \textit{new equation 3 = equation 3 + (−1)equation 1} \end{cases}$

Eliminate the y-term from equation 3.

$$\begin{array}{r} 10y-8z=10 \\ -10y+8z=-10 \\ \hline 0=0 \end{array}$$

$\begin{cases} x+3y-2z=2 & \textit{Write equation 1.} \\ \quad -5y+4z=-5 & \textit{Write equation 2.} \\ \quad 0=0 & \textit{new equation 3 = equation 3 + (−2)equation 2} \end{cases}$

This final step eliminates both the y- and z-terms of equation 3. Since equation 3 is a true statement, there are infinitely many solutions to the system.

When a system has infinitely many solutions, use back-substitution to determine an expression for the value of each variable.

TIP

The variable a, called a parameter, represents any real number. Any letter (other than x, y, or z) can be used here to represent any real number.

❷ Let $z = a$ (where a represents any real number). Using back-substitution, solve for x and y in terms of a.

Substitute $z = a$ into equation 2, and solve for y.

$$\begin{aligned} -5y+4a &= -5 \\ -5y &= -4a-5 \\ y &= \frac{-4a-5}{-5} \\ y &= \frac{4}{5}a+1 \end{aligned}$$

Substitute $z = a$ and $y = \frac{4}{5}a+1$ into equation 1 to solve for x.

$$\begin{aligned} x+3\left(\frac{4}{5}a+1\right)-2a &= 2 \\ x+\frac{12}{5}a+3-2a &= 2 \\ x+\frac{2}{5}a+3 &= 2 \\ x &= -\frac{2}{5}a-1 \end{aligned}$$

The solution is the set of all points such that $\left(-\frac{2}{5}a-1,\ \frac{4}{5}a+1,\ a\right)$, where a is any real number.

Note that the system in **Example 4** has infinitely many solutions. These solutions follow the specific form defined by $\left(-\frac{2}{5}a-1,\ \frac{4}{5}a+1,\ a\right)$, where a is any real number. For example, one possible solution is found by letting $a = 0$. In this case, the solution is $(-1, 1, 0)$. Another possible solution is found by letting $a = 5$, which yields the solution $(-3, 5, 5)$.

8.4.5 Nonsquare Systems

The systems examined so far have had an equal number of equations and variables. When the number of equations in a system is equal to the number of variables in the system, the system is called a **square system**. **Nonsquare systems of equations** have fewer equations than variables (or fewer variables than equations). It is not possible to find a unique solution for a nonsquare system of equations; it will either have no solution or infinitely many solutions.

EXAMPLE 5 Solving a Nonsquare System of Equations

NOTICE THAT

The system is nonsquare, so it does not have a unique solution.

Solve the system of equations. $\begin{cases} x + y + 2z = 0 \\ -x + y - z = 2 \end{cases}$

SOLUTION

Solve the system using Gaussian elimination.

❶ Eliminate the x-term from equation 2.

$$\begin{cases} x + y + 2z = 0 & \textit{Write equation 1.} \\ \quad 2y + z = 2 & \textit{new equation 2 = equation 1 + equation 2} \end{cases}$$

❷ Let $z = a$ (where a represents any real number). Using back-substitution, solve for x and y in terms of a.

Substitute $z = a$ into equation 2, and solve for y.

$$2y + a = 2$$
$$2y = 2 - a$$
$$y = -\frac{1}{2}a + 1$$

Substitute $z = a$ and $y = -\frac{1}{2}a + 1$ into equation 1 to solve for x.

$$x + \left(-\frac{1}{2}a + 1\right) + 2a = 0$$
$$x + \frac{3}{2}a + 1 = 0$$
$$x = -\frac{3}{2}a - 1$$

TIP

These points lie on the line of intersection of the two planes described by the equations in the system.

The solution is the set of all points such that $\left(-\frac{3}{2}a - 1,\ -\frac{1}{2}a + 1,\ a\right)$, where a is any real number.

EXAMPLE 6 Solving a Nonsquare System of Equations

NOTICE THAT

The system is nonsquare, so it does not have a unique solution.

Solve the system of equations. $\begin{cases} 2x - 3y - z = 5 \\ 6x - 9y - 3z = 2 \end{cases}$

SOLUTION

Multiply equation 1 by –3, add equations 1 and 2, and replace equation 2 with that sum.

$$\begin{cases} 2x - 3y - z = 5 & \textit{Write equation 1.} \\ \qquad\quad 0 = -13 & \textit{new equation 2 = equation 2 + (–3)equation 1} \end{cases}$$

This step eliminates the x-, y- and z-terms of equation 2. Equation 2 is now a false statement. Therefore, there is no solution to the system of equations.

8.4.6 Modeling with Multivariable Linear Systems

EXAMPLE 7 Modeling with a Multivariable Linear System

A recycling company buys bulk recycling (paper, aluminum, plastic, glass, etc.) from three cities and then sells the three most popular materials (aluminum, plastic, and glass). The following table shows the percent of the bulk recycling purchased from each city that is aluminum, plastic, or glass (e.g., 30% of the total bulk recycling bought from city A is aluminum), and the total number of tons of each type of material that the company can sell per day. How many tons per day should the company buy from each city to match the demand for the materials?

	City A	City B	City C	Total Sold
Aluminum	30%	20%	30%	30 tons
Plastic	10%	10%	20%	15 tons
Glass	30%	40%	50%	50 tons

SOLUTION

TIP

Identify the unknowns and write a system of equations.

Let a, b, and c be the number of tons of recycling bought from cities A, B, and C, respectively.

$$\begin{cases} 0.3a + 0.2b + 0.3c = 30 \\ 0.1a + 0.1b + 0.2c = 15 \\ 0.3a + 0.4b + 0.5c = 50 \end{cases}$$

The total of 30% of the bulk recycling from City A, 20% of the bulk recycling from City B, and 30% of the bulk recycling from City C is 30 tons of aluminum.

Multiply each equation by 10 to eliminate the decimals (so the equations will be more manageable) and switch equations 1 and 2 (so that the coefficient of x in equation 1 will be 1).

$$\begin{cases} a + b + 2c = 150 \\ 3a + 2b + 3c = 300 \\ 3a + 4b + 5c = 500 \end{cases}$$

Switch equation 1 and equation 2.

TIP

Use the allowable operations to write an equivalent system in triangular form.

Eliminate the a-term from equation 3.

$$\begin{cases} a + b + 2c = 150 & \text{Write equation 1.} \\ 3a + 2b + 3c = 300 & \text{Write equation 2.} \\ 2b + 2c = 200 & \text{new equation 3 = equation 3 + (–1)equation 2} \end{cases}$$

Eliminate the a-term from equation 2.

$$\begin{cases} a + b + 2c = 150 & \text{Write equation 1.} \\ -b - 3c = -150 & \text{new equation 2 = equation 2 + (–3)equation 1} \\ 2b + 2c = 200 & \text{Write equation 3.} \end{cases}$$

Eliminate the b-term from equation 3.

$$\begin{cases} a + b + 2c = 150 & \text{Write equation 1.} \\ -b - 3c = -150 & \text{Write equation 2.} \\ -4c = -100 & \text{new equation 3 = equation 3 + (2)equation 2} \end{cases}$$

TIP

Solve the triangular system using back-substitution.

Solve equation 3 for c by dividing both sides by –4. $-4c = -100 \Rightarrow c = 25$

Substitute the value for c into equation 2 and solve for b.

$$-b - 3(25) = -150 \Rightarrow b = 75$$

Substitute the values for b and c into equation 1 to solve for a.

$$3a + 2(75) + 3(25) = 300 \Rightarrow a = 25$$

The solution for the system of equations is (25, 75, 25). The company can meet its demand by purchasing 25 tons from city A, 75 tons from city B, and 25 tons from city C.

SECTION 8.4 EXERCISES

Warm Up

True or False?

1. A solution to $\begin{cases} 3x - y = 7 \\ x = 3 \end{cases}$ is (2, 3).

2. A solution to $\begin{cases} x - 2y = -15 \\ x = 20 - 5y \end{cases}$ is (−5, 5).

3. The following system has an infinite number of solutions.

$$\begin{cases} 4x + 9y = 15 \\ x + 2.25y = 3.75 \end{cases}$$

Just the Facts

Fill in the blanks.

4. $x + y = 1$ is the equation of a(n) ______, while $x + y + z = 1$ is the equation of a(n) ______.

5. Two different planes that are not parallel intersect at a(n) ______.

6. A system with more variables than equations is called a(n) ______.

7. The first step of Gaussian elimination is to manipulate a system into ______ using the allowable operations.

8. A nonsquare system of equations may have no solution or ______.

9. In a real-world situation where many variables simultaneously satisfy a set of relationships, ______ are used to find the values of the variables.

10. Suppose data is given from two power plants concerning temperature, pressure, and volume. The corresponding multivariable linear system is ______.

Essential Skills

In Exercises 1–8, solve each system.

1. $\begin{cases} -2x + y + z = -5 \\ 3y + z = 13 \\ z = 4 \end{cases}$

2. $\begin{cases} 2x - 3y + 4z = -2 \\ 2y + 5z = -1 \\ \frac{2}{3}z = -6 \end{cases}$

3. $\begin{cases} x + y + z = -6 \\ -2x - y + z = -2 \\ x - 2y - z = 1 \end{cases}$

4. $\begin{cases} 3x + y + 2z = 3 \\ 2x - 3y - z = -3 \\ x + 2y + z = 4 \end{cases}$

5. $\begin{cases} 7x - 4y - 2z = 7 \\ 8x - 2y - 8z = 1 \\ x + 2y - 6z = -6 \end{cases}$

6. $\begin{cases} 2x + y + 5z = 4 \\ 3x - 2y + 2z = 2 \\ 5x - 8y - 4z = 1 \end{cases}$

7. $\begin{cases} 9x + 5y + 9z = 3 \\ -45x + 10y + 27z = -6 \end{cases}$

8. $\begin{cases} 6x - 4y + 2z = 5 \\ 3x - 2y + z = 7 \end{cases}$

9. The table shows the numbers of pastels, brushes, and canvases bought at an art supply store and the amount of money spent. Identify the price of each item.

Pastels	Brushes	Canvases	Total Spent
3	4	5	\$22
2	5	7	\$23
4	3	6	\$24

10. The table shows the numbers of different types of drinks bought at a charity event, and the amount of money spent. Identify the price of each kind of drink.

Soda	Lemonade	Water	Total Spent
5	4	6	\$43
2	5	4	\$34
3	7	2	\$41

Extensions

11. Solve the system. $\begin{cases} x - y + 2z = \dfrac{16}{3} \\ 2x + y - z = -\dfrac{2}{3} \\ 3x - 2y - 3z = 0 \end{cases}$

12. Determine the value for k so that $\left(-1, \frac{4}{5}, \frac{1}{5}\right)$ is the solution for the following system.

$$\begin{cases} 5x - 5y + 10z = 1 \\ 10x + ky - 5z = -15 \\ 15x + 15y + 5z = -26 \end{cases}$$

13. The perimeter of a triangle is 47 inches. The longest side is 4 inches shorter than twice the shortest side. The sum of the lengths of the two shorter sides is 7 inches more than the length of the longest side. Find the lengths of the sides of the triangle.

14. Last Friday, Texas Cinemas sold a total of 8500 movie tickets. Proceeds totaled $64,600. Tickets are sold in one of three ways: adult admission costs $8, senior admission costs $7, and youth admission costs $6. How many of each type of ticket was sold if twice as many youth tickets were sold as senior tickets?

15. Find the equation of the circle $x^2 + y^2 + Dx + Ey + F = 0$ that passes through (2, 3), (5, 0), and (−1, 0).

16. In the 2011–12 NBA regular season, Kevin Durant scored 1850 points by hitting 1207 of his 1-point, 2-point, and 3-point attempts. He made 86% of his 501 free-throw (1-point) attempts. How many 1-, 2-, and 3-point baskets did Mr. Durant complete?

8.5 PARTIAL FRACTIONS

OBJECTIVES

- Find partial fraction decompositions when denominators have distinct linear factors.
- Find partial fraction decompositions when denominators have repeated linear factors.
- Find partial fraction decompositions when denominators have distinct linear factors and quadratic factors.
- Find partial fraction decompositions when denominators have repeated quadratic factors.

PREREQUISITE VOCABULARY TERMS

linear factor

linear system of equations

quadratic equation

rational expression

8.5.1 Partial Fraction Decomposition

Recall that the sum of two rational expressions can be found by first writing each expression with a common denominator and then adding the numerators. It is also possible to reverse this process and separate a rational expression into a sum of simpler rational expressions, or **partial fractions**. This process is called **partial fraction decomposition**.

The steps for finding a partial fraction decomposition depend on the denominator of the rational expression being decomposed. The simplest case of partial fraction decomposition involves a rational expression with two distinct (i.e., different) linear factors in the denominator.

Steps for Finding a Partial Fraction Decomposition of a Rational Expression: A Denominator with Two Distinct Linear Factors

❶ Factor the denominator.

❷ Write the rational expression as a sum of two partial fractions where the numerators are the unknown constants A and B, and the denominators are each of the linear factors from the original rational expression.

$$\frac{ex+f}{(ax+b)(cx+d)} = \frac{A}{(ax+b)} + \frac{B}{(cx+d)}$$

❸ Add the partial fractions and simplify the numerator.

❹ Write a system of two linear equations.

- Equation 1: Set the coefficient of the x-term in the numerator of the original rational expression equal to the coefficient of the x-term in the numerator of the summed partial fractions.
- Equation 2: Set the constant term in the numerator of the original rational expression equal to the constant term in the numerator of the summed partial fractions.

❺ Substitute the values of A and B into the original partial fractions.

IMPORTANT

For partial fraction decomposition to work, the degree of the numerator must be less than the degree of the denominator.

EXAMPLE 1 Finding a Partial Fraction Decomposition with Two Distinct Linear Factors in the Denominator

Find the partial fraction decomposition. $\dfrac{13x-74}{(x-5)(x-6)}$

SOLUTION

❶ The denominator is factored.

❷ Write the sum of the two partial fractions. $\dfrac{13x-74}{(x-5)(x-6)}=\dfrac{A}{(x-5)}+\dfrac{B}{(x-6)}$

❸ Add the partial fractions.

$$\frac{A}{(x-5)}+\frac{B}{(x-6)}=\frac{A(x-6)}{(x-5)(x-6)}+\frac{B(x-5)}{(x-5)(x-6)}$$ *Multiply to obtain a common denominator.*

$$=\frac{Ax-6A+Bx-5B}{(x-5)(x-6)}$$ *Add. Simplify the numerator.*

$$=\frac{(A+B)x+(-6A-5B)}{(x-5)(x-6)}$$ *Factor out x. Group the constants.*

TIP

Write the numerator in the form ____x + ____.

Therefore, $\dfrac{13x-74}{(x-5)(x-6)}=\dfrac{(A+B)x+(-6A-5B)}{(x-5)(x-6)}$. *The denominators are equal, so the numerators are also equal.*

❹ Write a system of linear equations.

TIP

Equation 1 sets the coefficients of x on each side equal to each other, while equation 2 sets the constants on each side equal to each other.

$$\begin{cases} A+B=13 \\ -6A-5B=-74 \end{cases}$$

The coefficients of x in the numerators are 13 and (A + B).
The constant terms in the numerators are −74 and (−6A − 5B).

Solve the system using substitution to find A and B.

$A=13-B$ *Solve equation 1 for A.*

$-6(13-B)-5B=-74$ *Substitute the expression for A into equation 2.*

$B=4$ *Simplify.*

$A=13-4=9$ *Substitute B = 4 into the expression for A.*

❺ Substitute $B = 4$ and $A = 9$ into the partial fractions, $\dfrac{A}{(x-5)}+\dfrac{B}{(x-6)}$.

$$\frac{13x-74}{(x-5)(x-6)}=\frac{9}{(x-5)}+\frac{4}{(x-6)}$$

TIP

Check the result by adding the partial fractions.

8.5.2 Repeated Linear Factors

When a rational expression has a denominator that is a linear factor repeated k times, the procedure for writing the partial fractions is a little different. Instead of always having two partial fractions, the number of partial fractions will be equal to the exponent k of the linear factor in the denominator. Each partial fraction will have a denominator equal to a power of the linear factor, where the powers run from 1 to k.

For example, if the original rational expression's denominator is $(x+4)^3$, then there will be three partial fractions (because the exponent is 3) and their denominators will be $(x+4)$, $(x+4)^2$, and $(x+4)^3$. In this case, the numerators of the partial fractions will be A, B, and C.

If the denominator contains a linear factor repeated k times and additional distinct linear factors, then there will be a partial fraction for each linear factor. For example, if the original rational expression's denominator is $x(x + 4)^3$, then there will be four partial fractions and their denominators will be x, $(x + 4)$, $(x + 4)^2$, and $(x + 4)^3$. In this case, the numerators of the partial fractions will be A, B, C, and D.

EXAMPLE 2 Finding a Partial Fraction Decomposition with Repeated Linear Factors in the Denominator

Find the partial fraction decomposition. $\dfrac{3x-2}{x(2x+1)^2}$

SOLUTION

Write the sum of the partial fractions.

$$\frac{3x-2}{x(2x+1)^2} = \frac{A}{x} + \frac{B}{(2x+1)} + \frac{C}{(2x+1)^2}$$

TIP

Write one partial fraction for the distinct factor x, and two partial fractions for the twice-repeated factor $(2x + 1)$.

Add the partial fractions.

$$\frac{A}{x} + \frac{B}{(2x+1)} + \frac{C}{(2x+1)^2} = \frac{A(2x+1)^2}{x(2x+1)^2} + \frac{Bx(2x+1)}{x(2x+1)^2} + \frac{Cx}{x(2x+1)^2}$$

Multiply to obtain a common denominator.

$$= \frac{4Ax^2 + 4Ax + A + 2Bx^2 + Bx + Cx}{x(2x+1)^2}$$

Add. Simplify the numerator.

$$= \frac{(4A+2B)x^2 + (4A+B+C)x + A}{x(2x+1)^2}$$

Factor out x^2 and x.

Therefore, $\dfrac{3x-2}{x(2x+1)^2} = \dfrac{(4A+2B)x^2 + (4A+B+C)x + A}{x(2x+1)^2}$.

The denominators are equal, so the numerators are also equal.

TIP

Write the numerator in the form

____x^2 + ____x + ____.

Write a system of linear equations.

$$\begin{cases} 4A + 2B = 0 \\ 4A + B + C = 3 \\ A = -2 \end{cases}$$

The coefficients of x^2 in the numerators are 0 and $(4A + 2B)$.
The coefficients of x in the numerators are 3 and $(4A + B + C)$.
The constant terms in the numerators are -2 and A.

TIP

Since there is no x^2-term in the left-side expression, $4A + 2B$ must be equal to 0.

Solve the system using back-substitution. Equation 3 is already solved for A.

Substitute $A = -2$ into equation 1, and solve for B.

$$4(-2) + 2B = 0$$
$$-8 + 2B = 0$$
$$B = 4$$

Substitute $A = -2$ and $B = 4$ into equation 2, and solve for C.

$$4(-2) + 4 + C = 3$$
$$-4 + C = 3$$
$$C = 7$$

Substitute $A = -2$, $B = 4$, and $C = 7$ into the partial fractions.

$$\frac{3x-2}{x(2x+1)^2} = \frac{-2}{x} + \frac{4}{(2x+1)} + \frac{7}{(2x+1)^2}$$

8.5.3 Distinct Linear and Quadratic Factors

If the denominator of a rational expression contains a quadratic factor, then the numerator of the partial fraction with the quadratic factor in the denominator will include a linear expression of the form $Ax + B$. This expression is one degree lower than the irreducible quadratic factor in the denominator.

EXAMPLE 3 Finding a Partial Fraction Decomposition with Distinct Linear and Quadratic Factors in the Denominator

Find the partial fraction decomposition. $\dfrac{6x^2+2x+30}{x^3+6x}$

SOLUTION

NOTICE THAT

The denominator is a product of a linear factor and an irreducible quadratic factor.

Factor the denominator. $\dfrac{6x^2+2x+30}{x^3+6x}=\dfrac{6x^2+2x+30}{x(x^2+6)}$

Write the sum of the partial fractions. $\dfrac{6x^2+2x+30}{x(x^2+6)}=\dfrac{A}{x}+\dfrac{Bx+C}{x^2+6}$

Add the partial fractions.

$$\frac{A}{x}+\frac{Bx+C}{x^2+6}=\frac{A(x^2+6)}{x(x^2+6)}+\frac{(Bx+C)x}{x(x^2+6)}$$ *Multiply to obtain a common denominator.*

$$=\frac{Ax^2+6A+Bx^2+Cx}{x(x^2+6)}$$ *Add. Simplify the numerator.*

$$=\frac{(A+B)x^2+Cx+6A}{x(x^2+6)}$$ *Factor out x^2.*

TIP

Write the numerator in the form ____x^2 + ____x + ____.

Therefore, $\dfrac{6x^2+2x+30}{x(x^2+6)}=\dfrac{(A+B)x^2+Cx+6A}{x(x^2+6)}$. *The denominators are equal, so the numerators are also equal.*

Write a system of linear equations.

$$\begin{cases} A+B=6 \\ C=2 \\ 6A=30 \end{cases}$$

The coefficients of x^2 in the numerators are 6 and (A + B).
The coefficients of x in the numerators are 2 and C.
The constant terms in the numerators are 30 and 6A.

Solve the system using back-substitution. Equation 2 is already solved for C.

Solve equation 3 for A.

$$6A=30$$
$$A=5$$

Substitute $A = 5$ into equation 1, and solve for B.

$$5+B=6$$
$$B=1$$

Substitute $A = 5$, $B = 1$, and $C = 2$ into the partial fractions.

$$\frac{6x^2+2x+30}{x(x^2+6)}=\frac{5}{x}+\frac{x+2}{x^2+6}$$

TIP

Check the result by adding the partial fractions.

8.5.4 Repeated Quadratic Factors

If the denominator of a rational expression contains repeated irreducible quadratic factors, then the numerators of the partial fraction decomposition for these factors will be linear expressions of the form $Ax + B$.

EXAMPLE 4 **Finding a Partial Fraction Decomposition with Repeated Quadratic Factors in the Denominator**

NOTICE THAT

The denominator is a power of an irreducible quadratic polynomial.

Find the partial fraction decomposition. $\dfrac{x^3+2x^2}{(x^2+1)^2}$

SOLUTION

Write the sum of the partial fractions. $\dfrac{x^3+2x^2}{(x^2+1)^2} = \dfrac{Ax+B}{(x^2+1)} + \dfrac{Cx+D}{(x^2+1)^2}$

Find the sum of the partial fractions.

$$\frac{x^3+2x^2}{(x^2+1)^2} = \frac{(Ax+B)(x^2+1)}{(x^2+1)(x^2+1)} + \frac{Cx+D}{(x^2+1)^2}$$

Multiply to obtain a common denominator.

$$= \frac{Ax^3+Ax+Bx^2+B+Cx+D}{(x^2+1)^2}$$

Add. Simplify the numerator.

$$= \frac{Ax^3+Bx^2+(A+C)x+(B+D)}{(x^2+1)^2}$$

Factor out x. Group the constants.

TIP

Write the numerator in the form

___x^3 + ___x^2 + ___x + ___.

Therefore, $\dfrac{x^3+2x^2}{(x^2+1)^2} = \dfrac{Ax^3+Bx^2+(A+C)x+(B+D)}{(x^2+1)^2}$.

The denominators are equal, so the numerators are also equal.

Write a system of linear equations.

$$\begin{cases} A = 1 \\ B = 2 \\ A+C = 0 \\ B+D = 0 \end{cases}$$

The coefficients of x^3 in the numerators are 1 and A.
The coefficients of x^2 in the numerators are 2 and B.
The coefficients of x in the numerators are 0 and (A + C).
The constant terms in the numerators are 0 and (B + D).

TIP

Since there is no x-term or constant term in the left-side expression, A + C and B + D must be equal to 0.

Solve the system using back-substitution.

Equation 1 is already solved for A and equation 2 is already solved for B.

Substitute $A = 1$ into equation 3, and solve for C.

$$1 + C = 0$$
$$C = -1$$

Substitute $B = 2$ into equation 4, and solve for D.

$$2 + D = 0$$
$$D = -2$$

Substitute $A = 1$, $B = 2$, $C = -1$, and $D = -2$ into the partial fractions.

$$\frac{x^3+2x^2}{(x^2+1)^2} = \frac{x+2}{(x^2+1)} + \frac{-x-2}{(x^2+1)^2} = \frac{x+2}{(x^2+1)} - \frac{x+2}{(x^2+1)^2}$$

WHAT'S THE BIG IDEA?

Write a rational expression with repeated linear factors and repeated quadratic factors in the denominator. Then write the form of the partial fraction decomposition.

SECTION 8.5 EXERCISES

Warm Up

Simplify.

1. $\frac{2}{x}+\frac{3}{x+1}$

2. $\frac{-2x+1}{x}-\frac{4}{5}$

3. $\frac{-x+7}{4}-\frac{3}{x-6}$

Just the Facts

Fill in the blanks.

4. Multiple rational expressions can be added together to form a single rational expression. Similarly, a single rational expression can be ______ into multiple rational expressions, which are called _______ fractions.

5. A rational expression with denominator $2x^2+11x+15$ can be decomposed into ________ fractions with denominators of ______ and _______.

6. When decomposing a rational expression, it is often necessary to ______ the denominator.

7. Two rational expressions can be combined when they have a common ______.

8. For the decomposition of $\frac{5x+7}{(x-1)^7}$, there will be _____ partial fractions.

9. For the decomposition of $\frac{6x-2}{(2x+5)(3x^2+2)}$, there will be ______ partial fractions.

10. If the numerator of a partial fraction is of the form $ax + b$, then its denominator must be _______.

Essential Skills

In Exercises 1–8, find each partial fraction decomposition.

1. $\frac{15x-21}{(x-2)(x-1)}$

2. $\frac{3x-87}{x^2-3x-54}$

3. $\frac{x-6}{x^2(x+3)}$

4. $\frac{-x^2+2x-14}{(x+3)(x^2-4x+4)}$

5. $\frac{-3x^2-3x-10}{x(x^2+2x+5)}$

6. $\frac{3x^2-x-4}{x^3-2x}$

7. $\frac{2x^3-x^2+9x-1}{(x^2+4)^2}$

8. $\frac{3x^3+x^2+22x+3}{(x^2+7)^2}$

Extensions

In Exercises 9–14, find each partial fraction decomposition.

9. $\frac{11x^2+141x+378}{(x+6)(x+3)(x+12)}$

10. $\frac{5x^3+53x-8}{(x^2+11)^2}$

11. $\frac{\frac{3}{4}x^4+4x^3+\frac{23}{2}x^2+2x+\frac{23}{4}}{(x+1)^2(x^2+3)^2}$

12. $\frac{10x^2-31x-3}{8x^3-27}$

13. $\frac{5}{a^2b^2-x^2}$

14. $\frac{-5x^3-6x^2-13x-34}{x^4+14x^2+45}$

8.6 SYSTEMS OF INEQUALITIES AND LINEAR PROGRAMMING

OBJECTIVES

- Graph linear inequalities.
- Graph nonlinear inequalities.
- Graph systems of inequalities.
- Optimize linear functions using linear programming.

PREREQUISITE VOCABULARY TERMS

system of equations
vertex (of a parabola)
x-intercept
y-intercept

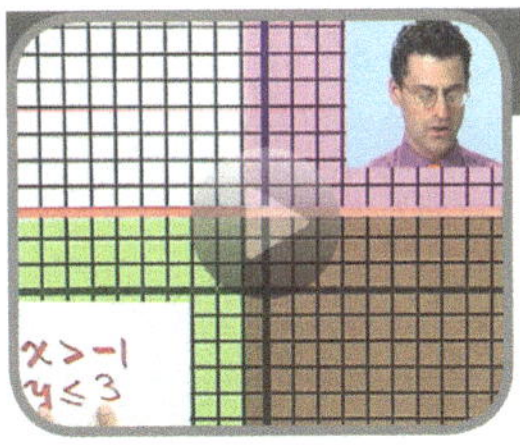

8.6.1 An Introduction to Graphing Linear Inequalities

Recall that the graph of a linear equation, for example, $y = 4$, is made up of an infinite number of points that are the solutions to the equation, and these solutions form a line. The graph of a linear inequality, for example, $y \geq 4$, is also made up of an infinite number of points that are the solutions to the inequality. However, the solutions for a linear inequality form an entire region of the coordinate plane (not just a line) called the **solution region** or **solution set** for the inequality. An inequality is graphed by shading its solution region.

To graph a linear inequality, begin by graphing the corresponding linear equation. For example, to graph the inequality $y \geq 4$, first graph the line $y = 4$. The points on this line (the graph of the corresponding linear equation) are included in the solution region when the inequality symbol is $\geq$ or $\leq$. To indicate that the line is also part of the solution region, use a solid line (as usual).

The points on the line are not included in the solution region when the inequality symbol is $>$ or $<$, that is, when the inequality is a **strict inequality**. In this case, use a dashed line to indicate that the points along the line are not included in the solution region.

DEFINITION

A **strict inequality** excludes the endpoints of the interval defined by the inequality.

TIP

Strict inequalities use the signs $<$ and $>$, meaning "strictly less than" and "strictly greater than."

The graph of the corresponding linear equation divides the coordinate plane into two regions, one of which is the solution region for the inequality. To determine which region is the solution region, choose a test point from either region (but not on the line) and substitute its coordinates into the inequality. If the resulting inequality is true, then the region containing the test point is the solution region for the inequality. If the resulting inequality is false, then the region that does not contain the test point is the solution region for the inequality. Shade the solution region to complete the graph of the inequality.

Steps for Graphing a Linear Inequality

❶ Graph the line for the equation corresponding to the inequality.
(Draw the line dashed for a strict inequality; otherwise, solid.)

❷ Test a point from one of the regions defined by the line by substituting its coordinates into the inequality. If the resulting inequality is
- true, then shade the region that contains the point.
- false, then shade the region that does not contain the point.

EXAMPLE 1 Graphing a Linear Inequality in One Variable

Graph the inequality. $y \geq -2$

SOLUTION

TIP

The graph of $y = -2$ is a horizontal line passing through $(0, -2)$.

❶ Graph the line $y = -2$ using a solid line (since $y \geq -2$ is not a strict inequality).

❷ Choose any point not on the line $y = -2$ to test in the inequality, such as (0, 0). Substitute the coordinates (in this case, only the y-coordinate) from (0, 0) into the inequality.

$0 \geq -2$ *This is a true statement.*

IMPORTANT

Any point on the plane can be tested except the points on the line $y = -2$.

The result of substituting with (0, 0) is a true statement, so the solution set is the region that includes (0, 0). Complete the graph of the inequality by shading the region that includes (0, 0), which is the region *above* the solid line $y = -2$.

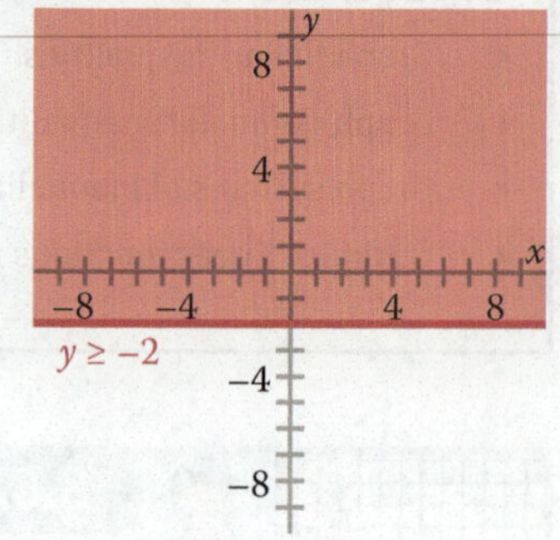

EXAMPLE 2 Graphing a Strict Linear Inequality in One Variable

Graph the inequality. $x < 1$

SOLUTION

TIP

The graph of $x = 1$ is a vertical line passing through $(1, 0)$.

❶ Graph the line $x = 1$ using a dashed line (since $x < 1$ is a strict inequality).

❷ Choose any point not on the line $x = 1$ to test in the inequality, such as (0, 0). Substitute the coordinates (in this case, only the x-coordinate) from (0, 0) into the inequality.

$0 < 1$ *This is a true statement.*

The result of substituting with (0, 0) is a true statement, so the solution set is the region that includes (0, 0). Complete the graph of the inequality by shading the region to the left of the dashed line $x = 1$.

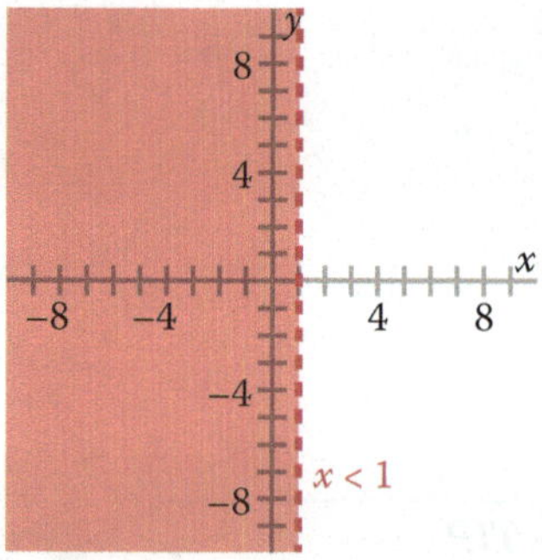

EXAMPLE 3 Graphing Multiple Linear Inequalities in One Variable

Graph the inequalities. $x \geq 2$ and $y < 4$

SOLUTION

Alternative Method

An alternative method for graphing linear inequalities in one variable is demonstrated in this example.

The solution set of $x \geq 2$ includes all points where the x-coordinate is greater than or equal to 2, which is all points on, or to the right of, the vertical line $x = 2$. So, graph $x \geq 2$ by sketching a solid line at $x = 2$ and shading the region to the right of that line.

The solution set of $y < 4$ includes all points where the y-coordinate is less than 4, which is all points below the horizontal line $y = 4$. So, graph $y < 4$ by sketching a dashed line at $y = 4$ and shading the region below that line.

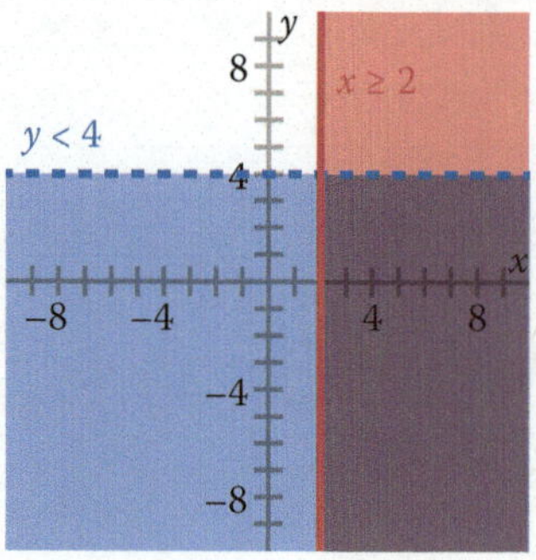

IMPORTANT

The solution set for the system of inequalities is the overlapping region that is included in the solution set for both inequalities.

8.6.2 Graphing Linear and Nonlinear Inequalities

When graphing a linear inequality in two variables, graph the line of the related equation by using the intercepts or by using the slope and y-intercept.

EXAMPLE 4 Graphing a Linear Inequality in Two Variables

Graph the inequality. $-2x + 4y \le 4$

SOLUTION

❶ Graph the line $-2x + 4y = 4$ using a solid line (since the inequality is not strict).

The graph of $-2x + 4y = 4$ is a line with intercepts at (0, 1) and (−2, 0).

REMEMBER

To find the x-intercept, substitute 0 for y. To find the y-intercept, substitute 0 for x.

❷ Choose any point not on the line to test in the inequality, such as (0, 0).

$-2(0) + 4(0) \le 4$ *Substitute the coordinates from (0, 0).*

$0 \le 4$ *This is a true statement.*

The result of substituting with (0, 0) is a true statement, so the solution set is the region that includes (0, 0). Complete the graph by shading the region below the solid line $-2x + 4y = 4$.

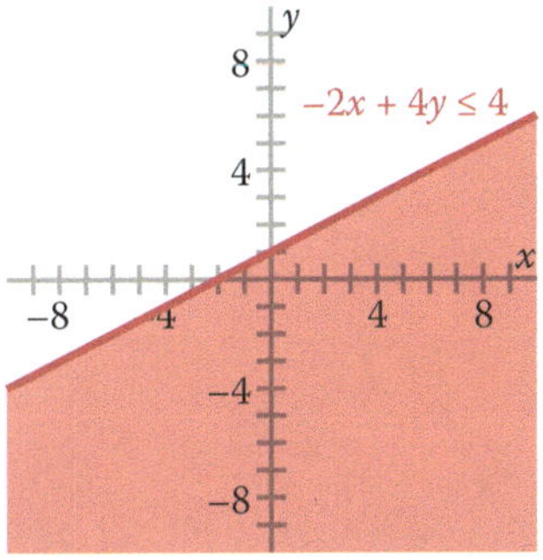

EXAMPLE 5 Graphing a Strict Linear Inequality in Two Variables

Graph the inequality. $x + 3y > 2$

SOLUTION

❶ Graph the line $x + 3y = 2$ using a dashed line (since the inequality is strict).

The graph of $x + 3y = 2$ is a line with intercepts at (0, 2/3) and (2, 0).

❷ Choose any point not on the line to test in the inequality, such as (0, 0).

$0 + 3(0) > 2$ *Substitute the coordinates from (0, 0).*

$0 > 2$ *This is a false statement.*

The result of substituting with (0, 0) is a false statement, so the solution set is the region that does not include (0, 0). Complete the graph by shading the region above the dashed line $x + 3y = 2$.

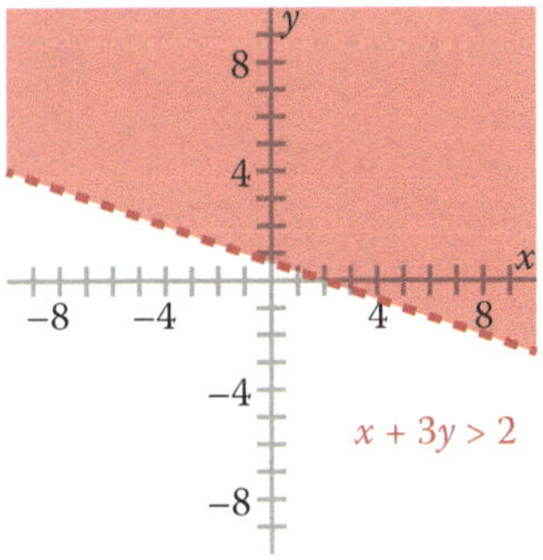

Graphing Nonlinear Inequalities

The steps for graphing a nonlinear inequality are the same as for graphing a linear inequality, except that the graph from step ❶ will not be a line. The graph of the nonlinear equation, such as a parabola or circle, will still divide the plane into two regions, one of which will be the solution set for the inequality. For example, a parabola divides the coordinate plane into two regions: the region inside of the parabola and the region outside of the parabola.

EXAMPLE 6 Graphing a Nonlinear Inequality

NOTICE THAT

The inequality is quadratic, so the graph in step ∑ will be a parabola.

Graph the inequality. $x^2 + y > 2$

SOLUTION

❶ Graph $x^2 + y = 2$. Use a dashed line for the parabola, since the inequality is strict.

The graph of $y = -x^2 + 2$ is a parabola that opens downward, with vertex at (0, 2).

❷ Choose any point not on the parabola to test in the inequality, such as (0, 0).

$(0)^2 + 0 > 2$ *Substitute the coordinates from (0, 0).*

$0 > 2$ *This is a false statement.*

The result of substituting with (0, 0) is a false statement, so the solution set is the region that does not include (0, 0), which is the region outside of the parabola. Complete the graph by shading the region outside of the parabola.

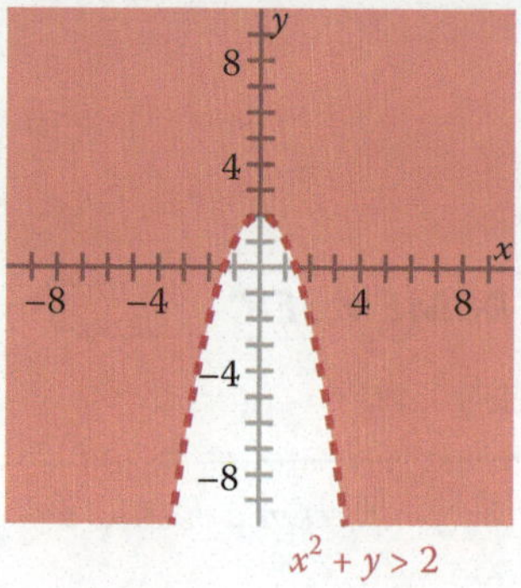

EXAMPLE 7 Graphing a Nonlinear Inequality

Graph the inequality. $y \geq 2x^2$

SOLUTION

❶ Graph $y = 2x^2$. Use a solid line for the parabola, since the inequality is not strict.

The graph of $y = 2x^2$ is a parabola that opens upward, with vertex at (0, 0).

CAUTION

The point (0, 0) cannot be chosen for testing because (0, 0) is on the parabola.

❷ Choose any point not on the parabola to test in the inequality, such as (0, 1).

$1 \geq 2(0)^2$ *Substitute the coordinates from (0, 1).*

$1 \geq 0$ *This is a true statement.*

The result of substituting with (0, 1) is a true statement, so the solution set is the region that includes (0, 1), which is the region inside of the parabola. Complete the graph by shading the region inside of the parabola.

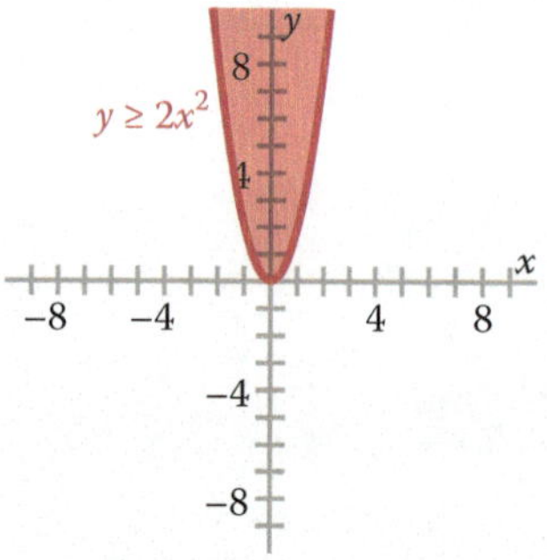

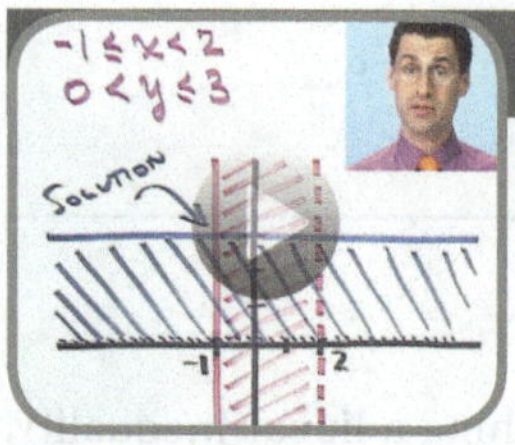

8.6.3 Graphing the Solution Set of a System of Inequalities

As with a system of equations, it is possible to define a **system of inequalities** that consists of two or more inequalities. A solution set for a system of inequalities is made up of all points that satisfy both inequalities. To graph a system of inequalities, graph each inequality in the system. If their solution regions overlap, this overlapping region is the solution region for the system of inequalities. If their solution regions do not overlap, then the system of inequalities has no solutions.

EXAMPLE 8 Graphing a System of Linear Inequalities in One Variable

Graph the system of inequalities. $\begin{cases} 1 < x \leq 4 \\ 2 \leq y < 3 \end{cases}$

SOLUTION

To graph $1 < x \leq 4$, sketch the dashed line $x = 1$ and the solid line $x = 4$, and then shade the region between these lines.

To graph $2 \leq y < 3$, sketch the solid line $y = 2$ and the dashed line $y = 3$, and then shade the region between these lines.

The solution set for the system of inequalities is the region where the two shadings overlap.

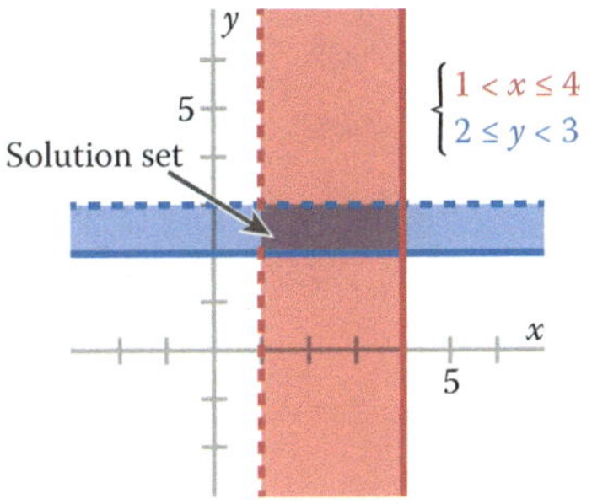

REMEMBER

The compound inequality $1 < x \leq 4$ can be written as $x > 1$ and $x \leq 4$. Similarly, $2 \leq y < 3$ can be written as $y \geq 2$ and $y < 3$.

NOTICE THAT

The system's solution region is a rectangle with vertices at (1, 3), (4, 3), (4, 2), and (1, 2). The vertex at (4, 2) is included in the system's solution region. The vertices at (1, 3), (4, 3), and (1, 2) are not included in the system's solution region.

EXAMPLE 9 Graphing a System of Linear Inequalities in Two Variables

Graph the system of inequalities. $\begin{cases} 2x - y \leq 1 \\ x + y > 3 \end{cases}$

SOLUTION

To graph $2x - y \leq 1$, sketch the solid line $2x - y = 1$, and then shade the region above the line.

To graph $x + y > 3$, sketch the dashed line $x + y = 3$, and then shade the region above the line.

The solution set for the system of inequalities is the region where the two shadings overlap.

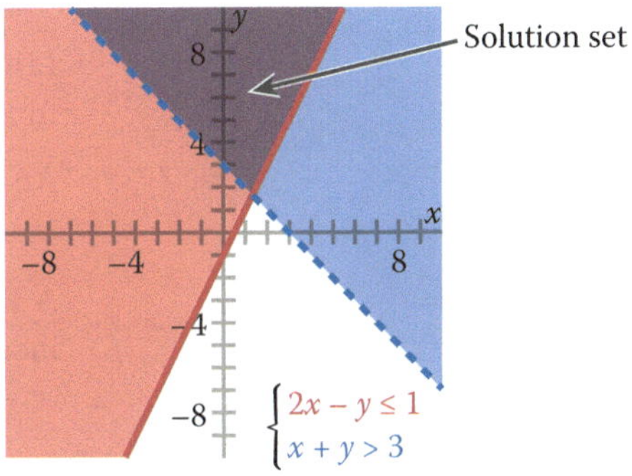

TIP

Check a solution set by testing any point in that region in each of the system's inequalities. If each inequality in the system is satisfied, then the point is in the solution region (as are all points in that region). For example, testing (0, 4) results in two true statements, $2(0) - 4 \leq 1$ and $0 + 4 > 3$. Therefore, the region containing (0, 4) is the solution set.

EXAMPLE 10 Graphing a Nonlinear System of Inequalities

Graph the system of inequalities. $\begin{cases} x^2 + y^2 \leq 25 \\ -2x + y > 2 \end{cases}$

SOLUTION

To graph $x^2 + y^2 \leq 25$, sketch the circle (using a solid line) with center (0, 0) and radius 5, and then shade the region inside the circle.

To graph $-2x + y > 2$, sketch the dashed line $-2x + y = 2$, and then shade the region above the line.

The solution set for the system of inequalities is the region where the two shadings overlap (all points on the circle, and inside of the circle, that are above the line $-2x + y = 2$).

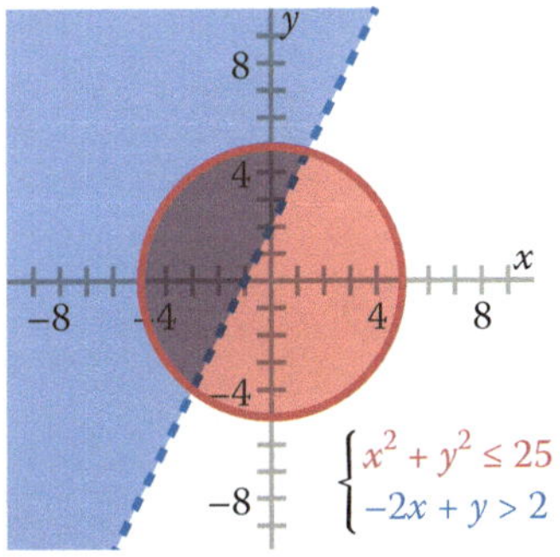

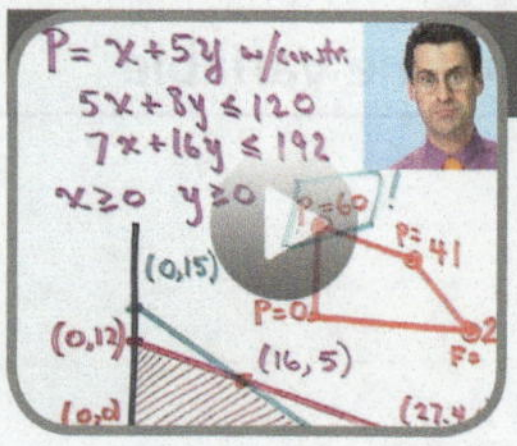

8.6.4 Solving for Maxima–Minima

An important application of systems of inequalities is called **linear programming**. Linear programming finds maximum or minimum solutions to linear functions given constraints on those functions as defined by a system of linear inequalities. Linear programming provides a systematic method for identifying the optimal solution(s), which is the best solution(s) given the constraints.

To find an optimal solution using linear programming, first use the information available to

- define variables for the unknowns in the system,
- write a linear function, called the **objective function**, and
- write a system of linear inequalities (each inequality is a constraint on the objective function).

Then graph the system of linear inequalities. The system's solution region is called the **feasible region** for the objective function. This feasible region contains all solutions (x, y) that satisfy every constraint (i.e., every inequality in the system).

When the feasible region is the shape of a polygon, the vertices of this region exist, and the objective function will always be maximized and minimized at these vertices. Identify any maxima or minima for the objective function by substituting the coordinates from the vertices into the objective function.

TIP

"Maxima" is the plural form of maximum, and "minima" is the plural form of minimum.

DEFINITION

Linear programming finds an optimal solution (minimum or maximum) to a linear function given constraints on the function as defined by a system of linear inequalities.

An **objective function** is a linear function that relates input variables to a quantity to be optimized.

A **feasible region** is the solution region of a system of linear inequalities, where each inequality defines constraints on an objective function.

Steps for Linear Programming

❶ Identify the unknowns, write the objective function, and write the system of inequalities defining the constraints.

❷ Graph the feasible region.

❸ Identify the vertices of the feasible region.

❹ Substitute the vertices into the objective function, and identify the minimum or maximum value.

EXAMPLE 11 Finding the Minimum or Maximum of a Linear Function Given Constraints

TIP

These examples include the constraints $x \geq 0$ and $y \geq 0$, limiting the solution to the first quadrant, which is typically the case in models of real-world situations.

Maximize the function $P = 3x + 2y$, given the constraints. $\begin{cases} x-1 \leq y \\ y \leq 2x \\ 2y \leq 10 - x \\ x \geq 0 \\ y \geq 0 \end{cases}$

SOLUTION

❶ The constraints (i.e., the inequalities) and the objective function ($P = 3x + 2y$) are given.

❷ Graph the feasible region defined by the constraints.

Each inequality is not a strict inequality, so all lines will be solid.

$x - 1 \leq y$: shade above a solid line with slope 1 and y-intercept at (0, –1)

$y \leq 2x$: shade below a solid line with slope 2 and y-intercept at (0, 0)

$2y \leq 10 - x$: shade below a solid line with slope –1/2 and y-intercept at (0, 5)

$x \geq 0$: shade to the right of the y-axis

$y \geq 0$: shade above the x-axis

The feasible region is the area of the graph in which the shaded regions overlap.

IMPORTANT

A feasible region's vertices will not always be easily readable from the graph. In this case, solve the system composed of the equations for the two intersecting lines. This will allow you to find the coordinates of the point of intersection (i.e., the feasible region's vertex).

❸ It is apparent from the graph that two of the feasible region's vertices are (0, 0) and (1, 0). Solve the related systems of equations (using substitution) to find the other two vertices.

$\begin{cases} x - 1 = y \\ 2y = 10 - x \end{cases}$

Substitute the expression for y from equation 1 into equation 2 and solve for x.

$$2(x-1) = 10 - x$$
$$2x - 2 = 10 - x$$
$$x = 4$$

Substitute $x = 4$ into either equation (here equation 1 is used) to find the corresponding y-coordinate.

$$4 - 1 = y \Rightarrow 3 = y$$

So, (4, 3) is a vertex of the feasible region.

$\begin{cases} y = 2x \\ 2y = 10 - x \end{cases}$

Substitute the expression for y from equation 1 into equation 2 and solve for x.

$$2(2x) = 10 - x$$
$$4x = 10 - x$$
$$x = 2$$

Substitute $x = 2$ into either equation (here equation 1 is used) to find the corresponding y-coordinate.

$$y = 2(2) = 4$$

So, (2, 4) is a vertex of the feasible region.

❹ Substitute the coordinates from the vertices into the objective function, $P = 3x + 2y$, to find the point that maximizes the function.

vertex (0, 0): $P = 3(0) + 2(0) = 0$

vertex (1, 0): $P = 3(1) + 2(0) = 3$

vertex (4, 3): $P = 3(4) + 2(3) = 18$ ✓

vertex (2, 4): $P = 3(2) + 2(4) = 14$

The greatest P-value is 18.

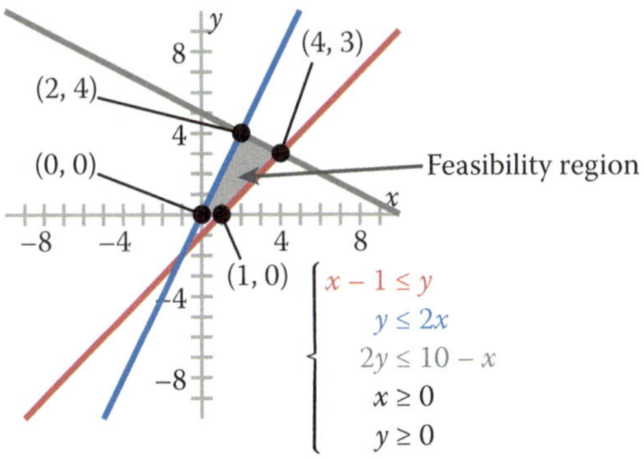

Therefore, $P = 3x + 2y$ is maximized when $x = 4$ and $y = 3$, and that maximum value is $P = 18$.

8.6.5 Applying Linear Programming

EXAMPLE 12 Finding a Maximum Using Linear Programming

A picture-framing business designs two kinds of pictures: matted and unmatted. The company employees can complete 36 pictures each day using up to 80 total person-hours of labor. It takes 4 person-hours to complete one matted picture and 2 person-hours to complete one unmatted picture. How many of each kind of picture should be made daily to maximize the company's profits if the profit on a matted picture is \$40 and the profit on an unmatted picture is \$35?

SOLUTION

TIP

When graphing, plot m along the horizontal axis and u along the vertical axis.

TIP

"Up to" translates into "less than or equal to;" "at least" translates into "greater than or equal to."

❶ Let m be the number of matted pictures and let u be the number of unmatted pictures.

Objective Function: $P = 40m + 35u$ — *The objective function is the equation for the profit. The profit is \$40 for each matted picture and \$35 for each unmatted picture.*

Constraints:

$$\begin{cases} m + u \le 36 \\ 4m + 2u \le 80 \\ m \ge 0 \\ u \ge 0 \end{cases}$$

The company can frame up to 36 pictures a day.

It takes 4 person-hours to make each matted picture, 2 person-hours to make each unmatted picture, and the company uses up to 80 total person-hours a day.

The company will make at least 0 matted pictures and at least 0 unmatted pictures each day.

❷ Graph the feasible region defined by the constraints.

$m + u \le 36$: shade below a solid line with intercepts at (36, 0) and (0, 36)

$4m + 2u \le 80$: shade below a solid line with intercepts at (20, 0) and at (0, 40)

$m \ge 0$ and $u \ge 0$: the feasible region is confined to quadrant I

❸ It is apparent from the graph that three of the feasible region's vertices are (0, 0), (0, 36), and (20, 0). Solve the related system of equations (using elimination) to find the fourth vertex.

$$\begin{cases} m + u = 36 \\ 4m + 2u = 80 \end{cases} \qquad \begin{aligned} -2m - 2u &= -72 \\ +\ (4m + 2u) &= 80 \\ \hline 2m &= 8 \\ m &= 4 \end{aligned}$$

Multiply equation 1 by −2 and add to eliminate the u-term.

Solve for m.

Substitute $m = 4$ into either equation (here equation 1 is used) to find the corresponding u-coordinate.

$4 + u = 36 \Rightarrow u = 32$ So, (4, 32) is a vertex of the feasible region.

❹ Substitute the coordinates from the vertices into the objective function, $P = 40m + 35u$, to find the point that maximizes the function.

vertex (0, 0): $P = 40(0) + 35(0) = 0$

vertex (0, 36): $P = 40(0) + 35(36) = 1260$

vertex (20, 0): $P = 40(20) + 35(0) = 800$

vertex (4, 32): $P = 40(4) + 35(32) = 1280$

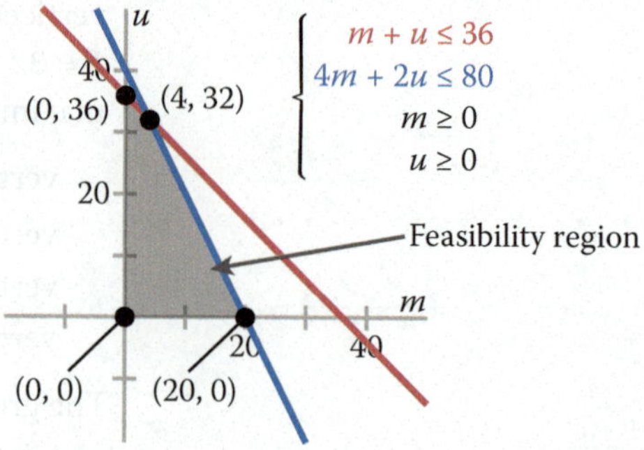

The greatest P-value, 1280, occurs at (4, 32).

Therefore, the company's profit is maximized when it produces 4 matted pictures and 32 unmatted pictures each day.

SECTION 8.6 EXERCISES

Warm Up

True or False? (–3, 3) is a solution to each inequality.

1. $5x \le -15$
2. $y > 4$
3. $2x - 3y < 10$

Just the Facts

Fill in the blanks.

4. When graphing linear inequalities, use a test point for the regions. If the test point produces a(n) _____ statement, then shade in that region.
5. The graph of a function defined by the inequality symbols of ≥ or ≤ would be represented with a(n) _____ line. The strict inequality symbols of < or > would be represented with a(n) _____ line. The graph of $y > 4$ is represented with a(n) _____ line and has shading _____ the line.
6. Solving a system of linear or nonlinear equations involves looking for the _____ of intersection. Solving a system of linear or nonlinear inequalities involves looking for the region in the graph that has _____ shading. No solution would mean that the shading does not _____.
7. The most important application of systems of inequalities is called _____. It finds _____ or _____ values of the objective function.
8. The function that is being maximized or minimized in linear programming is called the _____ function. Use the _____ of the feasible region to determine the value that will achieve this.
9. A linear inequality that is graphed as a solid vertical line at $x = 5$ and shaded to the left of the line is written _____.
10. If a system of inequalities is composed of $x \ge 0$ and $y \ge 0$, then the solution set is quadrant _____.

Essential Skills

In Exercises 1–20, graph.

1. $x \le -2$
2. $y > 2$
3. $y \le 3$
4. $x \ge -5$
5. $x \ge 1$ and $y \le 0$
6. $x < 3$ and $y > -4$
7. $x < -2$ and $y \ge 2$
8. $x \le -1$ and $y < 4$
9. $2x + 3y \le 6$
10. $3x - 5y > -15$
11. $x - 2y \ge -2$
12. $6x - 4y < -24$
13. $y \ge -3x^2$
14. $y < -(x - 2)^2$
15. $y + 4 \le x^2$
16. $y > 3x^2 + 5x + 3$
17. $\begin{cases} x^2 + y \le 3 \\ x - y < 2 \end{cases}$
18. $\begin{cases} (x+2)^2 + y^2 \le 9 \\ 3y \le 2x - 3 \end{cases}$
19. $\begin{cases} 4x + 2y < 6 \\ x^2 + y^2 > 16 \end{cases}$
20. $\begin{cases} y \ge (x-5)^2 + 3 \\ y \le 5x + 3 \end{cases}$

In Exercises 21–22, graph each system and maximize or minimize each objective function.

21. Maximize $C = x + y$ given the following constraints.

$$\begin{cases} y \ge 0 \\ x \ge 0 \\ 4x + 2y \le 8 \\ 2x - y \le 0 \end{cases}$$

22. Minimize $C = 2x + y$ given the following constraints.

$$\begin{cases} 0 \le x \le 6 \\ 0 \le y \le 6 \\ -6x + 3y \text{ ″ } 12 \\ 2x + y \ge 4 \end{cases}$$

In Exercises 23–28, maximize or minimize each objective function.

23. Minimize $C = 4x + 6y$ given the following constraints.

$$\begin{cases} x + y \ge 3 \\ 7x - 2y \le 21 \\ 5y \le 4x + 15 \end{cases}$$

24. Maximize $C = 3x + 4y$ given the following constraints.

$$\begin{cases} 0 \le x \le 6 \\ 0 \le y \le 6 \\ -4x + 2y \le 8 \\ 2x + y \ge 4 \end{cases}$$

25. Maximize $P = 4x + 7y$ given the following constraints.

$$\begin{cases} x + y \geq 2 \\ 6x - 2y \leq 12 \\ y \leq x + 2 \end{cases}$$

26. Minimize $C = 5x + y$ given the following constraints.

$$\begin{cases} 0 \leq x \leq 8 \\ 0 \leq y \leq 8 \\ -8x + 4y \leq 24 \\ 2x + y \geq 6 \end{cases}$$

27. Maximize $P = 4x - 3y$ given the following constraints.

$$\begin{cases} x \geq 0 \\ y \geq 0 \\ 3x - 6y \leq 6 \\ -x + 8y \leq 16 \end{cases}$$

28. Minimize $C = 3x + 7y$ given the following constraints.

$$\begin{cases} 8x - 3y \leq 16 \\ 5y \leq 6x + 10 \\ x + y \geq 2 \end{cases}$$

29. Family Foods makes jumbo cookies and regular cookies. The oven can make at most 200 cookies per day. Each jumbo cookie requires 2 ounces of flour, and each regular cookie requires 1 ounce of flour. There are 250 ounces of flour available. The income from each jumbo cookie is 11 cents and from each regular cookie is 5 cents. How many of each size cookie should be made to maximize income?

30. A florist designs two high-profit arrangements—a bridal centerpiece and a funeral wreath. The company's employees can complete up to 18 arrangements each day using up to 24 total person-hours of labor. It takes 4 person-hours to complete 1 funeral wreath, and 1 person-hour to complete 1 bridal centerpiece. How many of each type of arrangement should the florist produce daily for maximum profit, if the profit on a funeral wreath is $75 and the profit on a bridal centerpiece is $40?

31. An office supply company manufactures desks and chairs. In a given week, the company can assemble 500 units. They must assemble at least 100 desks and 150 chairs to meet their usual demands. Their profit is $30 per desk and $20 per chair. How many units of each should they assemble to maximize their profit?

32. A florist designs two high-profit arrangements—a funeral wreath and a bridal centerpiece. The company's employees can complete up to 17 arrangements each day using up to 26 total person-hours of labor. It takes 4 person-hours to complete 1 funeral wreath, and 1 person-hour to complete 1 bridal centerpiece. How many of each type of arrangement should the florist produce daily for maximum profit, if the profit on a funeral wreath is $60 and the profit on a bridal centerpiece is $44?

Extensions

33. Angus Farms encompasses 240 acres. The owner will plant wheat and oats. The profit per acre of wheat is $40, and the profit per acre of oats is $30. He has workers who will provide 320 hours of labor. Each acre of wheat requires 2 hours of labor, and each acre of oats requires 1 hour of labor. How many acres of each should he plant to maximize his profit?

34. Write a system of nonlinear inequalities consisting of two circles, centered at the origin with no solution.

35. Graph the solution region. $\begin{cases} y \leq \sqrt{x+3} \\ y \geq -\sqrt{-x+3} \\ y \leq \sqrt{-x+3} \\ y \geq -\sqrt{x+3} \end{cases}$

36. Write the system of nonlinear inequalities whose solution is the shaded region.

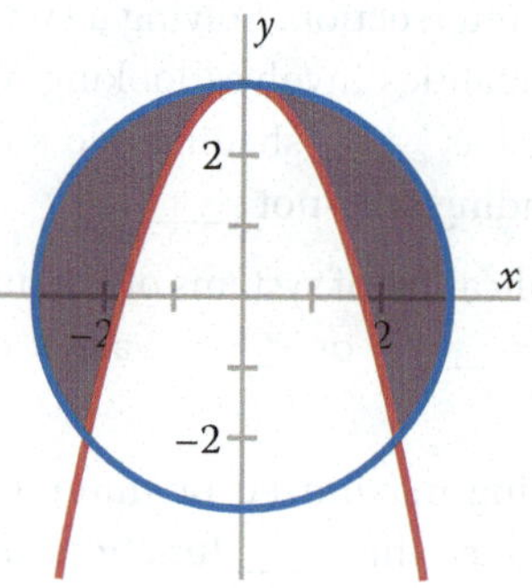

37. For a concert event, there are $50 reserved seat tickets and $25 general admission tickets. There are 3000 reserved seats available and a limit of 5000 paid ticket holders. The production company wants to take in at least $90,000 in ticket sales. Find and graph a system of inequalities describing the different numbers of tickets that can be sold.

38. An office supply company sells two models of laser printers at costs of $150 and $200. The $150 model yields a profit of $35 per unit and the $200 model yields a profit of $40 per unit. The manager estimates that the total monthly demand will not exceed 175 units. The manager wants to keep no more than $30,000 in inventory for these products. What is the minimum inventory level for these printers? What is the maximum profit?

CHAPTER 8 REVIEW EXERCISES

Section 8.1 Solving Systems of Two Linear Equations in Two Variables

1. Solve the system by graphing. $\begin{cases} 3x + y = 2 \\ -5x + y = -14 \end{cases}$

2. Solve the system by substitution. $\begin{cases} -3x + y = 5 \\ x + 4y = 7 \end{cases}$

In Exercises 3-4, solve each system by elimination.

3. $\begin{cases} 2x - 3y = 1 \\ 3x - 3y = 4 \end{cases}$

4. $\begin{cases} -6x + y = 7 \\ 3x - 2y = -5 \end{cases}$

5. Without solving, identify the system as consistent and independent, consistent and dependent, or inconsistent.
$\begin{cases} 2x + 4y = 26 \\ 6x + 12y = 77 \end{cases}$

Section 8.2 Nonlinear Systems

6. Solve the system by graphing. $\begin{cases} y = (x - 2)^2 \\ y = -2x + 7 \end{cases}$

In Exercises 7-8, solve each system by substitution.

7. $\begin{cases} 2x - y = 2 \\ y = -(x + 3)^2 \end{cases}$

8. $\begin{cases} 2x^2 + y^2 = 16 \\ y = -\frac{1}{9}x^2 + \frac{8}{3} \end{cases}$

In Exercises 9-10, solve each system by elimination.

9. $\begin{cases} y = x^2 - 5x + 6 \\ -2x - 3y = 8 \end{cases}$

10. $\begin{cases} x^2 + y^2 = 36 \\ 25x^2 + y^2 = 156 \end{cases}$

Section 8.3 Modeling with Systems

11. A businesswoman invests a total of \$8787 in two savings accounts. One account yields 8.5% simple interest and the other 9.5% simple interest. She earns a total of \$758.58 interest for the year. How much was invested in each account?

12. An airplane travels 1800 miles from Los Angeles to Chicago in 3.6 hours with the wind. On its return trip, against the wind, it takes 4 hours. Find the speed of the airplane with no wind and find the speed of the wind. (Assume that both speeds remain constant.)

13. Gold is alloyed with different metals to make it hard enough to be used in jewelry. The amount of gold present in a gold alloy is measured in 24ths called karats. 24-karat gold is 24/24 or 100% gold. Similarly, 18-karat gold is 18/24 or 75% gold. How many ounces of 10-karat gold should be added to an amount of 18-karat gold to make 6 ounces of 14-karat gold?

14. An elevator begins to ascend at a rate of 8 feet per second, and at the same time, a ball is thrown straight up into the air near the elevator. The following system of equations describes the height of the elevator (equation 1) and of the ball (equation 2), where the height is given in feet and the time is given in seconds.
$\begin{cases} h = 8t \\ h = -16t^2 + 20t \end{cases}$
To the nearest hundredth of a second, how long does it take for the ball and the elevator to reach the same height?

15. An asteroid is traveling toward the moon on a parabolic path that can be modeled by the equation $y - 3 = \frac{1}{5}x^2$. On the same plane, an unmanned research satellite orbits the moon on a path that can be modeled by the equation $36x^2 + 25y^2 = 900$. Is there a danger of collision? If so, find the coordinates of the points where the satellite and the asteroid might collide.

Section 8.4 Multivariable Linear Systems

In Exercises 16-19, solve each system.

16. $\begin{cases} 2x - 2y + 4z = 10 \\ 2y + 2z = 1 \\ 12z = 12 \end{cases}$

17. $\begin{cases} 4x - 4y + 8z = 10 \\ 8x + 4y - 4z = -9 \\ 12x + 12y - 4z = -16 \end{cases}$

18. $\begin{cases} x + 2y + z = 9 \\ 2x + 2y - z = 18 \\ x - y - z = 4 \end{cases}$

19. $\begin{cases} -x + \frac{1}{3}y + \frac{2}{3}z = 9 \\ 6x - 2y - 4z = 27 \end{cases}$

20. The table shows the number of hamburgers, orders of fries, and medium-sized drinks purchased at a restaurant, as well as the total amount spent on these items by three families. Find the price of each item.

	Hamburger	Fries	Drink	Total Spent
Family 1	2	1	2	\$13.25
Family 2	3	3	2	\$21.50
Family 3	5	2	5	\$32

Section 8.5 Partial Fractions

In Exercises 21-25, find each partial fraction decomposition.

21. $\dfrac{-x-9}{(x-1)(x-3)}$

22. $\dfrac{10x+81}{x^2+15x+54}$

23. $\dfrac{-2x^2+x-21}{(x-1)(x+3)^2}$

24. $\dfrac{9x^2-3x+4}{2x^3-x^2+x}$

25. $\dfrac{3x^3-x^2+13x-1}{(x^2+5)^2}$

Section 8.6 Systems of Inequalities and Linear Programming

In Exercises 26-29, graph.

26. $x < 4$ and $y \geq -1$

27. $-x - 4y < 8$

28. $y > (x + 3)^2$

29. $\begin{cases} -2x > 5y + 10 \\ x^2 + (y+2)^2 \geq 16 \end{cases}$

30. A car dealer has room for 40 vehicles for a clearance sale. There are a total of 186 person-hours available to prepare vehicles for the sale. It takes 6 person-hours to prepare a truck for sale and 3 person-hours to prepare a car for sale. How many trucks and cars should the dealer prepare for the sale if the profit for selling each truck is \$500 and the profit for selling each car is \$400?

PRECALCULUS

MATRICES AND DETERMINANTS

CHAPTER

9

9.1 MATRICES AND SYSTEMS OF EQUATIONS

OBJECTIVES

- Write an augmented matrix for a linear system.
- Solve linear systems of equations by using elementary row operations.
- Solve linear systems of equations by using Gaussian elimination.
- Identify inconsistent linear systems and dependent linear systems.

PREREQUISITE VOCABULARY TERMS

consistent system
dependent system
inconsistent system
independent system
triangular form (of a linear system of equations)

9.1.1 An Introduction to Matrices

A **matrix** is a way of displaying numbers systematically in rows and columns (i.e., in a rectangular array). The plural of matrix is matrices.

DEFINITION

If m and n are positive integers, then an $\boldsymbol{m \times n}$ **matrix** is a rectangular array with m rows and n columns. There are n **entries** (or elements) in each row, and m entries in each column. Each entry a_{ij} in the matrix is named by its position in the ith row and jth column. A matrix containing m rows and n columns is said to be of **order** $m \times n$ (or has dimensions $m \times n$). **Equivalent matrices** have exactly the same dimensions and equal corresponding entries.

For example, matrix A (matrices are typically named with an uppercase letter) has 2 rows and 3 columns, so it is a 2×3 matrix (i.e., a matrix of order 2×3).

$$A = \begin{bmatrix} a_{11} & a_{12} & a_{13} \\ a_{21} & a_{22} & a_{23} \end{bmatrix} \begin{matrix} \leftarrow \text{row } 1 \\ \leftarrow \text{row } 2 \end{matrix}$$

$$\begin{matrix} \uparrow & \uparrow & \uparrow \\ \text{column 1} & \text{column 2} & \text{column 3} \end{matrix}$$

A matrix with the same number of rows and columns is called a **square matrix**.

The rows of a matrix can be identified with an uppercase R and a subscripted number, indicating the row's position in the matrix. For example, row 2 can be referred to as R_2.

EXAMPLE 1 Finding the Order of a Matrix

Determine the order of each matrix.

A. $\begin{bmatrix} 5 & -9 \\ 0 & 5 \end{bmatrix}$ **B.** $\begin{bmatrix} 1 & 6 \\ 0.5 & 10 \\ -3 & 8 \\ 16 & 8 \end{bmatrix}$ **C.** $\begin{bmatrix} 10 & -5 & 0 & 1 & 0 \end{bmatrix}$

SOLUTION

A. The matrix has two rows and two columns, so the matrix is of order 2×2.

B. The matrix has four rows and two columns, so the matrix is of order 4×2.

C. The matrix has one row and five columns, so the matrix is of order 1×5.

9.1.2 Augmented Matrices

Recall from Chapter 8 that a system of equations is just a group of equations, typically with common variables. Consider the Gaussian elimination method for solving a system of linear equations discussed in Chapter 8. This method involved lining up the equations in the system so that the terms with like variables aligned vertically, and the constant terms aligned vertically. Then, using operations on the equations, the x- and y- terms were eliminated in order to solve for z. This was accomplished by focusing on the coefficients.

In this section, you will use the coefficients and constant terms from a system of linear equations as the entries in a matrix, and then apply operations to the rows of that matrix to solve the corresponding system. In this way, the system will be solved without writing the variables.

A matrix that lists the coefficients and constant terms of a system of linear equations is called an **augmented matrix**. In an augmented matrix, each row corresponds to one equation in the system, and each column contains the coefficients of one of the variables in the system, where the last column contains the constant terms. A dashed line is used to separate the columns that correspond to the coefficients from the column that corresponds to the constant terms.

A Linear System of Two Equations in Two Variables

An augmented matrix for a system of two linear equations in two variables, x and y, will contain the coefficients of x in the first column, the coefficients of y in the second column, and the constant terms in the third column. Therefore, an augmented matrix for a system of two linear equations in two variables will always be a 2×3 matrix. An example of a system of two linear equations in two variables and its corresponding augmented matrix follows.

TIP

Equation 1, $2x - y = 4$, is represented by the entries in R_1 of the augmented matrix, 2, −1, and 4. Equation 2, $x + y = 2$, is represented by the entries in R_2 of the augmented matrix, 1, 1, and 2.

$$\begin{cases} 2x - y = 4 \\ x + y = 2 \end{cases}$$

System of Two Linear Equations in Two Variables

$$\begin{array}{c} \left[\begin{array}{cc:c} 2 & -1 & 4 \\ 1 & 1 & 2 \end{array}\right] \begin{array}{l} \leftarrow \text{equation 1} \\ \leftarrow \text{equation 2} \end{array} \\ \begin{array}{ccc} \uparrow & \uparrow & \uparrow \\ x & y & \text{constants} \end{array} \end{array}$$

Augmented 2 × 3 Matrix

A Linear System of Three Equations in Three Variables

An augmented matrix that represents a linear system of three equations in three variables, x, y, and z, will contain the coefficients of x, y, and z in the first three columns respectively, and the constant terms in the fourth column. Therefore, an augmented matrix for a system of three linear equations in three variables will always be a 3×4 matrix. An example of a system of three linear equations in three variables and its corresponding augmented matrix follows.

$$\begin{cases} 2x + y - z = 0 \\ x + 3z = 4 \\ -x + 4y + z = 2 \end{cases}$$

System of Three Linear Equations in Three Variables

$$\begin{array}{c} \left[\begin{array}{ccc:c} 2 & 1 & -1 & 0 \\ 1 & 0 & 3 & 4 \\ -1 & 4 & 1 & 2 \end{array}\right] \begin{array}{l} \leftarrow \text{equation 1} \\ \leftarrow \text{equation 2} \\ \leftarrow \text{equation 3} \end{array} \\ \begin{array}{cccc} \uparrow & \uparrow & \uparrow & \uparrow \\ x & y & z & \text{constants} \end{array} \end{array}$$

Augmented 3 × 4 Matrix

Equation 2, $x + 3z = 4$, is in terms of x and z only, so only two coefficients appear in the equation. The corresponding row in the augmented matrix, R_2, must contain an entry in each column, so some number must be used as the y-coefficient (in other words, that entry cannot be left blank in the augmented matrix). Since $x + 3z = 4$ is equal to $x + 0y + 3z = 4$, the y-term's coefficient in R_2 of the augmented matrix is 0.

EXAMPLE 2 Writing an Augmented Matrix

Write the augmented matrix for the system of linear equations.

$$\begin{cases} z - 2y = 1 \\ x + 3z - w = 0 \\ z = 2 \\ 2x + y + 5w = 6 \end{cases}$$

SOLUTION

Identify the order of the augmented matrix.

The system contains four equations, so the augmented matrix will contain four rows. The system's equations include four variables, x, y, z, and w, so the augmented matrix will contain five columns (one for each variable, plus another column for the constant terms). Therefore, the augmented matrix will be of order 4×5.

NOTICE THAT

The system contains four variables, x, y, z, and w, but no single equation in the system contains all four variables.

Write the system with the variables vertically aligned.

Write the 4×5 augmented matrix using the coefficients and constant terms as the entries. Use 0 as the coefficient for all missing terms.

$$\begin{cases} \quad - 2y + z \qquad = 1 \\ x \qquad + 3z - w = 0 \\ \qquad\qquad z \qquad = 2 \\ 2x + y \qquad + 5w = 6 \end{cases} \qquad \left[\begin{array}{cccc|c} 0 & -2 & 1 & 0 & 1 \\ 1 & 0 & 3 & -1 & 0 \\ 0 & 0 & 1 & 0 & 2 \\ 2 & 1 & 0 & 5 & 6 \end{array}\right]$$

TIP

The variables x, y, z, and w typically correspond to columns 1, 2, 3, and 4 respectively in an augmented matrix.

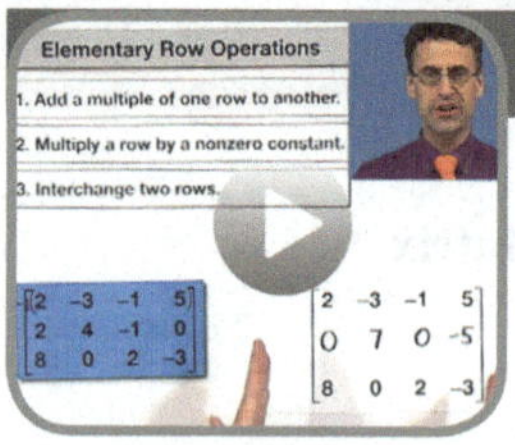

9.1.3 Elementary Row Operations

In Chapter 8 you saw that the equations in a system of linear equations could be manipulated and written as equal systems by using the operations listed in the following table.

Operations That Yield Equivalent Systems of Linear Equations
1. Add a nonzero multiple of one equation to another.
2. Multiply an equation by a nonzero constant.
3. Interchange the position of two equations.

These operations can also be applied to augmented matrices. Since a row of an augmented matrix corresponds to an equation in a linear system, the word "equation" is replaced by "row" when discussing these operations in reference to matrices. For matrices, these operations are referred to as **elementary row operations**.

Elementary Row Operations
1. Add a nonzero multiple of one row to another.
2. Multiply a row by a nonzero constant.
3. Interchange the position of two rows.

The operations for matrices are not referred to as "operations that yield equivalent matrices" because the original augmented matrix and the updated augmented matrix are no longer equal after an operation is performed (because the corresponding entries are no longer the same). However, the corresponding systems are still equal. When two augmented matrices have equal corresponding systems, those matrices are said to be **row-equivalent**. So, any matrix obtained by using elementary row operations is row-equivalent to the original matrix.

These elementary row operations are demonstrated in the following example. Later in this section, you will see how to use elementary row operations on an augmented matrix to solve the corresponding system of linear equations.

EXAMPLE 3 Using Elementary Row Operations

Write the resulting matrix after the stated row operation is applied to the given matrix.

A. Interchange R_2 and R_3.

$$\begin{bmatrix} -1 & 5 & 0 & 2 \\ 8 & 0 & 1 & 9 \\ 1 & 3 & 7 & 3 \end{bmatrix}$$

B. Multiply R_1 by -1 and multiply R_3 by $-\frac{1}{3}$.

$$\begin{bmatrix} 6 & -3 & 0 & -9 \\ 5 & 5 & 2 & 0 \\ -6 & 3 & -1 & 5 \end{bmatrix}$$

C. Replace R_3 with $-R_2 + R_3$.

$$\begin{bmatrix} 1 & 0 & -4 & 8 \\ -3 & 2 & -1 & 0 \\ -3 & 2 & 7 & 9 \end{bmatrix}$$

SOLUTION

A. To interchange two rows, swap the entries.

Original Matrix

$$\begin{bmatrix} -1 & 5 & 0 & 2 \\ 8 & 0 & 1 & 9 \\ 1 & 3 & 7 & 3 \end{bmatrix}$$

Row-Equivalent Matrix

$$\begin{bmatrix} -1 & 5 & 0 & 2 \\ 1 & 3 & 7 & 3 \\ 8 & 0 & 1 & 9 \end{bmatrix} \begin{matrix} R_1 \text{ does not change.} \\ R_3 \\ R_2 \end{matrix}$$

NOTICE THAT

Multiplying a row by $-1/3$ is equivalent to dividing the row by -3.

B. To multiply a row by a nonzero constant, multiply each entry in that row by the constant.

Original Matrix

$$\begin{bmatrix} 6 & -3 & 0 & -9 \\ 5 & 5 & 2 & 0 \\ -6 & 3 & -1 & 5 \end{bmatrix}$$

Row-Equivalent Matrix

$$\begin{bmatrix} -6 & 3 & 0 & 9 \\ 5 & 5 & 2 & 0 \\ 2 & -1 & \frac{1}{3} & -\frac{5}{3} \end{bmatrix} \begin{matrix} -R_1 \\ R_2 \text{ does not change.} \\ -\frac{1}{3}R_3 \end{matrix}$$

NOTICE THAT

-1 times a row plus another row is equivalent to subtracting the rows, where the row multiplied by -1 is the row being subtracted.

C. Replace R_3 with $-R_2 + R_3$.

Original Matrix

$$\begin{bmatrix} 1 & 0 & -4 & 8 \\ -3 & 2 & -1 & 0 \\ -3 & 2 & 7 & 9 \end{bmatrix}$$

Row-Equivalent Matrix

$$\begin{bmatrix} 1 & 0 & -4 & 8 \\ -3 & 2 & -1 & 0 \\ 0 & 0 & 8 & 9 \end{bmatrix} \begin{matrix} R_1 \text{ does not change.} \\ R_2 \text{ does not change.} \\ -R_2 + R_3 \end{matrix}$$

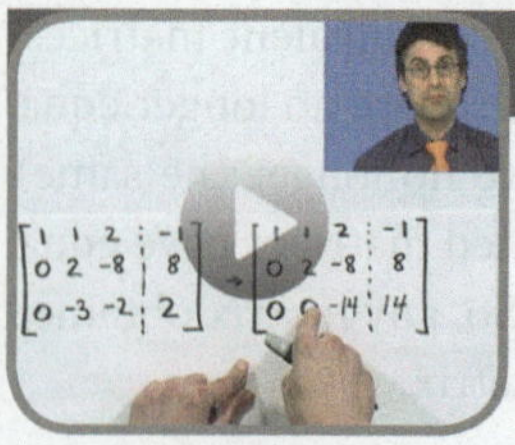

9.1.4 Gauss-Jordan Elimination

The following 3 × 4 augmented matrix is said to be in **reduced row-echelon form** because the first nonzero entry in each row is a 1, called a leading 1, and the entries above and below each leading 1 are 0.

Augmented Matrix in Reduced Row-Echelon Form

$$\left[\begin{array}{ccc|c} 1 & 0 & 0 & a \\ 0 & 1 & 0 & b \\ 0 & 0 & 1 & c \end{array}\right]$$

a, b, and c are some real numbers.

When an augmented matrix is in reduced row-echelon form, the solution of the corresponding linear system of equations can be read directly from the matrix.

Corresponding System of Linear Equations

$$\begin{cases} x = a \\ y = b \\ z = c \end{cases}$$

REMEMBER

In a square system of linear equations, the number of equations is equal to the number of variables.

Gauss-Jordan elimination is a method for solving a square system of linear equations by manipulating the corresponding augmented matrix using the elementary row operations until it is in reduced row-echelon form, at which time the system's solution can be read from the augmented matrix.

TIP

The entries can be turned into the 0s or 1s in any order. However, this systematic method can be helpful.

1. 0s in column 1
2. 0s in column 2
3. 0s in column 3

Then the 1s are obtained by multiplying each row by the reciprocal of the leading coefficient.

Steps for Solving a System of Linear Equations by Using Gauss-Jordan Elimination

1. Write the augmented matrix that corresponds to the linear system of equations.
2. Apply elementary row operations to the augmented matrix until the matrix is in reduced row-echelon form.
3. Read the system's solution from the reduced row-echelon form matrix.

EXAMPLE 4 Using Gauss-Jordan Elimination to Solve a Linear System

Use Gauss-Jordan elimination to solve the system.
$$\begin{cases} x + 2y + z = 5 \\ x - 2y + 2z = 6 \\ x + 2y + 4z = -4 \end{cases}$$

SOLUTION

1. Write the corresponding augmented matrix.

$$\left[\begin{array}{ccc|c} 1 & 2 & 1 & 5 \\ 1 & -2 & 2 & 6 \\ 1 & 2 & 4 & -4 \end{array}\right]$$

The goal is to write the augmented matrix in reduced row-echelon form.

$$\left[\begin{array}{ccc|c} 1 & 0 & 0 & a \\ 0 & 1 & 0 & b \\ 0 & 0 & 1 & c \end{array}\right]$$

> ***TIP***
>
> *The process shown here is just one of many ways to write the augmented matrix in reduced row-echelon form.*

❷ Use elementary row operations to write the augmented matrix in reduced row-echelon form.
Get the 0s in column 1 (rows 2 and 3).

$$\left[\begin{array}{ccc|c} 1 & 2 & 1 & 5 \\ 0 & -4 & 1 & 1 \\ 0 & 0 & 3 & -9 \end{array}\right] \quad \begin{array}{l} \\ -R_1 + R_2 \\ -R_1 + R_3 \end{array}$$

Get the 0s in column 2 (rows 1 and 3).

$$\left[\begin{array}{ccc|c} 1 & 0 & \frac{3}{2} & \frac{11}{2} \\ 0 & -4 & 1 & 1 \\ 0 & 0 & 3 & -9 \end{array}\right] \quad \begin{array}{l} \frac{1}{2}R_2 + R_1 \\ \\ \textit{The row 3 column 2 entry is already a 0.} \end{array}$$

Get the 0s in column 3 (rows 1 and 2).

$$\left[\begin{array}{ccc|c} 1 & 0 & 0 & 10 \\ 0 & -4 & 0 & 4 \\ 0 & 0 & 3 & -9 \end{array}\right] \quad \begin{array}{l} -\frac{1}{2}R_3 + R_1 \\ -\frac{1}{3}R_3 + R_2 \\ \\ \end{array}$$

Change the leading entry in each row to a 1.

$$\left[\begin{array}{ccc|c} 1 & 0 & 0 & 10 \\ 0 & 1 & 0 & -1 \\ 0 & 0 & 1 & -3 \end{array}\right] \quad \begin{array}{l} \textit{Row 1's leading entry is already a 1.} \\ -\frac{1}{4}R_2 \\ \frac{1}{3}R_3 \end{array}$$

❸ Write the corresponding system of equations. $\begin{cases} x = 10 \\ y = -1 \\ z = -3 \end{cases}$

Therefore, the system's solution is (10, –1, –3).

9.1.5 Gaussian Elimination

Recall that Gaussian elimination with systems of linear equations is a method used to solve the system by using operations to write the system in triangular form, and then using back-substitution to find the values of the variables.

System of Three Linear Equations in Triangualr Form

$$\begin{cases} \_\_x + \_\_y + \_\_z = \_\_ \\ \qquad\;\; \_\_y + \_\_z = \_\_ \\ \qquad\qquad\quad \_\_z = \_\_ \end{cases}$$

Gaussian elimination can also be used after a system of linear equations is written as an augmented matrix. With Gaussian elimination, the augmented matrix is not written in reduced row-echelon form (as it was with Gauss-Jordan elimination). Instead, the augmented matrix is written in a **row-echelon form**, which means that the matrix has a corresponding system that is in triangular form where the leading coefficient in each row is a 1.

Steps for Solving a System of Linear Equations by Using Gaussian Elimination on the Augmented Matrix

1. Write the augmented matrix that corresponds to the linear system of equations.
2. Apply elementary row operations to the augmented matrix until the matrix is in row-echelon form (i.e., until the matrix corresponds to a system that is in triangular form where the leading coefficient in each row is a 1).
3. Write the row-echelon form matrix as its corresponding linear system.
4. Use back-substitution to find the values of the variables.

EXAMPLE 5 Using Gaussian Elimination to Solve a Linear System

Use Gaussian elimination to solve the system. $\begin{cases} x+3y+z=4 \\ -x-3y+2z=5 \\ x-6y-2z=4 \end{cases}$

SOLUTION

1. Write the corresponding augmented matrix. $\left[\begin{array}{ccc|c} 1 & 3 & 1 & 4 \\ -1 & -3 & 2 & 5 \\ 1 & -6 & -2 & 4 \end{array}\right]$

IMPORTANT

The goal is to write the augmented matrix in row-echelon form.

$\left[\begin{array}{ccc|c} 1 & d & e & a \\ 0 & 1 & f & b \\ 0 & 0 & 1 & c \end{array}\right]$

2. Use elementary row operations to write the augmented matrix in row-echelon form.
Get the 0s in column 1 (rows 2 and 3).

$$\left[\begin{array}{ccc|c} 1 & 3 & 1 & 4 \\ 0 & 0 & 3 & 9 \\ 0 & -9 & -3 & 0 \end{array}\right] \begin{array}{l} \\ R_1+R_2 \\ -R_1+R_3 \end{array}$$

Get the 0 in column 2 (row 3).

$$\left[\begin{array}{ccc|c} 1 & 3 & 1 & 4 \\ 0 & -9 & -3 & 0 \\ 0 & 0 & 3 & 9 \end{array}\right] \begin{array}{l} \\ R_3 \\ R_2 \end{array}$$

Change the leading entry in each row to a 1.

$$\left[\begin{array}{ccc|c} 1 & 3 & 1 & 4 \\ 0 & 1 & \frac{1}{3} & 0 \\ 0 & 0 & 1 & 3 \end{array}\right] \begin{array}{l} \text{Row 1's leading entry is already a 1.} \\ -\frac{1}{9}R_2 \\ \frac{1}{3}R_2 \end{array}$$

3. Write the corresponding system of linear equations. $\begin{cases} x+3y+z=4 \\ y+\frac{1}{3}z=0 \\ z=3 \end{cases}$

4. Use back-substitution to find the values of the variables.

Find y.

Substitute $z=3$ into equation 2.

$$y+\tfrac{1}{3}(3)=0 \Rightarrow y=-1$$

Find x.

Substitute $z=3$ and $y=-1$ into equation 1.

$$x+3(-1)+3=4 \Rightarrow x=4$$

Therefore, the system's solution is $(4, -1, 3)$.

9.1.6 Inconsistent and Dependent Systems

Independent Systems

Recall that a system with no solution is called an inconsistent system. An inconsistent system is easily recognizable because during the elimination process you will obtain a false statement, such as $0 = 5$.

A false statement in an augmented matrix will appear as a row with all 0s on the left of the dashed line and a nonzero number on the right of the dashed line (in the constant column). So, if the elimination process yields a row with all 0 entries except for the last entry, then the corresponding system must be inconsistent.

EXAMPLE 6 **Solving an Inconsistent System**

Solve the system. $\begin{cases} -x - 2y + 3z = 1 \\ x - 3y + z = 1 \\ 2x - y - 2z = 2 \end{cases}$

SOLUTION

Write the corresponding augmented matrix. $\left[\begin{array}{ccc:c} -1 & -2 & 3 & 1 \\ 1 & -3 & 1 & 1 \\ 2 & -1 & -2 & 2 \end{array}\right]$

Use elementary row operations.

$$\left[\begin{array}{ccc:c} -1 & -2 & 3 & 1 \\ 0 & -5 & 4 & 2 \\ 0 & -5 & 4 & 4 \end{array}\right] \begin{array}{l} \\ R_1 + R_2 \\ 2R_1 + R_3 \end{array}$$

$$\left[\begin{array}{ccc:c} -1 & -2 & 3 & 1 \\ 0 & -5 & 4 & 2 \\ 0 & 0 & 0 & 2 \end{array}\right] \begin{array}{l} \\ \\ -R_2 + R_3 \end{array}$$

The corresponding equation for R_3 is $0x + 0y + 0z = 2$, or $0 = 2$.

Since $0 = 2$ is a false statement, the system is inconsistent and has no solution.

Dependent Systems

The complete solution for a dependent system can be expressed by solving for the **leading variable** (a variable that corresponds to a leading entry in the row-echelon form of the augmented matrix) in terms of the **free variables** (a variable that is not a leading variable). Each free variable can take on the value of any real number, as demonstrated in the following example.

EXAMPLE 7 Solving a Dependent System

Solve the system. $\begin{cases} -x+7z=5 \\ -x-y+10z=4 \\ -3x-5y+36z=10 \end{cases}$

SOLUTION

Write the corresponding augmented matrix. $\left[\begin{array}{ccc|c} -1 & 0 & 7 & 5 \\ -1 & -1 & 10 & 4 \\ -3 & -5 & 36 & 10 \end{array}\right]$

Use elementary row operations.

$$\left[\begin{array}{ccc|c} -1 & 0 & 7 & 5 \\ 0 & -1 & 3 & -1 \\ 0 & -5 & 15 & -5 \end{array}\right] \begin{array}{l} \\ -R_1+R_2 \\ -3R_1+R_3 \end{array}$$

$$\left[\begin{array}{ccc|c} -1 & 0 & 7 & 5 \\ 0 & 1 & -3 & 1 \\ 0 & 0 & 0 & 0 \end{array}\right] \begin{array}{l} \\ -R_2 \\ -5R_2+R_3 \end{array}$$

NOTICE THAT

The matrix contains a row with all 0 entries. So, the corresponding system now contains two equations and three variables. Since the system has fewer equations than variables, the system must be either inconsistent or dependent.

The corresponding equation for R_3 is $0x + 0y + 0z = 0$, or $0 = 0$. This is a true statement, and this row of zeros indicates the presence of a free variable.

Write the corresponding system. $\begin{cases} -x+7z=5 \\ y-3z=1 \end{cases}$

The leading variables are x and y, and z is the free variable.

So, let $z = a$, where a represents any real number.

Use the two equations in the system to express x and y in terms of a.

Substitute $z = a$ into equation 1 and solve for x.

$-x + 7a = 5$

$x = 7a - 5$

Substitute $z = a$ into equation 2 and solve for y.

$y - 3a = 1$

$y = 3a + 1$

So, the complete solution of the dependent system is the set of all points of the form $(7a - 5, 3a + 1, a)$, where a is any real number.

IMPORTANT

The system has an infinite number of solutions. Each is of the form $(7a - 5, 3a + 1, a)$, where a is any real number (e.g., $(-5, 1, 0)$, $(2, 4, 1)$, $(9, 7, 2)$, and $(16, 10, 3)$ are solutions to the system).

SECTION 9.1 EXERCISES

Warm Up

Write each equation in the standard form of a linear equation, $Ax + By = C$.

1. $-4y + 2 = 3x + 7$
2. $x - 5 = 3(y + 8)$
3. $-(-6 - y) = 9x$

Just the Facts

Fill in the blanks.

4. An $m \times n$ matrix for which $m = n$ is called a(n) _____ matrix.
5. A(n) _____ matrix can be used to represent both the _____ and _____ terms of a system of linear equations.
6. In a(n) _____ matrix representing a system of linear equations, R_i represents the _____ equation.
7. If an operation can be applied to a linear system of equations without changing the solution, that same operation applied to a(n) _____ matrix is called a(n) _____ row operation.
8. Gauss-Jordan elimination uses _____ row operations to rewrite a matrix in _____ row echelon form. A system of linear equations represented by a matrix in this form can be solved easily.
9. Gaussian elimination combines _____ operations with the _____ method for solving a system of equations.
10. If a row of the _____ matrix representing a system of linear equations contains all zeros, that system is _____.

Essential Skills

In Exercises 1–6, determine the order of each matrix.

1. $\begin{bmatrix} 2 & -5 & 0 \end{bmatrix}$

2. $\begin{bmatrix} 1 & 2 \\ 4 & 0 \\ -1 & 3 \\ 2 & 7 \\ -3 & 0 \end{bmatrix}$

3. $\begin{bmatrix} 5 \\ 0 \\ 9 \\ 2 \end{bmatrix}$

4. $\begin{bmatrix} 4 & 5 & 3 \\ 8 & 4 & 8 \\ 4 & 19 & 0 \\ 0 & 8 & 0 \end{bmatrix}$

5. $\begin{bmatrix} -2 & 0 \\ 8 & -5 \\ 1 & 6 \end{bmatrix}$

6. $\begin{bmatrix} 7 & 4 & -3 & 4 & 7 \\ 4 & 4 & 0 & 7 & 15 \end{bmatrix}$

In Exercises 7–12, write the augmented matrix for each system of linear equations.

7. $\begin{cases} x - 2y + 5z = -12 \\ 3x + 4y + z = 1 \\ -6x - y - 3z = -8 \end{cases}$

8. $\begin{cases} 2x + y = 4 \\ 7x - y - 3z = -4 \end{cases}$

9. $\begin{cases} x + 6y = 12 \\ -9x - 4y = 5 \end{cases}$

10. $\begin{cases} 2x + y - 2z + 7w = 1 \\ 2x + y + z - w = 1 \\ 7x - y - 7z + 2w = 0 \end{cases}$

11. $\begin{cases} 3x + 7y - 6z + 7w = 4 \\ 6x + 20y - 10z + w = 26 \\ -4x + 2y - 3z - 8w = -67 \end{cases}$

12. $\begin{cases} 6x + y + z = 0 \\ 8x - 8z = -1 \\ x + 8y + 6z = -8 \end{cases}$

In Exercises 13–20, write the resulting matrix after the stated row operation is applied to each given matrix.

13. Interchange R_1 and R_3. $\begin{bmatrix} 2 & -1 & 5 & 0 \\ 4 & 8 & -7 & 1 \\ 3 & 0 & -2 & 9 \end{bmatrix}$

14. Multiply R_2 by -1 and R_3 by 4. $\begin{bmatrix} 2 & 0 & -1 & 1 \\ 4 & -5 & 1 & 0 \\ -1 & 3 & 0 & 9 \end{bmatrix}$

15. Replace R_1 with $R_2 + R_1$. $\begin{bmatrix} 5 & -2 & 6 \\ 0 & 1 & 7 \end{bmatrix}$

16. Replace R_2 with $(-3)R_1 + R_2$. $\begin{bmatrix} 1 & 2 & -4 & 0 \\ 3 & -2 & -1 & 5 \\ -8 & 1 & -2 & -6 \end{bmatrix}$

17. Replace R_1 with $(4)R_2 + R_1$. $\begin{bmatrix} 1 & 5 & -2 \\ 3 & 0 & 7 \\ 2 & 1 & -4 \end{bmatrix}$

18. Multiply R_1 by 2 and R_3 by 5. $\begin{bmatrix} 2 & 3 & 1 & 5 \\ 8 & -4 & 1 & 0 \\ 2 & 3 & 0 & 12 \end{bmatrix}$

19. Multiply R_3 by 3 and R_2 by $-1/2$. $\begin{bmatrix} -2 & 5 \\ 0 & -6 \\ 7 & 3 \end{bmatrix}$

20. Replace R_3 with $R_1 + (-2)R_3$. $\begin{bmatrix} 5 & 0 & 5 & 2 \\ 4 & 2 & 2 & 0 \\ 2 & 3 & 0 & 5 \end{bmatrix}$

In Exercises 21–38, use Gauss-Jordan elimination or Gaussian elimination to solve each system. If the system is dependent, write the complete solution, letting the free variable(s) be any real number.

21. $\begin{cases} x + y - z = 9 \\ y + 3z = 3 \\ -x - 2z = 2 \end{cases}$

22. $\begin{cases} x + y - z = 3 \\ x + 2y + 3z = 5 \\ -2x - y + z = -4 \end{cases}$

23. $\begin{cases} x + y = 10 \\ 2x + y + z = 21 \\ y + 2z = -1 \end{cases}$

24. $\begin{cases} x + z = 11 \\ 2x + y = 15 \\ x + 2z = 19 \end{cases}$

25. $\begin{cases} -6x - 3y + 3z = 6 \\ 2x + 2y - 2z = 2 \\ x + 2y + 3z = 5 \end{cases}$

26. $\begin{cases} x + z = 0 \\ 5x - y + z = 10 \\ x + y = 5 \end{cases}$

27. $\begin{cases} x - 2y + 3z = 9 \\ -x + 3y = -4 \\ 2x - 5y + 5z = 17 \end{cases}$

28. $\begin{cases} x - y + 2z = 7 \\ -x + 3y + 2z = -3 \\ -3x + y + 5z = -10 \end{cases}$

29. $\begin{cases} 2x + z = -1 \\ 8x + y - z = -7 \\ 4x + y + z = -1 \end{cases}$

30. $\begin{cases} x + 2y = 0 \\ 6x + 5y + 2z = 3 \\ -28y + z = 26 \end{cases}$

31. $\begin{cases} -9y + z = -11 \\ 4x + 8y = 32 \\ 4x + 5y + 2z = 9 \end{cases}$

32. $\begin{cases} 3x + 12y + z = 24 \\ 2x + y + z = 10 \\ x + 2y = 5 \end{cases}$

33. $\begin{cases} 2x + y - 4z = -4 \\ y + 2z = 6 \\ 4x + 3y - 6z = 1 \end{cases}$

34. $\begin{cases} x - 2y - z = 0 \\ 2x + y - z = 1 \\ 3x + 4y - z = 2 \end{cases}$

35. $\begin{cases} 8x + 4z = -4 \\ 16x + y - z = 6 \\ 8x + y + z = 4 \end{cases}$

36. $\begin{cases} 2x + z = 0 \\ 8x + y - z = -27 \\ 6x + y + z = -9 \end{cases}$

37. $\begin{cases} x - 2y - z = 0 \\ -5x + y - z = 4 \\ -11x + 4y - z = 8 \end{cases}$

38. $\begin{cases} 8x + y - z = 7 \\ 4x + z = 10 \\ 12x + y + z = 23 \end{cases}$

Extensions

39. True or False? Using the Gauss-Jordan method of solving systems of equation, the reduced row-echelon form of the system $\begin{cases} 4x+5y=\dfrac{25}{8} \\ x-y=\dfrac{5}{8} \end{cases}$ is $\left[\begin{array}{cc|c} 1 & 0 & \frac{3}{4} \\ 0 & 1 & \frac{1}{8} \end{array}\right]$.

40. Use Gaussian elimination with back substitution to solve the system. $\begin{cases} -x-3y+2z=5 \\ x+4y+2z=9 \\ -2x+3y+z=19 \end{cases}$

41. The sum of three numbers, x, y, and z, is 24. The third number, z, is four less than four times the second number, y. The third number is also twice the first number, x, increased by four times the second number. Write a system of equations and solve the system using Gaussian elimination.

42. True or False? Using the Gauss-Jordan method of solving systems of equation, the reduced row-echelon form of the system $\begin{cases} 2x+3y+2z=-4 \\ 4x+3y+4z=-3 \\ x-y+2z=6 \end{cases}$ is $\left[\begin{array}{ccc|c} 1 & 0 & 0 & -\frac{10}{3} \\ 0 & 1 & 0 & -\frac{5}{3} \\ 0 & 0 & 1 & \frac{23}{6} \end{array}\right]$.

43. Solve the system. $\begin{cases} 2x+y-2z-w=5 \\ x+2y+z+2w=-7 \\ -x-2y-2z-3w=10 \\ z+w=-3 \end{cases}$

44. Solve the system. $\begin{cases} A+3B-3C=4 \\ -3A+6B+9C=-7 \\ -2A+B-2C=-3 \end{cases}$

9.2 OPERATIONS WITH MATRICES

OBJECTIVES

- Identify equal matrices.
- Add and subtract matrices.
- Multiply a matrix by a scalar.
- Solve matrix equations.
- Multiply matrices.

PREREQUISITE VOCABULARY TERMS

augmented matrix
equivalent matrices
matrix
order (of a matrix)
row-equivalent matrices

9.2.1 Equality of Matrices

In Section 9.1, a specific type of matrix, an augmented matrix, was used as a way to express a system of linear equations. However, matrices can express all types of information, not only systems of equations. In this section, you will use matrices that are not augmented (so there will be no vertical dashed line).

Recall that equal matrices have exactly the same order (i.e., dimensions) and equal corresponding entries.

Equal Matrices

Two $m \times n$ matrices A and B are equal if each entry of A, A_{ij}, is equal to each entry of B, B_{ij}.

To determine whether two matrices are equal, check that the matrices have

1. the same order (i.e., the same number of rows and the same number of columns), and
2. the same entries in all the same positions.

For example, consider matrices A and B.

$$A = \begin{bmatrix} -1 & \frac{1}{2} & 1 & -5 \\ 9 & 2 & 0 & 4 \end{bmatrix} \qquad B = \begin{bmatrix} -1 & 0.5 & 1 & -5 \\ 3^2 & 2 & 0 & 4 \end{bmatrix}$$

Check the order and entries of A and B to determine whether the matrices are equal.

Both A and B are 2×4 matrixes (because both have 2 rows and 4 columns), so they have equal dimensions (i.e., they have the same order).

REMEMBER

The A_{12} entry is the entry in the first row, second column of matrix A.

Note that the A_{12} entry is $\frac{1}{2}$, and the B_{12} entry is 0.5. These entries do not contain the same digits, but the values of the entries are equal. Similarly, $A_{21} = 9 = 3^2 = B_{21}$, so the A_{21} and B_{21} entries are also equal. The rest of the corresponding entries are also equal.

$$A_{11} = -1 = B_{11} \quad A_{13} = 1 = B_{13} \quad A_{14} = -5 = B_{14}$$

$$A_{22} = 2 = B_{22} \quad A_{23} = 0 = B_{23} \quad A_{24} = 4 = B_{24}$$

Each entry in A is equal to the corresponding entry in B. Therefore, $A = B$.

EXAMPLE 1 Using Equal Matrices

Find p, q, and r given that $M = \begin{bmatrix} -3 & 2q \\ p+1 & 2 \\ -1 & 0 \\ r-p & 8 \end{bmatrix}$, $N = \begin{bmatrix} -3 & 10 \\ 15 & 2 \\ -1 & 0 \\ 0 & 8 \end{bmatrix}$, and $M = N$.

SOLUTION

The matrices are equal, so use the corresponding entries (that is, entries in the same positions) to write equations for p, q, and r.

$2q = 10$, so $q = 5$. $\quad p + 1 = 15$, so $p = 14$. $\quad r - p = 0 \Rightarrow r - 14 = 0$, so $r = 14$.

9.2.2 The Arithmetic of Matrices

Matrices can be added and subtracted when they have the same dimensions. To add two matrices A and B, add each entry in A to the corresponding entry in B, and place that sum in the corresponding position in the matrix $A + B$ (the matrix representing the sum of A and B). Matrix $A + B$ will have the same dimensions as the original matrices. Subtraction of two matrices is similar, except that the corresponding entries are subtracted instead of added.

EXAMPLE 2 Adding and Subtracting Matrices

Complete each operation given $C = \begin{bmatrix} 5 & 9 & -11 \\ -3 & 0 & 4 \end{bmatrix}$ and $D = \begin{bmatrix} -12 & 1 & 0 \\ 4 & 6 & -1 \end{bmatrix}$, if possible.

A. $C + D$

B. $D - C$

SOLUTION

The sum and difference can be found because C and D have the same dimensions (both are 2×3 matrices).

A. $C + D = \begin{bmatrix} 5 & 9 & -11 \\ -3 & 0 & 4 \end{bmatrix} + \begin{bmatrix} -12 & 1 & 0 \\ 4 & 6 & -1 \end{bmatrix} = \begin{bmatrix} -7 & 10 & -11 \\ 1 & 6 & 3 \end{bmatrix}$

B. $D - C = \begin{bmatrix} -12 & 1 & 0 \\ 4 & 6 & -1 \end{bmatrix} - \begin{bmatrix} 5 & 9 & -11 \\ -3 & 0 & 4 \end{bmatrix} = \begin{bmatrix} -17 & -8 & 11 \\ 7 & 6 & -5 \end{bmatrix}$

9.2.3 Multiplying Matrices by a Scalar

A matrix of any size can be multiplied by a number, called a **scalar**. To multiply a matrix A by a scalar, multiply each entry in A by that scalar and place each product in the corresponding position of the **scalar product** (the matrix representing the product). The scalar product matrix will have the same dimensions as the original matrix.

EXAMPLE 3 Multiplying a Matrix by a Scalar

Given that $A = \begin{bmatrix} 8 & 1 \\ 0 & -3 \end{bmatrix}$, find $-2A$.

SOLUTION

Multiply each entry in A by -2. $-2A = \begin{bmatrix} -16 & -2 \\ 0 & 6 \end{bmatrix}$

The properties of addition and scalar multiplication of matrices are similar to the properties of real numbers. The following table lists these properties where A and B are $m \times n$ matrices, and c and d are scalars.

Properties of Addition and Scalar Multiplication of Matrices	
Commutative Property of Matrix Addition	$A + B = B + A$
Associative Property of Matrix Addition	$(A + B) + C = A + (B + C)$
Associative Property of Scalar Multiplication	$c(dA) = (cd)A$
Distributive Property of Scalar Multiplication	$(c + d)A = cA + dA$ $c(A + B) = cA + cB$

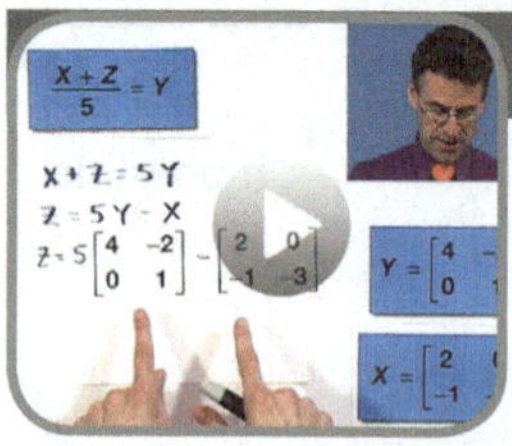

9.2.4 Solving a Matrix Equation

A matrix equation contains at least one matrix. For example, if A, B, and C are matrices, then $5A + B = C$ is a matrix equation. The process for solving a matrix equation for an unknown matrix is demonstrated in the following example.

EXAMPLE 4 Solving a Matrix Equation

Solve $2P + Q = R$ for P if $Q = \begin{bmatrix} 8 & 2 & -6 \\ 0 & 11 & 1 \end{bmatrix}$ and $R = \begin{bmatrix} -2 & 0 & 4 \\ 1 & 3 & -5 \end{bmatrix}$.

SOLUTION

Solve the equation $2P + Q = R$ for P.

$$2P + Q = R \Rightarrow 2P = R - Q \Rightarrow P = \frac{1}{2}(R - Q)$$

Substitute the given matrices for R and Q and simplify.

$$P = \frac{1}{2}\left(\begin{bmatrix} -2 & 0 & 4 \\ 1 & 3 & -5 \end{bmatrix} - \begin{bmatrix} 8 & 2 & -6 \\ 0 & 11 & 1 \end{bmatrix}\right) = \frac{1}{2}\begin{bmatrix} -10 & -2 & 10 \\ 1 & -8 & -6 \end{bmatrix} = \begin{bmatrix} -5 & -1 & 5 \\ \frac{1}{2} & -4 & -3 \end{bmatrix}$$

TIP

Check the answer by simplifying $2P + Q$, and then comparing this matrix to R. The two matrices must be equal.

9.2.5 Multiplying Matrices

Multiplication of matrices is not as straightforward as addition and subtraction of matrices. The product of two matrices AB can be found only when the number of *columns* in A is equal to the number of *rows* in B. So, if A is an $m \times n$ matrix and B is an $n \times p$ matrix, then AB can be found. Furthermore, and the product matrix AB is an $m \times p$ matrix. For example, consider the following matrices A and B.

$$A = \begin{bmatrix} a & b & c \\ d & e & f \end{bmatrix} \qquad B = \begin{bmatrix} g & h \\ i & j \\ k & l \end{bmatrix}$$

2 × 3 matrix — *3 × 2 matrix*

The product AB is defined because the number of columns in A is equal to the number of rows in B.

The product matrix AB will be a 2×2 matrix.

The entries in the 2×2 product matrix AB are found by multiplying each entry in a row from A by each corresponding entry in a column from B, then adding those individual products. Specifically, the combination of the first row in A and the first column in B gives the AB_{11} entry. The combination of the first row in A and the second column in B gives the AB_{12} entry, and so on. Generally, the combination of the ith row in A and the jth column in B gives the AB_{ij} entry. The procedure for multiplying matrices is demonstrated in the following example.

EXAMPLE 5 Multiplying Matrices

Complete each operation given $C = \begin{bmatrix} -5 & 0 \\ -1 & 2 \\ 4 & 1 \end{bmatrix}$ and $D = \begin{bmatrix} 1 & -3 \\ 0 & 2 \end{bmatrix}$, if possible.

A. CD **B.** DC

SOLUTION

A. C is a 3×2 matrix and D is a 2×2 matrix, so CD is defined, and CD is a 3×2 matrix.

$$CD = \begin{bmatrix} -5 & 0 \\ -1 & 2 \\ 4 & 1 \end{bmatrix} \begin{bmatrix} 1 & -3 \\ 0 & 2 \end{bmatrix} = \begin{bmatrix} \square & \square \\ \square & \square \\ \square & \square \end{bmatrix}$$

Find the entries in CD by completing the row-column multiplication using the rows in C and the columns in D.

$$CD = \begin{bmatrix} -5 & 0 \\ -1 & 2 \\ 4 & 1 \end{bmatrix} \begin{bmatrix} 1 & -3 \\ 0 & 2 \end{bmatrix} = \begin{bmatrix} -5(1)+0(0) & -5(-3)+0(2) \\ -1(1)+2(0) & -1(-3)+2(2) \\ 4(1)+1(0) & 4(-3)+1(2) \end{bmatrix} = \begin{bmatrix} -5 & 15 \\ -1 & 7 \\ 4 & -10 \end{bmatrix}$$

B. The product DC is not defined because the number of columns in D is not equal to the number of rows in C.

TIP

The product matrix CD is defined because the number of columns in matrix C is equal to the number of rows in matrix D.

Commutativity of Matrix Multiplication

In **Example 5,** the product CD is defined, but the product DC is not defined. Therefore, matrix multiplication is not commutative. Generally, even with matrices where both CD and DC are defined, the products are still not necessarily equal.

SECTION 9.2 EXERCISES

Warm Up

State the order of each matrix.

1. $\begin{bmatrix} 1 & 2 & -8 & 7 & 0 \\ 9 & -2 & 6 & 3 & -4 \\ 12 & 8 & 10 & 6 & 1 \end{bmatrix}$

2. $\begin{bmatrix} -3 & 2 & 0 \\ 4 & 8 & 7 \\ 6 & -7 & 1 \\ 0 & 6 & 5 \end{bmatrix}$

3. $\begin{bmatrix} 7 & 3 & 67 & 45 & 13 \end{bmatrix}$

Just the Facts

Fill in the blanks.

4. If a matrix A contains the fraction _____ for every entry and matrix B contains 0.2 for every entry, the two matrices are equal.

5. If two matrices have the same dimensions, they can be _____ or _____. For _____, the commutative property holds as it would for real numbers; and for _____, it does not hold, just as for real numbers.

6. A nonsquare matrix can be _____ by a scalar.

7. An equation which contains at least one matrix is called a(n) _____ equation.

8. The commutative property does not hold for matrix _____.

9. Each row of a matrix A must be multiplied by each _____ of a matrix B to get the product AB.

10. Multiplying two matrices A and B to get the product AB is only valid when the number of _____ in A is equal to the number of _____ in B.

Essential Skills

1. Find a, b, and c given that $P = Q$,

$P = \begin{bmatrix} 2 & a & -11 \end{bmatrix}$, and $Q = \begin{bmatrix} b & 4 & c+2 \end{bmatrix}$.

2. Find a, b, and c given that $X = Y$,

$X = \begin{bmatrix} 0 & -2 & 5 \\ 2 & 1 & -7 \end{bmatrix}$, and $Y = \begin{bmatrix} 0 & -2 & 2a+b \\ a & 1 & 1-c \end{bmatrix}$.

3. Find a, b, and c given that $R = S$,

$R = \begin{bmatrix} -2 & 2a-b & 6 & c \\ a+c & 5 & -7 & 8 \end{bmatrix}$, and

$S = \begin{bmatrix} -2 & 10 & 6 & -5 \\ -7 & 5 & -7 & 8 \end{bmatrix}$.

4. Find a, b, and c given that $M = N$,

$M = \begin{bmatrix} -5 & 8 \\ 10 & 0 \\ -6 & 3 \\ 12 & 9 \end{bmatrix}$, and $N = \begin{bmatrix} -5 & a+3b \\ 10 & c-a \\ b & 3 \\ 12 & 9 \end{bmatrix}$.

5. Find a, b, and c given that $P = Q$,

$P = \begin{bmatrix} a^2-1 & 8 & -4 \\ 6 & 0 & b+c \\ 5 & a+b & 13 \end{bmatrix}$, and $Q = \begin{bmatrix} 48 & 8 & -4 \\ 6 & 0 & 20 \\ 5 & -4 & 13 \end{bmatrix}$.

6. Find a, b, and c given that $X = Y$,

$X = \begin{bmatrix} 2 & 1 & -4 \\ b+a & 2 & 2 \\ 0 & c^2 & 8 \end{bmatrix}$, and $Y = \begin{bmatrix} 2 & a & -4 \\ 7 & 2 & 2 \\ 0 & 16 & 8 \end{bmatrix}$.

In Exercises 7–18, complete each operation, if possible.

7. Find $D + E$ and $D - E$ given that

$D = \begin{bmatrix} 33 & -19 \\ 2 & -34 \end{bmatrix}$ and $E = \begin{bmatrix} -15 & -35 \\ 9 & 3 \end{bmatrix}$.

8. Find $A + B$ and $B - A$ given that

$A = \begin{bmatrix} 5 & -1 & 0 & -2 \\ 2 & 1 & -3 & 4 \\ 1 & 0 & 2 & -2 \end{bmatrix}$ and $B = \begin{bmatrix} -3 & 0 & 2 & -6 \\ 1 & -1 & 5 & -2 \\ 0 & 8 & -7 & 1 \end{bmatrix}$.

9. Find $H + G$ and $G - H$ given that

$H = \begin{bmatrix} -7 & 8 & 0 \\ 12 & -5 & 6 \end{bmatrix}$ and $G = \begin{bmatrix} 4 & -8 & 11 \\ -5 & 0 & -4 \end{bmatrix}$.

10. Find $X + Y$ and $Y - X$ given that

$X = \begin{bmatrix} 13 & -1 & 7 \\ 7 & 8 & 0 \\ 1 & 3 & 10 \end{bmatrix}$ and $Y = \begin{bmatrix} 5 & 1 & 3 \\ 3 & -1 & 11 \\ 8 & 17 & -2 \end{bmatrix}$.

11. Find $J + K$ and $J - K$ given that

$$J = \begin{bmatrix} 3 & -6 \\ 7 & 0 \\ 23 & 10 \\ -8 & -3 \end{bmatrix} \text{ and } K = \begin{bmatrix} 1 & 4 \\ -6 & 7 \\ -5 & 6 \\ 11 & 4 \end{bmatrix}.$$

12. Find $F + G$ and $G - F$ given that

$$F = \begin{bmatrix} 5 & -1 & 9 \\ 6 & 10 & 5 \end{bmatrix} \text{ and } G = \begin{bmatrix} 5 & 7 \\ 4 & -1 \end{bmatrix}.$$

13. $-5\begin{bmatrix} 5 & 10 & -4 \\ 7 & 2 & 8 \end{bmatrix}$

14. $-\frac{1}{2}\begin{bmatrix} 6 & -4 & 8 \\ 20 & -2 & 10 \\ 2 & 12 & 14 \end{bmatrix}$

15. $-\frac{1}{4}\begin{bmatrix} -20 & 4 \\ 16 & -36 \\ 12 & 32 \\ -4 & -8 \end{bmatrix}$

16. $-4\begin{bmatrix} -3 & 3 & -4 \\ -7 & 5 & -4 \\ -1 & -2 & -7 \\ -1 & -4 & -7 \end{bmatrix}$

17. $6\begin{bmatrix} -2 & 0 \\ 1 & -6 \\ 7 & 4 \end{bmatrix}$

18. $\frac{1}{3}\begin{bmatrix} 9 & -15 & 12 \\ 6 & -3 & 15 \\ -3 & 18 & 21 \end{bmatrix}$

In Exercises 19–24, solve each equation for Z if

$$X = \begin{bmatrix} 5 & -2 \\ 0 & 1 \\ -1 & 4 \end{bmatrix} \text{ and } Y = \begin{bmatrix} -3 & 0 \\ -1 & -5 \\ 2 & 1 \end{bmatrix}.$$

19. $X - Z = Y$

20. $Y - \frac{1}{2}Z = X$

21. $2X - Y = 2Z$

22. $Z + 3X = 2Y$

23. $0.5Z + X = 3Y$

24. $2Y - Z = 4X$

In Exercises 25–32, state whether the products AB and BA are defined. If the product is defined, give the dimensions of the product matrix.

25. $A = \begin{bmatrix} 4 & -1 \\ 3 & 7 \end{bmatrix}$ and $B = \begin{bmatrix} 3 & 0 \\ 5 & -8 \end{bmatrix}$.

26. $A = \begin{bmatrix} 4 & 5 \\ -1 & 0 \\ 1 & 2 \end{bmatrix}$ and $B = \begin{bmatrix} 10 & 0 & 1 & -4 \\ -1 & 5 & 7 & -2 \end{bmatrix}$

27. $A = \begin{bmatrix} -1 & 9 & 6 & -7 \end{bmatrix}$ and $B = \begin{bmatrix} 2 & -5 \\ 0 & 1 \\ -9 & 12 \\ 3 & 10 \end{bmatrix}$

28. $A = \begin{bmatrix} 8 & 0 & -1 \\ 2 & -3 & 1 \end{bmatrix}$ and $B = \begin{bmatrix} 2 & -1 & -3 & 0 \\ 1 & 5 & 0 & -2 \end{bmatrix}$

29. $A = \begin{bmatrix} 2 & -5 & 8 \\ 10 & 3 & -81 \\ -9 & 4 & 0 \end{bmatrix}$ and $B = \begin{bmatrix} 3 & -9 & 0 \\ 1 & 7 & 3 \\ 5 & 8 & -8 \\ 10 & 4 & 6 \end{bmatrix}$

30. $A = \begin{bmatrix} 6 & 8 \\ -1 & 0 \\ 1 & 3 \end{bmatrix}$ and $B = \begin{bmatrix} 10 & 0 & 3 & 3 \\ 0 & 8 & 8 & -2 \\ 2 & 5 & 7 & 0 \end{bmatrix}$

31. $A = \begin{bmatrix} -7 & 5 \\ 7 & 3 \end{bmatrix}$ and $B = \begin{bmatrix} -8 & 7 & 0 \\ 3 & 16 & 8 \\ -5 & 0 & 7 \end{bmatrix}$

32. $A = \begin{bmatrix} 6 & 6 & 4 \\ 3 & 0 & 4 \end{bmatrix}$ and $B = \begin{bmatrix} 10 & 0 \\ 1 & 11 \end{bmatrix}$

In Exercises 33–38, find each product, if possible.

33. PQ if $P = \begin{bmatrix} 9 & -6 & -1 \\ 2 & -7 & -3 \end{bmatrix}$ and $Q = \begin{bmatrix} 1 \\ -9 \\ 7 \end{bmatrix}$

34. NM if $N = \begin{bmatrix} 0 & 3 & 1 \\ 3 & -1 & 0 \end{bmatrix}$ and $M = \begin{bmatrix} 1 & -6 \\ 0 & 1 \\ 1 & -1 \end{bmatrix}$

35. YX if $X = \begin{bmatrix} 3 & 0 \\ -1 & 5 \\ 7 & 2 \\ -4 & 6 \end{bmatrix}$ and $Y = \begin{bmatrix} -8 \\ 2 \end{bmatrix}$

36. SR if $R = \begin{bmatrix} 6 & 1 \\ 0 & 8 \end{bmatrix}$ and $S = \begin{bmatrix} 8 & -5 \\ 5 & -6 \end{bmatrix}$

37. JK if $K = \begin{bmatrix} 1 & 9 & -8 \\ 0 & 5 & 1 \\ 4 & -3 & 2 \end{bmatrix}$ and $J = \begin{bmatrix} -5 & 6 & 0 \\ 7 & 1 & -2 \end{bmatrix}$

38. NM if $M = \begin{bmatrix} 1 & -6 \\ 0 & 1 \\ 1 & -1 \end{bmatrix}$ and $N = \begin{bmatrix} 0 & 3 & 1 \\ 3 & -1 & 0 \end{bmatrix}$

Extensions

In Exercises 39–46, use the following information.

The Texas Window Decor company produces horizontal and vertical blinds and operates two factories in Houston and one in Dallas. Matrix *A* shows the production for week 1, matrix *B* for week 2, and matrix *C* shows the combined production for the third and fourth weeks. In each matrix, row 1 represents horizontal blinds, row 2 represents vertical blinds, column 1 represents the Dallas factory, and column 2 represents the Houston factory.

$$A = \begin{bmatrix} 200 & 80 \\ 1000 & 420 \end{bmatrix} \quad B = \begin{bmatrix} 160 & 80 \\ 880 & 640 \end{bmatrix} \quad C = \begin{bmatrix} 200 & 100 \\ 1200 & 420 \end{bmatrix}$$

39. How many blinds did the Houston factory produce in the first week?

40. How many horizontal blinds did the Dallas factory produce in the first week?

41. How many blinds did the Houston factory produce in the third and fourth weeks?

42. How many vertical blinds did the Houston factory produce in the third and fourth weeks?

43. How many vertical blinds were produced in the second week?

44. How many horizontal blinds were produced in the second week?

45. Write a matrix that describes the total production for the four weeks.

46. How many horizontal blinds were produced in the Dallas factory during the four-week period?

In Exercises 47–50, use the following information.

A sporting goods company has two stores. For store A, the June and July incomes were \$178,000 and \$166,500, and the June and July profits were \$28,500 and \$22,700. For store B, the June and July incomes were \$122,550 and \$145,000, and the June and July profits were \$18,600 and \$22,800.

47. Write a matrix that shows the incomes for the two stores for June and July.

48. Create a second matrix for the profits.

49. Write a matrix equation for the expenses using the matrices from Exercises 47 and 48.

50. What were the expenses of store A in June? in July?

In Exercises 51–57, evaluate each expression using the following matrices.

$$A = \begin{bmatrix} 2.2 & 2.4 \\ 12 & 0.8 \\ 8.8 & 6 \end{bmatrix} \quad C = \begin{bmatrix} 9 & 5.2 \\ 14 & 7 \\ 7.6 & 2.9 \end{bmatrix}$$

$$B = \begin{bmatrix} 16 & 4.4 & 12 \\ 8.8 & 3.9 & 11 \end{bmatrix} \quad D = \begin{bmatrix} 4 & 2.2 \\ 3.2 & 6.9 \end{bmatrix}$$

51. $8A$

52. $A + B$

53. $2A + 5C$

54. $2.2C - 1.5A$

55. $4.2(C - A)$

56. BD

57. DB

In Exercises 58–60, use the following information.

Two manufacturing plants are able to produce the following numbers of pet toys (in thousands) each day.

Item	Plant A	Plant B
dog toy	24	15
cat toy	35	12
hamster ball	18	8

58. Write a matrix representing five days of production.

59. Write a matrix to represent n days of production.

60. If a dog toy costs \$0.75 to make, a cat toy costs \$0.50, and the production of one hamster ball costs \$2.04, create a matrix to determine the production costs for each item in one day and then one week.

61. Given $M = N$, solve for the variables.

$$M = \begin{bmatrix} -5 & 2x-5 & 13 & y-11 \\ 3z+7 & 0 & -8 & 2a \\ 16 & 0.5b & 9 & 2-c \\ 45 & 1.2d+8 & 7 & 3 \end{bmatrix}$$

$$N = \begin{bmatrix} f & 12.6 & 13 & 2-y \\ 13 & g-6 & -8 & 15 \\ -h+19 & 67 & 9 & 2c+1 \\ j & 10 & w & 3 \end{bmatrix}$$

62. Given $A = \begin{bmatrix} 5 & -6 & 1 \\ -12 & 3 & 4 \end{bmatrix}$ and $F = \begin{bmatrix} -2 & 4 & -3 \\ 16 & 7 & 0 \end{bmatrix}$, determine matrix E so $2A - E = F$.

9.3 DETERMINANTS AND CRAMER'S RULE

OBJECTIVES

- Find determinants for 2 × 2 matrices.
- Find determinants of matrices by converting to triangular form.
- Find determinants of 3 × 3 matrices by expanding by cofactors.
- Find areas of polygons using determinants.
- Solve systems of linear equations in two variables and in three variables using Cramer's Rule.

PREREQUISITE VOCABULARY TERMS

collinear
linear system of equations
polygon
row operations
square matrix
triangular form (of a matrix)

9.3.1 Evaluating 2 × 2 Determinants

The **determinant** of a matrix is a number associated with any square matrix that is useful, among other things, for solving systems of linear equations. Several methods for finding the determinant of a matrix will be discussed in this section, but the first method works only for 2 × 2 matrices.

Determinant of a 2 × 2 Matrix

For a 2 × 2 matrix, $A = \begin{bmatrix} a & b \\ c & d \end{bmatrix}$, the determinant is $\det(A) = |A| = \begin{vmatrix} a & b \\ c & d \end{vmatrix} = ad - bc$.

TIP

Two vertical bars denote a determinant in this context, not the absolute value.

EXAMPLE 1 Finding the Determinant of a 2 × 2 Matrix

Find the determinant of each matrix.

A. $A = \begin{bmatrix} 3 & 2 \\ 6 & 3 \end{bmatrix}$ **B.** $B = \begin{bmatrix} 8 & 4 \\ -1 & 1 \end{bmatrix}$ **C.** $C = \begin{bmatrix} 2 & 4 \\ 6 & 12 \end{bmatrix}$

SOLUTION

NOTICE THAT

The three notations for determinants are used here.

Apply the formula for the determinant of a 2 × 2 matrix to each matrix.

A. $\det(A) = (3)(3) - (2)(6) = 9 - 12 = -3$

B. $|B| = (8)(1) - (4)(-1) = 8 + 4 = 12$

C. $\begin{vmatrix} 2 & 4 \\ 6 & 12 \end{vmatrix} = (2)(12) - (4)(6) = 24 - 24 = 0$

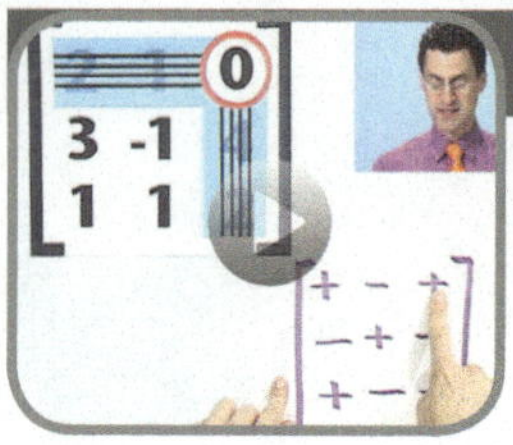

9.3.2 Finding a Determinant Using Expanding by Cofactors

The method demonstrated in the previous topic for finding the determinant of a 2 × 2 matrix is a simple calculation. The calculations for determinants of matrices that are larger than 2 × 2 can be less simple. Three methods for finding the determinants will be introduced in this section.

Finding the Determinant of a Square Matrix

Method 1: The Diagonal Method

Method 2: Using Cofactors

Method 3: Using Elementary Row Operations

Method 1: The Diagonal Method

The diagonal method is demonstrated in **Example 2**.

EXAMPLE 2 Finding the Determinant of a 3 × 3 Matrix

Find the determinant of the matrix. $A = \begin{bmatrix} 1 & 3 & 4 \\ -2 & 6 & 8 \\ 0 & 5 & -4 \end{bmatrix}$

SOLUTION

Multiply the elements along each of the three right diagonals.

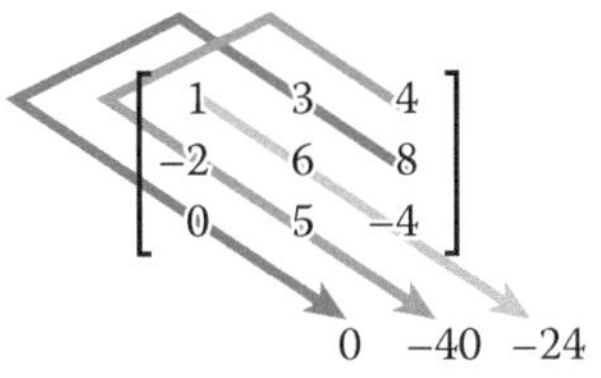

Multiply the elements along each of the three left diagonals.

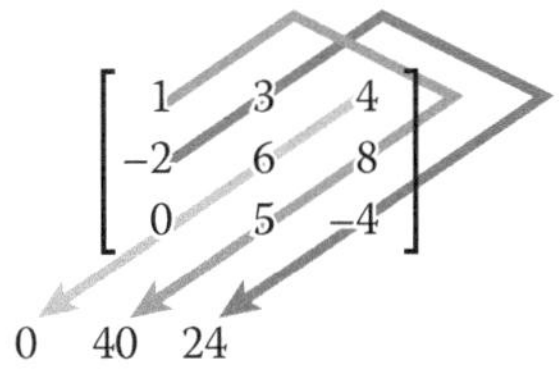

Add the three right-diagonal products.

$$0 + (-40) + (-24) = -64$$

Add the three left-diagonal products.

$$0 + 40 + 24 = 64$$

Subtract the sum of the left-diagonal products from the sum of the right-diagonal products.

$$-64 - 64 = -128$$

The determinant of A found using the diagonal method is −128.

The diagonal method is very simple in the case of a matrix in upper or lower triangular form because the only nonzero product is from the matrix's main diagonal. So, the determinant of a matrix in upper or lower triangular form is simply the product of the entries on the main diagonal.

Method 2: Using Cofactors

A second method for finding the determinant of a 3 × 3 matrix involves breaking the matrix down into three 2 × 2 matrices, and then finding the determinant of each 2 × 2 matrix. The determinants of these 2 × 2 matrices are called **minors**, denoted M_{ij}.

DEFINITION

The **minor** M_{ij} of a matrix A is the determinant of the matrix formed by disregarding the ith row and jth column of A.

The process of identifying the minors for a particular row (or column) is called expanding on a row (or column). For example, the minors calculated by expanding on row 1 are shown here.

TIP

The expansion here is along row 1, but any row or column can be chosen to calculate the determinant with this method.

$$\begin{vmatrix} E_{11} & E_{12} & E_{13} \\ E_{21} & E_{22} & E_{23} \\ E_{31} & E_{32} & E_{33} \end{vmatrix} \qquad \begin{vmatrix} E_{11} & E_{12} & E_{13} \\ E_{21} & E_{22} & E_{23} \\ E_{31} & E_{32} & E_{33} \end{vmatrix} \qquad \begin{vmatrix} E_{11} & E_{12} & E_{13} \\ E_{21} & E_{22} & E_{23} \\ E_{31} & E_{32} & E_{33} \end{vmatrix}$$

$$M_{11} = E_{22}E_{33} - E_{23}E_{32} \qquad M_{12} = E_{21}E_{33} - E_{23}E_{31} \qquad M_{13} = E_{21}E_{32} - E_{22}E_{31}$$

When minor M_{ij} is multiplied by an appropriate sign (+ or –), this product is called **cofactor** C_{ij}. There are two ways to determine the appropriate sign for a cofactor C_{ij}.

NOTICE THAT

$(-1)^{i+j}$ (from the cofactor formula) gives the appropriate sign, which is the same as the sign given in the sign matrix. For example, the E_{13} sign from the sign matrix is +, and $(-1)^{1+3} = (-1)^4 = 1$, meaning that the cofactor's sign is positive.

1. Use a sign matrix.
2. Use the formula $(-1)^{i+j}$.

3 × 3 Sign Matrix

$$\begin{bmatrix} + & - & + \\ - & + & - \\ + & - & + \end{bmatrix}$$

The appropriate sign for cofactor C_{ij} is given by the E_{ij} element of the corresponding sign matrix.

DEFINITION

The **cofactor** C_{ij} of the element E_{ij} in a matrix is the product of $(-1)^{i+j}$ and minor M_{ij}.

The minors are used to find the cofactors, and then the cofactors are used to find the determinant of the matrix. To find the determinant of a matrix A, multiply each of a row's (or column's) cofactors C_{ij} by the value of element E_{ij}, from A, and then add these products.

Calculating the Determinant of Matrix A Using Cofactors

If A is any $n \times n$ matrix ($n \geq 3$), then $|A|$ is the sum of the products of the elements of any row or column and their cofactors. The determinant can be calculated using

- the rth row of A: $|A| = E_{r1}C_{r1} + E_{r2}C_{r2} + \ldots + E_{rn}C_{rn}$.
- the cth column of A: $|A| = E_{1c}C_{1c} + E_{2c}C_{2c} + \ldots + E_{nc}C_{nc}$.

EXAMPLE 3 Finding the Determinant of a 3 × 3 Matrix Using Cofactors

$$A = \begin{bmatrix} 1 & 3 & 4 \\ -2 & 6 & 8 \\ 0 & 5 & -4 \end{bmatrix}$$

Expand on row 1 to find the determinant of matrix A from Example 2.

SOLUTION

Find the minors using row 1.

$$M_{11} = \begin{vmatrix} 6 & 8 \\ 5 & -4 \end{vmatrix} = -64 \qquad M_{12} = \begin{vmatrix} -2 & 8 \\ 0 & -4 \end{vmatrix} = 8 \qquad M_{13} = \begin{vmatrix} -2 & 6 \\ 0 & 5 \end{vmatrix} = -10$$

Find the cofactors using row 1.

$C_{11} = (-64)(1) = -64$ — *The appropriate sign for M_{11} is +1.*

$C_{12} = (8)(-1) = -8$ — *The appropriate sign for M_{11} is −1.*

$C_{13} = (-10)(1) = -10$ — *The appropriate sign for M_{11} is +1.*

Find the determinant of A using the formula for expanding on a row.

From matrix A, $E_{11} = 1$, $E_{12} = 3$, and $E_{13} = 4$.

$$|A| = E_{11}C_{11} + E_{12}C_{12} + E_{13}C_{13} = (1)(-64) + (3)(-8) + (4)(-10) = -128$$

The determinant is −128.

9.3.3 Evaluating a Determinant Using Elementary Row Operations

Recall that the determinant of a matrix in upper or lower triangular form can easily be found because it is equal to the product of the elements along the main diagonal. The third method for finding the determinant of a square matrix is to use row operations to transform a square matrix to one in triangular form. However, applying the elementary row operations to a matrix can change the determinant of a matrix, so you need to keep track of what row operations are performed and the effect of each on the determinant.

REMEMBER

A triangular-form matrix has 0s for entries above or below the diagonal. Upper triangular form means the 0s are below the diagonal. Lower triangular form means the 0s are above the diagonal.

The following examples show the effects of the row operations on the determinant.

EXAMPLE 4 The Determinant of a 2 × 2 Matrix After a Row Interchange

Find the determinant of $A = \begin{bmatrix} 3 & 5 \\ 4 & 4 \end{bmatrix}$ before and after a row interchange.

SOLUTION

Find the determinant of A. $\quad \det(A) = (3)(4) - (5)(4) = -8$

Find the determinant after a row interchange.

Write the matrix after switching rows 1 and 2. $\quad \begin{bmatrix} 4 & 4 \\ 3 & 5 \end{bmatrix}$ *Interchange R_1 and R_2.*

Find the determinant of the new matrix. $\quad \begin{vmatrix} 4 & 4 \\ 3 & 5 \end{vmatrix} = (4)(5) - (4)(3) = 8$

Thus, the determinant of A is -8, and the determinant of A after a row interchange is 8.

NOTICE THAT

The determinants have the same absolute value, but opposite signs.

EXAMPLE 5 The Determinant of a 2 × 2 Matrix After Multiplication

Find the determinant of $A = \begin{bmatrix} 3 & 5 \\ 4 & 4 \end{bmatrix}$ before and after multiplying row 2 by a factor of 2.

SOLUTION

From **Example 4**, $\det(A) = -8$.

Find the determinant after multiplying row 2 by 2.

Write the matrix after multiplying row 2 by 2. $\quad \begin{bmatrix} 3 & 5 \\ 8 & 8 \end{bmatrix}$ $2R_2$

Find the determinant of the new matrix. $\quad \begin{vmatrix} 3 & 5 \\ 8 & 8 \end{vmatrix} = (3)(8) - (5)(8) = -16$

Thus, the determinant of A is -8, and the determinant of A after multiplying a row by 2 is -16.

NOTICE THAT

The determinant of A after multiplying a row by 2 is twice the determinant of A.

EXAMPLE 6 The Determinant of a 2 × 2 Matrix After a Row Addition

Find the determinant of $A = \begin{bmatrix} 3 & 5 \\ 4 & 4 \end{bmatrix}$ before and after replacing R_2 with $3R_1 + R_2$.

SOLUTION

From **Example 2**, $\det(A) = -8$. Find the determinant after replacing R_2 with $3R_1 + R_2$.

Write the new matrix. $\begin{bmatrix} 3 & 5 \\ 13 & 19 \end{bmatrix} \; 3R_1 + R_2$

Find the determinant of the new matrix. $\begin{vmatrix} 3 & 5 \\ 13 & 19 \end{vmatrix} = (3)(19) - (5)(13) = -8$

Thus, the determinant of A is -8, and the determinant of A after replacing R_2 with $3R_1 + R_2$ is -8.

NOTICE THAT

The determinant of A after replacing a row with a sum of rows is equal to the determinant of A.

When carrying out row operations on a matrix, keep track of the operations needed to correct the determinant with each step. In addition, write down the operation itself to make it easier to check work once the answer is obtained.

Effects of Row Operations on the Determinant of a Matrix
Interchanging two rows in a matrix reverses the sign of the determinant.
Multiplying a row by a factor also multiplies the determinant by that factor.
Adding a multiple of another row to a row does not affect the value of the determinant.

EXAMPLE 7 Finding the Determinant of a 3 × 3 Matrix

Find the determinant of the matrix by writing the matrix in triangular form. $\begin{bmatrix} 2 & -1 & 2 \\ 0 & 2 & -1 \\ 2 & 0 & 2 \end{bmatrix}$

SOLUTION

Use row operations to write the matrix in triangular form.

$$\det \begin{bmatrix} 2 & -1 & 2 \\ 0 & 2 & -1 \\ 2 & 0 & 2 \end{bmatrix} = \begin{vmatrix} 2 & -1 & 2 \\ 0 & 2 & -1 \\ 0 & 1 & 0 \end{vmatrix} \; R_3 + -R_1$$ *Determinant is unchanged.*

$$= \left(-\frac{1}{2}\right) \begin{vmatrix} 2 & -1 & 2 \\ 0 & 2 & -1 \\ 0 & -2 & 0 \end{vmatrix} \; -2R_3$$ *Divide the determinant by −2.*

$$= \left(-\frac{1}{2}\right) \begin{vmatrix} 2 & -1 & 2 \\ 0 & 2 & -1 \\ 0 & 0 & -1 \end{vmatrix} \; R_3 + R_2$$ *Determinant is unchanged.*

$$= \left(-\frac{1}{2}\right)(2)(2)(-1)$$ *Multiply the elements on the main diagonal.*

$$= 2$$

IMPORTANT

Keep track of the necessary corrections of the determinant caused by the row operations.

TIP

Multiply the determinant by −1/2 (i.e., divide by −2) to correct for multiplying a row by −2.

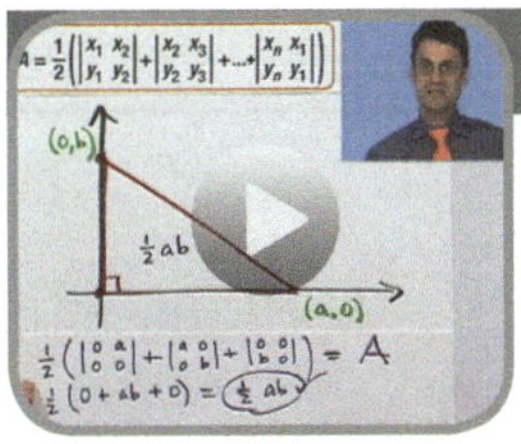

9.3.4 Applying Determinants

One interesting application of determinants is their use in finding the area of a polygon described in the Cartesian coordinate plane. Any polygon can be divided into triangles and the method works by finding and summing the areas of the triangles that make up the polygon.

For any polygon, the area is equal to one half the sum of the determinants of the matrices produced by listing adjacent coordinates of the polygon. The coordinates must be listed in counterclockwise order to produce a positive result.

NOTICE THAT

The second column of each determinant is the first column of the next determinant.

Area of a Polygon Using Determinants

For any polygon with vertices $(x_1, y_1), (x_2, y_2), (x_3, y_3), \ldots, (x_n, y_n)$, the area of the polygon is

$$A = \frac{1}{2}\left(\begin{vmatrix} x_1 & x_2 \\ y_1 & y_2 \end{vmatrix} + \begin{vmatrix} x_2 & x_3 \\ y_2 & y_3 \end{vmatrix} + \begin{vmatrix} x_3 & x_4 \\ y_3 & y_4 \end{vmatrix} + \ldots + \begin{vmatrix} x_n & x_1 \\ y_n & y_1 \end{vmatrix}\right).$$

The vertices should be listed in counterclockwise order to produce a positive result.

The method can be demonstrated by finding the area of a triangle. The area of a triangle is one half the base length multiplied by the height of the triangle. Consider the triangle shown in Figure 9.3a.

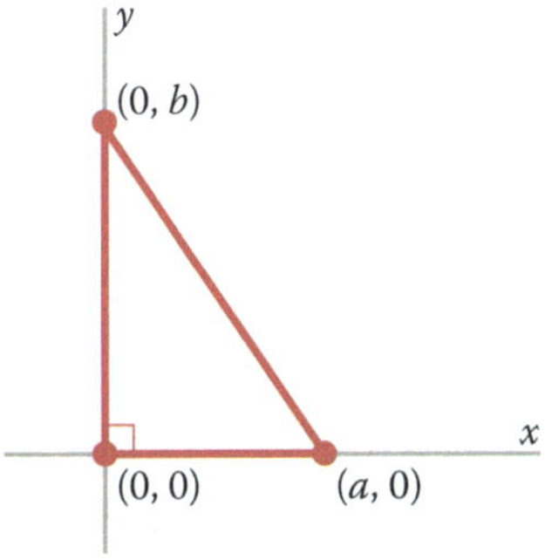

Figure 9.3a

Using the coordinates of the vertices, the area of this right triangle can be written fairly simply.

$$A = \frac{1}{2}(a-0)(b-0) = \frac{1}{2}ab$$

To calculate the area using the determinants method, first list the coordinates in a counterclockwise order beginning arbitrarily at the origin.

$$(0, 0), (a, 0), (0, b)$$

Then substitute the values for the coordinates into the formula for the area of a polygon and solve.

$$A = \frac{1}{2}\left(\begin{vmatrix} 0 & a \\ 0 & 0 \end{vmatrix} + \begin{vmatrix} a & 0 \\ 0 & b \end{vmatrix} + \begin{vmatrix} 0 & 0 \\ b & 0 \end{vmatrix}\right)$$

$$= \frac{1}{2}([(0)(0)-(a)(0)]+[(a)(b)-(0)(0)]+[(0)(0)-(0)(b)])$$ *Find the determinants.*

$$= \frac{1}{2}(0+ab+0)$$ *Multiply and subtract.*

$$= \frac{1}{2}ab$$ *Simplify.*

This result is the same as the result from the traditional formula.

EXAMPLE 8 Finding the Area of a Triangle Using Determinants

Find the area of the triangle with vertices (3, 0), (6, 2), (1, 3).

SOLUTION

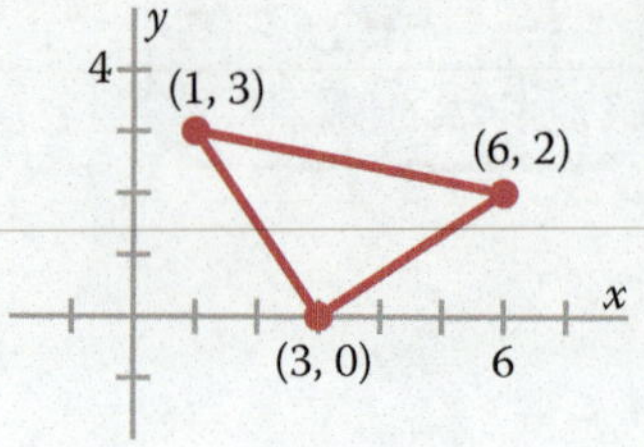

Begin by verifying the coordinates are listed in counterclockwise order. Draw a graph of the triangle.

The coordinates are listed in counterclockwise order.

Substitute the coordinates into the formula and solve.

$$A = \frac{1}{2}\left(\begin{vmatrix} 3 & 6 \\ 0 & 2 \end{vmatrix} + \begin{vmatrix} 6 & 1 \\ 2 & 3 \end{vmatrix} + \begin{vmatrix} 1 & 3 \\ 3 & 0 \end{vmatrix}\right)$$ *Substitute the coordinates into the formula for area.*

$$= \frac{1}{2}([(3)(2)-(6)(0)]+[(6)(3)-(1)(2)]+[(1)(0)-(3)(3)])$$ *Find the determinants.*

$$= \frac{1}{2}(6+16-9)$$ *Multiply and subtract.*

$$= \frac{1}{2}(13)$$ *Add.*

$$= \frac{13}{2}$$ *Simplify.*

The area of the triangle is $\frac{13}{2}$, or $6\frac{1}{2}$, square units.

If three points lie on one line, then the area of the "triangle" they form would be equal to zero. For this reason, the formula for the area of a polygon is also a test for collinearity of points.

WHAT'S THE BIG IDEA?

Use the formula for the area of a polygon to find the area of a rectangle in the Cartesian coordinate plane, and compare the result with length-by-width calculations.

EXAMPLE 9 Testing for Collinearity Using Determinants

Determine if the points are collinear using determinants. (−6, −2), (0, 4), (−3, 1)

SOLUTION

TIP

The test for collinearity does not require the points to be in any particular order.

Substitute the coordinates into the formula and solve for the area.

$$A = \frac{1}{2}\left(\begin{vmatrix} -6 & 0 \\ -2 & 4 \end{vmatrix} + \begin{vmatrix} 0 & -3 \\ 4 & 1 \end{vmatrix} + \begin{vmatrix} -3 & -6 \\ 1 & -2 \end{vmatrix}\right)$$ *Substitute into the formula for area.*

$$= \frac{1}{2}([(-6)(4)-(0)(-2)]+[(0)(1)-(-3)(4)]+[(-3)(-2)-(-6)(1)])$$ *Find determinants.*

$$= \frac{1}{2}(-24+12+12)$$ *Multiply and subtract.*

$$= \frac{1}{2}(0)$$ *Add.*

$$= 0$$ *Simplify.*

The area is zero, meaning that the three points are collinear.

9.3.5 Using Cramer's Rule

A matrix with a determinant equal to 0 is said to be a **singular matrix**. If the matrix represents a system of linear equations, then that system has a unique solution if and only if the matrix is nonsingular. If the matrix is singular, the system of equations has no solution or infinitely many solutions.

When a square matrix represents a system of linear equations, the system can be solved using determinants. The method used is **Cramer's Rule**. Consider the following system of linear equations.

$$\begin{cases} a_1x + b_1y = c_1 \\ a_2x + b_2y = c_2 \end{cases}$$

The system's **coefficient matrix** is the first two columns from the corresponding augmented matrix. The determinant of the coefficient matrix is called D.

$$\begin{bmatrix} a_1 & b_1 \\ a_2 & b_2 \end{bmatrix} \qquad D = \begin{vmatrix} a_1 & b_1 \\ a_2 & b_2 \end{vmatrix}$$

Coefficient Matrix Determinant of the Coefficient Matrix

Two other determinants can be defined by replacing the x or y coefficients with the appropriate constant terms from the system.

$$D_x = \begin{vmatrix} c_1 & b_1 \\ c_2 & b_2 \end{vmatrix} \qquad D_y = \begin{vmatrix} a_1 & c_1 \\ a_2 & c_2 \end{vmatrix}$$

Provided D is not equal to zero, the solution to the system of equations can be found using the determinants.

$$x = \frac{D_x}{D} \qquad y = \frac{D_y}{D}$$

Solution to the System of Linear Equations

TIP

Cramer's Rule cannot be used when $D = 0$ because division by 0 is undefined.

Why Does Cramer's Rule Work?

To see why Cramer's Rule produces a solution to a system of linear equations, consider the general expression for a system of two equations.

$$\begin{cases} a_1x + b_1y = c_1 \\ a_2x + b_2y = c_2 \end{cases}$$

This system can be solved by substitution (or some other method) to give expressions for x and y.

$$x = \frac{c_1b_2 - c_2b_1}{a_1b_2 - a_2b_1} \qquad y = \frac{a_1c_2 - a_2c_1}{a_1b_2 - a_2b_1}$$

NOTICE THAT

The expressions in the numerators and denominators are in the form of the expression for the determinant of a 2×2 matrix.

The expressions for x and y can be written as ratios of determinants.

$$x = \frac{\begin{vmatrix} c_1 & b_1 \\ c_2 & b_2 \end{vmatrix}}{\begin{vmatrix} a_1 & b_1 \\ a_2 & b_2 \end{vmatrix}} \qquad y = \frac{\begin{vmatrix} a_1 & c_1 \\ a_2 & c_2 \end{vmatrix}}{\begin{vmatrix} a_1 & b_1 \\ a_2 & b_2 \end{vmatrix}}$$

These expressions are the same as the solutions given by Cramer's Rule.

$$x = \frac{D_x}{D} \qquad y = \frac{D_y}{D}$$

EXAMPLE 10 Using Cramer's Rule to Solve a System of Linear Equations

Solve the system of linear equations using Cramer's Rule. $\begin{cases} 2x - 7y = 0 \\ x - 4y = -1 \end{cases}$

SOLUTION

$D = \begin{vmatrix} a_1 & b_1 \\ a_2 & b_2 \end{vmatrix}$ ▶ Find D and proceed if $D \neq 0$.

$$D = \begin{vmatrix} 2 & -7 \\ 1 & -4 \end{vmatrix} = (2)(-4) - (-7)(1) = -1$$

The value of D is nonzero, therefore the system has a unique solution which can be found using Cramer's Rule.

Find the values of D_x and D_y.

$D_x = \begin{vmatrix} c_1 & b_1 \\ c_2 & b_2 \end{vmatrix}$ ▶

$$D_x = \begin{vmatrix} 0 & -7 \\ -1 & -4 \end{vmatrix} = (0)(-4) - (-7)(-1) = -7$$

$D_y = \begin{vmatrix} a_1 & c_1 \\ a_2 & c_2 \end{vmatrix}$ ▶

$$D_y = \begin{vmatrix} 2 & 0 \\ 1 & -1 \end{vmatrix} = (2)(-1) - (0)(1) = -2$$

Use the determinants to find the solution to the system of linear equations.

$$x = \frac{D_x}{D} = \frac{-7}{-1} = 7 \qquad y = \frac{D_y}{D} = \frac{-2}{-1} = 2$$

The solution to the system of linear equations is (7, 2).

9.3.6 Using Cramer's Rule in a 3 × 3 Matrix

Cramer's Rule can be used to solve square systems of equations with more than two equations and variables. The addition of each new variable requires the calculation of a new determinant. The determinants can be found by the method of expansion using cofactors.

Consider the general form of a system of three linear equations with three variables. There are four determinants defined for this system.

$$\begin{cases} a_1x + b_1y + c_1z = d_1 \\ a_2x + b_2y + c_2z = d_2 \\ a_3x + b_3y + c_3z = d_3 \end{cases}$$

$$D = \begin{vmatrix} a_1 & b_1 & c_1 \\ a_2 & b_2 & c_2 \\ a_3 & b_3 & c_3 \end{vmatrix} \quad D_x = \begin{vmatrix} d_1 & b_1 & c_1 \\ d_2 & b_2 & c_2 \\ d_3 & b_3 & c_3 \end{vmatrix} \quad D_y = \begin{vmatrix} a_1 & d_1 & c_1 \\ a_2 & d_2 & c_2 \\ a_3 & d_3 & c_3 \end{vmatrix} \quad D_z = \begin{vmatrix} a_1 & b_1 & d_1 \\ a_2 & b_2 & d_2 \\ a_3 & b_3 & d_3 \end{vmatrix}$$

The solution to the system of three linear equations is found in a way analogous to the two-equation system. If D is nonzero, then the system has a unique solution that can be found using Cramer's Rule.

$$x = \frac{D_x}{D} \qquad y = \frac{D_y}{D} \qquad z = \frac{D_z}{D}$$

EXAMPLE 11 Using Cramer's Rule with a System of Three Linear Equations

Find the solution to the system of linear equations using Cramer's Rule. $\begin{cases} x + 2z = -12 \\ 3y - z = 14 \\ 5x - 8y = -34 \end{cases}$

SOLUTION

REMEMBER

When a variable does not appear in one of the equations, its coefficient is 0.

REMEMBER

When choosing a row or column to expand around, the one with the most 0s will have the simplest solution. In this case, there is not a row or column with more than one 0, so any one can be used.

Write the determinant of the coefficient matrix D.

$$D = \begin{vmatrix} 1 & 0 & 2 \\ 0 & 3 & -1 \\ 5 & -8 & 0 \end{vmatrix}$$

Find the determinant D using expansion about row 1.

$$D = \begin{vmatrix} 1 & 0 & 2 \\ 0 & 3 & -1 \\ 5 & -8 & 0 \end{vmatrix} = (1)(1)(-8) + (0)(-1)(5) + (2)(1)(-15) = -8 - 30 = -38$$

The value of D is nonzero, therefore the system has a unique solution which can be found using Cramer's Rule.

Solve for D_x, D_y, and D_z using expansion. In each case, the matrix is expanded around row 1.

$$D_x = \begin{vmatrix} -12 & 0 & 2 \\ 14 & 3 & -1 \\ -34 & -8 & 0 \end{vmatrix} = (-12)(1)(-8) + (0)(-1)(-34) + (2)(1)(-10) = 96 - 20 = 76$$

$$D_y = \begin{vmatrix} 1 & -12 & 2 \\ 0 & 14 & -1 \\ 5 & -34 & 0 \end{vmatrix} = (1)(1)(-34) + (-12)(-1)(5) + (2)(1)(-70) = -34 + 60 - 140 = -114$$

$$D_z = \begin{vmatrix} 1 & 0 & -12 \\ 0 & 3 & 14 \\ 5 & -8 & -34 \end{vmatrix} = (1)(1)(10) + (0)(-1)(-70) + (-12)(1)(-15) = 10 + 180 = 190$$

Use the determinants to find the solution to the system of linear equations.

$$x = \frac{D_x}{D} = \frac{76}{-38} = -2 \qquad y = \frac{D_y}{D} = \frac{-114}{-38} = 3 \qquad z = \frac{D_z}{D} = \frac{190}{-38} = -5$$

The solution to the system of linear equations is $(-2, 3, -5)$.

SECTION 9.3 EXERCISES

Warm Up

Write the augmented matrix for each system of equations.

1. $\begin{cases} -3x - 5y = 2 \\ x + y = 15 \end{cases}$

2. $\begin{cases} -2x + y = 7 \\ 2x + 8y = -20 \end{cases}$

3. $\begin{cases} -3x + y + 2z = 7 \\ x + 9z = 2 \\ -5y - z = 8 \end{cases}$

Just the Facts

Fill in the blanks.

4. A 2 × 2 matrix $\begin{bmatrix} a & b \\ c & d \end{bmatrix}$ is _____ if ad = _____.

5. The determinant of a matrix in _____ form can be computed by multiplying the entries on the diagonal of the matrix.

6. The only elementary row operation that does not change the value of the determinant is _____ one row to another.

7. If an elementary row operation is applied to a matrix with a determinant equal to 2 and the resulting matrix has a determinant of 4, then a row of the matrix was _____.

8. If A is a 2 × 2 matrix and $|A| \neq 0$, then the matrix is _____.

9. The formula for finding the area of a polygon only gives a positive result if the _____ are input in _____ order.

10. Cramer's Rule can be used to solve a _____.

Essential Skills

In Exercises 1–2, find the determinant of each matrix.

1. $B = \begin{bmatrix} -2 & -5 \\ -3 & 7 \end{bmatrix}$

2. $A = \begin{bmatrix} -\frac{2}{3} & \frac{1}{4} \\ \frac{2}{3} & \frac{1}{2} \end{bmatrix}$

In Exercises 3–4, expand along the given row to find the determinant of each matrix.

3. Row 1; $E = \begin{bmatrix} 1 & 3 & 4 \\ -2 & 6 & 8 \\ 0 & 5 & -4 \end{bmatrix}$

4. Row 3; $A = \begin{bmatrix} 10 & 1 & 7 \\ -3 & 8 & 3 \\ 2 & 6 & 9 \end{bmatrix}$

In Exercises 5–6, use row operations to find the determinant of each matrix.

5. $A = \begin{bmatrix} 3 & -4 & -2 \\ 1 & -2 & 5 \\ -1 & 1 & 2 \end{bmatrix}$

6. $D = \begin{bmatrix} 1 & 2 & 1 \\ 2 & 2 & 0 \\ -1 & 0 & -1 \end{bmatrix}$

In Exercises 7–8, using matrices, find the area of each polygon.

7. a triangle with vertices (–3, 0), (2, 1), and (–4, 4)

8. a quadrilateral with the following vertices listed counterclockwise around the perimeter (–2, –1), (–5, –4), (–2, –7), and (1, –4)

In Exercises 9–10, use determinants to determine if the points are collinear.

9. (4, –5), (7, 0), and (5, –1)

10. (0, 2), (–3, 4), and (2, 2/3)

In Exercises 11–14, use Cramer's Rule to solve each system of equations. If $D = 0$, use another method to solve.

11. $\begin{cases} 2x - y = -5 \\ 2x + 2y = 16 \end{cases}$

12. $\begin{cases} 5x + 2y = -3 \\ -y - 3x = 1 \end{cases}$

13. $\begin{cases} 3x + 5y = 27 \\ 2x + 7z = 22 \\ 4y + 3z = 18 \end{cases}$

14. $\begin{cases} x + y + 2z = 3 \\ x + y - 2z = -5 \\ x - 2y + z = 7 \end{cases}$

Extensions

15. Use determinants to determine if the points $\left(3, \frac{2}{3}\right)$, $\left(1, -\frac{2}{3}\right)$, and $\left(-4, -\frac{8}{3}\right)$ are collinear.

16. Using matrices, find the area of the pentagon with vertices (0, 2), (−2, −1), (0, −2), (1, −2), and (3, 0).

17. Given the following system, find the value of D_z as used in Cramer's Rule.

$$\begin{cases} 5x + 2y - z = -3 \\ x - y - 3z = 1 \\ x + 2z = 0 \end{cases}$$

18. Find the determinant of matrix B using row operations.

$$B = \begin{bmatrix} \frac{2}{3} & \frac{4}{3} & 1 \\ 0 & \frac{5}{3} & -1 \\ \frac{4}{3} & 1 & -3 \end{bmatrix}$$

19. Using $J = \begin{bmatrix} 1 & -8 & 6 \\ 7 & -8 & 2 \\ 0 & 4 & 5 \end{bmatrix}$, a student expanded around row 2 of matrix J to find the determinant and got $(7)(-1)(-40 - 24) + (-8)(1)(5 - 0) + (2)(-1)(4 - 8)$. Where is the error in her computation?

9.4 INVERSES OF MATRICES

OBJECTIVES

- Find inverses of 2 × 2 matrices.
- Find inverses of 2 × 2 matrices using the general method.
- Find inverses of 3 × 3 matrices using the general method.
- Find inverses of 3 × 3 matrices using row operations.
- Solve systems of linear equations using inverse matrices.

PREREQUISITE VOCABULARY TERMS

augmented matrix
linear system of equations
singular matrix
square matrix

9.4.1 Finding the Inverse of a 2 × 2 Matrix

The Identity Matrix

The square-matrix equivalent of the number 1 is the **identity matrix *I***. Multiplying the $n \times n$ identity matrix I_n by any matrix of the same dimensions results in the original matrix, just as multiplying any number by 1 gives that number. So, if A is a square matrix, then $A \cdot I = A = I \cdot A$, where I has the same dimensions as A.

The identity matrix is a square matrix where each of the elements along the **main diagonal** are 1, and all other elements are 0.

$$I_3 = \begin{bmatrix} 1 & 0 & 0 \\ 0 & 1 & 0 \\ 0 & 0 & 1 \end{bmatrix}$$

I_3 is the identity matrix for a 3 × 3 matrix.

Inverse Matrices

REMEMBER

Matrix A is nonsingular, if its determinant is not equal to 0.

It can be shown that any nonsingular square matrix has an **inverse matrix**. The inverse matrix of the nonsingular square matrix A is denoted A^{-1} (read as "A inverse"). The product of a matrix and its inverse is the identity matrix, $A \cdot A^{-1} = I$. A singular matrix (i.e., a matrix where its determinant is 0) does not have an inverse matrix.

DEFINITION

The **main diagonal** of a matrix is the diagonal running from left to right that begins at element E_{11}; it is the set of elements E_{ij} for which $i = j$.

The **identity matrix *I*** is a square matrix with the main diagonal elements equal to 1 and all other elements equal to 0.

The **inverse matrix** A^{-1} of nonsingular square matrix A is the matrix with the same dimensions as A such that $A \cdot A^{-1} = I = A^{-1} \cdot A$.

IMPORTANT

If matrix A is singular (i.e., its determinant is 0), then A^{-1} does not exist.

The procedure for finding the inverse of a nonsingular 2 × 2 matrix is relatively simple.

The Inverse of a Nonsingular 2 × 2 Matrix

If $A = \begin{bmatrix} a & b \\ c & d \end{bmatrix}$ and $\det(A) \neq 0$, then $A^{-1} = \dfrac{1}{|A|}\begin{bmatrix} d & -b \\ -c & a \end{bmatrix}$.

EXAMPLE 1 Finding and Verifying the Inverse of a 2 × 2 Matrix

If $A = \begin{bmatrix} 2 & 1 \\ 3 & -2 \end{bmatrix}$, find A^{-1}. Verify that the result is the inverse.

SOLUTION

Find the determinant of A. $\det(A) = (2)(-2) - (1)(3) = -7$

Use the determinant of A to find the inverse of A. $A^{-1} = -\frac{1}{7}\begin{bmatrix} -2 & -1 \\ -3 & 2 \end{bmatrix} = \begin{bmatrix} \frac{2}{7} & \frac{1}{7} \\ \frac{3}{7} & -\frac{2}{7} \end{bmatrix}$

> **TIP**
>
> *From matrix A,*
> $a = 2$,
> $b = 1$,
> $c = 3$, *and*
> $d = -2$.
> *Therefore,*
> $\begin{bmatrix} d & -b \\ -c & a \end{bmatrix} = \begin{bmatrix} -2 & -1 \\ -3 & 2 \end{bmatrix}$.

CHECK

Confirm that $AA^{-1} = I$ to verify that A^{-1} is the inverse of A.

$$AA^{-1} = \begin{bmatrix} 2 & 1 \\ 3 & -2 \end{bmatrix}\begin{bmatrix} \frac{2}{7} & \frac{1}{7} \\ \frac{3}{7} & -\frac{2}{7} \end{bmatrix} = \begin{bmatrix} \frac{4}{7} + \frac{3}{7} & \frac{2}{7} + \left(-\frac{2}{7}\right) \\ \frac{6}{7} + \left(-\frac{6}{7}\right) & \frac{3}{7} + \frac{4}{7} \end{bmatrix} = \begin{bmatrix} 1 & 0 \\ 0 & 1 \end{bmatrix} \checkmark$$

WHAT'S THE BIG IDEA?

Demonstrate algebraically that $AA^{-1} = I$ for $A = \begin{bmatrix} a & b \\ c & d \end{bmatrix}$.

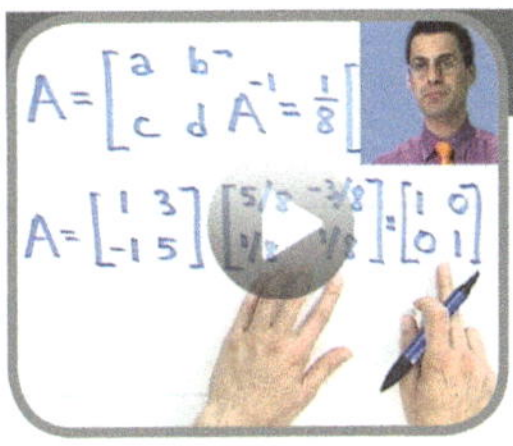

9.4.2 Finding the Inverse of a 2 × 2 Matrix: Another Example

There is a general method for finding the inverse of a matrix that can be applied to square nonsingular 2 × 2 matrices, as well as square nonsingular matrices with dimensions higher than 2 × 2. The method utilizes determinants, minors, and cofactors.

Recall from Section 9.3 that a cofactor C_{ij} is the factor $(-1)^{i+j}$ times the minor M_{ij}, and the minor M_{ij} is the determinant of the matrix formed by ignoring the ith row and the jth column.

Steps for Using Cofactors to Find the Inverse of a Matrix A

❶ Find the determinant of matrix A.

❷ Construct a matrix N with elements C_{ij}, where $C_{ij} = (-1)^{i+j}M_{ij}$.

❸ Reflect the elements in matrix N about the main diagonal.

❹ Multiply the matrix from step ❸ by the scalar $\frac{1}{\det(A)}$, resulting in A^{-1}.

EXAMPLE 2 Finding and Verifying the Inverse of a 2 × 2 Matrix

If $A = \begin{bmatrix} 8 & -3 \\ -4 & 2 \end{bmatrix}$, find A^{-1}. Verify that the result is the inverse.

SOLUTION

❶ Find the determinant of A. $\det(A) = (8)(2) - (-3)(-4) = 4$

❷ Construct a matrix N with elements C_{ij}. $N = \begin{bmatrix} (-1)^2(2) & (-1)^3(-4) \\ (-1)^3(-3) & (-1)^4(8) \end{bmatrix} = \begin{bmatrix} 2 & 4 \\ 3 & 8 \end{bmatrix}$

❸ Reflect the elements in matrix N about the main diagonal. $\begin{bmatrix} 2 & 3 \\ 4 & 8 \end{bmatrix}$

❹ Multiply the matrix from step ❸ by $\frac{1}{4}$. $A^{-1} = \frac{1}{4}\begin{bmatrix} 2 & 3 \\ 4 & 8 \end{bmatrix} = \begin{bmatrix} \frac{1}{2} & \frac{3}{4} \\ 1 & 2 \end{bmatrix}$

CHECK

Confirm that $AA^{-1} = I$ to verify that A^{-1} is the inverse of A.

$$AA^{-1} = \begin{bmatrix} 8 & -3 \\ -4 & 2 \end{bmatrix}\begin{bmatrix} \frac{1}{2} & \frac{3}{4} \\ 1 & 2 \end{bmatrix} = \begin{bmatrix} 4+-3 & 6+-6 \\ -2+2 & -3+4 \end{bmatrix} = \begin{bmatrix} 1 & 0 \\ 0 & 1 \end{bmatrix} \checkmark$$

REMEMBER

In the case of a 2 × 2 matrix, the minors M_{ij} are simply the values of the individual elements remaining when the ith row and jth column are removed.

IMPORTANT

AA^{-1} is the identity matrix I_2, therefore A^{-1} is the inverse of A.

9.4.3 Finding the Inverse of an $n \times n$ Matrix

The method for finding the inverse of a square matrix that was demonstrated in the previous topic is applied to a 3 × 3 matrix in the following example.

EXAMPLE 3 Finding the Inverse of a 3 × 3 Matrix

If $A = \begin{bmatrix} 2 & -3 & 4 \\ 2 & -1 & 0 \\ 4 & -2 & 3 \end{bmatrix}$, find A^{-1}.

SOLUTION

❶ Find the determinant of the matrix. $|A| = 12$

❷ Construct a matrix N.

$$N = \begin{bmatrix} +\begin{vmatrix} -1 & 0 \\ -2 & 3 \end{vmatrix} & -\begin{vmatrix} 2 & 0 \\ 4 & 3 \end{vmatrix} & +\begin{vmatrix} 2 & -1 \\ 4 & -2 \end{vmatrix} \\ -\begin{vmatrix} -3 & 4 \\ -2 & 3 \end{vmatrix} & +\begin{vmatrix} 2 & 4 \\ 4 & 3 \end{vmatrix} & -\begin{vmatrix} 2 & -3 \\ 4 & -2 \end{vmatrix} \\ +\begin{vmatrix} -3 & 4 \\ -1 & 0 \end{vmatrix} & -\begin{vmatrix} 2 & 4 \\ 2 & 0 \end{vmatrix} & +\begin{vmatrix} 2 & -3 \\ 2 & -1 \end{vmatrix} \end{bmatrix} = \begin{bmatrix} -3 & -6 & 0 \\ 1 & -10 & -8 \\ 4 & 8 & 4 \end{bmatrix}$$

❸ Reflect the matrix N about the main diagonal.

$$\begin{bmatrix} -3 & 1 & 4 \\ -6 & -10 & 8 \\ 0 & -8 & 4 \end{bmatrix}$$

❹ Multiply the matrix from step ❸ by $\frac{1}{12}$.

$$A^{-1} = \frac{1}{12}\begin{bmatrix} -3 & 1 & 4 \\ -6 & -10 & 8 \\ 0 & -8 & 4 \end{bmatrix} = \begin{bmatrix} -\frac{1}{4} & \frac{1}{12} & \frac{1}{3} \\ -\frac{1}{2} & -\frac{5}{6} & \frac{2}{3} \\ 0 & -\frac{2}{3} & \frac{1}{3} \end{bmatrix}$$

TIP

Any method for finding the determinant of a 3 × 3 matrix can be used.

9.4.4 Finding the Inverse of an $n \times n$ Matrix Using Row Operations

Another approach to finding the inverse of an $n \times n$ nonsingular matrix A involves creating an $n \times 2n$ augmented matrix with A on the left side and the identity matrix on the right side.

> **Steps for Using Elementary Row Operations to Find the Inverse of a Matrix A**
>
> ❶ Construct the augmented matrix $[A : I]$.
>
> ❷ Use row operations to reduce A in $[A : I]$ to I. The resulting augmented matrix is in the form $[I : A^{-1}]$ (where A^{-1} is the inverse of A).
>
> ❸ Write the inverse matrix A^{-1} from the result of step ❷.

EXAMPLE 4 Finding the Inverse of a 3×3 Matrix Using Row Operations

If $A = \begin{bmatrix} 2 & -3 & 4 \\ 2 & -1 & 0 \\ 4 & -2 & 3 \end{bmatrix}$, find A^{-1}.

SOLUTION

❶ Create the augmented matrix.

$$\left[\begin{array}{ccc|ccc} 2 & -3 & 4 & 1 & 0 & 0 \\ 2 & -1 & 0 & 0 & 1 & 0 \\ 4 & -2 & 3 & 0 & 0 & 1 \end{array}\right]$$

❷ Use row operations to reduce A in $[A : I]$ to I.

$$\begin{array}{r} \\ -R_1 + R_2 \\ -2R_2 + R_3 \end{array} \left[\begin{array}{ccc|ccc} 2 & -3 & 4 & 1 & 0 & 0 \\ 0 & 2 & -4 & -1 & 1 & 0 \\ 0 & 0 & 3 & 0 & -2 & 1 \end{array}\right]$$

Get 0s in A_{21}, A_{31}, and A_{32}.

$$\begin{array}{r} R_1 + R_2 \\ \frac{1}{2}R_2 \\ \frac{1}{3}R_3 \end{array} \left[\begin{array}{ccc|ccc} 2 & -1 & 0 & 0 & 1 & 0 \\ 0 & 1 & -2 & -\frac{1}{2} & \frac{1}{2} & 0 \\ 0 & 0 & 1 & 0 & -\frac{2}{3} & \frac{1}{3} \end{array}\right]$$

Get 0 in A_{13}.
Get 1s in A_{22} and A_{33}.

$$\begin{array}{r} R_1 + R_2 \\ \\ \\ \end{array} \left[\begin{array}{ccc|ccc} 2 & 0 & -2 & -\frac{1}{2} & \frac{3}{2} & 0 \\ 0 & 1 & -2 & -\frac{1}{2} & \frac{1}{2} & 0 \\ 0 & 0 & 1 & 0 & -\frac{2}{3} & \frac{1}{3} \end{array}\right]$$

Get 0 in A_{12}.

$$\begin{array}{r} R_1 + 2R_3 \\ R_2 + 2R_3 \\ \\ \end{array} \left[\begin{array}{ccc|ccc} 2 & 0 & 0 & -\frac{1}{2} & \frac{1}{6} & \frac{2}{3} \\ 0 & 1 & 0 & -\frac{1}{2} & -\frac{5}{6} & \frac{2}{3} \\ 0 & 0 & 1 & 0 & -\frac{2}{3} & \frac{1}{3} \end{array}\right]$$

Get 0s in A_{13} and A_{23}.

$$\begin{array}{r} \frac{1}{2}R_1 \\ \\ \\ \end{array} \left[\begin{array}{ccc|ccc} 1 & 0 & 0 & -\frac{1}{4} & \frac{1}{12} & \frac{1}{3} \\ 0 & 1 & 0 & -\frac{1}{2} & -\frac{5}{6} & \frac{2}{3} \\ 0 & 0 & 1 & 0 & -\frac{2}{3} & \frac{1}{3} \end{array}\right]$$

Get 1 in A_{11}.

❸ Therefore, $A^{-1} = \begin{bmatrix} -\frac{1}{4} & \frac{1}{12} & \frac{1}{3} \\ -\frac{1}{2} & -\frac{5}{6} & \frac{2}{3} \\ 0 & -\frac{2}{3} & \frac{1}{3} \end{bmatrix}$.

IMPORTANT

The left side of the augmented matrix is the identity matrix I_3, so the process is complete and the right side is the inverse.

9.4.5 Solving a System of Equations with Inverses

A system of linear equations can be represented as a matrix equation, as shown here.

$$\begin{cases} a_1x + b_1y = c_1 \\ a_2x + b_2y = c_2 \end{cases} \Rightarrow \begin{bmatrix} a_1 & b_1 \\ a_2 & b_2 \end{bmatrix}\begin{bmatrix} x \\ y \end{bmatrix} = \begin{bmatrix} c_1 \\ c_2 \end{bmatrix}$$

Matrix inverses can be used to solve a matrix equation.

Let the coefficient matrix be A. $$A\begin{bmatrix} x \\ y \end{bmatrix} = \begin{bmatrix} c_1 \\ c_2 \end{bmatrix}$$

Multiply both sides of the equation by A^{-1} and simplify. $$A^{-1}A\begin{bmatrix} x \\ y \end{bmatrix} = A^{-1}\begin{bmatrix} c_1 \\ c_2 \end{bmatrix}$$

$$I\begin{bmatrix} x \\ y \end{bmatrix} = A^{-1}\begin{bmatrix} c_1 \\ c_2 \end{bmatrix}$$

Therefore, the solution to the system of linear equations is the product of the inverse of the coefficient matrix and the matrix of constant terms. $$\begin{bmatrix} x \\ y \end{bmatrix} = A^{-1}\begin{bmatrix} c_1 \\ c_2 \end{bmatrix}$$

EXAMPLE 5 **Solving a System of Linear Equations Using Inverse Matrices**

Solve the system of equations using the inverse of the coefficient matrix. $\begin{cases} -x - 3y = -2 \\ 4x + 5y = 8 \end{cases}$

SOLUTION

Write the coefficient matrix from the system. $A = \begin{bmatrix} -1 & -3 \\ 4 & 5 \end{bmatrix}$

Write the matrix equation using the constants from the system (−2 and 8). $\begin{bmatrix} x \\ y \end{bmatrix} = A^{-1}\begin{bmatrix} -2 \\ 8 \end{bmatrix}$

$|A| = (-1)(5) - (-3)(4) = 7$ ▶

Find A^{-1} using $A^{-1} = \frac{1}{|A|}\begin{bmatrix} d & -b \\ -c & a \end{bmatrix}$. $$A^{-1} = \frac{1}{7}\begin{bmatrix} 5 & 3 \\ -4 & -1 \end{bmatrix} = \begin{bmatrix} \frac{5}{7} & \frac{3}{7} \\ -\frac{4}{7} & -\frac{1}{7} \end{bmatrix}$$

Substitute A^{-1} into the matrix equation and solve. $$\begin{bmatrix} x \\ y \end{bmatrix} = \begin{bmatrix} \frac{5}{7} & \frac{3}{7} \\ -\frac{4}{7} & -\frac{1}{7} \end{bmatrix}\begin{bmatrix} -2 \\ 8 \end{bmatrix} = \begin{bmatrix} 2 \\ 0 \end{bmatrix}$$

The solution to the system of linear equations is (2, 0).

TIP

Substitute $x = 2$ and $y = 0$ into each of the original equations to check the system's solution.

SECTION 9.4 EXERCISES

Warm Up

Find the determinant of each matrix.

1. $\begin{bmatrix} 0 & 4 \\ -3 & 2 \end{bmatrix}$

2. $\begin{bmatrix} -5 & 1 \\ 2 & 3 \end{bmatrix}$

3. $\begin{bmatrix} 3 & -8 \\ 0 & 7 \end{bmatrix}$

Just the Facts

Fill in the blanks.

4. If A^{-1} exists, then det(A) _____.

5. If a square matrix B has no inverse, then det(B) must be _____.

6. The cofactor C_{ij} is _____ by E_{ij} in the process of finding the inverse of a matrix A.

7. There is a simple formula for calculating the inverse of a(n) _____ matrix.

8. To calculate the inverse of square matrix F, the _____ matrix $[F : I]$ can be used, where I represents the _____ matrix.

9. The product of the inverse of the _____ matrix and the matrix of the _____ terms gives the solution to a system of linear equations.

10. The final step in finding the inverse of a matrix is to _____ by the _____ of the original matrix.

Essential Skills

In Exercises 1–8, find the inverse of each matrix, if it exists.

1. $B = \begin{bmatrix} -3 & 1 \\ 5 & -2 \end{bmatrix}$

2. $A = \begin{bmatrix} 2 & -1 \\ 4 & 3 \end{bmatrix}$

3. $C = \begin{bmatrix} 4 & 1 \\ 5 & 2 \end{bmatrix}$

4. $A = \begin{bmatrix} 12 & 6 \\ -4 & -10 \end{bmatrix}$

5. $M = \begin{bmatrix} 1 & 1 & 1 \\ 2 & 1 & -1 \\ 1 & -1 & 1 \end{bmatrix}$

6. $A = \begin{bmatrix} -4 & 2 & -3 \\ -2 & 3 & 1 \\ 4 & 0 & -1 \end{bmatrix}$

7. $E = \begin{bmatrix} 3 & 1 & 3 \\ 1 & 2 & -1 \\ 2 & -1 & 3 \end{bmatrix}$

8. $A = \begin{bmatrix} 5 & 2 & 1 \\ 2 & -1 & 1 \\ 4 & 3 & 2 \end{bmatrix}$

In Exercises 9–10, use the inverse of the coefficient matrix to find the solution to the system of linear equations.

9. $\begin{cases} -x - 3y = -2 \\ 4x + 5y = 8 \end{cases}$

10. $\begin{cases} \dfrac{x}{2} + y = -4 \\ 2x - y = 9 \end{cases}$

Extensions

11. Write the coefficient matrix and the constant matrix for the following system. Solve the system by using the inverse of the coefficient matrix. $\begin{cases} \frac{1}{2}y + 2x + 6 = 15 \\ 3x + 5y - 50 = 6 \end{cases}$

12. A group of 26 friends is going on a road trip to visit national parks. They are taking 5 vehicles: compact cars that seat 4 people and minivans that seat 7 people.

A. Write a system of equations that represents this situation. Let x be the number of compact cars and y be the number of minivans.

B. Write the coefficient matrix.

C. Write the constant matrix.

D. Find the inverse of the coefficient matrix.

E. Use the inverse of the coefficient matrix to find how many of each vehicle type the group took on this trip.

13. A game show host says that he has \$10,000 in \$100 bills and \$20 bills, and he will give you the \$10,000 if you can tell him how many of each type of bill he has. You get a hint that there are 300 bills total. Use an inverse matrix to find how many bills of each type he has.

14. For what values of a, b, c, and d will matrix $\begin{bmatrix} a & b \\ c & d \end{bmatrix}$ be its own inverse?

15. Find the missing element of matrix $F = \begin{bmatrix} 5 & 10 \\ 6 & ? \end{bmatrix}$ so that it has no inverse.

16. The Texas Country Store is selling gift packages containing Texas pecans, walnuts, and chocolate candy. The small package contains $\frac{1}{2}$ pound of each and costs \$7. The medium package has 2 pounds of pecans, 1 pound of walnuts, and 1 pound of candy, and is priced at \$18.50. Their large package contains 3 pounds of pecans, 2 pounds of walnuts, and 3 pounds of candy for \$38. Write the matrix equation and solve to find the cost per pound of the pecans, walnuts, and candy.

CHAPTER 9 REVIEW EXERCISES

Section 9.1 Matrices and Systems of Equations

1. Write the augmented matrix for the following system of linear equations.

$$\begin{cases} 3x - 2y + 5z = 30 \\ 5x + 4z = 35 \\ 3y - 2z = -4 \end{cases}$$

2. Write the resulting matrix after multiplying R_2 by -1 and interchanging R_1 and R_3.

$$\begin{bmatrix} -4 & -7 & 0 & 1 \\ 3 & -5 & 6 & 2 \\ 0 & 2 & 8 & -9 \end{bmatrix}$$

In Exercises 3–5, use Gauss-Jordan elimination or Gaussian elimination to solve each system.

3. $\begin{cases} 3x + y + z = 20 \\ -2x - y + z = -7 \\ x + 2y + 3z = 19 \end{cases}$

4. $\begin{cases} 2x - 2y + z = 1 \\ 2x - y + 2z = 10 \\ 2x + z = 5 \end{cases}$

5. $\begin{cases} -x + 3y + 2z = -3 \\ x - y + 2z = 7 \\ -3x + y + 5z = -10 \end{cases}$

Section 9.2 Operations with Matrices

6. Find m, n, and p given that $X = Y$,

$$X = \begin{bmatrix} -2 & m & mp \\ 0 & 7 & 6 \end{bmatrix}, \text{ and } Y = \begin{bmatrix} -2 & -2 & 5 \\ n+m & 7 & 6 \end{bmatrix}.$$

7. If possible, find $F + X$ and $F - X$ given that

$$F = \begin{bmatrix} -2 & 7 & 1 \\ -5 & 0 & 9 \\ 3 & -1 & 12 \end{bmatrix} \text{ and } X = \begin{bmatrix} 1 & 0 & -8 \\ 4 & -4 & 7 \\ 2 & -1 & 5 \end{bmatrix}.$$

8. Solve $N - 2P = M$ for P, given that

$$M = \begin{bmatrix} -4 & 0 & 7 \\ 8 & 1 & 9 \end{bmatrix} \text{ and } N = \begin{bmatrix} 1 & -5 & 6 \\ 0 & 2 & -3 \end{bmatrix}.$$

9. State whether the products CD and DC are defined, and if so, give the dimensions of the product matrix.

$$C = \begin{bmatrix} 3 & 6 \\ -4 & 1 \end{bmatrix} \text{ and } D = \begin{bmatrix} 0 & 1 & -8 & 10 \\ 6 & 0 & 5 & 9 \end{bmatrix}$$

10. Find SR, if possible, given the following matrices.

$$R = \begin{bmatrix} -2 & 1 & 7 \\ 6 & 0 & 1 \end{bmatrix} \text{ and } S = \begin{bmatrix} 1 & 0 \\ 6 & 2 \\ -1 & -5 \\ 9 & 8 \end{bmatrix}$$

Section 9.3 Determinants and Cramer's Rule

11. Find the determinant of matrix K. $K = \begin{bmatrix} \frac{5}{2} & -\frac{3}{2} \\ \frac{1}{2} & -\frac{1}{3} \end{bmatrix}$

12. Find the determinant of matrix J using row operations.

$$J = \begin{bmatrix} 2 & -4 & -2 \\ 1 & -1 & 5 \\ -1 & 2 & 6 \end{bmatrix}$$

13. Expand along R_2 to find $|G|$. $G = \begin{bmatrix} -1 & 2 & -2 \\ 0 & 5 & 0 \\ 1 & 0 & 1 \end{bmatrix}$

14. Use matrices to find the area of the polygon with the following vertices (listed counterclockwise around the perimeter). (3, 8), (0, 5), (1, 3), (4, 3), (5, 5)

15. Use Cramer's Rule to solve the system of equations. If $D = 0$, use another method to solve.

$$\begin{cases} x - y + z = 4 \\ x - y - z = -6 \\ 3x + y - 2z = 3 \end{cases}$$

Section 9.4 Inverses of Matrices

In Exercises 16–19, find the inverse of each matrix, if it exists.

16. $Q = \begin{bmatrix} 5 & 1 \\ 3 & -1 \end{bmatrix}$

17. $W = \begin{bmatrix} -2 & 3 \\ 1 & -1 \end{bmatrix}$

18. $Z = \begin{bmatrix} -2 & 1 & 0 \\ 3 & -5 & 1 \\ 4 & -1 & 0 \end{bmatrix}$

19. $P = \begin{bmatrix} 0 & -1 & 2 \\ 3 & 7 & 0 \\ 1 & 6 & -4 \end{bmatrix}$

20. Use the inverse of the coefficient matrix to find the solution to the system of linear equations. $\begin{cases} -2x - 5y = -1 \\ 3x + 7y = 3 \end{cases}$

Analytic Geometry

Chapter

10

10.1 POLAR COORDINATES

OBJECTIVES

- Plot a point in the plane using polar coordinates.
- Find polar coordinates that identify a given point in the plane.
- Convert from polar coordinates to rectangular coordinates.
- Convert from rectangular coordinates to polar coordinates.

PREREQUISITE VOCABULARY TERMS

inverse trigonometric functions
Pythagorean Theorem
radians
trigonometric functions

10.1.1 An Introduction to Polar Coordinates

Recall that each point on a Cartesian plane can be identified by an ordered pair (x, y) where the x-coordinate represents the horizontal distance between the point and the origin, and the y-coordinate represents the vertical distance between the point and the origin (Figure 10.1a). The x- and y-coordinates of an ordered pair on a Cartesian plane are also called **rectangular coordinates**.

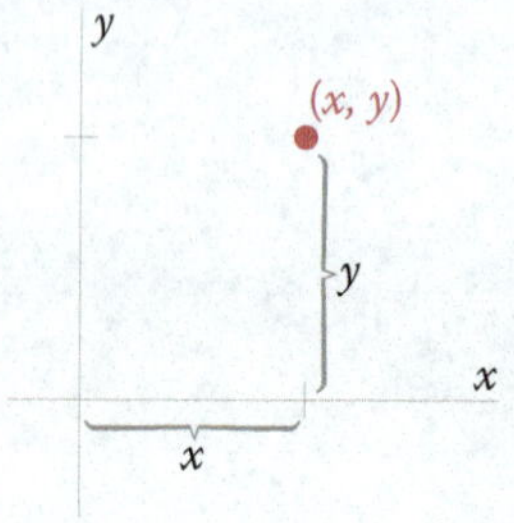

Figure 10.1a

The rectangular coordinates of a point (x, y) can be considered as directions for navigating from the origin to the point (x, y). For example, the directions for navigating to the point (2, 2) are to first move 2 units right from the origin, then 2 units up. However, there are other directions from the origin that will result in the same position in the plane. An alternative way to navigate to a point in the plane, and to identify a given point, will be introduced in this chapter.

The point with rectangular coordinates (2, 2) can also be reached by moving $2\sqrt{2}$ units from the origin on the ray that is the terminal side of an angle measuring $\pi/4$ radians (Figure 10.1b).

Thus, an alternative method for navigating to a point in the plane is given by moving a distance r from the origin on the terminal side of an angle θ, rotated from the positive x-axis.

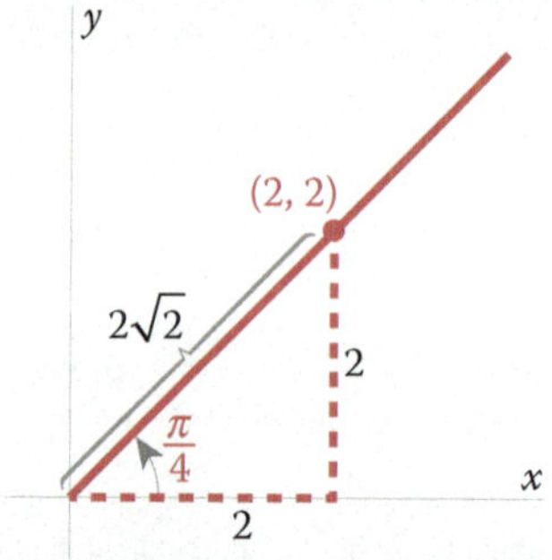

Figure 10.1b

TIP

The distance from the origin to the point (2, 2), $2\sqrt{2}$, can be found using the Pythagorean Theorem or the Distance Formula.

Polar Coordinates

If P is a point in the plane, then the **polar coordinates** of P are (r, θ), where $|r|$ is the distance from the origin to P, and θ is the angle between the positive x-axis and the ray through P.

The positive x-axis is also called the **polar axis**, and the origin is also called the **pole**.

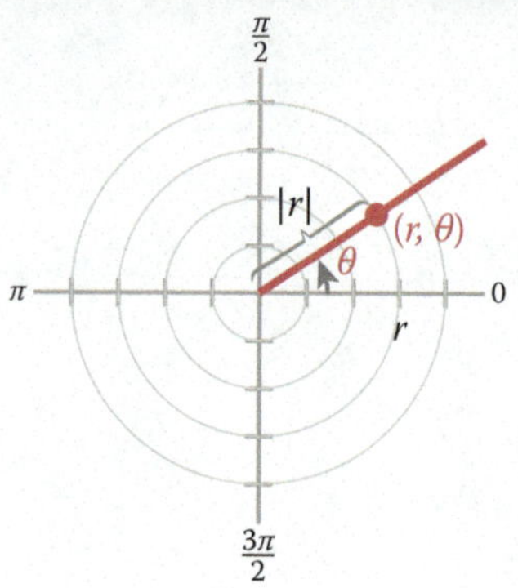

IMPORTANT

If $\theta > 0$, then the ray containing P is rotated counterclockwise from the polar axis. If $\theta < 0$, then the ray containing P is rotated clockwise from the polar axis.

Polar Coordinates When $r < 0$

For rectangular coordinates (x, y), r represents the distance from the origin to (x, y), so r is always positive. However, for polar coordinates (r, θ), r is allowed to be negative because r represents not only the distance from the pole to (r, θ), but also the direction from the pole to (r, θ).

If P is the point (r, θ) such that $r < 0$, then P is $|r|$ units from the origin and the ray containing P is in the direction opposite to the terminal side of θ (Figure 10.1c). Specifically, the ray containing P is rotated π radians (clockwise or counterclockwise) from θ.

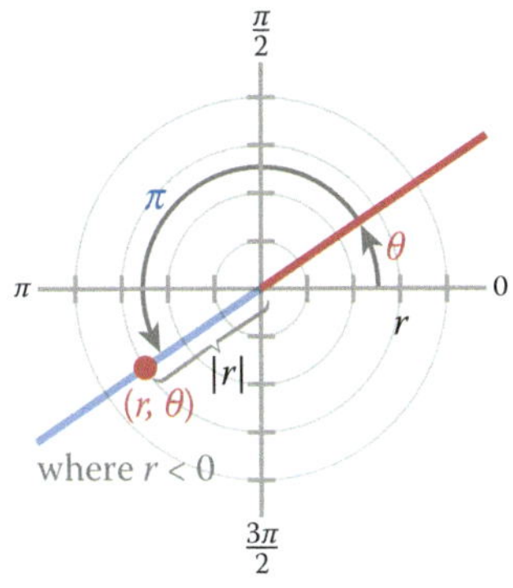

Figure 10.1c

NOTICE THAT

Rotating a ray by π radians in the clockwise direction results in a ray that is the same position as if the original ray was rotated by π radians in the counterclockwise direction.

EXAMPLE 1 Plotting Polar Coordinates

Plot each point using the given polar coordinates.

A. $P(3, \pi)$ **B.** $Q(-2, 3\pi/4)$ **C.** $R(-5, -\pi/6)$

SOLUTION

A. $P(3, \pi) \Rightarrow r = 3$ and $\theta = \pi$

P is 3 units from the origin on the ray rotated π radians counterclockwise from the polar axis.

So, P is located at -3 on the negative x-axis.

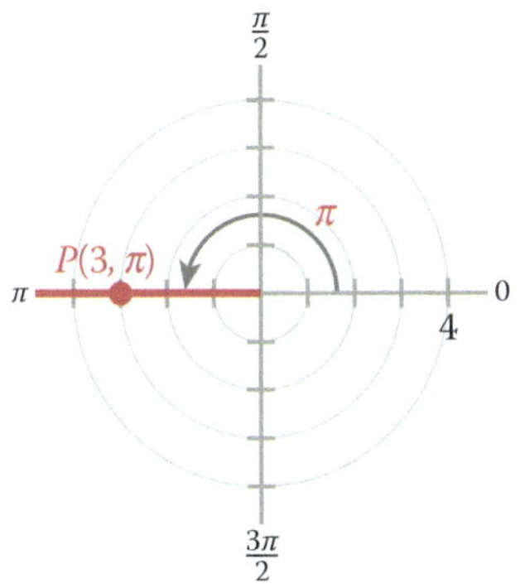

NOTICE THAT

P(3, π) is at the same location on the plane as the point (–3, 0). So, the rectangular coordinates of P are (–3, 0).

B. $Q(-2, 3\pi/4) \Rightarrow r = -2$ and $\theta = 3\pi/4$

Since $r < 0$, the ray containing Q is rotated $3\pi/4$ radians counterclockwise from the polar axis (in quadrant II), and then π radians counterclockwise. Thus, the ray containing Q is in quadrant IV.

Q is 2 units from the origin on this ray.

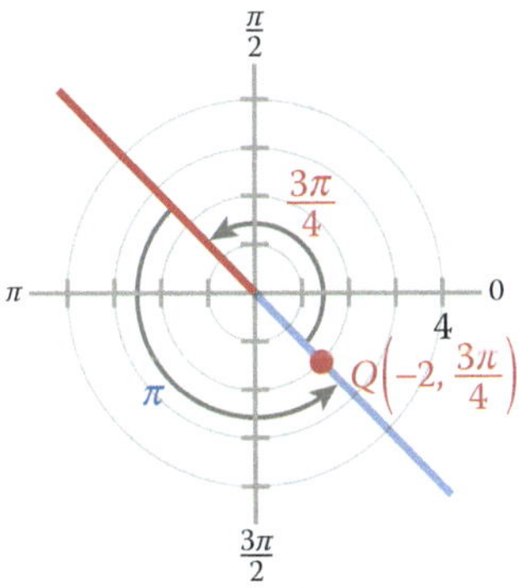

TIP

Any point 2 units from the origin (on any ray) is on a circle centered at the origin with radius 2.

C. $R(-5, -\pi/6) \Rightarrow r = -5$ and $\theta = -\pi/6$

The ray containing R is rotated $\pi/6$ radians clockwise from the polar axis (clockwise because $\theta < 0$), and then π radians clockwise (because $r < 0$). Thus, the ray containing R is in quadrant II.

R is 5 units from the origin on this ray.

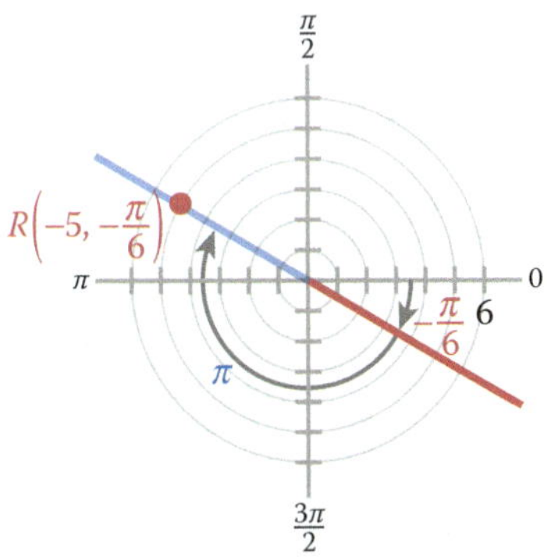

IMPORTANT

When r < 0, the ray is rotated θ + π or θ + π radians from the polar axis (i.e., π radians clockwise or counterclockwise from θ).

With rectangular coordinates, every point in the plane is represented by a unique ordered pair (x, y). However, this is not the case with polar coordinates. Each point can be represented by an infinite number of polar coordinates. For example, the polar coordinates (r, θ) and $(r, \theta + 2\pi n)$, where n is any integer, represent the same point in the plane because the angles θ and $\theta + 2\pi n$ have the same terminal side. Furthermore, the polar coordinates $(-r, \theta \pm \pi)$ also represent the same point in the plane as (r, θ) and $(r, \theta + 2\pi n)$.

EXAMPLE 2 Identifying Polar Coordinates from a Graph

Write two pairs of polar coordinates to represent each point.

A.

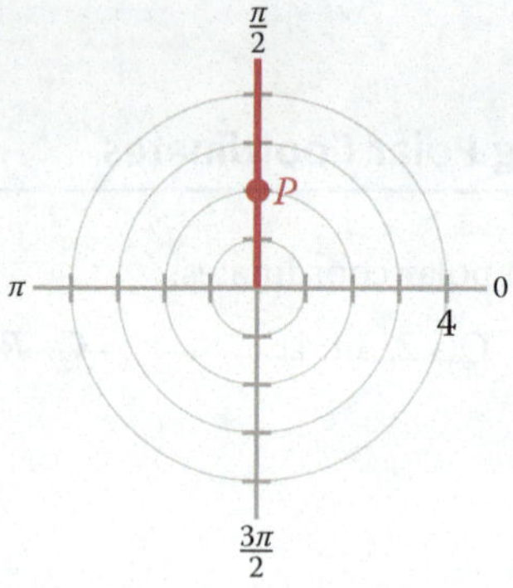

B.

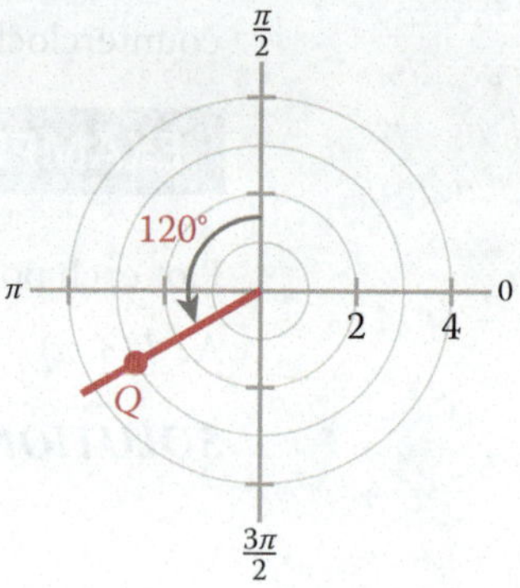

SOLUTION

> ***NOTICE THAT***
>
> *P is 2 units from the origin on the positive y-axis. So, the rectangular coordinates of P are (0, 2).*

A. The angle from the polar axis (i.e., the positive x-axis) to the ray containing P is $\pi/2$ radians, so $\theta = \pi/2$. P is 2 units from the origin, so $|r| = 2$. Thus, polar coordinates of P are $(2, \pi/2)$.

Alternatively, if $r = -2$, then the ray containing P can be rotated π radians from θ, so $\theta = 3\pi/2$. Thus, polar coordinates of P are also $(-2, 3\pi/2)$.

B. Since the point Q is located 120° counterclockwise from the positive y-axis, the angle from the polar axis to the ray containing Q is 210°, that is, $7\pi/6$ radians. So, $\theta = 7\pi/6$. Since Q is 3 units from the origin, $|r| = 3$. Thus, polar coordinates of Q are $(3, 7\pi/6)$.

> ***REMEMBER***
>
> *When $r < 0$, the rotation of π radians can be clockwise or counterclockwise.*

Alternatively, if $r = -3$, then the ray containing Q can be rotated π radians from θ, so $\theta = \pi/6$. Thus, polar coordinates of Q are also $(-3, \pi/6)$.

10.1.2 Converting between Polar and Rectangular Coordinates

We have seen that a point in the plane can be represented by a unique ordered pair of rectangular coordinates (x, y) or a pair of polar coordinates (r, θ). When the point is on an axis, then θ is a quadrantal angle and the polar coordinates, as well as the rectangular coordinates, are easily identified. For example, the point in Figure 10.1d has polar coordinates $(1, \pi/2)$ and rectangular coordinates $(0, 1)$. It is important to be able to convert between polar and rectangular coordinates even when the point is not on an axis.

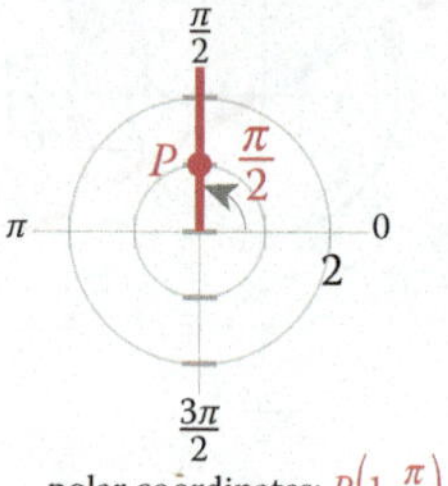

polar coordinates: $P\left(1, \frac{\pi}{2}\right)$
rectangular coordinates: $P(0, 1)$

Figure 10.1d

Converting from Polar to Rectangular Coordinates

Suppose P is a point in the plane with polar coordinates (r, θ) and rectangular coordinates (x, y). Trigonometric functions can be used to determine the x- and y-values when the r- and θ-values are known.

Recall that if (x, y) is a point on the terminal side of θ, then $\cos\theta = x/r$ and $\sin\theta = y/r$, where $r = \sqrt{x^2 + y^2}$. Solving the trigonometric functions for x and y respectively gives $x = r\cos\theta$ and $y = r\sin\theta$. Therefore, if the polar coordinates of a point P are (r, θ), then the rectangular coordinates of P are $(r\cos\theta, r\sin\theta)$.

IMPORTANT

If a point in the plane has polar coordinates (r, θ) and rectangular coordinates (x, y), then $x = r\cos\theta$ and $y = r\sin\theta$.

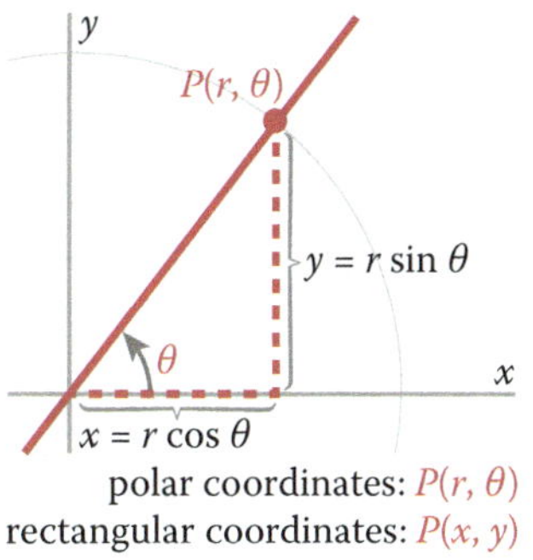

Figure 10.1e

EXAMPLE 3 Converting from Polar Coordinates to Rectangular Coordinates

Convert each pair of polar coordinates (r, θ) to rectangular coordinates (x, y).

A. $P(1, 5\pi/4)$ **B.** $Q(1.5, 2\pi/3)$

SOLUTION

Plot each point to visualize its location on the plane.

P is in quadrant III.

Q is in quadrant II.

Use $x = r\cos\theta$ and $y = r\sin\theta$ to find the rectangular coordinates (x, y) of P and Q.

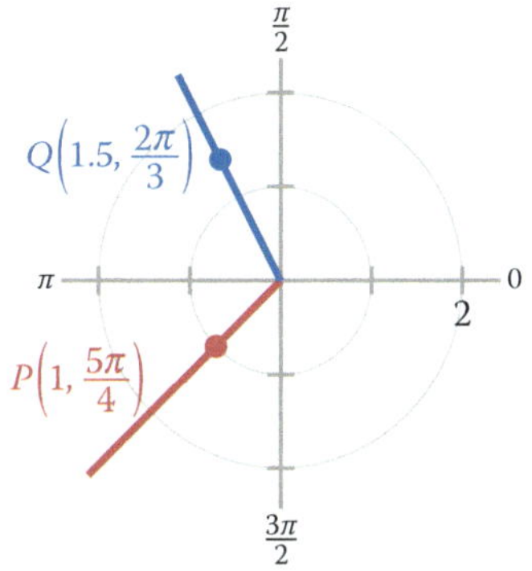

A. $P(1, 5\pi/4) \Rightarrow r = 1$ and $\theta = 5\pi/4$

$$x = r\cos\theta = (1)\cos\frac{5\pi}{4} = -\frac{\sqrt{2}}{2}$$

$$y = r\sin\theta = (1)\sin\frac{5\pi}{4} = -\frac{\sqrt{2}}{2}$$

B. $Q(1.5, 2\pi/3) \Rightarrow r = 1.5$ and $\theta = 2\pi/3$

$$x = r\cos\theta = (1.5)\cos\frac{2\pi}{3} = (1.5)\left(-\frac{1}{2}\right) = -\frac{3}{4}$$

$$y = r\sin\theta = (1.5)\sin\frac{2\pi}{3} = (1.5)\left(\frac{\sqrt{3}}{2}\right) = \frac{3\sqrt{3}}{4}$$

Therefore, the rectangular coordinates are $P\left(-\frac{\sqrt{2}}{2}, -\frac{\sqrt{2}}{2}\right)$ and $Q\left(-\frac{3}{4}, \frac{3\sqrt{3}}{4}\right)$.

CHECK

Find the location of each point on the plane using decimal approximations for the x- and y-coordinates.

Since $-\frac{\sqrt{2}}{2} \approx -0.7$, the approximate rectangular coordinates of P are $(-0.7, -0.7)$.

Since $\frac{3\sqrt{3}}{4} \approx 1.3$, the approximate rectangular coordinates of Q are $(-0.75, 1.3)$.

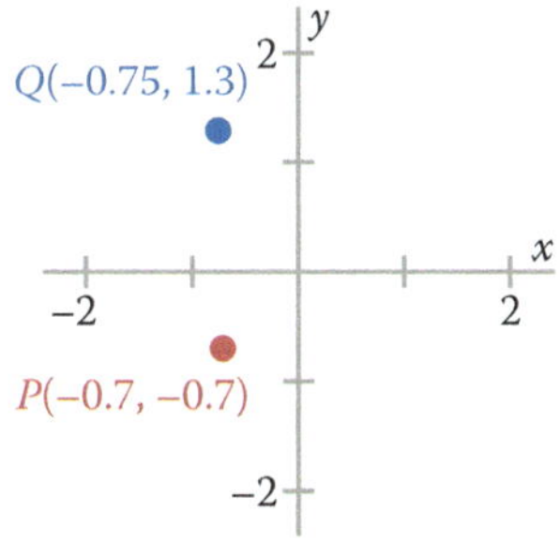

The rectangular coordinates of P and Q appear to correspond to the same positions on the plane as the polar coordinates of P and Q.

Converting from Rectangular to Polar Coordinates

A similar procedure can be used to convert from rectangular coordinates to polar coordinates. In this case, we use expressions for r and θ in terms of x and y.

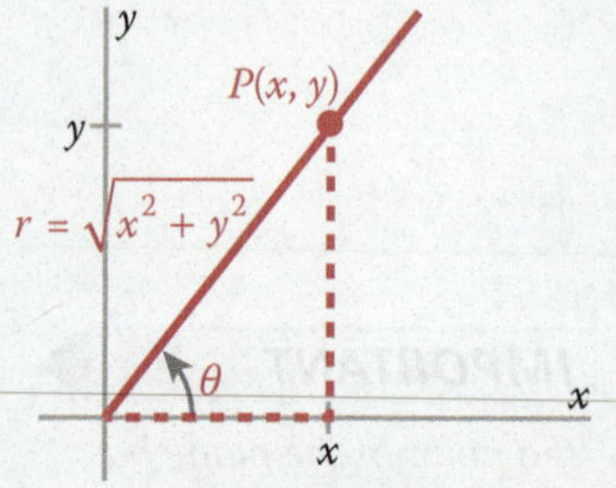

Figure 10.1f

IMPORTANT

The r given here is always positive. The ordered pair (r, θ) gives one possible representation of the point in terms of polar coordinates, however, there can always be other possible representations.

Again, recall that if (x, y) is a point on the terminal side of θ, then $\sin\theta = y/r$, $\cos\theta = x/r$, and $\tan\theta = y/x$, where $r = \sqrt{x^2 + y^2}$. So, the r-value in the polar coordinates (r, θ) of a point P with rectangular coordinates (x, y) is given by $r = \sqrt{x^2 + y^2}$.

Since $\tan\theta = y/x$, the θ-value in the polar coordinates (r, θ) of a point P with rectangular coordinates (x, y) is given by $\theta = \tan^{-1}(y/x)$, but only when θ is in the range of the inverse tangent function $(-\pi/2, \pi/2)$, that is, only when θ is in quadrant I or IV. If θ is in quadrant II or III, then $\tan^{-1}(y/x)$ can still be used to find θ, though in this case, θ will differ from $\tan^{-1}(y/x)$ by π, that is, $\theta = \tan^{-1}(y/x) \pm \pi$.

EXAMPLE 4 Converting from Rectangular Coordinates to Polar Coordinates

Convert each pair of rectangular coordinates (x, y) to polar coordinates (r, θ).

A. $P\left(\sqrt{3}, 3\right)$ **B.** $Q(-6, 6)$ **C.** $R(5, -3)$ **D.** $S(-9, -2)$

IMPORTANT

When a point is in quadrant II or III, θ is not in the range of the inverse tangent function.

SOLUTION

Use $r = \sqrt{x^2 + y^2}$ and $\tan^{-1}\dfrac{y}{x}$ to find the polar coordinates (r, θ) of P, Q, R, and S.

A. Since $x = \sqrt{3}$ and $y = 3$, $r = \sqrt{\sqrt{3}^2 + 3^2} = \sqrt{12} = 2\sqrt{3}$ and $\tan^{-1}\dfrac{3}{\sqrt{3}} = \tan^{-1}\sqrt{3} = \dfrac{\pi}{3}$.

P is in quadrant I, so $\theta = \pi/3$. Thus, the polar coordinates of P are $\left(2\sqrt{3}, \dfrac{\pi}{3}\right)$.

B. Since $x = -6$ and $y = 6$, $r = \sqrt{(-6)^2 + 6^2} = 6\sqrt{2}$ and $\tan^{-1}\dfrac{6}{-6} = \tan^{-1}(-1) = -\dfrac{\pi}{4}$.

Q is in quadrant II, so $\theta = -\pi/4 + \pi = 3\pi/4$. Thus, the polar coordinates of Q are $\left(6\sqrt{2}, \dfrac{3\pi}{4}\right)$.

IMPORTANT

Since −3/5 is not a multiple of π/6 or π/4, use a calculator to evaluate the inverse tangent function of −3/5.

C. Since $x = 5$ and $y = -3$, $r = \sqrt{5^2 + (-3)^2} = \sqrt{34}$ and $\tan^{-1}\dfrac{-3}{5} \approx -0.54$.

R is in quadrant IV, so $\theta \approx -0.54$.

Thus, the polar coordinates of R are approximately $\left(\sqrt{34}, -0.54\right)$.

D. Since $x = -9$ and $y = -2$, $r = \sqrt{(-9)^2 + (-2)^2} = \sqrt{85}$ and $\tan^{-1}\dfrac{-2}{-9} = \tan^{-1}\dfrac{2}{9} \approx 0.22$.

S is in quadrant III, so $\theta \approx 0.22 + \pi \approx 3.36$.

Thus, the polar coordinates of S are approximately $\left(\sqrt{85}, 3.36\right)$.

SECTION 10.1 EXERCISES

Warm Up

Determine the quadrant containing the terminal side of each angle.

1. $2\pi/3$
2. $-150°$
3. $-\pi/6$

Just the Facts

Fill in the blanks.

4. In general, if Q is a point in the plane, then the rectangular coordinates of Q are given by _____, and the polar coordinates of Q are given by _____.
5. The polar coordinates $(3, 5\pi/6)$ describe the point in the plane that is located _____ units from the origin on the ray rotated _____ radians counterclockwise from the _____, which, in rectangular coordinates, is called the positive x-axis.
6. To convert from polar coordinates to rectangular coordinates, use $r\cos\theta$ for _____ and $r\sin\theta$ for _____.
7. To convert from rectangular coordinates to polar coordinates, use $r =$ _____ and $\tan\theta =$_____.
8. If P is a point with polar coordinates (r, θ) and $r < 0$, then the ray containing P is in the _____ direction of θ.
9. Every point in the plane can be represented by a unique pair of _____ coordinates (x, y). When using _____ coordinates, each point can be represented by multiple pairs of coordinates such as (r, θ), $(-r, \theta \pm \pi)$, $(-r, \theta \pm 2\pi n)$, and _____.
10. When $\theta < 0$, the ray containing the point with polar coordinates (r, θ) is rotated _____ from the polar axis.

Essential Skills

In Exercises 1–8, plot each point using the given polar coordinates.

1. $(3, 2\pi)$
2. $(4, \pi/2)$
3. $(5, 7\pi/6)$
4. $(2, \pi/4)$
5. $(-3, 4\pi/3)$
6. $(-4, \pi/9)$
7. $(-1, -\pi/4)$
8. $(-6.5, -5\pi/6)$

In Exercises 9–14, write two pairs of polar coordinates (r, θ) to represent each point, where θ is in the interval $[0, 2\pi)$.

9.
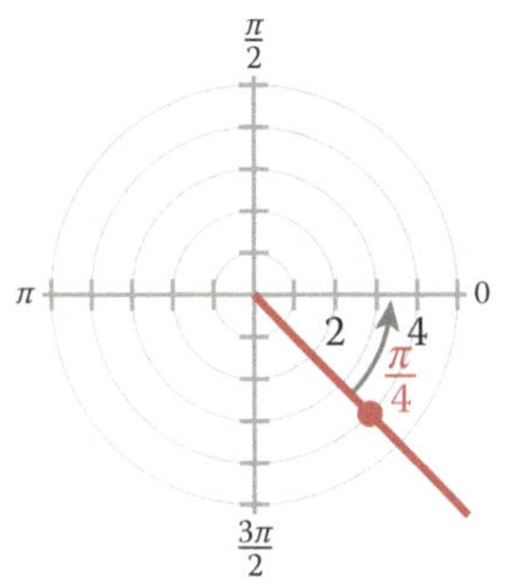

10.
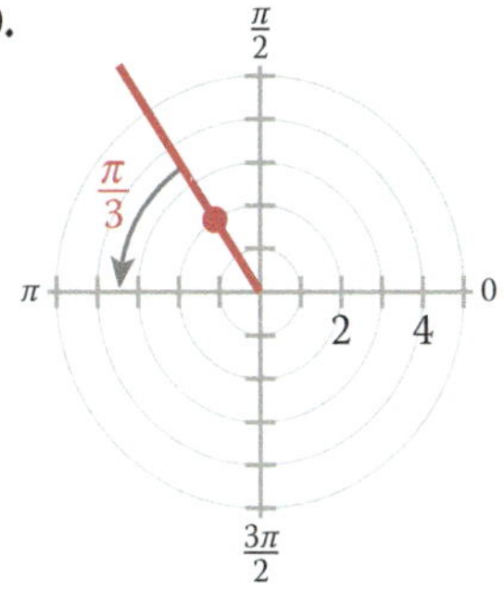

11. a point 2 units from the origin on a ray that forms an angle of $4\pi/3$ radians counterclockwise with the polar axis
12. a point 6 units from the origin on a ray that forms an angle of $11\pi/6$ radians counterclockwise with the polar axis
13. a point 5 units from the origin on a ray that forms a 300° angle counterclockwise with the polar axis
14. a point 3.8 units from the origin on a ray that forms a 225° angle counterclockwise with the polar axis

In Exercises 15–20, convert each pair of polar coordinates (r, θ) to rectangular coordinates (x, y).

15. $(4, -\pi/6)$
16. $(-3, 2\pi/3)$
17. $(-5, 315°)$
18. $(1, -5\pi/6)$
19. $(6.5, -150°)$
20. $(-4, 3\pi/4)$

In Exercises 21–26, convert each pair of rectangular coordinates (x, y) to polar coordinates (r, θ).

21. $\left(-\sqrt{3}, 1\right)$
22. $(4, -4)$
23. $\left(2\sqrt{2}, -2\sqrt{2}\right)$
24. $\left(-2\sqrt{3}, -2\right)$
25. $\left(\sqrt{6}, -\sqrt{2}\right)$
26. $\left(-5, 5\sqrt{3}\right)$

In Exercises 27–32, convert each pair of rectangular coordinates (x, y) to polar coordinates (r, θ). Round θ to the nearest hundredth.

27. $(2, -5)$

28. $(-1, -3)$

29. $(-4, -6)$

30. $(3, 5)$

31. $(-7, 8)$

32. $(-3, 2)$

Extensions

In Exercises 33–38, for each pair of rectangular coordinates (x, y) find every pair of polar coordinates (r, θ) that satisfy $-2\pi < \theta < 2\pi$.

33. $\left(-2\sqrt{2}, 2\sqrt{2}\right)$

34. $\left(-3\sqrt{6}, 3\sqrt{2}\right)$

35. $\left(5\sqrt{2}, -5\sqrt{6}\right)$

36. $\left(-\sqrt{3}, -1\right)$

37. $(-4, 7)$

38. $(4, -3)$

In Exercises 39–44, for each pair of rectangular coordinates (x, y) find every pair of polar coordinates (r, θ) that satisfy $0 < \theta < 4\pi$.

39. $\left(-3\sqrt{3}, 3\right)$

40. $\left(-4\sqrt{2}, -4\sqrt{2}\right)$

41. $\left(5, -5\sqrt{3}\right)$

42. $(7, 7)$

43. $(-4, -3)$

44. $(5, -12)$

45. Determine the polar coordinates of the vertices of a square that has its center at the origin, two sides parallel to the horizontal axis, and side length s. Use angles within $[0, 2\pi]$ and $r > 0$.

10.2 POLAR EQUATIONS

OBJECTIVES

- Write a polar equation in rectangular form.
- Write a rectangular equation in polar form.
- Graph a polar equation by converting to a rectangular equation.
- Graph simple polar equations.
- Graph special polar equations.

PREREQUISITE VOCABULARY TERMS

circle
inverse trigonometric functions
polar coordinates
radius
rectangular coordinates

10.2.1 Writing a Polar Equation in Rectangular Form

In the last section, we converted between rectangular coordinates and polar coordinates. Similar methods can be used to convert between rectangular equations (in terms of x and y) and **polar equations** (in terms of r and θ). To convert equations between polar form and rectangular form, make substitutions using the following equations.

$x = r\cos\theta$	$r^2 = x^2 + y^2$
$y = r\sin\theta$	$\tan\theta = y/x$

IMPORTANT

Squaring both sides of $r = \sqrt{x^2 + y^2}$ gives $r^2 = x^2 + y^2$.

EXAMPLE 1 Writing a Polar Equation in Rectangular Form

IMPORTANT

There are often several ways in which a polar equation can be converted to a rectangular equation, but each will result in the same general form equation (if completed correctly).

Write each polar equation as a rectangular equation in general form.

A. $r = 3\cos\theta$ **B.** $r = -2\tan\theta$

SOLUTION

A. Notice that the right side contains $\cos\theta$. If the right side contained $r\cos\theta$, then we could substitute x for $r\cos\theta$. So, begin by multiplying both sides of the equation by r. Then substitute using the equations for converting between polar coordinates and rectangular coordinates.

$r = 3\cos\theta$

$r^2 = r(3\cos\theta)$ *Multiply each side by r.*

$r^2 = 3r\cos\theta$ *Commutative Property*

$x^2 + y^2 = 3x$ *Substitute $x^2 + y^2$ for r^2 and x for $r\cos\theta$.*

$x^2 + y^2 - 3x = 0$ *Write the equation in general form.*

REMEMBER

An equation in general form has 0 on one side of the equation.

ALTERNATIVE METHOD

Begin by substituting x/r for $\cos\theta$.

$r = 3\cos\theta$

$r = 3x/r$ *Substitute x/r for $\cos\theta$.*

$r^2 = 3x$ *Multiply each side by r.*

$x^2 + y^2 = 3x$ *Substitute $x^2 + y^2$ for r^2.*

$x^2 + y^2 - 3x = 0$ *Write the equation in general form.*

B. Square both sides of the equation so that the left side is r^2. Then substitute using the equations for converting between polar coordinates and rectangular coordinates.

$r = -2 \tan \theta$	
$r^2 = (-2 \tan \theta)^2$	*Square each side.*
$x^2 + y^2 = (-2(y/x))^2$	*Substitute $x^2 + y^2$ for r^2 and y/x for $\tan \theta$.*
$x^2 + y^2 = 4y^2/x^2$	*Simplify.*
$x^2(x^2 + y^2) = 4y^2$	*Multiply each side by x^2.*
$x^4 + x^2y^2 - 4y^2 = 0$	*Write the equation in general form.*

TIP

$$\left(\frac{-2y}{x}\right)^2 = \frac{(-2)^2(y)^2}{(x)^2} = \frac{4y^2}{x^2}$$

10.2.2 Writing a Polar Equation in Rectangular Form: Another Example

Any of the six definitions of the trigonometric functions of θ can be used when converting between polar and rectangular form. The same is true for any of the trigonometric identities.

EXAMPLE 2 Writing a Polar Equation in Rectangular Form

Write each polar equation as a rectangular equation in general form.

A. $5 \sec \theta = 4r$ **B.** $r = 4$ **C.** $2\theta - 3\pi/2 = 0$

SOLUTION

IMPORTANT

Use trigonometric identities to write the equations in terms of sine, cosine, or tangent, as needed.

A. Use the reciprocal identity $\sec \theta = \dfrac{1}{\cos \theta}$ to write the equation in terms of cosine.

$5 \sec \theta = 4r$	
$5/(\cos \theta) = 4r$	*Reciprocal Identity*
$5 = 4r \cos \theta$	*Multiply each side by $\cos \theta$.*
$5 = 4x$	*Substitute x for $r \cos \theta$.*
$4x - 5 = 0$	*Write the equation in general form.*

B. Square both sides of the equation.

$r = 4$	
$r^2 = 16$	*Square each side.*
$x^2 + y^2 = 16$	*Substitute $x^2 + y^2$ for r^2.*
$x^2 + y^2 - 16 = 0$	*Write the equation in general form.*

NOTICE THAT

The graph of the rectangular equation is a circle centered at the origin, and with radius 4.

C. Add $3\pi/2$ to each side and then divide each side by 2 to solve the equation for θ.

$2\theta - 3\pi/2 = 0$	
$\theta = 3\pi/4$	*Solve for θ.*
$\tan \theta = \tan (3\pi/4)$	*Take the tangent of each side.*
$y/x = -1$	*Evaluate $\tan (3\pi/4)$ and substitute y/x for $\tan \theta$.*
$y = -x$	*Multiply each side by x.*
$x + y = 0$	*Write the equation in general form.*

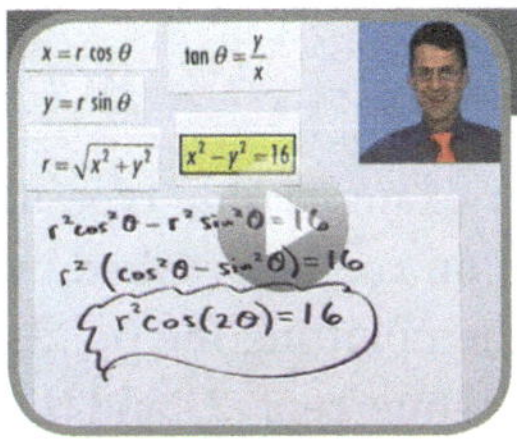

10.2.3 Writing a Rectangular Equation in Polar Form

A polar equation is written in terms of r and θ. To write a rectangular equation (an equation in terms of x and y) in polar form, again use the formulas to convert between rectangular coordinates and polar coordinates.

EXAMPLE 3 Writing a Rectangular Equation in Polar Form

Write each rectangular equation in polar form.

A. $y = 5$ **B.** $y = 3x + 1$ **C.** $y^2 - 6x^2 = 0$

IMPORTANT

Use trigonometric identities to write the equations in terms of sine, cosine, or tangent, as needed.

SOLUTION

Manipulate each equation as needed, then substitute using the equations for converting between polar coordinates and rectangular coordinates.

A. Use $y = r \sin\theta$ to write the rectangular equation in terms of r and θ.

$$y = 5$$

$$r\sin\theta = 5 \qquad \textit{Substitute } r\sin\theta \textit{ for } y.$$

$$r = 5/\sin\theta \qquad \textit{Divide each side by } \sin\theta \textit{ to solve for } r.$$

$$r = 5\csc\theta \qquad \textit{Reciprocal Identity}$$

A polar form of the rectangular equation $y = 5$ is $r = 5\csc\theta$.

TIP

The equation $r \sin\theta = 5$ is in polar form because it is in terms of r and θ.

B. Use $x = r\cos\theta$ and $y = r\sin\theta$ to write the rectangular equation in terms of r and θ.

$$y = 3x + 1$$

$$r\sin\theta = 3(r\cos\theta) + 1 \qquad \textit{Substitute } r\cos\theta \textit{ for } x \textit{ and } r\sin\theta \textit{ for } y.$$

$$r\sin\theta - 3r\cos\theta = 1 \qquad \textit{Subtract } 3r\cos\theta \textit{ from each side.}$$

$$r(\sin\theta - 3\cos\theta) = 1 \qquad \textit{Factor } r \textit{ from each term.}$$

$$r = \frac{1}{\sin\theta - 3\cos\theta} \qquad \textit{Solve for } r.$$

A polar form of the rectangular equation $y = 3x + 1$ is $r = \dfrac{1}{\sin\theta - 3\cos\theta}$.

TIP

Isolate the r-terms on one side of the equation so that the equation can be solved for r.

C. Manipulate the equation so that it contains y/x, then use $\tan\theta = y/x$ to write the rectangular equation in terms of r and θ.

$$y^2 - 6x^2 = 0$$

$$y^2 = 6x^2 \qquad \textit{Add } 6x^2 \textit{ to each side.}$$

$$y^2/x^2 = 6 \qquad \textit{Divide each side by } x^2.$$

$$(y/x)^2 = 6 \qquad \textit{Power of a Quotient Property}$$

$$(\tan\theta)^2 = 6 \qquad \textit{Substitute } \tan\theta \textit{ for } y/x.$$

$$\tan^2\theta = 6 \qquad \textit{Simplify.}$$

A polar form of the rectangular equation $y^2 - 6x^2 = 0$ is $\tan^2\theta = 6$.

10.2.4 Graphing Simple Polar Equations

We have seen several methods for graphing rectangular equations. For example, one type of rectangular equation is a linear equation, such as $y = x + 2$. This equation can be graphed by making a table showing the corresponding x- and y-coordinates, and then plotting the points from that table. Additionally, this equation can be graphed by using the y-intercept, 2, and the slope, 1.

There are also several methods for graphing polar equations. One method is to first convert the polar equation to rectangular form and then use known techniques for graphing rectangular equations, as shown in **Example 4**.

EXAMPLE 4 Graphing the Rectangular Form of a Polar Equation

Graph the polar equation $r = 4 \sin \theta$.

SOLUTION

Write the equation in rectangular form.

$$r = 4 \sin \theta$$

$$r^2 = 4r \sin \theta \qquad \textit{Multiply each side by } r.$$

$$x^2 + y^2 = 4y \qquad \textit{Substitute } x^2 + y^2 \textit{ for } r^2 \textit{ and } y \textit{ for } r \sin \theta.$$

$$x^2 + y^2 - 4y = 0 \qquad \textit{Write the equation in general form.}$$

$$x^2 + (y^2 - 4y + 4) = 4 \qquad \textit{Complete the square.}$$

$$x^2 + (y - 2)^2 = 4 \qquad \textit{Factor.}$$

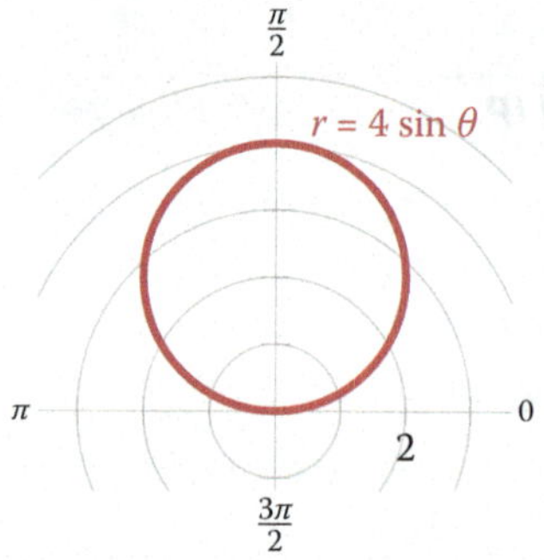

REMEMBER

The standard form equation for a circle with center (h, k) and radius r is $(x - h)^2 + (y - k)^2 = r$.

The resulting equation $x^2 + (y - 2)^2 = 4$ describes a circle with center (0, 2) and radius 2.

Therefore, the graph of $r = 4 \sin \theta$ is a circle with center (0, 2) and radius 2.

Comparing the Graphs of $x = 9$ and $r = 9$

Rectangular equations and polar equations that appear very similar can have very different graphs. Consider the rectangular equation $x = 9$ and the polar equation $r = 9$, which appear to be similar.

We know that the graph of $x = 9$ is a vertical line that passes through 9 on the x-axis. Converting to polar form gives $r \cos \theta = 9$, that is, $r = 9/(\cos \theta)$. Thus, the graph of the polar equation $r = 9/(\cos \theta)$ is a vertical line that passes through 9 on the horizontal axis.

Now consider the graph of the polar equation $r = 9$. Converting $r = 9$ to rectangular form gives $x^2 + y^2 = 81$, which is a circle with center at the origin and radius 9. Thus, the graph of the polar equation $r = 9$ is a circle with center at the origin and radius 9.

So, the graph of $x = 9$ (i.e., $r = 9/(\cos \theta)$) is a vertical line and the graph of $r = 9$ is a circle, as shown on a polar graph in Figure 10.2a.

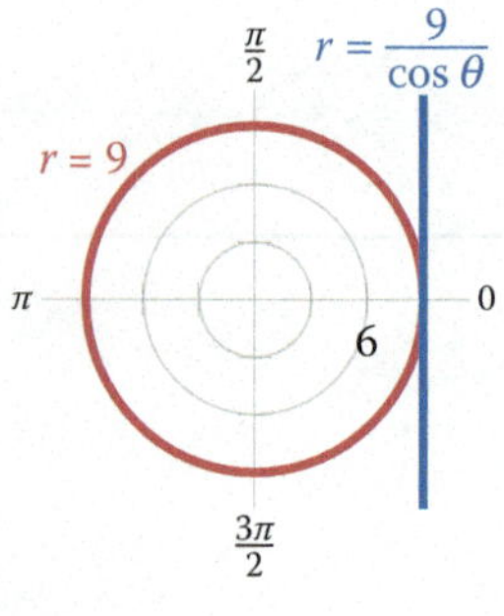

Figure 10.2a

Generalizing the Polar Graphs of $r = c$, $\theta = c$, and $r = \theta$

Suppose c is some real number. Recall that the graph of

- $x = c$ is all points in the plane with x-coordinate c (i.e., a vertical line),
- $y = c$ is all points in the plane with y-coordinate c (i.e., a horizontal line), and
- $y = x$ is all points in the plane with equal x- and y-coordinates (a diagonal line).

In a similar way, the graphs of simple polar equations in which c is a real number such that $c > 0$ can be generalized.

Graphs of Simple Polar Equations

$r = c$	$\theta = c$	$r = \theta$
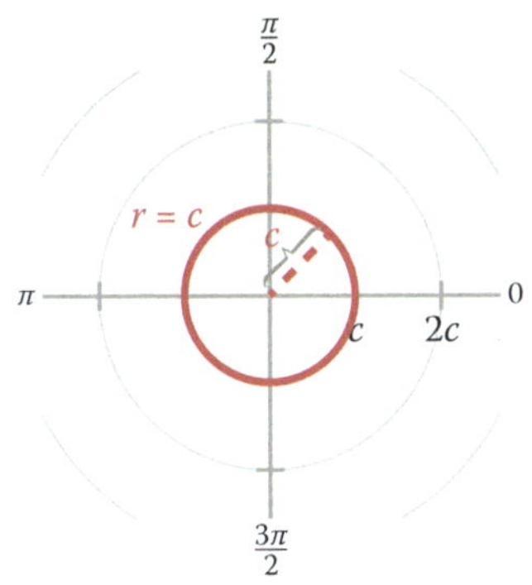	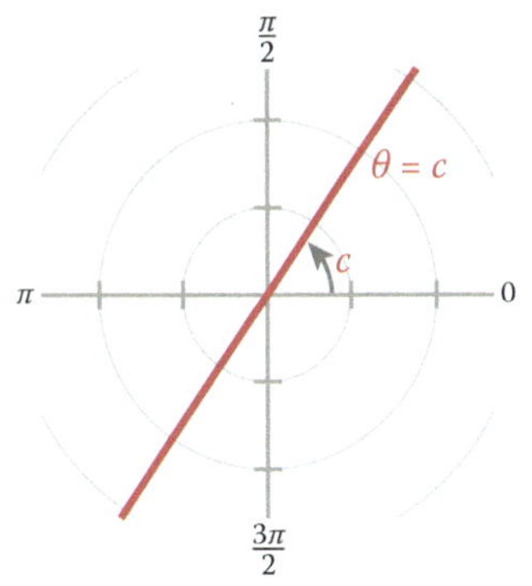	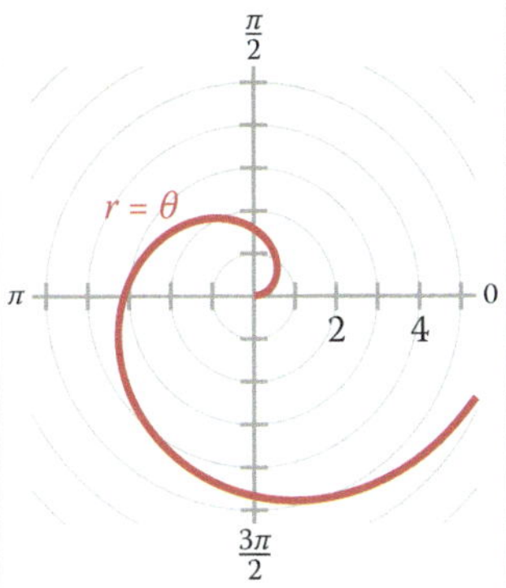
The graph is all points on the polar plane with r-coordinate c (i.e., all points c units from the origin).	*The graph is all points on the polar plane with θ-coordinate c (i.e., a diagonal line formed by all points along the rays θ and θ + π).*	*The graph is all points on the polar plane with equal r- and θ-coordinates (i.e., a spiral starting at the origin).*

NOTICE THAT

The graph of polar equation $r = c$ is a circle with center $(0, 0)$ and radius c.

The graph of $r = c$ is a circle because the value of r (i.e., the distance from the origin) is constant. The graph of $\theta = c$ is a line because the value of θ (i.e., the angle between the polar axis and the ray containing the point) is constant. The graph of $r = \theta$ is a spiral because as r increases, θ also increases.

EXAMPLE 5 Graphing a Simple Polar Equation

Graph each polar equation.

A. $r = 7$ **B.** $\theta = \pi/9$

SOLUTION

A. The graph of $r = 7$ is all points on the polar plane with r-coordinate 7. Since θ is not specified in the equation, θ can take on any value.

So, the graph of $r = 7$ is a circle with center at the origin and radius 7. In other words, the graph of $r = 7$ is all points on the polar plane that are 7 units from the origin.

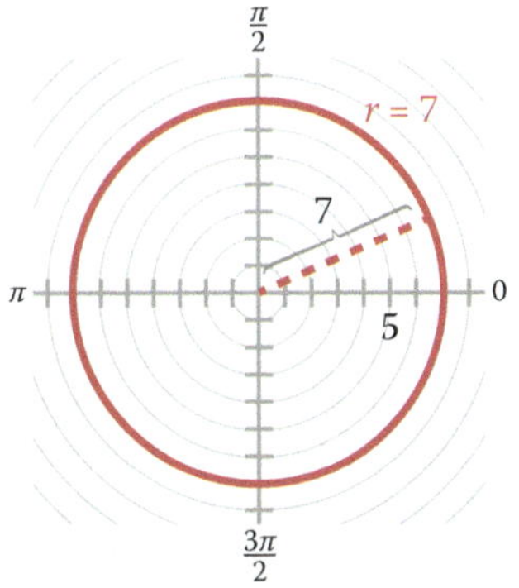

CHECK

Write the equation in rectangular form.

$$r = 7$$

$$r^2 = 49 \qquad \textit{Square each side.}$$

$$x^2 + y^2 = 49 \qquad \textit{Substitute } x^2 + y^2 \textit{ for } r^2.$$

The resulting equation $x^2 + y^2 = 49$ describes a circle with center (0, 0) and radius 7. ✔

B. The graph of $\theta = \pi/9$ is all points on the polar plane with θ-coordinate $\pi/9$. Since r is not specified in the equation, r can take on any value.

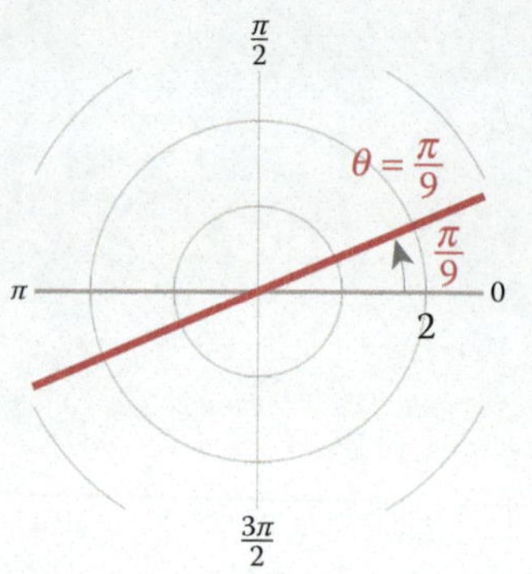

TIP

π/9 radians is 20°.

Recall that θ is the angle of rotation from the positive x-axis (or polar axis) to a ray containing a polar point. So, $\theta = \pi/9$ is the graph of all points on the ray that is rotated $\pi/9$ radians (counterclockwise) from the polar axis.

Furthermore, since r can be any positive or negative value, the ray extends out from the origin in both directions.

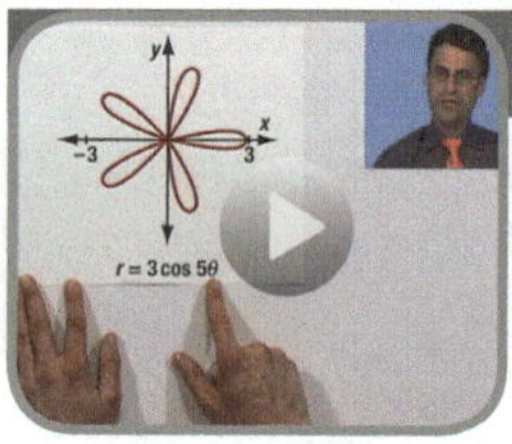

10.2.5 Graphing Special Polar Equations

Polar Equations of Circles

The equation of a circle appears simpler in polar form than it does in rectangular form. For example, the polar form of the rectangular equation $x^2 + y^2 = c^2$ (a circle centered at the origin and radius c) is simply $r = c$.

The polar equations $r = a \cos \theta$ and $r = a \sin \theta$ also define circles. However, these circles are not centered at the origin, but instead pass through the origin. The centers, in rectangular coordinates, are at $(a/2, 0)$ and $(0, a/2)$, respectively. Thus, each circle's radius is $|a|/2$. The attributes of the polar graphs of $r = a \cos \theta$ and $r = a \sin \theta$ where $a > 0$ are summarized in the following table.

Polar Equations of Circles: $r = a \cos \theta$ and $r = a \sin \theta$, $a > 0$	
$r = a \cos \theta$	$r = a \sin \theta$
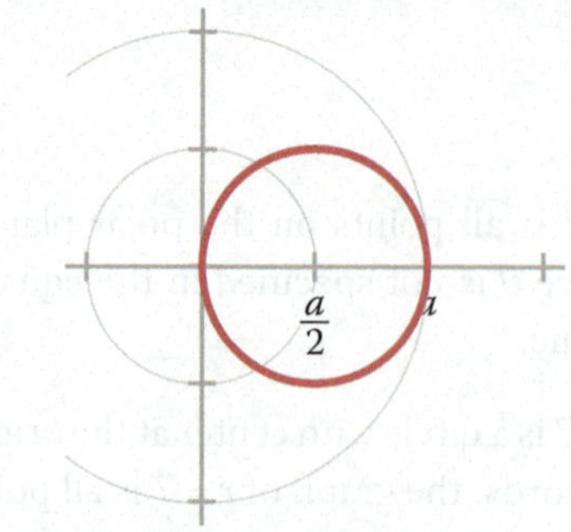	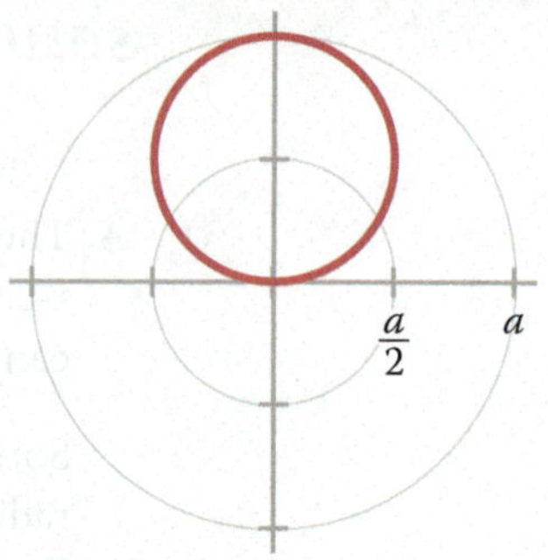
center at $(a/2, 0)$ *on the horizontal axis*	center at $(a/2, \pi/2)$ *on the vertical axis*

TIP

When $a < 0$, the graph of $r = a \cos \theta$ is the reflection across the vertical axis of the graph of $r = |a| \cos \theta$. Similarly, when $a < 0$, the graph of $r = a \sin \theta$ is the reflection across the horizontal axis of the graph of $r = |a| \sin \theta$.

Also, note that the graph of $r = a \cos \theta$ is symmetric about the horizontal axis, and the graph of $r = a \sin \theta$ is symmetric about the vertical axis, as expected because cosine is an even function and sine is an odd function.

EXAMPLE 6 Graphing the Polar Equation of a Circle

Identify the center, radius, and symmetry of each circle, and then sketch the circle's graph.

A. $r = 9 \cos \theta$ **B.** $r = 6 \sin \theta$

SOLUTION

A. The polar graph of an equation of the form $r = a \cos \theta$, where $a = 9$, is a circle symmetric about the horizontal axis that passes through the origin. The circle is centered at $(9/2, 0)$ and its radius is $9/2$. The circle also passes through $(9, 0)$.

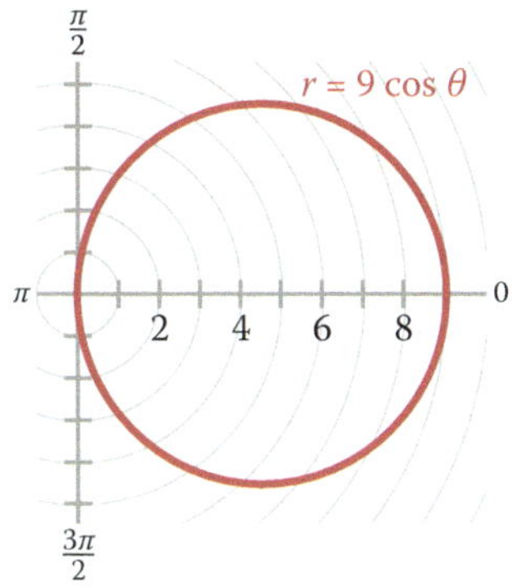

TIP

The circle is centered at (4.5, 0), which is the point 4.5 units from the origin on the polar axis. In rectangular coordinates, the center is at (4.5, 0), which is on the x-axis.

B. The polar graph of an equation of the form $r = a \sin \theta$, where $a = 6$, is a circle symmetric about the vertical axis that passes through the origin. The circle is centered at $(3, \pi/2)$ and its radius is 3. The circle also passes through $(6, \pi/2)$.

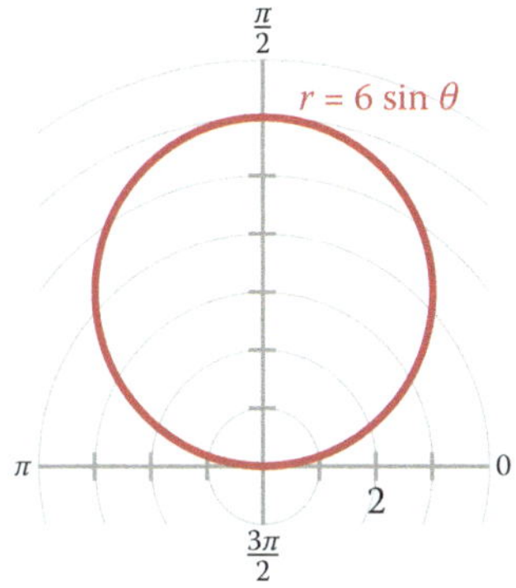

TIP

The circle is centered at $(3, \pi/2)$, which is the point 3 units from the origin on the ray rotated $\pi/2$ radians from the pole (i.e., the vertical axis). In rectangular coordinates, the center is at (0, 3), which is on the y-axis.

Polar Equations of Rose Curves

If n is an integer and $n > 1$, then the graph of a polar equation of the form $r = a \sin n\theta$ or $r = a \cos n\theta$, called a **rose curve**, is composed of evenly spaced **petals**, as shown in the table below. The **radius of a petal** is the maximum r-value on the rose curve, which occurs at each point that defines the tip of a petal.

TIP

Like the equation of a circle, the equation of a rose curve appears much simpler in polar form than it does in rectangular form.

Examples of Rose Curves

$r = 3 \cos 5\theta$	$r = 2 \cos 4\theta$	$r = 4 \sin 7\theta$	$r = 4 \sin 6\theta$
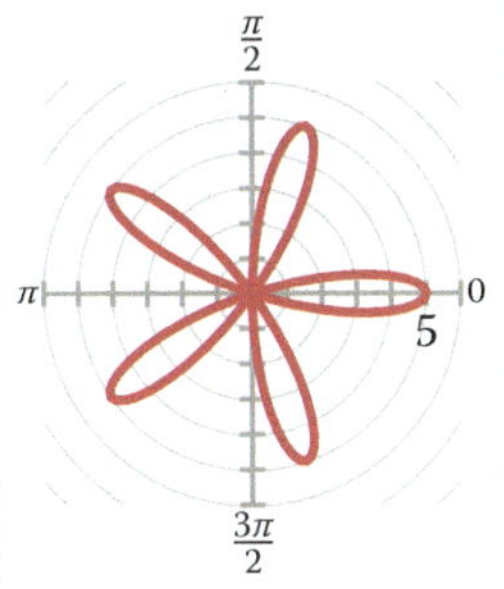	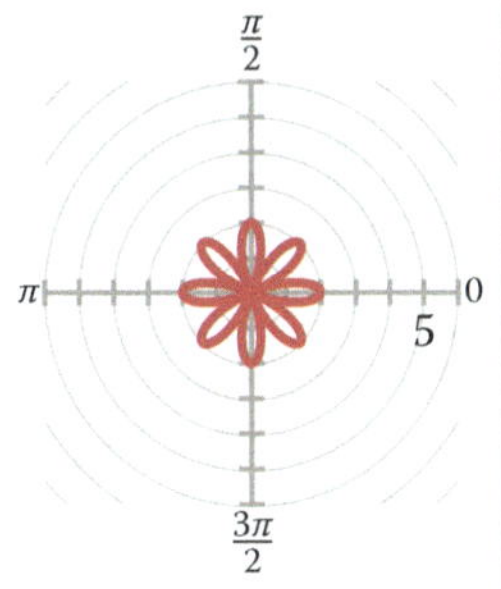	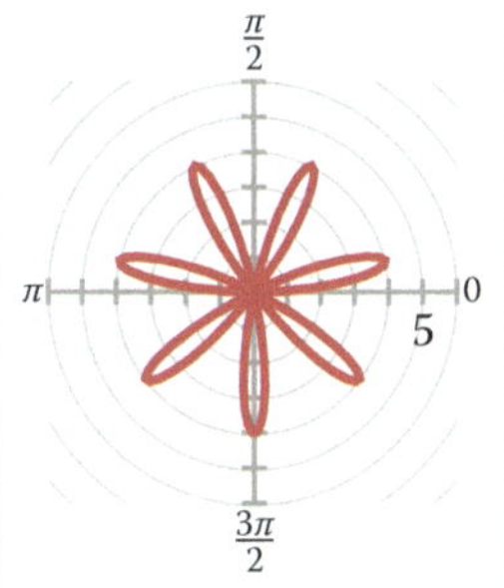	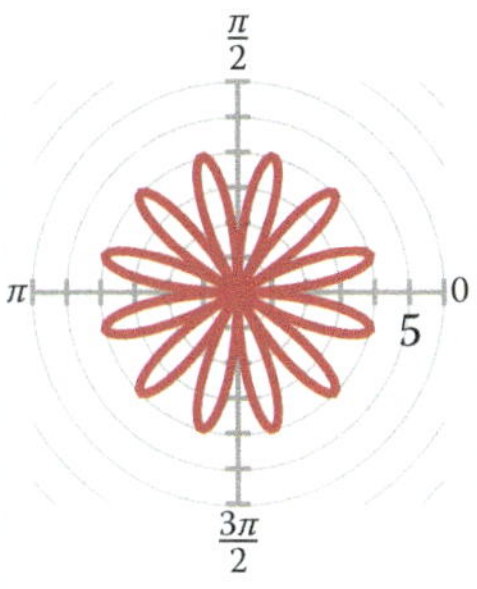
5 petals, radius 5	*8 petals, radius 2*	*7 petals, radius 4*	*12 petals, radius 4*

The n-value and the a-value in the equation of a rose curve can be used to determine the number of petals, the radius of the petals, and the angle between the petals of a rose curve.

- A rose curve has n petals when n is odd, and $2n$ petals when n is even.
- The angle between pairs of consecutive petals is equal to 2π divided by the number of petals.
- The radius of the petals is equal to $|a|$.

Notice that if $\theta = 0$, then $r = a \cos n\theta = a \cos 0 = a(1) = a$ (which is also the petal radius). Therefore, in the graph of a rose curve defined by $r = a \cos n\theta$, the point at the tip of one of the petals is always at $(a, 0)$ on the horizontal axis.

Additionally, if $\theta = \pi/(2n)$, then $r = a \sin n\theta = a \sin(\pi/2) = a(1) = a$. Therefore, in the graph of a rose curve defined by $r = a \sin n\theta$, the point at the tip of one of the petals is always at $\left(a, \dfrac{\pi}{2n}\right)$.

The attributes of the polar graphs of $r = a \cos n\theta$ and $r = a \sin n\theta$, where $a > 0$ and $n > 1$, are summarized in the following table.

Polar Equations of Rose Curves: $r = a \cos n\theta$ and $r = a \sin n\theta$, $a > 0$ and $n > 1$

	$r = a \cos n\theta$		$r = a \sin n\theta$	
	n is odd	**n is even**	**n is odd**	**n is even**
Symmetric About	the horizontal axis	the vertical and horizontal axes	the vertical axis	the vertical and horizontal axes
Number of Petals	n	$2n$	n	$2n$
Angle Between the Petals	$2\pi/n$	π/n	$2\pi/n$	π/n
Tip of One Petal	$(a, 0)$		$\left(a, \dfrac{\pi}{2n}\right)$	

If $a < 0$, then the graph of $r = a \cos n\theta$ is a reflection over the vertical axis of the graph of $r = |a| \cos n\theta$, and the graph of $r = a \sin n\theta$ is a reflection over the horizontal axis of the graph of $r = |a| \sin n\theta$.

Figure 10.2b shows the graph of $r = 2 \cos 3\theta$, which is a rose curve with 3 petals with radius 2, where the angle between each pair of consecutive petals is $2\pi/3$.

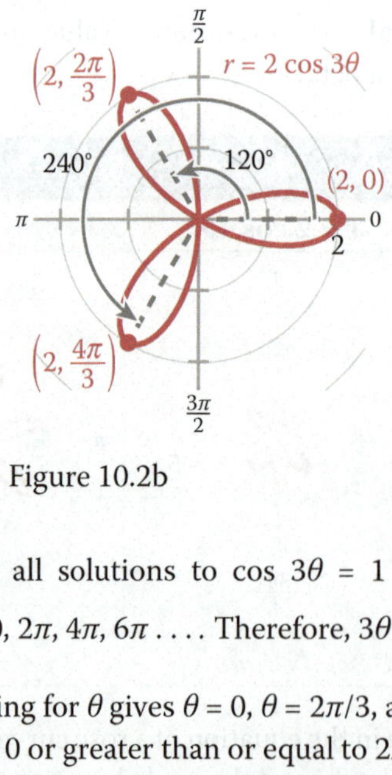

Figure 10.2b

Since the radius of each petal is also the r-coordinate at the point that defines the tip of each petal, the θ-coordinate of the tip of each petal can be found by substituting the radius in the equation of the rose curve for r, and then solving for θ.

For example, since the graph of $r = 2 \cos 3\theta$ has petals with radius 2, solving the equation $2 = 2 \cos 3\theta$ for θ gives the θ-coordinate of the tip of each petal.

$$2 = 2 \cos 3\theta \ \Rightarrow \ \cos 3\theta = 1$$

Find all solutions to $\cos 3\theta = 1$ on the interval $[0, 2\pi)$. We know that $\cos x = 1$ when $x = 0, 2\pi, 4\pi, 6\pi \ldots$. Therefore, $3\theta = 0, 3\theta = 2\pi, 3\theta = 4\pi \ldots$.

TIP

Because this rose curve has 3 petals, it is necessary to find the polar coordinates for three points.

Solving for θ gives $\theta = 0$, $\theta = 2\pi/3$, and $\theta = 4\pi/3$. (All other solutions of $\cos 3\theta = 1$ are either less than 0 or greater than or equal to 2π.) Therefore, the polar coordinates of the point that defines the tip of the 3 petals in the graph of $r = 2 \cos 3\theta$ are $(2, 0)$, $(2, 2\pi/3)$, and $(2, 4\pi/3)$, as shown in Figure 10.2b.

EXAMPLE 7 Graphing the Polar Equation of a Rose Curve

Identify the number of petals, petal radius, and symmetry (if any) of each rose curve, and then sketch its graph.

A. $r = 3\cos 5\theta$ **B.** $r = 2\sin 3\theta$

SOLUTION

A. The equation $r = 3\cos 5\theta$ is of the form $r = a\cos n\theta$, where $a = 3$ and $n = 5$.

Therefore:

- The graph has 5 petals (since $n = 5$, and n is odd), each with a radius of 3 (since $r = 3$).
- The graph is symmetric about the horizontal axis (since the equation is of the form $r = a\cos n\theta$, where n is odd).
- The tip of one petal is at (3, 0) because $\theta = 0$ when $r = 3$.

As with every rose curve, the petals of the graph of $r = 3\cos 5\theta$ are evenly spaced. Since there are 5 petals, the angle between the petals is $2\pi/5$.

Plot the points for the tip of each petal: $(3, 0)$, $(3, 2\pi/5)$, $(3, 4\pi/5)$, $(3, 6\pi/5)$, and $(3, 8\pi/5)$.

Then sketch the petals from the origin to each petal tip and back.

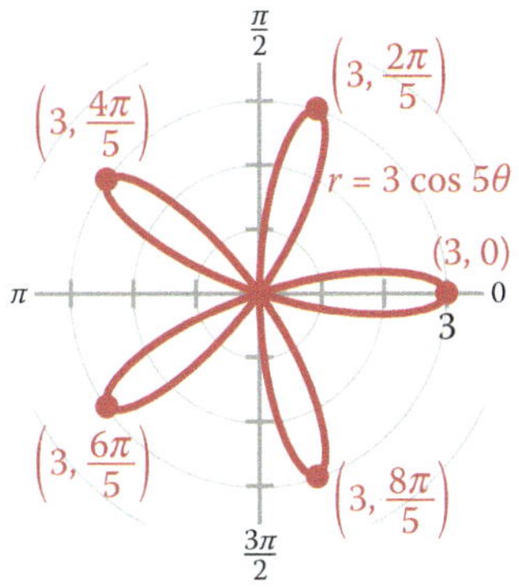

B. The equation $r = 2\sin 3\theta$ is of the form $r = a\sin n\theta$, where $a = 2$ and $n = 3$.

Therefore:

- The graph has 3 petals (since $n = 3$, and n is odd), each with a radius of 2 (since $r = 2$).
- The graph is symmetric about the vertical axis (since the equation is of the form $r = a\sin n\theta$, where n is odd).
- The tip of one petal is at $(2, \pi/6)$ because $\theta = \pi/6$ when $r = 2$.

As with every rose curve, the petals of the graph of $r = 2\sin 3\theta$ are evenly spaced. Since there are 3 petals, the angle between the petals is $2\pi/3$.

Plot the points for the tip of each petal: $(2, \pi/6)$, $(2, 5\pi/6)$, and $(2, 3\pi/2)$.

Then sketch the petals from the origin to each petal tip and back.

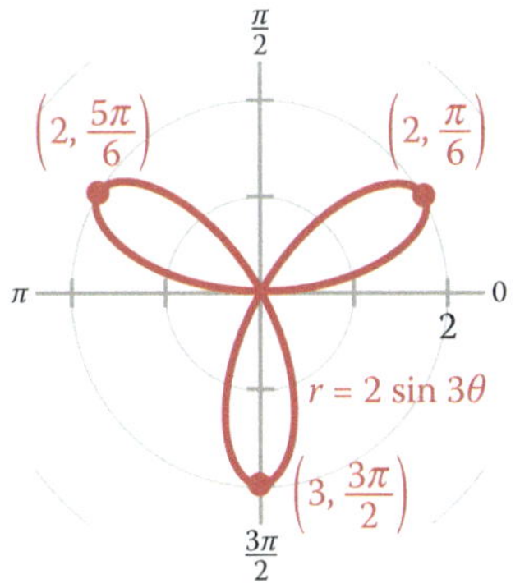

Polar Equations of Limaçons

Another type of polar equation that appears much simpler in polar form than it does in rectangular form describes a graph called a **limaçon**. The shape of a limaçon is somewhat circular, and includes either a dimple or an inner loop. Examples of limaçons are shown in the following table.

Examples of Limaçons

$r = 5 - 4\cos\theta$	$r = 1 - 3\cos\theta$	$r = 4 - 3\sin\theta$	$r = 3 - 6\sin\theta$
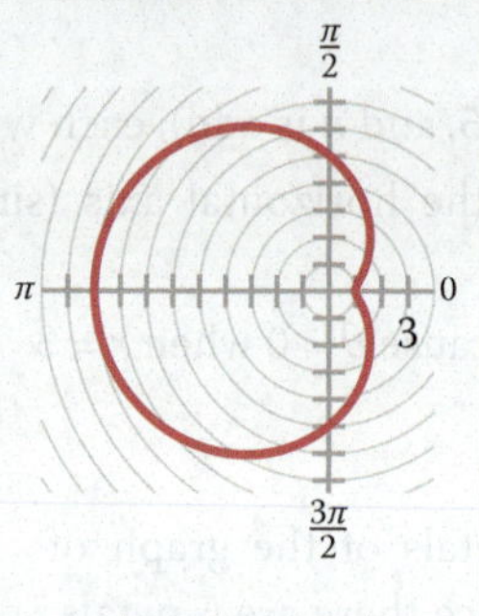	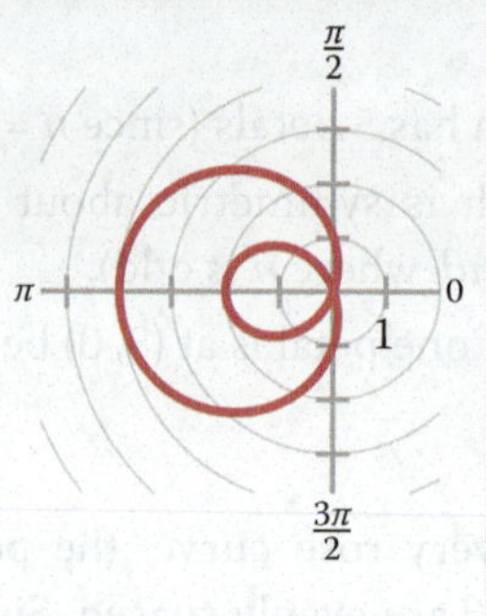	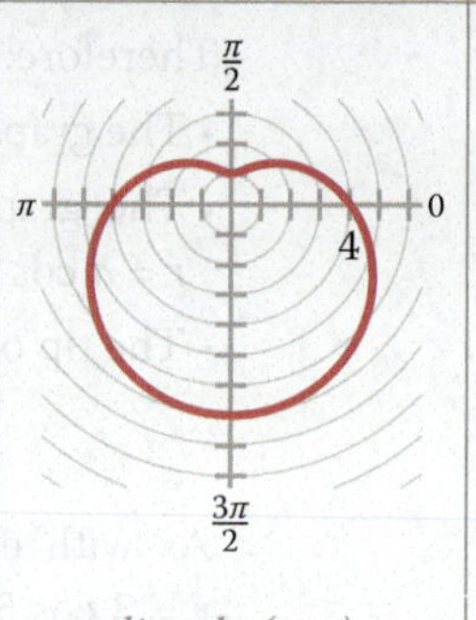	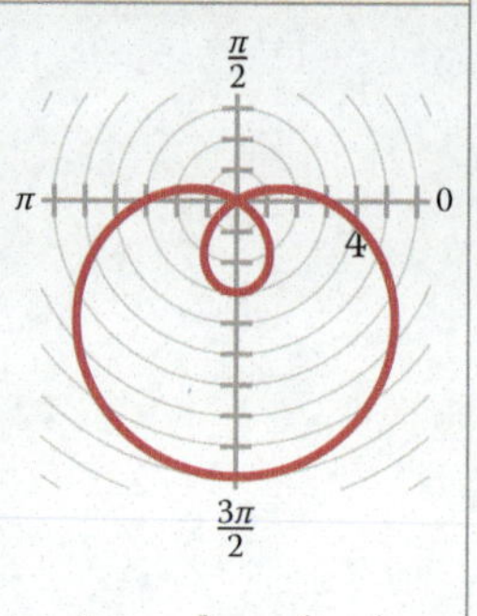
dimple (right)	*inner loop (right)*	*dimple (top)*	*inner loop (top)*

A **horizontal limaçon** is symmetric about the horizontal axis. Its equation is of the form $r = a - b\cos\theta$ and its dimple or inner loop is on the left or right part of the curve (depending on the signs of a and b).

A **vertical limaçon** is symmetric about the vertical axis. Its equation is of the form $r = a - b\sin\theta$ and its dimple or inner loop is on the top or bottom part of the curve (depending on the signs of a and b).

TIP

On a polar graph, the origin is the point where $r = 0$ for any value of θ.

Limaçons have an inner loop when $|a| < |b|$. Every limaçon with an inner loop passes through the origin. Limaçons have a dimple (but no inner loop) when $|a| \geq |b|$. If $|a| \leq |b|$, then the limaçon passes through the origin.

The following table summarizes the attributes of the polar graphs of $r = a - b\cos\theta$ and $r = a - b\sin\theta$, where $a > 0$ and $b > 0$.

Polar Equations of Limaçons: $r = a - b\cos\theta$ and $r = a - b\sin\theta$, $a > 0$ and $b > 0$

	$r = a - b\cos\theta$		$r = a - b\sin\theta$	
	$a > b$	$a < b$	$a > b$	$a < b$
Graph contains	dimple (right)	inner loop (right)	dimple (top)	inner loop (top)
Symmetric about	the horizontal axis		the vertical axis	
nonzero horizontal axis intercepts	$(a + b, \pi)$ and $(a - b, 0)$		(a, π) and $(a, 0)$	
nonzero vertical axis intercepts	$(a, \pi/2)$ and $(a, 3\pi/2)$		$(a - b, \pi/2)$ and $(a + b, 3\pi/2)$	

IMPORTANT

The graph of a limaçon passes through the origin when $|a| \leq |b|$.

If $b < 0$, then the graph of $r = a - b\cos\theta$ is the reflection over the vertical axis of the graph of $r = a - |b|\cos\theta$, and the graph of $r = a - b\sin\theta$ is the reflection over the horizontal axis of the graph of $r = a - |b|\sin\theta$.

The case where $a = b$ is a special type of limaçon called a **cardioid**. The graph of a cardioid is similar to the dimpled limaçon, but it is more heart-shaped. The tip of a cardioid's dimple is always at the origin. Examples of cardioids are shown in the following table.

Examples of Cardioids			
$r = 5 - 5\cos\theta$	$r = 1 - \cos\theta$	$r = 4 - 4\sin\theta$	$r = 5 - 5\sin\theta$
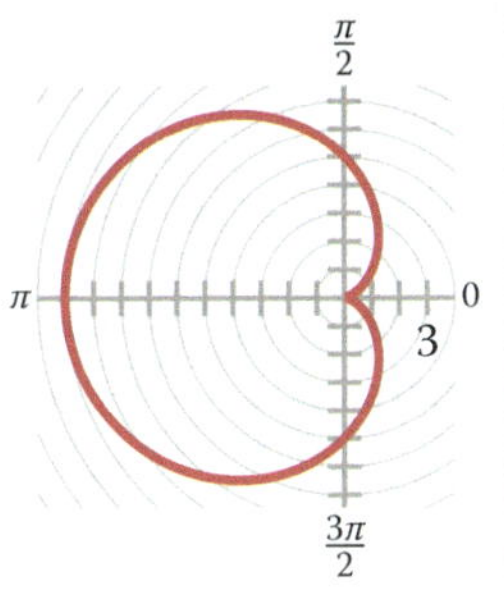	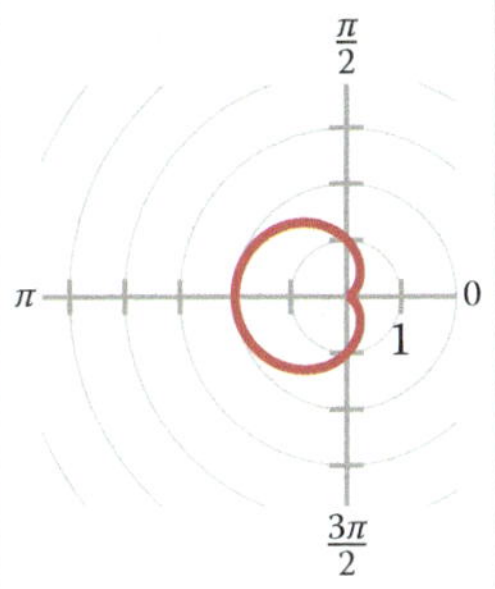	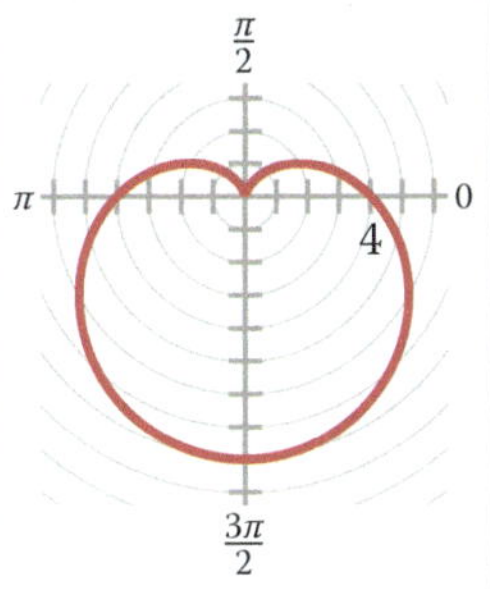	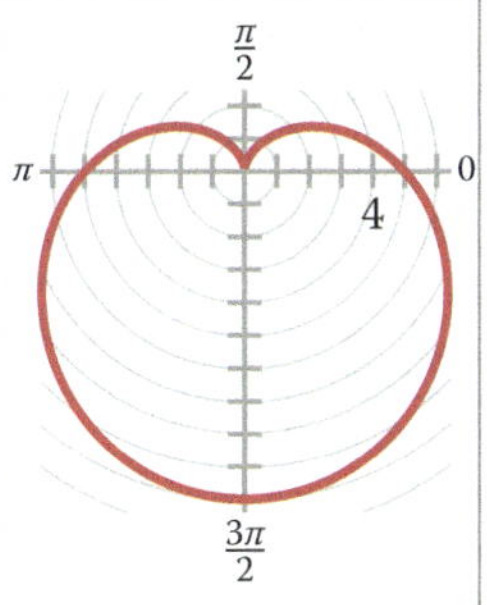
dimple (right)	*dimple (right)*	*dimple (top)*	*dimple (top)*

EXAMPLE 8 Graphing the Polar Equation of a Limaçon

Graph each polar equation.

A. $r = 2 - \cos\theta$ **B.** $r = 1 - 3\sin\theta$

SOLUTION

TIP

This limaçon will not pass through the origin because $|a| > |b|$.

A. The equation is of the form $r = a - b\cos\theta$, where $a = 2$ and $b = 1$. So, the graph is a horizontal limaçon symmetric about the horizontal axis with a dimple on the right. Find the polar coordinates of the horizontal and vertical axis intercepts.

$\theta = 0 \Rightarrow r = 2 - \cos 0 = 2 - 1 = 1$

$\theta = \pi \Rightarrow r = 2 - \cos\pi = 2 - (-1) = 3$

$\theta = \pi/2 \Rightarrow r = 2 - \cos(\pi/2) = 2 - 0 = 2$

$\theta = 3\pi/2 \Rightarrow r = 2 - \cos(3\pi/2) = 2 - 0 = 2$

Plot the four intercepts, $(1, 0)$, $(3, \pi)$, $(2, \pi/2)$ and $(2, 3\pi/2)$, then sketch the dimpled limaçon through the intercepts.

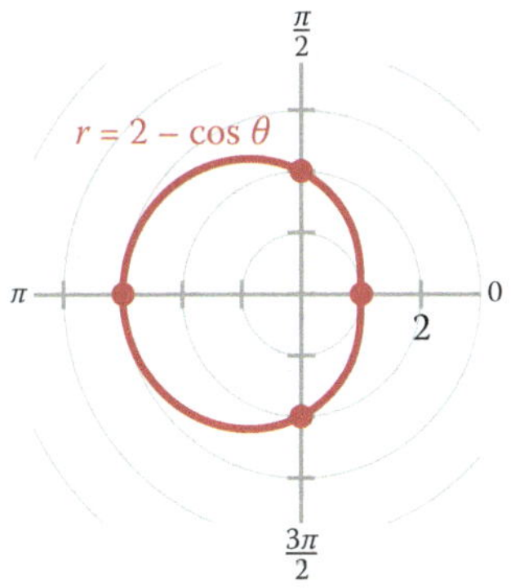

B. The equation is of the form $r = a - b\sin\theta$, where $a = 1$ and $b = 3$. Therefore, the graph is a vertical limaçon symmetric about the vertical axis with an inner loop on the top.

Find the polar coordinates of the horizontal and vertical axis intercepts. Since $|a| \le |b|$, the limaçon passes through the origin.

$\theta = 0 \Rightarrow r = 1 - 3\sin 0 = 1 - 3(0) = 1$

$\theta = \pi \Rightarrow r = 1 - 3\sin\pi = 1 - 3(0) = 1$

$\theta = \pi/2 \Rightarrow r = 1 - 3\sin(\pi/2) = 1 - 3(1) = -2$

$\theta = 3\pi/2 \Rightarrow r = 1 - 3\sin(3\pi/2) = 1 - 3(-1) = 4$

Plot the five intercepts, $(0, 0)$, $(1, 0)$, $(1, \pi)$, $(-2, \pi/2)$ and $(4, 3\pi/2)$, then sketch the limaçon through the intercepts with an inner loop at the top.

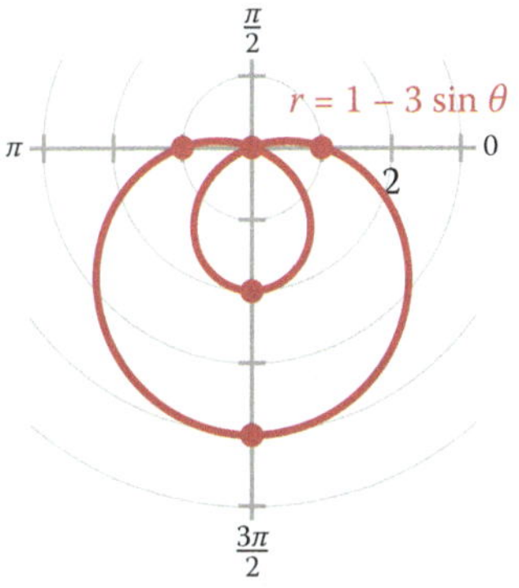

SECTION 10.2 EXERCISES

Warm Up

Write the equation of each conic in standard form by completing the square.

1. $x^2 - 6x + y^2 = 0$
2. $x^2 - y^2 + 8y = 0$
3. $x^2 - 10x + y^2 + 2y = 0$

Just the Facts

Fill in the blanks.

4. To convert a polar equation to a rectangular equation, substitute x for _____, y for _____, y/x for _____, and $x^2 + y^2$ for _____.
5. When substituting for csc θ, sec θ, and cot θ in the conversion from a polar equation to a rectangular equation, use the _____ identities.
6. The graph of $x = c$, where c is a real number, is the graph of a(n) _____, whereas the graph of $r = c$, where c is a real number, is the graph of a(n) _____.
7. The graph of a diagonal line formed by all points along the rays θ and $\theta \pm \pi$ has the equation _____.
8. An equation in the form $r = \theta$ is the graph of a _____ starting at the origin.
9. A rose curve graph has a polar equation of the form _____ or _____.
10. The number of petals in a rose curve graph can be determined by _____ in the equation $r = a \sin n\theta$.

Essential Skills

In Exercises 1–18, write each polar equation as a rectangular equation in general form.

1. $r = 2 \cos \theta$
2. $r - 3 \sin \theta = 0$
3. $r + \sin \theta = 0$
4. $r - 5 \cos \theta = 0$
5. $r - \tan \theta = 0$
6. $r = 4 \tan \theta$
7. $r = -4 \sec \theta$
8. $-5 \cot \theta = 10r$
9. $r = 2 \csc \theta$
10. $-8 \sec \theta = 2r$
11. $6r = 2 \cot \theta$
12. $-42 \csc \theta = 7r$
13. $\theta = 5\pi/3$
14. $0 = \theta - 11\pi/6$
15. $4\theta + \pi = 0$
16. $0 = \theta - 5\pi/6$
17. $r = 2$
18. $r = -5$

In Exercises 19–28, write each rectangular equation in polar form.

19. $x = 8$
20. $y = -3$
21. $2x + 5y = 3$
22. $y = -4x + 2$
23. $-3x + 2y = 9$
24. $y = 4x - 2$
25. $x^2 + y^2 = 16$
26. $2x^2 = -2y^2 + 8$
27. $x^2 = 5y$
28. $3x^2 + 3y^2 = 27$

In Exercises 29–36, graph each polar equation.

29. $r = 5$
30. $r = 3$
31. $\theta = 5\pi/4$
32. $\theta = -\pi/6$
33. $r = 2 \cos \theta$
34. $r = -6 \sin \theta$
35. $r = 4 \sin \theta$
36. $r = -3 \cos \theta$

In Exercises 37–42, identify the center and the radius of each circle.

37. $r = 5 \sin \theta$
38. $r = 11 \cos \theta$
39. $r = 4 \cos \theta$
40. $r = -3 \sin \theta$
41. $r = -4 \sin \theta$
42. $r = 7 \cos \theta$

In Exercises 43–48, identify the number of petals and petal radius of each rose curve. Then state whether the graph has a vertical, horizontal, or no axis of symmetry.

43. $r = 6 \cos 3\theta$

44. $r = 3 \sin 4\theta$

45. $r = 5 \sin 7\theta$

46. $r = 7 \cos 6\theta$

47. $r = 4 \cos 2\theta$

48. $r = 8 \sin 5\theta$

In Exercises 49–60, graph each polar equation.

49. $r = 5 \cos 3\theta$

50. $r = 4 \sin 5\theta$

51. $r = 7 \sin 2\theta$

52. $r = 6 \cos 4\theta$

53. $r = 8 \sin 6\theta$

54. $r = 2 \cos 7\theta$

55. $r = 1 - 3 \cos \theta$

56. $r = 2 - 4 \sin \theta$

57. $r = 3 - 5 \sin \theta$

58. $r = 6 - 5 \cos \theta$

59. $r = 2 - \sin \theta$

60. $r = 4 - 5 \cos \theta$

Extensions

In Exercises 61–64, graph each pair of polar equations and determine the point(s) of intersection.

61. $r = -4 \cos \theta$; $r = -2 \sec \theta$

62. $r = 16 \sin \theta$; $r = 4 \csc \theta$

63. $r = 5 + 2 \sin \theta$; $r = 4$

64. $r = -\sqrt{3} + 4 \cos \theta$; $r = \sqrt{3}$

In Exercises 65–66, convert each polar equation to a rectangular equation.

65. $r \cos (\theta - 7\pi/4) = 5$

66. $r \sin (\theta + 4\pi/3) = 4$

In Exercises 67–68, graph each polar equation.

67. $r = 3 - 3 \sin \theta$

68. $r = 2 - 2 \cos \theta$

In Exercises 69–74, write a polar equation whose graph has the given characteristics.

69. a circle with radius 8 and center $(8, 3\pi/2)$

70. a line that passes through $(7, 4\pi/3)$

71. a rose curve with 5 petals, vertical axis of symmetry, and a petal tip at $(4, \pi/2)$

72. a rose curve with 7 petals, horizontal axis of symmetry, and petal radius 6

73. a horizontal limaçon with intercepts $(6, 0)$, $(-10, 0)$, $(8, \pi/2)$, and $(-8, \pi/2)$

74. a vertical limaçon with intercepts $(4, 0)$, $(4, \pi)$, $(1, \pi/2)$, and $(7, -\pi/2)$

10.3 ROTATION OF AXES

OBJECTIVES

- Use the rotation of axes formula to find the coordinates of a point after being rotated about the origin through an angle of ϕ.
- Identify the type of conic by its equation.
- Graph a rotated conic.

PREREQUISITE VOCABULARY TERMS

axis of symmetry
conic section
Sum of Angles Formulas
trigonometric functions

10.3.1 The Rotation of Axes Formulas

We have seen that various transformations, such as translations or reflections, can be applied to the graph of an equation.

Another type of transformation, called a **rotation**, occurs when a graph or figure is rotated about a fixed point on a plane. The fixed point is called the **center of rotation**. The **angle of rotation**, typically represented by ϕ (the Greek letter "phi"), defines the amount of rotation.

Figure 10.3a

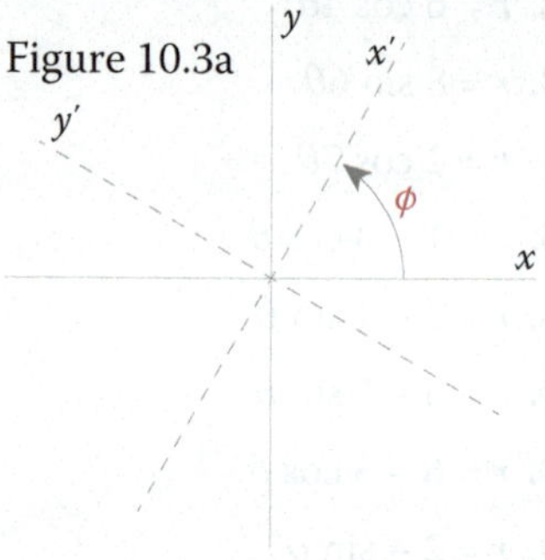

If the x- and y-axes are rotated by a given angle ϕ, then the resulting axes are referred to as the x'-axis and the y'-axis (the x-prime axis and the y-prime axis). For example, a counterclockwise rotation about the origin of the x- and y-axes by an angle of ϕ is shown in Figure 10.3a.

IMPORTANT

A rotation "about the origin" means that the center of rotation is the point (0, 0).

When a point $P(x, y)$ is rotated about the origin by angle ϕ, then the image of $P(x, y)$ after the rotation is referred to as $P'(x', y')$. In this case, if P is on the terminal side of standard position angle θ, then $P'(x', y')$ is on the terminal side of a standard position angle measuring $\theta + \phi$ (Figure 10.3b). Furthermore, if the distance between the origin and P is r, where $r = \sqrt{x^2 + y^2}$, it follows that the distance from the origin to P' is also r, and the relationship between x', y', and r is given by $r = \sqrt{(x')^2 + (y')^2}$.

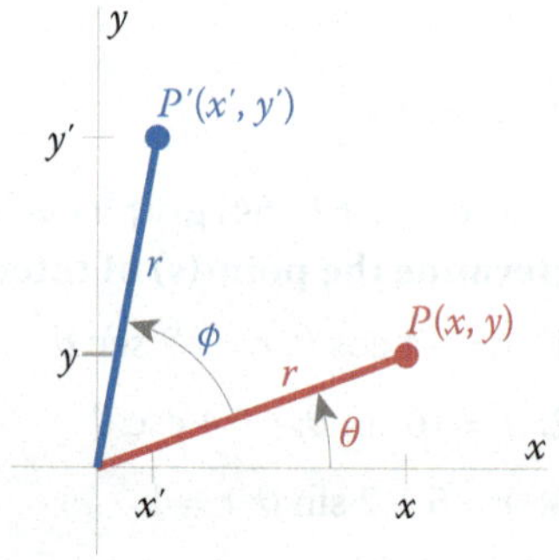

Figure 10.3b

IMPORTANT

The distance from the origin to P is equal to the distance from the origin to P′ when the center of rotation is the origin.

Additionally, since $x = r\cos\theta$ and $y = r\sin\theta$ (Figure 10.3c), it follows that $x' = r\cos(\theta + \phi)$ and $y' = r\sin(\theta + \phi)$ (Figure 10.3d).

NOTICE THAT

The measure of the standard position angle with terminal side through P' is $\theta + \phi$, where ϕ is the angle of rotation.

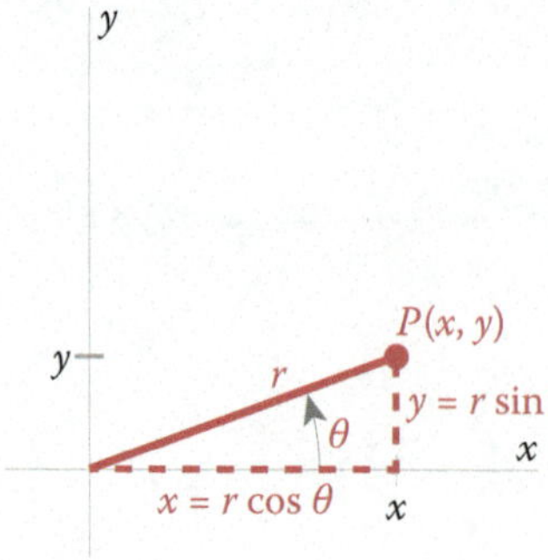

Figure 10.3c

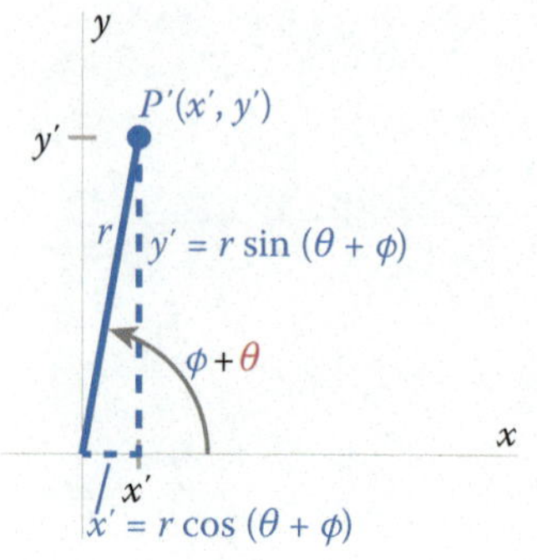

Figure 10.3d

The sum of angles formulas for sine and cosine (from Section 6.3) can be used to simplify $x' = r\cos(\theta + \phi)$ and $y' = r\sin(\theta + \phi)$.

$$\begin{aligned} x' &= r\cos(\theta + \phi) \\ &= r(\cos\theta\cos\phi - \sin\theta\sin\phi) && \textit{Sum of Angles Formulas} \\ &= r\cos\theta\cos\phi - r\sin\theta\sin\phi && \textit{Distribute } r. \\ &= x\cos\phi - y\sin\phi && \textit{Substitute } x \textit{ for } r\cos\theta \textit{ and } y \textit{ for } r\sin\theta. \end{aligned}$$

$$\begin{aligned} y' &= r\sin(\theta + \phi) \\ &= r(\sin\theta\cos\phi + \cos\theta\sin\phi) \\ &= r\sin\theta\cos\phi + r\cos\theta\sin\phi \\ &= y\cos\phi + x\sin\phi \end{aligned}$$

The **rotation of axes formulas** follow from this result.

Rotation of Axes Formulas

If a point $P(x, y)$ on the x- and y-axes is rotated by an angle ϕ counterclockwise about the origin then the image of $P(x, y)$ is $P(x', y')$ where

$$x' = x\cos\phi - y\sin\phi \quad \text{and} \quad y' = y\cos\phi + x\sin\phi.$$

EXAMPLE 1 Using the Rotation of Axes Formulas

Suppose a point $P(4, 1)$ is rotated by an angle of $\pi/6$ counterclockwise about the origin. Find the coordinates of P', the image of P after the rotation.

SOLUTION

Use the rotation of axes formula where $x = 4$, $y = 1$, and $\phi = \pi/6$ to find x' and y'.

$$x' = x\cos\phi - y\sin\phi = 4\cos\frac{\pi}{6} - 1\sin\frac{\pi}{6} = 4\left(\frac{\sqrt{3}}{2}\right) - 1\left(\frac{1}{2}\right) = \frac{4\sqrt{3} - 1}{2}$$

$$y' = y\cos\phi + x\sin\phi = 1\cos\frac{\pi}{6} + 4\sin\frac{\pi}{6} = 1\left(\frac{\sqrt{3}}{2}\right) + 4\left(\frac{1}{2}\right) = \frac{\sqrt{3} + 4}{2}$$

Therefore, P' has coordinates $\left(\frac{4\sqrt{3} - 1}{2}, \frac{\sqrt{3} + 4}{2}\right)$.

EXAMPLE 2 Using the Rotation of Axes Formulas

IMPORTANT

If the center of rotation is not stated, then the center of rotation should be assumed to be the origin.

Suppose $P'(x', y')$ is the image of $P(x, y)$ after being rotated by an angle of ϕ counterclockwise about the origin. Find y' if $\phi = 3\pi/2$, $x = 5$, and $x' = -1$.

SOLUTION

Use the rotation of axes formula for x' where $x = 5$, $x' = -1$, and $\phi = 3\pi/2$ to find y.

$$\begin{aligned} x' &= x\cos\phi - y\sin\phi \\ -1 &= 5\cos(3\pi/2) - y\sin(3\pi/2) \\ -1 &= 5(0) - y(-1) \\ -1 &= y \end{aligned}$$

Now use the rotation of axes formula for y' where $x = 5$, $y = -1$, and $\phi = 3\pi/2$ to find y'.

$$y' = y\cos\phi + x\sin\phi = (-1)\cos(3\pi/2) + 5\sin(3\pi/2) = (-1)(0) + 5(-1) = -5$$

So, $y' = -5$.

10.3.2 Rotating the Axes to Graph Conics

Recall from Chapter 3 that a conic section is a figure formed by the intersection of a cone with a plane. Parabolas, ellipses (including circles), and hyperbolas are all conic sections.

The equation of any conic can be written in the general form, $Ax^2 + Bxy + Cy^2 + Dx + Ey + F = 0$, where A, B, and C are not all equal to 0. The discriminant, $B^2 - 4AC$, and coefficients of a general form equation can be used to identify the type of conic, as summarized below.

Classifying Conics: $Ax^2 + Bxy + Cy^2 + Dx + Ey + F = 0$	
Circle	$B^2 - 4AC < 0$, $B = 0$, and $A = C$
Ellipse (non-circular)	$B^2 - 4AC < 0$, and $B \neq 0$ or $A \neq C$
Hyperbola	$B^2 - 4AC > 0$
Parabola	$B^2 - 4AC = 0$

If the equation of a conic has no B-term (i.e., $B = 0$), then the conic has an axis of symmetry that is either parallel to, or coincides with, the x- or y-axis. (This was the case for all conics in Chapter 3.)

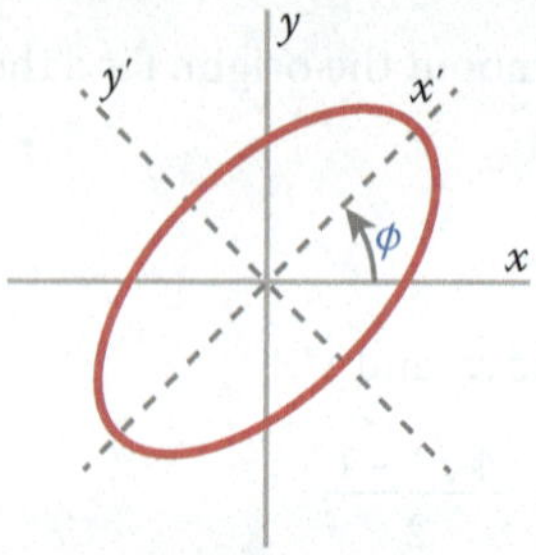

Figure 10.3e

A **rotated conic** is a conic in which the axes of symmetry have been rotated about the origin by some angle of rotation ϕ. An example of a rotated ellipse is shown in Figure 10.3e. The equation of a rotated conic will have a B-term (i.e., $B \neq 0$).

Given the equation for a conic and an angle of rotation, the rotation of axes formulas can be used to find the equation of the resulting rotated conic, as demonstrated in **Example 3**.

EXAMPLE 3 Writing the Equation of a Rotated Conic

Write the equation of the resulting conic if the graph of $y = x^2$ is rotated counterclockwise by an angle of $\pi/2$ radians about the origin. Then graph both conics.

NOTICE THAT

Writing $y = x^2$ in general form gives $y - x^2 = 0$. So, $A = 1$, $B = 0$, and $C = 0$. Therefore, since $B^2 - 4AC = 0$, the graph of $y = x^2$ is a parabola.

SOLUTION

First, use the rotation of axes formulas to find expressions for x' and y' in terms of x and y. Then solve each equation for x and y, and substitute to write the equation of the rotated conic.

$$x' = x\cos\phi - y\sin\phi = x\cos(\pi/2) - y\sin(\pi/2) = x(0) - y(1) = -y$$

$$y' = y\cos\phi + x\sin\phi = y\cos(\pi/2) + x\sin(\pi/2) = y(0) + x(1) = x$$

So, $x' = -y$ and $y' = x$.

Solve for x and y.

$$y' = x \Rightarrow x = y' \quad \text{and} \quad x' = -y \Rightarrow y = -x'$$

Substitute $x = y'$ and $y = -x'$ into the original equation.

$$y = x^2 \Rightarrow -x' = y'^2 \Rightarrow y'^2 = -x'$$

The graph of $y'^2 = -x'$ is a horizontal parabola with vertex at (0, 0), opening to the left.

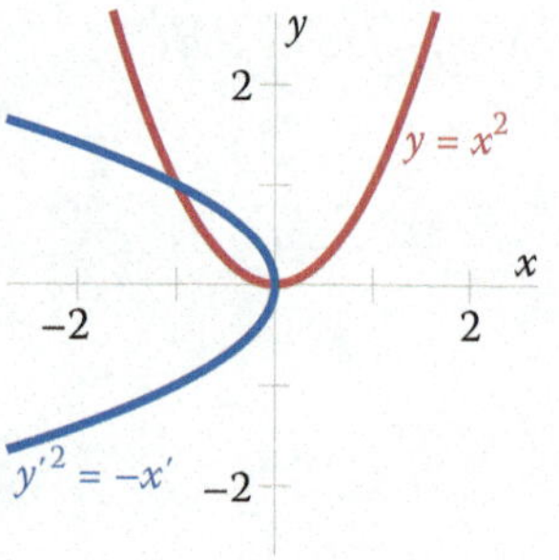

NOTICE THAT

The graph of $y'^2 = -x'$ is the image of the graph of $y = x^2$ after a counterclockwise rotation of 90° about the origin.

EXAMPLE 4 Identifying the Type of Conic

Identify the type of conic, if any, given by each equation.

A. $x^2 + 3xy + 2y^2 + x - 3y + 9 = 0$ **B.** $3x^2 - 2xy + y^2 + 1 = y - 7x$

SOLUTION

Identify A, B, and C from each equation (in general form), then find the discriminant.

A. Evaluate the discriminant using $A = 1$, $B = 3$, and $C = 2$. $B^2 - 4AC = 3^2 - 4(1)(2) = 1$

Since $B^2 - 4AC > 0$, the equation is a hyperbola.

REMEMBER

Write the equation in general form before identifying the values of the coefficients.

B. The general form equation is $3x^2 - 2xy + y^2 + 7x - y + 1 = 0$.

Evaluate the discriminant using $A = 3$, $B = -2$, and $C = 1$. $B^2 - 4AC = (-2)^2 - 4(3)(1) = -8$

Since $B^2 - 4AC < 0$ and $A \neq C$, the equation is an ellipse, but not a circle.

Graphing $Ax^2 + Bxy + Cy^2 + Dx + Ey + F = 0$ When $B \neq 0$

Each of the conics graphed in Chapter 3 has an equation that can be written in the general form $Ax^2 + Bxy + Cy^2 + Dx + Ey + F = 0$ where $B = 0$. When $B \neq 0$, a conic with an equation of the form $Ax^2 + Bxy + Cy^2 + Dx + Ey + F = 0$ can be graphed using the following procedure.

Rotated Conics

The graph of $Ax^2 + Bxy + Cy^2 + Dx + Ey + F = 0$ (where $B \neq 0$) is the graph of $A'(x')^2 + C'(y')^2 + D'x' + E'y' + F' = 0$ rotated counterclockwise by an angle ϕ such that $\cot 2\phi = \dfrac{A-C}{B}$ and $0 < 2\phi < \pi$.

The coefficients A', C', D', E', and F' are given by

$A' = A\cos^2\phi + B\cos\phi\sin\phi + C\sin^2\phi$,
$C' = A\sin^2\phi - B\cos\phi\sin\phi + C\cos^2\phi$,
$D' = D\cos\phi + E\sin\phi$,
$E' = -D\sin\phi + E\cos\phi$, and
$F' = F$.

NOTICE THAT

The formulas for A' and C' are very similar, and the formulas for D' and E' are very similar.

Steps for Graphing a Conic $Ax^2 + Bxy + Cy^2 + Dx + Ey + F = 0$ Where $B \neq 0$

1. Use the discriminant to determine the type of conic given by the original equation.
2. Calculate the angle of rotation ϕ.
3. Calculate the coefficients A', C', D', E', and F'.
4. Write and graph the new equation $A'(x')^2 + C'(y')^2 + D'x' + E'y' + F' = 0$.
5. Rotate the graph counterclockwise by angle ϕ about the origin to obtain the graph of $Ax^2 + Bxy + Cy^2 + Dx + Ey + F = 0$.

IMPORTANT

The graph of the equation in terms of x' and y' is the pre-image of the rotated conic.

EXAMPLE 5 Graphing a Conic

Graph. $\sqrt{3}x^2 + 3xy + 2\sqrt{3}y^2 - 12 = 0$

SOLUTION

Identify the type of conic. $B^2 - 4AC = 3^2 - 4\left(\sqrt{3}\right)\left(2\sqrt{3}\right) = 9 - 24 = -15$

Since $B^2 - 4AC < 0$ and $A \neq C$, the conic is an ellipse, but not a circle.

Calculate the angle of rotation ϕ.

$$\cot 2\phi = \frac{\sqrt{3} - 2\sqrt{3}}{3} \qquad \textit{Substitute.}$$

$$\cot 2\phi = -\frac{\sqrt{3}}{3} \qquad \textit{Simplify.}$$

$$2\phi = \frac{2\pi}{3} \qquad \textit{Take the inverse cotangent of each side.}$$

$$\phi = \frac{\pi}{3} \qquad \textit{Solve for } \phi.$$

IMPORTANT

Since cotangent is negative in the quadrants II and IV, and $0 < 2\phi < \pi$, it follows that 2ϕ is in quadrant II.

Calculate the coefficients A', C', D', E', and F'.

$$A' = \sqrt{3}\cos^2\frac{\pi}{3} + 3\cos\frac{\pi}{3}\sin\frac{\pi}{3} + 2\sqrt{3}\sin^2\frac{\pi}{3} = \sqrt{3}\left(\frac{1}{2}\right)^2 + 3\left(\frac{1}{2}\right)\left(\frac{\sqrt{3}}{2}\right) + 2\sqrt{3}\left(\frac{\sqrt{3}}{2}\right)^2 = \frac{5\sqrt{3}}{2}$$

$$C' = \sqrt{3}\sin^2\frac{\pi}{3} - 3\cos\frac{\pi}{3}\sin\frac{\pi}{3} + 2\sqrt{3}\cos^2\frac{\pi}{3} = \sqrt{3}\left(\frac{\sqrt{3}}{2}\right)^2 - 3\left(\frac{1}{2}\right)\left(\frac{\sqrt{3}}{2}\right) + 2\sqrt{3}\left(\frac{1}{2}\right)^2 = \frac{\sqrt{3}}{2}$$

NOTICE THAT

Since $D = 0$ and $E = 0$, $D' = 0$ and $E' = 0$.

$$D' = D\cos\phi + E\sin\phi = 0\cos(\pi/3) + 0\sin(\pi/3) = 0$$

$$E' = -D\sin\phi + E\cos\phi = -(0)\sin(\pi/3) + 0\cos(\pi/3) = 0$$

$$F' = F = -12$$

Write the equation of the conic in terms of x' and y' by substituting the values of the coefficients into $A'(x')^2 + C(y')^2 + D'x' + E'y' + F' = 0$.

$$\frac{5\sqrt{3}}{2}(x')^2 + \frac{\sqrt{3}}{2}(y')^2 - 12 = 0$$

Write the equation in the standard form of an ellipse.

$$\frac{5\sqrt{3}}{2}(x')^2 + \frac{\sqrt{3}}{2}(y')^2 = 12 \Rightarrow \frac{5\sqrt{3}}{24}(x')^2 + \frac{\sqrt{3}}{24}(y')^2 = 1 \Rightarrow \frac{y'^2}{24/\sqrt{3}} + \frac{x'^2}{24/(5\sqrt{3})} = 1$$

REMEMBER

The standard form equation of a vertical ellipse is $y^2/a^2 + x^2/b^2 = 1$, where $a \geq b$.

So, $a = \sqrt{24/\sqrt{3}} \approx 3.7$ and $b = \sqrt{24/(5\sqrt{3})} \approx 1.7$.

Therefore, the ellipse's intercepts are at approximately $(0, \pm 3.7)$ and $(\pm 1.7, 0)$. Use the intercepts to graph the ellipse.

Then rotate this graph counterclockwise about the origin $\pi/3$ radians to make the graph of $\sqrt{3}x^2 + 3xy + 2\sqrt{3}y^2 - 12 = 0$.

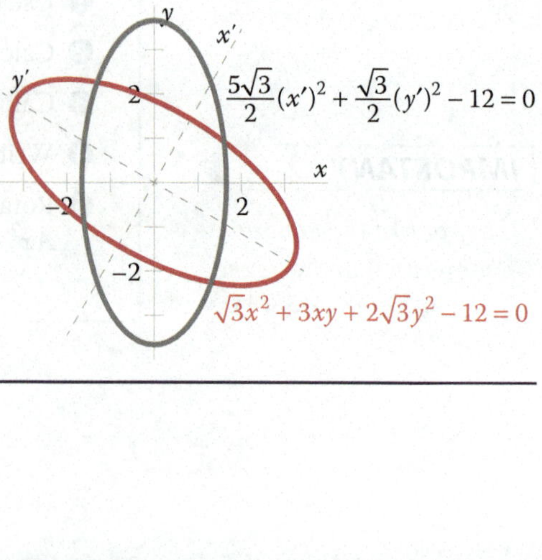

SECTION 10.3 EXERCISES

Warm Up

Graph each conic.

1. $y^2 - 9x^2 = 36$
2. $x^2 + 25y^2 = 225$
3. $y = -x^2 - 6x$

Just the Facts

Fill in the blanks.

4. A transformation involving a graph being turned about a fixed point on a plane is called a(n) _____.
5. When the x- and y-axes are rotated by some given angle, the resulting axes are referred to as the _____-axis and the _____-axis.
6. If a point $P(x, y)$ is rotated by an angle ϕ counterclockwise about the origin to $P(x', y')$, then $x' =$ _____ and $y' =$ _____.
7. To identify the type of conic given by an equation, use the _____, which is $B^2 - 4AC$.
8. The equations for ellipses, hyperbolas, and parabolas have discriminant values that are _____ 0, _____ 0, and _____ 0, respectively.
9. The angle of rotation of a conic is found using $\cot 2\phi =$ _____.
10. When the general form equation of a conic, $Ax^2 + Bxy + Cy^2 + Dx + Ey + F = 0$, has no B-term, then the conic has an axis of symmetry that is _____ to the x- or y-axis.

Essential Skills

In Exercises 1–4, find the coordinates of P' if P' is the image of P after a counterclockwise rotation $\pi/4$ about the origin.

1. $P(0, 6)$
2. $P(-2, 5)$
3. $P(-4, 0)$
4. $P(-2, -8)$

In Exercises 5–8, find the coordinates of P' if P' is the image of P after a counterclockwise rotation $\pi/6$ about the origin.

5. $P(-2, 0)$
6. $P(0, -14)$
7. $P(-6, 2)$
8. $P(-3, -5)$

In Exercises 9–12, $P'(x', y')$ is the point that results when $P(x, y)$ is rotated by an angle of ϕ counterclockwise about the origin. Use the given information to find the missing coordinate.

9. $\phi = \pi/6$, $x = -4$, and $x' = -2\sqrt{3} + 10$, find y'.
10. $\phi = \pi/6$, $y = -10$, and $y' = 4 - 5\sqrt{3}$, find x'.
11. $\phi = \pi/3$, $y = 8$, and $y' = 3\sqrt{3} + 4$, find x'.
12. $\phi = \pi/3$, $x = 12$, and $x' = 6 - \sqrt{3}$, find y'.

In Exercises 13–18, identify the type of conic, if any, given by the equation.

13. $2x^2 + 3x - 4xy = y - 3y^2$
14. $y^2 - 2xy = 2y - x^2 - 6x$
15. $y + xy - y^2 - x^2 = 5x + x^2 - y^2$
16. $1 - y^2 + 3y = x - x^2 + xy$
17. $3y^2 = 5 - 2x - y$
18. $2y - 2y^2 - x^2 = 1 - 3x - xy$

In Exercises 19–24, graph each conic.

19. $xy + 2 = 0$
20. $xy = 1$
21. $x^2 + 4xy + y^2 - 3 = 0$
22. $x^2 + 2xy + y^2 + x - y = 0$
23. $5x^2 + 6xy + 5y^2 - 8 = 0$
24. $2\sqrt{3}x^2 - 6xy + \sqrt{3}x + 3y = 0$

Extensions

In Exercises 25–28, for each conic below, determine the angle of rotation from the original x- and y-axes to the axes of symmetry. Round to the nearest degree.

25. $9x^2 - 24xy + 16y^2 - 400x - 300y = 0$
26. $8x^2 + 12xy + 13y^2 - 34 = 0$
27. $4xy + 3y^2 + 4x + 6y = 0$
28. $x^2 + 2xy + y^2 + x + y - 2 = 0$

In Exercises 29–32, if P' is the image of P, find the coordinates of P' for each angle of rotation counterclockwise about the origin: $\pi/2$, π, and $3\pi/2$.

29. $P(5, 3)$
30. $P(-4, 1)$
31. $P(8, -6)$
32. $P(-7, -2)$

In Exercises 33–34, find the equation of the conic after each given conic is rotated counterclockwise $\pi/2$ about the origin.

33. $\dfrac{x^2}{4} + \dfrac{y^2}{9} = 1$

34. $\dfrac{x^2}{25} - \dfrac{y^2}{16} = 1$

In Exercises 35–36, find the equation of the conic after each given conic is rotated counterclockwise $3\pi/2$ about the origin.

35. $y^2 = x$

36. $\dfrac{x^2}{81} + \dfrac{y^2}{36} = 1$

In Exercises 37–38, find the equation of the conic after each given conic is rotated counterclockwise π about the origin.

37. $(x + 3)^2 - 3(y + 2)^2 = 9$

38. $6(x - 5)^2 + (y + 1)^2 = 42$

10.4 POLAR EQUATIONS OF CONICS

OBJECTIVES

- Define a conic section geometrically.
- Write the polar equation of a conic.
- Graph the polar equation of a conic.

PREREQUISITE VOCABULARY TERMS

conic section
directrix
focus
polar coordinates
transverse axis

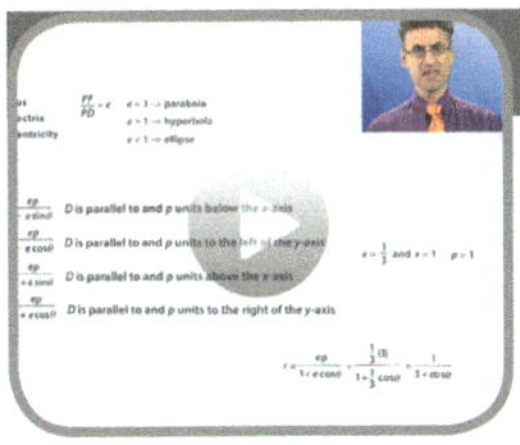

10.4.1 Polar Equations of Conics

In Chapter 3, a parabola was defined as the set of all points that are equidistant from a fixed point (called the parabola's focus, F) and a fixed line (called the parabola's directrix).

Suppose P is a point on the parabola and D is a point on the parabola's directrix such that PD is perpendicular to the directrix, as shown in the table below. We can then define a parabola as the set of all points $P(x, y)$ in a plane such that $PF = PD$. Notice that, in this case, the ratio of PF to PD is 1 (i.e., $PF/PD = 1$).

All points P on a hyperbola or non-circular ellipse can also be defined in terms of the ratio of PF to PD. The following geometric definition applies to conics (parabolas, non-circular ellipses, and hyperbolas) with a focus at the origin.

IMPORTANT

P is any point on the conic. PD is perpendicular to the directrix.

Geometric Definition of a Conic with Focus at the Origin

Suppose a plane contains a point F (called the focus) at the origin, a fixed line (called the directrix) that does not pass through F, and a point D on the directrix. A conic is the set of all points $P(x, y)$ in the plane such that if PD is perpendicular to the directrix, the ratio of PF to PD is constant. This constant, represented by e, is a positive real number called the **eccentricity of the conic.**

$$\frac{PF}{PD} = e$$

PF is the distance between P and F.
PD is the distance between P and D.

Parabola if $e = 1$ | Ellipse if $e < 1$ | Hyperbola if $e > 1$

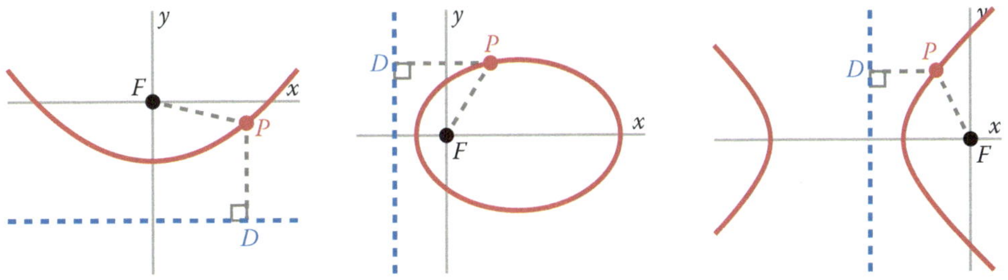

Standard form polar equations of certain conics follow from the fact that $PF/PD = e$.

Polar Equations of Conics

A polar equation for a conic with its focus at the origin and its directrix parallel to either the horizontal or vertical axes (i.e., the x- or y-axis for a rectangular plane) can be derived from the geometric definition of a conic.

Suppose the directrix of a conic is a line parallel to, and p units below, the horizontal axis (Figure 10.4a). It follows that the vertical distance between any point on the directrix and the horizontal axis is p, and $PD = p + r \sin \theta$. Additionally, because the point $P(r, \theta)$ is r units from the origin (the focus of the conic), $PF = r$.

By substituting these expressions for PF and PD into the equation $PF/PD = e$, we see that the polar graph of a conic with focus at the origin and directrix parallel to, and p units below, the horizontal axis is the set of all points $P(r, \theta)$, such that $r/(p + r \sin \theta) = e$.

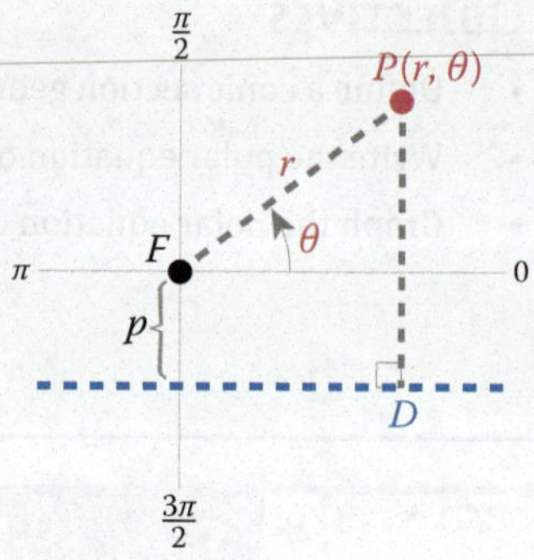

Figure 10.4a

IMPORTANT

If the directrix of a conic is parallel to, and p units below, the horizontal axis (i.e., the x-axis), then it is a horizontal line for which the rectangular equation is $y = -p$.

Solving $r/(p + r \sin \theta) = e$ for r gives the standard form polar equation for this type of conic.

$r/(p + r \sin \theta) = e$	
$r = e(p + r \sin \theta)$	*Multiply both sides by the denominator.*
$r = ep + er \sin \theta$	*Distribute e.*
$r - er \sin \theta = ep$	*Subtract er sin θ from both sides.*
$r(1 - e \sin \theta) = ep$	*Factor r from each term.*
$r = ep/(1 - e \sin \theta)$	*Divide each side by 1 − e sin θ to solve for r.*

Therefore, the polar equation for this type of conic (i.e., with focus at the origin and directrix parallel to, and p units below, the horizontal axis) is given by $r = \dfrac{ep}{1 - e \sin \theta}$, where e is the eccentricity of the conic.

The equations for the three other types of conics with focus at the origin and directrix parallel to the horizontal or vertical axes (which can be derived in a similar way) are given in the following table.

Polar Equations of Conics: F(0, 0) and Directrix Parallel to the Horizontal or Vertical Axes	
$r = \dfrac{ep}{1 - e \sin \theta}$	The rectangular equation of the directrix is $y = -p$. *The directrix is parallel to, and p units below, the horizontal axis.*
$r = \dfrac{ep}{1 - e \cos \theta}$	The rectangular equation of the directrix is $x = -p$. *The directrix is parallel to, and p units to the left of, the vertical axis.*
$r = \dfrac{ep}{1 + e \sin \theta}$	The rectangular equation of the directrix is $y = p$. *The directrix is parallel to, and p units above, the horizontal axis.*
$r = \dfrac{ep}{1 + e \cos \theta}$	The rectangular equation of the directrix is $x = p$. *The directrix is parallel to, and p units to the right of, the vertical axis.*

EXAMPLE 1 Identifying a Conic by Its Polar Equation

Identify the type of conic given by each polar equation, then write the equation of its directrix.

A. $r = \dfrac{2}{4 - \cos\theta}$ **B.** $r = \dfrac{9}{3 + 5\sin\theta}$

SOLUTION

A. Write the equation in the standard form of a conic with its focus at the origin.

$$r = \frac{2}{4 - \cos\theta} = \frac{2}{4\left(1 - \frac{1}{4}\cos\theta\right)} = \frac{\frac{1}{2}}{1 - \frac{1}{4}\cos\theta}$$

Thus, $e = 1/4$. Since $e < 1$, the conic is an ellipse.

Because the ellipse's equation is of the form $r = \dfrac{ep}{1 - e\cos\theta}$, its directrix is the vertical line for which the rectangular equation is $x = -p$.

From the numerator of the equation of the ellipse, $ep = 1/2$. Substituting $e = 1/4$ into this equation gives $(1/4)p = 1/2$. Therefore, $p = 2$, meaning the directrix is the vertical line $x = -2$.

B. Write the equation in the standard form of a conic with its focus at the origin.

$$r = \frac{9}{3 + 5\sin\theta} = \frac{9}{3\left(1 + \frac{5}{3}\sin\theta\right)} = \frac{3}{1 + \frac{5}{3}\sin\theta}$$

Therefore, $e = 5/3$. Since $e > 1$, the conic is a hyperbola.

Because the hyperbola's equation is of the form $r = \dfrac{ep}{1 + e\sin\theta}$, its directrix is the horizontal line for which the rectangular equation is $y = p$.

From the numerator of the equation of the hyperbola, $ep = 3$. Substituting $e = 5/3$ into this equation gives $(5/3)p = 3$. Therefore, $p = 9/5$, meaning the directrix is the vertical line $y = 9/5$.

EXAMPLE 2 Writing the Polar Equation of a Conic

Write the polar equation of a conic with focus at the origin if its directrix is the line $x = -2$ and its eccentricity is 1.

SOLUTION

Since the directrix is the line $x = -2$, the directrix is a vertical line parallel to and 2 units to the left of the vertical axis. So, the conic is given by an equation of the form $r = \dfrac{ep}{1 - e\cos\theta}$.

Since p is the distance between the directrix and the parallel axis, $p = 2$.

Substitute $e = 1$ and $p = 2$ into the standard form equation to write the equation of the conic.

$$r = \frac{ep}{1 - e\cos\theta} = \frac{1 \cdot 2}{1 - 1\cos\theta} = \frac{2}{1 - \cos\theta}$$

10.4.2 Graphing Polar Equations of Conics

To graph the polar equation of a conic with its directrix, first write the equation in one of the four standard polar forms of a conic with focus at the origin. Next, identify e and p from the equation. Use e to determine whether the conic is an ellipse (if $e < 1$), a parabola (if $e = 1$), or a hyperbola (if $e > 1$). Then, using the information regarding the directrix given from the standard form equation and p, graph the directrix.

We can use a conic's directrix to determine whether the conic is horizontal or vertical because the directrix of an ellipse or a hyperbola is perpendicular to its transverse axis, and because the directrix of a parabola is perpendicular to its axis of symmetry. So, if a conic's directrix is horizontal, the conic is vertical, and if a conic's directrix is vertical, the conic is horizontal.

Make a table to identify some of the points on the conic. Then, plot the points from the table and sketch the conic through those points, keeping in mind the type and direction of the conic.

EXAMPLE 3 Graphing the Polar Equation of a Conic

IMPORTANT

If a polar equation can be written in one of the four standard forms of a conic, then the conic's focus is at the origin.

Graph the conic and its directrix. $r(5 - 10 \sin \theta) = 8$

SOLUTION

Write the equation in the standard form of a conic with focus at the origin. Begin by dividing both sides by $5 - 10 \sin \theta$ in order to solve the equation for r.

$$r = \frac{8}{5 - 10 \sin \theta} = \frac{8}{5(1 - 2 \sin \theta)} = \frac{\frac{8}{5}}{1 - 2 \sin \theta}$$

Therefore, $e = 2$. Since $e > 1$, the conic is a hyperbola.

The directrix of a conic of the form $r = \dfrac{ep}{1 - e \sin \theta}$ is the horizontal line $y = -p$.

Since $e = 2$ and $ep = 8/5$, it follows that $2p = 8/5$, and $p = 4/5$. Thus, the rectangular equation of the directrix is $y = -4/5$.

REMEMBER

A hyperbola's transverse axis is perpendicular to its directrix.

Plot the focus at the origin and sketch the directrix (the line $y = -4/5$). The conic is a vertical hyperbola because the directrix is horizontal. Since the focus is at the origin and the hyperbola is vertical, the hyperbola's vertices must be on the vertical axis (i.e., the y-axis).

Make a table to identify some of the points on the hyperbola.

r	$8/5$	$-8/5$	$8/5$	$8/15$
θ	0	$\pi/2$	π	$3\pi/2$

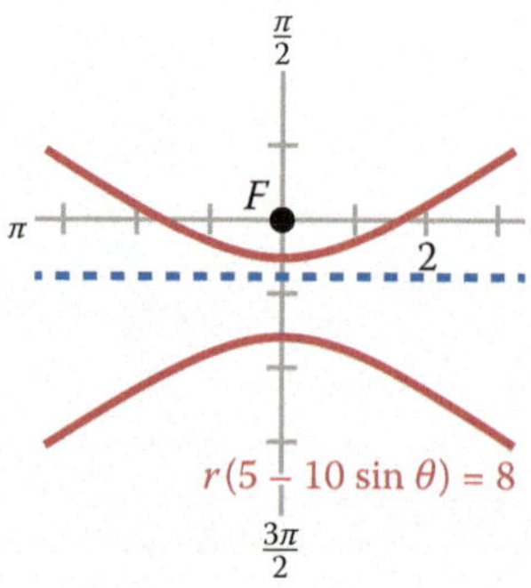

Sketch the vertical hyperbola through these points. The vertices are on the vertical axis, so the polar coordinates of the hyperbola's vertices are $(-8/5, \pi/2)$ and $(8/15, 3\pi/2)$. Note that both of these points are below the horizontal axis.

SECTION 10.4 EXERCISES

Warm Up

1. Find the focus and directrix of the graph of $2y^2 = 6x$.
2. Find the foci and eccentricity of the graph of $100x^2 + 50y^2 - 300 = 0$.
3. Find the foci and vertices of the graph of $\frac{y^2}{9} - x^2 = 1$.

Just the Facts

Fill in the blanks.

4. If the eccentricity of a conic _____ 1, then the conic is a parabola. If the eccentricity of a conic _____ 1, then the conic is an ellipse. If the eccentricity of a conic _____ 1, then the conic is a hyperbola.
5. The polar equation of a conic with a focus at the origin and a directrix parallel to, and p units to the left of, the vertical axis is _____.
6. The polar equation of a conic with a focus at the origin and a directrix parallel to, and p units to the right of, the vertical axis is _____.
7. The polar equation of a conic with a focus at the origin and a directrix parallel to, and p units above, the horizontal axis is _____.
8. The polar equation of a conic with focus at the origin and a directrix parallel to, and p units below, the horizontal axis is _____.
9. A horizontal directrix is either _____ to the horizontal axis or it coincides with the horizontal axis.
10. The major axis of an ellipse is _____ to its directrix.

Essential Skills

In Exercises 1–6, identify the type of conic given by each polar equation. Write the rectangular equation of its directrix and state its eccentricity.

1. $r = \dfrac{5}{10 - 15\sin\theta}$
2. $r = \dfrac{9}{2 + 3\cos\theta}$
3. $r = \dfrac{4}{6 - \cos\theta}$
4. $r = \dfrac{4}{1 - 3\sin\theta}$
5. $r = \dfrac{3}{2 + 6\sin\theta}$
6. $r = \dfrac{2}{1 - \cos\theta}$

In Exercises 7–12, write the polar equation of each conic (focus at the origin) with the given directrix and eccentricity.

7. $x = 3$, $e = 2/3$
8. $y = 3$, $e = 4$
9. $y = 2$, $e = 1$
10. $x = -3$, $e = 4/3$
11. $x = -1$, $e = 3/2$
12. $y = -3$, $e = 1$

In Exercises 13–18, graph each conic. Write the rectangular equation of its directrix and state its eccentricity.

13. $r = \dfrac{1}{1 + \sin\theta}$
14. $2r - 4r\cos\theta = 3$
15. $3r - 2r\sin\theta = 6$
16. $4r + 8r\sin\theta = 12$
17. $r(3 - 9\sin\theta) = 6$
18. $5r + 4r\cos\theta = 10$

Extensions

In Exercises 19–22, write each polar equation as a rectangular equation in general form.

19. $r = \dfrac{12}{1 + 4\sin\theta}$
20. $r = \dfrac{2}{1 - \cos\theta}$
21. $r = \dfrac{9}{2 + 3\cos\theta}$
22. $r = \dfrac{5}{10 - 15\sin\theta}$

In Exercises 23–26, find a polar equation for a parabola with focus at the pole and the given polar coordinates as its vertex.

23. $(6, \pi)$
24. $(3, 0)$
25. $(2, \pi/2)$
26. $(5, 3\pi/2)$

In Exercises 27–30, find a polar equation for each ellipse with a focus at the pole and the given polar coordinates as the endpoints of its major axis.

27. $(2, 0)$ and $(8, \pi)$

28. $(1, \pi/2)$ and $(7, 3\pi/2)$

29. $(3, \pi/2)$ and $(6, -\pi/2)$

30. $(4, 0)$ and $(2, \pi)$

In Exercises 31–34, find a polar equation for each hyperbola with a focus at the pole and the given polar coordinates as endpoints of its transverse axis.

31. $(4, \pi/2)$ and $(-8, 3\pi/2)$

32. $(1, \pi)$ and $(-5, 2\pi)$

33. $(2, 0)$ and $(-12, \pi)$

34. $(3, -\pi/2)$ and $(-6, \pi/2)$

In Exercises 35–38, graph each conic. Write the rectangular equation of its directrix and state its eccentricity.

35. $r = \dfrac{6 \sec \theta}{2 \sec \theta - 1}$

36. $r = \dfrac{12 \csc \theta}{3 \csc \theta + 2}$

37. $r = \dfrac{4 \csc \theta}{2 \csc \theta + 3}$

38. $r = \dfrac{\sec \theta}{\sec \theta - 2}$

10.5 PLANE CURVES AND PARAMETRIC EQUATIONS

OBJECTIVES

- Graph parametric equations.
- Write the rectangular form equation of a plane curve.
- Find parametric equations that describe the graph of a rectangular equation.

PREREQUISITE VOCABULARY TERMS

absolute value function
parabola
Pythagorean identities
trigonometric functions

10.5.1 Graphing Plane Curves

The equations modeling real world situations in the preceding chapters are written in terms of either one or two variables, typically x and y (or r and θ for polar equations). For example, consider a soccer ball kicked into the air. The path the ball takes can be described by an equation written in terms of x and y, where x is the ball's horizontal position relative to its starting point (i.e., the distance the ball has traveled along the ground from its starting point), and y is the ball's vertical position relative to its starting point (i.e., the distance between the ball and the ground).

However, both variables (x and y) are functions of time t because any point along the path of the ball is defined by the ball's horizontal position x at time t and its vertical position y at that same time t. So, the ball's path can be described by a pair of functions, $f(t)$ and $g(t)$, where $f(t)$ is the ball's horizontal distance from the starting point at time t and $g(t)$ is the ball's vertical height above the ground at time t. These two functions of the same variable t, $f(t)$ and $g(t)$, are called **parametric equations**.

IMPORTANT

$f(t)$ and $g(t)$ are functions of the same variable, t. They can also be expressed using other letters, such as $x(t)$ and $y(t)$.

Parametric Equations

Let $f(t)$ and $g(t)$ be continuous functions on some interval I of real numbers. A **plane curve** is the set of all ordered pairs $(f(t), g(t))$. The equations $x = f(t)$ and $y = g(t)$ are the parametric equations for the plane curve, where t (the independent variable) is called the **parameter**.

TIP

If the parametric equations $x = f(t)$ and $y = g(t)$ describe a ball's horizontal and vertical position, then the plane curve represents the path of the ball.

One method for graphing a plane curve given by a pair of parametric equations involves making a table of ordered pairs $(f(t), g(t))$. Since the parametric equations $x = f(t)$ and $y = g(t)$ are in terms of three variables, a table of ordered pairs $(f(t), g(t))$ will have three columns. The first column contains the t-values, the second column contains the $f(t)$-values, and the third column contains the $g(t)$-values.

After making the table, plot the ordered pairs $(f(t), g(t))$ as points on an xy-plane. Then draw the plane curve through the points. If the interval I is not all real numbers, then the plane curve will have endpoints.

Arrowheads drawn along the curve indicate the **orientation** of the plane curve, which is the direction of the curve as the values of the parameter increase.

EXAMPLE 1 Graphing a Plane Curve

Graph the plane curve defined by each pair of parametric equations.

A. $x(t) = t + 3$
$y(t) = 4t - 4$
$0 \le t \le 7$

B. $x(t) = 2t^2 - 4$
$y(t) = 2t$
$-2 \le t \le 3$

C. $x(t) = \sin t$
$y(t) = \cos t$
$-\pi/2 \le t \le \pi/2$

SOLUTION

A. The plane curve defined by $x(t) = t + 3$ and $y(t) = 4t - 4$ is all points $(x(t), y(t))$ such that $0 \le t \le 7$. Make a table of ordered pairs $(x(t), y(t))$ using t-values such that $0 \le t \le 7$.

TIP

The t-values chosen for the table do not have to be integers, but the t-values must be in the given interval. The interval's endpoints should always be chosen for the table.

Substitute a t-value into $x(t) = t + 3$ to find the $x(t)$-coordinate of a point on the plane curve. Then substitute that same t-value into $y(t) = 4t - 4$ to find the corresponding $y(t)$-coordinate of the point on the plane curve.

Repeat the process for each t-value in the table. From the table, the plane curve passes through the points (3, –4), (4, 0), (5, 4), (6, 8), (7, 12), (8, 16), (9, 20), and (10, 24).

t	$x(t)$	$y(t)$
0	3	–4
1	4	0
2	5	4
3	6	8
4	7	12
5	8	16
6	9	20
7	10	24

Furthermore, since the endpoints of the interval are where $t = 0$ and $t = 7$, the endpoints of the plane curve are (3, –4) and (10, 24).

IMPORTANT

Since this plane curve is all points (x(t), y(t)) such that 0 ≤ t ≤ 7, the plane curve is a continuous curve, not a set of discrete points.

Sketch the plane curve through the points and include the endpoints (3, –4) and (10, 24). Draw an arrowhead pointing to the second endpoint to indicate the plane curve's orientation.

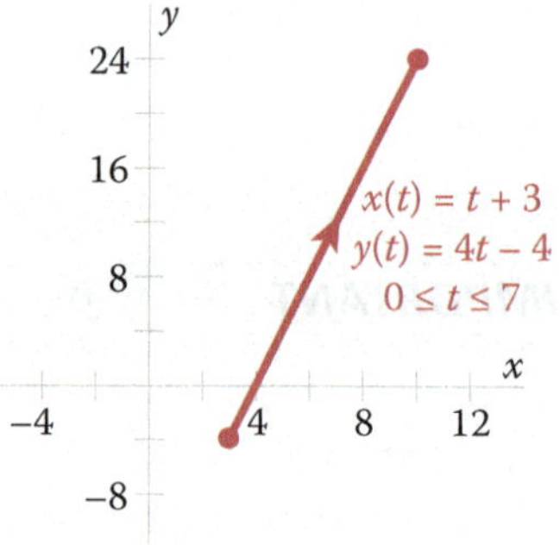

B. The plane curve defined by $x(t) = 2t^2 - 4$ and $y(t) = 2t$ is all points $(x(t), y(t))$ such that $-2 \le t \le 3$. Make a table of ordered pairs $(x(t), y(t))$ using t-values such that $-2 \le t \le 3$.

Substitute a t-value into $x(t) = 2t^2 - 4$ to find the $x(t)$-coordinate of a point on the plane curve. Then substitute that same t-value into $y(t) = 2t$ to find the corresponding $y(t)$-coordinate of the point on the plane curve.

Repeat the process for each t-value in the table. From the table, the plane curve passes through the points (4, –4), (–2, –2), (–4, 0), (–2, 2), (4, 4), and (14, 6).

t	$x(t)$	$y(t)$
–2	4	–4
–1	–2	–2
0	–4	0
1	–2	2
2	4	4
3	14	6

NOTICE THAT

This plane curve does not pass the Vertical Line Test. Therefore, a plane curve given by parametric equations is not always the graph of a function.

Furthermore, since the endpoints of the interval are where $t = -2$ and $t = 3$, the endpoints of the plane curve are (4, –4) and (14, 6).

Sketch the plane curve through the points and include the endpoints (4, –4) and (14, 6). Draw arrowheads pointing to the second endpoint to indicate the plane curve's orientation.

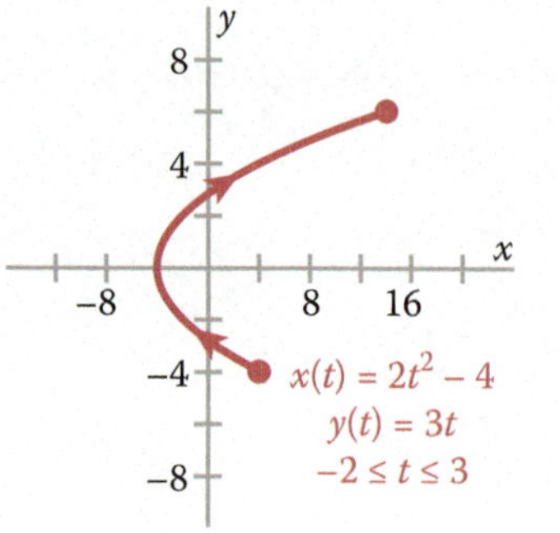

C. The plane curve defined by $x(t) = \sin t$ and $y(t) = \cos t$ is all points $(x(t), y(t))$ such that $-\pi/2 \le t \le \pi/2$. Make a table of ordered pairs $(x(t), y(t))$ using t-values such that $-\pi/2 \le t \le \pi/2$.

Substitute a t-value into $x(t) = \sin t$ to find the $x(t)$-coordinate of a point on the plane curve. Then substitute that same t-value into $y(t) = \cos t$ to find the corresponding $y(t)$-coordinate of the point on the plane curve.

Repeat the process for each t-value in the table.

t (rad)	$x(t)$	$y(t)$
$-\pi/2$	-1	0
$-\pi/4$	$-\sqrt{2}/2$	$\sqrt{2}/2$
0	0	1
$\pi/4$	$\sqrt{2}/2$	$\sqrt{2}/2$
$\pi/2$	1	0

From the table, the plane curve passes through the points $(-1, 0)$, $\left(-\sqrt{2}/2, \sqrt{2}/2\right)$, $(0, 1)$, $\left(\sqrt{2}/2, \sqrt{2}/2\right)$, and $(1, 0)$.

The endpoints of the interval are where $t = -\pi/2$ and $t = \pi/2$, so the endpoints of the plane curve are $(-1, 0)$ and $(1, 0)$.

Sketch the plane curve through the points and include the endpoints $(-1, 0)$and $(1, 0)$. Draw arrowheads pointing to the second endpoint to indicate the plane curve's orientation.

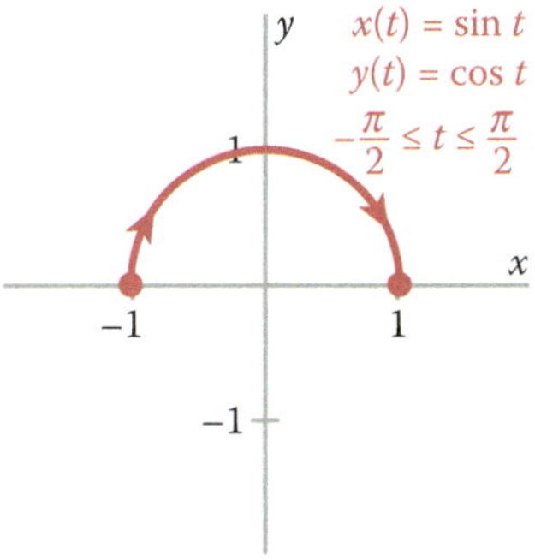

TIP

Indicate the plane curve's orientation using one arrowhead in the interval where the curve is increasing and another in the interval where the curve is decreasing.

10.5.2 Eliminating the Parameter

Figure 10.5a shows the plane curve for the parametric equations ($x(t) = t + 3$ and $y(t) = 4t - 4$), which are given in **Example 1a**, with no restrictions on the interval for t (i.e., $-\infty < t < \infty$). Notice that the unrestricted plane curve is a line with slope 4 and y-intercept -16. So, the rectangular equation of this line in terms of x and y is $y = 4x - 16$.

Generally, a plane curve can be described by parametric equations, or it can be described by a rectangular equation, which does not contain the parameter t.

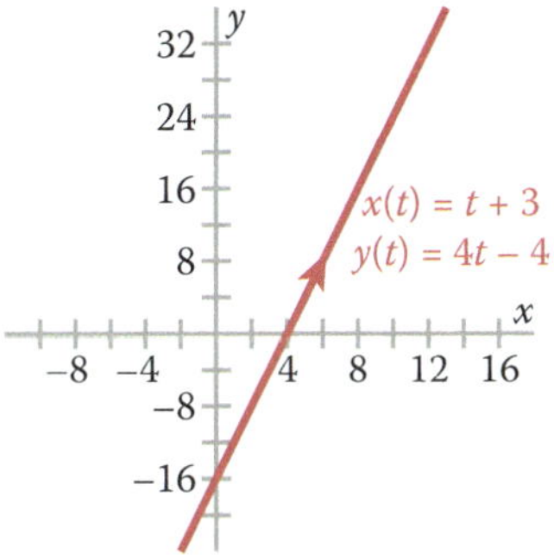

Figure 10.5a

REMEMBER

The equation of a line can be found by substituting its slope and y-intercept into slope-intercept form, $y = mx + b$.

A technique called **eliminating the parameter** is a process by which two parametric equations are written as a single rectangular equation which does not contain the parameter.

Steps for Eliminating the Parameter from Parametric Equations

❶ Solve either one of the parametric equations for the parameter t.

❷ Substitute the expression for t from step ❶ into the other parametric equation and simplify (eliminating the variable t).

TIP

The equations $x(t) = t + 3$ and $y(t) = 4t - 4$ can also be written as $x = t + 3$ and $y = 4t - 4$, respectively.

For example, to eliminate the parameter for the parametric equations $x(t) = t + 3$ and $y(t) = 4t - 4$, solve $x(t) = t + 3$ for t, then substitute the expression for t into $y(t) = 4t - 4$, as follows.

$x(t) = t + 3 \Rightarrow t = x - 3$ *Solve for t.*

$y = 4(x - 3) - 4 = 4x - 12 - 4 = 4x - 16$ *Substitute x – 3 for t and simplify.*

Therefore, the parametric equations can be expressed as the rectangular equation $y = 4x - 16$.

EXAMPLE 2 Eliminating the Parameter

Eliminate the parameter in each pair of parametric equations.

A. $x(t) = \sqrt{t}$, $y(t) = t - 2$

B. $x(t) = |t + 2|$, $y(t) = t - 4$

SOLUTION

TIP

Start by solving either equation for t.

A. Solve $x(t) = \sqrt{t}$ for t. $x(t) = \sqrt{t} \Rightarrow t = x^2$ when $t \geq 0$

Substitute $t = x^2$ into the second equation $y(t) = t - 2$. $y = t - 2 = x^2 - 2$

Therefore, $y = x^2 - 2$ for all $x \geq 0$.

ALTERNATIVE METHOD

First, solve $y(t)$ for t. $y(t) = t - 2 \Rightarrow t = y + 2$

IMPORTANT

The square root expression must be greater than or equal to 0, so $x \geq 0$.

Substitute $t = y + 2$ into the first equation. $x = \sqrt{t} = \sqrt{y+2}$

Square both sides to solve for y. Thus, $y = x^2 - 2$ for all $x \geq 0$.

TIP

Solving $y(t) = t - 4$ for t is easier than solving $x(t) = |t + 2|$ for t, so start with $y(t) = t - 4$.

B. Solve $y(t) = t - 4$ for t. $y(t) = t - 4 \Rightarrow t = y + 4$

Substitute $t = y + 4$ into the first equation $x(t) = |t + 2|$.

$$x = |t + 2| = |y + 4 + 2| = |y + 6|$$

Therefore, the rectangular equation for the plane curve described by the given parametric equations is $x = |y + 6|$.

When the parametric equations involve trigonometric functions, it is sometimes easier to use a trigonometric identity to eliminate the parameter rather than solving for the parameter in one equation and substituting that expression into the remaining equation. This process is demonstrated in **Example 3**.

EXAMPLE 3 Eliminating the Parameter

Eliminate the parameter in each pair of parametric equations.

A. $x(t) = \sin t$
$y(t) = \cos t$

B. $x = 5 \tan t$
$y = 5 \sec t$

SOLUTION

A. Since the Pythagorean identity $\sin^2 t + \cos^2 t = 1$ gives a relationship between $\sin t$ and $\cos t$ (the two trigonometric functions in the given parametric equations), use $\sin^2 t + \cos^2 t = 1$ to write a relationship between x and y.

$\sin^2 t + \cos^2 t = 1$ *Pythagorean Identity*

$x^2 + y^2 = 1$ *Substitute x for sin t and y for cos t.*

NOTICE THAT

The plane curve is a circle centered at the origin with radius 1.

B. Since the Pythagorean identity $1 + \tan^2 t = \sec^2 t$ gives a relationship between $\tan t$ and $\sec t$ (the two trigonometric functions in the given parametric equations), use $1 + \tan^2 t = \sec^2 t$ to write a relationship between x and y. Note that solving $x = 5 \tan t$ for $\tan t$ gives $\tan t = x/5$, and solving $y = 5 \sec t$ for $\sec t$ gives $\sec t = y/5$.

$1 + \tan^2 t = \sec^2 t$ *Pythagorean Identity*

$1 + (x/5)^2 = (y/5)^2$ *Substitute x/5 for tan t and y/5 for sec t.*

$1 + x^2/25 = y^2/25$ *Simplify.*

$1 = y^2/25 - x^2/25$ *Subtract $x^2/25$ from both sides.*

NOTICE THAT

The plane curve is a vertical hyperbola centered at the origin with vertices at (0, ±5) and co-vertices at (±5, 0).

Eliminating the Parameter to Graph a Plane Curve

Consider again the plane curve for the parametric equations from **Example 1a**. As shown in Figure 10.5b, $x(t) = t + 3$ and $y(t) = 4t - 4$ when $0 \le t \le 7$.

Recall that the plane curve for parametric equations $x(t) = t + 3$ and $y(t) = 4t - 4$ can also be described by the rectangular equation $y = 4x - 16$, which is the result of eliminating the parameter from the original parametric equations.

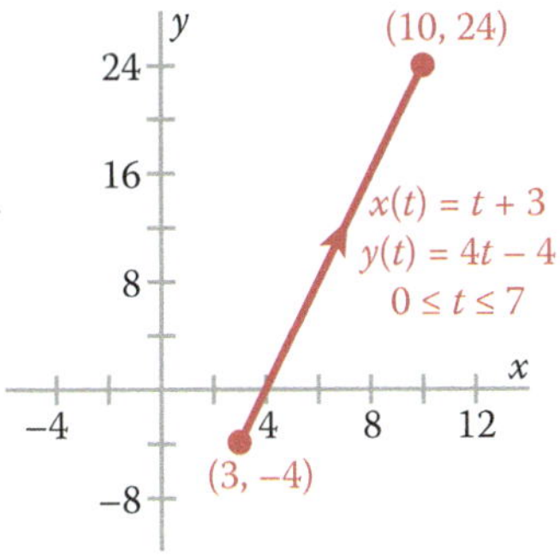

Figure 10.5b

However, the interval for x must be restricted in some way in order for $y = 4x - 16$ to represent the plane curve for $x(t) = t + 3$ and $y(t) = 4t - 4$ when $0 \le t \le 7$. If the $x(t)$-values are strictly increasing or decreasing as t increases (for all t within the interval), then the interval for x is given by writing the interval for t in terms of x. Specifically, since $x = t + 3$, the interval $0 \le t \le 7$ can be written in terms of x by adding 3 to each part of the inequality.

$0 \le t \le 7 \Rightarrow 0 + 3 \le t + 3 \le 7 + 3 \Rightarrow 3 \le x \le 10$ *Simplify and substitute x for t + 3.*

Therefore, the plane curve for $x(t) = t + 3$ and $y(t) = 4t - 4$ when $0 \le t \le 7$ can also be represented by $y = 4x - 16$ when $3 \le x \le 10$. Notice that the graph shown in Figure 10.5b confirms the interval $3 \le x \le 10$, as the x-coordinates of the left and right endpoints are 3 and 10 respectively.

Steps for Graphing a Plane Curve by Eliminating the Parameter

❶ Write the parametric equations as a single rectangular equation in terms of x and y by eliminating the parameter t from the parametric equations.

❷ Find the interval for x (or y).

❸ Graph the rectangular equation over the interval for x (or y).

TIP

The interval for x and y can sometimes be found by making a table of t-, $x(t)$-, and $y(t)$-values.

EXAMPLE 4 Graphing a Plane Curve by Eliminating the Parameter

Graph the plane curve described by the parametric equations.

A. $x(t) = \sqrt{t}$
$y(t) = t - 2$
$0 \le t \le 16$

B. $x(t) = |t + 2|$
$y(t) = t - 4$
$-4 \le t \le 8$

SOLUTION

A. From **Example 2A**, we know that the plane curve described by the given parametric equations has rectangular equation $y = x^2 - 2$. The graph of $y = x^2 - 2$ is a vertical parabola opening upward. It is the image of $y = x^2$ after a vertical translation down 2 units, so the vertex is at $(0, -2)$.

Since $x(t) = \sqrt{t}$ is strictly increasing for all t such that $0 \le t \le 16$, use the interval for t, $0 \le t \le 16$, and the equation for x, $x(t) = \sqrt{t}$, to find the interval for x.

$\sqrt{0} \le \sqrt{t} \le \sqrt{16}$ *Take the square root of 0, t, and 16.*

$0 \le x \le 4$ *Substitute x for $\sqrt{t}$ and simplify.*

So, the plane curve of the given parametric equations is the parabola described by $y = x^2 - 2$ when $0 \le x \le 4$.

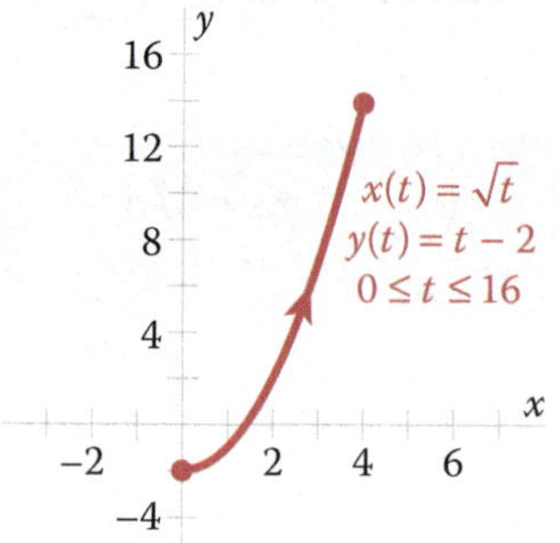

CHECK

Confirm several points on the plane curve (including the endpoints) by making a table for the parametric equations. Since $x(t) = \sqrt{t}$, choose perfect-square t-values from the interval $0 \le t \le 16$ for the table so that the $x(t)$-coordinates are whole numbers.

From the table, the plane curve passes through the points $(0, -2)$, $(1, -1)$, $(2, 2)$, $(3, 7)$, and $(4, 14)$, which confirms the plane curve.

t	0	1	4	9	16
$x(t)$	0	1	2	3	4
$y(t)$	−2	−1	2	7	14

B. In **Example 2B** it was shown that the plane curve described by the given parametric equations has the rectangular equation $x = |y + 6|$. The graph of $x = |y + 6|$ is an absolute value graph opening to the right. It is the image of $x = |y|$ after a vertical translation down 6 units.

Since $x(t) = |t + 2|$ decreases and then increases as t increases for all t such that $-4 \le t \le 8$, but $y(t) = t - 4$ only increases as t increases for all t such that $-4 \le t \le 8$, use the interval for t, $-4 \le t \le 8$, and the equation for y, $y = t - 4$, to find the interval for y.

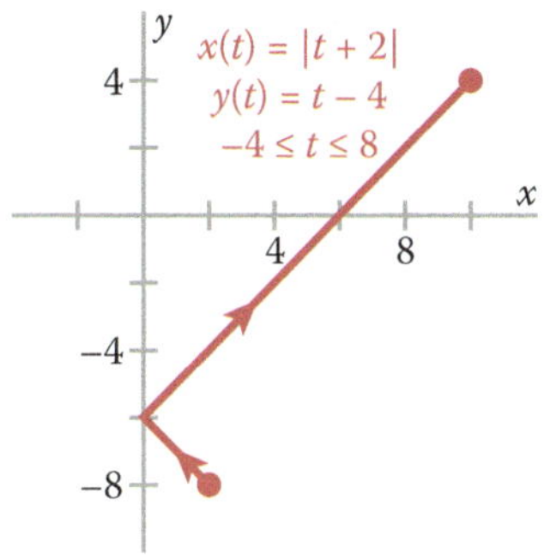

$-4 - 4 \le t - 4 \le 8 - 4$ *Subtract 4 from −4, t, and 8.*

$-8 \le y \le 4$ *Substitute y for t − 4 and simplify.*

Therefore, the plane curve of the given parametric equations is the absolute value graph described by $x = |y + 6|$ when $-8 \le y \le 4$.

TIP

Since the rectangular equation is solved for x, find the interval for y instead of x.

ALTERNATIVE METHOD

Make a table for all integer t-values such that $-4 \le t \le 8$.

t	−4	−3	−2	−1	0	1	2	3	4	5	6	7	8
$x(t)$	2	1	0	1	2	3	4	5	6	7	8	9	10
$y(t)$	−8	−7	−6	−5	−4	−3	−2	−1	0	1	2	3	4

From the table, the least $x(t)$-value is 0 and the greatest $x(t)$-value is 10. Therefore, the interval for x is $0 \le x \le 10$, which is confirmed by the plane curve.

The interval for y, $-8 \le y \le 4$, is confirmed by the table because the least $y(t)$-value is −8 and the greatest $y(t)$-value is 4.

10.5.3 Writing Parametric Equations

Just as a plane curve for a given pair of parametric equations can also be described by a rectangular equation, the graph of any given rectangular equation $y = f(x)$ can be described by a pair of parametric equations in terms of a third variable, t.

Unlike rectangular equations which have only one simplified form, there are an infinite number of possible pairs of parametric equations that represent a given plane curve. One such pair of parametric equations can be determined by letting x be equal to the parameter t. In this case, $x(t) = t$, and so $y(t) = f(t)$.

NOTICE THAT

Substituting $t = x$ into $y = f(t)$ confirms that $y = f(x)$, as claimed.

EXAMPLE 5 Writing Parametric Equations

Find parametric equations $x(t)$ and $y(t)$ to represent the graph of each rectangular equation.

A. $y = 3x - 1$ **B.** $x^2/4 + y^2/9 = 1$

SOLUTION

A. Let $x(t) = t$. Substituting t for x in the rectangular equation $y = 3x - 1$ gives $y = 3t - 1$.

So, the parametric equations $x(t) = t$ and $y(t) = 3t - 1$ represent the graph of $y = 3x - 1$.

Additional pairs of parametric equations can be determined by choosing another expression for x in terms of t, such as $x(t) = 2t$. Substituting $2t$ for x in the rectangular equation $y = 3x - 1$ gives $y = 3(2t) - 1 = 6t - 1$.

Thus, the following pairs of parametric equations each represent the graph of $y = 3x - 1$.

$x(t) = t$	$x(t) = 2t$
$y(t) = 3t - 1$	$y(t) = 6t - 1$

B. Let $x(t) = t$. Since $x^2/4 + y^2/9 = 1$ does not give y explicitly as a function of x, solve the equation for y, then find $y(t)$ by substituting t for x.

$$\frac{y^2}{9} = 1 - \frac{x^2}{4}$$ *Isolate the y^2-term.*

$$y = \pm\sqrt{9\left(1 - \frac{x^2}{4}\right)}$$ *Multiply both sides by 9 and take ± the square root of each side.*

$$y = \pm 3\sqrt{1 - \frac{x^2}{4}}$$ *Simplify.*

IMPORTANT

Since y can take on both positive and negative values, both the positive and negative square roots must be included.

Substituting t for x in the rectangular equation gives $y = \pm 3\sqrt{1 - \frac{t^2}{4}}$.

So, $x(t) = t$ and $y(t) = \pm 3\sqrt{1 - \frac{t^2}{4}}$ represent the graph of $x^2/4 + y^2/9 = 1$.

A second pair of parametric equations can be found using the Pythagorean identity $\sin^2 t + \cos^2 t = 1$. Choose another expression for x in terms of t, such as $x(t) = 2 \sin t$.

NOTICE THAT

The coefficient of $\sin t$ is chosen to be 2 so that the denominator of $x^2/4$ will cancel out after $2 \sin t$ is squared, thus simplifying the process. However, the coefficient does not have to be 2. For example, letting $x(t) = \sin t$ gives yet another pair of parametric equations for $x^2/4 + y^2/9 = 1$.

Substitute $2 \sin t$ for x in the rectangular equation $x^2/4 + y^2/9 = 1$.

$(2 \sin t)^2/4 + y^2/9 = 1$ *Substitute.*

$(4 \sin^2 t)/4 + y^2/9 = 1$ *Power of a Product Property*

$\sin^2 t + y^2/9 = 1$ *Simplify.*

$y^2/9 = 1 - \sin^2 t$ *Isolate the y^2-term.*

$y^2/9 = \cos^2 t$ *Pythagorean Identity*

$y = \pm\sqrt{9 \cos^2 t}$ *Multiply each side by 9 and then take ± the square root.*

$y = \pm 3 \cos t$ *Simplify.*

Thus, the following pairs of parametric equations each represent the graph of $x^2/4 + y^2/9 = 1$.

$x(t) = t$	$x(t) = 2 \sin t$
$y(t) = \pm 3\sqrt{1 - \frac{t^2}{4}}$	$y(t) = \pm 3 \cos t$

SECTION 10.5 EXERCISES

Warm Up

Evaluate each function for domain values {−2, −1, 0, 1, 2}.

1. $y = x^2 - 1$
2. $y = |3x + 4|$
3. $y = \sqrt{x+5}$

Just the Facts

Fill in the blanks.

4. The set of all ordered pairs $(f(t), g(t))$ for all t in the interval I is called a _____.
5. If $x = f(t)$ and $y = g(t)$ are parametric equations for a plane curve, then t, the independent variable, is called the _____.
6. Eliminating the parameter is a process by which two parametric equations are written as a single _____ equation.
7. The graph of any given rectangular equation $y = f(x)$ can be described by a pair of parametric equations in terms of a _____ variable.
8. Arrowheads drawn along a plane curve indicate the _____ of the plane curve.
9. To eliminate the parameter described by the parametric equations $x(t) = \sin t$ and $y(t) = \cos t$, it is easier to use _____ identities than it is to solve for x or y.
10. Some plane curves do not pass the Vertical Line Test. Such plane curves are given by parametric equations, but they are not the graph of a(n) _____.

Essential Skills

In Exercises 1–6, use the given t-values to make a table for each pair of parametric equations.

1. $x(t) = t^2 - 4$; $t = -3, -2, 1, 6$
   $y(t) = \sqrt{t+3}$
2. $x(t) = |1 - t|$; $t = -1, 0, 1, 2$
   $y(t) = \dfrac{1+t^2}{2}$
3. $x(t) = \dfrac{t+2}{t+1}$; $t = -2, 0, 1, 2$
   $y(t) = t^3 + 1$
4. $x(t) = -t^3 + t^2$; $t = -5, 0, 3, 4$
   $y(t) = \sqrt{4-t}$
5. $x(t) = t^2 + 3t$; $t = -1, 0, 1, 2$
   $y(t) = |t + 5|$
6. $x(t) = \dfrac{2}{t-2}$; $t = -1, 0, 1, 3$
   $y(t) = t^2 + 3$

In Exercises 7–18, graph the plane curve defined by each pair of parametric equations.

7. $x(t) = t + 2$
   $y(t) = t^2 - 1$
   $-3 \le t \le 1$
8. $x(t) = \dfrac{t^2}{t+2}$
   $y(t) = \sqrt{1+t}$
   $-1 \le t \le 3$
9. $x(t) = \sqrt{t+1}$
   $y(t) = t - 2$
   $-1 \le t \le 15$
10. $x(t) = \dfrac{t^3}{t+5}$
    $y(t) = \sqrt{1+t}$
    $-1 \le t \le 2$
11. $x(t) = |2t - 1|$
    $y(t) = t^2$
    $-2 \le t \le 2$
12. $x(t) = \dfrac{t^2}{t+4}$
    $y(t) = \sqrt{t+2}$
    $-1 \le t \le 5$
13. $x(t) = 2\cos t$
    $y(t) = 4\sin t$
    $0 \le t \le 3\pi/2$
14. $x(t) = 2\cos t$
    $y(t) = -\dfrac{1}{2}\sin t$
    $-\pi/2 \le t \le \pi/2$
15. $x(t) = \sin t + 1$
    $y(t) = -3\cos t$
    $0 \le t \le \pi$
16. $x(t) = -2\sin t$
    $y(t) = \cos t$
    $0 \le t \le 3\pi/2$

17. $x(t) = 5\cos t - 2$

$y(t) = 3 - \sin t$

$-\pi \le t \le \pi/2$

18. $x(t) = -\cos t$

$y(t) = -\sin t$

$-\pi \le t \le \pi/2$

In Exercises 19–30, eliminate the parameter to find the rectangular equation for the plane curve described by the parametric equations. State the domain of the rectangular equation for the plane curve.

19. $x(t) = \sqrt{t+2}$

$y(t) = t - 3$

20. $x(t) = 4 + 6t$

$y(t) = |2t|$

21. $x(t) = t - 5$

$y(t) = t^2 - 1$

22. $x(t) = |5t|$

$y(t) = 4 - 15t$

23. $x(t) = t^2 - 1$

$y(t) = 5t + 4$

24. $x(t) = 3 + 28t$

$y(t) = |4t|$

25. $x(t) = 2\cos t$

$y(t) = \sin t$

26. $x(t) = -\cos t$

$y(t) = 3\sin t$

27. $x(t) = -4\csc t$

$y(t) = 2\cot t$

28. $x(t) = 3\tan t$

$y(t) = -\sec t$

29. $x(t) = \dfrac{\sec t}{2}$

$y(t) = \dfrac{\tan t}{3}$

30. $x(t) = -5\cot t$

$y(t) = -2\csc t$

In Exercises 31–36, graph the plane curve described by the parametric equations.

31. $x(t) = 4t + 1$

$y(t) = -3t + 2$

$0 \le t \le 4$

32. $x(t) = \dfrac{1}{2}t + 3$

$y(t) = t + 1$

$-2 \le t \le 2$

33. $x(t) = |-2t| + 4$

$y(t) = t + 3$

$-3 \le t \le 3$

34. $x(t) = 0.2t - 5$

$y(t) = -7t - 0.6$

$-4 \le t \le 4$

35. $x(t) = -2t + 3$

$y(t) = \sqrt{t+2}$

$-2 \le t \le 2$

36. $x(t) = -t/3 + 2$

$y(t) = 4t - 0.7$

$-2 \le t \le 2$

In Exercises 37–42, find parametric equations $x(t)$ and $y(t)$ to represent the graph of each rectangular equation.

37. $y = -x^2$

38. $y = |2 - x|$

39. $y = x^2 + 3$

40. $y = 4x^2 - 7$

41. $x^2/9 + y^2 = 1$

42. $y = |15 - 3x|$

Extensions

In Exercises 43–47, describe each plane curve for parametric equations $x(t)$ and $y(t)$. Assume that $x(t)$ and $y(t)$ are defined for all real numbers.

43. $x(t) = y(t)$

44. $x(t) = -y(t)$

45. $x(t) = 2(y(t))$

46. $-3(x(t)) = 4(y(t))$

47. $2(x(t)) = y(t)$

In Exercises 48–52, graph the plane curve described by the parametric equations.

48. $x(t) = -\cos 2t$

$y(t) = \sin t$

$-\pi/4 \le t \le \pi/4$

49. $x(t) = \cos t$

$y(t) = \sin 2t$

$-\pi/2 \le t \le \pi/4$

50. $x(t) = \sin 3t$

$y(t) = \cos 2t$

$-\pi/2 \le t \le \pi/6$

51. $x(t) = -\sin 4t$

$y(t) = \cos 2t$

$-\pi/8 \le t \le \pi/2$

52. $x(t) = -\cos 2t$

$y(t) = -2 \sin t$

$-\pi/2 \le t \le \pi/2$

53. True or False? The two sets of parametric equations $x(t) = 5t$, $y(t) = t^3$ and $x(t) = t$, $y(t) = t^3/125$ have the same rectangular equation.

In Exercises 54–55, eliminate the parameter to find the rectangular equation for the plane curve described by the parametric equations. State the domain of the rectangular equation for the plane curve.

54. $x(t) = \sin t + 1$

$y(t) = -3 \cos t$

55. $x(t) = 5 \cos t - 2$

$y(t) = 3 - \sin t$

CHAPTER 10 REVIEW EXERCISES

Section 10.1 Polar Coordinates

In Exercises 1–2, represent the point by writing four pairs of polar coordinates in which θ is in the interval $(-2\pi, 2\pi)$.

1. a point 4.5 units from the origin on a ray that forms an angle of $3\pi/4$ radians counterclockwise from the polar axis
2. a point 7 units from the origin on a ray that forms a 210° angle counterclockwise from the polar axis
3. Convert $(-5.5, -4\pi/3)$ to rectangular coordinates (x, y).
4. Convert $(-8\sqrt{3}, -8)$ to polar coordinates (r, θ).
5. Convert $(-3.6, -2.1)$ to polar coordinates (r, θ). Round θ to the nearest hundredth.

Section 10.2 Polar Equations

6. Write $-12 \csc \theta = 4r$ in rectangular form.
7. Write $2x + 6y^2 = 0$ in polar form.

In Exercises 8–10, graph the polar equation.

8. $\theta = 5\pi/3$
9. $r = -5 \cos \theta$
10. $r = 5 \sin 4\theta$

Section 10.3 Rotation of Axes

11. For coordinate $P(7, -3)$, find the coordinates of P' when the angle of rotation is $\pi/3$.
12. For coordinate $P(-1, -4)$, find the coordinates of P' when the angle of rotation is $11\pi/4$.
13. $P'(x', y')$ is the point that results when $P(x, y)$ is rotated by an angle of ϕ counterclockwise about the origin. Use $\phi = 5\pi/4$, $x = 6$, and $x' = -4$ to find y'.
14. Identify the type of conic, if any, given by $-2y^2 - 2x^2 = -5 - 5x - 4xy$.
15. Graph $30y + 73y^2 + 52x^2 = 75 + 40x - 72xy$.

Section 10.4 Polar Equation of Conics

In Exercises 16–17, identify the conic given by the polar equation, then determine its directrix and eccentricity.

16. $r = \dfrac{6}{3 - 2\sin\theta}$
17. $r = \dfrac{12}{4 + 8\cos\theta}$

In Exercises 18–19, use the given directrix and eccentricity to write the polar equation of the conic (with focus at the origin).

18. directrix $x = 6$, $e = 1/4$
19. directrix $y = -4$, $e = 2$
20. Graph the equation $3r + 3r\cos\theta = 8$ and find its directrix and eccentricity.

Section 10.5 Plane Curves and Parametric Equations

In Exercises 21–22, graph the plane curve defined by the pair of parametric equations.

21. $x(t) = t^3$
    $y(t) = |0.5t - 8|$
    $-2 \le t \le 2$
22. $x(t) = -2\cos t$
    $y(t) = 3\sin t - 1$
    $0 \le t \le 3\pi/2$

In Exercises 23–24, eliminate the parameter to find the rectangular equation for the plane curve described by the parametric equations. State the domain of the rectangular equation for the plane curve.

23. $x(t) = 2 - t$
    $y(t) = 3t^2 - 1$
24. $x(t) = -4\sin t$
    $y(t) = \cos t$
25. Graph $x(t) = \sqrt{3t + 3}$.
    $y(t) = t^2 - 4$
    $-1 \le t \le 2$

PRECALCULUS

SEQUENCES, SERIES, AND PROBABILITY

CHAPTER

11

11.1 SEQUENCES AND SERIES

OBJECTIVES

- Find terms of sequences.
- Find general functions for sequences given their terms.
- Find terms of recursive sequences.
- Evaluate series.
- Find summations of series.

PREREQUISITE VOCABULARY TERMS

domain

function

natural numbers

11.1.1 Introduction to Sequences

A **sequence** is a function that has natural numbers (i.e., positive integers) as its domain. The function's values are the **terms of the sequence**. These terms are typically written as an ordered list of numbers where the position of the term in the list is the term's **index**.

$a_1, a_2, a_3, \ldots$ *a_1 is the sequence's first term (index is 1), a_2 is the sequence's second term (index is 2), and so on.*

DEFINITION

An **infinite sequence** is a function whose domain is the set of all natural numbers with terms $a_1, a_2, a_3, \ldots$.

A **finite sequence** is a function whose domain is a set of the first n natural numbers with terms $a_1, a_2, a_3, \ldots, a_n$.

A sequence can be written in a general form. For example, the following general function (i.e., formula) represents a sequence where the terms are the multiples of 4.

$a_n = 4n$ *The terms of the sequence are the multiples of 4: 4, 8, 12, ...*

The terms of the sequence $a_n = 4n$ can be generated by evaluating the function $f(n) = 4n$ using the term's index n (i.e., the position of the term in the list). For example, to find a_1 (the first term of the sequence), evaluate $f(1)$, and to find a_2 (the second term of the sequence), evaluate $f(2)$, and so on.

$$a_1 = 4(1) = 4 \qquad a_2 = 4(2) = 8 \qquad a_3 = 4(3) = 12 \qquad a_4 = 4(4) = 16 \qquad \ldots$$

EXAMPLE 1 Finding Terms of a Sequence

Find the first 3 terms and the 10th term of the sequence $a_n = \dfrac{3n+1}{n}$.

SOLUTION

Find the indicated terms by evaluating the function for each index value (1, 2, 3, and 10).

$$a_1 = \frac{3(1)+1}{1} = 4 \qquad a_2 = \frac{3(2)+1}{2} = \frac{7}{2} \qquad a_3 = \frac{3(3)+1}{3} = \frac{10}{3} \qquad a_{10} = \frac{3(10)+1}{10} = \frac{31}{10}$$

1st term *2nd term* *3rd term* *10th term*

The first 3 terms of the sequence are 4, $\frac{7}{2}$, and $\frac{10}{3}$, and the 10th term is $\frac{31}{10}$.

11.1.2 Finding the *n*th Term of a Sequence

It is often possible to write a general function (or formula) for a sequence by observing a pattern in the sequence's terms, and then relating the index value for each term to the value of the term. Some sequences can be described by more than one general function.

EXAMPLE 2 Finding the General Function for a Sequence from Its Terms

Find a general function for each sequence.

A. 3, 6, 9, 12, . . .

B. $\frac{1}{2}, \frac{2}{3}, \frac{3}{4}, \frac{4}{5}, \ldots$

SOLUTION

Look for a pattern in the terms that relates each term's index *n* to the value of the term.

A. The first four terms of the sequence are shown.

Terms	$a_1 = 3$	$a_2 = 6$	$a_3 = 9$	$a_4 = 12$
Index	$n = 1$	$n = 2$	$n = 3$	$n = 4$

Each term is 3 times its index (n).

Write a general function to describe this pattern. $a_n = 3n$

TIP

Check the general function by substituting each index value for n.

B. The first four terms of the sequence are shown.

Terms	$a_1 = \frac{1}{2}$	$a_2 = \frac{2}{3}$	$a_3 = \frac{3}{4}$	$a_4 = \frac{4}{5}$
Index	$n = 1$	$n = 2$	$n = 3$	$n = 4$

Each term's numerator is equal to the index (n) and each term's denominator is equal to 1 more than the index (n + 1).

Write a general function to describe this pattern. $a_n = \dfrac{n}{n+1}$

11.1.3 Recursive Sequences

Consider the sequence 5, 10, 15, 20, ... The terms in this sequence are the multiples of 5, and the general function for this sequence is $a_n = 5n$. Each of the terms in this sequence can be generated using only the index values. There are some sequences where the terms are generated using not only the index value, but also the value of one (or more) preceding terms. This type of sequence is called a **recursive sequence**.

DEFINITION

In a **recursive sequence**, the value of each term depends on one or more of its preceding terms.

An example of a recursive sequence is the **Fibonacci sequence**, where both a_1 and a_2 are 1, and then each successive term a_n is the sum of its two preceding terms.

TIP

Find each term (other than the first two terms, which are both 1) by adding the two preceding terms.

1, 1, 2, 3, 5, 8, 13, . . .

Fibonacci sequence: 1, 1, 2, 3, 5, 8, . . . , $a_n = a_{n-1} + a_{n-2}$

So, the first two terms of a Fibonacci sequence are defined explicitly, and each subsequent term is defined by the general function $a_n = a_{n-1} + a_{n-2}$. As with any sequence, any term in a Fibonacci sequence can be found by replacing *n* with the index of the term. For example, the 6th term is given by $a_6 = a_{6-1} + a_{6-2} = a_5 + a_4 = 5 + 3 = 8$.

EXAMPLE 3 Finding the Terms of a Recursive Sequence

Write the first 5 terms of the sequence if $a_1 = 1$ and $a_n = 2a_{n-1}$.

SOLUTION

The first term a_1 is given to be 1. Generate the next 4 terms of the sequence evaluating the function using $n = 2$, $n = 3$, $n = 4$, and $n = 5$.

$a_2 = 2a_{2-1} = 2a_1 = 2(1) = 2$ *Substitute n = 2, then use the value of a_1, 1.*

$a_3 = 2a_{3-1} = 2a_2 = 2(2) = 4$ *Substitute n = 3, then use the value of a_2, 2.*

$a_4 = 2a_{4-1} = 2a_3 = 2(4) = 8$ *Substitute n = 4, then use the value of a_3, 4.*

$a_5 = 2a_{5-1} = 2a_4 = 2(8) = 16$ *Substitute n = 5, then use the value of a_4, 8.*

The first 5 terms are 1, 2, 4, 8, and 16.

WHAT'S THE BIG IDEA?

Write a formula for a recursive sequence where the terms are only two values and $a_1 = 2$.

11.1.4 Summation Notation and Finite Series

A **series** is a sum of the terms in a sequence. A series is represented using **summation notation**, which identifies the index values used to create the terms of the series. Summation notation makes use of the Greek capital letter sigma, Σ, to signify a sum. Below the sigma is the name of the index and its starting value. Above the sigma is the ending value of the index.

DEFINITION

A **series** is the sum of the terms in a sequence. A series can be described using **summation notation**.

$$\sum_{n=j}^{k} a_n = a_j + a_{j+1} + a_{j+2} + \ldots + a_k$$

TIP

The left side of the equation is read as "the summation of a sub n from n equals j to n equals k."

EXAMPLE 4 Finding the Value of a Summation

Expand each summation and find its value.

A. $\displaystyle\sum_{n=1}^{3} (4n+2)$

B. $\displaystyle\sum_{n=5}^{8} 3n$

SOLUTION

A. Evaluate $4n + 2$ for $n = 1$, $n = 2$, and $n = 3$, then add those terms.

$$\sum_{n=1}^{3} (4n+2) = [4(1)+2] + [4(2)+2] + [4(3)+2]$$
$$= 6 + 10 + 14$$
$$= 30$$

B. Evaluate $3n$ for $n = 5$, $n = 6$, $n = 7$, and $n = 8$, then add those terms.

$$\sum_{n=5}^{8} 3n = 3(5) + 3(6) + 3(7) + 3(8)$$
$$= 15 + 18 + 21 + 24$$
$$= 78$$

EXAMPLE 5 Writing a Summation from a Series

Write an expression using summation notation that represents the series $\frac{1}{2}+\frac{1}{4}+\frac{1}{6}+\frac{1}{8}$.

SOLUTION

Look for a pattern in the series. Notice that the successive denominators are multiples of 2.

$$\frac{1}{2}+\frac{1}{4}+\frac{1}{6}+\frac{1}{8}=\frac{1}{2(1)}+\frac{1}{2(2)}+\frac{1}{2(3)}+\frac{1}{2(4)}$$

Each denominator is the product of 2 and the term's index, where the index begins with 1 and ends with 4.

So, this series can be translated into summation notation using the values in parentheses as the index values.

$$\frac{1}{2}+\frac{1}{4}+\frac{1}{6}+\frac{1}{8}=\sum_{n=1}^{4}\frac{1}{2n}$$

SECTION 11.1 EXERCISES

Warm Up

Evaluate each function for the domain values {1, 2, 3, 4, 5}.

1. $g(a) = 2a - 1$
2. $f(b) = b^2$
3. $p(c) = -3c + 4$

Just the Facts

Fill in the blanks.

4. A(n) _____ is a function whose domain is the set of integers beginning with 1.
5. The first term in a sequence is denoted as _____. The nth term is denoted as _____.
6. The sum of the terms in a sequence is called a(n) _____.
7. Σ is a Greek symbol used to represent the _____ of a series.
8. A(n) _____ sequence is a sequence in which the terms of the sequence depend on one or more previous terms.
9. $a_6 = 24$ states that the _____ term of the sequence is _____.
10. The summation from n equals 4 to 20 of $9n - 5$ would be written in summation notation as _____.

Essential Skills

In Exercises 1–2, find the indicated terms of each sequence.

1. the 2nd and 15th terms of the sequence $a_n = 7n$
2. the 1st and 10th terms of the sequence $a_n = \frac{n+1}{2n}$

In Exercises 3–6, find a general function for each sequence.

3. 20, 17, 14, 11, 8, . . .
4. $\frac{3}{1}, \frac{4}{2}, \frac{5}{3}, \frac{6}{4}, \frac{7}{5}, \ldots$
5. −1, −2, −3, −4, −5, . . .
6. 2, 4, 6, 8, 10, . . .

In Exercises 7–8, find the first 5 terms for each recursive sequence.

7. $a_1 = 5; a_n = a_{n-1} - 6$
8. $a_1 = 2; a_n = 3a_{n-1} + 1$

In Exercises 9–10, find the value of each summation.

9. $\sum_{n=3}^{7} -2n$

10. $\sum_{n=1}^{4} (3n+2)$

In Exercises 11–12, write a summation that represents each series.

11. 2 + 4 + 6 + 8 + 10
12. $\frac{2}{3} + \frac{2}{6} + \frac{2}{9} + \frac{2}{12}$

Extensions

13. Write a general function for $\sqrt{3}, \sqrt{6}, 3, 2\sqrt{3}, \sqrt{15}, \ldots$.
14. Express the series −3 + 9 − 27 + 81 − 243 using sigma notation.
15. True or False? 4, 9, 14, 18, 23 are terms of a sequence.
16. Find the first 5 terms of the recursive sequence defined by $a_1 = 4$, $a_2 = 7$, and $a_n = 4a_{n-1} - 6a_{n-2}$.
17. Suppose that $2500 is invested at 5.2% compounded annually. The value of the investment after n years is given by the general function $a_n = 2500(1.052)^n$. Find the value of the investment after each of the first 7 years. Round to the nearest cent.
18. Write the general function for 1, 0, 1, 4, 9, 16, . . . and find a_{50}.

11.2 ARITHMETIC SEQUENCES

OBJECTIVES

- Identify arithmetic sequences.
- Find particular terms in arithmetic sequences.
- Find general functions for arithmetic sequences given two terms.
- Find sums of finite arithmetic sequences.

PREREQUISITE VOCABULARY TERMS

index (of a sequence)
sequence
series
summation
terms (of a sequence)

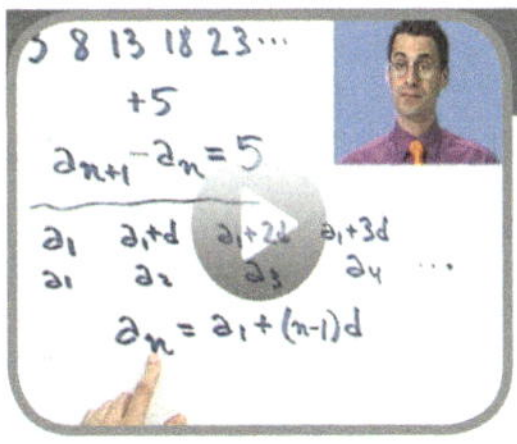

11.2.1 Introduction to Arithmetic Sequences

A sequence that increases (or decreases) by the same amount between consecutive terms is called an **arithmetic sequence**. Therefore, the difference between any term in an arithmetic sequence and its immediately preceding term is the same for all terms in the sequence. This difference is called the **common difference**.

IMPORTANT

If the terms are increasing, then $d > 0$. If the terms are decreasing, then $d < 0$.

IMPORTANT

Subtract any two consecutive terms in an arithmetic sequence to find the common difference, but be sure to subtract the preceding term, not the following term.

DEFINITION

In an **arithmetic sequence**, the difference d (called the **common difference**) between any two consecutive terms is the same.

Consider the terms of the following sequence. 2, 8, 14, 20, 26, 32, 38, . . .

Each term is 6 more than the preceding term. Therefore, the sequence is arithmetic and $d = 6$. A general pattern found in this arithmetic sequence is examined in the following table.

	a_1	a_2	a_3	a_4
Term	2	8	14	20
Pattern	2 + 0	2 + 6	2 + 2(6)	2 + 3(6)
General Pattern	$a_1 + 0d$	$a_1 + d$	$a_1 + 2d$	$a_1 + 3d$

This pattern suggests that each term in the sequence is the sum of the first term a_1 and a multiple of d, where consecutive terms are given by consecutive multiples of d. Specifically, the first term a_1 is $a_1 + 0d$. The second term a_2 is $a_1 + 1d$. The third term a_3 is $a_1 + 2d$, and so on. It follows that the nth term is given by $a_1 + (n - 1)d$.

The nth Term of an Arithmetic Sequence

If a_1 is the first term and d is the common difference between consecutive terms in an arithmetic sequence, then the nth term of the sequence is given by $a_n = a_1 + (n - 1)d$.

When numeric values of a_1 and d are substituted into the formula for the nth term of the arithmetic sequence, $a_n = a_1 + (n - 1)d$, the resulting equation is called the **general function for the sequence**. For example, because $a_1 = 2$ and $d = 6$, the general function for the arithmetic sequence 2, 8, 14, 20, 26, 32, 38, . . . is $a_n = 2 + 6(n - 1)$.

EXAMPLE 1 Finding a Term of an Arithmetic Sequence Given a_1 and d

Find a_{10} for the arithmetic sequence with $a_1 = 2$ and $d = 5$.

SOLUTION

Substitute the given values into the formula for the *n*th term of an arithmetic sequence.

Arithmetic Sequence ▶ $a_n = a_1 + (n - 1)d$

$$a_{10} = 2 + (10 - 1)5 = 2 + (9)5 = 47$$

11.2.2 Finding Terms of an Arithmetic Sequence

If two terms of an arithmetic sequence are consecutive, then *d* is the difference between the two terms. If two given terms are not consecutive, then the formula for the *n*th term of an arithmetic sequence can be used to find *d*, as demonstrated in **Example 2**. Finding unknown terms in an arithmetic sequence often requires finding the value of *d* first.

EXAMPLE 2 Finding Missing Terms for an Arithmetic Sequence Given a_1 and Another Term

Find the missing terms of the arithmetic sequence.

5, __, __, __, __, 20

SOLUTION

From the given arithmetic sequence, $a_1 = 5$ and $a_6 = 20$.

Substitute the values for a_1 and a_6 into the formula for the *n*th term of an arithmetic sequence, and then solve for *d*.

Arithmetic Sequence ▶ $a_n = a_1 + (n - 1)d$

$$20 = 5 + (6 - 1)d \Rightarrow 20 = 5 + 5d \Rightarrow 3 = d$$ *Simplify and solve for d.*

Since $a_1 = 5$ and $d = 3$, the general function of the sequence is $a_n = 5 + 3(n - 1)$.

Use the general function or the common difference to find the missing terms.

Method 1: Use the general function.	**Method 2: Add the common difference.**
$a_2 = 5 + (2 - 1)3 = 5 + 3 = 8$	$a_2 = a_1 + 3 = 5 + 3 = 8$
$a_3 = 5 + (3 - 1)3 = 5 + 6 = 11$	$a_3 = a_2 + 3 = 8 + 3 = 11$
$a_4 = 5 + (4 - 1)3 = 5 + 9 = 14$	$a_4 = a_3 + 3 = 11 + 3 = 14$
$a_5 = 5 + (5 - 1)3 = 5 + 12 = 17$	$a_5 = a_4 + 3 = 14 + 3 = 17$

The complete sequence is 5, 8, 11, 14, 17, 20.

CHECK

As a check, use each method to verify the final value in the sequence, 20.

Method 1: $a_6 = 5 + (6 - 1)3 = 5 + 15 = 20$ ✓ **Method 2:** $a_6 = a_5 + 3 = 17 + 3 = 20$ ✓

11.2.3 Using Two Terms to Find an Arithmetic Sequence

We have seen that an arithmetic sequence can be written as a list of terms (e.g., 2, 8, 14, . . .) or represented by a general function of the form $a_n = a_1 + (n - 1)d$, where a_1 and d are specific values corresponding to the sequence.

For example, the general function for the sequence 2, 8, 14, . . . is $a_n = 2 + 6(n - 1)$ because $a_1 = 2$ and $d = 6$. This general function can be simplified and written as $a_n = 6n - 4$.

NOTICE THAT

The function $a_n = 6n - 4$ is linear and the slope is equal to the common difference. This is always true for an arithmetic sequence.

EXAMPLE 3 Finding a General Function for an Arithmetic Sequence Given Two Terms

If $a_7 = 26$ and $a_9 = 18$, find a_1 and d. Then write a general function for the arithmetic sequence.

SOLUTION

Write two different equations by substituting the given values into the formula for the nth term of an arithmetic sequence.

$26 = a_1 + 6d$ *Since $a_7 = 26$, substitute 26 for a_n and substitute 7 for n. Simplify.*

$18 = a_1 + 8d$ *Since $a_9 = 18$, substitute 18 for a_n and substitute 9 for n. Simplify.*

Arithmetic Sequence
$a_n = a_1 + (n - 1)d$

The two equations form a system of linear equations. $\begin{cases} 26 = a_1 + 6d \\ 18 = a_1 + 8d \end{cases}$

Use the elimination method to solve the system for a_1 and d.

$$\begin{aligned} 26 &= a_1 + 6d \\ +(-18 &= -a_1 - 8d) && \textit{Multiply equation 2 by } -1. \\ 8 &= -2d && \textit{Add the equations.} \\ -4 &= d && \textit{Solve for } d. \end{aligned}$$

NOTICE THAT

The common difference is negative as expected since $a_9 < a_7$ (i.e., the terms are decreasing).

Substitute $d = -4$ into either of the equations in the system and solve for a_1.

$18 = a_1 + 8(-4) \Rightarrow a_1 = 50$

Substitute the values of a_1 and d into the formula for the nth term of an arithmetic sequence.

$a_n = 50 + (n - 1)(-4) = 50 - 4(n - 1) = -4n + 54$ *Simplify.*

So, the general function for an arithmetic sequence where $a_7 = 26$ and $a_9 = 18$ is $a_n = -4n + 54$.

CHECK

The general function can be checked by using $a_n = -4n + 54$ to find the values of a_7 and a_9.

$a_7 = -4(7) + 54 = 26$ ✓ $a_9 = -4(9) + 54 = 18$ ✓

ALTERNATIVE METHOD

If the first term is 50 and the common difference is −4, then the arithmetic sequence is 50, 46, 42, 38, 34, 30, 26, 22, 18, 14, 10, ... So, the 7th term is 26 and the 9th term is 18, as expected.

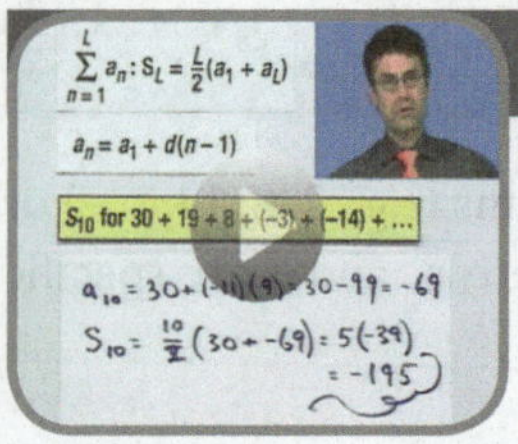

11.2.4 Finding the Sum of an Arithmetic Sequence

The sum of a finite arithmetic sequence can be found easily using a formula attributed to Carl Friedrich Gauss. The finite arithmetic sequence 3, 8, 13, 18, 23, 28, where S is the sum of the sequence, can be used to demonstrate the basis for this formula. If we first write this sequence as a series with its terms in order ($S = 3 + 8 + 13 + 18 + 23 + 28$), and then write the series again with the terms in the reverse order, we can find the sum of the sequence by lining up the terms, adding the equations, and then solving for S.

$$S = 3 + 8 + 13 + 18 + 23 + 28 \quad \textit{Write the terms of the sequence as a series.}$$

$$S = 28 + 23 + 18 + 13 + 8 + 3 \quad \textit{Write the terms in the series in reverse order.}$$

$$2S = 31 + 31 + 31 + 31 + 31 + 31 \quad \textit{Add the equations.}$$

$$2S = 6 \cdot 31$$

$$S = \frac{6}{2} \cdot 31 \quad \textit{Solve for S.}$$

Notice that S is equal to 6 (the number of terms in the sequence) times 31 (the sum of the first and last terms) divided by 2.

More generally, the sum of any finite arithmetic sequence can be found by adding just the first term and the last term, multiplying that sum by the number of terms in the sequence, and then dividing that product by 2. This pattern can be used to write a formula for the sum of a finite arithmetic sequence.

Sum of a Finite Arithmetic Sequence

If a_1 and a_L are the first and last terms of an arithmetic sequence, respectively, then the sum of the first L terms of the sequence is given by $S_L = \frac{L}{2}(a_1 + a_L)$.

IMPORTANT

This formula only works for arithmetic sequences.

EXAMPLE 4 Calculating the Sum of a Finite Arithmetic Sequence

Find S_{10}, the sum of the first 10 terms, for the arithmetic sequence 13, 8, 3, −2, −7, ...

SOLUTION

The formula $S_L = \frac{L}{2}(a_1 + a_L)$ can be used to find the sum because the sequence is arithmetic.

The sum includes the first 10 terms, so $L = 10$. The first term of the sequence is given, $a_1 = 13$.

To find the 10th term, substitute $a_1 = 13$, $d = -5$, and $n = 10$ into the formula for the nth term of an arithmetic sequence.

Arithmetic Sequence ▶ $a_n = a_1 + (n - 1)d$

$$a_{10} = 13 + (10 - 1)(-5) = 13 + 9(-5) = -32$$

Substitute the values into the formula for the sum of a finite arithmetic sequence.

Sum of an Arithmetic Sequence ▶ $S_L = \frac{L}{2}(a_1 + a_L)$

$$S_{10} = \frac{10}{2}(13 + (-32)) = 5(13 - 32) = -95$$

The sum of the first 10 terms of the sequence is −95.

EXAMPLE 5 Calculating the Sum of a Finite Arithmetic Sequence

Find the sum. $\sum_{n=1}^{5}(6-4n)$

SOLUTION

IMPORTANT

$6-4n$ is not the general form of the sequence. The coefficient of n is equal to d, but $a_1 \neq 6$.

The sequence given by this summation is arithmetic with a common difference of –4. This is apparent because as n is replaced in $6-4n$ with successive values of the index, the value of each term will change by the common difference of –4.

The summation $\sum_{n=1}^{5}(6-4n)$ indicates that the first 5 terms will be summed. So, $L = 5$.

Use $a_n = 6-4n$ to find a_1 and a_5. $a_1 = 6-4(1) = 2$ $a_5 = 6-4(5) = -14$

Substitute the values into the formula.

Sum of an Arithmetic Sequence $S_L = \frac{L}{2}(a_1 + a_L)$

$$S_5 = \frac{5}{2}(2+(-14)) = \frac{5}{2}(-12) = -30$$

The sum is –30.

SECTION 11.2 EXERCISES

Warm Up

Determine whether each sequence is arithmetic.

1. 0, 4, 8, 12, 16, . . .
2. 1, −2, 4, −8, 16, . . .
3. −7, −6, −5, −4, −3, . . .

Just the Facts

Fill in the blanks.

4. A sequence is _____ if the difference between consecutive terms is constant.
5. In the arithmetic sequence 2, 8, 14, 20, 26, . . . , the common difference is _____.
6. The formula for the *n*th term of an arithmetic sequence is _____, where the first term is _____ and _____ is the common difference.
7. The *n*th term of an arithmetic sequence is represented by _____.
8. $S_L = \frac{L}{2}(a_1 + a_L)$ is used to find the _____ of a(n) _____ sequence where a_L is the _____ of the *L*th term and *L* is the _____ of terms in the sequence.
9. S_8 is used to denote the _____ of the first _____ terms of an arithmetic sequence.
10. To write a general function of an arithmetic sequence, only _____ pieces of information are needed.

Essential Skills

In Exercises 1–2, find the indicated term for each arithmetic sequence.

1. a_{15} if $a_1 = 7$ and $d = -6$
2. a_9 if $a_1 = 3$ and $d = -2$

In Exercises 3–6, find the general function for each arithmetic sequence.

3. $a_1 = 7$ and $a_5 = 47$
4. $a_1 = 36$ and $a_9 = 4$
5. $a_3 = 45$ and $a_6 = 96$
6. $a_6 = 20$ and $a_{12} = 8$

In Exercises 7–8, find the indicated sum for each arithmetic sequence.

7. S_9 for 5, 7, 9, 11, . . .
8. S_{11} for 10, 7, 4, 1, −2, −5, −8, . . .

In Exercises 9–10, find each sum.

9. $\sum_{n=1}^{5} 4n$

10. $\sum_{n=1}^{6} (2+3n)$

Extensions

11. Find the sum. $\sum_{j=7}^{49} (30 - 0.2j)$
12. Determine the seating capacity of a concert venue with 70 rows of seats if the number of seats in each row forms an arithmetic sequence and there are 50 seats in the first row, 54 seats in the second row, 58 seats in the third row, and so on.
13. Find the sum of the sequence 0.8, 0.6, 0.4, 0.2, . . . if $L = 50$.
14. A woman borrowed $1800 to purchase a used car. She agreed to pay back the loan with monthly payments of $160 plus 2% interest on the unpaid balance.
    - **A.** Find the first 5 monthly payments and the unpaid balance after each month.
    - **B.** Find the total amount of interest paid over the term of the loan.
15. The sum of the first 25 terms of an arithmetic sequence is 720, and $a_1 = 7.2$. Find a_n.
16. Find the missing terms of the arithmetic sequence 2, 9, _____, _____, _____, _____, _____, _____, 58. Write a recursive rule for the sequence.
17. At a charity golf tournament, the 12 golfers with the lowest scores are awarded cash prizes to be given to their favorite charities. First place receives $5000, second place receives $4900, third place receives $4800, and so on, following an arithmetic sequence. What is the total amount of prize money?

11.3 GEOMETRIC SEQUENCES

OBJECTIVES

- Identify geometric sequences.
- Find particular terms in geometric sequences.
- Find general functions for geometric sequences given two terms.
- Find sums of finite geometric sequences.
- Find sums of infinite geometric sequences.
- Write repeating decimals as fractions.

PREREQUISITE VOCABULARY TERMS

index (of a sequence)
ratio
repeating decimal
sequence
series
summation

11.3.1 Introduction to Geometric Sequences

A sequence where consecutive terms form a **common ratio** r, rather than a common difference, is called a **geometric sequence**. So, if the sequence $a_1, a_2, a_3, a_4, \ldots$ is geometric, then $\frac{a_2}{a_1} = \frac{a_3}{a_2} = \frac{a_4}{a_3} = \frac{a_{n+1}}{a_n} = r$ (where a_n and a_{n+1} are two consecutive terms of the sequence).

DEFINITION

In a **geometric sequence**, the ratios of consecutive terms are all equal to a **common ratio** r.

IMPORTANT

Each term in a geometric sequence is the product of the previous term and a constant.

For example, consider the following sequence. $2, 6, 18, 54, 162, 486, \ldots$

The sequence is geometric because the ratios of consecutive terms are equal.

$$\frac{6}{2} = \frac{18}{3} = \frac{54}{18} = \frac{162}{54} = \frac{486}{162}$$

The sequence's common ratio is 3.

The terms in this sequence increase quickly as each term in the sequence is 3 times the preceding term. A general pattern found in this geometric sequence is examined in the following table.

	a_1	a_2	a_3	a_4
Term	2	6	18	54
Pattern	$2 \cdot 3^0$	$2 \cdot 3^1$	$2 \cdot 3^2$	$2 \cdot 3^3$
General Pattern	$a_1 r^0$	$a_1 r^1$	$a_1 r^2$	$a_1 r^3$

Each term in the sequence is the product of the first term a_1 and a power of r, where consecutive terms are given by consecutive powers of r. Specifically, $a_1 = a_1 r^0$, $a_2 = a_1 r^1$, $a_3 = a_1 r^2$, and so on. It follows that the nth term is given by $a_n = a_1 r^{n-1}$.

The nth Term of a Geometric Sequence

The nth term of a geometric sequence, where a_1 is the first term of the sequence and r is the common ratio between consecutive terms, is given by $a_n = a_1 r^{n-1}$.

EXAMPLE 1 Finding a Term of a Geometric Sequence

What is the 5th term of the geometric sequence for which $a_1 = 2$ and $a_2 = 10$?

SOLUTION

Find the ratio for the sequence. $r = \frac{a_{n+1}}{a_n} = \frac{a_2}{a_1} = \frac{10}{2} = 5$

Substitute $a_1 = 2$, $r = 5$, and $n = 5$ into the formula for the nth term of a geometric sequence.

Geometric Sequence ▶ $a_n = a_1 r^{n-1}$

$a_5 = 2(5)^{5-1} = 2(5)^4 = 2(625) = 1250$

The fifth term of the sequence is 1250.

EXAMPLE 2 Finding the General Function for a Geometric Sequence Given Two Terms

Find the general function for the geometric sequence for which $a_4 = 9$ and $a_9 = \frac{1}{27}$.

SOLUTION

IMPORTANT

To write the general function of a geometric sequence, substitute the values of a_1 and r into the formula for the nth term of a geometric sequence.

Substitute the values given for the nth term into the formula of a geometric sequence to give two equations.

$$a_4 = 9 \Rightarrow a_1 r^{4-1} = 9 \Rightarrow a_1 r^3 = 9$$

$$a_9 = \frac{1}{27} \Rightarrow a_1 r^{9-1} = \frac{1}{27} \Rightarrow a_1 r^8 = \frac{1}{27}$$

The two equations form a system of equations. $\begin{cases} a_1 r^3 = 9 \\ a_1 r^8 = \frac{1}{27} \end{cases}$

Use the substitution method to solve the system.

TIP

Equation 2 could also have been chosen.

Solve equation 1 for a_1. $9 = a_1 r^3 \Rightarrow a_1 = \frac{9}{r^3}$

Substitute the expression for a_1 into equation 2 and solve for r.

$\left(\frac{9}{r^3}\right) r^8 = \frac{1}{27}$ *Substitute $9/r^3$ for a_1 in equation 2.*

$9r^5 = \frac{1}{27}$ *Simplify.*

$r^5 = \frac{1}{243}$ *Divide each side by 9.*

$r = \frac{1}{3}$ *Take the 5th root of each side and simplify the radical.*

TIP

Equation 2 could also have been chosen.

Find a_1. Substitute $r = \frac{1}{3}$ into equation 1 and simplify. $a_1 = \frac{9}{r^3} = \frac{9}{\left(\frac{1}{3}\right)^3} = \frac{9}{\frac{1}{27}} = 243$

Substitute the values of a_1 and r into $a_n = a_1 r^{n-1}$. $a_n = 243\left(\frac{1}{3}\right)^{n-1}$

11.3.2 Finding the Sum of a Geometric Sequence

The sum of a finite set of terms in a geometric sequence can be found using the formula derived below if r does not equal 1.

Let the sum of the finite sequence with n terms be S_n.

$$S_n = a_1 + a_1r + a_1r^2 + \ldots + a_1r^{n-1}$$ *Write the sum of the first n terms.*

$$rS_n = a_1r + a_1r^2 + \ldots + a_1r^{n-1} + a_1r^n$$ *Write the equation again, multiplied by r.*

$$S_n - rS_n = a_1 - a_1r^n$$ *Subtract the equations.*

$$(1-r)S_n = a_1(1-r^n)$$ *Factor.*

$$S_n = \frac{a_1(1-r^n)}{1-r},\ r \neq 1$$ *Solve for S_n, the sum of the terms.*

NOTICE THAT

All but the first and last terms cancel out when the right sides are subtracted.

Sum of a Finite Geometric Sequence

If a_1 is the first term and r is the common ratio between consecutive terms of a geometric sequence, then the sum of the first n terms of the sequence is given by $S_n = \dfrac{a_1(1-r^n)}{1-r}$.

NOTICE THAT

The assumption that $r \neq 1$ was necessary because the formula would be undefined for this value, since the denominator would equal 0.

EXAMPLE 3 Calculating the Sum of a Geometric Sequence

Find the sum of the first 10 terms of the geometric sequence for which $a_1 = 4$ and $r = 2$.

SOLUTION

Substitute the given values into the formula for the sum.

$$S_{10} = \frac{4(1-2^{10})}{1-2} = \frac{4(1-1024)}{-1} = 4092$$

Sum of a Finite Geometric Series

$$S_n = \frac{a_1(1-r^n)}{1-r}$$

Infinite Geometric Series

If the value of r is constrained such that $|r| < 1$, that is, $-1 < r < 1$, then as n gets larger and larger, the size of the term gets smaller and smaller. This is called a **converging series**, because higher terms in the series become less and less important, and the sum of the sequence converges to a particular value. This convergence is reflected in the formula for the sum of a geometric sequence. As n becomes larger and larger, the numerator of the sum of the geometric series gets closer and closer to a_1, and as n approaches ∞, r^n approaches 0 and the numerator approaches a_1. When n approaches ∞, the series is an **infinite series**, which is expressed in the following way.

$$\sum_{n=1}^{\infty} a_n = a_1 + a_2 + \ldots$$

Despite the fact that it has an infinite number of terms, when an infinite series is converging, it is possible to calculate its sum.

IMPORTANT

Many infinite series do not have a finite sum. For example, the series $1 + 2 + 3 + \ldots$ does not converge to any number. It just grows without bound.

DEFINITION

An **infinite series** is the sum of an infinite sequence of terms.

Sum of an Infinite Geometric Sequence

The sum S of a convergent infinite geometric sequence, where a_1 is the first term and r is the common ratio between consecutive terms, is $S = \dfrac{a_1}{1-r}$, where $|r| < 1$.

EXAMPLE 4 Calculating the Sum of an Infinite Geometric Sequence

Find the sum of the infinite geometric sequence in which $a_1 = 100$ and $r = \dfrac{1}{2}$.

SOLUTION

Since $|r| < 1$, the formula can be used to calculate the sum.

Substitute the given values into the formula for the sum of an infinite geometric sequence and simplify.

Sum of an Infinite Geometric Sequence ▶

$S = \dfrac{a_1}{1-r}$, *where* $|r| < 1$

$$S = \frac{100}{1-\frac{1}{2}} = \frac{100}{\frac{1}{2}} = 200$$

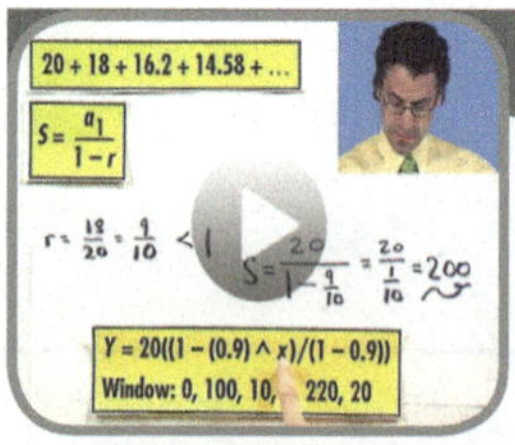

11.3.3 Finding the Sum of an Infinite Geometric Sequence

EXAMPLE 5 Finding the Sum of an Infinite Geometric Sequence

Find the sum of the infinite geometric sequence. 50, 30, 18, 10.8, 6.48, . . .

SOLUTION

The series appears to be converging because the terms in the sequence are steadily decreasing in value.

Find the value of r by finding the ratio between any two consecutive terms of the series.

$$r = \frac{a_2}{a_1} = \frac{30}{50} = \frac{3}{5}$$

Substitute the values for a_1 and r into the formula for the sum of an infinite geometric sequence and simplify to find the sum of the sequence.

$$S = \frac{50}{1-\frac{3}{5}} = \frac{50}{\frac{2}{5}} = 125$$

The sum of the infinite sequence is 125.

EXAMPLE 6 Evaluating a Geometric Series

Evaluate. $\sum_{n=1}^{\infty} \frac{2}{3^{n-1}}$

SOLUTION

First find the value of r by finding the values of a_1 and a_2 and then their ratio.

$$a_1 = \frac{2}{3^{1-1}} = \frac{2}{3^0} = \frac{2}{1} = 2 \quad a_2 = \frac{2}{3^{2-1}} = \frac{2}{3^1} = \frac{2}{3}$$

The first term in the series is 2. The second term in the series is $\frac{2}{3}$.

Find the ratio of the two consecutive terms. $\frac{a_2}{a_1} = \frac{\frac{2}{3}}{2} = \frac{2}{3} \cdot \frac{1}{2} = \frac{1}{3}$

The constant ratio for this series is $\frac{1}{3}$. Since $|r| < 1$, the sum of the infinite sequence exists.

To find the sum, substitute the values for a_1 and r into the formula. $S = \frac{2}{1 - \frac{1}{3}} = \frac{2}{\frac{2}{3}} = 3$

The sum of the infinite geometric series is 3.

11.3.4 Writing a Repeated Decimal as a Fraction

Geometric series provide a useful way to find the fraction equal to a particular repeated decimal.

EXAMPLE 7 Using the Sum of an Infinite Geometric Sequence to Write a Repeating Decimal as a Fraction

Write $0.\overline{36}$ as a fraction in simplest form.

SOLUTION

Express the decimal as an infinite geometric series.

$$0.\overline{36} = 0.363636\ldots = 0.36 + 0.0036 + 0.000036 + \ldots$$

Use the first two terms to find the common ratio r.

$$r = \frac{0.0036}{0.36} = 0.01$$ The common ratio r is 0.01.

Substitute the values for a_1 and r into the formula and simplify.

Sum of an Infinite Geometric Sequence ▶

$S = \frac{a_1}{1-r}$, *where* $|r| < 1$

$$S = \frac{0.36}{1 - 0.01} = \frac{0.36}{0.99} = \frac{36}{99} = \frac{4}{11}$$

Therefore, $0.\overline{36} = \frac{4}{11}$.

SECTION 11.3 EXERCISES

Warm Up

Simplify.

1. $\dfrac{\frac{2}{5}}{\frac{1}{2}}$ 2. $\dfrac{-4}{1-\frac{4}{7}}$ 3. $\dfrac{5}{\left(\frac{2}{3}\right)^2}$

Just the Facts

Fill in the blanks.

4. A(n) _____ sequence is a sequence in which each consecutive term is a multiple of the previous term.
5. The formula for the nth term of a geometric sequence is _____, in which a_1 is the _____ term and _____ is the common ratio.
6. $r = \dfrac{a_{n+1}}{a_n}$ is the _____ of any two _____ terms in a geometric sequence.
7. For a geometric sequence to be convergent, the _____ value must be between _____ and _____.
8. $S_n = \dfrac{a_1(1-r^n)}{1-r}$ is the formula for the sum of a(n) _____ set of terms in a(n) _____ sequence.
9. A(n) _____ geometric series where n goes from 1 to _____ is denoted by $\sum_{n=1}^{\infty} a_n = a_1 + a_2 + \ldots$. If the series is convergent, _____ can be used to calculate its sum.
10. To write a repeating decimal as a fraction, use the formula _____.

Essential Skills

In Exercises 1–2, find the indicated term of each geometric sequence.

1. 4th term, if $a_1 = 9$ and $a_2 = 18$
2. 8th term, if $a_1 = 3$ and $a_2 = 9$

In Exercises 3–4, find the general function for each geometric sequence.

3. $a_3 = 32$ and $a_7 = 512$
4. $a_5 = 2$ and $a_9 = 1/8$

In Exercises 5–6, find the indicated sum of each geometric sequence.

5. the first 5 terms; $a_1 = 3$ and $r = 2$
6. the first 6 terms; $a_1 = 2$ and $r = 3$

In Exercises 7–12, find the sum of each infinite geometric sequence. Round the answer to the nearest tenth, if needed.

7. $a_1 = 7$ and $r = 4/7$
8. $a_1 = 300$ and $r = 3/5$
9. 2, 0.8, 0.64, 0.512, 0.4096, . . .
10. 300, 225, 168.75, 126.5625, 94.921875, . . .
11. $\sum_{n=1}^{\infty}\left(\frac{2}{3}\right)^n$
12. $\sum_{n=1}^{\infty}\frac{2}{5^{n-1}}$

In Exercises 13–14, write each repeating decimal as a fraction in simplest form.

13. 0.35353535 . . .
14. 0.42424242 . . .

Extensions

15. Use summation notation to write the sum 20 + 40 + 80 + . . . + 20,480. Find the sum.
16. $100 is invested on the first day of each month in an account that pays 5% interest compounded monthly. What is the balance after 3 years?
17. Find the sum of the infinite series 35 + 14 + 5.6 +2.24 + ... Write using summation notation.
18. Find the 30th term of the geometric sequence in which $a_3 = 40$ and $a_{10} = -5120$.
19. Write 3.1414141414 . . . as a fraction in simplest form.
20. True or False? The sum of any infinite geometric series can be found when the r-value and n are known.
21. Evaluate $\sum_{n=0}^{51}(-2)^n$. Round to the nearest ten million.

11.4 MATHEMATICAL INDUCTION

OBJECTIVES

- Identify the steps of a proof by mathematical induction.
- Prove a formula for the sum of a finite sequence using mathematical induction.

PREREQUISITE VOCABULARY TERMS

sequence
series
summation

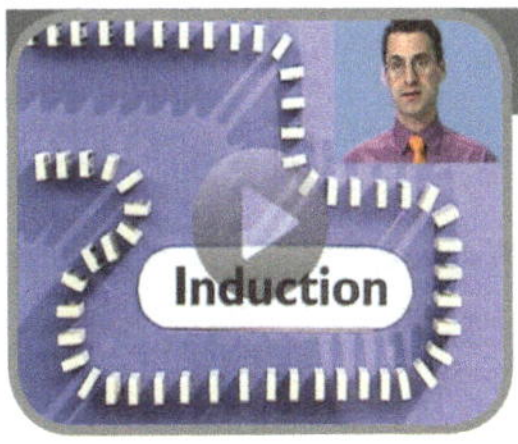

11.4.1 Introduction to Proof by Induction

At the heart of mathematics are logical proofs that prove that some general statement, such as a formula, is true for all cases. Unlike empirical evidence, in which many cases are tested, a proof demonstrates that *all* cases are always true. **Mathematical induction** is one such method of proof that is often used to prove that a certain formula holds for every natural number.

Consider the formula for the sum of the first n natural numbers, $1 + 2 + \ldots + n = \dfrac{n(n+1)}{2}$. How can we prove that this formula is true for all natural numbers n?

Using an empirical method of proof, we can show that this formula holds for specific cases by substituting values of n into the formula. For example, substituting $n = 1$ in the formula yields the true statement $1 = 1$. So, we know that the formula is true when $n = 1$. The following calculations prove that the formula is also true for $n = 2$ and $n = 3$.

NOTICE THAT

This process is the same as checking the solution to an equation.

$$n = 2: \quad 1+2 \stackrel{?}{=} \frac{2(2+1)}{2} \qquad 3 \stackrel{?}{=} \frac{6}{2} \qquad 3 = 3 \checkmark$$

$$n = 3: \quad 1+2+3 \stackrel{?}{=} \frac{3(3+1)}{2} \qquad 6 \stackrel{?}{=} \frac{12}{2} \qquad 6 = 6 \checkmark$$

Substituting $n = 2$ and $n = 3$ results in true statements, so the formula holds for $n = 2$ and $n = 3$.

NOTICE THAT

Using this empirical method (substitution) to prove that the formula is true for all natural numbers would take an infinite amount of time (since there are an infinite number of natural numbers). So, another method of proof must be used.

From this evidence, we know that the formula holds for the natural numbers 1, 2, and 3. However, we cannot assume from this evidence that it is true for *all* natural numbers because we've only shown that it is true for the first three natural numbers.

Mathematical induction can be used to prove that the formula is true for all natural numbers at once. Induction has two parts:

❶ Prove the base case.
So, prove that the statement is true for the least value of the variable (usually $n = 1$).

❷ Prove that
- if we assume that the statement is true for $n = k$ (that is, a particular value of n),
- then the statement is true for $n = k + 1$ (that is, for the very next value of n).

Why do these two parts prove the original statement? Think of the proof as an infinitely tall ladder, where each of the ladder's rungs represents a different natural number. The *base case* (step ❶) gets us on the first rung of the ladder. Step ❷ allows us to climb from one rung of the ladder to the next. Given our starting point and the ability to climb, we are able to hit every rung of the ladder. In other words, we are able to prove the statement for every single natural number.

Note that without a starting point (the base case), the second step does not actually prove the original statement for any actual values of n. Without the ability to prove the statement for one value of n, based on the truth of the statement for the preceding value of n, we have only proven the statement for just the one natural number.

11.4.2 Proving with Induction

Steps for Proof by Mathematical Induction

1. Prove the base case (typically where $n = 1$).
2. Prove that if the formula holds for an arbitrary case ($n = k$), then the formula holds for the next case ($n = k + 1$).

The formula for the sum of squares of the first n natural numbers is proven for all natural numbers in **Example 1**.

EXAMPLE 1 Prove a Formula with Induction

Prove that $1+4+9+16+\ldots+n^2 = \dfrac{n(n+1)(2n+1)}{6}$ for all natural numbers n.

SOLUTION

IMPORTANT

In step 1, show that the base case is true by substituting n = 1 into the given formula.

1. Prove the base case, in which $n = 1$.

When $n = 1$, the left side of the formula is 1. Show that the right side is also 1.

$$\frac{(1)(1+1)(2(1)+1)}{6} = \frac{(1)(2)(3)}{6} = \frac{6}{6} = 1 \checkmark$$ *Substitute n = 1 in the formula and simplify.*

So, the formula holds true for the base case.

IMPORTANT

In step 2, make the statement by substituting k for n. Assume the statement is true and then substitute using the n = k in the n = k + 1 case.

2. Prove that if the formula holds for $n = k$, then the formula holds for $n = k + 1$.

Assume the formula holds when $n = k$.

$$1+4+9+16+\ldots+k^2 = \frac{k(k+1)(2k+1)}{6}$$ *Substitute k for n in the formula.*

Prove that the formula holds for $n = k + 1$.

$$\text{Prove: } 1+4+9+16+\ldots+k^2+(k+1)^2 = \frac{(k+1)((k+1)+1)(2(k+1)+1)}{6}$$

Substitute n = k + 1 into the original formula to determine the statement to be proven.

Manipulate the expression on the left side until it is in the same form as the right side.

IMPORTANT

Keep in mind that the goal of the algebraic manipulation in step 2 is not to simplify. Instead, the goal is to write the expression in the same form as the right side of the n = k + 1 equation. Since the first factor in the right-side expression is (k + 1), factor (k + 1) from each term to match that expression.

$$1+4+9+16+\ldots+k^2+(k+1)^2$$

$$= \frac{k(k+1)(2k+1)}{6}+(k+1)^2$$ *Substitute the right side of the n = k equation for $1 + 4 + 9 + 16 + \ldots + k^2$.*

$$= \frac{k(k+1)(2k+1)}{6}+\frac{6(k+1)^2}{6}$$ *Find a common denominator.*

$$= \frac{k(k+1)(2k+1)+6(k+1)^2}{6}$$ *Add.*

$$= \frac{(k+1)[k(2k+1)+6(k+1)]}{6}$$ *Factor (k + 1) from each term in the numerator.*

$$= \frac{(k+1)[2k^2+7k+6)]}{6}$$ *Distribute k and combine the like terms.*

$$= \frac{(k+1)(k+2)(2k+3)}{6}$$ *Factor the trinomial.*

$$= \frac{(k+1)((k+1)+1)(2(k+1)+1)}{6}$$ *Write the sums in terms of (k + 1).*

Therefore, since the $k + 1$ case is true assuming the kth case is true, and the base case is true, then the formula holds for all natural numbers by induction.

11.4.3 Proving with Induction: Another Example

The formula for the sum of cubes of the first n natural numbers is proven for all natural numbers in **Example 2**.

EXAMPLE 2 Proving a Formula with Induction

Prove that $\displaystyle\sum_{r=1}^{n} r^3 = \frac{n^2(n+1)^2}{4}$ for all natural numbers n.

SOLUTION

❶ Prove the base case, in which $n = 1$.

When $n = 1$, the left side of the formula is 1. Show that the right side is also 1.

$$\frac{(1)^2(1+1)^2}{4} = \frac{(1)(2)^2}{4} = \frac{4}{4} = 1 \checkmark$$ *Substitute $n = 1$ into the formula and simplify.*

So, the formula holds true for the base case.

TIP

The left side is equal to 1 when $n = 1$ because $1^3 = 1$.

❷ Prove that if the formula holds for $n = k$, then the formula holds for $n = k + 1$.
Assume the formula holds when $n = k$.

$$1 + 8 + 27 + \ldots + k^3 = \frac{k^2(k+1)^2}{4}$$ *Substitute k for n in the formula.*

TIP

The left side of the equation shows $\displaystyle\sum_{r=1}^{n} r^3$ which can be written as a sum by substituting the index values for r.

Prove that the formula holds for $n = k + 1$.

$$\text{Prove: } 1 + 8 + 27 + \ldots + k^3 + (k+1)^3 = \frac{(k+1)^2((k+1)+1)^2}{4}$$

Substitute $n = k + 1$ into the original formula to determine the statement to be proven.

Manipulate the expression on the left side until it is in the same form as the right side.

$$1 + 8 + 27 + \ldots + k^3 + (k+1)^3 = \frac{k^2(k+1)^2}{4} + (k+1)^3$$ *Substitute the right side of the $n = k$ equation for $1 + 8 + 27 + \ldots + k^3$.*

$$= \frac{k^2(k+1)^2}{4} + \frac{4(k+1)^3}{4}$$ *Find a common denominator.*

$$= \frac{k^2(k+1)^2 + 4(k+1)^3}{4}$$ *Add.*

$$= \frac{(k+1)^2[k^2 + 4(k+1)]}{4}$$ *Factor $(k + 1)^2$ from each term in the numerator.*

$$= \frac{(k+1)^2[k^2 + 4k + 4)]}{4}$$ *Distribute the 4.*

$$= \frac{(k+1)^2(k+2)^2}{4}$$ *Factor the trinomial.*

$$= \frac{(k+1)^2((k+1)+1)^2}{4}$$ *Write $(k + 2)$ in terms of $(k + 1)$.*

IMPORTANT

Keep in mind that the goal of the algebraic manipulation in step ❷ is not to simplify. Instead, the goal is to write the expression in the same form as the right side of the $n = k + 1$ equation.

Therefore, since the $k + 1$ case is true assuming the kth case is true, and the base case is true, then the formula holds for all natural numbers by induction.

SECTION 11.4 EXERCISES

Warm Up

Determine whether each sequence is arithmetic, geometric, or neither.

1. 20, 24, 28, 33, 37, . . .
2. 5, –10, 20, –40, 80, . . .
3. –7, –9, –11, –13, –15, . . .

Just the Facts

Fill in the blanks.

4. A(n) _____ is used in mathematics to show that formulas are true for _____ cases.
5. The first step in a proof by _____ is to prove the _____ case to be true where $n =$ _____.
6. Assuming an arbitrary case to be true is the _____ step in proof by induction.
7. Once it is proven that the formula holds true for _____, then the proof by induction is finished.
8. The $k + 1$ statement for the inductive proof of $1 + 2 + 3 + 4 + \ldots + n = 0.5n(n + 1)$ is written _____.
9. The proofs by induction in this section were used to prove statements about _____ numbers.
10. If the base case is not true, then the _____ fails.

Essential Skills

In Exercises 1–10, use mathematical induction to prove each formula for all natural numbers *n*.

1. $2 + 10 + 18 + \ldots + (8n - 6) = 2n(2n - 1)$
2. $1 + 3 + 5 + \ldots + (2n - 1) = n^2$
3. $\sum_{k=1}^{n} (6k - 7) = 3n^2 - 4n$
4. $\sum_{k=1}^{n} (2k)^3 = 2n^2(n+1)^2$
5. $\sum_{k=1}^{n} 10^{k-1} = \frac{1}{9}(10^n - 1)$
6. $1 + 3 + 9 + \ldots + 3^{n-1} = \frac{1}{2}(3^n - 1)$
7. $1 \cdot 2 + 2 \cdot 2^2 + 3 \cdot 2^3 + \ldots + n \cdot 2^n = 2[1 + (n - 1)2^n]$
8. $1 + 8 + 64 + \ldots + 8^{n-1} = \frac{1}{7}(8^n - 1)$
9. $\sum_{k=1}^{n} \frac{1}{k(k+1)} = \frac{n}{n+1}$
10. $1^3 + 3^3 + 5^3 + 7^3 + \ldots + (2n - 1)^3 = n^2(2n^2 - 1)$

Extensions

In Exercises 11–12, use mathematical induction to prove each statement for all natural numbers *n*.

11. $8^n - 3^n$ is divisible by 5
12. $7^n - 4^n$ is divisible by 3

11.5 COUNTING PRINCIPLES

OBJECTIVES

- Use the Fundamental Counting Principle and the Factorial Rule of Counting.
- Find numbers of permutations and distinguishable permutations.
- Find numbers of combinations.

PREREQUISITE VOCABULARY TERMS

factorial

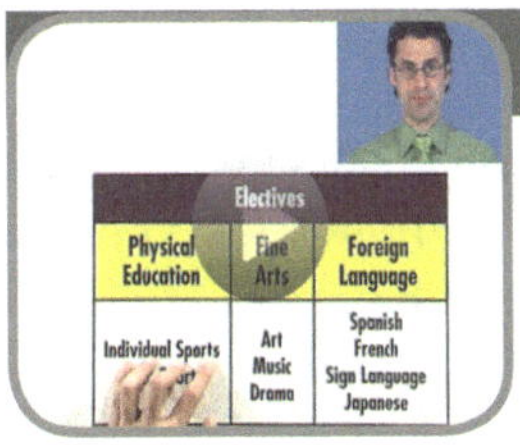

11.5.1 Using the Fundamental Counting Principle

Counting seems like a simple activity, when, in fact, counting can be challenging. The challenge with counting is to make sure that all items are counted and that they are *only counted once*. Simple enumeration or listing procedures are often not reasonable counting methods because the counts can reach high numbers.

The most basic counting tool is the **Fundamental Counting Principle**. The Fundamental Counting Principle calculates the number of ways there are of combining a set of items.

TIP

Use the Fundamental Counting Principle when counting the number of ways of combining a series of items, each of which has several varieties.

Fundamental Counting Principle

If there are n ways to choose a first item and m ways to choose a second item after the first item, then there are $n \cdot m$ total ways to choose both items.

EXAMPLE 1 Applying the Fundamental Counting Principle

How many different schedules are possible if a student must include 1 science class, 1 math class, and 1 social studies class in her schedule, and there are 3 science courses, 4 math courses, and 2 social studies courses from which to choose?

TIP

A tree diagram is a helpful way to visualize a counting example when the numbers are small. The order that the items are chosen for a tree diagram does not matter.

SOLUTION

Method 1: The Tree Method

The number of possible schedules can be found using a tree diagram. List the course possibilities for the first item (subject), science (labeled S1, S2, S3). Then, under each of the science courses, draw a branch to each possibility for math (labeled M1, M2, M3, M4). Finally, under each science and math branch, draw a branch to each possibility for social studies (labeled T1, T2).

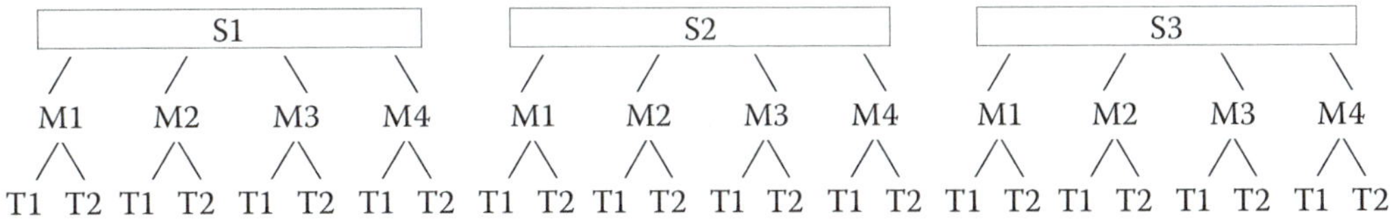

IMPORTANT

In a tree diagram, the total number of combinations is the number of branches at the final level.

Fundamental Counting Principle ▶

$n_1 \cdot n_2 \cdot n_3 \cdot \ldots$

There are 24 branches in the third level of the tree diagram, so there are 24 possible schedules.

Method 2: The Fundamental Counting Principle

There are 3 items (classes) and 3, 4, and 2 choices within each item. To apply the Fundamental Counting Principle to find the total number of ways these items can be combined, multiply the number of choices in each.

$3 \cdot 4 \cdot 2 = 24$ *Apply the Fundamental Counting Principle.*

There are 24 different possible schedules.

Counts can get very large. Methods like the Fundamental Counting Principle were developed to count large numbers of things without actually having to identify each one individually as in the tree method.

EXAMPLE 2 Applying the Fundamental Counting Principle

TIP

The tree method, in which every combination is listed, would be unreasonable with this example.

Canadian postal codes use an alternating letter and number system in the format A1A 1A1. In the number positions, any of the digits from 0 to 9 may be used. In the letter positions, all of the letters may be used except D, F, I, O, Q, and U, because the machines that read the handwritten postal codes have trouble distinguishing these from other characters. Given the postal code format and the numbers of digits and letters available, how many unique postal codes are possible?

SOLUTION

TIP

Use the Fundamental Counting Principle when counting the combinations produced by a series of choices.

There are 6 items to choose (3 letter slots in the postal code and 3 number slots in the postal code). The letters have 26 – 6 = 20 possible choices and the numbers have 10 possible choices. Apply the Fundamental Counting Principle to find the total number of combinations.

$$20 \cdot 10 \cdot 20 \cdot 10 \cdot 20 \cdot 10 = 10^3 \cdot 20^3 = 8{,}000{,}000$$

There are 8 million different possible codes.

11.5.2 Permutations

Sets of items are often ordered for some purpose. For example, when the three candidates in an election are ranked by the number of votes they received, the items in the set (i.e., the three candidates) are ordered by first place, second place, and third place. When the items in a set are ordered (or arranged) in a particular way, that arrangement of the items is called a **permutation**.

Counting Permutations Using a Whole Set of Items

There are many methods for counting the number of permutations of a set of items. In other words, there are many methods to count the number of ways in which the items in a set can be ordered. The **Factorial Rule of Counting** can be used to find the number of permutations of the set when *all* of the items in a set are being ordered.

Factorial Rule of Counting

The number of ways of ordering n different items is $n!$.

IMPORTANT

$n! = n(n-1)(n-2)\cdots(1)$ and by convention, $0! = 1$.

EXAMPLE 3 Counting Permutations

A professor has prepared a 4-question test for her class. She would like to make the test each student receives different by providing each student in the class with the same questions, but in a different order. How many tests can be made using all 4 questions?

SOLUTION

TIP

Use the Factorial Rule of Counting if the order of the items matters and all of the items are being counted.

Use the Factorial Rule of Counting to find the total number of permutations because all 4 of the questions are being used and the questions will be delivered in a particular order.

$4! = 4(3)(2)(1) = 24$ *There are 4 possibilities for the first question, 3 for the second question, 2 for the third question, and 1 for the fourth question.*

There are 24 different possible tests.

Counting Permutations Using a Subset of Items

When selecting r items to order from a larger set of n items, the number of ways of ordering those r items is found with the general permutations formula.

> ***Permutations***
>
> The number of ways of ordering r items selected from a total of n different items is
>
> $$_nP_r = \frac{n!}{(n-r)!}.$$

NOTICE THAT

When $n = r$, meaning the whole set is being ordered, the denominator becomes $0! = 1$, and the formula is equivalent to the Factorial Rule of Counting.

Notice that the numerator of the permutations formula is equal to the number of ways to order all n items, $n!$. The denominator is the number of ways to order the items *not* chosen, $(n - r)!$. Dividing the number of ways to order *all* of the items by the number of ways to order the *not chosen* items leaves the number of ways to order the chosen items.

EXAMPLE 4 Counting Permutations

Competing in a 100-meter race are 9 sprinters. How many possible outcomes are there for gold, silver, and bronze medals in the race?

SOLUTION

The selections in this situation are ordered, so the permutations formula can be used.

Identify n and r.

$n = 9$ *The total number of items (runners) is 9.* $r = 3$ *The number of selected runners is 3.*

Substitute the values into the permutation formula to find the number of ways the medals can be awarded.

Permutations ▶ $_nP_r = \frac{n!}{(n-r)!}$

$$_9P_3 = \frac{9!}{(9-3)!} = \frac{9!}{6!} = \frac{9 \cdot 8 \cdot 7 \cdot 6!}{6!} = \frac{9 \cdot 8 \cdot 7 \cdot \cancel{6!}}{\cancel{6!}} = 504$$ *Use the permutations formula.*

There are 504 different ways the gold, silver, and bronze medals can be awarded to 3 of the 9 runners.

> **WHAT'S THE BIG IDEA?**
>
> As the size r of the subset selected from a set of items gets larger, what happens to the number of permutations?

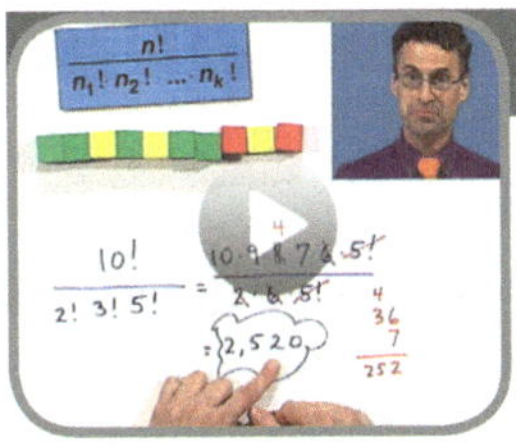

11.5.3 Distinguishable Permutations

Sometimes a set of n items will have duplications. For example, consider the number of ways to order all of the letters in the word *Mississippi*. In this word, there is one of the letter m, two of the letter p, four of the letter i, and four of the letter s. Some of the orderings will have the first i in the first position and some will have the second, third, or fourth i in the first position. These selections cannot be distinguished from each other.

Finding the number of **distinguishable permutations** requires a variation of the permutations formula.

TIP

Count using distinguishable permutations if the order of the n items matters and there are duplicates among the items.

Distinguishable Permutations

The number of ways of ordering n items in which there are n_1 of one kind, n_2 of another kind, and so on, is $\dfrac{n!}{n_1! \cdot n_2! \cdot \ldots \cdot n_k!}$, where $n_1 + n_2 + \ldots + n_k = n$.

EXAMPLE 5 Counting Distinguishable Permutations

In a word game, a player has 7 letters: *E, X, M, E, A, E, M*, and they would like to use all 7 of the letters in making their word. How many distinguishable permutations are there for the 7 letters?

SOLUTION

There are 7 total letters, so $n = 7$. Out of those 7 letters, there are 4 different letters: *E, M, X*, and *A*. So, define the values of n_1, n_2, n_3, and n_4. It does not matter which letter is associated with n_1, n_2, and so on.

$n_1 = 3$ *The player has 3 of the letter E.* $n_2 = 2$ *The player has 2 of the letter M.*

$n_3 = 1$ *The player has 1 of the letter X.* $n_4 = 1$ *The player has 1 of the letter A.*

Substitute the values into the formula to find the number of distinguishable permutations for 7-letter words.

Distinguishable Permutations ▶ $\dfrac{n!}{n_1! \cdot n_2! \cdot \ldots \cdot n_k!}$

$$\frac{7!}{3!2!1!1!} = \frac{7 \cdot 6 \cdot 5 \cdot 4 \cdot 3!}{3! \cdot 2} = \frac{7 \cdot 6 \cdot 5 \cdot 4 \cdot \cancel{3!}}{\cancel{3!} \cdot 2} = 420$$ *Use the distinguishable permutations formula.*

There are 420 distinguishable letter orderings possible.

11.5.4 Combinations

In many cases, the order of a selection of items does not matter. For example, when dealing a hand of cards from a deck of playing cards, the order in which the cards are dealt does not matter, only the identity of each of the cards. The number of ways of selecting r items from a set of n items when order does *not* matter is called the number of **combinations** of r items chosen from n items.

TIP

$C(n, r)$, ${}_nC_r$ and $\binom{n}{r}$ are read as "n choose r."

Combinations

The number of ways of choosing r items from n items when order does not matter is

$$C(n, r) = {}_nC_r = \binom{n}{r} = \frac{n!}{r!(n-r)!}.$$

Any of the three notations for a combination can be used.

IMPORTANT

Count using combinations if a subset of items is being selected from the whole and the order of the items does not matter.

The formula for the number of combinations can be derived intuitively from the formula for the number of permutations. The only difference between the two is a factor of $r!$ in the denominator of the combinations formula. According to the Factorial Counting Rule, this is the number of ways of ordering r items. Since counting combinations ignores order, dividing by $r!$ factors out the number of different orders from the total to give just the number of ways r items can be selected from n.

EXAMPLE 6 Counting Using Combinations

At a rowing regatta, 9 rowers in individual boats compete in a heat from which the first 3 rowers to finish are chosen to advance to the final race. How many possible outcomes are there for rowers chosen to advance to the final?

SOLUTION

This is a combination because the order of the 3 chosen rowers does not matter (there are no medals given for a heat, so the rowers all have equal rank). Identify the total number of items, n, and the number of items chosen, r.

$n = 9$ *There are 9 total rowers.* $r = 3$ *The number of chosen rowers is 3.*

Substitute the values into the formula to find the number of combinations.

Combinations ▶ $_nC_r = \dfrac{n!}{r!(n-r)!}$

$$\binom{9}{3} = \frac{9!}{3!(9-3)!} = \frac{9!}{3!6!} = \frac{9\cdot 8\cdot 7\cdot 6!}{3\cdot 2\cdot 1\cdot 6!} = \frac{9\cdot 8\cdot 7\cdot \cancel{6!}}{3\cdot 2\cdot 1\cdot \cancel{6!}} = \frac{504}{6} = 84$$ *Use the combinations formula.*

There are 84 different 3-rower finishes possible.

There are many fewer combinations than there were permutations for a similar situation (i.e., in **Example 4**). This makes sense because the ordering creates distinctions among the selections that are not being considered when order is ignored.

WHAT'S THE BIG IDEA?

As the size r of the subset selected from a set of items gets larger, what happens to the number of combinations?

SECTION 11.5 EXERCISES

Warm Up

Simplify.

1. $6!$ **2.** $\frac{5!}{3!}$ **3.** $\frac{7!5!}{6!}$

Just the Facts

Fill in the blanks.

4. The _____ states that if there are n ways to choose a first item and m ways to choose a second item after the first item, then there are $n \cdot m$ ways to choose both items.

5. $n!$ is the number of _____ possible for all _____ items.

6. If the order of the items matters when subsets of the items are being selected, count using _____.

7. If the order of the items does not matter when subsets of the items are being selected, count using _____.

8. A _____ diagram is a visual way to answer questions involving the Fundamental Counting Principle where the number of combinations is the number of _____ at the last level.

9. $\frac{n!}{n_1! \cdot n_2! \cdot \ldots \cdot n_k!}$ is used when counting the number of _____ of n objects.

10. ${}_nC_r$ is read _____ choose _____.

Essential Skills

1. At the local ice cream store, there are many choices for a sundae. There are 16 flavors of ice cream, 4 liquid toppings, 33 dry toppings, and with or without whipped cream. How many different sundae varieties are there if 1 flavor of ice cream, 1 liquid topping, 1 dry topping, and whipped cream are chosen?

2. A sandwich bar allows customers to construct their own sandwiches by choosing from 2 kinds of buns, 2 kinds of condiments, 4 kinds of veggies, and 3 kinds of meats. How many different sandwiches consisting of 1 bun type, 1 condiment type, 1 veggie type, and 1 meat type can be created?

3. A phone banking password consists of 2 letters followed by 4 digits. The letters B and X are not used, and the last digit cannot be 0. How many different passwords are possible?

4. A personal identification code consists of 3 letters followed by 3 digits. The letters X, Y, and Z are not used, and the last digit cannot be 0 or 9. How many different codes are possible?

In Exercises 5–6, determine the number of arrangements in each situation.

5. There are 8 different coffee mugs that have to be arranged in a showcase.

6. There are 5 children who have to form a line.

In Exercises 7–8, determine the number of ways first, second, and third place finishers can be chosen in each situation.

7. There are 8 dogs running in a race at a track.

8. There are 20 cars participating in a race.

In Exercises 9–10, determine the number of distinguishable permutations for the letters in each word.

9. fantastic

10. intelligent

11. A raffle contains 12 different tickets. In how many ways can 3 different tickets be chosen?

12. Students taking an online math test are randomly assigned 4 questions out of a set of 10 different questions. How many different sets of questions (i.e., tests) are possible for the students?

Extensions

13. Find the number of distinguishable permutations of the letters in "supercalifragilisticexpialidocious." Write your answer in scientific notation.

14. In the Texas Mega Millions lottery game, a player chooses 5 distinct numbers from the first field of 56 numbers and 1 Mega Ball number from the second field of 46 numbers. In how many ways can a player choose the first 5 numbers?

15. In a typical football team, a coach will have the following number and types of players on the offensive roster: 4 quarterbacks, 11 wide receivers, 3 centers, 3 fullbacks, 4 tight ends, 5 running backs, 7 offensive tackles, and 5 guards. If a play calls for 1 quarterback, 2 wide receivers, 1 center, 1 fullback, 1 tight end, 1 running back, 2 offensive tackles, and 2 guards, how many different offensive groups can the coach pick?

16. Should permutations, combinations, or the Fundamental Counting Principle be used to solve each situation?

A. arrangements of used cars
B. a student council committee
C. a motorcycle license plate
D. a poker hand

17. In a class with 13 girls and 10 boys, 8 will be selected to meet with the dean of student affairs. How many groups of 8 can be chosen that have the following?

A. all boys
B. all girls
C. half boys
D. half girls
E. 5 boys and 3 girls
F. no more than 2 boys

18. When ordering a pizza, there are 2 different crust options and 15 different toppings to choose from. How many combinations of pizza can be ordered with just 3 toppings?

11.6 PROBABILITY

OBJECTIVES

- Calculate probabilities of events.
- Use counting methods to find probabilities.
- Calculate joint probabilities of independent events.
- Find probabilities of inclusive events.
- Find combined probabilities of mutually exclusive events.
- Use the complement to find probabilities.

PREREQUISITE VOCABULARY TERMS

Fundamental Counting Principle

permutation

11.6.1 The Probability of an Event

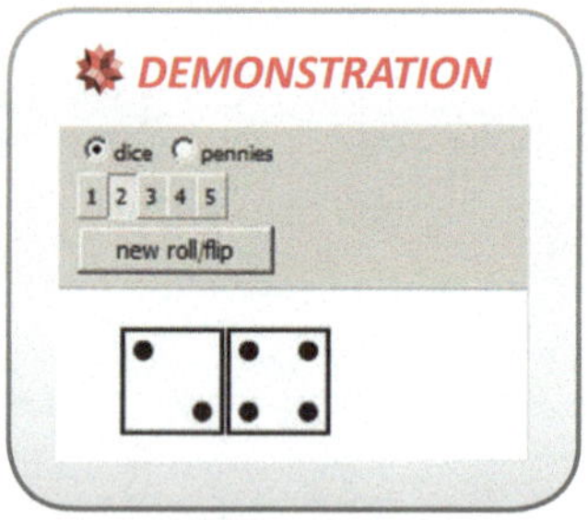

Dice Rolls and Coin Flips

Introduction to Probability: Experiments and Sample Space

The **probability** of something occurring is a measure of the likelihood that it will occur. Probabilities occur in the context of what mathematicians call "probability **experiments**," which are the procedures followed to obtain some **outcome**. An outcome, or a set of outcomes, is called an **event**, and the set of all outcomes possible from an experiment is the experiment's **sample space**. Events are often symbolized by capital letters such as A or B, and the probability that event A will occur is symbolized as $P(A)$.

Consider the experiment where one coin is flipped. This experiment's sample space contains two outcomes: the coin landing on heads, H, and the coin landing on tails, T. These outcomes can be represented by {H, T}. If two coins are flipped in the experiment, then there are four possible outcomes in the sample space.

HH	*Both coins land on heads.*
HT	*The first coin lands on heads, and the second coin lands on tails.*
TH	*The first coin lands on tails, and the second coin lands on heads.*
TT	*Both coins land on tails.*

So, the sample space for flipping two coins can be represented by {HH, HT, TH, TT}.

IMPORTANT

The sample space for flipping one coin, {H, T}, could be represented equivalently as {T, H}. The sample space for flipping two coins, {HH, HT, TH, TT}, can be represented equivalently as {HT, TH, TT, HH}, {TH, TT, HH, HT}, {TT, HH, HT, TH}, . . . or with the outcomes in any order.

Calculating the Probability of an Event

The **probability of an event** can be determined when all of the outcomes in an experiment have the same probability of occurring (i.e., each outcome is equally likely). Experiments where all of the outcomes are equally likely include rolling a die, choosing a card from a deck of standard playing cards, or playing a lottery.

Probability of an Event

The probability of an event A is the number of successful outcomes divided by the total number of possible outcomes in an experiment's sample space, where all of the outcomes in the sample space are equally likely.

$$P(A) = \frac{\text{number of equally likely successful outcomes}}{\text{number of equally likely outcomes in sample space}}$$

Consider again the experiment where one coin is flipped. In this experiment, the two possible outcomes, a heads and a tails, have the same probability of occurring (assuming the coin flip is fair). So, you can calculate the probability of the coin landing on heads (or of the coin landing on tails). To find the probability of the coin landing on heads, $P(H)$, divide the number of successful outcomes in the sample space by the total number of possible outcomes in the sample space.

Experiment: Flipping One Coin

Number of outcomes in the sample space = 2
The coin could land on heads or tails.

Number of successes = 1
The successful outcome is heads.

$$P(\text{H}) = \frac{\text{number of successes}}{\text{total number of outcomes}} = \frac{1}{2}$$

TIP

A probability of 1/2 can also be expressed as 0.5 or 50%.

Therefore, the probability of getting heads on one coin flip is 1/2.

The probability of an event that is certain to occur is 100%, or 1. The probability of an event that will never occur is 0. It follows that the probability of any event will always be a number between 0 and 1.

EXAMPLE 1 Calculating Probability: Coins

Suppose an experiment consists of flipping three coins. Find each probability.

A. exactly one tails **B.** exactly two heads **C.** at least one heads

SOLUTION

Describe the experiment's sample space (i.e., all of the possible outcomes when three coins are flipped).

TIP

The individual coin flips are not counted by themselves, even though they are individually distinguishable. The possible outcomes formed by all three occurring together are counted.

Sample space: {HHH, HHT, HTH, THH, HTT, THT, TTH, TTT}

So, there are 8 outcomes in the sample space.

To find each probability, find the number of ways to obtain a success, and then divide the number of successes by 8 (the total number of outcomes in the experiment's sample space).

A. The successes are the outcomes where there is "exactly one tails." From the 8 possible outcomes, the ways to get one tails are HHT, HTH, and THH. So, the number of successes is 3.

$$P(\text{exactly one tails}) = \frac{\text{number of successes}}{\text{total number of outcomes}} = \frac{3}{8}$$

The probability of getting one tails when three coins are flipped is 3/8, or 0.375.

B. The successes are the outcomes where there are "exactly two heads." From the 8 possible outcomes, the ways to get two heads are HHT, HTH, and THH. So, the number of successes is 3.

$$P(\text{exactly two heads}) = \frac{\text{number of successes}}{\text{total number of outcomes}} = \frac{3}{8}$$

TIP

The probability of getting one tails is equal to the probability of getting two heads.

The probability of getting two heads when three coins are flipped is 3/8, or 0.375.

C. The successes are the outcomes where there are "at least one heads." From the 8 possible outcomes, the ways to get at least one heads are HHH, HHT, HTH, THH, HTT, THT, and TTH. So, the number of successes is 7.

$$P(\text{at least one heads}) = \frac{\text{number of successes}}{\text{total number of outcomes}} = \frac{7}{8}$$

The probability of getting at least one heads when three coins are flipped is 7/8, or 0.875.

EXAMPLE 2 Calculating Probability: Dice

Find the probability of obtaining a sum of 6 from rolling two dice.

SOLUTION

Make a table for all possible outcomes in the sample space (i.e., all possible sums from two dice).

Second Die \ First Die	1	2	3	4	5	6
1	2	3	4	5	6	7
2	3	4	5	6	7	8
3	4	5	6	7	8	9
4	5	6	7	8	9	10
5	6	7	8	9	10	11
6	7	8	9	10	11	12

NOTICE THAT

It matters which number appears on which die, since there are different ways of getting the same outcome, so 1 + 2 and 2 + 1 are counted as distinct outcomes.

There are 36 cells in the table showing a sum, so the number of outcomes in the sample space for this experiment is 36.

Use the table to find the number of ways to obtain a success (a sum of 6 on the two dice).

$5 + 1$ $\quad 4 + 2$ $\quad 3 + 3$ $\quad 2 + 4$ $\quad 1 + 5$

So, there are 5 ways to get a sum of 6.

Divide the number of successes (the ways to get a sum of 6 from two dice) by the total number of outcomes in the sample space to find the probability.

$$P(\text{sum of } 6) = \frac{\text{number of successes}}{\text{total number of outcomes}} = \frac{5}{36}$$

Using the Fundamental Counting Principle to Find the Number of Outcomes in a Sample Space

The sample space for the experiments in **Examples 1** and **2** were found by listing all of the possible outcomes. This method is useful for finding the total number of outcomes in a sample space when the total number of possible outcomes is not very large, or when there is a systematic way to list the outcomes.

The number of outcomes in a sample space can often be found by applying the Fundamental Counting Principle.

For example, in **Example 1**, there are 2 outcomes for the first coin, 2 for the second, and 2 for the third. So, by the Fundamental Counting Principle, there are $2 \cdot 2 \cdot 2 = 2^3 = 8$ possible outcomes in the sample space.

The number of outcomes in the sample space for **Example 2** can also be found by using the Fundamental Counting Principle, where the number of outcomes is the product of the number of outcomes from one die (6) and the number of outcomes from the second die (6), $6 \cdot 6 = 36$.

The Fundamental Counting Principle is extremely useful when the number of outcomes for an experiment is too large to list them all.

WHAT'S THE BIG IDEA?

Describe a method other than the method shown in **Example 2** to calculate the probability that a roll of one die lands on a 3 is 1/6.

11.6.2 The Probability of an Event: Another Example

Experiments with dice and coins are commonly seen applications of probability. Another common application of probability concerns a standard deck of cards. Note that a standard deck of playing cards contains 52 cards in four suits: hearts, diamonds, clubs, and spades (jokers are not counted). There are 13 cards in each suit: the numbers 2 through 10, a jack, a queen, a king, and an ace. The spades and clubs are "black" cards, and the hearts and diamonds are "red" cards. So, a deck of cards contains 26 black cards and 26 red cards.

It can be assumed that the cards in a deck have been well shuffled and that any card chosen from a deck is picked "at random," meaning that the probability of selecting each particular card is equal. Therefore, it is possible to calculate a probability involving choosing a card (or several cards) from a deck.

EXAMPLE 3 Calculating Probability

Find the probability of selecting each type of card from a standard deck of cards.

A. a queen **B.** a club **C.** a red 5

SOLUTION

Find the total number of possible outcomes in the sample space.

Number of outcomes in sample space = 52 *There are 52 cards in a deck.*

Find the number of ways to obtain a success for each situation, then divide the number of successes by 52 to find the probability.

A. A "success" is choosing a queen. There are 4 queens in a deck of cards, so the number of successful outcomes is 4.

P(selecting a queen) = 4/52 = 1/13 *The probability is approximately 8%.*

The probability of selecting a queen is 4 in 52, or 1 in 13 (a relatively rare event).

B. A "success" is choosing a club. There are 13 clubs in a deck of cards, so the number of successful outcomes is 13.

P(selecting a club) = 13/52 = 1/4 *The probability is 25%.*

The probability of selecting a club is 13 in 52, or 1 in 4.

C. A "success" is choosing a red 5. There are two red 5s: the 5 of hearts and the 5 of diamonds. So, the number of successful outcomes is 2.

P(selecting a red 5) = 2/52 = 1/26 *The probability is approximately 4%.*

The probability of selecting a red 5 is 2 in 52, or 1 in 26.

TIP

This is a fairly likely event. If cards were selected from a deck (at random) over and over (and returned to the deck), a club would get selected every 1 in 4 cards on average.

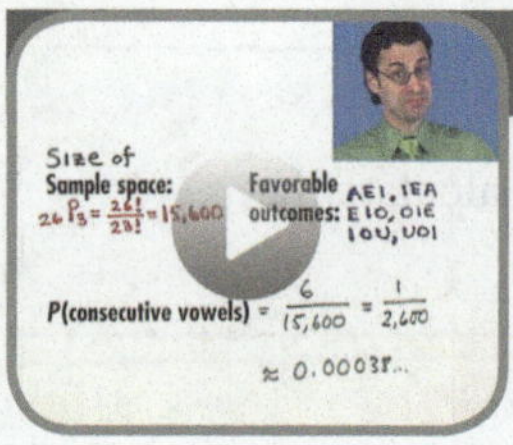

11.6.3 Calculating Probability by Counting

The experiments examined until now have been situations in which the sample space was small enough to count, although the Fundamental Counting Principle could also be used to find the number of outcomes in the sample space. In addition, the number of successes could be obtained by listing or enumerating the outcomes. Often, probabilities must be calculated for events that come from large sample spaces and involve large numbers of successes—too many to easily list. In these cases, counting methods can be used to find probabilities.

EXAMPLE 4 Calculating Probabilities Using Permutations

A 3-letter code is constructed from any of the 26 letters of the alphabet, but each letter can only be used once. What is the probability that the code consists of only vowels?

SOLUTION

TIP

If order matters, as when choices have specific positions, use the permutations formula.

Constructing the code in this example can be thought of as a permutation, in that 3 letters are being selected from the 26 letters of the alphabet without replacement. The order in which the letters are arranged matters, since different codes would arise from different orderings of the same set of 3 letters. Therefore, the sample space can be counted using the formula for permutations.

Find the total number of possible outcomes in the sample space.

There are 26 letters in the alphabet, with 3 letters being selected each time. The number of possible orderings is

Permutations ▶ $_nP_r = \dfrac{n!}{(n-r)!}$

$$_{26}P_3 = \frac{26!}{(26-3)!} = \frac{26 \cdot 25 \cdot 24 \cdot 23!}{23!} = 15{,}600\,.$$

Find the number of ways to obtain a success.

A "success" in this example is an ordering that is constructed only of vowels. There are 5 vowels in the alphabet: A, E, I, O, and U. The number of successes is equal to the number of orderings of selections of 3 of those vowels at a time (without replacement). This number is also a permutation.

$$_5P_3 = \frac{5!}{(5-3)!} = \frac{5 \cdot 4 \cdot 3 \cdot 2!}{2!} = 60$$

Divide the number of successes by the sample space to find the probability.

$$P(\text{a code with three vowels}) = \frac{_5P_3}{_{26}P_3} = \frac{60}{15{,}600} = \frac{1}{260}$$

TIP

1/260 is less than 1%.

The probability of all vowels is 1/260, so the event is not very likely.

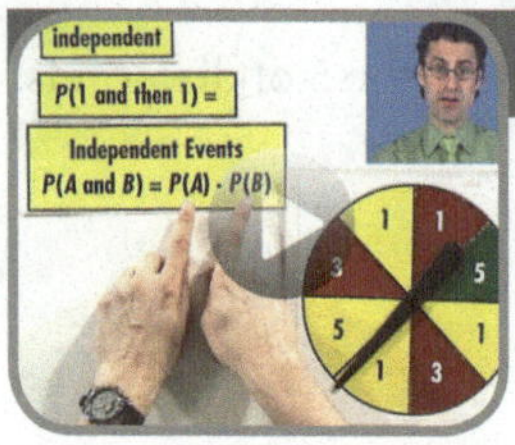

11.6.4 Independent Events

An event in a probability experiment is a possible outcome or group of outcomes of the experiment. Events can be classified as **simple events**, which are single, uncombined outcomes of an experiment, and **compound events**, which are events made up of more than one simple event. Examples of simple events are the outcome of a coin flip or the outcome of a card drawn from a deck of playing cards. Examples of compound events are the outcome of three coin flips or the outcome of five cards dealt from a deck of cards.

Some compound events are joint events in that they co-occur and their probability is described as "the probability of event *A and* event *B*." The word *and* means that both events occur together. When two events are independent, the occurrence of one event has no effect on the likelihood that the other event will occur. For example, when two people flip a coin, the outcome of one of the coin flips does not affect the probability of a "heads" or "tails" from the second coin flip. The coin flips are **independent events**.

DEFINITION

A **simple event** is an uncombined outcome of a probability experiment.

A **compound event** is an outcome of a probability experiment made up of more than one simple event.

Two events are **independent events** when the outcome of one event does not affect the probability of the outcome of the second event.

The joint probability of two independent events is equal to the product of their individual probabilities.

Joint Probability of Independent Events

$P(A \text{ and } B) = P(A) \cdot P(B)$, where A and B are independent events.

EXAMPLE 5 Calculating Probabilities of Joint Independent Events

IMPORTANT

Each sector is the same size, so the probability of the arrow landing on each sector is equal. Therefore, it is possible to calculate probabilities using the number of sectors.

Find each probability of a compound event on the spinner.

A. spinning a red and then a 5

B. spinning a 4, then a 2, and then a blue

SOLUTION

The outcome of one spin will not affect the outcome of subsequent spins, so each spin is an independent event. First, find the probabilities of the individual events that make up the compound event. Then multiply these probabilities to find the joint probability. The number of outcomes in the sample space for the spinner is 12.

A. There are 4 red sectors and 4 sectors labeled with a "5."

$$P(\text{red}) = \frac{4}{12} = \frac{1}{3} \qquad P(5) = \frac{4}{12} = \frac{1}{3}$$

The joint probability is the product of the individual probabilities.

$$P(\text{red and } 5) = P(\text{red}) \cdot P(5) = \frac{1}{3} \cdot \frac{1}{3} = \frac{1}{9}$$

Joint Probability of Independent Events $P(A \text{ and } B) = P(A) \cdot P(B)$

The probability of spinning a red and then a 5 is 1/9, or about 0.1. With a large number of pairs of spins, 1 in 9 is expected to come up a red first and then a 5.

B. There are 2 sectors labeled with a "4," 2 sectors labeled with a "2," and 2 blue sectors.

$$P(4) = \frac{2}{12} = \frac{1}{6} \qquad P(2) = \frac{2}{12} = \frac{1}{6} \qquad P(\text{blue}) = \frac{2}{12} = \frac{1}{6}$$

The joint probability is the product of the individual probabilities.

$$P(4 \text{ and } 2 \text{ and blue}) = P(4) \cdot P(2) \cdot P(\text{blue}) = \frac{1}{6} \cdot \frac{1}{6} \cdot \frac{1}{6} = \frac{1}{216}$$

TIP

Find the joint probability of any number of independent events by multiplying the individual probabilities.

The joint probability of the three spins coming up as 4, 2, and then blue are quite low.

WHAT'S THE BIG IDEA?

When selecting a sequence of cards from a regular deck of playing cards without replacing them, are the cards selected independent events? Explain.

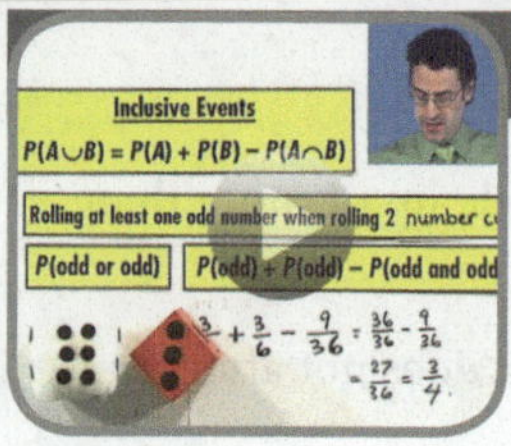

11.6.5 Inclusive Events

There are many circumstances in which it is important to know a combined probability, or a probability of **inclusive events**. In general, these situations can be recognized by the use of the word "or" in the description of the probability. For example, what is the probability of having a birthday in March *or* April? The word "or" implies that the two individual probabilities should be combined, or added, to find the total probability.

TIP

Mutually exclusive events cannot occur together.

Inclusive events are complicated when there is overlap between the two events, that is, when they are not mutually exclusive. It is not possible to have a birthday in both March and April, but what if the question asked, "What is the probability of having a birthday on a Friday or in March?" In this case, there are a number of ways of being born on a Friday during a year and a number of ways of being born in March. It is also possible to be born in March and on a Friday because the two events are not mutually exclusive. The combined probability is obtained by adding the probability of a Friday birthday to the probability of a March birthday, but the result will include March and Friday birthdays twice (once from the Friday probability, once from the March probability). This can be corrected for by subtracting the probability of joint occurrences from the sum of the probabilities. When the events cannot occur together, meaning there is never any overlap, this probability is just 0.

TIP

The formula for inclusive probability can be written using the intersection and union symbols.
$P(A \cup B) =$
$P(A) + P(B) - P(A \cap B)$

Inclusive Probability

The probability of inclusive events A or B occurring is given by

$$P(A \text{ or } B) = P(A) + P(B) - P(A \text{ and } B).$$

EXAMPLE 6 Calculating Probabilities for Inclusive Events

When rolling a die, what is the probability of rolling a 3 or a multiple of 3?

SOLUTION

TIP

Use a sum of probabilities when "or" is used; use a product of probabilities when "and" is used.

In describing the probability, the word "or" is used, so the events are inclusive. Find the individual probabilities by dividing the sample space of a single roll of a die, which is 6, by the possible number of successful outcomes.

$P(3) = \frac{1}{6}$ *There is 1 way to roll a 3.* $P(\text{multiple of } 3) = \frac{2}{6}$ *There are 2 ways to roll a multiple of 3 (3 or 6).*

Can these two events overlap, or happen together? Yes: rolling a 3 is both rolling a 3 and rolling a multiple of 3. There is 1 way within the sample space that the events occur together.

$$P(\text{rolling a 3 and rolling a multiple of 3}) = 1/6$$

Use the formula for inclusive probability.

$$P(3 \text{ or multiple of } 3) = P(3) + P(\text{multiple of } 3) - P(3 \text{ and a multiple of } 3) = \frac{1}{6} + \frac{2}{6} - \frac{1}{6} = \frac{1}{3}$$

The probability of rolling a 3 or a multiple of 3 is 1/3.

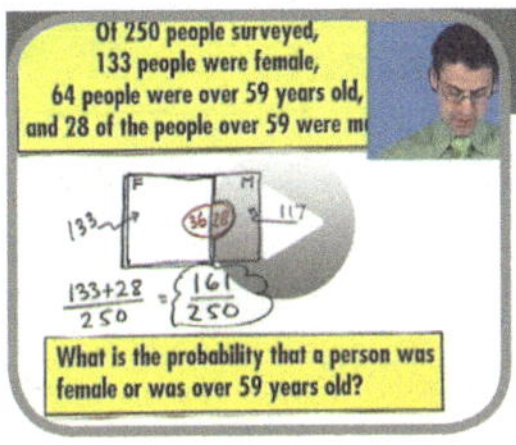

11.6.6 Inclusive Events: Another Example

EXAMPLE 7 Calculating Probabilities for Inclusive Events

Of 250 people surveyed, 133 were female, 64 were over 59 years old, and 28 of those over 59 years old were male. To the nearest percent, what is the probability that a randomly chosen survey respondent was female or over 59 years old?

SOLUTION

In describing the probability, the word "or" is used. The probability is about inclusive events and not about joint events (which would have used the word "and" in the description). Find the individual probabilities by dividing the possible number of successful outcomes by the sample space (250).

> **TIP**
>
> *A Venn diagram can also be used to answer this question. The solution shown here uses the formula for the probability of inclusive events directly.*

Given Information	Probabilities
Number of females: 133	$P(\text{female}) = \dfrac{133}{250}$
Number of respondents over 59 years old: 64	$P(>59) = \dfrac{64}{250}$
Number of respondents over 59 years old and male: 28	$P(>59 \text{ and male}) = \dfrac{28}{250}$

Is there a way these two events (female and over 59 years old) can overlap? Yes: not all of the over-59 individuals are male, so the remaining ones are female. These individuals are both over 59 and female. Find the probability of a respondent being a female over 59.

$$P(>59 \text{ and female}) = \frac{64-28}{250} = \frac{36}{250}$$

There are 64 respondents over 59, 28 of which are male.

Use the formula for inclusive probability.

$$P(\text{female or} > 59) = P(\text{female}) + P(>59) - P(>59 \text{ and female})$$

$$= \frac{133}{250} + \frac{64}{250} - \frac{36}{250} = \frac{161}{250} \approx 0.64$$

The probability that a randomly chosen survey respondent is female or over 59 years old is approximately 64%.

11.6.7 Mutually Exclusive Events

To find the combined probability of two mutually exclusive events, the individual probabilities of the events are simply added. The principle is the same as finding the combined probability of two events that are not mutually exclusive (that might occur together), except that the probability they occur together is 0, so there is nothing to subtract from the sum of the individual probabilities.

> ***Combined Probability of Mutually Exclusive Events***
>
> The probability of mutually exclusive events A or B occurring is given by
> $P(A \text{ or } B) = P(A) + P(B)$.

EXAMPLE 8 Combined Probability of Mutually Exclusive Events

NOTICE THAT

The given information is the same as in ***Example 7****.*

Of 250 people surveyed, 133 respondents were female, 64 respondents were over 59 years old, and 28 of those respondents over 59 years old were male. What is the probability that a randomly chosen respondent was a female under 59 years old or a male over 59 years old?

SOLUTION

Is there a way these two events (female under 59 and male over 59) can overlap? No: in this example, the two events cannot overlap, because an individual cannot be male and female at the same time. These are mutually exclusive events. The number of males over 59 is given, 28.

Find the number of females over 59, and then use that number to find the number of females under 59.

$$\text{Females over } 59 = 64 - 28 = 36 \qquad \text{Females under } 59 = 133 - 36 = 97$$

Use the formula for mutually exclusive events to find the combined probability.

$$P(\text{female under 59 or male over 59}) = P(\text{female under 59}) + P(\text{male over 59})$$

$$= \frac{97}{250} + \frac{28}{250} = \frac{125}{250} = 0.5$$

TIP

For the individual events, divide the number of successes by the sample space (250).

The probability that a randomly chosen survey respondent is a female under 59 or a male over 59 years old is 50%.

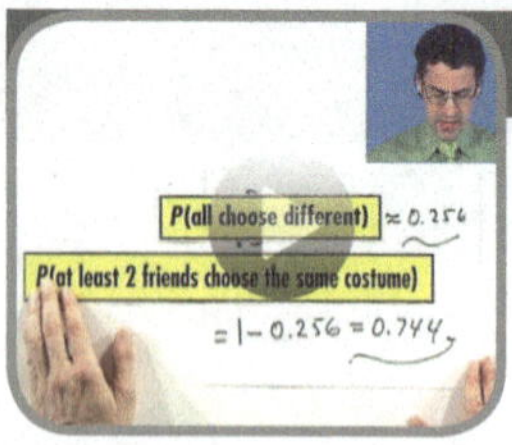

11.6.8 Using the Complement

The sum of the probabilities for all of the outcomes in a sample space is always equal to 1. For example, consider the experiment of flipping one coin. There are two outcomes in the sample space, H and T, and for each the probability is 1/2.

$$P(\text{H}) + P(\text{T}) = \frac{1}{2} + \frac{1}{2} = 1$$

So, the sum of the probabilities for all of the outcomes in the sample space is 1.

A consequence of this is that the sum of the probability of an event A and the probability of all events in the sample space that are not A is equal to 1. The event "not A" is called the **complement** of A, and it is represented as A', which is read as "A-prime."

TIP

Event A and all events that are not A comprise the complete sample space.

Sample space = {A, A'}

$$P(A) + P(A') = 1$$

For example, if the experiment is rolling two dice and the event is "the sum of the dice is a number greater than 4," then the complement of the event is "the sum of the dice is a number less than or equal to 4." Since the sum of the two dice must be a number that is either greater than 4 or less than or equal to 4, the sum of the probabilities of those events must be 1.

By subtracting $P(A)$ from each side of the equation $P(A) + P(A') = 1$, we see that the probability of the complement of A, $P(A')$, is equal to 1 minus $P(A)$.

The Probability of the Complement of an Event

The probability of the complement of event A is equal to 1 minus the probability of A.

$$P(A') = 1 - P(A)$$

This relationship can be exploited in situations in which finding $P(A)$ is difficult but finding $P(A')$ might be very simple. Typically, these situations involve phrases like "at least" or "one or more," which describe a lot of possibilities.

EXAMPLE 9 Calculating Probability Using the Complement

Suppose there are 6 students in a study group. What is the probability (to the nearest thousandth) that at least 2 of the students will have a birthday in the same month?

SOLUTION

Solving this example directly is difficult, because there are many ways the birthdays could happen that constitute a success. For example, 2 students could share a birthday in any of the 12 months, but there are many combinations of 2 students. Then 3-student combinations would have to be considered, and 4-student combinations, and so on, since the event is "at least 2 students."

The question includes the phrase "at least," so look to see if the probability of the complement event is easier to calculate. The complement of the event *at least 2 students have the same birth month* is *no students have the same birth month,* which is an event that is easier to count.

So, find the probability that no students have the same birth month.

The number of outcomes in the sample space can be calculated using the Fundamental Counting Principle. It is the product of the number of choices for the first student, 12 months, and the number of choices for the second student, 12 months, and so on. The number of outcomes in the sample space is equal to 12^6.

TIP

Alternatively, the first student has 12 months to choose from, the second student has 11 months, and so on; thus the number is $12 \cdot 11 \cdot 10 \cdot 9 \cdot 8 \cdot 7$. This is the same number as the permutation.

If no student can have the same birth month, this is like choosing 6 birth months out of the 12 months without replacement and with order mattering (since the students have different identities). The count is therefore a permutation: ${}_{12}P_6$.

$$P(\text{no students have same birth month}) = \frac{{}_{12}P_6}{12^6}$$

Use the complement to find the probability that at least 2 students will have a birthday in the same month.

$$\begin{aligned} P(\text{at least 2 students have same birth month}) &= 1 - P(\text{no students have same birth month}) \\ &= 1 - \frac{{}_{12}P_6}{12^6} \\ &= 1 - \frac{665{,}280}{2{,}985{,}984} \\ &\approx 0.777 \end{aligned}$$

The probability that at least 2 students in the group share a birth month is approximately 0.777.

SECTION 11.6 EXERCISES

Warm Up

Determine whether each situation represents a permutation, a combination, or neither.

1. the number of ways the letters in the word SKEEBALL can be arranged
2. the number of groups of 4 dogs that can be chosen from a pack of 20
3. the number of ways a president, vice president, and secretary can be chosen from a committee of 12

Just the Facts

Fill in the blanks.

4. The _____ of an event is the likelihood of that event occurring.
5. The set of all outcomes possible from an experiment is the experiment's _____.
6. When all of the _____ in an experiment have the same probability of occurring, the probability of an event A is the number of successful outcomes divided by the _____ number of possible outcomes in an experiment's sample space.
7. The probability of any event will always be a number between 0 and _____.
8. $P(A \text{ and } B) = P(A) \cdot P(B)$ is the formula for joint probability of _____ events.
9. A(n) _____ event is an uncombined outcome of a probability experiment. A(n) _____ event is an outcome of a probability experiment made up of more than one _____ event.
10. The event "not A" is called the _____ of A and is represented as A'.

Essential Skills

In Exercises 1–6, find each probability.

1. at least 2 tails in 3 coin flips
2. exactly 1 heads in 3 coin flips
3. a sum of 4 from rolling 2 dice
4. a sum of 10 from rolling 2 dice
5. rolling an even number on a die
6. drawing a black king from a standard deck of 52 playing cards

7. In a quiz show, 3 questions on Sports, 3 questions on General Knowledge, and 4 questions on Science are printed separately on 10 cards and placed upside down. A contestant is asked to select 2 cards at random. What is the probability of the contestant selecting 2 questions on Science?
8. Party guests are each given a number at the door when they come in. During the party, 3 numbers will be chosen at random from a hat to win door prizes. There are 20 people at the party, of whom 10 are women and 10 are men. What is the probability that all 3 prizes go to women?

In Exercises 9–10, use the spinner to find each probability.

9. the probability of spinning a yellow and then a 4
10. the probability of spinning a 5, then a green, then a 2

11. From the set of all whole numbers less than 25, what is the probability of getting a number divisible by 5 or divisible by 3?
12. Find the probability of drawing a red card or a 5 from a standard deck of 52 playing cards.
13. Of 1000 students surveyed, 530 were boys and 180 had bicycles. Of those who had bicycles, 30 were girls. What is the probability that a student surveyed was a boy or had a bicycle?
14. Of 800 people surveyed, 420 were male and 325 had cell phones. Of those with cell phones, 200 were female. What is the probability that a person surveyed was male or had a cell phone?
15. Rolling a die, what is the probability of rolling an odd number or a 4?
16. A card is drawn from a standard deck of 52 playing cards. What is the probability of drawing a black card or a red card?
17. Each of 5 boys randomly chooses a watch from 12 different styles. What is the probability that at least 2 boys choose the same type of watch?
18. Each of 6 women randomly chooses a woolen overcoat from 15 different styles. What is the probability that at least 2 women choose the same type of overcoat?

Extensions

19. A restaurant gives a free fortune cookie to every guest. The restaurant claims that there are 100 different equally likely messages inside the fortune cookies. What is the probability that a group of 5 people receives at least 2 cookies with the same message inside?

20. A police lineup was composed of 5 randomly arranged men, where 2 of the 5 men were undercover detectives. Assume that 1 of the 5 men was guilty, but the guilty man was not one of the detectives.

A. What is the probability that the 2 detectives stood next to each other?

B. What is the probability that the guilty person stood in the far left position?

C. What is the probability that the guilty person stood in the middle position?

21. If a 6-volume set of books is placed on a shelf at random, what is the probability that the books will be arranged in either correct order (beginning with Volume 1) or reverse order (beginning with Volume 6)?

22. Out of 200 students in a senior class, 113 students are either varsity athletes or on the honor roll. There are 74 seniors who are varsity athletes and 51 seniors who are on the honor roll. Find the probability that a randomly selected senior:

A. is a varsity athlete but not on the honor roll.

B. is on the honor roll.

C. is neither a varsity athlete nor on the honor roll.

D. is both a varsity athlete and on the honor roll.

23. At a local diner, the probability that a patron will order pancakes is 0.3 (event P) and the probability that she will order bacon is 0.45 (event B). The probability that she will order both pancakes and bacon is 0.135.

A. Are P and B independent events?

B. What is the probability that she will order neither pancakes nor bacon?

C. What is the probability that she will order either bacon or pancakes?

D. What is the probability that she will order bacon but not pancakes?

24. A hand of 5 cards is dealt from a well-shuffled standard deck of 52 cards. Find the probability that the hand contains:

A. 4 of a kind (for example, 4 kings)

B. at least 2 hearts

C. 5 cards of the same color

11.7 THE BINOMIAL THEOREM

OBJECTIVES

- Use the Binomial Theorem and Pascal's Triangle to expand powers of binomials.
- Use the combinations formula to find binomial coefficients.
- Find specific terms of a binomial expansion.

PREREQUISITE VOCABULARY TERMS

binomial
combination
factorial
polynomial

11.7.1 Using the Binomial Theorem

Recall that a power of a binomial can be expanded by multiplying.

$$(x + y)^2 = x^2 + 2xy + y^2$$
$$(x + y)^3 = x^3 + 3x^2y + 3xy^2 + y^3$$
$$(x + y)^4 = x^4 + 4x^3y + 6x^2y^2 + 4xy^3 + y^4$$
$$(x + y)^5 = x^5 + 5x^4y + 10x^3y^2 + 10x^2y^3 + 5xy^4 + y^5$$

TIP

x and y are the terms of the binomial.

Close examination of these expansions reveals several patterns. Let the binomial's power be n. Notice that the coefficients when $n = 2$ are 1, 2, and 1, and when $n = 3$ the coefficients are 1, 3, 3, and 1. Clearly, the coefficients begin and end with 1 in each expansion. So that you can observe this pattern, the coefficients from example expansions are listed in Figure 11.7a.

When the coefficients are listed in this triangular form, each coefficient between the 1s is the sum of the two coefficients diagonally above it. This pattern can be continued, giving the coefficients for the binomial expansion when $n = 6$, $n = 7$, and so on.

$n = 2$:	1 2 1
$n = 3$:	1 3 3 1
$n = 4$:	1 4 6 4 1
$n = 5$:	1 5 10 10 5 1

Figure 11.7a

The coefficients follow the pattern seen in **Pascal's Triangle** (Figure 11.7b). Pascal's Triangle is created by writing a 1, then two 1s diagonally below the first, then a 1, 2, and 1 below those. Each subsequent row begins and ends with a 1, and each entry between the 1s is the sum of the two entries diagonally above it.

Pascal's Triangle

Level 0	1
Level 1	1 1
Level 2	1 2 1
Level 3	1 3 3 1
Level 4	1 4 6 4 1
Level 5	1 5 10 10 5 1
Level 6	1 6 15 20 15 6 1
⋮	⋮

Figure 11.7b

The entries in level 2 (1, 2, and 1) are the coefficients of a binomial raised to the power of 2 (i.e., where $n = 2$): $(x + y)^2 = x^2 + 2xy + y^2$. Notice also that the entries in level 1 (1 and 1) are the coefficients of a binomial raised to the power of 1: $(x + y)^1 = x + y$. Furthermore, the entry in level 0 (1) is the coefficient of a binomial raised to the 0 power: $(x + y)^0 = 1$. Generally, the coefficients in the expansion of $(x + y)^n$, the **binomial coefficients**, are the entries in the nth level of Pascal's Triangle.

The coefficients in the expansion of $(x + y)^n$, called the **binomial coefficients**, are the entries in the nth level of Pascal's Triangle.

Further examination of the expansions of $(x + y)^2$, $(x + y)^3$, $(x + y)^4$, and $(x + y)^5$ reveals additional patterns.

- The number of terms in each expansion is equal to $n + 1$.
- The power of the first and last term is n.
- The powers of x (the first term in the binomial) begin with n and decrease by 1 in each successive term.
- The powers of y (the second term in the binomial) increase by 1 in each successive term and end with n.
- The sum of the powers in each term is equal to n.

These patterns are summarized in the **Binomial Theorem**, which is a formula for expanding a power of a binomial.

IMPORTANT

x and y represent the first and last terms in a binomial.

Binomial Theorem

$$(x + y)^n = {}_nC_0 x^n y^0 + {}_nC_1 x^{n-1} y^1 + {}_nC_2 x^{n-2} y^2 + \ldots + {}_nC_{n-1} x^1 y^{n-1} + {}_nC_n x^0 y^n,$$

where n is a natural number.

The notation ${}_nC_0$, ${}_nC_1$, . . . refers to the binomial coefficients, which can be evaluated in several ways. First, as seen previously, the ${}_nC_0$ coefficient is the first entry from the nth level of Pascal's Triangle. The first and last entry of each level in Pascal's Triangle is 1, so ${}_nC_0 = 1$ and ${}_nC_n = 1$. (A second method for evaluating binomial coefficients is demonstrated in the following topic.)

The expansion in the Binomial Theorem can be simplified by substituting 1 for the first and last binomial coefficients, and simplifying the factors in each term.

$$(x + y)^n = x^n + {}_nC_1 x^{n-1} y + {}_nC_2 x^{n-2} y^2 + \ldots + {}_nC_{n-1} x y^{n-1} + y^n$$

EXAMPLE 1 Finding a Binomial Expansion Using Pascal's Triangle

Use the Binomial Theorem to expand each binomial.

A. $(a + b)^4$

B. $(3x + 2y)^6$

SOLUTION

Write the terms of the expansion by substituting the coefficients from level n of Pascal's Triangle and the values of the two terms of the binomial into the Binomial Theorem. Then simplify the resulting polynomial.

A. The power of the binomial is 4, so $n = 4$ and the binomial coefficients are the values from level 4 of Pascal's Triangle: 1, 4, 6, 4, and 1.

The first term of the binomial is a and the second term is b. Therefore, $x = a$ and $y = b$.

Substitute $n = 4$, $x = a$, $y = b$, and the binomial coefficients into the Binomial Theorem.

$(a + b)^4 = a^4 + 4a^3b + 6a^2b^2 + 4ab^3 + b^4$ *Apply the Binomial Theorem.*

B. The power of the binomial is 6, so $n = 6$ and the binomial coefficients are the values from level 6 of Pascal's Triangle: 1, 6, 15, 20, 15, 6, and 1.

The first term of the binomial is $3x$ and the second term is $2y$. So, substitute 6 for n, $3x$ for x, $2y$ for y, and the binomial coefficients into the Binomial Theorem.

IMPORTANT

Enclose 3x and 2y within parentheses and be sure to apply each power to both factors in each expression.

Note that the terms in the binomial, $3x$ and $2y$, are not single variables (or numeric values). So, the Power of a Product Property, $(ab)^n = a^n b^n$, must be used when simplifying.

$$(3x + 2y)^6$$

$$= (3x)^6 + 6(3x)^5(2y) + 15(3x)^4(2y)^2 + 20(3x)^3(2y)^3 + 15(3x)^2(2y)^4 + 6(3x)(2y)^5 + (2y)^6$$

$$= 3^6x^6 + 6(3^5x^5)(2y) + 15(3^4x^4)(2^2y^2) + 20(3^3x^3)(2^3y^3) + 15(3^2x^2)(2^4y^4) + 6(3x)(2^5y^5) + 2^6y^6$$

$$= 729x^6 + 6(243x^5)(2y) + 15(81x^4)(4y^2) + 20(27x^3)(8y^3) + 15(9x^2)(16y^4) + 6(3x)(32y^5) + 64y^6$$

$$= 729x^6 + 2{,}916x^5y + 4{,}860x^4y^2 + 4{,}320x^3y^3 + 2{,}160x^2y^4 + 576xy^5 + 64y^6$$

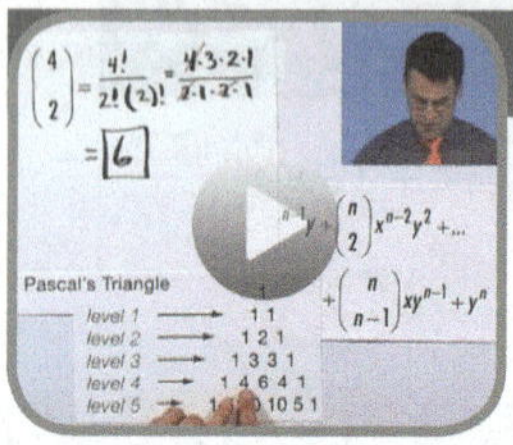

11.7.2 Binomial Coefficients

In **Example 1**, the binomial coefficients were found using Pascal's Triangle. However, when the power of a binomial is large, using Pascal's Triangle is unwieldy. The binomial coefficients can also be evaluated using the formula for combinations (which is also called the formula for binomial coefficients).

Binomial Coefficients (Combinations)

For a binomial raised to the nth power,

$$(x + y)^n = x^n + {}_nC_1x^{n-1}y + {}_nC_2x^{n-2}y^2 + \ldots + {}_nC_{n-1}xy^{n-1} + y^n,$$

the binomial coefficient of $x^{n-m}y^m$ is given by $C(n, m) = {}_nC_m = \binom{n}{m} = \dfrac{n!}{m!(n-m)!}$.

Any of the three notations for a combination can be used. Each is read as "n choose m."

REMEMBER

$n! = n(n-1)(n-2)\ldots(2)(1)$

Thus, the binomial coefficient of the second term in the expansion of $(x + y)^n$ is given by $C(n, 1)$, the binomial coefficient of the third term in the expansion of $(x + y)^n$ is given by $C(n, 2)$, and so on. Note that the coefficient of a term in the *simplified* expansion of $(x + y)^n$ will be equal to its corresponding binomial coefficient only when the coefficient of the x- and y-terms is 1.

A simplified version of the Binomial Theorem is shown below, using the $\binom{n}{m}$ notation for the binomial coefficients.

Binomial Theorem—Alternate Form

$$(x + y)^n = x^n + \binom{n}{1}x^{n-1}y + \binom{n}{2}x^{n-2}y^2 + \ldots + \binom{n}{n-1}xy^{n-1} + y^n$$

where n is a natural number.

EXAMPLE 2 Finding a Binomial Expansion

Use the Binomial Theorem and the formula for binomial coefficients to expand each binomial.

A. $(a-3b)^3$ **B.** $\left(b^3+\sqrt{3}a\right)^3$

SOLUTION

The power of each binomial is 3. So, the binomial coefficients will be the same in each expansion. First, evaluate the binomial coefficients using the formula for binomial coefficients where $n = 3$.

m	Binomial Coefficient
0	$\binom{3}{0}=\frac{3!}{0!(3-0)!}=\frac{3\cdot2\cdot1}{1(3\cdot2\cdot1)}=1$
1	$\binom{3}{1}=\frac{3!}{1!(3-1)!}=\frac{3\cdot2\cdot1}{1(2\cdot1)}=3$
2	$\binom{3}{2}=\frac{3!}{2!(3-2)!}=\frac{3\cdot2\cdot1}{2\cdot1(1)}=3$
3	$\binom{3}{3}=\frac{3!}{3!(3-3)!}=\frac{3\cdot2\cdot1}{3\cdot2\cdot1(1)}=1$

NOTICE THAT

The binomial coefficient for the first and last term (i.e., the 0th and 3rd term) is 1.

Write the terms of each expansion using these coefficients and the appropriate powers of the two terms in each binomial. Then simplify the resulting polynomials, paying special attention to negative signs.

A. The binomial can be written as $(a + (-3b))$. So, substitute a for x and $-3b$ for y. Since $-3b$ is a product, enclose this expression within parentheses and apply the Power of Product Property when simplifying.

$$(a-3b)^3 = a^3 + 3a^2(-3b) + 3a(-3b)^2 + (-3b)^3 = a^3 - 9a^2b + 27ab^2 - 27b^3$$

B. Substitute b^3 for x and $\sqrt{3}a$ for y. Since $\sqrt{3}a$ is a product, enclose this expression within parentheses and apply the Power of Product Property when simplifying. Additionally, since b^3 is a power, apply the Power of a Power Property, $(a^m)^n = a^{m\cdot n}$, when simplifying.

$$\begin{aligned}\left(b^3+\sqrt{3}a\right)^3 &= (b^3)^3+3(b^3)^2\left(\sqrt{3}a\right)+3(b^3)\left(\sqrt{3}a\right)^2+\left(\sqrt{3}a\right)^3\\ &= b^{3\cdot3}+3(b^{3\cdot2})\left(\sqrt{3}a\right)+3(b^3)\left(\sqrt{3}\right)^2(a^2)+\left(\sqrt{3}\right)^3(a^3)\\ &= b^9+3\sqrt{3}ab^6+9a^2b^3+3\sqrt{3}a^3\end{aligned}$$

11.7.3 Finding a Term of a Binomial Expansion

The Binomial Theorem can also be used to find the value for any individual term in an expansion of $(x + y)^n$. Looking at the expansion, a pattern emerges in which the rth term's coefficient is $\binom{n}{r-1}$, the power of x is $n - r + 1$, and the power of y is $r - 1$.

TIP

Another way of writing the formula that may be easier to remember is $\binom{n}{r-1}x^{n-(r-1)}y^{r-1}$.

***r*th Term of a Binomial Expansion of $(x + y)^n$**

The rth term of a binomial $(x + y)$ raised to the nth power is $\binom{n}{r-1}x^{n-r+1}y^{r-1}$.

EXAMPLE 3 **Finding a Specific Term in a Binomial Expansion**

A. Find the 12th term of $(a^2 + b)^{15}$. **B.** Find the 6th term of $(a - 2b)^{12}$.

SOLUTION

Determine n, r, and the terms of the binomial. Then substitute the values into the formula for the rth term.

TIP

n is the exponent of the binomial, and r is the number of the term.

A. $n = 15$, $r = 12$, $x = a^2$, $y = b$

$$\binom{15}{11}(a^2)^{15-12+1}b^{12-1} = \frac{15!}{11!4!}(a^2)^4b^{11} = 1365a^8b^{11}$$

The 12th term of the expansion is $1365a^8b^{11}$.

B. $n = 12$, $r = 6$, $x = a$, $y = -2b$

$$\binom{12}{5}a^{12-6+1}(-2b)^{6-1} = \frac{12!}{5!7!}a^7(-2b)^5$$

$$= \frac{12!}{5!7!}(-2)^5a^7b^5$$

$$= -25{,}344a^7b^5$$

The 6th term of the expansion is $-25{,}344a^7b^5$.

SECTION 11.7 EXERCISES

Warm Up

Simplify.

1. $(a^7d^4)^3$
2. $(-6mn^3)^5$
3. $\left(5\sqrt{6}f^3g\right)^4$

Just the Facts

Fill in the blanks.

4. The _____ is a formula for expanding the power of a binomial.
5. Binomial coefficients can be denoted as _____, _____, or _____. They can be found using combinations or with _____ Triangle.
6. To find the 20th term of a binomial expansion raised to the 80th power, it is more efficient to use the _____ rather than Pascal's Triangle.
7. The expression $\binom{8}{5-1}(3x)^{8-5+1}(-2y)^{5-1}$ is used to find the _____ term of a binomial expression raised to the _____ power, where the first term in the binomial is _____ and the second term is _____.
8. Level 8 in Pascal's Triangle provides the _____ for a binomial expansion where n = _____. These coefficients are _____.
9. The binomial coefficients of the first and last terms of a binomial expansion will always be _____.
10. The first binomial coefficient is the number of combinations for _____ choose _____.

Essential Skills

In Exercises 1–8, expand each binomial.

1. $(x+8)^4$
2. $(a+b)^6$
3. $(2g+h)^3$
4. $(3a+4b)^4$
5. $(a-2b)^4$
6. $(2x-5y)^5$
7. $\left(p+\sqrt{2q}\right)^4$
8. $\left(a^2+\sqrt{5b}\right)^3$

In Exercises 9–10, find the indicated term in each binomial expansion.

9. 5th term of $(2a-b^2)^9$
10. 4th term of $(c^5-2d^4)^7$

Extensions

11. Expand. $\left(5x-2\sqrt{y}\right)^7$
12. Find the 13th term of the expansion of $(3m+n^4)^{15}$.
13. 1024 is the binomial coefficient for which term in the expansion of $(2-3x)^{10}$?

In Exercises 14–16, expand each binomial.

14. $(3-2i)^5$
15. $\left(\frac{2}{x}-3y\right)^4$
16. $\left(f^{\frac{3}{4}}+g^{\frac{1}{4}}\right)^5$

CHAPTER 11 REVIEW EXERCISES

Section 11.1 Sequences and Series

1. Find the 5th and 14th terms of the sequence. $a_n = \frac{n^2 - 11}{3n - 5}$

2. Find a general function for the sequence. $-7, 1, 9, 17, \ldots$

3. Find the first 5 terms for the recursive sequence if $a_1 = 3$ and $a_n = -2a_{n-1} + 4$.

4. Find the value of the summation. $\sum_{n=3}^{6} (2^n - 10)$

5. Write a summation that represents the series.
$-7 - 15 - 23 - 31 - 39$

Section 11.2 Arithmetic Sequences

6. Find a_{26} for the arithmetic sequence if $a_1 = 11$ and $d = 3$.

7. Find the missing terms in the arithmetic sequence.
3, __, __, __, 23

8. Find a general function for the arithmetic sequence if $a_4 = 74$ and $a_9 = 144$.

In Exercises 9–10, find the indicated sum for each arithmetic sequence.

9. Find S_{18} for the arithmetic sequence.
$-3, 5, 13, 21, 29, \ldots$

10. $\sum_{n=6}^{18} (38 - 5n)$

Section 11.3 Geometric Sequences

11. Find the 10th term of the geometric sequence if $a_1 = 8$ and $a_2 = 16$.

12. Find the general function for the geometric sequence if $a_3 = 3$ and $a_6 = -1/9$.

13. Find the sum of the first 8 terms of the geometric sequence if $a_1 = 6$ and $r = -1/2$.

14. Find the sum of the infinite geometric sequence.
$-3, 2, -\frac{4}{3}, \frac{8}{9}, \ldots$

15. Write 0.88888888 . . . as a fraction in simplest form.

Section 11.4 Mathematical Induction

16. Write the $n = k + 1$ statement for the inductive proof of $2 + 4 + 6 + \ldots + 2n = n^2 + n$ (for all natural numbers n).

Section 11.5 Counting Principles

17. On a "build a stuffed animal" website, there are 41 choices for the type of animal, 84 outfit choices, and 21 footwear choices available for purchase. How many different stuffed-animal varieties are there if 1 animal, 1 outfit, and 1 style of footwear are chosen?

18. How many different ways can you arrange 6 family members in a row for a portrait?

19. Determine the number of ways first, second, and third place finishers can be chosen from 12 teams.

20. Determine the number of distinguishable permutations of the letters in the word "calculus."

Section 11.6 Probability

21. Find the probability of pulling a jack or a club from a standard deck of 52 playing cards.

22. A box contains 4 white ribbons and 8 pink ribbons. Determine the probability of picking a white ribbon and then another white ribbon (without replacing the first white ribbon).

23. In a large high school, 55% of the students are boys, 60% of the students play sports, and 40% of the girls play sports. If a student is chosen at random, what is the probability that the student is female and plays sports?

24. Seven cards are dealt from a well-shuffled standard deck of cards. Find the probability of getting at least one ace.

25. The probability of rain on a certain day is 65% in Houston and 60% in Mexico City. Find the probability that it will not rain in either city.

Section 11.7 The Binomial Theorem

In Exercises 26–28, expand each binomial.

26. $(2 - y)^5$

27. $(5a - 2b)^3$

28. $(x^2 - \sqrt{6}y)^4$

In Exercises 29–30, find the indicated term in each binomial expansion.

29. 5th term of $(v^3 + 2w^2)^7$

30. 11th term of $(3x + 2)^{10}$

ANSWERS

SECTION P.1 ANSWERS

Warm Up/Just the Facts

1. −11/24

$$-\frac{5}{8}-\left(-\frac{1}{6}\right)$$
$$= -\frac{5}{8}+\frac{1}{6}$$ *Write subtraction of a negative as addition.*
$$= -\frac{15}{24}+\frac{4}{24}$$ *Write each term as an equivalent fraction with a common denominator.*
$$= \frac{-15+4}{24}$$ *Combine the numerators.*
$$= -\frac{11}{24}$$ *Simplify the numerator.*

2. $-6k^2 + 8k - 4$

$$7k - 4 + k^2 + k - 7k^2 = k^2 - 7k^2 + 7k + k - 4$$
$$= -6k^2 + 8k - 4$$

3. The values of x include all numbers between −10 and 3, including 3 but not −10.

4. irrational numbers

5. whole numbers

6. Rational; integers; ≠ 0

7. −100

8. includes b but not a; includes a but not b

9. closed; open

10. distance; 0

Essential Skills/Extensions

1. $-3, 2.09, \sqrt{10}, \frac{11}{3}, 3.72$

3. $-4.11, -4.1, 1/4, 12/5, \sqrt{13}$

5. whole number, integer, rational number, real number

7. rational number, real number

9. natural number, whole number, integer, rational number, real number

11. 60

13. −1

15. 11/4

17. −100

19. −3/4; 4/3

21. 8.1; −10/81

23. Associative Property of Addition

25. Identity Property of Multiplication

27. $(0, \infty)$

29. $(-\infty, 10]$

31. $(0, 3.2]$

33. $(7, 20]$

35. $x \le 4$

37. $x \ge -1.5$

39. $5 \le x \le 10$

41. $-3.8 \le x < 12$

43. $-4 < x < 0$; $(-4, 0)$

45. $(-\infty, 1) \cup (1, \infty)$

47. $-2 \le x \le 4$; $[-2, 4]$; $\{x \mid -2 \le x \le 4\}$

49. {1, 2, 3, 4, . . . }

51. {2, 4, 6, 8, . . . }

53. 0

55. 3

57. when $x \ge 0$, $-|x| = -x$; when $x < 0$, $-|x| = x$

59. when $x \ge 0$, $-4x + |x| = -3x$; when $x < 0$, $-4x + |x| = -5x$

61. 15 units

63. 5.75 units

65. −10/3, 0.15, 4/5 (possible answers)

67. 0, −4, −9 (possible answers)

69. 0

71. The student substituted the value for n incorrectly in the denominator, which should be 5 + 1 − (−3).

73. $A \cup B = [-5, \infty)$, $A \cap B = [-1, 0]$

75. $A \cup B = (-\infty, -4) \cup (-3, \infty)$, $A \cap B = \varnothing$ (empty set)

77. $A \cup B = (-\infty, \infty)$, $A \cap B = [10]$

79. true

81. true

83. yes; $AB = CD = AD = BC = 11$

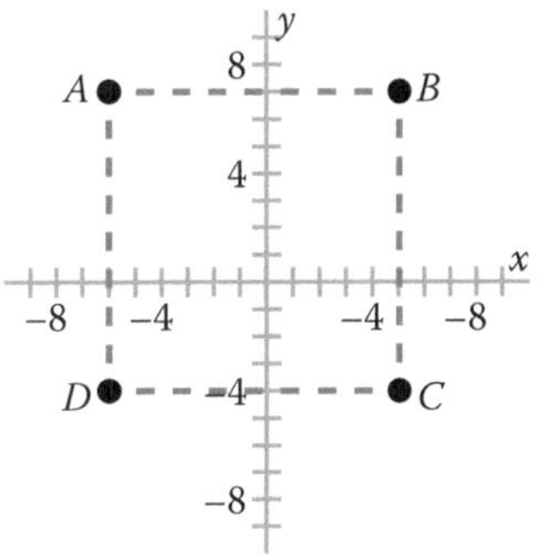

SECTION P.2 ANSWERS

Warm Up/Just the Facts

1. 125

$5^3 = 5 \cdot 5 \cdot 5 = 125$ *In 5^3, 5 is a factor 3 times.*

2. 3/7

$$\left(\frac{3}{7}\right)^2 = \left(\frac{3}{7}\right)\left(\frac{3}{7}\right) = \frac{3 \cdot 3}{7 \cdot 7} = \frac{9}{49}$$

In $\left(\frac{3}{7}\right)^2$, $\frac{3}{7}$ is a factor twice.

3. 8^4

$8 \cdot 8 \cdot 8 \cdot 8 = 8^4$ *In 8^4, 8 is a factor 4 times.*

4. nonzero; base; exponent, power

5. like bases; added

6. base; multiplied

7. Quotient of Powers; like bases; subtracted

8. Zero Exponent; zero

9. greater than or equal to; less than; power

10. negative; 1

Essential Skills/Extensions

1. 49

3. −32

5. −32

7. −1

9. x^8

11. simplified

13. m^{11}

15. $w^{13}x^6$

17. $2b^5c^3d^{14}$

19. $-16g^{12}h^5k^8$

21. d^{10}

23. h^{32}

25. $25p^{16}q^2$

27. $12g^{10}h^8k$

29. h^8

31. $1/m^7$

33. $1/j$

35. 16

37. 1/729

39. −125/8

41. 6.02×10^5

43. 3.0×10^{-5}

45. 4.69×10^9

47. 710,000

49. 0.00052

51. 0.00000018

53. 4.8×10^{10}

55. 3.0×10^{-7}

57. 9.8×10^4

59. 9.72×10^{-4}

61. $\frac{d^6}{2}$; Power of a Power, Product of Powers, Quotient of Powers

63. $-\frac{1}{27n^6}$; Zero Exponent, Negative Exponent, Power of a Product, Power of a Power, Quotient of Powers

65. $3x^0$; $\frac{6x^0}{2}$ (possible answers)

67. $\frac{4x^4}{2x^6}$; $\frac{6x^{10}}{3x^{12}}$ (possible answers)

69. $(2x^3y^2z)^4$; $(4x^6y^4z^2)^2$ (possible answers)

71. The error is in $2^{-3}c^{-5(-3)} = -8c^{15}$ because $2^{-3} \neq -8$.

Correction: $2^{-3}c^{-5(-3)} = \left(\frac{1}{2^3}\right)c^{15} = \frac{c^{15}}{8}$

SECTION P.3 ANSWERS

Warm Up/Just the Facts

1. 64
   $2^6 = 64$ *2 is a factor 6 times.*
2. $6x + 2y - 6$
   $5x - 7 - y + x + 3y + 1 = 5x + x - y + 3y - 7 + 1$
   $= 6x + 2y - 6$
3. k^{11}
   $k^3(k^4)^2 = k^3(k^8) = k^{11}$
4. p^n
5. square; 3
6. Even
7. −512
8. perfect squares
9. like radicals; radicand, index
10. $p - q\sqrt{n}$; conjugate

Essential Skills/Extensions

1. true
3. true
5. false
7. true
9. $5\sqrt{7}$
11. $2\sqrt[3]{2}$
13. $-24\sqrt{2}$
15. $\frac{8\sqrt[3]{3}}{3}$
17. $6\sqrt[3]{6}$
19. $-2\sqrt{2} + 12\sqrt{3}$
21. $3\sqrt{3}$
23. $\sqrt{2}$
25. $-\frac{\sqrt{3}}{3}$
27. $\frac{4\sqrt{14}}{7}$
29. $\frac{\sqrt[3]{75}}{5}$
31. $\frac{5\sqrt[3]{9}}{3}$
33. $-2\sqrt{7} + 6$
35. $\frac{5 + \sqrt{3}}{11}$
37. $\frac{8 - \sqrt{10}}{9}$
39. 125
41. 81
43. 32
45. $\sqrt{x}$
47. $2\sqrt[6]{32}$
49. $y\sqrt{y}$
51. 1/256
53. 1/343
55. 1/8
57. $3\sqrt{3}$
59. 2
61. 9
63. $3u^3v\sqrt{uz}$
65. $6g^2h^3j\sqrt{2j}$
67. $2n^4p^5\sqrt[3]{2mp}$
69. $\frac{w\sqrt{v}}{2uv}$
71. $\frac{4t^3}{y^3}$
73. $mn^3p\sqrt{7}$
75. 5 and 6; 5.6
77. 2 and 3; 2.7
79. $\frac{5\sqrt{2x}}{2}$
81. $\frac{2\sqrt{1+y}}{1+y}$
83. E only
85. $2\sqrt[4]{2}$
87. **A:** Writing the radical with a rational exponent gives $4^{\frac{1}{4}}$. Since $\frac{1}{4} = \frac{1}{2} \cdot \frac{1}{2}$, by the Power of a Power Property, the power can be written as $(4^{\frac{1}{2}})^{\frac{1}{2}}$. Simplifying the power within parentheses gives $2^{\frac{1}{2}}$, which is equal to $\sqrt{2}$ when written in radical form.

    B: Writing the radical with a rational exponent and 4 as a power gives $(2^2)^{\frac{1}{4}}$. By the Power of a Power Property, $(2^2)^{\frac{1}{4}} = 2^{2 \cdot \frac{1}{4}} = 2^{\frac{1}{2}}$ which is equal to $\sqrt{2}$ when written in radical form.

SECTION P.4 ANSWERS

Warm Up/Just the Facts

1. $15x^2 - 35x + 10$

   $5(3x^2 - 7x + 2) = 5(3x^2) - 5(7x) + 5(2) = 15x^2 - 35x + 10$

2. $18y + 2$

   $$\begin{aligned} 2(4y + 1) + 10y &= 2(4y) + 2(1) + 10y \\ &= 8y + 2 + 10y \\ &= 18y + 2 \end{aligned}$$

3. $-54p^{13}qr^4$

   $$\begin{aligned} -54p^{13}qr^4 = -6p^5qr^2(3p^4r)^2 &= -6p^5qr^2(3)^2(p^4)^2(r)^2 \\ &= -6p^5qr^2(9)(p^8)(r^2) \\ &= -6(9)p^{5+8}qr^{2+2} \\ &= -54p^{13}qr^4 \end{aligned}$$

4. monomial
5. exponents
6. difference; monomials (or terms)
7. 2; 3
8. quadratic
9. degree; leading coefficient
10. multiplying; binomials

Essential Skills/Extensions

1. $k^2 - 5k + 12$; leading coefficient: 1; quadratic trinomial
3. $-5m^5 - 6m^3 + 3m^2 + 16m$; leading coefficient: -5; quintic polynomial
5. $b^4 + 4b^3 + 2b^2 - b$; leading coefficient: 1; quartic polynomial
7. $-6x^3 + x + 6$
9. $4y^3 - 3y^2 - 7y + 1$
11. $7m^3 + 5m^2 - 12m + 6$
13. $-5x^2 + 38x - 48$
15. $6a^3 - 15a^2 + 8a - 20$
17. $-7m^4 - 19m^2 + 6$
19. $3x^4 - 8x^3 + 11x^2 - 7x + 1$
21. $2x^4 + x^3 + 5x^2 - 13x + 5$
23. $36y^6 - 24y^4z - 24y^3z^2 + 4y^2z^2 + 8yz^3 + 4z^4$
25. $9x^2 + 42x + 49$
27. $125a^6 + 150a^4bc + 60a^2b^2c^2 + 8b^3c^3$
29. $4x^2 - 25$
31. $2\sqrt{2}y^3 - 42y^2z^3 + 147\sqrt{2}yz^6 - 343z^9$
33. $81n^4 - n^2$
35. $9b^4 - 19b^2 + 1$
37. $64r^6 - 48r^5 + 108r^4 - 49r^3 + 54r^2 - 12r + 8$
39. $(3x - 1)(2x - 10)$, $(3x - 1)(2x + 10)$, $(3x - 5)(2x - 2)$, $(6x - 5)(x - 2)$, $(6x - 5)(x + 2)$, $(6x + 10)(x - 1)$ (possible answers)
41. $(3x + 10)(3x - 10)$ (possible answer)
43. $(5x^2 + 1)(4x + 1)$, $(4x^2 + 1)(5x + 1)$, $(2x^2 + 1)(10x + 1)$, $(10x^2 + 1)(2x + 1)$, $(20x^2 + 1)(x + 1)$, $(x^2 + 1)(20x + 1)$ (possible answers)
45. $(3x + 1)(x + 10) - (x + 2)^2 = 2x^2 + 27x + 6$

SECTION P.5 ANSWERS

Warm Up/Just the Facts

1. $16m^4 + 2m^3 - 8m^2$

   $$\begin{aligned} 2m^2(8m^2 + m - 4) &= 2m^2(8m^2) + 2m^2(m) - 2m^2(4) \\ &= 16m^4 + 2m^3 - 8m^2 \end{aligned}$$

2. $b^2 + 11b + 18$

   $(b + 2)(b + 9) = b^2 + 9b + 2b + 18 = b^2 + 11b + 18$

3. $4p^2 - 25$

   $(2p + 5)(2p - 5) = 4p^2 - 10p + 10p - 25 = 4p^2 - 25$

4. product
5. xy
6. $6mn$
7. $5y$
8. multiplying
9. $(x - y)$
10. difference

Essential Skills/Extensions

1. $2x^3y^2(y^6 + 3x + 5x^2y^8)$

3. $(2x + 5y)(3x^2 + 7y^2)$

5. $7b(c + 4)(3b^2 + 6b - 2)$

7. $2a(b + 5)(4a^2 + a - 11)$

9. $(2x + 3y)(5x + y)$

11. $4(a + 6)(a + 2b)$

13. $5(x - 3)(x - 3y)$

15. $(k + 12)(k + 2)$

17. $(b + 15)(b - 2)$

19. $(a - 6)(a + 1)$

21. $(3h + 1)(h + 6)$

23. $(3x - 1)(2x - 3)$

25. $(b + 3)(3b - 2)$

27. $(2x + y)(7x + 4y)$

29. $(3p^2 - 2q)(p^2 + 2q)$

31. $(5w^3 - 3x)(w^3 + 3x)$

33. $4(c - 5)(2c - 3)$

35. $3(5m - 2)(4m + 5)$

37. $4(5x - 2)(4x - 5)$

39. $(n + 10)^2$

41. $(1 - 5z^2)^2$ or $(5z^2 - 1)^2$

43. $(7x - 5y)^2$

45. $(3x + 7)(3x - 7)$

47. not factorable

49. $4(6x + 5y)(6x - 5y)$

51. $(3p + 2)(9p^2 - 6p + 4)$

53. $(u^2 - 2)(u^4 + 2u^2 + 4)$

55. $(7 - 4y)(49 + 28y + 16y^2)$

57. $(q^3 + r^2)(q^6 - q^3r^2 + r^4)$

59. $((x + 2y)^2 - 2)((x + 2y)^4 + 2(x + 2y)^2 + 4)$

61. $((m + n)^3 - 3)((m + n)^6 + 3(m + n)^3 + 9)$

63. $\pi r^2 h\left(1 + \frac{4}{3}rh + \frac{1}{3}h^2\right)$

65. $-7n(3p - z^2)(4p + z)$

67. $(10y^n + 3)(2y^n + 1)$

69. $(x + a)^2 - (y - b)^2$

71. $(8(x - 5y) + 3z)(64(x - 5y)^2 - 24z(x - 5y) + 9z^2)$

73. $(m^4 + 3)(2 - a)$

75. c can be any perfect square.

SECTION P.6 ANSWERS

Warm Up/Just the Facts

1. 3/5

$$\frac{24}{40} = \frac{\cancel{8}\cdot 3}{\cancel{8}\cdot 5} = \frac{3}{5}$$

2. 3/25

$$\frac{9}{10} \div \frac{15}{2} = \frac{9}{10}\cdot\frac{2}{15} = \frac{9\cdot 2}{10\cdot 15} = \frac{\cancel{3}\cdot 3\cdot \cancel{2}}{\cancel{2}\cdot 5\cdot \cancel{3}\cdot 5} = \frac{3}{25}$$

3. 9/8

$$\frac{5}{8}+\left(\frac{2}{3}-\frac{1}{6}\right) = \frac{5}{8}+\left(\frac{4}{6}-\frac{1}{6}\right) = \frac{5}{8}+\frac{3}{6} = \frac{5}{8}+\frac{1}{2} = \frac{5}{8}+\frac{4}{8} = \frac{9}{8}$$

4. polynomials; rational expressions

5. allowable

6. all real numbers

7. 3; domain

8. common factors

9. $x + 1$

10. $x^6(x + 1)^2(x + 2)$

Essential Skills/Extensions

1. $(-\infty, \infty)$

3. $\{x \mid x \neq 0 \text{ and } x \neq -3/4\}$

5. $\{x \mid x \neq 2 \text{ and } x \neq 3\}$

7. $\frac{x+3}{x-1}$, $x \neq 4$ and $x \neq 1$

9. $\frac{a+5}{a+3}$, $a \neq 6$ and $a \neq -3$

11. $\frac{4}{x+8}$, $x \neq 5$ and $x \neq -8$

13. $\frac{2x}{x-3}$, $x \neq 3$ and $x \neq -3$

15. $\frac{5a+10}{a+8}$, $a \neq 2$ and $a \neq -8$

17. $\frac{b-3}{6b}$, $b \neq 0$ and $b \neq -6$

19. $\frac{-x-1}{x-2}$, $x \neq 2$ and $x \neq 1$

21. $\frac{-x}{2x+5}$, $x \neq 5/2$ and $x \neq -5/2$

23. $\frac{-a-6}{a+5}$, $a \neq 6$ and $a \neq -5$

25. $\frac{15x-10}{2}$

27. $\frac{4}{3x}$

29. $\frac{8z+8}{z^2+1}$

31. $\frac{-x-6}{16}$

33. $\frac{20b^2-12b}{-2b^2-9b+5}$

35. $\frac{2k^2-2k}{-4k^2-17k+15}$

37. $\frac{3x+20}{10x^2+5x}$

39. $\frac{2x-1}{x^2-x-2}$

41. $\frac{-2x-3}{x^2+2x+1}$

43. $\frac{x^2+13x-16}{3x^2-3x}$

45. $\frac{-a^2+4a+2}{a^3}$

47. $\frac{7m^2-2m-4}{2m^2+3m}$

49. $3/x$

51. -4

53. $18/13$

55. $\frac{n}{n^2-2}$

57. $\frac{1}{x-3}$

59. $\frac{3x}{-x+4}$

61. $\frac{3x^2-5x+2}{15x-20}$; $\{x \mid x \neq -1 \text{ and } x \neq 4/3\}$

63. $\frac{1-3y}{2y^2-y-3}$; $\{x \mid x \neq -1 \text{ and } x \neq 3/2\}$

65. $\frac{5x+3}{3x+2}$

67. $\frac{2b\sqrt{y^2-b^2}}{y^2-b^2}$

69. $\frac{m^2-4m-13}{m^2-7m-9}$

71. perimeter: $\frac{16w+28}{15}$; area: $\frac{w^2+2w-8}{15}$

SECTION P.7 ANSWERS

Warm Up/Just the Facts

1. $-5x-5$

$10-5(x+3) = 10-5x-15$ *Distribute.*

$= -5x-5$ *Combine like terms.*

2. $17m+7$

$8m+11-(4-2m)+7m$

$= 8m+11-4+2m+7m$ *Distribute.*

$= 17m+7$ *Combine like terms.*

3. $8\sqrt{5}$

$\sqrt{125}+\sqrt{45}$

$= \sqrt{25\cdot 5}+\sqrt{9\cdot 5}$

$= 5\sqrt{5}+3\sqrt{5}$ *Simplify the radicals.*

$= 8\sqrt{5}$ *Combine like terms.*

4. addition; division

5. LCM

6. extraneous

7. Completing the square; Quadratic Formula

8. square roots

9. Complex numbers

10. negative

Essential Skills/Extensions

1. no solution

3. $-10/11$

5. −3

7. 4/5

9. 1/2

11. 7/4

13. −36/13

15. 25/28

17. −20

19. 31/3

21. −5

23. −7/10

25. 11

27. −9

29. 11

31. −7, 5

33. 0, 5

35. 1, 9

37. −7/2, 3

39. −5/2, 5/4

41. −5, 7/2

43. $3\sqrt{3}+4i\sqrt{3}$

45. $-4\sqrt{7}-3i\sqrt{5}$

47. $\sqrt{2}+3i\sqrt{11}$

49. $-3\pm 3\sqrt{3}$

51. $-4\pm 3\sqrt{5}$

53. $-3\pm 5i\sqrt{7}$

55. $-6\pm 2\sqrt{11}$

57. $\dfrac{-42\pm 6\sqrt{7}}{7}$

59. $-3\pm\dfrac{3i}{2}$

61. −1, −1/11

63. $\dfrac{-3\pm\sqrt{17}}{6}$

65. $\dfrac{3\pm i\sqrt{23}}{8}$

67. 14 in. × 6 in.

69. 16 m × 7 m

71. $P = 16, 32$

73. $s = 2.5, 5$

75. yes

77. \$413.10, \$648, \$52.20

79. 25%

81. \$109.41

83. The first piece is 60 m, the second piece is 64 m, and the third piece is 48 m.

85. 4.4

87. 80 mi/h

89. Yes. The student should have subtracted $32x$ from both sides to make the right side of the equation equal to 0. Factoring out an x shows that there are two solutions: $x = 0$ and $x = -4$.

91. $x^2 - 4x + 1 = 0$ (possible answer)

SECTION P.8 ANSWERS

Warm Up/Just the Facts

1. −17

$-|8-3(-5+2)| = -|8-3(-3)| = -|8+9| = -|17| = -17$

2. 37/10

$\dfrac{3(1)+4}{1+1}+\dfrac{1}{5}=\dfrac{7}{2}+\dfrac{1}{5}=\dfrac{35+2}{10}=\dfrac{37}{10}$

3. $(2x+1)(3x-4)$

4. polynomial

5. squared

6. radical

7. extraneous

8. rational; denominator of the rational exponent

9. absolute value; $A = -B$

10. multiplying the equation by a common denominator

Essential Skills/Extensions

1. −1/9, 0

3. −7, 3, 0

5. −4, −2, 0

7. −3, −1/3, 3

9. −5, 2/3, 5

11. −5, 3/4, 5

13. −6

15. 13/5

17. 1/2

19. 4

21. 12

23. 10, 46

25. 3/4

27. 1

29. −1/2

31. 4

33. 2

35. −8

37. −2

39. −1/3, 1/3

41. −2/5, 2/5

43. 1

45. 4, 11

47. −1, 254

49. no solution

51. −6, 1

53. 8/5, 4

55. −5/2, 1/2

57. −20/3, −16/3

59. −37, 27

61. 1, 2, $\frac{1}{2}\left(5-\sqrt{5}\right)$

63. 1/4, −3

65. −2/5 , 3/2

67. $y = -2 \pm \sqrt{3x-5}$

69. $p = 0, p = c/a$

71. 1 mi/h

SECTION P.9 ANSWERS

Warm Up/Just the Facts

1. $(5, \infty)$

2. $(-\infty, 0) \cup (0, \infty)$

3. $[-2, 70]$

4. inequality; multiplied, divided; negative

5. $x \geq y$

6. $x \leq y$

7. compound inequality

8. and; or

9. <, >

10. ≤, ≥

Essential Skills/Extensions

1. $(-\infty, -6]$

3. $(-1/7, \infty)$

5. $[1, \infty)$

7. $(-\infty, 2] \cup [7, \infty)$

9. $(-\infty, \infty)$

11. $(-\infty, -5) \cup [3, \infty)$

13. $[2, 5]$

15. $(6, 12]$

17. $(-7, -4]$

19. $(-\infty, -1) \cup (4, \infty)$

21. $(-6, -5)$

23. $[-5, -3]$

25. $(2, 6)$

27. $(-\infty, -8] \cup [-4, \infty)$

29. $(-\infty, 2) \cup (5, \infty)$

31. $(-\infty, 4/3] \cup [4, \infty)$

33. $(-\infty, \infty)$

35. $[-4/3, 6]$

37. $(2/3, 4)$

39. $[-8, 1/5]$

41. $(-\infty, -5/2] \cup [4/3, \infty)$

43. $(-\infty, 3/4] \cup [4, \infty)$

45. $(-\infty, -9/2] \cup [3, \infty)$

47. $(-\infty, 3/4] \cup [4, \infty)$

49. $(-\infty, -3)$

51. $(-\infty, 2]$

53. $(8, \infty)$

55. $(-\infty, -1] \cup (7/2, \infty)$

57. $[-3, 5/2)$

59. $(-\infty, -7/3) \cup (4/5, \infty)$

61. $(1, 8/3)$

63. $(-5, -2] \cup [2, \infty)$

65. $(-\infty, -2) \cup [5/7, \infty)$

67. $(-\infty, -11) \cup (1, \infty)$

69. $\left(-3, 1-\sqrt{10}\right) \cup \left(3, 1+\sqrt{10}\right)$

71. $\left(0, \dfrac{3+\sqrt{21}}{2}\right)$

73. 18 lawns

75. 1800 boxes

CHAPTER P.10 REVIEW ANSWERS

1. -96

2. $(-10, -7]$

3. $x \geq -1$

4. 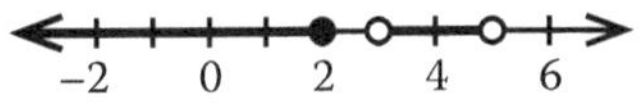

5. -2

6. $-45a^{10}b^3c$

7. $\dfrac{1}{4g}$

8. $9/128$

9. 4.7×10^{-6}

10. 2.56×10^6

11. $12\sqrt{2} - \sqrt{15}$

12. $\dfrac{5\sqrt{6} - 35}{43}$

13. $16/9$

14. $2bcd^5\sqrt{3bd}$

15. $\dfrac{5xz^3\sqrt{xy}}{y}$

16. $-5x^3 + 10x^2 - 21x - 6$; leading coefficient: -5; cubic polynomial

17. $6x^2 + x - 15$; leading coefficient: 6; quadratic trinomial

18. $x^3 - x^2 - 21x + 5$; leading coefficient: 1; cubic polynomial

19. $c^4 - 4d^2$

20. $216x^3 - 540x^2y + 450xy^2 - 125y^3$

21. $2(a - 1)(3a + 5)$

22. $(5n^3 - 1)(5n^3 + 1)$

23. $(3x - 2y)(x^2 + 4)$

24. $(2cd^4 + 1)(4c^2d^8 - 2cd^4 + 1)$

25. $4(z - 1)^2(z + 1)(z^2 + z + 1)$

26. $x + 5, x \neq 7$

27. $\dfrac{x^2 - 3x + 9}{x - 4}, x \neq -3, 4$

28. $\dfrac{2k - 2k^2}{4k^2 + 17k - 15}$

29. $-\dfrac{2}{x^2 + 2x - 3}$

30. $\dfrac{5}{-6x + 12}$

31. $2/3$

32. 4

33. 0

34. 4/7

35. 7/2, –5/4

36. $3 \pm \sqrt{11}$

37. $\dfrac{\sqrt{7} \pm i\sqrt{65}}{12}$

38. 130 mi

39. –19/23, 37/7

40. no real solution

41. 3

42. $-1, \pm\dfrac{\sqrt{2}}{2}$

43. [–21, 5)

44. $\left(-\infty, -\dfrac{23}{3}\right] \cup \left[\dfrac{25}{3}, \infty\right)$

45. $\left[-\dfrac{2}{3}, 3\right]$

46. $\left(\dfrac{1}{5}, \dfrac{4}{3}\right]$

SECTION 1.1 ANSWERS

Warm Up/Just the Facts

1. $\sqrt{113}$
2. (–1, –3)
3. $(x - 2)^2$
4. axes; four; origin
5. I; II; III
6. distance; (x_1, y_1), (x_2, y_2)
7. midpoint; *A*, *B*
8. collinear
9. *y*; *x*; 0; *x*; *y*
10. circle; (0, 0); 10

Essential Skills/Extensions

1. quadrant I

3. on the *y*-axis

5. quadrant IV

7.

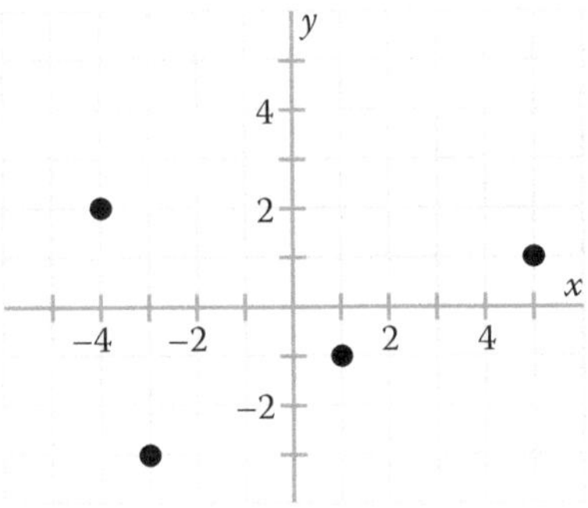

9. 9.43

11. 25.55

13. 5.83

15. (–1, 13/2)

17. (1, –13/2)

19. (3/40, 1/16)

21. (–10, 1)

23. (–13, –7)

25. (–10, –11.5)

27. not collinear

29. collinear

31. isosceles triangle

33. scalene triangle

35. equilateral triangle

37. straight line

39.

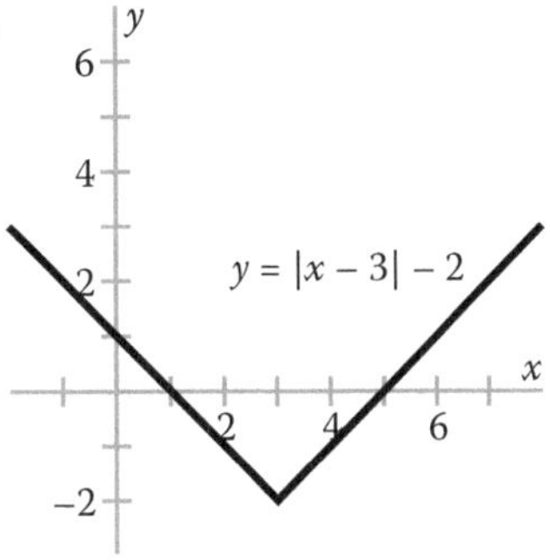

41.

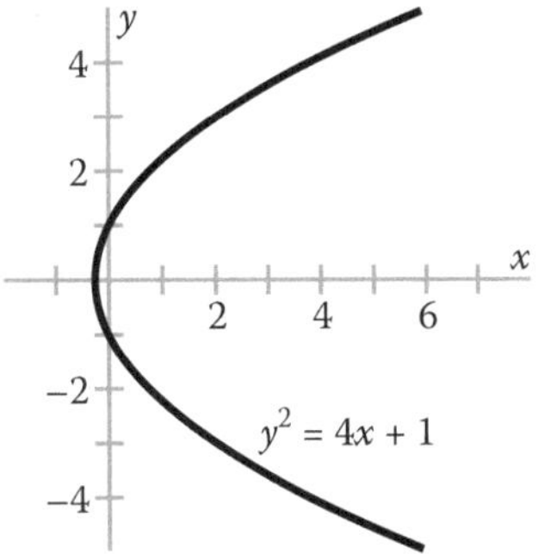

43.

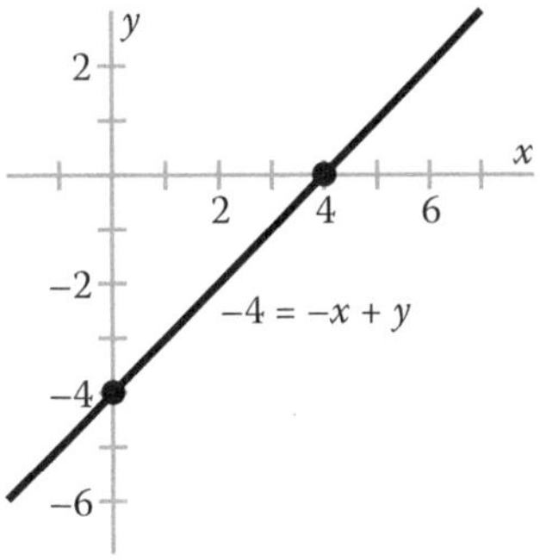

45.

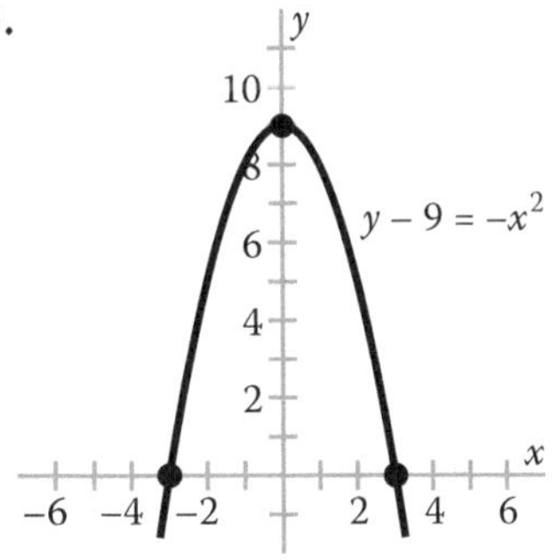

47. $x^2 + y^2 = 49$

49. $(x + 4)^2 + (y + 3)^2 = 36$

51. $(x + 7)^2 + (y - 2)^2 = 1/4$

53. center (0, –4); radius 5

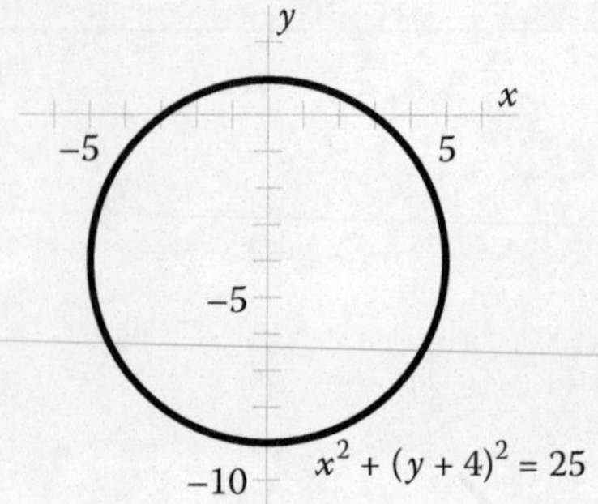

55. $(x - 1)^2 + (y + 5)^2 = 196$; center (1, –5); radius 14

57. $(x + 6)^2 + (y - 7)^2 = 289$; center (–6, 7); radius 17

59. $(x - 4)^2 + (y + 8)^2 = 64$; center (4, –8); radius 8

61. $(x + 1)^2 + (y - 5)^2 = 17$

63. $(x - 3)^2 + (y - 2)^2 = 89$

65. $(x - 2)^2 + (y - 1)^2 = 34$

67. with respect to the *y*-axis

69. with respect to the *y*-axis

71.

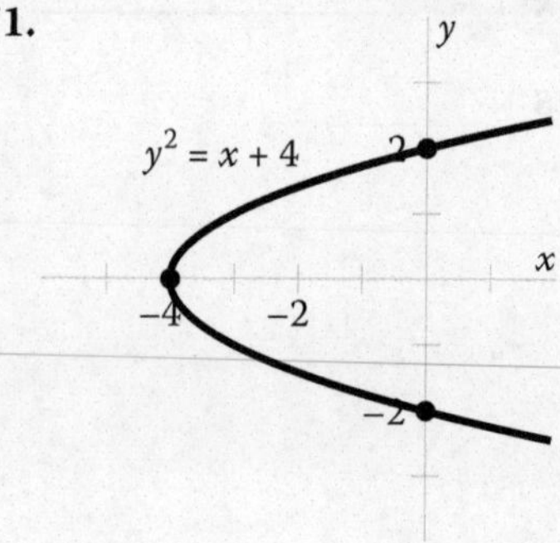

73. $(x + 3\sqrt{2})^2 + (y + 2\sqrt{2})^2 = 28$

75. true

77. $m = 0$; n can be any real number

79. false

81. $(x - 1)^2 + (y + 2)^2 = 6$

SECTION 1.2 ANSWERS

Warm Up/Just the Facts

1. $\dfrac{-4-6}{7-(-1)} = \dfrac{-10}{8} = \dfrac{-5}{4}$

2. $y = -13x + 26$

3. –3/2

4. slope; vertical; horizontal

5. $m = \dfrac{y_2 - y_1}{x_2 - x_1}$; (x_1, y_1), (x_2, y_2)

6. slope-intercept; point-slope; slope; *y*-intercept; point

7. undefined; 0; horizontal; vertical

8. $Ax + By = C$

9. same; perpendicular; opposite reciprocals; parallel; ⊥

10. q; (r, p)

Essential Skills/Extensions

1. 8/3

3. –2/3

5. –2/3

7. 0

9.

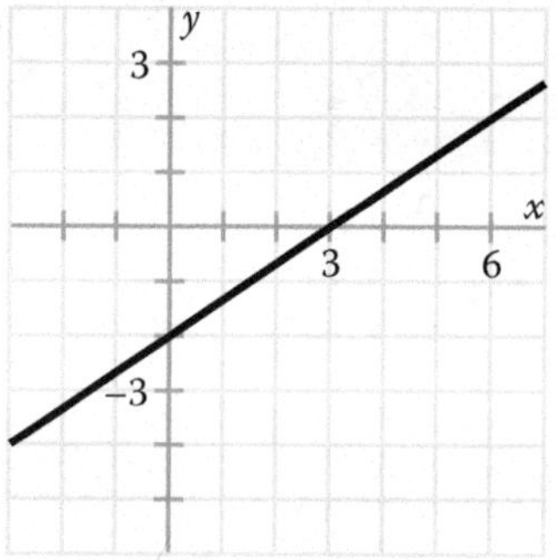

11. $y = 2x + 1$

13. $y = \dfrac{3}{5}x + 4$

15. $y = \dfrac{9}{4}x$

17.

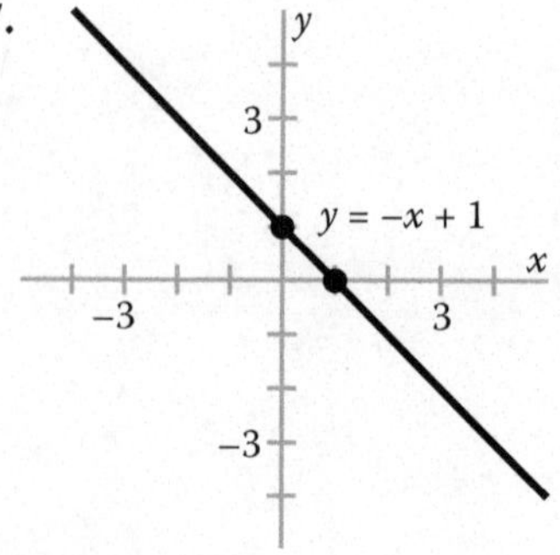

19. $y = x - 11$

21. $y = \dfrac{1}{2}x + 5$

23. $y = 22x + 29$

25. $y = -\dfrac{11}{4}x + \dfrac{13}{4}$

27. $y = -\dfrac{1}{3}x + \dfrac{17}{3}$

29. $y + 3 = 4(x - 5)$; $y = 4x - 23$

31. $y - 5 = \dfrac{1}{2}(x + 6)$; $y = \dfrac{1}{2}x + 8$

33. $y - 3 = 6(x - 8)$; $y = 6x - 45$

35. $y - 7 = -\dfrac{4}{5}(x - 10)$; $y = -\dfrac{4}{5}x + 15$

37. $y = -5$

39. $y = 9$

41. $x = 2$

43. slope: 3/2; y-intercept at (0, –4)

45. slope: –3/4; y-intercept at (0, 3/4)

47. slope: 2/3; y-intercept at (0, –5/3)

49. $y = -x + 13$

51. $y = -5x + 26$

53. $y = \dfrac{3}{4}x - \dfrac{29}{4}$

55. $y = -\dfrac{2}{3}x - \dfrac{5}{3}$

57. $y = -\dfrac{1}{2}x - \dfrac{1}{2}$

59. $y = \dfrac{1}{3}x - 3$

61. $y = 5x - 3$

63. $y = -\dfrac{6}{5}x + \dfrac{41}{5}$

65. $y = 73x$

67. 10 h 58 min

69. The slope is 30, so the ant travels at an average rate of 30 meters per minute.

71. yes

73. $W = 0.05S + 2250$ for $S < 10{,}000$, $W = 0.1S + 2250$ for 10,000 " S " 14,999 , $W = 0.12S + 2250$ for 15,000 " S " 17,999 ; $4170

75. $y = 39.25x - 74.75$

77. Approximately 279 stores; Yes, because 2009 is within the years that we are assuming linear growth.

79. $R = 23.75t$

81. 2489 h

SECTION 1.3 ANSWERS

Warm Up/Just the Facts

1. 16

2. 10/63

3. 45

4. decreases; increases

5. direct, inverse, joint, combined

6. direct; origin; k

7. $y = \dfrac{k}{x}$; x and k

8. krm^3

9. joint; jointly; m, n; k

10. combined

Essential Skills/Extensions

1. $d = \dfrac{8g}{7}$; 21.875

3. $t = 4s/7$; $t = 12$; $s = 35$

5. $p = 12q/7$; $q = 2.1$

7. $u = -10v/7$; $v = -6.44$; $u = -7$

9. $w = -28r/5$; $w = 56/3$; $r = -5/16$

11. $l = \dfrac{16w}{13}$; 78 cm

13. $l = \dfrac{34w}{25}$; 3.75 ft

15. $38,080

17. $x = 8.4/y$; 2.1

19. $a = 22.875/b$; $b = 549/8 = 68.625$, $a = 61/40 = 1.525$

21. $y = -\dfrac{9}{x}$; $y = -33/2 = -16.5$; $x = 6$

23. $c = 2.1/d$; $d = -7/3$; $c = 28/5 = 5.6$

25. 39 days

27. 15 days

29. 8.5 ohms

31. 9.4 cd

33. 13.9 cd

35. The frequency is 1/3 the frequency of the original.

37. 20 in.

39. 155 square inches

41. 9 in.

43. $E = \dfrac{53.1}{d^2}$

45. 58 kg

47. 15 kg

49. 1129 ft

51. yes; under 5.2 lb/in^2; B. 243 lb

53. 243 lb

55. It becomes 32 times larger.

57. 11,025 lb; $F = \dfrac{3ws^2}{r}$

SECTION 1.4 ANSWERS

Warm Up/Just the Facts

1. –9
2. 26
3. 50
4. function; y-value
5. vertical line; once; function
6. y; $f(x) = 2x + 1$
7. equal
8. all real numbers
9. verbally (words), algebraically (equation), visually (graph), numerically (table); numerically
10. piecewise; domain; simplify

Essential Skills/Extensions

1. a function

3. a function

5. a function

7. a function

9. not a function

11. 28; –7

13. 2; –10

15. –1; 56

17. $|4m - 31|$

19. $-3r^2 + 32z - 85$

21. $2z^2 + 13z + 25$

23. 25; –1

25. 144; –4

27. 222; 2

29. 8/7

31. –21

33. –3/4; 5

35. –3/5; 2, 0

37. 5/6

39. –3; 7

41. all real numbers

43. –3; 0; 6

45. $p(h) = 9h$

h	0	5	15	35	60	100
$p(h)$	0	45	135	315	540	900

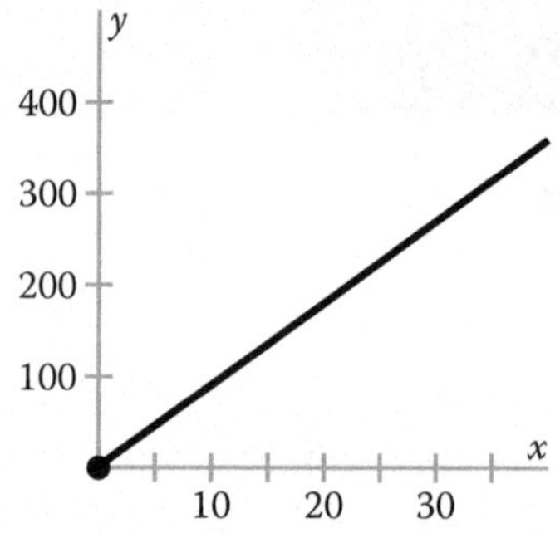

47. 18 ft

49. 71 ft

51. $f(s) = \dfrac{36 - 24s + 4s^2}{\pi}$;

$g(r) = \dfrac{36 - 12\pi r + \pi^2 r^2}{4}$

53. $\{-3, 0, 2, 4\}$

55. $(-\infty, 9]$

57. $\{x \mid x \neq -2, x \neq 12\}$

59. $\{x \mid x \neq -6, x \neq 6\}$

61. $(-\infty, -3] \cup [3, 6) \cup (6, \infty)$

63. $2x + 3 + h$

65. yes

SECTION 1.5 ANSWERS

Warm Up/Just the Facts

1. V-shaped
2. U-shaped
3. linear
4. open
5. $f(x) = [[x]]$; step
6. zeros; x-intercepts
7. >
8. maximum, minimum
9. closed; open
10. strictly increasing, strictly decreasing; V

Essential Skills/Extensions

1. domain: $[-2, 2)$; range: $[-4, 4)$

3. domain: $(-\infty, \infty)$; range: $\{y \mid y = 2\}$

5. domain: $(-\infty, \infty)$; range: $(-\infty, -4]$

7. domain: $(-\infty, \infty)$; range: $(-\infty, \infty)$

9. 8; 1

11. 2; −2

13. 0; 3

15.

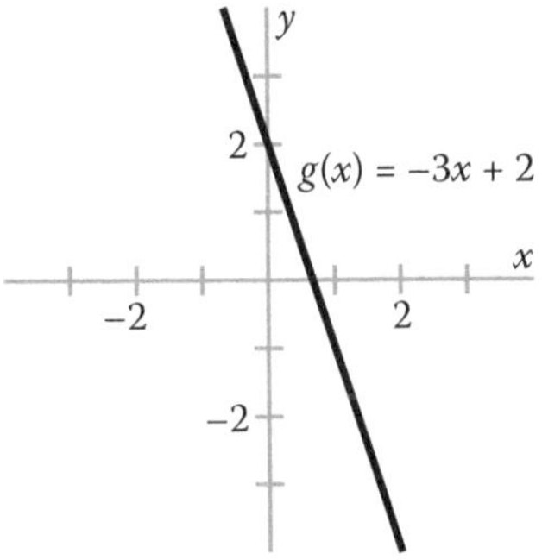

17.

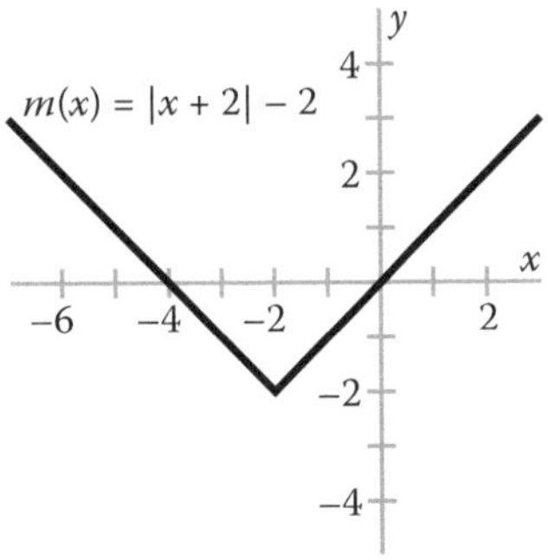

19.

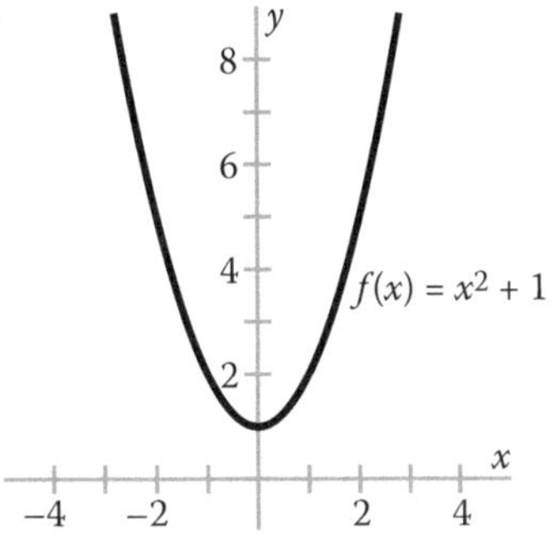

21.

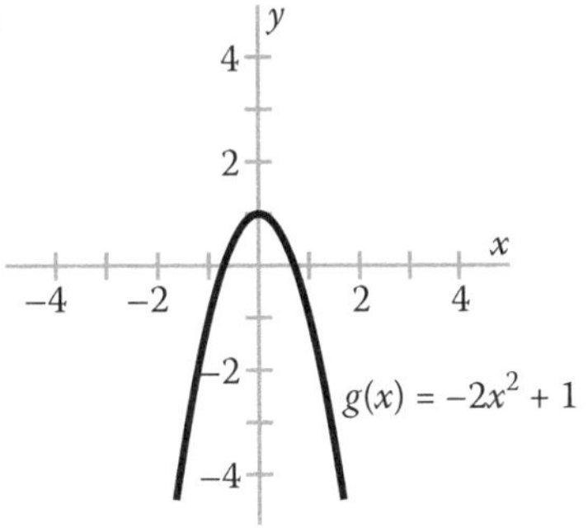

23. $p(x) = \begin{cases} x & \text{if } x < 0 \\ 3 & \text{if } 0 \le x < 2 \\ -x + 4 & \text{if } x \ge 2 \end{cases}$

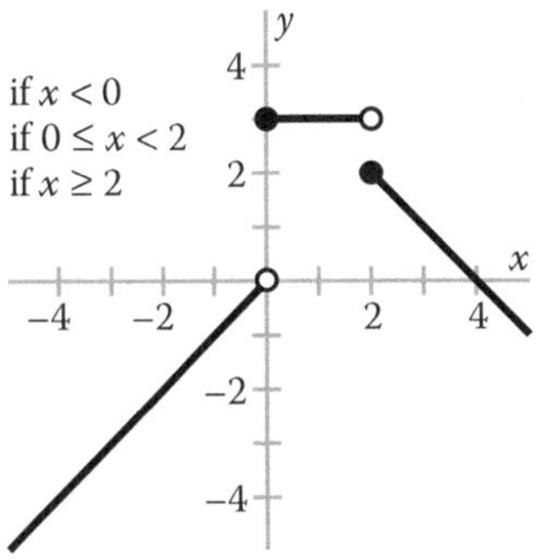

25.

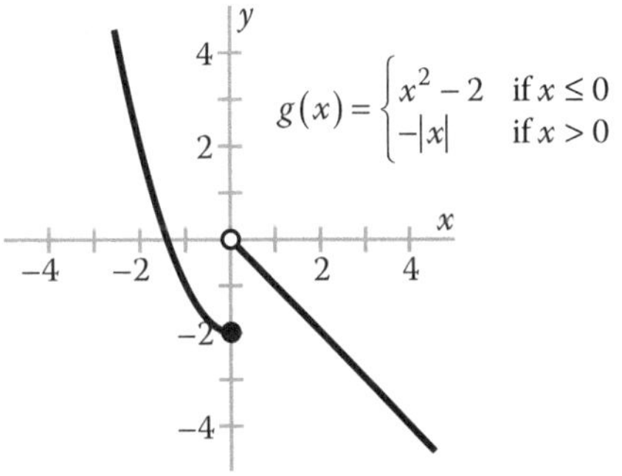

27. $f(m) = \begin{cases} 40 & \text{if } 0 \le m \le 450 \\ 60 & \text{if } 450 < m \le 900 \\ 100 & \text{if } m > 900 \end{cases}$

29. $f(m) = \begin{cases} 25 \text{ if } 0 \le m \le 200 \\ 35 \text{ if } 200 < m \le 500 \\ 60 \text{ if } m > 500 \end{cases}$

31. $f(x) = \begin{cases} 20x \text{ if } 0 < x \le 6 \\ 10x + 60 \text{ if } 6 < x \le 22 \\ 5x + 170 \text{ if } x > 22 \end{cases}$

33. 14

35. 39

37. –22

39. 4

41.

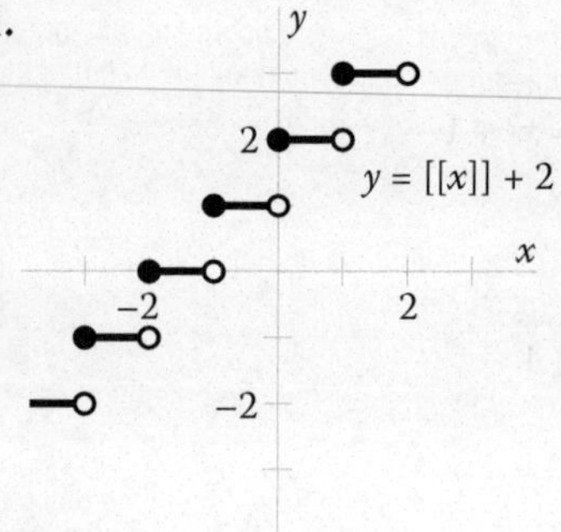

43. –6, –1

45. 0, 5

47. 5/6, –4

49. no real zeros

51. 0, 2, –2

53. 2

55. no real zeros

57. 9, –9

59. 4/5

61. –4/5, 1/3

63. strictly increasing: (–∞, 0); strictly decreasing: (0, ∞)

65. strictly increasing: (–2, ∞)

67. strictly decreasing: (–∞, 1); constant: (1, ∞)

69. constant: (0, 2); strictly increasing: (2, 4); strictly decreasing: (4, ∞)

71. relative minimum at (1, –1); relative maximum at (0, 0)

73. domain: (–∞, ∞); range: (–∞, ∞); strictly increasing: (–∞, ∞)

75. domain: (–∞, ∞); range: (–2, ∞); strictly decreasing: (–∞, 3/4); strictly increasing: (3/4, ∞)

77.

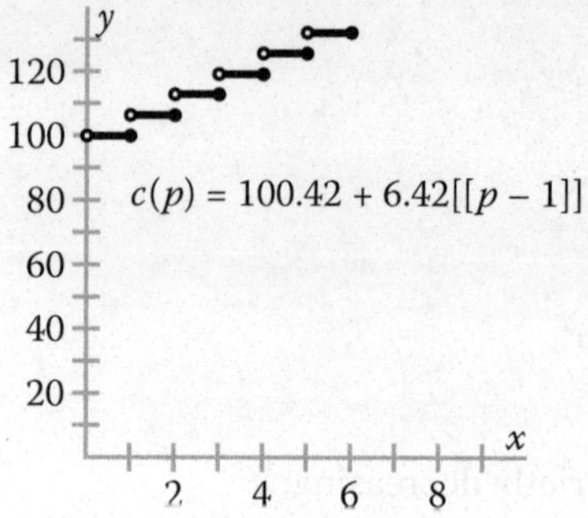

79. $p(x) = \begin{cases} x & \text{if } 0 < x < 20 \\ 20 & \text{if } 20 \le x < 1000 \\ 50 & \text{if } 1000 \le x < 3000 \\ 100 & \text{if } x \ge 3000 \end{cases}$

81. 5, $\pm i\frac{\sqrt{35}}{7}$; 1 real zero

83.

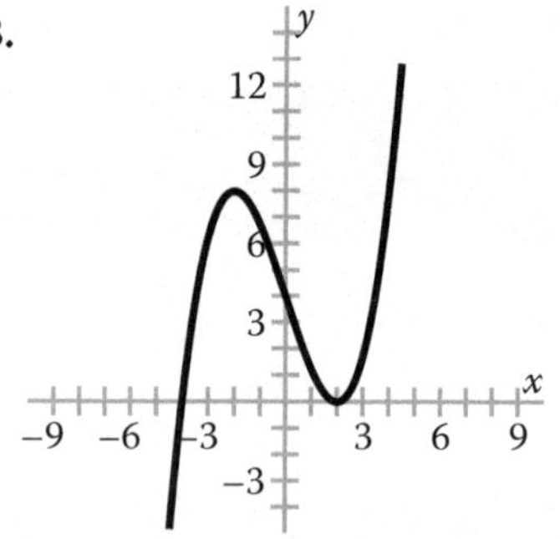

SECTION 1.6 ANSWERS

Warm Up/Just the Facts

1. True. $f(-x) = f(x)$, which proves that the graph of $f(x)$ is symmetric with respect to the y-axis.
2. True. The expression inside the radicand must be set greater than or equal to zero to determine the domain, but the domain might be all real numbers (e.g., when the radicand is x^2).
3. False. Zeros are x-intercepts.
4. Translations, reflections; doesn't; nonrigid
5. c units left
6. reflection; x-axis
7. wider
8. (–2, –4)
9. x^2; left; down
10. even

Essential Skills/Extensions

1.

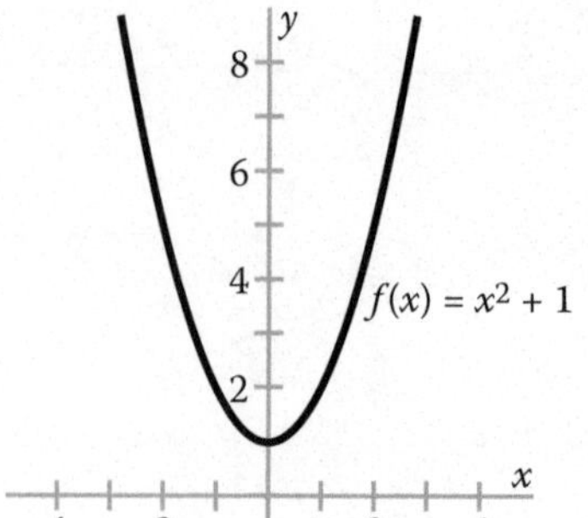

3.

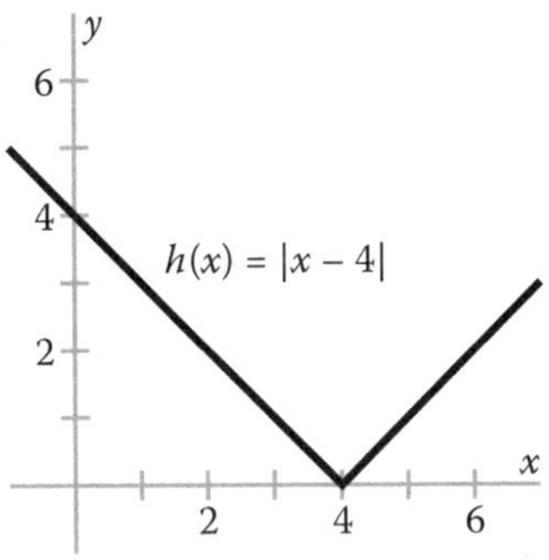

5.

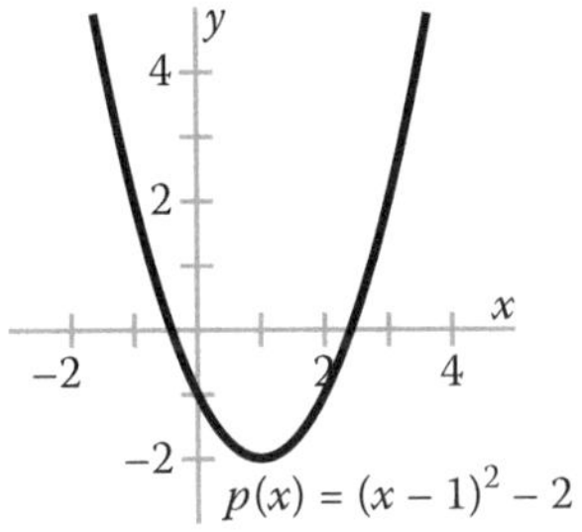

7.

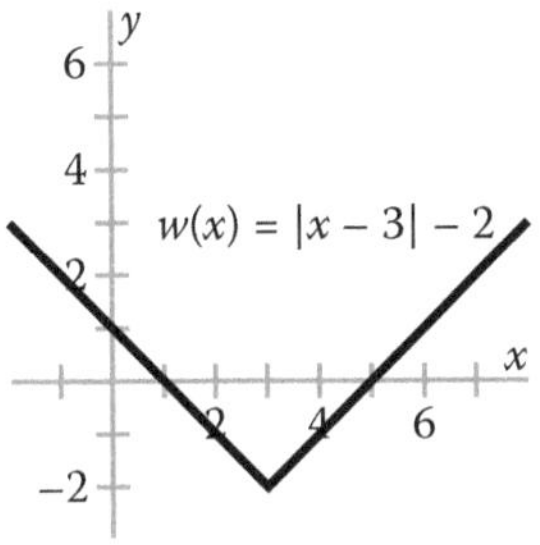

9.

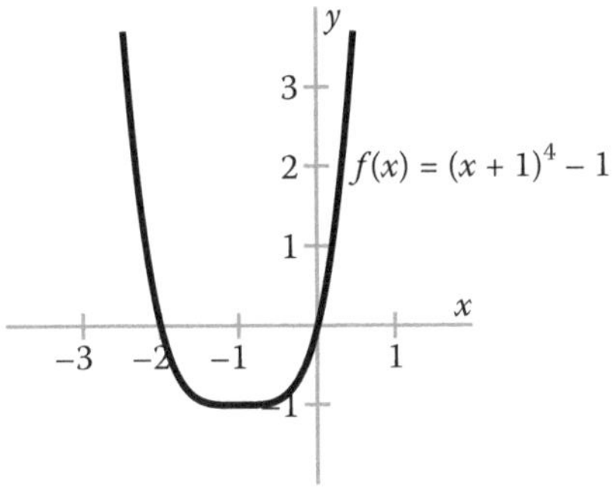

11.

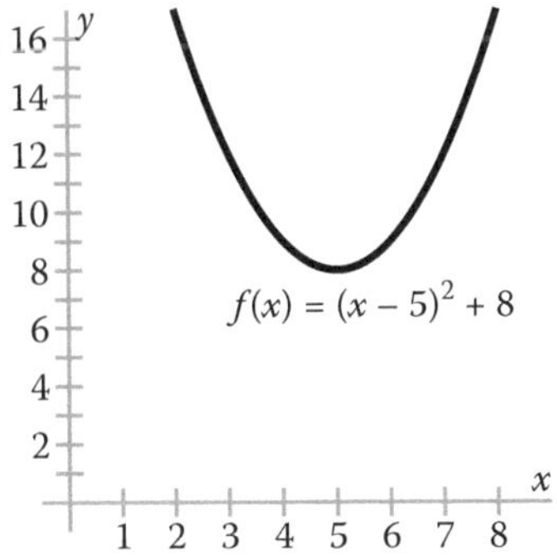

13.

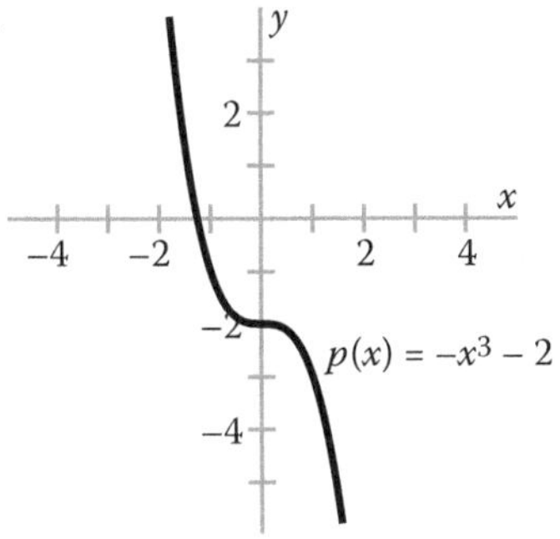

15.

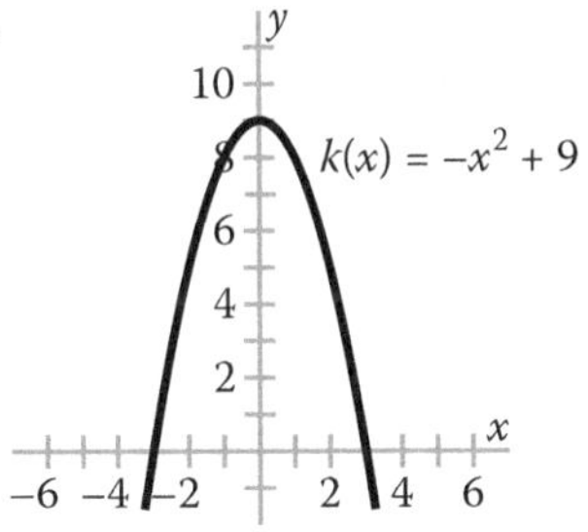

17.

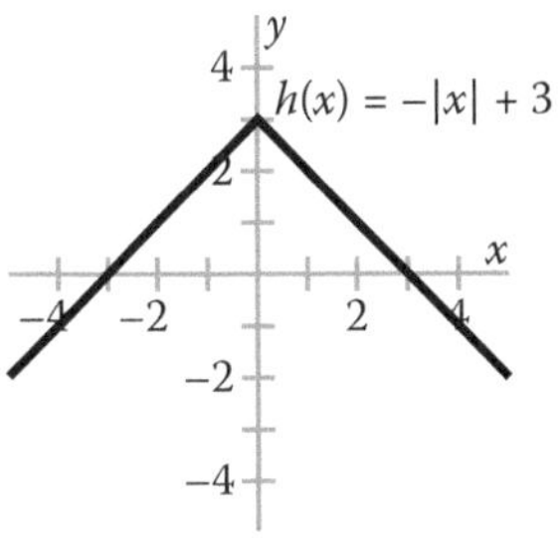

19.

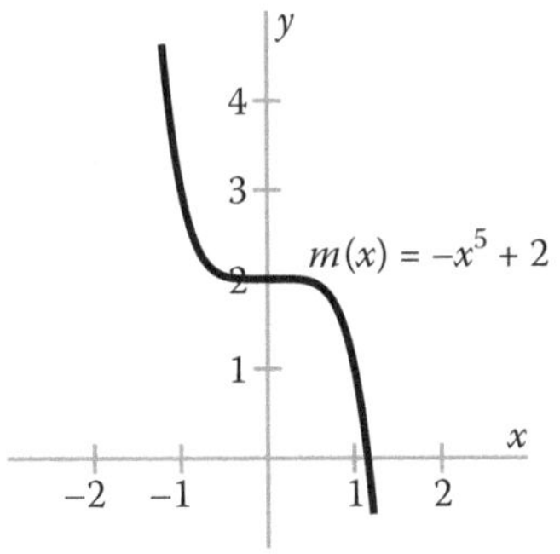

21.

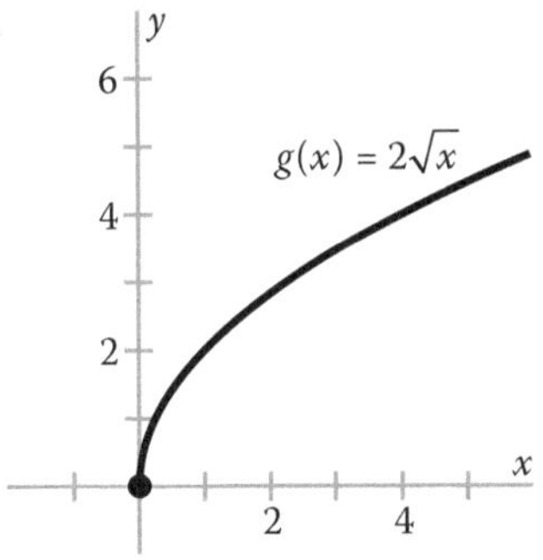

23.

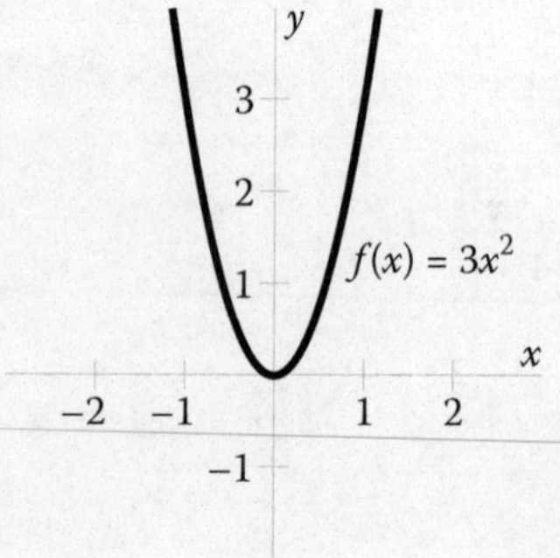

25.

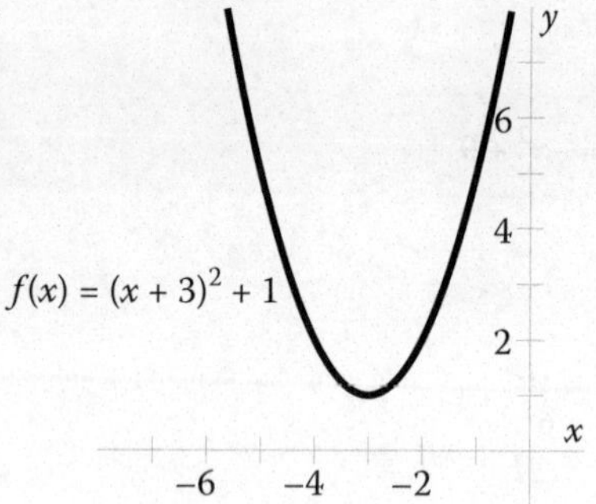

27.

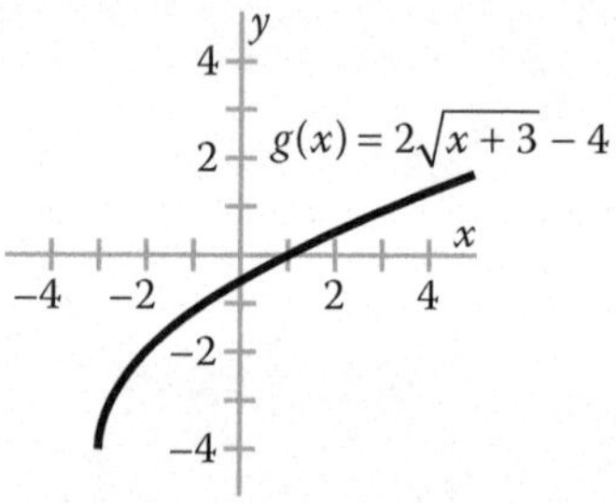

29.

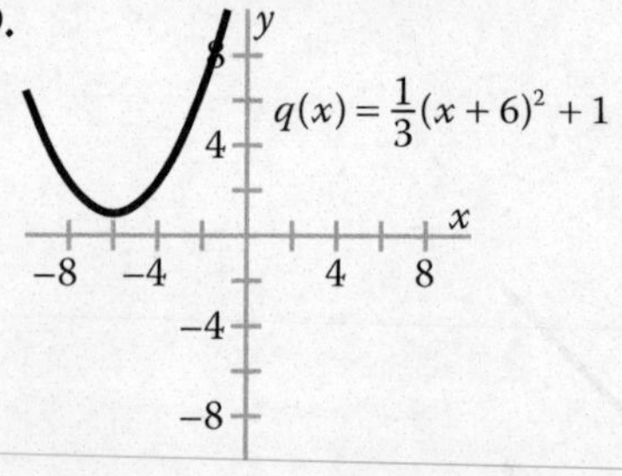

31. even

33. even

35. neither

37. odd

39. neither

41. even

43. $h(x) = -|x + 1| + 2$

45. false

SECTION 1.7 ANSWERS

Warm Up/Just the Facts

1. x-intercept: (1/2, 0); y-intercept: (0, –1)

2. x-intercepts: (–3, 0), (3, 0); y-intercept: (0, –27)

3. x-intercept: (–1, 0); y-intercept: (0, 1)

4. addition, subtraction, multiplication, division

5. composition

6. g, f

7. m

8. difference quotient; h

9. Decomposing

10. $h(x)$

Essential Skills/Extensions

1. $-x + 4$

3. $2x^2 + 6$

5. $7x^2 - 11x + 15$

7. $-4x^2 - 2x + 7$

9. $\frac{5x - 2}{x^2 - x}$

11. $\frac{x^2 + 4x + 2}{x^2 - 25}$

13. $-2x^2 + 15x + 8$

15. $\frac{1}{x + 7}$

17. $9x^3 - 24x^2 + 24x - 64$

19. $5x - 10$

21. $\frac{-5x^2 - 10x}{x^2 + x - 12}$

23. $\frac{2x^2 - 8x + 6}{x + 4}$

25. –112

27. 7

29. –18

31. −10

33. −3

35. $-4a - 2h$

37. $12a + 6h - 4$

39. $3a^2 + 3ah + h^2 - 5$

41. 82

43. −52

45. 8/7

47. 11

49. 13

51. 5/4

53. $(g \circ f)(x) = 20x + 20$; $(f \circ g)(x) = 20x - 14$

55. $(g \circ f)(x) = x$; $(f \circ g)(x) = x$

57. $(g \circ f)(x) = 25x^2 - 65x + 42$; $(f \circ g)(x) = 5x^2 - 25x + 26$

59. $(g \circ f)(x) = 4x + 3$; $(f \circ g)(x) = 2\sqrt{x^2 + 3}$

61. $f(x) = -3x - 10$ (possible answer)

63. $f(x) = 2x + 6$ (possible answer)

65. $g(x) = 3x - 1$ (possible answer)

67. $g(x) = x^3$ (possible answer)

69. $p(m) = 0.80m$; $c(m) = m - 10$; $(p \circ c)(m) = 0.80m - 8$; $(p \circ c)(m)$ represents the cost of the merchandise when the \$10-off coupon is applied first and then the 20% employee discount is applied.

71. −1

73. −6

75. −1/2

77. 0

79. 2

79. 31

81. 4

83. A. $t(m) = (f + p)(m)$; B. the total number of employees in July

85. $(h \circ g)(x) = \dfrac{2}{-3x+3}$, all real numbers except 1;

$(g \circ h)(x) = \dfrac{8x - 46}{x - 5}$, all real numbers except 5

87. $(f - g)(x) = 2x^2$; graph is a vertical stretch of $y = x^2$ by a factor of 2

89. $(fg)(x) = 3x^4$; graph is a vertical stretch of $y = x^4$ by a factor of 3

91. $(g/f)(x) = 1/3$; graph is a horizontal line, $y = 1/3$, hole at $x = 0$

SECTION 1.8 ANSWERS

Warm Up/Just the Facts

1. $(-\infty, \infty)$
2. $[-3, 3]$
3. $(-\infty, -3) \cup (-3, \infty)$
4. $y = x$
5. (y, x)
6. Horizontal Line Test; inverse
7. $\{y \mid y > 9\}$
8. inverse
9. Horizontal Line Test
10. subtraction

Essential Skills/Extensions

1. 3

3. cannot be determined

5. −0.6

7. yes

9. no

11. yes

13. no

15. no

17. yes

19. yes

21. no

23.

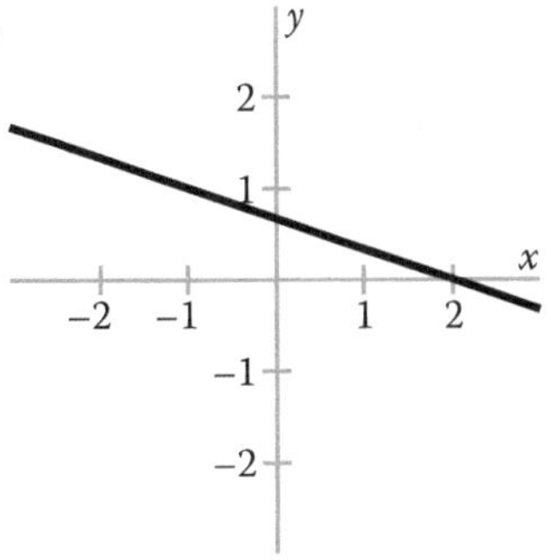

25. $g^{-1}(x) = -x + 3$

27. $p^{-1}(x) = \dfrac{x+1}{6}$

29. $t^{-1}(x) = \dfrac{2x-11}{9}$

31. $g^{-1}(x) = x^2 - 4, x \geq 0$

33. $h^{-1}(x) = \dfrac{x^2+1}{2}; x \geq 0$

35. inverse does not exist

37. $k^{-1}(x) = \dfrac{1}{x} - 6$

39. $m^{-1}(x) = \dfrac{4x-7}{x-2}$

41. $s^{-1}(x) = \dfrac{x^3-7}{2}$

43. $(-\infty, 0]$, $[0, \infty)$ (possible answers)

45. $-9/2$

47. -1

49. $(g^{-1} \circ f^{-1})(x) = \dfrac{x-4}{2}$

51. 2

53. $g^{-1}(x) = \sqrt[3]{x+27}$

CHAPTER 1 REVIEW ANSWERS

1. $AB = 6.3$; $AC = 6.3$; $BC = 12.6$; collinear

2. center $(-2, 5)$; radius 4

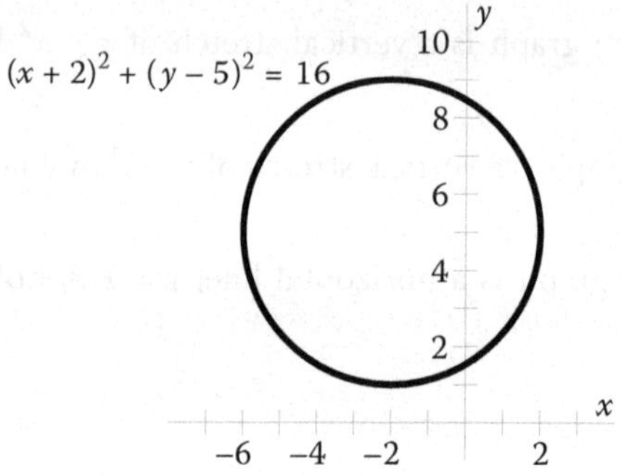

3.

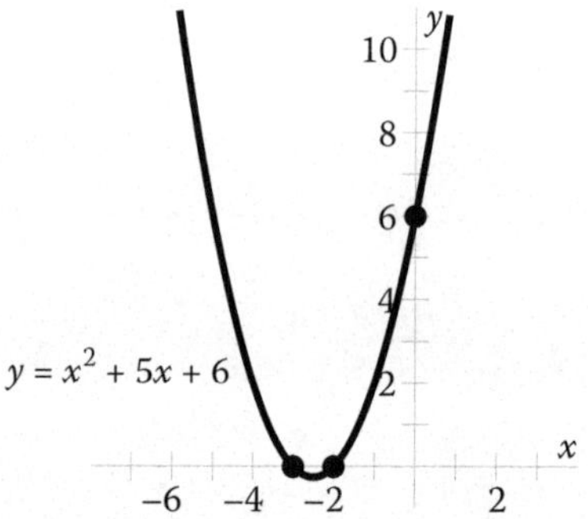

4. $(x - 2)^2 + (y + 3)^2 = 16$

5.

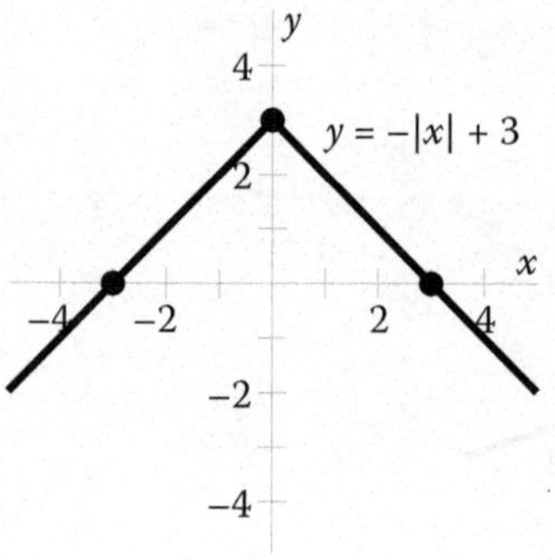

6.

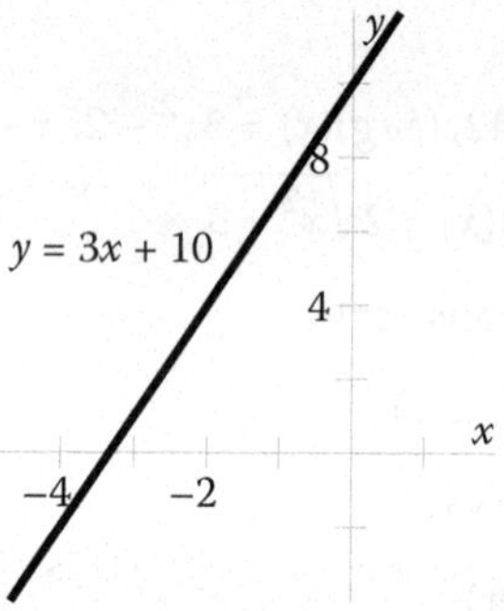

7. $y = 22x + 29$

8. slope: $-5/2$; y-intercept: -2

9. $y + 2 = \dfrac{3}{4}(x - 7)$

10. $y = -\dfrac{3}{4}x + 1$

11. 8 N/ft^2

12. 156

13. 32.7 h

14. 3/20 Hz

15. 882.315 N

16. not a function

17. -5; 51; $a^2 - 9a + 15$

18. $A = \dfrac{1}{4}s^2\sqrt{3}$

19. -3, 1

20. $(-\infty, -1) \cup (-1, 3) \cup (3, \infty)$

21. domain: $[-2, \infty)$; range: $[0, \infty)$; 1; 2; -2

22. $k(x) = \begin{cases} -x - 4 & \text{if } x \le -1 \\ 2x + 1 & \text{if } -1 < x \le 2 \\ -4 & \text{if } x > 2 \end{cases}$

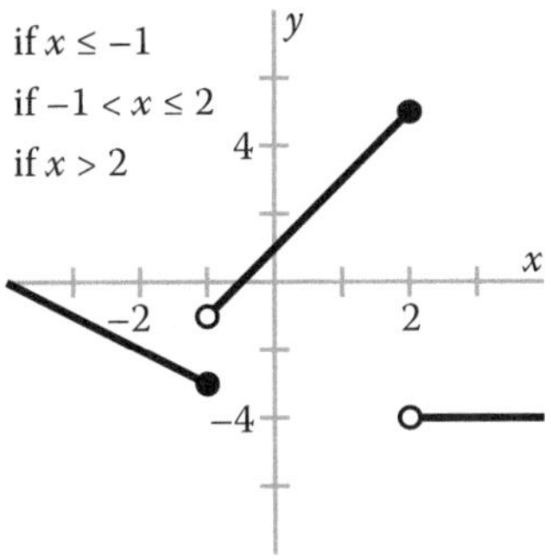

23.

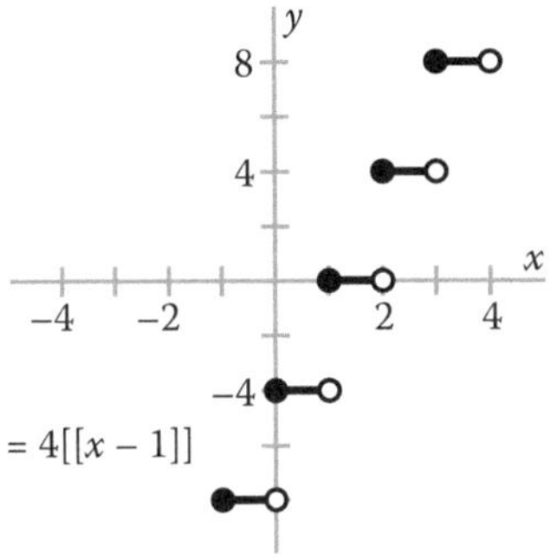

24. −6, −2, 6

25. strictly increasing: (−1, 2); strictly decreasing: (−4, −1); constant: (2, 4)

26.

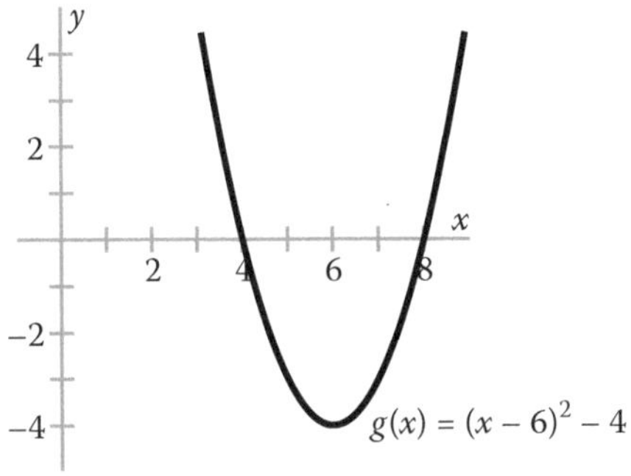

27.

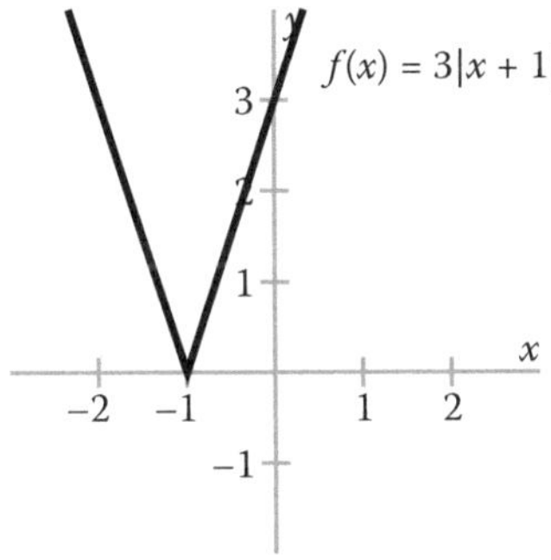

28.

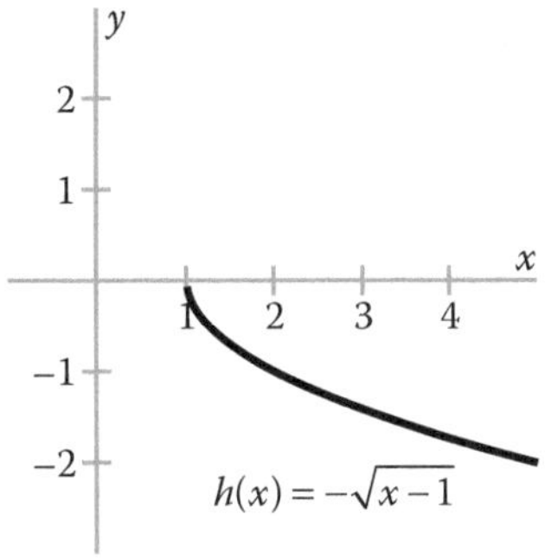

29. even

30. neither

31. $3x^2 - x - 2$; 68

32. $x^2 - 7x - 8$; $\dfrac{x-8}{x+1}$

33. $-6x - 3h + 5$

34. $4x^2 - 6x - 3$; 15

35. $7x - 4$

36. −5; 6; not possible

37. yes

38. yes

39. does not exist

40. $k^{-1}(x) = \dfrac{x^2 + 45}{9}$, $x \ge 0$

SECTION 2.1 ANSWERS

Warm Up/Just the Facts

1. −3, 15
2. −9, 3
3. −3, 4
4. *a*-value; upward, downward; wider
5. vertex; maximum, minimum
6. $x = -\dfrac{b}{2a}$
7. *k*; *h* (or $-\dfrac{b}{2a}$)
8. =; >; <
9. symmetry
10. $f(x) = x^2$; right; up

Essential Skills/Extensions

1.

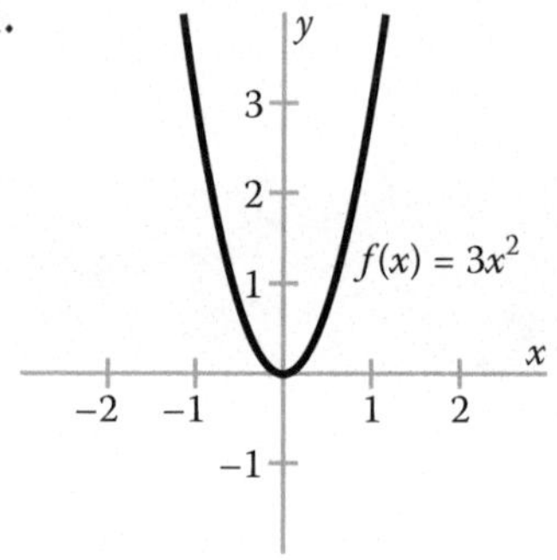

3. (6, −2); $x = 6$; minimum

5. (−1, 0); $x = -1$; maximum

7. (5, 4); $x = 5$; minimum

9. (−2, −10)

11. (−5, 1)

13. (5, 5)

15. (−4, −28)

17. (0, 6)

19. (0, −2)

21. (−2, −3)

23.

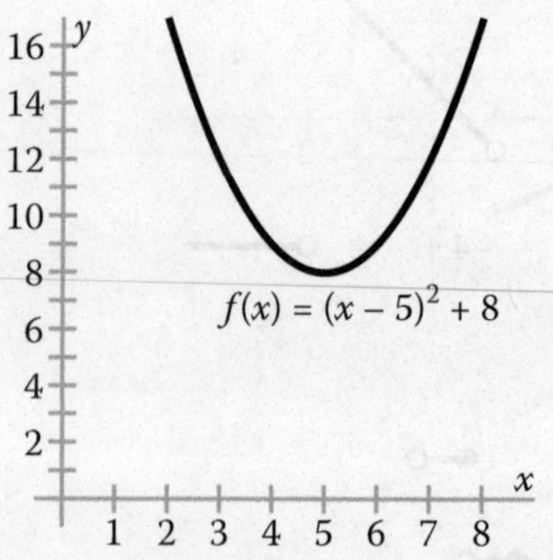

25. $f(x) = 3(x + 11)^2 - 108$

27. $f(x) = -2(x + 5)^2 + 25$

29. The graph has two *x*-intercepts.

31. The graph has one *x*-intercept.

33. The graph has no *x*-intercepts.

35.

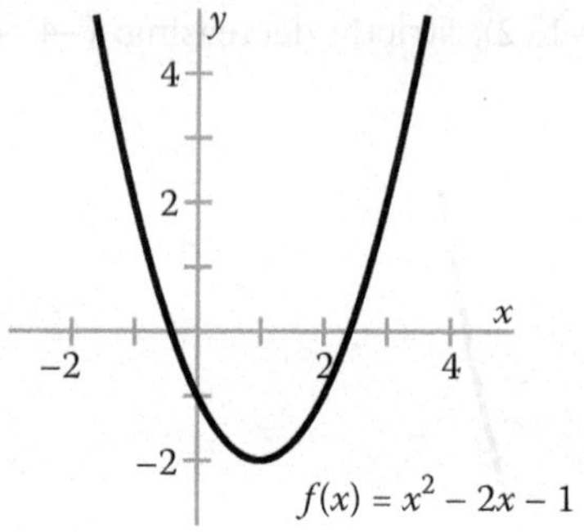

37. $y = 3x^2 + 6x + 3$

39. $y = -2x^2 + 12x - 19$

41. $y = 2x^2 + 4x + 4$

43. minimum: $y = -20$

45. minimum: $y = 8/3$

47. maximum: $y = -7/2$

49. 900 ft^2

51. 26,406.25 ft^2

53. $y = -0.12x^2 + 3$

55. $y = 1.25x^2 - 5$ and $y = -1.25x^2 + 5$,
$y = 0.5x^2 - 2$ and $y = -0.5x^2 + 2$ (possible answers)

57. true

SECTION 2.2 ANSWERS

Warm Up/Just the Facts

1. $9x^4 + 5x^3 + 4x^2 - 12x$; 4; 9
2. $-4.4x^7 + 2x^5 - 11x - 10$; 7; -4.4
3. $-4x^3 + x^2 + 6x + 30$; 3; -4
4. degree, leading coefficient
5. odd; positive
6. multiplicity
7. four
8. Intermediate Value Theorem; zero
9. even
10. up; left; $f(x) = x^3$

Essential Skills/Extensions

1. As $x \to -\infty, f(x) \to \infty$; and as $x \to \infty, f(x) \to \infty$.

3. As $x \to -\infty, f(x) \to \infty$; and as $x \to \infty, f(x) \to -\infty$.

5. As $x \to -\infty, f(x) \to -\infty$; and as $x \to \infty, f(x) \to \infty$.

7.

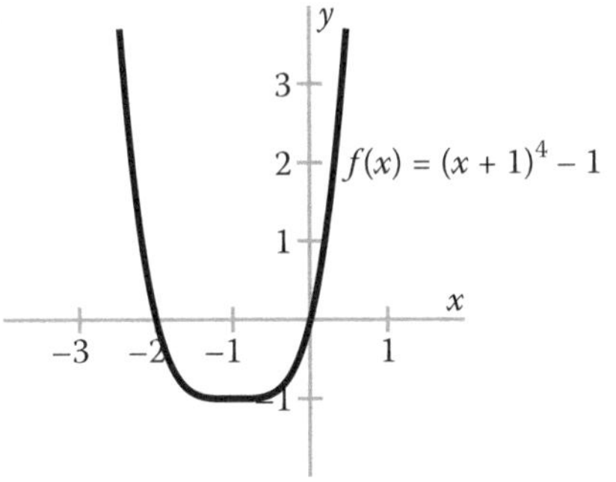

9.

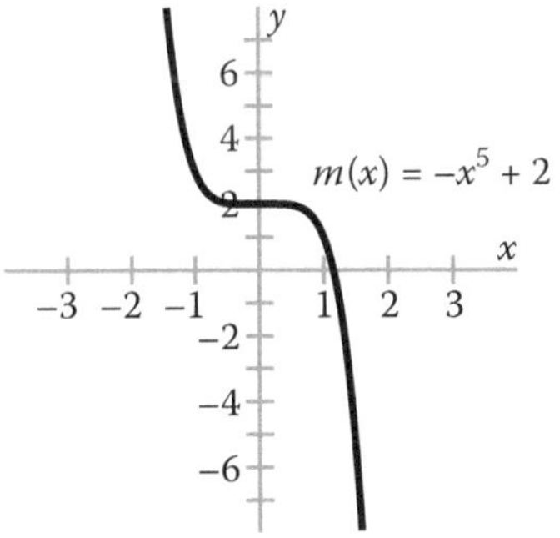

11. 0: multiplicity 1; –4: multiplicity 4; 1: multiplicity 2

13. 0: multiplicity 2; –5: multiplicity 6; –8: multiplicity 5

15. –1: multiplicity 2; –9: multiplicity 1; 7: multiplicity 3

17. 0: multiplicity 1; –2: multiplicity 2

19. 0: multiplicity 4; 4: multiplicity 1; –6: multiplicity 1

21. 0: multiplicity 2; 5: multiplicity 1; –3: multiplicity 1

23.

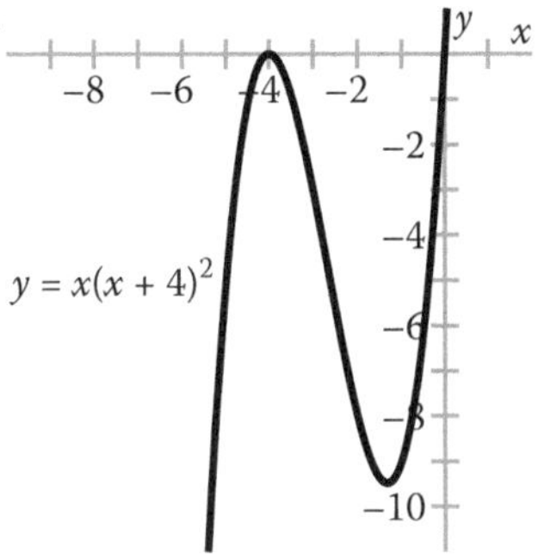

25.

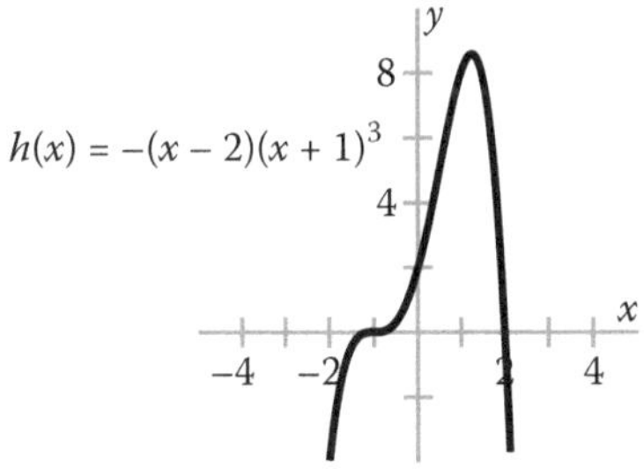

27. one local maximum between –3 and –1; one local minimum between 0 and 2

29. one local maximum between –1 and 1; one local minimum between 0 and 2

31. two local maximums: one local maximum between –4 and –3 and one between –1 and 1; two local minimums: one local minimum between –2 and –1 and one between 0 and 2

33. one local maximum between 1 and 2; two local minimums: one local minimum between –1 and 0 and one at 3.

35. $f(x) = -0.2(x + 2)^2(x - 3)^2(x - 1)^3$

37. $f(x) = -(x + 1)(x - 2)(x - 6)$, $g(x) = (x + 1)(x - 2)(x - 6)$; $f(x) = 2(x + 1)(x - 2)(x - 6)$, $g(x) = -2(x + 1)(x - 2)(x - 6)$ (possible answers)

39. The signs of the zeros were incorrect: the x-intercepts should be (1, 0) and (–4, 0), rather than (–1, 0) and (4, 0).

41. $f(x) = 2(x - 1)^3(x - 3)^2(x + 1)^4$

43. $f(x) = -4(x - 5)(x + 0.5)^3(x - 6)$

45. $f(x) = -x^3 + 2x^2$

SECTION 2.3 ANSWERS

Warm Up/Just the Facts

1. $\dfrac{x^2-25}{x+5} = \dfrac{\cancel{(x+5)}(x-5)}{\cancel{(x+5)}}$
$= x-5$

2. $\dfrac{2x^2+28x+98}{x+7} = \dfrac{2(x^2+14x+49)}{x+7}$
$= \dfrac{2\cancel{(x+7)}(x+7)}{\cancel{(x+7)}}$
$= 2(x+7)$

3. $\dfrac{2x^2-18}{x^2+6x+9} = \dfrac{2(x^2-9)}{(x+3)(x+3)}$
$= \dfrac{2\cancel{(x+3)}(x-3)}{\cancel{(x+3)}(x+3)}$
$= \dfrac{2(x-3)}{x+3}$

4. long, synthetic
5. coefficients; zero
6. zero
7. b; $f(b)$
8. dividend; divisor; $q(x)$; $r(x)$
9. quotient; divisor
10. –5

Essential Skills/Extensions

1. $2x^2+x+3+\dfrac{1}{x+1}$

3. $x^2+2x+8+\dfrac{9}{x-2}$

5. $2x^2+2x+1+\dfrac{1}{x+2}$

7. $3x^3-4x^2+8x-7+\dfrac{10}{x+2}$

9. $6x^2+29x+109+\dfrac{448}{x-4}$

11. $3x^3-4x^2+6x+2+\dfrac{10}{x+2}$

13. $4x^3+12x^2-10x-35-\dfrac{210}{x-3}$

15. $4x^3-20x^2+97x-478+\dfrac{2390}{x+5}$

17. The remainder must be –11 when $h(x)$ is divided by $x - 6$.

19. The remainder must be 10 when $f(x)$ is divided by $x - 3$.

21. The remainder must be –7 when $m(x)$ is divided by $x + 2$.

23. The remainder must be 0 when $q(x)$ is divided by $x + 9$.

25. $g(-1) = 4$

27. $h(8) = 0$

29. $k(-9) = 12$

31. $j(10) = -23$

33. $x + 1$ is not a factor of $g(x)$.

35. $t - 6$ is a factor of $t(x)$.

37. $p + 2$ is a factor of $p(x)$.

39. $r - 4$ is not a factor of $r(x)$.

41. 5

43. 4046

45. yes

47. $x^2 + 5x - 1$; all real numbers except –3 and 4

49. not possible

51. $x^3-x^2+2x-2+\dfrac{3}{x+1}$

53. $(x - 3)$, $(x - 5)$, $(x + 2)$, $(x + 3/4)$

55. $(x + 4)$, $(x - 5)$, $(x + 3/2)$

SECTION 2.4 ANSWERS

Warm Up/Just the Facts

1. True. The Factor Theorem states that the binomial $(x - k)$ is a factor of a polynomial $f(x)$ if and only if $f(k) = 0$.

2. False. $x^2 - 4$ is the difference of squares and can be factored to $(x + 2)(x - 2)$.

3. False. Every polynomial function has a y-intercept.

4. four

5. Descartes' Rule of Signs

6. $(x + 4)$; $(x - 7)$; $(x - 12)$

7. Rational Zero Theorem; leading; constant

8. lower; upper

9. upper; lower

10. 0

Essential Skills/Extensions

1. $g(x) = (x + 4)(x + 5)(x - 1)$; −4, −5, 1

3. $h(x) = (x + 4)(x - 2)^3$; −4, 2

5. $m(x) = (x + 3)(x - 5)^3$; −3, 5

7. $k(x) = (x - 2)(x + 2)(x - 3)(x - 6)$; −2, 2, 3, 6

9. $f(x) = x^3 + 3x^2 - 10x$

11. $f(x) = x^3 - 8x^2 + 17x - 10$

13. $f(x) = x^3 + 10x^2 + 27x + 18$

15. $f(x) = x^3 - 7x^2 + 14x - 8$

17. $p(x) = -2x^2 + 24x - 54$

19. $g(x) = x^3 + 8x^2 + 11x - 20$

21. $h(x) = x^3 + 4x^2 + x - 6$

23. $m(x) = x^3 + x^2 - 64x - 64$

25. ±1, ±2, ±5, ±10; 1, 2, 5

27. ±1/6, ±1/3, ±1/2, ±2/3, ±1, ±2; −2, −1/3, 1/2

29. ±1/2, ±1, ±3/2, ±3, ±5, ±5/2, ±15/2, ±15; 5

31. ±1/2, ±1, ±2; 1/2, 1, 2

33. −1/2, 0, 4

35. −2, 5/6, 3, 4

37. −5, −4, −3/4, 3

39. −4 (multiplicity 2), −3/2, 4

41. 4, 2, or 0 positive zeros; 0 negative zeros

43. 3 or 1 positive zeros; 1 negative zero

45. 3 or 1 positive zeros; 2 or 0 negative zeros

47. 3 or 1 positive zeros; 2 or 0 negative zeros

49. −11 is a lower bound for the real zeros.

51. 7 is a real zero.

53. 8 is an upper bound for the real zeros.

55. −2 is a lower bound for the real zeros.

57. $f(x) = x^5 + 3x^2 - 1$

59. $p(x) = 2(x + 1)(x + 6)(x - 9)(x - 5)$

61. false

63. −0.5, 0.3, 4.5

SECTION 2.5 ANSWERS

Warm Up/Just the Facts

1. $$\frac{4}{4-\sqrt{5}} = \frac{4}{4-\sqrt{5}} \cdot \frac{4+\sqrt{5}}{4+\sqrt{5}} = \frac{16+4\sqrt{5}}{16-5} = \frac{16+4\sqrt{5}}{11}$$

2. $$\frac{7}{5+\sqrt{2}} = \frac{7}{5+\sqrt{2}} \cdot \frac{5-\sqrt{2}}{5-\sqrt{2}} = \frac{35-7\sqrt{2}}{25-2} = \frac{35-7\sqrt{2}}{23}$$

3. $4 + \left(4+\sqrt{3}\right)\left(4-\sqrt{3}\right) = 16 - 3 = 13$

4. imaginary; i

5. complex; real; i

6. $a - bi$

7. 1, −1, i, $-i$

8. i^2

9. conjugate

10. complex; one

Essential Skills/Extensions

1. −1

3. i

5. $-i$

7. $-1 - 4i$

9. $5 - 6i$

11. $-20 + 5i$

13. $58 + 50i$

15. $-5 + 10i$

17. $7 - 3i$

19. $29 + 63i$

21. $40 + 32i$

23. $\frac{1}{37} + \frac{6}{37}i$

25. $\frac{1}{13} + \frac{5i}{13}$

27. $\frac{36}{25} - \frac{2i}{25}$

29. $-\frac{22}{65} - \frac{71i}{65}$

31. $3, -2, 1-\sqrt{2}, 1+\sqrt{2}$

33. $-4/3, 1, 2+\sqrt{5}, 2-\sqrt{5}$

35. $-3/2, 3, 3+\sqrt{5}, 3-\sqrt{5}$

37. $-5/2, 5, 4+\sqrt{5}, 4-\sqrt{5}$

39. $-2, 3, 3i, -3i$

41. $-1, 3, 2i, -2i$

43. $-1, 3, 3i, -3i$

45. $-2, 3, 4i, -4i$

47. $1, 5i, -5i$

49. $-7, 1 - 5i, 1 + 5i$

51. $3, 4 + 2i, 4 - 2i$

53. $-4, 3 + 3i, 3 - 3i$

55. $i, 2 - 3i, \frac{1+i}{3}$

57. $-\frac{83}{85} + \frac{9}{85}i$

59. $-71 + 11i$

61. $\frac{3}{2} - \frac{i}{2}$

63. $\frac{293}{26} + \frac{17i}{26}$

65. $f(x) = x^7 - 14x^6 + 91x^5 - 344x^4 + 779x^3 - 1006x^2 + 689x - 676$,
$f(x) = 2x^7 - 28x^6 + 182x^5 - 688x^4 + 1558x^3 - 2012x^2 + 1378x - 1352$
(possible answers)

CHAPTER 2 REVIEW ANSWERS

1. (0, 1); $x = 0$; maximum

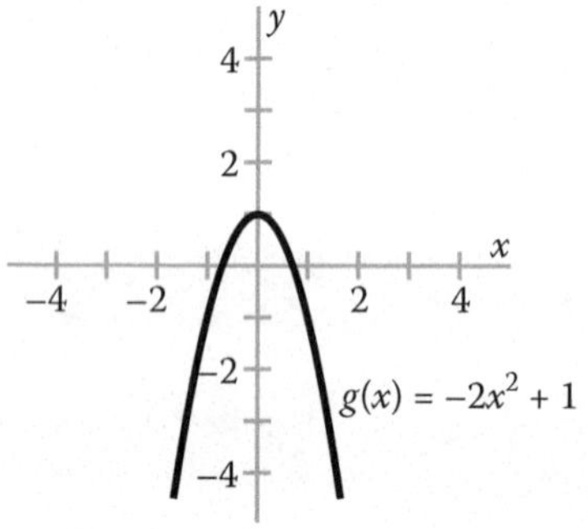

2. (−1, 3)

3.

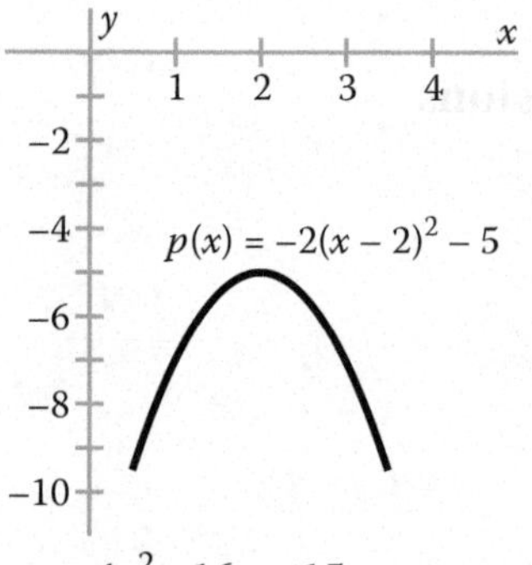

4. $y = -4x^2 - 16x - 15$

5. maximum: −131/20

6. As $x \to -\infty$, $d(x) \to \infty$, and as $x \to \infty$, $d(x) \to -\infty$.

7.

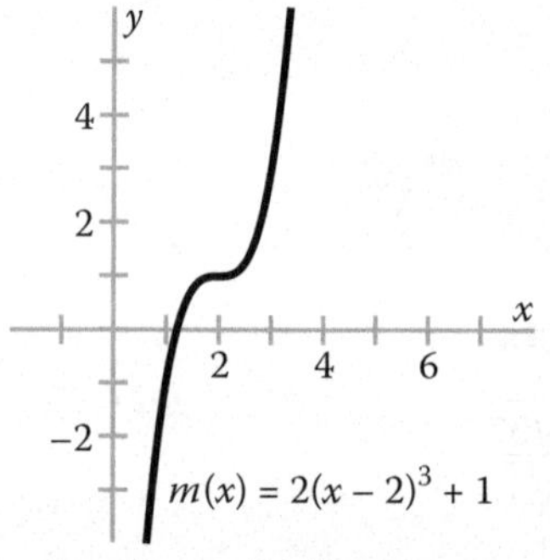

8. −5, −1/4: multiplicity 1; −4: multiplicity 2

9.

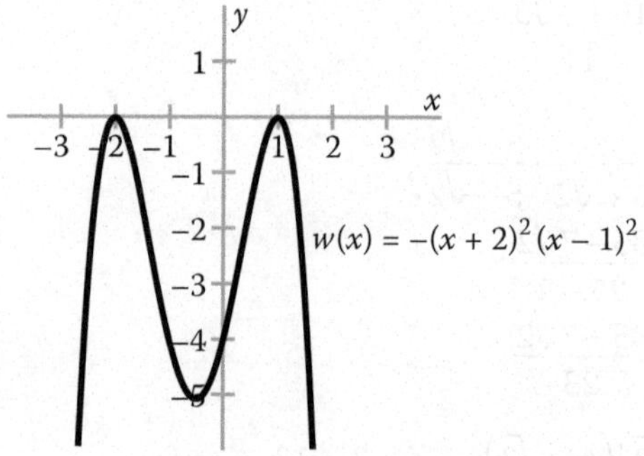

10. one local maximum between 1 and 2, and one local minimum between −1 and 1

11. $5x^2 + 14x + 56 + \frac{232}{x-4}$

12. $x^3 - x^2 - 8x + 12$

13. $3x^2+6x+8+\dfrac{26}{x-2}$

14. $f(-9) = 12$

15. $x - 3$ is a factor of $g(x)$.

16. $h(x) = (x + 3)(x + 5)(x - 4)$; $x = -3, -5, 4$

17. $m(x) = x^3 + 6x^2 + 11x + 6$

18. ±1, ±5, ±7, ±35, ±1/2, ±5/2, ±7/2, ±35/2; −1, −7/2, 5

19. $n(x) = (x + 2)(x - 5)(x - 3)$; $x = -2, 3, 5$

20. 4, 2, or 0 positive zeros; 1 negative zero

21. $11 + 13i$

22. $\dfrac{19}{145}-\dfrac{112}{145}i$

23. $4, -1, -2+\sqrt{6}, -2-\sqrt{6}$

24. $3, -1, 4i, -4i$

25. $5, -1, 2i, -2i$

SECTION 3.1 ANSWERS

Warm Up/Just the Facts

1. $\dfrac{x+1}{x^2-x}$; $x \neq -3$

$$\frac{x^2+4x+3}{x^3+2x^2-3x} = \frac{\cancel{(x+3)}(x+1)}{x\cancel{(x+3)}(x-1)} = \frac{x+1}{x(x-1)} = \frac{x+1}{x^2-x}$$

2. 0, −1, and 1

$$3x^3 - 3x = 0$$
$$3x(x^2 - 1) = 0$$
$$3x(x+1)(x-1) = 0$$
$$3x = 0 \quad \text{or} \quad x+1 = 0 \quad \text{or} \quad x-1 = 0$$
$$x = 0 \qquad x = -1 \qquad x = 1$$

3.

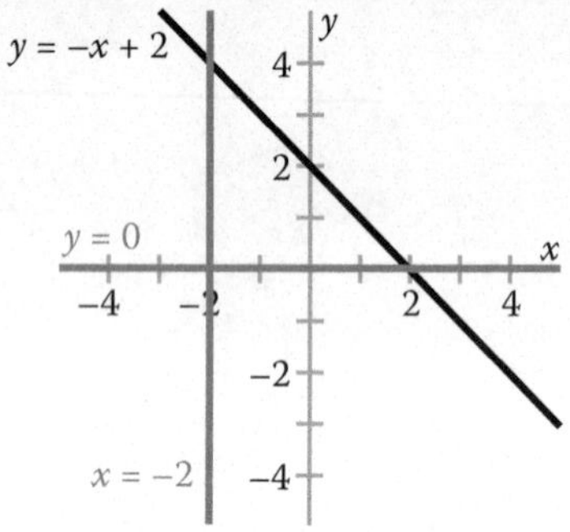

4. quotients
5. $y = 0$; $x = 0$
6. horizontal, oblique
7. horizontal
8. horizontal; $y = 0$
9. long division; 1 plus
10. oblique; $y = 2x + 5$

Essential Skills/Extensions

1. $x = -4$; $y = 0$

3. D: $(-\infty, 1) \cup (1, \infty)$; R: $(-\infty, 0) \cup (0, \infty)$

5.

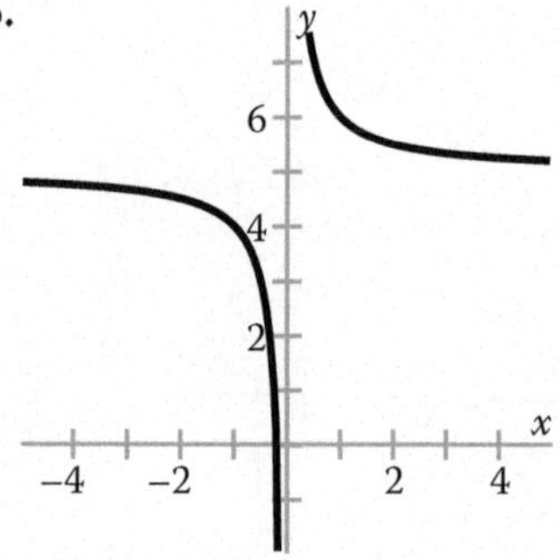

7.

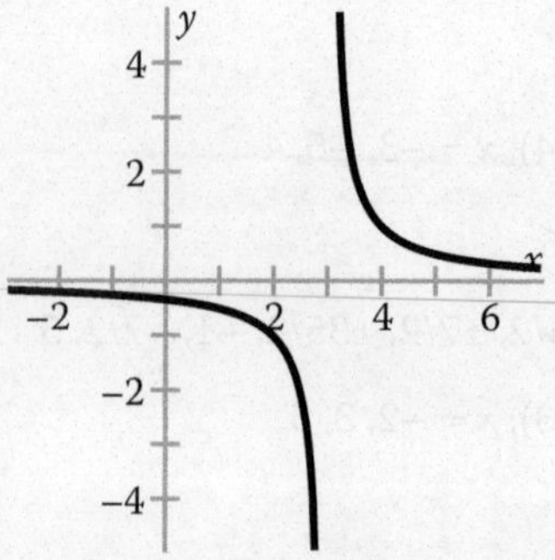

9.

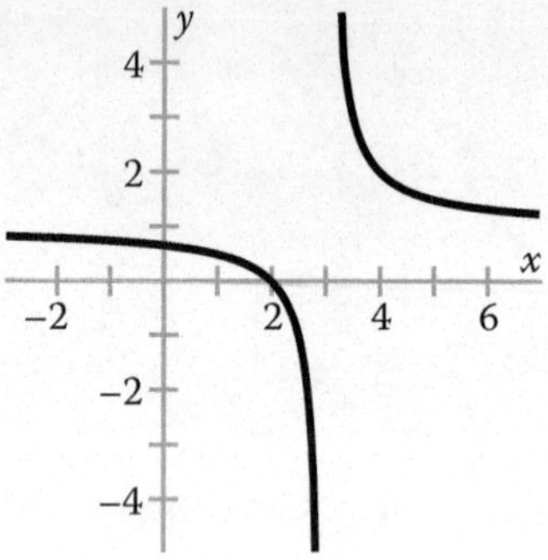

11. $x = 2$

13. $x = -6$, $x = 6$

15. $x = -1/2$

17. $x = 0$, $x = 2$, $x = 1/2$

19.

x	−3	−2	−1	0	1
$f(x)$	−8	und	0	−1/2	0

21. zeros: $x = -2/3$, $x = -1/2$; vertical asymptote: $x = 1$

23.

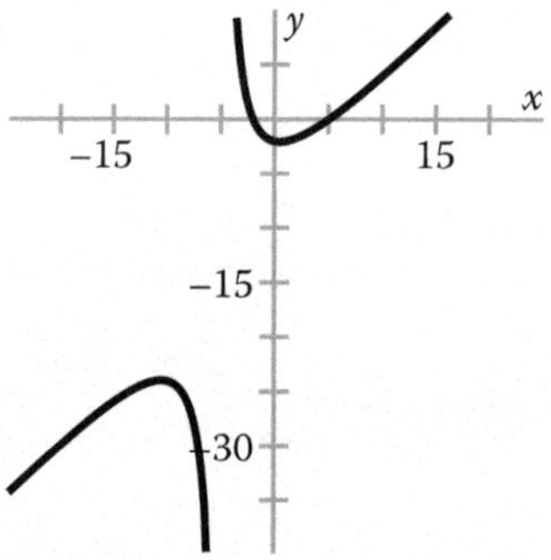

25. B

27. no horizontal asymptote

29. $y = 0$

31. $y = 4$

33.

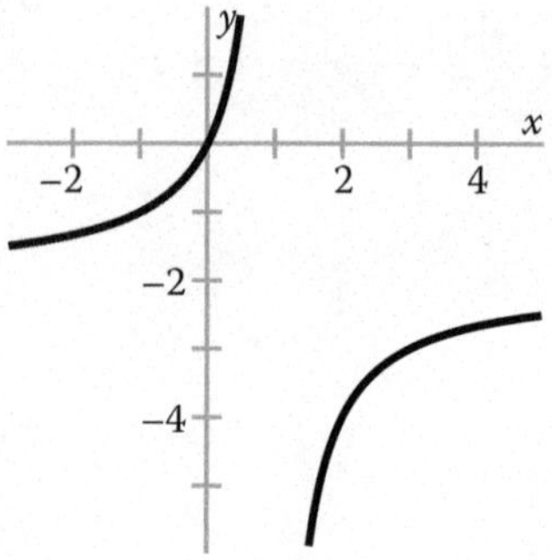

35.

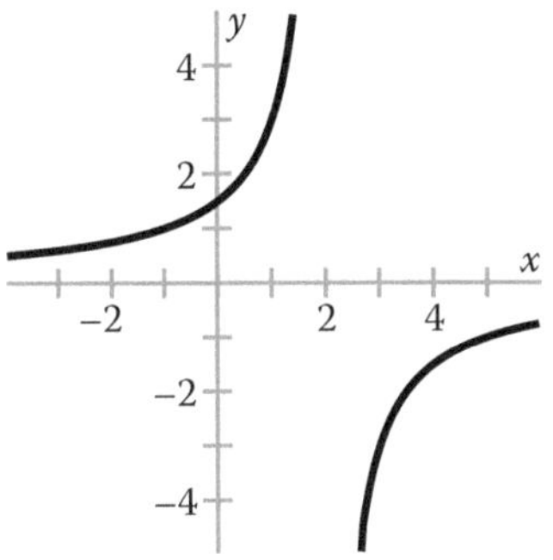

37. vertical asymptote: $x = -3$; horizontal asymptote: none; oblique asymptote: $y = -x + 3$

39. D

41. no hole

43.

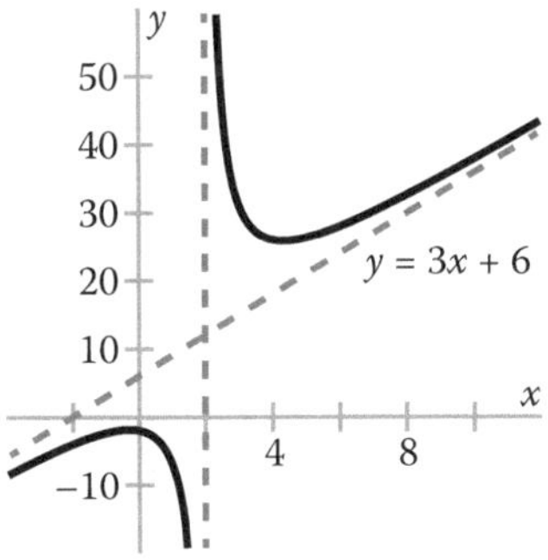

45. vertical asymptote: $x = -5$; horizontal asymptote: none; oblique asymptote: $y = x - 5$

47. (5, 7.5)

49.

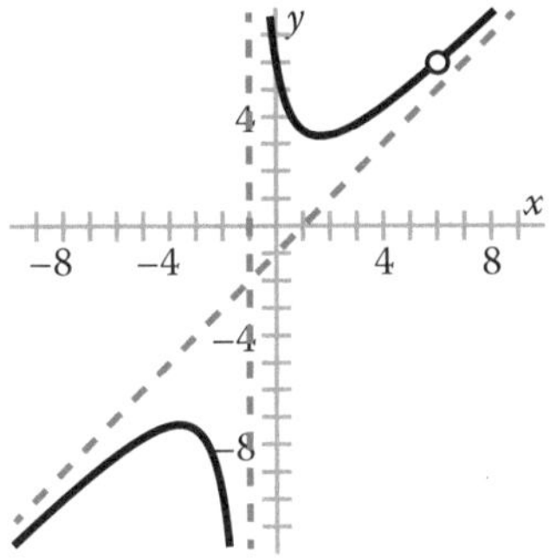

51. Errors:

a. "x-intercept at $x = 2$"

correction: x-intercept at $x = 0$

b. "one vertical asymptote, at the line $x = 0$"

correction: one vertical asymptote, at the line $x = 2$

53. no errors

55. The function has no horizontal asymptote. It has an oblique asymptote.

57. $g(x) = \dfrac{2}{x^2} + 6$

59. $g(x) = \dfrac{2}{(x-2)^2} - 3$

61. Answers vary, for example $f(x) = \dfrac{x}{x^2 - 2x - 3}$.

The expression in the denominator must have the factors $(x - 3)$ and $(x + 1)$, while the expression in the numerator does not have either of those factors.

63. Answers vary, for example $f(x) = \dfrac{(x+1)(x-3)}{x-3}$.

The expression in the numerator and the expression in the denominator must both have the factor $(x - 3)$. The simplified function must not contain a variable in the denominator (e.g., linear or quadratic function), and when 3 is substituted into the simplified function for x, the result must be $y = 4$.

65. Answers vary, for example $f(x) = \dfrac{x^2 + x - 1}{x - 1}$.

The expression in the denominator must have the factor $(x - 1)$, while the expression in the numerator does not have that factor. The degree of the numerator must be 1 greater than the degree of the denominator, and the quotient must be $x + 2 + \dfrac{R}{x-1}$, where the remainder is any nonzero number.

SECTION 3.2 ANSWERS

Warm Up/Just the Facts

1. vertex form: $f(x) = (x - 0)^2 + 0 \Rightarrow h = 0, k = 0$
vertex: (0, 0); axis of symmetry: $x = 0$
Find several points on the parabola with x-values near the vertex.

x	−2	−1	1	2
$f(x)$	4	1	1	4

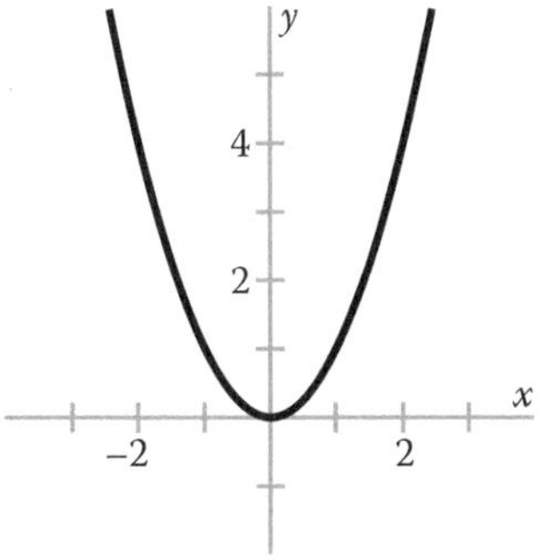

2. vertex form: $f(x) = (x - 3)^2 + 1 \Rightarrow h = 3, k = 1$
vertex: (3, 1); axis of symmetry: $x = 3$
Find several points on the parabola with x-values near the vertex.

x	1	2	4	5
$f(x)$	−3	0	0	−3

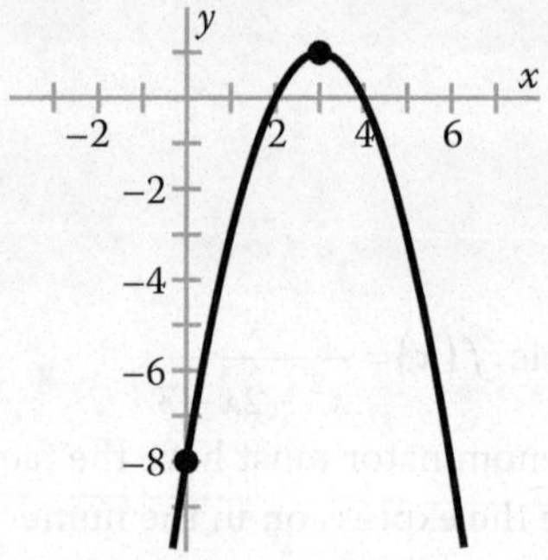

3. vertex form: $f(x) = (x + 2)^2 + 0 \Rightarrow h = -2, k = 0$
vertex: (−2, 0); axis of symmetry: $x = -2$
$f(0) = (0)^2 + 4(0) + 4 = 4$: y-intercept: (0, 4)
Find several points on the parabola with x-values near the vertex.

x	−4	−3	−1
$f(x)$	4	1	1

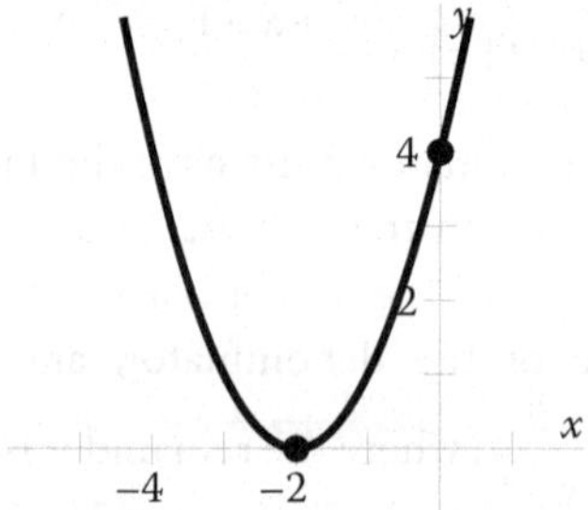

4. cone (double cone), plane
5. point; line; directrix
6. parabolas; down; axis of symmetry; y-axis; focus; $y = -p$
7. parabolas; to the right; axis of symmetry; x-axis; vertex; $x = -p$
8. vertex; focus
9. vertex; (0, 0)
10. right, left

Essential Skills/Extensions

1. focus: (0, 5/8); directrix: $y = -5/8$
3. right

5.

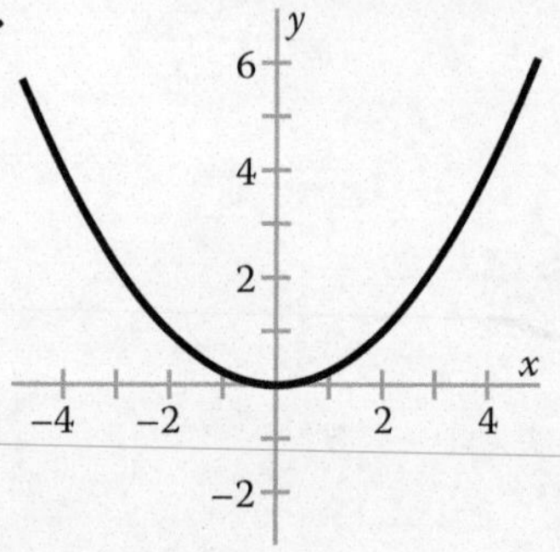

7. $y^2 = \dfrac{x}{2}$

9. $x^2 = \dfrac{10}{3}y$

11. $x^2 = -y$

13. Errors:

a. *"the focus is at (20, 0)"*
correction: the focus is at (1/20, 0)
b. *"the equation of the directrix is y = −20"*
correction: the equation of the directrix is x = −1/20

15. Errors:

a. *"The axis of symmetry is at y = 0"*
correction: the axis of symmetry is at x = 0
b. *"the parabola opens to the left"*
correction: the parabola opens downward
c. *"the focus is at (−3/2, 0)"*
correction: the focus is at (0, −3/2)
d. *"the equation of the directrix is x = 3/2"*
correction: the equation of the directrix is y = 3/2

17. The equation $x^2 = 4py$ is a function because it describes a parabola with a vertical axis of symmetry. A parabola that opens vertically (opening upward or downward) will intersect any vertical line no more than one time. Therefore, the parabola passes the Vertical Line Test and is a function. The equation $y^2 = 4px$ is not a function because it describes a parabola with a horizontal axis of symmetry. A parabola that opens horizontally (opening left or right) will intersect a vertical line more than one time (when the vertical line passes through the parabola, not at the vertex). Therefore, the parabola does not pass the Vertical Line Test and is not a function.

19. A parabola described by an equation of the form $x^2 = 4py$ cannot have a directrix at $y = x$ because the parabola's directrix must be at $y = -p$, where p is a constant. Additionally, the axis of symmetry is perpendicular to the directrix, and since the axis of symmetry is vertical, the directrix must be horizontal. The equation $y = x$ does not define a horizontal line. Therefore, $y = x$ cannot be the equation of the directrix.

21. $x^2 = y$ and $y^2 = 8x$

23. 1/2

25. 12

SECTION 3.3 ANSWERS

Warm Up/Just the Facts

1. $c^2 = a^2 - b^2$

   $c^2 = 5^2 - 1^2$ *Substitute.*

   $c^2 = 25 - 1$ *Simplify the powers.*

   $c^2 = 24$ *Subtract.*

   $c = \pm\sqrt{24}$ *Take ± the square root of each side.*

   $c = \pm 2\sqrt{6}$ *Simplify the radical.*

2. $c^2 = a^2 - b^2$

   $20 = 70 - b^2$ *Substitute.*

   $b^2 = 50$ *Solve for b^2.*

   $b = \pm\sqrt{50}$ *Take ± the square root of each side.*

   $b = \pm 5\sqrt{2}$ *Simplify the radical.*

3. $\frac{2}{5\sqrt{3}} = \frac{2}{5\sqrt{3}} \cdot \frac{\sqrt{3}}{\sqrt{3}} = \frac{2\sqrt{3}}{5\sqrt{9}} = \frac{2\sqrt{3}}{15}$
4. sum; foci
5. x; y
6. y; x
7. vertices (or y-intercepts), foci, major axis
8. co-vertices (or x-intercepts), minor axis
9. minor axis
10. major axis
11. horizontal; origin; $(\pm a, 0)$; $(0, \pm b)$
12. vertical; origin; $(0, \pm a)$; $(\pm b, 0)$

Essential Skills/Extensions

1. $\frac{y^2}{64} + \frac{x^2}{16} = 1$

3. $\frac{x^2}{9} + y^2 = 1$

5. $\frac{y^2}{16} + x^2 = 1$

7. $\frac{y^2}{169} + \frac{x^2}{144} = 1$

9. $\frac{x^2}{100} + \frac{y^2}{36} = 1$

11. $(\pm 11, 0)$ and $(0, \pm 20)$

13. major axis: $6\sqrt{2}$; minor axis: $2\sqrt{6}$

15. 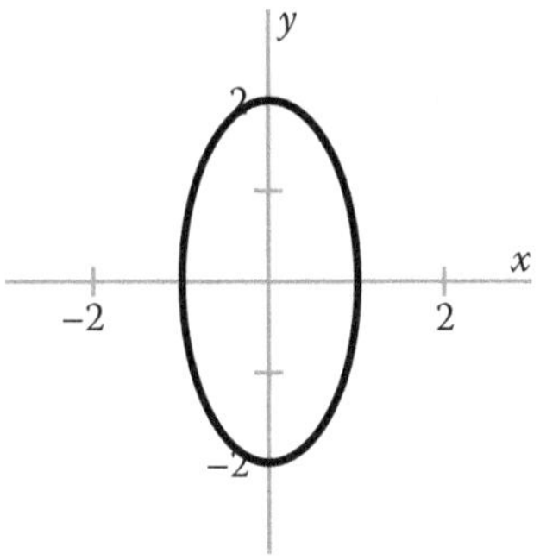

17. 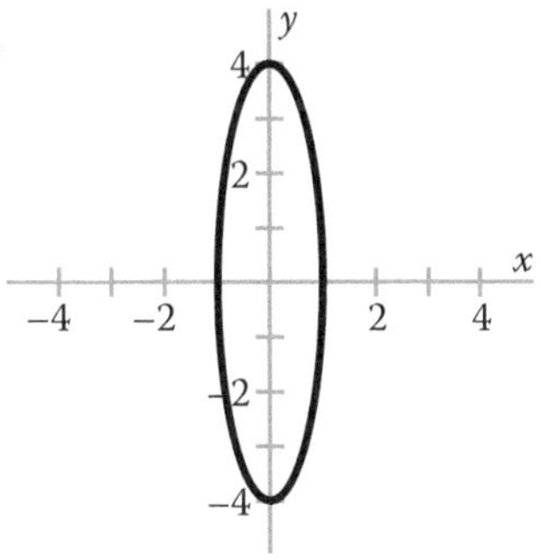

19. $(0, \pm 2\sqrt{5})$

21. $\frac{2\sqrt{2}}{3}$

23. $\frac{3\sqrt{2}}{5}$

25. $\left(0, \pm\frac{6}{5}\right)$

27. $\frac{y^2}{400} + \frac{x^2}{396} = 1$

29. center: (0, 0); foci: $(0, \pm 3\sqrt{3})$; vertices: $(0, \pm 6)$, co-vertices: $(\pm 3, 0)$, major axis: 12 units; minor axis: 6 units; eccentricity: $\frac{\sqrt{3}}{2}$

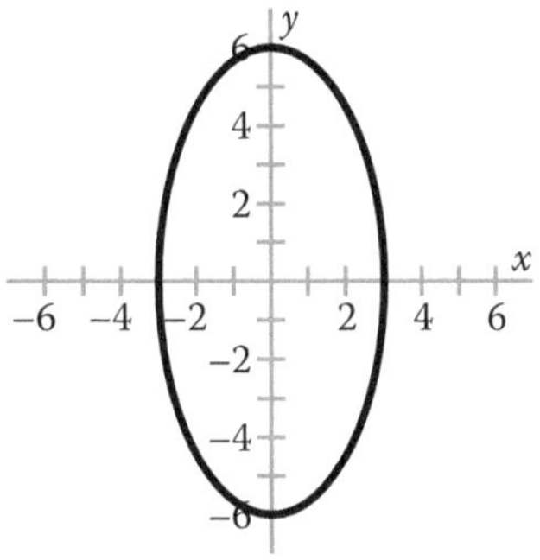

31. $y = \sqrt{\ }(18(1 - X \wedge 2/7)$ and $y = -\sqrt{\ }(18(1 - X \wedge 2/7)$

33. $\frac{x^2}{36} + \frac{y^2}{16} = 1$; $\frac{y^2}{36} + \frac{x^2}{9} = 1$ (possible answers)

35. $\frac{x^2}{49} + \frac{y^2}{16} = 1$; $\frac{y^2}{49} + \frac{x^2}{16} = 1$ (possible answers)

37. $\dfrac{y^2}{10}+x^2=1$; $\dfrac{y^2}{2}+x^2=1$ (possible answers)

39. $\dfrac{x^2}{16}+\dfrac{y^2}{9}=1$; $\dfrac{x^2}{16}+\dfrac{y^2}{4}=1$ (possible answers)

41. $\dfrac{y^2}{25}+\dfrac{x^2}{16}=1$; $\dfrac{x^2}{25}+\dfrac{y^2}{16}=1$ (possible answers)

43. $\dfrac{y^2}{49}+\dfrac{x^2}{25}=1$; $\dfrac{y^2}{36}+\dfrac{x^2}{25}=1$ (possible answers)

45. $\dfrac{x^2}{36}+\dfrac{y^2}{32}=1$; $\dfrac{y^2}{36}+\dfrac{x^2}{32}=1$ (possible answers)

47. B, A, C

SECTION 3.4 ANSWERS

Warm Up/Just the Facts

1. $\dfrac{a}{45}=\dfrac{4}{9}$

$9a = 180$ — *Cross multiply.*

$a = 20$ — *Divide each side by 9.*

2. $\dfrac{x^2}{10}-\dfrac{y^2}{12}=1$

$-\dfrac{y^2}{12}=1-\dfrac{x^2}{10}$ — *Subtract the x^2-term from each side.*

$y^2=-12\left(1-\dfrac{x^2}{10}\right)$ — *Multiply each side by –12.*

$y=\pm\sqrt{-12\left(1-\dfrac{x^2}{10}\right)}$ — *Take ± the square root of each side.*

3. $\sqrt{54}=\sqrt{9\cdot 6}=3\sqrt{6}$

4. foci; difference

5. branches

6. vertical

7. horizontal

8. x-intercepts, transverse axis, foci, vertices; x

9. y-intercepts, transverse axis, foci, vertices; y

10. vertices

11. $2x$

12. false

Essential Skills/Extensions

1. vertices: $(\pm 6, 0)$; co-vertices: $(0, \pm 7)$

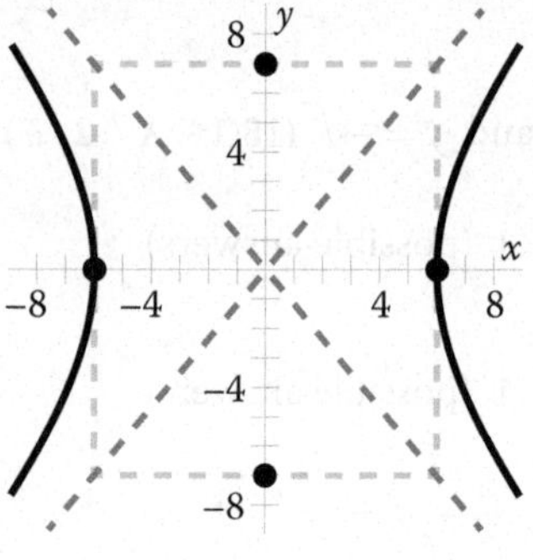

3. $y^2-\dfrac{x^2}{25}=1$

5. $\dfrac{y^2}{225}-\dfrac{x^2}{175}=1$

7. $\dfrac{y^2}{100}-\dfrac{x^2}{21}=1$

9. $\dfrac{x^2}{81}-\dfrac{y^2}{2025}=1$

11. $\dfrac{y^2}{9}-x^2=1$

13. vertices: $(0, \pm 1)$; foci: $(0, \pm 5\sqrt{2})$; asymptotes: $y=\pm\dfrac{1}{7}x$

15.

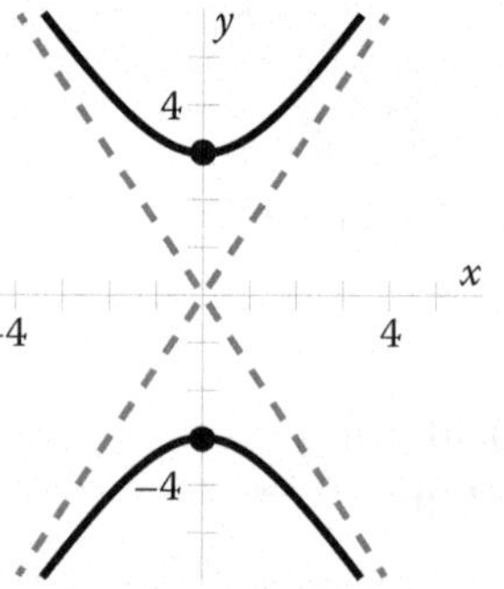

17.

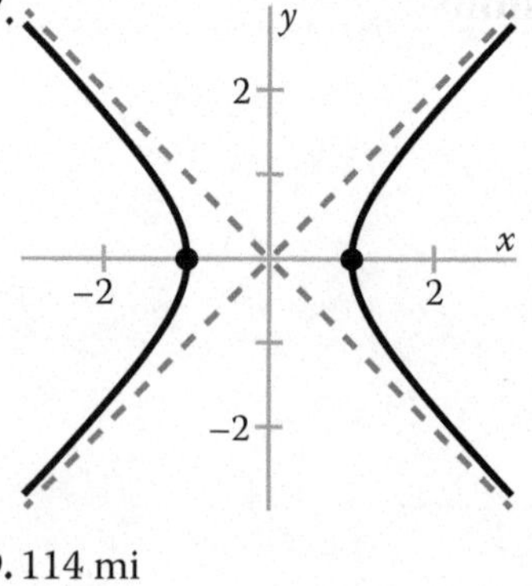

19. 114 mi

21. $\dfrac{y^2}{3600}-\dfrac{x^2}{3456}=1$

23. $\dfrac{y^2}{36}-\dfrac{x^2}{16}=1$; $\dfrac{x^2}{36}-\dfrac{y^2}{49}=1$ (possible answers)

25. $\frac{y^2}{4} - \frac{x^2}{100} = 1$; $\frac{x^2}{121} - \frac{y^2}{100} = 1$ (possible answers)

27. $\frac{y^2}{16} - \frac{x^2}{9} = 1$; $\frac{x^2}{16} - \frac{y^2}{9} = 1$ (possible answers)

29. $\frac{y^2}{4} - x^2 = 1$; $\frac{x^2}{16} - \frac{y^2}{64} = 1$ (possible answers)

31. When $a = b$, the vertices and co-vertices are equidistant from the origin and the asymptotes are at $y = \pm x$.

33. Errors:
a. "of length $2n$ when $n > m$, and $2m$ when $m > n$"
correction: of length 2m
b. "co-vertices are at $(0, \pm n)$"
correction: co-vertices are at $(\pm n, 0)$
c. "foci are at $(\pm(m^2 + n^2), 0)$"
correction: foci are at (0, $\left(0, \pm\sqrt{m^2 + n^2}\right)$*)*

SECTION 3.5 ANSWERS

Warm Up/Just the Facts

1.

$x^2 + 10x - 8 = 1$	*Isolate the constant terms on the right side.*
$x^2 + 10x = 9$	
$x^2 + 10x + 25 = 9 + 25$	*Add $(10/2)^2$ to both sides.*
$(x + 5)^2 = 34$	*Factor.*

2.
$$2x^2 + 4x - 3y^2 + 24y + 10 = 0$$
$$2(x^2 + 2x) - 3(y^2 - 8y) = -10$$
$$2(x^2 + 2x + 1) - 3(y^2 - 8y + 16) = -10 + 2 - 48$$
$$2(x + 1)^2 - 3(y - 4)^2 = -56$$

3.

$7 + 5x^2 - x = 3x$	*Write the equation in the form $ax^2 + bx + c = 0$, then identify a, b, and c from the equation.*
$5x^2 - 4x + 7 = 0$	
$a = 5, b = -4, c = 7$	
$b^2 - 4ac = (-4)^2 - 4(5)(7)$	*Substitute the values of a, b, and c into the formula for the discriminant, $b^2 - 4ac$.*
$= 16 - 140$	
$= -124$	

4. up, down; major, minor

5. transverse

6. (–1, 2)

7. horizontal parabola; to the left

8. (7, 6); vertical; $x = 7$

9. (–3, 0); 3; horizontal; $y = 0$

10. C; equal to 0

Essential Skills/Extensions

1. opens to the right; $(y - k)^2 = 4p(x - h)$; $p > 0$

3. $(x + 2)^2 = -16(y - 10)$

5. (–4, 3)

7. horizontal major axis: $4\sqrt{10}$ units; vertical minor axis: 2 units

9.

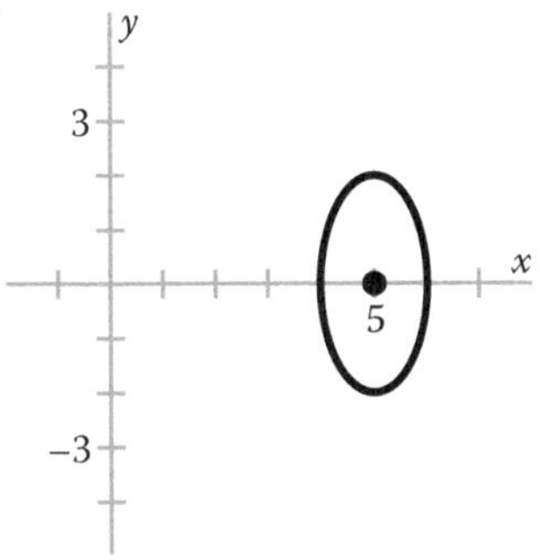

11. (10, –6)

13. (–2, 18) and (–2, 22)

15. $y - 9 = \pm 2(x + 16)$

17.

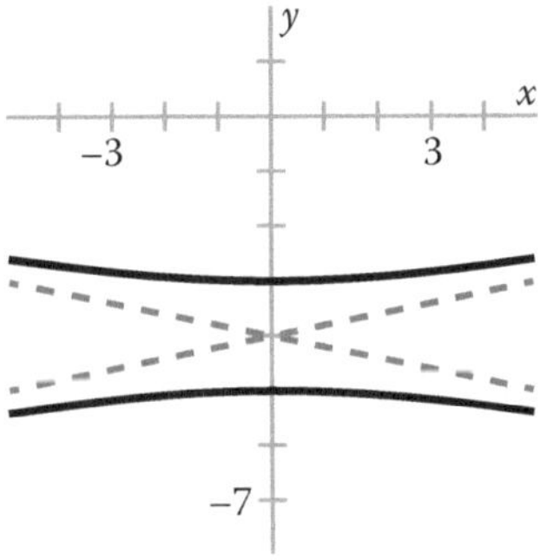

19. parabola

21. not a conic

23. ellipse (not a circle)

25. circle

27. not a conic

29. parabola

31. $(x - 15)^2 - y^2 = 1$ (hyperbola)

33. $\frac{(x+1)^2}{9} + (y-4)^2 = 1$ (ellipse)

35. $x^2 + 2x + 2y + 21 = 0$;
$A = 1, B = 0, C = 0, D = 2, E = 2, F = 21$

37. parabola; $B^2 - 4AC = (-6)2 - 4(9)(1) = 0$

39. circle; $B^2 - 4AC = (0)^2 - 4(-1)(-1) = -4 < 0$, $B = 0$, and $A = C$

41. $x^2 = -6(y - 4)$

43. $(x - 3)^2 = -6(y + 1)$

45. $\dfrac{(y-2)^2}{6} + (x-7)^2 = 1$

47. $\dfrac{(x+m)^2}{a^2} + \dfrac{(y+n)^2}{b^2} = 1$

49. $\dfrac{(x-2)^2}{12} - \dfrac{(y+1)^2}{8} = 1$

51. $B = (-6, 0) \cup (0, 6)$

53. $B = \pm 6$

CHAPTER 3 REVIEW ANSWERS

1. D: $(-\infty, 4) \cup (4, \infty)$; **R:** $(-\infty, -2) \cup (-2, \infty)$

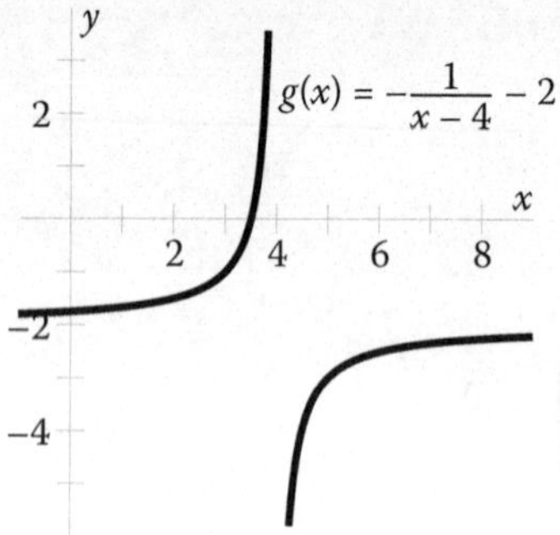

2. $x = 1, x = 3, x = -3$

3.

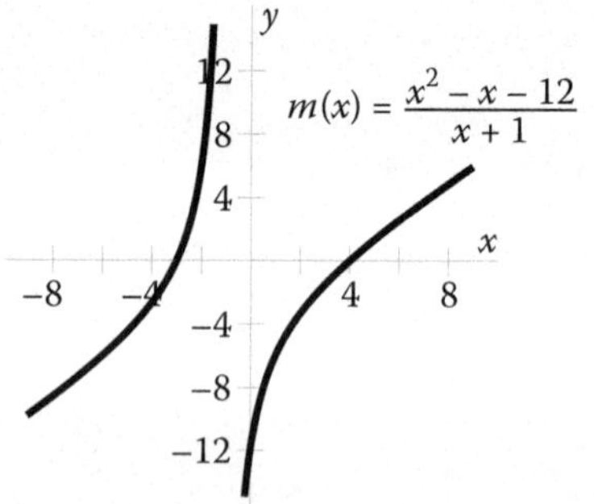

4. $y = 3x + 10$

5. hole: (3, 2.7)

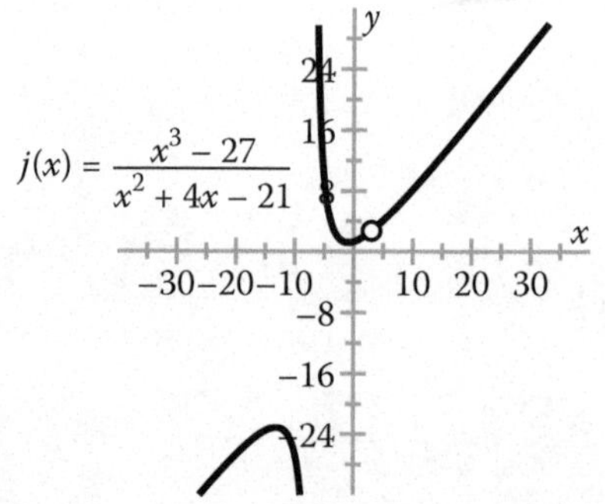

6. focus: $(0, -3/4)$; directrix: $y = 3/4$

7. down

8.

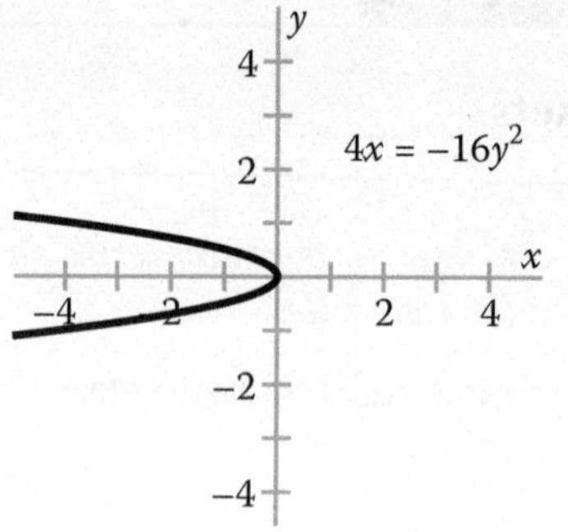

9. $8y = x^2$

10. $2y = x^2$

11. $\dfrac{x^2}{25} + \dfrac{y^2}{4} = 1$

12. $\dfrac{y^2}{16} + \dfrac{x^2}{4} = 1$

13. intercepts: $(\pm 10, 0)$ and $(0, \pm 6)$; foci: $(\pm 8, 0)$

14. $e = 2/3$

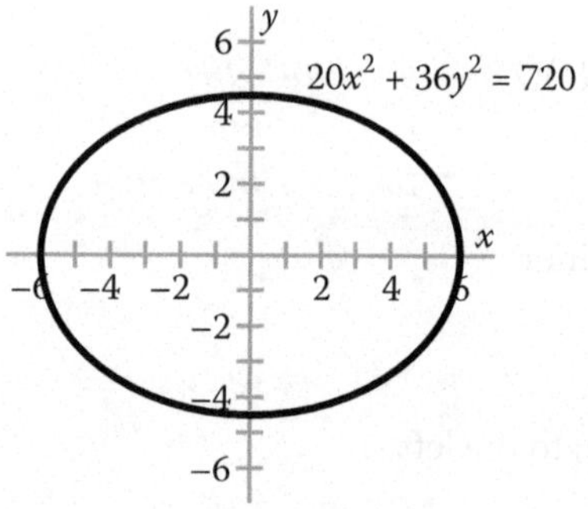

15. $\dfrac{x^2}{81} + \dfrac{y^2}{72} = 1$

16. $\dfrac{y^2}{25} - \dfrac{x^2}{36} = 1$

17. $\dfrac{x^2}{49} - \dfrac{y^2}{23} = 1$

18. vertices: $(\pm 2, 0)$; co-vertices: $(0, \pm 3)$; foci: $(\pm\sqrt{13}, 0)$; asymptotes: $y = \pm\dfrac{3}{2}x$

19.

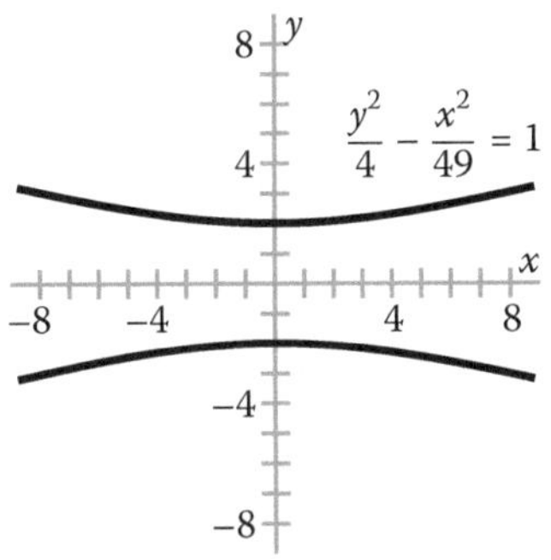

20. 146 nautical mi

21. $(y + 5)^2 = -16(x + 3)$

22. center: (−4, 2)

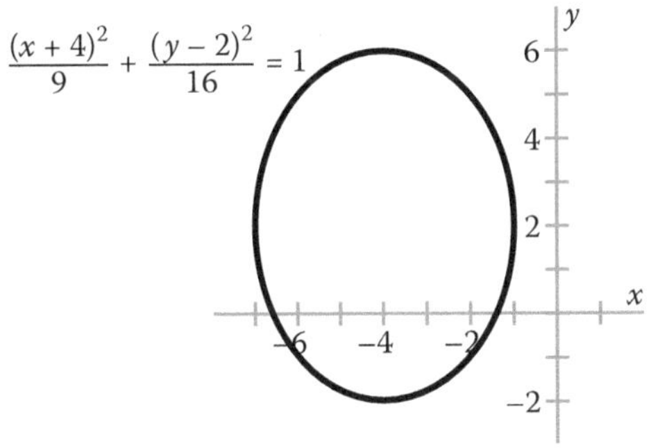

23. center: (2, 3)

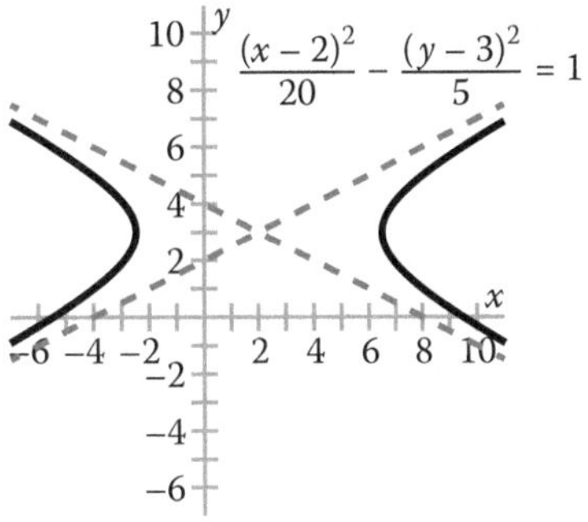

24. $-x^2 + 4y^2 - 6x - 8y - 21 = 0$;
$A = -1, B = 0, C = 4, D = -6, E = -8, F = -21$

25. ellipse

SECTION 4.1 ANSWERS

Warm Up/Just the Facts

1. translation 4 units down
2. reflection across the x-axis, vertical stretch by a factor of 5
3. translation left 3 units and up 1 unit
4. positive, real; 0
5. 1; 0
6. b^{-x} or $\left(\frac{1}{b}\right)^x$; $-b^x$
7. steeper
8. \$102.50; \$100
9. taking the 40th root
10. natural

Essential Skills/Extensions

1. $g(x) = 5^x$, $h(x) = 2^{3x}$, $q(x) = 8^x - 3$

3. $g(x) = 2^{x+3}$, $m(x) = 7\pi^x$, $q(x) = -5^x - 9$, $f(x) = 8^{\frac{1}{2}x}$

5. 16, 1, 128

7. −1728, −1/12, $-2\sqrt{3}$

9. 1/243, 27, $\frac{\sqrt{3}}{9}$

11. 1/2401, 7, 1/343

13. 729, $3\sqrt{3}$,1/9

15. 8000, 1/400, $40\sqrt{5}$

17. 144, 1/12, $2\sqrt{3}$

19. 243/1024, 256/81, $\frac{2\sqrt{3}}{3}$

21. 1/4096, 512, 1/32

23. 1, 9/4, 243/32

25. $f(x) = 5^x$ and $g(x) = \left(\frac{1}{5}\right)^{-x}$

27. $d(x) = 8^{-x}$ and $m(x) = \left(\frac{1}{8}\right)^x$; $p(x) = \left(\frac{1}{8}\right)^{-x}$ and $h(x) = 8^x$

29. $h(x) = \left(\frac{3}{5}\right)^{-x}$ and $g(x) = \left(\frac{5}{3}\right)^x$; $r(x) = -\left(\frac{5}{3}\right)^x$ and $w(x) = -\left(\frac{3}{5}\right)^{-x}$; $f(x) = -\left(\frac{3}{5}\right)^x$ and $p(x) = -\left(\frac{5}{3}\right)^{-x}$

31.

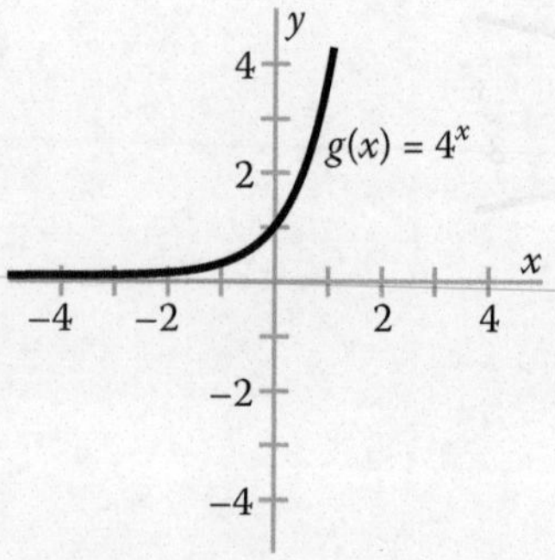

33.

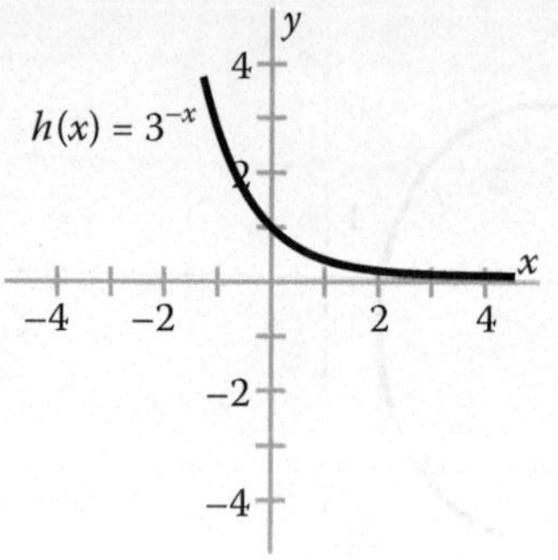

35.

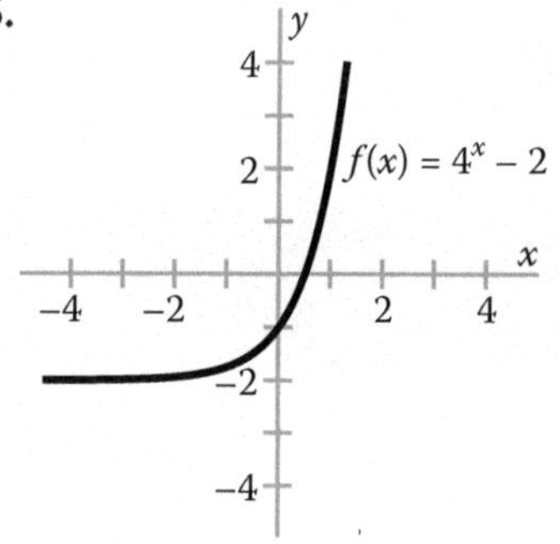

37. \$34,023.01

39. \$9410.96

41. \$22,562.94

43. \$28,652.93

45. \$4752.24

47. 5.45%

49. 6.99%

51. 0.41, 1.1, 3, 8.15, 22.17

53. −0.27, −0.74, −2, −5.44, −14.78

55. 10.34, 3.81, 1.4, 0.52, 0.19

57. 0.04, 0.12, 0.325, 0.88, 2.4

59.

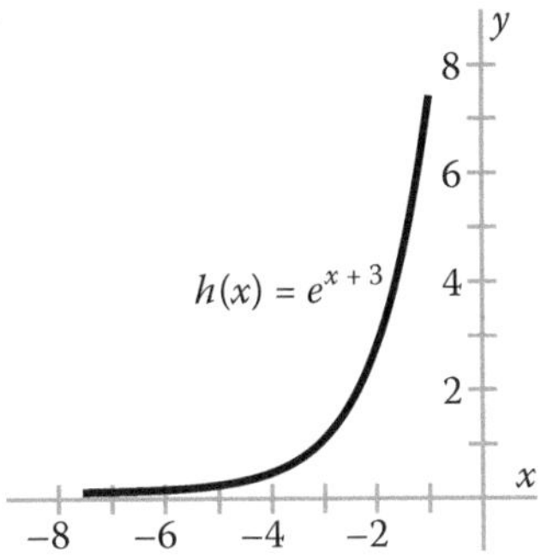

61. 4.9 lb/in^2

63. 6.8 lb/in^2

65. 4616

67. 2729

69. 8.8 lb/in^2

71. $g(x) = 2^{x-3} + 6$

73. $g(x) = \left(\frac{1}{2}\right)^{-x} - 5$

75. False. When $x = 0$, y is approximately 20.

77. True.

SECTION 4.2 ANSWERS

Warm Up/Just the Facts

1. 0

2. 1/9

3. 8

4. logarithmic; b

5. 0; zero-exponent

6. negative

7. exponential; table of values

8. base; 10

9. $\ln x$; e

10. x

Essential Skills/Extensions

1. 2

3. 3

5. 4

7. 5

9. $\log_3 81 = 4$

11. $\log_{\frac{2}{5}} \frac{25}{4} = -2$

13. $\log_9 \frac{1}{81} = -2$

15. $\log_{\frac{4}{11}} \frac{121}{16} = -2$

17. $\log_{27} 243 = \frac{5}{3}$

19. $\log_7 \frac{1}{2401} = -4$

21. $\log_{\frac{2}{7}} \frac{4}{49} = 2$

23. $6^3 = 216$

25. $\left(\frac{1}{5}\right)^{-4} = 625$

27. $7^5 = 16{,}807$

29. $\left(\frac{4}{5}\right)^3 = \frac{64}{125}$

31. $27^{\frac{4}{3}} = 81$

33. $\left(\frac{4}{3}\right)^2 = \frac{16}{9}$

35. $8^{\frac{7}{3}} = 128$

37. 2

39. 9

41. 4

43. 1/3

45. 3/4

47. 2/3

49. 1

51. 5

53. 0

55. 64

57.

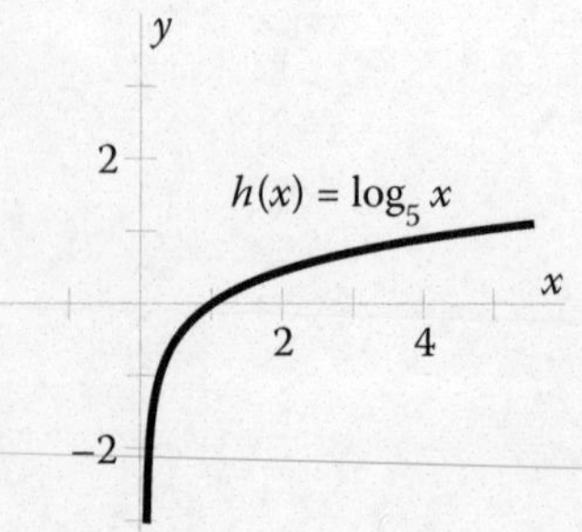

59.

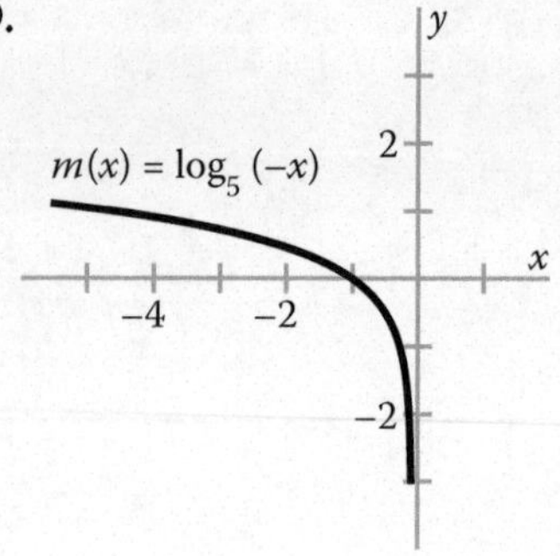

61. e

63. 4/7

65. 10

67. 28

69. −0.98

71. 0.19

73. 3.8

75. 46.81 dB

77. 59.03 dB

79. 4.88

81. (7, ∞)

83. (−3/4, ∞)

85. (1/3, ∞)

87. 31

89. $y = \log 1$

91. b

93. y

95. neither

SECTION 4.3 ANSWERS

Warm Up/Just the Facts

1. Power Property
2. Product Property
3. Quotient Property
4. sum; x, y
5. difference; x; y
6. $\frac{1}{2}\log 20$
7. expand, combine
8. base
9. Change of Base Formula
10. equal

Essential Skills/Extensions

1. 5/6

3. 5/4

5. 3/8

7. −2

9. −2

11. 5/3

13. −3/4

15. 3

17. −4

19. 2

21. $5\log_4 x + \frac{1}{9}\log_4 y$

23. $\log_6 a + \frac{1}{2}\log_6 c - 2\log_6 b$

25. $2\ln x + \ln y + \frac{1}{3}\ln z$

27. $\ln p + 5\ln(q + 4) - \ln r$

29. $2\log r - \log w - \frac{1}{2}\log(x+1)$

31. $\frac{1}{5}\ln(6+x) - \frac{4}{5}\ln y$

33. $\frac{1}{3}\log_2(2+t) - \frac{4}{3} - \frac{1}{3}\log_2 s$

35. $\log_6 \frac{x^4}{\sqrt[5]{y}}$

37. $\log_7 \frac{8m^{12}}{\sqrt[6]{2x+1}}$

39. $\log_5 \dfrac{x^3z^4}{y^2}$

41. $\log_b \dfrac{8x^3y^5}{z^6}$

43. $\log_8 \dfrac{16r^4\sqrt{3w-2}}{(x+8)^2}$

45. $\log_2 \dfrac{\sqrt[4]{(6-a)^3}}{(b-4)^{12}c^3}$

47. $\log_5 \sqrt[3]{\dfrac{(5-m)^2}{16n^4k^8}}$

49. –2.32

51. 0.47

53. 1.18

55. $\log_2(1-x)+\log_2(3+x)-\log_2(2-x)$

57. $\frac{3}{2}\log_a 5+3\log_a x+\frac{1}{2}\log_a(1-4x)-\frac{3}{2}\log_a 7-3\log_a(x+1)$

59. $\log_4 \dfrac{10+x}{4-x}+\log_5(3-2x)$

61. They both are correct.

63. 1.06

65. –7.13

67. –0.3

SECTION 4.4 ANSWERS

Warm Up/Just the Facts

1. $5^4 = 625$
2. $2^6 = 64$
3. $3^5 = 243$
4. One-to-One; $b > 0$, $b \neq 1$
5. Logarithms
6. Natural
7. isolate; exponential
8. parsecs
9. extraneous; logarithmic
10. Quadratic

Essential Skills/Extensions

1. 26/21

3. 11/12

5. 10/9

7. 31/25

9. 13/4

11. 16/17

13. 13/9

15. 1/9

17. 2.26

19. 0.77

21. –1.46

23. –1.37

25. –38.55

27. –31.14

29. 3.26

31. 1.65

33. 0.44

35. 0.37

37. 2.61

39. 0.05

41. ln 3 or 1.1

43. 0.04

45. –0.1, 0.39

47. 0.05, 0.22

49. 4.95

51. 50.6

53. 149.41

55. 409.43

57. 23.25

59. 2046.5

61. 19

63. 210

65. 8.87 parsecs

67. 29.92 parsecs

69. 8

71. 2

73. 4

75. 0

77. 8.1 yr (approximately 8 years 1 month)

79. 29.4 yr (approximately 29 years 5 months)

81. 0.684

83. −10

85. 2.65 parsecs, 8.76 light years

SECTION 4.5 ANSWERS

Warm Up/Just the Facts

1. −2.457

2. −2.824

3. 0.057

4. logarithmic; three

5. decay; greater; decrease

6. 1

7. medication

8. 1; 30 min

9. initial; surroundings; *t*

10. more

Essential Skills/Extensions

1. -5.011×10^{-6} mol/L

3. 2021

5. 2 h

7. 56.9 min

9. 23.2 yr

11. 18

13. 21.8 h

15. $500

17. 12.05 V

19. 10.352 yr

21. 7:26 p.m.

CHAPTER 4 REVIEW ANSWERS

1.

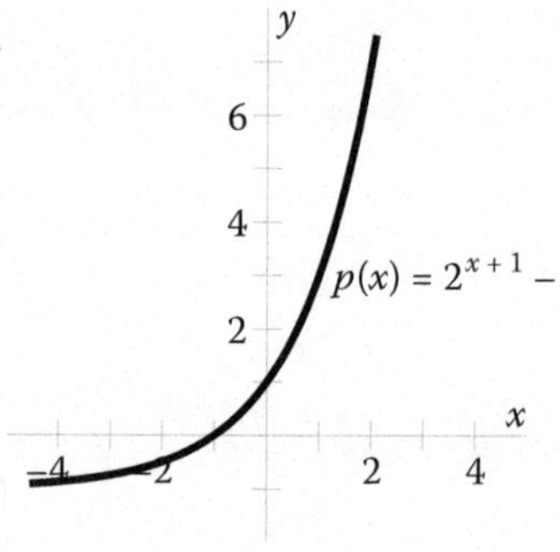

2.

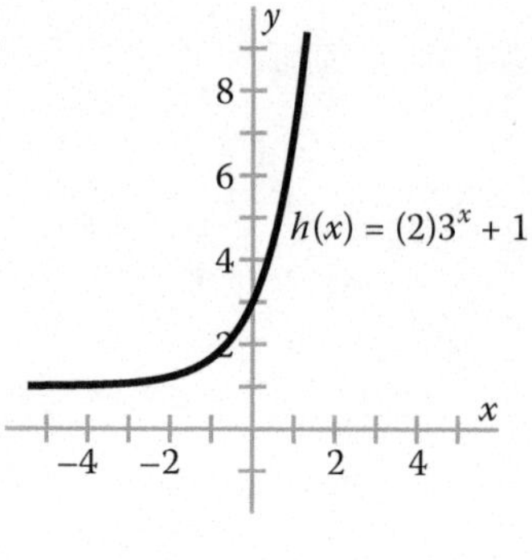

3.

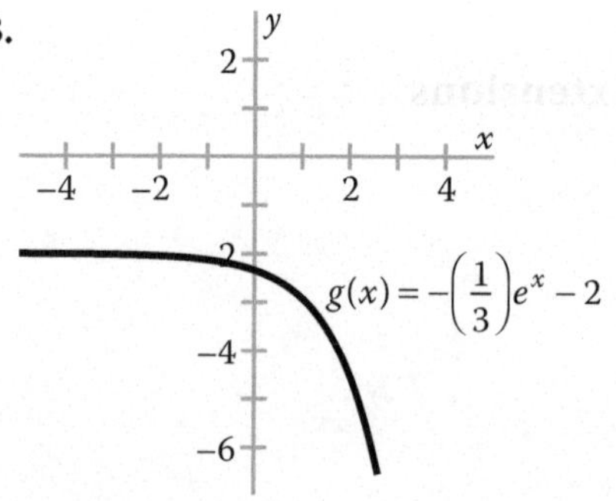

4. $10,509.13

5. 64,000; 107,460,590

6. 10/3

7. −1

8.

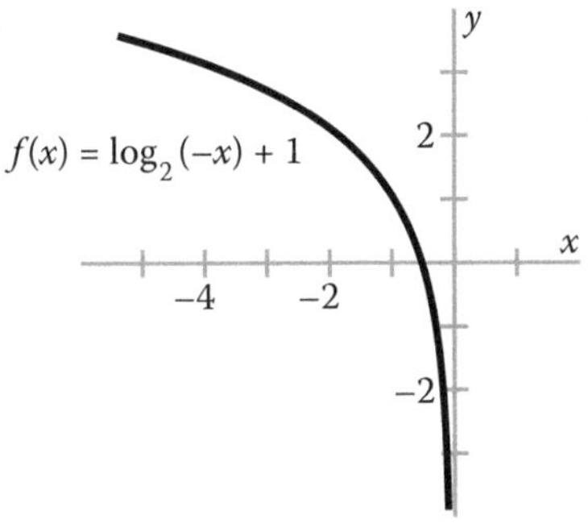

9.

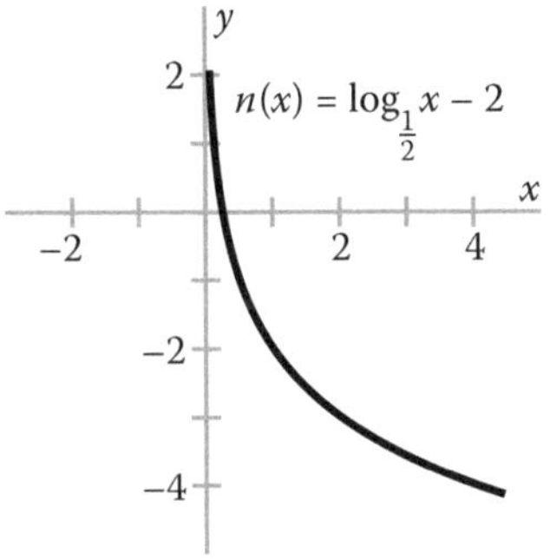

10. $(-\infty, 8)$

11. 2

12. −2

13. $3 + 3\log_8 y + 3\log_8(x-5) - \frac{3}{4}\log_8 x$

14. $\log_b \frac{(x+2)^4 \sqrt[3]{y^2}}{3}$

15. 0.62

16. 1/9

17. 0.35

18. 0

19. 6

20. 7.7 yr

21. 0.064 W

22. 15.3 days

23. 17.3 yr

24. 2.5 h

25. 12.9 yr

SECTION 5.1 ANSWERS

Warm Up/Just the Facts

1. $C \approx 94.2$ cm, $A \approx 706.9\text{ cm}^2$

Substitute the radius into the appropriate formula and simplify.

$C = 2\pi r$	*Circumference Formula*
$= 2\pi(15)$	*Substitute.*
≈ 94.2 cm	*Simplify.*
$A = \pi r^2$	*Area Formula*
$= \pi(15)^2$	*Substitute.*
$\approx 706.9\text{ cm}^2$	*Simplify.*

2. II; I; positive x-axis; IV; III

3. $158\text{ in.} \cdot \dfrac{1\text{ ft}}{12\text{ in.}} = \dfrac{158}{12}\text{ ft} = 13\dfrac{1}{6}\text{ ft}$

4. x-axis; vertex; clockwise; positive

5. 180°; right; acute

6. coterminal

7. III; coterminal; $25\pi/18$ rad

8. linear speed

9. 45° or $\pi/4$ rad; 90° or $\pi/2$ rad

10. $\dfrac{180^\circ}{\pi\text{ rad}}$; –600°; II

Essential Skills/Extensions

1.

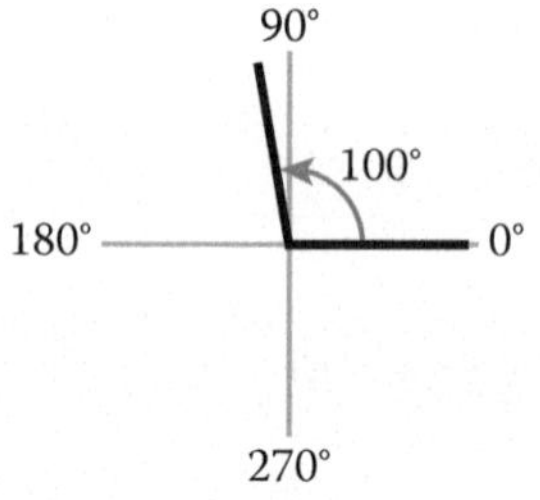

3.

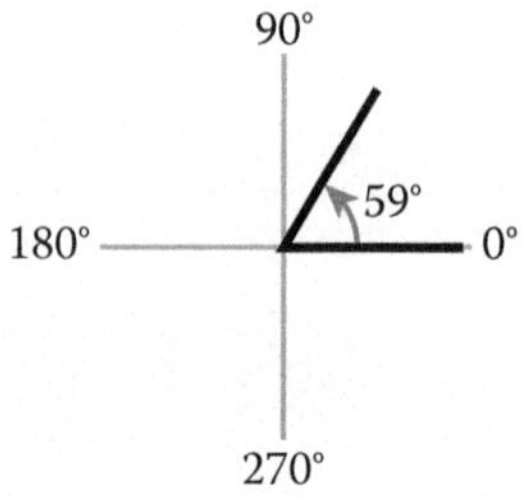

5.

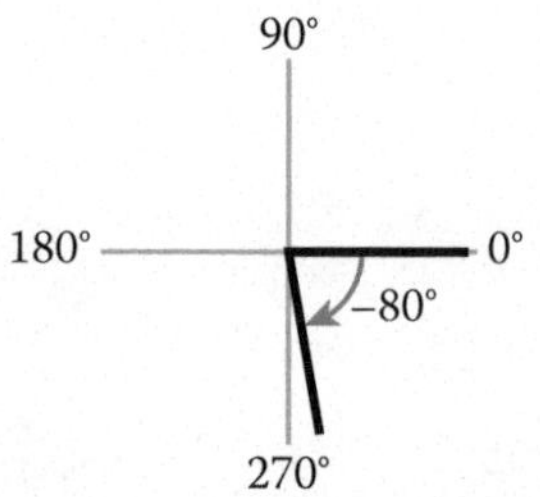

7.

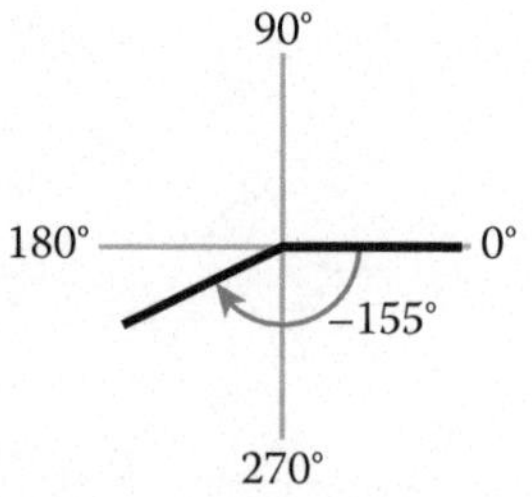

9.

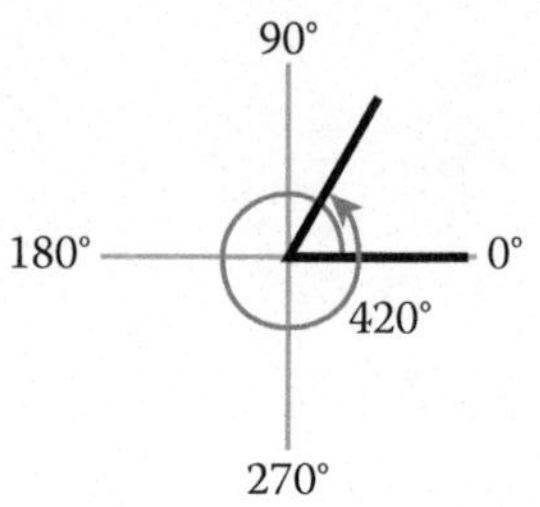

11.

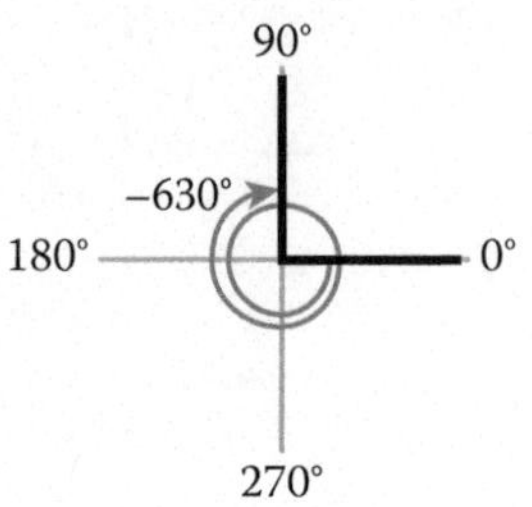

13. 195°

15. 455°, 175°, –181°

17. quadrant I

19. quadrant III

21. quadrant IV

23. quadrant II

25. B, C, E

27. A, D, E

29. B, D, E

31. 600°, –120°

33. 300°, –60°

35. 518°, –202°

37. 190°, –530°

39. 78°, –282°

41. 47°, –313°

43. 83°

45. no complement

47. 105°

49. 36°

51. $\pi/4$ rad

53. $-67\pi/60$ rad

55. $-71\pi/36$ rad

57. 30°

59. –660°

61. 269°

63. 1.75 m

65. 266.9 ft

67. 14.05 in.

69. 9 units

71. 16.6 in.

73. 123.5 m

75. 3 cm/s

77. 3.8 mm/s

79. 7 in./s

81. $\pi/5$ rad/min

83. 15.2 rad/s

85. 155,167.3 rad/h

87. 51 mm^2

89. 887 mm^2

91. 325.2 in^2

93. 11 cm

95. 9.02 in.

97. 29.5 m

99. 4°

101. 6426 mi

103. 90.9 ft/s

105. 12.1 rev/s

107. 1.2 rad. 1.2 rad is approximately 68.754935°, whereas 68° 45′ 13″ is approximately 68.753611°.

109. 3.8 m/s

111. 19.635 ft

113. 10.4 in.

115. 6.6 in./s

SECTION 5.2 ANSWERS

Warm Up/Just the Facts

1. obtuse; acute; right; straight; acute

150° 15′ 22″ is greater than 90° and less than 180°; therefore, the angle is obtuse. 32° is less than 90° and greater than 0°; therefore, the angle is acute. 90° is a right angle. 180° is a straight angle. 47.325° is less than 90° and greater than 0°; therefore, the angle is acute.

2. $10\sqrt{2}$ ft

Since an isosceles right triangle is a special right triangle (45°-45°-90°), the side lengths are a, a, and $a\sqrt{2}$, where a is the length of the leg. Therefore, the length of the hypotenuse is $10\sqrt{2}$ ft.

3. $12\sqrt{3}$ cm

The height divides the equilateral triangle into two 30°-60°-90° triangles, each with side lengths related by a, $a\sqrt{3}$, and $2a$, where $2a$ is the hypotenuse and $a\sqrt{3}$ is the leg that is also the equilateral triangle's height. Since the hypotenuse (i.e., a side of the equilateral triangle) is 24, $2a = 24$ and so $a = 12$. Thus, the height of the equilateral triangle is $12\sqrt{3}$ cm.

4. largest; hypotenuse; right or 90°; legs

5. a, $a\sqrt{3}$, $2a$; a, a, $a\sqrt{2}$

6. tangent, cotangent

7. cosine; 0

8. secant; tangent; sine

9. rationalizing

10. elevation; depression; down

Essential Skills/Extensions

1. $\sin\theta = \frac{\sqrt{7}}{4}$, $\cos\theta = \frac{3}{4}$, $\tan\theta = \frac{\sqrt{7}}{3}$,

$\csc\theta = \frac{4\sqrt{7}}{7}$, $\sec\theta = \frac{4}{3}$, $\cot\theta = \frac{3\sqrt{7}}{7}$

3. $\sin\theta = \frac{7}{9}$, $\cos\theta = \frac{4\sqrt{2}}{9}$, $\tan\theta = \frac{7\sqrt{2}}{8}$,

$\csc\theta = \frac{9}{7}$, $\sec\theta = \frac{9\sqrt{2}}{8}$, $\cot\theta = \frac{4\sqrt{2}}{7}$

5. $\sin\theta = \frac{5\sqrt{89}}{89}$, $\cos\theta = \frac{8\sqrt{89}}{89}$, $\tan\theta = \frac{5}{8}$,

$\csc\theta = \frac{\sqrt{89}}{5}$, $\sec\theta = \frac{\sqrt{89}}{8}$, $\cot\theta = \frac{8}{5}$

7. $\sin\theta = \frac{\sqrt{11}}{6}$, $\cos\theta = \frac{5}{6}$, $\tan\theta = \frac{\sqrt{11}}{5}$

9. $\sin\theta = 4/5$, $\cos\theta = 3/5$, $\tan\theta = 4/3$

11. $\sin\theta = \frac{4\sqrt{3}}{7}$, $\cos\theta = \frac{1}{7}$, $\tan\theta = 4\sqrt{3}$

13. $\csc\theta = 5/3$, $\sec\theta = 5/4$, $\cot\theta = 4/3$

15. $\csc\theta = \frac{13}{9}$, $\sec\theta = \frac{13\sqrt{22}}{44}$, $\cot\theta = \frac{2\sqrt{22}}{9}$

17. $\csc\theta = \frac{\sqrt{137}}{4}$, $\sec\theta = \frac{\sqrt{137}}{11}$, $\cot\theta = \frac{11}{4}$

19. $\sin 45° = \frac{\sqrt{2}}{2}$, $\cos 45° = \frac{\sqrt{2}}{2}$, $\tan 45° = 1$,

$\csc 45° = \sqrt{2}$, $\sec 45° = \sqrt{2}$, $\cot 45° = 1$

21. $\sin\theta = \frac{\sqrt{119}}{12}$, $\tan\theta = \frac{\sqrt{119}}{5}$, $\csc\theta = \frac{12\sqrt{119}}{119}$,

$\sec\theta = \frac{12}{5}$, $\cot\theta = \frac{5\sqrt{119}}{119}$

23. $\sin\theta = \dfrac{2\sqrt{85}}{85}$, $\tan\theta = \dfrac{2}{9}$, $\csc\theta = \dfrac{\sqrt{85}}{2}$,

$\sec\theta = \dfrac{\sqrt{85}}{9}$, $\cot\theta = \dfrac{9}{2}$

25. $\sin\theta = \dfrac{\sqrt{102}}{51}$, $\cos\theta = \dfrac{7\sqrt{51}}{51}$, $\tan\theta = \dfrac{\sqrt{2}}{7}$,

$\csc\theta = \dfrac{\sqrt{102}}{2}$, $\cot\theta = \dfrac{7\sqrt{2}}{2}$

27. $\pi/6$ rad, 30°

29. $\pi/4$ rad, 45°

31. $\pi/3$ rad, 60°

33. $\pi/4$ rad, 45°

35. 0.239

37. −1.019

39. 1.477

41. 8.55

43. 4.66

45. 40.96

47. 8.23

49. 85.82

51. 18.7

53. 1155 ft

55. 82.2 yd

57. 344 ft

59. 57.9555 ft

61. 1.01 mi

63. 56.3 ft

65. 44.55 ft

67. 11,322.5 ft

69. 91.7 ft

71. 210 ft

73. 50/17 in^2

75. 1008/53 units^2

77. 1853.6 ft

79. 45°

81. 8.1 units

83. no

85. 148,513,292 km

87. 442.2 m; 542.7 m; 1199.7 m

89. 277.9 km

SECTION 5.3 ANSWERS

Warm Up/Just the Facts

1. $3\pi/4$ rad; $4\pi/3$ rad; $\pi/2$ rad; $11\pi/6$ rad

Multiply each degree measure by $\dfrac{\pi \text{ rad}}{180°}$ to convert to radians.

$$135°\cdot\frac{\pi \text{ rad}}{180°} = \frac{135\pi}{180}\text{ rad} = \frac{3\pi}{4}\text{ rad;}$$

$$240°\cdot\frac{\pi \text{ rad}}{180°} = \frac{240\pi}{180}\text{ rad} = \frac{4\pi}{3}\text{ rad;}$$

$$90°\cdot\frac{\pi \text{ rad}}{180°} = \frac{90\pi}{180}\text{ rad} = \frac{\pi}{2}\text{ rad;}$$

$$330°\cdot\frac{\pi \text{ rad}}{180°} = \frac{330\pi}{180}\text{ rad} = \frac{11\pi}{6}\text{ rad}$$

2. 120°; 225°; 36°; 270°

Multiply each radian measure by $\dfrac{180°}{\pi \text{ rad}}$ to convert to degrees.

$$\frac{2\pi}{3}\text{ rad}\cdot\frac{180°}{\pi \text{ rad}} = \frac{360°}{3} = 120°;$$

$$\frac{5\pi}{4}\text{ rad}\cdot\frac{180°}{\pi \text{ rad}} = \frac{900°}{4} = 225°;$$

$$\frac{\pi}{5}\text{ rad}\cdot\frac{180°}{\pi \text{ rad}} = \frac{180°}{5} = 36°;$$

$$\frac{3\pi}{2}\text{ rad}\cdot\frac{180°}{\pi \text{ rad}} = \frac{540°}{2} = 270°$$

3. Start at the positive x-axis and rotate in a counterclockwise direction.

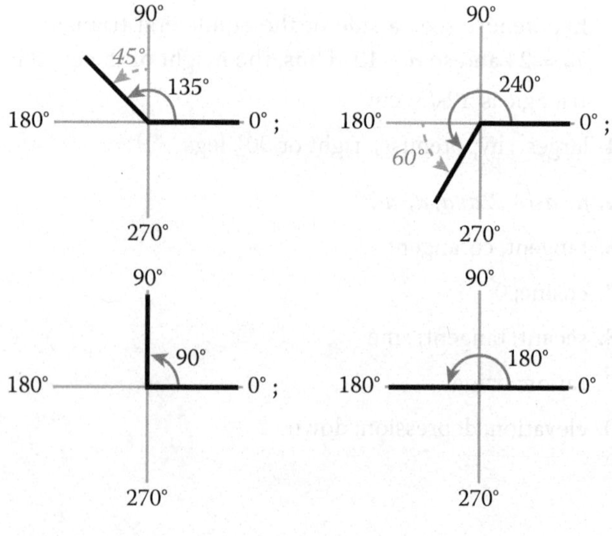

4. Pythagorean

5. cosine; tangent; cosecant; positive; tangent, cotangent

6. II

7. θ'; acute; terminal; x-axis

8. $\theta' = 35°$

9. positive; sine, cosecant; "students"

10. fourth

Essential Skills/Extensions

1. $\sin\theta = \frac{7\sqrt{65}}{65}$, $\cos\theta = \frac{4\sqrt{65}}{65}$, $\tan\theta = \frac{7}{4}$,

$\csc\theta = \frac{\sqrt{65}}{7}$, $\sec\theta = \frac{\sqrt{65}}{4}$, $\cot\theta = \frac{4}{7}$

3. $\sin\theta = \frac{2\sqrt{5}}{5}$, $\cos\theta = \frac{\sqrt{5}}{5}$, $\tan\theta = 2$,

$\csc\theta = \frac{\sqrt{5}}{2}$, $\sec\theta = \sqrt{5}$, $\cot\theta = \frac{1}{2}$

5. $\sin\theta = \frac{5\sqrt{34}}{34}$, $\cos\theta = -\frac{3\sqrt{34}}{34}$, $\tan\theta = -\frac{5}{3}$,

$\csc\theta = \frac{\sqrt{34}}{5}$, $\sec\theta = -\frac{\sqrt{34}}{3}$, $\cot\theta = -\frac{3}{5}$

7. $\sin\theta = \frac{\sqrt{5}}{5}$, $\cos\theta = -\frac{2\sqrt{5}}{5}$, $\tan\theta = -\frac{1}{2}$,

$\csc\theta = \sqrt{5}$, $\sec\theta = -\frac{\sqrt{5}}{2}$, $\cot\theta = -2$

9. $\sin\theta = -\frac{2\sqrt{5}}{5}$, $\cos\theta = -\frac{\sqrt{5}}{5}$, $\tan\theta = 2$,

$\csc\theta = -\frac{\sqrt{5}}{2}$, $\sec\theta = -\sqrt{5}$, $\cot\theta = \frac{1}{2}$

11. $\sin\theta = -\frac{3\sqrt{10}}{10}$, $\cos\theta = -\frac{\sqrt{10}}{10}$, $\tan\theta = 3$,

$\csc\theta = -\frac{\sqrt{10}}{3}$, $\sec\theta = -\sqrt{10}$, $\cot\theta = \frac{1}{3}$

13. $\sin\theta = -12/13$, $\cos\theta = 5/13$, $\tan\theta = -12/5$, $\csc\theta = -13/12$, $\sec\theta = 13/5$, $\cot\theta = -5/12$

15. $\sin\theta = -\frac{5.1\sqrt{37.57}}{37.57}$, $\cos\theta = \frac{3.4\sqrt{37.57}}{37.57}$, $\tan\theta = -\frac{3}{2}$,

$\csc\theta = -\frac{\sqrt{37.57}}{5.1}$, $\sec\theta = \frac{\sqrt{37.57}}{3.4}$, $\cot\theta = -\frac{2}{3}$

17. 1/2

19. −1

21. $\sqrt{3}$

23. $-\sqrt{3}$

25. −2

27. $-\frac{\sqrt{3}}{2}$

29. $\sec\theta = -13/12$, $\cot\theta = 12/5$

31. $\sin\theta = 7/25$, $\cot\theta = -24/7$

33. $\cos\theta = \frac{2\sqrt{11}}{7}$, $\csc\theta = \frac{7\sqrt{5}}{5}$

35. $\sin\theta = \frac{\sqrt{33}}{9}$, $\tan\theta = -\frac{\sqrt{11}}{4}$

37. $\cos\theta = \frac{\sqrt{3}}{3}$, $\cot\theta = -\frac{\sqrt{2}}{2}$

39. $\cos\theta = \frac{3\sqrt{13}}{13}$, $\cot\theta = -\frac{3}{2}$

41. $\sec\theta = -\frac{\sqrt{74}}{7}$, $\csc\theta = -\frac{\sqrt{74}}{5}$

43. $\cos\theta = \frac{9\sqrt{97}}{97}$, $\sin\theta = -\frac{4\sqrt{97}}{97}$

45. $\tan\theta = \frac{\sqrt{159}}{4}$, $\csc\theta = -\frac{5\sqrt{1113}}{159}$

47. $\sqrt{3}$

49. 1/2

51. 0

53. $\frac{\sqrt{2}}{2}$

55. $\frac{\sqrt{3}}{3}$

57. $\frac{\sqrt{3}}{2}$

59. 2

61. $\sqrt{2}$

63. 0

65. 15/4

67. 5/2

69. $-\frac{1}{\sqrt{1-a^2}}$

71. $x = 2\sqrt{3} + \sqrt{221}$ in., $x' = -2 - \frac{4\sqrt{663}}{221}$ in./rad

73. $x = \sqrt{2} + \sqrt{23}$ in., $x' = -\sqrt{2} - \frac{2\sqrt{23}}{23}$ in./rad

75. $8\sqrt{2}$ ft/s^2

77. $\pi/6$ rad

79. $\pi/3$ rad

81. A reference angle is a positive acute angle formed by the terminal side of θ and the x-axis. If θ is a quadrantal angle, then the positive angle formed by its terminal side and the x-axis is not acute.

SECTION 5.4 ANSWERS

Warm Up/Just the Facts

1. $\sqrt{3}$; $\frac{\sqrt{2}}{2}$; $\frac{\sqrt{3}}{2}$; $\frac{2\sqrt{3}}{3}$; 1

From the chart in 5.3.5, $\tan 60° = \sqrt{3}$, $\cos 45° = \frac{\sqrt{2}}{2}$,

$\sin 60° = \frac{\sqrt{3}}{2}$, $\sec 30° = \frac{2\sqrt{3}}{3}$, and $\cot 45° = 1$.

2. II

3. 5/7

4. $x^2 + y^2 = 1$; (0, 0); 1

5. $\left(\frac{\sqrt{2}}{2}, -\frac{\sqrt{2}}{2}\right)$; $4\pi/3$; π

6. [–1, 1]; –1; 1

7. periodic; period

8. $f(x)$; $-f(x)$; even

9. 1; equal

10. 2π; clockwise, counterclockwise; 2π

Essential Skills/Extensions

1. $\left(-\frac{\sqrt{2}}{2}, \frac{\sqrt{2}}{2}\right)$

3. $\left(-\frac{1}{2}, \frac{\sqrt{3}}{2}\right)$

5. $\left(-\frac{\sqrt{3}}{2}, \frac{1}{2}\right)$

7. (–1, 0)

9. $\left(\frac{\sqrt{3}}{2}, \frac{1}{2}\right)$

11. $\left(-\frac{1}{2}, -\frac{\sqrt{3}}{2}\right)$

13. (0, 1)

15. $\left(\frac{\sqrt{2}}{2}, -\frac{\sqrt{2}}{2}\right)$

17. $\left(-\frac{\sqrt{3}}{2}, -\frac{1}{2}\right)$

19. $\sin\frac{4\pi}{3} = -\frac{\sqrt{3}}{2}$, $\cos\frac{4\pi}{3} = -\frac{1}{2}$, $\tan\frac{4\pi}{3} = \sqrt{3}$,

$\csc\frac{4\pi}{3} = -\frac{2\sqrt{3}}{3}$, $\sec\frac{4\pi}{3} = -2$, $\cot\frac{4\pi}{3} = \frac{\sqrt{3}}{3}$

21. $\sin\frac{5\pi}{6} = \frac{1}{2}$, $\cos\frac{5\pi}{6} = -\frac{\sqrt{3}}{2}$, $\tan\frac{5\pi}{6} = -\frac{\sqrt{3}}{3}$,

$\csc\frac{5\pi}{6} = 2$, $\sec\frac{5\pi}{6} = -\frac{2\sqrt{3}}{3}$, $\cot\frac{5\pi}{6} = -\sqrt{3}$

23. $\sin\frac{11\pi}{4} = \frac{\sqrt{2}}{2}$, $\cos\frac{11\pi}{4} = -\frac{\sqrt{2}}{2}$, $\tan\frac{11\pi}{4} = -1$,

$\csc\frac{11\pi}{4} = \sqrt{2}$, $\sec\frac{11\pi}{4} = -\sqrt{2}$, $\cot\frac{11\pi}{4} = -1$

25. $\sin\pi = 0$, $\cos\pi = -1$, $\tan\pi = 0$, $\csc\pi$ is undefined,

$\sec\pi = -1$, $\cot\pi$ is undefined

27. $\sin\frac{2\pi}{3} = \frac{\sqrt{3}}{2}$, $\cos\frac{2\pi}{3} = -\frac{1}{2}$, $\tan\frac{2\pi}{3} = -\sqrt{3}$,

$\csc\frac{2\pi}{3} = \frac{2\sqrt{3}}{3}$, $\sec\frac{2\pi}{3} = -2$, $\cot\frac{2\pi}{3} = -\frac{\sqrt{3}}{3}$

29. $\sin\frac{\pi}{2} = 1$, $\cos\frac{\pi}{2} = 0$, $\tan\frac{\pi}{2}$ is undefined,

$\csc\frac{\pi}{2} = 1$, $\sec\frac{\pi}{2}$ is undefined, $\cot\frac{\pi}{2} = 0$

31. $\sin\left(-\frac{\pi}{6}\right) = -\frac{1}{2}$, $\cos\left(-\frac{\pi}{6}\right) = \frac{\sqrt{3}}{2}$, $\tan\left(-\frac{\pi}{6}\right) = -\frac{\sqrt{3}}{3}$,

$\csc\left(-\frac{\pi}{6}\right) = -2$, $\sec\left(-\frac{\pi}{6}\right) = \frac{2\sqrt{3}}{3}$, $\cot\left(-\frac{\pi}{6}\right) = -\sqrt{3}$

33. $\sin\left(-\frac{5\pi}{4}\right) = \frac{\sqrt{2}}{2}$, $\cos\left(-\frac{5\pi}{4}\right) = -\frac{\sqrt{2}}{2}$, $\tan\left(-\frac{5\pi}{4}\right) = -1$,

$\csc\left(-\frac{5\pi}{4}\right) = \sqrt{2}$, $\sec\left(-\frac{5\pi}{4}\right) = -\sqrt{2}$, $\cot\left(-\frac{5\pi}{4}\right) = -1$

35. $\sin\left(-\frac{2\pi}{3}\right) = -\frac{\sqrt{3}}{2}$, $\cos\left(-\frac{2\pi}{3}\right) = -\frac{1}{2}$, $\tan\left(-\frac{2\pi}{3}\right) = \sqrt{3}$,

$\csc\left(-\frac{2\pi}{3}\right) = -\frac{2\sqrt{3}}{3}$, $\sec\left(-\frac{2\pi}{3}\right) = -2$, $\cot\left(-\frac{2\pi}{3}\right) = \frac{\sqrt{3}}{3}$

37. $\sin\left(-\frac{\pi}{2}\right) = -1$, $\cos\left(-\frac{\pi}{2}\right) = 0$, $\tan\left(-\frac{\pi}{2}\right)$ is undefined,

$\csc\left(-\frac{\pi}{2}\right) = -1$, $\sec\left(-\frac{\pi}{2}\right)$ is undefined, $\cot\left(-\frac{\pi}{2}\right) = 0$

39. $\sin\left(-\frac{3\pi}{4}\right) = -\frac{\sqrt{2}}{2}$, $\cos\left(-\frac{3\pi}{4}\right) = -\frac{\sqrt{2}}{2}$, $\tan\left(-\frac{3\pi}{4}\right) = 1$,

$\csc\left(-\frac{3\pi}{4}\right) = -\sqrt{2}$, $\sec\left(-\frac{3\pi}{4}\right) = -\sqrt{2}$, $\cot\left(-\frac{3\pi}{4}\right) = 1$

41. $\sin\left(-\frac{5\pi}{6}\right) = -\frac{1}{2}$, $\cos\left(-\frac{5\pi}{6}\right) = -\frac{\sqrt{3}}{2}$, $\tan\left(-\frac{5\pi}{6}\right) = \frac{\sqrt{3}}{3}$,

$\csc\left(-\frac{5\pi}{6}\right) = -2$, $\sec\left(-\frac{5\pi}{6}\right) = -\frac{2\sqrt{3}}{3}$, $\cot\left(-\frac{5\pi}{6}\right) = \sqrt{3}$

43. $\cot\left(-\frac{5\pi}{3}\right) = \frac{\sqrt{3}}{3}$, $\tan\left(-\frac{5\pi}{3}\right) = \sqrt{3}$

45. $\sin\left(-\frac{5\pi}{6}\right) = -\frac{1}{2}$, $\cos\left(-\frac{5\pi}{6}\right) = -\frac{\sqrt{3}}{2}$

47. $\csc -\frac{3\pi}{4} = -\sqrt{2}$, $\sec -\frac{3\pi}{4} = -\sqrt{2}$

49. $\sin(-t) = 4/9$, $\csc(-t) = 9/4$

51. $\tan(-t) = -3/5$, $\cot(-t) = -5/3$

53. $\csc(-t) = 5/4$, $\sin(-t) = 4/5$

55. $\pi/6$

57. −3.3

59. 0

61. 2/3; 0; −2/3

63. −11/3; −22/3; −11

65. 2; 0; −2

67. true

69. true

71. false

SECTION 5.5 ANSWERS

Warm Up/Just the Facts

1. $\dfrac{\pi}{2} + \pi n$

The values of t where $\cos t = 0$ occur at $\pi/2$ and $3\pi/2$. Since cosine has a period of 2π, the values occur at regular intervals of 2π. Therefore, the solutions are $\dfrac{\pi}{2} + 2\pi n$ and $\dfrac{3\pi}{2} + 2\pi n$ or simply $\dfrac{\pi}{2} + \pi n$.

2. $\dfrac{\pi}{2} + 2\pi n$

The value of t where $\sin t = 1$ occurs at $\pi/2$. Since sine has a period of 2π, the values occur at regular intervals of 2π. Therefore, the solutions are $\dfrac{\pi}{2} + 2\pi n$.

3. $\pi + 2\pi n$

The value of t where $\cos t = -1$ occurs at π. Since cosine has a period of 2π, the values occur at regular intervals of 2π. Therefore, the solutions are $\pi + 2\pi n$.

4. 2; amplitude

5. cosine; zeros; maximums

6. [−10, 10]

7. amplitude

8. $\pi/2$; $y = a \cos x$

9. $\dfrac{2\pi}{|b|}$; 8π; 4

10. $c > 0$; $c < 0$

Essential Skills/Extensions

1. 1/2

3. 5

5.

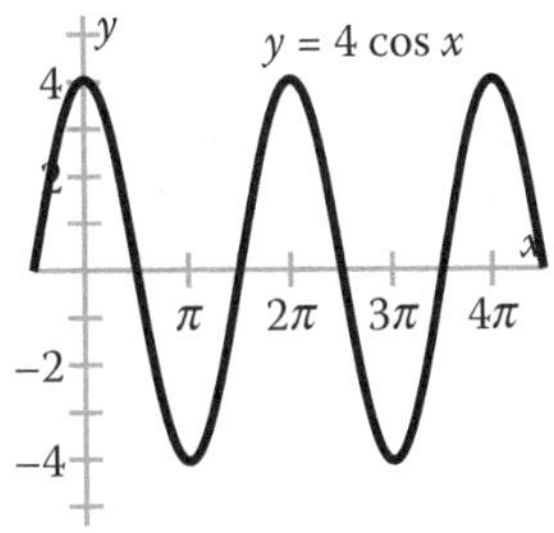

7.

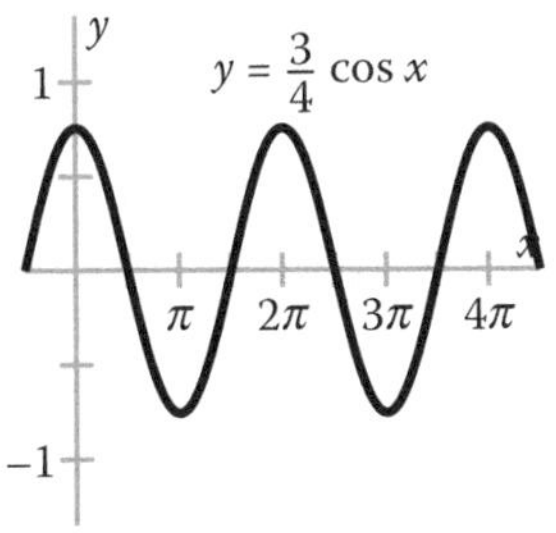

9.

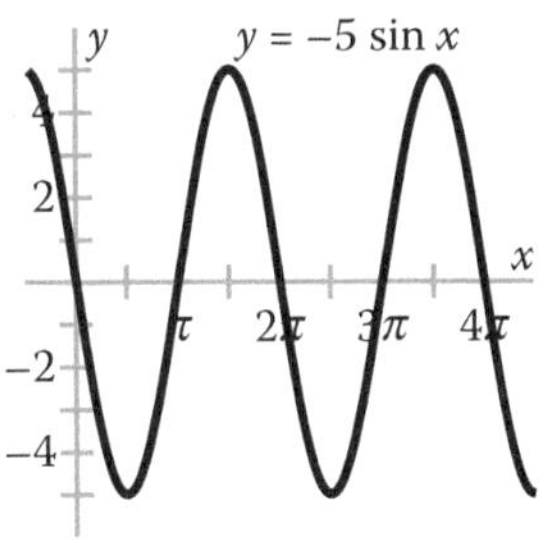

11.

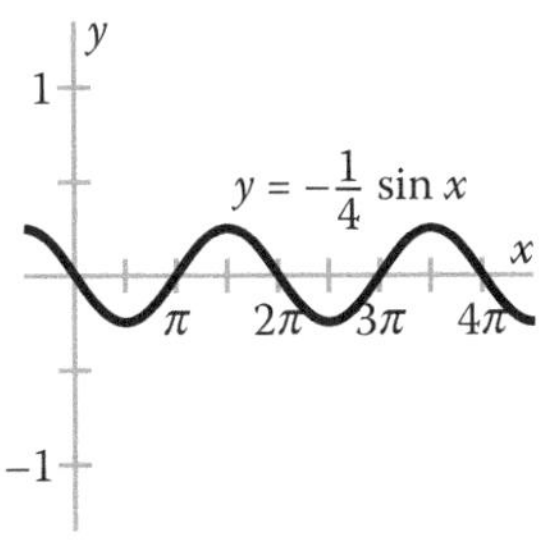

13. 2

15. $8\pi/3$

17.

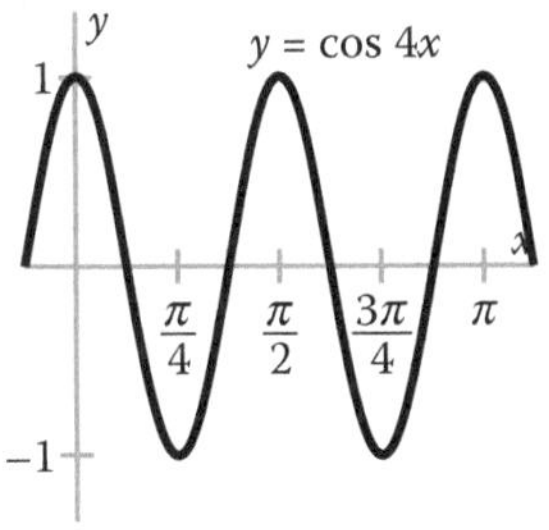

19.

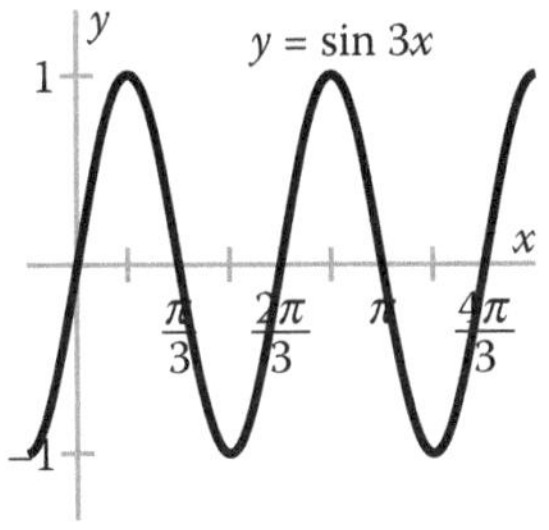

21.

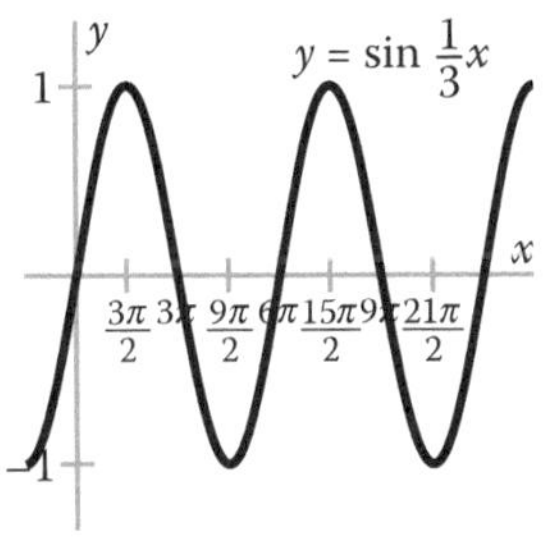

23.

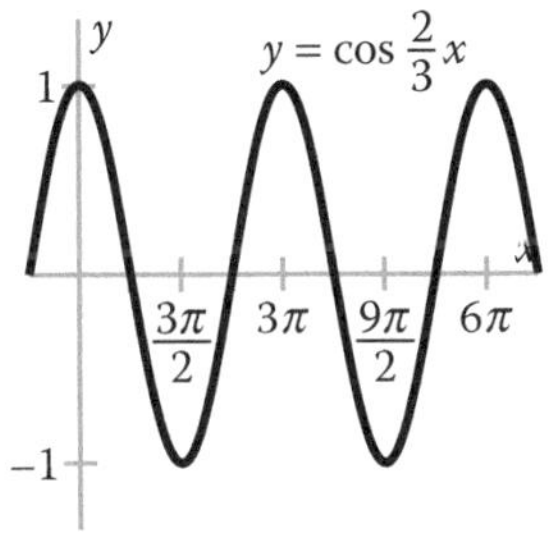

25. period: $2\pi/3$; amplitude: 4

27. period: 8; amplitude: 1/2

29.

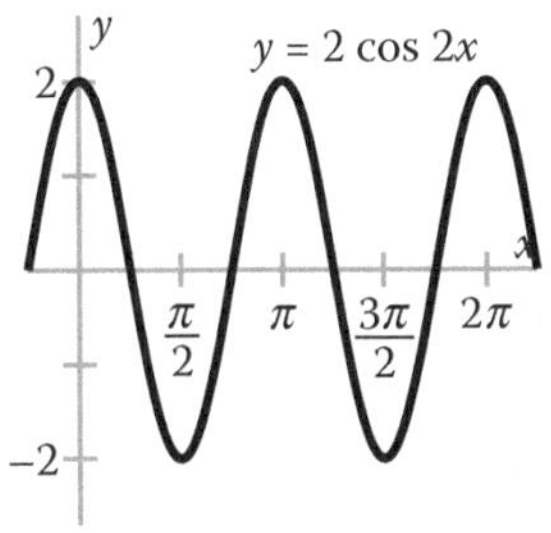

31.

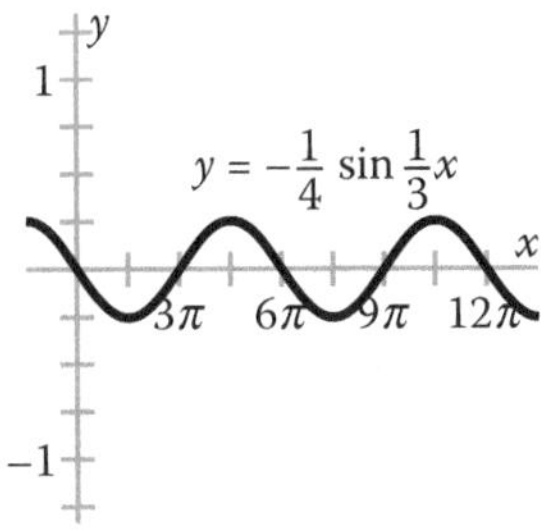

33.

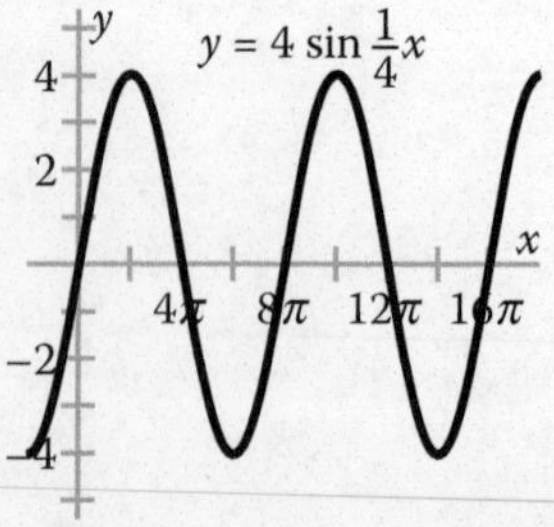

35.

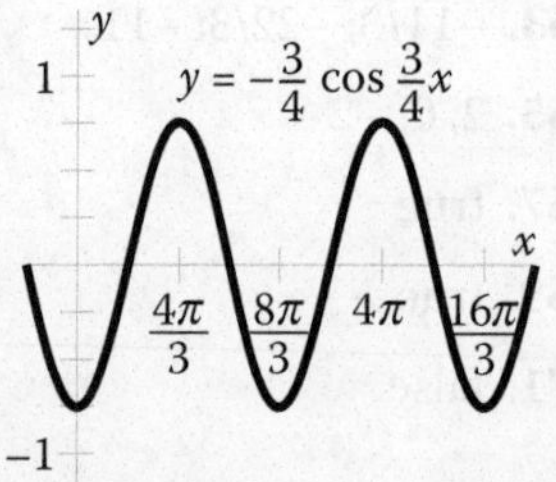

37. maximums: $x = \pi + 4\pi n$; minimums: $x = 3\pi + 4\pi n$; zeros: $x = 2\pi n$ (n is an integer)

39. maximums: $x = \frac{3\pi}{8} + \frac{\pi}{2}n$; minimums: $x = \frac{\pi}{8} + \frac{\pi}{2}n$; zeros: $x = \frac{\pi}{4}n$ (n is an integer)

41. maximums: $x = \pi n$; minimums: $x = \pi/2 + \pi n$; zeros: $x = \frac{\pi}{4} + \frac{\pi}{2}n$ (n is an integer)

43. maximums: $x = 3\pi + 6\pi n$; minimums: $x = 6\pi n$; zeros: $x = \frac{3\pi}{2} + 3\pi n$ (n is an integer)

45. zeros: $x = \pi n$ (n is an integer)

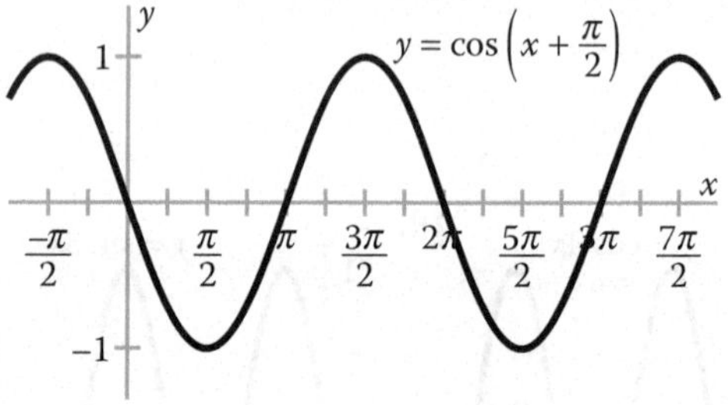

47. zeros: $x = -\frac{2\pi}{3} + \pi n$ (n is an integer)

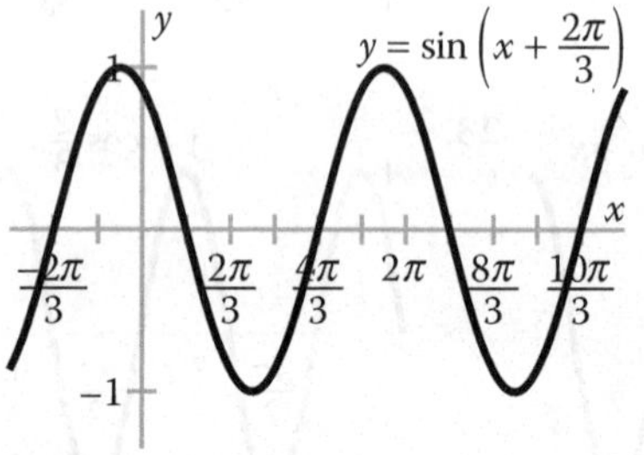

49. amplitude: 2; period: 2π; phase shift: right $\pi/3$ units; vertical shift: 1 unit upward

51. amplitude: 3/4; period: 2π; phase shift: left π units; vertical shift: 5 units downward

53.

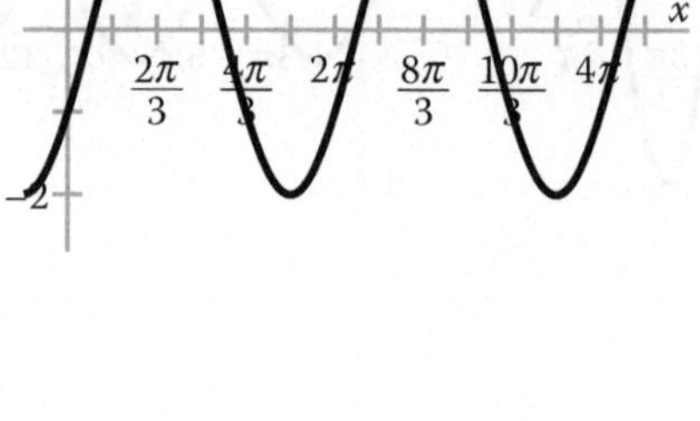

55.

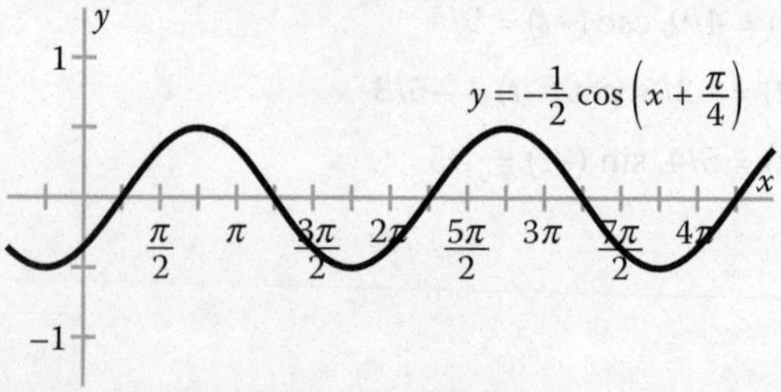

57.

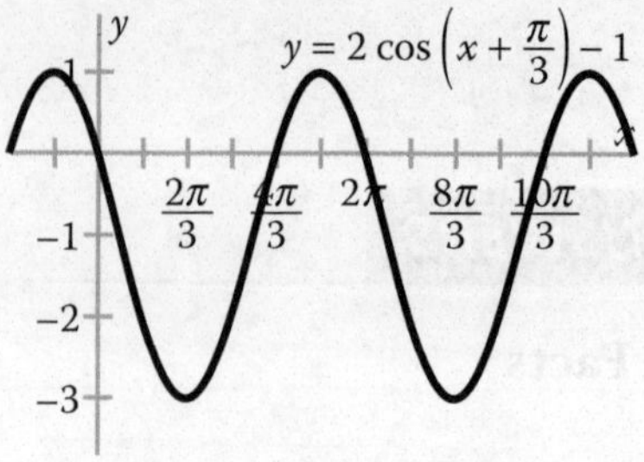

59.

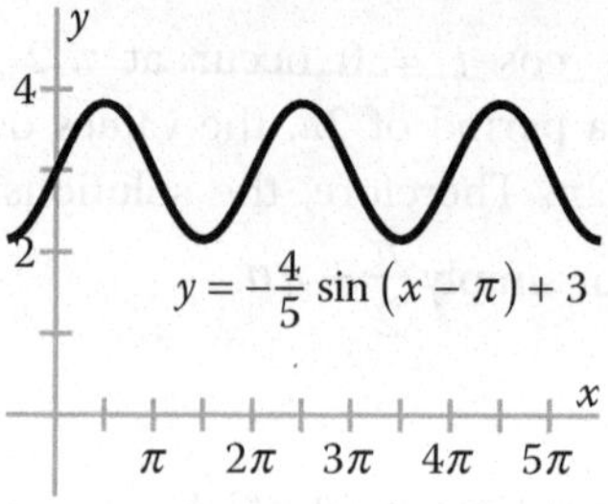

61. amplitude: 2; period: 2π; phase shift: right $\pi/3$ units; vertical shift: 1 units upward

63. amplitude: 1/4; period: 4π; phase shift: left π units; vertical shift: 1 unit downward

65.

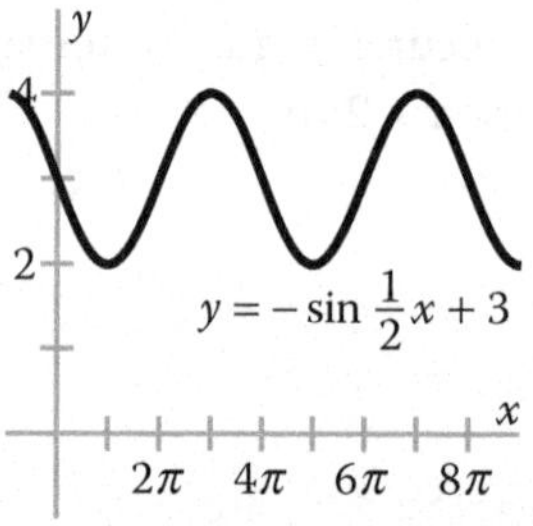

67.

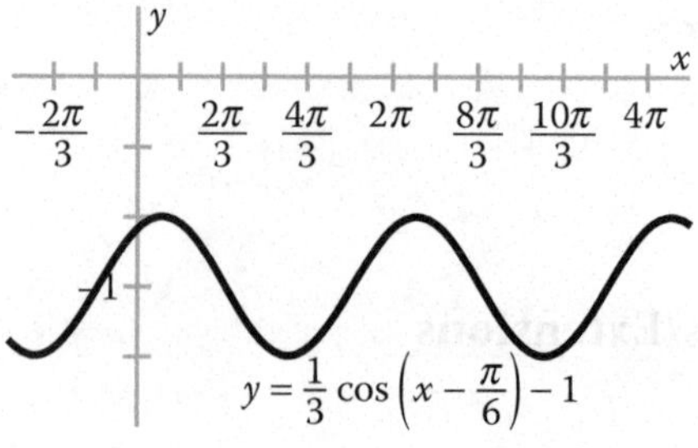

69.

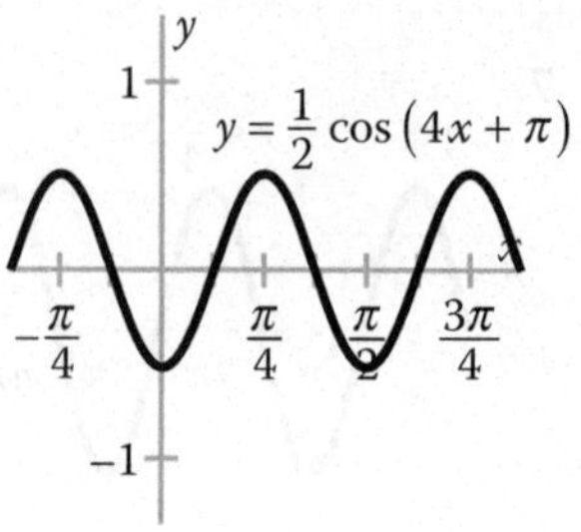

71.

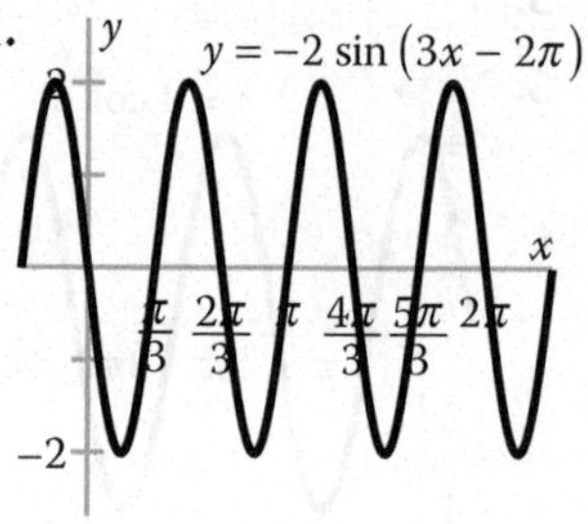

73.

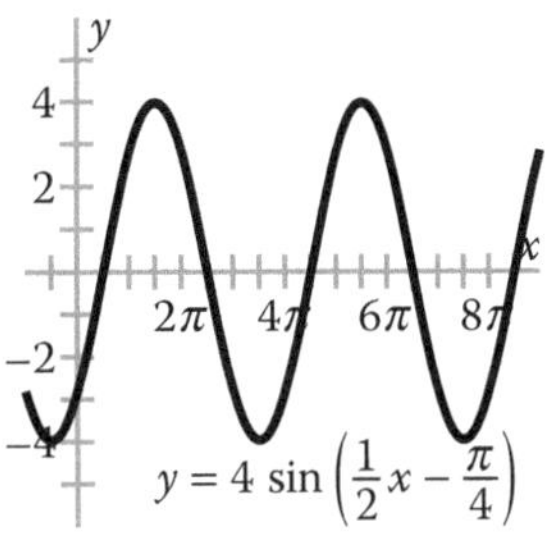

75.

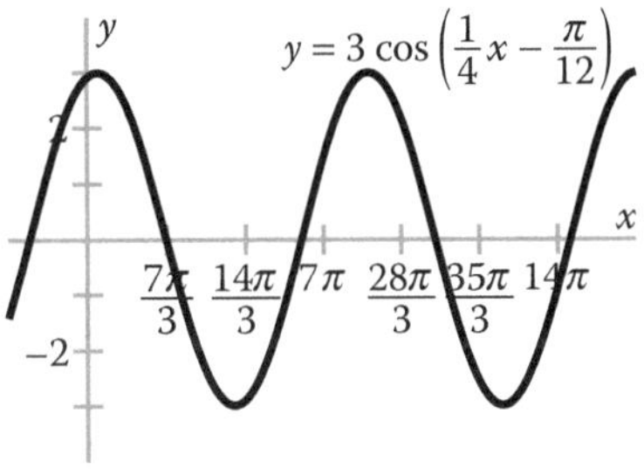

77. maximum: 65 ft

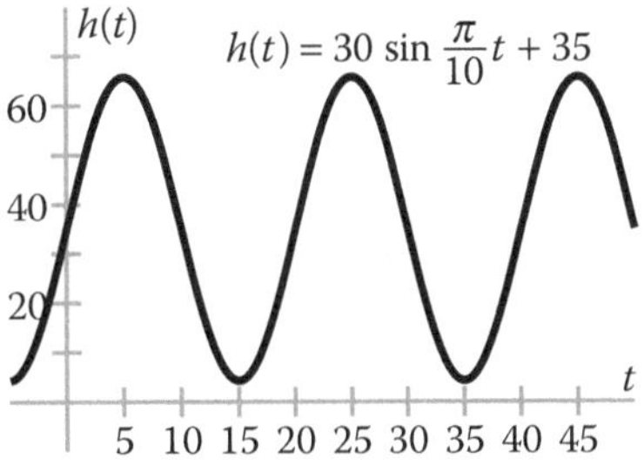

79. maximum: 105 mm HG

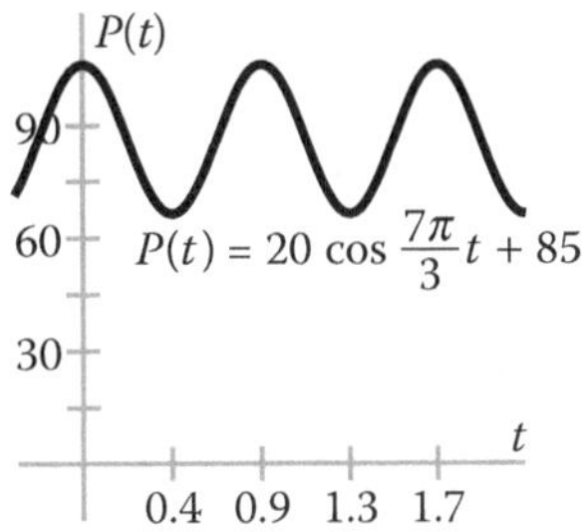

81. maximum: 200 amp

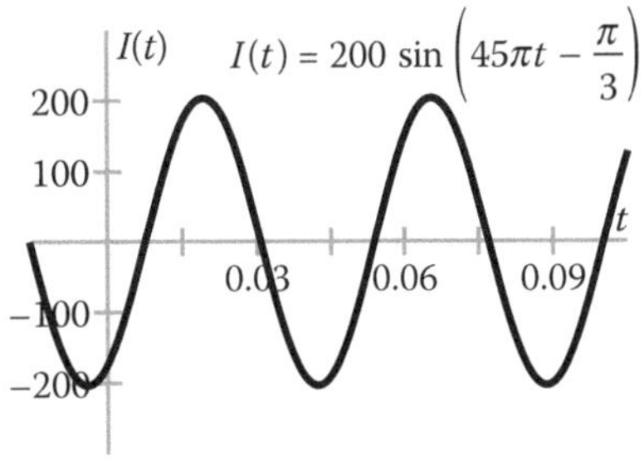

83. 35 white spotted owls

85. True

87. False. The period is 6π, which is three times the period of the corresponding basic function.

89. 98°

91. $y = 4 \sin \frac{\pi}{3}(x - 3) + 7$

93. $y = 15 \sin 6\left(x + \frac{4\pi}{5}\right) - 30$

95.

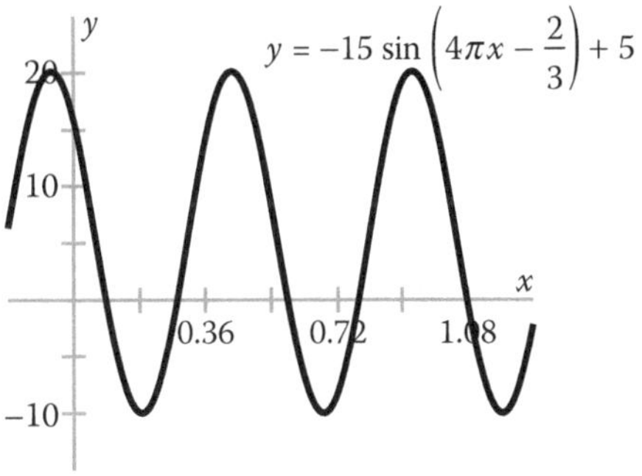

97.

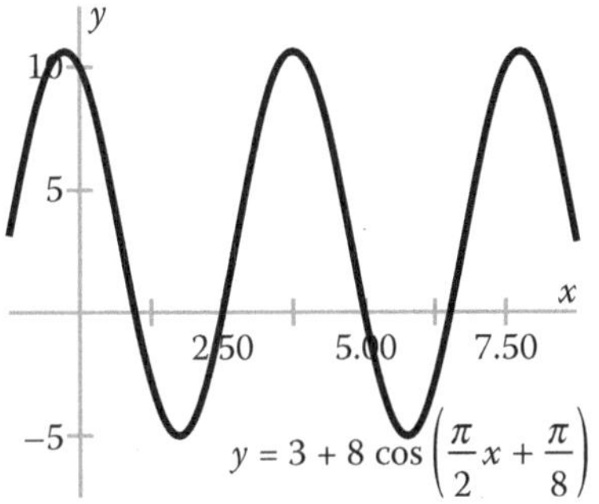

SECTION 5.6 ANSWERS

Warm Up/Just the Facts

1. $(-\infty, \infty)$; $[-1, 1]$

The domain of $y = \cos x$ is $(-\infty, \infty)$ and the range is $[-1, 1]$.

2. πn

The zeros of $y = \sin x$ occur at $0, \pi, 2\pi, 3\pi, \ldots$, which can be written as πn, where n is an integer.

3. $\frac{\square}{2} + \square n$

The tangent function $y = \tan x$ is undefined for $x = \frac{\square}{2} + \square n$ because $\tan \frac{\square}{2} = \frac{y}{x} = \frac{1}{0}$, which is undefined. Since the period of tangent is π, the undefined values occur at intervals of π, so the answer is $\frac{\square}{2} + \square n$.

4. $y = \tan x$; vertical; n

5. cosecant; cosine

6. tangent; π; vertical; a; $\pi/|b|$; a

7. $\pi/2$; left, right; x-axis

8. secant; vertical; secant

9. $c > 0$; 2π

10. b-value

Essential Skills/Extensions

1. vertical asymptotes:
$x = \frac{\square}{2} + \square n$;
zeros: $x = \pi n$

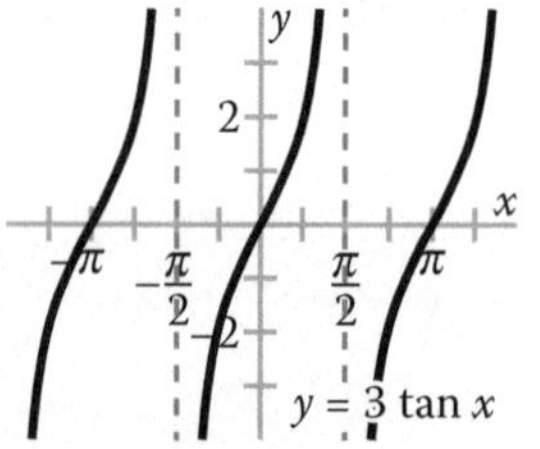

3. vertical asymptotes:
$x = \frac{\square}{2} + \square n$;
zeros: $x = \pi n$

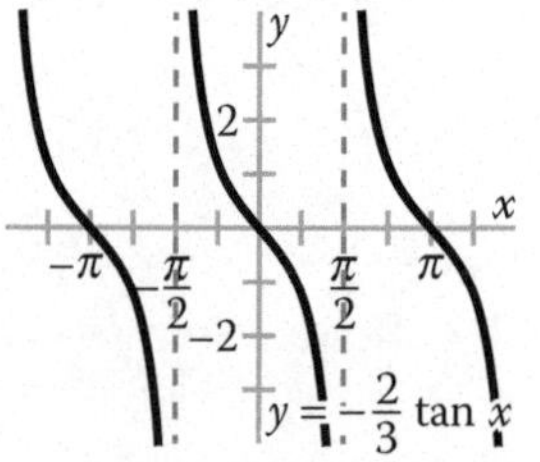

5. vertical asymptotes:
$x = \frac{\square}{6} + \frac{\square}{3} n$;
zeros: $x = \frac{\square}{3} n$

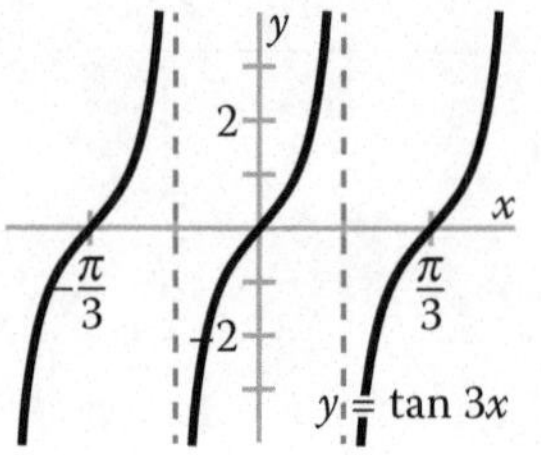

7. vertical asymptotes:
$x = 2\pi + 4\pi n$;
zeros: $x = 4\pi n$

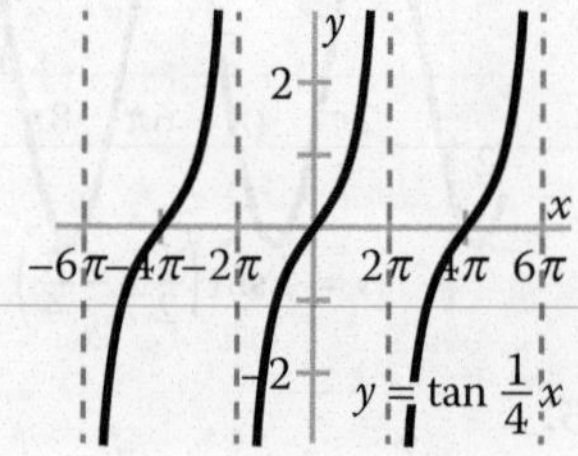

9. vertical asymptotes:
$x = \frac{\square}{4} + \frac{\square}{2} n$;
zeros: $x = \frac{\square}{2} n$

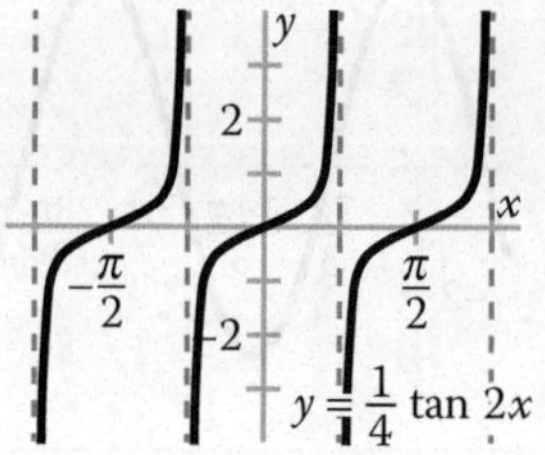

11. vertical asymptotes:
$x = \frac{3\square}{2} + 3\square n$;
zeros: $x = 3\pi n$

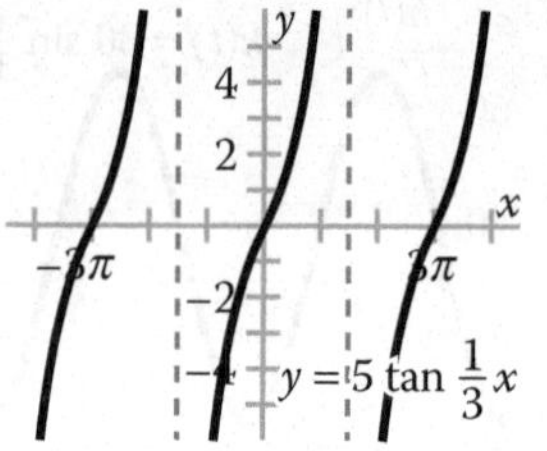

13. vertical asymptotes:
$x = \frac{\square}{8} + \frac{\square}{4} n$;
zeros: $x = \frac{\square}{4} n$

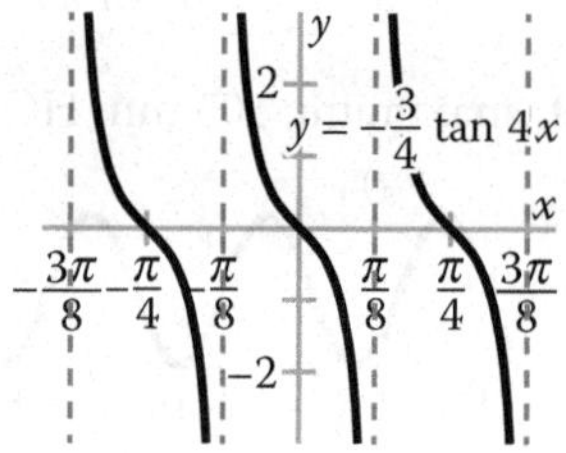

15. vertical asymptotes:
$x = 2\pi n$;
zeros: $x = \pi + 2\pi n$

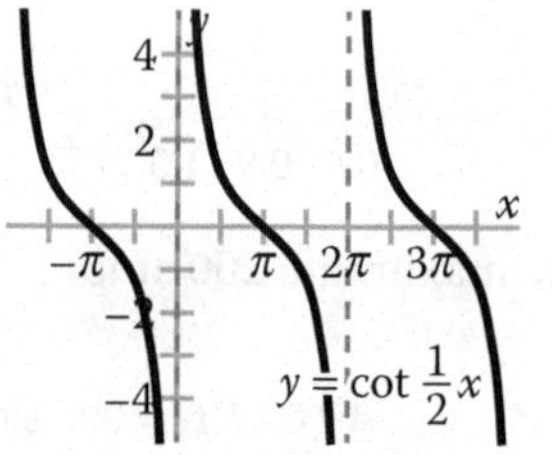

17. vertical asymptotes:
$x = \frac{\square}{2} n$;
zeros: $x = \frac{\square}{4} + \frac{\square}{2} n$

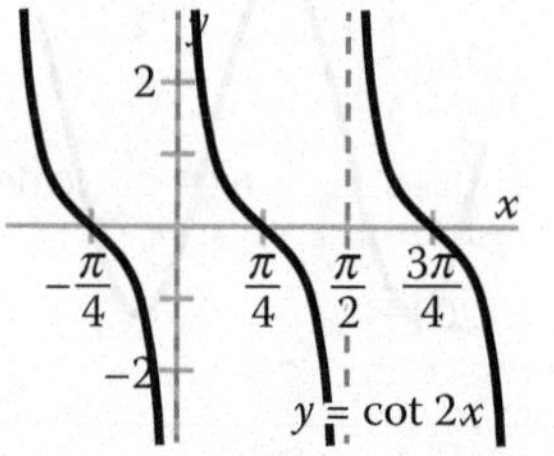

19. vertical asymptotes:
$x = \frac{\square}{4} n$;
zeros: $x = \frac{\square}{8} + \frac{\square}{4} n$

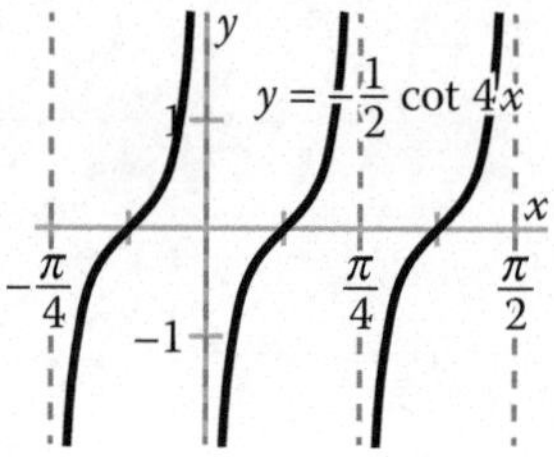

21. vertical asymptotes:
$x = \frac{\square}{3}n$;
zeros: $x = \frac{\square}{6} + \frac{\square}{3}n$

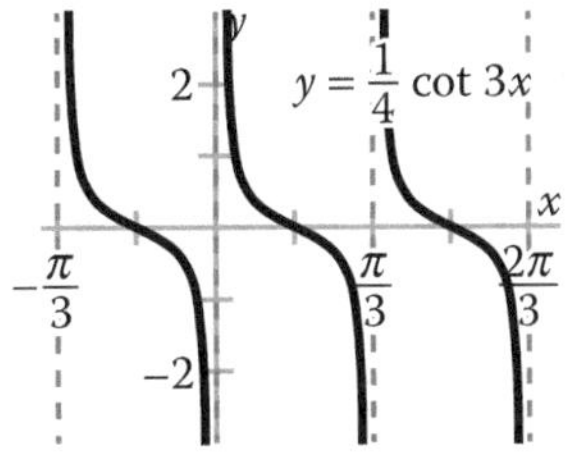

23. vertical asymptotes:
$x = 2\pi n$;
zeros: $x = \pi + 2\pi n$

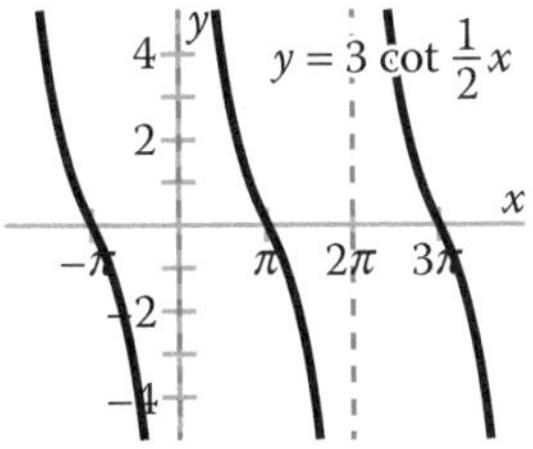

25. vertical asymptotes:
$x = 2\pi n$;
zeros: $x = \pi + 2\pi n$

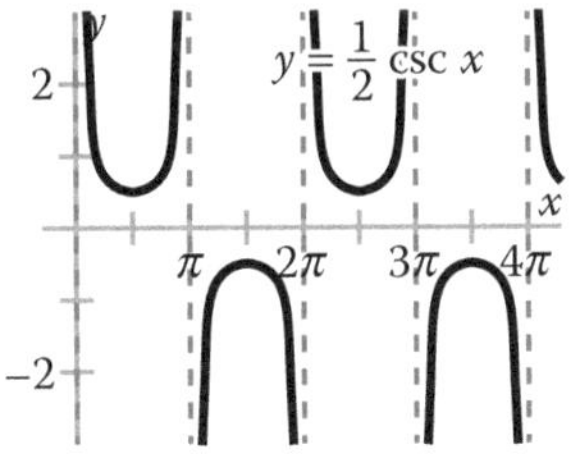

27. vertical asymptotes:
$x = \pi n$;
zeros: none

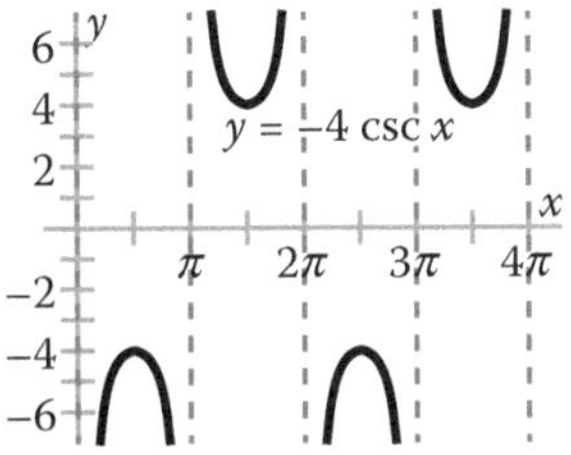

29. vertical asymptotes:
$x = \frac{\square}{2} + \square n$;
zeros: none

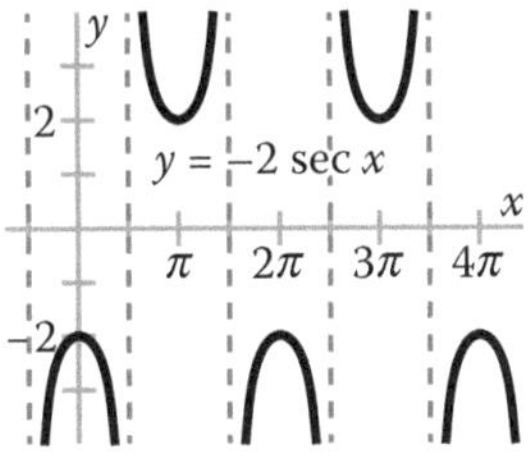

31. vertical asymptotes:
$x = \frac{\square}{2} + \square n$;
zeros: none

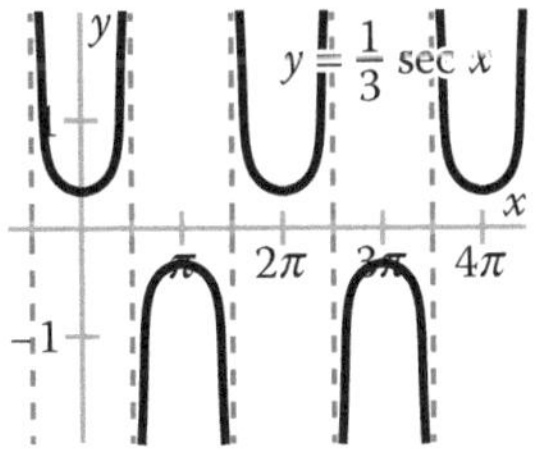

33. 4π

35. 5π

37. vertical asymptotes: $x = \frac{\square}{3}n$; zeros: none

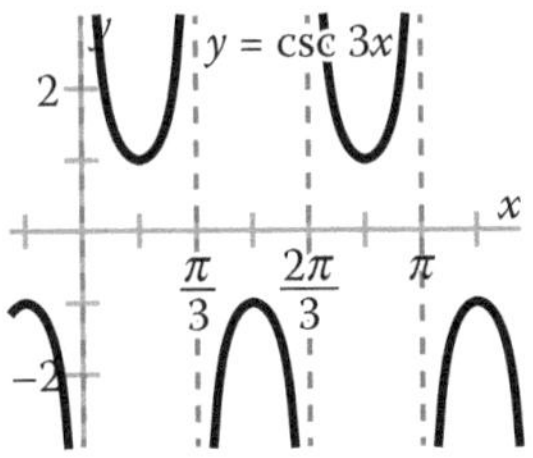

39. vertical asymptotes: $x = \frac{\square}{4} + \frac{\square}{2}n$; zeros: none

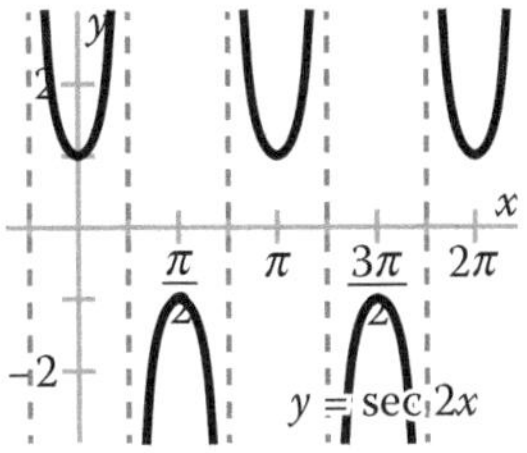

41. vertical asymptotes: $x = \frac{3\square}{2} + 3\square n$; zeros: none

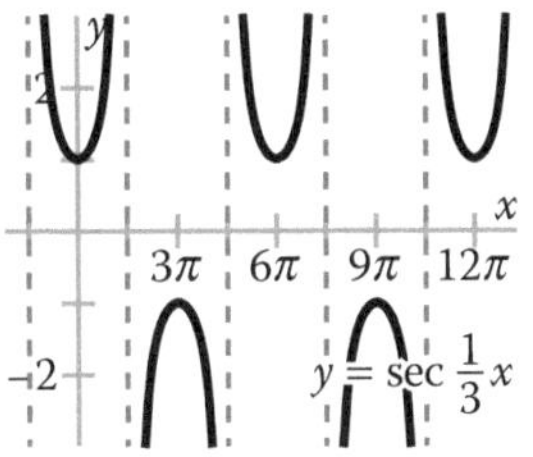

43. vertical asymptotes: $x = \frac{4\square}{3}n$; zeros: none

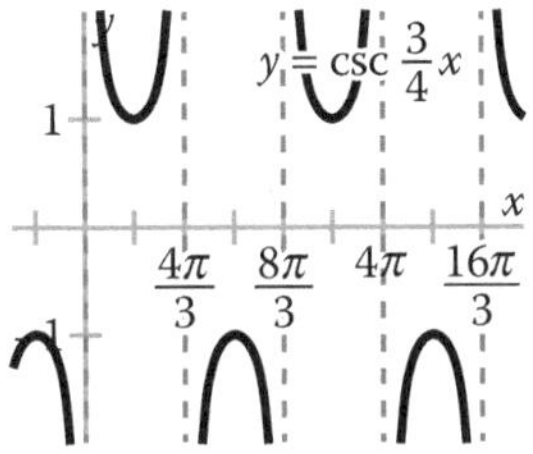

45. vertical asymptotes: $x = \frac{\square}{3}n$; zeros: none

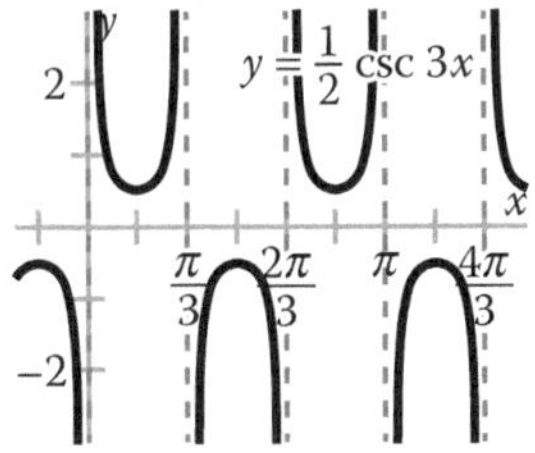

47. vertical asymptotes: $x = \frac{\square}{4} + \frac{\square}{2}n$; zeros: none

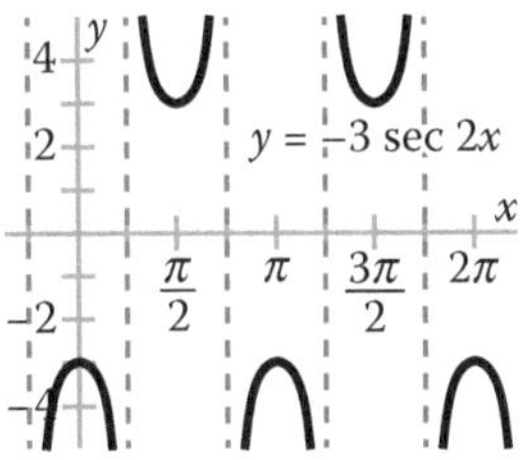

49. vertical asymptotes: $x = 2\pi n$; zeros: none

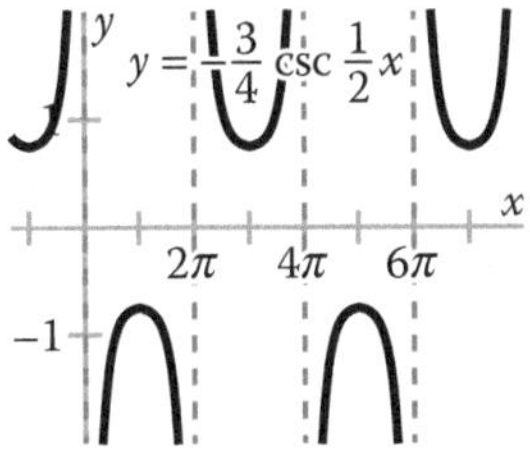

51. vertical asymptotes: $x = \frac{5\square}{6} + \square n$; zeros: $x = \frac{\square}{3} + \square n$

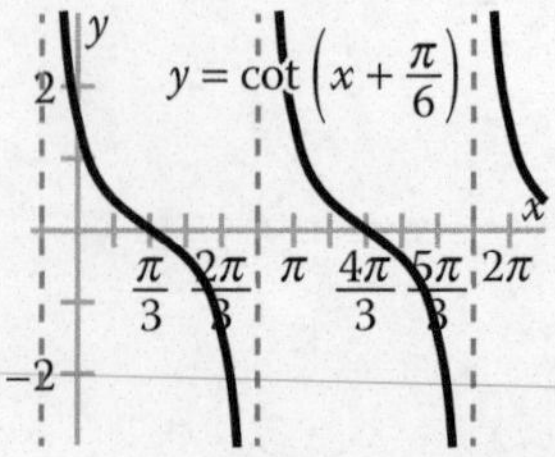

53. vertical asymptotes: $x = \frac{\square}{4} + \square n$; zeros: $x = \frac{3\square}{4} + \square n$

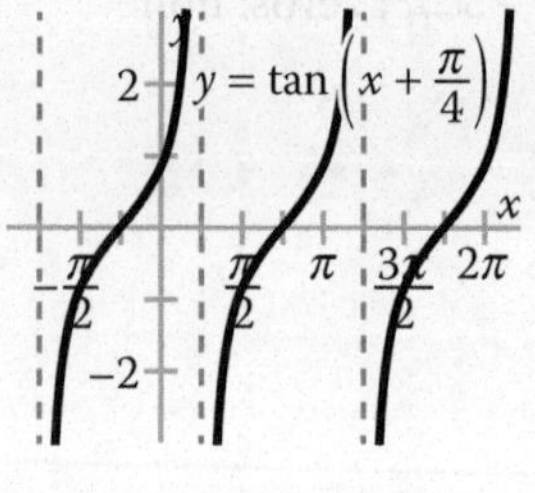

55. vertical asymptotes: $x = \frac{\square}{4} + \square n$; zeros: none

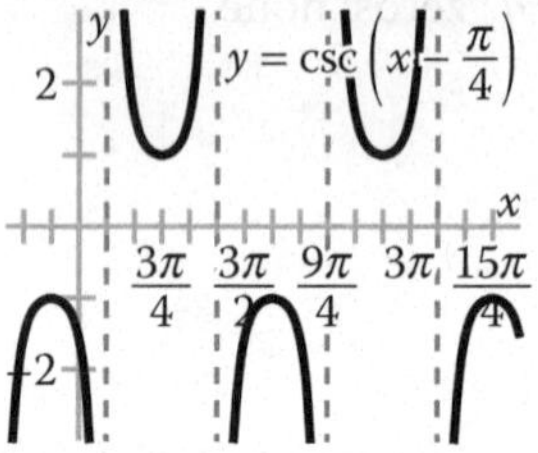

57. vertical asymptotes: $x = \frac{2\square}{3} + \square n$; zeros: none

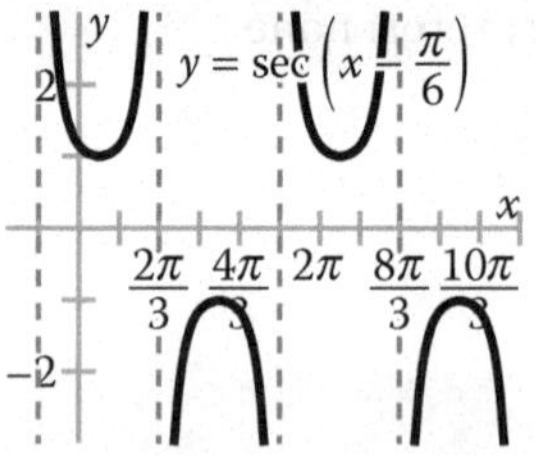

59. vertical asymptotes: $x = \frac{\square}{6} + \frac{\square}{3}n$; zeros: $x = \frac{\square}{3}n$

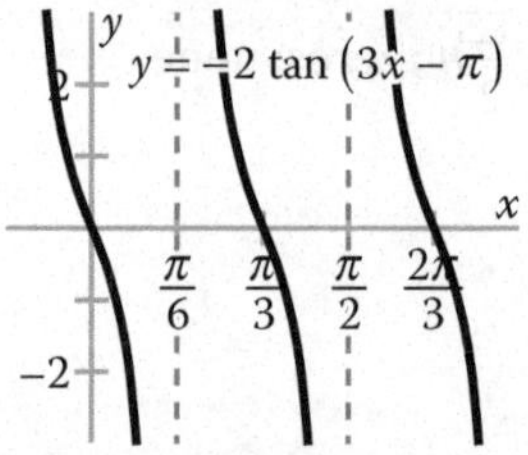

61. vertical asymptotes: $x = \frac{\square}{4}n$; zeros: $x = \frac{\square}{8} + \frac{\square}{4}n$

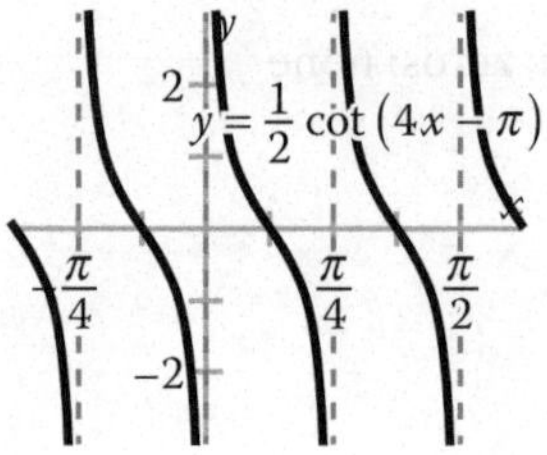

63. vertical asymptotes: $x = \frac{\square}{12} + \frac{\square}{2}n$; zeros: none

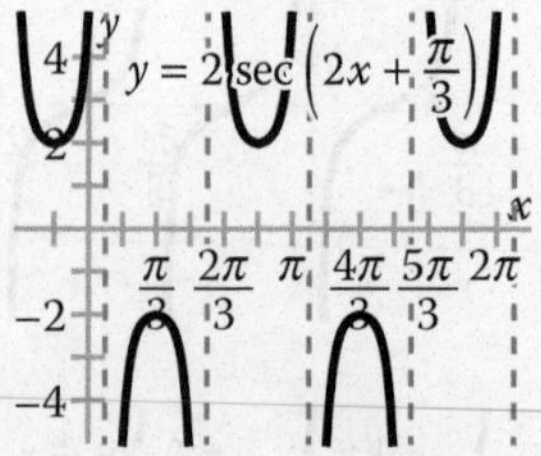

65. vertical asymptotes: $x = \frac{\square}{3}n$; zeros: none

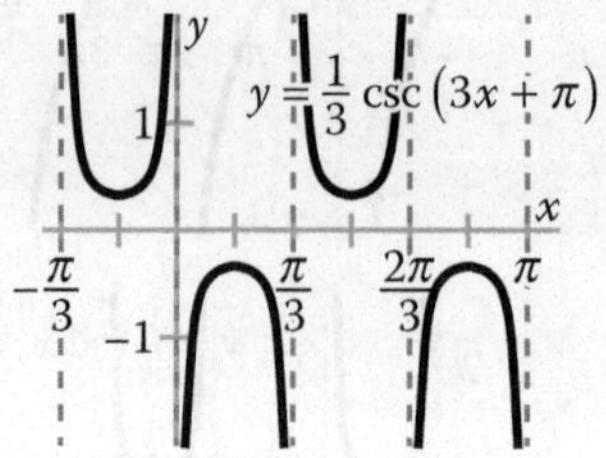

67. A

69. B

71. E

73. A

75. A

77. C

79. The graph of $y = 2 \cot \pi x$ has a vertical stretch of 2, a period of 1, vertical asymptotes at −1, 0, 1, 2, . . . or n (n is an integer), and zeros at −1/2, 1/2, 3/2, 5/2, . . . or $1/2 + n$ (n is an integer). The graph of $y = \cot x$ has no vertical stretch, a period of π, vertical asymptotes at $-\pi$, 0, π, 2π, . . . or πn (n is an integer), and zeros at $-\pi/2$, $\pi/2$, $3\pi/2$, $5\pi/2$, . . . or $\pi/2 + \pi n$ (n is an integer).

81. true

83. $y = -5 \tan\left(4x + \frac{4\square}{5}\right)$ or $y = -5 \tan 4\left(x + \frac{\square}{5}\right)$

85. $y = 4 \tan 3x$

87. $y = -5 \cot \frac{2}{3}\left(x - \frac{\square}{6}\right)$

89. $y = -\frac{1}{2} \csc \frac{2}{3}\left(x + \frac{\square}{6}\right)$

91. $y = -8 \sec 8\left(x - \frac{2\square}{5}\right)$

SECTION 5.7 ANSWERS

Warm Up/Just the Facts

1. $f^{-1}(x) = \dfrac{x+1}{3}$

$f(x) = 3x - 1$ is a linear equation, so the graph is a line. The line is not horizontal because the equation contains x. Any nonhorizontal line passes the Horizontal Line Test, so f is a one-to-one function and has an inverse function. To find the inverse function, interchange x and y, then solve for y.

$f(x) = 3x - 1$

$y = 3x - 1$ *Replace f(x) with y.*

$x = 3y - 1$ *Interchange x and y.*

$x + 1 = 3y$ *Add 1 to both sides.*

$\dfrac{x+1}{3} = y$ *Divide both sides by 3.*

$f^{-1}(x) = \dfrac{x+1}{3}$ *Replace y with $f^{-1}(x)$.*

2. No inverse function.

$y = x^2$ is a quadratic function, so the graph is a U-shaped graph and does not pass the Horizontal Line Test. Therefore, it does not have an inverse function.

3. $h^{-1}(x) = \dfrac{x^2-1}{2},\ x \ge 0$

$h(x) = \sqrt{2x+1}$ is a square root function, so the graph passes the Horizontal Line Test, is a one-to-one function, and therefore has an inverse function. The domain of h is the interval $[-1/2, \infty)$, so the range of h^{-1} must also be $[-1/2, \infty)$. Furthermore, the range of h is the interval $[0, \infty)$, so the domain of h^{-1} must also be $[0, \infty)$. To find the inverse function of h, replace $h(x)$ with y, interchange x and y, then solve for y.

$h(x) = \sqrt{2x+1}$

$y = \sqrt{2x+1}$ *Replace h(x) with y.*

$x = \sqrt{2y+1}$ *Interchange x and y.*

$x^2 = 2y + 1$ *Square both sides.*

$\dfrac{x^2-1}{2} = y$ *Subtract 1 and divide by 2.*

$h^{-1}(x) = \dfrac{x^2-1}{2},\ x \ge 0$ *Replace y with $h^{-1}(x)$.*

4. horizontal; once
5. reflections; $y = x$
6. six; Horizontal Line Test; inverses
7. $\left[-\dfrac{□}{2}, \dfrac{□}{2}\right]$; one-to-one; $\left[-\dfrac{□}{2}, \dfrac{□}{2}\right]$; domain
8. $y = \arccos x$
9. $[0, \pi]$
10. cosecant; 0; cosecant; restricted

Essential Skills/Extensions

1. $\sin^{-1}\dfrac{\sqrt{2}}{2} = \dfrac{□}{4}$, $\cos^{-1}\dfrac{\sqrt{2}}{2} = \dfrac{□}{4}$, $\tan^{-1}\dfrac{\sqrt{2}}{2}$ cannot be determined without a calculator
3. $\sin^{-1} 0 = 0$, $\cos^{-1} 0 = \pi/2$, $\tan^{-1} 0 = 0$
5. $\sin^{-1}(-1) = -\pi/2$, $\cos^{-1}(-1) = \pi$, $\tan^{-1}(-1) = -\pi/4$
7. $\sin^{-1}\dfrac{\sqrt{3}}{2} = \dfrac{□}{3}$, $\cos^{-1}\dfrac{\sqrt{3}}{2} = \dfrac{□}{6}$, $\tan^{-1}\dfrac{\sqrt{3}}{2}$ cannot be determined without a calculator
9. 1.362 rad
11. 0.177 rad
13. not possible
15. 0.846 rad
17. 0.658 rad
19. −0.068 rad
21. $-\pi/4$
23. $-\pi/3$
25. $5\pi/6$
27. $\pi/7$
29. $\pi/6$
31. $-\pi/4$
33. $\dfrac{\sqrt{21}}{2}$
35. $\dfrac{\sqrt{22}}{5}$
37. $\dfrac{\sqrt{66}}{22}$
39. $-15/17$
41. $3/5$
43. $\dfrac{5\sqrt{29}}{29}$
45. $-\dfrac{5\sqrt{26}}{26}$
47. $\csc^{-1}\dfrac{2}{\sqrt{3}} = \dfrac{□}{3}$, $\sec^{-1}\dfrac{2}{\sqrt{3}} = \dfrac{□}{6}$, $\text{arccot}\,\dfrac{2}{\sqrt{3}}$ cannot be determined without a calculator
49. $\csc^{-1} 2 = \pi/6$, $\sec^{-1} 2 = \pi/3$, $\cot^{-1} 2$ cannot be determined without a calculator
51. $\dfrac{5\sqrt{21}}{21}$

53. $-\frac{\sqrt{65}}{4}$

55. 13/5

57. $\left[-\frac{1}{2}, \frac{1}{2}\right]$

59. $\theta = \arctan \frac{10.32}{x}$

61. false

63. $-\frac{\sqrt{2}}{2}$

65. $\pi/4$

SECTION 5.8 ANSWERS

Warm Up/Just the Facts

1. N 60° W

Since 150° is in quadrant II, the north and west bearing are used. Subtract 90° from 150° to determine the acute angle θ that is rotated from the north-south line. 150° – 90° = 60°; therefore, 150° as a bearing is N 60° W.

2. N 50° E

Since 40° is in quadrant I, the north and east bearing are used. Subtract 40° from 90° to determine the acute angle θ that is rotated from the north-south line. 90° – 40° = 50°; therefore, 40° as a bearing is N 50° E.

3. S 20° E

Since 290° is in quadrant IV, the south and east bearing are used. Subtract 270° from 290° to determine the acute angle θ that is rotated from the north-south line. 290° – 270° = 20°; therefore, 290° as a bearing is S 20° E.

4. subtracting; 180°

5. subtracting; 90°

6. sine, cosine

7. acute; N, S; north, south

8. equilibrium

9. simple; 9/2; 2/9

10. coefficients; *a*; ke^{-ct}

Essential Skills/Extensions

1. $A = 55°$, $a \approx 9.8$, $b \approx 6.9$

3. $B = 47°$, $a \approx 6.7$, $c \approx 9.8$

5. $C = 8°$, $b \approx 172.9$, $c \approx 174.6$

7. 133.9 ft

9. 559.8 ft

11. 11.9 ft

13. 44.8 ft

15. 85.1 ft

17. bearing = N 50.12° E, distance ≈ 99.7 NM

19. bearing = S 45.03° E, distance ≈ 131.87 NM

21. bearing = N 44.18° E, distance ≈ 100.51 NM

23. 1/8 cycle per unit of time

25. 3/14 cycle per unit of time

27. 4/17 cycle per unit of time

29. $d = -12 \cos \frac{\square}{8} t$

31. $d = 23 \sin \frac{2\square}{11} t$

33. $d = 22 \sin \frac{\square}{3} t$

35. $d = -26 \cos \frac{\square}{21} t$

37. $d = 0.3e^{-2t} \cos 523.26\pi t$

39. $d = 0.38e^{-2t} \cos 587.33\pi t$

41. 82.54 ft

43. 0.00006 s

45. –12,285,622.4 mm/s^2

47. –569.3 mm/s

49. $d = 15 \cos 3\pi t, d = -15 \cos 3\pi t, d = 15 \sin 3\pi t, d = -15 \sin 3\pi t$ (possible answers)

51. $c = -\frac{\ln\left(\frac{d}{k \sin \square t}\right)}{t}$

53. $d = 2.25 \cos \frac{5\square}{6} t$

CHAPTER 5 REVIEW ANSWERS

1. I; 405°; –315°; 45°; 135°; $\pi/4$

2. II; 520°; –200°; no complement; 20°; $8\pi/9$

3. IV; 355°; –5°; no complement; no supplement; $143\pi/36$

4. IV; 320°; –40°; no complement; no supplement; $-20\pi/9$

5. 492.8 ft^2; 53.3 ft; 165°

6. $\sin\theta = \frac{7}{10}$, $\cos\theta = \frac{\sqrt{51}}{10}$, $\tan\theta = \frac{7\sqrt{51}}{51}$,
$\csc\theta = \frac{10}{7}$, $\sec\theta = \frac{10\sqrt{51}}{51}$, $\cot\theta = \frac{\sqrt{51}}{7}$

7. 46.4

8. 6.1

9. 5.6

10. 330.2

11. $\sin\theta = \frac{5\sqrt{26}}{26}$, $\cos\theta = -\frac{\sqrt{26}}{26}$, $\tan\theta = -5$,
$\csc\theta = \frac{\sqrt{26}}{5}$, $\sec\theta = -\sqrt{26}$, $\cot\theta = -\frac{1}{5}$

12. $\sin\theta = -\frac{3\sqrt{58}}{58}$, $\cos\theta = -\frac{7\sqrt{58}}{58}$, $\tan\theta = \frac{3}{7}$,
$\csc\theta = -\frac{\sqrt{58}}{3}$, $\sec\theta = -\frac{\sqrt{58}}{7}$, $\cot\theta = \frac{7}{3}$

13. $\sin(-225°) = \frac{\sqrt{2}}{2}$, $\cos(-225°) = -\frac{\sqrt{2}}{2}$, $\tan(-225°) = -1$,
$\csc(-225°) = \sqrt{2}$, $\sec(-225°) = -\sqrt{2}$, $\cot(-225°) = -1$

14. $\sin\frac{7\square}{3} = \frac{\sqrt{3}}{2}$, $\cos\frac{7\square}{3} = \frac{1}{2}$, $\tan\frac{7\square}{3} = \sqrt{3}$,
$\csc\frac{7\square}{3} = \frac{2\sqrt{3}}{3}$, $\sec\frac{7\square}{3} = 2$, $\cot\frac{7\square}{3} = \frac{\sqrt{3}}{3}$

15. $\sin\theta = -\frac{2\sqrt{29}}{29}$; $\cos\theta = \frac{5\sqrt{29}}{29}$

16. $\sin\frac{7\square}{6} = -\frac{1}{2}$, $\cos\frac{7\square}{6} = -\frac{\sqrt{3}}{2}$, $\tan\frac{7\square}{6} = \frac{\sqrt{3}}{3}$,
$\csc\frac{7\square}{6} = -2$, $\sec\frac{7\square}{6} = -\frac{2\sqrt{3}}{3}$, $\cot\frac{7\square}{6} = \sqrt{3}$

17. $\sin(-\pi) = 0$, $\cos(-\pi) = -1$, $\tan(-\pi) = 0$,
$\csc(-\pi)$ is undefined, $\sec(-\pi) = -1$, $\cot(-\pi)$ is undefined

18. $\sin\frac{9\square}{2} = 1$, $\cos\frac{9\square}{2} = 0$, $\tan\frac{9\square}{2}$ is undefined,
$\csc\frac{9\square}{2} = 1$, $\sec\frac{9\square}{2}$ is undefined, $\cot\frac{9\square}{2} = 0$

19. $\csc(-t) = 6/5$, $\sin(-t) = 5/6$

20. $\tan t = -7/6$ $\cot t = -6/7$

21. amplitude: 2, period: 2, phase shift: none, vertical shift: 3 units upward, maximums: $x = 2n + 1$, minimums: $x = 2n$, zeros: none (n is an integer)

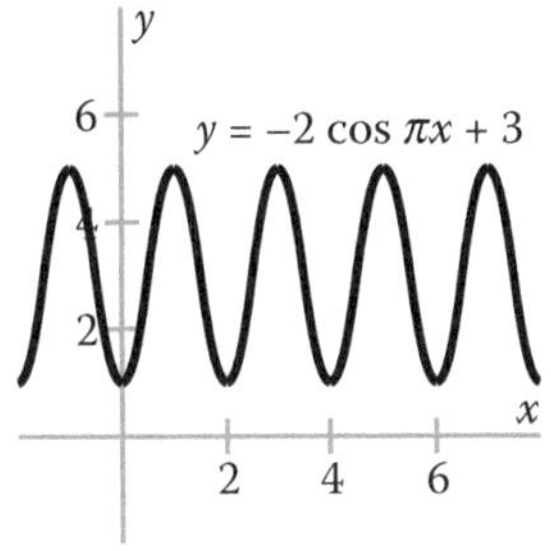

22. amplitude: 1/2, period: 2π, phase shift: left $2\pi/3$ units, vertical shift: 1 unit upward, maximums: $x = 5\pi/6 + 2\pi n$, minimums: $x = 11\pi/6 + 2\pi n$, zeros: none (n is an integer)

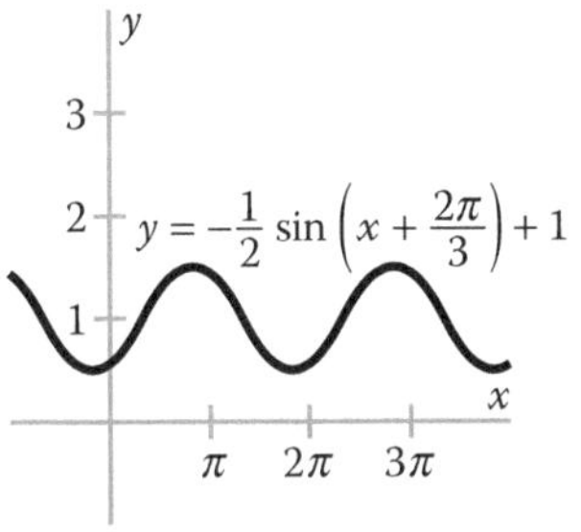

23. amplitude: 2/3, period: $2\pi/3$, phase shift: right $2\pi/3$ units, vertical shift: none, maximums: $x = \frac{2\square}{3}n$, minimums: $x = \frac{\square}{3} + \frac{2\square}{3}n$, zeros: $x = \frac{\square}{6} + \frac{\square}{3}n$ (n is an integer)

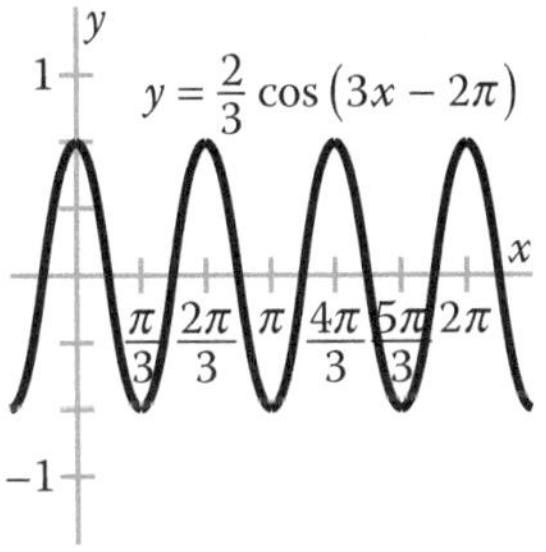

24. amplitude: 3, period: 4, phase shift: none, vertical shift: 1 unit downward, maximums: $x = 1 + 4n$, minimums: $x = 3 + 4n$ (n is an integer)

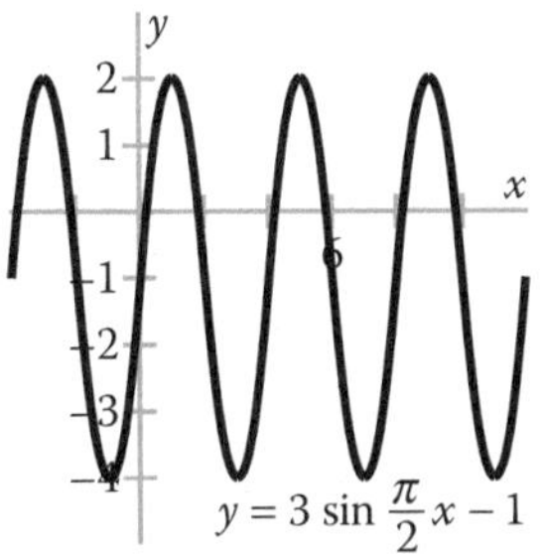

25. maximum: 80 ft

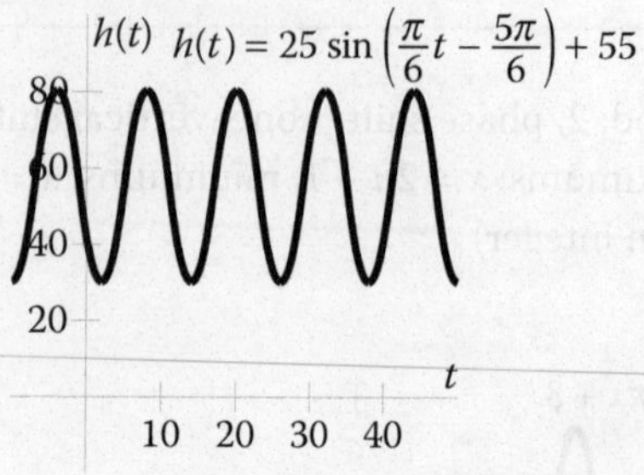

26. period: $3\pi/2$; phase shift: right $3\pi/8$ units; vertical shift: none; vertical asymptotes: $x = \frac{9\square}{8} + \frac{3\square}{2}n$; zeros: $x = \frac{3\square}{8} + \frac{3\square}{2}n$ (n is an integer)

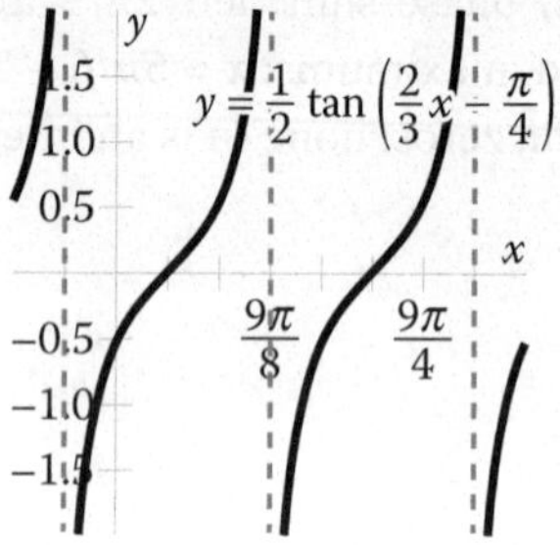

27. period: 2π; phase shift: left π units; vertical shift: none; vertical asymptotes: $x = \pi + 2\pi n$; zeros: $x = 2\pi n$ (n is an integer)

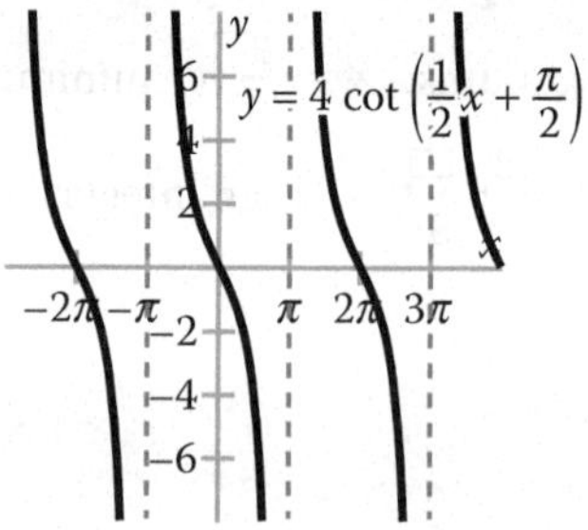

28. period: 2; phase shift: left 2 units; vertical shift: 2 units downward; vertical asymptotes: $x = n$ (n is an integer)

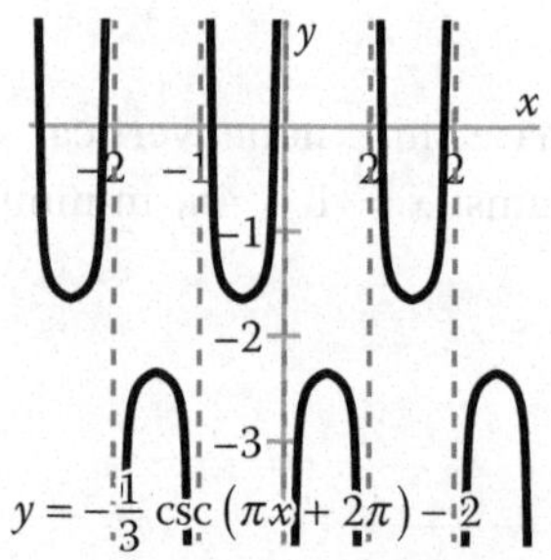

29. period: π; phase shift: right $\pi/4$ units; vertical shift: 1 unit upward; vertical asymptotes: $x = \pi/2 + \pi n$ (n is an integer)

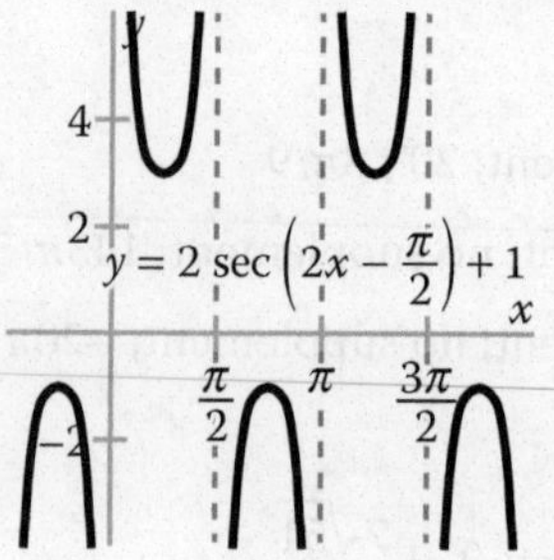

30. period: $\pi/2$; phase shift: left π units; vertical shift: 1 unit downward; vertical asymptotes: $x = \frac{\square}{2}n$ (n is an integer)

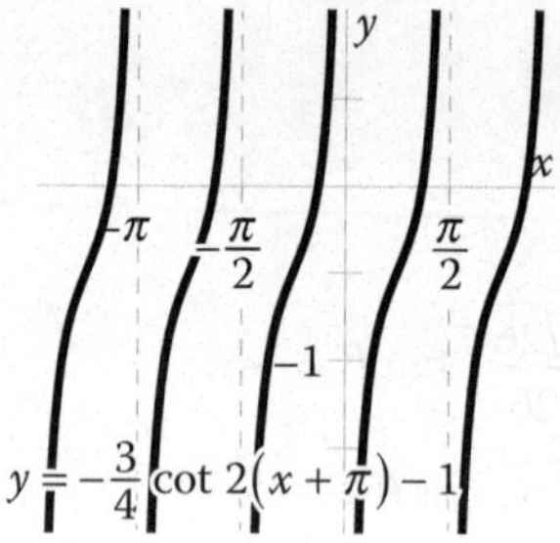

31. $\pi/6$

32. 2.65 rad

33. $2\pi/3$

34. $\frac{7\sqrt{55}}{55}$ rad

35. $-4/3$ rad

36. $A = 53.8°$, $b \approx 8.7$, $c \approx 14.7$

37. 71.1°

38. 360.73 ft

39. N 46.7° W

40. $d = 20\sin\frac{4\square}{3}t$; 2/3

SECTION 6.1 ANSWERS

Warm Up/Just the Facts

1. True. 45° corresponds to the point $\left(\frac{\sqrt{2}}{2}, \frac{\sqrt{2}}{2}\right)$ on the unit circle, $\cot\theta = x/y$, and $\frac{\sqrt{2}}{2} \div \frac{\sqrt{2}}{2} = 1$.
2. True. 240° and 120° have the same reference angle of 60°. Since secant is negative in quadrants II and III, the values are the same for sec 240° and sec 120°.
3. False. sin 150° = 1/2 and 2(1/2) ≠ −1.
4. identity
5. even-odd
6. Pythagorean
7. cofunction
8. both; equal; variables
9. $\frac{\sin\theta}{\cos\theta}$; $\sin\theta$
10. counterexample

Essential Skills/Extensions

1. $\cos\theta = 12/13$, $\tan\theta = -5/12$, $\csc\theta = -13/5$, $\sec\theta = 13/12$, $\cot\theta = -12/5$

3. $\sin\theta = \frac{4\sqrt{2}}{9}$, $\tan\theta = \frac{4\sqrt{2}}{7}$, $\csc\theta = \frac{9\sqrt{2}}{8}$, $\sec\theta = \frac{9}{7}$, $\cot\theta = \frac{7\sqrt{2}}{8}$

5. $\cos\theta = -\frac{\sqrt{5}}{3}$, $\tan\theta = \frac{2\sqrt{5}}{5}$, $\csc\theta = -\frac{3}{2}$, $\sec\theta = -\frac{3\sqrt{5}}{5}$, $\cot\theta = \frac{\sqrt{5}}{2}$

7. $\sin\theta - \frac{7\sqrt{65}}{65}$, $\cos\theta = \frac{4\sqrt{65}}{65}$, $\csc\theta = \frac{\sqrt{65}}{7}$, $\sec\theta = \frac{\sqrt{65}}{4}$, $\cot\theta = \frac{4}{7}$

9. $\sin\theta = -\frac{\sqrt{10}}{10}$, $\cos\theta = \frac{3\sqrt{10}}{10}$, $\tan\theta = -\frac{1}{3}$, $\csc\theta = -\sqrt{10}$, $\sec\theta = \frac{\sqrt{10}}{3}$

11. $\sin\theta = \frac{8\sqrt{65}}{65}$, $\cos\theta = -\frac{\sqrt{65}}{65}$, $\csc\theta = \frac{\sqrt{65}}{8}$, $\sec\theta = -\sqrt{65}$, $\cot\theta = -\frac{1}{8}$

13. $\sin\theta = -\frac{\sqrt{5}}{3}$, $\cos\theta = -\frac{2}{3}$, $\tan\theta = \frac{\sqrt{5}}{2}$, $\csc\theta = -\frac{3\sqrt{5}}{5}$, $\cot\theta = \frac{2\sqrt{5}}{5}$

15. $\sin\theta = -24/25$, $\cos\theta = -7/25$, $\tan\theta = 24/7$, $\sec\theta = -25/7$, $\cot\theta = 7/24$

17. $\sin\theta = -\frac{\sqrt{105}}{11}$, $\cos\theta = \frac{4}{11}$, $\tan\theta = -\frac{\sqrt{105}}{4}$, $\csc\theta = -\frac{11\sqrt{105}}{105}$, $\cot\theta = -\frac{4\sqrt{105}}{105}$

19. $\sec^3 x$
21. $\cos^2 x$
23. $\cos x$
25. $\sin^2 x$
27. $\cos^2 x$
29. 1
31. $\cot^3\theta$
33. $2\cot x$
35. $2\sec x$
37. 1
39. $\tan^2 x - \sin^2 x$
41. $9\cot^2 x$
43. $\csc^2 x\sec^2 x$
45. $2\sin x\cos x + 1$
47. $\cot^2\theta$
49. 1
51. $\sin^2\theta$
53. $(\tan x + 5)(\tan x + 1)$
55. $(2\cos\theta - 3)(\cos\theta + 4)$
57. $(3\sin t + 7\cos t)^2$
59. $\sin^5 t\cos^2 t$
61. $\cos^4 t$
63. $$\begin{aligned}(\cos x - \sec x)^2 &= \cos^2 x - \cos x\sec x - \sec x\cos x + \sec^2 x\\ &= \cos^2 x - 2\cos x\sec x + \sec^2 x\\ &= 1 - \sin^2 x - 2(1) + \sec^2 x\\ &= -\sin^2 x - 1 + \tan^2 x + 1\\ &= \tan^2 x - \sin^2 x\end{aligned}$$

(possible answer)

65. $\dfrac{1-\sin t}{\cos t}+\dfrac{\cos t}{1-\sin t}=\dfrac{1-\sin t}{\cos t}\cdot\dfrac{1-\sin t}{1-\sin t}+\dfrac{\cos t}{1-\sin t}\cdot\dfrac{\cos t}{\cos t}$

$=\dfrac{1-\sin t-\sin t+\sin^2 t+\cos^2 t}{(\cos t)(1-\sin t)}$

$=\dfrac{1-2\sin t+1}{(\cos t)(1-\sin t)}$

$=\dfrac{2-2\sin t}{(\cos t)(1-\sin t)}$

$=\dfrac{2(1-\sin t)}{(\cos t)(1-\sin t)}$

$=\dfrac{2}{\cos t}$

$=2\sec t$

(possible answer)

67. $\dfrac{1+2\cot^2 x+\cot^4 x}{1-\cot^2 x}=\dfrac{(1+\cot^2 x)(1+\cot^2 x)}{1-\cot^2 x}$

$=\dfrac{(\csc^2 x)(\csc^2 x)}{1-\cot^2 x}$

$=\dfrac{\csc^4 x}{1-\cot^2 x}$

(possible answer)

69. not an identity

71. 125

73. 12

75. $\dfrac{1+\sin x+\cos x}{1-\sin x+\cos x}=\dfrac{1+\sin x+\cos x}{1-\sin x+\cos x}\cdot\dfrac{1+\sin x-\cos x}{1+\sin x-\cos x}$

$=\dfrac{1+2\sin x+\sin^2 x-\cos^2 x}{1+2\cos x\sin x-\sin^2 x-\cos^2 x}$

$=\dfrac{1+2\sin x+\sin^2 x-(1-\sin^2 x)}{1+2\cos x\sin x-(1-\cos^2 x)-\cos^2 x}$

$=\dfrac{1+2\sin x+\sin^2 x-1+\sin^2 x}{1+2\cos x\sin x-1+\cos^2 x-\cos^2 x}$

$=\dfrac{2\sin x+2\sin^2 x}{2\cos x\sin x}$

$=\dfrac{(2\sin x)(1+\sin x)}{2\cos x\sin x}$

$=\dfrac{1+\sin x}{\cos x}$

(possible answer)

77. $\dfrac{\sec^3\theta-\cos^3\theta}{\sec\theta-\cos\theta}=\dfrac{(\sec\theta-\cos\theta)(\sec^2\theta+\sec\theta\cos\theta+\cos^2\theta)}{\sec\theta-\cos\theta}$

$=\sec^2\theta+\cos^2\theta+1$

(possible answer)

79. $(2\sin x+3\cos x)^2+(3\sin x-2\cos x)^2=4\sin^2 x+12\cos x\sin x+9\cos^2 x+9\sin^2 x-12\cos x\sin x+4\cos^2 x$

$=13\sin^2 x+13\cos^2 x$

$=13(\sin^2 x+\cos^2 x)$

$=13$

(possible answer)

81. $\tan^2 x-1$

SECTION 6.2 ANSWERS

Warm Up/Just the Facts

1. $x = 1/3$ and $x = -3/2$

$6x^2 + 7x = 3$

$6x^2 + 7x - 3 = 0$ *Write the equation in general form.*

$(3x - 1)(2x + 3) = 0$ *Factor.*

$3x - 1 = 0$ or $2x + 3 = 0$ *Set each factor equal to 0.*

$x = 1/3$ $x = -3/2$ *Solve each equation.*

2. $x = 0$ and $x = 5$

$20x = 4x^2$

$4x^2 - 20x = 0$ *Write the equation in general form.*

$4x(x - 5) = 0$ *Factor.*

$4x = 0$ or $x - 5 = 0$ *Set each factor equal to 0.*

$x = 0$ $x = 5$ *Solve each equation.*

3. $x = -2/5$ and $x = 3$

$13x = -6 + 5x^2$

$5x^2 - 13x - 6 = 0$ *Write the equation in general form.*

$(5x + 2)(x - 3) = 0$ *Factor.*

$5x + 2 = 0$ or $x - 3 = 0$ *Set each factor equal to 0.*

$x = -2/5$ $x = 3$ *Solve each equation.*

4. true
5. periodic
6. add; period
7. inverse
8. no solution
9. $-\tan x$
10. mode

Essential Skills/Extensions

1. $\pi/3 + 2\pi n, 2\pi/3 + 2\pi n$
3. $3\pi/4 + \pi n$
5. $5\pi/6 + \pi n$
7. $2\pi n$
9. $\pi/2 + \pi n$
11. $2\pi/3 + 2\pi n, 4\pi/3 + 2\pi n$
13. $3\pi/4 + \pi n$
15. $\pi/3 + 2\pi n, 2\pi/3 + 2\pi n$
17. $\pi/3 + \pi n, 2\pi/3 + \pi n$
19. $\pi/3 + \pi n, 2\pi/3 + \pi n$
21. $\frac{\pi}{4} + \frac{\pi}{2}n$
23. $\frac{\pi}{4} + \frac{\pi}{2}n$
25. 1.181, 5.102
27. 1.042, 2.1
29. $0, \pi$
31. $0, \pi/3, 2\pi/3, \pi, 4\pi/3, 5\pi/3$
33. $0, 3\pi/4, \pi, 7\pi/4$
35. $\pi/6, 5\pi/6, 3\pi/2$
37. no real solution
39. $\pi/3, \pi, 5\pi/3$
41. $\pi/6, 5\pi/6, \pi/2$
43. $\pi/3, \pi, 5\pi/3$
45. $0, \pi/3, 2\pi/3, \pi, 4\pi/3, 5\pi/3$
47. $x = \pi/3$
49. $x = \pi/6, 5\pi/6, 7\pi/6, 11\pi/6$
51. $3\pi/2$
53. $\pi/12, 5\pi/12, 7\pi/12, 11\pi/12, 13\pi/12, 17\pi/12, 19\pi/12, 23\pi/12$
55. $\pi/6, \pi/3, 2\pi/3, 5\pi/6, 7\pi/6, 4\pi/3, 5\pi/3, 11\pi/6$
57. $\pi/8, 3\pi/8, 5\pi/8, 7\pi/8, 9\pi/8, 11\pi/8, 13\pi/8, 15\pi/8$
59. $0, \pi/2, \pi, 3\pi/2$
61. $0, 2\pi/3, 4\pi/3$
63. $0, \pi/6, \pi/2, 5\pi/6, \pi, 7\pi/6, 3\pi/2, 11\pi/6$
65. $\pi/4, \pi/3, 2\pi/3, 3\pi/4, 5\pi/4, 4\pi/3, 5\pi/3, 7\pi/4$
67. $\pi/6, \pi/3, 2\pi/3, 5\pi/6, 7\pi/6, 4\pi/3, 5\pi/3, 11\pi/6$
69. 1.179, 5.104
71. 0.666, 2.475
73. 0.196, 1.375, 3.338, 4.516
75. no solution in $[0, 2\pi)$
77. 1.231, 5.052
79. $\pi/2, 3\pi/2$
81. $\pi/2$
83. $\pi/6, 5\pi/6$
85. $0, \pi$
87. no solution
89. 0.214 s

91. 0.117 s

93. 0.681 s

95. $\pi/48$ s

97. 0.004 s

99. 0, $4\pi/3$, $8\pi/3$, 4π, $16\pi/3$, $20\pi/3$, 8π, $28\pi/3$

101. 6.29°

103. $\tan 3x - \sec 3x = \tan 3 \cdot \frac{2\pi}{3} - \sec 3 \cdot \frac{2\pi}{3}$
$= \tan 2\pi - \sec 2\pi$
$= 0 - 1$
$= -1$

SECTION 6.3 ANSWERS

Warm Up/Just the Facts

1. $\pi/3 + \pi/4$

$7\pi/12 = 4\pi/12 + 3\pi/12 = \pi/3 + \pi/4$

2. 45° + 30°

75° = 45° + 30°

3. −60° + −45°

−105° = −60° + −45°

4. identity

5. False

6. sum; $\dfrac{\tan 60° + \tan 45°}{1 - \tan 60° \tan 45°}$

7. sum; $\cos(A + B)$

8. difference; $\sin(A - B)$

9. extraneous

10. difference; $\pi/4$

Essential Skills/Extensions

1. sin 120° cos 45° − cos 120° sin 45°,

cos 120° cos 45° + sin 120° sin 45°, $\dfrac{\tan 120° - \tan 45°}{1 + \tan 120° \tan 45°}$

(possible answers)

3. $\sin\frac{7\pi}{4}\cos\frac{5\pi}{6} - \cos\frac{7\pi}{4}\sin\frac{5\pi}{6}$,

$\cos\frac{7\pi}{4}\cos\frac{5\pi}{6} + \sin\frac{7\pi}{4}\sin\frac{5\pi}{6}$, $\dfrac{\tan\frac{7\pi}{4} - \tan\frac{5\pi}{6}}{1 + \tan\frac{7\pi}{4}\tan\frac{5\pi}{6}}$

(possible answers)

5. $\dfrac{-\sqrt{2} - \sqrt{6}}{4}$

7. $\dfrac{-\sqrt{6} + \sqrt{2}}{4}$

9. $\dfrac{-\sqrt{2} - \sqrt{6}}{4}$

11. $\dfrac{\sqrt{6} + \sqrt{2}}{4}$

13. $\dfrac{1 - \sqrt{3}}{1 + \sqrt{3}}$

15. $\dfrac{1 - \sqrt{3}}{1 + \sqrt{3}}$

17. $-\dfrac{\sqrt{2}}{2}$

19. $\sqrt{3}$

21. $-\dfrac{\sqrt{2}}{2}$

23. 1/2

25. $-\sin\theta$

27. $-\tan x$

29. $-\cos x$

31. not possible

33. $\sin\left(\frac{\pi}{6} + x\right) = \sin\frac{\pi}{6}\cos x + \cos\frac{\pi}{6}\sin x$
$= \frac{1}{2} \cdot \cos x + \frac{\sqrt{3}}{2} \cdot \sin x$
$= \frac{1}{2}\left(\cos x + \sqrt{3}\sin x\right)$

(possible answer)

35. $\pi/2$

37. $5\pi/4$, $7\pi/4$

39. 16/65

41. 63/65

43. 16/63

45. 4/5

47. 3/5

49. 4/3

51. $\dfrac{-\sqrt{2} - \sqrt{6}}{4}$

53. $\dfrac{1 - \sqrt{3}}{1 + \sqrt{3}}$

55. not possible

SECTION 6.4 ANSWERS

Warm Up/Just the Facts

1. $-6\sin^2 x$

$3\cos^2 x - 3\sin^2 x - 3 = 3(1 - \sin^2 x) - 3\sin^2 x - 3$

$= 3 - 3\sin^2 x - 3\sin^2 x - 3 = -6\sin^2 x$

2. -1

$\tan^2 x - \sec^2 x = \tan^2 x - (1 + \tan^2 x)$

$= \tan^2 x - 1 - \tan^2 x = -1$

3. 5

$5\csc^2 x - 5\cot^2 x = 5(1 + \cot^2 x) - 5\cot^2 x$

$= 5 + 5\cot^2 x - 5\cot^2 x = 5$

4. $2\sin\theta\cos\theta$

5. $\cos 2\theta$

6. $\dfrac{2\tan\theta}{1-\tan^2\theta}$

7. $\tan^2\theta$

8. $\pm\sqrt{\dfrac{1-\cos\theta}{2}}$

9. $\cos\theta/2$

10. $\sin^2\theta$

Essential Skills/Extensions

1. $-\sqrt{3}$

3. 0

5. $10\sin 2\theta$

7. $-10\cos 2\theta$

9. $10\tan 2\theta$

11. $\sin 2\theta = 336/625$, $\cos 2\theta = 527/625$, $\tan 2\theta = 336/527$, $\csc 2\theta = 625/336$, $\sec 2\theta = 625/527$, $\cot 2\theta = 527/336$

13. $\sin 2\theta = 4/5$, $\cos 2\theta = -3/5$, $\tan 2\theta = -4/3$, $\csc 2\theta = 5/4$, $\sec 2\theta = -5/3$, $\cot 2\theta = -3/4$

15. $\sin 2\theta = -24/25$, $\cos 2\theta = -7/25$, $\tan 2\theta = 24/7$, $\csc 2\theta = -25/24$, $\sec 2\theta = -25/7$, $\cot 2\theta = 7/24$

17. $\sin 2\theta = -112/113$, $\cos 2\theta = 15/113$, $\tan 2\theta = -112/15$, $\csc 2\theta = -113/112$, $\sec 2\theta = 113/15$, $\cot 2\theta = -15/112$

19. $\sin 2\theta = -120/169$, $\cos 2\theta = 119/169$, $\tan 2\theta = -120/119$, $\csc 2\theta = -169/120$, $\sec 2\theta = 169/119$, $\cot 2\theta = -119/120$

21. $\pi/4, \pi/2, 3\pi/4, 5\pi/4, 3\pi/2, 7\pi/4$

23. $\pi/2, 7\pi/6, 3\pi/2, 11\pi/6$

25. $7\pi/6, 11\pi/6$

27. $\pi/6, \pi/2, 5\pi/6, 7\pi/6, 3\pi/2, 11\pi/6$

29. $\dfrac{3+\cos 4t+4\cos 2t}{128}$

31. $\cos 2\theta$

33. $-2-\sqrt{3}$

35. $\dfrac{-\sqrt{6}+\sqrt{2}}{4}$

37. $\sqrt{2}-1$

39. $\sin\theta/2 = 3/5$, $\cos\theta/2 = 4/5$, $\tan\theta/2 = 3/4$, $\csc\theta/2 = 5/3$, $\sec\theta/2 = 5/4$, $\cot\theta/2 = 4/3$

41. $\sin\dfrac{\theta}{2} = \dfrac{\sqrt{5}}{5}$, $\cos\dfrac{\theta}{2} = \dfrac{2\sqrt{5}}{5}$, $\tan\dfrac{\theta}{2} = \dfrac{1}{2}$, $\csc\dfrac{\theta}{2} = \sqrt{5}$, $\sec\dfrac{\theta}{2} = \dfrac{\sqrt{5}}{2}$, $\cot\dfrac{\theta}{2} = 2$

43. $\dfrac{5\sqrt{26}}{26}$

45. $\dfrac{-8+\sqrt{39}}{5}$

47. $\dfrac{1}{2}\sin 27x + \dfrac{1}{2}\sin 13x$

49. $\dfrac{1}{2}\cos 2x + \dfrac{1}{2}\cos 6x$

51. $-2\sin 8x\sin 5x$

53. $2\cos 4x\cos x$

55. $2\sin 3x\cos 2x$

57. $\pi/2$

59. $A = 2r^2\sin 2\theta$

61. $A = \dfrac{c^2\sin 2A}{4}$

63. $\tan\theta\cdot\dfrac{1-\cos 2\theta}{1+\cos 2\theta}$

65. $3\pi/2$

67. $\pi/8, 3\pi/8, 5\pi/8, 7\pi/8, 9\pi/8, 11\pi/8, 13\pi/8, 15\pi/8$

69. $\pi/8, 3\pi/8, 5\pi/8, 7\pi/8, 9\pi/8, 11\pi/8, 13\pi/8, 15\pi/8$

71. $50\pi - 100\sin 2\theta$

73. 0.84, 2.3, 3.98, 5.44

75. 0, 2, 3.14, 4.28

CHAPTER 6 REVIEW ANSWERS

1. $\sin\theta = -\frac{5}{12}$, $\cos\theta = -\frac{\sqrt{119}}{12}$, $\tan\theta = \frac{5\sqrt{119}}{119}$,

$\sec\theta = -\frac{12\sqrt{119}}{119}$, $\cot\theta = \frac{\sqrt{119}}{5}$

2. –sin x

3. 1

4.
$$\frac{\cos^2 x + \tan^2 x - 1}{\sin^2 x} = \frac{1-\sin^2 x + \tan^2 x - 1}{\sin^2 x}$$
$$= \frac{\tan^2 x - \sin^2 x}{\sin^2 x}$$
$$= \frac{\tan^2 x}{\sin^2 x} - \frac{\sin^2 x}{\sin^2 x}$$
$$= \sec^2 x - 1$$
$$= \tan^2 x$$

(possible answer)

5.
$$\frac{\sec x + \csc x}{\cot x + \tan x} = \frac{\frac{1}{\cos x} + \frac{1}{\sin x}}{\frac{\cos x}{\sin x} + \frac{\sin x}{\cos x}}$$
$$= \frac{\frac{\sin x + \cos x}{\cos x \sin x}}{\frac{\cos^2 x + \sin^2 x}{\sin x \cos x}}$$
$$= \frac{\sin x + \cos x}{\cos x \sin x} \cdot \frac{\sin x \cos x}{1}$$
$$= \cos x + \sin x$$

(possible answer)

6. $3\pi/2 + 2\pi n$

7. $\pi/6 + 2\pi n$, $5\pi/6 + 2\pi n$

8. $\pi/3 + \pi n$, $2\pi/3 + \pi n$, $3\pi/4 + \pi n$

9. 2.203, 4.081

10. 0.743, 2.927, 3.884, 6.069

11. $\frac{1+\sqrt{3}}{\sqrt{3}-1}$

12. $\frac{\sqrt{6}-\sqrt{2}}{4}$

13. $-\cos x$

14. $\sin(x+y) - \sin(x-y)$

$= \sin x \cos y + \cos x \sin y - (\sin x \cos y - \cos x \sin y)$

$= \sin x \cos y + \cos x \sin y - \sin x \cos y + \cos x \sin y$

$= 2\cos x \sin y$

(possible answer)

15.
$$\frac{\sin(x+y) - \sin(x-y)}{\cos(x+y) + \cos(x-y)}$$
$$= \frac{\sin x \cos y + \cos x \sin y - \sin x \cos y + \cos x \sin y}{\cos x \cos y - \sin x \sin y + \cos x \cos y + \sin x \sin y}$$
$$= \frac{2\cos x \sin y}{2\cos x \cos y}$$
$$= \frac{\sin y}{\cos y}$$
$$= \tan y$$

(possible answer)

16. $\sin 2\theta = \frac{12\sqrt{10}}{49}$, $\cos 2\theta = \frac{31}{49}$, $\tan 2\theta = \frac{12\sqrt{10}}{31}$,

$\csc 2\theta = \frac{49\sqrt{10}}{120}$, $\sec 2\theta = \frac{49}{31}$, $\cot 2\theta = \frac{31\sqrt{10}}{120}$,

$\sin\frac{\theta}{2} = \sqrt{\frac{7-2\sqrt{10}}{14}}$, $\cos\frac{\theta}{2} = \sqrt{\frac{7+2\sqrt{10}}{14}}$, $\tan\frac{\theta}{2} = \frac{7-2\sqrt{10}}{3}$,

$\csc\frac{\theta}{2} = \sqrt{\frac{14}{7-2\sqrt{10}}}$, $\sec\frac{\theta}{2} = \sqrt{\frac{14}{7+2\sqrt{10}}}$, $\cot\frac{\theta}{2} = \frac{3}{7-2\sqrt{10}}$

17. $\sin 2\theta = -36/85$, $\cos 2\theta = -77/85$, $\tan 2\theta = 36/77$,

$\csc 2\theta = -85/36$, $\sec 2\theta = -85/77$, $\cot 2\theta = 77/36$,

$\sin\frac{\theta}{2} = \sqrt{\frac{85+2\sqrt{85}}{170}}$, $\cos\frac{\theta}{2} = \sqrt{\frac{85-2\sqrt{85}}{170}}$, $\tan\frac{\theta}{2} = \frac{2+\sqrt{85}}{9}$,

$\csc\frac{\theta}{2} = \sqrt{\frac{170}{85+2\sqrt{85}}}$, $\sec\frac{\theta}{2} = \sqrt{\frac{170}{85-2\sqrt{85}}}$, $\cot\frac{\theta}{2} = \frac{9}{2+\sqrt{85}}$

18. $\pi/6$, $5\pi/6$, $7\pi/6$, $11\pi/6$

19. $\frac{30\cos x + 15\cos 3x + 3\cos 5x}{16}$

20. $2\sin 5x \sin x$

SECTION 7.1 ANSWERS

Warm Up/Just the Facts

1. 99.92

   $\sin 15° = 25.86/c$ — *Sine trigonometric function.*

   $c = 25.86/\sin 15°$ — *Solve for c.*

2. 8.08

   $\sin 21° = b/22.55$ — *Sine trigonometric function.*

   $b = 22.55(\sin 21°)$ — *Solve for b.*

3. 53.87°

   $\sin \alpha = 73.5/91$ — *Sine trigonometric function.*

   $\alpha = \sin^{-1} \frac{73.5}{91}$ — *Solve for α.*

4. oblique
5. $\frac{\sin \alpha}{a}, \frac{\sin \beta}{b}, \frac{\sin \gamma}{c}$
6. ambiguous
7. zero
8. one
9. two
10. $\frac{1}{2}ca \sin \beta$

Essential Skills/Extensions

1. $b \approx 2.87, c \approx 7.78, \beta = 14°$
3. $a \approx 72.92, b \approx 199.58, \beta = 142°$
5. $a \approx 23.98, b \approx 12.87, \alpha = 81°$
7. $a \approx 13.55, b \approx 15, \gamma = 60°$
9. $a \approx 10.87, c \approx 33.55, \beta = 38°$
11. $b \approx 32.78, c \approx 40.15, \alpha = 15°$
13. $a \approx 372, \alpha \approx 44.14°, \beta \approx 13.86°$
15. $c \approx 135.49, \beta \approx 30.55°, \gamma \approx 110.45°$
17. two
19. zero
21. first triangle: $\beta \approx 49.89°, \gamma \approx 95.11°, c \approx 10.42$, second triangle: $\beta \approx 130.11°, \gamma \approx 14.89°, c \approx 2.69$
23. first triangle: $\alpha \approx 57.7°, \beta \approx 97.3°, b \approx 2.35$, second triangle: $\alpha \approx 122.3°, \beta \approx 32.7°, b \approx 1.28$
25. zero triangle: $b > 17.11$, one oblique triangle: $0 \leq b \leq 11$, two triangles: $11 < b < 17.11$
27. zero triangle: $b > 8\sqrt{2}$, one oblique triangle: $0 \leq b \leq 8$, two triangles: $8 < b < 8\sqrt{2}$
29. 7377.3 in^2
31. 218.5 in^2
33. 2130.8 in^2
35. 20.2 in^2
37. 35 m
39. 148 ft
41. 14.4 mi
43. 227.45 ft
45. 374.82 ft
47. 12.98 m
49. 0.99 km
51. 22.9
53. angle 1: 109°, angle 2: 71°

SECTION 7.2 ANSWERS

Warm Up/Just the Facts

1. $\gamma = \cos^{-1}\left(-\frac{c^2 - a^2 - b^2}{2ab}\right)$

   $c^2 = a^2 + b^2 - 2ab \cos \gamma$

   $c^2 - a^2 - b^2 = -2ab \cos \gamma$ — *Subtract a^2 and b^2.*

   $-\frac{c^2 - a^2 - b^2}{2ab} = \cos \gamma$ — *Divide each side by -2ab.*

   $\gamma = \cos^{-1}\left(-\frac{c^2 - a^2 - b^2}{2ab}\right)$ — *Take the inverse cosine.*

2. α: the smallest angle is opposite the shortest side.

   Since side a is the shortest side, α is the smallest angle.

3. β: the largest angle is opposite the longest side.

   Since side b is the longest side, β is the largest angle.

4. Cosines
5. $b^2 + c^2 - 2bc \cos \alpha$; $a^2 + c^2 - 2ac \cos \beta$; $a^2 + b^2 - 2ab \cos \gamma$
6. Sines
7. Cosines
8. Sines
9. Heron's
10. $\sqrt{s(s-a)(s-b)(s-c)}$; $\frac{1}{2}(a+b+c)$

Essential Skills/Extensions

1. 92.123°
3. 98.213°
5. 20.742°
7. 43.49°
9. 150°
11. 90.27°
13. 36.1°
15. 42 mi
17. 6.8 mi
19. 3460 m
21. 9.6 in^2
23. 764.9 ft^2
25. 2.9 in^2
27. 3.83 ft
29. 2.9 ft
31. 5.99 in
33. 6.6
35. α: 54.5, β: 44.2, and γ: 81.3
37. 42 cm^2
39. 70.2 m^2
41. 2.9 ft

SECTION 7.3 ANSWERS

Warm Up/Just the Facts

1. $3\sqrt{10}$

$d = \sqrt{(-1-2)^2 + (4-(-5))^2}$ *Substitute the coordinates into the distance formula.*

$= 3\sqrt{10}$ *Simplify.*

2. $2\sqrt{41}$

$d = \sqrt{(7-(-3))^2 + (-2-6)^2}$ *Substitute the coordinates into the distance formula.*

$= 2\sqrt{41}$ *Simplify.*

3. 15

$d = \sqrt{(1-(-8))^2 + (9-(-3))^2}$ *Substitute the coordinates into the distance formula.*

$= 15$ *Simplify.*

4. vector
5. length; Pythagorean
6. angle
7. $\| \vec{v} \|$
8. unit
9. $\vec{v} / \| \vec{v} \|$
10. $r \cos \theta$; $r \sin \theta$

Essential Skills/Extensions

1. $\langle 3, -5 \rangle$
3. $\langle -6, 7 \rangle$
5. $\langle -10, -15 \rangle$
7. $\langle -4, -3 \rangle$
9. $\langle 7, -3 \rangle$
11. $\langle -5, 7 \rangle$
13. 2, 150°
15. $4\sqrt{2}$, 225°
17. 4, 300°
19. $3\sqrt{5}$, 63.43°
21. $\sqrt{65}$, 330.26°
23. $\sqrt{26}$, 168.69°
25. $\langle -4, 10 \rangle$
27. $\langle 41, -6.5 \rangle$
29. $\langle 2.5, 3.75 \rangle$
31.

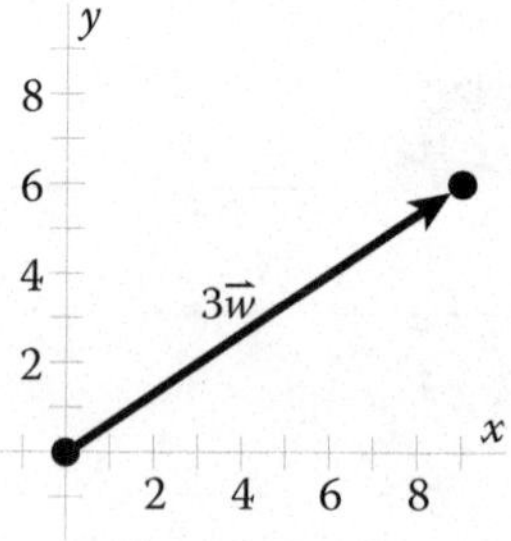

33.

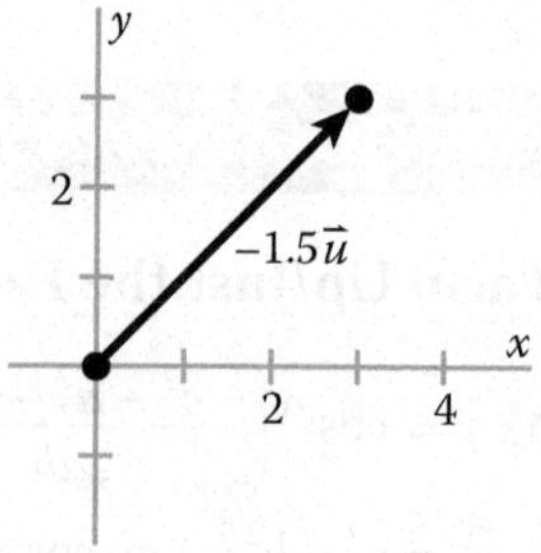

35.

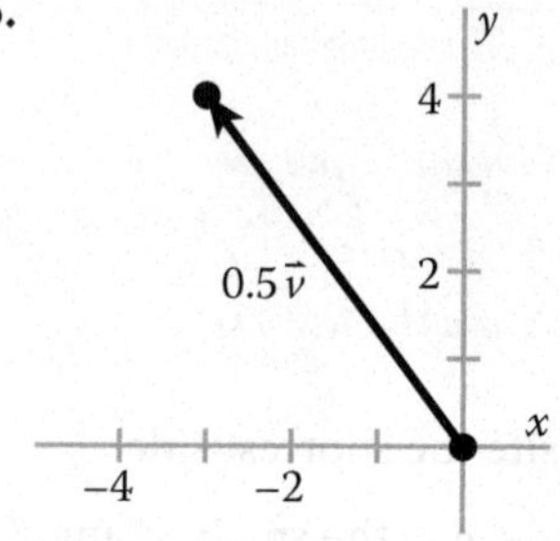

37. $\langle 1, -1 \rangle$

39. $\langle -4.6, -14.6 \rangle$

41.

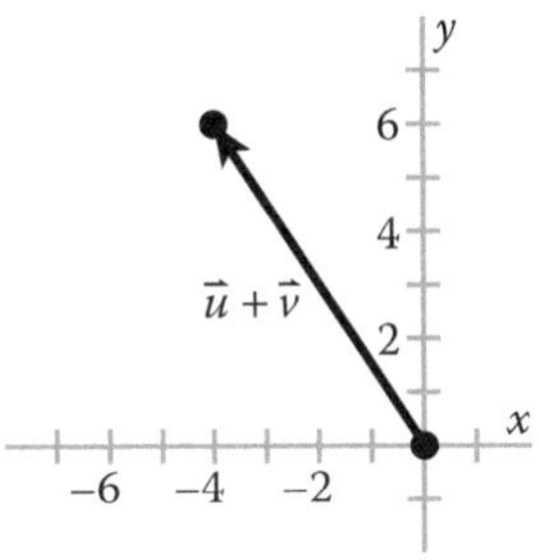

43.

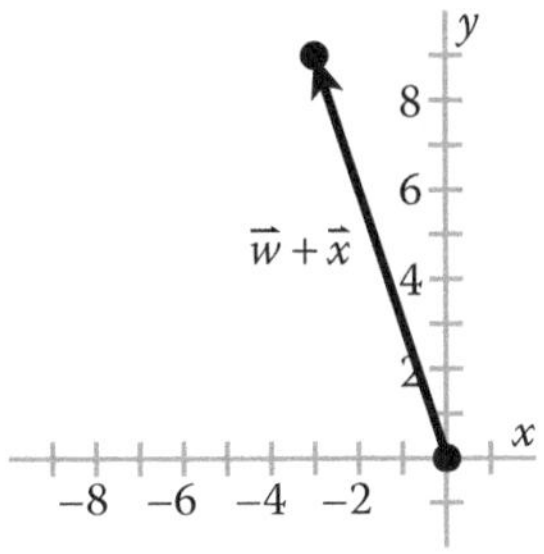

45. $\langle -9.3, 6.3 \rangle$

47. $\langle 3.5, -12 \rangle$

49. $\langle 11.7, 9.8 \rangle$

51.

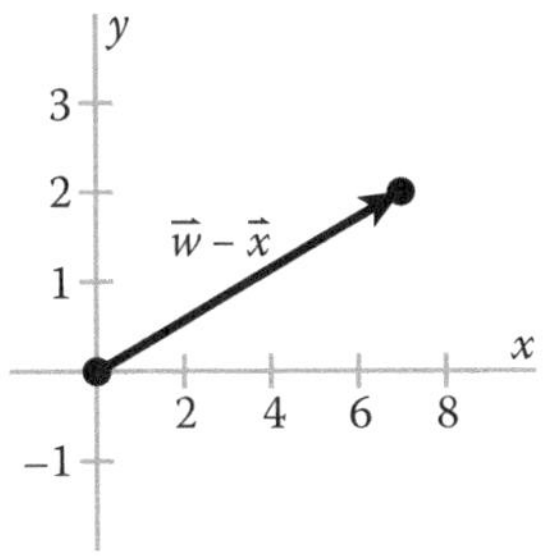

53.

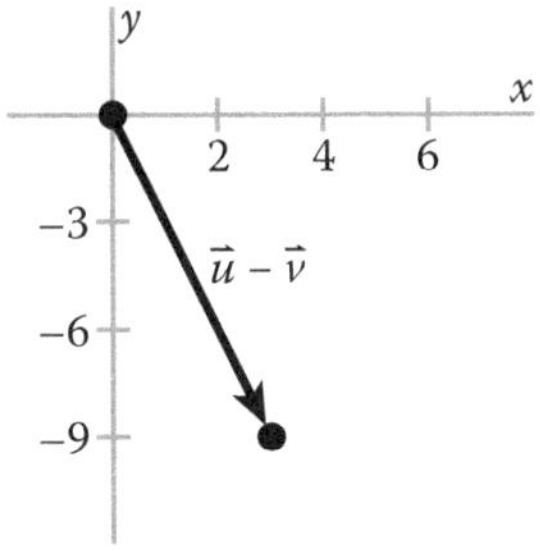

55.

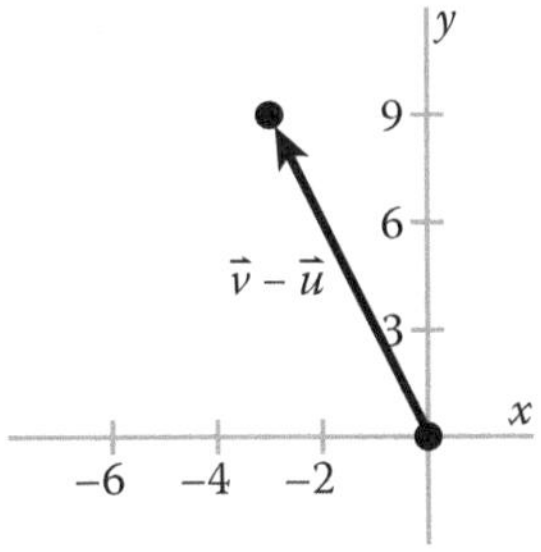

57. $\langle 5\sqrt{3}, 5 \rangle$

59. $\langle 9, 9 \rangle$

61. $\langle -3\sqrt{3}, 3 \rangle$

63. $\langle 3, -3 \rangle$

65. $\langle -5, -5\sqrt{3} \rangle$

67. $\langle -3.6, 2.3 \rangle$

69. $\langle -3.3, -1.8 \rangle$

71. $\langle 1.2, -5.5 \rangle$

73. $\left\langle \frac{3\sqrt{34}}{34}, \frac{5\sqrt{34}}{34} \right\rangle$

75. $\left\langle -\frac{\sqrt{5}}{5}, -\frac{2\sqrt{5}}{5} \right\rangle$

77. $\left\langle \frac{2\sqrt{13}}{13}, -\frac{3\sqrt{13}}{13} \right\rangle$

79. $\langle 35, 35\sqrt{3} \rangle$

81. $\langle -30\sqrt{3}, 30 \rangle$

83. $\langle -27\sqrt{2}, -27\sqrt{2} \rangle$

85. $4\vec{i} + 9\vec{j}$

87. $5\vec{i} - 5\vec{j}$

89. 45°

91. 330°

93. $2\pi/3$

95. 153.4°

97. 341.6°

99. 209.7°

101. $\langle 13.2, 2.6 \rangle$

103. $\langle 12.5, -10 \rangle$

105. $\langle 35.36, 35.36 \rangle$

107. $\langle 36.25, 28.32 \rangle$

109. $\langle -5/13, -12/13 \rangle$; $\langle -5/13, -12/13 \rangle$; $\langle -5/13, -12/13 \rangle$

111. 3.9 s

SECTION 7.4 ANSWERS

Warm Up/Just the Facts

1. Distributive Property

2. Commutative Property of Multiplication

3. Zero Product Property

4. $v_1 w_1 + v_2 w_2$

5. $\dfrac{\vec{u} \cdot \vec{v}}{\|\vec{u}\| \cdot \|\vec{v}\|}$

6. 0

7. 180°

8. $\left(\dfrac{\vec{u} \cdot \vec{v}}{\|\vec{v}\|^2} \right) \vec{v}$

9. scalar

10. magnitude

Essential Skills/Extensions

1. −18.3

3. 14.2

5. −48.7

7. 10; 68; 88

9. 7; 28.25; 0.8

11. −17.65; 41.29; −76.6

13. $2\sqrt{10}$

15. $\sqrt{146}$

17. $\sqrt{25.81}$

19. $11\pi/21$

21. $19\pi/24$

23. 66°

25. 141°

27. 31.3°

29. 122°

31. 98.4°

33. 96.3°

35. 90.6°

37. 85.6°

39. neither; orthogonal; parallel

41. orthogonal; parallel; neither

43. $\langle -4, 8\rangle$

45. $\langle -3, 3\rangle$

47. $\vec{w}_1 = \langle -5/2, 5/2\rangle$; $\vec{w}_2 = \langle 9/2, 9/2\rangle$

49. $\vec{w}_1 = \langle -18/13, -27/13\rangle$; $\vec{w}_2 = \langle -60/13, 40/13\rangle$

51. 169.7°, 7.1°, 3.2°

53. 123.7°, 45°, 11.3°

55. $\dfrac{3\sqrt{65} - 2\sqrt{37}}{6}$

57. $\dfrac{\sqrt{37} - 12}{3}$

59. −41/3

SECTION 7.5 ANSWERS

Warm Up/Just the Facts

1. $1 - 3i$. The complex conjugate of $a + bi$ is $a - bi$.

2. $-5 - 4i$. The complex conjugate of $a + bi$ is $a - bi$.

3. $-2 + 6i$. The complex conjugate of $a - bi$ is $a + bi$.

4. real; imaginary

5. modulus

6. $r(\cos\theta + i\sin\theta)$

7. $r_1 r_2 (\cos(\theta_1 + \theta_2) + i\sin(\theta_1 + \theta_2))$

8. $r_1 / r_2 (\cos(\theta_1 - \theta_2) + i\sin(\theta_1 - \theta_2))$

9. $z^n (\cos n\theta + i\sin n\theta)$

10. nth

Essential Skills/Extensions

1.

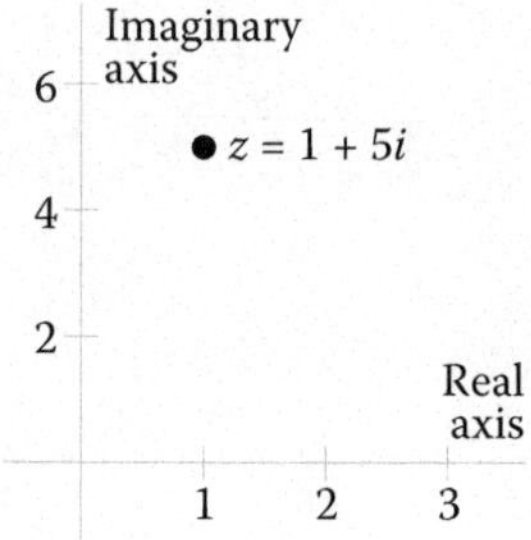

3.

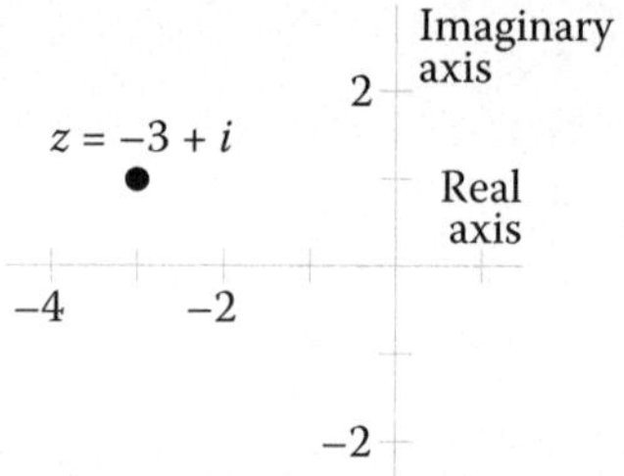

5.

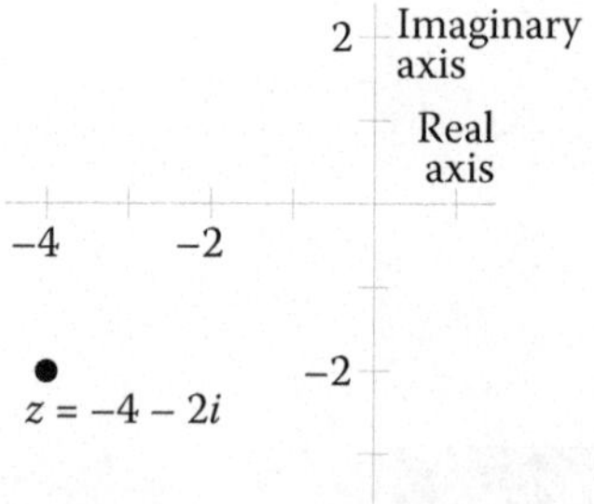

7. $2\sqrt{5}$

9. $3\sqrt{5}$

11. $\sqrt{17}$

13. $4\left(\cos\dfrac{\pi}{3} + i\sin\dfrac{\pi}{3}\right)$

15. $3\sqrt{2}\left(\cos\dfrac{7\pi}{4} + i\sin\dfrac{7\pi}{4}\right)$

17. $12\left(\cos\dfrac{5\pi}{6} + i\sin\dfrac{5\pi}{6}\right)$

19. $3\sqrt{5}\left(\cos 63° + i \sin 63°\right)$

21. $\sqrt{5}\left(\cos 117° + i \sin 117°\right)$

23. $\sqrt{17}\left(\cos 194° + i \sin 194°\right)$

25. $6(\cos 120° + i \sin 120°)$

27. $32(\cos 160° + i \sin 160°)$

29. $18(\cos 290° + i \sin 290°)$

31. $2(\cos 70° + i \sin 70°)$

33. $9(\cos 20° + i \sin 20°)$

35. $4(\cos 120° + i \sin 120°)$

37. $343\left(\cos \frac{\pi}{2} + i \sin \frac{\pi}{2}\right)$

39. $324(\cos \pi + i \sin \pi)$

41. $64\left(\cos \frac{2\pi}{3} + i \sin \frac{2\pi}{3}\right)$

43. $-16 + 16i$

45. $-648 - 648\sqrt{3}i$

47. $1 + \sqrt{3}i$

49. $-8i$

51. -256

53. $\frac{3125\sqrt{3}}{2} + \frac{3125}{2}i$

55. $\pm\sqrt{3}, \pm i\sqrt{3}$

57. $\pm\sqrt[3]{2}, \frac{\sqrt[3]{2}(-1 \pm \sqrt{3}i)}{2}, \frac{\sqrt[3]{2}(1 \pm \sqrt{3}i)}{2}$

59. $\pm 3, \pm 3i$

61. $\pm 3, \frac{3 \pm 3\sqrt{3}i}{2}, \frac{-3 \pm 3\sqrt{3}i}{2}$

63. $\pm 2, \pm 2i$

65. $6, -3 \pm 3\sqrt{3}i$

67. $\frac{\sqrt{3} + i}{2}, \frac{-\sqrt{3} + i}{2}, -i$

69. $2\sqrt[6]{2}\left(\cos \frac{5\pi}{12} + i \sin \frac{5\pi}{12}\right), 2\sqrt[6]{2}\left(\cos \frac{13\pi}{12} + i \sin \frac{13\pi}{12}\right),$
$2\sqrt[6]{2}\left(\cos \frac{7\pi}{4} + i \sin \frac{7\pi}{4}\right)$

71. $3125(\cos 175° + i \sin 175°)$

73. $272\sqrt{34}(\cos 3° + i \sin 3°)$

75. $1 + \sqrt{3}i$, $-1.34 + 1.48i$, $-1.83 - 0.81i$, $0.21 - 1.99i$, $1.96 - 0.42i$

77. $1.29 + 4.83i$, $-4.83 - 1.29i$, $3.53 - 3.54i$

CHAPTER 7 REVIEW ANSWERS

1. 54.2°, 87.8°, 17.9, $A \approx 79.7$; 125.8°, 16.2°, 5, $A \approx 22.3$

2. 40.6°, 9.4°, 4.3, $A \approx 27.8$

3. 34.4°, 99.6°, 57.6, $A \approx 683.3$

4. 18.8°, 20.2°, 10.9, $A \approx 35$

5. 10°, 87.9, 96.1, $A \approx 733.5$

6. 88.9°, 56.4°, 34.7°, $A \approx 1229.8$, $h \approx 60$

7. 74.9°, 55.8°, 49.3°, $A \approx 63.7$, $h \approx 11.6$

8. 49.3°, 39.7°, 25.1, $A \approx 152$, $h \approx 19$

9. 76.9°, 27.1°, 91.6, $A \approx 1919.2$, $h \approx 89.3$

10. 26°, 22°, 47.5, $A \approx 249.7$, $h \approx 20.8$

11. $\langle 6, -2\rangle$; $2\sqrt{10}$; 341.6°; $\left\langle \frac{3\sqrt{10}}{10}, -\frac{\sqrt{10}}{10} \right\rangle$; $6\vec{i} - 2\vec{j}$

12. $\langle -9, 9\rangle$; $9\sqrt{2}$; 135°; $\left\langle -\frac{\sqrt{2}}{2}, \frac{\sqrt{2}}{2} \right\rangle$; $-9\vec{i} + 9\vec{j}$

13. $\langle 7.2, 1.8\rangle$; $18\sqrt{0.17}$; 14°; $\left\langle \frac{40\sqrt{0.17}}{17}, \frac{10\sqrt{0.17}}{17} \right\rangle$; $7.2\vec{i} + 1.8\vec{j}$

14. $\langle -11.2, 2.6\rangle$

15. $\langle -32.6, 4\rangle$

16. -25.8; 12.24; 61.2; $\sqrt{13}$

17. -20.2; 31.76; 9.6; $\sqrt{61}$

18. -6.7; 18.64; 45.9; $\sqrt{49.46}$

19. neither; $\langle 5/17, -3/17\rangle$; $\vec{w}_1 = \langle 5/17, -3/17\rangle$; $\vec{w}_2 = \langle -39/17, -65/17\rangle$

20. parallel; $\langle -1, 8\rangle$; $\vec{w}_1 = \langle -1, 8\rangle$; $\vec{w}_2 = \langle 0, 0\rangle$

21. $4(\cos 120° + i \sin 120°)$

22. $-6 - 6\sqrt{3}i$; $0.7 + 0.26i$

23. $50{,}000\sqrt{3} - 50{,}000i$; $100{,}000(\cos 330° + i \sin 330°)$

24. $\pm\sqrt[6]{23}, \frac{\sqrt[6]{23}(1 \pm \sqrt{3}i)}{2}, \frac{\sqrt[6]{23}(-1 \pm \sqrt{3}i)}{2}$

25. $\sqrt[8]{72}\left(\cos \frac{7\pi}{16} + i \sin \frac{7\pi}{16}\right), \sqrt[8]{72}\left(\cos \frac{15\pi}{16} + i \sin \frac{15\pi}{16}\right),$
$\sqrt[8]{72}\left(\cos \frac{23\pi}{16} + i \sin \frac{23\pi}{16}\right), \sqrt[8]{72}\left(\cos \frac{31\pi}{16} + i \sin \frac{31\pi}{16}\right)$

SECTION 8.1 ANSWERS

Warm Up/Just the Facts

1. True. (4) = 7(2) – 10
2. True. 3(–4) + 5(–1) = –17
3. False. The x-intercept is 1/3.
4. system
5. no; infinitely many
6. graphing, substitution, elimination
7. substitution
8. consistent, independent
9. no; inconsistent
10. elimination

Essential Skills/Extensions

1.

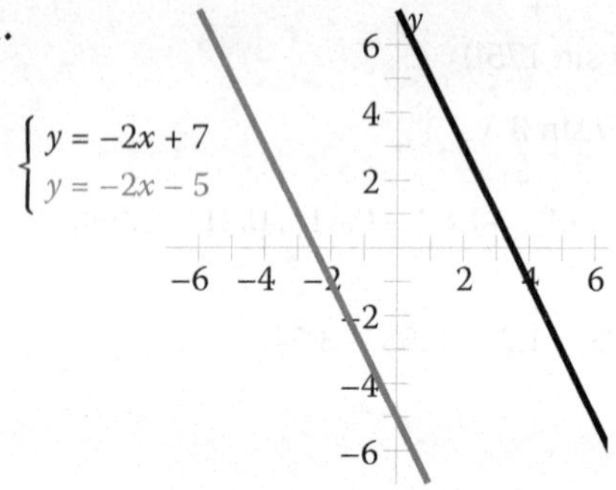

3.

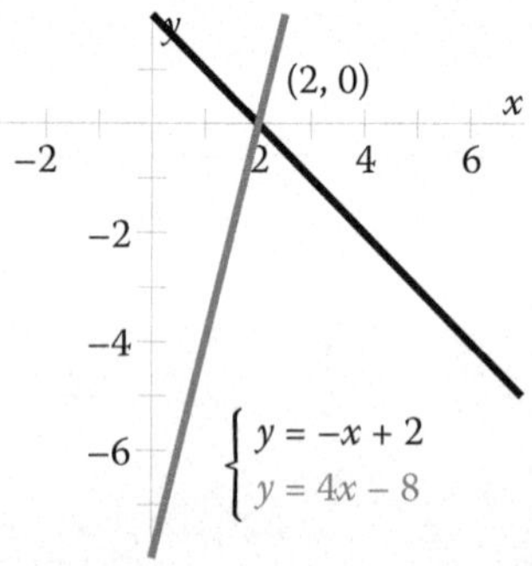

5. (–3, 4)
7. (–2, –3)
9. (–3, –1)
11. (–4, 2)
13. (1, 0)
15. (1, –2)
17. (6, 0)
19. (7, –1)
21. (2, 2)
23. (0, –2)
25. (–3, –1)
27. (2, 11/2)
29. infinite number of solutions
31. (47, –61/2)
33. (83/5, –34)
35. (3, –4)
37. (11/15, –14/15)
39. (14/23, 6/23)
41. consistent and dependent
43. consistent and independent
45. inconsistent
47. consistent and independent
49. 7/3
51. (–5850, 1850)
53. $\left(\frac{3}{2}, 1\right)$
55. $y = 3x + 2$ and $y = 5x - 2$, $y = 9x + 8$ and $4x + 7y = 6$ (possible answers)
57. (1, 2)

SECTION 8.2 ANSWERS

Warm Up/Just the Facts

1. $x = \frac{y - 14}{5}$
2. $y = 12 - 6x^2$
3. $x^2 = 20 - 5y^2$
4. nonlinear
5. one; two; zero; tangent
6. graphing, substitution, elimination
7. no
8. four
9. variable
10. elimination

Essential Skills/Extensions

1.

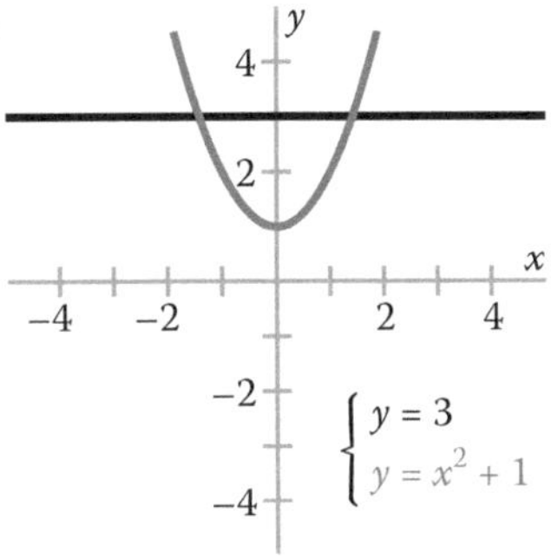

3.

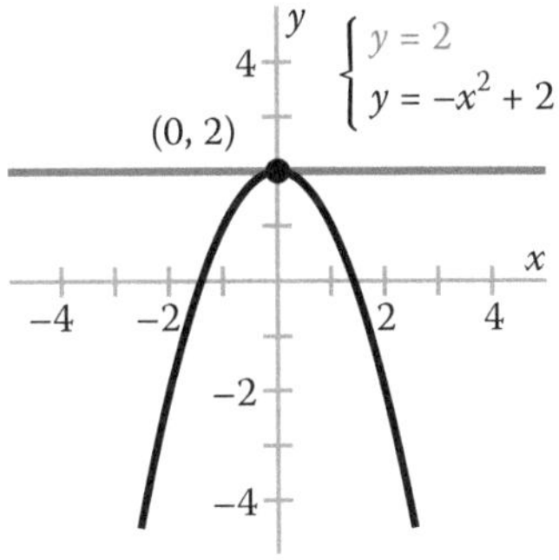

5. $(-7, -16), (-1, -4)$

7. $(0, 9)$

9. $(1, -1), (-1/2, 5/4)$

11. $(0, -4), (\sqrt{7}, 3), (-\sqrt{7}, 3)$

13. no solution

15. $(2\sqrt{6}, -1), (-2\sqrt{6}, -1)$

17. $(0, 4), (-1, 3)$

19. $(-5, -3), (1, 3)$

21. $(-4, -1), (2, 11)$

23. $(3, 1), (-1, 9)$

25. $(-4, -2), (5/4, -11/16)$

27. $(1, -2), (-7/2, -17/4)$

29. $(3, 0), (-3, 0)$

31. $(0, 5), (0, -5)$

33. $\left(\frac{5\sqrt{69}}{12}, \frac{65\sqrt{3}}{12}\right), \left(\frac{5\sqrt{69}}{12}, -\frac{65\sqrt{3}}{12}\right), \left(-\frac{5\sqrt{69}}{12}, \frac{65\sqrt{3}}{12}\right), \left(-\frac{5\sqrt{69}}{12}, -\frac{65\sqrt{3}}{12}\right)$

35.

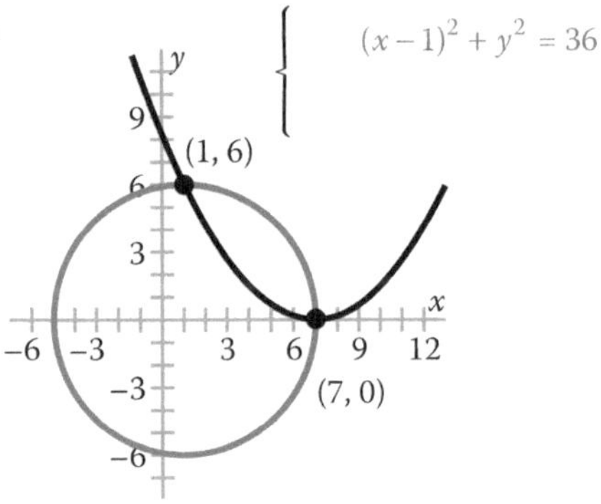

37. 5, 11

39.

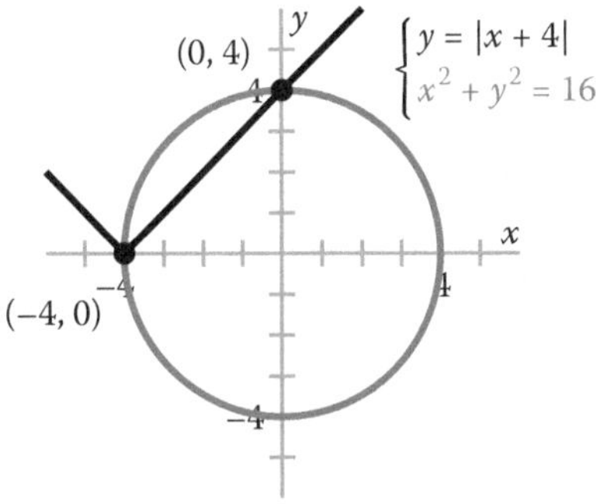

41. $(0, 4), (0, -4)$

SECTION 8.3 ANSWERS

Warm Up/Just the Facts

1. $x + x + 2 = 18$
2. $P(0.06)(3) = 45$
3. $d = 50h$
4. $x + y$
5. decimals
6. unknowns
7. graphing, substitution, elimination
8. table
9. $0.4x + 0.15y$
10. substitution; variable

Essential Skills/Extensions

1. \$2346

3. 5 mi/h; 3 mi/h

5. espresso: \$7/lb; vanilla: \$7.50/lb

7. 1.63 s

9. yes; $(0, -3)$

11. 9995 adult, 3005 children

13. yes; $(0, -15), (-9, -12)$

15. 12 oz of solution A, 18 oz of solution B

SECTION 8.4 ANSWERS

Warm Up/Just the Facts

1. False. The solution is (3, 2).
2. True. Substitute $x = -5$ and $y = 5$ into each equation.
3. True. Multiplying equation 2 by 4 yields equation 1. Therefore, they are equations of the same line.
4. line; plane
5. line
6. nonsquare system
7. triangular form
8. infinitely many solutions
9. multivariable linear systems
10. nonsquare

Essential Skills/Extensions

1. (6, 3, 4)
3. (−2, 1, −5)
5. infinitely many solutions
7. infinitely many solutions
9. pastel: $3, brush: $2, canvas: $1
11. $\left(1, -1, \frac{5}{3}\right)$
13. 12 in., 15 in., 20 in.
15. $x^2 + y^2 - 4x - 5 = 0$

SECTION 8.5 ANSWERS

Warm Up/Just the Facts

1.
$$\begin{aligned}\frac{2}{x}+\frac{3}{x+1} &= \frac{2(x+1)}{x(x+1)}+\frac{3(x)}{x(x+1)} && \textit{Write with a common denominator.}\\ &= \frac{2(x+1)+3(x)}{x(x+1)} && \textit{Add.}\\ &= \frac{2x+2+3x}{x(x+1)} && \textit{Distribute.}\\ &= \frac{5x+2}{x(x+1)} && \textit{Combine the like terms.}\end{aligned}$$

2.
$$\begin{aligned}\frac{-2x+1}{x}-\frac{4}{5} &= \frac{(-2x+1)(5)}{x(5)}-\frac{4(x)}{5(x)} && \textit{Write with a common denominator.}\\ &= \frac{(-2x+1)(5)-4(x)}{5x} && \textit{Subtract.}\\ &= \frac{-10x+5-4x}{5x} && \textit{Distribute.}\\ &= \frac{-14x+5}{5x} && \textit{Combine the like terms.}\end{aligned}$$

3.
$$\begin{aligned}\frac{-x+7}{4}-\frac{3}{x-6} &= \frac{(-x+7)(x-6)}{4(x-6)}-\frac{3(4)}{4(x-6)} && \textit{Write with a common denominator.}\\ &= \frac{(-x+7)(x-6)-3(4)}{4(x-6)} && \textit{Subtract.}\\ &= \frac{-x^2+6x+7x-42-12}{4(x-6)} && \textit{FOIL}\\ &= \frac{-x^2+13x-54}{4(x-6)} && \textit{Combine the like terms.}\end{aligned}$$

4. decomposed; partial
5. partial; $2x + 5$, $x + 3$
6. factor
7. denominator
8. 7

9. 2

10. quadratic

Essential Skills/Extensions

1. $\dfrac{9}{x-2}+\dfrac{6}{x-1}$

3. $-\dfrac{1}{x+3}+\dfrac{1}{x}-\dfrac{2}{x^2}$

5. $-\dfrac{2}{x}+\dfrac{-x+1}{x^2+2x+5}$

7. $\dfrac{2x-1}{x^2+4}+\dfrac{x+3}{(x^2+4)^2}$

9. $\dfrac{4}{x+6}+\dfrac{2}{x+3}+\dfrac{5}{x+12}$

11. $\dfrac{4x-1}{(x^2+3)^2}+\dfrac{\frac{3}{4}}{(x+1)^2}$

13. $\dfrac{5}{2ab(x+ab)}-\dfrac{5}{2ab(x-ab)}$

SECTION 8.6 ANSWERS

Warm Up/Just the Facts

1. True. $5(-3) \le -15 \Rightarrow -15 \le -15$, which is a true statement.
2. False. $3 > 4$, which is not a true statement.
3. True. $2(-3) - 3(3) < 10 \Rightarrow -6 - 9 < 10 \Rightarrow -15 < 10$, which is a true statement.
4. true
5. solid; dashed; dashed; above
6. point(s); overlapping; overlap
7. linear programming; maximum, minimum
8. objective; vertices
9. $x \le 5$
10. I

Essential Skills/Extensions

1.

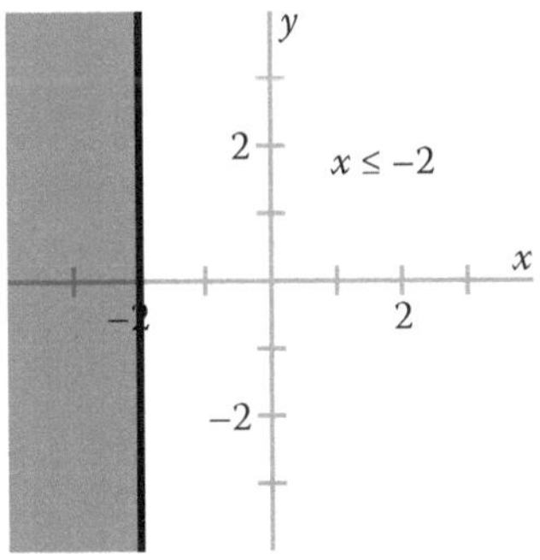

3.

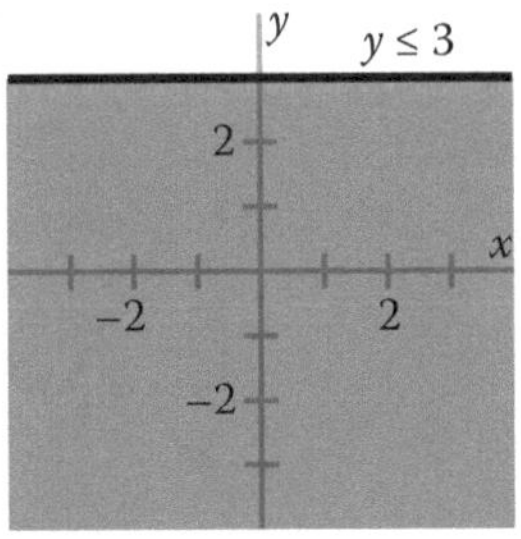

5.

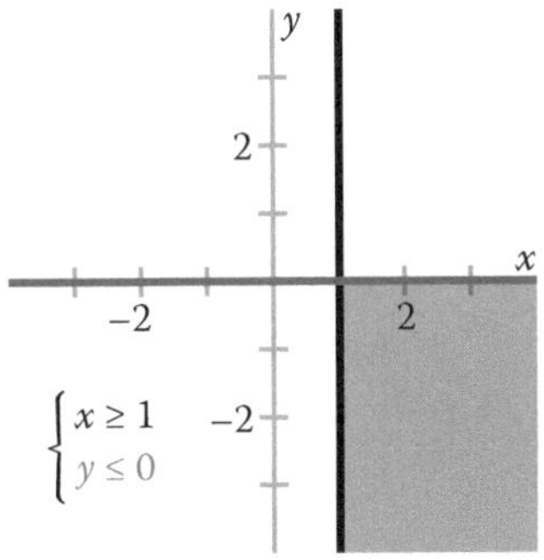

7.

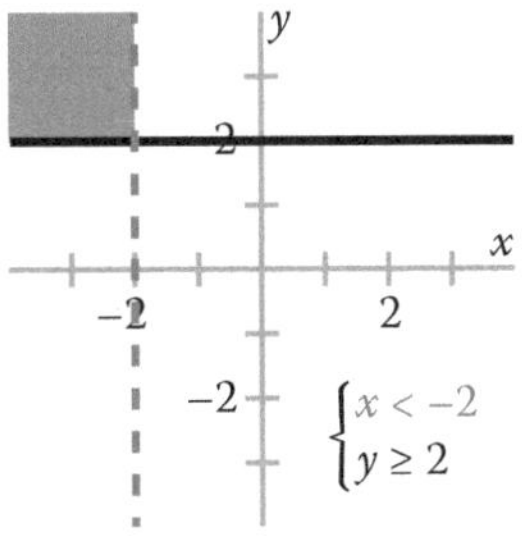

9.

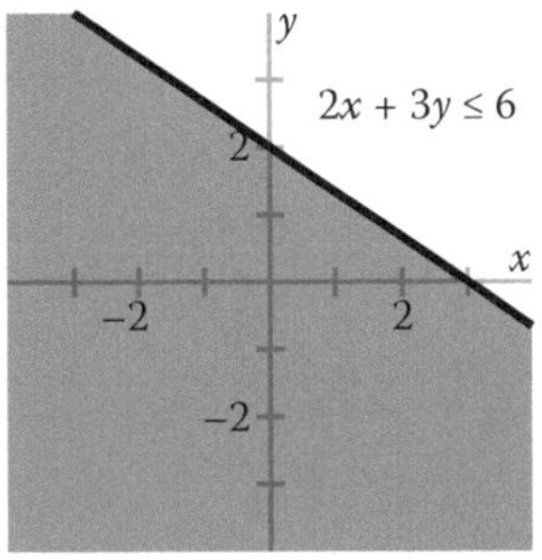

11.

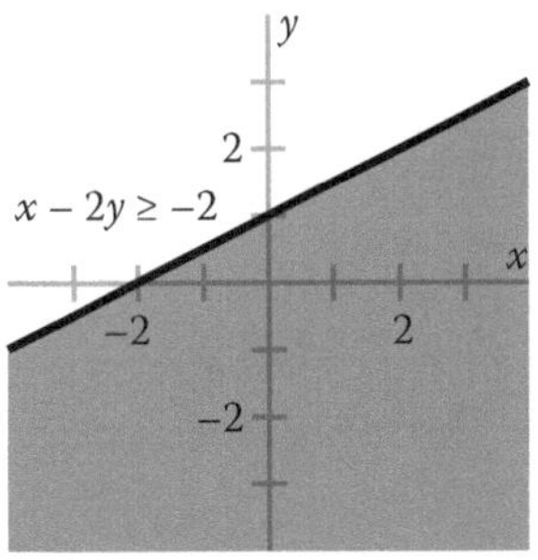

13.

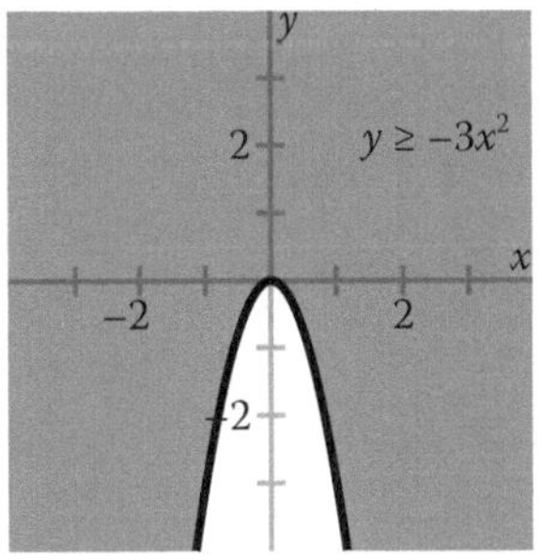

15.

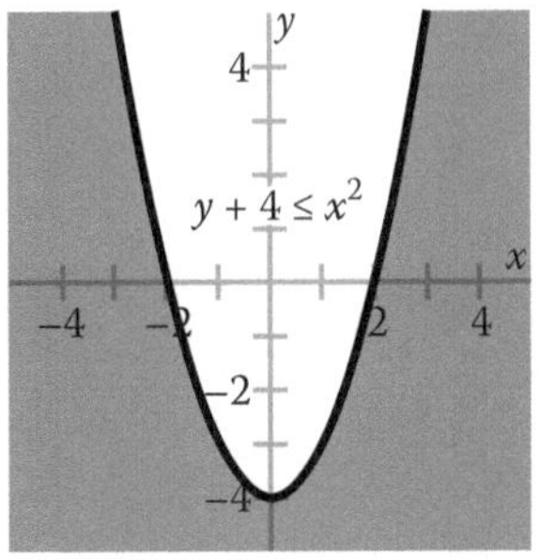

17.

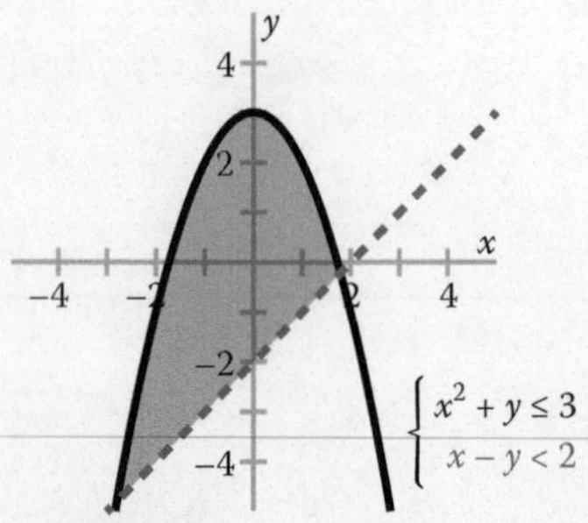

19.

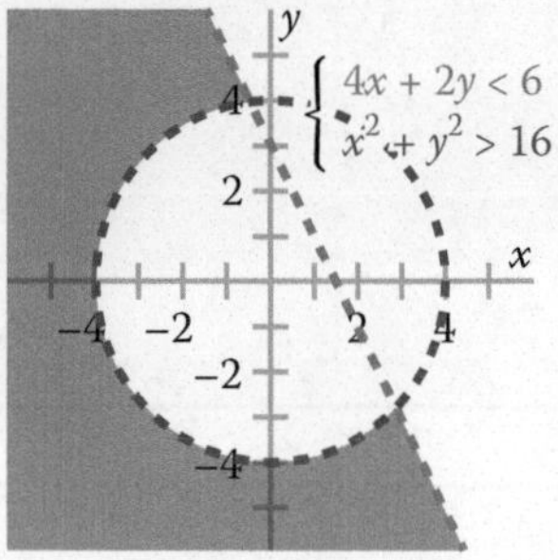

21.

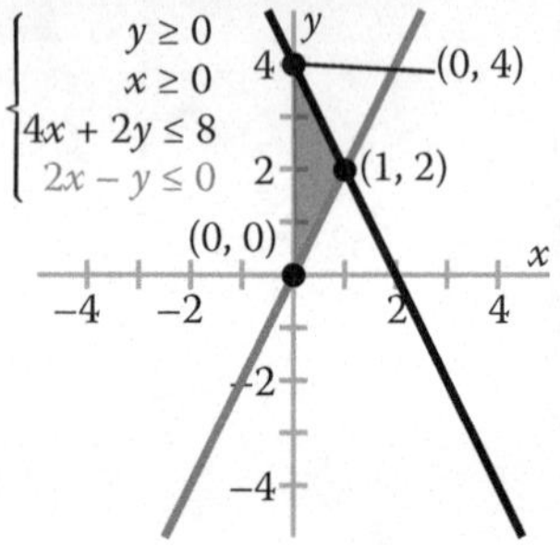

23. The minimum value is 12 at (3, 0).

25. The maximum value is 58 at (4, 6).

27. The maximum value is 23 at (8, 3).

29. 125 jumbo and no regular cookies

31. 350 desks, 150 chairs

33. 80 acres of wheat, 160 acres of oats

35.

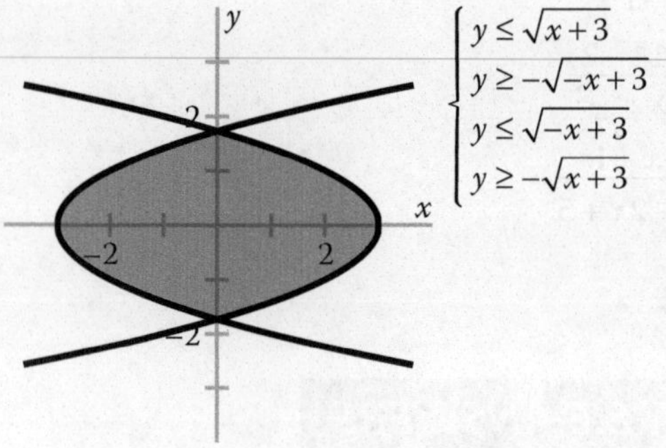

37.

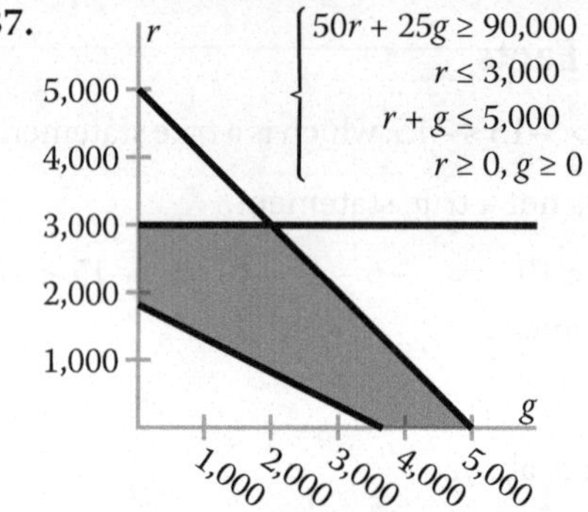

CHAPTER 8 REVIEW ANSWERS

1.

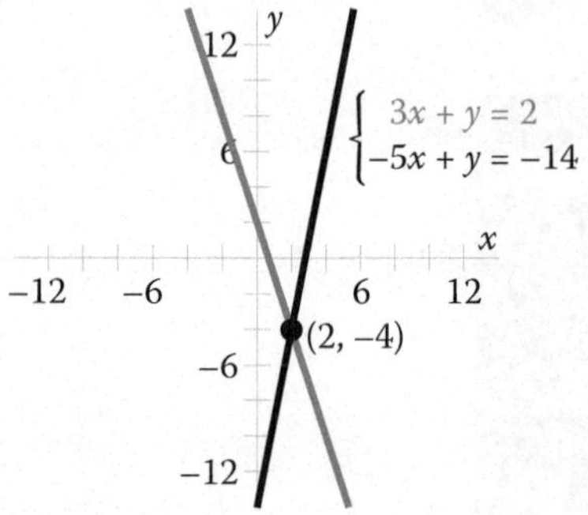

2. (−1, 2)

3. $\left(3, \dfrac{5}{3}\right)$

4. (−1, 1)

5. inconsistent

6. (−1, 9), (3, 1)

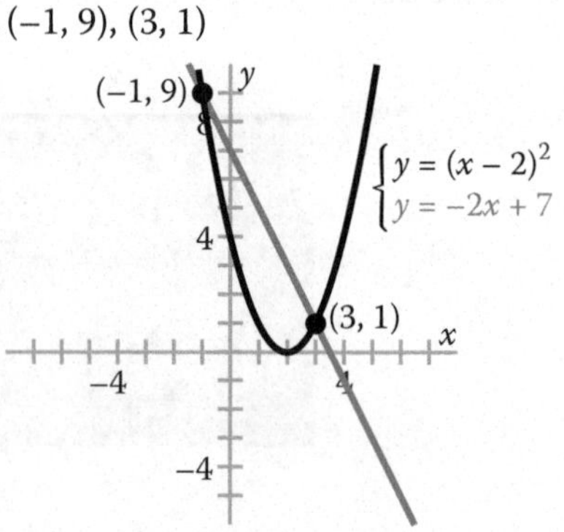

7. (−7, −16), (−1, −4)

8. $\left(\sqrt{6}, 2\right), \left(-\sqrt{6}, 2\right)$

9. no solution

10. $\left(\sqrt{5}, \sqrt{31}\right), \left(-\sqrt{5}, \sqrt{31}\right), \left(\sqrt{5}, -\sqrt{31}\right), \left(-\sqrt{5}, -\sqrt{31}\right)$

11. 8.5% account: \$7618.50, 9.5% account: \$1168.50

12. airplane: 475 mi/h, wind: 25 mi/h

13. 3 oz

14. 0.75 s

15. yes; (3, 24/5), (−3, 24/5)

16. $\left(\frac{5}{2}, -\frac{1}{2}, 1\right)$

17. $\left(-\frac{1}{4}, -\frac{3}{4}, 1\right)$

18. (5, 3, −2)

19. no solution

20. hamburger: \$3.75, fries: \$2.25, drink: \$1.75

21. $-\frac{6}{x-3}+\frac{5}{x-1}$

22. $\frac{3}{x+9}+\frac{7}{x+6}$

23. $-\frac{\frac{11}{8}}{(x-1)}-\frac{\frac{5}{8}}{(x+3)}+\frac{\frac{21}{2}}{(x+3)^2}$

24. $\frac{4}{x}+\frac{x+1}{2x^2-x+1}$

25. $\frac{3x-1}{x^2+5}-\frac{2x-4}{(x^2+5)^2}$

26.

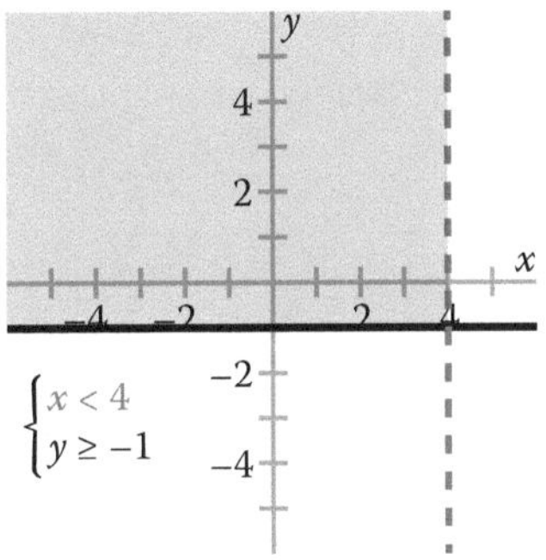

27.

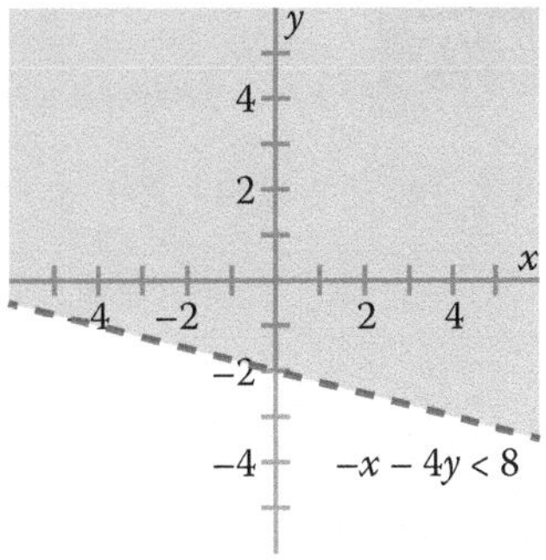

28.

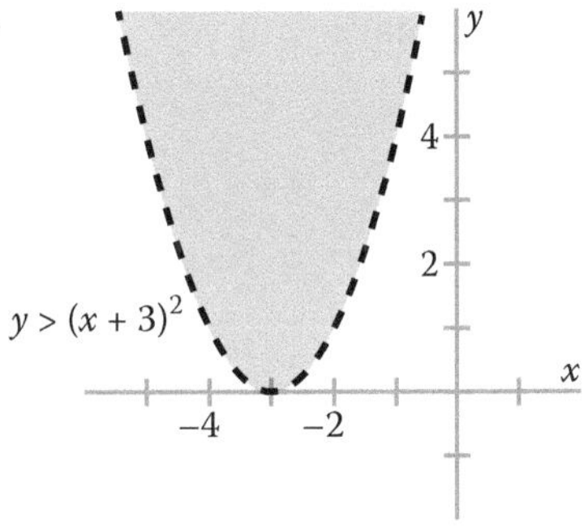

29.

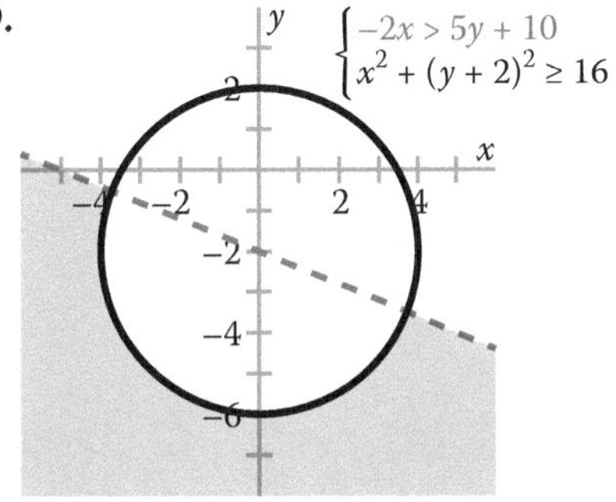

30. 22 trucks, 18 cars

SECTION 9.1 ANSWERS

Warm Up/Just the Facts

1. $3x + 4y = -5$
2. $x - 3y = 29$
3. $9x - y = 6$
4. square
5. augmented; coefficients; constant
6. augmented; *i*th
7. augmented; elementary
8. elementary; reduced
9. matrix; substitution
10. augmented; dependent

Essential Skills/Extensions

1. 1 × 3

3. 4 × 1

5. 3 × 2

7. $\left[\begin{array}{rrr|r} 1 & -2 & 5 & -12 \\ 3 & 4 & 1 & 1 \\ -6 & -1 & -3 & -8 \end{array}\right]$

9. $\left[\begin{array}{rr|r} 1 & 6 & 12 \\ -9 & -4 & 5 \end{array}\right]$

11. $\left[\begin{array}{rrrr|r} 3 & 7 & -6 & 7 & 4 \\ 6 & 20 & -10 & 1 & 26 \\ -4 & 2 & -3 & -8 & -67 \end{array}\right]$

13. $\begin{bmatrix} 3 & 0 & -2 & 9 \\ 4 & 8 & -7 & 1 \\ 2 & -1 & 5 & 0 \end{bmatrix}$

15. $\begin{bmatrix} 5 & -1 & 13 \\ 0 & 1 & 7 \end{bmatrix}$

17. $\begin{bmatrix} 13 & 5 & 26 \\ 3 & 0 & 7 \\ 2 & 1 & -4 \end{bmatrix}$

19. $\begin{bmatrix} -2 & 5 \\ 0 & 3 \\ 21 & 9 \end{bmatrix}$

21. $\left(\frac{2}{3}, 7, -\frac{4}{3}\right)$

23. (11, –1, 0)

25. (–3, 4, 0)

27. (1, –1, 2)

29. (–1, 2, 1)

31. (122/15, –1/15, –58/5)

33. no solution, inconsistent

35. (0, 5, –1)

37. $\left(\frac{8+3a}{9}, -\frac{4+6a}{9}, a\right)$

39. false

41. $\begin{cases} x + y + z = 24 \\ 4y - z = 4 \\ 2x + 4y - z = 0 \end{cases}$, (–2, 6, 20)

43. $\left(\frac{2-a}{3}, \frac{-a-7}{3}, -a-3, a\right)$, where a is any real number.

SECTION 9.2 ANSWERS

Warm Up/Just the Facts

1. 3 × 5
2. 4 × 3
3. 1 × 5
4. 1/5
5. added, subtracted; addition; subtraction
6. multiplied
7. matrix
8. multiplication
9. column
10. columns; rows

Essential Skills/Extensions

1. $a = 4, b = 2, c = -13$

3. $a = -2, b = -14, c = -5$

5. $a = 7, b = -11, c = 31$ and $a = -7, b = 3, c = 17$

7. $D + E = \begin{bmatrix} 18 & -54 \\ 11 & -31 \end{bmatrix}$, $D - E = \begin{bmatrix} 48 & 16 \\ -7 & -37 \end{bmatrix}$

9. $H + G = \begin{bmatrix} -3 & 0 & 11 \\ 7 & -5 & 2 \end{bmatrix}$,

$G - H = \begin{bmatrix} 11 & -16 & 11 \\ -17 & 5 & -10 \end{bmatrix}$

11. $J + K = \begin{bmatrix} 4 & -2 \\ 1 & 7 \\ 18 & 16 \\ 3 & 1 \end{bmatrix}$, $J - K = \begin{bmatrix} 2 & -10 \\ 13 & -7 \\ 28 & 4 \\ -19 & -7 \end{bmatrix}$

13. $\begin{bmatrix} -25 & -50 & 20 \\ -35 & -10 & -40 \end{bmatrix}$

15. $\begin{bmatrix} 5 & -1 \\ -4 & 9 \\ -3 & -8 \\ 1 & 2 \end{bmatrix}$

17. $\begin{bmatrix} -12 & 0 \\ 6 & -36 \\ 42 & 24 \end{bmatrix}$

19. $\begin{bmatrix} 8 & -2 \\ 1 & 6 \\ -3 & 3 \end{bmatrix}$

21. $\begin{bmatrix} 6.5 & -2 \\ 0.5 & 3.5 \\ -2 & 3.5 \end{bmatrix}$

23. $\begin{bmatrix} -28 & 4 \\ -6 & -32 \\ 14 & -2 \end{bmatrix}$

25. AB and BA are defined, both products are 2×2 matrices.

27. AB is defined, $AB = 1 \times 2$

29. BA is defined, $BA = 4 \times 3$

31. Neither products are defined.

33. $\begin{bmatrix} 56 \\ 44 \end{bmatrix}$

35. not possible

37. $\begin{bmatrix} -5 & -15 & 46 \\ -1 & 74 & -59 \end{bmatrix}$

39. 500

41. 520

43. 1520

45. $\begin{bmatrix} 560 & 260 \\ 3080 & 1480 \end{bmatrix}$

47. Using column 1 for company A, column 2 for company B, row 1 for June, and row 2 for July yields $\begin{bmatrix} 178{,}000 & 122{,}550 \\ 166{,}500 & 145{,}000 \end{bmatrix}$.

49. $\begin{bmatrix} 178{,}000 & 122{,}550 \\ 166{,}500 & 145{,}000 \end{bmatrix} - \begin{bmatrix} 28{,}500 & 18{,}600 \\ 22{,}700 & 22{,}800 \end{bmatrix} =$
$\begin{bmatrix} 149{,}500 & 103{,}950 \\ 143{,}800 & 122{,}200 \end{bmatrix}$

51. $\begin{bmatrix} 17.6 & 19.2 \\ 96 & 6.4 \\ 70.4 & 48 \end{bmatrix}$

53. $\begin{bmatrix} 49.4 & 30.8 \\ 94 & 36.6 \\ 55.6 & 26.5 \end{bmatrix}$

55. $\begin{bmatrix} 28.56 & 11.76 \\ 8.4 & 26.04 \\ -5.04 & -13.02 \end{bmatrix}$

57. $\begin{bmatrix} 83.36 & 26.18 & 72.2 \\ 111.92 & 40.99 & 114.3 \end{bmatrix}$

59. $\begin{bmatrix} 24n & 15n \\ 35n & 12n \\ 18n & 8n \end{bmatrix}$ (possible answer)

61. $a = 7.5, b = 134, c = 1/3, d = 5/3, f = -5, g = 6, h = 3, j = 45, x = 8.8, y = 6.5, z = 2$

SECTION 9.3 ANSWERS

Warm Up/Just the Facts

1. $\left[\begin{array}{cc|c} -3 & -5 & 2 \\ 1 & 1 & 15 \end{array}\right]$

2. $\left[\begin{array}{cc|c} -2 & 1 & 7 \\ 2 & 8 & -20 \end{array}\right]$

3. $\left[\begin{array}{ccc|c} -3 & 1 & 2 & 7 \\ 1 & 0 & 9 & 2 \\ 0 & -5 & -1 & 8 \end{array}\right]$

4. singular; *bc*
5. triangular
6. adding
7. multiplied by 2
8. nonsingular
9. vertices; counterclockwise
10. system of linear equations

Essential Skills/Extensions

1. −29
3. −128
5. 3
7. 21/2 units^2
9. The points are not collinear.
11. (1, 7)
13. (4, 3, 2)
15. The points are not collinear.
17. −1
19. (4 – 8) should be (4 – 0).

SECTION 9.4 ANSWERS

Warm Up/Just the Facts

1. (0)(2) – (4)(–3) = 0 – (–12) = 12
2. (–5)(3) – (1)(2) = –15 – 2 = –17
3. (3)(7) – (–8)(0) = 21 – 0 = 21
4. does not equal 0
5. 0
6. not multiplied
7. 2 × 2
8. augmented; identity
9. coefficient; constant
10. divide; determinant

Essential Skills/Extensions

1. $\begin{bmatrix} -2 & -1 \\ -5 & -3 \end{bmatrix}$

3. $\begin{bmatrix} \frac{2}{3} & -\frac{1}{3} \\ -\frac{5}{3} & \frac{4}{3} \end{bmatrix}$

5. $\begin{bmatrix} 0 & \frac{1}{3} & \frac{1}{3} \\ \frac{1}{2} & 0 & -\frac{1}{2} \\ \frac{1}{2} & -\frac{1}{3} & \frac{1}{6} \end{bmatrix}$

7. $\begin{bmatrix} -1 & \frac{6}{5} & \frac{7}{5} \\ 1 & -\frac{3}{5} & -\frac{6}{5} \\ 1 & -1 & -1 \end{bmatrix}$

9. (2, 0)

11. $\begin{bmatrix} 2 & \frac{1}{2} \\ 3 & 5 \end{bmatrix}$; $\begin{bmatrix} 9 \\ 56 \end{bmatrix}$; (2, 10)

13. 50 \$100 bills, 250 \$20 bills

15. 12

CHAPTER 9 REVIEW ANSWERS

1. $\left[\begin{array}{ccc|c} 3 & -2 & 5 & 30 \\ 5 & 0 & 4 & 35 \\ 0 & 3 & -2 & -4 \end{array}\right]$

2. $\begin{bmatrix} 0 & 2 & 8 & -9 \\ -3 & 5 & -6 & -2 \\ -4 & -7 & 0 & 1 \end{bmatrix}$

3. (5, 1, 4)

4. (−1, 2, 7)

5. (5, 0, 1)

6. $m = -2$, $n = 2$, $p = -5/2$

7. $F + X = \begin{bmatrix} -1 & 7 & -7 \\ -1 & -4 & 16 \\ 5 & -2 & 17 \end{bmatrix}$; $F - X = \begin{bmatrix} -3 & 7 & 9 \\ -9 & 4 & 2 \\ 1 & 0 & 7 \end{bmatrix}$

8. $P = \begin{bmatrix} \frac{5}{2} & -\frac{5}{2} & -\frac{1}{2} \\ -4 & \frac{1}{2} & -6 \end{bmatrix}$

9. CD is defined, $CD = 2 \times 4$; DC is not defined.

10. $SR = \begin{bmatrix} -2 & 1 & 7 \\ 0 & 6 & 44 \\ -28 & -1 & -12 \\ 30 & 9 & 71 \end{bmatrix}$

11. −1/12

12. 10

13. 5

14. 31/2 units^2

15. (3, 4, 5)

16. $\begin{bmatrix} \frac{1}{8} & \frac{1}{8} \\ \frac{3}{8} & -\frac{5}{8} \end{bmatrix}$

17. $\begin{bmatrix} 1 & 3 \\ 1 & 2 \end{bmatrix}$

18. $\begin{bmatrix} \frac{1}{2} & 0 & \frac{1}{2} \\ 2 & 0 & 1 \\ \frac{17}{2} & 1 & \frac{7}{2} \end{bmatrix}$

19. $\begin{bmatrix} -\frac{14}{5} & \frac{4}{5} & -\frac{7}{5} \\ \frac{6}{5} & -\frac{1}{5} & \frac{3}{5} \\ \frac{11}{10} & -\frac{1}{10} & \frac{3}{10} \end{bmatrix}$

20. (8, −3)

SECTION 10.1 ANSWERS

Warm Up/Just the Facts

1. II

2. III

3. IV

4. (x, y); (r, θ)

5. 3; $5\pi/6$; polar axis

6. x; y

7. $\sqrt{x^2 + y^2}$; y/x

8. opposite

9. rectangular; polar; $(r, \theta + 2\pi n)$

10. clockwise

Essential Skills/Extensions

1.

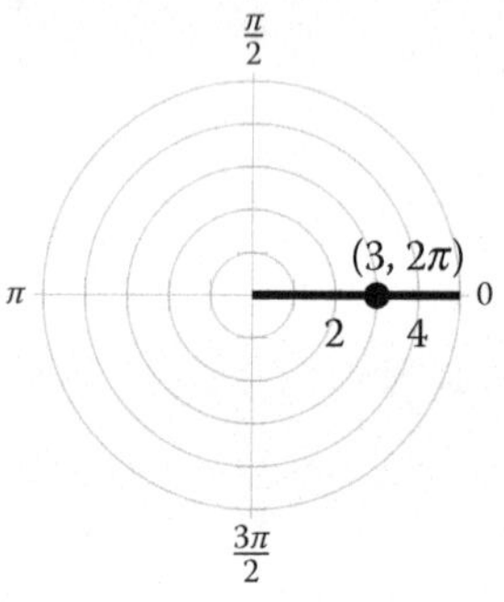

3.

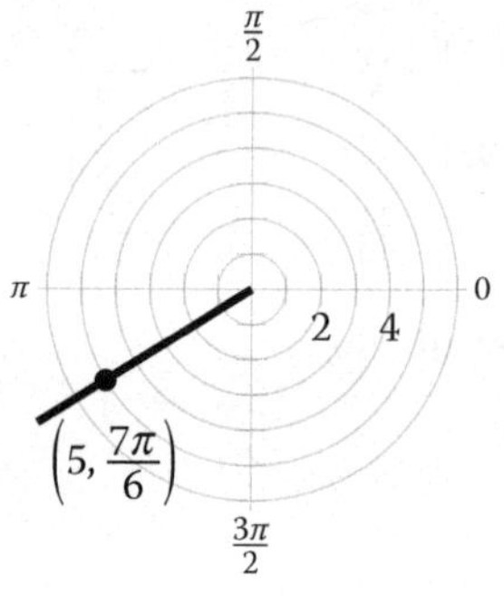

5.

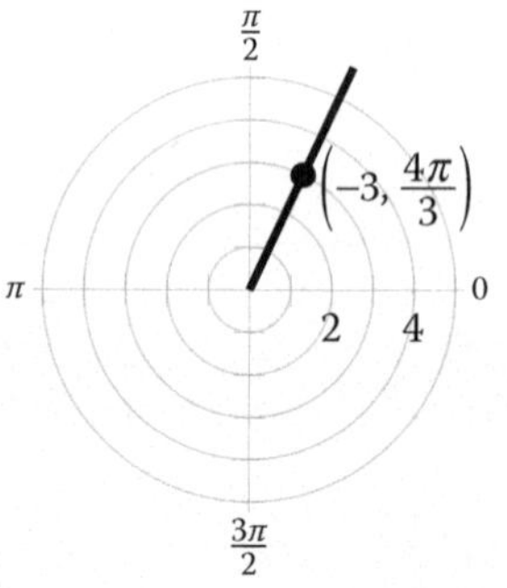

7.

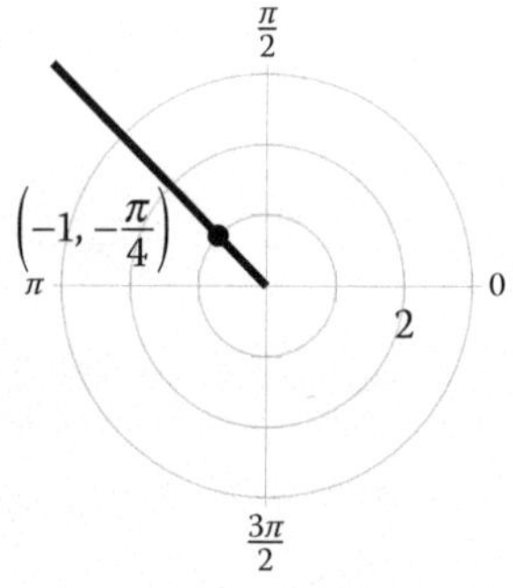

9. $(4, 7\pi/4)$, $(-4, 3\pi/4)$

11. $(2, 4\pi/3)$, $(-2, \pi/3)$

13. $(5, 5\pi/3)$, $(-5, 2\pi/3)$

15. $\left(2\sqrt{3}, -2\right)$

17. $\left(-\dfrac{5\sqrt{2}}{2}, \dfrac{5\sqrt{2}}{2}\right)$

19. $\left(-\dfrac{13\sqrt{3}}{4}, -\dfrac{13}{4}\right)$

21. $(2, 5\pi/6)$

23. $(4, -\pi/4)$

25. $\left(2\sqrt{2}, -\dfrac{\pi}{6}\right)$

27. $\left(\sqrt{29}, -1.19\right)$

29. $\left(2\sqrt{13}, 4.12\right)$

31. $\left(\sqrt{113}, 2.29\right)$

33. $(4, -5\pi/4)$, $(-4, -\pi/4)$, $(4, 3\pi/4)$, $(-4, 7\pi/4)$

35. $\left(-10\sqrt{2}, -\dfrac{4\pi}{3}\right)$, $\left(10\sqrt{2}, -\dfrac{\pi}{3}\right)$, $\left(-10\sqrt{2}, \dfrac{2\pi}{3}\right)$, $\left(10\sqrt{2}, \dfrac{5\pi}{3}\right)$

37. $\left(\sqrt{65}, -4.19\right)$, $\left(-\sqrt{65}, -1.05\right)$, $\left(\sqrt{65}, 2.09\right)$, $\left(-\sqrt{65}, 5.23\right)$

39. $(6, 5\pi/6)$, $(-6, 11\pi/6)$, $(6, 17\pi/6)$, $(-6, 23\pi/6)$

41. $(-10, 2\pi/3)$, $(10, 5\pi/3)$, $(-10, 8\pi/3)$, $(10, 11\pi/3)$

43. $(-5, 0.64)$, $(5, 3.79)$, $(-5, 6.93)$, $(5, 10.07)$

45. $\left(\dfrac{s\sqrt{2}}{2}, \dfrac{\pi}{4}\right)$, $\left(\dfrac{s\sqrt{2}}{2}, \dfrac{3\pi}{4}\right)$, $\left(\dfrac{s\sqrt{2}}{2}, \dfrac{5\pi}{4}\right)$, $\left(\dfrac{s\sqrt{2}}{2}, \dfrac{7\pi}{4}\right)$

SECTION 10.2 ANSWERS

Warm Up/Just the Facts

1. $(x-3)^2 + y^2 = 9$

$x^2 - 6x + y^2 = 0$

$x^2 - 6x + 9 + y^2 = 9$ — *Complete the square.*

$(x-3)^2 + y^2 = 9$ — *Factor.*

2. $(y-4)^2 - x^2 = 16$

$x^2 - y^2 + 8y = 0$

$x^2 - (y^2 - 8y + 16) = -16$ — *Complete the square.*

$x^2 - (y-4)^2 = -16$ — *Factor.*

$-x^2 + (y-4)^2 = 16$ — *Divide by −1.*

$(y-4)^2 - x^2 = 16$ — *Write in standard form.*

3. $(x-5)^2+(y+1)^2=26$

$$x^2-10x+y^2+2y=0$$

$$x^2-10x+25+y^2+2y+1=25+1 \quad \textit{Complete the square.}$$

$$(x-5)^2+(y+1)^2=26 \quad \textit{Factor.}$$

4. $r\cos\theta$; $r\sin\theta$; $\tan\theta$; r^2

5. reciprocal

6. vertical line; circle

7. $\theta = c$

8. spiral

9. $r = a\sin n\theta$, $r = a\cos n\theta$

10. n

Essential Skills/Extensions

1. $x^2+y^2-2x=0$

3. $x^2+y^2+y=0$

5. $x^4+x^2y^2-y^2=0$

7. $x+4=0$

9. $y-2=0$

11. $9y^4+9x^2y^2-x^2=0$

13. $\sqrt{3}x+y=0$

15. $x+y=0$

17. $x^2+y^2-4=0$

19. $r=8\sec\theta$

21. $r=\dfrac{3}{2\cos\theta+5\sin\theta}$

23. $r=\dfrac{9}{2\sin\theta-3\cos\theta}$

25. $r=4$

27. $r=5\tan\theta\sec\theta$

29.

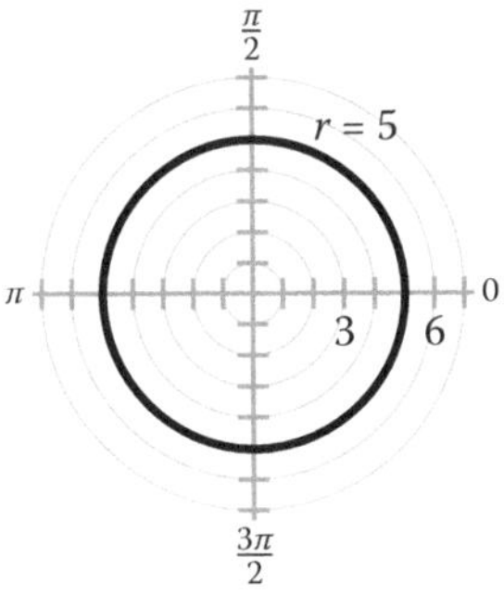

31.

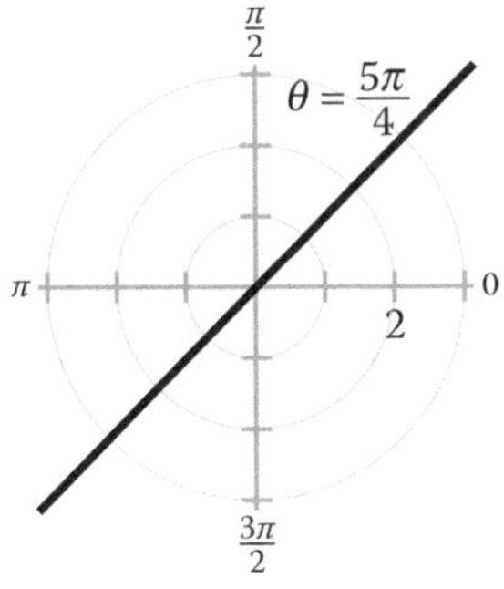

33.

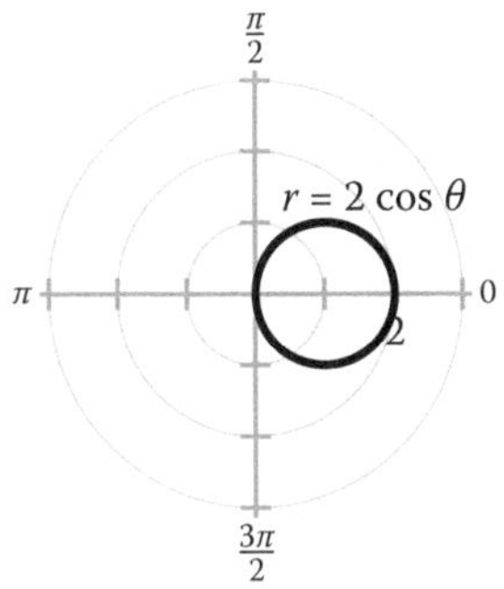

35.

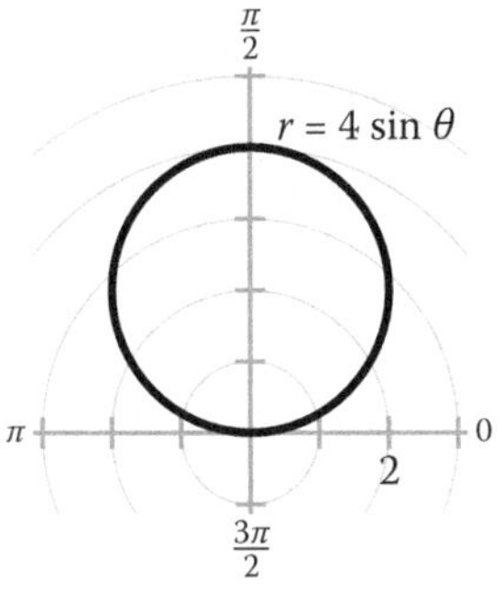

37. center $(5/2, \pi/2)$; radius 2.5

39. center $(2, 0)$; radius 2

41. center $(2, 3\pi/2)$; radius 2

43. 3 petals; petal radius 6; horizontal axis of symmetry

45. 7 petals; petal radius 5; vertical axis of symmetry

47. 4 petals; petal radius 4; horizontal axis of symmetry and vertical axis of symmetry

49.

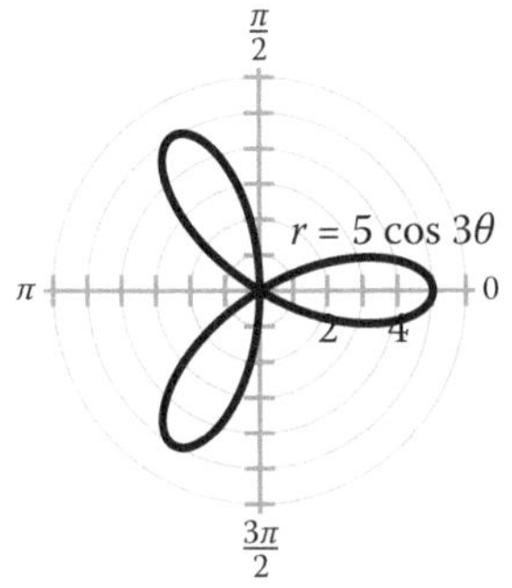

51.

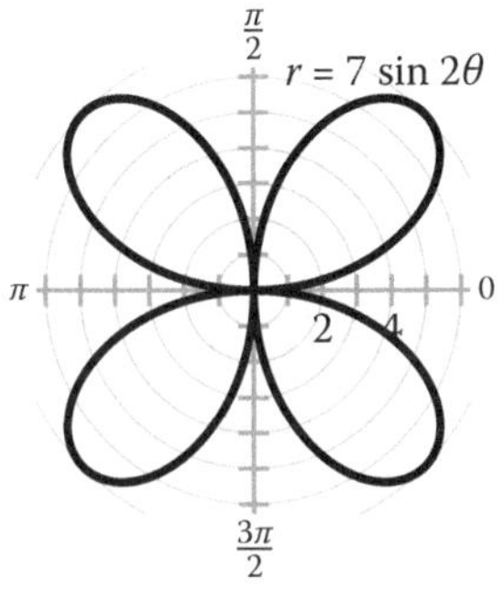

53.

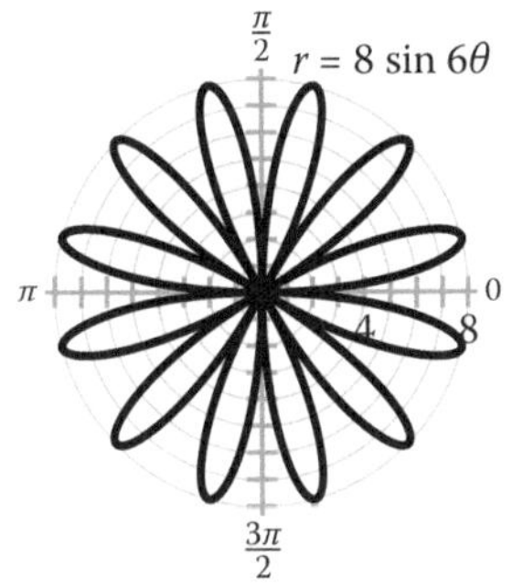

55.

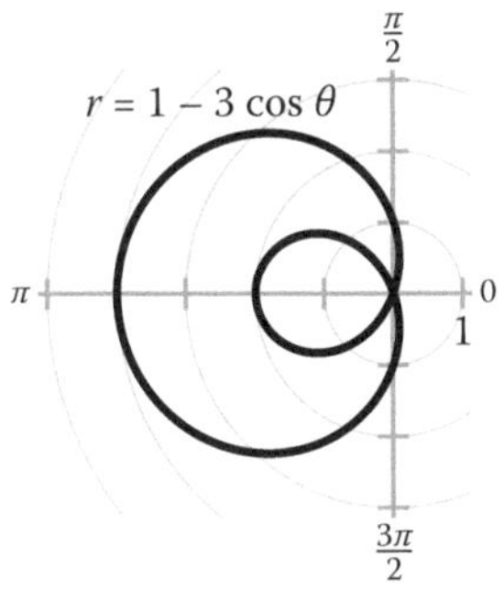

57.

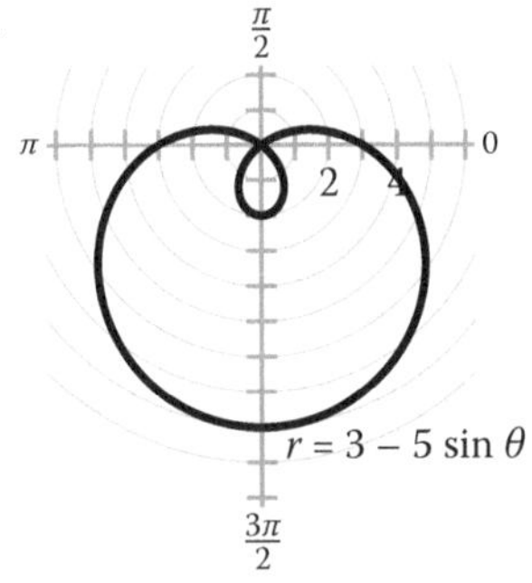

59.

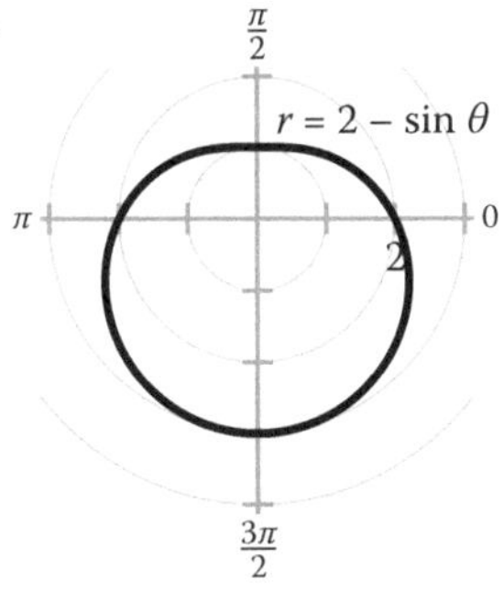

61.

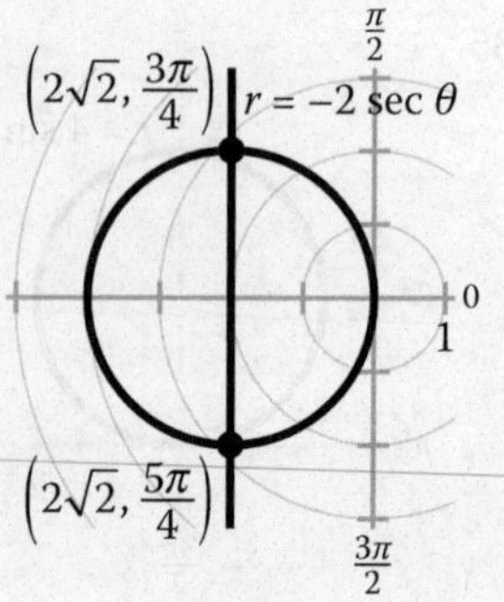

63.

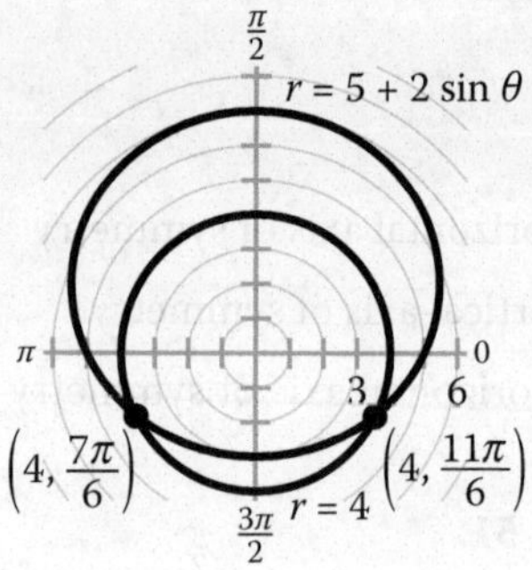

65. $\sqrt{2}x - \sqrt{2}y - 10 = 0$

67.

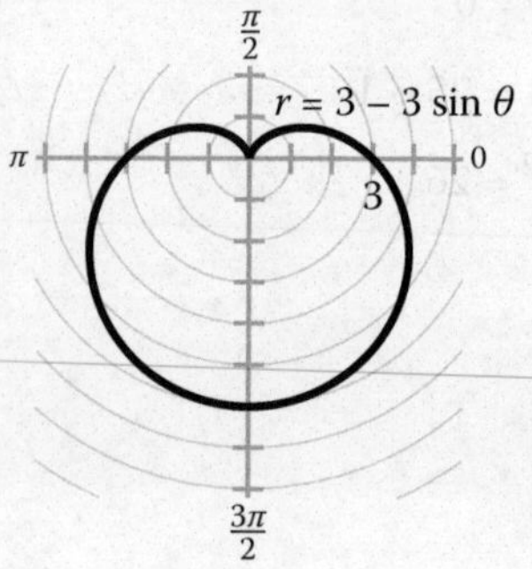

69. $r = -16\sin\theta$

71. $r = 4\sin 5\theta$

73. $r = 8 - 2\cos\theta$

SECTION 10.3 ANSWERS

Warm Up/Just the Facts

1.

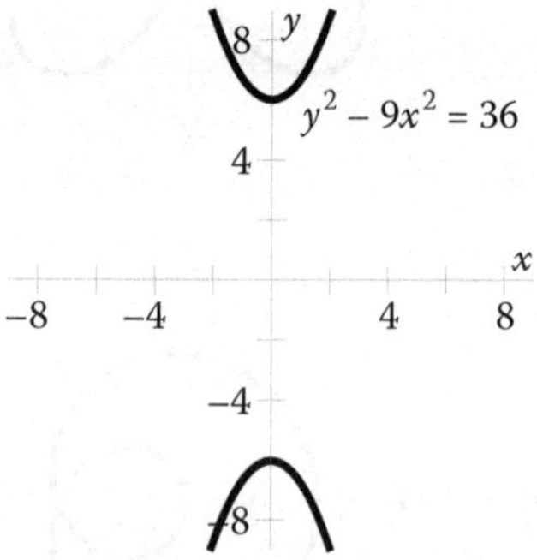

2.

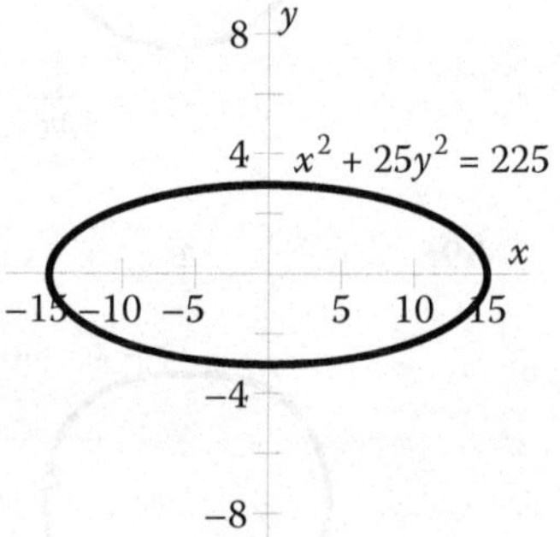

3.

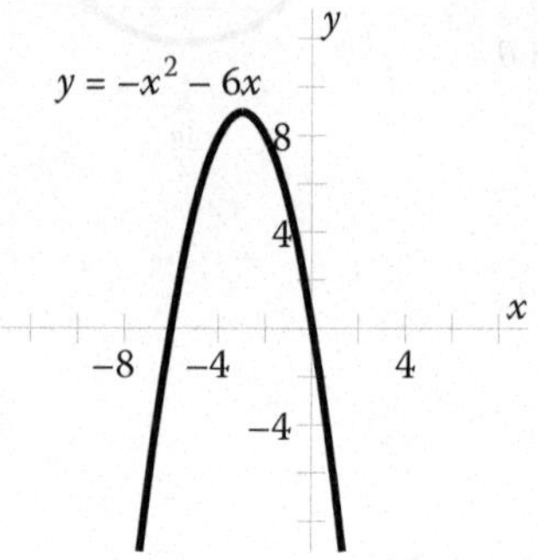

4. rotation

5. x', y'

6. $x\cos\phi - y\sin\phi$; $y\cos\phi + x\sin\phi$

7. discriminant

8. <; >; =

9. $\dfrac{A - C}{B}$

10. parallel

Essential Skills/Extensions

1. $\left(-3\sqrt{2}, 3\sqrt{2}\right)$

3. $\left(-2\sqrt{2}, -2\sqrt{2}\right)$

5. $\left(-\sqrt{3}, -1\right)$

7. $\left(-3\sqrt{3} - 1, -3 + \sqrt{3}\right)$

9. $-10\sqrt{3} - 2$

11. $-4\sqrt{3} + 3$

13. ellipse

15. hyperbola

17. parabola

19.

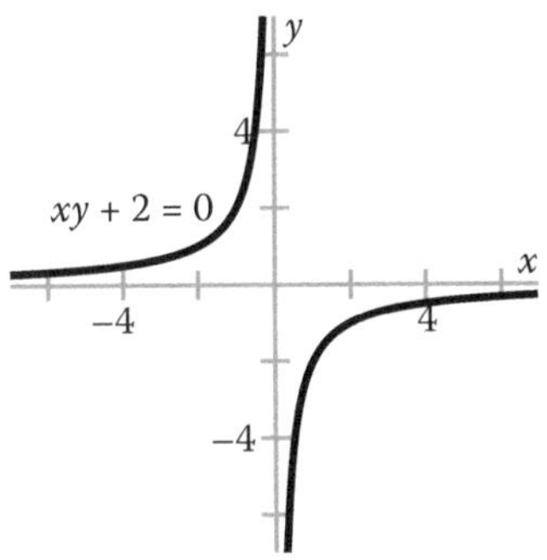

21.

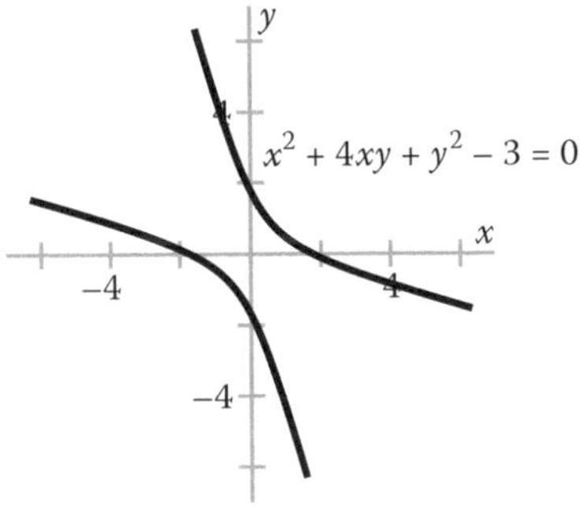

23.

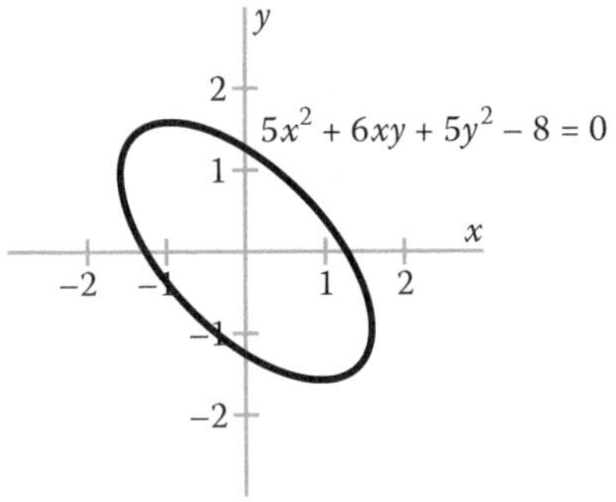

25. 37°

27. 63°

29. (−3, 5); (−5, −3); (3, −5)

31. (6, 8); (−8, 6); (−6, −8)

33. $\frac{x'^2}{9} + \frac{y'^2}{4} = 1$

35. $y' = -x'^2$

37. $(3 - x')^2 - 3(2 - y')^2 = 9$

SECTION 10.4 ANSWERS

Warm Up/Just the Facts

1. focus: (3/4, 0)

directrix: $x = -3/4$

$2y^2 = 6x$

$y^2 = 3x$ *Write the equation in standard form $y^2 = 4px$.*

Use the value for p to find the focus $(p, 0)$ and the directrix $x = -p$. The value for p is 3/4, therefore, the focus is at (3/4, 0) and the directrix is $x = -3/4$.

2. foci: $\left(0, \pm\sqrt{3}\right)$

eccentricity: $\frac{\sqrt{2}}{2}$

$100x^2 + 50y^2 - 300 = 0$

$\frac{y^2}{6} + \frac{x^2}{3} = 1$ *Write the equation in standard form.*

The equation is that of an ellipse in $\frac{y^2}{a^2} + \frac{x^2}{b^2} = 1$ form. The eccentricity e is given by $e = c/a$ where $c^2 = a^2 - b^2$ and the foci are given by $(0, \pm c)$. Since the value for c is $\sqrt{3}$, the eccentricity is $\frac{\sqrt{2}}{2}$ and the foci are $\left(0, \pm\sqrt{3}\right)$.

3. foci: $\left(0, \pm\sqrt{10}\right)$

vertices: (0, ±3)

The equation $\frac{y^2}{9} - x^2 = 1$ is in the standard form of a hyperbola $\frac{y^2}{a^2} - \frac{x^2}{b^2} = 1$ with foci $(0, \pm c)$, where $c^2 = a^2 + b^2$ and the vertices are $(0, \pm a)$. Since the value for c is $\sqrt{10}$, the foci are $\left(0, \pm\sqrt{10}\right)$. The value for a is 3, therefore, the vertices are (0, ±3).

4. =; <; >

5. $r = \frac{ep}{1 - e\cos\pi}$

6. $r = \frac{ep}{1 + e\cos\pi}$

7. $r = \frac{ep}{1 + e\sin\pi}$

8. $r = \frac{ep}{1 - e\sin\pi}$

9. parallel; coincides

10. perpendicular

Essential Skills/Extensions

1. hyperbola; directrix: $y = -1/3$; eccentricity: 3/2

3. ellipse; directrix: $x = -4$; eccentricity: 1/6

5. hyperbola; directrix: $y = 1/2$; eccentricity: 3

7. $r = \dfrac{6}{3 + 2\cos\pi}$

9. $r = \dfrac{2}{1 + \sin\pi}$

11. $r = \dfrac{3}{2 - 3\cos\pi}$

13. directrix: $y = 1$; $e = 1$

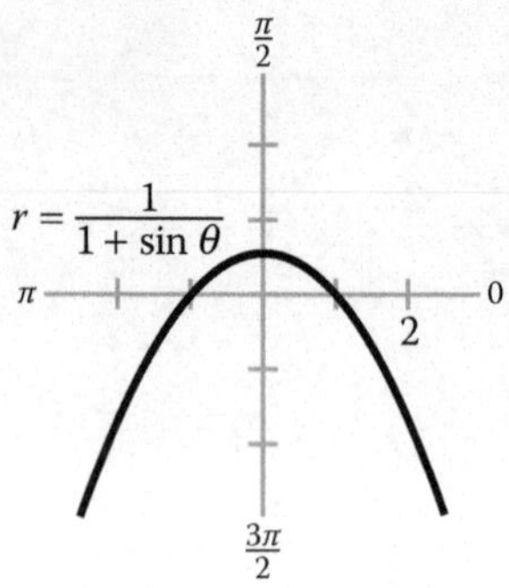

15. directrix: $y = -3$; $e = 2/3$

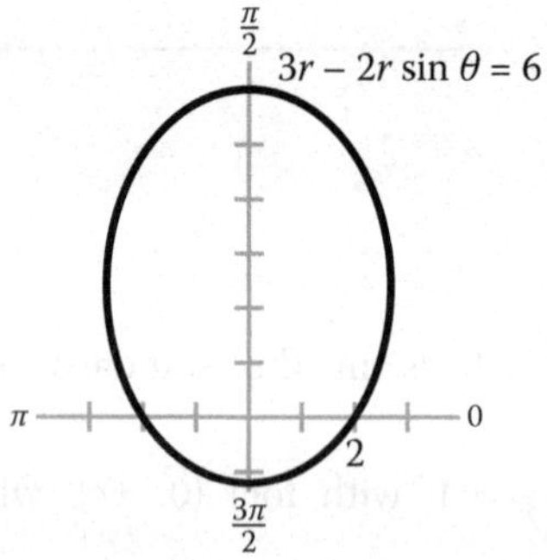

17. directrix: $y = -2/3$; $e = 3$

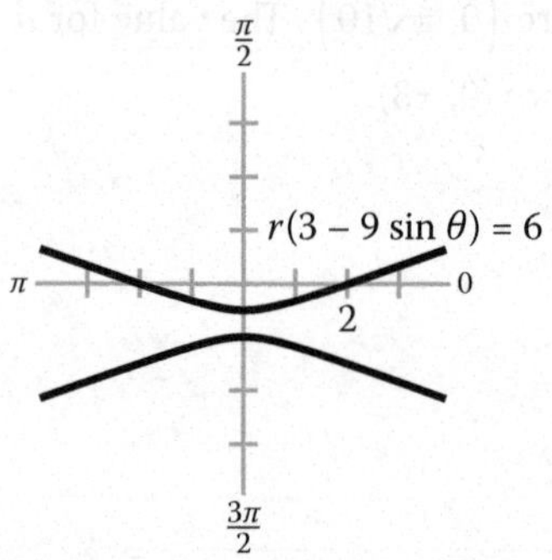

19. $x^2 - 15y^2 + 96y - 144 = 0$

21. $5x^2 - 4y^2 - 54x + 81 = 0$

23. $r = \dfrac{12}{1 - \cos\pi}$

25. $r = \dfrac{4}{1 + \sin\pi}$

27. $r = \dfrac{48}{15 + 9\cos\pi}$

29. $r = \dfrac{12}{3 + \sin\pi}$

31. $r = \dfrac{16}{1 + 3\sin\pi}$

33. $r = \dfrac{24}{5 + 7\cos\pi}$

35. directrix: $x = -6$; $e = 1/2$

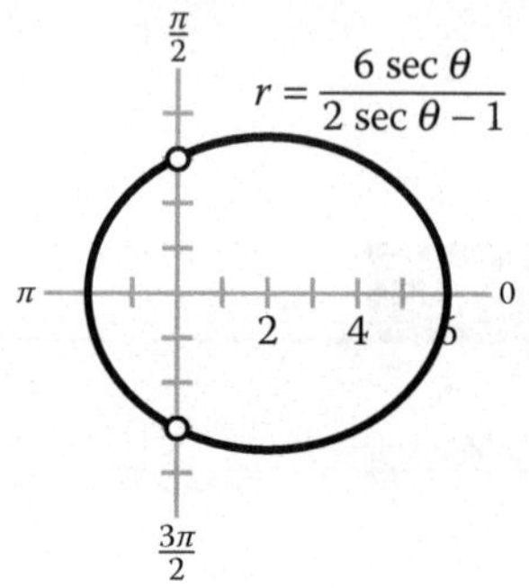

37. directrix: $y = 4/3$; $e = 3/2$

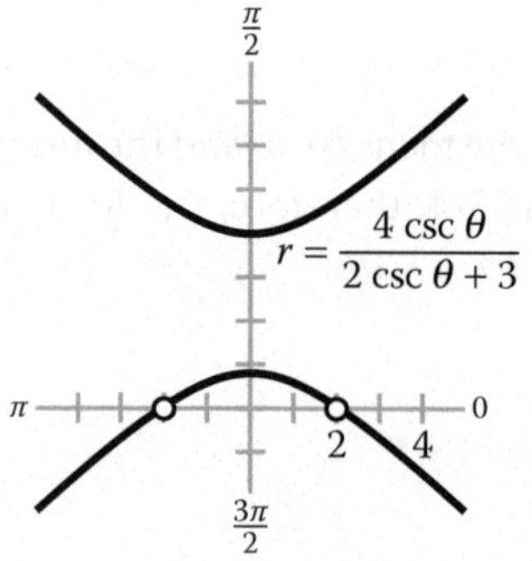

SECTION 10.5 ANSWERS

Warm Up/Just the Facts

1. $y = (-2)^2 - 1 = 3$, $y = (-1)^2 - 1 = 0$, $y = (0)^2 - 1 = -1$, $y = (1)^2 - 1 = 0$, $y = (2)^2 - 1 = 3$
   Substitute each x-value in $y = x^2 - 1$ to find the corresponding y-value.
2. $y = |3(-2) + 4| = 2$, $y = |3(-1) + 4| = 1$, $y = |3(0) + 4| = 4$, $y = |3(1) + 4| = 7$, $y = |3(2) + 4| = 10$
   Substitute each x-value in $y = |3x + 4|$ to find the corresponding y-value.
3. $y = \sqrt{-2+5} = \sqrt{3}$, $y = \sqrt{-1+5} = 2$, $y = \sqrt{0+5} = \sqrt{5}$, $y = \sqrt{1+5} = \sqrt{6}$, $y = \sqrt{2+5} = \sqrt{7}$
   Substitute each x-value in $y = \sqrt{x+5}$ to find the corresponding y-value.
4. plane curve
5. parameter
6. rectangular
7. third
8. orientation
9. trigonometric
10. function

Essential Skills/Extensions

1.

t	−3	−2	1	6
$x(t) = t^2 - 4$	5	0	−3	32
$y(t) = \sqrt{t+3}$	0	1	2	3

3.

t	−2	0	1	2
$x(t) = \dfrac{x+2}{x+1}$	0	2	3/2	4/3
$y(t) = t^3 + 1$	−7	1	2	9

5.

t	−1	0	1	2
$x(t) = t^2 + 3t$	−2	0	4	10
$y(t) = \|t + 5\|$	4	5	6	7

7.

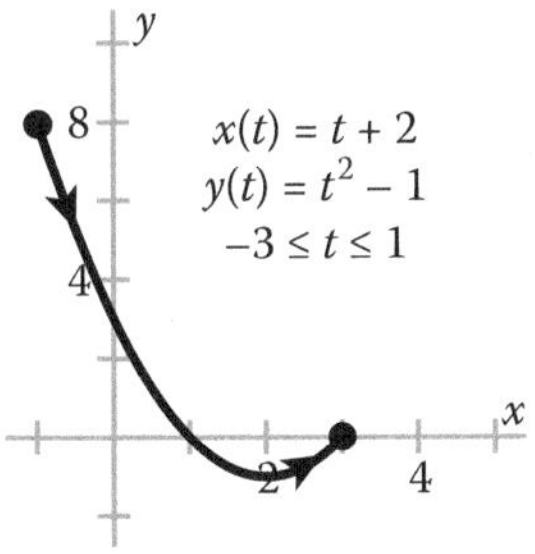

9.

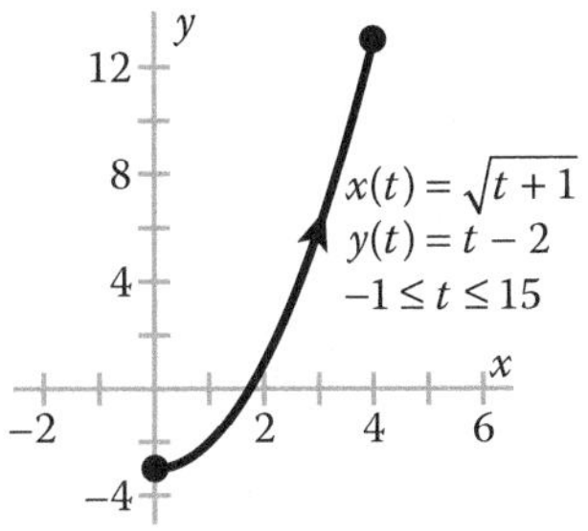

11.

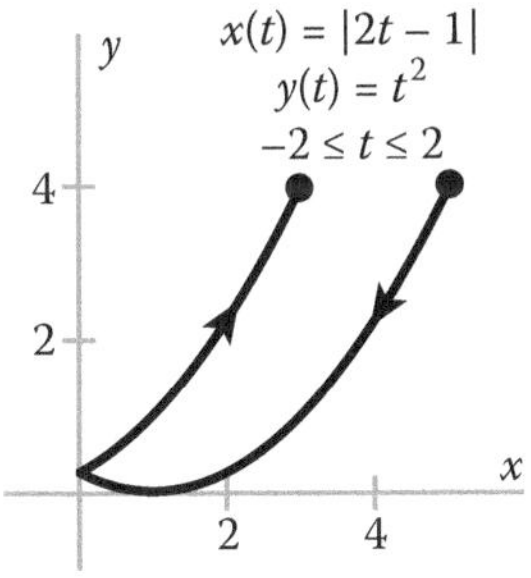

13.

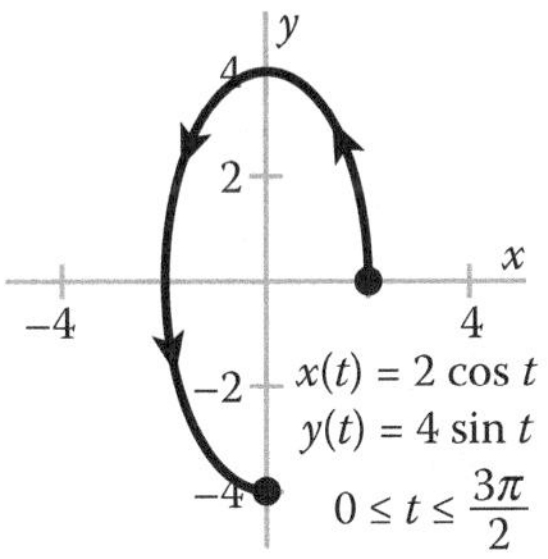

15.

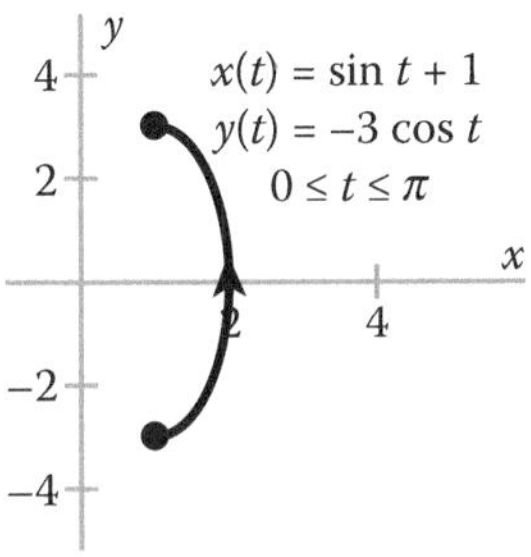

17.

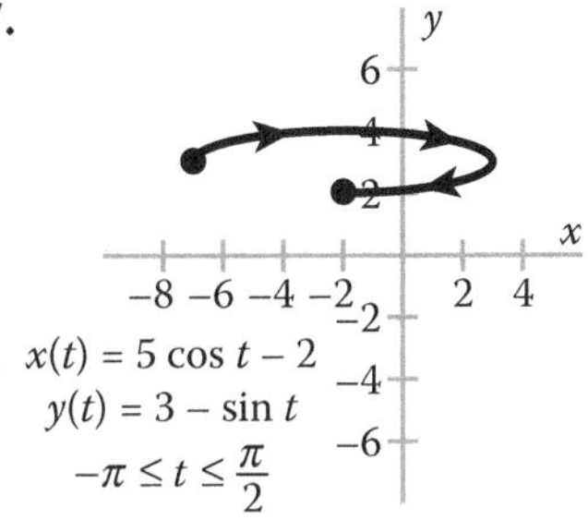

19. $y = x^2 - 5$, $[0, \infty)$

21. $y = x^2 + 10x + 24$, $(-\infty, \infty)$

23. $x = \dfrac{(y-4)^2 - 25}{25}$, $[-1, \infty)$

25. $x^2/4 + y^2 = 1$, $[-2, 2]$

27. $\dfrac{x^2}{16} - \dfrac{y^2}{4} = 1$, $(-\infty, -4] \cup [4, \infty)$

29. $4x^2 - 9y^2 = 1$, $(-\infty, -1/2] \cup [1/2, \infty)$

31.

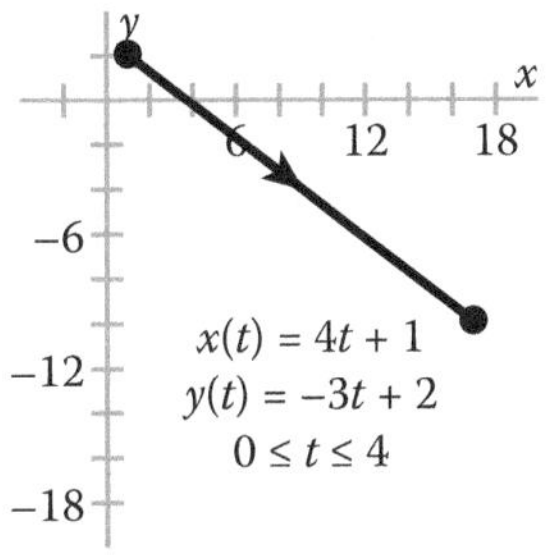

33.

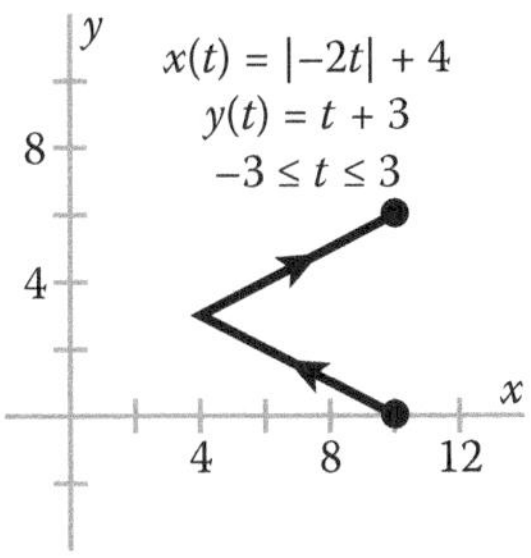

35.

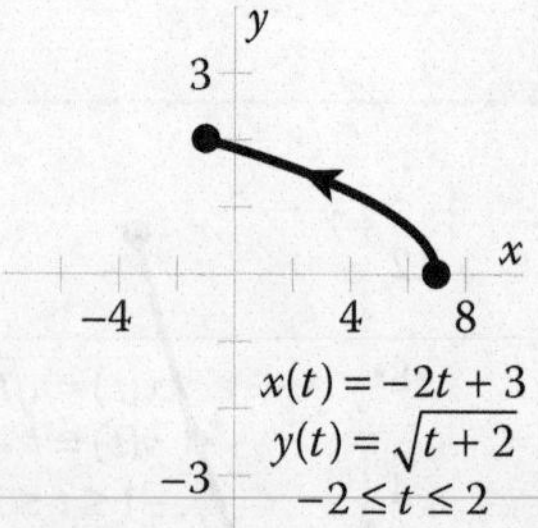

37. $x(t) = t, y(t) = -t^2$; $x(t) = -t, y(t) = t^2$ (possible answers)

39. $x(t) = t, y(t) = t^2 + 3$; $x(t) = t^2, y(t) = t + 3$ (possible answers)

41. $x(t) = t,\ y(t) = \pm\sqrt{1 - \dfrac{t^2}{9}}$; $x(t) = 3 \cos t, y(t) = \sin t$ (possible answers)

43. The plane curve is a line with slope 1 and y-intercept 0 (the line $y = x$).

45. The plane curve is a line with slope 2 and y-intercept 0 (the line $y = 2x$).

47. The plane curve is a line with slope 1/2 and y-intercept 0 (the line $y = x/2$).

49.

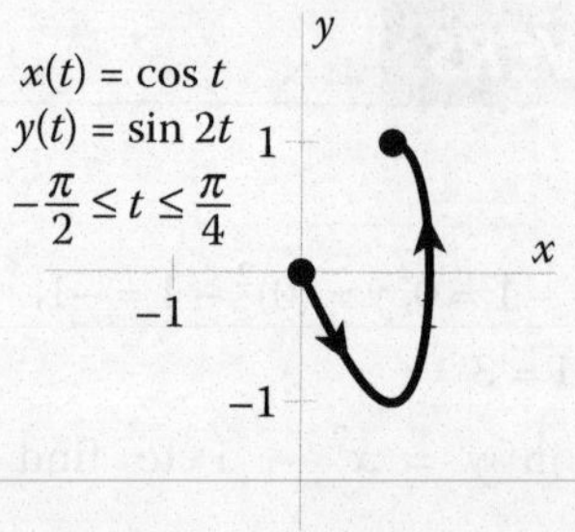

51.

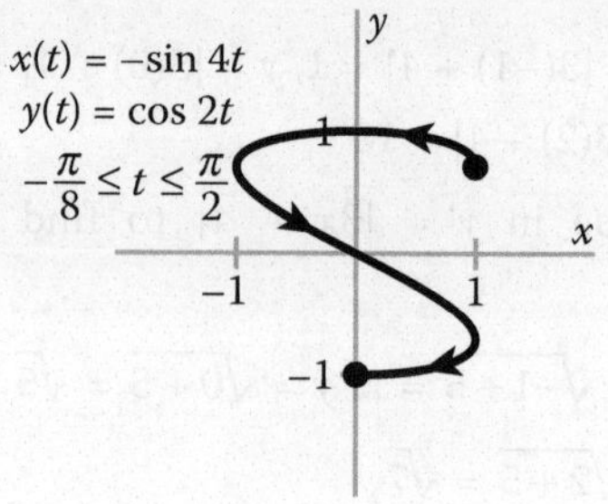

53. True. $y = x^3/125$ is the rectangular equation for both pairs of parametric equations.

55. $(x + 2)^2/25 + (3 - y)^2 = 1$

CHAPTER 10 REVIEW ANSWERS

1. $(4.5, 3\pi/4)$, $(-4.5, 7\pi/4)$, $(-4.5, -\pi/4)$, $(4.5, -5\pi/4)$

2. $(-7, \pi/6)$, $(7, 7\pi/6)$, $(7, -5\pi/6)$, $(-7, -11\pi/6)$

3. $(2.75, -2.75\sqrt{3})$

4. $(16, 7\pi/6)$

5. $(\sqrt{17.37}, 3.67)$

6. $y = -3$

7. $r = -\dfrac{\cot\theta\csc\theta}{3}$

8.

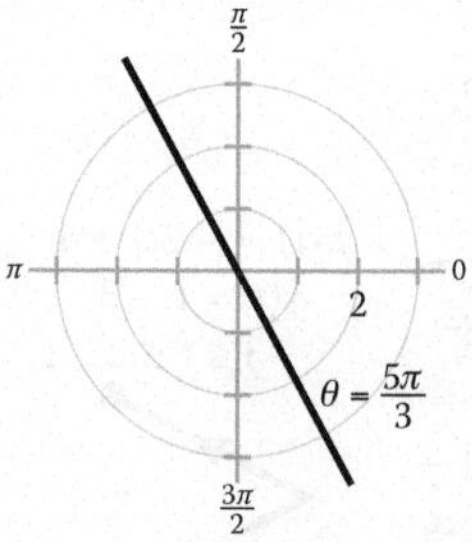

9.

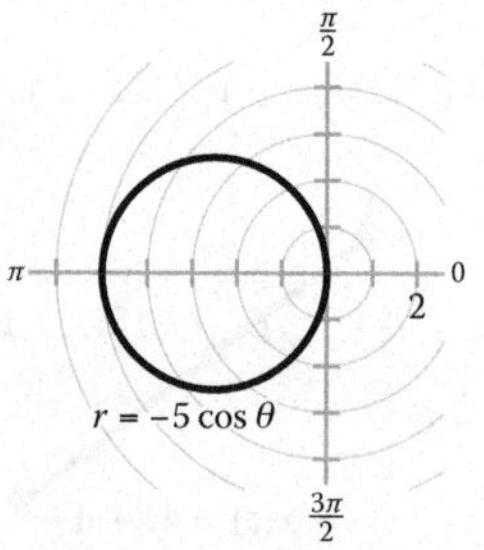

10.

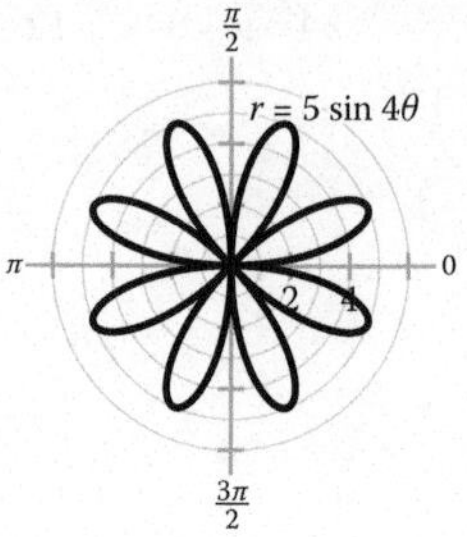

11. $\left(\dfrac{7+3\sqrt{3}}{2}, \dfrac{-3+7\sqrt{3}}{2}\right)$

12. $\left(\dfrac{5\sqrt{2}}{2}, \dfrac{3\sqrt{2}}{2}\right)$

13. $4 - 6\sqrt{2}$

14. parabola

15.

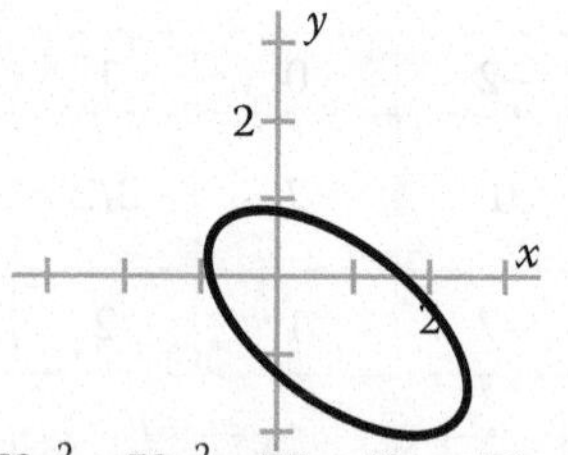

$30y + 73y^2 + 52x^2 = 75 + 40x - 72xy$

16. ellipse; $y = -3$; $e = 2/3$

17. hyperbola; $x = 3/2$; $e = 2$

18. $r = \dfrac{6}{4 + \cos\theta}$

19. $r = \dfrac{8}{1 - 2\sin\theta}$

20. $x = 8/3$; $e = 1$

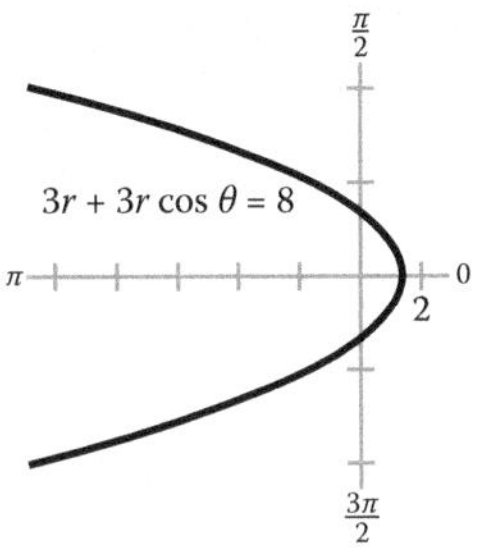

21.

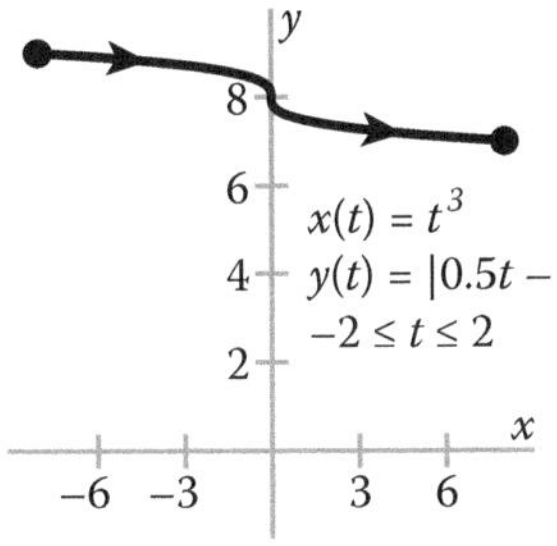

22.

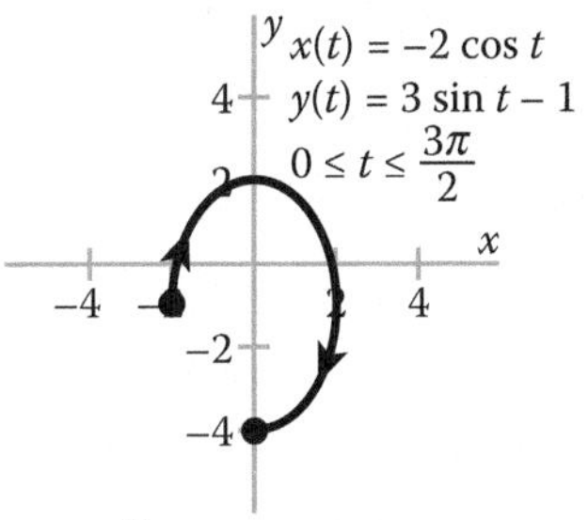

23. $y = 3x^2 - 12x + 11$, $(-\infty, \infty)$

24. $x^2/16 + y^2 = 1$, $[-4, 4]$

25.

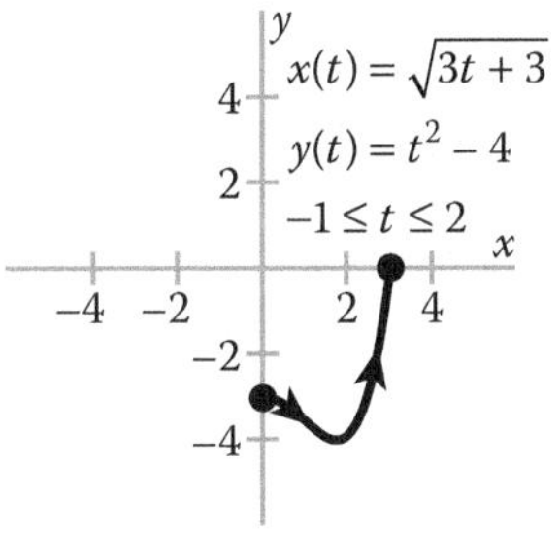

SECTION 11.1 ANSWERS

Warm Up/Just the Facts

1. 1, 3, 5, 7, 9

2. 1, 4, 9, 16, 25

3. 1, –2, –5, –8, –11

4. sequence

5. a_1; a_n

6. series

7. summation

8. recursive

9. 6th; 24

10. $\sum_{n=4}^{20}(9n-5)$

Essential Skills/Extensions

1. 14; 105

3. $a_n = 23 - 3n$

5. $a_n = -n$

7. 5, –1, –7, –13, –19

9. –50

11. $\sum_{n=1}^{5} 2n$

13. $a_n = \sqrt{3n}$

15. False. There is no function to represent the values.

17. \$2630, \$2766.76, \$2910.63, \$3061.98, \$3221.21, \$3388.71, \$3564.92

SECTION 11.2 ANSWERS

Warm Up/Just the Facts

1. Yes. The values increase by 4 each time.

2. No. There is no common difference.

3. Yes. The values decrease by 1 each time.

4. arithmetic

5. 6

6. $a_n = a_1 + (n-1)d$; a_1; d

7. a_n

8. sum; arithmetic; value; number

9. sum; 8

10. 2

Essential Skills/Extensions

1. –77

3. $a_n = 7 + (n-1)10$

5. $a_n = 11 + (n-1)17$

7. 117

9. 60

11. 1049.2

13. –205

15. 50.4

17. \$53,400

SECTION 11.3 ANSWERS

Warm Up/Just the Facts

1. 4/5 $\quad \dfrac{\frac{2}{5}}{\frac{1}{2}} = \dfrac{2}{5}\cdot\dfrac{2}{1} = \dfrac{4}{5}$

2. –28/3 $\quad \dfrac{-4}{1-\frac{4}{7}} = \dfrac{-4}{\frac{3}{7}} = -4\cdot\dfrac{7}{3} = -\dfrac{28}{3}$

3. 45/4 $\quad \dfrac{5}{\left(\frac{2}{3}\right)^2} = \dfrac{5}{\frac{4}{9}} = 5\cdot\dfrac{9}{4} = \dfrac{45}{4}$

4. geometric

5. $a_n = a_1 r^{n-1}$; first; r

6. ratio; consecutive

7. r; –1, 1

8. finite; geometric

9. infinite; ∞; $S = \dfrac{a_1}{1-r}$

10. $S = \dfrac{a_1}{1-r}$

Essential Skills/Extensions

1. 72

3. $a_n = 8 \cdot 2^{n-1}$

5. 93

7. 49/3

9. 10/3

11. 2

13. 35/99

15. $\sum_{n=1}^{11} 20 \cdot 2^{n-1}$; 40,940

17. $\sum_{n=1}^{\infty} 35 \cdot 0.4^{n-1} = 58.\overline{33}$

19. 311/99

21. –1,501,199,880,000,000

SECTION 11.4 ANSWERS

Warm Up/Just the Facts

1. neither

2. geometric

3. arithmetic

4. proof; all

5. induction; base; 1

6. second

7. $n = k + 1$

8. $1 + 2 + 3 + 4 + \ldots + k + (k + 1) = 0.5(k + 1)((k + 1) + 1)$

9. natural

10. proof

Essential Skills/Extensions

1. When $n = 1$, the left side is 2 and the right side is $2(1)(2(1) - 1) = 2(2 - 1) = 2$. So, the base case is true.

Assume that the statement is true when $n = k$. $2 + 10 + 18 + \ldots + (8k - 6) = 2k(2k - 1)$.

Prove:

$$\boxed{2 + 10 + 18 + \ldots + (8k - 6) + (8(k + 1) - 6) = 2(k + 1)(2(k + 1) - 1)}$$

$$\begin{aligned} 2 + 10 + 18 + \ldots + (8k - 6) + (8(k + 1) - 6) &= 2k(2k - 1) + (8(k + 1) - 6) \\ &= 4k^2 - 2k + (8k + 8 - 6) \\ &= 4k^2 + 6k + 2 \\ &= 2(2k^2 + 3k + 1) \\ &= 2(k + 1)(2k + 1) \\ &= 2(k + 1)(2(k + 1) - 1) \end{aligned}$$

By the principle of mathematical induction, $2 + 10 + 18 + \ldots + (8n - 6) = 2n(2n - 1)$ for all natural numbers n.

3. When $n = 1$, the left side is -1 and the right side is $3(1)^2 - 4(1) = 3 - 4 = -1$. So, the base case is true.

Assume that the statement is true for $n = k$. $-1 + 5 + 11 + \ldots + 6k - 7 = 3k^2 - 4k$.

Prove:

$$\boxed{-1 + 5 + 11 + \ldots + 6k - 7 + 6(k + 1) - 7 = 3(k + 1)^2 - 4(k + 1)}$$

$$\begin{aligned} -1 + 5 + 11 + \ldots + 6k - 7 + 6(k + 1) - 7 &= 3k^2 - 4k + 6(k + 1) - 7 \\ &= 3k^2 - 4k + 6k + 6 - 7 \\ &= 3k^2 + 2k - 1 \\ &= (k + 1)(3k - 1) \\ &= 3(k + 1)^2 - 4(k + 1) \end{aligned}$$

By the principle of mathematical induction, $-1 + 5 + 11 + \ldots + 6n - 7 = 3n^2 - 4n$ for all natural numbers n.

5. When $n = 1$, the left side is 1 and the right side is $\frac{1}{9}(10^1 - 1) = \frac{1}{9}(9) = 1$. So, the base case is true.

Assume that the statement is true for $n = k$. $1 + 10 + 100 + \ldots + 10^{k-1} = \frac{1}{9}(10^k - 1)$.

Prove:

$$\boxed{1 + 10 + 100 + \ldots + 10^{k-1} + 10^{((k+1)-1))} = \frac{1}{9}(10^{k+1} - 1)}$$

$$\begin{aligned} 1 + 10 + 100 + \ldots + 10^{k-1} + 10^{((k+1)-1))} &= \frac{1}{9}(10^k - 1) + 10^{((k+1)-1))} \\ &= \frac{1}{9}(10^k - 1) + 10^k \\ &= \frac{1}{9}10^k - \frac{1}{9} + 10^k \\ &= 10^k\left(\frac{1}{9} + 1\right) - \frac{1}{9} \\ &= 10^k\left(\frac{10}{9}\right) - \frac{1}{9} \\ &= \left(\frac{10^{k+1}}{9}\right) - \frac{1}{9} \\ &= \frac{1}{9}(10^{k+1} - 1) \end{aligned}$$

By the principle of mathematical induction, $1 + 10 + 100 + \ldots + 10^{n-1} = \frac{1}{9}(10^n - 1)$ for all natural numbers n.

7. When $n = 1$, the left side is 2 and the right side is $2[1 + (1 - 1)2^1] = 2$. So, the base case is true.

Assume that the statement is true for $n = k$. $1 \cdot 2 + 2 \cdot 2^2 + 3 \cdot 2^3 + \ldots + k \cdot 2^k = 2[1 + (k - 1)2^k]$

Prove:

$$\boxed{1 \cdot 2 + 2 \cdot 2^2 + 3 \cdot 2^3 + \ldots + k \cdot 2^k + (k + 1)2^{k+1} = 2[1 + ((k + 1) - 1)2^{k+1}]}$$

$$\begin{aligned} 2 + 2 \cdot 2^2 + 3 \cdot 2^3 + \ldots + k \cdot 2^k + (k + 1)2^{k+1} &= 2[1 + (k - 1)2^k] + (k + 1)2^{k+1} \\ &= 2 + (k - 1)2^{k+1} + (k + 1)2^{k+1} \\ &= 2 + 2k \cdot 2^{k+1} \\ &= 2[1 + ((k + 1) - 1)2^{k+1}] \end{aligned}$$

By the principle of mathematical induction, $1 \cdot 2 + 2 \cdot 2^2 + 3 \cdot 2^3 + \ldots + n \cdot 2^n = 2[1 + (n - 1)2^n]$ for all natural numbers n.

9. When $n = 1$, the left side is 1/2 and the right side is $1/(1+1) = 1/2$. So, the base case is true.

Assume that the statement is true when $n = k$. $1/2 + 1/6 + 1/12 + 1/20 + \ldots + \frac{1}{k(k+1)} = k/(k+1)$

Prove:

$$1/2 + 1/6 + 1/12 + \ldots + \frac{1}{k(k+1)} + \frac{1}{(k+1)((k+1)+1)} = \frac{k+1}{((k+1)+1)}$$

$$\begin{aligned} 1/2 + 1/6 + 1/12 + \ldots + \frac{1}{k(k+1)} + \frac{1}{(k+1)((k+1)+1)} &= \frac{k}{k+1} + \frac{1}{(k+1)((k+1)+1)} \\ &= \frac{k((k+1)+1)+1}{(k+1)((k+1)+1)} \\ &= \frac{k^2+2k+1}{(k+1)((k+1)+1)} \\ &= \frac{(k+1)^2}{(k+1)((k+1)+1)} \\ &= \frac{k+1}{((k+1)+1)} \end{aligned}$$

By the principle of mathematical induction, $1/2 + 1/6 + 1/12 + 1/20 + \ldots + \frac{1}{n(n+1)} = n/(n+1)$ for all natural numbers n.

11. When $n = 1$, $8^1 - 3^1 = 5$ which is divisible by 5. So, the base case is true.

Assume that the statement is true for $n = k$. $8^k - 3^k$ is divisible by 5.

Prove:

$8^{k+1} - 3^{k+1}$ is divisible by 5.

$$\begin{aligned} 8^{k+1} - 3^{k+1} &= 8 \cdot 8^k - 3 \cdot 3^k \\ &= 8 \cdot 8^k - (8-5) \cdot 3^k \\ &= 8 \cdot 8^k - 8 \cdot 3^k + 5 \cdot 3^k \\ &= 8(8^k - 3^k) + 5 \cdot 3^k \end{aligned}$$

The first term, $8^k - 3^k$, can be written as a product of a whole number and 5 because (by the induction hypothesis) $8^k - 3^k$ is divisible by 5. The second term, $5 \cdot 3^k$, is the product of the whole number 3^k and 5. Therefore, 5 can be factored from each term in $8(8^k - 3^k) + 5 \cdot 3$ and so $8(8^k - 3^k) + 5 \cdot 3$ is divisible by 5.

By the principle of mathematical induction, 8^n 3^n is divisible by 5 for all natural numbers n.

SECTION 11.5 ANSWERS

Warm Up/Just the Facts

1. 720 $6! = 6(5)(4)(3)(2)(1) = 720$

2. 20 $\frac{5!}{3!} = \frac{5 \cdot 4 \cdot 3!}{3!} = \frac{5 \cdot 4 \cdot \cancel{3!}}{\cancel{3!}} = 20$

3. 840

$$\frac{7!5!}{6!} = \frac{7 \cdot 6! \cdot 5 \cdot 4 \cdot 3 \cdot 2 \cdot 1}{6!} = \frac{7 \cdot \cancel{6!} \cdot 5 \cdot 4 \cdot 3 \cdot 2 \cdot 1}{\cancel{6!}} = 840$$

4. Fundamental Counting Principle

5. orderings; n

6. permutations

7. combinations

8. tree; branches

9. distinguishable permutations

10. n; r

Essential Skills/Extensions

1. 2112

3. 5,184,000

5. 40,320

7. 336

9. 90,720

11. 220

13. 1.4×10^{30}

15. 8,316,000

17. A. 45; B. 1287; C. 150,150; D. 150,150; E. 72,072; F. 95,667

SECTION 11.6 ANSWERS

Warm Up/Just the Facts

1. permutation

2. combination

3. permutation

4. probability

5. sample space

6. outcomes; total

7. 1

8. independent

9. simple; compound; simple

10. complement

Essential Skills/Extensions

1. 1/2

3. 1/12

5. 1/2

7. 2/15

9. 1/24

11. 12/25

13. 0.56

15. 2/3

17. ≈ 0.618

19. ≈ 0.097

21. ≈ 0.0028

23. A. yes; B. 0.385; C. 0.615; D. 0.315

SECTION 11.7 ANSWERS

Warm Up/Just the Facts

1. $a^{21}d^{12}$
2. $-7776m^5n^{15}$
3. $22{,}500f^{12}g^4$
4. Binomial Theorem
5. ${}_nC_m$, $\binom{n}{m}$, $\frac{n!}{m!(n-m)!}$, or $C(n, m)$; Pascal's
6. formula for the rth term of a binomial expansion: $\binom{n}{r-1}x^{n-r+1}y^{r-1}$
7. 5th; 8th; $3x$; $-2y$
8. coefficients; 7; 1, 7, 21, 35, 35, 21, 7, 1
9. 1
10. n; 0

Essential Skills/Extensions

1. $x^4 + 32x^3 + 384x^2 + 2048x + 4096$

3. $8g^3 + 12g^2h + 6gh^2 + h^3$

5. $a^4 - 8a^3b + 24a^2b^2 - 32ab^3 + 16b^4$

7. $p^4 + 4\sqrt{2}p^3q + 12p^2q^2 + 8\sqrt{2}pq^3 + 4q^4$

9. $4032a^5b^8$

11. $78{,}125x^7 - 218{,}750x^6\sqrt{y} + 262{,}500x^5y - 175{,}000x^4y\sqrt{y} + 70{,}000x^3y^2 - 16{,}800x^2y^2\sqrt{y} + 2240xy^3 - 128y^3\sqrt{y}$

13. 11th

15. $\frac{16}{x^4} - \frac{96y}{x^3} + \frac{216y^2}{x^2} - \frac{216y^3}{x} + 81y^4$

CHAPTER 11 REVIEW ANSWERS

1. 7/5, 5

2. $a_n = 8n - 15$

3. 3, −2, 8, −12, 28

4. 80

5. $\sum_{n=1}^{5}(-8n+1)$

6. 86

7. 8, 13, 18

8. $a_n = 14n + 18$

9. 1170

10. −286

11. 4096

12. $a_n = 27\left(-\frac{1}{3}\right)^{n-1}$

13. 255/64

14. −9/5

15. 8/9

16. $2 + 4 + 6 + \ldots + 2k + 2(k+1) = (k+1)^2 + (k+1)$

17. 72,324

18. 720

19. 1320

20. 5040

21. 4/13

22. 1/11

23. 0.18

24. 0.45

25. 0.14

26. $32 - 80y + 80y^2 - 40y^3 + 10y^4 - y^5$

27. $125a^3 - 150a^2b + 60ab^2 - 8b^3$

28. $x^8 - 4\sqrt{6}x^6y + 36x^4y^2 - 24\sqrt{6}x^2y^3 + 36y^4$

29. $560v^9w^8$

30. 1024

INDEX

A

B

C

D

E

F

G

H

I

S

T

U

V

W

X

Y

Z

APPENDIX A: COMMON FORMULAS AND PROPERTIES

Algebra

Properties of Powers and Radicals

Product of Powers	$a^m a^n = a^{m+n}$	Power of a Power	$(a^m)^n = a^{m \cdot n}$
Power of a Product	$(ab)^n = a^n b^n$	Zero Exponent	$a^0 = 1$
Negative Exponent	$a^{-m} = \dfrac{1}{a^m}$	Rational Exponent	$a^{\frac{1}{n}} = \sqrt[n]{a}\ ;\ a^{\frac{m}{n}} = \left(\sqrt[n]{a}\right)^m = \sqrt[n]{a^m}$
Quotient of Powers	$\dfrac{a^m}{a^n} = a^{m-n}$	Power of a Quotient	$\left(\dfrac{a}{b}\right)^n = \dfrac{a^n}{b^n}$
Product Property of Radicals	$\sqrt[n]{ab} = \sqrt[n]{a} \cdot \sqrt[n]{b}$	Quotient Property of Radicals	$\sqrt[n]{\dfrac{a}{b}} = \dfrac{\sqrt[n]{a}}{\sqrt[n]{b}}$

The Quadratic Formula

If $ax^2 + bx + c = 0$, then $x = \dfrac{-b \pm \sqrt{b^2 - 4ac}}{2a}$.

Special Products of Binomials

Square of a Sum	$(A + B)^2 = A^2 + 2AB + B^2$
Square of a Difference	$(A - B)^2 = A^2 - 2AB + B^2$
Cube of a Sum	$(A + B)^3 = A^3 + 3A^2B + 3AB^2 + B^3$
Cube of a Difference	$(A - B)^3 = A^3 - 3A^2B + 3AB^2 - B^3$

Special Factoring

Perfect Square	$A^2 + 2AB + B^2 = (A + B)^2$
Difference of Two Squares	$A^2 - B^2 = (A + B)(A - B)$
Sum of Two Cubes	$A^3 + B^3 = (A + B)(A^2 - AB + B^2)$
Difference of Two Cubes	$A^3 - B^3 = (A - B)(A^2 + AB + B^2)$

Binomial Theorem

$$(x + y)^n = x^n + \binom{n}{1}x^{n-1}y + \binom{n}{2}x^{n-2}y^2 + \ldots + \binom{n}{n-1}xy^{n-1} + y^n \text{ where } \binom{n}{m} = \frac{n!}{m!(n-m)!}$$

Geometric Formulas

Formulas for Perimeter P, Area A, Circumference C, Surface Area S, and Volume V

Square

$P = 4s$

$A = s^2$

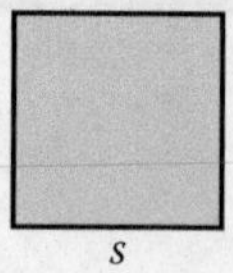

Cube

$S = 6s^2$

$V = s^3$

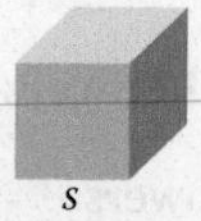

Rectangle

$P = 2l + 2w$

$A = lw$

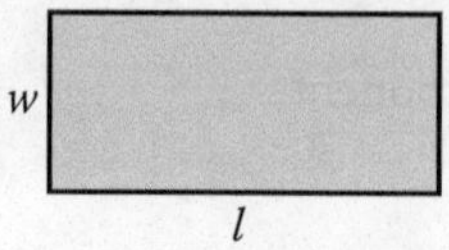

Rectangular Prism

$S = 2lw + 2lh + 2hw$

$V = lwh$

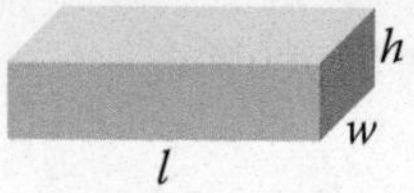

Triangle

$A = \frac{1}{2}bh$

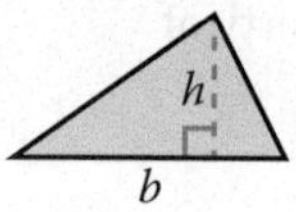

Right Triangle

$A = \frac{1}{2}ab$

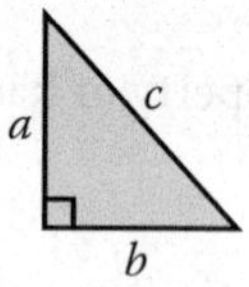

Circle

$C = 2\pi r$

$A = \pi r^2$

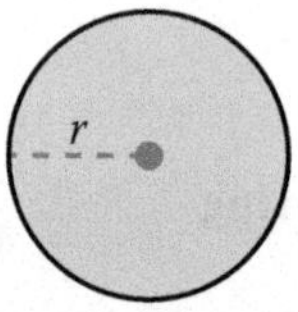

Sphere

$S = 4\pi r^2$

$V = \frac{4}{3}\pi r^3$

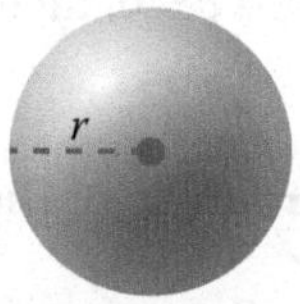

Cylinder

$S = 2\pi rh + 2\pi r^2$

$V = \pi r^2 h$

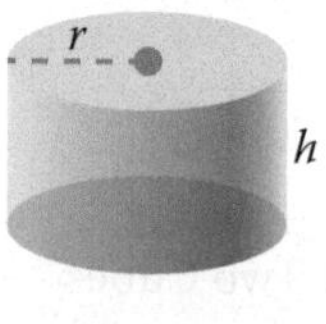

Cone

$S = \pi r\sqrt{r^2 + h^2} + \pi r^2$

$V = \frac{1}{3}\pi r^2 h$

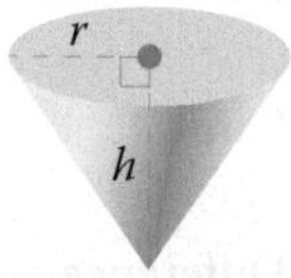

The Pythagorean Theorem

$a^2 + b^2 = c^2$

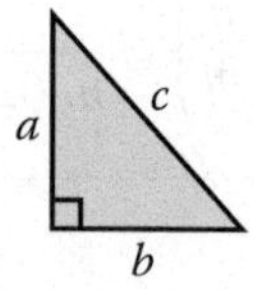

Example: Equilateral Triangle

$h = \sqrt{s^2 - \left(\frac{s}{2}\right)^2} = \frac{s\sqrt{3}}{2}$

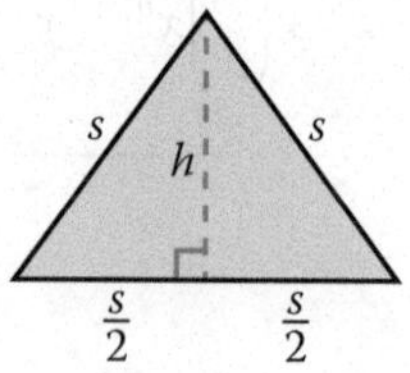

Distance and Midpoint *between points (x_1, y_1) and (x_2, y_2)*

Distance $\quad d = \sqrt{(x_2 - x_1)^2 + (y_2 - y_1)^2}$

Midpoint $\quad M = \left(\frac{x_1 + x_2}{2}, \frac{y_1 + y_2}{2}\right)$

Lines *through points (x_1, y_1) and (x_2, y_2) with slope m and y-intercept b*

Slope $\quad m = \frac{y_2 - y_1}{x_2 - x_1}$

Slope-Intercept Form $\quad y = mx + b$

Point-Slope Form $\quad y - y_1 = m(x - x_1)$

Interest Formulas

Simple	$I = Prt$	*I is the amount of simple interest paid (or earned) when a principal P is borrowed (or invested) for t years at an interest rate r.*
Compound	$A = P\left(1+\frac{r}{m}\right)^{mt}$	*A is the balance in an account after a principal P is borrowed (or invested) for t years at an annual interest rate r compounded m times per year.*
Continuously compounded	$A = Pe^{rt}$	*A is the balance in an account after a principal P is borrowed (or invested) for t years at an annual interest rate r compounded continuously.*

Logarithms (where a and b are positive real numbers such that $a \neq 1$ and $b \neq 1$)

	Logarithm with Base b	**Common Logarithm**	**Natural Logarithm**
Logarithmic Form	$y = \log_b x$	$y = \log x$	$y = \ln x$
Exponential Form	$b^y = x$	$10^y = x$	$e^y = x$
Properties of Logs	$\log_b 1 = 0$	$\log 1 = 0$	$\ln 1 = 0$
	$\log_b b = 1$	$\log 10 = 1$	$\ln e = 1$
	$\log_b b^x = x$	$\log 10^x = x$	$\ln e^x = x$
	$b^{\log_b x} = x$	$10^{\log x} = x$	$e^{\ln x} = x$
Product Property	$\log_b (xy) = \log_b x + \log_b y$	$\log (xy) = \log x + \log y$	$\ln (xy) = \ln x + \ln y$
Quotient Property	$\log_b\left(\frac{x}{y}\right) = \log_b x - \log_b y$	$\log\left(\frac{x}{y}\right) = \log x - \log y$	$\ln\left(\frac{x}{y}\right) = \ln x - \ln y$
Power Property	$\log_b x^y = y \log_b x$	$\log x^y = y \log x$	$\ln x^y = y \ln x$

Change of Base Formula

$$\log_b x = \frac{\log_a x}{\log_a b} = \frac{\log x}{\log b} = \frac{\ln x}{\ln b}$$

Sequences, Series, and Probability

Sequences and Series

General Formula for an Arithmetic Sequence	$a_n = a_1 + (n-1)d$	*a_n is the nth term of an arithmetic sequence where a_1 is the first term and d is the common difference.*
Sum of an Arithmetic Sequence	$S_L = \frac{L}{2}(a_1 + a_L)$	*S_L is the sum of the first L terms of an arithmetic sequence where a_1 is the first term.*
General Formula for a Geometric Sequence	$a_n = a_1 r^{n-1}$	*a_n is the nth term of a geometric sequence where a_1 is the first term and r is the common ratio.*
Sum of a Finite Geometric Sequence	$S_n = \frac{a_1(1-r^n)}{1-r}$	*S_n is the sum of the first n terms of a geometric sequence where a_1 is the first term and r is the common ratio.*
Sum of an Infinite Geometric Sequence	$S = \frac{a_1}{1-r}$	*S is the sum of a convergent infinite geometric sequence where a_1 is the first term and r is the common ratio.*

Counting

Permutations	${}_nP_r = \frac{n!}{(n-r)!}$	*${}_nP_r$ is the number of ways of ordering r items selected from a total of n different items.*
Distinguishable Permutations	$\frac{n!}{n_1! \cdot n_2! \cdot \ldots \cdot n_k!}$	*ways of ordering n items when there are n_1 of one kind, n_2 of another, and so on, where $n_1 + n_2 + \ldots + n_k = n$*
Combinations	${}_nC_r = \binom{n}{r} = \frac{n!}{r!(n-r)!}$	*${}_nC_r$ is the number of ways of choosing r items from n items when order does not matter.*

Probability

Probability of an Event	$P(A) = \frac{\text{number of successful outcomes}}{\text{number of outcomes in sample space}}$	*P(A) is the probability of event A occurring when all successful outcomes and all outcomes in the sample space are equally likely.*
Independent Events	$P(A \text{ and } B) = P(A) \cdot P(B)$	*P(A and B) is the probability of events A and B occurring when A and B are independent events.*
Mutually Exclusive Events	$P(A \text{ or } B) = P(A) + P(B)$	*P(A or B) is the probability of events A or B occurring when A and B are mutually exclusive events.*
Inclusive Events	$P(A \text{ or } B) = P(A) + P(B) - P(A \text{ and } B)$	*P(A or B) is the probability of events A or B occurring when A and B are inclusive events.*

The Trigonometric Functions of Acute Angle θ

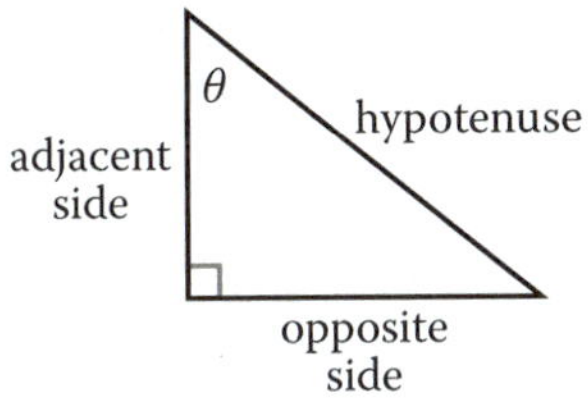

$\text{sine } \theta = \sin\theta = \dfrac{\text{opp}}{\text{hyp}}$

$\text{cosine } \theta = \cos\theta = \dfrac{\text{adj}}{\text{hyp}}$

$\text{tangent } \theta = \tan\theta = \dfrac{\text{opp}}{\text{adj}}$

$\text{cosecant } \theta = \csc\theta = \dfrac{\text{hyp}}{\text{opp}}$

$\text{secant } \theta = \sec\theta = \dfrac{\text{hyp}}{\text{adj}}$

$\text{cotangent } \theta = \cot\theta = \dfrac{\text{adj}}{\text{opp}}$

The Trigonometric Functions of Angle θ

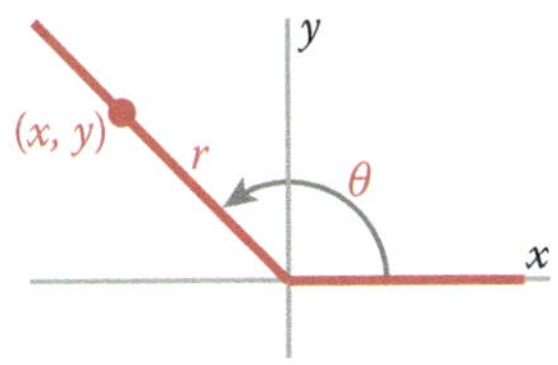

$r = \sqrt{x^2 + y^2} \neq 0$

$\sin\theta = y/r$

$\cos\theta = x/r$

$\tan\theta = y/x,\ x \neq 0$

$\csc\theta = r/y,\ y \neq 0$

$\sec\theta = r/x,\ x \neq 0$

$\cot\theta = x/y,\ y \neq 0$

The Trigonometric Functions of Real Number t on the Unit Circle

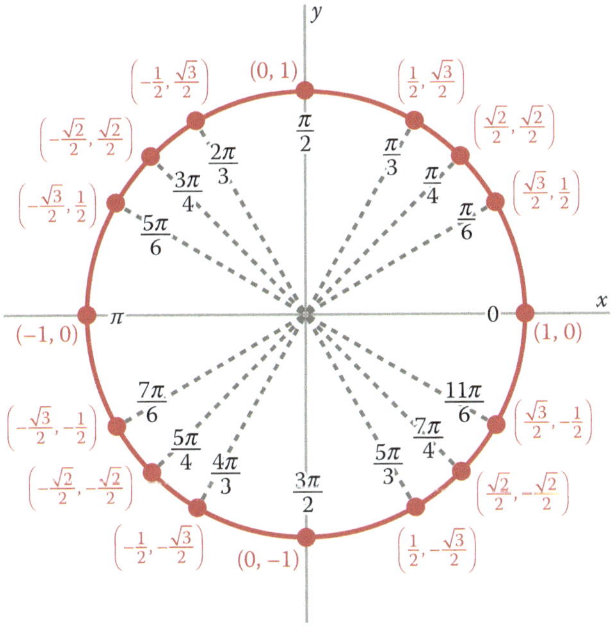

$\sin t = y$

$\cos t = x$

$\tan t = y/x,\ x \neq 0$

$\csc t = 1/y,\ y \neq 0$

$\sec t = 1/x,\ x \neq 0$

$\cot t = x/y,\ y \neq 0$

The Trigonometric Functions of Important Angles

θ (degrees)	θ (radians)	sin θ	cos θ	tan θ
0°	0	0	1	0
30°	$\pi/6$	1/2	$\sqrt{3}/2$	$1/\sqrt{3}$
45°	$\pi/4$	$\sqrt{2}/2$	$\sqrt{2}/2$	1
60°	$\pi/3$	$\sqrt{3}/2$	1/2	$\sqrt{3}$
90°	$\pi/2$	1	0	und

The Law of Sines and the Law of Cosines

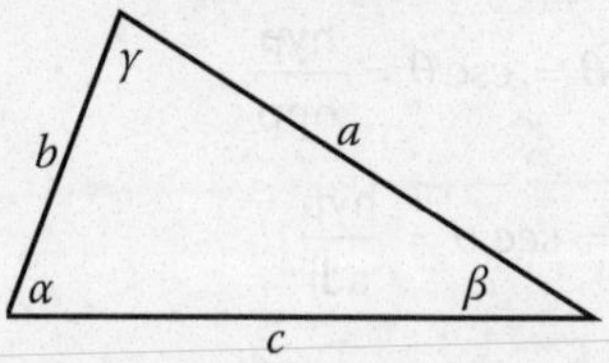

$$\frac{\sin \alpha}{a} = \frac{\sin \beta}{b} = \frac{\sin \gamma}{c}$$

$$\frac{a}{\sin \alpha} = \frac{b}{\sin \beta} = \frac{c}{\sin \gamma}$$

$$a^2 = b^2 + c^2 - 2bc \cos \alpha$$

$$b^2 = a^2 + c^2 - 2ac \cos \beta$$

$$c^2 = a^2 + b^2 - 2ab \cos \gamma$$

Trigonometric Identities

Even-Odd Identities

$\sin(-\theta) = -\sin \theta$ $\cos(-\theta) = \cos \theta$ $\tan(-\theta) = -\tan \theta$

$\csc(-\theta) = -\csc \theta$ $\sec(-\theta) = \sec \theta$ $\cot(-\theta) = -\cot \theta$

Reciprocal Identities

$\sin \theta = \dfrac{1}{\csc \theta}$ $\cos \theta = \dfrac{1}{\sec \theta}$ $\tan \theta = \dfrac{1}{\cot \theta}$

$\csc \theta = \dfrac{1}{\sin \theta}$ $\sec \theta = \dfrac{1}{\cos \theta}$ $\cot \theta = \dfrac{1}{\tan \theta}$

Quotient Identities

$\tan \theta = \dfrac{\sin \theta}{\cos \theta}$ $\cot \theta = \dfrac{\cos \theta}{\sin \theta}$

Pythagorean Identities

$\sin^2 \theta + \cos^2 \theta = 1$ $\tan^2 \theta + 1 = \sec^2 \theta$ $1 + \cot^2 \theta = \csc^2 \theta$

Cofunction Identities

$\sin\left(\dfrac{\pi}{2} - \theta\right) = \cos \theta$ $\cos\left(\dfrac{\pi}{2} - \theta\right) = \sin \theta$ $\tan\left(\dfrac{\pi}{2} - \theta\right) = \cot \theta$

$\csc\left(\dfrac{\pi}{2} - \theta\right) = \sec \theta$ $\sec\left(\dfrac{\pi}{2} - \theta\right) = \csc \theta$ $\cot\left(\dfrac{\pi}{2} - \theta\right) = \tan \theta$

Trigonometric Formulas

Sum and Difference Formulas

$\sin(\theta_1 + \theta_2) = \sin\theta_1 \cos\theta_2 + \cos\theta_1 \sin\theta_2$

$\sin(\theta_1 - \theta_2) = \sin\theta_1 \cos\theta_2 - \cos\theta_1 \sin\theta_2$

$\cos(\theta_1 + \theta_2) = \cos\theta_1 \cos\theta_2 - \sin\theta_1 \sin\theta_2$

$\cos(\theta_1 - \theta_2) = \cos\theta_1 \cos\theta_2 + \sin\theta_1 \sin\theta_2$

$$\tan(\theta_1 + \theta_2) = \frac{\tan\theta_1 + \tan\theta_2}{1 - \tan\theta_1 \tan\theta_2}$$

$$\tan(\theta_1 - \theta_2) = \frac{\tan\theta_1 - \tan\theta_2}{1 + \tan\theta_1 \tan\theta_2}$$

Power-Reducing Formulas

$$\sin^2\theta = \frac{1 - \cos 2\theta}{2} \qquad \cos^2\theta = \frac{1 + \cos 2\theta}{2} \qquad \tan^2\theta = \frac{1 - \cos 2\theta}{1 + \cos 2\theta}$$

Double-Angle Formulas

$\sin 2\theta = 2\sin\theta\cos\theta$

$$\cos 2\theta = \cos^2\theta - \sin^2\theta = 1 - 2\sin^2\theta = 2\cos^2\theta - 1$$

$$\tan 2\theta = \frac{2\tan\theta}{1 - \tan^2\theta}$$

Half-Angle Formulas

$$\sin\frac{\theta}{2} = \pm\sqrt{\frac{1 - \cos\theta}{2}} \qquad \cos\frac{\theta}{2} = \pm\sqrt{\frac{1 + \cos\theta}{2}} \qquad \tan\frac{\theta}{2} = \frac{1 - \cos\theta}{\sin\theta}$$

Product-to-Sum Formulas

$$\sin\theta_1 \sin\theta_2 = \frac{1}{2}[\cos(\theta_1 - \theta_2) - \cos(\theta_1 + \theta_2)]$$

$$\cos\theta_1 \cos\theta_2 = \frac{1}{2}[\cos(\theta_1 - \theta_2) + \cos(\theta_1 + \theta_2)]$$

$$\sin\theta_1 \cos\theta_2 = \frac{1}{2}[\sin(\theta_1 + \theta_2) + \sin(\theta_1 - \theta_2)]$$

$$\cos\theta_1 \sin\theta_2 = \frac{1}{2}[\sin(\theta_1 + \theta_2) - \sin(\theta_1 - \theta_2)]$$

The Graph of the Basic Sine Function: $y = \sin x$

x	0	$\pi/2$	π	$3\pi/2$	2π	$5\pi/2$	3π	$7\pi/2$	4π	$9\pi/2$
y	0	1	0	−1	0	1	0	−1	0	1

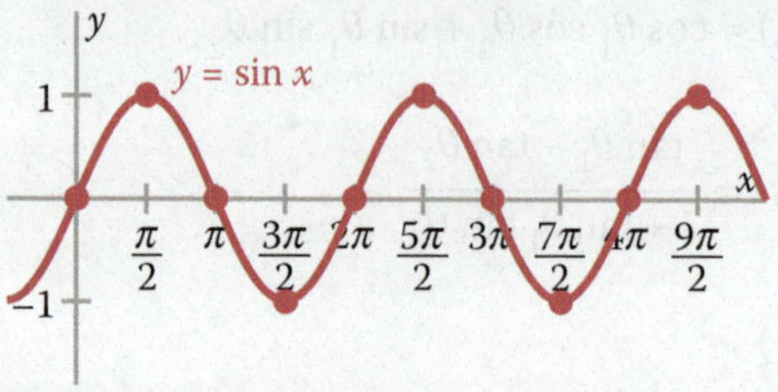

Domain: $(-\infty, \infty)$

Range: $[-1, 1]$

Period: 2π

The Graph of the Basic Cosine Function: $y = \cos x$

x	0	$\pi/2$	π	$3\pi/2$	2π	$5\pi/2$	3π	$7\pi/2$	4π	$9\pi/2$
y	1	0	−1	0	1	0	−1	0	1	0

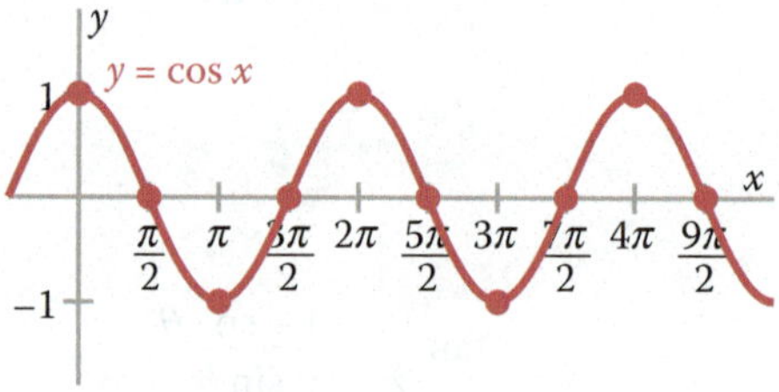

Domain: $(-\infty, \infty)$

Range: $[-1, 1]$

Period: 2π

The Graph of the Basic Tangent Function: $y = \tan x$

x	$-5\pi/4$	$-\pi$	$-3\pi/4$	$-\pi/4$	0	$\pi/4$	$3\pi/4$	π	$5\pi/4$	$7\pi/4$	2π	$9\pi/4$
y	−1	0	1	−1	0	1	−1	0	1	−1	0	1

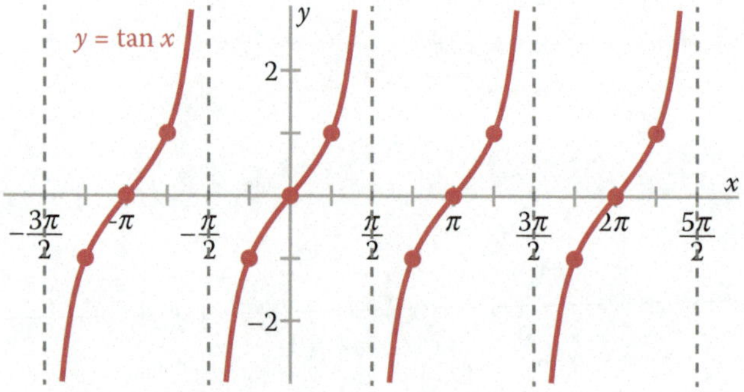

Domain: $\{x : x \neq \frac{\pi}{2} + \pi n\}$, where n is an integer

Range: $(-\infty, \infty)$

Vertical Asymptotes:
$x = \frac{\pi}{2} + \pi n$, where n is an integer

Period: π

The Graph of the Basic Cosecant Function: $y = \csc x$

x	$-3\pi/2$	$-\pi/2$	$\pi/2$	$3\pi/2$	$5\pi/2$	$7\pi/2$
y	1	−1	1	−1	1	−1

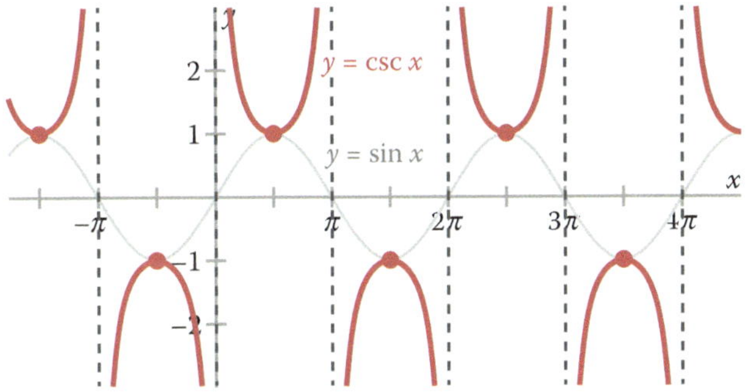

Domain: $\{x : x \neq \pi n\}$, where n is an integer

Range: $(-\infty, -1] \cup [1, \infty)$

Vertical Asymptotes: $x = \pi n$, where n is an integer

Period: 2π

The Graph of the Basic Secant Function: $y = \sec x$

x	$-\pi$	0	π	2π	3π	4π
y	−1	1	−1	1	−1	1

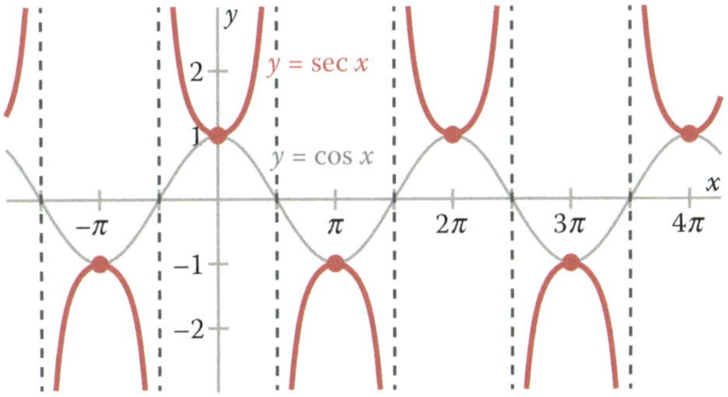

Domain: $\{x : x \neq \frac{\pi}{2} + \pi n\}$, where n is an integer

Range: $(-\infty, -1] \cup [1, \infty)$

Vertical Asymptotes:

$x = \frac{\pi}{2} + \pi n$, where n is an integer

Period: 2π

The Graph of the Basic Cotangent Function: $y = \cot x$

x	$-3\pi/2$	$-5\pi/4$	$-3\pi/4$	$-\pi/2$	$-\pi/4$	$\pi/4$	$\pi/2$	$3\pi/4$	$5\pi/4$	$3\pi/2$	$7\pi/4$	$9\pi/4$	$5\pi/2$
y	0	−1	1	0	−1	1	0	−1	1	0	−1	1	0

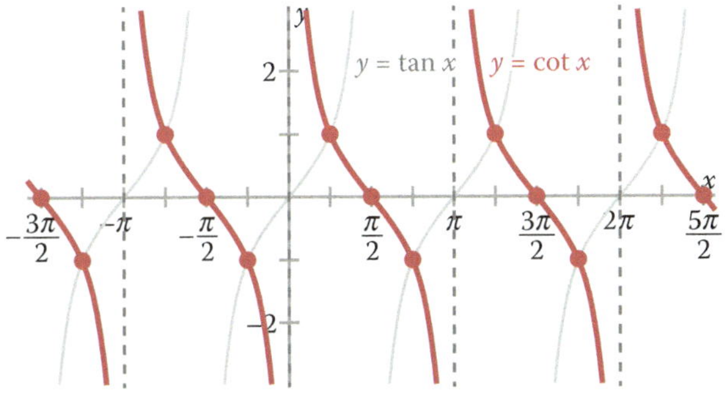

Domain: $\{x : x \neq \pi n\}$, where n is an integer

Range: $(-\infty, \infty)$

Vertical Asymptotes: $x = \pi n$, where n is an integer

Period: π

Vectors

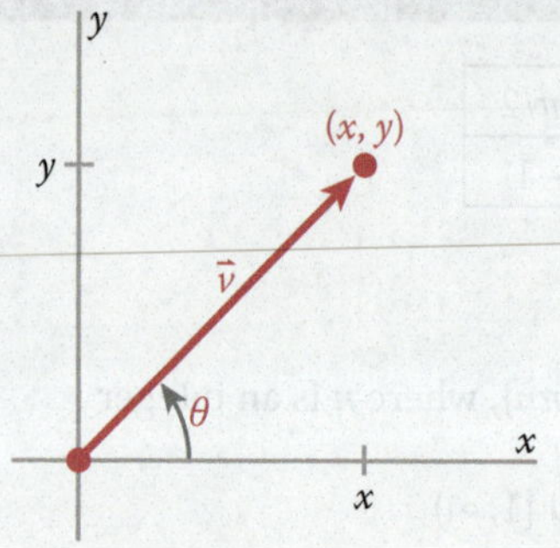

$\vec{v} = \langle x, y \rangle$

θ is the direction angle of $\vec{v}$.

Magnitude

If $\vec{v} = \langle x, y \rangle$, then the magnitude of $\vec{v}$ is $\|\vec{v}\| = \sqrt{x^2 + y^2}$.

Vector Addition and Subtraction

If $\vec{v} = \langle v_1, v_2 \rangle$ and $\vec{u} = \langle u_1, u_2 \rangle$, then $\vec{v} + \vec{u} = \langle v_1 + u_1, v_2 + u_2 \rangle$ and $\vec{v} - \vec{u} = \langle v_1 - u_1, v_2 - u_2 \rangle$.

Unit Vector

If $\vec{v} = \langle x, y \rangle$ and $\|\vec{v}\| \neq 0$, then the unit vector in the same direction as $\vec{v}$ is $\vec{u} = \frac{1}{\|\vec{v}\|}\vec{v} = \left\langle \frac{x}{\|\vec{v}\|}, \frac{y}{\|\vec{v}\|} \right\rangle$.

Dot Product

If $\vec{u} = \langle u_1, u_2 \rangle$ and $\vec{v} = \langle v_1, v_2 \rangle$, then the dot product of $\vec{u}$ and $\vec{v}$ is $\vec{u} \cdot \vec{v} = u_1 v_1 + u_2 v_2$.

APPENDIX B: ABBREVIATIONS

° C	degrees Celsius
° F	degrees Fahrenheit
A	ampere
cd	candela
cm	centimeter
ft	foot
gal	gallon
h	hour
Hz	hertz
in.	inch
K	Kelvin
kg	kilogram
km	kilometer
kPa	kilopascal
kW	kilowatt
L	liter
lb	pound
m	meter
mg	milligram
mi	mile
min	minute
mL	milliliter
N	Newton
oz	ounce
s	second
V	volt
W	watt
yd	yard
yr	year

APPENDIX C: SYMBOLS AND NOTATION

$=$	equal to	$\cup$	union
$\approx$	approximately equal to	$\cap$	disjunction
$<$	less than	$\varnothing$	empty set
$>$	greater than	$f(x)$	function notation
$\leq$	less than or equal to	$\pm$	plus or minus
$\geq$	greater than or equal to	%	percent
$\neq$	not equal to	$\parallel$	parallel
$+$	addition	$\perp$	perpendicular
$-$	subtraction	$\mathbb{R}$	the set of all real numbers
$\cdot$	multiplication	$\mathbb{Q}$	the set of all rational numbers
$\div$	division	Z	the set of all integers
/	division	N	the set of all natural numbers
$\sqrt{}$	square root	i	imaginary unit
$\sqrt[n]{}$	*n*th root	π	pi
$\lvert\ \rvert$	absolute value	∞	infinity
!	factorial	$\rightarrow$	approaches
[[ ]]	greatest integer	$\Leftrightarrow$	if and only if
$\circ$	composed with	$\Rightarrow$	implies
$\binom{n}{r}$	*n* choose *r*	$\begin{bmatrix} \square & \square \\ \square & \square \end{bmatrix}$	matrix
${}_nC_r$	*n* choose *r*	$n \times m$	matrix dimensions
$\sum$	summation	$\begin{vmatrix} \square & \square \\ \square & \square \end{vmatrix}$	determinant